In accordance with the latest syllabus prescribed by the
Central Board of Secondary Education, New Delhi

MOST LIKELY CBSE QUESTION BANK

CLASS XII

MATHEMATICS

Category Wise & Chapter Wise

By

A PANEL OF AUTHORS

AF517344

GURUKUL

EDITION
YEAR 2021

COPYRIGHT RESERVED
BY THE PUBLISHER

All rights reserved. No part of this publication may be reproduced in any form without the prior permission of the Oswal Publishers.

DISCLAIMER

With the ambition of providing standard academic resources, we have exercised extreme care in publishing the content. In case of any discrepancies in the matter, we request readers to excuse the unintentional lapse and not hold us liable for the same. Suggestions are always welcome.

ISBN : 978-93-91184-29-2

PUBLISHED BY

GURUKUL BOOKS & PACKAGING
(Unit of Oswal Printers & Publishers Pvt. Ltd.)

1/12, Sahitya Kunj, M.G. Road, Agra - 282002

(0562) 2527771-4, +91 7534077222

info@oswalpublishers.in

www.oswalpublishers.com

The cover of this book has been designed using resources from Freepik.com

PREFACE

Gurukul's CBSE Most Likely Question Bank series is updated as per the latest syllabus given by the Central Board of Secondary Education.

This title highlights the knowledge-based and skill-based goals of the Bloom's Taxonomy by making the students acquainted with relevant facts and concepts and also teaching them ways to apply their subjective knowledge into practise for desired academic results.

This book provides ample practice questions arranged section-wise so that the students can revise the whole syllabus in a comparatively lesser time and develop the ability of prioritising and categorising topics for effective learning. It covers all probable types of questions asked in the exams. The solutions have been prepared by eminent subject experts with explanations. Solutions follow the standard Marking Scheme of the CBSE Board.

In addition, questions from compartment paper, foreign paper and latest board paper have also been incorporated in the chapters.

The series is an attempt to instil confidence in students to face the board examination. The language used is simple, to the point and questions cover all the important topics as per the weightage given to them by the board.

We hope this book will be a valuable asset for the students. All further suggestions towards improving the series are welcome and would be incorporated in the future editions.

Note : Questions marked with :

'★' are board exam questions from previous years

'★★' are frequently asked board exam questions

—The Publisher

How to Prepare for Board Exams Using Gurukul's Most Likely Question Bank

The Most Likely Question Bank is a great learning tool for CBSE Examinations as it is the most comprehensive and concisely composed set of questions for students. In order to prepare well for their exams, students may follow the following approach..

- **Step 1 : Prepare Your Theory**

  Students must go through the theory of the chapters, from their textbooks covering the syllabus as per the Board Guidelines..

- **Step 2 : Understand the Categories**

  The contents have been arranged as per different 'Category' of Questions. Each category in turn covers topics from all the chapters of the syllabus. e.g., Short Answers category will cover the short answer type questions from all the chapters of the syllabus in that subject.

- **Step 3 : Complete Syllabus Revision**

  When students revise questions from any single category, they are revising the entire syllabus in different assemblages.

- **Step 4 : Customised Learning**

  Students with different learning calibers can customize their preparations. They can refer the objective questions first to get a basic understanding of the chapters and topics, and then they can take up the most complex categories.

- **Step 5 : Check Your Progress**

  Students can check their answers against the solutions given next to the questions. This will save time and help in faster revisions. Students can benefit tremendously from this series by using them optimally for their exam preparations

FINDING THE RIGHT BIN FOR MASKS?

Masks are a powerful shield against virus. The purpose behind wearing is to stop the spread of saliva and nasal discharge from infecting the other person.

Global use of disposable face masks and gloves is estimated to have been 100 billion pieces every month.

Masks are made up of a synthetic material called, TNT which takes more than 400 years to decompose.

Also masks come in a myriad of shades. The first step towards its healthy disposal is to choose which one to wear at the first place.

For casual use, choose tight cloth masks which can be reused over and over again. However, wearing it in vicinity of an infected person can be a wrong choice.

When visiting hospitals and other contaminated zones, choose loose fitting surgical masks which blocks droplets and splashes of saliva. However, they are ineffective against very small particles that come off sneezing and coughing in close vicinity.

N95 Masks are not for routine use and must be spared for medical experts who are in direct contact with the patient.

When it comes to disposing a mask, the very first step is to wash your hands before removing it from your face.

Wash your hands for atleast 20 seconds, using either an alcohol based sanitizer (having 60% alcohol) or using soap and water.

Remove your mask either using the mask strings or downside up. Don't touch the center of mask.

If you are wearing a cloth mask, put it separately into wash without mixing it with other clothes.

In case of a surgical mask, fold it inwards such that the droplets are not exposed to the air. Make it into a roll and wrap it up in two polythene bags or tissue bag, one inside the other.

Prefer cloth masks for casual use and avoid one time used masks.

CONTENTS

Note: Questions marked with :
'★' are board exam questions from previous years

SYLLABUS

COURSE STRUCTURE CLASS – XII

Time : 3 Hours **Max. Marks : 80**

No.	Units	No. of Periods	Marks
I	Relations and Functions	30	08
II	Algebra	50	10
III	Calculus	80	35
IV	Vectors and Three - Dimensional Geometry	30	14
V	Linear Programming	20	05
VI	Probability	30	08
	Total	240	80
	Internal Assessment		**20**

Unit - I : Relations and Functions

1. **Relations and Functions** **15 Periods**

Types of relations: reflexive, symmetric, transitive and equivalence relations. One to one and onto functions, composite functions, inverse of a function.

2. **Inverse Trigonometric Functions** **15 Periods**

Definition, range, domain, principal value branch. Graphs of inverse trigonometric functions. Elementary properties of inverse trigonometric functions.

Unit - II : Algebra

1. **Matrices** **25 Periods**

Concept, notation, order, equality, types of matrices, zero and identity matrix, transpose of a matrix, symmetric and skew symmetric matrices. Operation on matrices: Addition and multiplication and multiplication with a scalar. Simple properties of addition, multiplication and scalar multiplication. Non- commutativity of multiplication of matrices and existence of non-zero matrices whose product is the zero matrix (restrict to square matrices of order 2).Concept of elementary row and column operations. Invertible matrices and proof of the uniqueness of inverse, if it exists; (Here all matrices will have real entries).

2. **Determinants** **25 Periods**

Determinant of a square matrix (up to 3 x 3 matrices), properties of determinants, minors, cofactors and applications of determinants in finding the area of a triangle. Adjoint and inverse of a square matrix. Consistency, inconsistency and number of solutions of system of linear equations by examples, solving system of linear equations in two or three variables (having unique solution) using inverse of a matrix.

Unit - III : Calculus

1. **Continuity and Differentiability** **20 Periods**

Continuity and differentiability, derivative of composite functions, chain rule, derivatives of inverse trigonometric functions, derivative of implicit functions. Concept of exponential and logarithmic functions.

Derivatives of logarithmic and exponential functions. Logarithmic differentiation, derivative of functions expressed in parametric forms. Second order derivatives. Rolle's and Lagrange's Mean Value Theorems (without proof) and their geometric interpretation.

2. **Applications of Derivatives** **10 Periods**

Applications of derivatives : rate of change of bodies, increasing/decreasing functions, tangents and normals, use of derivatives in approximation, maxima and minima (first derivative test motivated geometrically and second derivative test given as a provable tool). Simple problems (that illustrate basic principles and understanding of the subject as well as real-life situations).

3. **Integrals** **20 Periods**

Integration as inverse process of differentiation. Integration of a variety of functions by substitution, by partial fractions and by parts, Evaluation of simple integrals of the following types and problems based on them.

$$\int \frac{dx}{x^2 \pm a^2}, \int \frac{dx}{\sqrt{x^2 \pm a^2}}, \int \frac{dx}{\sqrt{a^2 - x^2}}, \int \frac{dx}{ax^2 + bx + c}, \int \frac{dx}{\sqrt{ax^2 + bx + c}}$$

$$\int \frac{px + q}{ax^2 + bx + c} dx, \int \frac{px + q}{\sqrt{ax^2 + bx + c}} dx, \int \sqrt{a^2 + x^2}\, dx, \int \sqrt{x^2 - a^2}\, dx$$

$$\int \sqrt{ax^2 + bx + c}\, dx, \int (px + q) \sqrt{ax^2 + bx + c}\, dx$$

Definite integrals as a limit of a sum, Fundamental Theorem of Calculus (without proof). Basic properties of definite integrals and evaluation of definite integrals.

4. **Applications of the Integrals** **15 Periods**

Applications in finding the area under simple curves, especially lines, circles/ parabolas/ellipses (in standard form only), Area between any of the two above said curves (the region should be clearly identifiable).

5. **Differential Equations** **15 Periods**

Definition, order and degree, general and particular solutions of a differential equation, formation of differential equation whose general solution is given. Solution of differential equations by method of separation of variables, solutions of homogeneous differential equations of first order and first degree. Solutions of linear differential equation of the type :

$\frac{dy}{dx} + py = q$, where p and q are functions of x or constants.

$\frac{dx}{dy} + px = q$, where p and q are functions of y or constants.

Unit - IV : Vectors and Three-Dimensional Geometry

1. **Vectors** **15 Periods**

Vectors and scalars, magnitude and direction of a vector. Direction cosines and direction ratios of a vector. Types of vectors (equal, unit, zero, parallel and collinear vectors), position vector of a point, negative of a vector, components of a vector, addition of vectors, multiplication of a vector by a scalar, position vector of a point dividing a line segment in a given ratio. Definition, Geometrical Interpretation, properties and application of scalar (dot) product of vectors, vector (cross) product of vectors, scalar triple product of vectors.

2. **Three - dimensional Geometry** **15 Periods**

Direction cosines and direction ratios of a line joining two points. Cartesian equation and vector equation of a line, coplanar and skew lines, shortest distance between two lines. Cartesian and vector equation of a plane. Angle between (i) two lines, (ii) two planes, (iii) a line and a plane. Distance of a point from a plane.

Unit - V : Linear Programming

1. **Linear Programming** **20 Periods**

Introduction, related terminology such as constraints, objective function, optimization, different types of linear programming (L.P.) problems, mathematical formulation of L.P. problems, graphical method of solution for problems in two variables, feasible and infeasible regions (bounded or unbounded), feasible and infeasible solutions, optimal feasible solutions (up to three non-trivial constraints).

Unit - VI : Probability

1. **Probability** **30 Periods**

Conditional probability, multiplication theorem on probability, independent events, total probability, Bayes' theorem, Random variable and its probability distribution, mean and variance of random variable. Binomial probability distribution.

MATHEMATICS
(Code No. 041)
CLASS-XII

Time : 3 Hours **Max. Marks : 80**

S. No.	Typology of Questions	Total Marks	% Weightage
1.	**Remembering :** Exhibit memory of previously learned material by recalling facts, terms, basic concepts, and answers. **Understanding :** Demonstrate understanding of facts and ideas by organizing, comparing, translating, interpreting, giving descriptions, and stating main ideas	44	55
2.	**Applying :** Solve problems to new situations by applying acquired knowledge, facts, techniques and rules in a different way.	20	25
3.	**Analysing :** Examine and break information into parts by identifying motives or causes. Make inferences and find evidence to support generalizations. **Evaluating :** Present and defend opinions by making judgments about information, validity of ideas, or quality of work based on a set of criteria. **Creating :** Compile information together in a different way by combining elements in a new pattern or proposing alternative solutions.	16	20
	Total	80	100

1. No chapter wise weightage. Care to be taken to cover all the chapters.
2. Suitable internal variations may be made for generating various templates keeping the overall weightage to different form of questions and typology of questions same.

Choice (s) :

There will be no overall choice in the question paper.

However, 33% internal choices will be given.

Basic Concepts

Chapter 1. Relations and Functions

1. A **relation** from a set A to a set B is a subset of the Cartesian product $A \times B$. Thus, every subset including the empty set ϕ and a whole of $A \times B$ is a relation from A to B. Thus, if R is any relation from a set A to a set B, then R is a subset of $A \times B$.

If $(a, b) \in R$, we say that a is related to b and write aRb and if an ordered pair $(a, b) \notin R$, then 'a' is not related to 'b' and we write $a \not R b$.

2. A relation is **empty relation** if no element of A is related with any element of B.

3. A relation R is called **universal relation** if each element of A is related to every element of B.

4. The relation, $I_A = \{(a, a) : a \in A\}$ is called the **identity relation** on A.

5. The set $D = \{a : (a, b) \in R\}$ is called the **domain** of R *i.e.*, the set of first entries of the ordered pairs in relation is called the domain of the relation and the set $T = \{b : (a, b) \in R\}$ is called the **range** of R, *i.e.*, the set of second entries of the ordered pairs in a relation is called the range of the relation. If $A = B$ and R is a relation from A to B, we say that R is a relation on (or in) A.

6. A relation R on a set A is an **equivalence relation**, if it is **reflexive, symmetric and transitive.**

(i) Reflexive relation : For all $a \in A$, $(a, a) \in R$ or aRa holds.

(ii) Symmetric relation : For $a, b \in A$ if $(a, b) \in R$ then $(b, a) \in R$ or $aRb \Leftrightarrow bRa$.

(iii) Transitive relation : For $a, b, c \in A$ if $(a, b) \in R$ and $(b, c) \in R$, then $(a, c) \in R$ or aRb and $bRc \Leftrightarrow aRc$.

7. All **functions** $f : A \to B$ are relations from A to B such that for each $a \in A$, there is an unique $b \in B$.

8. A function f from a non-empty set A to a non-empty set B is a relation from A to B which satisfies the following conditions :

(i) For each $a \in A$, there exists $b \in B$ such that $(a, b) \in f$.

(ii) Whenever $(a, b) \in f$ and $(a, c) \in f$, then $b = c$.

In other words, $f : A \to B$ is a function if for each $a \in A$, there is a unique $b \in B$ such that $(a, b) \in f$.

9. The element b is called the image of a under f and generally, we write $f(a) = b$.

The set A is called the domain of f and B is called the co-domain of f.

Also, the subset $f(A) = \{b : (a, b) \in f\}$ of B is called the range of f.

10. In general, the range of a function is a subset of the co-domain.

11. All functions are relations but the converse is not true.

Let $A = \{1, 2, 3\}$ and $B = \{a, b\}$.

Then the relation $R_1 = \{(1, a), (2, b), (3, a)\}$ defines a function from A to B.

But the relation $R_2 = \{(1, a), (2, a), (3, a), (3, b)\}$ does not define a function.

12. The relation $R_3 = \{(1, a), (2, a), (3, a)\}$ defines a function whose range is $\{a\}$. Such a function whose range consists of a single element is called a **constant function.**

13. $f : A \to B$ is a **one-one or injective function** if for $a_1, a_2 \in A$, $f(a_1) = f(a_2) \Rightarrow a_1 = a_2$

or $\quad a_1 \neq a_2 \Rightarrow f(a_1) \neq f(a_2)$.

Otherwise f is called **Many-one function.**

For example :

(i) The constant function $f : A \to B$ defined as $f(a) = b$ for each $a \in A$ and a fixed element b of B can never be one to one because for $a_1 \neq a_2 \in A$.

$\Rightarrow \quad f(a_1) = b, f(a_2) = b$

$\therefore \quad f(a_1) = f(a_2)$

(if $A = \{a\}$, the constant function is one to one)

(ii) Let $A = \{1, 2, 3, 4\}$ and $B = \{a, b, c, d\}$

The function $f = \{(1, a), (2, b), (3, c), (4, d)\}$ is a one-to-one function.

But the function $g = \{(1, a), (2, b), (3, a), (4, c)\}$ is not a one-to-one function.

(iii) The function $f : R \to R$ defined as $f(x) = \sin(x)$ is not one-to-one, since the real numbers $x, x + 2\pi, x + 4\pi, \ldots, x + 2n\pi$ all have the same value.

14. $f : A \to B$ is called an **onto or surjective function** if $f(a) = b$, *i.e.*, for each $b \in B$, there is at least one $a \in A$ such that $f(a) = b$.

15. A function, $f : A \to B$ that is both one-one and onto is called a **bijective function**. It has one to one correspondence of elements from set A to set B.

16. If there is a bijection between two sets A and B, then the two sets are equivalent *i.e.*, they have the same number of elements.

For example, suppose you wish to find out if the number of students in the class is more than the chairs. You can count both students and chairs. Another way is to ask students to sit down. If all chairs are taken but students are left standing, then the number of students is more. If no student is standing but some chairs are empty, the number of chairs is more. However, if all students are sitting and no chair is left then we have no one to one correspondence and the number of chairs is equal to the number of students. The same idea is used in counting the number of elements of a set.

17. A function $f: A \to B$ is onto if and only if Range of $f = B$.

18. If $f: A \to B$ and $g: B \to C$ are two functions, then $h: A \to C$ defined as $h(a) = g(f(a))$ is called the **composite function** of f and g denoted by gof.

Alternatively, let f and g be two functions. Then the composition of the function f with g [denoted by $fog(x)$] is defined as $fog(x) = f(g(x))$, provided the range of g is in the domain of f. To evaluate $fog(x)$, the sequence of operations may be represented as :

$$x \xrightarrow{g} g(x) \xrightarrow{f} f(g(x))$$

19. A function $f: A \to B$ will have an inverse defined as $g: B \to A$, provided f is a bijective function *i.e.*, it is both one-one and onto.

Let $f: A \to B$ be a function and $g: B \to A$ be a function, such that

$$fog(x) = I_B(x) = x, \text{ for all } x \in B$$
$$gof(x) = I_A(x) = x, \text{ for all } x \in A$$

Then g is called the **inverse of f** and is denoted by

$$g = f^{-1}$$

If f be a function, then its inverse exists if and only if f is a bijection. Further the inverse is unique.

Chapter 2. Inverse Trigonometric Functions

1. Let $f: X \to Y$ be an bijective function *i.e.,* both one-one and onto. Then for each $y \in Y$, there exists a unique $x \in X$, such that, $f(x) = y$. Thus, we can talk about a function $g: Y \to X$ defined as $g(y) = x$, such that, $f(x) = y$.

In this case, g is called the inverse of f and is denoted by f^{-1}. Thus, g is the inverse of f iff f is the inverse of g. Thus, $f(f^{-1}(y)) = y$ and $f^{-1}(f(x)) = x$ for all $x \in X$ and $y \in Y$.

2. Set of all values that a function will take for different values of x are called **principal value.**

3. The **inverse of the sine function** is the $\sin^{-1}$ function. The domain of the function $x \to \sin x$ is the set of all real numbers and its range is $(-1, 1)$. It increases strictly as x varies from $-\frac{\pi}{2}$ to $\frac{\pi}{2}$ and decreases strictly as x increases from $\frac{\pi}{2}$ to $\frac{3\pi}{2}$.

4. We have restricted the sine function to the interval $\left[-\frac{\pi}{2}, \frac{\pi}{2}\right]$ to make it a bijection.

In fact, if we restrict the domain of $\sin\theta$ to one of the intervals $\left[-\frac{\pi}{2}, \frac{\pi}{2}\right], \left[\frac{3\pi}{2}, \frac{5\pi}{2}\right], \left[-\frac{5\pi}{2}, -\frac{3\pi}{2}\right]$ etc., $\sin\theta$ is one-one onto function with range $(-1, 1)$.

We therefore, conclude that in each of these intervals, we can define the inverse of the sine function.

Thus, $\sin^{-1} x$ is a function with domain $(-1, 1)$ and range $\left[-\frac{\pi}{2}, \frac{\pi}{2}\right]$ or $\left[-\frac{5\pi}{2}, -\frac{3\pi}{2}\right]$ or $\left[\frac{3\pi}{2}, \frac{5\pi}{2}\right]$ and so on.

5. The **inverse of the cosine function** is the $\cos^{-1}$ function. The domain of the function $x \to \cos x$ is R and its range is $(-1, 1)$. Also, the function decreases strictly from 1 to -1 as x increases from 0 to π, and it increases strictly from -1 to $+1$ as x increases from π to 2π and so on.

6. We have restricted the cosine function to the interval $[0, \pi]$ to make it a bijection. We conclude that $\cos^{-1}x$ is a function with domain $[-1, 1]$ and range $[0, \pi]$, $[\pi, 0]$ or $[\pi, 2\pi]$ and so on.

7. The **inverse of the tan function** is the $\tan^{-1}$ function. The domain of the function $x \to \tan x$ is

$$R - \left\{\pm\frac{\pi}{2}, \pm\frac{3\pi}{2}, \pm\frac{5\pi}{2}, \ldots\ldots\ldots\right\}$$

so that we exclude from R the number for which $\cos x = 0$. The range is $(-\infty, \infty)$. The function $x \to \tan x$ increases strictly from $-\infty$ to $+\infty$ as x increases in $\left(-\frac{\pi}{2}, \frac{\pi}{2}\right)$

8. As in the case of $\sin^{-1}x$, $\tan^{-1}x$ is a function with domain $(-\infty, \infty)$ and range $\left[-\frac{\pi}{2}, \frac{\pi}{2}\right]$ or $\left[-\frac{5\pi}{2}, -\frac{3\pi}{2}\right]$ or $\left[\frac{3\pi}{2}, \frac{5\pi}{2}\right]$ and so on.

9. The inverse of the cot function is the $\cot^{-1}$ function. The domain of the function $x \to \cot x$ is $R - (0, \pm\pi, \pm 2\pi, \pm 3\pi, \ldots)$ so that we exclude from R the number for which $\sin x = 0$. The range is $(-\infty, \infty)$. The function $x \to \cot x$ decreases strictly from $-\infty$ to $+\infty$ as x increases in $(0, \pi)$ so that the function is invertible in $(0, \pi)$.

10. $\cot^{-1}x$ is a function with domain $(-\infty, \infty)$ and range $[0, \pi]$ or $[-\pi, 0]$ or $[\pi, 2\pi]$ and so on.

11. The inverse of the secant function is the $\sec^{-1}$ function. The domain of the function $x \to \sec x$ is

$$R - \left\{\pm\frac{\pi}{2}, \pm\frac{3\pi}{2}, \pm\frac{5\pi}{2}, \ldots\ldots\ldots\right\}$$

so that we exclude from R the number for which $\cos x = 0$.

The range is

$$(-\infty, -1) \cup (1, \infty) = R \sim (-1, 1).$$

The function $x \to \sec x$ increases strictly from 1 to ∞ as x increases in $\left[0, \frac{\pi}{2}\right)$ and strictly increasing from $-\infty$ to -1 as x increases in $\left(\frac{\pi}{2}, \pi\right]$.

Thus, $x \to \sec x$ is invertible in the domain

$$\left(0, \frac{\pi}{2}\right) \cup \left(\pi, \frac{\pi}{2}\right) = [0, \pi] - \left(\frac{\pi}{2}\right).$$

12. The inverse of the cosecant function is the cosec^{-1} function. The domain of the function $x \to \text{cosec}\, x$ is

$$R - (0, \pm\pi, \pm 2\pi, \pm 3\pi, ...)$$

so that we exclude from R the set of number for which $\sin x = 0$. The range is

$$(-\infty, -1) \cup (1, \infty) = \text{R} - (-1, 1)$$

The function $x \to \text{cosec}\, x$ decreases strictly from -1 to $-\infty$ as x increases in $\left[-\frac{\pi}{2}, 0\right)$ and strictly decreases from $+\infty$ to 1 as x increases in $\left(0, \frac{\pi}{2}\right]$.

13. The function $x \to \text{cosec}\, x$ is thus invertible in the domain

$$\left[-\frac{\pi}{2}, 0\right) \cup \left(0, \frac{\pi}{2}\right] = \left[-\frac{\pi}{2}, \frac{\pi}{2}\right] \sim \{0\}$$

$\text{cosec}^{-1} x$ denotes the number which belongs to $\left[-\frac{\pi}{2}, 0\right) \cup \left(0, \frac{\pi}{2}\right]$ whose cosecant is x.

14. Domain and Principal Values of Trigonometric Functions

Function	Domain	Principal Value Branches
$\sin^{-1} x$	$[-1, 1]$	$\left[-\frac{\pi}{2}, \frac{\pi}{2}\right]$
$\cos^{-1} x$	$[-1, 1]$	$[0, \pi]$
$\tan^{-1} x$	R	$\left(-\frac{\pi}{2}, -\frac{\pi}{2}\right)$
$\text{cosec}^{-1} x$	$(-\infty, -1] \cup [1, \infty)$	$\left[-\frac{\pi}{2}, 0\right) \cup \left(0, \frac{\pi}{2}\right]$
$\sec^{-1} x$	$(-\infty, -1] \cup [1, \infty)$	$\left[0, \frac{\pi}{2}\right) \cup \left(\frac{\pi}{2}, \pi\right]$
$\cot^{-1} x$	R	$(0, \pi)$

15. Graphs of the Trigonometric functions and their respective inverse functions.

(i) The graphs of the function $x \to \sin x$ and its inverse $x \to \sin^{-1} x$.

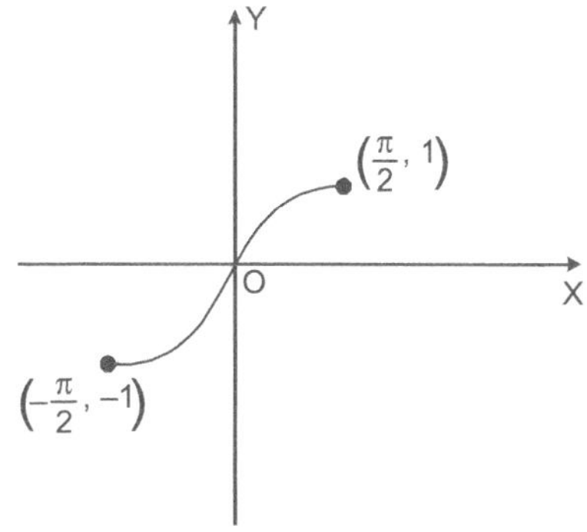

$y = \sin x$

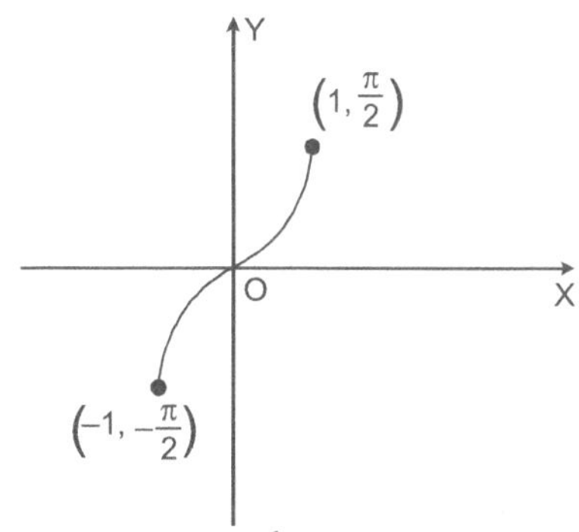

$y = \sin^{-1} x$

(ii) The graphs of the function $x \to \cos x$ and its inverse $x \to \cos^{-1} x$.

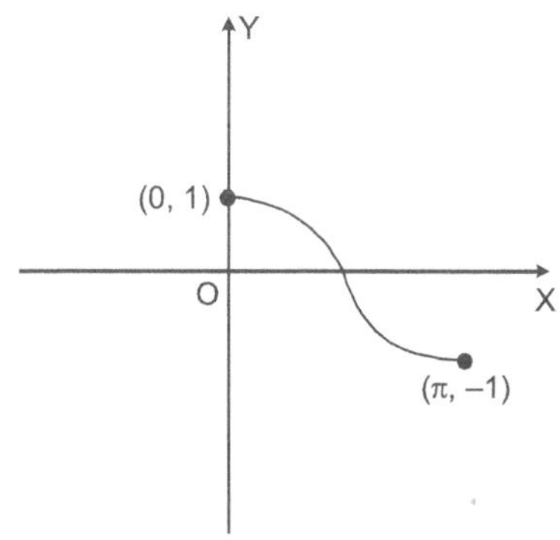

$y = \cos x$

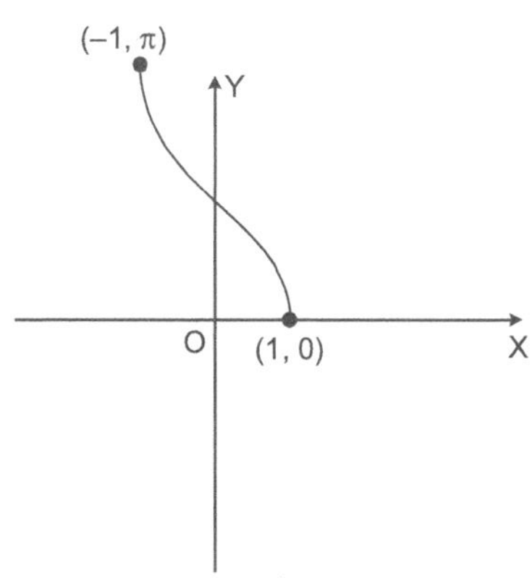

$y = \cos^{-1} x$

(iii) The graphs of the function $x \to \tan x$ and its inverse $x \to \tan^{-1} x$.

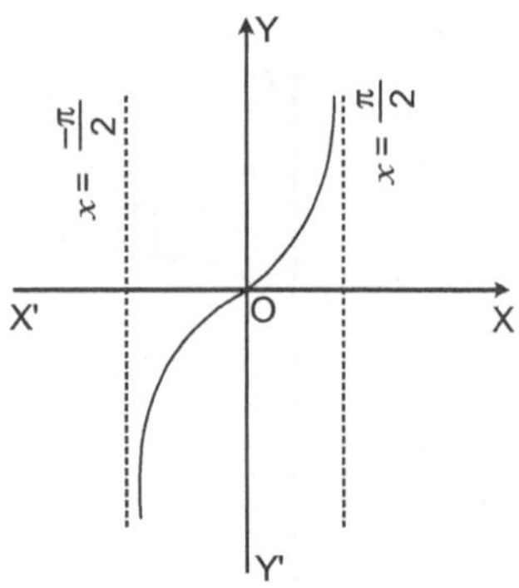

$y = \tan x$

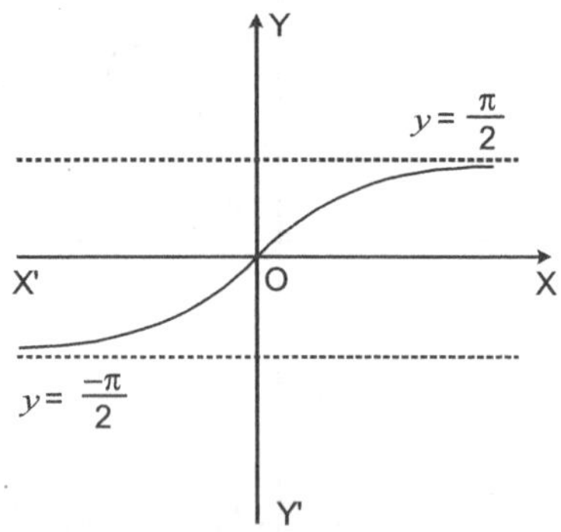

$y = \tan^{-1} x$

(iv) The graphs of the functions $x \to \cot x$ and its inverse $x \to \cot^{-1} x$.

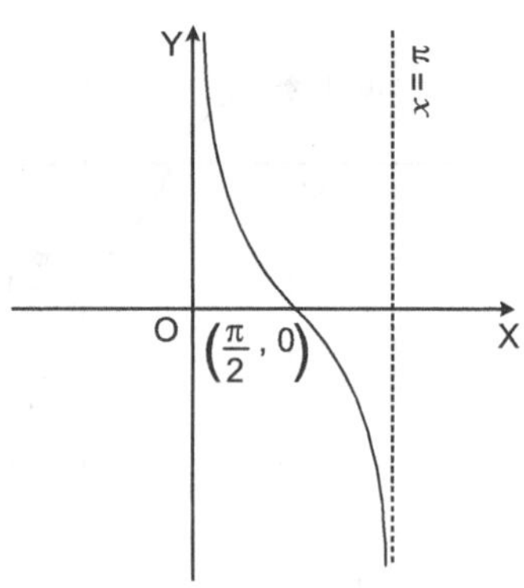

$y = \cot x$

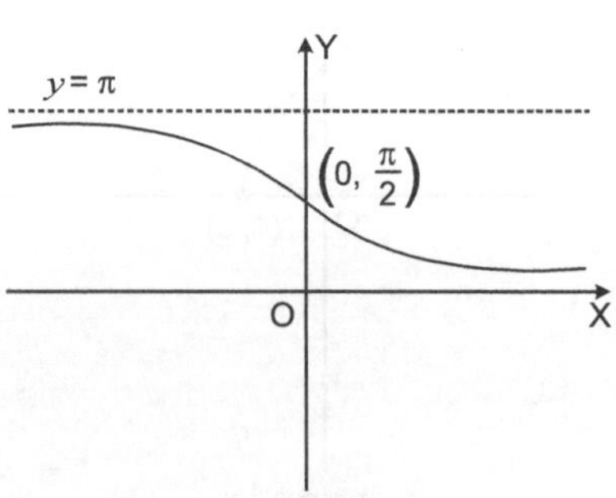

$y = \cot^{-1} x$

(v) The graphs of the functions $x \to \sec x$ and its inverse $x \to \sec^{-1} x$.

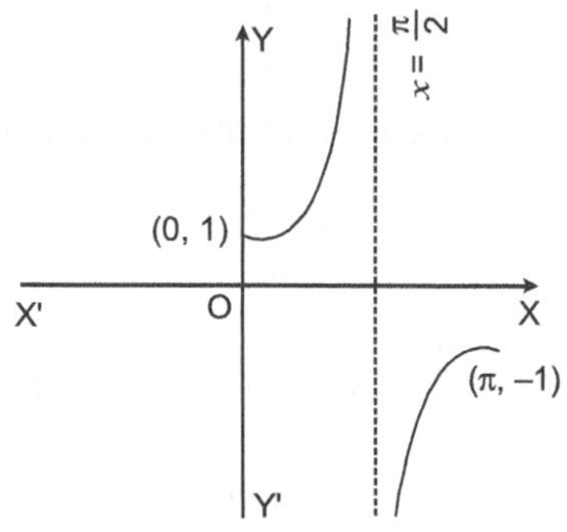

$y = \sec x$

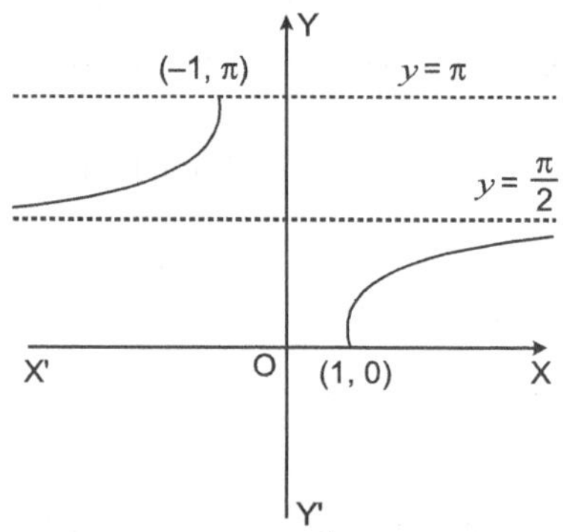

$y = \sec^{-1} x$

(vi) The graphs of the function $x \to \text{cosec } x$ and its inverse $x \to \text{cosec}^{-1} x$.

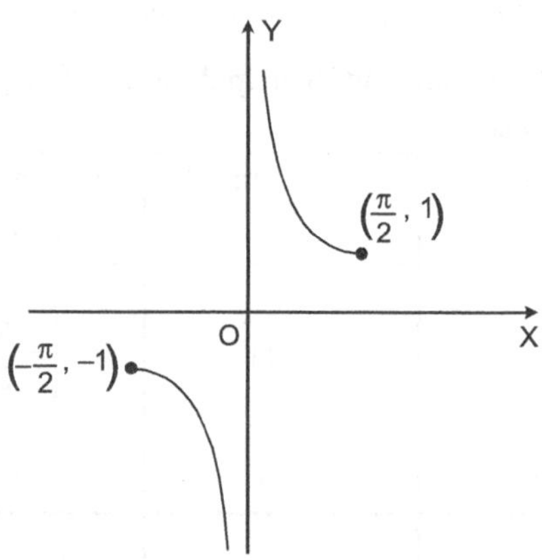

$y = \text{cosec } x$

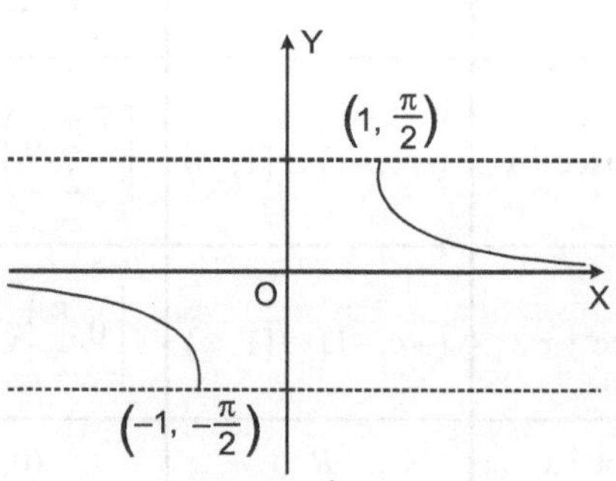

$y = \text{cosec}^{-1} x$

16. Properties of Inverse Trigonometric Functions

(i) $\sin^{-1}(\sin\theta) = \theta$ and $\sin(\sin^{-1}x) = x$, provided that $-\frac{\pi}{2} \le \theta \le \frac{\pi}{2}$ and $-1 \le x \le 1$.

(ii) $\cos^{-1}(\cos\theta) = \theta$ and $\cos(\cos^{-1}x) = x$, provided that $0 \le \theta \le \pi$ and $-1 \le x \le 1$.

(iii) $\tan^{-1}(\tan\theta) = \theta$ and $\tan(\tan^{-1}x) = x$, provided that $-\frac{\pi}{2} < \theta < \frac{\pi}{2}$ and $-\infty < x < \infty$.

(iv) $\cot^{-1}(\cot\theta) = \theta$ and $\cot(\cot^{-1}x) = x$, provided that $0 < \theta < \pi$ and $-\infty < x < \infty$.

(v) $\sec^{-1}(\sec\theta) = \theta$ and $\sec(\sec^{-1}x) = x$, provided that $0 \le \theta < \frac{\pi}{2}$ or $\frac{\pi}{2} < \theta \le \pi$ and $-\infty < x \le 1$ or $1 \le x < \infty$.

(vi) $\operatorname{cosec}^{-1}(\operatorname{cosec}\theta) = \theta$ and $\operatorname{cosec}(\operatorname{cosec}^{-1}x) = x$, provided that $-\frac{\pi}{2} \le \theta < 0$ or $0 < \theta \le \frac{\pi}{2}$ and $-\infty < x \le -1$ or $1 \le x < \infty$.

17. Properties Corresponding to Reciprocal Relationship Between Some Trigonometric Functions

(i) $\sin^{-1}x = \operatorname{cosec}^{-1}\left(\frac{1}{x}\right)$ and $\operatorname{cosec}^{-1}x = \sin^{-1}\left(\frac{1}{x}\right)$.

(ii) $\cos^{-1}x = \sec^{-1}\left(\frac{1}{x}\right)$ and $\sec^{-1}x = \cos^{-1}\left(\frac{1}{x}\right)$.

(iii) $\tan^{-1}x = \cot^{-1}\left(\frac{1}{x}\right)$ and $\cot^{-1}x = \tan^{-1}\left(\frac{1}{x}\right)$.

18. Properties Corresponding to the Change in Argument of Trigonometric Functions

(i) $\sin^{-1}(-x) = -\sin^{-1}x$

(ii) $\tan^{-1}(-x) = -\tan^{-1}x$

(iii) $\sec^{-1}(-x) = \pi - \sec^{-1}x$

(iv) $\cos^{-1}(-x) = \pi - \cos^{-1}x$

(v) $\operatorname{cosec}^{-1}(-x) = -\operatorname{cosec}^{-1}x$

(vi) $\cot^{-1}(-x) = -\cot^{-1}x$

(vii) $\sin^{-1}x + \cos^{-1}x = \frac{\pi}{2}$

(viii) $\sec^{-1}x + \operatorname{cosec}^{-1}x = \frac{\pi}{2}$

(ix) $\tan^{-1}x + \cot^{-1}x = \frac{\pi}{2}$

19. Properties of tan Inverse Function

(i) $\tan^{-1}x + \tan^{-1}y = \tan^{-1}\left(\frac{x+y}{1-xy}\right)$,

If $x > 0, y > 0$ and $xy < 1$.

(ii) $\tan^{-1}x + \tan^{-1}y = \pi + \tan^{-1}\left(\frac{x+y}{1-xy}\right)$,

If $x > 0,\ y > 0$ and $xy > 1$.

(iii) $\tan^{-1}x + \tan^{-1}y = -\pi + \tan^{-1}\left(\frac{x+y}{1-xy}\right)$.

If $x < 0, y < 0$ and $xy > 1$

(iv) $\tan^{-1}x - \tan^{-1}y = \tan^{-1}\left(\frac{x-y}{1+xy}\right)$. If $xy > -1$

(v) $\tan^{-1}x - \tan^{-1}y = \pi + \tan^{-1}\left(\frac{x-y}{1+xy}\right)$.

If $x > 0, y < 0$ and $xy < -1$

(vi) $\tan^{-1}x - \tan^{-1}y = -\pi + \tan^{-1}\left(\frac{x-y}{1+xy}\right)$.

If $x < 0, y > 0$ and $xy < -1$

(vii) $\tan^{-1}x + \tan^{-1}y + \tan^{-1}z = \tan^{-1}\left(\frac{x+y+z-xyz}{1-xy-yz-zx}\right)$

20. Sum of Two Inverse Sine Functions

(i) $\sin^{-1}x \pm \sin^{-1}y = \sin^{-1}\left(x\sqrt{1-y^2} \pm y\sqrt{1-x^2}\right)$.

If $-1 \le x, y \le 1$ and $x^2 + y^2 \le 1$.

(ii) $\sin^{-1}x \pm \sin^{-1}y = \pi - \sin^{-1}\left(x\sqrt{1-y^2} + y\sqrt{1-x^2}\right)$,

If $x > 0, y \le 1$ and $x^2 + y^2 > 1$.

(iii) $\sin^{-1}x + \sin^{-1}y$

$= -\pi - \sin^{-1}\left(x\sqrt{1-y^2} + y\sqrt{1-x^2}\right)$

If $x \ge -1, y < 0$ and $x^2 + y^2 > 1$

(iv) $\sin^{-1}x - \sin^{-1}y$

$= \pi$ □ $- \sin^{-1}\left(x\sqrt{1-y^2} - y\sqrt{1-x^2}\right)$

If $0 < x \le 1, -1 \le y \le 0$ and $x^2 + y^2 > 1$

(v) $\sin^{-1}x - \sin^{-1}y$

$= -\pi - \sin^{-1}\left(x\sqrt{1-y^2} - y\sqrt{1-x^2}\right)$

If $-1 \le x < 0, 0 < y \le 1$ and $x^2 + y^2 > 1$.

21. Sum of Two Inverse Cosine Functions

(i) $\cos^{-1}x + \cos^{-1}y = \cos^{-1}\left(xy - \sqrt{1-x^2}\sqrt{1-y^2}\right)$,

If $x > -1, y \leq 1$ and $x + y \geq 0$

(ii) $\cos^{-1}x + \cos^{-1}y$

$= 2\pi - \cos^{-1}\left(xy - \sqrt{1-x^2}\sqrt{1-y^2}\right)$

If $x > -1, y \leq 1$ and $x + y \leq 0$

(iii) $\cos^{-1} - \cos^{-1}y = \cos^{-1}\left(xy - \sqrt{1-x^2}\sqrt{1-y^2}\right)$

If $x \geq -1, y \leq 1$ and $x \leq y$

(iv) $\cos^{-1}x - \cos^{-1}y$

$= \cos^{-1}\left(xy + \sqrt{1-x^2}\sqrt{1-y^2}\right)$

If $0 < x \leq 1, -1 \leq y \leq 0$ and $x \geq y$

22. Formulae Corresponding to Half Angle

(i) $2\sin^{-1}x = \sin^{-1}\left(2x\sqrt{1-x^2}\right)$

(ii) $2\cos^{-1}x = \cos^{-1}(2x^2 - 1)$

(iii) $2\tan^{-1}x = \tan^{-1}\left(\frac{2x}{1-x^2}\right) = \sin^{-1}\left(\frac{2x}{1+x^2}\right)$

$= \cos^{-1}\left(\frac{1-x^2}{1+x^2}\right)$.

23. Formulae Corresponding to Triple Angle

(i) $3\sin^{-1}x = \sin^{-1}(3x - 4x^3)$

(ii) $3\cos^{-1}x = \cos^{-1}(4x^3 - 3x)$

(iii) $3\tan^{-1}x = \tan^{-1}\left(\frac{3x - x^3}{1-3x^2}\right)$

24. Formulae Corresponding to the Formulae Relating Sine to Cosine.

(i) $\sin^{-1}x = \cos^{-1}\sqrt{1-x^2} = \tan^{-1}\frac{x}{\sqrt{1-x^2}}$

$= \cot^{-1}\frac{\sqrt{1-x^2}}{x} = \sec^{-1}\left(\frac{1}{\sqrt{1-x^2}}\right) = \operatorname{cosec}^{-1}\left(\frac{1}{x}\right)$.

(ii) $\cos^{-1}x = \sin^{-1}\sqrt{1-x^2} = \tan^{-1}\frac{\sqrt{1-x^2}}{x}$

$= \cot^{-1}\left(\frac{x}{\sqrt{1-x^2}}\right) = \sec^{-1}\frac{1}{x} = \operatorname{cosec}^{-1}\left(\frac{1}{\sqrt{1-x^2}}\right)$.

(iii) $\tan^{-1}x = \sin^{-1}\left(\frac{x}{\sqrt{1+x^2}}\right) = \cos^{-1}\left(\frac{1}{\sqrt{1+x^2}}\right)$

$= \cot^{-1}\left(\frac{1}{x}\right) = \sec^{-1}\sqrt{1+x^2} = \operatorname{cosec}^{-1}\left(\frac{\sqrt{1+x^2}}{x}\right)$.

Chapter 3. Matrices

1. A **matrix** is a rectangular array of $m \times n$ numbers arranged in m rows and n columns. Matrix numbers are called the **elements or entries** of the matrix.

$$A = \begin{pmatrix} a_{11} & a_{12} & \dots & a_{1n} \\ a_{21} & a_{22} & \dots & a_{2n} \\ \vdots & \vdots & \dots & \vdots \\ a_{m1} & a_{m2} & & a_{mn} \end{pmatrix}$$

or $A = [a_{ij}]$,

where $i = 1, 2, \dots m$ and $j = 1, 2, \dots n$

2. The number of rows and columns that a matrix has is called its **order or its dimension.** By convention rows are listed first and columns second. Thus, if a matrix has m rows and n columns then its dimension is $m \times n$.

3. A matrix which has only one row is called **row matrix**.

For example : $A = [1\ \ 4\ \ 8\ \ 9]_{1 \times 4}$ is a row matrix of order 1×4.

In general, $A = [a_{ij}]_{1 \times n}$ is a row matrix of order $1 \times n$.

4. A matrix which has only one column is called **column matrix.**

For example, $A = \begin{bmatrix} 4 \\ 5 \\ 7 \\ 8 \end{bmatrix}_{4 \times 1}$ is a column matrix of order

4×1

In general, $A = [a_{ij}]_{m \times 1}$ is a column matrix of order $m \times 1$.

5. A matrix in which the number of rows are equal to the number of columns, is called a **square matrix.**

$$A = [a_{ij}]_{m \times m}$$

For example, $A = \begin{bmatrix} a & b & c \\ c & d & e \\ f & g & h \end{bmatrix}$

A is a square matrix of order 3×3.

6. A square matrix is called a **diagonal matrix** if all the elements, except the diagonal elements are zero.

$$A = [a_{ij}]_{m \times m}$$

where $a_{ij} = 0$, for $i \neq j$ and $a_{ij} \neq 0$, $i = j$

For example, $A = \begin{bmatrix} a & 0 & 0 \\ 0 & b & 0 \\ 0 & 0 & c \end{bmatrix}$

which is also denoted by $[a, b, c]$ and is called a diagonal matrix of order 3×3.

7. A square matrix is called **scalar matrix** if its all the elements, except diagonal elements are zero and diagonal elements are equal non-zero quantity.

$$A = [a_{ij}]_{n \times n},$$

where $a_{ij} = 0$, for $i \neq j$

and $a_{ij} = \alpha$, for $i = j$

where α is any real number.

In other words, a diagonal matrix in which all diagonal elements are equal is called a scalar matrix.

For example, $A = \begin{bmatrix} 5 & 0 & 0 \\ 0 & 5 & 0 \\ 0 & 0 & 5 \end{bmatrix}$

[A is a scalar matrix of order 3×3]

8. A square matrix in which all the non-diagonal elements are zero and diagonal elements are unity is called **identity or unit matrix.** In other words, the square matrix $A = [a_{ij}]_{m \times m}$ is an identity matrix if

$$a_{ij} = \begin{cases} 1, \text{ if } i = j \\ 0, \text{ if } i \neq j \end{cases}$$

For example, $[1]$, $\begin{bmatrix} 1 & 0 \\ 0 & 1 \end{bmatrix}$ and $\begin{bmatrix} 1 & 0 & 0 \\ 0 & 1 & 0 \\ 0 & 0 & 1 \end{bmatrix}$

are identity matrices of order 1×1, 2×2 and 3×3 respectively.

An n order identity matrix is represented by I_n.

9. A matrix in which all elements are zero is called a **null matrix or zero matrix.**

For example, $[0]$, $\begin{bmatrix} 0 & 0 \\ 0 & 0 \end{bmatrix}$, $\begin{bmatrix} 0 & 0 & 0 \\ 0 & 0 & 0 \\ 0 & 0 & 0 \end{bmatrix}$ are null matrices of order 1×1, 2×2 and 3×3 respectively.

10. Two matrices are said to be **equal** if they have same order and all their corresponding elements are equal.

For example, $\begin{bmatrix} 1 & 3 \\ 0 & 5 \end{bmatrix}$ and $\begin{bmatrix} 1 & 3 \\ 0 & 5 \end{bmatrix}$ are equal matrices

but $\begin{bmatrix} 1 & 7 \\ 5 & 9 \end{bmatrix}$ and $\begin{bmatrix} 2 & 1 \\ 0 & 3 \end{bmatrix}$ are not equal matrices.

11. A square matrix $A = [a_{ij}]$ is called an **upper triangular matrix** if $a_{ij} = 0$, for all $i > j$. Thus all elements below diagonal elements are zero in upper triangular matrix.

For example, $A = \begin{bmatrix} 2 & 7 & 5 \\ 0 & 8 & 6 \\ 0 & 0 & -4 \end{bmatrix}$

is an upper triangular matrix of order 3×3.

12. A square matrix $A = [a_{ij}]$ is called a **lower triangular matrix** if $a_{ij} = 0$ for all $i < j$.

Thus, in a lower triangular matrix, all elements above diagonal elements are zero.

For example, $A = \begin{bmatrix} 6 & 0 & 0 \\ 9 & 4 & 0 \\ 2 & 8 & 5 \end{bmatrix}$ is a lower triangular matrix

13. If A is the given matrix, then the matrix obtained by interchanging the rows and columns is called the **transpose of a matrix**. Transpose of matrix 'A' is denoted by A' or A^T.

14. Properties of Transpose : If A, B and C are matrices such that their sum and product are defined, then

(i) $(A^T)^T = A$

(ii) $(A \pm B)^T = A^T \pm B^T$

(iii) $kA^T = k \cdot A^T$ where k is a scalar

(iv) $(AB)^T = B^T A^T$

(v) $(ABC)^T = C^T B^T A^T$

15. A square matrix is said to be **symmetric** if $A = A^T$ *i.e.*, if $A = [a_{ij}]_{m \times m}$, then $a_{ij} = a_{ji}$ for all i, j. Also elements of the symmetric matrix are symmetric about the diagonal.

If $[a_{ij}] = [a_{ji}]$, then matrix A is said to be symmetric matrix for all possible values of i and j.

For example, $A = \begin{bmatrix} 2 & 7 & 5 \\ 7 & 3 & 9 \\ 5 & 9 & 4 \end{bmatrix}$

then $A^T = \begin{bmatrix} 2 & 7 & 5 \\ 7 & 3 & 9 \\ 5 & 9 & 4 \end{bmatrix}^T$

$$= \begin{bmatrix} 2 & 7 & 5 \\ 7 & 3 & 9 \\ 5 & 9 & 4 \end{bmatrix} = A$$

Hence, A is a symmetric matrix.

16. A square matrix is said to be **skew symmetric matrix** if $A' = -A$.

If $A = [a_{ij}]_{m \times m}$, then $a_{ij} = -a_{ji}$ for all i, j.

For example,

$$A=\begin{bmatrix}0&7&-9\\-7&0&3\\9&-3&0\end{bmatrix}$$

$$\Rightarrow \quad A'=\begin{bmatrix}0&7&-9\\-7&0&3\\9&-3&0\end{bmatrix}'$$

$$=\begin{bmatrix}0&-7&9\\7&0&-3\\-9&3&0\end{bmatrix}$$

$$=-\begin{bmatrix}0&7&-9\\-7&0&3\\9&-3&0\end{bmatrix}$$

$$=-A$$

Thus, A is a skew-symmetric matrix.

17. Diagonal elements of a skew-symmetric matrix are always zero.

18. Addition of Matrices : If A and B are two matrices of same order $m \times n$ such that $A=[a_{ij}]_{m\times n}$ and $B=[b_{ij}]_{m\times n}$ for all possible values of i and j then

$$A+B=C$$

where $C=[c_{ij}]_{m\times n}$ where $c_{ij}=a_{ij}+b_{ij}$

For example,

$$\begin{bmatrix}1&2\\-3&5\end{bmatrix}_{2\times2}+\begin{bmatrix}-2&-1\\1&3\end{bmatrix}_{2\times2}=\begin{bmatrix}-1&1\\-2&8\end{bmatrix}_{2\times2}$$

19. Properties of Matrix addition :

(i) Matrix addition is commutative.

$$A+B=B+A$$

(ii) Matrix addition is associative.

$$A+(B+C)=(A+B)+C$$

(iii) If O be a zero matrix of order $m\times n$, then

$$A+O=O+A=A$$

Then O is called additive identity.

(iv) If A be a given matrix then the matrix $-A$ is the additive inverse of A as $A+(-A)=O$ which is the additive identity matrix.

Then O is called additive identity.

20. Subtraction of Matrices : If $A=[a_{ij}]$ and $B=[b_{ij}]$ be the two matrices of order $m\times n$, then we define

$$A-B=A+(-B)$$

In other words, we subtract each element of B from each element of A.

or $\quad C=A-B=A+(-1)B$

For example, $\quad A=\begin{bmatrix}1&2\\3&4\end{bmatrix},\ B=\begin{bmatrix}2&3\\1&5\end{bmatrix}$

$$A-B=\begin{bmatrix}1&2\\3&4\end{bmatrix}-\begin{bmatrix}2&3\\1&5\end{bmatrix}$$

$$=\begin{bmatrix}1&2\\3&4\end{bmatrix}+(-1)\begin{bmatrix}2&3\\1&5\end{bmatrix}$$

$$=\begin{bmatrix}1&2\\3&4\end{bmatrix}+\begin{bmatrix}-2&-3\\-1&-5\end{bmatrix}$$

$$=\begin{bmatrix}-1&-1\\2&-1\end{bmatrix}$$

21. Multiplication of matrix by a scalar :

If $A=[a_{ij}]_{m\times n}$ be a matrix of order $m\times n$, then multiplication of A by a scalar k is matrix of order $m\times n$, such that each element of matrix A is multiplied by scalar k.

For example,

$$A=\begin{bmatrix}1&7\\-2&5\end{bmatrix} \text{ and } k=2$$

then $\quad kA=2A=2\begin{bmatrix}1&7\\-2&5\end{bmatrix}=\begin{bmatrix}2&14\\-4&10\end{bmatrix}$

22. Multiplication or product of Matrices : If A and B are two matrices, then multiplication of A and B *i.e.*, AB is defined, if

number of columns of A = number of rows of B

i.e., $\quad A=[a_{ij}]_{m\times n},\ B=[b_{jk}]_{n\times p}$

then $\quad AB=[c_{ik}]_{m\times p}$.

23. Properties of Multiplication of Matrices :

(i) Multiplication of matrices is not commutative *i.e.*,

$$AB\neq BA$$

(ii) Multiplication of matrices is associative *i.e.*,

$$\Rightarrow \quad (AB)C=A(BC)$$

(iii) Multiplication of matrices is distributive over addition *i.e.*,

$$A\cdot(B+C)=A\cdot B+A\cdot C$$

or $\quad (A+B)\cdot C=AC+BC$

(iv) For every square matrix A, there exists a multiplication identity matrix I of same order such that

$$AI=A=IA$$

(v) A matrix 'B' is said to be inverse of a matrix 'A' if,

$$AB=BA=I$$

Inverse of a matrix A is denoted by A^{-1}.

(vi) If product of two matrices is a zero matrix, then it is not necessary that one of the matrices is zero matrix.

For example,

$$A = \begin{bmatrix} 0 & 1 \\ 0 & 5 \end{bmatrix} \neq 0$$

$$B = \begin{bmatrix} 5 & 7 \\ 0 & 0 \end{bmatrix} \neq 0$$

$$\Rightarrow \quad AB = \begin{bmatrix} 0 & 1 \\ 0 & 5 \end{bmatrix}\begin{bmatrix} 5 & 7 \\ 0 & 0 \end{bmatrix} = \begin{bmatrix} 0 & 0 \\ 0 & 0 \end{bmatrix} = 0$$

24. A square matrix A of order m is said to be **invertible**, if there exists another matrix of order B of same order m, such that

$$AB = BA = I$$

where I is a unit matrix of same order m. The matrix B is called the inverse of matrix A and is represented as A^{-1}.

25. Properties of inverse of a matrix : If A and B are two invertible matrices of same order, then

(i) $(A^{-1})^{-1} = A$

(ii) $(AB)^{-1} = B^{-1}A^{-1}$

(iii) $(A^T)^{-1} = (A^{-1})^T$

(iv) $|A^{-1}| = |A|^{-1}$

(v) $AA^{-1} = A^{-1}A = I$

(vi) $(kA)^{-1} = kA^{-1}$

26. Some Important Theorems

(i) For a square matrix A with real numbers, $A + A'$ is a symmetric and $A - A'$ is a skew-symmetric matrix.

(ii) If A is a square matrix, then it can be represented as a sum of symmetric and skew-symmetric matrices.

i.e., $$A = \frac{1}{2}(A + A') + \frac{1}{2}(A - A').$$

(iii) All positive integral powers of a symmetric matrix are symmetric.

(iv) All odd integral powers of a skew-symmetric matrix are skew-symmetric.

Chapter 4. Determinants

1. Every square matrix can be associated to an expression or a number which is known as **determinant.**

2. Let $A = [a]$ be a matrix of order 1, then determinant of A is defined to be equal to a.

3. Let $A = \begin{bmatrix} a & b \\ c & d \end{bmatrix}$ be a matrix of order 2, then determinant of A is given by

$$|A| = \begin{vmatrix} a & b \\ c & d \end{vmatrix} = (ad - bc).$$

4. Let $A = \begin{bmatrix} a_1 & b_1 & c_1 \\ a_2 & b_2 & c_2 \\ a_3 & b_3 & c_3 \end{bmatrix}$ be a matrix of order 3, then determinant of A can be calculated by expanding it along R_1.

$$|A| = \begin{vmatrix} a_1 & b_1 & c_1 \\ a_2 & b_2 & c_2 \\ a_3 & b_3 & c_3 \end{vmatrix}$$

$$= a_1\begin{vmatrix} b_2 & c_2 \\ b_3 & c_3 \end{vmatrix} - b_1\begin{vmatrix} a_2 & c_2 \\ a_3 & c_3 \end{vmatrix} + c_1\begin{vmatrix} a_2 & b_2 \\ a_3 & b_3 \end{vmatrix}$$

$$= a_1(b_2c_3 - b_3c_2) - b_1(a_2c_3 - a_3c_2) + c_1(a_2b_3 - a_3b_2)$$

$$= a_1b_2c_3 - a_1b_3c_2 - a_2b_1c_3 + a_3b_1c_2 + a_2b_3c_1 - a_3b_2c_1$$

is the expansion or the value of the third order determinant.

5. The determinant of a square matrix of order $n \in N$ can be obtained by expanding it along any row or column.

6. Minors and Cofactors

Minor : Minor of an element is the determinant obtained by leaving the elements of the column and row containing that element.

Example : Let $\Delta = \begin{vmatrix} a_1 & b_1 & c_1 \\ a_2 & b_2 & c_2 \\ a_3 & b_3 & c_3 \end{vmatrix}$,

To get minor of a_3 we shall leave the entries of the column and the row of Δ containing a_3 and we shall keep the remaining entries in the same position.

So Minor of $a_3 = \begin{vmatrix} b_1 & c_1 \\ b_2 & c_2 \end{vmatrix}$

[**Note :** Minor of an element of a determinant of order n $(n \geq 2)$ is a determinant of order $(n - 1)$].

Cofactor : If the value of a minor is multiplied by $(-1)^{i+j}$, we get the cofactor of that element, where i is the number of the row and j is the number of the column containing that element.

If M_{ij} is the minor of an element a_{ij}, then cofactor of a_{ij},

$$c_{ij} = (-1)^{i+j} M_{ij}$$

Note : (i) On adding the products of the elements of any of its rows (columns) by their corresponding cofactors gives the value of the determinant.

i.e., $$\Delta = \begin{vmatrix} a_{11} & a_{12} & a_{13} \\ a_{21} & a_{22} & a_{23} \\ a_{31} & a_{32} & a_{33} \end{vmatrix}$$

or $\Delta = a_{11}c_{11} + a_{12}c_{12} + a_{13}c_{13}$

or $\Delta = a_{21}c_{21} + a_{22}c_{22} + a_{23}c_{23}$

or $\Delta = a_{31}c_{31} + a_{32}c_{32} + a_{33}c_{33}$

(ii) If all the elements of any row (column) are multiplied by the cofactors of the corresponding elements of other row (column) and the products are added, then the sum will be zero.

Example:

$$\Delta = a_{11}c_{21} + a_{12}c_{22} + a_{13}c_{23}$$

$$= a_{11}(-1)^{2+1}\begin{vmatrix} a_{12} & a_{13} \\ a_{32} & a_{33} \end{vmatrix} + a_{12}(-1)^{2+2}\begin{vmatrix} a_{11} & a_{13} \\ a_{31} & a_{33} \end{vmatrix} + a_{13}(-1)^{2+3}\begin{vmatrix} a_{11} & a_{12} \\ a_{31} & a_{32} \end{vmatrix}$$

$$= \begin{vmatrix} a_{11} & a_{12} & a_{13} \\ a_{11} & a_{12} & a_{13} \\ a_{31} & a_{32} & a_{33} \end{vmatrix} = 0 \quad [\because R_1 \text{ and } R_2 \text{ are identical}]$$

7. Adjoint of Matrix : Let $A = [a_{ij}]$ be a square matrix of order n and let c_{ij} be cofactor of a_{ij} in A. Then the transpose of the matrix of cofactors of elements of A is called the adjoint of A and is denoted by adj A.

$$\text{Adjoint of } A = \text{Adj } A = [c_{ij}]^T\,[A_{ij}]^T$$

$$A(\text{Adj } A) = (\text{Adj } A)A = |A|\,I_n$$

If $A = \begin{bmatrix} a & b \\ c & d \end{bmatrix}$, then

$$\text{adj }(A) = \begin{bmatrix} d & -b \\ -c & a \end{bmatrix}$$

8. Properties of adjoint of matrix :

(i) adj $(A^T) = [\text{adj }(A)]^T$

(ii) adj $(kA) = k^{n-1}$ (adj A), $k \in R$

(iii) adj $(AB) = [\text{adj }(B).\text{adj }(A)]$

(iv) $|\text{adj } A| = |A|^{n-1}$

(v) $|\text{adj (adj } (A))| = |A|^{(n-2)} \cdot A$

9. Singular Matrix : A square matrix 'A' of order 'n' is said to be singular if $|A| = 0$.

10. Non-singular Matrix : A square matrix 'A' of order 'n' is said to be non-singular matrix if $|A| \neq 0$.

11. Theorems on Determinants :

Theorem 1 : If all the rows of determinant (D) of a matrix are converted into the corresponding columns, the value of the determinant remains same.

[**Note :** If $R_i \to C_i$ is performed on D, then the value of determinant does not change.]

Example : $\begin{vmatrix} a_1 & b_1 & c_1 \\ a_2 & b_2 & c_2 \\ a_3 & b_3 & c_3 \end{vmatrix} = \begin{vmatrix} a_1 & a_2 & a_3 \\ b_1 & b_2 & b_3 \\ c_1 & c_2 & c_3 \end{vmatrix}$

Theorem 2 : If two rows (columns) of a determinant are interchanged then the value of the new determinant is the additive inverse of the value of the given determinant.

[**Note :** If R_{ij} (C_{ij}) is performed on D and we get new determinant D', then $D' = -D$]

Example :

$$\begin{vmatrix} a_1 & b_1 & c_1 \\ a_2 & b_2 & c_2 \\ a_3 & b_3 & c_3 \end{vmatrix} = (-1)\begin{vmatrix} a_2 & b_2 & c_2 \\ a_1 & b_1 & c_1 \\ a_3 & b_3 & c_3 \end{vmatrix} = (-1)^2\begin{vmatrix} a_2 & b_2 & c_2 \\ a_3 & b_3 & c_3 \\ a_1 & b_1 & c_1 \end{vmatrix}$$

Theorem 3 : If every element of any row or column of a determinant is multiplied by a constant k, then the value of new determinant is k times the value of original determinant.

[**Note :** If the result of $R_i(k)$ or $C(k)$ on D is D', then $D' = kD$; $(k \neq 0)$].

Example : $\begin{vmatrix} ka_1 & kb_1 & kc_1 \\ a_2 & b_2 & c_2 \\ a_3 & b_3 & c_3 \end{vmatrix} = k\begin{vmatrix} a_1 & b_1 & c_1 \\ a_2 & b_2 & c_2 \\ a_3 & b_3 & c_3 \end{vmatrix}$

Theorem 4 : If all the entries of any two rows or columns of a determinant are identical, then the value of the determinant is zero.

i.e., $\begin{vmatrix} a_1 & b_1 & c_1 \\ a_1 & b_1 & c_1 \\ a_3 & b_3 & c_3 \end{vmatrix} = 0$

Theorem 5 : If some or all elements of any row or column of a determinant are expressed as sum of two or more terms, then such determinant can be expressed as sum of two or more determinants of the same order.

$$\begin{vmatrix} a_1+d_1 & b_1+e_1 & c_1+f_1 \\ a_2 & b_2 & c_2 \\ a_3 & b_3 & c_3 \end{vmatrix} = \begin{vmatrix} a_1 & b_1 & c_1 \\ a_2 & b_2 & c_2 \\ a_3 & b_3 & c_3 \end{vmatrix} + \begin{vmatrix} d_1 & e_1 & f_1 \\ a_2 & b_2 & c_2 \\ a_3 & b_3 & c_3 \end{vmatrix}$$

Theorem 6 : If all the elements of any two rows or columns of a determinant are proportional, then the value of such determinant becomes zero.

$$\begin{vmatrix} a_1 & b_1 & c_1 \\ ka_1 & kb_1 & kc_1 \\ a_3 & b_3 & c_3 \end{vmatrix} = k\begin{vmatrix} a_1 & b_1 & c_1 \\ a_1 & b_1 & c_1 \\ a_3 & b_3 & c_3 \end{vmatrix} = 0$$

Theorem 7 : If all the elements of any row or column of a determinant are zero, then the value of such determinant becomes zero.

Example : $\begin{vmatrix} 0 & 0 & 0 \\ a & b & c \\ d & e & f \end{vmatrix} = 0$

Theorem 8 : If the elements of any row or column are added k times, the corresponding elements of any other row or column of the determinant, then the value of determinant remains same if we apply the equation

$$R_i \to R_i + kR_j \text{ or } C_i \to C_i + kC_j$$

Example :

$$\begin{vmatrix} a_1 & b_1 & c_1 \\ a_2 & b_2 & c_2 \\ a_3 & b_3 & c_3 \end{vmatrix} = \begin{vmatrix} a_1 & b_1 & c_1 \\ a_2 + ka_1 & b_2 + kb_1 & c_2 + kc_1 \\ a_3 & b_3 & c_3 \end{vmatrix}$$

Theorem 9 : Let A be a square matrix and let A' be the transpose matrix of A then

$$|A| = |A'|.$$

12. Applications of Determinants to Coordinate Geometry :

(i) Area of a triangle : If the vertices of a ΔABC are $A(x_1, y_1)$, $B(x_2, y_2)$ and $C(x_3, y_3)$, then area of ΔABC can be expressed in the form of determinant as

$$\text{Area of triangle} = \frac{1}{2}\begin{vmatrix} x_1 & y_1 & 1 \\ x_2 & y_2 & 1 \\ x_3 & y_3 & 1 \end{vmatrix}$$

$$= \frac{1}{2}|x_1(y_2 - y_3) + x_2(y_3 - y_1) + x_3(y_1 - y_2)|$$

Since area is always a positive quantity, so we take the absolute value of the above determinant.

[**Note : (a)** When area is given, then take both positive and negative values of determinant for further calculation.

(b) On shifting the origin, the area of triangle remains same]

(ii) Collinearity of points : Three points (x_1, y_1), (x_2, y_2) and (x_3, y_3) are said to be collinear if

$$\begin{vmatrix} x_1 & y_1 & 1 \\ x_2 & y_2 & 1 \\ x_3 & y_3 & 1 \end{vmatrix} = 0$$

(iii) Equation of a line passing through two given points : Cartesian equation of a line passing through two distinct points (x_1, y_1) and (x_2, y_2) is given by

$$\begin{vmatrix} x & y & 1 \\ x_1 & y_1 & 1 \\ x_2 & y_2 & 1 \end{vmatrix} = 0$$

where (x, y) is the point on the line.

13. Applications of determinant in solving a system of linear equations

Consistent and Inconsistent System : A system of equation is said to be consistent if it has one or more solutions and a system of equations is said to be inconsistent if it has no solution.

Let the system of linear equations be

$$a_1x + b_1y + c_1z = d_1$$
$$a_2x + b_2y + c_2z = d_2$$
and $$a_3x + b_3y + c_3z = d_3$$

Then these equations can be represented in the matrix form as :

Let $A = \begin{bmatrix} a_1 & b_1 & c_1 \\ a_2 & b_2 & c_2 \\ a_3 & b_3 & c_3 \end{bmatrix}$, $X = \begin{bmatrix} x \\ y \\ z \end{bmatrix}$ and $B = \begin{bmatrix} d_1 \\ d_2 \\ d_3 \end{bmatrix}$

$$\Rightarrow \quad AX = B$$

Case 1 : If A is a non-singular matrix *i.e.*, $|A| \neq 0$, then its inverse exists.

$$\therefore \quad AX = B \quad \ldots(i)$$

Multiply both sides of equation (i) by A^{-1}

$$A^{-1}AX = A^{-1}B$$
$$(A^{-1}A)X = A^{-1}B$$
$$IX = A^{-1}B \quad [\because A^{-1}A = I]$$
$$\Rightarrow \quad X = A^{-1}B \quad [\because IX = X]$$

Since inverse of the matrix is unique so it will have unique solution.

Case 2 : If A is a singular matrix *i.e.*, $|A| = 0$.

$\Rightarrow$ Its inverse does not exist.

(i) When $(\text{adj } A)B = 0$

$\Rightarrow$ The given system of equations is consistent and has infinitely many solutions.

(ii) When $(\text{adj } A)B \neq 0$

$\Rightarrow$ The given system of equations is inconsistent and has no solution.

Chapter 5. Continuity and Differentiability

1. Let f be a function with domain $D \subseteq R$, the set of real numbers and let $a \in D$. Then, 'f' is said to be **continuous** at 'a' if $\lim\limits_{x \to a} f(x)$ exists and its value is equal to $f(a)$.

In other words, 'f' is continuous at a if the following three conditions are satisfied :

(i) $f(a)$ exists, that is f is defined at $x = a$.

(ii) The limit $\lim\limits_{x \to a} f(x)$ exists *i.e.*,

$$\lim_{x \to a^-} f(x) = \lim_{x \to a^+} f(x) \text{ and}$$

(iii) $\lim\limits_{x \to a} f(x) = f(a)$

2. A function f is said to be **discontinuous** at a point, if it is not continuous at that point. This might happen in one of the following ways :

Case 1 : If both $\lim_{x\to a^+} f(x)$ and $\lim_{x\to a^-} f(x)$ exists but are not equal.

Consider the function :

$$f(x)=\begin{cases}\dfrac{|x|}{x}, & \text{if } x\neq 0\\ 0, & \text{if } x=0\end{cases}$$

The graph of the above function is as follows :

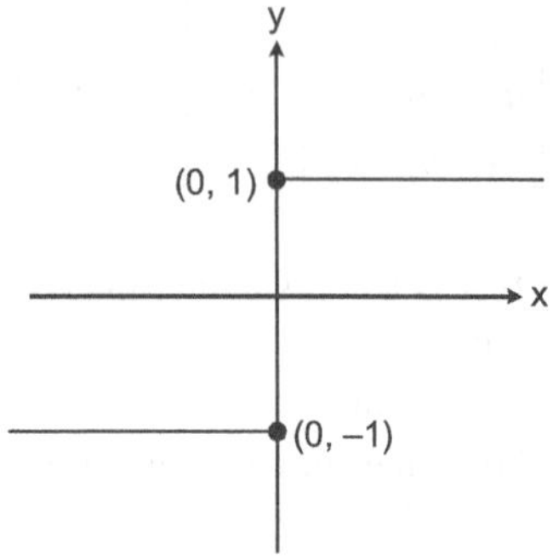

Then, $\lim_{x\to 0^-} f(x) = -1 \neq \lim_{x\to 0^+} f(x) = 1$

Thus, f is a discontinuous function at $x = 0$.

Case 2 : If either or both of $\lim_{x\to a^-} f(x)$ and $\lim_{x\to a^+} f(x)$ does not exist.

For example, consider the function :

$$f(x)=\begin{cases}\sin\dfrac{1}{x}, & \text{if } x>0\\ 0, & \text{if } x\leq 0\end{cases}$$

Then, $\lim_{x\to 0^-} f(x)$ and $\lim_{x\to 0^+} f(x)$ both do not exist.

Thus, f is a discontinuous function at $x = 0$.

3. A function f is said to have **removable discontinuity** if $\lim_{x\to a} f(x)$ exists (*i.e.*, both $\lim_{x\to a^-} f(x)$ and $\lim_{x\to a^+} f(x)$ exist and are equal), then it is possible to make the function continuous at $x = a$ by redefining the value of the function at that point a and putting it as equal to the value of $\lim_{x\to a} f(x)$.

For example, consider the function

$$f(x)=\begin{cases}1, & \text{if } x\neq 0\\ 2, & \text{if } x=0\end{cases}$$

Then, $\lim_{x\to a} f(x) = 1 \neq 2 = f(1)$

Thus, f has a removable discontinuity at $x = 0$, since we can redefine $f(0) = 1$ and make it continuous.

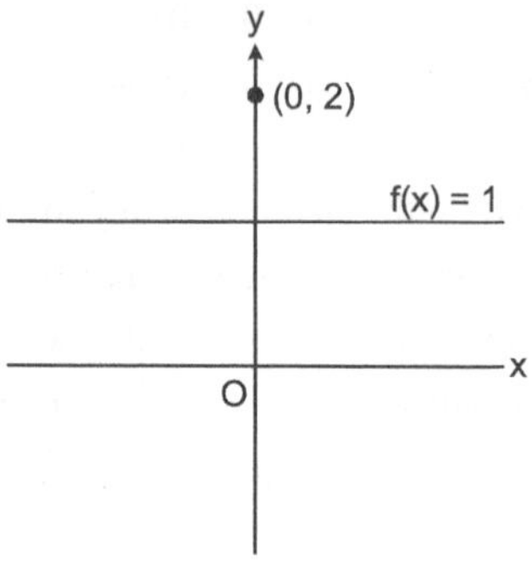

4. Continuity over an Interval : Consider a function f defined on an interval I of real numbers. At the interior points of I, the continuity of f can be discussed as given in the definition.

Assuming, $I = [2, 3)$, the left hand limit of f at $x = 2$ cannot be found as f is not defined for values of $x < 2$. In this case, f is said to be continuous at $x = 2$ if the right hand limit, $\lim_{x\to 2^+} f(x)$ exists and is equal to $f(2)$.

Similarly, if $I = (2, 3]$, then the continuity at $x = 3$ can be defined similarly.

5. Continuity over an open interval : Let $I = (a, b)$ be an open interval.

Consider $f : I \to R$ to be a function.

Then, f is continuous on I, if for each $y \in I$,

$$\lim_{x\to y^-} f(x) = \lim_{x\to y^+} f(x) = f(y)$$

6. Continuity at the left end point : Let $I = [a, b)$ and $f : I \to R$ be a function.

Then, f is continuous at the left end point $x = a$, if

$\lim_{x\to a^+} f(x) = f(a)$.

7. Continuity at the right end point : Let $I = (a, b]$ and $f : I \to R$ be a function.

Then, f is continuous at the right end point $x = b$, if

$$\lim_{x\to b^-} f(x) = -f(b)$$

8. Continuity over a closed interval : Let $I = [a, b]$ and $f : I \to R$ be a function.

Then, f is continuous on I if f is continuous over the open interval (a, b) and f is also continuous at the left end point a and the right end point b.

In other words, for each $a < y < b$

$$\lim_{x\to y^-} f(x) = \lim_{x\to y^+} f(x) = f(y)$$

$$\lim_{x\to a^+} f(x) = f(a) \text{ and } \lim_{x\to b^-} f(x) = f(b)$$

For example, let

$$f(x) = \begin{cases} 2x-1 \text{ if } x<0 \\ 2x+1 \text{ if } x \geq 0 \end{cases}$$

The function $f(x)$ is continuous in the closed interval $[1, 2]$.

$f(x) = 2x + 1$, being a polynomial is continuous at each point of $x > 0$.

Note that the function is continuous at every point except 0. It is continuous on every interval that does not contain 0.

9. Combining Continuous Functions : Continuous functions defined on the same domain can be combined in many ways (for example, by addition, subtraction, multiplication, division etc.) to get new continuous functions.

10. Sum of continuous functions : Let f and g be two continuous functions defined on the same domain. Then $f + g$ is a continuous function with domain D. (Similarly, $f - g$ is also a continuous with domain D)

For example, the function $f : R \to R$ defined by $f(x) = \sin x + e^x - x$ being the sum of three continuous functions $\sin x$, e^x and $-x$ is a continuous function.

11. Scalar multiple of a continuous function : Let f be a continuous function defined on $D \subseteq R$ and α be a real number. Then, the function αf defined by

$(\alpha f)(x) = \alpha f(x), \forall x \in D$ is a continuous function with domain D.

For example, $f(x) = 9x$, $g(x) = 5 \sin x$ and $h(x) = 7e^x$ are continuous functions on R.

12. Product of continuous functions : Let f and g be two continuous functions both defined on $D \subseteq R$.

Then, the product function fg defined by $(f.g)(x) = f(x).g(x), \forall x \in D$ is a continuous function on D.

For example, $f(x) = x \sin x$, $g(x) = e^x \cos x$ are continuous functions on R.

In particular, if n is a natural number then, $h(x) = x^n$ is a continuous function on R.

13. Quotient of a continuous function : Let f and g be two continuous functions with the same domain D such that $g(x) \neq 0$ for all $x \in D$.

Then, the quotient $\frac{f}{g}$ defined by $\frac{f}{g}(x) = \frac{f(x)}{g(x)}$, $\forall x \in D$ is a continuous function with domain D.

For example, consider the function $f(x) = \frac{\sin x}{e^x}$, $\forall x \in R$.

Since e^x is continuous and $e^x \neq 0, \forall x \in R$ and $\sin x$ is also continuous for all $x \in R$, therefore $\frac{\sin x}{e^x}$ $\forall x \in R$ is continuous.

14. Some Important Limits :

$$\lim_{x\to a} \frac{x^n - a^n}{x - a} = na^{n-1}$$

$$\lim_{x\to 0} \frac{\sin x}{x} = 1 = \lim_{x\to a} \frac{x}{\sin x}$$

$$\lim_{x\to 0} \frac{\tan x}{x} = 1 = \lim_{x\to a} \frac{x}{\tan x}$$

$$\lim_{x\to 0} \frac{e^x - 1}{x} = 1$$

$$\lim_{x\to\infty} \frac{1}{x^a} = 0, \forall a \in (0, \infty)$$

$$\lim_{x\to 0} \frac{a^x - 1}{x} = \log_e a$$

$$\lim_{x\to\infty} \left(1 + \frac{1}{x}\right)^x = e$$

$$\lim_{x\to 0} \frac{\log(1+x)}{x} = 1$$

$$\lim_{x\to 0} (1+x)^{1/x} = e$$

$$\lim_{x\to\infty} \left(1 + \frac{a}{x}\right)^x = e^a$$

$$\lim_{x\to e} \log_e x = 1.$$

15. Derivative of a function at a Point : Let f be a function.

The derivative of a function at the point $x = c$ (denoted by $f'(c)$) is defined as

$$f'(c) = \lim_{h\to 0} \frac{f(c+h) - f(c)}{h}$$

provided the limit exists and is finite.

If the limit exists and is finite, we say that the function f is differentiable at c.

16. Left Hand and Right Hand Derivatives : Left hand and right hand derivatives correspond to taking left hand and right hand limits respectively.

The limit $\lim_{h \to a^-} \frac{f(a+h)-f(a)}{h}$, if it exists is called the left-hand derivatives of f at $x = a$ and is denoted by $Lf'(a)$ or LHD.

The limit $\lim_{h \to a^+} \frac{f(a+h)-f(a)}{h}$, if it exists is called the right-hand derivative of f at $x = a$ and is denoted by $Rf'(a)$ and RHD.

In an open interval, a function is differentiable only if the left hand and right hand limits exist are finite and equal to each other.

However, let f be a function defined on $[a, b)$.

Then, f is said to be differentiable at the left end point 'a' if $Rf'(a)$ exists.

Similarly, let f be a function defined on $(a, b]$.

Then, f is said to be differentiable at the right end point 'b' if $Lf'(a)$ exists.

17. All differentiable functions are continuous but all continuous functions are not differentiable *i.e.,* if f is differentiable at $x = a$, then it is also continuous at that point.

Conclusion 1 : If a function is not continuous at a point then it is not differentiable at that point.

The converse of the above is not true, *i.e.*, all continuous functions are not differentiable.

For example : Modulus function $f(x) = | x |$ is not differentiable at $x = 0$ but it is continuous at $x = 0$.

Conclusion 2 : Since differentiability implies continuity, if a function is not continuous, then it is not differentiable.

18. Differentiation of Composite Functions [Chain Rule] : The Chain Rule is the method to differentiate a function which is a function of another function.

That is, if y is a function of t and t is a function of x, then we can write $y = f(t)$ where $t = g(x)$.

Then $\frac{d}{dx}(f(g(x))) = f'(g(x))g'(x)$ (Chain rule)

In other words if $y = f(g(x))$, then

$$\frac{dy}{dx} = \frac{dy}{dt} \cdot \frac{dt}{dx}$$

Yet another form of the chain rule is

Put $g(x) = t$

Then $\frac{d}{dx}(f(t)) = \frac{d}{dt}(f(t)) \cdot \frac{dt}{dx}$

19. Product Rule : Let u and v be two functions of x then

$$\frac{d}{dx}(uv) = u\frac{dv}{dx} + v\frac{du}{dx}$$

i.e., $\frac{d}{dx}$(Product of two functions) = first function $\frac{d}{dx}$(second function) + second function $\frac{d}{dx}$(first function).

20. Differentiation of Implicit Functions : Implicit functions are equations of the type $f(x, y) = 0$ in which neither y can be written as a function of x nor x can be written as a function of y.

Or, they may be equations (such as given an implicit function $f(x, y) = 0$ to evaluate $\frac{dy}{dx}$ we consider y as a dependent variable and x as an independent variable.

21. Differentiation of Logarithmic Functions : If a function is given in the exponential form, *i.e.*, $y = f(x)^{g(x)}$, then we apply the logarithmic method.

Here, $y = f(x)^{g(x)}$

Taking the logarithm on both sides, we get

$$\log y = g(x) \log (f(x))$$

Differentiating both sides with respect to x, we get

$$\frac{1}{y}\frac{dy}{dx} = \frac{d}{dx}(g(x) \log f(x))$$

$$= g(x)\frac{d}{dx}\log f(x) + \left(\log f(x)\frac{d}{dx}g(x)\right)$$

$$\Rightarrow \frac{dy}{dx} = y\left[\frac{g(x)}{f(x)}\frac{d}{dx}f(x) + \left(\log f(x)\frac{d}{dx}g(x)\right)\right]$$

$$\Rightarrow \frac{dy}{dx} = f(x)^{g(x)}\left[\frac{g(x)}{f(x)}\frac{d}{dx}f(x) + \left(\log f(x)\frac{d}{dx}g(x)\right)\right]$$

We can also apply this method when $y = f_1(x).f_2(x) \ldots f_n(x)$, where n is a large number.

22. Parametric Functions : Curves in a plane are described by a function in two variables, say x and y.

However, by introducing a new variable say t, we may write x and y as functions of t.

Functions of this form are known as parametric functions.

For example, $x = f(t), y = g(t), t \in [a, b]$ is a parametric function with parameter t.

For example,

(i) A circle with centre at origin and radius r is given by $x^2 + y^2 = r^2$.

It has the following parametric form :

$$x = r\cos\theta, y = r\sin\theta, \theta \in [0, 2\pi].$$

(ii) $x = at^2, y = 2at$ $t \in R$ is the parametric form of the parabola $y^2 = 4ax$.

23. Differentiation of Parametric Functions : We use a form of chain rule to differentiate parametrically defined functions.

Let x and y be functions of a parameter t.

Then, $$\frac{dy}{dx} = \frac{\left(\frac{dy}{dt}\right)}{\left(\frac{dx}{dt}\right)}$$

This rule follows from chain rule by considering y as a function of x and x as a function of t.

24. Second Order Derivatives :

Let $$y = f(x)$$

Then, $\frac{dy}{dx}$ is also a function of x.

Thus, we can take the derivative of $\frac{dy}{dx}$ *i.e.*, $\frac{d}{dx}\left(\frac{dy}{dx}\right)$.

We write, $$\frac{d^2y}{dx^2} = \frac{d}{dx}\left(\frac{dy}{dx}\right)$$

In this case, $\frac{d^2y}{dx^2}$ is called the second order derivative of $y = f(x)$.

Going on, in this way we can define the derivative as :

$$\frac{d^3y}{dx^3} = \frac{d}{dx}\left(\frac{d^2y}{dx^2}\right) = \frac{d}{dx}\left(\frac{d}{dx}\left(\frac{dy}{dx}\right)\right) \ldots \text{ and so on.}$$

In general, $$\frac{d^ny}{dx^n} = \frac{d}{dx}\left(\frac{d^{n-1}y}{dx^{n-1}}\right)$$

$$= \frac{d}{dx}\left(\frac{d}{dx}\left(\frac{d^{n-2}y}{dx^{n-2}}\right)\right) = \ldots$$

$$= \frac{d}{dx}\left(\frac{d}{dx} \cdots \frac{d}{dx}\left(\frac{dy}{dx}\right)\right)$$

is called the nth order derivative of $y = f(x)$.

Here, $n = 1, 2, 3, \ldots$

We denote $\frac{d^ny}{dx^n}$ by y^n also.

Thus, $$y_1 = \frac{dy}{dx}, y_2 = \frac{d^2y}{dx^2}, y_3 = \frac{d^3y}{dx^3} \ldots$$

25. Rolle's Theorem : Let f be a real function defined in the closed interval $[a, b]$ such that :

(i) f is continuous in $[a, b]$

(ii) f is differentiable in the open interval (a, b)

(iii) $f(a) = f(b)$

Then there exists a point c in the open interval (a, b) such that $f'(c) = 0$.

In other words, Rolle's theorem says that if f is a differentiable function and if it agrees at two distinct points a and b then its derivative must vanish at least once between a and b.

Note that : The point c at which the derivative vanishes need not be a unique point. There may be one or more finite and infinitely many such points.

Geometrically $f'(c) = 0$ means that the tangent drawn to the graph of $f(x)$ at the point $x = c$ is parallel to the X-axis.

As $f(a) = f(b)$, the chord drawn to the graph, which passes through points $(a, f(a))$ and $(b, f(b))$ is also parallel to the X-axis. Thus, in particular, this means that the tangent at $x = c$ is parallel to the chord joining $(a, f(a))$ and $(b, f(b))$.

Physically, the fact that $f'(c) = 0$ means that the rate of change of f is zero at $x = c$.

[Rolle's Theorem is an example of a mean value theorem]

26. Lagrange's Mean Value Theorem : Lagrange's Mean Value Theorem is a generalization of Rolle's Theorem.

Let f be a real function continuous on the closed interval $[a, b]$ and differentiable in the open interval (a, b), then there is a point c at which $f'(c) = \frac{f(b) - f(a)}{b - a}$.

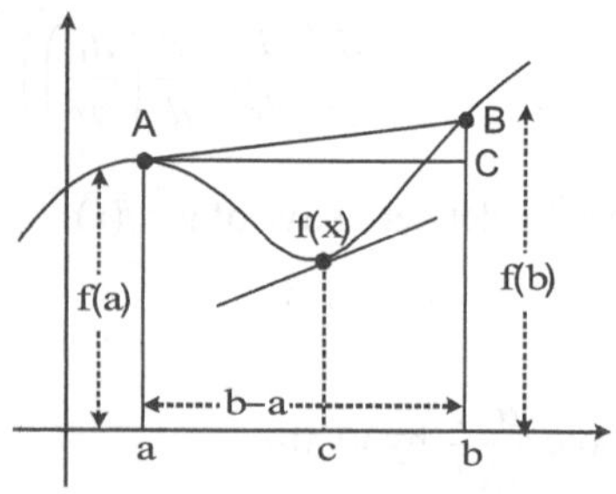

For an explanation of this have a look at the above figure.

Geometrically the quantity $\frac{f(b)-f(a)}{b-a}$ only gives the slope of the chord AB that joins the points $A(a, f(a))$ and $B(b, f(b))$.

Note : If you put $f(a) = f(b)$ in the Mean Value Theorem, we can say that there is some c between a and b and $f'(c)$ $= \frac{f(b)-f(a)}{b-a} = 0$ which is the assertion of Rolle's theorem.

27. Important Series which are used in Limits :

(i) $(1+x)^n = 1 + nx + \frac{n(n-1)x^2}{2!} + \frac{n(n-1)(n-2)x^3}{3!} + ...$

(ii) $e^x = 1 + \frac{x}{1!} + \frac{x^2}{2!} + \frac{x^3}{3!} + ... \infty$

and $e^{-x} = 1 - \frac{x}{1!} + \frac{x^2}{2!} - \frac{x^3}{3!} + ...$.

(iii) $a^x = 1 + x \log_e a + \frac{x^2}{2!}(\log_e a)^2 + ...$

and $\log|1+x| = x - \frac{x^2}{2} + \frac{x^3}{3} ....$

(iv) $\log|1-x| = -x - \frac{x^2}{2} - \frac{x^3}{3}$

and $\sin x = x - \frac{x^3}{3!} + \frac{x^5}{5!} - \frac{x^7}{7!} + ...$

(v) $\cos x = 1 - \frac{x^2}{2!} + \frac{x^4}{4!} - \frac{x^6}{6!}$

and $\tan x = x + \frac{x^3}{3!} + \frac{x^5}{5!} + ...$

28. Derivatives of Some Elementary Functions :

$$\frac{d}{dx}(x^n) = nx^{n-1}$$

$$\frac{d}{dx}(e^x) = e^x$$

$$\frac{d}{dx}(\ln x) = \frac{1}{x}$$

$$\frac{d}{dx}(a^x) = a^x \cdot \ln a$$

$$\frac{d}{dx}(\sin x) = \cos x$$

$$\frac{d}{dx}(\cos x) = -\sin x$$

$$\frac{d}{dx}(\tan x) = \sec^2 x$$

$$\frac{d}{dx}(\cot x) = -\text{cosec}^2 x$$

$$\frac{d}{dx}(\sec x) = \sec x \tan x$$

$$\frac{d}{dx}(\text{cosec } x) = -\text{cosec } x \cot x$$

$$\frac{d}{dx}(\sin^{-1} x) = \frac{1}{\sqrt{1-x^2}},$$

$$\frac{d}{dx}(\cos^{-1} x) = -\frac{1}{\sqrt{1-x^2}}$$

$$\frac{d}{dx}(\tan^{-1} x) = \frac{1}{1+x^2},$$

$$\frac{d}{dx}(\cot^{-1} x) = -\frac{1}{1+x^2}$$

$$\frac{d}{dx}(\sec^{-1} x) = \frac{1}{|x|\sqrt{x^2-1}},$$

$$\frac{d}{dx}(\text{cosec}^{-1} x) = -\frac{1}{|x|\sqrt{x^2-1}}$$

29. Evaluation of Limits :

(i) Direct substituting method : We substitute the value of the point in the given expression and if we get a finite number, then this number is the limit of the given function.

(ii) Factorisation method : On substituting $x = a$ in the given expression, if we get $\frac{0}{0}, \frac{\infty}{\infty}$ ...., etc. form, then we factorize the numerator and denominator and take $(x - a)$ as a common factor from numerator and denominator. After cancelling out $(x - a)$, we put $x = a$. If we get a finite number, then it is the required value otherwise repeat the step till we get a finite number.

(iii) Rationalisation method : Rationalisation method is applicable when

(a) numerator, denominator or both are in square root

or

(b) after substituting the value of limit if we get the negative number in square root. Hence, after simplifying in both the cases, we get the required value.

(iv) L'Hospital Rule : With the help of this rule, if we have to evaluate $\lim_{x\to x}\frac{f(x)}{g(x)}$ such that it takes indeterminate form, *i.e.*, $\frac{0}{0}$ or $\frac{\infty}{\infty}$ then we differentiate numerator and denominator to get $\lim_{x\to c}\frac{f(x)}{g(x)}$ if it is determinate form then it is required value, otherwise repeat the step till we get a determinate form and thus required value.

Note : According to L' Hospital rule

$$\lim_{x\to c}\frac{f(x)}{g(x)} = \lim_{x\to\infty}\frac{f'(x)}{g'(x)}, \text{where } g'(x)\neq 0\ \forall x\in Z \text{ with } x\neq c$$

Chapter 6. Applications of Derivatives

1. **Rate of change :** Let $y=f(x)$ be a relation between two variables x and y.

   Then $\frac{dy}{dx}$ [or $f'(x)$] represents the rate of change of y with respect to x and $\left[\frac{dy}{dx}\right]_{x\,=\,x_0}$ [or $f'(x_0)$] represents the rate of change of y with respect to $x=x_0$.

2. Let $y = f(x)$ be any function of x which is **differentiable** in (a, b). The derivative of this function at some point x of (a, b) is given by the relation

$$\frac{dy}{dx} = \lim_{\Delta x\to 0}\frac{\Delta y}{\Delta x}$$

$$\lim_{\Delta x\to 0}\frac{f(x+\Delta x)-f(x)}{\Delta x} = f'(x)$$

$$\Rightarrow \quad \frac{dy}{dx} = f'(x)$$

$$\Rightarrow \quad dy = f'(x)\,dx$$

where dy is the differential of the function.

The error Δx in x is called the **absolute error** in x.

If Δx is an error in x then $\frac{\Delta x}{x}$ is called the **relative error** in x.

If Δx is an error in x then $\frac{\Delta x}{x}\times 100$ is called **percentage error** in x.

3. If $y=f(x)$ and $P(x, y)$ be any point of $y=f(x)$ then

$$\left(\frac{dy}{dx}\right)_P = \text{slope of tangent to } y=f(x) \text{ at point } P(x, y)$$

$$\frac{-1}{\left(\frac{dy}{dx}\right)_P} = \text{slope of the normal to } y=f(x) \text{ at point } P(x, y)$$

4. If the tangent is parallel to X-axis, then $\frac{dy}{dx}=0$

   If the tangent is parallel to Y-axis, then $\frac{dx}{dy}=0$.

5. The equation of tangent to the curve $y=f(x)$ at point $P(x_1, y_1)$ is given by

$$y-y_1 = \left(\frac{dy}{dx}\right)_{(x_1,\,y_1)}(x-x_1)$$

The equation of normal to the curve $y=f(x)$ at point $P(x_1, y_1)$ is given by

$$y-y_1 = \frac{-1}{\left(\frac{dy}{dx}\right)_{(x_1,\,y_1)}}(x-x_1)$$

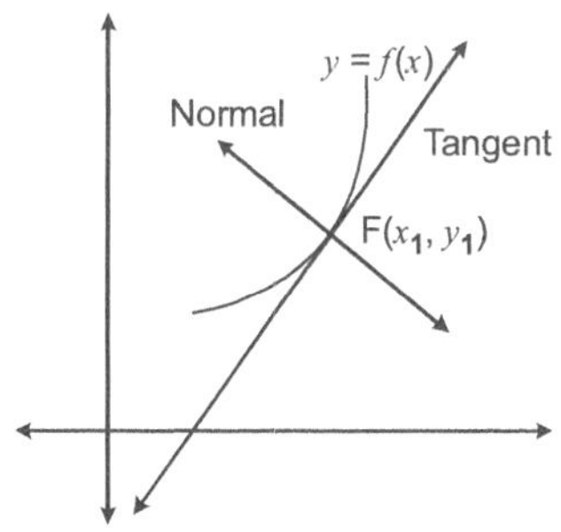

6. If θ is the angle between two given curves $y=f_1(x)$ and $y=f_2(x)$ at point (x_1, y_1) then θ is given by

$$\tan\theta = \left(\frac{m_1-m_2}{1+m_1m_2}\right)$$

where (x_1, y_1) is the point of intersection of two curves and m_1, m_2 are slopes of their tangents at point (x_1, y_1).

7. **Increasing function :** (i) A function $f(x)$ is an increasing function in an interval (a, b), if

$$x_1 < x_2 \Rightarrow f(x_1)\le f(x_2)$$

for all $x_1, x_2\in(a, b)$.

(ii) A function $f(x)$ is said to be strictly increasing in (a, b) when if

$$x_1 < x_2 \Rightarrow f(x_1) < f(x_2)$$

for all $x_1, x_2 \in (a, b)$.

8. **Decreasing function :** (i) A function $f(x)$ is said to be decreasing function in an interval (a, b),

It $\qquad x_1 < x_2 \Rightarrow f(x_1) \geq f(x_2)$

for all $x_1, x_2 \in (a, b)$.

(ii) A function $f(x)$ is said to be strictly decreasing in (a, b) when

$$x_1 < x_2 \Rightarrow f(x_1) > f(x_2)$$

for all $x_1, x_2 \in (a, b)$.

9. A function is said to be **monotonic** in an interval if it is either increasing or decreasing in the given interval.

10. A function $f(x)$ is said to attain a **local maxima** at $x = a$, if there exists a neighbourhood $(a - \delta, a + \delta)$ of a such that

$$f(x) < f(a) \text{ for all } x \in (a - \delta, a + \delta), x \neq a$$

then $f(a)$ is the local maximum value of $f(x)$ at $x = a$.

11. A function $f(x)$ is said to attain a **local minima** at $x = a$ if there exists a neighbourhood $(a - \delta, a + \delta)$ of a such that $f(x) > f(a)$ for all $x \in (a - \delta, a + \delta), x \neq a$ then $f(a)$ is called the local minimum value of $f(x)$ at $x = a$.

12. Local maximum and minimum values of a function are its respective largest and smallest values of the given function over given interval.

13. A point $x = c$ is called **critical point** of the function $f(x)$, if $f(c)$ exists and either $f'(c) = 0$ or $f'(c) = \infty$.

14. **First derivative test :**

(i) Given $y = f(x)$

(ii) Find $f'(x)$

(iii) Now find critical points $x = c_1, c_2, \ldots, c_n$ by putting $f'(x) = 0$.

(iv) Now find sign of $f''(x)$ at critical points : consider $x = c_1$.

If $\frac{dy}{dx}$ changes its sign from positive to negative as x increases through c_1, then the function attains its local maximum at $x = c_1$.

If $\frac{dy}{dx}$ changes its sign from negative to positive as x increases through c_1, then the function attains a local minimum at $x = c_1$.

If $\frac{dy}{dx}$ does not change sign as x increases through c_1, then $x = c_1$ is neither a point of local maximum nor a point of local minimum. In this case $x = c_1$ is a point of inflection.

15. **Higher order derivative test.**

Let f be a differentiable function defined on an interval I also let c be an interior point of I such that :

(a) $f'(c) = f''(c) = f'''(c) = \ldots = f^{n-1}(c) = 0$

(b) $f^n(c)$ exists and is non-zero.

Then, if n is even and $f^n(c) > 0 \Rightarrow x = c$ is a point of local minimum.

If n is even and $f^n(c) < 0 \Rightarrow x = c$ is a point of local maximum.

If n is odd, $x = c$ is neither a point of local maximum nor a point of local minimum.

16. If $f(x)$ is a function and $x = c$ is critical point then $x = c$ is called **point of inflection** if

(i) $f'(c) = 0$

(ii) $f''(c) = 0$

(iii) $f'''(c) \neq 0$.

Chapter 7. Integrals

1. Integration is the inverse process of differentiation. If we are given the derivative of a function and asked to find the original function, such a process is called **Integration.**

2. Some important formulae of Integration :

(i) $\int [f(x) + g(x)]\, dx = \int f(x)\, dx + \int g(x)\, dx$

(ii) $\int Af(x)\, dx = A\int f(x)\, dx$, where A is any real number

(iii) $\int x^n\, dx = \frac{x^{n+1}}{n+1} + c \left[\int x^n\, dx = \frac{x^{n+1}}{n+1} + c\right]$ is valid only for those algebraic expressions in which $n \neq -1$ and base is always linear

Where c is the constant of integration.

(iv) $\int 1\, dx = x + c$

(v) $\int e^x\, dx = e^x + c$

(vi) $\int \frac{1}{x}\, dx = \log x + c$

(vii) $\int a^x\, dx = \frac{a^x}{\log a} + c$

(viii) $\int \cos x\, dx = \sin x + c$

(ix) $\int \sin x\, dx = -\cos x + c$

(x) $\int \tan x\, dx = \log|\sec x| + c$ or $-\log|\cos x| + c$

(xi) $\int \cot x\, dx = \log|\sin x| + c$ or $-\log|\text{cosec}\, x| + c$

(xii) $\int \sec x\, dx = \log|\sec x + \tan x| + c$

(xiii) $\int \text{cosec}\, x\, dx = \log|\text{cosec}\, x - \cot x| + c$

(xiv) $\int \sec x \tan x\, dx = \sec x + c$

(xv) $\int \text{cosec}\, x \cot x\, dx = -\text{cosec}\, x + c$

(xvi) $\int \sec^2 x\, dx = \tan x + c$

(xvii) $\int \text{cosec}^2 x\, dx = -\cot x + c$

(xviii) $\int \frac{1}{\sqrt{x^2 - a^2}}\, dx = \log\left|x + \sqrt{x^2 - a^2}\right| + c$

(xix) $\int \frac{1}{\sqrt{a^2 - x^2}}\, dx = \sin^{-1} \frac{x}{a} + c$

(xx) $\int \frac{1}{\sqrt{x^2 + a^2}}\, dx = \log\left|x + \sqrt{x^2 + a^2}\right| + c$

(xxi) $\int \frac{1}{x^2 + a^2}\, dx = \frac{1}{a} \tan^{-1} \frac{x}{a} + c$

(xxii) $\int \frac{1}{x\sqrt{x^2 - a^2}}\, dx = \frac{1}{a} \sec^{-1} \frac{x}{a} + c$

Note : While using formulae from (xviii) to (xxii), coefficient of x^2 must be unity.

3. Some important methods of integration :

(i) Integration by decomposition :

(a) $\int (ax^2 + bx + c)\, dx = \int ax^2\, dx + \int bx\, dx + \int c\, dx$

$$= a\int x^2\, dx + b\int x\, dx + c\int 1\, dx$$

$$= \frac{ax^3}{3} + \frac{bx^2}{2} + cx + d$$

(b) $\int \frac{1}{\sin^2 x \cos^2 x}\, dx = \int \frac{\sin^2 x + \cos^2 x}{\sin^2 x \cos^2 x} dx$

$$= \int \frac{\sin^2 x}{\sin^2 x \cos^2 x}\, dx + \int \frac{\cos^2 x}{\sin^2 x \cos^2 x}\, dx$$

$$= \int \frac{1}{\cos^2 x}\, dx + \int \frac{1}{\sin^2 x}\, dx$$

$$= \int \sec^2 x\, dx + \int \text{cosec}^2 x\, dx$$

$$= \tan x - \cot x + c$$

(ii) Integration by substitution : In this method we change the variable by substituting in place of the given variable.

Example : $\int \tan x\, dx = \int \frac{\sin x}{\cos x}\, dx$

Put $\cos x = t$

Differentiating w.r.t. x

$$-\sin x = \frac{dt}{dx}$$

$$\sin x\, dx = -dt$$

$$\therefore \quad \int \tan x\, dx = -\int \frac{dt}{t}$$

$$= -\log|t| + c$$

$$= -\log|\cos x| + c$$

$$= -\log\left|\frac{1}{\sec x}\right| + c$$

$$= -\{\log 1 - \log|\sec x|\} + c$$

$$\int \tan x\, dx = \log|\sec x| + c$$

(iii) Integration by parts : This method is used when the given function is the product of two different functions as follows :

$$\int \underset{\text{I}}{f(x)} \cdot \underset{\text{II}}{g(x)}\, dx = f(x)\left[\int g(x)\, dx\right] - \int\left[\frac{d}{dx}\{f(x)\} \cdot \int g(x)\, dx\right]$$

where I and II functions are taken according to

ILATE

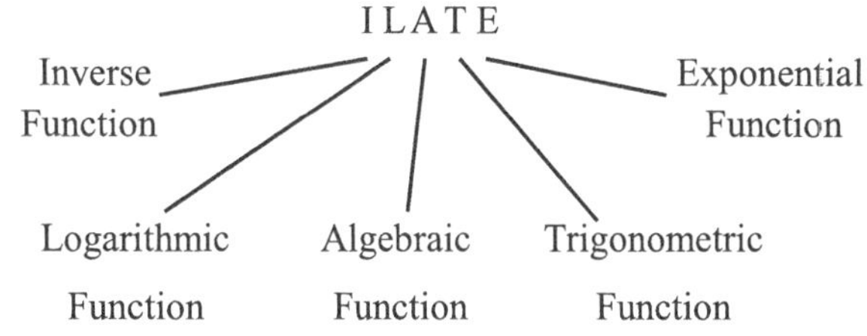

(iv) Integration by partial fraction : In this method the proper fraction can be expressed as the sum of partial functions. If the degree of numerator is greater than the degree of denominator then using long division method the given fraction can be expressed as proper fraction.

Example :

$$\frac{3x + 1}{(x-2)^2 (x+2)}\, dx = \frac{A}{x-2} + \frac{B}{(x-2)^2} + \frac{C}{x+2}$$

then by equating the coefficients find the value of A, B and C.

(v) Some important properties of definite integral :

(a) $\int_a^b f(x)\,dx = -\int_b^a f(x)\,dx$ or $\int_a^b f(t)\,dt$

(b) $\int_a^a f(x)\,dx = 0$

(c) $\int_a^b f(x)\,dx = \int_a^c f(x)\,dx + \int_c^b f(x)\,dx$

(d) $\int_a^b f(x)\,dx = \int_a^b f(a+b-x)\,dx$

(e) $\int_0^{2a} f(x)\,dx = \int_0^a f(x)\,dx + \int_0^a f(2a-x)\,dx$

(f) $\int_0^{2a} f(x)\,dx = \begin{cases} 2\int_0^a f(x)\,dx & \text{if } f(x) \text{ is even} \\ 0 & \text{if } f(x) \text{ is odd} \end{cases}$

$f(x)$ is even if $f(-x) = f(x)$

and $f(x)$ is odd if $f(-x) = -f(x)$

(g) $\int_{-a}^a f(x)\,dx$

$= \begin{cases} 0 & \text{if } f(x) \text{ is an odd function} \\ 2\int_0^a f(x)\,dx & \text{if } f(x) \text{ is an even function} \end{cases}$

(vi) Definite integral as the limit of sum :

$$I = \lim_{h\to 0} h\,[f(a) + f(a+h) + f(a+2h) + \ldots + f\{a+(n-1)h\}]$$

where $h = \dfrac{b-a}{n}$ or $b - a = nh$

4. Some important forms of integration :

(i) $\int \dfrac{f'(x)}{f(x)}\,dx = \log|f(x)| + c$

(ii) $\int [f(x)]^n \times f'(x)\,dx = \dfrac{[f(x)]^{n+1}}{n+1} + c$

(iii) $\int e^x [f(x) + f'(x)]\,dx = e^x f(x) + c$

(iv) $\int \sqrt{a^2 - x^2}\,dx = \dfrac{x}{2}\sqrt{a^2 - x^2} + \dfrac{a^2}{2}\sin^{-1}\dfrac{x}{a} + c$

(v) $\int \sqrt{x^2 + a^2}\,dx = \dfrac{x}{2}\sqrt{x^2 + a^2} + \dfrac{a^2}{2}\log|x + \sqrt{x^2 + a^2}| + c$

(vi) $\int \sqrt{x^2 - a^2}\,dx = \dfrac{x}{2}\sqrt{x^2 - a^2} - \dfrac{a^2}{2}\log|x + \sqrt{x^2 - a^2}| + c$

Chapter 8. Applications of the Integrals

By using the concepts of integration we will find the area bounded by curves.

Area under simple curves :

(i) The area enclosed by the curve $y = f(x)$ and the ordinates $x = a$, $x = b$ and X-axis is given by

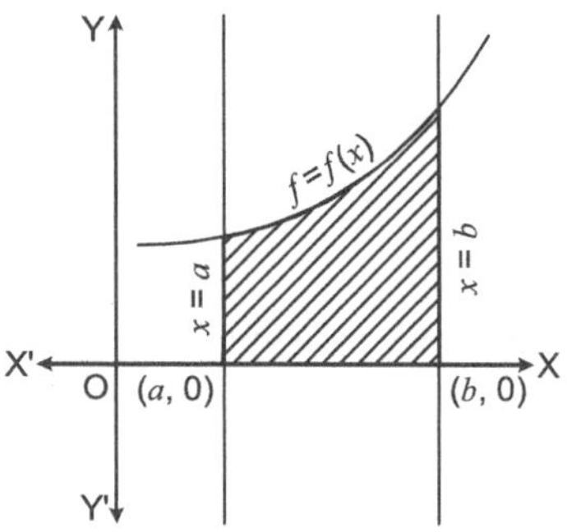

$$\text{Area} = \int_a^b y\,dx = \int_a^b f(x)\,dx$$

(ii) The area enclosed by the curve $x = f(y)$, Y-axis and abscissas $y = a$ and $y = b$ is given by

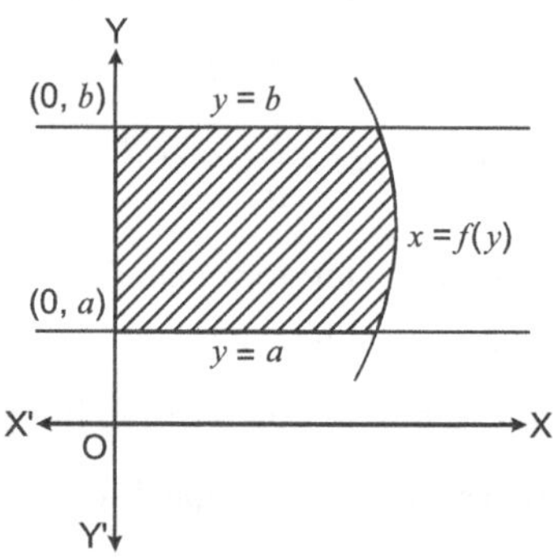

$$A = \int_a^b x\,dy = \int_a^b f(y)\,dy$$

(iii) When the curve $y = f(x)$ lies below the X-axis, then $y = f(x)$ is negative so the area under the curve is also negative so we take the modulus of the area *i.e.*, area will be given by

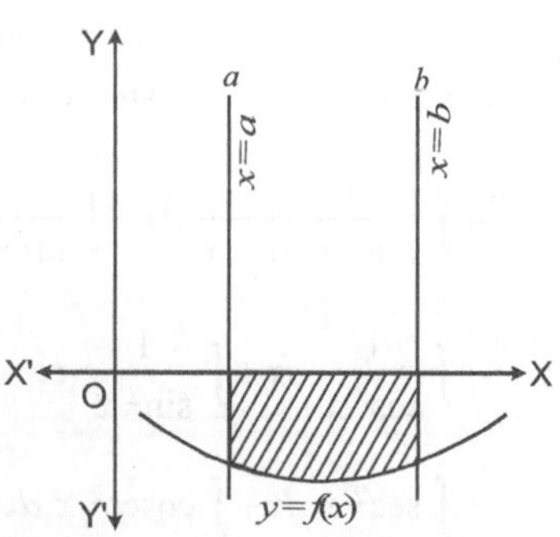

$$A = \left|\int_a^b f(x)\,dx\right|$$

(iv) To find the area of a circle $x^2 + y^2 = a^2$

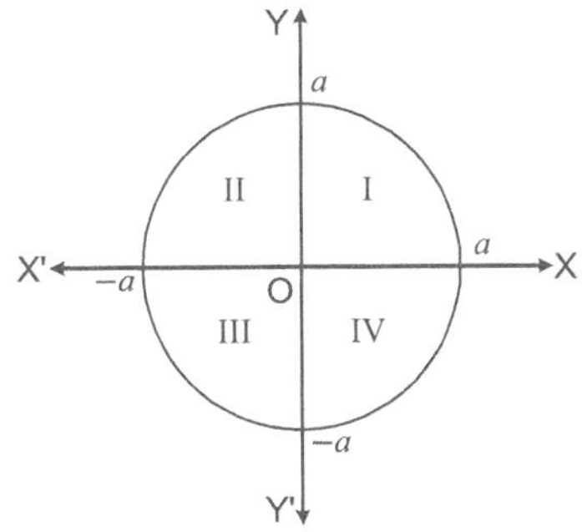

Area of the circle is divided into four equal parts I, II, III and IV

$$A = \int_a^b y\, dx$$

$$= \int_a^b \sqrt{a^2 - x^2}\, dx$$

(v) Area bounded by two curves : Let the given curves be $y = f(x)$ and $y = g(x)$. After drawing the sketch of these curves, find the point of intersection of the curves. Let (p, q) and (r, s) are points of intersection of the curves then required area will be

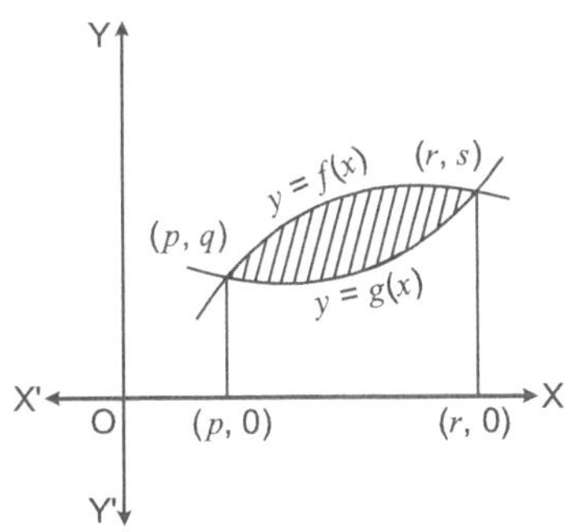

$$A = \int_p^r f(x)\, dx - \int_p^r g(x)\, dx$$

Similarly

$$A = \int_q^s f(y)\, dy - \int_q^s g(y)\, dy$$

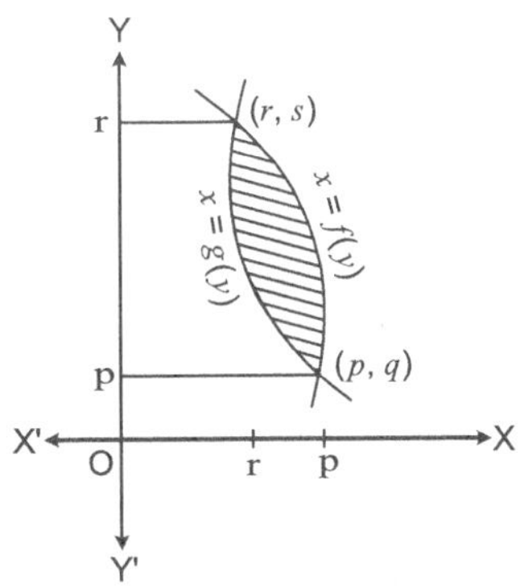

Example : To find the area of two curves one of them is circle $x^2 + y^2 = 16$ and other is parabola $x^2 = 6y$.

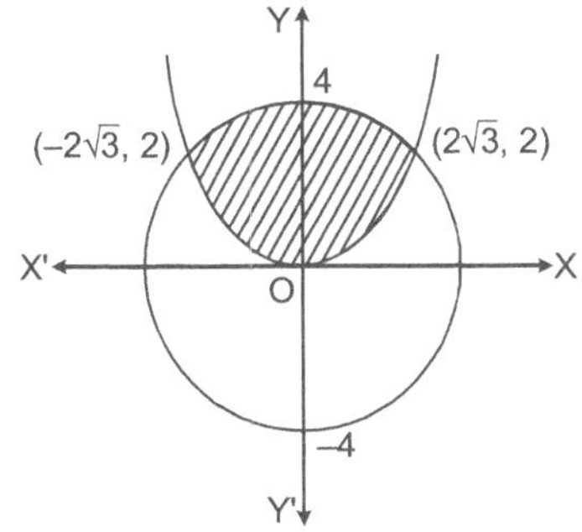

By solving these two equations of curves simultaneously, we get the points of intersections $(-2\sqrt{3}, 2)$ and $(2\sqrt{3}, 2)$

$$A = \int_{-2\sqrt{3}}^{2\sqrt{3}} y \text{ (circle) } dx - \int_{-2\sqrt{3}}^{2\sqrt{3}} y \text{ (parabola) } dx$$

OR

$$A = \int_0^2 x \text{ (parabola) } dy + \int_2^4 x \text{ (circle) } dy$$

Chapter 9. Differential Equations

1. **Differential equation** is an equation which involves unknown functions and their derivatives with respect to one or more independent variables.
2. **Ordinary differential equation** is an equation which involves only one independent variable and its derivatives with respect to that independent variable.
3. The **order** of the differential equation is the order of the highest order derivative, which occurs in the differential equation.
4. The **degree** of a differential equation is the index of the highest power of the highest order derivative when it has been made free from negative and fractional numbers, so as derivatives are concerned and it does not contain logarithmic, exponential and trigonometric functions of derivatives.

   For example :

   $$\frac{d^2y}{dx^2} = x^2 \sin^2 x + x^3 + y$$

   is ordinary differential equation whose order is 2 and degree is 1.
5. **Formation of differential equation :** To obtain differential equation which has independent variable x (say), dependent variable y (say), and n (say) arbitrary constant.

   (i) Differentiate the given equation n times with respect to x.

   (ii) Now, eliminate n arbitrary constant with the help of n obtained equations.

(iii) Now, the obtained equation is the desired differential equation.

6. **Solution of differential equation :** Solution of differential equation is a relation between the variables which satisfy the given differential equation.

7. The most general solution of an ordinary differential equation of first order contains one arbitrary constant, of second order contains two arbitrary constants and so on.

8. Particular solution of differential equation satisfies the differential equation but contains no arbitrary constant.

9. **Solving a differential equation with variable separable :** Equation whose variables are separable can be put into the given form :

$$f(x)\,dx + f(y)\,dy = 0$$

The general solution of this equation is obtained by integrating above equation.

$$\int f(x)\,dx + \int f(y)\,dy = c$$

where c is an arbitrary constant.

10. **Homogeneous equation** is the differential equation of the form

$$\frac{dy}{dx} = \frac{f(x, y)}{g(x, y)}.$$

Where $f(x, y)$ and $g(x, y)$ are homogeneous functions of same degree in x and y, *i.e.*, equation of the form $\frac{dy}{dx} = f\left(\frac{y}{x}\right)$, is called homogeneous differential equation.

11. Let $\frac{dy}{dx} = f\left(\frac{y}{x}\right)$ be the given homogeneous differential equation, to solve this we substitute

$$y = vx \Rightarrow \frac{dy}{dx} = v + x\frac{dv}{dx}$$

and proceed by substitution for $\frac{dy}{dx}$ and y in given equation.

12. **Linear differential equation** is the differential equation of the form

$$\frac{dy}{dx} + Py = Q \text{ or } \frac{dx}{dy} + P_1 x = Q_1$$

To solve linear differential equation we multiply both sides of the given differential equation by integrating factor of the given differential equation.

where $e^{\int P dx}$ is the integrating factor of equation $\frac{dy}{dx} + Py = Q$ and $e^{\int P_1 dx}$ is the integrating factor of

$$\frac{dy}{dx} + P_1 x = Q_1.$$

Chapter 10. Vectors

1. A **scalar quantity** has only magnitude but no direction, *e.g.*, mass, volume, temperature and real number. By a scalar, we always mean a real number.

2. A **vector quantity** has magnitude as well as direction, *e.g.*, force, velocity, displacement.

3. Given a line, there are two directions associated with it. Any line with one of these directions is called a **directed line**. Any segment of a directed line is called a **'directed line segment.'** Thus, a directed line segment has a magnitude and a direction and therefore is a vector.

4. The directed line, of which it is a part, is called its **support**.

5. Graphically, we represent vectors by directed line segments. A directed line segment $\vec{OP}$ has direction from O to P. P is called its **terminal point** (or terminus) and O is called its **initial point.**

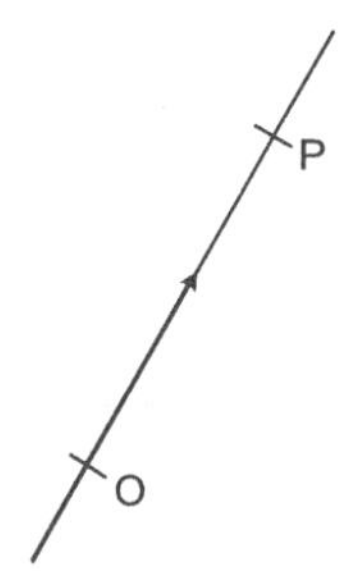

Analytically, a vector is represented by a letter with an arrow on top of it, like : $\vec{a}, \vec{b}, \vec{c}$ etc.

Note : Any vector has (i) magnitude, (ii) direction and (iii) support. The magnitude of a vector is a scalar associated with it. More precisely, we have :

The modulus or magnitude of a vector is the positive number which is the measure of its length. For vector $|\vec{A}|$, its magnitude is denoted by $|\vec{A}|$.

6. Let P be any point in space having coordinates (x, y, z) with respect to origin $O(0, 0, 0)$. Then the vector $\vec{OP}$ with O as its initial and P as its terminal point is called the **position vector** of the point P with respect to O.

$$|\vec{OP}| = \sqrt{x^2 + y^2 + z^2}$$

7. A vector whose initial and terminal points coincide is called a **zero vector**. Clearly, the modulus or magnitude of the zero vector is the real number 0, but its direction is indeterminate. Zero vector is also called the null vector.

8. Vectors having the same initial point are called **coinitial vectors** (see figure).

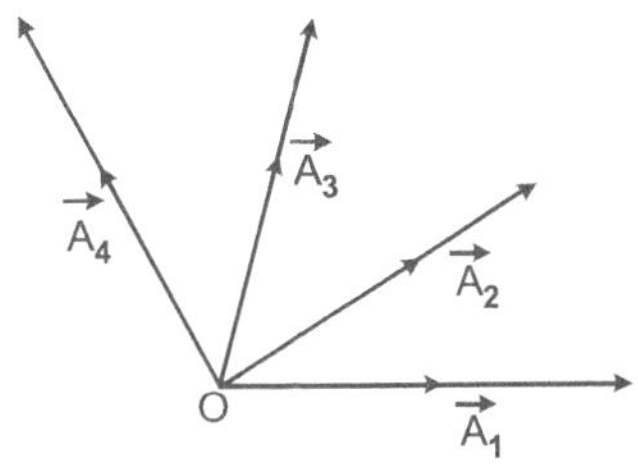

9. Vectors having the same terminal point are called **coterminous vectors** (see figure).

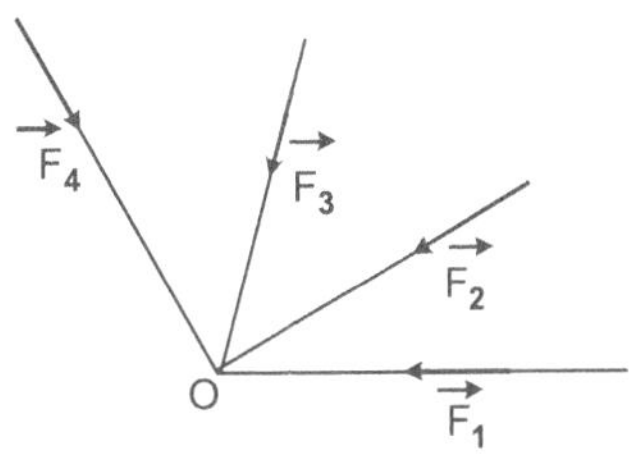

10. A vector whose magnitude is unity is called **unit vector**. The unit vector in the direction of given vector $\vec{a}$ is denoted by $\hat{a}$

11. Two vectors $\vec{a}$ and $\vec{b}$ are called **equal vectors** if they have same magnitude and direction irrespective of positions of their initial points.

12. A vector whose magnitude is the same as that of a given vector but direction opposite to that of it, is called **negative** of the given vector.

13. We can consider vectors to lie in 2- or 3-dimensional space, as required by the application. For example, 2-dimensional vectors are considered to be free vectors but with an extra condition imposed - their initial and terminal points lie in the *XY* plane. Or, if the application is such, we may consider vectors to be lying in a 3-dimensional space. If the context does not make the dimension of the underlying space clear, you should assume that we are in 3-dimensional space.

14. Vector Addition : Vector addition is defined as follows.

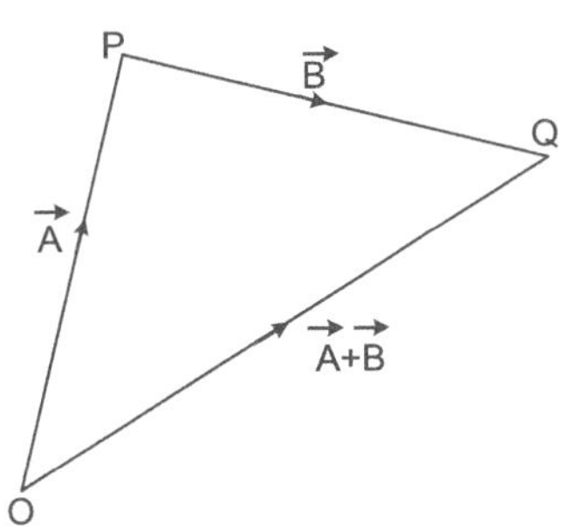

Let, $\vec{A}, \vec{B}$ be two given vectors.

Take any point O as the origin and let $\vec{OP} = \vec{A}, \vec{PQ} = \vec{B}$, so that the initial point of $\vec{B}$ is the terminal point of $\vec{A}$.

Then, the vector $\vec{OQ}$ is called the sum of the vectors $\vec{A}$ and $\vec{B}$ and we write $\vec{OQ} = \vec{OP} + \vec{PQ} = \vec{A} + \vec{B}$.

Note :

(i) This law is called the **triangle law of addition** : If any two vectors are represented completely by two sides of a triangle taken in the same order, then the resultant of the two vectors is represented both in magnitude and direction by the third side of the triangle taken in the opposite order.

(ii) This sum does not depend upon the choice of *O*, the origin.

(iii) The triangle law of addition of two vectors is equivalent to the following parallelogram law of addition :

$$\vec{OC} = \vec{a} + \vec{b}$$

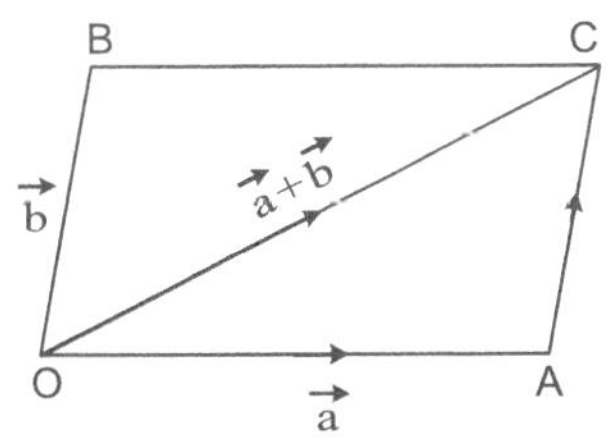

15. Laws of Vector Addition : The laws of addition of vectors are given by theorems 1 and 2 below.

Theorem 1 : If $\vec{a}, \vec{b}, \vec{c}$ are any three vectors, then the following holds:

(i) $\vec{a} + \vec{b} = \vec{b} + \vec{a}$, *i.e.*, vector addition is commutative.

(ii) $\vec{a} + \left(\vec{b} + \vec{c}\right) = \left(\vec{a} + \vec{b}\right) + \vec{c}$, *i.e.*, vector addition is associative.

Theorem 2 : For any vector $\vec{a}$,

(i) $\vec{a} + \vec{O} = \vec{O} + \vec{a} = \vec{a}$, where $\vec{O}$ denotes the zero vector and is called the additive identity.

(ii) $\vec{a} + \left(-\vec{a}\right) = \left(-\vec{a}\right) + \vec{a} = \vec{O}$ where $(-\vec{a})$ is called the additive inverse.

16. Vector Subtraction : The difference of vectors $\vec{a}$ and $\vec{b}$ denoted by $\vec{a} - \vec{b}$ is a vector, which when added to $\vec{b}$ gives the vector $\vec{a}$.

Graphically,

$$\vec{OP} = \vec{a},\ \vec{PQ} = \vec{b}, \vec{PQ'} = -\vec{b}$$

$$\vec{OQ'} = \vec{OP} + \vec{PQ'} = \vec{a} - \vec{b}$$

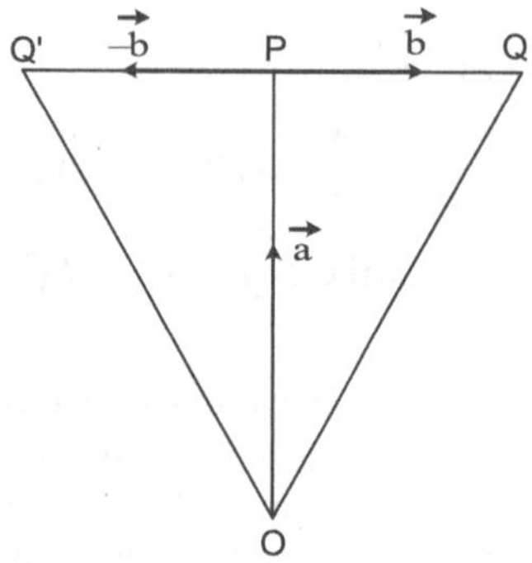

17. Scalar Multiplication of a Vector: The multiplication of a vector $\vec{a}$ by a scalar m is a vector $m\vec{a}$, whose

(i) magnitude is $|m|$ times that of the magnitude of $\vec{a}$

(ii) direction is the same as that of $\vec{a}$ if $m > 0$, and opposite to that of $\vec{a}$ if $m < 0$.

For example, $5\vec{a}$ is a vector whose magnitude is 5 times that of $\vec{a}$ and has the same direction as that of $\vec{a}$

18. Laws of Scalar Multiplication :

Theorem : Let $\vec{a}$ be any vector and m, n, any scalars. Then, the following holds :

(i) $\vec{OA} = \vec{O}$

(ii) $(-1)\vec{a} = -\vec{a}$

(iii) $m(-\vec{a}) = (-m)\vec{a} = -(m\vec{a})$

(iv) $m(n)\vec{a} = (mn)\vec{a}$

(v) $(m+n)\vec{a} = m\vec{a} + n\vec{a}$

19. $\vec{a}$ is parallel to $\vec{b}$ if $\vec{a} = \lambda\vec{b}$ where λ is a scalar.

20. Components of a Vector in Three-Dimensions :

Let OX, OY and OZ be three mutually perpendicular lines taken as the coordinate axes and let $P(x, y, z)$ be any point.

Through P, draw planes parallel to the YZ-plane, ZX-plane and XY-plane to meet the X-axis, Y-axis and Z-axis in L, M and N respectively.

If $\hat{i}, \hat{j}, \hat{k}$ are the unit vectors along OX, OY and OZ respectively, then $\vec{OL} = x\hat{i}$ and $\vec{OM} = y\hat{j}$ and $\vec{ON} = z\hat{k}$ are the components of $\vec{OP}$ along OX, OY and OZ respectively.

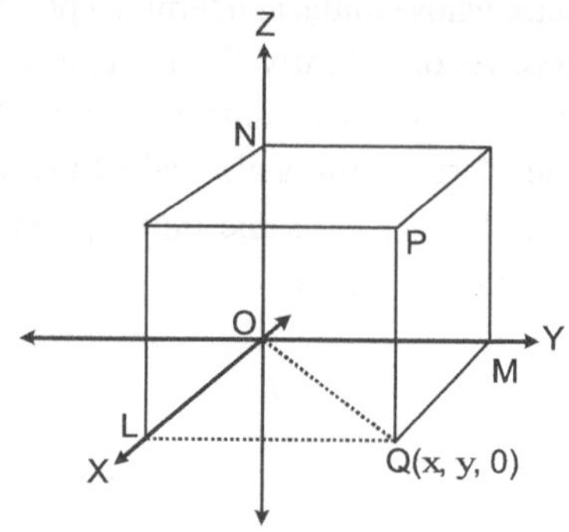

The magnitude of $\vec{OP}$ is

$$OP = \sqrt{x^2 + y^2 + z^2}$$

(using length of diagonal of cuboid)

Similarly, the components of $\vec{PQ}$, which is a join of $P(x_1, y_1, z_1)$ and $Q(x_2, y_2, z_2)$ are

$$(x_2 - x_1)\,\hat{i},\ (y_2 - y_1)\,\hat{j},\ (z_2 - z_1)\,\hat{k}$$

21. Section Formula for Internal Division : If P and Q be two points with position vectors $\vec{a}$ and $\vec{b}$ respectively and let R be a point dividing PQ internally in the ratio $m : n$. Then, $\vec{OR}$, the position vector of R, is given by $\vec{OR} = \dfrac{m\vec{b} + n\vec{a}}{m+n}$, where O is the origin.

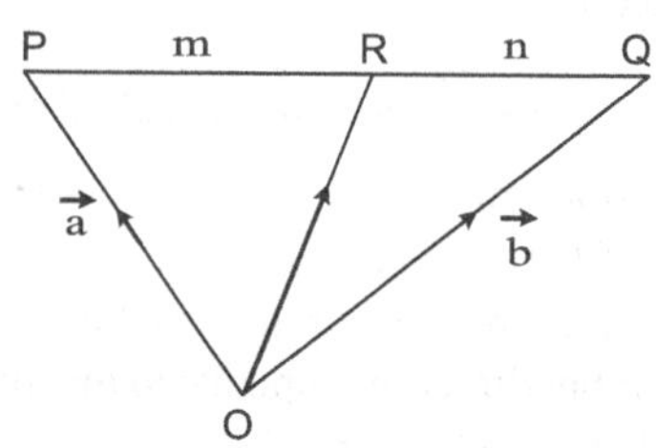

22. Section Formula for External Division : If $\vec{a}, \vec{b}$ be the position vectors of the points P and Q respectively and let R divide their join PQ in the ratio $m : n\ (m \neq n)$ externally. Then, the position vector of R is

$$\frac{m\vec{b} - n\vec{a}}{m-n}$$

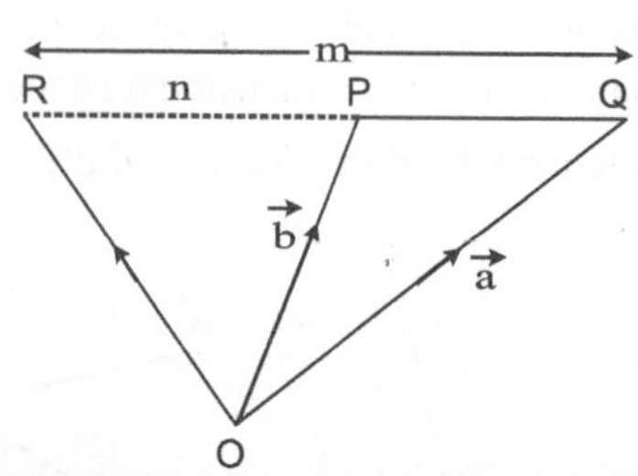

23. Scalar Product : For real numbers, multiplication is defined in a unique way, but for vectors, there are two ways to multiply — scalar (or dot) product and vector (or cross) product.

As is evident from the name, the former is a scalar quantity and the latter is a vector quantity. The scalar product is defined as follows :

Definition : The scalar product $\vec{a} \cdot \vec{b}$ of two vectors $\vec{a}$ and $\vec{b}$ is defined as $\vec{a} \cdot \vec{b} = ab \cos\theta$, where θ is the angle between $\vec{a}$ and $\vec{b}$.

This product is also called the dot product because of the notation involved in the multiplication process.

For example, if $\hat{i}$ and $\hat{j}$ are unit vectors along OX and OY in the Cartesian coordinate system, then,

$$\hat{i} \cdot \hat{j} = 1 \times 1 \times \cos 90° = 1 \times 0 = 0$$

similarly, $\hat{j} \cdot \hat{k} = \hat{k} \cdot \hat{i} = 0$

Note : From the definition of a scalar product, we make a few observations :

(i) Two vectors $\vec{a}$ and $\vec{b}$ are perpendicular if the angle θ between them is 90°.

In this case, $\vec{a} \cdot \vec{b} = 0$

Thus, $\vec{a}$ and $\vec{b}$ are perpendicular, if $\vec{a} \cdot \vec{b} = 0$

(ii) For like vectors $\vec{a}$ and $\vec{b}$, $\vec{a} \cdot \vec{b} = |\vec{a}||\vec{b}|$

For unlike vectors $\vec{a}$ and $\vec{b}$ $\cdot$ $\vec{a} \cdot \vec{b} = -|\vec{a}||\vec{b}|$ since the respective angles are 0° and π radians.

(iii) For any vector $\vec{a}$, $\vec{a} \cdot \vec{a} = |\vec{a}|^2$

(iv) If $\vec{a}$ and $\vec{b}$ are any vectors, then the angle θ between them is

$$\cos\theta = \frac{\vec{a} \cdot \vec{b}}{|\vec{a}||\vec{b}|} = \frac{\vec{a}}{|\vec{a}|} = \frac{\vec{b}}{|\vec{b}|}\, \hat{a} \cdot \hat{b}.$$

(v) If $\vec{a} = a_1\hat{i} + a_2\hat{j} + a_3\hat{k}$, $\vec{b} = b_1\hat{i} + b_2\hat{j} + b_3\hat{k}$, then $\vec{a} \cdot \vec{b} = a_1 b_1 + a_2 b_2 + a_3 b_3$.

(vi) Scalar product of two vectors is a real number.

(vii) The scalar product of vectors has the following properties :

Theorem : If $\vec{a}$, $\vec{b}$ and $\vec{c}$ are three vectors and n is a scalar, then

(i) $\vec{a} \cdot \vec{b} = \vec{b} \cdot \vec{a}$ (scalar product of vectors is commutative)

(ii) $\vec{a} \cdot \left(n\vec{b}\right) = \left(n\vec{a}\right) \cdot \vec{b} = n\left(\vec{a} \cdot \vec{b}\right)$

(iii) $\vec{a} \cdot \left(\vec{b} + \vec{c}\right) = \vec{a} \cdot \vec{b} + \vec{a} \cdot \vec{c}$

(scalar product distributes itself over the addition of vectors.)

24. Vector Product of Two Vectors : The vector product of two vectors is a vector, and it is defined as follows :

Definition : If $\vec{a}$ and $\vec{b}$ are two vectors, then their vector product (or cross product) $\vec{a} \times \vec{b}$ (taken in this order) is a vector $\vec{c}$ whose magnitude is $ab \sin\theta$, where θ is the angle between $\vec{a}$ and $\vec{b}$ and has a direction perpendicular to the plane of the vectors $\vec{a}$ and $\vec{b}$ such that $\vec{a}$, $\vec{b}$ and $\vec{c}$ form a right-handed system.

If $\hat{n}$ is the unit vector in the direction perpendicular to $\vec{a}$ and $\vec{b}$ so that $\vec{a}$, $\vec{b}$, $\hat{n}$ form a right-handed system, then,

$$\vec{c} = \vec{a} \times \vec{b} = |\vec{a}||\vec{b}| \sin\theta\, \hat{n},\ 0 < \theta < 2\pi.$$

If $\vec{a}$ and $\vec{b}$ are like vectors, then obviously, $\theta = 0°$ and if $\vec{a}$, $\vec{b}$ are unlike vectors, then $\theta = \pi$ in that case, $\vec{a} \times \vec{b} = 0$.

Same is the case, when $\vec{a}$ and $\vec{b}$ are collinear.

Therefore, if $\vec{a}$, $\vec{b}$ are collinear, then $\vec{a} \times \vec{b} = 0$

Note, that $\vec{a} \times \vec{b}$ can vanish in some other cases too.

Geometrically, $|\vec{a} \times \vec{b}|$ represents the area of a parallelogram with sides $\vec{a}$ and $\vec{b}$.

Let h be the height of the parallelogram with sides $\vec{a}$ and $\vec{b}$.

Then the area of the parallelogram = hA.

But $h = |\vec{b}| \sin\theta$ is the angle between $\vec{a}$ and $\vec{b}$

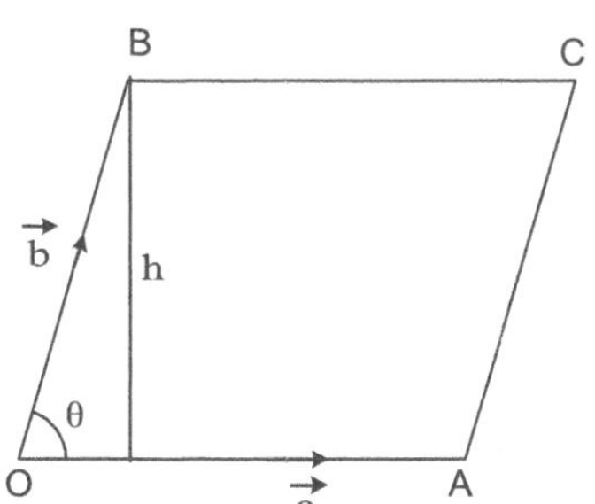

Therefore, the area of the parallelogram

$$= b(\sin\theta)\, a = ab \sin\theta = |\vec{a} \times \vec{b}|$$

25. We have the following relationship between scalar and vector for any two vectors $\vec{a}$ and $\vec{b}$

$$|\vec{a} \times \vec{b}|^2 + \left(\vec{a} \cdot \vec{b}\right)^2 = a^2b^2$$

26. Properties of Vector Product

The properties are given by the following theorems :

Theorem 1 : $\vec{a}$ and $\vec{b}$ are any two vectors and α is a scalar, then the following hold :

(i) $\vec{a}\times\vec{b} = -\vec{b}\times\vec{a}$, *i.e.,* vector product is not commutative.

(ii) $\alpha\left(\vec{a}\times\vec{b}\right)=\left(\alpha\vec{a}\right)\times\vec{b}=\vec{a}\times\left(\alpha\vec{b}\right)$

(iii) $\vec{a}\times\vec{a}=0$

Theorem 2 : Let $\vec{a},\vec{b},\vec{c}$ be any three vectors.

Then $\vec{a}\times\left(\vec{b}+\vec{c}\right)=\vec{a}\times\vec{b}+\vec{a}\times\vec{c}$

Corollary : If $\vec{a}=a_1\hat{i}+a_2\hat{j}+a_3\hat{k}$ and $b=b_1\hat{i}+b_2\hat{j}+b_3\hat{k}$, then,

$$\vec{a}\times\vec{b}=\begin{vmatrix}\hat{i} & \hat{j} & \hat{k}\\ a_1 & a_2 & a_3\\ b_1 & b_2 & b_3\end{vmatrix}$$

27. Two non-zero vectors $\vec{a}$ and $\vec{b}$ are parallel if $\vec{a}\times\vec{b}=0$.

28. Area of triangle whose sides are $\vec{a}$ and $\vec{b}$ is

$$\frac{1}{2}|\vec{a}\times\vec{b}|$$

29. If $\hat{i},\hat{j},\hat{k}$ are vectors along *X*-, *Y*- and *Z*-axes then $\hat{i}\times\hat{i}=0,\ \hat{j}\times\hat{j}=0,\ \hat{k}\times\hat{k}=0,\ \hat{i}\times\hat{j}=\hat{k},\ \hat{j}\times\hat{k}=\hat{i},\ \hat{k}\times\hat{i}=\hat{j}$

30. Unit vector $\hat{n}$ perpendicular to $\vec{a}$ and $\vec{b}$ is

$$\hat{n}=\frac{\vec{a}\times\vec{b}}{|\vec{a}\times\vec{b}|}$$

31. For vectors $\vec{a},\vec{b},\vec{c}$ scalar triple product is given by $\vec{a}\cdot(\vec{b}\times\vec{c})$ or $(\vec{a}\times\vec{b})\cdot\vec{c}$ denoted by $[a\,b\,c]$.

32. For vectors $\vec{a}\cdot\vec{b}=0\Leftrightarrow\vec{a}=0,\ \vec{b}=0$ or $\vec{a}\perp\vec{b}$.

33. Projection of $\vec{a}$ along $\vec{b}$ is $\dfrac{\vec{a}\cdot\vec{b}}{|\vec{b}|}$ and projection vector of $\vec{a}$ along $\vec{b}$ is $\left(\dfrac{\vec{a}\cdot\vec{b}}{|\vec{b}|}\right)\vec{b}$.

34. If $\hat{i},\hat{j},\hat{k}$ are unit vectors along *X*-, *Y*- and *Z*-axes then $\hat{i}^2=\hat{j}^2=\hat{k}^2=1,\ \hat{i}\cdot\hat{j}=0,\ \hat{j}\cdot\hat{k}=0,\ \hat{k}\cdot\hat{i}=0$.

Chapter 11. Three-Dimensional Geometry

1. The angles α, β and γ which a directed line *OA* makes with the positive direction of *X*-, *Y*- and *Z*-axes respectively are called the **direction angles** of the line *OA*.

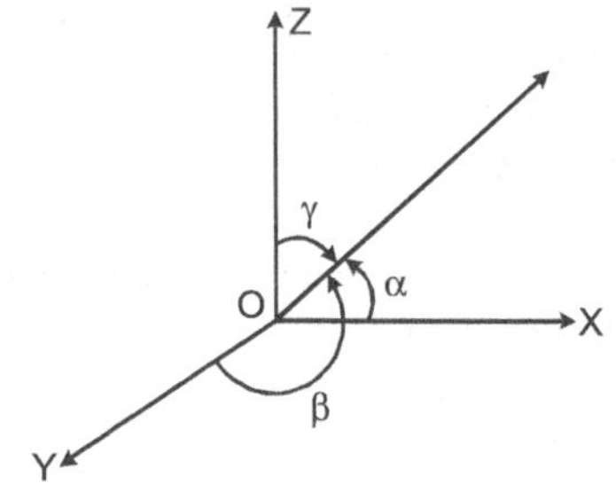

2. Cosines of the angles made by *OA* with the positive direction of the coordinate axes *i.e.,* $\cos\alpha$, $\cos\beta$ and $\cos\gamma$ are called the **direction cosines** of this line.

The direction cosines of a line are usually denoted by the letters *l, m, n*.

3. Direction cosines of the *X*-axis are (1, 0, 0) that of the *Y*-axis are (0, 1, 0) and that of the *Z*-axis are (0, 0, 1).

4. Relation between the Direction Cosines of a Line

$$l^2+m^2+n^2=1$$

Also $l=\cos\alpha$, $m=\cos\beta$ and $n=\cos\gamma$

From (i) and (ii), we can deduce :

$$\cos^2\alpha+\cos^2\beta+\cos^2\gamma=1$$

5. The coordinates of a point $A(x, y, z)$ in terms of the Direction Cosines

If *l, m, n* are the direction cosines of a line *OA* and $|OA|=r$.

Then $OA=r\cos\alpha$ or $x=lr$ $\quad(\because \cos\alpha=l)$

Similarly $y=mr$ and $z=nr$

Therefore, the coordinates of the point *A* can be given as $A(lr, mr, nr)$.

6. Direction cosines of a line joining two points $P(x_1, y_1, z_1)$ and $Q(x_2, y_2, z_2)$ are

$$\frac{x_2-x_1}{PQ},\ \frac{y_2-y_1}{PQ},\ \frac{z_2-z_1}{PQ}$$

where $PQ=\sqrt{(x_2-x_1)^2+(y_2-y_1)^2+(z_2-z_1)^2}$

7. A set of three numbers *a, b, c* which are proportional to the direction cosines of a line are called the **direction ratios** of the line.

If *l, m, n* are the direction cosines and *a, b, c* are the direction ratios of a line, then

(i) $l : m : n = a : b : c$

(ii) $l = \dfrac{a}{\sqrt{a^2+b^2+c^2}} : m = \dfrac{b}{\sqrt{a^2+b^2+c^2}} : n = \dfrac{c}{\sqrt{a^2+b^2+c^2}}$

(iii) The direction cosines of a line are unique as far as their magnitudes are concerned but a line may have an infinite number of direction ratios.

For example, the direction cosines of the X-axis are (1, 0, 0); those of the Y-axis are (0, 1, 0) and for the Z-axis are (0, 0, 1). These are unique, but (– 3, 0, 0); (4, 0, 0) or in general (k, 0, 0) (for any value of k) are direction ratios of the X-axis.

8. The vector equation of a line that passes through the given point whose position vector is $\vec{a}$ and is parallel to a given vector $\vec{b}$ is

$$\vec{r} = \vec{a} + \lambda \vec{b}$$

9. Equation of a line through a point (x, y, z) and having direction cosines l, m, n is

$$\frac{x-x_1}{l} = \frac{y-y_1}{m} = \frac{z-z_1}{n}$$

10. The vector equation of a line which passes through two points whose position vectors are $\vec{a}$ and $\vec{b}$ is $\vec{a} + \lambda(\vec{a} - \vec{b})$

11. Cartesian equation of a line that passes through two points (x_1, y_1, z_1) and (x_2, y_2, z_2) is

$$\frac{x-x_1}{x_2-x_1} = \frac{y-y_1}{y_2-y_1} = \frac{z-z_1}{z_2-z_1}$$

12. Angle between Two Lines : Let $\vec{R} = \vec{A_1} + t\vec{B_1}$ and $\vec{R} = \vec{A_2} + u\vec{B_2}$ be two given lines. Since these lines are parallel to the vectors $\vec{B_1}$ and $\vec{B_2}$ respectively, the angles between these lines is the angle between the vectors $\vec{B_1}$ and $\vec{B_2}$. Hence, if θ is the angle between $\vec{B_1}$ and $\vec{B_2}$, then the angle between the given lines is

$$\theta = \cos^{-1}\frac{(\vec{B_1}\cdot\vec{B_2})}{B_1B_2}$$

If $\vec{B_1} = x_1\hat{i} + y_1\hat{j} + z_1\hat{k}$ and $\vec{B_2} = x_2\hat{i} + y_2\hat{j} + z_2\hat{k}$, then

$$\theta = \cos^{-1}\left(\frac{x_1x_2+y_1y_2+z_1z_2}{\sqrt{x_1^2+y_1^2+z_1^2}\sqrt{x_2^2+y_2^2+z_2^2}}\right)$$

If (l_1, m_1, n_1) and (l_2, m_2, n_2) are actual direction cosines of the lines, then the angle θ between them is

$$\theta = \cos^{-1}(l_1l_2 + m_1m_2 + n_1n_2)$$

$$\theta = \sin^{-1}\left(\sqrt{(l_1m_2-l_2m_1)^2-(m_1n_2-m_2n_1)^2+(n_1l_2-n_2l_1)^2}\right)$$

Thus, the lines are perpendicular, if

$$l_1l_2 + m_1m_2 + n_1n_2 = 0.$$

13. Two lines with direction ratios (a_1, b_1, c_1) and (a_2, b_2, c_2) are

(i) perpendicular if $a_1a_2 + b_1b_2 + c_1c_2 = 0$

(ii) parallel if

$$\frac{a_1}{a_2} = \frac{b_1}{b_2} = \frac{c_1}{c_2}$$

14. Skew lines are lines in space which are neither parallel nor intersecting each other. They lie in different planes.

15. Shortest Distance between Two Skew Lines : Shortest distance between two skew lines is the line segment perpendicular to both the lines.

If $\vec{r} = \vec{a_1} + t\vec{b_1}$ and $\vec{r} = \vec{a_2} + t\vec{b_2}$ be two skew lines, then the shortest distance between them is given as

$$d = \left|\frac{(\vec{b_1}\times\vec{b_2})\cdot(\vec{a_2}-\vec{a_1})}{|\vec{b_1}\times\vec{b_2}|}\right|$$

From here we deduce the following :

(i) Two lines intersect if and only if $d = 0$ and only if $(\vec{b_1}\times\vec{b_2}).(\vec{a_2}-\vec{a_1}) = 0$.

(ii) If two lines are parallel then their equations can be taken as

$$\vec{r} = \vec{a_1} + t\vec{b_1} \text{ and } \vec{r} = \vec{a_2} + u\vec{b_1}$$

16. Shortest Distance between the Lines :

$$\frac{x-x_1}{a_1} = \frac{y-y_1}{b_1} = \frac{z-z_1}{c_1} \text{ and } \frac{x-x_2}{a_2} = \frac{y-y_2}{b_2} = \frac{z-z_2}{c_2}$$

is

$$\frac{\begin{vmatrix} x_2-x_1 & y_2-y_1 & z_2-z_1 \\ a_1 & b_1 & c_1 \\ a_2 & b_2 & c_2 \end{vmatrix}}{\sqrt{(b_1c_2-b_2c_1)^2+(c_1a_2-c_2a_1)^2+(a_1b_2-a_2b_1)^2}}$$

17. Distance between parallel lines $a_1 + \pi b$ and $a_2 + \mu b$ is $\left|\dfrac{\vec{b}\times(\vec{a_2}-\vec{a_1})}{|\vec{b}|}\right|$

18. A plane has the direction of its normal from origin as $\vec{n}$ and contains a point P with position vector $\vec{a}$. then the position vector $\vec{r}$ of any point in the plane is given by $(\vec{r}-\vec{a})\vec{n} = 0$.

19. General equation of a plane in vector form is $\vec{r} \cdot \vec{n} = d$, where $\vec{n}$ is unit normal vector to the plane and d is the perpendicular distance of the plane from the origin. If

$$\vec{r} = x\hat{i} + y\hat{j} + z\hat{k},\ \vec{n} = l\hat{i} + m\hat{j} + n\hat{k}$$

and then $lx + my + nz = d$ is the general equation of plane in Cartesian form.

20. The equation of a plane passing through a point whose position vector is $\vec{a}$ and perpendicular to the vector $\vec{n}$ is $(\vec{r} - \vec{a}) \cdot \vec{n} = 0$.

21. Cartesian form of equation of a plane passing through a point : Let the given point A be (x_1, y_1, z_1) and P be (x, y, z) and direction ratios of $\vec{n}$ are l, m, n.

So
$$\vec{a} = x_1\hat{i} + y_1\hat{j} + z_1\hat{k}$$
$$\vec{r} = x\hat{i} + y\hat{j} + z\hat{k}$$
$$\vec{n} = l\hat{i} + m\hat{j} + n\hat{k}$$

Substituting it in equation written in $(\hat{r} - \hat{a}) \cdot \hat{n} = 0$, we have

$$((x - x_1)\hat{i} + (y - y_1)\hat{j} + (z - z_1)\hat{k}) \cdot (l\hat{i} + m\hat{j} + n\hat{k}) = 0$$

which is required equation in Cartesian form.

22. Equation of a plane passing through a point A and parallel to two lines $\vec{r}$, $\vec{r'}$:

Let $\vec{a}$ be the position vector of point A.

Let the plane be parallel to lines $\vec{r_1} = \vec{a} + \lambda\vec{b}$ and $\vec{r_2} = \vec{a} + \mu\vec{b}$

Let P be a point with position vector $\vec{r}$, then P lies in the plane iff $\vec{AP}, \vec{b'}, \vec{b}$ are coplanar.

$$(\vec{r} - \vec{a}) \cdot (\vec{b} \times \vec{b'}) = 0$$

is the required vector equation of plane.

Cartesian form : Let the plane passes through the point $A(x_1, y_1, z_1)$, parallel to lines whose direction ratios are b_1, b_2, b_3 and b_1', b_2', b_3'.

Let $P(x, y, z)$ be any point on the plane with position vector $\vec{r}$. Then cartesian equation of plane is

$$\begin{vmatrix} x - x_1 & y - y_1 & z - z_1 \\ b_1 & b_2 & b_3 \\ b_1' & b_2' & b_3' \end{vmatrix} = 0$$

Note : The above formula holds even if the two lines lie in the plane.

23. Equation of plane in normal form is $lx + my + nz = d$, where l, m, n are direction cosines of perpendicular from origin and $d\ (d \neq 0)$ is the distance of the plane from origin.

24. Equation of a plane passing through three non-collinear (but co-planar) points in vector form is given by

$$(\vec{r} - \vec{a}) \cdot [(\vec{b} - \vec{a}) \times (\vec{c} - \vec{a})] = 0$$

Where $\vec{a}, \vec{b}, \vec{c}$ are position vectors of non-collinear points.

25. Equation of a plane passing through three non-collinear (but co-planar) points

If $A(x_1, y_1, z_1)$, $B(x_2, y_2, z_2)$ and $C(x_3, y_3, z_3)$ are three collinear points and the coordinates of $\vec{r}$ are (x, y, z) then the equation of plane is given as,

$$\begin{vmatrix} x - x_1 & y - y_1 & z - z_1 \\ x_2 - x_1 & y_2 - y_1 & z_2 - z_1 \\ x_3 - x_1 & y_3 - y_1 & z_3 - z_1 \end{vmatrix} = 0$$

26. Equation of a plane that cuts the coordinate axes at $(a, 0, 0)$, $(0, b, 0)$ and $(0, 0, c)$ is

$$\frac{x}{a} + \frac{y}{b} + \frac{z}{c} = 1.$$

27. Vector equation of a plane that passes through the intersection of planes $\vec{r} \cdot \vec{n_1} = d_1$ and $\vec{r} \cdot \vec{n_2} = d_2$ is

$$\vec{r} \cdot (\vec{n_1} + \lambda\vec{n_2}) = d_1 + \lambda d_2$$

In Cartesian form : Equation of a plane that passes through the intersection of two given planes

$$A_1x + B_1y + C_1z + D_1 = 0 \text{ and } A_2x + B_2y + C_2z + D_2 = 0$$

is

$$(A_1x + B_1y + C_1z + D_1) + \lambda(A_2x + B_2y + C_2z + D_2) = 0$$

28. Two lines are said to be **coplanar** if they lie in the same plane. Let the given lines be

$$\vec{r} = \vec{a_1} + \lambda\vec{b_1} \quad \ldots(1)$$

and

$$\vec{r} = \vec{a_2} + \lambda\vec{b_2} \quad \ldots(2)$$

The lines will be coplanar if

$$(\vec{a_2} - \vec{a_1}) \cdot (\vec{b_1} \times \vec{b_2}) = 0$$

In Cartesian form : If the coordinates of points $A(x_1, y_1, z_1)$ and $B(x_2, y_2, z_2)$ and a_1, b_1, c_1 and a_2, b_2, c_2 are the direction ratios of vector $\vec{b_1}$ and $\vec{b_2}$ respectively the lines will be coplanar iff

$$\begin{vmatrix} x_2 - x_1 & y_2 - y_1 & z_2 - z_1 \\ a_1 & b_1 & c_1 \\ a_2 & b_2 & c_2 \end{vmatrix} = 0$$

29. Angle between Two Planes : The angle between two planes is the angle between the normals to them from any point (in particular from the origin).

Let the two planes have equations $\vec{r} \cdot \vec{n_1} = d_1$ and $\vec{r} \cdot \vec{n_2} = d_2$ and θ be the angle between these planes.

Then, $\cos\theta = \dfrac{\vec{n_1} \cdot \vec{n_2}}{|\vec{n_1}||\vec{n_2}|}$ and the plane containing $\vec{n_1}$ and $\vec{n_2}$ is perpendicular to both the given planes.

Two planes are perpendicular, if $\vec{n_1} \cdot \vec{n_2} = 0$ and they are parallel, if $\vec{n_1} = \pm \vec{n_2}$

In Cartesian form : The angle between the planes $A_1x + B_1y + C_1z + D_1 = 0$ and $A_2x + B_2y + C_2z + D_2 = 0$ is given by

$$\cos\theta = \left| \frac{A_1A_2 + B_1B_2 + C_1C_2}{\sqrt{A_1^2 + B_1^2 + C_1^2}\sqrt{A_2^2 + B_2^2 + C_2^2}} \right|$$

30. The angle ϕ between the line and the plane :

Let the line be $\vec{r} = \vec{a} + \lambda \vec{b}$ and the plane be

$$\vec{r} \cdot \hat{n} = d$$

Then the angle ϕ between them is given by

$$\sin\phi = \left| \frac{\vec{b} \cdot \hat{n}}{|\vec{b}||\vec{n}|} \right|$$

31. Distance of a Point from a Plane :

Let $\vec{r} \cdot \vec{n} = d$ be the equation of plane and A be the point with position vector $\vec{a}$ on the plane then distance of the plane from the point is given by

$$\left| \frac{\vec{a} \cdot \vec{n} - d}{|n|} \right|$$

In Cartesian form : The distance of a point (x_1, y_1, z_1) from the plane $Ax + By + Cz + D = 0$ is

$$\left| \frac{Ax_1 + By_1 + Cz_1 + D}{\sqrt{A^2 + B^2 + C^2}} \right|$$

Chapter 12. Linear Programming

1. Linear Programming is a process by which real-life problem may be transformed into a mathematical model function and then solved using algorithms. The word 'linear' indicates that all the mathematical functions in the model must be linear and 'programming' means planning. The most common application of linear programming is allocating resources among activities in the best possible way.

2. Linear Programmig Model (LP - model) : The LP-model is formulated in terms of the general problem of allocating resources to activities.

Let us consider n number of activities.

Let $x_1, x_2 \ldots x_n$ be n(decision) variable and $c_1, c_2 \ldots c_n$ be n constants.

(i) The linear function $z = c_1x_1 + c_2x_2 + \ldots + c_nx_n$ which is to be maximized or minimized is called the **objective function**. It is always non-negative.

(ii) The restriction on the variable of the LP-model are called **constraints.**

(iii) The non-negative constraints of a LP-model are the inequalities of the form

$$x_i \geq 0,$$

where $\quad i = 1, 2, \ldots n$

i.e. $\quad x_1 \geq 0, x_2 \geq 0, \ldots, x_n \geq 0$

3. Types of Solutions of a Linear Programming Problem (LPP) : Possible solutions of the LPP are sets of values of the variables $x_1, x_2 \ldots x_n$ which are desirable or not desirable.

(i) A **feasible solution** is one for which all the constraints are satisfied. The collection of the feasible solutions gives the feasible region. The feasible region of an LPP is always a convex polygon, that is, a region bounded by straight lines such that the line segment joining any two arbitrary points of the region always lies entirely within the region.

A feasible region may be **bounded** or **unbounded**, depending on whether the variables are bounded or not in the region. There may be problem which have no feasible solutions. Here, we say that the feasible region is empty.

A **optimal solution** is a feasible solution that gives the most favourable value of the objective function. The most favourable value is the largest or the smallest value, depending on whether the objective function is to maximized or minimized.

An LPP may have more than one optimal solution (*i.e.*, multiple optimal solutions) or no optimal solution.

(ii) A **infeasible solution** is one, such that at least one constraint is violated.

4. Graphical Methods to Solve the LPP in Two Variables : There are two ways of solving the LPP in two variables using the graphical method :

(i) Corner point method where a 'corner point' solution is a solution that lies at a corner of the feasible region.

(ii) Iso-profit or Iso-cost method.

Theorem : Let R be the feasible region for linear programming problem and let $z = ax + by$ be the objective function, when z has an optimal value (maximum or minimum) where the variables x and y are subject to constraints described by linear inequalities, this optimal value must occur at a corner point of the feasible region.

Theorem : Let R be the feasible region for a LPP and let $z = ax + by$ be the objective function. If R is bounded then the objective function, z has both a maximum and a minimum value on R and each of these occurs at a corner point of R.

Note that if R is unbounded then a maximum or a minimum value of the objective function may not exist. However if it exists, it must occur at a corner point of R.

Chapter 13. Probability

1. **Probability** is a measure of uncertainty of events in a random experiment.
2. In probability, a numerical value is assigned to the chance of happening of an event.
3. **Experiment** is an operation which can produce some well defined outcomes.
4. If all possible outcomes of an experiment are known but the exact output cannot be predicted in advance then the experiment is called a **random experiment.**
5. The set of all possible outcomes of an experiment is called the **sample space.**
6. Any subset of a sample space is called an **event**.
7. The probability of occurrence of an event E is defined as

$$P(E) = \frac{n(E)}{n(S)}$$

where $n(E)$ = No. of elements in the event E

$n(S)$ = No. of elements of the sample space S

8. Two or more events associated with a random experiment are **mutually exclusive** if no two or more of them can occur simultaneously in the same trial.
9. Two or more events associated with a random experiment are said to be **exhaustive** if their union is the sample space.
10. (i) The probability of a certain event *i.e.*, the event which is likely to happen is 1.

(ii) The probability of any event E must be greater than or equal to 0 and less than or equal to 1 *i.e.*,

$$0 \le P(E) \le 1.$$

(iii) The probability of uncertain event *i.e.*, event which can not be possible is 0.

11. For any events A and B,

$$P(A \cup B) = P(A) + P(B) - P(A \cap B).$$

12. If $\overline{A}$ or A' or A^c denotes (not A) then

$$P(\overline{A}) = P(A') = P(A^c) = 1 - P(A).$$

13. If A and B are two independent events, then

$$P(A \cap B) = P(A) \cdot P(B).$$

14. If A and B are two mutually exclusive events, then

$$A \cap B = \phi \Rightarrow P(A \cap B) = 0.$$

15. Two events A and B are said to be **independent** if the occurrence of one does not depend upon the occurrence of the other.
16. If two events are not independent, they are said to be **dependent**.
17. **Addition theorem : (i)** When A and B are mutually exclusive events then probability of occurrence of either A or B is given by,

$$P(A \cup B) = P(A) + P(B).$$

(ii) When A and B are not mutually exclusive events then probability of occurrence either A or B is given by

$$P(A \cup B) = P(A) + P(B) - P(A \cap B)$$

For three mutually exclusive events A, B and C :

$$P(A \cup B \cup C) = P(A) + P(B) + P(C) - P(A \cap B) - P(B \cap C) - P(C \cap A) + P(A \cap B \cap C)$$

18. If A and B are two events associated with the same sample space of a random experiment then the conditional probability of the event A given that B has already occurred *i.e.*, $P(A/B)$ is given by

$$P(A/B) = \frac{P(A \cap B)}{P(B)} \text{ provided } P(B) \ne 0$$

Similarly, $$P(B/A) = \frac{P(A \cap B)}{P(A)} \text{ provided } P(A) \ne 0.$$

19. Properties of conditional probability are as follows :

(i) If A and B are the events of a sample space S of an experiment then

$$P(A/A) = P(B/B) = 1$$

(ii) If A and B are any two events of a sample space S and F is an event of S such that $P(F) \ne 0$, then

$$P[(A \cup B)/F] = P(A/F) + P(B/F) - P[(A \cap B)/F]$$

(c) $P(A'/B) = 1 - P(A/B)$

20. Let A and B be two events associated with a sample space S.

Then $$P(A \cap B) = P(A) \cdot P(B/A) = P(B) \cdot P(A/B)$$

provided $P(A) \ne 0$ and $P(B) \ne 0$

This result is known as the **multiplication theorem of probability.**

21. **Multiplication theorem of probability for more than two events :** If A, B and C are three events of sample space S, then we have

$$P(A \cap B \cap C) = P(A) \cdot P(B/A) \cdot P(C/(A \cap B)]$$
$$= P(A) \cdot P(B/A) \cdot P(C/AB)$$

22. Two events A and B are said to be independent, if

$$P(B/A) = P(B), \text{ provided } P(A) \neq 0$$
and $$P(A/B) = P(A), \text{ provided } P(B) \neq 0$$

i.e., occurrence or non-occurrence of one event does not affect the probability of the occurrence or non-occurrence of other event.

23. Three events A, B and C are said to be mutually independent, if

$$P(A \cap B) = P(A) \cdot P(B)$$
$$P(A \cap C) = P(A) \cdot P(C)$$
$$P(B \cap C) = P(B) \cdot P(C)$$
$$P(A \cap B \cap C) = P(A) \cdot P(B) \cdot P(C)$$

24. If the events A and B are independent, then
(i) $\overline{A}$ and B are independent.
(ii) A and $\overline{B}$ are independent.
(iii) $\overline{A}$ and $\overline{B}$ are independent.

25. Let E_1, E_2, ..., E_n be a mutually exclusive and exhaustive events associated with a random experiment and S be the sample space.
Suppose that each of the events E_1, E_2, ..., E_n has non-zero probability of occurrence. Let A be any event associated with S, then

$$P(A) = P(E_1) \cdot P(A/E_1) + P(E_2) \cdot P(A/E_2) + P(E_3) \cdot P(A/E_3) + \ldots + P(E_n) P(A/E_n)$$

$$P(A) = \sum_{j=1}^{n} P(E_j) \cdot P(A/E_j)$$

This result is called the **law of total probability.**

26. **Bayes' Theorem :** Let S be the sample space and let $E_1, E_2, \ldots E_n$ be n mutually exclusive and exhaustive events associated with a random experiment.
If A is any event of non-zero probability which occurs with E_1 or E_2 or ... or E_n, then

$$P(E_i/A) = \frac{P(E_i) \cdot P(A/E_i)}{\sum_{i=1}^{n} P(E_i) \cdot P(A/E_i)} \text{ for any } i = 1, 2, 3, \ldots, n$$

27. A random variable is a real valued function whose domain is the sample space of a random experiment.

28. A description giving the values of the random variable along with the corresponding probabilities is called the **probability distribution** of the random variable X.

29. The probability distribution of a random variable X is the system of numbers.

X:	x_1	x_2	...	x_n
$P(X)$:	p_1	p_2	...	p_n

where $p_i > 0$, $\sum_{i=1}^{n} p_i = 1, i = 1, 2, 3, \ldots, n$

where x_i is the possible values of the random variable X.

p_i is the probability of random variable X taking the value x_i.

30. **Mean** is a measure of location or central tendency in the sense that it roughly locates a middle or average value of the random variable.

31. The mean of expectation of a random variable X (*i.e.*, $x_1, x_2, \ldots, x_n$) is the sum of the products of all possible values of X by their respective probabilities.

Thus, $E(X) = \mu = \sum_{i=1}^{n} x_i p_i = x_1p_1 + x_2p_2 + \ldots + x_np_n$.

32. Let X be a random variable whose possible values x_1, $x_2, \ldots, x_n$ occur with probabilities $p(x_1), p(x_2), \ldots p(x_n)$ respectively.
Let $\mu = E(X)$ be the mean of X. The variance of X, denoted by var (X) or σ_x^2 is defined as

$$\sigma_x^2 = \text{var}(X) = \sum_{i=1}^{n} (x_i - \mu)^2 \cdot p(x_i)$$

33. The non-negative number $\sigma_x = \sqrt{\sum_{i=1}^{n} (x_i - \mu)^2 \, p(x_i)}$ is called the **standard deviation** of the random variable 'X'.

34. Trials of a random experiment are called **Bernoulli trials** if they satisfy the following conditions :
(i) There should be a finite no. of trials
(ii) The trials should be independent of each other.
(iii) Each tiral has exactly two outcomes : success or failure.
(iv) The probability of success remains the same in each trial.

35. In an experiment of n-Bernoulli's trials, the probabilities of 0, 1, 2, ..., n success can be obtained as 1st, 2nd, ... $(n+1)$th terms in the expansion of $(q+p)^n$.

36. The probability of x success in n-Bernoulli trials is ${}^nC_x p^x q^{n-x}$, where $x = 0, 1, 2, \ldots, n$ and $q = 1 - p$.

37. A **binomial distribution** with n-Bernoulli trials and probability of success in each trial as p, is denoted by $B(n, p)$. The probability function of the binomial distribution is

$$P(x) = {}^nC_x p^x q^{n-x}, x = 0, 1, 2, \ldots, n.$$

38. The mean and variance of a binomial variate with parameters n and p are np and npq, $(q = 1 - p)$ respectively.

39. $(p+q)^n = {}^nC_0 p^0 q^{n-0} + {}^nC_1 p^1 q^{n-1} + \ldots + {}^nC_n p^n q^{n-n} = 1.$

Multiple Choice Questions

Set 1

Chapter 1. Relations and Functions

Q. 1. Let R be a relation on the set N of natural numbers defined by *nRm* if *n* divides *m*. Then R is

(A) Reflexive and symmetric

(B) Transitive and symmetric

(C) Equivalence

(D) Reflexive, transitive but not symmetric

Sol. Since n divides n, $\forall\ n \in$ N, R is reflexive. R is not symmetric since for 3, 6 $\in$ N, 3 R 6 $\neq$ 6 R 3.

R is transitive since for n, m, r whenever n/m and m/r $\Rightarrow n/r$, *i.e.*, n divides m and m divides r, then n will devide r.

So, the correct option is (D).

Q. 2. Let L denote the set of all straight lines in a plane. Let a relation R be defined by *l*R*m* if and only if *l* is perpendicular to m $\forall$ $l, m \in$ L. Then R is

(A) reflexive **(B)** symmetric

(C) transitive **(D)** none of these

Sol. The correct option is (B).

Q. 3. Let N be the set of natural numbers and the function f : N $\to$ N be defined by $f(n) = 2n + 3\ \forall \in$ N. Then f is

(A) surjective **(B)** injective

(C) bijective **(D)** none of these

Sol. The correct option is (B).

Q. 4. Set A has 3 elements and the set B has 4 elements. Then the number of injective mappings that can be defined from A to B is

(A) 144 **(B)** 12

(C) 24 **(D)** 64

Sol. The total number of injective mappings from the set containing 3 elements into the set containing 4 elements is ${}^4P_3 = 4! = 24$.

So the correct option is (C).

Q. 5. Let f : R $\to$ R be defined by $f(x)$ = sin x and g : R $\to$ R be defined by $g(x) = x^2$, then *fog* is

(A) $x^2 \sin x$ **(B)** $(\sin x)^2$

(C) $\sin x^2$ **(D)** $\dfrac{\sin x}{x^2}$

Sol.
$$fog = f(g(x)) = f(x^2) = \sin x^2$$

So the correct option is (C).

Q. 6. Let f : R $\to$ R be defined by $f(x) = 3x - 4$. Then $f^{-1}(x)$ is given by

(A) $\dfrac{x+4}{3}$ **(B)** $\dfrac{x}{3} - 4$

(C) $3x + 4$ **(D)** none of these

Sol. $f(x) = 3x - 4$

Let $y = f(x)$

$$y = 3x - 4$$
$$y + 4 = 3x$$
$$\frac{y+4}{3} = x$$

$\therefore$ $f^{-1}(x) = \dfrac{x+4}{3}$

So the correct option is (A).

Q. 7. Let f : R $\to$ R be defined by $f(x) = x^2 + 1$. Then, pre-images of 17 and – 3, respectively, are

(A) ϕ, {4, – 4} **(B)** {3, – 3}, ϕ

(C) {4, – 4}, ϕ **(D)** {4, – 4}, {2, – 2}

Sol. Since for $f^{-1}(17) = x$

$\Rightarrow$ $f(x) = 17$ or $x^2 + 1 = 17$

$\Rightarrow$ $x = \pm 4$

or $f^{-1}(17) = \{4, -4\}$

and for $f^{-1}(-3) = x$

$\Rightarrow$ $f(x) = -3$

$\Rightarrow$ $x^2 + 1 = -3$

$\Rightarrow$ $x^2 = -4$

Hence $f^{-1}(-3) = \phi$.

So the correct option is (C).

Q. 8. For real numbers *x* and *y*, define *xRy* if and only if $x - y + \sqrt{2}$ is an irrational number. Then the relation R is

(A) reflexive **(B)** symmetric

(C) transitive **(D)** none of these

Sol. The correct option is (A).

Chapter 2. Inverse Trigonometric Functions

Q. 1. Which of the following corresponds to the principal value branch of $\tan^{-1}$?

(A) $\left(-\dfrac{\pi}{2}, \dfrac{\pi}{2}\right)$ **(B)** $\left[-\dfrac{\pi}{2}, \dfrac{\pi}{2}\right]$

(C) $\left(-\dfrac{\pi}{2}, \dfrac{\pi}{2}\right) - \{0\}$ **(D)** $(0, \pi)$

Sol. The correct option is (A).

Q. 2. The principal value branch of $\sec^{-1}$ is

(A) $\left[-\frac{\pi}{2}, \frac{\pi}{2}\right] - \{0\}$ (B) $[0, \pi] - \left\{\frac{\pi}{2}\right\}$

(C) $(0, \pi)$ (D) $\left(-\frac{\pi}{2}, \frac{\pi}{2}\right)$

Sol. The correct option is (B).

Q. 3. The principal value of the expression $\cos^{-1}[\cos(-680°)]$ is :

(A) $\frac{2\pi}{9}$ (B) $\frac{-2\pi}{9}$

(C) $\frac{34\pi}{9}$ (D) $\frac{\pi}{9}$

Sol. $\cos^{-1}[\cos(-680°)] = \cos^{-1}[\cos(720° - 40°)]$

$= \cos^{-1}[\cos(-40°)]$

$= \cos^{-1}[\cos(40°)]$

$= 40° = \frac{2\pi}{9}.$

So the correct option is (A).

Q. 4. The value of $\cot(\sin^{-1} x)$ is

(A) $\frac{\sqrt{1+x^2}}{x}$ (B) $\frac{x}{\sqrt{1+x^2}}$

(C) $\frac{1}{x}$ (D) $\frac{\sqrt{1-x^2}}{x}$.

Sol. Let $\sin^{-1} x = \theta$,

then $\sin\theta = x$

$\Rightarrow \quad \operatorname{cosec}\theta = \frac{1}{x}$

$\Rightarrow \quad \operatorname{cosec}^2\theta = \frac{1}{x^2}$

$\Rightarrow \quad 1 + \cot^2\theta = \frac{1}{x^2}$

$\Rightarrow \quad \cot\theta = \frac{\sqrt{1-x^2}}{x}$

$\Rightarrow \quad \cot(\sin^{-1} x) = \frac{\sqrt{1-x^2}}{x}$

So the correct option is (D).

Q. 5. If $\tan^{-1} x = \frac{\pi}{10}$ for some $x \in R$, then the value of $\cot^{-1} x$ is

(A) $\frac{\pi}{5}$ (B) $\frac{2\pi}{5}$

(C) $\frac{3\pi}{5}$ (D) $\frac{4\pi}{5}$

Sol. We know that

$$\tan^{-1} x + \cot^{-1} x = \frac{\pi}{2}.$$

$\therefore \quad \cot^{-1} x = \frac{\pi}{2} - \frac{\pi}{10}$

$\Rightarrow \quad \cot^{-1} x = \frac{\pi}{2} - \frac{\pi}{10}$

$= \frac{2\pi}{5}.$

So the correct option is (B).

Q. 6. The domain of $\sin^{-1} 2x$ is

(A) $[0, 1]$ (B) $[-1, 1]$

(C) $\left[-\frac{1}{2}, \frac{1}{2}\right]$ (D) $[-2, 2]$

Sol. Let $\sin^{-1} 2x = \theta$

So that $2x = \sin\theta$.

Now, $1 \leq \sin\theta \leq 1$, *i.e.*, $-1 \leq 2x \leq 1$ which gives

$-\frac{1}{2} \leq x \leq \frac{1}{2}$.

So the correct option is (C).

Q. 7. The principal value of $\sin^{-1}\left(\frac{-\sqrt{3}}{2}\right)$ is

(A) $-\frac{2\pi}{3}$ (B) $-\frac{\pi}{3}$

(C) $\frac{4\pi}{3}$ (D) $\frac{5\pi}{3}$.

Sol. $\sin^{-1}\left(\frac{-\sqrt{3}}{2}\right) = \sin^{-1}\left(-\sin\frac{\pi}{2}\right)$

$= -\sin^{-1}\left(\sin\frac{\pi}{3}\right)$

$= -\frac{\pi}{3}.$

So the correct option is (B).

Q. 8. The greatest and least values of $(\sin^{-1} x)^2 + (\cos^{-1} x)^2$ are respectively

(A) $\frac{5\pi^2}{4}$ and $\frac{\pi^2}{8}$ (B) $\frac{\pi}{2}$ and $\frac{-\pi}{2}$

(C) $\frac{\pi^2}{4}$ and $\frac{-\pi^2}{4}$ (D) $\frac{\pi^2}{4}$ and 0.

Sol. We have

$(\sin^{-1} x)^2 + (\cos^{-1} x)^2$

$= (\sin^{-1} x + \cos^{-1} x)^2 - 2\sin^{-1} x \cos^{-1} x$

$= \frac{\pi^2}{4} - 2\sin^{-1} x\left(\frac{\pi}{2} - \sin^{-1} x\right)$

$$= \frac{\pi^2}{4} - \pi \sin^{-1} x + 2(\sin^{-1} x)^2$$

$$= 2\left[(\sin^{-1} x)^2 - \frac{\pi}{2}\sin^{-1} x + \frac{\pi^2}{8}\right]$$

$$= 2\left[\left(\sin^{-1} x - \frac{\pi}{4}\right)^2 + \frac{\pi^2}{16}\right].$$

Thus, the least value is $2\left(\frac{\pi^2}{16}\right)$, *i.e.,* $\frac{\pi^2}{8}$ and the greatest value is $2\left[\left(\frac{-\pi}{2} - \frac{\pi}{4}\right)^2 + \frac{\pi^2}{16}\right]$, *i.e.,* $\frac{5\pi^2}{4}$.

So the correct option is (A).

Q. 9. The domain of $y = \cos^{-1}(x^2 - 4)$ is

(A) $[3, 5]$ **(B)** $[0, \pi]$

(C) $[-\sqrt{5}, -\sqrt{3}] \cap [-\sqrt{5}, \sqrt{3}]$

(D) $[-\sqrt{5}, -\sqrt{3}] \cup [\sqrt{3}, \sqrt{5}]$

Sol. $y = \cos^{-1}(x^2 - 4)$

$\Rightarrow \cos y = x^2 - 4$

i.e., $-1 \le x^2 - 4 \le 1$ (since $-1 \le \cos y \le 1$)

$\Rightarrow 3 \le x^2 \le 5$

$\Rightarrow \sqrt{3} \le |x| \le \sqrt{5}$

$\Rightarrow x \in [-\sqrt{5}, -\sqrt{3}] \cup [\sqrt{3}, \sqrt{5}]$

So the correct option is (D).

Q. 10. The domain of the function defined by $f(x) = \sin^{-1} x + \cos x$ is

(A) $[-1, 1]$ **(B)** $[-1, \pi + 1]$

(C) $(-\infty, \infty)$ **(D)** ϕ

Sol. The domain of cos is R and the domain of $\sin^{-1}$ is $[-1, 1]$.

$\therefore$ The domain of $\cos x + \sin^{-1} x$ is $R \cap [-1, 1]$, *i.e.,* $[-1, 1]$.

So the correct option is (A).

Q. 11. The value of $\sin[2\sin^{-1}(0.6)]$ is

(A) ·48 **(B)** ·96

(C) 1·2 **(D)** sin 1·2

Sol. Let $\sin^{-1}(\cdot 6) = \theta$, *i.e.,* $\sin\theta = \cdot 6$.

Now, $\sin(2\theta) = 2\sin\theta\cos\theta = 2(\cdot 6)(\cdot 8) = \cdot 96$

So the correct option is (B).

Q. 12. If $\sin^{-1} x + \sin^{-1} y = \frac{\pi}{2}$, then value of $\cos^{-1} x + \cos^{-1} y$ is

(A) $\frac{\pi}{2}$ **(B)** π

(C) 0 **(D)** $\frac{2\pi}{3}$

Sol. Given that $\sin^{-1} x + \sin^{-1} y = \frac{\pi}{2}$.

$$\therefore \left(\frac{\pi}{2} - \cos^{-1} x\right) + \left(\frac{\pi}{2} - \cos^{-1} y\right) = \frac{\pi}{2}$$

$$\Rightarrow \cos^{-1} x + \cos^{-1} y = \frac{\pi}{2}.$$

So the correct option is (A).

Q. 13. The equation $\tan^{-1} x - \cot^{-1} x = \tan^{-1}\left(\frac{1}{\sqrt{3}}\right)$ has

(A) no solution

(B) unique solution

(C) infinite number of solutions

(D) two solutions

Sol. We have

$\tan^{-1} x - \cot^{-1} x = \frac{\pi}{6}$ and $\tan^{-1} x + \cot^{-1} x = \frac{\pi}{2}$

Adding them, we get

$$2\tan^{-1} x = \frac{2\pi}{3}$$

$$\Rightarrow \tan^{-1} x = \frac{\pi}{3}$$

i.e., $x = \sqrt{3}$.

So the correct option is (B).

Q. 14. If $\alpha \le 2\sin^{-1} x + \cos^{-1} x \le \beta$, then

(A) $\alpha = \frac{-\pi}{2}, \beta = \frac{\pi}{2}$ **(B)** $\alpha = 0, \beta = \pi$

(C) $\alpha = \frac{-\pi}{2}, \beta = \frac{3\pi}{2}$ **(D)** $\alpha = 0, \beta = 2\pi$

Sol. We have

$$\frac{-\pi}{2} \le \sin^{-1} x \le \frac{\pi}{2}$$

$$\Rightarrow \frac{-\pi}{2} + \frac{\pi}{2} \le \sin^{-1} x + \frac{\pi}{2} \le \frac{\pi}{2} + \frac{\pi}{2}$$

$$\Rightarrow 0 \le \sin^{-1} x + (\sin^{-1} x + \cos^{-1} x) \le \pi$$

$$\Rightarrow 0 \le 2\sin^{-1} x + \cos^{-1} x \le \pi$$

So the correct option is (B).

Q. 15. The value of $\tan^2(\sec^{-1} 2) + \cot^2(\text{cosec}^{-1} 3)$ is

(A) 5 **(B)** 11

(C) 13 **(D)** 15

Sol. $\tan^2(\sec^{-1} 2) + \cot^2(\text{cosec}^{-1} 3)$

$= \sec^2(\sec^{-1} 2) - 1 + \text{cosec}^2(\text{cosec}^{-1} 3) - 1$

$= 2^2 \times 1 + 3^2 - 2 = 11.$

So the correct option is (B).

Q. 16. The value of $\sin^{-1}\left(\cos\frac{3\pi}{5}\right)$ is :★

(A) $\frac{\pi}{10}$ (B) $\frac{3\pi}{5}$

(C) $\frac{-\pi}{10}$ (D) $\frac{-3\pi}{5}$

Sol. $\sin^{-1}\left(\cos\frac{3\pi}{5}\right) = \frac{\pi}{2} - \cos^{-1}\left(\cos\frac{3\pi}{5}\right)$

$$= \frac{\pi}{2} - \frac{3\pi}{5}$$

$$= \frac{5\pi - 6\pi}{10}$$

$$= \frac{-\pi}{10}$$

So the correct option is (C).

Q. 17. The value of $\tan\left[\frac{1}{2}\cos^{-1}\left(\frac{\sqrt{5}}{3}\right)\right]$ is :★

(A) $\frac{3+\sqrt{5}}{2}$ (B) $\frac{3-\sqrt{5}}{2}$

(C) $\frac{-3+\sqrt{5}}{2}$ (D) $\frac{-3-\sqrt{5}}{2}$

Sol. Let $y = \tan\left[\frac{1}{2}\cos^{-1}\left(\frac{\sqrt{5}}{3}\right)\right]$

Putting, $x = \cos^{-1}\left(\frac{\sqrt{5}}{3}\right)$

$\Rightarrow \cos x = \frac{\sqrt{5}}{3}$ ...(i)

Now, $y = \tan\left(\frac{1}{2}\times x\right)$

$y = \tan\left(\frac{x}{2}\right)$

$\therefore y = \sqrt{\frac{1-\cos(x)}{1+\cos(x)}}$

$\therefore y = \sqrt{\frac{1-\sqrt{5}/3}{1+\sqrt{5}/3}}$ [From (i)]

$\therefore y = \sqrt{\frac{3-\sqrt{5}}{3+\sqrt{5}}} = \sqrt{\frac{(3-\sqrt{5})(3-\sqrt{5})}{(3+\sqrt{5})(3-\sqrt{5})}}$

$\therefore y = \sqrt{\frac{(3-\sqrt{5})^2}{9-5}} = \frac{3-\sqrt{5}}{\sqrt{4}} = \frac{3-\sqrt{5}}{2}$

$\therefore y = \frac{3-\sqrt{5}}{2}$

i.e., $\tan\left[\frac{1}{2}\cos^{-1}\left(\frac{\sqrt{5}}{3}\right)\right] = \frac{3-\sqrt{5}}{2}$

So the correct option is (B).

Chapter 3. Matrices

Q. 1. If A and B are square matrices of the same order, then (A+B) (A–B) is equal to

(A) $A^2 - B^2$

(B) $A^2 - BA - AB - B^2$

(C) $A^2 - B^2 + BA - AB$

(D) $A^2 - BA + B^2 + AB$

Sol. $(A+B)(A-B) = A(A-B) + B(A-B)$

$= A^2 - AB + BA - B^2$

So the correct option is (C).

Q. 2. If $A = \begin{bmatrix} 2 & -1 & 3 \\ -4 & 5 & 1 \end{bmatrix}$ and $B = \begin{bmatrix} 2 & 3 \\ 4 & -2 \\ 1 & 5 \end{bmatrix}$, then

(A) only AB is defined

(B) only BA is defined

(C) AB and BA both are defined

(D) AB and BA both are not defined.

Sol. Let $A = [a_{ij}]_{2\times 3}$ and $B = [b_{ij}]_{3\times 2}$

Both AB and BA are defined.

So the correct option is (C).

Q. 3. The matrix $A = \begin{bmatrix} 0 & 0 & 5 \\ 0 & 5 & 0 \\ 5 & 0 & 0 \end{bmatrix}$ is a

(A) scalar matrix **(B)** diagonal matrix

(C) unit matrix **(D)** square matrix

Sol. The correct option is (D).

Q. 4. If A and B are symmetric matrices of the same order, then (AB′ – BA′) is a

(A) Skew symmetric matrix (B) Null matrix

(C) Symmetric matrix (D) None of these

Sol. Since $(AB' - BA') = (AB')' - (BA')'$

$= (BA' - AB')$

$= -(AB' - BA')$

So the correct option is (A).

★ are board exam questions from previous years

Q. 5. If A = [2 – 3 4], B = $\begin{bmatrix} 3 \\ 2 \\ 2 \end{bmatrix}$, X = [1 2 3] and Y = $\begin{bmatrix} 2 \\ 3 \\ 4 \end{bmatrix}$, then AB + XY equals :★

(A) [28] **(B)** [24]
(C) 28 **(D)** 24

Sol. $AB + XY = [2\ -3\ 4]\begin{bmatrix} 3 \\ 2 \\ 2 \end{bmatrix} + [1\ 2\ 3]\begin{bmatrix} 2 \\ 3 \\ 4 \end{bmatrix}$

$= [8] + [20] = [28]$

So the correct option is (A).

Q. 6. If $[x\quad 1]\begin{bmatrix} 1 & 0 \\ -2 & 0 \end{bmatrix} = 0$, then x equals :★

(A) 0 **(B)** –2
(C) –1 **(D)** 2

Sol. Given :

$$[x\quad 1]\begin{bmatrix} 1 & 0 \\ -2 & 0 \end{bmatrix} = 0$$

$\Rightarrow \quad [x-2\quad 0] = [0\quad 0]$

$\Rightarrow \quad x - 2 = 0$

$\Rightarrow \quad x = 2$

So the correct option is (D).

Chapter 4. Determinants

Q. 1. Let $\Delta = \begin{vmatrix} Ax & x^2 & 1 \\ By & y^2 & 1 \\ Cz & z^2 & 1 \end{vmatrix}$ and $\Delta_1 = \begin{vmatrix} A & B & C \\ x & y & z \\ zy & zx & xy \end{vmatrix}$, then

(A) $\Delta_1 = -\Delta$ **(B)** $\Delta \neq \Delta_1$
(C) $-\Delta_1 = 0$ **(D)** None of these

Sol. Since $\Delta_1 = \begin{vmatrix} A & B & C \\ x & y & z \\ zy & zx & xy \end{vmatrix} = \begin{vmatrix} A & x & yz \\ B & y & zx \\ C & z & xy \end{vmatrix}$

$$= \frac{1}{xyz}\begin{vmatrix} Ax & x^2 & xyz \\ By & y^2 & xyz \\ Cz & z^2 & xyz \end{vmatrix}$$

$$= \frac{xyz}{xyz}\begin{vmatrix} Ax & x^2 & 1 \\ By & y^2 & 1 \\ Cz & z^2 & 1 \end{vmatrix} = \Delta$$

So the correct option is (C).

★ are board exam questions from previous years

Q. 2. If $x, y \in$ R, then the determinant $\Delta = \begin{vmatrix} \cos x & -\sin x & 1 \\ \sin x & \cos x & 1 \\ \cos(x+y) & -\sin(x+y) & 0 \end{vmatrix}$ lies in the interval

(A) $[-\sqrt{2}, \sqrt{2}]$ **(B)** $[-1, 1]$
(C) $[-\sqrt{2}, 1]$ **(D)** $[-1 -\sqrt{2}]$

Sol. Applying $R_3 \rightarrow R_3 - \cos y\, R_1 + \sin y\, R_2$, we get

$$\Delta = \begin{vmatrix} \cos x & -\sin x & 1 \\ \sin x & \cos x & 1 \\ 0 & 0 & \sin y - \cos y \end{vmatrix}$$

Expanding along R_3, we have

$\Delta = (\sin y - \cos y)(\cos^2 x + \sin^2 x)$

$= (\sin y - \cos y)$

$= \sqrt{2}\left[\frac{1}{\sqrt{2}}\sin y - \frac{1}{\sqrt{2}}\cos y\right]$

$= \sqrt{2}\left[\cos\frac{\pi}{4}\sin y - \sin\frac{\pi}{4}\cos y\right]$

$= \sqrt{2}\sin\left(y - \frac{\pi}{4}\right)$

$\therefore \quad -\sqrt{2} \leq \Delta \leq \sqrt{2}$.

So the correct option is (A).

Q. 3. If $\begin{vmatrix} 2 & 3 & 2 \\ x & x & x \\ 4 & 9 & 1 \end{vmatrix} + 3 = 0$, then the value of x is :★

(A) 3 **(B)** 0
(C) –1 **(D)** 1

Sol. Given : $\begin{vmatrix} 2 & 3 & 2 \\ x & x & x \\ 4 & 9 & 1 \end{vmatrix} + 3 = 0$

$\Rightarrow 2(x - 9x) - 3(x - 4x) + 2(9x - 4x) + 3 = 0$

$\Rightarrow -16x + 9x + 10x + 3 = 0$

$\Rightarrow \quad 3x = -3$

$\Rightarrow \quad x = -1$

So the correct option is (C).

Q. 4. Let A = $\begin{bmatrix} 200 & 50 \\ 10 & 2 \end{bmatrix}$ and B = $\begin{bmatrix} 50 & 40 \\ 2 & 3 \end{bmatrix}$, then | AB | is equal to :★

(A) 460 **(B)** 2000
(C) 3000 **(D)** –7000

Sol. $|AB| = |A||B|$
$= (400 - 500)(150 - 80)$
$= (-100)(70)$
$= -7000$

So the correct option is (D).

Q. 5. If $A = \begin{bmatrix} a & 0 & 0 \\ 0 & a & 0 \\ 0 & 0 & a \end{bmatrix}$**, then det (adj A) equals★**

(A) a^{27} **(B)** a^9

(C) a^6 **(D)** a^2

Sol. $|A| = a^3$

$\therefore$ $|\text{adj } A| = |A|^2 = (a^3)^2 = a^6$

So the correct option is (C).

Q. 6. If A is a square matrix of order 3, such that A (adj A) = 10 I, then | adj A | is equal to :★

(A) 1 **(B) 10**

(C) 100 **(D) 101**

Sol. Given : A (adj A) = 10I

We know, $A.\text{adj } A = |A|.I$

$\therefore$ $|A| = 10$

$\because$ $|\text{adj } A| = |A|^{n-1}$

$\therefore$ $|\text{adj } A| = |A|^{3-1} = 10^2 = 100$

So the correct option is (C).

Q. 7. If A is a 3 × 3 matrix such that | A | = 8, then | 3A | equals.★

(A) 8 **(B) 24**

(C) 72 **(D) 216**

Sol. Given : $|A| = 8$

We know, $|kA| = k^3|A|$, where k is constant

$\therefore$ $|3A| = 3^3|A|$
$= 27 \times 8 = 216$

So the correct option is (D).

Q. 8. If A is a skew symmetric matrix of order 3, then the value of | A | is :★

(A) 3 **(B) 0**

(C) 9 **(D) 27**

Sol. Since, A is a skew symmetric matrix.

$\therefore$ $A^T = -A$

$\Rightarrow$ $|A^T| = |-A|$

$\Rightarrow$ $|A| = |-1|^3|A|$

$\Rightarrow$ $|A| = -|A|$

$\Rightarrow$ $2|A| = 0$

$\Rightarrow$ $|A| = 0$

So the correct option is (B).

★ are board exam questions from previous years

Chapter 5. Continuity and Differentiability

Q. 1. Check whether the function $f(x) = 3x - 5$ is continuous at $x = 0, x = -5$ and at $x = 3$.

(A) $f(x)$ is continuous at $x = 0$

(B) $f(x)$ is not continuous at $x = 0$

(C) $f(x)$ is not continuous at $x = -5$

(D) $f(x)$ is not continuous at $x = 3$

Sol. $f(x) = 3x - 5$

At $x = 0$,

$f(0) = 3 \times 0 - 5 = -5$

$\lim_{x \to 0^+} f(x) = \lim_{x \to 0}(3x - 5) = 3 \times 0 - 5 = -5$

$\therefore$ $\lim_{x \to 0^+} f(x) = f(0)$

Therefore, f is continuous at $x = 0$.

At $x = -5$,

$f(-5) = 3 \times (-5) - 5 = -20$

$\lim_{x \to -5^+} f(x) = \lim_{x \to -5}(3x - 5) = 3 \times (-5) - 5 = -20$

$\therefore$ $\lim_{x \to -5} f(x) = f(-5)$

Therefore, f is continuous at $x = -5$.

At $x = 3$, $f(3) = 3 \times (3) - 5 = 4$

$\lim_{x \to 3^+} f(x) = \lim_{x \to 3}(3x - 5) = 3 \times (3) - 5 = 4$

$\therefore$ $\lim_{x \to 3} f(x) = f(3)$

Therefore, f is continuous at $x = 3$

So, the correct option is (D).

Q. 2. Find the continuity of $f(x) = x$ at $x = k$, k be any positive value.

(A) $f(x)$ is continuous at $x = k$

(B) $f(x)$ is not continuous at $x = k$

(C) $f(x)$ is continuous at $x = 0$

(D) None of the above.

Sol. The given function is,

$f(x) = x$

At $x = 0$,

$f(k) = k$

$\lim_{x \to k} f(x) = \lim_{x \to k}(x) = k$

$\therefore$ $\lim_{x \to k} f(x) = f(k)$

$\therefore f(x)$ is continuous at $x = k$.

So the correct option is (A).

Q. 3. Find the continuity of $f(x) = \dfrac{x^2 - 16}{x + 4}, x \neq -4$ at $x = k$, k be any positive value.

(A) $f(x)$ is not continuous at $x = k$

(B) $f(x)$ is continuous at $x = k$

(C) $f(x)$ is continuous at $x = -4$

(D) $f(x)$ is not continuous at $x = -4$

Sol. The given function is, $f(x) = \dfrac{x^2-16}{x+4}$

At $x = k$,

$$f(k) = \frac{x^2-16}{k+4} = \frac{(k+4)(k-4)}{k+4} = k-4$$

$$\lim_{x\to k} f(x) = \lim_{x\to k}\left(\frac{x^2-16}{x+4}\right) = \frac{(k+4)(k-4)}{k+4} = k-4$$

$$\therefore \lim_{x\to k} f(x) = f(k)$$

So the correct option is (B).

Q. 4. Find all points of discontinuous of f, where f is defined by

$$f(x) = x+3, \text{ if } x \le 3$$
$$f(x) = x-3, \text{ if } x > 3$$

(A) f is discontinuous at $x = 3$

(B) f is continuous at $x = 3$

(C) f is continuous at $x = -3$

(D) f is discontinuous at $x = -3$

Sol. The given function f is

$$f(x) = x+3, \text{ if } x \le 3$$
$$f(x) = x-3, \text{ if } x > 3$$

(i) $c < 3$

(ii) $c = 3$

(iii) $c > 3$

where c be a point on the real line. Then these three cases are valid

Case (i) : $c < 3$

then $f(x) = x+3$

$f(c) = c+3$

$$\lim_{x\to c} f(x) = \lim_{x\to c}(x+3) = c+3$$

$$\lim_{x\to c} f(x) = f(c)$$

Therefore, f is continuous at all points x, such that $x < 3$

Case (ii) : $c = 3$

then $f(x) = x+3$

$$\lim_{x\to c} f(x) = \lim_{x\to c}(x+3) = 3+3 = 6$$

The value of the function f at $c = 3$ is 6.

Case (iii) : $c > 3$

then $f(x) = x-3$

$f(c) = c-3$

$$\lim_{x\to c} f(x) = \lim_{x\to c}(x-3) = c-3$$

$$\lim_{x\to c} f(x) = f(c)$$

The given function f is discontinuous at point $c = 3$.

So the correct option is (A).

Q. 5. Find all points of discontinuous of f, where f is defined by

$$f(x) = x^2+3, \text{ if } x \le 3$$
$$f(x) = x^2-3, \text{ if } x > 3$$

(A) f is discontinuous at $x = -3$

(B) f is discontinuous at $x = 3$

(C) f is continuous at $x = -3$

(D) f is continuous at $x = 3$

Sol. The given function f is

$$f(x) = x^2+3, \text{ if } x \le 3$$
$$f(x) = x^2-3, \text{ if } x > 3$$

(i) $c < 3$

(ii) $c = 3$

(iii) $c > 3$

where c be a point on the real line. Then these three cases are valid

Case (i) : $c < 3$

then $f(x) = x^2+3$

$f(c) = c^2+3$

$$\lim_{x\to c} f(x) = \lim_{x\to c}(x^2+3) = c^2+3$$

$$\lim_{x\to c} f(x) = f(c)$$

Therefore, f is continuous at all points x, such that $x < 3$

Case (ii) : $c = 3$

then $f(x) = x^2+3$

$$\lim_{x\to c} f(x) = \lim_{x\to c}(x^2+3) = 3^2+3 = 12$$

The value of the function f at $c = 3$ is 12.

Case (iii) : $c > 3$

then $f(x) = x^2-3$

$f(c) = c^2-3$

$$\lim_{x\to c} f(x) = \lim_{x\to c}(x^2-3) = c^2-3$$

$$\lim_{x\to c} f(x) = f(c)$$

The given function f is discontinuous at point $c = 3$.

So the correct option is (B).

Q. 6. Find all points of discontinuous of f, where f is defined by

$$f(x) = x^2+x+3, \text{ if } x < 0$$
$$f(x) = 0, \quad \text{if } 0 \le x \le 1$$
$$f(x) = x^2-x+3, \text{ if } x > 1$$

(A) f is discontinuous at all points of c

(B) f is continuous at all points of c

(C) f is discontinuous at $x = 0$

(D) f is continuous at $x = 0$

Sol. The given function f is

$$f(x) = x^2+x+3, \text{ if } x < 0$$

$$f(x) = 0, \quad \text{if } 0 \le x \le 1$$
$$f(x) = x^2 - x + 3, \text{if } x > 1$$

(i) $c < 0$

(ii) $c > 1$

(iii) $0 \le c \le 1$

where c be a point on the real line. Then these three cases are valid.

Case (i) : $c < 0$

then $f(x) = x^2 + x + 3$

$f(c) = c^2 + c + 3$

$$\lim_{x\to c} f(x) = \lim_{x\to c} (x^2 + x + 3) = c^2 + c + 3$$

$$\lim_{x\to c} f(x) = f(c)$$

Therefore, f is continuous at all points x, such that $x < 0$

Case (ii) : $c > 1$

then $f(x) = x^2 - x + 3$

$$\lim_{x\to c} f(x) = \lim_{x\to c} (x^2 - x + 3) = c^2 - c + 3$$

Therefore, f is continuous at all points x, such that $x > 1$.

Case (iii) : $0 \le c \le 1$

then $f(x) = 0$

$f(c) = 0$

$$\lim_{x\to c} f(x) = \lim_{x\to c} (x^2 - x + 3) = c^2 - c + 3$$

$$\lim_{x\to c} f(x) = f(c)$$

Therefore, f is continuous at all points x, such that $0 \le x \le 1$.

But, the given function f is discontinuous at all points of c.

So the correct option is (A).

Q. 7. Find all points of discontinuous of f, where f is defined by

$$f(x) = 3, \quad \text{if } x \le -1$$
$$f(x) = 3x, \text{ if } -1 < x \le 1$$
$$f(x) = 3, \quad \text{if } x > 1$$

(A) f is discontinuous at all points of c

(B) f is continuous at all points of c

(C) f is discontinuous at $x = 0$

(D) f is continuous at $x = 0$

Sol. The given function f is

$$f(x) = 3, \quad \text{if } x \le -1$$
$$f(x) = 3x, \text{ if } -1 < x \le 1$$
$$f(x) = -3, \text{ if } x > 1$$

(i) $c \le -1$

(ii) $c > 1$

(iii) $-1 < c \le 1$

where c be a point on the real line. Then these three cases are valid.

Case (i) : $c \le -1$

then $f(x) = 3$

$f(c) = 3$

$$\lim_{x\to c} f(x) = \lim_{x\to c} 3 = 3$$

$$\lim_{x\to c} f(x) = f(c)$$

Therefore, f is continuous at all points x, such that $x \le -1$

Case (ii) : $c > 1$

then $f(x) = -3$

$$\lim_{x\to c} f(x) = \lim_{x\to c} -3 = -3$$

Therefore, f is continuous at all points x, such that $x > 1$.

Case (iii) : $-1 < c \le 1$

then $f(x) = 3x$

$f(c) = 3c$

$$\lim_{x\to c} f(x) = \lim_{x\to c} (3x) = 3c$$

$$\lim_{x\to c} f(x) = f(c)$$

Therefore, f is continuous at all points x, such that $-1 < c \le 1$.

But, the given function f is discontinuous at all points of c.

So the correct option is (A).

Q. 8. If $y = Ae^{5x} + Be^{-5x}$, then $\frac{d^2y}{dx^2}$ is equal to :★

(A) $25\,y$ **(B)** $5\,y$

(C) $-25\,y$ **(D)** $15\,y$

Sol. Given: $y = Ae^{5x} + Be^{-5x}$

$$\therefore \quad \frac{dy}{dx} = 5Ae^{5x} - 5Be^{-5x}$$

$$\Rightarrow \quad \frac{d^2y}{dx^2} = 25Ae^{5x} + 25Be^{-5x}$$

$$\Rightarrow \quad \frac{d^2y}{dx^2} = 25y$$

So the correct option is (A).

Q. 9. If $y = \log_e\left(\frac{x^2}{e^2}\right)$, then $\frac{d^2y}{dx^2}$ equals :

(A) $-\frac{1}{x}$ **(B)** $-\frac{1}{x^2}$

(C) $\frac{2}{x^2}$ **(D)** $-\frac{2}{x^2}$

★ are board exam questions from previous years

Sol. Given : $y = \log\left(\frac{x^2}{e^2}\right)$

$$\therefore \quad \frac{dy}{dx} = \frac{1}{\frac{x^2}{e^2}} \times \frac{1}{e^2}(2x)$$

$$\Rightarrow \quad \frac{dy}{dx} = \frac{2}{x}$$

$$\Rightarrow \quad \frac{d^2y}{dx^2} = \frac{-2}{x^2}$$

So the correct option is (D).

Chapter 6. Applications of Derivatives

Q. 1. The abscissa of the point on the curve $3y = 6x - 5x^3$, at which the normal passes through origin is :

(A) 1 (B) $\frac{1}{3}$

(C) 2 (D) $\frac{1}{2}$

Sol. Let (x_1, y_1) be the point on the given curve $3y = 6x - 5x^3$ at which the normal passes through the origin. Then we have $\left(\frac{dy}{dx}\right)_{(x_1, y_1)} = 2 - 5x_1^2$. Again, the equation of the normal of (x_1, y_1) passing through the origin gives $2 - 5x_1^2 = \frac{-x_1}{y_1} = \frac{-3}{6-5x_1^2}$.

Since $x_1 = 1$ satisfies the equation.

So the correct option is (A).

Q. 2. The two curves $x^3 - 3xy^2 + 2 = 0$ and $3x^2y - y^3 = 2$

(A) touch each other (B) cut at right angle

(C) cut at an angle $\frac{\pi}{3}$ (D) cut at an angle $\frac{\pi}{4}$

Sol. From first equation of the curve, we have

$$3x^2 - 3y^2 - 6xy \frac{dy}{dx} = 0$$

$$\Rightarrow \quad \frac{dy}{dx} = \frac{x^2 - y^2}{2xy} = (m_1) \text{ say}$$

and second equation of the curve gives

$$6xy + 3x^2 \frac{dy}{dx} - 3y^2 \frac{dy}{dx} = 0$$

$$\Rightarrow \quad \frac{dy}{dx} = \frac{-2xy}{x^2 - y^2} = (m_2) \text{ say}$$

Since $m_1 \cdot m_2 = -1$.

So the correct option is (B).

Q. 3. The tangent to the curve given by $x = e^t . \cos t, y = e^t . \sin t$ at $t = \frac{\pi}{4}$ makes with x-axis an angle:

(A) π (B) $\frac{\pi}{4}$

(C) $\frac{\pi}{3}$ (D) $\frac{\pi}{2}$

Sol. $\frac{dx}{dt} = -e^t . \sin t + e^t \cos t, \; \frac{dy}{dt} = e^t \cos t + e^t \sin t$

$$\therefore \left(\frac{dy}{dx}\right)_{t=\frac{\pi}{4}} = \frac{\cos t + \sin t}{\cos t - \sin t} = \frac{\sqrt{2}}{0}$$

So the correct option is (D).

Q. 4. The equation of the normal to the curve $y = \sin x$ at (0, 0) is :

(A) $x = 0$ (B) $y = 0$

(C) $x + y = 0$ (D) $x - y = 0$

Sol. $\frac{dy}{dx} = \cos x.$

$$\therefore \text{Slope of normal} = \left(\frac{-1}{\cos x}\right)_{x=0} = -1.$$

Hence, the equation of normal is

$y - 0 = -1(x + 0)$

or $x + y = 0$

So the correct option is (C).

Q. 5. The point on the curve $y^2 = x$, where the tangent makes an angle of $\frac{\pi}{4}$ with x-axis is

(A) $\left(\frac{1}{2}, \frac{1}{4}\right)$ (B) $\left(\frac{1}{4}, \frac{1}{2}\right)$

(C) (4, 2) (D) (1, 1)

Sol. $\frac{dy}{dx} = \frac{1}{2y} = \tan\frac{\pi}{4} = 1$

$$\Rightarrow y = \frac{1}{2} \Rightarrow x = \frac{1}{4}$$

So the correct option is (B).

Chapter 7. Integrals

Q. 1. By the inspection method, evaluate :

$$\int (au^2 + bu + c)\, du.$$

(A) $a\left(\frac{u^3}{3}\right) + b\left(\frac{u^2}{2}\right) + cu + C$

(B) $-a\left(\frac{u^3}{3}\right) + b\left(\frac{u^2}{2}\right) - cu + C$

(C) $a\left(\frac{u^3}{3}\right) + b\left(\frac{u^2}{2}\right) - cu + C$

(D) $a\left(\frac{u^3}{3}\right)-b\left(\frac{u^2}{2}\right)-cu+C$

Sol. The given integral is $\int (au^2+bu+c)\,du$.

$$\int (au^2+bu+c)\,du = \int au^2du+\int budu+\int cdu$$
$$= a\int u^2du+b\int udu+c\int du$$
$$= a\left(\frac{u^3}{3}\right)+b\left(\frac{u^2}{2}\right)+cu+C$$

So the correct option is (A).

Q. 2. Find the integral of $\sqrt{u}+\frac{1}{\sqrt{u}}$.

(A) $\frac{2}{3}u^{3/2}+2u^{1/2}+c$ **(B)** $\frac{2}{3}u^{3/2}-2u^{1/2}+c$

(C) $-\frac{2}{3}u^{3/2}-2u^{1/2}+c$ **(D)** $-\frac{2}{3}u^{3/2}+2u^{1/2}+c$

Sol. $\int\left(\sqrt{u}+\frac{1}{\sqrt{u}}\right)du = \int(\sqrt{u})\,du+\int\left(\frac{1}{\sqrt{u}}\right)du$

$$= \int(u^{1/2})\,du+\int\left(\frac{1}{u^{1/2}}\right)du$$
$$= \frac{u^{3/2}}{\frac{3}{2}}+\frac{u^{1/2}}{\frac{1}{2}}+C$$
$$= \frac{2}{3}u^{3/2}+2u^{1/2}+C$$

So the correct option is (A).

Q. 3. Evaluate : $\int \frac{10u^9+10^u\log_e 10}{u^{10}+10^u}du$.

(A) $\log(10^u+u^{10})+C$ **(B)** $(10^u-u^{10})+C$

(C) $10^u+u^{10}+C$ **(D)** $-10^u+u^{10}+C$

Sol. The given integral is $\int\frac{10u^9+10^u\log_e 10}{u^{10}+10^u}du$.

Let $u^{10}+10^u = v$

Differentiating both sides, we get

$$\frac{d}{du}(u^{10}+10^u) = \frac{dv}{du}$$
$$10u^9+10^u\log_e 10 = \frac{dv}{du}$$

$(10u^9+10^u\log_e 10)\,du = dv$

Substitue v for $u^{10}+10^n$ and dv for $(10u^9+10^u\log_e 10)$ du in the given integral.

$$\int\frac{10u^9+10^u\log_e 10}{u^{10}+10^u}du = \int\frac{dv}{v}$$
$$= \int\frac{1}{v}dv$$
$$= \log v+C$$
$$= \log(u^{10}+10^u)+C$$
$$(\because v = u^{10}+10^u)$$

So the correct option is (A).

Q. 4. Find the integral of sin *u* sin 2*u* sin 3*u*.

(A) $\frac{1}{4}[\cos 6u-\cos 4u-\cos 2u\}+C$

(B) $\frac{1}{4}[\cos 4-\cos 4u-\cos 2u]+C$

(C) $\frac{1}{8}\left[\frac{\cos 6u}{3}-\frac{\cos 4u}{2}-\cos 2u\right]+C$

(D) $\frac{1}{8}\left[\frac{\cos 6u}{3}-\frac{\cos 4u}{2}+\cos 2u\right]+C$

Sol. The given integral is,

$\int \sin u\sin 2u\sin 3u du.$

We know that,

$$\sin A\sin B = \frac{1}{2}[\cos(A-B)-\cos(A+B)]$$

Using the above formula for sin 2*u* sin 3*u* in the given integral, we get

$$\int \sin u\sin 2u\sin 3u du$$
$$= \int \sin u\cdot\frac{1}{2}[\cos(2u-3u)-\cos(2u+3u)]\,du$$
$$= \frac{1}{2}\int[\sin u\cos(-u)-\sin u\cos 5u]\,du$$
$$= \frac{1}{2}\int(\sin u\cos u-\sin u\cos 5u)\,du$$

We know that,

$2\sin A\cos A = \sin 2A$ and $\sin A\cos B$

$$= \frac{1}{2}[\sin(A+B)+\sin(A-B)].$$

Using the above two formulas in the integral, we get

$$\frac{1}{2}\int(\sin u\cos u-\sin u\cos 5u)\,du = \frac{1}{2}\int\frac{2\sin u\cos u}{2}du$$
$$-\frac{1}{2}\int\left[\frac{1}{2}\sin(u+5u)+\sin(u-5u)\right]du$$
$$= \frac{1}{4}\int\sin 2udu-\frac{1}{4}\int[\sin 6u+\sin(-4u)]\,du$$
$$= \frac{1}{4}\int\sin 2udu-\left[\int\sin 6udu-\int\sin 4udu\right]$$
$$= \frac{1}{4}\left[\frac{-\cos 2u}{2}-\left(\frac{-\cos 6u}{6}+\frac{\cos 4u}{4}\right)\right]+C$$

$$= \frac{1}{4} \times \frac{1}{2}\left[-\cos 2u + \frac{\cos 6u}{3} - \frac{\cos 4u}{2}\right] + C$$

$$= \frac{1}{8}\left[\frac{\cos 6u}{3} - \frac{\cos 4u}{2} - \cos 2u\right] + C$$

So the correct option is (C).

Q. 5. Integrate : $\frac{u^2}{1-u^6}$.

(A) $\frac{1}{6}\log\left|\frac{1+u^3}{1-u^3}\right| + C$ **(B)** $\log\left|\frac{1+u^3}{1-u^3}\right| + C$

(C) $\log\left|\frac{1+u}{1-u}\right| + C$ **(D)** $\log|1+u^3| + C$

Sol. The given integral is $\int \frac{u^2}{1-u^6} du$

$$\int \frac{u^2}{1-u^6} du = \int \frac{u^2}{1-(u^3)^2} du$$

Substitute z for u^3 and differentiate w.r.t. x,

$$u^3 = z$$

$$\frac{d}{du}u^3 = \frac{dz}{du}$$

$$3u^2 = \frac{dz}{du}$$

$$3u^2\, du = dz$$

$$\int \frac{u^2}{1-(u^3)^2} du = \frac{1}{3}\int \frac{dz}{1-z^2}$$

$$= \frac{1}{3}\left[\frac{1}{2}\log\left|\frac{1+z}{1-z}\right|\right] + C$$

$$= \frac{1}{3}\left[\frac{1}{2}\log\left|\frac{1+u^3}{1-u^3}\right|\right] + C \quad (\because z = u^3)$$

$$= \frac{1}{6}\log\left|\frac{1+u^3}{1-u^3}\right| + C$$

So the correct option is (A).

Q. 6. Evaluate $\int 1.\sin^{-1} x\, dx$.

(A) $x\sin^{-1} x - (1-x^2)^{-\frac{1}{2}} + C$

(B) $\sin^{-1} x - (1-x^2)^{-\frac{1}{2}} + C$

(C) $x\sin^{-1} x + (1-x^2)^{-\frac{1}{2}} + C$

(D) $x\sin^{-1} x - (1-x^2)^{\frac{1}{2}} + C$

Sol. Let, the given integral be $I = \int 1.\sin^{-1} x\, dx$

Integrating by parts,

$$I = \sin^{-1} x \int 1.dx - \int\left\{\left(\frac{d}{dx}\sin^{-1} x\right)\int 1.dx\right\}dx$$

$$= \sin^{-1} x.x - \int \frac{1}{\sqrt{1-x^2}}.x\, dx$$

Substitute $1 - x^2 = t$ in the given integral and differentiate it.

$$\frac{d}{dx}(1-x^2) = \frac{dt}{dx}$$

$$-2x = \frac{dt}{dx}$$

$$xdx = -\frac{1}{2}dt$$

$$= x\sin^{-1} x - \int -\frac{1}{2}\frac{dt}{\sqrt{t}}$$

$$= x\sin^{-1} x + \frac{1}{2}t^{\frac{1}{2}-1} \times (-2)$$

$$= x\sin^{-1} x - (1-x^2)^{-\frac{1}{2}} + C$$

$$(\because t = 1-x^2)$$

So the correct option is (A).

Q. 7. Find the integral of $x^2 . \sin^{-1} x^2\, dx$

(A) $\frac{1}{3}x^2\sin x^2 - \frac{1}{6}(1-x^2)^{-\frac{1}{2}} + C$

(B) $\frac{1}{3}x^2\sin^{-1} x^2 - \frac{1}{6}(1-x^2)^{-\frac{1}{2}} + C$

(C) $\frac{1}{3}x^2\sin^{-1} x^2 - \frac{1}{6}(1-x^2) + C$

(D) $\frac{1}{3}x^2\sin^{-1} x^2 + \frac{1}{6}(1-x^2)^{-\frac{1}{2}} + C$

Sol. The given integral is,

$$I = \int x^2.\sin^{-1} x^2 dx.$$

Integrating by parts,

$$I = \sin^{-1} x^2 \int x^2.dx - \int\left\{\left(\frac{d}{dx}\sin^{-1} x^2\right)\int x^2.dx\right\}dx$$

$$= \frac{1}{3}\sin^{-1} x^2.x^3 - \frac{1}{3}\int \frac{1}{\sqrt{1-x^4}}.x^3 dx$$

Substitute t for $1 - x^4$ and differentiate w.r.t. to x.

$$\frac{d}{dx}(1-x^4) = \frac{dt}{dx}$$

$$-4x^3 = \frac{dt}{dx}$$

$$x^3\, dx = -\frac{1}{4}dt$$

Substitute the values, $1 - x^4 = t$ and $x^3\, dx = -\frac{1}{4}dt$ in the integral.

$$= \frac{1}{3}x^3\sin^{-1} x^2 - \frac{1}{3}\int -\frac{1}{4}\frac{dt}{\sqrt{t}}$$

$$= \frac{1}{3}x^3 \sin^{-1} x + \frac{1}{12} t^{\frac{1}{2}-1} \times (-2)$$

$$= \frac{1}{3}x^2 \sin^{-1} x^2 - \frac{1}{6}(1-x^2)^{-\frac{1}{2}} + C$$

So the correct option is (B).

Q. 8. Evaluate the integral :

$$\int e^x (\cos x - \sin x) dx.$$

(A) $e^x \sin x + C$ **(B)** $\cos x + C$

(C) $e^x \cos x + C$ **(D)** $e^{-x} \cos x + C$

Sol. The given integral is,

$$I = \int e^x (\cos x - \sin x)\, dx$$

Let, $f(x) = \cos x$

Differentiating $f(x)$,

$$f'(x) = -\sin x$$

Substitute $f(x)$ for $\cos x$ and $f'(x)$ for $-\sin x$ in the given integral.

$$I = \int e^x (\cos x - \sin x)\, dx$$

$$= \int e^x \{f(x) - f'(x)\}\, dx$$

We know that $\int e^x \{f(x) - f'(x)\}\, dx = e^x f(x) + C$

$\therefore$ $I = e^x \cos x + C$

So the correct option is (C).

Q. 9. Find the value of :

$$\int e^x \left(\frac{1}{x^2} - \frac{1}{3x^3}\right) dx.$$

(A) $e^x \frac{1}{x^2} + C$ **(B)** $e^x x^2 + C$

(C) $e^x x + C$ **(D)** $e^x + C$

Sol. The given integral is,

$$I = \int e^x \left(\frac{1}{x^2} - \frac{1}{3x^3}\right) dx$$

Let, $f(x) = \frac{1}{x^2}$

Differentiating $f(x)$,

$$f'(x) = -\frac{1}{3x^3}$$

Substituting $f(x)$ for $\frac{1}{x^2}$ and $f'(x)$ for $-\frac{1}{3x^3}$ in the given integral, we get

$$I = \int e^x \left(\frac{1}{x^2} - \frac{1}{3x^3}\right) dx$$

$$= \int e^x \{f(x) + f'(x)\}\, dx$$

We know that

$$\int e^x \{f(x) + f'(x)\}\, dx = e^x f(x) + C.$$

★ are board exam questions from previous years

$\therefore$ $I = e^x \frac{1}{x^2} + C$

So the correct option is (A).

Q. 10. $\int_0^{\pi/8} \tan^2 (2x)\, dx$ **is equal to★**

(A) $\frac{4-\pi}{8}$ **(B)** $\frac{4+\pi}{8}$

(C) $\frac{4-\pi}{4}$ **(D)** $\frac{4-\pi}{2}$

Sol. $\int_0^{\pi/8} \tan^2(2x)dx = \int_0^{\pi/8} \left(\sec^2(2x) - 1\right) dx$

$$= \left[\frac{\tan(2x)}{2} - x\right]_0^{\pi/8}$$

$$= \frac{1}{2} - \frac{\pi}{8} - 0$$

$$= \frac{4-\pi}{8}$$

So the correct option is (A).

Q. 11. $\int x^2 e^{x^3}\, dx$ **equals :★**

(A) $\frac{1}{3}e^{x^3} + C$ **(B)** $\frac{1}{3}e^{x^4} + C$

(C) $\frac{1}{2}e^{x^3} + C$ **(D)** $\frac{1}{2}e^{x^2} + C$

Sol. Let $I = \int x^2 e^{x^3} dx$

Put $x^3 = t$

$\Rightarrow$ $3x^2 dx = dt$

$\Rightarrow$ $x^2 dx = \frac{1}{3} dt$

$\therefore$ $I = \frac{1}{3}\int e^t dt$

$$I = \frac{1}{3}e^t + C$$

$$I = \frac{1}{3}e^{x^3} + C$$

So the correct option is (A).

Q. 12. $\int 4^x 3^x\, dx$ **equals :★**

(A) $\frac{12^x}{\log 12} + C$ **(B)** $\frac{4^x}{\log 4} + C$

(C) $\left(\frac{4^x \cdot 3^x}{\log 4 \cdot \log 3}\right)+C$ (D) $\frac{3^x}{\log 3}+C$

Sol. Let $I = \int 4^x 3^x \, dx$

$\Rightarrow \quad I = \int 12^x \, dx$

$\Rightarrow \quad I = \frac{12^x}{\log 12}+C$

So the correct option is (A).

Chapter 8. Applications of the Integrals

Q. 1. The area enclosed by the circle $x^2 + y^2 = 2$ is equal to

(A) 4π sq units (B) $2\sqrt{2}\,\pi$ sq units

(C) $4\pi^2$ sq units (D) 2π sq units

Sol. $\because$ Area $= 4\int_0^{\sqrt{2}} \sqrt{2-x^2}$

$$= 4\left(\frac{x}{2}\sqrt{2-x^2} + \sin^{-1}\frac{x}{\sqrt{2}}\right)_0^{\sqrt{2}}$$

$= 2\pi$ sq units

So the correct option is (D).

Q. 2. The area enclosed by the ellipse $\frac{x^2}{a^2}+\frac{y^2}{b^2}=1$ is equal to

(A) $\pi^2 ab$ (B) πab

(C) $\pi a^2 b$ (D) πab^2

Sol. Since, Area $= 4\int_0^a \frac{b}{a}\sqrt{a^2-x^2}\,dx$

$$= \frac{4b}{a}\left[\frac{\pi}{2}\sqrt{a^2-x^2} + \frac{a^2}{2}\sin^{-1}\frac{x}{a}\right]_0^a$$

$= \pi ab$

So the correct option is (B).

Chapter 9. Differential Equations

Q. 1. The degree of the differential equation

$$\left(1+\frac{dy}{dx}\right)^3 = \left(\frac{d^2y}{dx^2}\right)^2 \text{ is}$$

(A) 1 (B) 2

(C) 3 (D) 4

Sol. The correct option is (B).

Q. 2. The degree of the differential equation

$$\frac{d^2y}{dx^2}+3\left(\frac{dy}{dx}\right)^2 = x^2\log\left(\frac{d^2y}{dx^2}\right) \text{ is}$$

(A) 1 (B) 2

(C) 3 (D) Not defined

Sol. The given differential equation is not a polynomial equation in terms of its derivatives, so its degree is not defined.

So the correct option is (D).

Q. 3. The order and degree of the differential equation

$$\left[1+\left(\frac{dy}{dx}\right)^2\right]^2 = \frac{d^2y}{dx^2} \text{ respectively, are}$$

(A) 1, 2 (B) 2, 2 (C) 2, 1 (D) 4, 2

Sol. The correct option is (C).

Q. 4. The order of the differential equation of all circles of given radius *a* is :

(A) 1 (B) 2 (C) 3 (D) 4

Sol. Let the equation of given family be $(x - h)^2 + (y - k)^2 = a^2$. It has two arbitrary constants h and k. Therefore, the order of the given differential equation will be 2.

So the correct option is (B).

Q. 5. The solution of the differential equation $2x.\frac{dy}{dx} - y = 3$ represents a family of :

(A) straight lines (B) circles

(C) parabolas (D) ellipses

Sol. Given equation can be written as

$$\frac{2dy}{y+3} = \frac{dx}{x} \Rightarrow 2\log(y+3) = \log x + \log c$$

$\Rightarrow (y + 3)^2 = cx$ which represents the family of parabolas.

So the correct option is (C).

Q. 6. The integrating factor of the differential equation

$$\frac{dy}{dx}(x\log x) + y = 2\log x \text{ is}$$

(A) e^x (B) $\log x$

(C) $\log(\log x)$ (D) x

Sol. Given equation can be written as

$$\frac{dy}{dx}+\frac{y}{x\log x} = \frac{2}{x}.$$

$$\therefore \quad \text{I.F.} = e^{\int \frac{1}{x\log x}dx} = e^{\log(\log x)} = \log x.$$

So the correct option is (B).

Q. 7. Solution of the differential equation

$$\frac{dx}{x}+\frac{dy}{y} = 0 \text{ is}$$

(A) $\frac{1}{x}+\frac{1}{y}=c$ (B) $\log x.\log y=c$

(C) $xy=c$ (D) $x+y=c$

Sol. From the given equation, we get

$$\log x+\log y=\log c$$

giving $xy=c$.

So the correct option is (C).

Q. 8. The solution of the differential equation

$$x\frac{dy}{dx}+2y=x^2 \text{ is}$$

(A) $y=\frac{x^2+c}{4x^2}$ (B) $y=\frac{x^2}{4}+c$

(C) $y=\frac{x^4+c}{x^2}$ (D) $y=\frac{x^4+c}{4x^2}$

Sol. I.F. $= e^{\int\frac{2}{x}dx}=e^{2\log x}=e^{\log x^2}=x^2$.

$\therefore$ The solution is

$$y.x^2=\int x^2.xdx=\frac{x^4}{4}+k$$

i.e., $$y=\frac{x^4+c}{4x^2}.$$

So the correct option is (D).

Chapter 10. Vectors

Q. 1. Find the magnitudes of the vectors given below:

$$\vec{a}=2\hat{i}+3\hat{j}+\hat{k},$$
$$\vec{b}=\hat{i}-2\hat{j}-3\hat{k},$$
$$\vec{c}=5\hat{i}+2\hat{j}+4\hat{k}$$

(A) $\sqrt{3},\sqrt{15},1$ (B) $\sqrt{35},\sqrt{2},9$

(C) 1, 1, 1 (D) $\sqrt{14},\sqrt{14},\sqrt{45}$

Sol. The given vectors are,

$$\vec{a}=2\hat{i}+3\hat{j}+\hat{k}$$
$$\vec{b}=\hat{i}-2\hat{j}-3\hat{k}$$
$$\vec{c}=5\hat{i}+2\hat{j}+4\hat{k}$$

Magnitude of $\vec{a}$,

$$|\vec{a}|=\sqrt{2^2+3^2+1^2}=\sqrt{4+9+1}=\sqrt{14}$$

Magnitude of $\vec{b}$,

$$|\vec{b}|=\sqrt{1^2+(-2)^2+(-3)^2}=\sqrt{1+4+9}=\sqrt{14}$$

Magnitude of $\vec{c}$,

$$|\vec{c}|=\sqrt{5^2+2^2+4^2}=\sqrt{25+4+16}=\sqrt{45}$$

Hence, the required magnitudes of the given vectors are $\sqrt{14}$, $\sqrt{14}$ and $\sqrt{45}$.

So the correct option is (D).

Q. 2. Find the vector components of a vector having the initial point (6, 4) and the terminal point (– 2, 7) :

(A) $2\hat{i},\hat{j}$ (B) $\hat{i},-7\hat{j}$

(C) $-8\hat{i},3\hat{j}$ (D) $-3\hat{i},8\hat{j}$

Sol. Let, the initial point of the vector is $A(6, 4)$ and the terminal point of the vector is $B(-2, 7)$.

$\therefore$ The required vector is,

$$\vec{AB}=(-2-6)\hat{i}+(7-4)\hat{j}=-8\hat{i}+3\hat{j}$$

Hence, the required vector components of the vector are $-8\hat{i}+3\hat{j}$.

So the correct option is (C).

Q. 3. Find the scalar components of a vector having the initial point (5, 9) and (– 7, 4).

(A) 12, – 5 (B) – 12, 5

(C) – 12, – 5 (D) 12, 5

Sol. Let, the initial point of the vector is $A(5, 9)$ and the terminal point of the vector is $B(-7, 4)$.

$\therefore$ The required vector is,

$$\vec{AB}=(-7-5)\hat{i}+(4-9)\hat{j}=-12\hat{i}-5\hat{j}$$

Hence, the required scalar components of the vector are – 12 and – 5.

So the correct option is (C).

Q. 4. The unit vector in the direction of the vector $\vec{a}=2\hat{i}+\hat{j}+3\hat{k}$ is :

(A) $\frac{2}{\sqrt{7}}\hat{i}+\frac{1}{\sqrt{7}}\hat{j}+\frac{1}{\sqrt{7}}\hat{k}$

(B) $\frac{2}{\sqrt{14}}\hat{i}+\frac{1}{\sqrt{14}}\hat{j}+\frac{3}{\sqrt{14}}\hat{k}$

(C) $\frac{1}{\sqrt{14}}\hat{i}+\frac{1}{\sqrt{14}}\hat{j}+\frac{1}{\sqrt{14}}\hat{k}$

(D) $\frac{4}{\sqrt{14}}\hat{i}+\frac{1}{\sqrt{14}}\hat{j}+\frac{1}{\sqrt{14}}\hat{k}$

Sol. Let, the unit vector in the direction of the vector

$$\vec{a} = 2\hat{j}+\hat{j}+3\hat{k} \text{ is } \hat{a}.$$

The unit vector $\hat{a}$ of the vector $\vec{a}$ is defined by

$$\hat{a} = \frac{\vec{a}}{|\vec{a}|}.$$

So, the magnitude of $\vec{a}$ is,

$$|\vec{a}| = \sqrt{2^2+1^2+3^2}$$
$$= \sqrt{4+1+9}$$
$$= \sqrt{14}$$

∴ The required unit vector is,

$$\hat{a} = \frac{\vec{a}}{|\vec{a}|}$$
$$= \frac{2\hat{i}+\hat{j}+3\hat{k}}{\sqrt{14}}$$
$$= \frac{2\hat{i}}{\sqrt{14}}+\frac{\hat{j}}{\sqrt{14}}+\frac{3\hat{k}}{\sqrt{14}}$$
$$= \frac{2}{\sqrt{14}}\hat{i}+\frac{1}{\sqrt{14}}\hat{j}+\frac{3}{\sqrt{14}}\hat{k}.$$

So the correct option is (B).

Q. 5. What is the unit vector in the direction of the vector $\vec{AB}$, where A and B are the points (2, 1, 2) and (1, 1, 1) respectively.

(A) $-\frac{1}{\sqrt{2}}\hat{i}-\frac{1}{\sqrt{2}}\hat{k}$ (B) $\frac{1}{\sqrt{2}}\hat{i}+\frac{1}{\sqrt{2}}\hat{k}$

(C) $\frac{1}{\sqrt{2}}\hat{i}-\frac{1}{\sqrt{2}}\hat{k}$ (D) $-\frac{1}{\sqrt{2}}\hat{i}+\frac{1}{\sqrt{2}}\hat{k}$

Sol. Given, $A(2, 1, 2)$ and $B(1, 1, 1)$ are the points of the vector $\vec{AB}$.

The unit vector in the direction of the vector $\vec{AB}$ is

$$\frac{\vec{AB}}{|\vec{AB}|}.$$

So, the vector $\vec{AB}$ is,

$$\vec{AB} = (1-2)\hat{i}+(1-1)\hat{j}+(1-2)\hat{k}$$
$$= -\hat{i}+0\cdot\hat{j}-\hat{k}$$
$$= -\hat{i}-\hat{k}$$

Magnitude of the vector $\vec{AB}$,

$$|\vec{AB}| = \sqrt{(-1)^2+(-1)^2}$$
$$= \sqrt{1+1}$$
$$= 2$$

Hence, the required unit vector is,

$$\frac{\vec{AB}}{|\vec{AB}|} = \frac{-\hat{i}-\hat{k}}{\sqrt{2}}$$
$$= \left(\frac{-\hat{i}}{\sqrt{2}}\right)+\left(\frac{-\hat{k}}{\sqrt{2}}\right)$$
$$= -\frac{1}{\sqrt{2}}\hat{i}-\frac{1}{\sqrt{2}}\hat{k}.$$

So the correct option is (A).

Q. 6. The direction cosines of the vector $\vec{a} = 5\hat{i}+7\hat{j}+3\hat{k}$ are :

(A) $\frac{1}{\sqrt{83}},\frac{3}{\sqrt{83}},\frac{9}{\sqrt{83}}$ (B) $\frac{5}{\sqrt{83}},\frac{7}{\sqrt{83}},\frac{3}{\sqrt{83}}$

(C) $\frac{5}{\sqrt{83}},\frac{1}{\sqrt{83}},\frac{-3}{\sqrt{83}}$ (D) $\frac{5}{\sqrt{83}},\frac{-7}{\sqrt{83}},\frac{-3}{\sqrt{83}}$

Sol. The given vector is

$$\vec{a} = 5\hat{i}+7\hat{j}+3\hat{k}.$$

The magnitude of the vector $\vec{a}$,

$$|\vec{a}| = \sqrt{5^2+7^2+3^2}$$
$$= \sqrt{25+49+9}$$
$$= \sqrt{83}$$

Hence, the direction cosines of the vector $\vec{a}$ are

$$\left(\frac{5}{\sqrt{83}},\frac{7}{\sqrt{83}},\frac{3}{\sqrt{83}}\right).$$

So the correct option is (B).

Q. 7. The direction cosines of the vector joining the points $L(2, 2, 3)$ and $M(-3, 1, 1)$ directed from L and M are :

(A) $\left(-\frac{5}{\sqrt{30}},-\frac{1}{\sqrt{30}},-\frac{2}{\sqrt{30}}\right)$

(B) $\left(\frac{5}{\sqrt{30}},\frac{1}{\sqrt{30}},\frac{2}{\sqrt{30}}\right)$

(C) $\left(\frac{5}{\sqrt{30}}, -\frac{1}{\sqrt{30}}, -\frac{2}{\sqrt{30}}\right)$

(D) $\left(-\frac{5}{\sqrt{30}}, -\frac{1}{\sqrt{30}}, \frac{2}{\sqrt{30}}\right)$

Sol. The given points are $L(2, 2, 3)$ and $M(-3, 1, 1)$

Therefore, the vector $\vec{LM}$ is,

$$\vec{LM} = (-3-2)\hat{i} + (1-2)\hat{j} + (1-3)\hat{k}$$
$$= -5\hat{i} - \hat{j} - 2\hat{k}$$

The magnitude of the vector $\vec{LM}$,

$$|\vec{LM}| = \sqrt{(-5)^2 + (-1)^2 + (-2)^2}$$
$$= \sqrt{25+1+4}$$
$$= \sqrt{30}$$

Hence, the direction cosines of the vector $\vec{LM}$ are,

$$\left(-\frac{5}{\sqrt{30}}, -\frac{1}{\sqrt{30}}, -\frac{2}{\sqrt{30}}\right).$$

So the correct option is (A).

Q. 8. Find the angle between the vectors $\vec{a}$ and $\vec{b}$ whose magnitudes are $\sqrt{3}$ and $\sqrt{3}$, where $\vec{a}.\vec{b} = 3$.

(A) 0° (B) $\frac{\pi}{2}$

(C) $\frac{\pi}{4}$ (D) $\frac{\pi}{3}$

Sol. Given, the magnitude of $\vec{a}$, $|\vec{a}| = \sqrt{3}$ and the magnitude of $\vec{b}$, $|\vec{b}| = \sqrt{3}$.

Also, $\vec{a}\cdot\vec{b} = 3$.

We know that

$$\vec{a}\cdot\vec{b} = |\vec{a}|\cdot|\vec{b}|\cos\theta$$
$$3 = \sqrt{3}\times\sqrt{3}\times\cos\theta$$
$$\cos\theta = \frac{3}{\sqrt{3}\times\sqrt{3}}$$
$$\cos\theta = \frac{3}{3}$$
$$\cos\theta = 1$$
$$\theta = 0°.$$

Hence, the angle between the vectors $\vec{a}$ and $\vec{b}$ is 0°.

So the correct option is (A).

Q. 9. Find the value of $(\vec{a} - 5\vec{b})\cdot(2\vec{a} + \vec{b})$.

(A) $2|\vec{a}|^2 - 9\vec{a}\cdot\vec{b} - 5|\vec{b}|^2$

(B) $|\vec{a}|^2 + 5\vec{a}\cdot\vec{b} - 9|\vec{b}|^2$

(C) $|\vec{a}|^2 + 5\vec{a}\cdot\vec{b} + |\vec{b}|^2$

(D) $|\vec{a}|^2 + \vec{a}\cdot\vec{b} + |\vec{b}|^2$

Sol.

$$(\vec{a} - 5\vec{b})\cdot(2\vec{a} + \vec{b}) = \vec{a}\cdot2\vec{a} + \vec{a}\cdot\vec{b} - 5\vec{b}\cdot2\vec{a} - 5\vec{b}\cdot\vec{b}$$
$$= 2\vec{a}\cdot\vec{a} + \vec{a}\cdot\vec{b} - 10\vec{a}\cdot\vec{b} - 5\vec{b}\cdot\vec{b}$$
$$= 2|\vec{a}|^2 - 9\vec{a}\cdot\vec{b} - 5|\vec{b}|^2$$

So the correct option is (A).

Q. 10. Find the unit vector which is perpendicular to both the vectors $\vec{a} + \vec{b}$ and $\vec{a} - \vec{b}$, where

$$\vec{a} = \hat{i} + 2\hat{j} + \hat{k}$$
$$\vec{b} = 2\hat{i} - \hat{j} + \hat{k}.$$

(A) $\mp\frac{1}{\sqrt{6}}\hat{i} \mp \frac{1}{\sqrt{6}}\hat{j} \pm \frac{2}{\sqrt{6}}\hat{k}$

(B) $\mp\frac{3}{\sqrt{6}}\hat{i} \mp \frac{1}{\sqrt{6}}\hat{j} \pm \frac{2}{\sqrt{6}}\hat{k}$

(C) $\mp\frac{3}{\sqrt{6}}\hat{i} \mp \frac{4}{\sqrt{6}}\hat{j} \pm \frac{2}{\sqrt{6}}\hat{k}$

(D) $\mp\frac{3}{\sqrt{6}}\hat{i} \mp \frac{5}{\sqrt{6}}\hat{j} \pm \frac{2}{\sqrt{6}}\hat{k}$

Sol. The given vectors are,

$$\vec{a} = \hat{i} + 2\hat{j} + \hat{k}$$
$$\vec{b} = 2\hat{i} - \hat{j} + \hat{k}.$$

So, $$\vec{a} + \vec{b} = \hat{i} + 2\hat{j} + \hat{k} + 2\hat{i} - \hat{j} + \hat{k}$$
$$= 3\hat{i} + \hat{j} + 2\hat{k}$$

Also, $$\vec{a} - \vec{b} = (\hat{i} + 2\hat{j} + \hat{k}) - (2\hat{i} - \hat{j} + \hat{k})$$
$$= \hat{i} + 2\hat{j} + \hat{k} - 2\hat{i} + \hat{j} - \hat{k}$$
$$= -\hat{i} + 3\hat{j}$$

$$\therefore (\vec{a} + \vec{b}) \times (\vec{a} - \vec{b}) = \begin{vmatrix} \hat{i} & \hat{j} & \hat{k} \\ 3 & 1 & 2 \\ -1 & 3 & 0 \end{vmatrix}$$
$$= \hat{i}(1-6) - \hat{j}(3+2) + \hat{k}(9+1)$$
$$= -5\hat{i} - 5\hat{j} + 10\hat{k}$$

$$\left|(\vec{a} + \vec{b}) \times (\vec{a} - \vec{b})\right| = \sqrt{(-5)^2 + (-5)^2 + 10^2}$$
$$= \sqrt{25+25+100}$$

$= \sqrt{150}$

$= 5\sqrt{6}$

Hence, the unit vector perpendicular to both the vectors $\vec{a}+\vec{b}$ and $\vec{a}-\vec{b}$ is,

$$\pm\frac{(\vec{a}+\vec{b})\times(\vec{a}-\vec{b})}{|(\vec{a}+\vec{b})\times(\vec{a}-\vec{b})|} = \pm\frac{-5\hat{i}-5\hat{j}+10\hat{k}}{5\sqrt{6}}$$

$$= \mp\frac{5}{5\sqrt{6}}\hat{i}\mp\frac{5}{5\sqrt{6}}\hat{j}\pm\frac{10}{5\sqrt{6}}\hat{k}$$

$$= \mp\frac{1}{\sqrt{6}}\hat{i}\mp\frac{1}{\sqrt{6}}\hat{j}\pm\frac{2}{\sqrt{6}}\hat{k}$$

So the correct option is (A).

Q. 11. If $\vec{a}\cdot\vec{b}=\frac{1}{2}|\vec{a}||\vec{b}|$, then the angle between $\vec{a}$ and $\vec{b}$ is:★

(A) 0°　　**(B)** 30°

(C) 60°　　**(D)** 90°

Sol. $\vec{a}.\vec{b} = \frac{1}{2}|\vec{a}||\vec{b}| = |\vec{a}||\vec{b}|\cos\theta$

$\Rightarrow \quad \cos\theta = \frac{1}{2}$

$\Rightarrow \quad \theta = 60°$

So the correct option is (C).

Q. 12. Let $\vec{a} = \hat{i} - 2\hat{j} + 3\hat{k}$. If $\vec{b}$ is a vector such that $\vec{a}.\vec{b} = |\vec{b}|^2$ and $\vec{a}.\vec{b} = |\vec{b}|^2$, then $|\vec{b}|$ equals★

(A) 7　　**(B)** 14

(C) $\sqrt{7}$　　**(D)** 21

Sol. $\quad |\vec{a}-\vec{b}|^2 = |\vec{a}|^2 + |\vec{b}|^2 - 2\vec{a}.\vec{b}$

$\Rightarrow \quad (\sqrt{7})^2 = (1+4+9)+|\vec{b}|^2 - 2|\vec{b}|^2$

$\Rightarrow \quad 7 - 14 = -|\vec{b}|^2$

$\Rightarrow \quad |\vec{b}|^2 = 7$

$\therefore \quad |\vec{b}| = \sqrt{7}$

So the correct option is (C).

Q. 13. If $\hat{i}, \hat{j}, \hat{k}$ are unit vectors along three mutually perpendicular directions, then★

(A) $\hat{i}.\hat{j} = 1$　　**(B)** $\hat{i}\times\hat{j} = 1$

(C) $\hat{i}.\hat{k} = 0$　　**(D)** $\hat{i}\times\hat{k} = 0$

Sol. The correct option is (C).

★ are board exam questions from previous years

Q. 14. ABCD is a rhombus whose diagonals intersect at E. Then $\vec{EA}+\vec{EB}+\vec{EC}+\vec{ED}$ equals :[H]

(A) $\vec{0}$　　**(B)** $\vec{AD}$

(C) $2\vec{BC}$　　**(D)** $2\vec{AD}$

Sol. The correct option is (A).

Chapter 11. Three-Dimensional Geometry

Q. 1. The coordinates of the foot of the perpendicular drawn from the point (2, 5, 7) on the *x*-axis are given by :

(A) (2, 0, 0)　　**(B)** (0, 5, 0)

(C) (0, 0, 7)　　**(D)** (0, 5, 7)

Sol. The correct option is (A).

Q. 2. P is a point on the line segment joining the points (3, 2, – 1) and (6, 2, – 2). If *x* coordinate of P is 5, then its *y* coordinate is :

(A) 2　　**(B)** 1　　**(C)** – 1　　**(D)** – 2

Sol. Let P divides the line segment in the ratio of λ : 1, *x* - coordinate of the point P may be expressed as

$$x = \frac{6\lambda+3}{\lambda+1}$$

giving $\frac{6\lambda+3}{\lambda+1} = 5$

so that $\lambda = 2$.

Thus, *y*-coordinate of P is $\frac{2\lambda+2}{\lambda+1} = 2$.

So the correct option is (A).

Q. 3. If α, β, γ are the angles that a line makes with the positive direction of *x*, *y*, *z* axis, respectively, then the direction cosines of the line are.

(A) sin α, sin β, sin γ　　(B) cos α, cos β, cos γ

(C) tan α, tan β, tan γ　　(D) $\cos^2\alpha, \cos^2\beta, \cos^3\gamma$

Sol. The correct option is (B).

Q. 4. The distance of a point P (*a*, *b*, *c*) from *x*-axis is

(A) $\sqrt{a^2+c^2}$　　**(B)** $\sqrt{a^2+b^2}$

(C) $\sqrt{b^2+c^2}$　　**(D)** b^2+c^2

Sol. The required distance is the distance of P (*a*, *b*, *c*) from Q (*a*, 0, 0), which is $\sqrt{b^2+c^2}$.

So the correct option is (C).

Q. 5. The equations of *x*-axis in space are :

(A) $x=0, y=0$　　**(B)** $x=0, z=0$

(C) $x=0$　　**(D)** $y=0, z=0$

Sol. On *x*-axis the *y* -coordinate and *z* - coordinates are zero.

So the correct option is (D).

Q. 6. A line makes equal angles with coordinate axis. Direction cosines of this line are

(A) $\pm(1, 1, 1)$ (B) $\pm\left(\frac{1}{\sqrt{3}}, \frac{1}{\sqrt{3}}, \frac{1}{\sqrt{3}}\right)$

(C) $\pm\left(\frac{1}{3}, \frac{1}{3}, \frac{1}{3}\right)$ (D) $\pm\left(\frac{1}{\sqrt{3}}, \frac{-1}{\sqrt{3}}, \frac{-1}{\sqrt{3}}\right)$

Sol. Let the line make angle α with each of the axis. Then, its direction cosines are $\cos\alpha$, $\cos\alpha$, $\cos\alpha$. Since $\cos^2\alpha + \cos^2\alpha + \cos^2\alpha = 1$.

$$\therefore \quad \cos\alpha = \pm\frac{1}{\sqrt{3}}$$

The correct option is (B).

Q. 7. The two lines $x = ay + b$, $z = cy + d$; and $x = a'y + b'$, $z = c'y + d'$ are perpendicular to each other, if*

(A) $\frac{a}{a'} + \frac{c}{c'} = 1$ (B) $\frac{a}{a'} + \frac{c}{c'} = -1$

(C) $aa' + cc' = 1$ (D) $aa' + cc' = -1$

So the correct option is (D).

Q. 8. The two planes $x - 2y + 4z = 10$ and $18x + 17y + kz = 50$ are perpendicular, if k is equal to :*

(A) –4 (B) 4

(C) 2 (D) –2

Sol. $18 - 34 + 4k = 0$

{using $A_1A_2 + B_1B_2 + C_1C_2 = 0$}

$\Rightarrow \quad 4k = 16$

$\Rightarrow \quad k = 4$

So the correct option is (B).

Q. 9. The line $\frac{x-2}{3} = \frac{y-3}{4} = \frac{z-4}{5}$ is parallel to the plane*

(A) $2x + 3y + 4z = 0$ (B) $3x + 4y - 5z = 7$

(C) $2x + y - 2z = 0$ (D) $x - y + z = 2$

Sol. The correct option is (C).

Q. 10. The lines $\frac{x-2}{1} = \frac{y-3}{1} = \frac{4-z}{k}$ and $\frac{x-1}{k} = \frac{y-4}{2} = \frac{z-5}{-2}$ are mutually perpendicular if the value of k is :*

(A) $-\frac{2}{3}$ (B) $\frac{2}{3}$

(C) –2 (D) 2

Sol. First line is,

$$\frac{x-2}{1} = \frac{y-3}{1} = \frac{z-4}{-k}$$

$\therefore$ Direction ratios of first line = $\langle 1, 1, -k\rangle$

Second line is,

$$\frac{x-1}{k} = \frac{y-4}{2} = \frac{z-5}{-2}$$

$\therefore$ Direction ratios of second line = $\langle k, 2, -2\rangle$

We know, two lines are perpendicular to each other, if the dot product of their direction ratios is zero.

i.e., $1 \times k + 1 \times 2 + (-k) \times (-2) = 0$

$\Rightarrow \quad k + 2 + 2k = 0$

$\Rightarrow \quad 3k = -2$

$\Rightarrow \quad k = -\frac{2}{3}$

So the correct option is (A).

Q. 11. The distance of the origin (0, 0, 0) from the plane $-2x + 6y - 3z = -7$ is :*

(a) 1 unit (B) $\sqrt{2}$ units

(C) $2\sqrt{2}$ units (D) 3 units

Sol. Given : Equation of plane is,

$-2x + 6y - 3z = -7$

$\Rightarrow \; -2x + 6y - 3z + 7 = 0$

We know distance of point (x_1, y_1, z_1) from a plane $Ax + By + Cz + D = 0$ is given as

$$\text{Distance} = \left|\frac{Ax_1 + By_1 + Cz_1 + D}{\sqrt{A^2 + B^2 + C^2}}\right|$$

$$\therefore \quad \text{Distance} = \left|\frac{-2(0) + 6(0) + (-3)(0) + 7}{\sqrt{4 + 36 + 9}}\right|$$

$$= \left|\frac{7}{\sqrt{49}}\right| = \left|\frac{7}{7}\right| = 1 \text{ unit}$$

So the correct option is (A).

Chapter 12. Linear Programming

Q. 1. In an LPP, if the objective function $z = ax + by$ has the same maximum value on two corner points of the feasible region, then the number of points at which z_{max} occurs is*

(A) 0 (B) 2

(C) finite (D) infinite

Sol. The correct option is (D).

Q. 2. The graph of the inequality $2x + 3y > 6$ is*

(A) half plane that contains the origin

(B) half plane that neither contains the origin nor the points of the line $2x + 3y = 6$

(C) whole XOY–plane excluding the points on the line $2x + 3y = 6$

(D) entire XOY plane.

Sol. The correct option is (B).

Chapter 13. Probability

Q. 1. Twelve cards numbered from 1 to 12 are placed in pack, mixed up throughly then one card is drawn randomly. If it is known the number on each drawn card is more than 6. What is the probability that it is an even number?

(A) $\frac{1}{2}$ (B) $\frac{3}{4}$

(C) $\frac{1}{6}$ (D) $\frac{3}{5}$

* are board exam questions from previous years

Sol. $A = \{2, 4, 6, 8, 10, 12\}$

$B = \{7, 8, 9, 10, 11, 12\}$

$P(A) = \frac{6}{12}, P(B) = \frac{6}{12},$

$A \cap B = \{8, 10, 12\}, P(A \cap B) = \frac{3}{12}$

$$P(A/B) = \frac{P(A \cap B)}{P(B)}$$

$$= \frac{\frac{3}{12}}{\frac{6}{12}} = \frac{3}{6} = \frac{1}{2}$$

So the correct option is (A).

Q. 2. A dice is thrown twice and sum of the number appearing is observed to be 4. What is the conditional probability that number 2 atleast appear once?

(A) 0.22 **(B)** 0.54

(C) 0.33 **(D)** 0.66

Sol. $F = \{(1, 3), (2, 2), (3, 1)\}$

$E = \{(2, 2)\}$

$(E \cap F) = \{(2, 2)\}$

$$P(E \cap F) = \frac{1}{36}$$

$$P(E/F) = \frac{P(E \cap F)}{P(F)}$$

$$= \frac{\frac{1}{36}}{\frac{3}{36}} = \frac{1}{3} = 0.33$$

So the correct option is (C).

Q. 3. Given, $P(A) = \frac{5}{12}$, $P(B) = \frac{5}{12}$, $P(A/B) = \frac{2}{5}$, find $P(A \cup B)$.

(A) $\frac{4}{3}$ **(B)** $\frac{3}{4}$

(C) $\frac{1}{3}$ **(D)** $\frac{3}{5}$

Sol. $P(A \cap B) = P(A/B) \times P(B)$

$$= \frac{2}{5} \times \frac{5}{12}$$

$$= \frac{2}{12} = \frac{1}{6}$$

$\because \quad P(A \cup B) = P(A) + P(B) - P(A \cap B)$

$$= \frac{5}{12} + \frac{5}{12} - \frac{2}{12}$$

$$= \frac{10-2}{12} = \frac{3}{4}$$

So the correct option is (B).

Q. 4. A bag contains 8 black and 5 blue balls three balls are drawn at random without replacements. What is the probability that all drawn balls are black colours?

(A) 0.122 **(B)** 0.195

(C) 0.132 **(D)** 0.022

Sol. 1st black ball drawn at random the probability

$$= \frac{8}{13}$$

2nd black ball drawn at random the probability

$$= \frac{7}{12}$$

3rd black ball drawn at random the probability

$$= \frac{6}{11}$$

$$P = \frac{8}{13} \times \frac{7}{12} \times \frac{6}{11}$$

So the total probability $= \frac{28}{13 \times 11}$

$= 0.195$

So the correct option is (B).

Q. 5. If $P(A) = \frac{2}{3}$ and $P(B) = \frac{2}{5}$, find $P(A \cap B)$ if A and B are independent events :

(A) $\frac{4}{15}$ **(B)** $\frac{3}{7}$

(C) $\frac{2}{15}$ **(D)** $\frac{1}{14}$

Sol. $P(A \cap B) = P(A) \times P(B)$

$$= \frac{2}{3} \times \frac{2}{5}$$

$$= \frac{4}{15}$$

So the correct option is (A).

Q. 6. Probability of obtaining an odd prime number on each die, when a die is rolled :

(A) $\frac{2}{5}$ **(B)** $\frac{1}{2}$

(C) $\frac{1}{3}$ **(D)** $\frac{1}{4}$

Sol. $E = \{(1, 1), (3, 3), (5, 5), (1, 3), (1, 5), (3, 1), (5, 1), (3, 5), (5, 3)\}$

Total number of events $= 6 \times 6$

$= 36$

Number of events occurs $= 9$

$$P = \frac{9}{36} = \frac{1}{4}$$

So, the probability is $\frac{1}{4}$

So the correct option is (D).

Q. 7. Three cards are drawn successively without replacements from a pack of 52 well suffled cards. What is the probability the first two cards are queen and third card drawn is a king?

(A) $\frac{5}{5525}$ (B) $\frac{3}{5525}$

(C) $\frac{2}{5525}$ (D) $\frac{1}{5525}$

Sol. Let Q denote the event that the cards drawn is queen and K is the event that the card drawn is an king, clearly we have to find $P(QQK)$.

Probability that the first card is queen, $P(Q) = \frac{4}{52}$

Also, $P(Q / Q)$ is the probability of second queen with the conditon that one queen has already been drawn. Now there will be three queens in 51 cards.

Probability that the second card is queen,

$$P(Q/Q) = \frac{3}{51}$$

Lastly, $P(K/QQ)$ is the probability of third drawn card to be a king, with the condition that two queens have already been drawn, now there are four kings left in 50 cards.

Therefore, $P(K/QQ) = \frac{4}{50}$

By multiplication law of probability, we have

$$P(QQK) = P(Q) \times P(Q/Q) \times P(K/QQ)$$

$$= \frac{4}{52} \times \frac{3}{51} \times \frac{4}{50} = \frac{2}{5525}$$

So the correct option is (C).

Q. 8. Nine cards numbered from 1 to 9 are placed in pack, mixed up thoroughly then one card is drawn randomly. If it is known that the number on each drawn card is more than 6. What is the probability that it is an even number ?

(A) $\frac{1}{2}$ (B) $\frac{3}{4}$

(C) 1 (D) $\frac{3}{6}$

Sol.
$$A = \{2, 4, 6, 8\}$$
$$B = \{8\}$$

$$P(A) = \frac{4}{9}, P(B) = \frac{1}{9}, A \cap B = \{8\}, P(A \cap B) = \frac{1}{9}$$

$$P(A/B) = \frac{P(A \cap B)}{P(B)}$$

$$= \frac{\frac{1}{9}}{\frac{1}{9}} = 1$$

So the correct option is (C).

Q. 9. From the set {1, 2, 3, 4, 5}, two numbers *a* and *b* ($a \neq b$) are chosen at random. The probability that is an integer is★

(A) $\frac{1}{3}$ (B) $\frac{1}{4}$

(C) $\frac{1}{2}$ (D) $\frac{3}{5}$

Sol. Total possible outcomes of $(a, b) = \{(1, 2), (1, 3), (1, 4), (1, 5), (2, 1), (2, 3), (2, 4), (2, 5), (3, 1), (3, 2), (3, 4), (3, 5), (4, 1), (4, 2), (4, 3), (4, 5), (5, 1), (5, 2), (5, 3), (5, 4)\}$

$\Rightarrow$ Number of possible outcomes = 20

Also, favourable outcomes of $(a, b) = \{(2, 1), (3, 1), (4, 1), (4, 2), (5, 1)\}$

$\Rightarrow$ Number of possible outcomes = 5

$$\therefore \quad P\left(\frac{a}{b} \text{ is integer}\right) = \frac{5}{20} = \frac{1}{4}$$

So the correct option is (B).

Q. 10. A bag contains 3 white, 4 black and 2 red balls. If 2 balls are drawn at random (without replacement), then the probability that both the balls are white is :★

(A) $\frac{1}{18}$ (B) $\frac{1}{36}$

(C) $\frac{1}{12}$ (D) $\frac{1}{24}$

Sol. Required probability $= \frac{{}^3C_2}{{}^9C_2} = \frac{1}{12}$

So the correct option is (C).

Q. 11. Three dice are thrown simultaneously. The probability of obtaining a total score of 5 is★

(A) $\frac{5}{216}$ (B) $\frac{1}{6}$

(C) $\frac{1}{36}$ (D) $\frac{1}{49}$

Sol. When three dice are thrown :

$$n(S) = 6 \times 6 \times 6 = 216$$

Let A be the event of getting total score 5.

Then A = {(1, 1, 3), (1, 3, 1), (3, 1, 1), (2, 2, 1), (2, 1, 2), (1, 2, 2)}

$$\therefore \quad n(A) = 6$$

$$\therefore \quad P(A) = \frac{n(A)}{n(S)}$$

$$= \frac{6}{216} = \frac{1}{36}$$

So the correct option is (C).

★ are board exam questions from previous years

Q. 12. A card is picked at random from a pack of 52 playing cards. Given that the picked card is a queen, the probability of this card to be a card of spade is :★

(A) $\frac{1}{3}$ **(B)** $\frac{4}{13}$

(C) $\frac{1}{4}$ **(D)** $\frac{1}{2}$

Sol. Let event A = Picked card is a queen. and event B = card of spade is picked.

$$\therefore \quad P(A) = \frac{4}{52}, P(B) = \frac{13}{52}$$

and, $P(A \cap B) = \frac{1}{52}$ [$\because A \cap B$ = Queen of space]

Now, required probability,

$$P(B/A) = \frac{P(A \cap B)}{P(A)} = \frac{1/52}{4/52} = \frac{1}{4}$$

So the correct option is (C).

Q. 13. A die is thrown once. Let A be the event that the number obtained is greater than 3. Let B be the event that the number obtained is less than 5. Then P(A∪B) is :★

(A) $\frac{2}{5}$ **(B)** $\frac{3}{5}$

(C) 0 **(D)** 1

Sol. $P(A) = \frac{3}{6}$

$$P(B) = \frac{4}{6}$$

and $P(A \cap B) = \frac{1}{6}$

We know, $P(A \cup B) = P(A) + P(B) - P(A \cap B)$

$$= \frac{3}{6} + \frac{4}{6} - \frac{1}{6}$$

$$= \frac{6}{6} = 1$$

So the correct option is (D).

Q. 14. A number is chosen randomly from numbers 1 to 60. The prabability that the chosen number is multiple of 2 or 5 is★

(A) $\frac{2}{5}$ **(B)** $\frac{3}{5}$

(C) $\frac{7}{10}$ **(D)** $\frac{9}{10}$

Sol. Let A = Numbers multiple of 2

and B = Numbers multiple of 5

$$\therefore \quad P(A \cup B) = P(A) + P(B) - P(A \cap B)$$

$$= \frac{30}{60} + \frac{12}{60} - \frac{6}{60}$$

$$= \frac{36}{60} = \frac{3}{5}$$

So the correct option is (B).

★ are board exam questions from previous years

Very Short Questions | Set 2 |

Chapter 1. Relations and Functions

Q. 1. Let $A = \{1, 2, 3\}$ and $B = \{a, b, c, d\}$. Then of the given relations which relation from A to B is not a function ?

$$R_1 = \{(1, a), (2, b), (3, c)\}$$
$$R_2 = \{(1, a), (2, b), (3, d)\}$$
$$R_3 = \{(1, a), (1, b), (2, c), (3, d)\}$$

Sol. R_3 is not a function since $1 \in A$ has two images $a, b \in B$. R_1 and R_2 are functions since in these relations, every element of A has a unique image in B.

Q. 2. Show that the relation $R = \{(a, a), (a, b), (a, c), (b, b), (b, c), (c, a), (c, b), (c, c)\}$ on the set $A = \{a, b, c\}$ is only reflexive and neither transitive nor symmetric.

Sol. R is reflexive, since $(a, a), (b, b), (c, c) \in R$

R is not symmetric, since $(a, b) \in R$ but $(b, a) \notin R$

R is not transitive, since $(b, c) \in R$ and $(c, a) \in R$ but $(b, a) \notin R$.

Thus, R is only reflexive.

Q. 3. State whether the function $f: N \rightarrow N$ given by $f(x) = 5x$ is injective, surjective or both.★

Sol. Here, $f(x_1) = f(x_2)$

$\Rightarrow \quad 5x_1 = 5x_2$

$\Rightarrow \quad x_1 = x_2, \forall\, x_1, x_2 \in N$

$\therefore f(x)$ is injective.

Now, the range of $f(n) = 5n, n \in N$

but the codomain of $f(n) = N$.

Since, range $\neq$ codomain,

$f(x)$ is not surjective.

Q. 4. Let $A = \{1, 2, 3\}$, $B = \{4, 5, 6, 7\}$ and let $f = \{(1, 4), (2, 5), (3, 6)\}$ be a function from A to B. State whether f is one-one or not.★

Sol. Here, $f: A \rightarrow B$ is defined as $\{(1, 4), (2, 5), (3, 6)\}$.

Since the images of distinct elements of A under f are distinct as :

$$f(1) = 4, f(2) = 5 \text{ and } f(3) = 6$$

From above it is evident that $x_1 \neq x_2$ and $f(x_1) \neq f(x_2)$.

$\therefore f: A \rightarrow B$ is one-one.

Q. 5. Let $f: A \rightarrow B$ be defined by $f(x) = x^3$ where $A = \{1, 3, 5\}$ and $B = \{1, 27, 125, 343\}$. State whether f^{-1} would exist or not.

Sol. Since $f: A \rightarrow B$ is defined by $f(x) = x^3$, we have

$f(1) = (1)^3 = 1$,

$f(3) = (3)^3 = 27$ and $f(5) = (5)^3 = 125$

$\therefore \quad f = \{(1, 1), (3, 27), (5, 125)\}$

Here, f is not onto, since $343 \in B$ does not have a preimage in A.

Thus, f^{-1} does not exist.

Q. 6. "If f and g are two functions such that the range of f is contained in the domain of g, then both fog and gof exist."

Is the above statement correct ? If not, rewrite the correct statement.

Sol. No. The correct statement would be :

If f and g are two functions such that the range of f is contained in the domain of g, then gof exists. (or any statement consistent to this)

Range of f is $f(A)$ and $f(A) \subset B$, where $f: A \rightarrow B$ and $g : B \rightarrow C$. Pictorially, we can depict this as

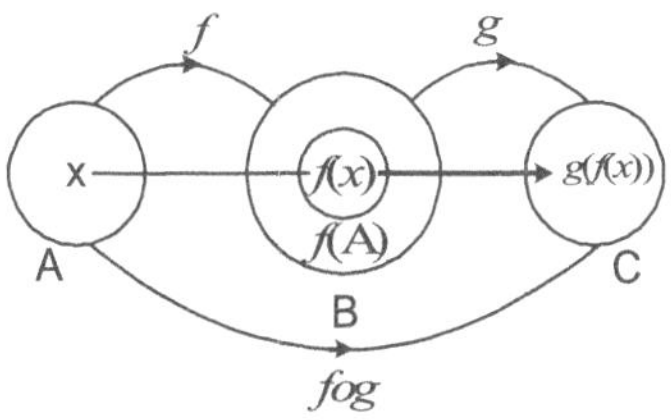

Q. 7. Let $f: R \rightarrow R, g: R \rightarrow R, h: R \rightarrow R$ such that $f(x) = x^2$, $g(x) = x + 1$ and $h(x) = 2x$. Find the value of $(fog)oh$.

Sol. We have $f, g, h: R \rightarrow R$

$$f(x) = x^2, g(x) = x + 1, h(x) = 2x$$
$$(fog)(x) = f(g(x)) = f(x + 1) = (x + 1)^2$$

$(fog)oh : R \rightarrow R$ is defined by

$[(fog)oh](x) = (fog)[h(x)] = fog(2x) = (2x + 1)^2$

Q. 8. If $f: R \rightarrow R$ is defined by $f(x) = x + 2$ and $g: R \rightarrow R$ is defined by $g(x) = x^2$, then verify that neither fog nor gof is a one-one function.

Sol. We have, $f(x) = x + 2$ and $g(x) = x^2$

$$(fog)(x) = f(g(x)) = f(x^2) = x^2 + 2$$
$$(gof)(x) = g(f(x)) = g(x + 2) = (x + 2)^2$$

Let $\quad (fog)(x) = (fog)(y)$

$\Rightarrow \quad x^2 + 2 = y^2 + 2$

$\Rightarrow \quad x^2 = y^2$

$\Rightarrow \quad x = \pm y$

$\therefore \quad (fog)(x) = (fog)(y)$

does not give the unique solution $x = y$ and hence is not one-one.

Let $\quad (gof)(x) = (gof)(y)$

$\Rightarrow \quad (x + 2)^2 = (y + 2)^2$

★ are board exam questions from previous years

$\Rightarrow \quad x+2=\pm(y+2)$

$\Rightarrow \quad$ either $x=y$ or $x=-y-4$

$\therefore$ *gof* is also not one-one.

Q. 9. Write *fog* if $f: R \to R$ and $g: R \to R$ are given by $f(x) = 8x^3$ and $g(x) = x^{1/3}$.★

Sol. Here, $\quad fog = f(g(x)) = f(x^{1/3})$

$\Rightarrow \quad fog = 8(x^{1/3})^3 = 8x$

Q. 10. If $f(x)$ is defined on [0, 1] by the rule

$$f(x) = \begin{cases} x, & \text{if } x \text{ is rational} \\ 1-x, & \text{if } x \text{ is irrational} \end{cases}$$

then find $f(f(x))$ for all $x \in [0, 1]$.

Sol. We have, $f(x) = \begin{cases} x, & \text{if } x \text{ is rational} \\ 1-x, & \text{if } x \text{ is irrational} \end{cases}$

If x is rational, then

$$f(x) = x$$

Therefore, $\quad f(f(x)) = f(x) = x$

If x is irrational, then

$$f(x) = 1-x$$

Therefore, $\quad f(f(x)) = f(1-x) = 1-(1-x) = x$

Thus, $\quad f(f(x)) = x$ for all $x \in [0, 1]$.

Q. 11. Write *fog* if $f: R \to R$ and $g: R \to R$ is given by $f(x) = |x|$ and $g(x) = |5x-2|$.★

Sol. Given,

$$f(x) = |x|, g(x) = |5x-2|$$

Now, $\quad (fog)(x) = f(g(x)) = f\{|5x-2|\}$

$$= ||5x-2||$$

$$= |5x-2|.$$

Q. 12. Write *fog* if $f: R \to R$ and $g: R \to R$ are given by $f(x) = 2x^2$ and $g(x) = x^{1/2}$.★

Sol. Given that

$$f(x) = 2x^2 \text{ and } g(x) = x^{1/2}$$

Now, $\quad fog(x) = f(g(x)) = f(x^{1/2}) = 2(x^{1/2})^2 = 2x$

Q. 13. State the reason for the relation R in the set {1, 2, 3} given by $R = \{(1, 2), (2, 1)\}$ not to be transitive.★

Sol. As $(1, 2) \in R$ and $(2, 1) \in R$ then for a relation to be transitive $(1, 1)$ should belong to R but $(1, 1) \notin R$. Hence, R is not transitive.

Q. 14. If $f: R \to R$ and $g: R \to R$ are given by $f(x) = \sin x$ and $g(x) = 5x^2$, find $gof(x)$.★

Sol. Given that $f(x) = \sin x$ and $g(x) = 5x^2$, then

$$gof(x) = g(f(x)) = g(\sin x)$$

$$= 5(\sin x)^2$$

$$= 5\sin^2 x.$$

Q. 15. If $f: R \to R$ is defined by $f(x) = \dfrac{3x+5}{2}$ is an invertible function, find f^{-1}.★

Sol. Given function is $f(x) = \dfrac{3x+5}{2}$ and is invertible.

Let $\quad y = \dfrac{3x+5}{2}$

$\Rightarrow \quad 2y = 3x+5$

$\Rightarrow \quad 3x = 2y-5$

$\Rightarrow \quad x = \dfrac{2y-5}{3}$

$\therefore \quad f^{-1}(y) = \dfrac{2y-5}{3}$

$\Rightarrow \quad f^{-1}(x) = \dfrac{2x-5}{3}$

Q. 16. If $f: R \to R$ defined by $f(x) = \dfrac{2x-7}{4}$ is an invertible function, find f^{-1}.★

Sol. Given that

$$f(x) = \frac{2x-7}{4}$$

$$y = \frac{2x-7}{4}$$

Let $\quad 4y = 2x-7$

$\Rightarrow \quad 2x = 4y+7$

$$x = \frac{4y+7}{2}$$

or $\quad f^{-1}(y) = \dfrac{4y+7}{2}$

$\Rightarrow \quad f^{-1}(x) = \dfrac{4x+7}{2}$

Q. 17. Find the value of the parameter α for which the function $f(x) = 1 + \alpha x, \alpha \neq 0$ is the inverse of itself.

Sol. Let $\quad f(x) = y$

Then $\quad 1 + \alpha x = y$

$\Rightarrow \quad x = \dfrac{y-1}{\alpha}$

$\therefore \quad f^{-1}(y) = \dfrac{y-1}{\alpha}$

Now, $\quad f(x) = f^{-1}(x)$

$\Rightarrow \quad 1 + \alpha x = \dfrac{x-1}{\alpha}$

★ are board exam questions from previous years

$$\Rightarrow \quad 1+\alpha x = \frac{1}{\alpha}x - \frac{1}{\alpha}$$

$$\Rightarrow \quad \alpha = \frac{1}{\alpha} \text{ and } 1 = \frac{-1}{\alpha}$$

[comparing the coefficients]

$$\Rightarrow \quad \alpha = -1$$

Q. 18. Find the inverse of the function $f: R \to R$ such that $f(x) = \cos x$.

Sol. Given, $f: R \to R$ is defined by $f(x) = \cos x$.

Clearly, f is not one-one.

as $\quad \cos\left(\frac{n\pi}{2}\right) = 0 \ \forall \ n \in N$

Thus, f is not bijective and f^{-1} does not exist.

Fill in the blanks :

Q. 19. The range of the function $f(x) = \frac{|x-1|}{(x-1)}$, $x \neq 1$ is ———— .

Sol. {1,–1}

Here, $\quad f(x) = \frac{|x-1|}{x-1}, x \neq 1$

$$\Rightarrow \quad f(x) = \frac{(x-1)}{(x-1)}, \ \forall x > 1$$

and $\quad f(x) = \frac{-(x-1)}{(x-1)}, \ \forall x < 1$

i.e., the range of $f(x)$ is $\{1, -1\}$.

Q. 20. If $f: R \to R$ be defined by $f(x) = (3 - x^3)^{1/3}$, then the value of fof is ———— .

Sol. x

Here, $\quad fof = f(f(x)) = f[(3-x^3)^{1/3}]$

$$\Rightarrow \quad fof = [3 - \{(3-x^3)^{1/3}\}^3]^{1/3}$$

$$\Rightarrow \quad fof = [3 - (3-x^3)]^{1/3}$$

$$\Rightarrow \quad fof = (x^3)^{1/3}$$

$$\Rightarrow \quad fof = x$$

Q. 21. If f is an invertible function, defined as $f(x) = \frac{3x-4}{5}$, then the value of $f^{-1}(x)$ is ———— .

Sol. $\frac{5x+4}{3}$

Given that $\quad f(x) = \frac{3x-4}{5}$

Let $\quad y = \frac{3x-4}{5}$

$$\Rightarrow \quad 5y = 3x - 4$$

$$\Rightarrow \quad 3x = 5y + 4$$

$$\Rightarrow \quad x = \frac{5y+4}{3}$$

$$\therefore \quad f^{-1}(y) = \frac{5y+4}{3}$$

$$\Rightarrow \quad f^{-1}(x) = \frac{5x+4}{3}$$

Q. 22. If $R = \{(x, y) : x + 2y = 8\}$ is a relation on N, then the range of R is ———.

Sol. $\{1, 2, 3\}$

The given relation on N is

$$R = \{(x, y) : x + 2y = 8\}$$

Since both $x, y \in N$,

x can take value 2, 4, 6, for all $x, y \in N$.

For $x = 2$, $\quad 2 + 2y = 8$

$\Rightarrow \quad y = 3$

For $x = 4$, $\quad 4 + 2y = 8$

$\Rightarrow \quad y = 2$

For $x = 6$, $\quad 6 + 2y = 8$

$\Rightarrow \quad y = 1$

$$R = \{(2, 3), (4, 2), (6, 1)\}$$

The range of R = Set of second elements

$= \{1, 2, 3\}$.

Q. 23. Let A and B be two non-empty finite sets with $n(A) = p$ and $n(B) = q$. Then, the number of relations from set A to set B is ———.

Sol. 2^{pq}

Number of elements of an n-element set is 2^n.

Since, the relations from A to B will be subsets of $(A \times B)$ and since, $n(A \times B) = n(A).n(B) = p.q$, there will be 2^{pq} relations from set A to set B.

Q.24. Let f be a function defined by $f(x) = \frac{x^2 + 2x + 1}{x^2 - 8x + 12}$ · Then, the domain of f is ———.

Sol. R – {2, 6}

$f(x)$ will be defined for all values of x except when $x = 2$ or $x = 6$ as $(x^2 - 8x + 12)$ has its two roots 2 and 6, where the value of the expression in the denominator will become zero making the function as not defined as these values.

$\therefore$ Domain of $f(x) = R - \{2, 6\}$.

Q. 25. A relation in a set A is called ______ relation, if each element of A is related to itself.★

Sol. Identity.

Q. 26. A relation R on a set A is called ________, if $(a_1, a_2) \in R$ and $(a_2, a_3) \in R$ implies that $(a_1, a_3) \in R$ for $a_1, a_2, a_3 \in A$.★

Sol. Transitive

★ are board exam questions from previous years

Chapter 2. Inverse Trigonometric Functions

Q. 1. Find the principal value of $\cos^{-1}\left(\frac{1}{\sqrt{2}}\right)$.

Sol. Let

$$\cos^{-1}\left(\frac{1}{\sqrt{2}}\right) = \theta$$

where $0 \le \theta \le \pi$

$$\Rightarrow \quad \cos\theta = \frac{1}{\sqrt{2}}$$

$$\Rightarrow \quad \cos\theta = \frac{1}{\sqrt{2}} = \cos\frac{\pi}{4}$$

$$\Rightarrow \quad \theta = \frac{\pi}{4}$$

Q. 2. If $\tan^{-1}x + \tan^{-1}y = \tan^{-1}\left[\frac{x+y}{1-xy}\right]$ then what are the constraints on the values of x, y and xy ?

Sol. $x, y > 0$,

$xy < 1$

Q. 3. Find the general solution of the equation :

$2\tan^{-1}(\cos x) = \tan^{-1}(2\operatorname{cosec} x)$.

Sol. Given,

$$2\tan^{-1}(\cos x) = \tan^{-1}(2\operatorname{cosec} x)$$

$$\Rightarrow \tan^{-1}\left(\frac{2\cos x}{1-\cos^2 x}\right) = \tan^{-1}(2\operatorname{cosec} x)$$

$$\left[\because 2\tan^{-1}x = \tan^{-1}\left(\frac{2x}{1-x^2}\right)\right]$$

$$\Rightarrow \quad \frac{2\cos x}{1-\cos^2 x} = 2\operatorname{cosec} x$$

$$\Rightarrow \quad \frac{2\cos x}{\sin^2 x} = \frac{2}{\sin x}$$

$$\Rightarrow \quad \frac{\cos x}{\sin x} = 1$$

$$\Rightarrow \quad \cot x = 1 = \cot\frac{\pi}{4}$$

$$\Rightarrow \quad x = n\pi + \frac{\pi}{4}$$

Q. 4. Write the principal value of

$$\cos^{-1}\left(\frac{1}{2}\right) - 2\sin^{-1}\left(-\frac{1}{2}\right).\star$$

Sol. $\cos^{-1}\left(\frac{1}{2}\right) - 2\sin^{-1}\left(-\frac{1}{2}\right)$

$$= \cos^{-1}\left(\cos\frac{\pi}{3}\right) - 2\sin^{-1}\left(\sin -\frac{\pi}{6}\right)$$

$$\left[\begin{array}{l}\because \text{principal value of } \cos^{-1}x \text{ is } [0,\pi] \\ \text{and that of } \sin^{-1}x \text{ is } \left[-\frac{\pi}{2},\frac{\pi}{2}\right]\end{array}\right]$$

$$= \frac{\pi}{3} - 2\left(-\frac{\pi}{6}\right)$$

$$= \frac{\pi}{3} + \frac{\pi}{3} = \frac{2\pi}{3}$$

Q.5. Find principal value of

$\tan^{-1}\sqrt{3} - \sec^{-1}(-2)$.★

Sol. We know that principal value of $\tan^{-1}x$ is $\left(-\frac{\pi}{2},\frac{\pi}{2}\right)$ and that of $\sec^{-1}x$ is $[0,\pi] - \left\{\frac{\pi}{2}\right\}$.

So,

$\tan^{-1}\sqrt{3} - \sec^{-1}(-2)$

$$= \tan^{-1}\left(\tan\frac{\pi}{3}\right) - \sec^{-1}\left(\sec\frac{2\pi}{3}\right)$$

$$\left[\because \tan\frac{\pi}{3} = \sqrt{3} \text{ and } \sec\frac{2\pi}{3} = -2\right]$$

$$= \frac{\pi}{3} - \frac{2\pi}{3} = \frac{-\pi}{3}$$

Q. 6. Write the value of $\sin\left[\frac{\pi}{3} - \sin^{-1}\left(-\frac{1}{2}\right)\right]$.★

Sol. We are given that $\sin\left[\frac{\pi}{3} - \sin^{-1}\left(-\frac{1}{2}\right)\right]$

$$= \sin\left[\frac{\pi}{3} + \sin^{-1}\left(\frac{1}{2}\right)\right]$$

$$[\because \sin^{-1}(-\theta) = -\sin^{-1}\theta]$$

$$= \sin\left[\frac{\pi}{3} + \sin^{-1}\left(\sin\frac{\pi}{6}\right)\right]$$

$$\left[\because \sin\frac{\pi}{6} = \frac{1}{2}\right]$$

$$= \sin\left[\frac{\pi}{3} + \frac{\pi}{6}\right]$$

$$= \sin\frac{\pi}{2} = 1$$

Q. 7. Write the value of $\cos^{-1}\left(\cos\frac{7\pi}{6}\right)$.★★

Sol. We know that the principal value of $\cos^{-1}\theta$ is $[0,\pi]$

★ are board exam questions from previous years
★★ are frequently asked board exam questions

$\therefore \quad \cos^{-1}\left(\cos\frac{7\pi}{6}\right) \neq \frac{7\pi}{6}$ as $\frac{7\pi}{6} \notin [0, \pi]$

$$\therefore \quad \cos^{-1}\left(\cos\frac{7\pi}{6}\right) = \cos^{-1}\left[\cos\left(2\pi - \frac{5\pi}{6}\right)\right]$$

$$= \cos^{-1}\left[\cos\frac{5\pi}{6}\right]$$

$$\left[\because \cos(2\pi - \theta) = \cos\theta\right]$$

$$= \frac{5\pi}{6} \in [0, \pi]$$

Q. 8. What is the principal value of tan^{-1}(– 1)?★★

Sol. We know that principal value of $\tan^{-1}(\theta)$ is $\left(-\frac{\pi}{2}, \frac{\pi}{2}\right)$.

$$\therefore \quad \tan^{-1}(-1) = \tan^{-1}\left(-\tan\frac{\pi}{4}\right) \quad \left\{\because \tan\frac{\pi}{4} = 1\right\}$$

$$= \tan^{-1}\tan\left(-\frac{\pi}{4}\right)$$

$$\left[\because -\tan\theta = \tan(-\theta)\right]$$

$$= -\frac{\pi}{4} \in \left(-\frac{\pi}{2}, \frac{\pi}{2}\right)$$

Q. 9. Using principal values, write the value of

$$\sin^{-1}\left(-\frac{\sqrt{3}}{2}\right).★$$

Sol. We know that principal value of $\sin^{-1}\theta$ is $\left[-\frac{\pi}{2}, \frac{\pi}{2}\right]$.

$$\therefore \quad \sin^{-1}\left(-\frac{\sqrt{3}}{2}\right) = \sin^{-1}\left(-\sin\frac{\pi}{3}\right)$$

$$= \sin^{-1}\left[\sin\left(-\frac{\pi}{3}\right)\right]$$

$$= -\frac{\pi}{3} \in \left[-\frac{\pi}{2}, \frac{\pi}{2}\right]$$

$$\therefore \quad \sin^{-1}\left(-\frac{\sqrt{3}}{2}\right) = -\frac{\pi}{3}$$

Q. 10. What is the principal value of sec^{-1}(– 2) ?★

Sol. We know that principal value of

$$\sec^{-1}\theta \text{ is } [0, \pi] - \left\{-\frac{\pi}{2}\right\}$$

$$\therefore \quad \sec^{-1}(-2) = \sec^{-1}\left(-\sec\frac{\pi}{3}\right)$$

$$= \frac{-\pi}{3}$$

But $\frac{-\pi}{3} \notin [0, \pi] - \left\{-\frac{\pi}{2}\right\}$

$$\therefore \quad \sec^{-1}(-2) = \sec^{-1}\left[\sec\left(\pi - \frac{\pi}{3}\right)\right]$$

$$\left[\because \sec(\pi - \theta) = -\sec\theta\right]$$

$$= \sec^{-1}\left(\sec\frac{2\pi}{3}\right)$$

$$= \frac{2\pi}{3} \in [0, \pi] - \left\{-\frac{\pi}{2}\right\}$$

Q. 11. Using principal value, find the value of

$$\cos^{-1}\left(\cos\frac{13\pi}{6}\right).★$$

Sol. As principal value of $\cos^{-1}\theta$ is $[0, \pi]$

$$\therefore \cos^{-1}\left(\cos\frac{13\pi}{6}\right) = \cos^{-1}\left[\cos\left(2\pi + \frac{\pi}{6}\right)\right]$$

$$= \cos^{-1}\left(\cos\frac{\pi}{6}\right)$$

$$\left[\because \cos(2\pi + \theta) = \cos\theta\right]$$

$$= \frac{\pi}{6} \in [0, \pi]$$

$$\therefore \cos^{-1}\left(\cos\frac{13\pi}{6}\right) = \frac{\pi}{6}$$

Q. 12. If $\tan^{-1}(\sqrt{3}) + \cot^{-1}(x) = \frac{\pi}{2}$, find x.★

Sol. Given that $\tan^{-1}(\sqrt{3}) + \cot^{-1}x = \frac{\pi}{2}$

$$\tan^{-1}\sqrt{3} = \frac{\pi}{2} - \cot^{-1}x$$

$$\Rightarrow \quad \tan^{-1}\sqrt{3} = \tan^{-1}x$$

$$\left[\because \tan^{-1}x + \cot^{-1}x = \frac{\pi}{2}\right]$$

$$\therefore \quad x = \sqrt{3}$$

Q. 13. Write principal value of $\sin^{-1}\left(\sin\frac{3\pi}{5}\right)$.★

Sol. As we know that principal value of $\sin^{-1}\theta$ is $\left[-\frac{\pi}{2}, \frac{\pi}{2}\right]$.

★ are board exam questions from previous years

★★ are frequently asked board exam questions

$$\sin^{-1}\left(\sin\frac{3\pi}{5}\right) \neq \frac{3\pi}{5} \qquad \left(\because \frac{3\pi}{5} \notin \left[-\frac{\pi}{2},\frac{\pi}{2}\right]\right)$$

$$\therefore \quad \sin^{-1}\left(\sin\frac{3\pi}{5}\right) = \sin^{-1}\sin\left(\pi-\frac{2\pi}{5}\right)$$

$$= \sin^{-1}\left(\sin\frac{2\pi}{5}\right)$$

$$\left[\because \sin(\pi-\theta) = \sin\theta\right]$$

$$= \frac{2\pi}{5} \in \left[-\frac{\pi}{2},\frac{\pi}{2}\right]$$

Q. 14. Using principal values, evaluate

$$\tan^{-1}(1) + \sin^{-1}\left(-\frac{1}{2}\right).\star$$

Sol. Principal value of $\tan^{-1}\theta$ is $\left(-\frac{\pi}{2},\frac{\pi}{2}\right)$ and that of $\sin^{-1}\theta$ is $\left[-\frac{\pi}{2},\frac{\pi}{2}\right]$.

$$\therefore \tan^{-1}(1) + \sin^{-1}\left(-\frac{1}{2}\right)$$

$$= \tan^{-1}\left(\tan\frac{\pi}{4}\right) + \sin^{-1}\left(-\sin\frac{\pi}{6}\right)$$

$$= \tan^{-1}\tan\frac{\pi}{4} + \sin^{-1}\sin\left(-\frac{\pi}{6}\right)$$

$$= \frac{\pi}{4} - \frac{\pi}{6} = \frac{\pi}{12}$$

Q. 15. If $\cos^{-1}p + \cos^{-1}q + \cos^{-1}r = \pi$, then prove that:

$$p^2 + q^2 + r^2 + 2pqr = 1$$

Sol. Given, $\cos^{-1}p + \cos^{-1}q + \cos^{-1}r = \pi$,

$$\Rightarrow \cos^{-1}\left[pq - \sqrt{1-p^2}\sqrt{1-q^2}\right] = \pi - \cos^{-1}r$$

$$\Rightarrow \cos^{-1}\left[pq - \sqrt{1-p^2}.\sqrt{1-q^2}\right] = \cos^{-1}(-r)$$

$$\Rightarrow \quad pq - \sqrt{1-p^2}\sqrt{1-q^2} = -r$$

$$\Rightarrow \quad (pq+r)^2 = (1-p^2)(1-q^2)$$

$$\Rightarrow \quad p^2q^2 + r^2 + 2pqr = 1 - p^2 - q^2 + p^2q^2$$

$$\Rightarrow \quad p^2 + q^2 + r^2 + 2pqr = 1$$

Q. 16. If $\sin^{-1}x + \sin^{-1}y + \sin^{-1}z = \frac{3\pi}{2}$, then find the value of $x^{100} + y^{100} + z^{100} - \frac{9}{x^{101} + y^{101} + z^{101}}$. [HOTS]

Sol. Since, we know that $\left|\sin^{-1}x\right| \leq \frac{\pi}{2}$

Therefore, each of $\sin^{-1}x$, $\sin^{-1}y$ and $\sin^{-1}z = \frac{\pi}{2}$.

as $\sin^{-1}x + \sin^{-1}y + \sin^{-1}z = \frac{3\pi}{2} = \frac{\pi}{2} + \frac{\pi}{2} + \frac{\pi}{2}$

$\therefore$ each of x, y and $z = \sin\frac{\pi}{2} = 1$

Therefore, the value of the given expression will be equal to $3 - \frac{9}{3} = 0$

Q. 17. Find the real solutions of the equation :

$$\tan^{-1}(1-x) + \tan^{-1}(1+x) = \tan^{-1}2x.$$

Sol. We know that

$$\tan^{-1}x + \tan^{-1}y = \tan^{-1}\left(\frac{x+y}{1-xy}\right)$$

Since we have,

$$\tan^{-1}\left(\frac{(1-x)+(1+x)}{1-(1-x^2)}\right) = \tan^{-1}2x$$

$$\Rightarrow \quad \frac{(1-x)+(1+x)}{1-(1-x^2)} = 2x$$

$$\Rightarrow \quad \frac{2}{x^2} = 2x$$

$$\Rightarrow \quad x^3 = 1$$

Therefore, the only real solution to the equation is $x = 1$.

Q. 18. Find the value of $\tan^{-1}\sqrt{3} - \cot^{-1}(-\sqrt{3})$. ⋆

Sol. We have, $\tan^{-1}\sqrt{3} - \cot^{-1}(-\sqrt{3})$

$$= \tan^{-1}\left(\tan\frac{\pi}{3}\right) - \cot^{-1}\left(\cot\frac{5\pi}{6}\right)$$

$$= \frac{\pi}{3} - \frac{5\pi}{6} = \frac{2\pi - 5\pi}{6}$$

$$= \frac{-3\pi}{6} = \frac{-\pi}{2}.$$

Fill in the blanks :

Q. 19. If $\sin^{-1}(1-x^2) + \cos^{-1}(1-x^2) = \frac{\pi}{2}$, then the range of x is ———.

Sol. $\left[-\sqrt{2},\sqrt{2}\right]$

We should have,

$$-1 \leq (1-x^2) \leq 1$$

$$-2 \leq -x^2 \leq 0$$

$$\Rightarrow \quad 0 \leq x^2 \leq 2$$

$$\Rightarrow \quad -\sqrt{2} \leq x \leq \sqrt{2}$$

$$\Rightarrow \quad x \in \left[-\sqrt{2},\sqrt{2}\right]$$

⋆ are board exam questions from previous years

Q. 20. The value of $\tan^{-1}\left(\tan\frac{3\pi}{4}\right)$ is ———.

Sol. $\frac{-\pi}{4}$

Principal value of $\tan^{-1}\theta$ is $\left(-\frac{\pi}{2},\frac{\pi}{2}\right)$

$\therefore$ The principal value of $\tan^{-1}\left(\tan\frac{3\pi}{4}\right)$

$$= \tan^{-1}\left[\tan\left(\pi-\frac{\pi}{4}\right)\right]$$

$$= \tan^{-1}\left(-\tan\frac{\pi}{4}\right)$$

$$[\because \tan(\pi-\theta) = -\tan\theta]$$

$$= \tan^{-1}\tan\left(-\frac{\pi}{4}\right)$$

$$[\because -\tan\theta = -\tan(-\theta)]$$

$$= -\frac{\pi}{4} \in \left(-\frac{\pi}{2},\frac{\pi}{2}\right)$$

Chapter 3. Matrices

Q. 1. Construct a matrix of order 2 × 2, whose elements are determined by $a_{ij} = (2i-j)^2$.

Sol. Let A be a matrix of order 2 × 2.

$$A = [a_{ij}]_{2\times 2} = \begin{bmatrix} a_{11} & a_{12} \\ a_{21} & a_{22} \end{bmatrix}$$

Given, $a_{ij} = (2i-j)^2$

For 2 × 2 order matrix,

$i = 1, 2$ and $j = 1, 2$

$\therefore \quad a_{11} = (2(1)-1)^2 = 1, a_{12} = (2(1)-2)^2 = 0$

$a_{21} = (2(2)-1)^2 = 9, a_{22} = (2(2)-2)^2 = 4$

$\Rightarrow \quad A = \begin{bmatrix} 1 & 0 \\ 9 & 4 \end{bmatrix}.$

Q. 2. If a matrix has 10 elements, find the possible orders of the matrix.

Sol. The possible orders of a matrix having 10 elements are 1 × 10, 10 × 1, 2 × 5 and 5 × 2.

Q. 3. If

$$\begin{bmatrix} x+3 & z+4 & 2y-7 \\ -6 & a-1 & 0 \\ b-3 & -21 & 0 \end{bmatrix} = \begin{bmatrix} 0 & 6 & 3y-2 \\ -6 & -3 & 2c+2 \\ 2b+4 & -21 & 0 \end{bmatrix},$$ **find the values of a, b, c, x, y and z.** **[NCERT]**

Sol. As the given matrices are equal therefore, their corresponding elements must be equal.

Comparing the corresponding elements, we get

$$x+3 = 0, z+4 = 6, 2y-7 = 3y-2$$

$$a-1 = -3, 0 = 2c+2, b-3 = 2b+4$$

Simplifying, we get

$$a = -2, b = -7, c = -1$$

$$x = -3, y = -5 \text{ and } z = 2.$$

Q. 4. Simplify :

$$\cos\theta\begin{bmatrix} \cos\theta & \sin\theta \\ -\sin\theta & \cos\theta \end{bmatrix} + \sin\theta\begin{bmatrix} \sin\theta & -\cos\theta \\ \cos\theta & \sin\theta \end{bmatrix}$$

[NCERT]

Sol. We have,

$$\cos\theta\begin{bmatrix} \cos\theta & \sin\theta \\ -\sin\theta & \cos\theta \end{bmatrix} + \sin\theta\begin{bmatrix} \sin\theta & -\cos\theta \\ \cos\theta & \sin\theta \end{bmatrix}$$

$$= \begin{bmatrix} \cos^2\theta & \sin\theta\cos\theta \\ -\sin\theta\cos\theta & \cos^2\theta \end{bmatrix} + \begin{bmatrix} \sin^2\theta & -\sin\theta\cos\theta \\ \sin\theta\cos\theta & \sin^2\theta \end{bmatrix}$$

$$= \begin{bmatrix} \cos^2\theta+\sin^2\theta & \sin\theta\cos\theta-\sin\theta\cos\theta \\ -\sin\theta\cos\theta+\sin\theta\cos\theta & \cos^2\theta+\sin^2\theta \end{bmatrix}$$

$$= \begin{bmatrix} 1 & 0 \\ 0 & 1 \end{bmatrix} \quad [\because \sin^2\theta+\cos^2\theta = 1]$$

Q. 5. Find XY if $X = \begin{bmatrix} -2 & 3 & 4 \\ 5 & 7 & 8 \end{bmatrix}$ and $Y = \begin{bmatrix} 3 & 5 \\ 8 & 9 \end{bmatrix}$

Sol. Since the dimensions of X are 2 × 3 and the dimensions of Y are 2 × 2. As a result, number of columns of matrix $X \neq$ number of rows of matrix Y.

So we cannot calculate XY.

Q. 6. Find $A+B$ if $A = \begin{bmatrix} 1 & 2 \\ 5 & 6 \end{bmatrix}$ and $B = \begin{bmatrix} 5 & 3 \\ 2 & 5 \\ 1 & 8 \end{bmatrix}$.

Sol. For addition of two matrices, order of both the matrices should be same. Here order of A is 2 × 2 and the order of B is 3 × 2, which is not same.

Hence, addition of above matrices A and B is not possible.

Q. 7. Find a matrix C such that $A+2B+C = 0$

$$A = \begin{bmatrix} 1 & -2 \\ 3 & -4 \end{bmatrix}, B = \begin{bmatrix} -6 & 7 \\ 8 & -9 \end{bmatrix}$$

Sol. We have,

$$A+2B+C=0$$

$$\Rightarrow \quad C=-A-2B$$

$$=-\begin{bmatrix}1 & -2\\3 & -4\end{bmatrix}-2\begin{bmatrix}-6 & 7\\8 & -9\end{bmatrix}$$

$$=\begin{bmatrix}-1 & 2\\-3 & 4\end{bmatrix}+\begin{bmatrix}12 & -14\\-16 & 18\end{bmatrix}$$

$$=\begin{bmatrix}-1+12 & 2-14\\-3-16 & 4+18\end{bmatrix}=\begin{bmatrix}11 & -12\\-19 & 22\end{bmatrix}$$

Q. 8. Find the value of x and y, if

$$3\begin{bmatrix}x & 2\\-2 & y-1\end{bmatrix}+\begin{bmatrix}2 & 1\\-1 & 2\end{bmatrix}=\begin{bmatrix}8 & 7\\-7 & -13\end{bmatrix}$$

Sol. Given,

$$3\begin{bmatrix}x & 2\\-2 & y-1\end{bmatrix}+\begin{bmatrix}2 & 1\\-1 & 2\end{bmatrix}=\begin{bmatrix}8 & 7\\-7 & -13\end{bmatrix}$$

First multiply by scalar 3 to the corresponding matrix then add both the matrices and compare the value of x and y.

$$\begin{bmatrix}3x & 6\\-6 & 3(y-1)\end{bmatrix}+\begin{bmatrix}2 & 1\\-1 & 2\end{bmatrix}=\begin{bmatrix}8 & 7\\-7 & -13\end{bmatrix}$$

$$\begin{bmatrix}3x+2 & 7\\-7 & 3y-1\end{bmatrix}=\begin{bmatrix}8 & 7\\-7 & -13\end{bmatrix}$$

On comparing each elements of both matrices, we get

$$3x+2=8$$

$$\Rightarrow \quad x=2$$

and $\quad 3y-1=-13$

$$\Rightarrow \quad y=-4.$$

Q. 9. If $A=\begin{bmatrix}1 & 0\\0 & 1\end{bmatrix}$ and $B=\begin{bmatrix}a & c\\b & d\end{bmatrix}$ prove that $(AB)'$ $=B'A'$.

Sol. $\quad AB=\begin{bmatrix}1 & 0\\0 & 1\end{bmatrix}\begin{bmatrix}a & c\\b & d\end{bmatrix}=\begin{bmatrix}a & c\\b & d\end{bmatrix}$

$$(AB)'=\begin{bmatrix}a & b\\c & d\end{bmatrix} \quad \ldots(i)$$

$$A'=\begin{bmatrix}1 & 0\\0 & 1\end{bmatrix}, B'=\begin{bmatrix}a & b\\c & d\end{bmatrix}$$

$$B'A'=\begin{bmatrix}a & b\\c & d\end{bmatrix}\begin{bmatrix}1 & 0\\0 & 1\end{bmatrix}=\begin{bmatrix}a & b\\c & d\end{bmatrix} \quad \ldots(ii)$$

Now, by equations (i) and (ii)

$$(AB)'=B'A'.$$

Q. 10. If $A=\dfrac{1}{\pi}\begin{bmatrix}\sin^{-1}(x\pi) & \tan^{-1}\left(\dfrac{x}{\pi}\right)\\ \sin^{-1}\left(\dfrac{x}{\pi}\right) & \cot^{-1}(\pi x)\end{bmatrix}$

and $\quad B=\dfrac{1}{\pi}\begin{bmatrix}-\cos^{-1}(\pi x) & \tan^{-1}\left(\dfrac{x}{\pi}\right)\\ \sin^{-1}\left(\dfrac{x}{\pi}\right) & -\tan^{-1}(\pi x)\end{bmatrix}$,

then find $A-B$.

Sol. Given, $\quad A=\dfrac{1}{\pi}\begin{bmatrix}\sin^{-1}(x\pi) & \tan^{-1}\left(\dfrac{x}{\pi}\right)\\ \sin^{-1}\left(\dfrac{x}{\pi}\right) & \cot^{-1}(\pi x)\end{bmatrix}$

and $\quad B=\dfrac{1}{\pi}\begin{bmatrix}-\cos^{-1}(\pi x) & \tan^{-1}\left(\dfrac{x}{\pi}\right)\\ \sin^{-1}\left(\dfrac{x}{\pi}\right) & -\tan^{-1}(\pi x)\end{bmatrix}$

$$A-B=A+(-1)B$$

$$=\frac{1}{\pi}\begin{bmatrix}\sin^{-1}(x\pi) & \tan^{-1}\left(\dfrac{x}{\pi}\right)\\ \sin^{-1}\left(\dfrac{x}{\pi}\right) & \cot^{-1}(\pi x)\end{bmatrix}$$

$$+\frac{(-1)}{\pi}\begin{bmatrix}-\cos^{-1}(\pi x) & \tan^{-1}\left(\dfrac{x}{\pi}\right)\\ \sin^{-1}\left(\dfrac{x}{\pi}\right) & -\tan^{-1}(\pi x)\end{bmatrix}$$

$$=\frac{1}{\pi}\begin{bmatrix}\sin^{-1}(x\pi)+\cos^{-1}(\pi x) & \tan^{-1}\left(\dfrac{x}{\pi}\right)-\tan^{-1}\left(\dfrac{x}{\pi}\right)\\ \sin^{-1}\left(\dfrac{x}{\pi}\right)-\sin^{-1}\left(\dfrac{x}{\pi}\right) & \cot^{-1}(\pi x)+\tan^{-1}(\pi x)\end{bmatrix}$$

$$=\frac{1}{\pi}\begin{bmatrix}\pi/2 & 0\\0 & \pi/2\end{bmatrix} \quad \begin{bmatrix}\because \sin^{-1}x+\cos^{-1}x=\pi/2\\ \tan^{-1}x+\cot^{-1}x=\pi/2\end{bmatrix}$$

Q. 11. Construct a 3×2 matrix, whose elements are given by $a_{ij}=e^{ix}\sin jx$. **[NCERT]**

Sol. In general, let A be a matrix of order 3×2.

$$\therefore \quad A=[a_{ij}]_{3\times 2}=\begin{bmatrix}a_{11} & a_{12}\\a_{21} & a_{22}\\a_{31} & a_{32}\end{bmatrix} \quad \ldots(i)$$

where $\quad a_{ij}=e^{ix}\sin jx$

$\therefore \quad a_{11}=e^{x}\sin x$

$a_{12}=e^{x}\sin 2x$

$a_{21}=e^{2x}\sin x$

$a_{22}=e^{2x}\sin 2x$

$a_{31}=e^{3x}\sin x$

$a_{32}=e^{3x}\sin 2x$

On putting all the values in equation (i), we get

$$A=\begin{bmatrix}e^{x}\sin x & e^{x}\sin 2x\\ e^{2x}\sin x & e^{2x}\sin 2x\\ e^{3x}\sin x & e^{3x}\sin 2x\end{bmatrix}$$

Q. 12. Three students in a history class score marks in three tests as follows : Anuj 89, 73, 91; Chitra 92, 97, 87; Samarth 79, 89, 96. Display this information as a 3 × 3 matrix.

Sol. Students Anuj, Chitra and Samarth score marks respectively as

$$89, 73, 91$$
$$92, 97, 87$$
$$79, 89, 96$$

As a matrix this can be represented as

	Test 1	Test 2	Test 3
Anuj	89	73	91
Chitra	92	97	87
Samarth	79	89	96

Q. 13. If $X = \begin{bmatrix} 4 & 1 \\ -1 & 2 \end{bmatrix}$, show that $6X - X^2 = 9I$, where I is the unit matrix.

Sol. Given, $X = \begin{bmatrix} 4 & 1 \\ -1 & 2 \end{bmatrix}$

On putting the value of matrix X in equation $6X - X^2 = 9I$

$$\text{L.H.S.} = 6X - X^2$$

$$= 6\begin{bmatrix} 4 & 1 \\ -1 & 2 \end{bmatrix} - \begin{bmatrix} 4 & 1 \\ -1 & 2 \end{bmatrix}\begin{bmatrix} 4 & 1 \\ -1 & 2 \end{bmatrix}$$

$$= \begin{bmatrix} 24 & 6 \\ -6 & 12 \end{bmatrix} - \begin{bmatrix} 16-1 & 4+2 \\ -4-2 & -1+4 \end{bmatrix}$$

$$= \begin{bmatrix} 24 & 6 \\ -6 & 12 \end{bmatrix} - \begin{bmatrix} 15 & 6 \\ -6 & 3 \end{bmatrix}$$

$$= \begin{bmatrix} 24-15 & 6-6 \\ -6+6 & 12-3 \end{bmatrix} = \begin{bmatrix} 9 & 0 \\ 0 & 9 \end{bmatrix}$$

$$= 9\begin{bmatrix} 1 & 0 \\ 0 & 1 \end{bmatrix}$$

$$= \text{R.H.S.}$$

Q. 14. Find the value of x and y, if

$$\begin{bmatrix} 1 & 2 \\ 3 & 3 \end{bmatrix}\begin{bmatrix} x & 0 \\ 0 & y \end{bmatrix} = \begin{bmatrix} x & 0 \\ 9 & 0 \end{bmatrix}$$

Sol. Given,

$$\begin{bmatrix} 1 & 2 \\ 3 & 3 \end{bmatrix}\begin{bmatrix} x & 0 \\ 0 & y \end{bmatrix} = \begin{bmatrix} x & 0 \\ 9 & 0 \end{bmatrix}$$

$$\begin{bmatrix} 1\times x+2\times 0 & 1\times 0+2\times y \\ 3\times x+3\times 0 & 3\times 0+3\times y \end{bmatrix} = \begin{bmatrix} x & 0 \\ 9 & 0 \end{bmatrix}$$

$$\Rightarrow \quad \begin{bmatrix} x & 2y \\ 3x & 3y \end{bmatrix} = \begin{bmatrix} x & 0 \\ 9 & 0 \end{bmatrix}$$

From the above illustration it is clear that both the matrices are equal, which is possible only when each corresponding element is equal to each other.

$$\therefore \quad 3x = 9$$
$$\Rightarrow \quad x = 3$$
$$\text{and} \quad 3y = 0$$
$$\Rightarrow \quad y = 0.$$

Q. 15. Let $A = \begin{bmatrix} 1 & 0 \\ 2 & 1 \end{bmatrix}$ and $f(X) = X^2 + X - I$, then find $f(A)$.

Sol. We have $A = \begin{bmatrix} 1 & 0 \\ 2 & 1 \end{bmatrix}$

$$A^2 = A \cdot A = \begin{bmatrix} 1 & 0 \\ 2 & 1 \end{bmatrix}\begin{bmatrix} 1 & 0 \\ 2 & 1 \end{bmatrix}$$

$$= \begin{bmatrix} 1\times 1+0\times 2 & 1\times 0+0\times 1 \\ 2\times 1+1\times 2 & 2\times 0+1\times 1 \end{bmatrix}$$

$$= \begin{bmatrix} 1 & 0 \\ 4 & 1 \end{bmatrix}$$

$$\therefore \quad f(A) = A^2 + A - I$$

$$= \begin{bmatrix} 1 & 0 \\ 4 & 1 \end{bmatrix} + \begin{bmatrix} 1 & 0 \\ 2 & 1 \end{bmatrix} - \begin{bmatrix} 1 & 0 \\ 0 & 1 \end{bmatrix}$$

$$= \begin{bmatrix} 1+1-1 & 0+0-0 \\ 4+2-0 & 1+1-1 \end{bmatrix}$$

$$= \begin{bmatrix} 2 & 0 \\ 6 & 1 \end{bmatrix}.$$

Q. 16. By using elementary operations, find the inverse of $A = \begin{bmatrix} 1 & 4 \\ 3 & 2 \end{bmatrix}$.

Sol. We know that

$$A = IA$$

$$\therefore \quad \begin{bmatrix} 1 & 4 \\ 3 & 2 \end{bmatrix} = \begin{bmatrix} 1 & 0 \\ 0 & 1 \end{bmatrix}A$$

Applying $R_2 \to R_2 - 3R_1$

$$\begin{bmatrix} 1 & 4 \\ 0 & -10 \end{bmatrix} = \begin{bmatrix} 1 & 0 \\ -3 & 1 \end{bmatrix}A$$

Applying $R_2 \to \left(\frac{-1}{10}\right)R_2$

$$\begin{bmatrix} 1 & 4 \\ 0 & 1 \end{bmatrix} = \begin{bmatrix} 1 & 0 \\ 3/10 & -1/10 \end{bmatrix}A$$

Applying $R_1 \to R_1 - 4R_2$

$$\begin{bmatrix} 1 & 0 \\ 0 & 1 \end{bmatrix} = \begin{bmatrix} -2/10 & 4/10 \\ 3/10 & -1/10 \end{bmatrix}A$$

Thus $\quad A^{-1} = \begin{bmatrix} -1/5 & 2/5 \\ 3/10 & -1/10 \end{bmatrix}$

Q. 17. If A is a square matrix such that $A^2 = A$, then write the value of $7A - (I + A)^3$, where I is an identity matrix.★

Sol.
$$7A-(I+A)^3 = 7A-(I^3+A^3+3I^2A+3IA^2)$$
$$[\because (a+b)^3 = a^3+b^3+3a^2b+3ab^2]$$
$$= 7A-(I+A^2.A+3IA+3IA^2)$$
$$[\because I^n = I \,\forall\, n \in N]$$
$$= 7A-(I+A.A+3A+3IA)$$
$$[\because A^2 = A, IA = A]$$
$$= 7A-(I+A^2+3A+3A)$$
$$= 7A-(I+A+3A+3A)$$
$$= 7A-I-7A = -I.$$

Q. 18. Write the element a_{23} of a 3 × 3 matrix $A = (a_{ij})$ whose elements a_{ij} are given be $a_{ij} = \dfrac{|i-j|}{2}$.★

Sol. Given,
$$a_{ij} = \frac{|i-j|}{2}$$
$$\therefore \quad a_{23} = \frac{|2-3|}{2} = \frac{|-1|}{2} = \frac{1}{2}$$

Q. 19. Use elementary column operation $C_2 \to C_2 + 2C_1$ in the following matrix equation :
$$\begin{bmatrix} 2 & 1 \\ 2 & 0 \end{bmatrix} = \begin{bmatrix} 3 & 1 \\ 2 & 0 \end{bmatrix} \begin{bmatrix} 1 & 0 \\ -1 & 1 \end{bmatrix} \star$$

Sol. We have
$$\begin{bmatrix} 2 & 1 \\ 2 & 0 \end{bmatrix} = \begin{bmatrix} 3 & 1 \\ 2 & 0 \end{bmatrix} \begin{bmatrix} 1 & 0 \\ -1 & 1 \end{bmatrix}$$
$$P = AB \text{ (say)}$$
Applying $C_2 \to C_2 + 2C_1$ on P, we get
$$\begin{bmatrix} 2 & 5 \\ 2 & 4 \end{bmatrix} = Q \text{ (say)}$$
Applying $C_2 \to C_2 + 2C_1$ on B, we get
$$\begin{bmatrix} 1 & 2 \\ -1 & -1 \end{bmatrix} = C \text{ (say)}$$
Now,
$$AC = \begin{bmatrix} 3 & 1 \\ 2 & 0 \end{bmatrix} \begin{bmatrix} 1 & 2 \\ -1 & -1 \end{bmatrix} = Q$$
$$\therefore \quad \begin{bmatrix} 2 & 5 \\ 2 & 4 \end{bmatrix} = \begin{bmatrix} 3 & 1 \\ 2 & 0 \end{bmatrix} \begin{bmatrix} 1 & 2 \\ -1 & -1 \end{bmatrix}$$

Q. 20. Write the number of all possible matrices of order 2 × 2 with each entry 1, 2 or 3.★

Sol. Total number of all possible matrices of order 2 × 2 with each entry 1, 2 or 3 are 3^4 *i.e.*, 81.

Fill in the blanks :

Q. 21. If the matrix $A = \begin{bmatrix} 0 & a & -3 \\ 2 & 0 & -1 \\ b & 1 & 0 \end{bmatrix}$ is skew-symmetric, then the values of a and b are ——— and ——— respectively.

★ are board exam questions from previous years

Sol. – 2, 3

Given,
$$A = \begin{bmatrix} 0 & a & -3 \\ 2 & 0 & -1 \\ b & 1 & 0 \end{bmatrix}$$
A is given to be skew-symmetric matrix
$$\therefore \quad A^T = -A$$
$$\therefore \quad \begin{bmatrix} 0 & 2 & b \\ a & 0 & 1 \\ -3 & -1 & 0 \end{bmatrix} = -\begin{bmatrix} 0 & a & -3 \\ 2 & 0 & -1 \\ b & 1 & 0 \end{bmatrix}$$
$$\begin{bmatrix} 0 & 2 & b \\ a & 0 & 1 \\ -3 & -1 & 0 \end{bmatrix} = \begin{bmatrix} 0 & -a & 3 \\ -2 & 0 & 1 \\ -b & -1 & 0 \end{bmatrix}$$
On comparing, we get
$$-a = 2 \text{ and } -b = -3$$
$$\Rightarrow \quad a = -2 \text{ and } b = 3.$$

Q. 22. The value of :
$$\begin{bmatrix} -2\sin 30° & \operatorname{cosec} 30° \\ \cot 45° & \sin 90° \end{bmatrix} \begin{bmatrix} \tan 45° & \sin 90° \\ 2\sec 0° & \sec 60° \end{bmatrix} \begin{bmatrix} \operatorname{cosec} 90° \\ 2\cos 60° \end{bmatrix}$$
is ———.

Sol. $\begin{bmatrix} 6 \\ 6 \end{bmatrix}$

Given,
$$\begin{bmatrix} -2\sin 30° & \operatorname{cosec} 30° \\ \cot 45° & \sin 90° \end{bmatrix} \times \begin{bmatrix} \tan 45° & \sin 90° \\ 2\sec 0° & \sec 60° \end{bmatrix} \times \begin{bmatrix} \operatorname{cosec} 90° \\ 2\cos 60° \end{bmatrix}$$
$$= \begin{bmatrix} -2\times1/2 & 2 \\ 1 & 1 \end{bmatrix} \times \begin{bmatrix} 1 & 1 \\ 2\times1 & 2 \end{bmatrix} \times \begin{bmatrix} 1 \\ 2\times1/2 \end{bmatrix}$$
$$= \left(\begin{bmatrix} -1 & 2 \\ 1 & 1 \end{bmatrix} \times \begin{bmatrix} 1 & 1 \\ 2 & 2 \end{bmatrix} \right) \times \begin{bmatrix} 1 \\ 1 \end{bmatrix}$$
$$= \begin{bmatrix} -1\times1+2\times2 & -1\times1+2\times2 \\ 1\times1+1\times2 & 1\times1+1\times2 \end{bmatrix} \times \begin{bmatrix} 1 \\ 1 \end{bmatrix}$$
$$= \begin{bmatrix} -1+4 & -1+4 \\ 1+2 & 1+2 \end{bmatrix} \times \begin{bmatrix} 1 \\ 1 \end{bmatrix} = \begin{bmatrix} 3 & 3 \\ 3 & 3 \end{bmatrix} \begin{bmatrix} 1 \\ 1 \end{bmatrix}$$
$$= \begin{bmatrix} 3+3 \\ 3+3 \end{bmatrix}$$
$$= \begin{bmatrix} 6 \\ 6 \end{bmatrix}.$$

Q. 23. If A = diag (3 – 2 4) and B = diag (1 – 2 3). Then, $A - 2B$ is ———.

Sol. diag $(1, 2, -2)$

We have,

$$A = \text{diag}(3\ \ -2\ \ 4) \text{ and } B = \text{diag}(1\ \ -2\ \ 3)$$

$$\Rightarrow \quad A = \begin{bmatrix} 3 & 0 & 0 \\ 0 & -2 & 0 \\ 0 & 0 & 4 \end{bmatrix} \text{ and } B = \begin{bmatrix} 1 & 0 & 0 \\ 0 & -2 & 0 \\ 0 & 0 & 3 \end{bmatrix}$$

$$\therefore \quad A - 2B = \begin{bmatrix} 3 & 0 & 0 \\ 0 & -2 & 0 \\ 0 & 0 & 4 \end{bmatrix} - 2\begin{bmatrix} 1 & 0 & 0 \\ 0 & -2 & 0 \\ 0 & 0 & 3 \end{bmatrix}$$

$$= \begin{bmatrix} 3 & 0 & 0 \\ 0 & -2 & 0 \\ 0 & 0 & 4 \end{bmatrix} + \begin{bmatrix} -2 & 0 & 0 \\ 0 & 4 & 0 \\ 0 & 0 & -6 \end{bmatrix}$$

$$= \begin{bmatrix} 3-2 & 0 & 0 \\ 0 & -2+4 & 0 \\ 0 & 0 & 4-6 \end{bmatrix}$$

$$= \begin{bmatrix} 1 & 0 & 0 \\ 0 & 2 & 0 \\ 0 & 0 & -2 \end{bmatrix}$$

$$= \text{diag}(1\ \ 2\ \ -2).$$

Q. 24. If matrix $A = \begin{bmatrix} 5 & -5 \\ -5 & 5 \end{bmatrix}$ and $A^2 = \lambda A$, then the value of λ is ———.

Sol. 10

Given,
$$A = \begin{bmatrix} 5 & -5 \\ -5 & 5 \end{bmatrix} \quad \ldots\text{(i)}$$

$$A^2 = \begin{bmatrix} 5 & -5 \\ -5 & 5 \end{bmatrix}\begin{bmatrix} 5 & -5 \\ -5 & 5 \end{bmatrix}$$

$$= \begin{bmatrix} 25+25 & -25-25 \\ -25-25 & 25+25 \end{bmatrix}$$

$$= \begin{bmatrix} 50 & -50 \\ -50 & 50 \end{bmatrix}$$

$$= 10\begin{bmatrix} 5 & -5 \\ -5 & 5 \end{bmatrix}$$

$$= 10A$$

Now, $\quad A^2 = \lambda A$

On comparing, we get

$$\lambda = 10.$$

Q. 25. If $A^T = \begin{bmatrix} 3 & 4 \\ -1 & 2 \\ 0 & 1 \end{bmatrix}$ and $B = \begin{bmatrix} -1 & 2 & 1 \\ 1 & 2 & 3 \end{bmatrix}$, then the value of $A^T - B^T$ is ———.

Sol. $\begin{bmatrix} 4 & 3 \\ -3 & 0 \\ -1 & -2 \end{bmatrix}$

Here,
$$B = \begin{bmatrix} -1 & 2 & 1 \\ 1 & 2 & 3 \end{bmatrix}$$

$$\Rightarrow \quad B^T = \begin{bmatrix} -1 & 1 \\ 2 & 2 \\ 1 & 3 \end{bmatrix}$$

$$\therefore \quad A^T - B^T = \begin{bmatrix} 3 & 4 \\ -1 & 2 \\ 0 & 1 \end{bmatrix} - \begin{bmatrix} -1 & 1 \\ 2 & 2 \\ 1 & 3 \end{bmatrix}$$

$$= \begin{bmatrix} 3+1 & 4-1 \\ -1-2 & 2-2 \\ 0-1 & 1-3 \end{bmatrix} = \begin{bmatrix} 4 & 3 \\ -3 & 0 \\ -1 & -2 \end{bmatrix}$$

Q. 26. If $-A + 3B + X = 0$, where $A = \begin{bmatrix} -2 & 6 \\ 5 & 8 \end{bmatrix}$ and $B = \begin{bmatrix} 1 & 2 \\ -2 & 3 \end{bmatrix}$. Then, the matrix $X =$ ———.

Sol. $\begin{bmatrix} -5 & 0 \\ 11 & -1 \end{bmatrix}$

Given,
$$A = \begin{bmatrix} -2 & 6 \\ 5 & 8 \end{bmatrix} \text{ and } B = \begin{bmatrix} 1 & 2 \\ -2 & 3 \end{bmatrix}$$

$$-A + 3B + X = 0$$

$$X = A - 3B$$

$$= \begin{bmatrix} -2 & 6 \\ 5 & 8 \end{bmatrix} + (-3)\begin{bmatrix} 1 & 2 \\ -2 & 3 \end{bmatrix}$$

$$= \begin{bmatrix} -2 & 6 \\ 5 & 8 \end{bmatrix} + \begin{bmatrix} -3 & -6 \\ 6 & -9 \end{bmatrix}$$

$$= \begin{bmatrix} -2-3 & 6-6 \\ 5+6 & 8-9 \end{bmatrix}$$

$$= \begin{bmatrix} -5 & 0 \\ 11 & -1 \end{bmatrix}.$$

Q.27. If $[x\ \ 1]\begin{bmatrix} 1 & 0 \\ -2 & 0 \end{bmatrix} = 0$, then the value of x is ———.

Sol. 2

Given,

$$[x\ \ 1]\begin{bmatrix} 1 & 0 \\ -2 & 0 \end{bmatrix} = 0.$$

$$\Rightarrow \quad [x-2\ \ \ 0+0] = [0\ \ 0]$$

$$\Rightarrow \quad x - 2 = 0$$

$$\therefore \quad x = 2$$

Q. 28. If matrix $A = \begin{bmatrix} 0 & 2b & -2 \\ 3 & 1 & 3 \\ 3a & 3 & -1 \end{bmatrix}$ is given to be symmetric, then the values of a and b are ———.

Sol. $\frac{-2}{3}$ and $\frac{3}{2}$

We have,

$$A = \begin{bmatrix} 0 & 2b & -2 \\ 3 & 1 & 3 \\ 3a & 3 & -1 \end{bmatrix}$$

It is given that the matrix is symmetric.

$\therefore \quad A = A'$

$$\Rightarrow \begin{bmatrix} 0 & 2b & -2 \\ 3 & 1 & 3 \\ 3a & 3 & -1 \end{bmatrix} = \begin{bmatrix} 0 & 3 & 3a \\ 2b & 1 & 3 \\ -2 & 3 & -1 \end{bmatrix}$$

Now, by equality of matrices, we get

$2b = 3$

$\Rightarrow \quad b = \frac{3}{2}$

and $\quad 3a = -2$

$\Rightarrow \quad a = \frac{-2}{3}$

Therefore, $\quad a = \frac{-2}{3}$ and $b = \frac{3}{2}$

Q. 29. If $\begin{bmatrix} x+y & 7 \\ 9 & x-y \end{bmatrix} = \begin{bmatrix} 2 & 7 \\ 9 & 4 \end{bmatrix}$, then $x \cdot y =$ __★

Sol. –3.

Comparing elements of both matrices, we get,

$x + y = 2 \quad$ ...(i)

and $\quad x - y = 4 \quad$ ...(ii)

Adding (i) and (ii), we get

$2x = 6$

$\Rightarrow \quad x = 3$

From (i), $\quad y = 2 - x$

$= 2 - 3 = -1$

$\therefore \quad x.y = 3(-1) = -3$

Q. 30. If $A + B = \begin{bmatrix} 1 & 0 \\ 1 & 1 \end{bmatrix}$ and $A - 2B = \begin{bmatrix} -1 & 1 \\ 0 & -1 \end{bmatrix}$, then A = ________.★

Sol. $\begin{bmatrix} \frac{1}{3} & \frac{1}{3} \\ \frac{2}{3} & \frac{1}{3} \end{bmatrix}$

Given: $\quad A + B = \begin{bmatrix} 1 & 0 \\ 1 & 1 \end{bmatrix} \quad$ ...(i)

and $\quad A - 2B = \begin{bmatrix} -1 & 1 \\ 0 & -1 \end{bmatrix} \quad$ ...(ii)

$$\Rightarrow A - 2\left\{\begin{bmatrix} 1 & 0 \\ 1 & 1 \end{bmatrix} - A\right\} = \begin{bmatrix} -1 & 1 \\ 0 & -1 \end{bmatrix} \quad \text{[From (i)]}$$

$$\Rightarrow A - A\begin{bmatrix} 2 & 0 \\ 2 & 2 \end{bmatrix} + 2A = \begin{bmatrix} -1 & 1 \\ 0 & -1 \end{bmatrix}$$

$$3A = \begin{bmatrix} -1 & 1 \\ 0 & -1 \end{bmatrix} + \begin{bmatrix} 2 & 0 \\ 2 & 2 \end{bmatrix}$$

$$3A = \begin{bmatrix} 1 & 1 \\ 2 & 1 \end{bmatrix}$$

$$A = \frac{1}{3}\begin{bmatrix} 1 & 1 \\ 2 & 1 \end{bmatrix}$$

Chapter 4. Determinants

Q. 1. Prove that the determinant

$$\begin{vmatrix} x & \sin\theta & \cos\theta \\ -\sin\theta & -x & 1 \\ \cos\theta & 1 & x \end{vmatrix} \text{ is independent of } \theta.$$

[NCERT]

Sol. Let

$$\Delta = \begin{vmatrix} x & \sin\theta & \cos\theta \\ -\sin\theta & -x & 1 \\ \cos\theta & 1 & x \end{vmatrix}$$

$$= x\begin{vmatrix} -x & 1 \\ 1 & x \end{vmatrix} - \sin\theta\begin{vmatrix} -\sin\theta & 1 \\ \cos\theta & x \end{vmatrix} + \cos\theta\begin{vmatrix} -\sin\theta & -x \\ \cos\theta & 1 \end{vmatrix}$$

$= x(-x^2 - 1) - \sin\theta(-x\sin\theta - \cos\theta)$

$+ \cos(-\sin\theta + x\cos\theta)$

$= -x^3 - x + x\sin^2\theta + \sin\theta\cos\theta - \cos\theta\sin\theta + x\cos^2\theta$

$= -x^3 - x + x(\sin^2\theta + \cos^2\theta)$

$= -x^3 - x + x$

$= -x^3$

Hence, Δ is independent of θ.

Q. 2. Without expanding the determinant, prove that:

$$\begin{vmatrix} a & a^2 & bc \\ b & b^2 & ca \\ c & c^2 & ab \end{vmatrix} = \begin{vmatrix} 1 & a^2 & a^3 \\ 1 & b^2 & b^3 \\ 1 & c^2 & c^3 \end{vmatrix}$$

Sol. $\quad$ L.H.S. $= \begin{vmatrix} a & a^2 & bc \\ b & b^2 & ca \\ c & c^2 & ab \end{vmatrix}$

$$= \frac{1}{abc}\begin{vmatrix} a^2 & a^3 & abc \\ b^2 & b^3 & abc \\ c^2 & c^3 & abc \end{vmatrix}$$

Taking out factor abc from column C_3

$$= \frac{1}{abc} \cdot abc\begin{vmatrix} a^2 & a^3 & 1 \\ b^2 & b^3 & 1 \\ c^2 & c^3 & 1 \end{vmatrix}$$

★ are board exam questions from previous years

$$= \begin{vmatrix} a^2 & a^3 & 1 \\ b^2 & b^3 & 1 \\ c^2 & c^3 & 1 \end{vmatrix}$$

Applying $C_1 \leftrightarrow C_3$ and $C_2 \leftrightarrow C_3$

$$= (-1)^2 \begin{vmatrix} 1 & a^2 & a^3 \\ 1 & b^2 & b^3 \\ 1 & c^2 & c^3 \end{vmatrix}$$

$$= \begin{vmatrix} 1 & a^2 & a^3 \\ 1 & b^2 & b^3 \\ 1 & c^2 & c^3 \end{vmatrix}$$

$=$ R.H.S.

Hence, the given result is proved.

Q. 3. Write minors and cofactors of elements of determinant $\begin{vmatrix} a & c \\ b & d \end{vmatrix}$. [NCERT]

Sol. Given, $\begin{vmatrix} a & c \\ b & d \end{vmatrix}$

Minors $\quad M_{11} = d$ and $M_{12} = b$

$M_{21} = c$ and $M_{22} = a$

Cofactors are

$$C_{11} = (-1)^{1+1} d = d$$

$$C_{12} = (-1)^{1+2} b = -b$$

$$C_{21} = (-1)^{2+1} c = -c$$

$$C_{22} = (-1)^{2+2} a = a.$$

Q. 4. If area of a triangle is 35 sq. units with vertices (2, – 6), (5, 4) and (k, 4), then find the values of k. [NCERT]

Sol. Given,

$$\frac{1}{2}\begin{vmatrix} 2 & -6 & 1 \\ 5 & 4 & 1 \\ k & 4 & 1 \end{vmatrix} = \pm 35$$

$$\Rightarrow \begin{vmatrix} 2 & -6 & 1 \\ 5 & 4 & 1 \\ k & 4 & 1 \end{vmatrix} = \pm 70$$

$$\Rightarrow 2(4 \times 1 - 4 \times 1) + 6(5 \times 1 - k \times 1) + 1(5 \times 4 - k \times 4) = \pm 70$$

$$\Rightarrow 0 + 6(5-k) + (20-4k) = \pm 70$$

$$\Rightarrow 50 - 10k = \pm 70$$

On taking positive sign,

$$50 - 10k = 70$$

$$\Rightarrow k = -2$$

On taking negative sign,

$$50 - 10k = -70$$

$$\Rightarrow 10k = 120$$

$$\Rightarrow k = 12.$$

Thus, $k = -2, 12$.

Q.5. If $A = \begin{bmatrix} 1 & 0 & 1 \\ 0 & 1 & 2 \\ 0 & 0 & 4 \end{bmatrix}$, then show that $|3A| = 27|A|$.

Sol. Given $\quad A = \begin{bmatrix} 1 & 0 & 1 \\ 0 & 1 & 2 \\ 0 & 0 & 4 \end{bmatrix}$

It can be observed that in the first column two entries are zero. Thus, we expand along the first column (C_1) for easier calculation.

$$|A| = 1\begin{vmatrix} 1 & 2 \\ 0 & 4 \end{vmatrix} - 0\begin{vmatrix} 0 & 1 \\ 0 & 4 \end{vmatrix} + 0\begin{vmatrix} 0 & 1 \\ 1 & 2 \end{vmatrix}$$

$$= 1(4-0) - 0 + 0 = 4$$

$$\therefore \quad 27|A| = 27 \times 4 = 108 \quad \text{...(i)}$$

Now, $$3A = 3\begin{bmatrix} 1 & 0 & 1 \\ 0 & 1 & 2 \\ 0 & 0 & 4 \end{bmatrix} = \begin{bmatrix} 3 & 0 & 3 \\ 0 & 3 & 6 \\ 0 & 0 & 12 \end{bmatrix}$$

$$\therefore \quad |3A| = \begin{vmatrix} 3 & 0 & 3 \\ 0 & 3 & 6 \\ 0 & 0 & 12 \end{vmatrix}$$

On expanding along column C_1

$$= 3\begin{vmatrix} 3 & 6 \\ 0 & 12 \end{vmatrix} - 0\begin{vmatrix} 0 & 3 \\ 0 & 12 \end{vmatrix} + 0\begin{vmatrix} 0 & 3 \\ 3 & 6 \end{vmatrix}$$

$$= 3(36-0) - 0 + 0$$

$$= 108 \quad \text{...(ii)}$$

From equations (i) and (ii), we have

$$|3A| = 27|A|.$$

Q. 6. Using the properties of determinants and without expanding, prove that :

$$\begin{vmatrix} \boldsymbol{a-b} & \boldsymbol{b-c} & \boldsymbol{c-a} \\ \boldsymbol{b-c} & \boldsymbol{c-a} & \boldsymbol{a-b} \\ \boldsymbol{c-a} & \boldsymbol{a-b} & \boldsymbol{b-c} \end{vmatrix} = \mathbf{0}.$$

Sol. Let $\quad \Delta = \begin{vmatrix} a-b & b-c & c-a \\ b-c & c-a & a-b \\ c-a & a-b & b-c \end{vmatrix}$

Applying $R_1 \to R_1 + R_2$, we have

$$\Delta = \begin{vmatrix} a-c & b-a & c-b \\ b-c & c-a & a-b \\ -(a-c) & -(b-a) & -(c-b) \end{vmatrix}$$

$$= -\begin{vmatrix} a-c & b-a & c-b \\ b-c & c-a & a-b \\ a-c & b-a & c-b \end{vmatrix}$$

Here, the two rows R_1 and R_2 are identical.

$$\therefore \quad \Delta = 0.$$

Q. 7. Using the property of determinants and without expanding, prove that :

$$\begin{vmatrix} 1 & bc & a(b+c) \\ 1 & ca & b(c+a) \\ 1 & ab & c(a+b) \end{vmatrix} = 0.$$

Sol. Let $\Delta = \begin{vmatrix} 1 & bc & a(b+c) \\ 1 & ca & b(c+a) \\ 1 & ab & c(a+b) \end{vmatrix}$

Applying $C_3 \to C_3 + C_2$, we have

$$\Delta = \begin{vmatrix} 1 & bc & ab+bc+ca \\ 1 & ca & bc+ab+ca \\ 1 & ab & ca+bc+ab \end{vmatrix}$$

$$= (ab+bc+ca)\begin{vmatrix} 1 & bc & 1 \\ 1 & ca & 1 \\ 1 & ab & 1 \end{vmatrix}$$

Here, the two columns C_1 and C_3 are identical.

$\therefore \quad \Delta = 0.$

Q. 8. Find area of the triangle with vertices at the point given in each of the following :

(i) (6, 0), (4, 3), (2, 1)

(ii) (– 7, – 2), (– 3, – 5), (5, 2).

Sol. We know that area of triangle with vertices (x_1, y_1), (x_2, y_2) and (x_3, y_3) is given by

$$\Delta = \frac{1}{2}\begin{vmatrix} x_1 & y_1 & 1 \\ x_2 & y_2 & 1 \\ x_3 & y_3 & 1 \end{vmatrix}$$

(i) So, area of triangle with vertices (6, 0), (4, 3) and (2, 1) is given by

$$\Delta = \frac{1}{2}\begin{vmatrix} 6 & 0 & 1 \\ 4 & 3 & 1 \\ 2 & 1 & 1 \end{vmatrix}$$

$$= \frac{1}{2}\left|6\begin{vmatrix} 3 & 1 \\ 1 & 1 \end{vmatrix} - 0\begin{vmatrix} 4 & 1 \\ 2 & 1 \end{vmatrix} + 1\begin{vmatrix} 4 & 3 \\ 2 & 1 \end{vmatrix}\right|$$

$$= \frac{1}{2}|6(3\times 1 - 1\times 1) - 0 + 1(4\times 1 - 2\times 3)|$$

$$= \frac{1}{2}|6\times 2 - 0 - 2|$$

$\Delta = 5$ sq. units.

(ii) The area of the triangle with vertices (– 7, – 2), (– 3, – 5) and (5, 2) is given by

$$\Delta = \frac{1}{2}\begin{vmatrix} -7 & -2 & 1 \\ -3 & -5 & 1 \\ 5 & 2 & 1 \end{vmatrix}$$

$$= \frac{1}{2}\left|-7\begin{vmatrix} -5 & 1 \\ 2 & 1 \end{vmatrix} + 2\begin{vmatrix} -3 & 1 \\ 5 & 1 \end{vmatrix} + 1\begin{vmatrix} -3 & -5 \\ 5 & 2 \end{vmatrix}\right|$$

$$= \frac{1}{2}|-7(-5\times 1 - 2\times 1) + 2(-3\times 1 - 5\times 1) + 1(-3\times 2 - (-5)\times 5)|$$

$$= \frac{1}{2}|-7(-7) + 2(-8) + (19)|$$

$$= \frac{1}{2}|49 - 16 + 19|$$

$= 26$ sq. units.

Q. 9. Show that points *A* (*a*, *b* + *c*), *B* (*b*, *c* + *a*), *C* (*c*, *a*+*b*) are collinear.

Sol. Area of ΔABC is given by the relation

$$\Delta = \frac{1}{2}\begin{vmatrix} a & b+c & 1 \\ b & c+a & 1 \\ c & a+b & 1 \end{vmatrix}$$

Applying $R_2 \to R_2 - R_1$ and $R_3 \to R_3 - R_1$

$$= \frac{1}{2}\begin{vmatrix} a & b+c & 1 \\ b-a & a-b & 0 \\ c-a & a-c & 0 \end{vmatrix}$$

$$= \frac{1}{2}(a-b)(c-a)\begin{vmatrix} a & b+c & 1 \\ -1 & 1 & 0 \\ 1 & -1 & 0 \end{vmatrix}$$

Applying $R_3 \to R_3 + R_2$

$$= \frac{1}{2}(a-b)(c-a)\begin{vmatrix} a & b+c & 1 \\ -1 & 1 & 0 \\ 0 & 0 & 0 \end{vmatrix}$$

Since all the elements of R_3 are 0

$\therefore \quad \Delta = 0$

$\therefore$ Area of the triangle formed by points *A*, *B* and *C* is zero.

Hence, the points *A*, *B* and *C* are collinear.

Q. 10. Find the values of *k* if area of triangle is 4 square units and vertices are

(i) (*k*, 0), (4, 0), (0, 2)

(ii) (– 2, 0), (0, 4), (0, *k*).

Sol. We know that the area of triangle whose vertices are (x_1, y_1), (x_2, y_2) and (x_3, y_3) is the absolute value of the determinant (Δ),

where $\Delta = \begin{vmatrix} x_1 & y_1 & 1 \\ x_2 & y_2 & 1 \\ x_3 & y_3 & 1 \end{vmatrix}$

(i) The area of triangle with vertices (*k*, 0), (4, 0), (0, 2) is given by

$$\Delta = \frac{1}{2}\begin{vmatrix} k & 0 & 1 \\ 4 & 0 & 1 \\ 0 & 2 & 1 \end{vmatrix}$$

$$\pm 4 = \frac{1}{2}[k(0-2) - 0(4-0) + 1(8-0)]$$

$$\pm 4 = \frac{1}{2}[-2k+8]$$

$$= -k+4$$

When $-k+4=4 \Rightarrow k=0$

When $-k+4=-4 \Rightarrow k=8$.

Thus, $k = 0, 8$.

(ii) The area of the triangle with vertices $(-2, 0)$, $(0, 4)$, $(0, k)$ is given by the relation,

$$\Delta = \frac{1}{2}\begin{vmatrix} -2 & 0 & 1 \\ 0 & 4 & 1 \\ 0 & k & 1 \end{vmatrix}$$

$$= \frac{1}{2}\left|-2\begin{vmatrix} 4 & 1 \\ k & k \end{vmatrix}\right|$$

$$= \frac{1}{2}|-2(4-k)|$$

$$\pm 4 = k-4$$

When $k-4=4$

$k=8$

When $k-4=-4$

$k=0$.

Thus, $k = 8, 0$.

Q. 11. Find the equation of line joining (2, 3) and (2, 7) using determinants.

Sol. Let $P(x, y)$ be any point on the line joining points $A(2, 3)$ and $B(2, 7)$. Then, the points A, B and P are collinear. Therefore, the area of the triangle ABP will be zero.

$$\therefore \quad \frac{1}{2}\begin{vmatrix} 2 & 3 & 1 \\ 2 & 7 & 1 \\ x & y & 1 \end{vmatrix} = 0$$

$$\Rightarrow \quad \frac{1}{2}|2(7-y) - 3(2-x) + 1(2y-7x)| = 0$$

$$\Rightarrow \quad 14 - 2y - 6 + 3x + 2y - 7x = 0$$

$$\Rightarrow \quad -4x + 8 = 0$$

$$\Rightarrow \quad x - 2 = 0.$$

Q. 12. If $A = \begin{bmatrix} \alpha & \beta \\ \gamma & \delta \end{bmatrix}$, find the values of adj A.

Sol. Given, $A = \begin{bmatrix} \alpha & \beta \\ \gamma & \delta \end{bmatrix}$

$$|A| = \begin{vmatrix} \alpha & \beta \\ \gamma & \delta \end{vmatrix} = (\alpha\delta - \gamma\beta) \neq 0$$

So, A^{-1} exist.

Minors of matrix A

$$M_{11} = \delta, M_{12} = \gamma$$

$$M_{21} = \beta, M_{22} = \alpha$$

Cofactors of matrix A

$$C_{11} = \delta, C_{12} = -\gamma$$

$$C_{21} = -\beta, C_{22} = \alpha$$

$$\therefore \quad \text{adj } A = \begin{bmatrix} \delta & -\gamma \\ -\beta & \alpha \end{bmatrix}^T = \begin{bmatrix} \delta & -\beta \\ -\gamma & \alpha \end{bmatrix}$$

Q. 13. If A is a 3×3 invertible matrix, then what will be the value of k, if $\det(A^{-1}) = (\det A)^k$.★

Sol. Given, A is an invertible matrix.

$$\therefore \quad A^{-1} = \frac{\text{adj. } A}{|A|}$$

$$\Rightarrow \text{adj. } A = A^{-1}|A|$$

$$\Rightarrow |\text{adj. } A| = |A^{-1}||A| \quad \left[\begin{array}{l} \because |\text{adj. } A| = |A|^{n-1} \\ \text{and } |A^{-1}| = |A|^k \\ \hfill \text{(given)} \end{array}\right]$$

$$\Rightarrow \quad |A|^{n-1} = |A|^k|A|$$

$$\Rightarrow \quad |A|^{3-1} = |A|^k|A|$$

$$\Rightarrow \quad |A|^2 = |A|^k|A|$$

$$\Rightarrow \quad |A| = |A|^k$$

$$\Rightarrow \quad k = 1.$$

Q. 14. Examine the consistency of the system of equations $x + 2y = 2$ and $2x + 3y = 3$. [NCERT]

Sol. The given system of equations can be represented in the matrix form as

$$AX = B$$

where $A = \begin{bmatrix} 1 & 2 \\ 2 & 3 \end{bmatrix}$, $X = \begin{bmatrix} x \\ y \end{bmatrix}$ and $B = \begin{bmatrix} 2 \\ 3 \end{bmatrix}$

Now, $|A| = \begin{vmatrix} 1 & 2 \\ 2 & 3 \end{vmatrix}$

$$= 1 \times 3 - 2 \times 2$$

$$= -1 \neq 0$$

$\therefore$ A is a non-singular matrix.

As a result, its inverse can be calculated.

So, given system of equations is consistent.

Q. 15. Show that the following system of equations is consistent :

$$\mathbf{3x + 5y = 7}$$

$$\mathbf{2x - 5y = 9.}$$

Sol. The given system of equations can be represented in the matrix form as

$$AX = B$$

★ are board exam questions from previous years

Where $A = \begin{bmatrix} 3 & 5 \\ 2 & -5 \end{bmatrix}$, $X = \begin{bmatrix} x \\ y \end{bmatrix}$ and $B = \begin{bmatrix} 7 \\ 9 \end{bmatrix}$

Now, $|A| = \begin{vmatrix} 3 & 5 \\ 2 & -5 \end{vmatrix}$

$= 3 \times (-5) - 2 \times 5$

$= -15 - 10$

$= -25 \neq 0$

$\therefore A^{-1}$ can be calculated.

As a result, given system of equations is consistent.

Q. 16. Write the value of

$$\Delta = \begin{vmatrix} x+y & x+z & z+x \\ z & x & y \\ -3 & -3 & -3 \end{vmatrix}. \star$$

Sol. Let $\Delta = \begin{vmatrix} x+y & x+z & z+x \\ z & x & y \\ -3 & -3 & -3 \end{vmatrix}$

Applying $R_1 \to R_1 + R_2$, we get

$$\Delta = \begin{vmatrix} x+y+z & x+y+z & x+y+z \\ z & x & y \\ -3 & -3 & -3 \end{vmatrix}$$

Taking out $(x + y + z)$ and (-3) common from R_1 and R_3, respectively.

$$\Delta = -3(x+y+z)\begin{vmatrix} 1 & 1 & 1 \\ z & x & y \\ 1 & 1 & 1 \end{vmatrix}$$

$\Rightarrow \quad \Delta = -3(x+y+z).0$

[$\because R_1$ and R_3 are identical]

$\Rightarrow \quad \Delta = 0.$

Fill in the blanks :

Q. 17. If A is a 3×3 matrix, $|A| \neq 0$ and $|3A| = k|A|$, then the value of k is ——— .

Sol. 27

We have, $|3A| = k|A|$

If A is a $n \times n$ matrix, then

$|kA| = k^n |A|$

So, $|3A| = 3^3 |A|$

$= 27|A|$

$\Rightarrow \quad 27|A| = k|A|$

$\Rightarrow \quad k = 27.$

Q. 18. If $\begin{vmatrix} x & 2 \\ 18 & x \end{vmatrix} = \begin{vmatrix} 6 & 2 \\ 18 & 6 \end{vmatrix}$, then the value of x is ——— .

Sol. ± 6

Given, $\begin{vmatrix} x & 2 \\ 18 & x \end{vmatrix} = \begin{vmatrix} 6 & 2 \\ 18 & 6 \end{vmatrix}$

$\Rightarrow \quad x^2 - 18 \times 2 = 36 - 36$

$\Rightarrow \quad x^2 - 36 = 0$

$\Rightarrow \quad x^2 = 36$

$\Rightarrow \quad x = \pm 6.$

Q. 19. The value of

$$\begin{vmatrix} \cos\alpha\cos\beta & \cos\alpha\sin\beta & -\sin\alpha \\ -\sin\beta & \cos\beta & 0 \\ \sin\alpha\cos\beta & \sin\alpha\sin\beta & \cos\alpha \end{vmatrix} \text{ is ——— .}$$

Sol. 1

Let $\Delta = \begin{vmatrix} \cos\alpha\cos\beta & \cos\alpha\sin\beta & -\sin\alpha \\ -\sin\beta & \cos\beta & 0 \\ \sin\alpha\cos\beta & \sin\alpha\sin\beta & \cos\alpha \end{vmatrix}$

Expanding along C_3, we have

$$\Delta = -\sin\alpha \begin{vmatrix} -\sin\beta & \cos\beta \\ \sin\alpha\cos\beta & \sin\alpha\sin\beta \end{vmatrix} - 0\begin{vmatrix} \cos\alpha\cos\beta & \cos\alpha\sin\beta \\ \sin\alpha\cos\beta & \sin\alpha\sin\beta \end{vmatrix} + \cos\alpha \begin{vmatrix} \cos\alpha\cos\beta & \cos\alpha\sin\beta \\ -\sin\beta & \cos\beta \end{vmatrix}$$

$= -\sin\alpha(-\sin\alpha\sin^2\beta - \sin\alpha\cos^2\beta) - 0 + \cos\alpha(\cos\alpha\cos^2\beta + \cos\alpha\sin^2\beta)$

$= \sin^2\alpha(\sin^2\beta + \cos^2\beta) + \cos^2\alpha(\cos^2\beta + \sin^2\beta)$

$= \sin^2\alpha + \cos^2\alpha$

$= 1.$

Q. 20. If the matrix $\begin{bmatrix} k & 2 \\ 3 & 4 \end{bmatrix}$ has no inverse, then the value of k is ——— .

Sol. $\frac{3}{2}$

Let $A = \begin{bmatrix} k & 2 \\ 3 & 4 \end{bmatrix}$

We know that if the given matrix has no inverse, then

$|A| = 0$

$\begin{vmatrix} k & 2 \\ 3 & 4 \end{vmatrix} = 0$

$4k - 6 = 0$

$k = 3/2.$

Q. 21. If the matrix $\begin{bmatrix} 3-2x & x+1 \\ 2 & 4 \end{bmatrix}$ is a singular matrix, then the value of x is ——— .

Sol. 1

For a singular matrix,

$|A| = 0$

$$\therefore \quad \begin{vmatrix} 3-2x & x+1 \\ 2 & 4 \end{vmatrix} = 0$$

$$\Rightarrow \quad 4(3-2x)-2(x+1)=0$$

$$\Rightarrow \quad 12-8x-2x-2=0$$

$$\Rightarrow \quad x=1.$$

Q. 22. If $x \in N$ and $\begin{vmatrix} x+3 & -2 \\ -3x & 2x \end{vmatrix} = 8$, then the value of x is ———.

Sol. 2

We have, $\begin{vmatrix} x+3 & -2 \\ -3x & 2x \end{vmatrix} = 0$

$$(x+3)\times 2x-(-2)\times(-3x)=8$$

$$2x^2+6x-6x=8$$

$$2x^2=8$$

$$x^2=4$$

But $\quad x \neq -2$ as $x \in N$

$\therefore \quad x=2.$

Q. 23. The maximum value of $\begin{vmatrix} 1 & 1 & 1 \\ 1 & 1+\sin\theta & 1 \\ 1 & 1 & 1+\cos\theta \end{vmatrix}$ is ———.

Sol. $\frac{1}{2}$

Let $\quad \Delta = \begin{vmatrix} 1 & 1 & 1 \\ 1 & 1+\sin\theta & 1 \\ 1 & 1 & 1+\cos\theta \end{vmatrix}$

On applying $R_1 \to R_1 - R_2$ and $R_2 \to R_2 - R_3$, we get

$$\Delta = \begin{vmatrix} 0 & -\sin\theta & 0 \\ 0 & \sin\theta & -\cos\theta \\ 1 & 1 & 1+\cos\theta \end{vmatrix}$$

On expanding along R_1, we get

$$\Delta = 0+\sin\theta\,(0+\cos\theta)+0$$

$$= \sin\theta\cos\theta$$

$$= \frac{1}{2}\sin 2\theta$$

Now, as $-1 \le \sin 2\theta \le 1$ for all $\theta \in R$

$$\Rightarrow \quad -\frac{1}{2} \le \frac{1}{2}\sin 2\theta \le \frac{1}{2} \text{ for all } \theta \in R$$

Clearly, maximum value of Δ is $\frac{1}{2}$.

Q. 24. If A and B are square matrices each of order 3 and $|A| = 5$, $|B| = 3$, then the value of $|3AB|$ is ______.★

Sol. 405

$\because \quad |A| = 5, |B| = 3$

$\therefore \quad |3\,AB| = 3^3|A|\,|B|$

$$= 27 \times 5 \times 3 = 405$$

★ are board exam questions from previous years

Chapter 5. Continuity and Differentiability

Q. 1. Determine the value of 'k' for which the following function is continuous at $x = 3$.★

$$f(x) = \begin{cases} \dfrac{(x+3)^2-36}{x-3}, & x \neq 3 \\ k, & x = 3 \end{cases}$$

Sol. Given, $\quad f(x) = \begin{cases} \dfrac{(x+3)^2-36}{x-3}, & x \neq 3 \\ k, & x = 3 \end{cases}$

Since $f(x)$ is continuous at $x = 3$

$$\therefore \quad \lim_{x\to 3} f(x) = f(3)$$

$$\Rightarrow \quad \lim_{x\to 3} f(x) = k$$

$$\Rightarrow \quad \lim_{x\to 3} \frac{(x+3)^2-36}{x-3} = k$$

$$\Rightarrow \quad \lim_{x+3\to 6} \frac{(x+3)^2-6^2}{(x+3)-6} = k$$

$$\Rightarrow \quad \lim_{x\to 3} x+3+6 = k$$

$$\therefore \quad 12 = k$$

Thus, $f(x)$ is continuous at $x = 3$, if $k = 12$.

Q. 2. Find the derivative of $\dfrac{e^x}{\sin x}$.

Sol. Let $\quad y = \dfrac{e^x}{\sin x}$

$$\therefore \quad \frac{dy}{dx} = \frac{\sin x.e^x - e^x\cos x}{\sin^2 x}$$

[using quotient rule]

$$= \frac{e^x(\sin x - \cos x)}{\sin^2 x}$$

Q. 3. Find $\left.\dfrac{dy}{dx}\right|_{x=\frac{\pi}{2}}$, where $y = e^{\sin x}$.

Sol. We have $\; y = e^{\sin x}$

$$\Rightarrow \quad \frac{dy}{dx} = e^{\sin x}.\cos x$$

$$\therefore \quad \left.\frac{dy}{dx}\right|_{x=\frac{\pi}{2}} = e^1.0 = 0.$$

Q. 4. Evaluate: $\displaystyle\lim_{x\to\pi/2}\left[x\tan x - \frac{\pi}{2}\sec x\right]$

Sol. Consider, $\displaystyle\lim_{x\to\pi/2}\left[x\tan x - \frac{\pi}{2}\sec x\right]$

$$= \lim_{x\to\pi/2}\left[\frac{x\sin x}{\cos x} - \frac{\pi}{2}\times\frac{1}{\cos x}\right]$$

$$= \lim_{x \to \pi/2} \left[\frac{x \sin x - \frac{\pi}{2}}{\cos x} \right] \quad \left(\frac{0}{0} \text{ form} \right)$$

By L'Hospital's rule,

$$= \lim_{x \to \pi/2} \left[\frac{x \cos x + \sin x}{- \sin x} \right]$$

$$= -1.$$

Q. 5. Find $\frac{dy}{dx}$ when, $y = \frac{e^{2x} + e^{-2x}}{e^{2x} - e^{-2x}}$.

Sol. Given, $y = \frac{e^{2x} + e^{-2x}}{e^{2x} - e^{-2x}}$

$$= \frac{e^{4x} + 1}{e^{4x} - 1}$$

Differentiating w.r.t. x, we get

$$\frac{dy}{dx} = \frac{4e^{4x}(e^{4x} - 1) - 4e^{4x}(e^{4x} + 1)}{(e^{4x} - 1)^2}$$

$$= \frac{-8e^{4x}}{(e^{4x} - 1)^2}$$

Fill in the blanks :

Q. 6. If the function $f(x) = \begin{cases} \frac{kx}{|x|}, & \text{if } x < 0 \\ 3, & \text{if } x \geq 0 \end{cases}$ is continuous at $x = 0$, then the value of k is ———.

Sol. -3

Given that the function $f(x)$ is continuous at $x = 0$

$$\therefore \quad \lim_{x \to 0^-} f(x) = \lim_{x \to 0^+} f(x) = f(0) \quad \ldots\text{(i)}$$

Now, $\lim_{x \to 0^-} f(x) = \lim_{x \to 0^-} \frac{kx}{|x|}$

$$= k \lim_{x \to 0^-} \frac{x}{-x}$$

$$= k(-1) = -k$$

$$\lim_{x \to 0^+} f(x) = \lim_{x \to 0^+} 3 = 3$$

From equation (i)

$$-k = 3$$

or $k = -3.$

Q. 7. If $y = e^{\sin x^2}$, then the value of $\frac{dy}{dx}$ is ———.

Sol. $2x \cos x^2 \, e^{\sin x2}$

Given, $y = e^{\sin x^2}$

Differentiating both sides w.r.t. x, we get

$$\frac{dy}{dx} = e^{\sin x^2} \frac{d}{dx}(\sin x^2)$$

$$= e^{\sin x^2} \times \cos x^2 \times \frac{d}{dx}(x^2)$$

$$= e^{\sin x^2} \cos x^2 (2x)$$

$$\frac{dy}{dx} = 2x \cos x^2 \, e^{\sin x^2}$$

Q. 8. If $x = a \sin^3 t$ and $y = a \cos^3 t$, then the value of $\frac{dy}{dx}$ is ———.

Sol. $-\cot t$

Given,

$$x = a \sin^3 t$$

$$\therefore \quad \frac{dx}{dt} = 3a \sin^2 t . \cos t$$

Also, $y = a \cos^3 t$

$$\therefore \quad \frac{dy}{dt} = 3a \cos^2 t (-\sin t)$$

$$\therefore \quad \frac{dy}{dx} = \frac{-3a \cos^2 t . \sin t}{3a \sin^2 t . \cos t} \quad \left[\because \frac{dy}{dx} = \frac{dy/dt}{dx/dt} \right]$$

$$\Rightarrow \quad \frac{dy}{dx} = -\cot t.$$

Q. 9. The derivative of $f(x) = |x^3|$ at $x = 0$ is ——.

Sol. 0

We have,

$$f(x) = |x^3| = \begin{cases} x^3, & x \geq 0 \\ -x^3, & x < 0 \end{cases}$$

$$f'(0) = \lim_{x \to 0} \frac{f(x) - f(0)}{x - 0}$$

Now, $f'(0^-) = \lim_{x \to 0^-} \frac{f(x) - f(0)}{x - 0}$

$$= \lim_{x \to 0^-} \frac{-x^3 - 0}{x}$$

$$= -\lim_{x \to 0^-} x^2 = 0$$

and $f'(0^+) = \lim_{x \to 0^+} \frac{f(x) - f(0)}{x - 0}$

$$= \lim_{x \to 0^+} \frac{+x^3 - 0}{x}$$

$$= \lim_{x \to 0^+} x^2 = 0$$

Thus, $f'(0) = f'(0^-) = f'(0^+) = 0.$

Q. 10. The number of points of discontinuity of f defined by $f(x) = |x| - |x + 1|$ is ______.★

Sol. 0

Here $f(x) = |x| - |x + 1|$

Let $h(x) = |x|$

★ are board exam questions from previous years

and $g(x) = |x+1|$

Then $h(x)$ and $g(x)$ are both continuous functions as modulus function and polynomial function are continuous everywhere.

Also $(hog)(x) = h(g(x)) = h(x+1)$
$= |x+1|$

Since composite of two continuous functions is also continuous in its domain

$(hog)(x) = |x+1|$ is also continuous

Also difference of two continuous functions is continuous everywhere

$\therefore \quad f(x) = |x| - |x+1| = h(x) - (hog)(x)$

is also continuous everywhere.

Hence, the number of points of discontinuity of

$f(x) = |x| - |x+1|$ is zero.

Q. 11. If $f(x) = 2|x| + 3|\sin x| + 6$, then the right hand derivative $f(x)$ at $x = 0$ is ______________.★

Sol. 5 $\quad f(x) = 2|x| + 3|\sin x| + 6$

Right hand derivative of f at $x = 0$ is :

$$Rf ¢(0) = \lim_{h\to 0} \frac{f(0+h) - f(0)}{h}$$, where h is a very small positive real number.

$\therefore \quad Rf'(0)$

$$= \lim_{h\to 0} \frac{(2|h| + 3|\sin h| + 6) - (2|0| + 3\sin|0| + 6)}{h}$$

$$= \lim_{h\to 0} \frac{(2h + 3\sin h + 6) - (0+0+6)}{h}$$

$$= \lim_{h\to 0} \frac{2h + 3\sin h}{h}$$

$$= \lim_{h\to 0} \frac{2h}{h} + 3\lim_{h\to 0} \frac{\sin h}{h}$$

$= 2 + 3(1)$

$= 5$

Q. 12. If $f(x) = x\,|x|$, then $f'(x) =$ ____________.★

Sol. $2x, x \geq 0, -2x, x < 0$

$$f(x) = x|x|$$

$$= \begin{cases} x^2, & \text{if } x \geq 0 \\ -x^2, & \text{if } x < 0 \end{cases}$$

$$\therefore \quad f'(x) = \begin{cases} 2x, & \text{if } x \geq 0 \\ -2x, & \text{if } x < 0 \end{cases}$$

Chapter 6. Applications of Derivatives

Q. 1. Find the maximum value of the function

$$f(x) = x^3 + x^2 + x - 4.$$

★ are board exam questions from previous years

Sol. We have, $f(x) = x^3 + x^2 + x - 4$

Differentiating $f(x)$ with respect to x, we get

$$f'(x) = 3x^2 + 2x + 1$$

Clearly, $f'(x) \neq 0$ at real values of x, so $f(x)$ does not have maximum value.

Q. 2. A stone is dropped into a quiet lake and wave move in a circle at a speed of 3·5 cm/sec. At the instant when the radius of the circular wave is 7·5 cm, how fast is the enclosed area increasing ? [NCERT]

Sol. Let r be the radius and A be the area of the circular wave at any time t.

Then, area of circular wave

$$= A = \pi r^2 \text{ and } \frac{dr}{dt} = 3\cdot 5 \text{ cm/sec}$$

Now, $\quad A = \pi r^2$

Differentiate with respect to t, we get

$$\frac{dA}{dt} = \pi\left(2r\frac{dr}{dt}\right)$$

$$\Rightarrow \quad \frac{dA}{dt} = 2\pi r\frac{dr}{dt}$$

$$\Rightarrow \quad \frac{dA}{dt} = 2\pi r(3\cdot 5) = 7\pi r \quad \text{(given)}$$

$$\Rightarrow \quad \left(\frac{dA}{dt}\right)_{r=7\cdot 5} = 7\pi(7\cdot 5)$$

$$= 52\cdot 5\pi \text{ cm}^2/\text{sec.}$$

Q. 3. The total revenue in rupees received from the sale of x units of a product is given by $R(x) = 3x^2 + 36x + 5$. Find the marginal revenue, when $x = 5$, where by marginal revenue we mean the rate of change of total revenue with respect to the number of items sold at an instant. [NCERT]

Sol. Since the marginal revenue is the rate of change of total revenue with respect to the number of units sold, we have

$$R(x) = 3x^2 + 36x + 5$$

Differentiate with respect to x, we get

Marginal Revenue

$$(MR) = \frac{dR}{dx} = 6x + 36$$

When $x = 5$, $\quad MR = 6(5) + 36 = 66$

Hence, the required marginal revenue is ₹ 66.

Q. 4. The total cost $C(x)$ associated with the production of x units of an item is given by $C(x) = 0\cdot 005x^3 - 0\cdot 02x^2 + 30x + 5000$. Find the marginal cost when 3 units are produced, where by marginal cost we mean the instantaneous rate of change of total cost at any level of output.

Sol. Cost function is given as

$$C(x) = 0{\cdot}005x^3 - 0{\cdot}02x^2 + 30x + 5000$$

$$\text{Marginal cost} = \left[\frac{d}{dx}(C(x))\right]_{x=3}$$

$$= [0{\cdot}015x^2 - 0{\cdot}04x + 30]_{x=3}$$

$$= 0{\cdot}015(3)^2 - 0{\cdot}04(3) + 30$$

$$= 0{\cdot}135 - 0{\cdot}12 + 30$$

$$= 30{\cdot}015.$$

Q. 5. A man 2 metres high, walks at a uniform speed of 6 metres per minute away from a lamp-post, 5 metres high. Find the rate at which the length of his shadow increases.

Sol. Let *AB* be the lamp-post.

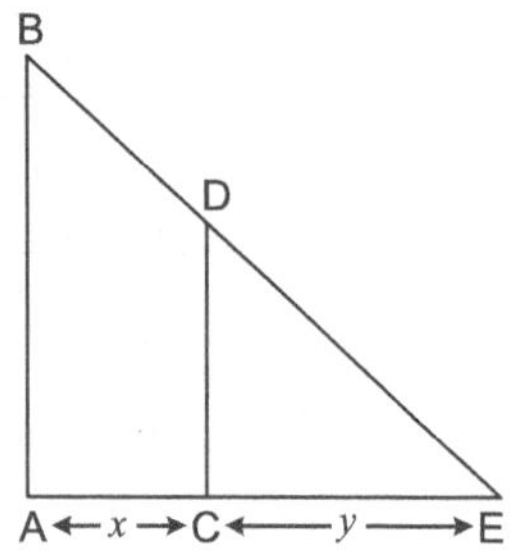

Let at any time, the man *CD* be at a distance *x* metres from the lamp-post and *y* metres be the length of his shadow *CE*.

Then, $\frac{dx}{dt} = 6$ metres/minute (given)

Clearly, triangles *ABE* and *CDE* are similar.

$$\therefore \quad \frac{AB}{CD} = \frac{AE}{CE}$$

$$\Rightarrow \quad \frac{5}{2} = \frac{x+y}{y}$$

$$\Rightarrow \quad 3y = 2x$$

Differentiating with respect to *t*, we get

$$3\frac{dy}{dt} = 2\frac{dx}{dt}$$

$$\Rightarrow \quad 3\frac{dy}{dt} = 2(6)$$

$$\Rightarrow \quad \frac{dy}{dt} = 4$$

Thus, the shadow increases at the rate of 4 metres/minute.

Q. 6. Show that the function *f* given by $f(x) = x^3 - 3x^2 + 4x, x \in R$ is strictly increasing on *R*.

Sol. We have,

$$f(x) = x^3 - 3x^2 + 4x, x \in R$$

Differentiate with respect to *x*, we get

$$f'(x) = 3x^2 - 6x + 4$$

$$= 3(x^2 - 2x + 1) + 1$$

$= 3(x-1)^2 + 1 > 0$, in every interval of *R*

Therefore, the function *f* is strictly increasing on *R*.

Q. 7. Find the points to the curve $\frac{x^2}{4} + \frac{y^2}{25} = 1$ at which the tangents are (i) parallel to *X*-axis, (ii) parallel to the *y*-axis.

Sol. Given, $\frac{x^2}{4} + \frac{y^2}{25} = 1$ ...(i)

Differentiating w.r.t. *x*, we get

$$\frac{2x}{4} + \frac{2y}{25}\frac{dy}{dx} = 0$$

$$\frac{dy}{dx} = \frac{-25x}{4y}$$

(i) For tangent parallel to *X*-axis

$$\frac{dy}{dx} = 0$$

$$\Rightarrow \quad x = 0$$

$\Rightarrow \quad y = \pm 5$ [From equation (i)]

$\Rightarrow$ Points are $(0, \pm 5)$

(ii) For tangent parallel to *Y*-axis, $\frac{dy}{dx}$ is not defined.

$$\Rightarrow \quad y = 0$$

$\Rightarrow \quad x = \pm 2$ [From equation (i)]

$\Rightarrow$ Points are $(\pm 2, 0)$.

Q. 8. Show that, the function $f(x) = \log(\cos x)$ is decreasing in $\left[0, \frac{\pi}{2}\right]$. **[HOTS]**

Sol. Given, $f(x) = \log(\cos x)$

Differentiate $f(x)$ w.r.t. *x*, we get

$$f'(x) = \frac{1}{\cos x}(-\sin x)$$

$$= -\tan x$$

But we know that $\tan x > 0$ for $\left[0, \frac{\pi}{2}\right]$

$$\Rightarrow \quad f'(x) < 0 \text{ for } x \in \left[0, \frac{\pi}{2}\right]$$

$\Rightarrow f(x)$ is decreasing in $\left[0, \frac{\pi}{2}\right]$.

Q. 9. The volume of a sphere is increasing at the rate of 8 cm³/s. Find the rate of which its surface area is increasing when the radius of the sphere is 12 cm.★

Sol. We have, volume of sphere $(V) = \frac{4}{3}\pi r^3$

Where, *r* is the radius of sphere.

$$\text{Now,} \quad \frac{dV}{dt} = \frac{4}{3}\pi\left(3r^2\frac{dr}{dt}\right) = 8 \text{ cm}^3/\text{s}$$

$$\Rightarrow \quad 4\pi r^2\frac{dr}{dt} = 8 \text{ cm}^3/\text{s}$$

★ are board exam questions from previous years

$$\Rightarrow \quad \frac{dr}{dt} = \frac{2}{\pi r^2} \text{ cm/s} \quad \ldots(i)$$

and surface area of sphere $(S) = 4\pi r^2$.

Then, $$\frac{dS}{dt} = 4\pi \cdot 2r\frac{dr}{dt}$$

$$\frac{dS}{dt} = 8\pi r \cdot \frac{2}{\pi r^2} \quad \text{[using (i)]}$$

$$\frac{dS}{dt} = \frac{16}{r} \text{ cm}^2\text{/s}$$

When, $r = 12$ cm,

$$\left[\frac{dS}{dt}\right]_{r=12} = \frac{16}{12} = \frac{4}{3} \text{ cm}^2\text{/s}$$

Hence, the surface area is increasing at the rate $\frac{4}{3}$ cm²/s, when radius of sphere is 12 cm.

Q. 10. The volume of a cube is increasing at the rate of 9 cm³/s. How fast is its surface area increasing when the length of an edge is 10 cm ?★

Sol. Let the side of the cube be a cm.

Then, volume of cube $(V) = a^3$

$$\frac{dV}{dt} = 3a^2\frac{da}{dt}$$

$$\Rightarrow \quad 3a^2\frac{da}{dt} = 9 \text{ cm}^3\text{/s}$$

$$\left[\text{Given, } \frac{dV}{dt} = 9 \text{ cm}^3 \text{/s}\right]$$

$$\frac{da}{dt} = \frac{3}{a^2} \text{ cm/s}$$

and surface area of cube $(S) = 6a^2$

$$\frac{dS}{dt} = 12a\frac{da}{dt}$$

$$\frac{dS}{dt} = 12a \times \frac{3}{a^2} = \frac{36}{a} = \text{cm}^2\text{/s}$$

$$\therefore \quad \left[\frac{dS}{dt}\right]_{a=10} = \frac{36}{10}$$

$$= 3{\cdot}6 \text{ cm}^2\text{/s.}$$

Q. 11. The length, x of a rectangle is decreasing at the rate of 5 cm/minute and the width, y is increasing at the rate of 4 cm/minute. When $x = 8$ cm and $y = 6$ cm, find the rate of change of the area of the rectangle.★

Sol. We have,

$$\frac{dx}{dt} = -5 \text{ cm/min} \quad \ldots(i)$$

and $$\frac{dy}{dt} = 4 \text{ cm/min} \quad \ldots(ii)$$

Now, area of the rectangle $(A) = xy$

$$\Rightarrow \quad \frac{dA}{dt} = x\frac{dy}{dt} + y\frac{dx}{dt}$$

$$\frac{dA}{dt} = x(4) + y(-5) \quad \text{[using (i) and (ii)]}$$

$$\therefore \quad \frac{dA}{dt} = 4x - 5y$$

When $x = 8$ cm and $y = 6$ cm,

$$\therefore \quad \left[\frac{dA}{dt}\right]_{\text{at } x=8,\ y=8} = 4(8) - 5(6)$$

$$= 32 - 30 = 2 \text{ cm}^2\text{/min}$$

Hence, the rate of change of the area of the rectangle is 2 cm²/min.

Fill in the blanks :

Q. 12. A balloon, which always remains spherical has a variable diameter $\frac{3}{2}(2x + 3)$. The rate of change of volume with respect to x is ———.

Sol. $\frac{27}{8}\pi(2x + 3)^2$

Let V be the volume of the balloon.

Radius of spherical balloon $= \frac{3}{4}(2x + 3)$

Since $$V = \frac{4\pi}{3}(\text{radius})^3$$

Then $$V = \frac{4\pi}{3}\left\{\frac{3}{4}(2x+3)\right\}^3$$

$$= \frac{9\pi}{16}(2x + 3)^3$$

Differentiate V with respect to x, we get

$$\frac{dV}{dx} = \frac{9\pi}{16}\, 3(2x + 3)^2 \frac{d}{dx}(2x + 3)$$

Hence $$\frac{dV}{dx} = \frac{27\pi}{8}(2x + 3)^2.$$

Q. 13. The slope of the tangent to the curve $x = t^2 + 3t - 8$, $y = 2t^2 - 2t - 5$ at $t = 2$ is ———.

Sol. $\frac{6}{7}$

Slope of the tangent

$$= \frac{dy}{dx} = \frac{\frac{dy}{dt}}{\frac{dx}{dt}}$$

$$= \frac{4t - 2}{2t + 3}$$

$$\Rightarrow \quad \left.\frac{dy}{dx}\right|_{t=2} = \left(\frac{4t-2}{2t+3}\right)_{t=2}$$

★ are board exam questions from previous years

$$= \frac{4\times2-2}{2\times2+3} = \frac{6}{7}$$

Q. 14. If $y = \log_e x$, then the value of Δy is ——— when $x = 3$ and $\Delta x = 0.03$.

Sol. 0.01

Given, $y = \log_e x$

$$\Delta y = \frac{dy}{dx}\Delta x$$

$$= \frac{1}{x}\Delta x$$

$$= \frac{0.03}{3}$$

$$\Delta y = 0.01.$$

Q. 15. The approximate change in the volume V of a cube of side x metres caused by increasing the side by 2% is ———. [NCERT]

Sol. $0.06x^3$ m^3

We know that,

$$V = x^3$$

or $$\Delta V = \left(\frac{dV}{dx}\right)\Delta x = (3x^2)\Delta x$$

$$= (3x^2)(0.02x)$$

$$= 0.06x^3 \text{ m}^3 \qquad [\text{As } 2\% \text{ of } x \text{ is } 0.02x]$$

Thus, the approximate change in volume is $0.06x^3$ m^3.

Q. 16. The slope of the tangent to the curve $y = x^3 - x$ at the point (2, 6) is ________.★

Sol. 11

$$y = x^3 - x$$

$$\Rightarrow \quad \frac{dy}{dx} = 3x^2 - 1$$

$\therefore$ Slope of tangent at (2, 6) = $3(2)^2 - 1 = 11$

Q. 17. The rate of change of the area of a circle with respect to its radius r, when r = 3 cm, is ______.★

Sol. 6π cm^2/cm

Area of circle, $A = \pi r^2$

$$\therefore \quad \frac{dA}{dr} = 2\pi r$$

$$\left[\frac{dA}{dr}\right]_{r = 3\text{cm}} = 2\times\pi\times3$$

$$= 6\pi \text{ cm}^2/\text{cm}$$

Q. 18. The least value of the function $f(x) = ax + \frac{b}{x}$ $(a > 0, b > 0, x > 0)$ is ———.★

Sol. $2\sqrt{ab}$

$$\because \quad f(x) = ax + \frac{b}{x}$$

$$\therefore \quad f'(x) = a - \frac{b}{x^2}$$

For maximum and minimum function,

Put $f'(x) = 0$

$$\Rightarrow \quad a - \frac{b}{x^2} = 0$$

$$\Rightarrow \quad ax^2 - b = 0$$

$$\Rightarrow \quad ax^2 = b$$

$$\Rightarrow \quad x = \sqrt{\frac{b}{a}} \qquad [\because x > 0]$$

Now, $$f''(x) = 0 + \frac{2b}{x^3}$$

$$= \frac{2b}{\left(\sqrt{b/a}\right)^3} > 0$$

$\therefore$ $f(x)$ is minimum at $x = \sqrt{\frac{b}{a}}$

$\therefore$ The least value of $f(x)$ is,

$$f\left(\sqrt{\frac{b}{a}}\right) = a\sqrt{\frac{b}{a}} + \frac{b}{\sqrt{\frac{b}{a}}}$$

$$= \sqrt{ab} + \sqrt{ab} = 2\sqrt{ab}$$

Chapter 7. Integrals

Q. 1. Find $\int \frac{x + \cos 6x}{3x^2 + \sin 6x} dx$

Sol. Let $I = \int \frac{x + \cos 6x}{3x^2 + \sin 6x} dx$

Multiply and divide by 6

$$\Rightarrow \quad I = \frac{1}{6}\int \frac{6x + 6\cos 6x}{3x^2 + \sin 6x} dx$$

$$= \frac{1}{6}\log|3x^2 + \sin 6x| + c$$

$$\left[\because \int \frac{f'(x)}{f(x)} dx = \log|f(x)|\right] + c$$

Q. 2. Find $\int \frac{2x}{1 + x^2} dx$

Sol. $\because \int \frac{f'(x)}{f(x)} dx = \log|f(x)| + c$

$$\therefore \quad \int \frac{2x}{1 + x^2} dx = \log|1 + x^2| + c$$

★ are board exam questions from previous years

Q. 3. Evaluate : $\int_0^2 \sqrt{4-x^2}\, dx$★

Sol. Given, $\int_0^2 \sqrt{2^2-x^2}\, dx$

$$\because \quad \int \sqrt{a^2-x^2}\,dx = \frac{x}{2}\sqrt{a^2-x^2}+\frac{a^2}{2}\sin^{-1}\frac{x}{a}$$

$$\therefore \quad \int_0^2 \sqrt{2^2-x^2}\, dx = \left[\frac{x}{2}\sqrt{4-x^2}+\frac{4}{2}\sin^{-1}\frac{x}{2}\right]_0^2$$

$$= \left(\frac{2}{2}\sqrt{4-4}+2\sin^{-1}\frac{2}{2}\right)-(0+2\sin^{-1}0)$$

$$= 0+2\sin^{-1}1-0$$

$$= 2\times\frac{\pi}{2}=\pi$$

Q. 4. Evaluate : $\int \sin^2(2x+5)\,dx$ **[NCERT]**

Sol. $\int \sin^2(2x+5)\,dx = \int \frac{1-\cos(4x+10)}{2}\,dx$

$$\left[\because \sin^2 x = \frac{1-\cos 2x}{2}\right]$$

$$= \frac{1}{2}\int 1\,dx - \frac{1}{2}\int \cos(4x+10)\,dx$$

$$= \frac{x}{2}-\frac{\sin(4x+10)}{8}+c$$

Q. 5. Find $\int \frac{x-1}{\sqrt{x^2-1}}\,dx$

Sol. We have, $\int \frac{x-1}{\sqrt{x^2-1}}\,dx$

$$= \int \frac{x}{\sqrt{x^2-1}}\,dx - \int \frac{1}{\sqrt{x^2-1}}\,dx$$

$$= \frac{1}{2}\int \frac{2x}{\sqrt{x^2-1}}\,dx - \int \frac{1}{\sqrt{x^2-1}}\,dx$$

$$= \frac{1}{2}\int (x^2-1)^{-1/2}.2x\,dx - \int \frac{1}{\sqrt{x^2-1}}\,dx$$

$$= \frac{1}{2}\frac{(x^2-1)^{1/2}}{1/2} - \log|x+\sqrt{x^2-1}|+c$$

$$= \sqrt{x^2-1}-\log|x+\sqrt{x^2-1}|+c$$

Q. 6. Find $\int \frac{x^3-x^2+x-1}{x-1}\,dx$

Sol. We have, $\int \frac{x^3-x^2+x-1}{x-1}\,dx$

$$= \int \frac{x^2(x-1)+1(x-1)}{x-1}\,dx = \int \frac{(x^2+1)(x-1)}{x-1}\,dx$$

$$= \frac{x^3}{3}+x+c$$

Q. 7. Find $\int (ax+b)^3\,dx$★

Sol. $\int (ax+b)^3\,dx = \frac{(ax+b)^{3+1}}{3+1}\times\frac{1}{a}+c$

$$= \frac{(ax+b)^4}{4a}+c$$

Q. 8. Evaluate $\int \frac{dx}{\sqrt{1-x^2}}$★

Sol. $\int \frac{dx}{\sqrt{1-x^2}} = \sin^{-1}x+c$

$$\left[\because \int \frac{dx}{\sqrt{a^2-x^2}} = \sin^{-1}\frac{x}{a}+c\right]$$

Q. 9. Find $\int \sec x(\sec x+\tan x)\,dx$

Sol. Let $I = \int \sec x(\sec x+\tan x)\,dx$

$$= \int (\sec^2 x+\sec x\tan x)\,dx$$

$$I = \tan x+\sec x+c$$

Q. 10. Find $\int \frac{2-3\sin x}{\cos^2 x}\,dx$★

Sol. Let $I = \int \frac{2-3\sin x}{\cos^2 x}\,dx$

$$= \int \left(\frac{2}{\cos^2 x}-\frac{3\sin x}{\cos x}\cdot\frac{1}{\cos x}\right)dx$$

$$= \int (2\sec^2 x-3\tan x\sec x)\,dx$$

$$= 2\tan x-3\sec x+c$$

Q. 11. $\int \frac{\log(\sin x)}{\tan x}\,dx$

Sol. Let $I = \int \frac{\log(\sin x)}{\tan x}\,dx$

$$= \int \log(\sin x)\cot x\,dx$$

Put $\log(\sin x) = t$

Differentiate w.r.t. x

$$\frac{1}{\sin x}\cos x = \frac{dt}{dx}$$

$$\cot x\,dx = dt$$

$$\therefore \quad I = \int t\,dt$$

$$= \frac{t^2}{2}+c$$

$$= \frac{(\log|\sin x|)^2}{2}+c$$

Fill in the blanks :

Q. 12. If $\int e^x(\tan x+1)\sec x\,dx = e^x f(x)+c$, **then the value of** $f(x)$ **satisfying the above is———.**

Sol. $\sec x$

$$\int e^x(\sec x\tan x+\sec x)\,dx = e^x f(x)+c$$

$$\because \quad \int e^x[g(x)+g'(x)]\,dx = e^x g(x)+c$$

$$\therefore \quad \int e^x(\sec x+\sec x\tan x)\,dx = e^x\sec x+c$$

$$\Rightarrow \quad e^x\sec x+c = e^x f(x)+c$$

$\therefore$ By comparing $\quad f(x) = \sec x.$

★ are board exam questions from previous years

Q. 13. The value of $\int \frac{3x^2}{x^6+1} dx$ is——.

Sol. $\tan^{-1}|x^3|+c$

$$\int \frac{3x^2}{x^6+1} dx = \int \frac{3x^2}{(x^3)^2+1} dx$$

Put $x^3 = t$

Differentiate both sides w.r.t. x

$$3x^2 = \frac{dt}{dx}$$

$$3x^2\, dx = dt$$

$$\therefore \int \frac{3x^2}{x^6+1} dx = \int \frac{dt}{t^2+1}$$

$$= \tan^{-1}|t|+c$$

$$= \tan^{-1}|x^3|+c$$

Q. 14. The value of $\int \frac{2\cos x}{3\sin^2 x} dx$ is——.

Sol. $\frac{-2}{3\sin x}+c$

Let $I = \int \frac{2\cos x}{3\sin^2 x} dx = \frac{2}{3}\int \frac{\cos x}{\sin^2 x} dx$

Put $\sin x = t$

$\cos x\, dx = dt$

$$\Rightarrow I = \frac{2}{3}\int \frac{dt}{t^2} = \frac{2}{3}\int t^{-2}\, dt$$

$$= \frac{2}{3}\left[\frac{t^{-1}}{-1}\right]+c$$

$$= -\frac{2}{3t}+c = -\frac{2}{3\sin x}+c$$

Q. 15*. The value of $\int \frac{(\log x)^2}{x} dx$ is——.

Sol. $\frac{(\log x)^3}{3}+c$

Let $I = \int \frac{(\log x)^2}{x} dx$

Put $\log x = t$

$\frac{1}{x} dx = dt$

$$\therefore I = \int t^2\, dt$$

$$= \frac{t^3}{3}+c$$

$$I = \frac{(\log x)^3}{3}+c$$

Q. 16. The value of $\int \frac{\sin^2 x - \cos^2 x}{\sin x \cos x} dx$ is——.

Sol. $-\log|\sin 2x|+c$

We have, $\int \frac{\sin^2 x - \cos^2 x}{\sin x \cos x} dx$

$$= -2\int \frac{\cos^2 x - \sin^2 x}{2\sin x \cos x} dx$$

$$= -2\int \frac{\cos 2x}{\sin 2x} dx$$

$$= -2\int \cot 2x\, dx$$

$$= -\log|\sin 2x|+c.$$

Q. 17. The value of $\int_{-\pi/4}^{\pi/4} \sin^3 x\, dx$ is——.

Sol. 0

$\int_{-\pi/4}^{\pi/4} \sin^3 x\, dx$

As $\sin^3 x$ is odd function

$\because \sin^3(-x) = (-\sin x)^3 = -\sin^3 x$

$$\therefore \int_{-\pi/4}^{\pi/4} \sin^3 x\, dx = 0 \quad \left[\because \int_{-a}^{a} f(x)\, dx = 0 \text{ if } f(x) \text{ is odd function}\right]$$

Chapter 8. Applications of the Integrals

Q. 1. Find the area of the region bounded by the curve $y = \frac{1}{x}$, X-axis and between $x = 1, x = 5$.

Sol. Given curve is $y = \frac{1}{x}$

$$\text{Required area} = \int_1^5 \frac{1}{x} dx$$

$$= [\log x]_1^5$$

$$= \log 5 - \log 1$$

$$= \log 5 \quad [\because \log 1 = 0]$$

Q. 2. Write an expression for finding the area bounded by the curves $y = \cos x$ and $y = \sin x$, between $x = 0, x = \frac{\pi}{2}$ and the X-axis.

Sol. Two curves intersect at $x = \frac{\pi}{4}$

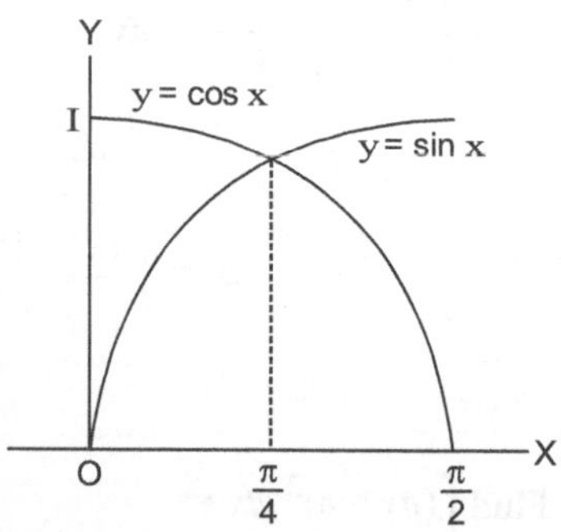

From graph,

$$\text{Required area} = \int_0^{\pi/4} \sin x\, dx + \int_{\pi/4}^{\pi/2} \cos x\, dx$$

Fill in the blanks :

Q. 3. The area bounded by the axes and the line $y = x+1$ is ———.

Sol. $\frac{1}{2}$ sq. unit

Q. 4. Area enclosed by the curve $y = \tan x$, the X-axis and the line $x = \frac{\pi}{3}$ is ———.

Sol. $\log_e 2$

Q. 5. Area bounded by the lines $x = 0$, $y = 0$ and $x + y = 1$ is ———.

Sol. $\frac{1}{2}$ sq. unit

Q. 6. Area bounded by the curves $y = e^x, x = 0, y = 0$ and $x = 1$ is ———.

Sol. $(e-1)$ *sq.* units

Q. 7. Area lying in the first quadrant and bounded by the circle $x^2 + y^2 = 4$ and the lines $x = 0$ and $x = 2$ is ———.

Sol. π sq. units

Q. 8. The area of the region bounded by the curve $x = y^2$, Y-axis and the line's $y = 3$ and $y = 4$ is ———.

Sol. $\frac{37}{3}$ sq. units

Q. 9. The area of the region bounded by the curve $y = x^2 + x$, X-axis and the lines $x = 2$ and $x = 5$ is equal to ———.

Sol. $\frac{297}{6}$ sq. units

Q. 10. The area of the region bounded by the circle $x^2 + y^2 = 1$ is ———.

Sol. π sq. units

Chapter 9. Differential Equations

Q. 1. What is the degree and order of the following differential equation ?

(i) $5x\left(\frac{dy}{dx}\right)^2 - \frac{d^2y}{dx^2} - 6y = \log x$

(ii) $\frac{d^4y}{dx^4} + \sin(y''') = 0$

(iii) $y' + 5y = 0$

(iv) $\left(\frac{ds}{dt}\right)^4 + 3s\frac{d^2s}{dt^2} = 0$

(v) $(y''')^2 + (y'')^3 + (y)^4 + y^5 = 0$

(vi) $y''' + 2y'' + y' = 0$

(vii) $y' + y = e^x$

(viii) $y'' + (y')^2 + 2y = 0$

(ix) $y'' + 2y' + \sin y = 0$

(x) $\left(\frac{d^2y}{dx^2}\right)^3 + \left(\frac{dy}{dx}\right)^2 + \sin\left(\frac{dy}{dx}\right) + 1 = 0$

(xi) $x\frac{dy}{dx} + y = 2$

(xii) $\frac{dy}{dx} + y\sin x = \sqrt{x\frac{dy}{dx} + 3}$

Sol. (i) Degree = 1, Order = 2

(ii) $\because$ It is not a polynomial in $\frac{dy}{dx}$

$\therefore$ degree is not defined, order = 4

(iii) order = 1 and degree = 1

(iv) order = 2, degree = 1

(v) order = 3, degree = 2

(vi) order = 3, degree = 1

(vii) order = 1, degree = 1

(viii) order = 2, degree = 1

(ix) order = 2, degree = 1

(x) order = 2, degree is not defined as it is not a polynomial in differential coefficients.

(xi) order = 1, degree = 1

(xii) Squaring both sides

$$\left(\frac{dy}{dx}\right)^2 + y^2\sin^2 x + 2y\sin x\frac{dy}{dx} = x\frac{dy}{dx} + 3$$

$\therefore$ order = 1, degree = 2.

Q. 2. Write integrating factor of the following differential equations :

(i) $\frac{dy}{dx} + \frac{1}{1+x^2}y = \sin x$

(ii) $x\frac{dy}{dx} + y\log x = x + y$

Sol. (i) Given differential equation is,

$$\frac{dy}{dx} + \frac{1}{1+x^2}y = \sin x$$

Integrating factor of above equation is given by

$$e^{\int \frac{1}{1+x^2}dx} = e^{\tan^{-1}x}$$

(ii) Given differential equation is,

$$x\frac{dy}{dx} + y\log x = x + y$$

$$\Rightarrow \quad \frac{dy}{dx} + \frac{\log x}{x}y = \frac{x+y}{x}$$

Integrating factor of above equation is given by

$$e^{\int \frac{\log x}{x} dx} = e^{\frac{(\log x)^2}{2}}$$

Q. 3. Find the differential equation representing the family of curves $v = \frac{A}{r} + B$, where A and B are arbitrary constants.★

Sol. We have,

$$v = \frac{A}{r} + B \quad \ldots\text{(i)}$$

Since, the given equation contains two arbitrary constants, we shall differentiate it two times. Now, differentiating (i) w.r.t. r, we get

$$\frac{dv}{dr} = -\frac{A}{r^2} + 0$$

$$\Rightarrow \quad r^2 \frac{dv}{dr} = -A \quad \ldots\text{(ii)}$$

Again, differentiating (ii) w.r.t. r, we get

$$r^2 \times \frac{d^2v}{dr^2} + 2r \times \frac{dv}{dr} = 0$$

$$\Rightarrow \quad r\frac{d^2v}{dr^2} + 2\frac{dv}{dr} = 0$$

This is the required differential equation representing the family of the given curve.

Q. 4. Find the differential equation representing the family of curves $y = ae^{bx+5}$, where a and b are arbitrary constants.★

Sol. Given curve is

$$y = ae^{bx+5} \quad \ldots\text{(i)}$$

Differentiate w.r.t. x, we get

$$\frac{dy}{dx} = ae^{bx+5} \cdot \frac{d}{dx}(bx+5)$$

$$\Rightarrow \quad \frac{dy}{dx} = ae^{bx+5} \cdot b \quad \ldots\text{(ii)}$$

$$\Rightarrow \quad \frac{dy}{dx} = y \cdot b \quad \text{[From (i)]}$$

Differentiating again w.r.t. x,

$$\frac{d^2y}{dx^2} = \frac{dy}{dx} \cdot b$$

$$\Rightarrow \quad \frac{d^2y}{dx^2} = \frac{dy}{dx}\left(\frac{dy}{dx} \cdot \frac{1}{y}\right) \quad \left[\because b = \frac{1}{y}\frac{dy}{dx}\right]$$

$$\Rightarrow \quad y\frac{d^2y}{dx^2} = \left(\frac{dy}{dx}\right)^2.$$

Q. 5. Find the integrating factor of the differential equation $\left(\frac{e^{-2\sqrt{x}}}{\sqrt{x}} - \frac{y}{\sqrt{x}}\right)\frac{dx}{dy} = 1.$★

Sol. We have,

$$\left(\frac{e^{-2\sqrt{x}}}{\sqrt{x}} - \frac{y}{\sqrt{x}}\right)\frac{dx}{dy} = 1$$

$$\frac{dy}{dx} = \frac{e^{-2\sqrt{x}}}{\sqrt{x}} - \frac{y}{\sqrt{x}}$$

$$\frac{dy}{dx} + \frac{y}{\sqrt{x}} = \frac{e^{-2\sqrt{x}}}{\sqrt{x}}$$

Which is a linear differential equation of the form

$$\frac{dy}{dx} + Py = Q$$

where $\quad P = \frac{1}{\sqrt{x}}$

and $\quad Q = \frac{e^{-2\sqrt{x}}}{\sqrt{x}}$

$$\therefore \quad \text{I.F.} = e^{\int P\,dx}$$

$$= e^{\int \frac{1}{\sqrt{x}} dx} = e^{2\sqrt{x}}$$

Q. 6. Write the sum of the order and degree of the following differential equation :

$$\frac{d}{dx}\left\{\left(\frac{dy}{dx}\right)^3\right\} = 0.\bigstar$$

Sol. Given,

$$\frac{d}{dx}\left\{\left(\frac{dy}{dx}\right)^3\right\} = 0$$

On differentiating w.r.t. x both sides, we get

$$3\left(\frac{dy}{dx}\right)^2 \times \frac{d^2y}{dx^2} = 0$$

Since the order and degree of the differential equation is 2 and 1 respectively,

So, the sum of the order and degree is 3.

Q. 7. Write the integrating factor of the following differential equation :

$$(1+y^2) + (2xy - \cot y)\frac{dy}{dx} = 0.\bigstar$$

Sol. Given,

$$(1+y^2) + (2xy - \cot y)\frac{dy}{dx} = 0$$

$$(1+y^2)\,dx + 2xy\,dy = \cot y\,dy$$

★ are board exam questions from previous years

$$\frac{dx}{dy}+\left(\frac{2y}{1+y^2}\right)x = \frac{\cot y}{1+y^2}$$

This is a linear differential equation of the form

$$\frac{dx}{dy}+Rx=S$$

where, $R=\dfrac{2y}{1+y^2}$

and $S=\dfrac{\cot y}{1+y^2}$

$\therefore$ Integrating factor $= e^{\int R\,dy} = e^{\int \frac{2y}{1+y^2}dy}$

$= e^{\log(1+y^2)}$

$= 1+y^2.$

Fill in the blanks :

Q. 8. The solution of the differential equation $(e^x+e^{-x})\,dy=(e^x-e^{-x})\,dx$ is ———.

Sol. $y=\log|e^x+e^{-x}|+c$

Given differential equation is

$$(e^x+e^{-x})\,dy=(e^x-e^{-x})\,dx$$

$$\Rightarrow \quad dy = \frac{(e^x-e^{-x})}{(e^x+e^{-x})}dx$$

On integrating both sides, we get

$$y=\int\frac{(e^x-e^{-x})}{(e^x+e^{-x})}dx$$

$$=\log|e^x+e^{-x}|+c.$$

Q. 9. The solution of differential equation $\dfrac{dy}{dx}=x^3+e^x+x^e$ is ———.

Sol. $y=\dfrac{x^4}{4}+e^x+\dfrac{x^{e+1}}{e+1}+c$

Given differential equation is

$$\frac{dy}{dx}=x^3+e^x+x^e$$

On integrating both sides, we get

$$y=\int x^3\,dx+\int e^x\,dx+\int x^e\,dx+c$$

$$=\frac{x^4}{4}+e^x+\frac{x^{e+1}}{e+1}+c$$

Q. 10. The order of differential equation $\left(\dfrac{d^2y}{dx^2}\right)^2+\cos\left(\dfrac{dy}{dx}\right)=0$ is ———.

Sol. 2

Q. 11. The integrating factor of $x\dfrac{dy}{dx}-3y=x^3$ is ———.

Sol. $\dfrac{1}{x^3}$

Q. 12. The integrating factor of $\dfrac{dy}{dx}+y\sec^2x=\sec x+\tan x$ is ———.

Sol. $e^{\tan x}$

Q. 13. The order and degree of differential equation $2x^2\dfrac{d^2y}{dx^2}-3\dfrac{dy}{dx}+y=0$ are —— and —— respectively.

Sol. 2 and 1.

Q. 14. The integrating factor of the differential equation $x\dfrac{dy}{dx}+2y=x^2$ is ———.★

Sol. x^2

$\because \quad x\dfrac{dy}{dx}+2y=x^2$

$\Rightarrow \quad \dfrac{dy}{dx}+\dfrac{2y}{x}=x$

Here, $P=\dfrac{2}{x}$, $Q=x$

$\therefore$ I.F. $=e^{\int P dx}=e^{\int\frac{2}{x}dx}$

$=e^{2\log x}=e^{\log x^2}$

$=x^2$

Q. 15. The degree of the differential equation $1+\left(\dfrac{dy}{dx}\right)^2=x$ is ———.★

Sol. 2

Chapter 10. Vectors

Q. 1. Find sum of vectors $\vec{a}=\hat{i}-2\hat{j}+\hat{k}$ and $\vec{b}=2\hat{i}-4\hat{j}+5\hat{k}$ and $\vec{c}=\hat{i}-6\hat{j}-7\hat{k}$.★

Sol. Sum of the vectors $\vec{a},\vec{b},\vec{c}$ is

$$\vec{a}+\vec{b}+\vec{c}=\left(\hat{i}-2\hat{j}+\hat{k}\right)+\left(2\hat{i}-4\hat{j}+5\hat{k}\right)+\left(\hat{i}-6\hat{j}-7\hat{k}\right)$$

$$=-4\hat{i}-12\hat{j}-\hat{k}$$

Q. 2. Find the scalar components of $\vec{AB}$ with initial point $A(2, 1)$ and terminal point $B(-5, 7)$.★

★ are board exam questions from previous years

Sol. Given, points $A(2, 1)$ and $B(-5, 7)$. Scalar component of $\overrightarrow{AB}$ are $x_2 - x_1$ and $y_2 - y_1$ *i.e.*, $-5-2$ and $7-1$ *i.e.*, -7 and 6.

Q. 3. For what value of '*a*', the vectors $2\hat{i} - 3\hat{j} + 4\hat{k}$ and $a\hat{i} + 6\hat{j} - 8\hat{k}$ are collinear ?★

Sol. Vectors $\vec{a}$ and $\vec{b}$ are said to be collinear, if

$$\vec{a} = k.\vec{b}, \text{ where } k = \text{scalar}$$

$$\therefore \quad 2\hat{i} - 3\hat{j} + 4\hat{k} = k(a\hat{i} + 6\hat{j} - 8\hat{k})$$

Above equation is satisfied when $a = -4$

$$\therefore \quad a = -4$$

Q. 4. Write the direction cosines of vector $-2\hat{i} + \hat{j} - 5\hat{k}$.★

Sol. Let $\vec{a} = -2\hat{i} + \hat{j} - 5\hat{k}$

$\therefore$ Direction cosines of $\vec{a}$ are

$$\frac{-2}{\sqrt{(-2)^2+(1)^2+(-5)^2}}, \frac{1}{\sqrt{(-2)^2+(1)^2+(-5)^2}}, \frac{-5}{\sqrt{(-2)^2+(1)^2+(-5)^2}} = \frac{-2}{\sqrt{30}}, \frac{1}{\sqrt{30}}, \frac{-5}{\sqrt{30}}.$$

Q. 5. Write the position vector of mid-point of the vector joining points $P(2, 3, 4)$ and $Q(4, 1, -2)$.★

Sol. Position vector of mid-point of a vector joining points $P(2, 3, 4)$ and $Q(4, 1, -2)$ is

$$\frac{OP+OQ}{2} = \frac{(2\hat{i}+3\hat{j}+4\hat{k})+(4\hat{i}+\hat{j}-2\hat{k})}{2}$$

$$= \frac{6\hat{i}+4\hat{j}+2\hat{k}}{2} = 3\hat{i}+2\hat{j}+\hat{k} = (3, 2, 1)$$

Q. 6. Write a unit vector in the direction of $\vec{a} = 2\hat{i} + \hat{j} + 2\hat{k}$.★

Sol. Unit vector in the direction of $\vec{a}$ is given as,

$$\hat{a} = \frac{\vec{a}}{|\vec{a}|}$$

$$= \frac{2\hat{i}+\hat{j}+2\hat{k}}{\sqrt{(2)^2+(1)^2+(2)^2}}$$

$$= \frac{2\hat{i}+\hat{j}+2\hat{k}}{\sqrt{9}}$$

$$= \frac{2}{3}\hat{i} + \frac{1}{3}\hat{j} + \frac{2}{3}\hat{k}$$

★ are board exam questions from previous years

Q. 7. Find the magnitude of the vector $\vec{a} = 3\hat{i} - 2\hat{j} + 6\hat{k}$.★

Sol. Magnitude of $\vec{a} = |\vec{a}|$

$$= \sqrt{(3)^2+(-2)+(6)^2}$$

$$= \sqrt{9+4+36}$$

$$= \sqrt{49} = 7$$

Q. 8. Find a unit vector in the direction of vector $\vec{a} = 2\hat{i} + 3\hat{j} + 6\hat{k}$.★

Sol. Unit vector in the direction of $\vec{a}$ is given as,

$$\hat{a} = \frac{\vec{a}}{|\vec{a}|}$$

$$= \frac{2\hat{i}+3\hat{j}+6\hat{k}}{\sqrt{(2)^2+(3)^2+(6)^2}}$$

$$= \frac{2\hat{i}+3\hat{j}+6\hat{k}}{\sqrt{49}}$$

$$= \frac{2}{7}\hat{i} + \frac{3}{7}\hat{j} + \frac{6}{7}\hat{k}$$

Q. 9. If *A*, *B* and *C* are the vertices of a triangle *ABC*, then what is the value of $\overrightarrow{AB} + \overrightarrow{BC} + \overrightarrow{CA}$?★

Sol. Let ΔABC be given triangle.

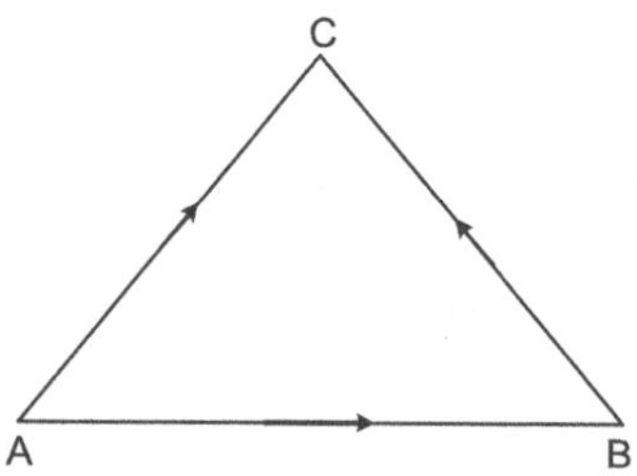

Now, by triangle law of vector addition, we have

$$\overrightarrow{AB} + \overrightarrow{BC} = \overrightarrow{AC}$$

$$\Rightarrow \quad \overrightarrow{AB} + \overrightarrow{BC} + \overrightarrow{CA} = \overrightarrow{CA} + \overrightarrow{AC}$$

$$\Rightarrow \quad \overrightarrow{AB} + \overrightarrow{BC} + \overrightarrow{CA} = \overrightarrow{CA} - \overrightarrow{CA} \quad [\because \overrightarrow{AC} = -\overrightarrow{CA}]$$

$$\therefore \quad \overrightarrow{AB} + \overrightarrow{BC} + \overrightarrow{CA} = 0$$

Q. 10. Find a vector in the direction of $\vec{a} = 2\hat{i} - \hat{j} + 2\hat{k}$ which has magnitude 6 units.★

Sol. First we find a unit vector in the direction of $\vec{a}$

$$\hat{a} = \frac{\vec{a}}{|\vec{a}|} = \frac{2\hat{i}-\hat{j}+2\hat{k}}{\sqrt{(2)^2+(-1)^2+(2)^2}}$$

$= \frac{2}{3}\hat{i} - \frac{1}{3}\hat{j} + \frac{2}{3}\hat{k}$

Now, vector of magnitude 6 units

$$= 6\left[\frac{2}{3}\hat{i} - \frac{1}{3}\hat{j} + \frac{2}{3}\hat{k}\right]$$

$$= 4\hat{i} - 2\hat{j} + 4\hat{k}$$

Q. 11. Find a position vector of mid-point of line segment *AB*, where *A* is point (3, 4, – 2) and *B* is point (1, 2, 4).★

Sol. Position vector of mid-point of line segment *AB*, where *A* is (3, 4, –2) and *B*(1, 2, 4) is given as follows:

Let $\vec{OA} = 3\hat{i} + 4\hat{j} - 2\hat{k}$

and $\vec{OB} = \hat{i} + 2\hat{j} + 4\hat{k}$

∴ Position vector of mid-point of $\vec{AB}$

$$= \frac{\vec{OA} + \vec{OB}}{2}$$

$$= \frac{(3\hat{i} + 4\hat{j} - 2\hat{k}) + (\hat{i} + 2\hat{j} + 4\hat{k})}{2}$$

$$= \frac{4\hat{i} + 6\hat{j} + 2\hat{k}}{2}$$

$$= 2\hat{i} + 3\hat{j} + \hat{k}$$

Q. 12. Write a vector of magnitude 15 units in the direction of vector $\hat{i} - 2\hat{j} + 2\hat{k}$.★

Sol. Let $\vec{a} = \hat{i} - 2\hat{j} + 2\hat{k}$

Unit vector in the direction of $\vec{a}$ is given as,

$$\hat{a} = \frac{\vec{a}}{|\vec{a}|} = \frac{\hat{i} - 2\hat{j} + 2\hat{k}}{\sqrt{(1)^2 + (-2)^2 + (2)^2}}$$

$$= \frac{\hat{i} - 2\hat{j} + 2\hat{k}}{\sqrt{9}}$$

$$= \frac{1}{3}\hat{i} - \frac{2}{3}\hat{j} + \frac{2}{3}\hat{k}$$

∴ Vector of magnitude 15 units

$$= 15\left(\frac{1}{3}\hat{i} - \frac{2}{3}\hat{j} + \frac{2}{3}\hat{k}\right)$$

$$= 5\hat{i} - 10\hat{j} + 10\hat{k}$$

Q. 13. What is the cosine of angle which the vector $\sqrt{2}\hat{i} + \hat{j} + \hat{k}$ makes with *Y*-axis?★

Sol. Let $\vec{a} = \sqrt{2}\hat{i} + \hat{j} + \hat{k}$

Now, unit vector in the direction of $\vec{a}$

$$= \frac{\vec{a}}{|\vec{a}|} = \frac{\sqrt{2}\hat{i} + \hat{j} + \hat{k}}{\sqrt{(\sqrt{2})^2 + (1)^2 + (1)^2}}$$

$$= \frac{\sqrt{2}\hat{i} + \hat{j} + \hat{k}}{2}$$

$$= \frac{\sqrt{2}}{2}\hat{i} + \frac{1}{2}\hat{j} + \frac{1}{2}\hat{k}$$

∴ Cosine of angle which the given vector makes with *Y*-axis is $\frac{1}{2}$.

Q. 14. If $\vec{a} = \hat{i} + 2\hat{j} - \hat{k}$ and $\vec{b} = 3\hat{i} + \hat{j} - 5\hat{k}$, find a unit vector in the direction of $\vec{a} - \vec{b}$.★

Sol. Here,

$$\vec{a} - \vec{b} = (\hat{i} + 2\hat{j} - \hat{k}) - (3\hat{i} + \hat{j} - 5\hat{k})$$

$$= -2\hat{i} + \hat{j} + 4\hat{k}$$

Let, $\vec{a} - \vec{b} = \vec{c} = -2\hat{i} + \hat{j} + 4\hat{k}$

Now, unit vector in the direction of $\vec{c}$

$$\frac{\vec{c}}{|\vec{c}|} = \frac{-2\hat{i} + \hat{j} + 4\hat{k}}{\sqrt{(-2)^2 + (1)^2 + (4)^2}}$$

$$= \frac{-2\hat{i} + \hat{j} + 4\hat{k}}{\sqrt{21}}$$

$$= -\frac{2}{\sqrt{21}}\hat{i} + \frac{1}{\sqrt{21}}\hat{j} + \frac{4}{\sqrt{21}}\hat{k}$$

Q. 15. Find λ when projection of $\vec{a} = \lambda\hat{i} + \hat{j} + 4\hat{k}$ on $\vec{b} = 2\hat{i} + 6\hat{j} + 3\hat{k}$ is 4 units.★

Sol. Given that projection of $\vec{a}$ over $\vec{b} = 4$

$$\Rightarrow \frac{\vec{a} \cdot \vec{b}}{|\vec{b}|} = 4$$

$$\left[\because \text{projection of } \vec{a} \text{ over } \vec{b} = \frac{\vec{a} \cdot \vec{b}}{|\vec{b}|}\right]$$

$$\Rightarrow \frac{(\lambda\hat{i} + \hat{j} + 4\hat{k}).(2\hat{i} + 6\hat{j} + 3\hat{k})}{\sqrt{(2)^2 + (6)^2 + (3)^2}} = 4$$

$$\Rightarrow \frac{2\lambda + 6 + 12}{\sqrt{49}} = 4$$

$$\Rightarrow \frac{2\lambda + 18}{7} = 4$$

$$\Rightarrow 2\lambda + 18 = 28$$

$$\Rightarrow 2\lambda = 10$$

$$\Rightarrow \lambda = 5$$

★ are board exam questions from previous years

Q. 16. If $\vec{a}\cdot\vec{a}=0$ and $\vec{a}\cdot\vec{b}=0$ then what can be concluded about the vector $\vec{b}$?★

Sol. $\vec{a}\cdot\vec{a}=0$

$\Rightarrow \quad |a|^2=0$

$\Rightarrow \quad |a|=0 \quad$ ...(i)

and $\quad \vec{a}\cdot\vec{b}=0$

$\Rightarrow \quad |\vec{a}||\vec{b}|\cos\theta=0 \quad$ ...(ii)

From equations (i) and (ii), it can be concluded that $\vec{b}$ is either zero or non-zero or perpendicular vector to $\vec{a}$.

Q. 17. Write the angle between vectors $\vec{a}$ and $\vec{b}$ with magnitudes $\sqrt{3}$ and 2 respectively, having $\vec{a}\cdot\vec{b}=\sqrt{6}$.★

Sol. Given $\quad |\vec{a}|=\sqrt{3}, |\vec{b}|=2$

and $\quad \vec{a}\cdot\vec{b}=\sqrt{6}$

Now, angle between $\vec{a}$ and $\vec{b}$ is given by

$$\cos\theta=\frac{\vec{a}\cdot\vec{b}}{|\vec{a}||\vec{b}|}=\frac{\sqrt{6}}{\sqrt{3}\times 2}$$

or $\quad \cos\theta=\frac{1}{\sqrt{2}}=\cos\frac{\pi}{4}$

$\therefore \quad \theta=\frac{\pi}{4} \quad \left[\because \cos\frac{\pi}{4}=\frac{1}{\sqrt{2}}\right]$

Q. 18. For what value of λ are the vectors $\hat{i}+2\lambda\hat{j}+\hat{k}$ and $2\hat{i}+\hat{j}-3\hat{k}$ perpendicular ?★

Sol. Given that the vectors are perpendicular, so their dot product is zero.

$\therefore \quad (\hat{i}+2\lambda\hat{j}+\hat{k}).(2\hat{i}+\hat{j}-3\hat{k})=0$

$\Rightarrow \quad 2+2\lambda-3=0$

$\Rightarrow \quad 2\lambda=1$

or $\quad \lambda=\frac{1}{2}$

Q. 19. If $|\vec{a}|=\sqrt{3}, |\vec{b}|=2$ and angle between $\vec{a}$ and $\vec{b}$ is 60°, find $\vec{a}\cdot\vec{b}$.★

Sol. We know that

$$\cos\theta=\frac{\vec{a}\cdot\vec{b}}{|\vec{a}|.|\vec{b}|}$$

Put $|\vec{a}|=\sqrt{3}, |\vec{b}|=\sqrt{2}$ and $\theta=60°$, we get

$$\cos 60°=\frac{\vec{a}\cdot\vec{b}}{\sqrt{3}\times 2}$$

or $\quad \vec{a}\cdot\vec{b}=\frac{1}{2}\times 2\sqrt{3}=\sqrt{3}$

$\therefore \quad \vec{a}\cdot\vec{b}=\sqrt{3}$

Q. 20. Find the value of λ, if the vectors $2\hat{i}+\lambda\hat{j}+3\hat{k}$ and $3\hat{i}+2\hat{j}-4\hat{k}$ are perpendicular to each other.★

Sol. Given that the vectors are perpendicular, so their dot product is zero.

$\therefore \quad (2\hat{i}+\lambda\hat{j}+3\hat{k}).(3\hat{i}+2\hat{j}-4\hat{k})=0$

$\Rightarrow \quad 6+2\lambda-12=0$

$\Rightarrow \quad 2\lambda=6$

$\Rightarrow \quad \lambda=3$

Q. 21. If $|\vec{a}|=2, |\vec{b}|=3$ and $\vec{a}\cdot\vec{b}=3$, find the projection of $\vec{b}$ on $\vec{a}$.★

Sol. Projection of $\vec{b}$ over $\vec{a}$

$$=\frac{\vec{b}\cdot\vec{a}}{|\vec{a}|}=\frac{\vec{a}\cdot\vec{b}}{|\vec{a}|} \quad [\because \vec{a}\cdot\vec{b}=\vec{b}\cdot\vec{a}]$$

$$=\frac{3}{2}$$

Q.22. Vectors $\vec{a}$ and $\vec{b}$ are such that $|\vec{a}|=\sqrt{3}, |\vec{b}|=\frac{2}{3}$ and $\vec{a}\times\vec{b}$ is a unit vector. Write the angle between $\vec{a}$ and $\vec{b}$.★

Sol. Given $\vec{a}\times\vec{b}$ is a unit vector.

$\therefore \quad |\vec{a}\times\vec{b}|=1$

$\Rightarrow \quad |\vec{a}||\vec{b}|\sin\theta\,\hat{n}=1$

$[\because \vec{a}\times\vec{b}=|\vec{a}||\vec{b}|\sin\theta\,\hat{n}]$

$\Rightarrow \quad \sqrt{3}\times\frac{2}{3}\times\sin\theta=1 \quad [\because |\hat{n}|=1]$

$\Rightarrow \quad \sin\theta=\frac{3}{2\sqrt{3}}$

$\Rightarrow \quad \sin\theta=\frac{(\sqrt{3})^2}{2\sqrt{3}}=\frac{\sqrt{3}}{2}$

$\Rightarrow \quad \sin\theta=\sin\frac{\pi}{3}$

★ are board exam questions from previous years

$\therefore \quad \theta = \frac{\pi}{3}$

$\therefore$ Angle between $\vec{a}$ and $\vec{b} = \frac{\pi}{3}$.

Q. 23. If $\vec{a}$ and $\vec{b}$ are two vectors such that $|\vec{a} \cdot \vec{b}| = |\vec{a} \times \vec{b}|$, then find angle between $\vec{a}$ and $\vec{b}$.★

Sol. Given that $|\vec{a} \cdot \vec{b}| = |\vec{a} \times \vec{b}|$

$\Rightarrow \quad |\vec{a}||\vec{b}| \cos\theta = |\vec{a}||\vec{b}| \sin\theta$

$[\because |\hat{n}| = 1]$

$\Rightarrow \quad \cos\theta = \sin\theta$

Divide both sides by $\cos\theta$, we get

$\tan\theta = 1$

$\therefore \quad \tan\theta = \tan\frac{\pi}{4} \quad \left[\because 1 = \tan\frac{\pi}{4}\right]$

$\therefore \quad \theta = \frac{\pi}{4}$

$\therefore$ Angle between $\vec{a}$ and $\vec{b} = \frac{\pi}{4}$

Q. 24. Find λ, if

$$(2\hat{i}+6\hat{j}+14\hat{k}) \times (\hat{i}-\lambda\hat{j}+7\hat{k}) = 0.★$$

Sol. Given that

$$(2\hat{i}+6\hat{j}+14\hat{k}) \times (\hat{i}-\lambda\hat{j}+7\hat{k}) = 0$$

$$\Rightarrow \quad \begin{vmatrix} \hat{i} & \hat{j} & \hat{k} \\ 2 & 6 & 14 \\ 1 & -\lambda & 7 \end{vmatrix} = 0$$

$\therefore \hat{i}(42+14\lambda) - \hat{j}(14-14) + \hat{k}(-2\lambda-6) = 0$

$\Rightarrow \quad \hat{i}(42+14\lambda) + \hat{k}(-2\lambda-6) = 0\hat{i}+0\hat{j}+0\hat{k}$

Comparing coefficients of $\hat{i}$ and $\hat{k}$ on both sides, we get

$42 + 14\lambda = 0$

$\Rightarrow \quad 14\lambda = -42$

$\Rightarrow \quad \lambda = -3$

and $\quad -2\lambda - 6 = 0$

$\Rightarrow \quad -2\lambda = 6$

$\therefore \quad \lambda = -3$

Q. 25. Find $\vec{a} \cdot \vec{b}$ if $\vec{a} = -\hat{i}+\hat{j}-2\hat{k}$ and $\vec{b} = 2\hat{i}+3\hat{j}-\hat{k}$.★

Sol. $\quad \vec{a} \cdot \vec{b} = (-\hat{i}+\hat{j}-2\hat{k})(2\hat{i}+3\hat{j}-\hat{k})$

$= -2+3+2 = 3$

Q. 26. Find the value of *P*, if

$$(2\hat{i}+6\hat{j}+27\hat{k}) \times (\hat{i}+3\hat{j}+P\hat{k}) = 0.★$$

Sol. Given that $(2\hat{i}+6\hat{j}+27\hat{k}) \times (\hat{i}+3\hat{j}+P\hat{k}) = 0$

$$\Rightarrow \quad \begin{vmatrix} \hat{i} & \hat{j} & \hat{k} \\ 2 & 6 & 27 \\ 1 & 3 & P \end{vmatrix} = 0$$

$\therefore \hat{i}(6P-81) - \hat{j}(2P-27) + \hat{k}(6-6) = 0$

$\Rightarrow \hat{i}(6P-81) - \hat{j}(2P-27) = 0\hat{i}+0\hat{j}+0\hat{k}$

On comparing coefficients of $\hat{i}$ and $\hat{j}$, we get

$6P - 81 = 0$

$\Rightarrow \quad 6P = 81$

$\Rightarrow \quad P = \frac{81}{6} = \frac{27}{2}$

and $\quad 2P - 27 = 0$

$\Rightarrow \quad 2P = 27$

$\therefore \quad P = \frac{27}{2}$

Q. 27. Find the angle between $\vec{a}$ and $\vec{b}$ with magnitude 1 and 2 respectively and when $|\vec{a} \times \vec{b}| = \sqrt{3}$.★

Sol. Given that $\quad |\vec{a} \times \vec{b}| = \sqrt{3}$

$\Rightarrow \quad |\vec{a}||\vec{b}| \sin\theta = \sqrt{3}$

$[\because \vec{a} \times \vec{b} = |\vec{a}||\vec{b}| \sin\theta . \hat{n}$ and $|\hat{n}| = 1]$

$\therefore \quad 1 \times 2 \times \sin\theta = \sqrt{3}$

$\therefore \quad \sin\theta = \frac{\sqrt{3}}{2} = \sin\frac{\pi}{3}$

$\therefore \quad \theta = \frac{\pi}{3}$

Angle between $\vec{a}$ and $\vec{b} = \frac{\pi}{3}$.

Q. 28. Write the value of *p* for which $\vec{a} = 3\hat{i}+2\hat{j}+9\hat{k}$ and $\vec{b} = \hat{i}+p\hat{j}+3\hat{k}$ are parallel vectors.★

★ are board exam questions from previous years

Sol. Since, $\vec{a}$ and $\vec{b}$ are parallel,

$\therefore \quad \vec{a} \times \vec{b} = 0$

$$\Rightarrow \begin{vmatrix} \hat{i} & \hat{j} & \hat{k} \\ 3 & 2 & 9 \\ 1 & p & 3 \end{vmatrix} = 0$$

$\therefore \hat{i}(6-9p) - \hat{j}(9-9) + \hat{k}(3p-2) = 0$

$\Rightarrow \hat{i}(6-9p) + \hat{k}(3p-2) = 0\hat{i} + 0\hat{j} + 0\hat{k}$

Comparing coefficient of $\hat{i}$ and $\hat{k}$ on both sides, we get

$6 - 9p = 0$

$\Rightarrow \quad 9p = 6$

$\therefore \quad p = \dfrac{6}{9} = \dfrac{2}{3}$

and $\quad 3p - 2 = 0$

$\Rightarrow \quad 3p = 2$

$\therefore \quad p = \dfrac{2}{3}$

Q. 29. Find the projection of $\vec{a}$ on $\vec{b}$ if $\vec{a} \cdot \vec{b} = 8$ and $\vec{b} = 2\hat{i} + 6\hat{j} + 3\hat{k}$.★

Sol. Projection of $\vec{a}$ on $\vec{b}$

$$= \frac{\vec{a} \cdot \vec{b}}{|\vec{b}|}$$

$$= \frac{8}{\sqrt{(2)^2 + (6)^2 + (3)^2}}$$

$$= \frac{8}{\sqrt{49}} = \frac{8}{7}$$

Q. 30. Find the value of the following :

$\hat{i} \cdot (\hat{j} \times \hat{k}) + \hat{j} \cdot (\hat{i} \times \hat{k}) + \hat{k} \cdot (\hat{i} \times \hat{j})$.★

Sol. $\hat{i} \cdot (\hat{j} \times \hat{k}) + \hat{j} \cdot (\hat{i} \times \hat{k}) + \hat{k}(\hat{i} \times \hat{j})$

$= \hat{i} \cdot \hat{i} + \hat{j} \cdot (-\hat{j}) + \hat{k} \cdot \hat{k}$

$= 1 - 1 + 1 = 1$

Q. 31. Find $|\vec{a} \times \vec{b}|$ if $\vec{a} = \hat{i} - 7\hat{j} + 7\hat{k}$ and $\vec{b} = 3\hat{i} - 2\hat{j} + 2\hat{k}$.★

Sol. Given, $\quad \vec{a} = \hat{i} - 7\hat{j} + 7\hat{k}$,

and $\quad \vec{b} = 3\hat{i} - 2\hat{j} + 2\hat{k}$.

$$\therefore \vec{a} \times \vec{b} = \begin{vmatrix} \hat{i} & \hat{j} & \hat{k} \\ 1 & -7 & 7 \\ 3 & -2 & 2 \end{vmatrix}$$

$= \hat{i}(-14 + 14) - \hat{j}(2 - 21) + \hat{k}(-2 + 21)$

$= 19\hat{j} + 19\hat{k}$

$\therefore \quad |\vec{a} \times \vec{b}| = \sqrt{(19)^2 + (19)^2}$

$= \sqrt{2(19)^2}$

$= 19\sqrt{2}$

Q. 32. If $|\vec{a}| = \sqrt{3}$, $|\vec{b}| = 2$ and $\vec{a} \cdot \vec{b} = 3$, find angle between $\vec{a}$ and $\vec{b}$.★

Sol. Angle between $\vec{a}$ and $\vec{b}$ is given by

$$\cos\theta = \frac{\vec{a} \cdot \vec{b}}{|\vec{a}||\vec{b}|}$$

$$= \frac{3}{\sqrt{3} \times 2}$$

$$= \frac{(\sqrt{3})^2}{\sqrt{3} \times 2}$$

$\therefore \quad \cos\theta = \dfrac{\sqrt{3}}{2}$

$\Rightarrow \quad \cos\theta = \cos\dfrac{\pi}{6}$

$\therefore \quad \theta = \dfrac{\pi}{6}$

Q. 33. Find $\vec{a} \cdot (\vec{b} \times \vec{c})$, if $\vec{a} = 2\hat{i} + \hat{j} + 3\hat{k}$, $\vec{b} = \hat{i} + 2\hat{j} + \hat{k}$ and $\vec{c} = 3\hat{i} + \hat{j} + 2\hat{k}$.★

Sol. Given $\quad \vec{a} = 2\hat{i} + \hat{j} + 3\hat{k}$

$\vec{b} = \hat{i} + 2\hat{j} + \hat{k}$

$\vec{c} = 3\hat{i} + \hat{j} + 2\hat{k}$

$$\therefore \quad (\vec{b} \times \vec{c}) = \begin{vmatrix} \hat{i} & \hat{j} & \hat{k} \\ 1 & 2 & 1 \\ 3 & 1 & 2 \end{vmatrix}$$

$= \hat{i}(4 - 1) - \hat{j}(2 - 3) + \hat{k}(1 - 6)$

$= 3\hat{i} + \hat{j} - 5\hat{k}$

So, $\quad \vec{a} \cdot (\vec{b} \times \vec{c}) = (2\hat{i} + \hat{j} + 3\hat{k}) \cdot (3\hat{i} + \hat{j} - 5\hat{k})$

$= 6 + 1 - 15$

$= -8.$

★ are board exam questions from previous years

Q. 34. Find the projection of the vector $\hat{i}+3\hat{j}+7\hat{k}$ on the vector $2\hat{i}-3\hat{j}+6\hat{k}$.★

Sol. Let $\vec{a}=\hat{i}+3\hat{j}+7\hat{k}$ and $\vec{b}=2\hat{i}-3\hat{j}+6\hat{k}$

$$\therefore \text{ Projection of } \vec{a} \text{ on } \vec{b}=\frac{\vec{a}\cdot\vec{b}}{|\vec{b}|}$$

$$=\frac{(\hat{i}+3\hat{j}+7\hat{k}).(2\hat{i}-3\hat{j}+6\hat{k})}{\sqrt{2^2+(-3)^2+6^2}}$$

$$=\frac{1\times 2+3\times(-3)+7\times 6}{\sqrt{4+9+36}}$$

$$=\frac{2-9+42}{\sqrt{49}}=\frac{35}{7}=5.$$

Q. 35. If $\vec{a}$ and $\vec{b}$ are two unit vectors such that $\vec{a}+\vec{b}$ is also a unit vector, then find the angle between $\vec{a}$ and $\vec{b}$.★

Sol. Given, $|\vec{a}|=1, |\vec{b}|=1$

and $|\vec{a}+\vec{b}|=1$

$\Rightarrow |\vec{a}+\vec{b}|^2=1$

$\Rightarrow (\vec{a}+\vec{b}).(\vec{a}+\vec{b})=1$

$\Rightarrow \vec{a}.\vec{a}+\vec{a}.\vec{b}+\vec{b}.\vec{a}+\vec{b}.\vec{b}=1$

$\Rightarrow |\vec{a}|^2+2\vec{a}.\vec{b}+|\vec{b}|^2=1$

$\Rightarrow 1+2\vec{a}.\vec{b}+1=1$

$\Rightarrow 2\vec{a}.\vec{b}=-1$

$\Rightarrow 2|\vec{a}|.|\vec{b}|\cos\theta=-1$

$\Rightarrow 2.1.1.\cos\theta=-1$

$\Rightarrow \cos\theta=-\frac{1}{2}$

$\Rightarrow \cos\theta=\cos 120°$

$\therefore \theta=120°$

Q. 36. If $\hat{a}, \hat{b}$ and $\hat{c}$ are mutually perpendicular unit vectors, then find the value of $|2\hat{a}+\hat{b}+\hat{c}|$.★

Sol. Let $x=|2\hat{a}+\hat{b}+\hat{c}|$

$\therefore x^2=|2\hat{a}+\hat{b}+\hat{c}|^2$

$=4|\hat{a}|^2+|\hat{b}|^2+|\hat{c}|^2+2(2\hat{a}.\hat{b}+\hat{b}.\hat{c}+2\hat{c}.\hat{a})$

Since $\hat{a}, \hat{b}$ and $\hat{c}$ are mutually perpendicular unit vectors

$\therefore \hat{a}.\hat{b}=\hat{b}.\hat{c}=\hat{c}.\hat{a}=0$

So, $x^2=4\times 1+1+1+2(2\times 0+0+2\times 0)$

$x^2=6$

$\therefore x=\sqrt{6}$

Hence, $|2\hat{a}+\hat{b}+\hat{c}|=\sqrt{6}$

Q. 37. Write a unit vector perpendicular to both the vectors $\vec{a}=\hat{i}+\hat{j}+\hat{k}$ and $\vec{b}=\hat{i}+\hat{j}$.★

Sol. A vector perpendicular to both $\vec{a}$ and $\vec{b}$ is $\vec{a}\times\vec{b}$.

$\therefore$ Unit vector perpendicular to $\vec{a}$ and $\vec{b}$

$$=\frac{\vec{a}\times\vec{b}}{|\vec{a}\times\vec{b}|}$$

Now, $$\vec{a}\times\vec{b}=\begin{vmatrix}\hat{i} & \hat{j} & \hat{k}\\ 1 & 1 & 1\\ 1 & 1 & 0\end{vmatrix}$$

$$=\hat{i}(0-1)-\hat{j}(0-1)+\hat{k}(1-1)$$

$$=-\hat{i}+\hat{j}$$

and $|\vec{a}\times\vec{b}|=\sqrt{(-1)^2+1^2}$

$=\sqrt{2}$

$\therefore$ Required unit vector $=\frac{-\hat{i}+\hat{j}}{\sqrt{2}}$

Q. 38. Find λ, if the vectors $\vec{a}=\hat{i}+3\hat{j}+\hat{k}$, $\vec{b}=2\hat{i}-\hat{j}-\hat{k}$ and $\vec{c}=\lambda\hat{j}+3\hat{k}$ are coplanar.★

Sol. Since, the given vectors are coplanar

$\therefore [\vec{a}\ \vec{b}\ \vec{c}]=0$

Here, $$[\vec{a}\ \vec{b}\ \vec{c}]=\begin{vmatrix}1 & 3 & 1\\ 2 & -1 & -1\\ 0 & \lambda & 3\end{vmatrix}=0$$

Expanding along R_3, we get

$0(-3+1)-\lambda(-1-2)+3(-1-6)=0$

$\Rightarrow 3\lambda=21$

$\lambda=7.$

★ are board exam questions from previous years

Q. 39. Write the position vector of the point which divides the join of points with position vectors $3\vec{a}-2\vec{b}$ and $2\vec{a}+3\vec{b}$ in the ratio 2 : 1.★

Sol. Let A and B be the given points with position vectors $3\vec{a}-2\vec{b}$ and $2\vec{a}+3\vec{b}$ respectively.

Let P and Q be the points dividing AB in the ratio 2 : 1 internally and externally respectively. Then

$$\text{Position vector of } P = \frac{1(3\vec{a}-2\vec{b})+2(2\vec{a}+3\vec{b})}{1+2} = \frac{7\vec{a}}{3}+\frac{4\vec{b}}{3}$$

$$\text{Position vector of } Q = \frac{1(3\vec{a}-2\vec{b})-2(2\vec{a}+3\vec{b})}{1-2} = \vec{a}+8\vec{b}$$

Q. 40. Write the number of vectors of unit length perpendicular to both the vectors $\vec{a}=2\hat{i}+\hat{j}+2\hat{k}$ and $\vec{b}=\hat{j}+\hat{k}$.★

Sol. We know that the unit vectors perpendicular to the plane $\vec{a}$ and $\vec{b}$ are $\pm\dfrac{\vec{a}\times\vec{b}}{|\vec{a}\times\vec{b}|}$

So,
$$\vec{a}\times\vec{b} = \begin{vmatrix}\hat{i} & \hat{j} & \hat{k}\\ 2 & 1 & 2\\ 0 & 1 & 1\end{vmatrix}$$

$$= (1-2)\hat{i}-(2-0)\hat{j}+(2-0)\hat{k}$$

$$= -\hat{i}-2\hat{j}+2\hat{k}$$

$$\Rightarrow \quad |\vec{a}\times\vec{b}| = \sqrt{(-1)^2+(-2)^2+(2)^2} = \sqrt{9} = 3$$

Hence, required vectors

$$= \pm\frac{1}{3}(-\hat{i}-2\hat{j}+2\hat{k}).$$

and number of vectors are 2.

Q. 41. Find the position vector of a point which divides the join of points with position vectors $\vec{a}-2\vec{b}$ and $2\vec{a}+\vec{b}$ externally in the ratio 2 : 1.★

Sol. Let A and B be the given points with position vectors $\vec{a}-2\vec{b}$ and $2\vec{a}+\vec{b}$ respectively. Let P be the point dividing AB in the ratio 2 : 1 externally.

∴ Position vector of P

$$= \frac{2\times(2\vec{a}+\vec{b})-1\times(\vec{a}-2\vec{b})}{2-1} = 3\vec{a}+4\vec{b}$$

Q. 42. The two vectors $\hat{j}+\hat{k}$ and $3\hat{i}-\hat{j}+4\hat{k}$ represent the two sides AB and AC, respectively of a ΔABC. Find the length of the median through A.★

Sol. In ΔABC

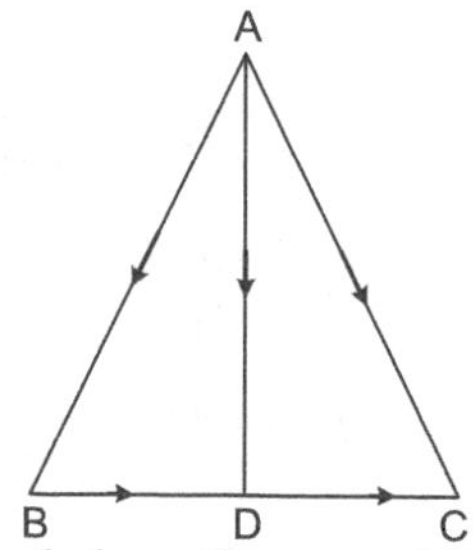

Using the triangle law of vector addition, we have

$$\vec{BC} = \vec{AC}-\vec{AB} = (3\hat{i}-\hat{j}+4\hat{k})-(\hat{j}+\hat{k}) = 3\hat{i}-2\hat{j}+3\hat{k}$$

$$\therefore \quad \vec{BD} = \frac{1}{2}\vec{BC} = \frac{3}{2}\hat{i}-\hat{j}+\frac{3}{2}\hat{k}$$

[∵ AD is the median]

In ΔABD, using the triangle law of vector addition, we have

$$\vec{AD} = \vec{AB}+\vec{BD} = (\hat{j}+\hat{k})+\left(\frac{3}{2}\hat{i}-\hat{j}+\frac{3}{2}\hat{k}\right) = \frac{3}{2}\hat{i}+0\hat{j}+\frac{5}{2}\hat{k}$$

$$\therefore \quad AD = \sqrt{\left(\frac{3}{2}\right)^2+0^2+\left(\frac{5}{2}\right)^2} = \frac{1}{2}\sqrt{34}$$

Hence, the length of the median through A is $\frac{1}{2}\sqrt{34}$ units.

Fill in the blanks :

Q. 43. If $\vec{a}=4\hat{i}+y\hat{j}-3\hat{k}$ and $|\vec{a}|=13$, then the value of y is ———.

Sol. ±2

Given, $|\vec{a}| = 13$

$$\Rightarrow \quad |4\hat{i}+y\hat{j}-3\hat{k}| = 13$$

★ are board exam questions from previous years

$\Rightarrow \quad \sqrt{(16+y^2+9)} = 13$

$\Rightarrow \quad y^2+25 = 169$

$\Rightarrow \quad y^2 = 144$

$\Rightarrow \quad y = \pm 12$

Q. 44. If $\vec{a} = (-1, 2)$, $\vec{b} = (3, -2)$, $\vec{c} = (0, 5)$, then the value of $\vec{a} + \vec{b} - 2\vec{c}$ is———.

Sol. (2, – 10)

$\vec{a} + \vec{b} - 2\vec{c}$

$= (-\hat{i}+2\hat{j}) + (3\hat{i}-2\hat{j}) - 2(0\hat{i}+5\hat{j})$

$= 2\hat{i} - 10\hat{j}$

$= (2, -10)$

Q. 45. If $\vec{a}$ and $\vec{b}$ are the position vectors of A and B respectively, then the position vector of a point C on the line AB, such that $\vec{AC} = 3\vec{AB}$ is———.

Sol. $3\vec{b} - 2\vec{a}$.

Let $\vec{r}$ be the position vector of C.

Then $\quad \vec{AC} = 3\vec{AB}$

$\Rightarrow \quad \vec{r} - \vec{a} = 3(\vec{b} - \vec{a})$

$\Rightarrow \quad \vec{r} = 3\vec{b} - 2\vec{a}$

Q. 46. If D, E, F are the mid-points of the sides BC, CA, AB respectively of the triangle ABC, then the value of $\vec{AD} + \vec{BE} + \vec{CF}$ is———.

Sol. 0

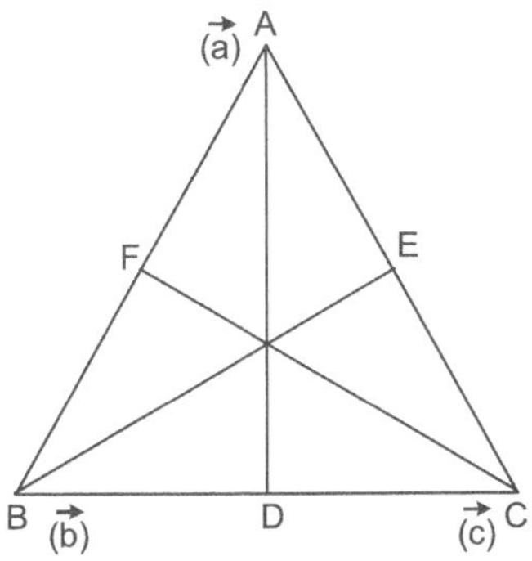

Let $\vec{a}, \vec{b}, \vec{c}$ be the position vectors of A, B, C of the triangle ABC.

Therefore, the position vector of

$$D = \frac{1}{2}(\vec{b} + \vec{c}),$$

the position vector of $E = \frac{1}{2}(\vec{c} + \vec{a})$

and the position vector of $F = \frac{1}{2}(\vec{a} + \vec{b})$

$\Rightarrow \vec{AD} + \vec{BE} + \vec{CF}$

$$= \left[\frac{\vec{b}+\vec{c}}{2} - \vec{a}\right] + \left[\frac{\vec{c}+\vec{a}}{2} - \vec{b}\right] + \left[\frac{\vec{a}+\vec{b}}{2} - \vec{c}\right]$$

$$= \frac{1}{2}\left[(\vec{b}+\vec{c}-2a) + (\vec{c}+\vec{a}-2b) + (\vec{a}+\vec{b}-2c)\right]$$

$$= \vec{0}$$

Q. 47. If $\hat{P}$ is a unit vector and $(\vec{x} - \vec{P}).(\vec{x} + \vec{P}) = 80$, then the value of $|\vec{x}|$ is———.

Sol. ± 9

Given that $\hat{P}$ is a unit vector

$\therefore \quad |\vec{P}| = 1$

Now, also given

$(\vec{x} - \vec{P}).(\vec{x} + \vec{P}) = 80$

$\therefore \quad \vec{x}.\vec{x} + \vec{x}.\vec{P} - \vec{P}.\vec{x} - \vec{P}.\vec{P} = 80$

$\Rightarrow \quad |x|^2 + \vec{x}.\vec{P} - \vec{x}.\vec{P} - |\vec{P}|^2 = 80$

$\Rightarrow \quad |x|^2 - |\vec{P}|^2 = 80$

$\Rightarrow \quad |x|^2 - 1 = 80$

$\therefore \quad |x|^2 = 81$

$\therefore \quad |x|^2 = \pm 9$

Q. 48. A unit vector in the direction of the sum of the vectors $\vec{a} = 2\hat{i} + 2\hat{j} - 5\hat{k}$ and $\vec{b} = 2\hat{i} + \hat{j} - 7\hat{k}$ is———.

Sol. $\frac{4}{13}\hat{i} + \frac{3}{13}\hat{j} - \frac{12}{13}\hat{k}$

Given, $\quad \vec{a} = 2\hat{i} + 2\hat{j} - 5\hat{k}$

$\vec{b} = 2\hat{i} + \hat{j} - 7\hat{k}$

$\Rightarrow \quad \vec{a} + \vec{b} = (2\hat{i}+2\hat{j}-5\hat{k}) + (2\hat{i}+\hat{j}-7\hat{k})$

$= 4\hat{i} + 3\hat{j} - 12\hat{k}$

and $\quad |\vec{a} + \vec{b}| = \sqrt{4^2 + 3^2 + (-12)^2}$

$= \sqrt{16+9+144}$

$= \sqrt{169}$

$= 13$

The unit vector in the direction of $\vec{a}+\vec{b}$ is

$$\frac{\vec{a}+\vec{b}}{|\vec{a}+\vec{b}|} = \frac{4\hat{i}+3\hat{j}-12\hat{k}}{13}$$

$$= \frac{4}{13}\hat{i}+\frac{3}{13}\hat{j}-\frac{12}{13}\hat{k}$$

Q.49. If $\vec{a}$ is a non-zero vector, then $(\vec{a}.\hat{i})\hat{i} + (\vec{a}.\hat{j})\hat{j} + (\vec{a}.\hat{k})\hat{k}$ equals ________.★

Sol. $\vec{a}$

Explanation Let $\vec{a} = a_1\hat{i}+b_1\hat{j}+c_1\hat{k}$

$\therefore (\vec{a}.\hat{i})\hat{i}+(\vec{a}.\hat{j})\hat{j}+(\vec{a}.\hat{k})\hat{k}$

$= a_1\hat{i}+b_1\hat{j}+c_1\hat{k}$

$= \vec{a}$

Q. 50. The projection of the vector $\hat{i}-\hat{j}$ on the vector $i+j$ is________.★

Sol. 0

Projection of $\hat{i}-\hat{j}$ on $\hat{i}+\hat{j}$

$$= \frac{(\hat{i}-\hat{j}).(\hat{i}+\hat{j})}{|\hat{i}+\hat{j}|} = 0$$

Chapter 11. Three-Dimensional Geometry

Q. 1. If P is a point in the space such that $OP = 12$ and $\vec{OP}$ is inclined at angle of 45° and 60° with OX and OY respectively, then find the position vector of P.

Sol. Let l, m, n be the direction cosines of $\vec{OP}$. Then, it is given that

$$l = \cos 45°, m = \cos 60°$$

$\Rightarrow \quad l = \frac{1}{\sqrt{2}}, m = \frac{1}{2}$

Since $\quad l^2+m^2+n^2 = 1$

$\Rightarrow \quad \frac{1}{2}+\frac{1}{4}+n^2 = 1$

$\Rightarrow \quad \frac{3}{4}+n^2 = 1$

$\Rightarrow \quad n^2 = \frac{1}{4}$

$\Rightarrow \quad n = \pm\frac{1}{2}$

Now $\quad \vec{r} = |\vec{r}|(l\hat{i}+m\hat{j}+n\hat{k})$

$\Rightarrow \quad \vec{r} = 12\left(\frac{1}{\sqrt{2}}\hat{i}+\frac{1}{2}\hat{j}\pm\frac{1}{2}\hat{k}\right)$

Therefore $\quad \vec{r} = 6\sqrt{2}\hat{i}+6\hat{j}\pm 6\hat{k}$

Q. 2. The direction cosines of a line are proportional to 2, – 3, 6. Find the actual values of the direction cosines.

Sol. If l, m, n are the direction cosines of the line then

$$l : m : n = 2 : -3 : 6$$

Hence $l = \frac{a}{\sqrt{a^2+b^2+c^2}}, m = \frac{b}{\sqrt{a^2+b^2+c^2}},$

$n = \frac{c}{\sqrt{a^2+b^2+c^2}}$

$l = \frac{2}{\sqrt{2^2+(-3)^2+6^2}}, m = \frac{-3}{\sqrt{2^2+(-3)^2+6^2}},$

$n = \frac{6}{\sqrt{2^2+(-3)^2+6^2}}$

i.e., $l = \frac{2}{7}, m = \frac{-3}{7}, n = \frac{6}{7}$

Q. 3. Find the distance between the points (2, 4, – 1) and (1, 3, 4) and then find the direction cosines of the line joining the points.

Sol. Let $P(2, 4, -1)$ and $Q(1, 3, 4)$ be the given points.

$\therefore \quad PQ = \sqrt{(2-1)^2+(4-3)^2+(-1-4)^2}$

$= \sqrt{1^2+1^2+5^2}$

$= \sqrt{27} = 3\sqrt{3}$

$\therefore \quad |\vec{QP}| = \sqrt{27} = |\vec{PQ}|$

The direction cosines of $\vec{PQ}$ are

$$\frac{1-2}{\sqrt{27}}, \frac{3-4}{\sqrt{27}}, \frac{4-(-1)}{\sqrt{27}}$$

i.e., $\quad \frac{-1}{\sqrt{27}}, \frac{-1}{\sqrt{27}}, \frac{5}{\sqrt{27}}$

Q. 4. If a line has direction ratios 2, – 1, – 2 then what are its direction cosines ?★

Sol. Given, direction ratios of line are 2, – 1, – 2.

$\therefore$ Direction cosines of above line are

$$\frac{2}{\sqrt{(2)^2+(-1)^2+(-2)^2}}, \frac{-1}{\sqrt{(2)^2+(-1)^2+(-2)^2}},$$

★ are board exam questions from previous years

$$\frac{-2}{\sqrt{(2)^2+(-1)^2+(-2)^2}}$$

$$=\frac{2}{\sqrt{4+1+4}},\frac{-1}{\sqrt{4+1+4}},\frac{-2}{\sqrt{4+1+4}}$$

$$=\frac{2}{\sqrt{9}},\frac{-1}{\sqrt{9}},\frac{-2}{\sqrt{9}}$$

$$=\frac{2}{3},\frac{-1}{3},\frac{-2}{3}.$$

Q. 5. What are the direction cosines of a line which makes equal angles with the coordinate axes ?**

Sol. Given that line makes equal angle with coordinate axes. Let α, β, γ be the angle made by the line with coordinate axes.

Given that $\alpha=\beta=\gamma$

$\Rightarrow$ $\cos\alpha=\cos\beta=\cos\gamma$

$\Rightarrow$ $l=m=n$ ...(i)

$[\because l=\cos\alpha, m=\cos\beta, n=\cos\gamma]$

We know that

$$l^2+m^2+n^2=1$$

$\therefore$ $l^2+l^2+l^2=1$

$\Rightarrow$ $3l^2=1$

$\Rightarrow$ $l^2=\frac{1}{3}$

$\Rightarrow$ $l=\pm\frac{1}{\sqrt{3}}$

$\therefore$ By equation (i), direction cosines of line are

$$\pm\frac{1}{\sqrt{3}},\pm\frac{1}{\sqrt{3}},\pm\frac{1}{\sqrt{3}}$$

Q. 6. Write the vector equation of the line given by $\frac{x-5}{3}=\frac{y+4}{7}=\frac{z-6}{2}$ **

Sol. Given equation of line in Cartesian form is

$$\frac{x-5}{3}=\frac{y+4}{7}=\frac{z-6}{2}$$

The point on line is $(5,-4,6)$ and direction ratios are $(3,7,2)$.

We know that vector equation of a line passing through $\vec{a}$ and parallel to $\vec{b}$ is

$$\vec{r}=\vec{a}+\lambda\vec{b}$$

Here, $\vec{a}=(5,-4,6)$ and $\vec{b}=(3,7,2)$

Its equation in vector form is

$$\vec{r}=(5\hat{i}-4\hat{j}+6\hat{k})+\lambda(3\hat{i}+7\hat{j}+2\hat{k})$$

Q. 7. Equation of line is $\frac{4-x}{2}=\frac{y+3}{2}=\frac{z+2}{1}$. Find the direction cosines of a line parallel to above line.*

Sol. Given equation of line can be written as

$$\frac{x-4}{-2}=\frac{y+3}{2}=\frac{z+2}{1}$$

$\therefore$ Direction cosines of line parallel to above line are given by

$$\frac{-2}{\sqrt{(-2)^2+(2)^2+(1)^2}},\frac{2}{\sqrt{(-2)^2+(2)^2+(1)^2}},\frac{1}{\sqrt{(-2)^2+(2)^2+(1)^2}}$$

$$=\frac{-2}{\sqrt{4+4+1}},\frac{2}{\sqrt{4+4+1}},\frac{1}{\sqrt{4+4+1}}$$

$$=\frac{-2}{\sqrt{9}},\frac{2}{\sqrt{9}},\frac{1}{\sqrt{9}}=-\frac{2}{3},\frac{2}{3},\frac{1}{3}$$

Hence, required direction cosines of a line parallel to the given line is $\left(-\frac{2}{3},\frac{2}{3},\frac{1}{3}\right)$.

Q. 8. If the equation of line AB is $\frac{3-x}{1}=\frac{y+2}{-2}=\frac{z-5}{4}$, write the direction ratios of line parallel to above line AB.*

Sol. Given equation of line can be written as

$$\frac{x-3}{-1}=\frac{y+2}{-2}=\frac{z-5}{4}$$

$\therefore$ Direction ratios of line parallel to above line are $-1,-2,4$ ($\because$ parallel lines have same direction ratios).

Q. 9. Find the value of λ such that the line $\frac{x-2}{6}=\frac{y-1}{\lambda}=\frac{z+5}{-4}$ is perpendicular to the plane $3x-y-2z=7$.*

Sol. Given that line $\frac{x-2}{6}=\frac{y-1}{\lambda}=\frac{z+5}{-4}$ is perpendicular to plane $3x-y-2z=7$

$\therefore$ $\frac{6}{3}=\frac{\lambda}{-1}=\frac{-4}{-2}$

[$\because$ when a line is perpendicular to a plane, their direction ratios are proportional]

$\Rightarrow$ $2=-\lambda=2$

$\Rightarrow$ $-\lambda=2$

$\Rightarrow$ $\lambda=-2$

Q. 10. Write the equation of line parallel to the line $\frac{x-2}{-3}=\frac{y+3}{2}=\frac{z+5}{6}$ and passing through point (1, 2, 3).*

* are board exam questions from previous years
** are frequently asked board exam questions

Sol. Given equation of line is written as

$$\frac{x-2}{-3} = \frac{y+3}{2} = \frac{z+5}{6} \quad \ldots\text{(i)}$$

Since, required line is parallel to given line, so direction ratios of required line will be proportional to – 3, 2 and 6.

∴ Equation of line passing through point (1, 2, 3) and parallel to equation (i) is

$$\frac{x-1}{-3} = \frac{y-2}{2} = \frac{z-3}{6} \cdot \left[\text{using } \frac{x-x_1}{a} = \frac{y-y_1}{b} = \frac{z-z_1}{c}\right]$$

Q. 11. If $P = (1, 5, 4)$ and $Q = (4, 1, -2)$, find the direction ratios of PQ.★

Sol. Given points are

$$P=(1,5,4) \text{ and } Q=(4,1,-2)$$

∴ Direction ratios of $PQ = 4 - 1, 1 - 5, -2 - 4 = 3, -4, -6$

[∵ Direction ratios of a line $P(x_1, y_1, z_1)$ and $Q(x_2, y_2, z_2)$ are $x_2-x_1, y_2-y_1, z_2-z_1$]

Q. 12. Find the Cartesian equation of the plane through the point with position vector $2\hat{i} - \hat{j} + \hat{k}$ and perpendicular to the vector $4\hat{i} + 2\hat{j} - 3\hat{k}$.

Sol. Plane is passing through the point A(say) having position vector

$$\vec{a} = 2\hat{i} - \hat{j} + \hat{k}$$

and is perpendicular to the vector

$$\vec{n} = 4\hat{i} + 2\hat{j} - 3\hat{k}$$

So, equation of plane is

$$(\vec{r} - \vec{a}).\vec{n} = 0 \text{ where}$$

$$\vec{r} = x\hat{i} + y\hat{j} + z\hat{k}$$

$$\Rightarrow [(x-2)\hat{i} + (y+1)\hat{j} + (z-1)\hat{k}].(4\hat{i} + 2\hat{j} - 3\hat{k}) = 0$$

$$\Rightarrow \quad 4(x-2) + 2(y+1) - 3(z-1) = 0$$

$\Rightarrow 4x + 2y - 3z - 3 = 0$ is the required equation of plane.

Q. 13. The position vector of two points P and Q are $3\hat{i} + \hat{j} + 2\hat{k}$ and $\hat{i} - 2\hat{j} - 4\hat{k}$ respectively. Find the equation of the plane through Q and perpendicular to PQ.

Sol. $\vec{PQ}$ = the position vector of Q – the position vector of P

$$= (\hat{i} - 2\hat{j} - 4\hat{k}) - (3\hat{i} + \hat{j} + 2\hat{k})$$

$$= -2\hat{i} - 3\hat{j} - 6\hat{k}$$

Therefore, the equation of the plane passing through the point Q and perpendicular to PQ is

$$[\vec{r} - (\hat{i} - 2\hat{j} - 4\hat{k})].(-2\hat{i} - 3\hat{j} - 6\hat{k}) = 0$$

$$[\text{Using } (\vec{r} - \vec{a}).\vec{n} = 0]$$

$$\Rightarrow \vec{r}.(-2\hat{i} - 3\hat{j} - 6\hat{k}) - (\hat{i} - 2\hat{j} - 4\hat{k}).(-2\hat{i} - 3\hat{j} - 6\hat{k}) = 0$$

$$\Rightarrow \vec{r}.(-2\hat{i} - 3\hat{j} - 6\hat{k}) - [1(-2) - 2(-3) - 4(-6)] = 0$$

$$\Rightarrow \vec{r}.(-2\hat{i} - 3\hat{j} - 6\hat{k}) - [-2 + 6 + 24] = 0$$

$$\Rightarrow \vec{r}.(-2\hat{i} - 3\hat{j} - 6\hat{k}) - 28 = 0$$

$$\Rightarrow \vec{r}.(2\hat{i} + 3\hat{j} + 6\hat{k}) + 28 = 0$$

This is the equation of the required plane.

Q. 14. Find the vector equation of the plane through the point $2\hat{i} - \hat{j} - 4\hat{k}$ and parallel to the plane $\vec{r}.(4\hat{i} - 12\hat{j} - 3\hat{k}) - 7 = 0$.

Sol. The equation of a plane parallel to the plane $\vec{r}.(4\hat{i} - 12\hat{j} - 3\hat{k}) - 7 = 0$ is

$$\vec{r}.(4\hat{i} - 12\hat{j} - 3\hat{k}) + \lambda = 0 \quad \ldots\text{(i)}$$

This plane passes through the point $2\hat{i} - \hat{j} - 4\hat{k}$.

Therefore, $(2\hat{i} - \hat{j} - 4\hat{k}).(4\hat{i} - 12\hat{j} - 3\hat{k}) + \lambda = 0$

$$\Rightarrow \quad 2(4) - 1(-12) - 4(-3) + \lambda = 0$$

$$\Rightarrow \quad 8 + 12 + 12 + \lambda = 0$$

$$\Rightarrow \quad \lambda = -32$$

Putting the value of λ in equation (i), we obtain the equation of the required plane as

$$\vec{r}.(4\hat{i} - 12\hat{j} - 3\hat{k}) - 32 = 0$$

$$\Rightarrow \quad \vec{r}.(4\hat{i} - 12\hat{j} - 3\hat{k}) = 32.$$

Q. 15. Find the vector equation of the plane passing through the point (2, 1, – 1) and passing through the line of intersection of the planes $\vec{r}.(\hat{i} + 3\hat{j} - \hat{k}) = 0$ and $\vec{r}.(\hat{j} + 2\hat{k}) = 0$.

Sol. The vector equation of a plane passing through the line of intersection of the planes $\vec{r}.(\hat{i} + 3\hat{j} - \hat{k}) = 0$ and $\vec{r}.(\hat{j} + 2\hat{k}) = 0$ can be written as

$$[\vec{r}.(\hat{i} + 3\hat{j} - \hat{k})] + \lambda[\vec{r}.(\hat{j} + 2\hat{k})] = 0$$

This plane passes through $2\hat{i} + \hat{j} - \hat{k}$

Therefore, $[(2\hat{i} + \hat{j} - \hat{k}).(\hat{i} + 3\hat{j} - \hat{k})] + \lambda[(2\hat{i} + \hat{j} - \hat{k}).(\hat{j} + 2\hat{k})] = 0$

★ are board exam questions from previous years

$$\Rightarrow [2(1)+1(3)-1(-1)]+\lambda[2(0)+1(1)-1(2)]=0$$

$$\Rightarrow \quad 6-\lambda=0$$

$$\Rightarrow \quad \lambda=6$$

Putting the value of λ in (i), we get the required equation of the plane as

$$[\vec{r}.(\hat{i}+3\hat{j}-\hat{k})]+6[\vec{r}.(\hat{j}+2\hat{k})]=0$$

$$\Rightarrow \quad \vec{r}.(\hat{i}+9\hat{j}+11\hat{k})=0.$$

Q. 16. Find the condition for the lines $\vec{r}=\vec{a}+\lambda(\vec{b}\times\vec{c})$ and $\vec{r}=\vec{b}+\mu(\vec{c}\times\vec{a})$ to intersect.

Sol. The lines $\vec{r}=\vec{a}+\lambda(\vec{b}\times\vec{c})$ and $\vec{r}=\vec{b}+\mu(\vec{c}\times\vec{a})$ passes through the points $\vec{a}$ and $\vec{b}$ respectively and are parallel to the vectors $\vec{b}\times\vec{c}$ and $\vec{c}\times\vec{a}$ respectively.

Therefore, they intersect if $\vec{a}-\vec{b}$, $\vec{b}\times\vec{c}$ and $\vec{c}\times\vec{a}$ are coplanar and so

$$[\vec{a}-\vec{b} \quad \vec{b}\times\vec{c} \quad \vec{c}\times\vec{a}]=0$$

$$\Rightarrow \quad (\vec{a}-\vec{b}).[(\vec{b}\times\vec{c})\times(\vec{c}\times\vec{a})]=0$$

$$\Rightarrow \quad (\vec{a}-\vec{b}).[(\vec{b}\times\vec{c}.\vec{a})\vec{c}-(\vec{b}\times\vec{c}.\vec{c})\vec{a}]=0$$

$$\Rightarrow \quad (\vec{a}-\vec{b}).[[\vec{b} \quad \vec{c} \quad \vec{a}]\vec{c}-0]=0$$

$$\Rightarrow \quad (\vec{a}-\vec{b}).[\vec{b} \quad \vec{c} \quad \vec{a}]\vec{c}=0$$

$$\Rightarrow \quad ((\vec{a}-\vec{b}).\vec{c})[\vec{b} \quad \vec{c} \quad \vec{a}]=0$$

$$\Rightarrow \quad (\vec{a}-\vec{b}).\vec{c}=0$$

$$\Rightarrow \quad \vec{a}.\vec{c}-\vec{b}.\vec{c}=0$$

$$\Rightarrow \quad \vec{a}.\vec{c}=\vec{b}.\vec{c}$$

Q. 17. Find the length of the perpendicular from the origin to the plane passing through the three non-collinear points $\vec{a}$, $\vec{b}$ and $\vec{c}$.

Sol. The vector equation of the plane passing through three non-collinear points $\vec{a}$, $\vec{b}$, $\vec{c}$ is

$$\vec{r}.[\vec{a}\times\vec{b}+\vec{b}\times\vec{c}+\vec{c}\times\vec{a}]-[\vec{a} \quad \vec{b} \quad \vec{c}]=0$$

Therefore, the length of the perpendicular from the origin to the plane is given by

$$\left|\frac{\vec{0}.(\vec{a}\times\vec{b}+\vec{b}\times\vec{c}+\vec{c}\times\vec{a})-[\vec{a} \quad \vec{b} \quad \vec{c}]}{|\vec{a}\times\vec{b}+\vec{b}\times\vec{c}+\vec{c}\times\vec{a}|}\right|$$

$$=\frac{[\vec{a} \quad \vec{b} \quad \vec{c}]}{|\vec{a}\times\vec{b}+\vec{b}\times\vec{c}+\vec{c}\times\vec{a}|}$$

Q. 18. Find the distance between the planes $x+y-z=3$ and $2x+2y-2z+10=0$.

Sol. Let (x_1, y_1, z_1) lie on the plane $x+y-z-3=0$

$$\Rightarrow \quad x_1+y_1-z_1-3=0$$

Now distance of (x_1, y_1, z_1) from the plane $2x+2y-2z+10=0$ is given by

$$d=\left|\frac{2x_1+2y_1-2z_1+10}{\sqrt{(2)^2+(2)^2+(-2)^2}}\right|$$

$$=\left|\frac{2(x_1+y_1-z_1)+10}{\sqrt{4+4+4}}\right|$$

$$\Rightarrow \quad d=\left|\frac{2(3)+10}{\sqrt{12}}\right|$$

$$\Rightarrow \quad d=\left|\frac{16}{\sqrt{12}}\right|$$

$$\Rightarrow \quad d=\frac{16}{2\sqrt{3}}$$

$$\Rightarrow \quad d=\frac{8}{\sqrt{3}}$$

$$\Rightarrow \quad d=\frac{8\sqrt{3}}{3}$$

Q. 19. If a line makes angles 90° and 60° respectively with the positive directions of X and Y axes, find the angle which it makes with the positive direction of Z-axis.*

Sol. We know,

$$l^2+m^2+n^2=1 \quad \ldots(i)$$

and $\quad l=\cos\alpha, m=\cos\beta, n=\cos\gamma$

Given $\quad \alpha=90°, \beta=60°$

$\therefore \quad \cos\alpha=\cos 90°=0$

and $\quad \cos\beta=\cos 60°=\frac{1}{2}$

From equation (i)

$$0^2+\left(\frac{1}{2}\right)^2+n^2=1$$

$$n^2=1-\frac{1}{4}=\frac{3}{4}$$

$$\Rightarrow \quad \cos^2\gamma=\frac{3}{4}$$

$$\text{or} \quad \cos\gamma=\pm\frac{\sqrt{3}}{2}$$

$$\Rightarrow \quad \gamma=\cos^{-1}\left(\frac{\sqrt{3}}{2}\right) \text{ or } \cos^{-1}\left(\frac{-\sqrt{3}}{2}\right)$$

$$\Rightarrow \quad \gamma=30° \text{ or } 150°.$$

* are board exam questions from previous years

Q. 20. Find distance of the plane $3x-4y+12z=3$ from the origin.★

Sol. Given equation of plane is $3x-4y+12z-3=0$ and the point is (0, 0, 0). We know that distance of the plane $Ax+By+Cz+D=0$ to the point (x_1, y_1, z_1) is

$$d=\frac{|Ax_1+By_1+Cz_1+D|}{\sqrt{A^2+B^2+C^2}}$$

Here, $x_1=y_1=z_1=0$ and $A=3, B=-4, C=12, D=-3$

$$\therefore \quad d=\frac{|3(0)-4(0)+12(0)-3|}{\sqrt{(3)^2+(-4)^2+(12)^2}}$$

$$d=\frac{|-3|}{\sqrt{9+16+144}}$$

$$=\frac{3}{\sqrt{169}}=\frac{3}{13}\text{ unit.}$$

Q. 21. Write the intercepts cut-off by plane $2x+y-z=5$ on X-axis.★

Sol. Given equation of plane is

$$2x+y-z=5$$

Divide both sides by 5, we get

$$\frac{2x}{5}+\frac{y}{5}-\frac{z}{5}=1$$

or $$\frac{x}{\left(\frac{5}{2}\right)}+\frac{y}{5}+\frac{z}{-5}=1$$

Comparing above equation of plane with the intercept form of equation of plane $\frac{x}{a}+\frac{y}{b}+\frac{z}{c}=1$, where a = x-intercept, b = y-intercept, c = z-intercept, we get

$$a=\frac{5}{2}$$

i.e., intercept cut-off on X-axis $=\frac{5}{2}$.

Q. 22. The equations of a line are $5x-3=15y+7=3-10z$. Write the direction cosines of the line.★

Sol. Given line is $5x-3=15y+7=3-10z$

Rewritting the equation in standard form :

$$5\left(x-\frac{3}{5}\right)=15\left(y+\frac{7}{15}\right)=10\left(z-\frac{3}{10}\right)$$

i.e., $$\frac{x-\frac{3}{5}}{\frac{1}{5}}=\frac{y+\frac{7}{15}}{\frac{1}{15}}=\frac{z-\frac{3}{10}}{\frac{1}{10}}$$

Thus, the direction ratios of the line are $\frac{1}{5}, \frac{1}{15}, \frac{-1}{10}$

i.e., $6, 2, -3$.

Hence, its direction cosines are

$$\pm\frac{6}{\sqrt{6^2+2^2+(-3)^2}}, \pm\frac{2}{\sqrt{6^2+2^2+(-3)^2}}, \pm\frac{-3}{\sqrt{6^2+2^2+(-3)^2}} \quad i.e., \pm\frac{6}{7}, \pm\frac{2}{7}, \mp\frac{3}{7}$$

i.e., $\frac{6}{7}, \frac{2}{7}, \frac{-3}{7}$ or $\frac{-6}{7}, \frac{-2}{7}, \frac{3}{7}$

Fill in the blanks :

Q. 23. If O is the origin, OP = 3 with direction ratios $-1, 2, -2$ then the coordinates of P are———.

Sol. $(-1, 2, -2)$

Direction ratios of $\vec{OP}$ are $-1, 2, -2$.

Therefore, direction cosines of $\vec{OP}$ are

$$\frac{-1}{\sqrt{(-1)^2+2^2+(-2)^2}}, \frac{2}{\sqrt{(-1)^2+2^2+(-2)^2}}, \frac{-2}{\sqrt{(-1)^2+2^2+(-2)^2}}$$

i.e., $\frac{-1}{3}, \frac{2}{3}, \frac{-2}{3}$

Hence $$\vec{OP}=|\vec{OP}|(l\hat{i}+m\hat{j}+n\hat{k})$$

$$=3\left(\frac{-1}{3}\hat{i}+\frac{2}{3}\hat{j}-\frac{2}{3}\hat{k}\right)$$

$$=-1\hat{i}+2\hat{j}-2\hat{k}$$

Therefore, coordinates of P are $(-1, 2, -2)$.

Q. 24. If α, β, γ are the angles which a directed line makes with the positive directions of the coordinates axes, then the value of $\sin^2\alpha+\sin^2\beta+\sin^2\gamma$ is———.

Sol. 2

Let l, m, n be the direction cosines of the line then

$$l=\cos\alpha, m=\cos\beta, n=\cos\gamma$$

Since $$l^2+m^2+n^2=1$$

$$\Rightarrow \quad \cos^2\alpha+\cos^2\beta+\cos^2\gamma=1$$

$$\Rightarrow \quad 1-\sin^2\alpha+1-\sin^2\beta+1-\sin^2\gamma=1$$

$$\Rightarrow \quad 3-(\sin^2\alpha+\sin^2\beta+\sin^2\gamma)=1$$

$$\Rightarrow \quad \sin^2\alpha+\sin^2\beta+\sin^2\gamma=3-1$$

$$\Rightarrow \quad \sin^2\alpha+\sin^2\beta+\sin^2\gamma=2$$

Q. 25. The equation of a line through the points $\hat{i}-2\hat{j}+\hat{k}$ and $3\hat{j}-2\hat{k}$ is———.

Sol. $-t\hat{i}+(3+5t)\hat{j}-(2+3t)\hat{k}$

★ are board exam questions from previous years

The line is parallel to $(3\hat{j}-2\hat{k})-(\hat{i}-2\hat{j}+\hat{k}) = -\hat{i}+5\hat{j}-3\hat{k}$ and passes through, say, $3\hat{j}-2\hat{k}$ (any of the two given points can be taken)

Hence the equation is

$$\vec{r} = 3\hat{j}-2\hat{k}+t(-\hat{i}+5\hat{j}-3\hat{k})$$

$$= -t\hat{i}+(3+5t)\hat{j}-(2+3t)\hat{k}$$

Q. 26. The angle between two lines with direction ratios as (1, 1, 2) and $(\sqrt{3}-1, -\sqrt{3}-1, 4)$ is———.

Sol. $\frac{\pi}{3}$

First we write their actual direction cosines.

They are $\frac{1}{\sqrt{6}}, \frac{1}{\sqrt{6}}, \frac{2}{\sqrt{6}}$ and $\frac{\sqrt{3}-1}{\sqrt{24}}, -\frac{\sqrt{3}+1}{\sqrt{24}}, \frac{4}{\sqrt{24}}$ respectively.

Thus if θ is the angle between them, then

$$\cos\theta = \frac{(\sqrt{3}-1)}{\sqrt{24}}\times\frac{1}{\sqrt{6}}+\frac{-(\sqrt{3}+1)}{\sqrt{24}}\times\frac{1}{\sqrt{6}} + \frac{4}{\sqrt{24}}\times\frac{2}{\sqrt{6}}$$

$$= \frac{6}{\sqrt{6}\sqrt{24}} = \frac{1}{2}$$

Therefore, $\theta = 60° = \frac{\pi}{3}$

Q. 27. The distance of point (2, 3, 4) from *X*-axis is———.

Sol. 5 units

Let any point on *X*-axis be $P(x, 0, 0)$. Then distance of point $B(2, 3, 4)$ from P is given as

$$d = \sqrt{(x_2-x_1)^2+(y_2-y_1)^2+(z_2-z_1)^2}$$

Here $(x_1, y_1, z_1) = (2, 0, 0)$

and $(x_2, y_2, z_2) = (2, 3, 4)$

$$\Rightarrow \quad d = \sqrt{(2-2)^2+(0-3)^2+(0-4)^2}$$

$$= \sqrt{9+16}$$

$$= \sqrt{25} = 5.$$

Q. 28. The direction cosines of the normal from the origin to the plane $\vec{r}.(\hat{i}+\hat{j}-\hat{k}) = 2$ is———.

Sol. $\frac{1}{\sqrt{3}}, \frac{1}{\sqrt{3}}, \frac{1}{\sqrt{3}}$

Equation of plane is given as,

$$\vec{r}.\hat{n} = d$$

So, $\quad \hat{n} = \hat{i}+\hat{j}-\hat{k}$

and $\quad d = 2$

Now, $|\hat{i}+\hat{j}-\hat{k}| = \sqrt{1+1+1} = \sqrt{3}$

Therefore, we can rewrite the equation of the plane as

$$\hat{r}\cdot\frac{(\hat{i}+\hat{j}-\hat{k})}{\sqrt{3}} = \frac{2}{\sqrt{3}}$$

Thus $\hat{n} = \frac{\hat{i}+\hat{j}-\hat{k}}{\sqrt{3}}$ and has direction cosines

$$\frac{1}{\sqrt{3}}, \frac{1}{\sqrt{3}}, \frac{1}{\sqrt{3}},$$

Q. 29. The angle between the planes $2x-y+z=6$ and $x+y+2z=7$ is———.

Sol. $\frac{\pi}{3}$

Let θ be the angle between the planes

$$\therefore \quad \cos\theta = \frac{2.1+(-1).1+1.2}{\sqrt{2^2+(-1)^2+(1)^2}\sqrt{1^2+1^2+2^2}}$$

$$= \frac{3}{\sqrt{6}\sqrt{6}}$$

$$= \frac{3}{6} = \frac{1}{2}$$

$$\Rightarrow \quad \cos\theta = \cos 60° = \frac{1}{2}$$

$$\Rightarrow \quad \theta = \frac{\pi}{3}$$

Q. 30. The perpendicular distance from the origin to the plane through the point (2, 3, – 1) and perpendicular to the vector $3\hat{i}-4\hat{j}+7\hat{k}$ is———.

Sol. $\frac{13}{\sqrt{74}}$ units.

The equation of the plane through (2, 3, – 1) and perpendicular to $3\hat{i}-4\hat{j}+7\hat{k}$ is

$$[\vec{r}-(2\hat{i}+3\hat{j}-\hat{k})].(3\hat{i}-4\hat{j}+7\hat{k}) = 0$$

[using $(\vec{r}-\vec{a}).\vec{n} = 0$]

$$\Rightarrow \vec{r}.(3\hat{i}-4\hat{j}+7\hat{k}) - (2\hat{i}+3\hat{j}-\hat{k}).(3\hat{i}-4\hat{j}+7\hat{k}) = 0$$

$$\Rightarrow \quad \vec{r}.(3\hat{i}-4\hat{j}+7\hat{k})-[6-12-7] = 0$$

$$\Rightarrow \quad \vec{r}.(3\hat{i}-4\hat{j}+7\hat{k})+13=0$$

Now the length of the perpendicular from the origin to this plane is given by

$$\left|\frac{\vec{0}.(3\hat{i}-4\hat{j}+7\hat{k})+13}{\sqrt{(3)^2+(-4)^2+(7)^2}}\right| = \left|\frac{13}{\sqrt{9+16+49}}\right| = \frac{13}{\sqrt{74}}$$

Q. 31. The angle between the line $\frac{x+1}{3}=\frac{y-1}{2}=\frac{z-2}{4}$ and the plane $2x+y-3z+4=0$ is ———.

Sol. $\sin^{-1}\left(\frac{-4}{\sqrt{406}}\right)$

Let the line makes an angle θ with the plane.

$$\therefore \quad \sin\theta = \frac{(2)(3)+(1)(2)+(-3)(4)}{\sqrt{(2)^2+(1)^2+(-3)^2}\sqrt{(3)^2+(2)^2+(4)^2}}$$

$$\Rightarrow \quad \sin\theta = \frac{6+2-12}{\sqrt{14}\sqrt{29}}$$

$$= \frac{-4}{\sqrt{406}}$$

Therefore, $\theta = \sin^{-1}\left(\frac{-4}{\sqrt{406}}\right)$

Q. 32. The vector equation of a line which passes through the points (3, 4, – 7) and (1, –1, 6) is ———.★

Sol. $(3\hat{i}+4\hat{j}-7\hat{k})+\lambda(-2\hat{i}-5\hat{j}+13\hat{k})$

$\because$ A = (3, 4, – 7) and B = (1, – 1, 6)

$\therefore$ Direction ratios of AB = < – 2, –5, 13 >

$\therefore$ Equation of required line is,

$$\vec{r} = (3\hat{i}+4\hat{j}-7\hat{k})+\lambda(-2\hat{i}-5\hat{j}+13\hat{k})$$

Q. 33 The line of shortest distance between two skew lines is ——— to both the lines.★

Sol. Perpendicular

Chapter 12. Linear Programming

Q. 1. A manufacturer produces two machine parts *A* and *B* in a certain time. Each of these machine parts undergoes four different operations O_1, O_2, O_3 and O_4. Given below are :

(a) The requirement (in hours) per unit of manufacturing of the machine.

(b) The available capacities of these operations (in hours) for the given time period.

(c) Profit on each unit of *A* and *B*.

Operation	Time (in hours) required for each unit of manufacturing		Available Capacities (hour)
	A	*B*	
O_1	1	2	30
O_2	3	1	60
O_3	4	3	200
O_4	5	4	200
Profit Per Unit (₹)	3	2	

Formulate the problem as an LP-model to maximize the profit.

Sol. Let x_1 and x_2 respectively be the number of units of machine part *A* and *B* that the company decides to produce.

The objective is to maximize the total profit that the company gets after selling *A* and *B*, which is given by the objective function :

$Z = 3x_1 + 2x_2$ [See profit per unit]

i.e., maximize $Z = 3x_1 + 2x_2$

For the required production, the number of hours taken by

Operation O_1 is : $x_1 + 2x_2$

Operation O_2 is : $3x_1 + x_2$

Operation O_3 is : $4x_1 + 3x_2$

Operation O_4 is : $5x_1 + 4x_2$

Now, the total time available for the four operations cannot exceed 30, 60, 200, 200 hours, respectively.

$\therefore$ We can write :

$$x_1 + 2x_2 \le 30$$
$$3x_1 + x_2 \le 60$$
$$4x_1 + 3x_2 \le 200$$
$$5x_1 + 4x_2 \le 200$$

As the number of machine parts produced will always non-negative, therefore, we have :

$$x_1 \ge 0,$$

and $x_2 \ge 0$ [non-negative constraints]

The LP-model of the given problem can then be written as :

Maximize $Z = 3x_1 + 2x_2$

subject to the constraints :

$$x_1 + 2x_2 \le 30$$
$$3x_1 + x_2 \le 60$$
$$4x_1 + 3x_2 \le 200$$
$$5x_1 + 4x_2 \le 200$$
$$x_1 \ge 0$$

and $x_2 \ge 0$

Q. 2. A jeweller makes gold necklaces and bracelets. Necklaces have a profit margin of ₹ 5,000 and bracelets ₹ 2,400. A necklace takes 2 hours of stone cutting, 7

★ are board exam questions from previous years

hours of setting and 6 hours for polishing, while a bracelet take 6 hours for stone cutting, 6 hours for setting and 3 hours for polishing. The jeweller has 40 hours for stone cutting, 70 hours for setting and 48 hours for polishing. Formulate the above data as a linear programming problem to maximize the profit.

Sol. Let x_1 and x_2 represent the number of necklaces and bracelets respectively. The given data can be summarized as follows :

	Necklaces (x_1)	Bracelets (x_2)	Maximum Available Capacity
Stone cutting (hour)	2	6	40
Setting (hour)	7	6	70
Polishing (hour)	6	3	48
Profit (in ₹)	5,000	2,400	

The linear programming problem, therefore, is

Maximize $Z = 5{,}000x_1 + 2{,}400x_2$

Subject to

$$2x_1 + 6x_2 \leq 40$$
$$7x_1 + 6x_2 \leq 70$$
$$6x_1 + 3x_2 \leq 48$$
$$x_1 \geq 0$$
$$x_2 \geq 0$$

Q. 3. Food *X* contains 6 units of Vitamin *A* per gram, and 7 units of vitamin *B* per gram and costs ₹ 2 per gram. Food *Y* contains 8 units of vitamin *A* per gram and 12 unit of vitamin *B* per gram and costs ₹ 2.50 per gram. The daily minimum requirement of vitamin *A* and vitamin *B* are 100 units and 120 units respectively. Formulate the above as a linear programming problem to minimize the cost.

Sol. The summary of the given data is as follows :

	Food *X*	Food *Y*	Daily minimum requirement
Vitamin *A* (per gram)	6	8	100
Vitamin *B* (per gram)	7	12	120
Cost (in ₹)	2	2·5	

Let x gram and y gram of food X and food Y respectively be needed for the minimum intake.

The linear programming problem, therefore, is

Minimize $Z = 2x + 2{\cdot}5y$

Subject to $6x + 8y \geq 100$

$$7x + 12y \geq 120$$
$$x \geq 0$$
$$y \geq 0$$

Q. 4. A firm produces the products, *A*, *B* and *C* at two plants I and II. The number of items produced and the operating cost per hour are as follows :

Plant	Product			Operating cost per hour (in ₹)
	A	*B*	*C*	
I	30	60	40	1,200
II	30	20	100	800

It is desired to produce at least 400 items of product *A*, 800 items of product *B* and 900 items of product *C*. Formulate the above as linear programming problem to minimize the operating cost per hour of the two plants.

Sol. Let x_1 and x_2 represent the number of hours available at the plants I and II respectively.

The linear programming problem is

Minimize $Z = 1200x_1 + 800x_2$

Subject to

$$30x_1 + 30x_2 \geq 400$$
$$60x_1 + 20x_2 \geq 800$$
$$40x_1 + 100x_2 \geq 900$$
$$x_1 \geq 0$$
$$x_2 \geq 0$$

Q. 5. A furniture firm manufactures chairs and tables, each requiring the use of three machines *A*, *B* and *C*. Production of one chair requires 2 hours on machine *A*, 1 hour on machine *B* and 1 hour on machine *C*. Each table requires 1 hour each on machines *A* and *B* and 3 hours on machine *C*. The profit realised by selling one chair is ₹ 30, while for a table the figure is ₹ 60. The total time available per week on machine *A* is 70 hour, on machine *B* is 40 hours and on machine *C* is 90 hours. Develop a mathematical formulation to maximize profit.

Sol. The given data may be summarized in the tabular form as follows :

Machine	Chair	Table	Available time per week (in hours)
A	2	1	70
B	1	1	40
C	1	3	90
Profit per unit (in ₹)	30	60	

Let x be the number of chairs to be produced per week and y be the number of table to be produced per week.

The mathematical formulation is :

Maximize $Z = 30x + 60y$

Subject to $2x + y \leq 70$

$$x + y \leq 40$$
$$x + 3y \leq 90$$
$$x, y \geq 0$$

Q. 6. The manager of an oil refinery must decide on the optimal mix of two possible blending processes of which the inputs and outputs per production sum are as follows :

	Input (units)		Output (units)	
Process	**Crude *A***	**Crude *B***	**Gasoline x**	**Gasoline y**
1	4	3	4	7
2	5	6	5	5

The maximum amounts available of crudes *A* and *B* are 225 units and 200 units respectively. Market demand shows that at least 150 units of gasoline x and 120 units of gasoline y must be produced. The profits per production seen from process 1 and process 2 are ₹ 200 and ₹ 300 respectively. Formulate the problem as a linear programming problem.

Sol. Let x_1 and x_2 denote the number of production runs of the two processes respectively. Then the objective is to maximize the total profit.

Maximize $\quad Z = 200x_1 + 300x_2$

Subject to

$$4x_1 + 5x_2 \leq 225$$
$$3x_1 + 6x_2 \leq 200$$
$$4x_1 + 5x_2 \geq 150$$
$$7x_1 + 5x_2 \geq 120$$
$$x_1, x_2 \geq 0$$

Q. 7. During the festival season, a company combines two items *A* and *B* to form gift packs. Each pack must weigh 5 kg and should contain at least 2 kg of *A* and not more than 4 kg of *B*. The net contribution to the company is ₹ 10 per kg of *A* and ₹ 12 per kg of *B*. Formulate this problem as a linear programming problem to maximize net contribution per pack.

Sol. Let x kg of item *A* and y kg of item *B* be mixed in each pack. Then the appropriate mathematical formulation of the given problem is

Maximize $\quad Z = 10x + 12y$

Subject to $\quad x + y = 5$

$$x \geq 2$$
$$y \leq 4$$
$$x, y \geq 0$$

Fill in the blanks :

Q. 8. The feasible region for any LPP is always a——polygon.

Sol. Convex.

Q. 9. In LPP, the objective function is always ———.

Sol. Iinear.

Q. 10. In LPP, the linear inequalities are called ———.

Sol. Constraints.

Q. 11. The function which has to be maximised or minimised a Linear Programming Problem is known as ———.

Sol. Objective function.

Q. 12. Any point in the feasible region that gives optimal value of the objective function is called an ———.

Sol. Optimal solution.

Q. 13. The maximum value of $Z = 3x + 2y$, subject to the constraints $x + 2y \leq 10$, $3x + y \leq 15$, $x, y \geq 0$ is ———.

Sol. 18

Q. 14. The maximum value of $Z = 4x + y$ subject to $x + y \leq 50$, $x, y \geq 0$

Sol. 200

Q. 15. The solution set of the inequation $x + 2y > 3$ is ——— (open half plane not contaning the origin/half plane contaning the origin).

Sol. Open half plane not contaning the origin.

Chapter 13. Probability

Q. 1. Let *A* and *B* be two events such that $P(A) = 3/8$ and $P(B) = 2/5$. Find :

(i) $P(A$ or $B)$, if *A* and *B* are mutually exclusive events.

(ii) $P(A$ and $B)$, if *A* and *B* are independent events.

Sol. (i) Let *A* and *B* be mutually exclusive events.

Then $\quad A \cap B = \phi$

$\therefore \quad P(A \text{ or } B) = P(A \cup B) = P(A) + P(B)$

$$= \frac{3}{8} + \frac{2}{5} = \frac{15+16}{40}$$

$$= \frac{31}{40}$$

(ii) Let *A* and *B* be two independent events. Then,

$$P(A \cap B) = P(A) \times P(B)$$

$$= \frac{3}{8} \times \frac{2}{5} = \frac{6}{40} = \frac{3}{20}$$

Q. 2. Let *A* and *B* be two events such that $P(A) = 0{\cdot}3$ and $P(A \cup B) = 0{\cdot}5$. Find $P(B)$ if :

(i) *A* and *B* are mutually exclusive

(ii) *A* and *B* are independent.

Sol. (i) Let *A* and *B* be mutually exclusive events.

$\Rightarrow \quad A \cap B = \phi$

$\therefore \quad P(A \cup B) = P(A) + P(B)$

$\Rightarrow \quad 0{\cdot}5 = 0{\cdot}3 + P(B)$

$\Rightarrow \quad P(B) = 0{\cdot}2$

(ii) Let *A* and *B* be independent events.Then

$$P(A \cap B) = P(A) \times P(B)$$

$\therefore \quad P(A \cup B) = P(A) + P(B) - P(A) \times P(B)$

$$0{\cdot}5 = 0{\cdot}3 + P(B)(1 - P(A))$$

$$0{\cdot}5 - 0{\cdot}3 = P(B)(1 - 0{\cdot}3)$$

$$0{\cdot}2 = P(B)(0{\cdot}7)$$

$$\therefore \quad P(B) = \frac{0{\cdot}2}{0{\cdot}7} = \frac{2}{7}$$

Q. 3. An urn contains 5 red balls, 6 green balls and 4 black balls. A ball is drawn at random from the urn. What is the probability that the ball drawn is either red or black ?

Sol. Given,

No. of red balls = 5

No. of green balls = 6

No. of black balls = 4

$\therefore$ Total no. of balls = 15 = $n(S)$

Let A be the event of drawing a red ball

$$\Rightarrow \quad P(A) = \frac{5}{15}$$

and B be the event of drawing a black ball

$$\Rightarrow \quad P(B) = \frac{4}{15}$$

Then $\quad A \cap B = \phi$

$$\therefore \quad P(A \cup B) = P(A \text{ or } B) = P(A) + P(B)$$

$$= \frac{5}{15} + \frac{4}{15}$$

$$= \frac{9}{15} = \frac{3}{5}$$

Hence, the probability that the ball drawn is either red or black is $\frac{3}{5}$.

Q. 4. A bag contains 4 blue balls and 6 red balls. Two balls are drawn at random. Find the probability that they are of same colour.

Sol. Let S be the sample space. Then

n(S) = Total number of ways of drawing 2 balls out of 10 balls

$$\therefore \quad n(S) = {}^{10}C_2 = \frac{10!}{2!(8!)} = \frac{10 \times 9 \times 8!}{1 \times 2 \times 8!}$$

$$= \frac{10 \times 9}{1 \times 2} = 45$$

Let A be the event of getting both balls of same colour.

Then, $n(A)$ = Total number of ways of drawing (2 balls out of 4 blue balls) or (2 balls out of 6 red balls)

i.e., $$n(A) = {}^4C_2 + {}^6C_2 = \frac{4!}{2!2!} + \frac{6!}{2!4!}$$

$$= \frac{4 \times 3}{1 \times 2} + \frac{6 \times 5}{1 \times 2}$$

$$= 6 + 15 = 21$$

$$\therefore \quad P(A) = \frac{n(A)}{n(S)} = \frac{21}{45} = \frac{7}{15}$$

Q. 5. The probability that a student is not an athlete is $\frac{1}{5}$. Then what is the probability that out of 5 students, 4 are athletes ?

Sol. Given, $P(\text{student is not an athlete}) = \frac{1}{5}$

i.e., $p = \frac{1}{5}$ then $q = 1 - \frac{1}{5} = \frac{4}{5}$

Number of students, $n = 5$

P(out of 5 students, 4 are athletes)

$$= {}^5C_4 \left(\frac{4}{5}\right)^4 \left(\frac{1}{5}\right)$$

$$= \frac{5!}{4!1!} \times \left(\frac{4}{5}\right)^4 \times \frac{1}{5} = \left(\frac{4}{5}\right)^4$$

Hence, the required probability is $\left(\frac{4}{5}\right)^4 = \frac{256}{625}$.

Q. 6. Find the value of k and its variance if X is a random variable with probability distribution as given below :

X	0	1	2	3
$P(X = x)$	$2k$	$3k$	$2k$	k

Sol. We know that sum of all $P(X = x) = 1$

$\therefore \quad 2k + 3k + 2k + k = 1$

$$8k = 1 \Rightarrow k = \frac{1}{8}$$

and $\quad$ variance, $\sigma^2 = \Sigma x_i^2 p_i - \mu^2 \quad$ (here, μ = mean)

Mean $= \Sigma x_i p_i$

$$= 0 \times 2\left(\frac{1}{8}\right) + 1 \times 3\left(\frac{1}{8}\right) + 2 \times 2\left(\frac{1}{8}\right) + 3 \times \frac{1}{8}$$

$$= \frac{3 + 4 + 3}{8} = \frac{10}{8}$$

$$= \frac{5}{4} = 1{\cdot}25$$

Variance, σ^2

$$= \left[\left(0 \times \frac{1}{4}\right) + \left(1 \times \frac{3}{8}\right) + \left(4 \times \frac{1}{4}\right) + \left(9 \times \frac{1}{8}\right)\right] - \left(\frac{5}{4}\right)^2$$

$$= \left(0 + \frac{3}{8} + 1 + \frac{9}{8}\right) - \left(\frac{5}{4}\right)^2$$

$$= \left(\frac{12}{8} + 1\right) - \frac{25}{16} = \frac{20}{8} - \frac{25}{16}$$

$$= \frac{40}{16} - \frac{25}{16} = \frac{15}{16}$$

Q. 7. If $P(X) = \frac{9}{10}$, $P(Y) = \frac{3}{10}$ and $P(X \cap Y) = \frac{7}{10}$. Find (i) $P(X/Y)$, (ii) $P(Y/X)$.

Sol. (i) We have

$$P(X/Y) = \frac{P(X \cap Y)}{P(Y)}$$

$$= \frac{7/10}{3/10} = \frac{7}{3}$$

(ii) We have,

$$P(Y/X) = \frac{P(X \cap Y)}{P(X)}$$

$$= \frac{7/10}{9/10} = \frac{7}{9}$$

Q. 8. If X and Y are events such that $P(X) = \frac{1}{3}$, $P(Y) = \frac{6}{5}$ and $P(X \cap Y) = \frac{2}{5}$. Are the events X and Y independent?

Sol. We know that,

$$P(X \cap Y) = \frac{2}{5} = \frac{1}{3} \times \frac{6}{5} = P(X)\,P(Y)$$

So, X and Y are independent events.

Q. 9. Two cards are drawn. What is the probability that an ace comes up given that a black card came up ?

Sol. Let A be the event of getting an ace and B be the event of getting a black card. Then,

$$P(A) = \frac{4}{52} = \frac{1}{13} \text{ and } P(B) = \frac{26}{52} = \frac{1}{2}$$

$$\therefore \quad P(A \cap B) = \frac{2}{52} = \frac{1}{26}$$

$\therefore$ Required probability

$$= P(A/B)$$

$$= \frac{P(A \cap B)}{P(B)} = \frac{1/26}{1/2} = \frac{1}{13}$$

Q. 10. Find the probability of getting a multiple of 2 if two dice are thrown and it is known that the second die always shows a prime number.

Sol. Let A be the event of getting a multiple of 2 on first die and B be the event of getting a prime number on second die. Then

$$A = \{2, 4, 6\} \text{ and } B = \{2, 3, 5\}$$

$\therefore$ Required probability $= P(A/B)$

$$= \frac{P(A \cap B)}{P(B)} = \frac{1/6}{3/6} = \frac{1}{3}$$

Q. 11. Five cards are drawn successively with replacement from the pack of 52 cards. What is the probability that all five cards are diamonds ?

Sol. Let $\quad n = 5$

x = no. of diamonds

p = probability of getting a diamond $= \frac{13}{52} = \frac{1}{4}$

q = probability of not getting a diamond

$$= 1 - \frac{1}{4} = \frac{3}{4}$$

$$P(X = 5) = {}^5C_5 \left(\frac{1}{4}\right)^5 \left(\frac{3}{4}\right)^0 = \frac{1}{1024}$$

Q. 12. In a box containing 200 bulbs, 20 are defective. Find the probability that out of a sample of 10 bulbs, 4 are defective.

Sol. Here $\quad n = 10$

X = no. of defective bulbs.

p = probability of getting a defective bulb

$$= \frac{20}{200} = \frac{1}{10}$$

q = probability of not getting a defective bulb

$$= 1 - \frac{1}{10} = \frac{9}{10}$$

Therefore, the probability of getting 4 defective bulbs

$$= {}^{10}C_4 \left(\frac{1}{10}\right)^4 \left(\frac{9}{10}\right)^6$$

Q. 13. A pair of dice is thrown 7 times. If getting a doublet is considered a success, find the probability of three successes.

Sol. Here, $n = 7$, Let p = probability of getting a doublet

$$= \frac{6}{36} = \frac{1}{6}$$

q = probability of not getting a doublet $= 1 - \frac{1}{6} = \frac{5}{6}$

If X = no. of successes, then

$$P(X = 3) = {}^7C_3 \left(\frac{1}{6}\right)^3 \left(\frac{5}{6}\right)^{7-3}$$

$$= 35 \left(\frac{1}{6}\right)^3 \left(\frac{5}{6}\right)^4$$

Q. 14. A die is rolled 20 times. Getting a number greater than 4 is a success. Find the mean and variance of the number of success.

Sol. Probability of getting a number greater than 4 in one throw of a die $= \frac{2}{6} = \frac{1}{3}$

Therefore, $P(\text{success}) = p = \frac{1}{3}$,

$P(\text{failure}) = q = 1 - p = \frac{2}{3}$ and $n = 20$.

$$\text{Mean} = np = 20 \times \frac{1}{3}$$

$$\text{Variance} = npq = 20 \times \frac{1}{3} \times \frac{2}{3}$$

$$= \frac{40}{9} = 4{\cdot}44$$

Q. 15. A salesman wants to know the average number of units he sells per sales call. His past record of sales has the following probability distribution :

Sales (in units)	0	1	2	3	4	5
Probability	0·15	0·20	0·10	0·05	0·30	0·20

What is the average number of units he sells per call ?

Sol. Let x denote the number of units he sells per sale call.

Therefore,

$$\bar{x} = \sum_{i=0}^{5} p_i x_i$$

$$= 0 \cdot 15 \times 0 + 0 \cdot 20 \times 1 + 0 \cdot 10 \times 2 + 0 \cdot 05 \times 3 + 0 \cdot 30 \times 4 + 0 \cdot 20 \times 5$$

$$= 0 \cdot 20 + 0 \cdot 20 + 0 \cdot 15 + 1 \cdot 20 + 1 \cdot 00$$

$$= 2.75$$

Thus, the average number of units he sells per call is 2·75.

Q. 16. Find the binomial distribution when the sum of its mean and variance for 5 trials is 1·8.

Sol. Given $n = 5, np + npq = 1 \cdot 8$

$\Rightarrow \quad 5[p + p(1-p)] = 1 \cdot 8$

$\Rightarrow \quad 5p + 5p(1-p) = 1 \cdot 8$

$\Rightarrow \quad 5p + 5p - 5p^2 = 1 \cdot 8$

$\Rightarrow \quad 5p^2 - 10p + 1 \cdot 8 = 0$

$\Rightarrow \quad p^2 - 2p + 0 \cdot 36 = 0$

$\Rightarrow (p - 0 \cdot 2)(p - 1 \cdot 8) = 0$

$\Rightarrow \quad p = 0.2, 1 \cdot 8$

$\because \quad p, q < 1 \Rightarrow p = 0 \cdot 2 = \frac{2}{10} = \frac{1}{5}$ and $q = \frac{4}{5}$

Therefore, the distribution is ${}^nC_r p^r q^{n-r}$

$$= {}^5C_r \left(\frac{1}{5}\right)^r \left(\frac{4}{5}\right)^{5-r}$$

Q. 17. The mean and standard deviation of a binomial distribution are 20 and 4 respectively. Find the value of *n* and *p*, the distribution parameters.

Sol. Given $np = 20, \sqrt{npq} = 4$

$\Rightarrow \quad \frac{npq}{np} = \frac{16}{20}$

$\Rightarrow \quad q = \frac{4}{5}$

$\Rightarrow \quad p = 1 - q = \frac{1}{5}$

Therefore, $n = \frac{20}{p} = 20 \times 5 = 100$

Fill in the Blanks :

Q. 18. If *A* and *B* are two events such that $P(A) \neq 0$ and $P(B/A) = 1$, then the relation between *A* and *B* is ———.

Sol. $A \subset B$

Given, $P(A) \neq 0$ and $P(B/A) = 1$

We know that,

$$P(B/A) = \frac{P(B \cap A)}{P(A)}$$

$\Rightarrow \quad 1 = \frac{P(B \cap A)}{P(A)}$

$\Rightarrow \quad P(A) = P(A \cap B)$

$\Rightarrow \quad A \subset B.$

Q. 19. If four cards are drawn at random from a pack of 52 cards, then the probability of getting all four cards of the same suit is ———.

Sol. $\frac{44}{4165}$

Total no. of cards = 52

No. of ways of drawing 4 cards = ${}^{52}C_4$

Number of suits = 4

Number of cards in each suit = $\frac{52}{4} = 13$

Number of ways of getting all 4 cards of same suit

$$= {}^{13}C_4 + {}^{13}C_4 + {}^{13}C_4 + {}^{13}C_4$$

$\therefore$ Required probability = $\frac{4({}^{13}C_4)}{{}^{52}C_4}$

$$= \frac{44}{4165}$$

Q. 20. If a coin is tossed and a die is thrown. The probability that the outcome will be a tail or a number greater than 3 or both is ———.

Sol. $\frac{3}{4}$

Let A = event of getting tail in a single toss of coin.

B = event of getting a number greater than 3, in a throw of die.

Then $P(A) = \frac{1}{2}$ and $P(B) = \frac{3}{6} = \frac{1}{2}$

$\therefore$ Required probability

$$= P(A \cup B)$$

$$= 1 - P(\bar{A}).P(\bar{B})$$

[$\because$ *A* and *B* are independent]

$$= 1 - \left(1 - \frac{1}{2}\right)\left(1 - \frac{1}{2}\right)$$

$$= 1 - \left(\frac{1}{2} \times \frac{1}{2}\right)$$

$$= 1 - \frac{1}{4} = \frac{3}{4}$$

Short Questions

| Set 3 |

Chapter 1. Relations and Functions

Q. 1. If the function $f : R \to R$ is given by $f(x) = x^2 + 3x + 1$ and $g : R \to R$ is given by $g(x) = 2x - 3$, find :**

(i) *fog* **(ii) *gof***

Sol.

(i) Finding *fog*

$$\begin{aligned} fog &= f(g(x)) \\ &= f(2x-3) \\ &= (2x-3)^2 + 3(2x-3) + 1 \\ &= 4x^2 + 9 - 12x + 6x - 9 + 1 \\ &= 4x^2 - 6x + 1 \end{aligned}$$

(ii) Finding *gof*

$$\begin{aligned} gof &= g(f(x)) \\ &= g(x^2 + 3x + 1) \\ &= [2(x^2 + 3x + 1)] - 3 \\ &= 2x^2 + 6x - 1. \end{aligned}$$

Q. 2. If $y = f(x) = \frac{x^2}{1+x^2}$, is the function one-one and onto provided $f : R \to R$?

Sol. $f(x)$ is not one-one (*i.e.*, injective)

$$\text{As } f(-x) = \frac{(-x)^2}{1+(-x)^2} = \frac{x^2}{1+x^2} = f(x)$$

but $x \neq -x$

For the function to be an onto function, let

$$y = f(x)$$

$$\Rightarrow \quad \frac{x^2}{1+x^2} = y$$

$$\Rightarrow \quad x^2 = \frac{y}{1-y}$$

$$\Rightarrow \quad \frac{y}{1-y} \geq 0, y \neq 1$$

$$\Rightarrow \quad 0 \leq y < 1$$

$\Rightarrow$ Range of $f(x)$ is not equal to its codomain.

Hence, it is not onto.

That is, the given function is neither one-one nor an onto function.

Q. 3. If the function $f : R \to R$ is given by $f(x) = x^2 + 2$ and $g : R \to R$ is given by $g(x) = \frac{1}{x-1}$, then show that the composition of functions is not commutative.

Sol. To show that the composition of functions is not commutative, we need to show that $fog \neq gof$.

$$\text{Here,} \quad fog = f(g(x)) = f\left(\frac{1}{x-1}\right) = \left(\frac{1}{x-1}\right)^2 + 2$$

$$gof = g(f(x)) = g(x^2 + 2)$$

$$= \frac{1}{x^2 + 2 - 1} = \frac{1}{x^2 + 1}$$

We can see that $fog \neq gof$ *i.e.*, composition of functions is not a commutative operation.

Q. 4. Let $f = \{(1, 3), (2, 6), (4, 12)\}$ and $g = \{(3, 7), (6, 8), (12, 9), (10, 13)\}$. Find *gof*.

Sol. Given, $f = \{(1, 3), (2, 6), (4, 12)\}$ and $g = \{(3, 7), (6, 8), (12, 9), (10, 13)\}$

Now, range of $f = \{3, 6, 12\}$ and domain of $g = \{3, 6, 12, 10)\}$

Therefore, range of f is a subset of domain of g.

Hence, *gof* exists.

$$\begin{aligned} \therefore \quad gof(1) &= g(f(1)) \\ &= g(3) \\ &= 7 \\ gof(2) &= g(f(2)) \\ &= g(6) \\ &= 8 \\ gof(4) &= g(f(4)) \\ &= g(12) \\ &= 9 \end{aligned}$$

$$\therefore \quad gof = \{(1, 7), (2, 8), (4, 9)\}.$$

Q. 5. Let f be the greatest integer function and g be the modulus function. Find the value of the following :

$$gof\left(\frac{-5}{3}\right) - fog\left(\frac{-5}{3}\right)$$

Sol. We have, $f(x) = [x]$ and $g(x) = |x|$

$$\text{For } gof\left(-\frac{5}{3}\right) - fog\left(-\frac{5}{3}\right)$$

$$= g\left[f\left(\frac{-5}{3}\right)\right] - f\left[g\left(\frac{-5}{3}\right)\right]$$

$$= g\left(\left[\frac{-5}{3}\right]\right) - f\left(\left|\frac{-5}{3}\right|\right)$$

** are frequently asked board exam questions

$$= g(-2) - f\left(\frac{5}{3}\right)$$

$$= |-2| - \left[\frac{5}{3}\right]$$

$$= 2 - 1 = 1.$$

Q. 6. If $A = \{1, 2, 3\}$ and $B = \{a, b, c\}$, define any two bijections from A to B. Also, find their inverse functions.

Sol. A bijection is one-one and onto function.

Let, $f: A \to B$ such that

$$f = \{(1, a), (2, b), (3, c)\}$$

[one-one and onto function]

Then, $f^{-1} = \{(a, 1), (b, 2), (c, 3)\}$

Let, $g: A \to B$, such that

$$g = \{(1, b), (2, a), (3, c)$$

[onto and one-one function]

Then, $g^{-1} = \{(b, 1), (a, 2), (c, 3)\}$

Q. 7. If $f: R \to R$ be defined such that $f(x) = x^3 - 3$, then prove that f^{-1} exists and find a formula for f^{-1}. Hence, find $f^{-1}(24)$ and $f^{-1}(5)$.

Sol. To prove that f is injective :

Since, $f(x) = f(y)$ for $x, y \in R$

$\Rightarrow$ $x^3 - 3 = y^3 - 3$

$\Rightarrow$ $x^3 = y^3$

$\Rightarrow$ $x = y$

Hence $f(x) = f(y)$

$\Rightarrow$ $x = y, \forall\, x, y \in R$

Thus, f is injective.

To prove that f is surjective :

Let $y \in R$

$\Rightarrow$ $f(x) = y$ *i.e.*, $x^3 - 3 = y$

$\Rightarrow$ $x^3 = y + 3$

$\Rightarrow$ $x = (y+3)^{1/3}$

For any $y \in R$, there exists an $x \in R$, such that

$f(x) = f[(y+3)^{1/3}] = ((y+3)^{1/3})^3 - 3 = (y+3) - 3 = y$

$f(x) = y$

Thus, f is surjective.

$\Rightarrow f$ is bijective and f^{-1} exists.

Now, $f^{-1}: R \to R$.

Let, $x \in R$ (domain) and $y \in R$ (codomain), such that

$f(x) = y$

$\Rightarrow$ $x^3 - 3 = y$

$\Rightarrow$ $x^3 = y + 3$

$\Rightarrow$ $x = (y+3)^{1/3}$

$f^{-1}(y) = (y+3)^{1/3}$

$\therefore$ $f^{-1}(x) = (x+3)^{1/3}$

$\Rightarrow$ $f^{-1}(24) = (24+3)^{1/3} = (27)^{1/3} = 3$

$\Rightarrow$ $f^{-1}(5) = (5+3)^{1/3} = 8^{1/3} = 2.$

Q. 8. Show that $f: N \to N$ given by

$$f(x) = \begin{cases} x+1, \text{ if } x \text{ is odd} \\ x-1, \text{ if } x \text{ is even} \end{cases}$$

is both one-one and onto.★ **[NCERT]**

Sol. Given function $f: N \to N$ such that

$$f(x) = \begin{cases} x+1, \text{ if } x \text{ is odd} \\ x-1, \text{ if } x \text{ is even} \end{cases}$$

One-one : From the given function, we observe that

Case I : When x is odd

Let $f(x_1) = f(x_2)$

$\Rightarrow$ $x_1 + 1 = x_2 + 1$

$\Rightarrow$ $x_1 = x_2$

$\because$ $f(x_1) = f(x_2)$

$\Rightarrow$ $x_1 = x_2, \forall\, x_1, x_2 \in N.$

so $f(x)$ is one-one.

Case II : When x is even.

Let $f(x_1) = f(x_2)$

$\Rightarrow$ $x_1 - 1 = x_2 - 1$

$\Rightarrow$ $x_1 = x_2$

$\because$ $f(x_1) = f(x_2)$

$\Rightarrow$ $x_1 = x_2, \forall\, x_1, x_2 \in N.$

So $f(x)$ is one-one.

Hence, from case I and case II, we observe that

$f(x_1) = f(x_2)$

$\Rightarrow$ $x_1 = x_2, \forall\, x_1, x_2 \in N$

$\because f(x)$ is a one-one function.

Onto : To show $f(x)$ is onto, we show that its range and codomain are same. From the definition of given function, we observe that

$$f(1) = 1 + 1 = 2$$

$$f(2) = 2 - 1 = 1$$

$$f(3) = 3 + 1 = 4$$

$$f(4) = 4 - 1 = 3 \text{ and so on}$$

So, we get set of natural numbers as the set of values of $f(x)$.

$\Rightarrow$ Range of $f(x) = N$

★ are board exam questions from previous years

Also, given that codomain =

$$N\left[\because \underset{\text{domain}}{f:N} \to \underset{\text{codomain}}{N}\right]$$

$\therefore$ Range = Codomain

$\therefore f(x)$ is an onto function.

Q. 9. Let $f: R \to R$ be defined as $f(x) = 10x + 7$, find the function $g: R \to R$, such that $gof = fog = I_R$.★

Sol. Given that

$$f(x) = 10x + 7$$

Let $y = 10x + 7$

$\Rightarrow 10x = y - 7$

$\Rightarrow x = \dfrac{y-7}{10}$

Now, let $g(x) = \dfrac{x-7}{10}$

Then, $gof(x)$ can be written as

$$gof(x) = g(f(x)) = g(10x+7)$$

$$= \frac{10x+7-7}{10} = x$$

Also, $fog(x)$ can be written as

$$fog(x) = f(g(x))$$

$$= f\left(\frac{x-7}{10}\right)$$

$$= 10\left(\frac{x-7}{10}\right) + 7$$

$\therefore fog(x) = x$

$\therefore gof = fog = I_R$

Hence, $g(x) = \dfrac{x-7}{10}$

Q. 10. Consider $f: R^+ \to [4, \infty)$ given by $f(x) = x^2 + 4$. Show that f is invertible with the inverse f^{-1} of f given by $f^{-1}(y) = \sqrt{y-4}$, where R is the set of all non-negative real numbers.★

Sol. Given that $f: R^+ \to [4, \infty)$ such that

$$f(x) = x^2 + 4$$

First we show that $f(x)$ is one-one function.

We know that $f: A \to B$, $f(x)$ is said to be one-one function, if

$$f(x_1) = f(x_2) \Leftrightarrow x_1 = x_2, \forall x_1, x_2 \in A$$

$\therefore$ for $x_1, x_2 \in R^+$

Let $f(x_1) = f(x_2)$

$\Rightarrow x_1^2 + 4 = x_2^2 + 4$

$\Rightarrow x_1^2 = x_2^2$

$\Rightarrow x_1^2 - x_2^2 = 0$

$\Rightarrow (x_1 - x_2)(x_1 + x_2) = 0$

$\Rightarrow$ Either $x_1 - x_2 = 0$

or $x_1 + x_2 = 0$

But $x_1 + x_2 = 0$ is not possible as $x_1, x_2 \in R$ *i.e.*, sum of two positive real numbers cannot be zero.

$\therefore x_1 - x_2 = 0$

$\Rightarrow x_1 = x_2$

Hence, $f(x)$ is a one-one function.

Now, we show that $f(x)$ is onto.

For onto, we prove that range = codomain

We find range of

$$f(x) = x^2 + 4$$

For that, let $y = x^2 + 4$

$\Rightarrow x^2 = y - 4$

$\Rightarrow x = \sqrt{y-4}$ ...(i)

It is clear from equation (i), that for all

$$y \in [4, \infty), x \in R$$

$\therefore$ Range = $[4, \infty)$

$\therefore$ Range = Codomain

Hence, f is an onto function.

Since, $f(x)$ is both one-one and onto, it is an invertible function with $f^{-1}(y) = \sqrt{y-4}$.

Q. 11. Show that the function $f: W \to W$ defined by

$$f(n) = \begin{cases} n+1, & \text{if } n \text{ is even} \\ n-1, & \text{if } n \text{ is odd} \end{cases}$$ **is bijective function.★**

Sol. One-one : Let

$$f(x_1) = f(x_2), \forall x_1, x_2 \in W$$

Case I : When x_1, x_2 both are even, then

Let $f(x_1) = f(x_2), \forall x_1, x_2 \in W$

$\Rightarrow x_1 + 1 = x_2 + 1$

$\Rightarrow x_1 = x_2$

$\therefore f(x_1) = f(x_2)$

$\Rightarrow x_1 = x_2, \forall x_1, x_2 \in W$

Case II : When both x_1, x_2 are odd, then

Let $f(x_1) = f(x_2), \forall x_1, x_2 \in W$

$\Rightarrow x_1 - 1 = x_2 - 1$

$\Rightarrow x_1 = x_2$

$\therefore f(x_1) = f(x_2)$

★ are board exam questions from previous years

$\Rightarrow \quad x_1 = x_2, \forall\, x_1, x_2 \in W$

Hence, from case I and case II, we observe that

$$f(x_1) = f(x_2)$$

$\Rightarrow \quad x_1 = x_2, \forall\, x_1, x_2 \in W$

$\therefore f(x)$ is a one-one function.

Onto :

Case I : Let $\quad n = 2m$

$\Rightarrow \quad f(2m) = 2m + 1 = n + 1$

Case II : Let $\quad n = 2m + 1$

$$f(2m+1) = (2m+1) - 1 = 2m$$

$\therefore \quad f(2m+1) = 2m = n \quad$ (even terms)

Also, $\quad f(1) = 0$

$\therefore f : W \to W$ is onto function.

Since, $f(x)$ is both one-one and onto function, so it is a bijective function.

Q. 12. Consider $f : R^+ \to [-5, \infty)$ given by $f(x) = 9x^2 + 6x - 5$, show that f is invertible with $f^{-1}(y) = \left(\dfrac{\sqrt{y+6} - 1}{3}\right)$.★ **[NCERT]**

Sol. Given function is $f : R^+ \to [-5, \infty)$ such that

$$f(x) = 9x^2 + 6x - 5$$

We shall show that it is both one-one and onto.

(i) For one-one : We know that a function $f : A \to B$ is said to be one-one function, if

$$f(x_1) = f(x_2)$$

$\Rightarrow \quad x_1 = x_2, \forall\, x_1, x_2 \in A$

So for $x_1, x_2 \in R^+$

Let $\quad f(x_1) = f(x_2)$

$\Rightarrow \quad 9x_1^2 + 6x_1 - 5 = 9x_2^2 + 6x_2 - 5$

$\Rightarrow \quad 9x_1^2 - 9x_2^2 + 6x_1 - 6x_2 = 0$

$\Rightarrow \quad 9(x_1^2 - x_2^2) + 6(x_1 - x_2) = 0$

$\Rightarrow \quad (x_1 - x_2)(9x_1 + 9x_2 + 6) = 0$

Now, either $\quad x_1 - x_2 = 0$

or $\quad 9x_1 + 9x_2 + 6 = 0$

But $9x_1 + 9x_2 + 6 = 0$ is not possible because $x_1, x_2 \in R^+$.

$\therefore \quad x_1 - x_2 = 0$

$\Rightarrow \quad x_1 = x_2$

$\therefore f(x)$ is a one-one function.

(ii) For onto : To show that $f(x)$ is onto, we show that range of $f(x)$ = codomain of $f(x)$.

Now, we find range of $f(x)$

Let, $\quad y = 9x^2 + 6x - 5$

$\Rightarrow \quad 9x^2 + 6x = y + 5$

$\Rightarrow \quad 9\left(x^2 + \dfrac{6x}{9}\right) = y + 5$

$\Rightarrow \quad 9\left(x^2 + \dfrac{2x}{3} + \dfrac{1}{9} - \dfrac{1}{9}\right) = y + 5$

$\Rightarrow \quad 9\left(x + \dfrac{1}{3}\right)^2 - 1 = y + 5$

$\Rightarrow \quad \left(x + \dfrac{1}{3}\right)^2 = \dfrac{y+6}{9}$

$\Rightarrow \quad x + \dfrac{1}{3} = \dfrac{\sqrt{y+6}}{3}$

or $\quad x = \dfrac{\sqrt{y+6} - 1}{3}$

From above equation, we get for every $y \in [-5, \infty)$ we have

$$x \in R$$

$\therefore \quad$ Range of $f(x) = [-5, \infty)$

Given, codomain of $f(x) = [-5, \infty)$

$\Rightarrow \quad$ Range of $f(x)$ = codomain of $f(x)$

$\therefore f(x)$ is onto function.

Since, $f(x)$ is both one-one and onto, so it is invertible function with

$$f^{-1}(y) = \frac{\sqrt{y+6} - 1}{3}$$

Q. 13. Show that the function $f : R \to R$ given by $f(x) = 4x^3 + 7$ for all $x \in R$ is bijective.★

Sol. $f : R \to R$ is given by

$$f(x) = 4x^3 + 7$$

For injective :

Let $\quad f(x) = f(y)$

$\Rightarrow \quad 4x^3 + 7 = 4y^3 + 7$

$\Rightarrow \quad x^3 = y^3$

$\Rightarrow \quad x^3 - y^3 = 0$

$\Rightarrow \quad (x - y)(x^2 + xy + y^2) = 0$

$\Rightarrow \quad x - y = 0$

(as the other term cannot be zero)

$\Rightarrow \quad x = y$

Thus, $\quad f(x) = f(y)$

$\Rightarrow \quad x = y \; \forall\, x, y \in R$

For surjective :

Let $\quad y \in R$

Then, $\quad f(x) = y$

$\Rightarrow \quad 4x^3 + 7 = y$

★ are board exam questions from previous years

$$\Rightarrow \quad x^3 = \frac{y-7}{4}$$

Thus, for any $y \in R, x = \left[\frac{y-7}{4}\right]^{1/3}$ is a real number or $x \in R$.

Since range of $f(x)$ and codomain of $f(x)$ are both real numbers, $f(x)$ is an onto function.

Therefore, f is bijective.

Q. 14. Let $A = R - \{3\}$ and $B = R - \{1\}$. Consider the function $f : A \to B$ defined by $f(x) = \frac{x-2}{x-3}$ · Show that f is one-one and onto and hence find f^{-1}.★

Sol.

(i) For one-one function.

Let $\quad f(x) = f(y) \ \forall \ x, y \in A$

$$\Rightarrow \quad \frac{x-2}{x-3} = \frac{y-2}{y-3}$$

$$\Rightarrow \quad (x-2)(y-3) = (x-3)(y-2)$$

$$\Rightarrow \quad xy - 3x - 2y + 6 = xy - 2x - 3y + 6$$

$$\Rightarrow \quad -x = -y \text{ or } x = y$$

$$\therefore \quad f(x) = f(y)$$

$$\Rightarrow \quad x = y, \ \forall \ x, y \in A$$

$\therefore f(x)$ is one-one function.

(ii) For onto function.

We have

$$f(x) = y = \frac{x-2}{x-3}$$

$$\Rightarrow \quad y(x-3) = (x-2)$$

$$\Rightarrow \quad xy - 3y = x - 2$$

$$\Rightarrow \quad xy - x = 3y - 2$$

$$\Rightarrow \quad x = \frac{3y-2}{y-1}$$

Since, $\quad x \in R - \{3\}, \forall \ y \in R,$

$\therefore$ Range of $f(x)$ is $R - \{1\}$.

We are also given the codomain of $f(x)$ as $R - \{1\}$.

Therefore, the range and codomain of R are same, hence $f(x)$ is onto function.

(iii) For inverse

$$x = f^{-1}(y) = \frac{3y-2}{y-1}, \ y \neq 1$$

$$\therefore \quad f^{-1}(x) = \frac{3x-2}{x-1}, \ x \neq 1.$$

Q. 15. If the function $f : R \to R$ is given by $f(x) = \frac{x+3}{3}$ and $g : R \to R$ is given by $g(x) = 2x - 3$. Find (i) fog, (ii) gof. Is $f^{-1} = g$?★

Sol. We have to find $fog(x)$ and $gof(x)$

(i) $\quad fog(x) = f(g(x))$

$$= f(2x-3) = \frac{(2x-3)+3}{3}$$

$$\therefore \quad fog(x) = \frac{2x}{3}$$

(ii) $\quad gof(x) = g(f(x))$

$$= g\left(\frac{x+3}{3}\right) = \left[2\left(\frac{x+3}{3}\right)\right] - 3$$

$$= \frac{2x+6}{3} - 3 = \frac{2x+6-9}{3}$$

$$\therefore \quad gof(x) = \frac{2x-3}{3}$$

Now, we find f^{-1}. For this let

$$y = \frac{x+3}{3}$$

$$\Rightarrow \quad 3y = x + 3$$

$$\Rightarrow \quad x = 3y - 3$$

$$\therefore \quad f^{-1}(y) = 3y - 3$$

But $\quad g(x) = 2x - 3$

$$\therefore \quad f^{-1}(x) = 3x - 3$$

$$\therefore \quad f^{-1} \neq g$$

Q. 16. Show that the relation S in the set R of real numbers defined as $S = \{(a, b) : a, b \in R \text{ and } a \leq b^3\}$ is neither reflexive nor symmetric nor transitive.★

Sol. Given relation is

$$S = \{(a, b) : a, b \in R \text{ and } a \leq b^3\}$$

(i) Reflexive : Take $\frac{1}{3} \in R$

As $\quad \frac{1}{3} \leq \left(\frac{1}{3}\right)^3$ is not true.

$$\Rightarrow \quad \left(\frac{1}{3}, \frac{1}{3}\right) \notin S$$

$\therefore S$ is not reflexive.

(ii) Symmetric : For $-2, 3 \in$ R

As $-2 \leq (3)^3$ is true but $3 \leq (-2)^3$ is not true

i.e., $\quad (-2, 3) \in S$ but $(3, -2) \notin S$

$\therefore S$ is not symmetric.

(iii) Transitive : For $3, \frac{3}{2}, \frac{4}{3} \in R$

As $3 \leq \left(\frac{3}{2}\right)^3$ and $\frac{3}{2} \leq \left(\frac{4}{3}\right)^3$ but $3 \leq \left(\frac{4}{3}\right)^3$ is not true.

★ are board exam questions from previous years

$\Rightarrow \quad \left(3, \frac{3}{2}\right) \in S$ and $\left(\frac{3}{2}, \frac{4}{3}\right) \in S$

But $\quad \left(3, \frac{4}{3}\right) \notin S$

$\therefore$ S is not transitive.

Hence, S is neither reflexive not symmetric nor transitive.

Q. 17. Show that the relation S in set $A = \{x \in Z : 0 \le x \le 12\}$ given by $S = \{(a, b) : a, b \in Z, |a - b|$ is divisible by 4} is an equivalence relation. Find the set of all elements related to 1.*

Sol. The given relation is

$S = \{(a, b) : a, b \in Z, |a - b|$ is divisible by 4} and $A = \{x : x \in Z : 0 \le x \le 12\}$

Now, A can be written as

$$A = \{0, 1, 2, 3, \ldots 12\}$$

(i) Reflexive : As for any $x \in A$ we get

$|x - x| = 0$, which is divisible by 4

$\Rightarrow \quad (x, x) \in S, \forall x \in A$

$\therefore$ S is reflexive.

(ii) Symmetric : As for any $(x, y) \in S$, we get

$|x - y|$ is divisible by 4 (by using definition of given relation)

$\Rightarrow \quad |x - y| = 4\lambda$ for some $\lambda \in Z$

$\Rightarrow \quad |y - x| = 4\lambda$ for some $\lambda \in Z$

$\Rightarrow \quad (y, x) \in S$

$\Rightarrow \quad (y, x) \in S, \forall x, y \in Z$

Hence, S is symmetric.

(iii) Transitive : For any $(x, y) \in S$ and $(y, z) \in S$ we get $|x - y|$ is divisible by 4 and $|y - z|$ is divisible by 4 (by using definition of given relation)

$\Rightarrow |x - y| = 4\lambda$ and $|y - z| = 4\mu$ for some $\lambda, \mu \in Z$

Now, $\quad x - z = (x - y) + (y - z)$

$\quad = \pm 4\lambda \pm 4\mu = \pm 4(\lambda + \mu)$

$\Rightarrow |x - z|$ is divisible by 4

$\Rightarrow \quad (x, z) \in S$

$\therefore$ We have shown that if $(x, y) \in S$ and $(y, z) \in S$

$\Rightarrow \quad (x, z) \in S, \forall x, y, z \in Z$

Hence, S is transitive.

Since, S is reflexive, symmetric and transitive, so it is equivalence relation.

Q. 18. Show that the relation S defined on set $N \times N$ by $(a, b)S(c, d) \Rightarrow a + d = b + c$ is an equivalence relation.*

Sol. The given relation is

$$(a, b)S(c, d) \Rightarrow a + d = b + c \text{ for } a, b, c, d \in N$$

We have to show that the above relation is an equivalence relation. For this, we show that it is reflexive, symmetric and transitive.

(i) Reflexive : As $(a, b)S(a, b)$

$\Rightarrow a + b = b + a, \forall a, b \in N \times N$ which is true.

$\therefore$ S is reflexive.

(ii) Symmetric : As $(a, b)S(c, d)$

$\Rightarrow \quad a + d = b + c, \forall (a, b), (c, d) \in N \times N$

$\Rightarrow \quad d + a = c + b$

$\Rightarrow \quad c + b = d + a$

$\Rightarrow (c, d)S(a, b) \forall (a, b), (c, d) \in N \times N$

$\therefore$ S is symmetric.

(iii) Transitive : Let $(a, b)S(c, d)$

$\Rightarrow \quad a + d \Rightarrow b + c, \forall (a, b), (c, d) \in N \times N \quad \ldots$(i)

and $(c, d)S(e, f)$

$\Rightarrow \quad c + f = d + e, \forall (c, d), (e, f) \in N \times N \quad \ldots$(ii)

On adding equations (i) and (ii), we get

$$(a + d) + (c + f) = (b + c) + (d + e)$$

$\Rightarrow \quad a + f = b + e$

$\Rightarrow (a, b)S(e, f)$

$\therefore$ We have shown that if $(a, b)S(c, d)$ and $(c, d)S(e, f)$

$\Rightarrow (a, b)S(e, f), \forall (a, b), (c, d), (e, f) \in N \times N.$

Hence, S is transitive.

Since, S is reflexive, symmetric and transitive. So it is an equivalence relation.

Q. 19. Show that relation R in the set of real numbers, defined as $R = \{(a, b) : a \le b^2\}$ is neither reflexive, nor symmetric nor transitive.* **[NCERT]**

Sol. The given relation is

$$R = \{(a, b) : a \le b^2\}$$

We have to show that R is neither reflexive nor symmetric nor transitive.

(i) Reflexive : Since,

$$\frac{1}{2} \le \left(\frac{1}{2}\right)^2 \text{ is not true}$$

$\Rightarrow \quad \left(\frac{1}{2}, \frac{1}{2}\right) \notin R$

$\therefore$ R is not reflexive.

(ii) Symmetric : Since, $\frac{1}{2} \le (1)^2$ is true but $1 \le \left(\frac{1}{2}\right)^2$ is not true.

$\Rightarrow \quad \left(\frac{1}{2}, 1\right) \in R$ but $\left(1, \frac{1}{2}\right) \notin R$

$\therefore$ R is not symmetric.

($\because$ R is said to be symmetric, if $(x, y) \in R \Rightarrow (y, x) \in R \; \forall x, y \in$ domain)

(iii) Transitive : Since, $2 \le (-3)^2$ and $(-3) \le (1)^2$ are both true, but $2 \le (1)^2$ is not true.

* are board exam questions from previous years

$\Rightarrow (2, -3)$ and $(-3, 1) \in R$ but $(2, 1) \notin R$

$\therefore$ R is not transitive.

($\because$ R is said to be transitive, if $(x, y) \in R$ and $(y, z) \in R \Rightarrow (x, z) \in R, \forall x, y, z \in$ domain)

Q. 20. If $f(x) = \tan^{-1} x$, then find the relation in $f(x), f(y)$ and $f(x+y)$.

Sol. We have

$$f(x) = \tan^{-1} x$$

$$\therefore \qquad x = \tan f(x) \qquad ...(1)$$

From equation (1),

$$y = \tan f(y)$$

and $$x + y = \tan f(x+y)$$

Hence, $\tan f(x) + \tan f(y) = \tan f(x+y)$ is the required relation.

Chapter 2. Inverse Trigonometric Functions

Q. 1. Find the principal value branch of $\cos^{-1} x$.

Sol. We define,

$$\cos^{-1} x : [-1, 1] \rightarrow [0, \pi]$$

For every $x \in [-1, 1]$ there exists a unique $\theta \in [0, \pi]$ such that $\cos\theta = x$. Thus θ is called a principal value of $\cos^{-1}x$. However we could also define.

$$\cos^{-1} x : [-1, 1] \rightarrow [-\pi, 0]$$

or $$\cos^{-1} x : [-1, 1] \rightarrow [\pi, 2\pi]$$

such that $\cos\theta = x$, Infact we can find infinitely many such intervals. But only the value of θ lying in $[0, \pi]$ is called the principal value.

That is, the principal value of $\cos^{-1} x$ is the numerically least among all the values of $\cos^{-1} x$. The principal value branch of $\cos^{-1}x$ is $[0, \pi]$.

Q. 2. Find the value of

$$\tan \frac{1}{2}\left[\sin^{-1}\frac{2x}{1+x^2} + \cos^{-1}\frac{1-y^2}{1+y^2}\right].$$

Sol. $$\tan\frac{1}{2}\left[\sin^{-1}\frac{2x}{1+x^2} + \cos^{-1}\frac{1-y^2}{1+y^2}\right]$$

Put $x = \tan\theta$

$y = \tan\phi$

$$= \tan\frac{1}{2}\left[\sin^{-1}\frac{2\tan\theta}{1+\tan^2\theta} + \cos^{-1}\frac{1-\tan^2\phi}{1+\tan^2\phi}\right]$$

$$= \tan\frac{1}{2}[\sin^{-1}(\sin 2\theta) + \cos^{-1}(\cos 2\phi)]$$

$$= \tan\frac{1}{2}[2\theta + 2\phi]$$

$$= \tan(\theta + \phi)$$

$$= \tan(\tan^{-1}x + \tan^{-1}y)$$

$$= \tan\left[\tan^{-1}\frac{x+y}{1-xy}\right]$$

$$= \frac{x+y}{1-xy}$$

Q. 3. Find the domain and range of $\sin^{-1}x$.

Sol. $\sin : R \rightarrow R$ such that

$\sin\theta = x$ for all $\theta \in R$ is a many-one into function. This function cannot have an inverse.

Therefore, we restrict the domain to the interval $\left[-\frac{\pi}{2}, \frac{\pi}{2}\right]$ and codomain to the interval $[-1, 1]$. Then,

$$\sin : \left[-\frac{\pi}{2}, \frac{\pi}{2}\right] \rightarrow [-1, 1] \text{ is a one-one, onto function}$$

and is therefore, invertible.

The inverse of the sine function is defined as

$$\sin^{-1} : [-1, 1] \rightarrow \left[-\frac{\pi}{2}, \frac{\pi}{2}\right]$$

such that $\quad \sin^{-1}x = \theta$

$\Leftrightarrow \quad \sin\theta = x$

Thus, the domain of the function is $[-1, 1]$ and the range is $\left[-\frac{\pi}{2}, \frac{\pi}{2}\right]$.

Q. 4. Prove that : $\sin^{-1}\frac{3}{5} + \sin^{-1}\frac{8}{17} = \sin^{-1}\frac{77}{85}$

Sol. Use $\sin^{-1}x + \sin^{-1}y$

$$= \sin^{-1}\left[x\sqrt{1-y^2} + y\sqrt{1-x^2}\right]$$

$$\text{L.H.S.} = \sin^{-1}\frac{3}{5} + \sin^{-1}\frac{8}{17}$$

$$= \sin^{-1}\left[\frac{3}{5}\sqrt{1-\left(\frac{8}{17}\right)^2} + \frac{8}{17}\sqrt{1-\left(\frac{3}{5}\right)^2}\right]$$

$$= \sin^{-1}\left[\frac{3}{5}\sqrt{\frac{225}{289}} + \frac{8}{17}\sqrt{\frac{16}{25}}\right]$$

$$= \sin^{-1}\left[\frac{3}{5}\times\frac{15}{17} + \frac{8}{17}\times\frac{4}{5}\right]$$

$$= \sin^{-1}\left[\frac{77}{85}\right]$$

= R.H.S.

Q. 5. Prove that :

$$\cot^{-1}\left(\frac{ab+1}{a-b}\right)+\cot^{-1}\left(\frac{bc+1}{b-c}\right)+\cot^{-1}\left(\frac{ca+1}{c-a}\right)=0$$

Sol. $\text{L.H.S.} = \cot^{-1}\left(\frac{ab+1}{a-b}\right)+\cot^{-1}\left(\frac{bc+1}{b-c}\right)+\cot^{-1}\left(\frac{ca+1}{c-a}\right)$

But we know that

$$\cot^{-1}x = \tan^{-1}\frac{1}{x} \text{ and } \tan^{-1}\frac{x-y}{1+xy} = \tan^{-1}x-\tan^{-1}y$$

$$\text{L.H.S} = \tan^{-1}\left(\frac{a-b}{1+ab}\right)+\tan^{-1}\left(\frac{b-c}{1+bc}\right)+\tan^{-1}\left(\frac{c-a}{1+ac}\right)$$

$$= \tan^{-1}a-\tan^{-1}b+\tan^{-1}b-\tan^{-1}c+\tan^{-1}c-\tan^{-1}a$$

$$=0$$

$$=\text{R.H.S.}$$

Q. 6. Prove that : $3\tan^{-1}x=\tan^{-1}\left(\frac{3x-x^3}{1-3x^2}\right)$

Sol. Let $\tan^{-1}x=\theta$

$\Rightarrow \tan\theta = x$

Now, $\tan 3\theta = \frac{3\tan\theta-\tan^3\theta}{1-3\tan^2\theta}$

$$=\frac{3x-x^3}{1-3x^2}$$

$\Rightarrow 3\theta = \tan^{-1}\left(\frac{3x-x^3}{1-3x^2}\right)$

$\Rightarrow 3\tan^{-1}x = \tan^{-1}\left(\frac{3x-x^3}{1-3x^2}\right)$

Q. 7. If $\tan^{-1}\left(\frac{\sqrt{1+x^2}-\sqrt{1-x^2}}{\sqrt{1+x^2}+\sqrt{1-x^2}}\right)=\alpha$, then prove that $x^2=\sin 2\alpha$.

Sol. We have,

$$\tan^{-1}\left(\frac{\sqrt{1+x^2}-\sqrt{1-x^2}}{\sqrt{1+x^2}+\sqrt{1-x^2}}\right)=\alpha$$

$\Rightarrow \tan\alpha = \frac{\sqrt{1+x^2}-\sqrt{1-x^2}}{\sqrt{1+x^2}+\sqrt{1-x^2}}$

Applying componendo and dividendo, we get

$$\frac{\tan\alpha+1}{\tan\alpha-1} = \frac{2\sqrt{1+x^2}}{-2\sqrt{1-x^2}}$$

$\Rightarrow \frac{\tan\alpha+1}{\tan\alpha-1} = -\frac{\sqrt{1+x^2}}{\sqrt{1-x^2}}$

$\Rightarrow \frac{\frac{\sin\alpha}{\cos\alpha}+1}{\frac{\sin\alpha}{\cos\alpha}-1} = -\frac{\sqrt{1+x^2}}{\sqrt{1-x^2}}$

$\Rightarrow \frac{\sin\alpha+\cos\alpha}{\sin\alpha-\cos\alpha} = -\frac{\sqrt{1+x^2}}{\sqrt{1-x^2}}$

Squaring both sides, we get

$$\left(\frac{\sin\alpha+\cos\alpha}{\sin\alpha-\cos\alpha}\right)^2 = \frac{1+x^2}{1-x^2}$$

$\Rightarrow \frac{1+2\sin\alpha\cos\alpha}{1-2\sin\alpha\cos\alpha} = \frac{1+x^2}{1-x^2}$

Applying componendo and dividendo, we get

$$\frac{2}{2\sin 2\alpha} = \frac{2}{2x^2}$$

$\Rightarrow \sin 2\alpha = x^2$

Q. 8. Draw the graph of $\tan^{-1}x$.

Sol. Table for $\tan^{-1}x$

x	$-\sqrt{3}$	-1	$\frac{-1}{\sqrt{3}}$	0	$\frac{1}{\sqrt{3}}$	1	$\sqrt{3}$
$\tan^{-1}x$	$\frac{-\pi}{3}$	$\frac{-\pi}{4}$	$\frac{-\pi}{6}$	0	$\frac{\pi}{6}$	$\frac{\pi}{4}$	$\frac{\pi}{3}$

Graph of $\tan^{-1}x$

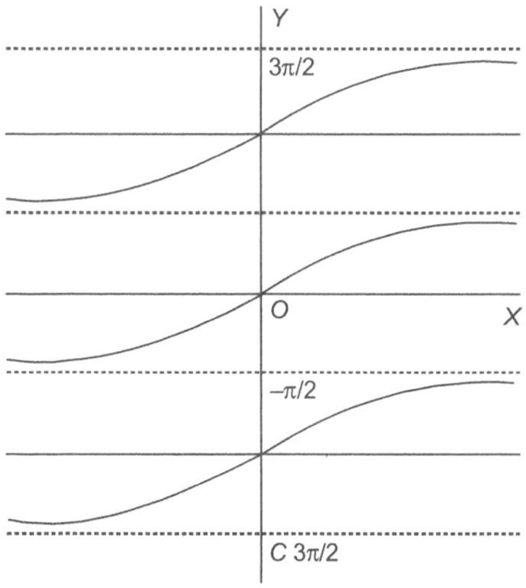

Q. 9. Write in the simplest form :

$$\tan^{-1}\left(\frac{x}{\sqrt{a^2-x^2}}\right)$$

Sol. Let $x = a\sin\theta$

$$\therefore \tan^{-1}\left(\frac{x}{\sqrt{a^2-x^2}}\right) = \tan^{-1}\left(\frac{a\sin\theta}{\sqrt{a^2-a^2\sin^2\theta}}\right)$$

$$= \tan^{-1}\left(\frac{a\sin\theta}{\sqrt{a^2\cos^2\theta}}\right)$$

$$= \tan^{-1}\left(\frac{a\sin\theta}{a\cos\theta}\right)$$

$$= \tan^{-1}(\tan\theta)$$

$$= \theta$$

$$\because \quad x = a\sin\theta$$

$$\Rightarrow \quad \sin\theta = \frac{x}{a}$$

$$\Rightarrow \quad \theta = \sin^{-1}\frac{x}{a}$$

$$\therefore \tan^{-1}\left(\frac{x}{\sqrt{a^2-x^2}}\right) = \sin^{-1}\frac{x}{a}$$

Q. 10. Solve the equation for x :★
$\sin^{-1}x + \sin^{-1}(1-x) = \cos^{-1}x$.

Sol. We have,

$$\sin^{-1}x + \sin^{-1}(1-x) = \cos^{-1}x$$

$$\sin^{-1}(x\sqrt{1-(1-x)^2} + (1-x)\sqrt{1-x^2}) = \cos^{-1}x$$

$$\sin^{-1}(x\sqrt{1-(1+x^2-2x)} + (1-x)\sqrt{1-x^2} = \cos^{-1}x$$

$$\sin^{-1}(x\sqrt{2x-x^2} + (1-x)\sqrt{1-x^2}) = \sin^{-1}(\sqrt{1-x^2})$$

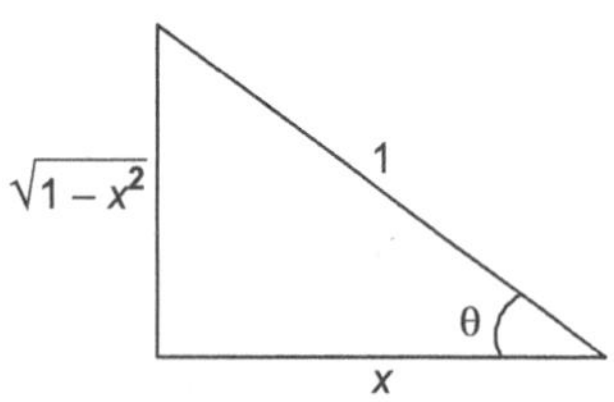

Let $\quad \cos^{-1}x = \theta$

$$x = \cos\theta$$

$$\sin\theta = \sqrt{1-x^2}$$

$$\theta = \sin^{-1}(\sqrt{1-x^2})$$

$$\Rightarrow x\sqrt{2x-x^2} + (1-x)\sqrt{1-x^2} = \sqrt{1-x^2}$$

$$\Rightarrow \quad x\sqrt{2x-x^2} = \sqrt{1-x^2}.(1-1+x)$$

$$\Rightarrow x\left[\sqrt{2x-x^2} - \sqrt{1-x^2}\right] = 0$$

$$\Rightarrow \quad x = 0$$

$$\text{or } \sqrt{2x-x^2} = \sqrt{1-x^2}$$

$$\Rightarrow \quad x = 0 \text{ or } 2x - x^2 = 1 - x^2$$

$$\Rightarrow \quad x = 0 \text{ or } x = \frac{1}{2}$$

Q. 11. Prove the following :★★

$$\cot^{-1}\left[\frac{\sqrt{1+\sin x}+\sqrt{1-\sin x}}{\sqrt{1+\sin x}-\sqrt{1-\sin x}}\right] = \frac{x}{2}; x \in \left(0, \frac{\pi}{4}\right).$$

Sol. L.H.S.

$$= \cot^{-1}\left[\frac{\sqrt{1+\sin x}+\sqrt{1-\sin x}}{\sqrt{1+\sin x}-\sqrt{1-\sin x}}\right]$$

$$= \cot^{-1}\left[\frac{\sqrt{\left(\cos\frac{x}{2}+\sin\frac{x}{2}\right)^2}+\sqrt{\left(\cos\frac{x}{2}-\sin\frac{x}{2}\right)^2}}{\sqrt{\left(\cos\frac{x}{2}+\sin\frac{x}{2}\right)^2}-\sqrt{\left(\cos\frac{x}{2}-\sin\frac{x}{2}\right)^2}}\right]$$

$$= \cot^{-1}\left[\frac{\cos\frac{x}{2}+\sin\frac{x}{2}+\cos\frac{x}{2}-\sin\frac{x}{2}}{\cos\frac{x}{2}+\sin\frac{x}{2}-\cos\frac{x}{2}+\sin\frac{x}{2}}\right]$$

$$= \cot^{-1}\left[\frac{2\cos\frac{x}{2}}{2\sin\frac{x}{2}}\right]$$

$$= \cot^{-1}\left(\cot\frac{x}{2}\right) = \frac{x}{2}$$

$$\therefore \quad \text{L.H.S.} = \text{R.H.S.}$$

Q. 12. Prove that :★

$$\tan^{-1}\left[\frac{\sqrt{1+x}-\sqrt{1-x}}{\sqrt{1+x}+\sqrt{1-x}}\right] = \frac{\pi}{4} - \frac{1}{2}\cos^{-1}x; -\frac{1}{\sqrt{2}} \le x \le 1.$$

Sol. L.H.S. $= \tan^{-1}\left[\frac{\sqrt{1+x}-\sqrt{1-x}}{\sqrt{1+x}+\sqrt{1-x}}\right]$

$$= \tan^{-1}\left[\frac{\sqrt{1+\cos 2\theta}-\sqrt{1-\cos 2\theta}}{\sqrt{1+\cos 2\theta}+\sqrt{1-\cos 2\theta}}\right]$$

(put $x = \cos 2\theta$)

$$= \tan^{-1}\left[\frac{\sqrt{2\cos^2\theta}-\sqrt{2\sin^2\theta}}{\sqrt{2\cos^2\theta}+\sqrt{2\sin^2\theta}}\right]$$

$$= \tan^{-1}\left[\frac{\sqrt{2}\cos\theta-\sqrt{2}\sin\theta}{\sqrt{2}\cos\theta+\sqrt{2}\sin\theta}\right]$$

$$= \tan^{-1}\left[\frac{\cos\theta-\sin\theta}{\cos\theta+\sin\theta}\right]$$

Dividing numerator and denominator by cos θ, we get

$$\text{L.H.S.} = \tan^{-1}\left[\frac{\frac{\cos\theta}{\cos\theta}-\frac{\sin\theta}{\cos\theta}}{\frac{\cos\theta}{\cos\theta}+\frac{\sin\theta}{\cos\theta}}\right]$$

$$= \tan^{-1}\left[\frac{1-\tan\theta}{1+\tan\theta}\right]$$

★ are board exam questions from previous years
★★ are frequently asked board exam questions

$$= \tan^{-1}\left[\frac{\tan\frac{\pi}{4} - \tan\theta}{1 + \tan\theta\tan\frac{\pi}{4}}\right]$$

$$= \tan^{-1}\tan\left(\frac{\pi}{4} - \theta\right)$$

$$= \frac{\pi}{4} - \theta$$

$$= \frac{\pi}{4} - \frac{1}{2}\cos^{-1}x \quad \left(\because\ \theta = \frac{1}{2}\cos^{-1}x\right)$$

= R.H.S.

L.H.S. = R.H.S.

Q. 13. Prove the following :**

$$\mathbf{2\tan^{-1}\left(\frac{1}{2}\right) + \tan^{-1}\left(\frac{1}{7}\right) = \tan^{-1}\left(\frac{31}{17}\right).}$$

Sol. L.H.S. $= 2\tan^{-1}\left(\frac{1}{2}\right) + \tan^{-1}\left(\frac{1}{7}\right)$

$$= \tan^{-1}\left[\frac{2\times\frac{1}{2}}{1-\left(\frac{1}{2}\right)^2}\right] + \tan^{-1}\left(\frac{1}{7}\right)$$

$$= \tan^{-1}\left[\frac{1}{1-\frac{1}{4}}\right] + \tan^{-1}\left(\frac{1}{7}\right)$$

$$= \tan^{-1}\left[\frac{1}{\frac{3}{4}}\right] + \tan^{-1}\left(\frac{1}{7}\right)$$

$$= \tan^{-1}\frac{4}{3} + \tan^{-1}\frac{1}{7}$$

$$= \tan^{-1}\left[\frac{\frac{4}{3}+\frac{1}{7}}{1-\frac{4}{3}\times\frac{1}{7}}\right]$$

$$= \tan^{-1}\left[\frac{\frac{28+3}{21}}{\frac{21-4}{21}}\right]$$

$$= \tan^{-1}\frac{31}{17}$$

= R.H.S.

Q. 14. Prove that :**

$$\mathbf{\tan^{-1}\left(\frac{1}{2}\right) + \tan^{-1}\left(\frac{1}{5}\right) + \tan^{-1}\left(\frac{1}{8}\right) = \frac{\pi}{4}.}$$

Sol. L.H.S. $= \tan^{-1}\left(\frac{1}{2}\right) + \tan^{-1}\left(\frac{1}{5}\right) + \tan^{-1}\left(\frac{1}{8}\right)$

$$= \tan^{-1}\left[\frac{\frac{1}{2}+\frac{1}{5}}{1-\frac{1}{2}\times\frac{1}{5}}\right] + \tan^{-1}\left(\frac{1}{8}\right)$$

$$= \tan^{-1}\left[\frac{5+2}{10-1}\right] + \tan^{-1}\left(\frac{1}{8}\right)$$

$$= \tan^{-1}\left(\frac{7}{9}\right) + \tan^{-1}\left(\frac{1}{8}\right)$$

$$= \tan^{-1}\left[\frac{\frac{7}{9}+\frac{1}{8}}{1-\frac{7}{9}\times\frac{1}{8}}\right]$$

$$= \tan^{-1}\left[\frac{56+9}{72-7}\right]$$

$$= \tan^{-1}\left[\frac{65}{65}\right]$$

$$= \tan^{-1}(1) = \frac{\pi}{4}$$

$\therefore$ L.H.S. = R.H.S.

Q. 15. Solve the following equation for x :**

$$\mathbf{\tan^{-1}\left(\frac{1-x}{1+x}\right) = \frac{1}{2}\tan^{-1}x;\ x > 0.}$$

Sol. Given equation is

$$\tan^{-1}\left(\frac{1-x}{1+x}\right) = \frac{1}{2}\tan^{-1}x;\ x > 0$$

$$\Rightarrow \quad 2\tan^{-1}\left(\frac{1-x}{1+x}\right) = \tan^{-1}x$$

$$\Rightarrow \quad \tan^{-1}\left[\frac{2\left(\frac{1-x}{1+x}\right)}{1-\left(\frac{1-x}{1+x}\right)^2}\right] = \tan^{-1}x$$

$$\Rightarrow \quad \tan^{-1}\left[\frac{2(1-x)(1+x)}{(1+x)^2-(1-x)^2}\right] = \tan^{-1}x$$

$$\Rightarrow \quad \tan^{-1}\left[\frac{2-2x^2}{4x}\right] = \tan^{-1}x$$

$$\Rightarrow \quad \frac{1-x^2}{2x} = x$$

** are frequently asked board exam questions

$$\Rightarrow \quad 1 - x^2 = 2x^2$$

$$\Rightarrow \quad 3x^2 = 1$$

$$\Rightarrow \quad x^2 = \frac{1}{3}$$

$$\therefore \quad x = \pm\frac{1}{\sqrt{3}}$$

But given $x > 0$,

$$\therefore \quad x = \frac{1}{\sqrt{3}}$$

Q. 16. Solve for x : $\cos(2\sin^{-1}x) = \frac{1}{9}$, $x > 0$.★

Sol. Given equation is $\cos(2\sin^{-1}x) = \frac{1}{9}, x > 0$ ...(i)

We put $\sin^{-1}x = y$

$$\Rightarrow \quad x = \sin y$$

Equation (i) becomes

$$\cos 2y = \frac{1}{9}$$

$$\Rightarrow \quad 1 - 2\sin^2 y = \frac{1}{9}$$

$$\Rightarrow \quad 2\sin^2 y = 1 - \frac{1}{9}$$

$$= \frac{8}{9}$$

$$\Rightarrow \quad \sin^2 y = \frac{4}{9}$$

$$\Rightarrow \quad x^2 = \frac{4}{9}$$

$$\therefore \quad x = \pm\frac{2}{3}$$

But given that $x > 0$

$$\therefore \quad x = \frac{2}{3}$$

Q. 17. Prove that $\tan^{-1}\sqrt{x} = \frac{1}{2}\cos^{-1}\left(\frac{1-x}{1+x}\right)$, $x \in (0, 1)$.★

Sol. $\text{R.H.S.} = \frac{1}{2}\cos^{-1}\left(\frac{1-x}{1+x}\right)$

$$= \frac{1}{2}\cos^{-1}\left(\frac{1-(\sqrt{x})^2}{1+(\sqrt{x})^2}\right)$$

Substituting $\sqrt{x} = \tan\theta$

$$= \frac{1}{2}\cos^{-1}\left(\frac{1-\tan^2\theta}{1+\tan^2\theta}\right)$$

$$= \frac{1}{2}\cos^{-1}(\cos 2\theta)$$

$\because$ The principal value of $\cos^{-1}x$ is $[0, \pi]$.

$$= \frac{1}{2}(2\theta)$$

$$= \theta$$

$$= \tan^{-1}\sqrt{x}$$

$$= \text{L.H.S.}$$

$$\therefore \quad \text{L.H.S.} = \text{R.H.S.}$$

Q. 18. Prove the following :★

$$\cos^{-1}\left(\frac{12}{13}\right) + \sin^{-1}\left(\frac{3}{5}\right) = \sin^{-1}\left(\frac{56}{65}\right).$$

Sol. Let $\cos^{-1}\left(\frac{12}{13}\right) = x$

$$\sin^{-1}\left(\frac{3}{5}\right) = y$$

so $\cos x = \frac{12}{13}$

and $\sin y = \frac{3}{5}$

$$\therefore \quad \sin^2 x = 1 - \cos^2 x$$

$$\Rightarrow \quad \sin^2 x = 1 - \frac{144}{169} = \frac{25}{169}$$

$$\sin x = \frac{5}{13}$$

and $\cos^2 y = 1 - \sin^2 y$

$$= 1 - \frac{9}{25} = \frac{16}{25}$$

and $\cos y = \frac{4}{5}$

Now, $\sin(x+y) = \sin x\cos y + \cos x\sin y$

$$\Rightarrow \quad \sin(x+y) = \left(\frac{5}{13}\times\frac{4}{5}\right) + \left(\frac{12}{13}\times\frac{3}{5}\right)$$

$$= \frac{20}{65} + \frac{36}{65} = \frac{56}{65}$$

$$\therefore \quad \sin(x+y) = \frac{56}{65}$$

$$\Rightarrow \quad x + y = \sin^{-1}\frac{56}{65}$$

★ are board exam questions from previous years

$\because \quad x = \cos^{-1}\frac{12}{13}$

and $\quad y = \sin^{-1}\frac{3}{5}$

$\therefore \quad \cos^{-1}\frac{12}{13} + \sin^{-1}\frac{3}{5} = \sin^{-1}\frac{56}{65}$

Q. 19. Prove that :*

$$\cos\left[\tan^{-1}\left\{\sin\left(\cot^{-1}x\right)\right\}\right] = \sqrt{\frac{1+x^2}{2+x^2}}$$

Sol. $\quad \text{L.H.S.} = \cos\left[\tan^{-1}\left\{\sin\left(\cot^{-1}x\right)\right\}\right]$

Let $\quad \cot^{-1}x = \theta$

$\Rightarrow \quad x = \cot\theta$

Then, given expression can be written as

$$\cos\left[\tan^{-1}\{\sin\theta\}\right] = \cos\left[\tan^{-1}\left(\frac{1}{\operatorname{cosec}\theta}\right)\right]$$

$$= \cos\left[\tan^{-1}\left(\frac{1}{\sqrt{1+\cot^2\theta}}\right)\right]$$

$(\because \operatorname{cosec}^2\theta = 1 + \cot^2\theta)$

$$= \cos\left[\tan^{-1}\left(\frac{1}{\sqrt{1+x^2}}\right)\right]$$

$$= \cos\phi \qquad ...(i)$$

where,

$$\tan^{-1}\left(\frac{1}{\sqrt{1+x^2}}\right) = \phi$$

or $\quad \tan\phi = \dfrac{1}{\sqrt{1+x^2}}$

From equation (i)

$$\cos[\tan^{-1}(\sin\theta)] = \frac{1}{\sec\phi} \qquad \left[\because \cos\phi = \frac{1}{\sec\phi}\right]$$

$$= \frac{1}{\sqrt{1+\tan^2\phi}}$$

$[\because \tan^2\phi + 1 = \sec^2\phi]$

$$= \frac{1}{\sqrt{1+\dfrac{1}{1+x^2}}}$$

$$= \frac{1}{\sqrt{\dfrac{1+x^2+1}{1+x^2}}}$$

$$= \sqrt{\frac{1+x^2}{2+x^2}}$$

$= \text{R.H.S.}$

$\therefore \quad \text{L.H.S.} = \text{R.H.S.}$

Q. 20. Prove that :**

$$\cos^{-1}\frac{4}{5} + \cos^{-1}\frac{12}{13} = \cos^{-1}\frac{33}{65}.$$

Sol. Let $x = \cos^{-1}\frac{4}{5}$ and $y = \cos^{-1}\frac{12}{13}$

$\Rightarrow \quad \cos x = \frac{4}{5}$ and $\cos y = \frac{12}{13}$

As we know that

$$\sin^2\theta = 1 - \cos^2\theta$$

$\Rightarrow \quad \sin\theta = \sqrt{1-\cos^2\theta}$

$\therefore \quad \sin x = \sqrt{1-\cos^2 x}$

$$= \sqrt{1-\frac{16}{25}} = \frac{3}{5}$$

and $\quad \sin y = \sqrt{1-\cos^2 y}$

$$= \sqrt{1-\frac{144}{169}}$$

$$= \frac{5}{13}$$

Now, using the identitiy

$$\cos(x+y) = \cos x\cos y - \sin x\sin y$$

We get,

$$\cos(x+y) = \left(\frac{4}{5}\times\frac{12}{13}\right) - \left(\frac{3}{5}\times\frac{5}{13}\right)$$

$$\cos(x+y) = \frac{48}{65} - \frac{15}{65}$$

$\Rightarrow \quad \cos(x+y) = \dfrac{33}{65}$

or $\quad x + y = \cos^{-1}\dfrac{33}{65}$

$\therefore \quad \cos^{-1}\frac{4}{5} + \cos^{-1}\frac{12}{13} = \cos^{-1}\frac{33}{65}$

$$\left[\because x = \cos^{-1}\frac{4}{5} \text{ and } y = \cos^{-1}\frac{12}{13}\right]$$

Q. 21. Solve for x :*

$2\tan^{-1}(\cos x) = \tan^{-1}(2\operatorname{cosec} x)$.

Sol. We have,

$$2\tan^{-1}(\cos x) = \tan^{-1}(2\operatorname{cosec} x)$$

* are board exam questions from previous years
** are frequently asked board exam questions

$$\Rightarrow \tan^{-1}\left(\frac{2\cos x}{1-\cos^2 x}\right) = \tan^{-1}\left(\frac{2}{\sin x}\right)$$

$$\left\{\because 2\tan^{-1} x = \tan^{-1}\left(\frac{2x}{1-x^2}\right)\right\}$$

$$\Rightarrow \frac{2\cos x}{\sin^2 x} = \frac{2}{\sin x} \quad ...(i)$$

$$\Rightarrow \sin x.\cos x - \sin^2 x = 0$$

$$\Rightarrow \sin x(\cos x - \sin x) = 0$$

$$\Rightarrow \sin x = 0 \text{ or } \cos x = \sin x$$

$$\Rightarrow \sin x = \sin 0 \text{ or } \cot x = 1 = \cot\frac{\pi}{4}$$

$$\Rightarrow x = 0 \text{ or } \frac{\pi}{4}$$

But here at $x = 0$, the given equation does not exist at equation (i)

$$\therefore \quad x = \frac{\pi}{4} \text{ is the only solution.}$$

$$\Rightarrow x = \frac{\pi}{4}.$$

Q. 22. Prove that :*

$$\sin^{-1}\left(\frac{4}{5}\right) + \sin^{-1}\left(\frac{5}{13}\right) + \sin^{-1}\left(\frac{16}{65}\right) = \frac{\pi}{2}.$$

Sol. L.H.S.

$$= \sin^{-1}\left(\frac{4}{5}\right) + \sin^{-1}\left(\frac{5}{13}\right) + \sin^{-1}\left(\frac{16}{65}\right)$$

$$= \sin^{-1}\left[\frac{4}{5}\sqrt{1-\frac{25}{169}} + \frac{5}{13}\sqrt{1-\frac{16}{25}}\right] + \sin^{-1}\left(\frac{16}{65}\right)$$

$$\left(\because \sin^{-1} x + \sin^{-1} y = \sin^{-1}\left[x\sqrt{1-y^2} + y\sqrt{1-x^2}\right]\right)$$

$$= \sin^{-1}\left[\frac{4}{5}\sqrt{\frac{144}{169}} + \frac{5}{13}\sqrt{\frac{9}{25}}\right] + \sin^{-1}\left(\frac{16}{65}\right)$$

$$= \sin^{-1}\left[\left(\frac{4}{5}\times\frac{12}{13}\right) + \left(\frac{5}{13}\times\frac{3}{5}\right)\right] + \sin^{-1}\left(\frac{16}{65}\right)$$

$$= \sin^{-1}\left(\frac{48}{65} + \frac{15}{65}\right) + \sin^{-1}\left(\frac{16}{65}\right)$$

$$= \sin^{-1}\frac{63}{65} + \sin^{-1}\frac{16}{65}$$

$$= \sin^{-1}\left[\frac{63}{65}\sqrt{1-\left(\frac{16}{65}\right)^2} + \frac{16}{65}\sqrt{1-\left(\frac{63}{65}\right)^2}\right]$$

$$= \sin^{-1}\left[\frac{63}{65}\sqrt{\frac{4225-256}{4225}} + \frac{16}{65}\sqrt{\frac{4225-3969}{4225}}\right]$$

$$= \sin^{-1}\left[\frac{63}{65}\times\sqrt{\frac{3969}{4225}} + \frac{16}{65}\times\sqrt{\frac{256}{4225}}\right]$$

$$= \sin^{-1}\left[\frac{63}{65}\times\frac{63}{65} + \frac{16}{65}\times\frac{16}{65}\right]$$

$$= \sin^{-1}\left(\frac{3969+256}{4225}\right)$$

$$= \sin^{-1}\left(\frac{4225}{4225}\right)$$

$$= \sin^{-1}(1)$$

$\because$ The principal value of $\sin^{-1} x$ is $\left[-\frac{\pi}{2}, \frac{\pi}{2}\right]$

$$= \sin^{-1}\left(\sin\frac{\pi}{2}\right) = \frac{\pi}{2}$$

= R.H.S.

$\therefore$ L.H.S. = R.H.S.

Q. 23. Solve for x :**

$$\tan^{-1} 3x + \tan^{-1} 2x = \frac{\pi}{4}$$

Sol. We have

$$\tan^{-1} 3x + \tan^{-1} 2x = \frac{\pi}{4}$$

$$\Rightarrow \tan^{-1}\left(\frac{2x+3x}{1-(2x)(3x)}\right) = \tan^{-1}(1)$$

$$\Rightarrow \tan^{-1}\left(\frac{5x}{1-6x^2}\right) = \tan^{-1}(1) \text{ if } 6x^2 < 1$$

$$\Rightarrow \frac{5x}{1-6x^2} = 1, \text{ if } 6x^2 < 1$$

$$\Rightarrow 5x = 1 - 6x^2 \text{ if } x^2 < \frac{1}{6}$$

$$\Rightarrow 6x^2 + 5x - 1 = 0, \text{ if } -\frac{1}{\sqrt{6}} < x < \frac{1}{\sqrt{6}}$$

$$\Rightarrow (x+1)(6x-1) = 0 \text{ if } -\frac{1}{\sqrt{6}} < x < \frac{1}{\sqrt{6}}$$

$$\Rightarrow x = -1 \text{ or } \frac{1}{6}, -\frac{1}{\sqrt{6}} < x < \frac{1}{\sqrt{6}}$$

$$\Rightarrow x = -1 \text{ is not root of given equation}$$

$$\therefore \quad x = \frac{1}{6}$$

** are frequently asked board exam questions

Q. 24. Prove that :★

$$\tan^{-1}\frac{3}{4}+\tan^{-1}\frac{3}{5}-\tan^{-1}\frac{8}{19}=\frac{\pi}{4}$$

Sol. L.H.S. $=\left[\tan^{-1}\frac{3}{4}+\tan^{-1}\frac{3}{5}\right]-\tan^{-1}\frac{8}{19}$

Applying $\tan^{-1}x+\tan^{-1}y=\tan^{-1}\left(\frac{x-y}{1+xy}\right)$.

we get

$$\tan^{-1}\left[\frac{\frac{3}{4}+\frac{3}{5}}{1-\frac{9}{20}}\right]-\tan^{-1}\frac{8}{19}=\tan^{-1}\left[\frac{\frac{27}{20}}{\frac{11}{20}}\right]-\tan^{-1}\frac{8}{19}$$

$$=\tan^{-1}\left(\frac{27}{11}\right)-\tan^{-1}\left(\frac{8}{19}\right)$$

Now, applying the identity $\tan^{-1}x-\tan^{-1}y$

$=\tan^{-1}\left(\frac{x-y}{1+xy}\right)$, we get

$$=\tan^{-1}\left[\frac{\frac{27}{11}-\frac{8}{19}}{1+\frac{27}{11}\times\frac{8}{19}}\right]$$

$$=\tan^{-1}\left[\frac{\frac{513-88}{209}}{\frac{209+216}{209}}\right]$$

$$=\tan^{-1}\left(\frac{425}{209}\times\frac{209}{425}\right)$$

$$=\tan^{-1}(1)$$

$\because$ The principal value of $\tan^{-1}x$ is $\left(-\frac{\pi}{2},\frac{\pi}{2}\right)$

$$=\tan^{-1}\left(\tan\frac{\pi}{4}\right)$$

$$=\left(\frac{\pi}{4}\right)$$

$$=\text{R.H.S.}$$

$\therefore$ L.H.S. = R.H.S.

Q. 25. Prove that :★★

$$\tan^{-1}\left(\frac{1}{3}\right)+\tan^{-1}\left(\frac{1}{5}\right)+\tan^{-1}\left(\frac{1}{7}\right)+\tan^{-1}\left(\frac{1}{8}\right)=\frac{\pi}{4}.$$

Sol. L.H.S.

$$=\tan^{-1}\left(\frac{1}{3}\right)+\tan^{-1}\left(\frac{1}{5}\right)+\tan^{-1}\left(\frac{1}{7}\right)+\tan^{-1}\left(\frac{1}{8}\right)$$

Applying the result $\tan^{-1}x+\tan^{-1}y=\tan^{-1}\left(\frac{x+y}{1-xy}\right)$

we get

$$\text{L.H.S.}=\tan^{-1}\left[\frac{\frac{1}{3}+\frac{1}{5}}{1-\frac{1}{15}}\right]+\tan^{-1}\left[\frac{\frac{1}{7}+\frac{1}{8}}{1-\frac{1}{56}}\right]$$

$$=\tan^{-1}\left[\frac{\frac{8}{15}}{\frac{14}{15}}\right]+\tan^{-1}\left[\frac{\frac{15}{56}}{\frac{55}{56}}\right]$$

$$=\tan^{-1}\left(\frac{4}{7}\right)+\tan^{-1}\left(\frac{3}{11}\right)$$

$$=\tan^{-1}\left[\frac{\frac{4}{7}+\frac{3}{11}}{1-\frac{12}{77}}\right]$$

$\left[\text{By applying } \tan^{-1}x+\tan^{-1}y=\tan^{-1}\left(\frac{x+y}{1-xy}\right)\right]$

$$=\tan^{-1}\left[\frac{\frac{65}{77}}{\frac{65}{77}}\right]$$

$$=\tan^{-1}(1)$$

$\because$ The principal value of $\tan^{-1}x$ is $\left(-\frac{\pi}{2},\frac{\pi}{2}\right)$

$$=\tan^{-1}\left(\tan\frac{\pi}{4}\right)$$

$$=\frac{\pi}{4}$$

$$=\text{R.H.S.}$$

$\therefore$ L.H.S. = R.H.S.

Q. 26. If $\tan^{-1}\frac{x-3}{x-4}+\tan^{-1}\frac{x+3}{x+4}=\frac{\pi}{4}$, then find the value of x.★

Sol. We have,

$$\tan^{-1}\left(\frac{x-3}{x-4}\right)+\tan^{-1}\left(\frac{x+3}{x+4}\right)=\frac{\pi}{4}$$

$$\Rightarrow \tan^{-1}\left(\frac{x-3}{x-4}\right)+\tan^{-1}\left(\frac{x+3}{x+4}\right)=\tan^{-1}1$$

$$\Rightarrow \tan^{-1}\left(\frac{x-3}{x-4}\right)=\tan^{-1}1-\tan^{-1}\left(\frac{x+3}{x+4}\right)$$

★ are board exam questions from previous years
★★ are frequently asked board exam questions

$$\Rightarrow \quad \tan^{-1}\left(\frac{x-3}{x-4}\right) = \tan^{-1}\left(\frac{1-\frac{x+3}{x+4}}{1+1\times\frac{x+3}{x+4}}\right)$$

$$\Rightarrow \quad \tan^{-1}\left(\frac{x-3}{x-4}\right) = \tan^{-1}\left(\frac{x+4-x-3}{x+4+x+3}\right)$$

$$\Rightarrow \quad \frac{x-3}{x-4} = \frac{1}{2x+7}$$

$$\Rightarrow \quad (2x+7)(x-3) = x-4$$

$$\Rightarrow \quad 2x^2-6x+7x-21 = x-4$$

$$\Rightarrow \quad 2x^2+x-21 = x-4$$

$$\Rightarrow \quad 2x^2 = 17$$

$$\Rightarrow \quad x = \pm\sqrt{\frac{17}{2}}.$$

Q. 27. Solve for x :★

$$\mathbf{\tan^{-1}\left(\frac{1+x}{1-x}\right) = \frac{\pi}{4} + \tan^{-1}x, 0 < x < 1.}$$

Sol. The given equation is

$$\tan^{-1}\left(\frac{1+x}{1-x}\right) = \frac{\pi}{4} + \tan^{-1}x, 0 < x < 1$$

$$\Rightarrow \quad \tan^{-1}\left(\frac{1+x}{1-x}\right) - \tan^{-1}x = \frac{\pi}{4}$$

$$\Rightarrow \quad \tan^{-1}\left[\frac{\frac{1+x}{1-x}-x}{1+\left(\frac{1+x}{1-x}\right)x}\right] = \frac{\pi}{4}$$

$$\left[\because \tan^{-1}x + \tan^{-1}y = \tan^{-1}\left(\frac{x+y}{1-xy}\right)\right]$$

$$\Rightarrow \quad \frac{1+x-x+x^2}{1-x+x+x^2} = \tan\frac{\pi}{4}$$

$$\Rightarrow \quad \frac{1+x^2}{1+x^2} = 1 \qquad \left[\because \tan\frac{\pi}{4} = 1\right]$$

$$\Rightarrow \quad 1 = 1$$

$\Rightarrow$ The given equation has many solutions as it is independent of x. Hence given equation will be satisfied for all $x \notin (0,1)$

Hence solution of

$$\tan^{-1}\left(\frac{1+x}{1-x}\right) = \frac{\pi}{4} + \tan^{-1}x, 0 < x < 1 \text{ is } x \in (0,1)$$

Q. 28. Prove that★

$$\mathbf{2\tan^{-1}\left(\frac{1}{5}\right) + \sec^{-1}\left(\frac{5\sqrt{2}}{7}\right) + 2\tan^{-1}\left(\frac{1}{8}\right) = \frac{\pi}{4}.}$$

★ **are board exam questions from previous years**

Sol. L.H.S.

$$= 2\tan^{-1}\left(\frac{1}{5}\right) + \sec^{-1}\left(\frac{5\sqrt{2}}{7}\right) + 2\tan^{-1}\left(\frac{1}{8}\right)$$

$$= 2\left[\tan^{-1}\left(\frac{1}{5}\right) + \tan^{-1}\left(\frac{1}{8}\right)\right] + \sec^{-1}\left(\frac{5\sqrt{2}}{7}\right)$$

$$= 2\tan^{-1}\left(\frac{\frac{1}{5}+\frac{1}{8}}{1-\frac{1}{5}\times\frac{1}{8}}\right) + \tan^{-1}\left(\sqrt{\left(\frac{5\sqrt{2}}{7}\right)^2 - 1}\right)$$

$$\left[\begin{aligned}\because \quad \tan^{-1}x + \tan^{-1}y &= \tan^{-1}\left(\frac{x+y}{1-xy}\right)\\ \because \quad \sec^{-1}x &= \tan^{-1}\sqrt{x^2-1}\end{aligned}\right]$$

$$= 2\tan^{-1}\left(\frac{13/40}{39/40}\right) + \tan^{-1}\left(\sqrt{\frac{1}{49}}\right)$$

$$= 2\tan^{-1}\left(\frac{13}{39}\right) + \tan^{-1}\left(\frac{1}{7}\right)$$

$$= 2\tan^{-1}\left(\frac{1}{3}\right) + \tan^{-1}\left(\frac{1}{7}\right)$$

$$= \tan^{-1}\left[\frac{2\times\frac{1}{3}}{1-\left(\frac{1}{3}\right)^2}\right] + \tan^{-1}\left(\frac{1}{7}\right)$$

$$\left[\because 2\tan^{-1}x = \tan^{-1}\left(\frac{2x}{1-x^2}\right)\right]$$

$$= \tan^{-1}\left(\frac{2/3}{8/9}\right) + \tan^{-1}\left(\frac{1}{7}\right)$$

$$= \tan^{-1}\left(\frac{3}{4}\right) + \tan^{-1}\left(\frac{1}{7}\right)$$

$$= \tan^{-1}\left(\frac{\frac{3}{4}+\frac{1}{7}}{1-\frac{3}{4}\times\frac{1}{7}}\right)$$

$$= \tan^{-1}\left(\frac{\frac{25}{28}}{\frac{25}{28}}\right)$$

$$= \tan^{-1}(1)$$

$$= \frac{\pi}{4} = \text{R.H.S.}$$

Q. 29. If $\sin[\cot^{-1}(x+1)] = \cos(\tan^{-1}x)$, then find x.★

Sol. Given, $\sin[\cot^{-1}(x+1)] = \cos(\tan^{-1}x)$

$$\sin\left\{\sin^{-1}\frac{1}{\sqrt{1+(x+1)^2}}\right\} = \cos\left\{\cos^{-1}\frac{1}{\sqrt{1+x^2}}\right\}$$

$$\left[\because \cot^{-1} x = \sin^{-1}\frac{1}{\sqrt{1+x^2}} \text{ and } \tan^{-1} x = \cos^{-1}\frac{1}{\sqrt{1+x^2}}\right]$$

$$\frac{1}{\sqrt{1+(x+1)^2}} = \frac{1}{\sqrt{1+x^2}}$$

$$\frac{1}{\sqrt{1+x^2+1+2x}} = \frac{1}{\sqrt{1+x^2}}$$

$$\frac{1}{\sqrt{x^2+2x+2}} = \frac{1}{\sqrt{1+x^2}}$$

$$\sqrt{1+x^2} = \sqrt{x^2+2x+2}$$

On squaring both sides, we get

$$1+x^2 = x^2+2x+2$$

$$\Rightarrow \quad 2x+2 = 1$$

$$x = \frac{-1}{2}$$

Q. 30. Prove that*

$$\tan\left\{\frac{\pi}{4}+\frac{1}{2}\cos^{-1}\frac{a}{b}\right\} + \tan\left\{\frac{\pi}{4}-\frac{1}{2}\cos^{-1}\frac{a}{b}\right\} = \frac{2b}{a}$$

Sol. Let $\frac{1}{2}\cos^{-1}\frac{a}{b} = A$

$$\text{L.H.S.} = \tan^{-1}\left(\frac{\pi}{4}+A\right)+\tan\left(\frac{\pi}{4}-A\right)$$

$$= \frac{\tan\frac{\pi}{4}+\tan A}{1-\tan\frac{\pi}{4}\tan A} + \frac{\tan\frac{\pi}{4}-\tan A}{1+\tan\frac{\pi}{4}\tan A}$$

$$= \frac{1+\tan A}{1-\tan A}+\frac{1-\tan A}{1+\tan A}$$

$$= \frac{(1+\tan A)^2+(1-\tan A)^2}{1-\tan^2 A}$$

$$= \frac{2+2\tan^2 A}{1-\tan^2 A}$$

$$= \frac{2(1+\tan^2 A)}{1-\tan^2 A} = 2\times\frac{1}{\cos 2A}$$

$$= 2\times\frac{1}{\cos 2\times\left(\frac{1}{2}\cos^{-1}\frac{a}{b}\right)}$$

$$= \frac{2}{\cos\left(\cos^{-1}\frac{a}{b}\right)} = \frac{2}{a/b}$$

$$= \frac{2b}{a}$$

* are board exam questions from previous years

Chapter 3. Matrices

Q. 1. Find the inverse of matrix A by using elementary row transformation, if $A = \begin{bmatrix} 2 & 6 \\ 4 & 14 \end{bmatrix}$.

Sol. Given, $\quad A = \begin{bmatrix} 2 & 6 \\ 4 & 14 \end{bmatrix}$

Matrix A can be further written as

$$A = IA$$

or $$\begin{bmatrix} 2 & 6 \\ 4 & 14 \end{bmatrix} = \begin{bmatrix} 1 & 0 \\ 0 & 1 \end{bmatrix} A$$

Apply $R_2 \to R_2 + (-2)R_1$,

$$\begin{bmatrix} 2 & 6 \\ 0 & 2 \end{bmatrix} = \begin{bmatrix} 1 & 0 \\ -2 & 1 \end{bmatrix} A$$

Apply $R_1 \to R_1 + (-3)R_2$

$$\begin{bmatrix} 2 & 0 \\ 0 & 2 \end{bmatrix} = \begin{bmatrix} 7 & -3 \\ -2 & 1 \end{bmatrix} A$$

Divide both sides by 2

$$\begin{bmatrix} 1 & 0 \\ 0 & 1 \end{bmatrix} = \frac{1}{2}\begin{bmatrix} 7 & -3 \\ -2 & 1 \end{bmatrix} A$$

Hence $$A^{-1} = \frac{1}{2}\begin{bmatrix} 7 & -3 \\ -2 & 1 \end{bmatrix}.$$

Q. 2. Find matrices X and Y if

$$2X+Y = \begin{bmatrix} 2 & -3 & 1 \\ 1 & 2 & 3 \end{bmatrix} \text{ and}$$

$$X-Y = \begin{bmatrix} 0 & 3 & 5 \\ -2 & -4 & 1 \end{bmatrix}$$

Sol. Given, $\quad 2X+Y = \begin{bmatrix} 2 & -3 & 1 \\ 1 & 2 & 3 \end{bmatrix}$

and $\quad X-Y = \begin{bmatrix} 0 & 3 & 5 \\ -2 & -4 & 1 \end{bmatrix}$

For finding the value of X, add $2X+Y$ and $X-Y$.

$$2X+Y+X-Y = \begin{bmatrix} 2 & -3 & 1 \\ 1 & 2 & 3 \end{bmatrix} + \begin{bmatrix} 0 & 3 & 5 \\ -2 & -4 & 1 \end{bmatrix}$$

$$3X = \begin{bmatrix} 2 & 0 & 6 \\ -1 & -2 & 4 \end{bmatrix}$$

$$\therefore \quad X = \frac{1}{3}\begin{bmatrix} 2 & 0 & 6 \\ -1 & -2 & 4 \end{bmatrix}$$

$$= \begin{bmatrix} 2/3 & 0 & 2 \\ -1/3 & -2/3 & 4/3 \end{bmatrix}$$

On putting the value of X in $(X-Y)$,

$$\begin{bmatrix} 2/3 & 0 & 2 \\ -1/3 & -2/3 & 4/3 \end{bmatrix} - Y = \begin{bmatrix} 0 & 3 & 5 \\ -2 & -4 & 1 \end{bmatrix}$$

$$Y = \begin{bmatrix} 2/3 & 0 & 2 \\ -1/3 & -2/3 & 4/3 \end{bmatrix} - \begin{bmatrix} 0 & 3 & 5 \\ -2 & -4 & 1 \end{bmatrix}$$

$$= \begin{bmatrix} 2/3 & -3 & -3 \\ 5/3 & 10/3 & 1/3 \end{bmatrix}$$

Hence, $$X = \begin{bmatrix} 2/3 & 0 & 2 \\ -1/3 & -2/3 & 4/3 \end{bmatrix}$$

and $$Y = \begin{bmatrix} 2/3 & -3 & -3 \\ 5/3 & 10/3 & 1/3 \end{bmatrix}$$

Q. 3. Find the value of λ, a non-zero scalar, if

$$\lambda \begin{bmatrix} 1 & 3 & 5 \\ 2 & 4 & 6 \end{bmatrix} + 2 \begin{bmatrix} 2 & 5 & 6 \\ 1 & 3 & 5 \end{bmatrix} = \begin{bmatrix} 7 & 19 & 27 \\ 8 & 18 & 28 \end{bmatrix}$$

Sol. Given,

$$\lambda \begin{bmatrix} 1 & 3 & 5 \\ 2 & 4 & 6 \end{bmatrix} + 2 \begin{bmatrix} 2 & 5 & 6 \\ 1 & 3 & 5 \end{bmatrix} = \begin{bmatrix} 7 & 19 & 27 \\ 8 & 18 & 28 \end{bmatrix}$$

Multiply by non-zero scalar λ to the corresponding matrix and then add to another matrix,

$$\begin{bmatrix} \lambda & 3\lambda & 5\lambda \\ 2\lambda & 4\lambda & 6\lambda \end{bmatrix} + \begin{bmatrix} 4 & 10 & 12 \\ 2 & 6 & 10 \end{bmatrix} = \begin{bmatrix} 7 & 19 & 27 \\ 8 & 18 & 28 \end{bmatrix}$$

$$\begin{bmatrix} \lambda & 3\lambda & 5\lambda \\ 2\lambda & 4\lambda & 6\lambda \end{bmatrix} = \begin{bmatrix} 7 & 19 & 27 \\ 8 & 18 & 28 \end{bmatrix} - \begin{bmatrix} 4 & 10 & 12 \\ 2 & 6 & 10 \end{bmatrix}$$

$$= \begin{bmatrix} 7 & 19 & 27 \\ 8 & 18 & 28 \end{bmatrix} + \begin{bmatrix} -4 & -10 & -12 \\ -2 & -6 & -10 \end{bmatrix}$$

$$= \begin{bmatrix} 7-4 & 19-10 & 27-12 \\ 8-2 & 18-6 & 28-10 \end{bmatrix}$$

$$= \begin{bmatrix} 3 & 9 & 15 \\ 6 & 12 & 18 \end{bmatrix}$$

On equating the corresponding elements, we get

$$\lambda = 3.$$

Q. 4. Let $A = \begin{bmatrix} 4 & 2 \\ -3 & 5 \\ 5 & 0 \end{bmatrix}$, $B = \begin{bmatrix} 1 & 1 & 1 \\ 3 & 3 & 3 \end{bmatrix}$ **and** $C = \begin{bmatrix} 3 & 1 \\ 5 & 2 \\ -2 & 4 \end{bmatrix}$, **verify that $BA = BC$.**

Sol. Given,

$$A = \begin{bmatrix} 4 & 2 \\ -3 & 5 \\ 5 & 0 \end{bmatrix}, B = \begin{bmatrix} 1 & 1 & 1 \\ 3 & 3 & 3 \end{bmatrix} \text{ and } C = \begin{bmatrix} 3 & 1 \\ 5 & 2 \\ -2 & 4 \end{bmatrix}$$

Here, number of columns of B = number of rows of A = 3.

So BA is defined.

$$BA = \begin{bmatrix} 1 & 1 & 1 \\ 3 & 3 & 3 \end{bmatrix} \begin{bmatrix} 4 & 2 \\ -3 & 5 \\ 5 & 0 \end{bmatrix}$$

$$= \begin{bmatrix} 4-3+5 & 2+5+0 \\ 12-9+15 & 6+15+0 \end{bmatrix} = \begin{bmatrix} 6 & 7 \\ 18 & 21 \end{bmatrix}$$

Similarly, number of columns of B = number of rows of $C = 3$, so BC is defined.

$$BC = \begin{bmatrix} 1 & 1 & 1 \\ 3 & 3 & 3 \end{bmatrix} \begin{bmatrix} 3 & 1 \\ 5 & 2 \\ -2 & 4 \end{bmatrix}$$

$$= \begin{bmatrix} 3+5-2 & 1+2+4 \\ 9+15-6 & 3+6+12 \end{bmatrix}$$

$$= \begin{bmatrix} 6 & 7 \\ 18 & 21 \end{bmatrix}$$

Hence $BA = BC$.

Q. 5. If A and B are symmetric matrices, then prove that $AB + BA$ is a symmetric matrix.

Sol. Given, A and B are symmetric matrices.

$\therefore$ $A' = A$ and $B' = B$

Now $(AB + BA)' = (AB)' + (BA)'$

$[\because (A+B)' = A' + B']$

$= B'A' + A'B'$ $[\because (AB)' = B'A']$

$= BA + AB$ $[\because A' = A$ and $B' = B]$

$= AB + BA$

So, $AB + BA$ is a symmetric matrix.

Q. 6. Show that all the diagonal elements of a skew symmetric matrix are zero.

Sol. Let $A = [a_{ij}]$, be a given matrix

Since it is skew symmetric $A = -A'$

$\therefore$ $a_{ij} = -a_{ji}$ For all i, j

$a_{ii} = -a_{ii}$ For all values of i

$\Rightarrow$ $2a_{ii} = 0$ For all values of i

$\Rightarrow$ $a_{ii} = 0$ For all values of i

$\Rightarrow$ $a_{11} = a_{22} = a_{33} = \ldots = a_{nn} = 0$

Hence all the diagonal elements of a skew symmetric matrix are zero as diagonal elements are

$$a_{11}; a_{22}; \ldots a_{nn}.$$

Q. 7. If $A_\alpha = \begin{bmatrix} \cos\alpha & \sin\alpha \\ -\sin\alpha & \cos\alpha \end{bmatrix}$, then prove that $A_\alpha \cdot A_\beta = A_{\alpha+\beta}$. **[NCERT]**

Sol. Given, $A_\alpha = \begin{bmatrix} \cos\alpha & \sin\alpha \\ -\sin\alpha & \cos\alpha \end{bmatrix}$ ...(i)

$$\therefore \quad A_\beta = \begin{bmatrix} \cos\beta & \sin\beta \\ -\sin\beta & \cos\beta \end{bmatrix} \quad \ldots(ii)$$

By equations (i) and (ii)

$$A_\alpha \cdot A_\beta = \begin{bmatrix} \cos\alpha & \sin\alpha \\ -\sin\alpha & \cos\alpha \end{bmatrix}\begin{bmatrix} \cos\beta & \sin\beta \\ -\sin\beta & \cos\beta \end{bmatrix}$$

$$= \begin{bmatrix} \cos\alpha\cos\beta - \sin\alpha\sin\beta & \cos\alpha\sin\beta + \sin\alpha\cos\beta \\ -\sin\alpha\cos\beta - \cos\alpha\sin\beta & -\sin\alpha\sin\beta + \cos\alpha\cos\beta \end{bmatrix}$$

$$= \begin{bmatrix} \cos(\alpha+\beta) & \sin(\alpha+\beta) \\ -\sin(\alpha+\beta) & \cos(\alpha+\beta) \end{bmatrix}$$

$$= A_{\alpha+\beta}.$$

Q. 8. If $A = \begin{bmatrix} 2 \\ -3 \\ 4 \end{bmatrix}$ and $B = [3 \ \ 6 \ \ 7]$, verify that $(AB)^T = B^T A^T$.

Sol. Given, $A = \begin{bmatrix} 2 \\ -3 \\ 4 \end{bmatrix}$ and $B = [3 \ 6 \ 7]$

Here, number of columns of matrix A = number of rows of matrix $B = 1$, so AB is defined.

$$\therefore \quad AB = \begin{bmatrix} 2 \\ -3 \\ 4 \end{bmatrix} [3 \ 6 \ 7]$$

$$= \begin{bmatrix} 6 & 12 & 14 \\ -9 & -18 & -21 \\ 12 & 24 & 28 \end{bmatrix}$$

$$\Rightarrow \quad (AB)^T = \begin{bmatrix} 6 & -9 & 12 \\ 12 & -18 & 24 \\ 14 & -21 & 28 \end{bmatrix}$$

Here $\quad B = [3 \ 6 \ 7]$

$$\therefore \quad B^T = \begin{bmatrix} 3 \\ 6 \\ 7 \end{bmatrix}$$

$$A = \begin{bmatrix} 2 \\ -3 \\ 4 \end{bmatrix}$$

$$\therefore \quad A^T = [2 \ -3 \ 4]$$

$$B^T A^T = \begin{bmatrix} 3 \\ 6 \\ 7 \end{bmatrix} [2 \ -3 \ 4]$$

$$= \begin{bmatrix} 6 & -9 & 12 \\ 12 & -18 & 24 \\ 14 & -21 & 28 \end{bmatrix}$$

Hence, $\quad (AB)^T = B^T A^T.$

Q. 9. Compute $A^2 - 4A - 5I$ where $A = \begin{bmatrix} 1 & 2 & 2 \\ 2 & 1 & 2 \\ 2 & 2 & 1 \end{bmatrix}$ and I is a unit matrix.

Sol. Given, $A = \begin{bmatrix} 1 & 2 & 2 \\ 2 & 1 & 2 \\ 2 & 2 & 1 \end{bmatrix}$

$$A^2 = A \cdot A = \begin{bmatrix} 1 & 2 & 2 \\ 2 & 1 & 2 \\ 2 & 2 & 1 \end{bmatrix}\begin{bmatrix} 1 & 2 & 2 \\ 2 & 1 & 2 \\ 2 & 2 & 1 \end{bmatrix}$$

$$= \begin{bmatrix} 1+4+4 & 2+2+4 & 2+4+2 \\ 2+2+4 & 4+1+4 & 4+2+2 \\ 2+4+2 & 4+2+2 & 4+4+1 \end{bmatrix}$$

$$= \begin{bmatrix} 9 & 8 & 8 \\ 8 & 9 & 8 \\ 8 & 8 & 9 \end{bmatrix}$$

$\therefore A^2 - 4A - 5I$

$$= \begin{bmatrix} 9 & 8 & 8 \\ 8 & 9 & 8 \\ 8 & 8 & 9 \end{bmatrix} - 4\begin{bmatrix} 1 & 2 & 2 \\ 2 & 1 & 2 \\ 2 & 2 & 1 \end{bmatrix} - 5\begin{bmatrix} 1 & 0 & 0 \\ 0 & 1 & 0 \\ 0 & 0 & 1 \end{bmatrix}$$

$$= \begin{bmatrix} 9-4-5 & 8-8 & 8-8 \\ 8-8 & 9-4-5 & 8-8 \\ 8-8 & 8-8 & 9-4-5 \end{bmatrix} = \begin{bmatrix} 0 & 0 & 0 \\ 0 & 0 & 0 \\ 0 & 0 & 0 \end{bmatrix}$$

$$= 0$$

Q. 10. If $A = \begin{bmatrix} \cos\alpha & \sin\alpha \\ -\sin\alpha & \cos\alpha \end{bmatrix}$, verify that $AA' = A'A = I$.

Sol. Given, $A = \begin{bmatrix} \cos\alpha & \sin\alpha \\ -\sin\alpha & \cos\alpha \end{bmatrix}$

$$\therefore \quad A' = \begin{bmatrix} \cos\alpha & -\sin\alpha \\ \sin\alpha & \cos\alpha \end{bmatrix}$$

$$\therefore \quad AA' = \begin{bmatrix} \cos\alpha & \sin\alpha \\ -\sin\alpha & \cos\alpha \end{bmatrix}\begin{bmatrix} \cos\alpha & -\sin\alpha \\ \sin\alpha & \cos\alpha \end{bmatrix}$$

$$= \begin{bmatrix} \cos^2\alpha + \sin^2\alpha & -\cos\alpha\sin\alpha + \sin\alpha\cos\alpha \\ -\sin\alpha\cos\alpha + \cos\alpha\sin\alpha & \sin^2\alpha + \cos^2\alpha \end{bmatrix}$$

$$= \begin{bmatrix} 1 & 0 \\ 0 & 1 \end{bmatrix} = I$$

$$A'A = \begin{bmatrix} \cos\alpha & -\sin\alpha \\ \sin\alpha & \cos\alpha \end{bmatrix}\begin{bmatrix} \cos\alpha & \sin\alpha \\ -\sin\alpha & \cos\alpha \end{bmatrix}$$

$$= \begin{bmatrix} \cos^2\alpha + \sin^2\alpha & \cos\alpha\sin\alpha - \sin\alpha\cos\alpha \\ \sin\alpha\cos\alpha - \cos\alpha\sin\alpha & \sin^2\alpha + \cos^2\alpha \end{bmatrix}$$

$$= \begin{bmatrix} 1 & 0 \\ 0 & 1 \end{bmatrix} = I$$

Hence $AA' = A'A = I.$

Q. 11. Find the matrix X so that

$$X\begin{bmatrix} 1 & 2 & 3 \\ 4 & 5 & 6 \end{bmatrix} = \begin{bmatrix} -7 & -8 & -9 \\ 2 & 4 & 6 \end{bmatrix}$$ **[NCERT]**

Sol. Here multiplier has 2×3 order matrix and product matrix has 2×3 order matrix.

Let $X = \begin{bmatrix} x & y \\ a & b \end{bmatrix}$

We have $\begin{bmatrix} x & y \\ a & b \end{bmatrix}\begin{bmatrix} 1 & 2 & 3 \\ 4 & 5 & 6 \end{bmatrix} = \begin{bmatrix} -7 & -8 & -9 \\ 2 & 4 & 6 \end{bmatrix}$

$$\Rightarrow \begin{bmatrix} x+4y & 2x+5y & 3x+6y \\ a+4b & 2a+5b & 3a+6b \end{bmatrix} = \begin{bmatrix} -7 & -8 & -9 \\ 2 & 4 & 6 \end{bmatrix}$$

Equating the corresponding elements, we get

$x + 4y = -7$ $a + 4b = 2$

$2x + 5y = -8$ and $2a + 5b = 4$

$3x + 6y = -9$ $3a + 6b = 6$

On solving these equations, we get

$$x = 1, y = -2$$
$$a = 2, b = 0$$

$$\Rightarrow X = \begin{bmatrix} 1 & -2 \\ 2 & 0 \end{bmatrix}$$

Q. 12. If $A = \begin{bmatrix} 1 & 0 \\ -1 & 7 \end{bmatrix}$ and $I = \begin{bmatrix} 1 & 0 \\ 0 & 1 \end{bmatrix}$, then find k so that $A^2 = 8A + kI$.

Sol. Given, $A = \begin{bmatrix} 1 & 0 \\ -1 & 7 \end{bmatrix}$

$$A^2 = A \cdot A = \begin{bmatrix} 1 & 0 \\ -1 & 7 \end{bmatrix}\begin{bmatrix} 1 & 0 \\ -1 & 7 \end{bmatrix}$$

$$= \begin{bmatrix} 1 & 0 \\ -8 & 49 \end{bmatrix}$$

and $8A + kI = 8\begin{bmatrix} 1 & 0 \\ -1 & 7 \end{bmatrix} + k\begin{bmatrix} 1 & 0 \\ 0 & 1 \end{bmatrix}$

$$= \begin{bmatrix} 8+k & 0 \\ -8 & 56+k \end{bmatrix}$$

Here $A^2 = 8A + kI$

$$\begin{bmatrix} 1 & 0 \\ -8 & 49 \end{bmatrix} = \begin{bmatrix} 8+k & 0 \\ -8 & 56+k \end{bmatrix}$$

On equating the corresponding elements, we get

$$1 = 8 + k \text{ and } 49 = 56 + k$$

$$\Rightarrow k = -7.$$

Q. 13. Find the value of x if

$$[1 \; x \; 1]\begin{bmatrix} 1 & 3 & 2 \\ 2 & 5 & 1 \\ 15 & 3 & 2 \end{bmatrix}\begin{bmatrix} 1 \\ 2 \\ x \end{bmatrix} = 0.$$

Sol. Given,

$$[1 \; x \; 1]\begin{bmatrix} 1 & 3 & 2 \\ 2 & 5 & 1 \\ 15 & 3 & 2 \end{bmatrix}\begin{bmatrix} 1 \\ 2 \\ x \end{bmatrix} = 0$$

$$\Rightarrow [1 \; x \; 1]\begin{bmatrix} 1+6+2x \\ 2+10+x \\ 15+6+2x \end{bmatrix} = 0$$

$$\Rightarrow [1 \; x \; 1]\begin{bmatrix} 7+2x \\ 12+x \\ 21+2x \end{bmatrix} = 0$$

$$\Rightarrow (7+2x) + x(12+x) + (21+2x) = 0$$

$$\Rightarrow 7 + 2x + 12x + x^2 + 21 + 2x = 0$$

$$\Rightarrow x^2 + 16x + 28 = 0$$

$$\Rightarrow (x+2)(x+14) = 0$$

or $x = -2, x = -14.$

Q. 14. Show that $A = \begin{bmatrix} 2 & -3 \\ 3 & 4 \end{bmatrix}$ satisfies the equation $x^2 - 6x + 17 = 0$. Hence, find A^{-1}.

Sol. We have, $A = \begin{bmatrix} 2 & -3 \\ 3 & 4 \end{bmatrix}$

$$\Rightarrow A^2 = A.A = \begin{bmatrix} 2 & -3 \\ 3 & 4 \end{bmatrix}\begin{bmatrix} 2 & -3 \\ 3 & 4 \end{bmatrix}$$

$$= \begin{bmatrix} 4-9 & -6-12 \\ 6+12 & -9+16 \end{bmatrix} = \begin{bmatrix} -5 & -18 \\ 18 & 7 \end{bmatrix}$$

$$6A = 6\begin{bmatrix} 2 & -3 \\ 3 & 4 \end{bmatrix} = \begin{bmatrix} 12 & -18 \\ 18 & 24 \end{bmatrix}$$

and $17I = 17\begin{bmatrix} 1 & 0 \\ 0 & 1 \end{bmatrix} = \begin{bmatrix} 17 & 0 \\ 0 & 17 \end{bmatrix}$

$$\Rightarrow A^2 - 6A + 17I_2 = \begin{bmatrix} -5 & -18 \\ 18 & 7 \end{bmatrix} - \begin{bmatrix} 12 & -18 \\ 18 & 24 \end{bmatrix} + \begin{bmatrix} 17 & 0 \\ 0 & 17 \end{bmatrix}$$

$$= \begin{bmatrix} -5-12+17 & -18+18+0 \\ 18-18+0 & 7-24+17 \end{bmatrix}$$

$$= \begin{bmatrix} 0 & 0 \\ 0 & 0 \end{bmatrix} = 0$$

Hence, matrix A satisfies the equation, $x^2-6x+17=0$.

Now, $A^2-6A+17I_2=0$

$\Rightarrow \quad A^2-6A=-17I_2$

Multiplying both sides by A^{-1}, we have

$$A-6I_2=-17A^{-1}$$

$$\Rightarrow \quad A^{-1}=\frac{1}{17}(6I_2-A)$$

$$=\frac{1}{17}\left\{\begin{bmatrix}6&0\\0&6\end{bmatrix}-\begin{bmatrix}2&-3\\3&4\end{bmatrix}\right\}$$

$$=\frac{1}{17}\begin{bmatrix}4&3\\-3&2\end{bmatrix}$$

Q. 15. If $A=\begin{bmatrix}\cos\theta & \sin\theta\\ -\sin\theta & \cos\theta\end{bmatrix}$, then prove that

$$A^n=\begin{bmatrix}\cos n\theta & \sin n\theta\\ -\sin n\theta & \cos n\theta\end{bmatrix}, n\in N.$$

Sol. Let the given statement be

$f(n)$: if $\quad A=\begin{bmatrix}\cos\theta & \sin\theta\\ -\sin\theta & \cos\theta\end{bmatrix}$

then $\quad A^n=\begin{bmatrix}\cos n\theta & \sin n\theta\\ -\sin n\theta & \cos n\theta\end{bmatrix}$

for $n=1 \quad f(1): A=\begin{bmatrix}\cos\theta & \sin\theta\\ -\sin\theta & \cos\theta\end{bmatrix}$

Then, $\quad A^1=\begin{bmatrix}\cos\theta & \sin\theta\\ -\sin\theta & \cos\theta\end{bmatrix}$

$\therefore f(1)$ is true.

Let $f(n)$ be true for $n=k$, so

$$f(k): A^k=\begin{bmatrix}\cos k\theta & \sin k\theta\\ -\sin k\theta & \cos k\theta\end{bmatrix}$$

Now, for $\quad n=k+1$

$$A^{k+1}=A^k\cdot A$$

$$=\begin{bmatrix}\cos k\theta & \sin k\theta\\ -\sin k\theta & \cos k\theta\end{bmatrix}\begin{bmatrix}\cos\theta & \sin\theta\\ -\sin\theta & \cos\theta\end{bmatrix}$$

$$=\begin{bmatrix}\cos k\theta\cos\theta-\sin k\theta\sin\theta & \cos k\theta\sin\theta+\sin k\theta\cos\theta\\ -\sin k\theta\cos\theta-\cos k\theta\sin\theta & -\sin k\theta\sin\theta+\cos k\theta\cos\theta\end{bmatrix}$$

$$=\begin{bmatrix}\cos(k\theta+\theta) & \sin(k\theta+\theta)\\ -\sin(k\theta+\theta) & \cos(k\theta+\theta)\end{bmatrix}$$

$$=\begin{bmatrix}\cos(k+1)\theta & \sin(k+1)\theta\\ -\sin(k+1)\theta & \cos(k+1)\theta\end{bmatrix}$$

Thus $f(k+1): A^{k+1}=\begin{bmatrix}\cos(k+1)\theta & \sin(k+1)\theta\\ -\sin(k+1)\theta & \cos(k+1)\theta\end{bmatrix}$

which is also true.

Hence, by the principle of mathematical induction,

$$A^n=\begin{bmatrix}\cos n\theta & \sin n\theta\\ -\sin n\theta & \cos n\theta\end{bmatrix} \text{ is true for all } n\in N.$$

Q. 16. If $f(x)=\begin{bmatrix}\cos x & -\sin x & 0\\ \sin x & \cos x & 0\\ 0&0&1\end{bmatrix}$, then show that $f(x)\cdot f(y)=f(x+y)$. **[NCERT]**

Sol. Given, $\quad f(x)=\begin{bmatrix}\cos x & -\sin x & 0\\ \sin x & \cos x & 0\\ 0&0&1\end{bmatrix}$

$\therefore \quad f(y)=\begin{bmatrix}\cos y & -\sin y & 0\\ \sin y & \cos y & 0\\ 0&0&1\end{bmatrix}$

L.H.S. $=f(x).f(y)$

$$=\begin{bmatrix}\cos x & -\sin x & 0\\ \sin x & \cos x & 0\\ 0&0&1\end{bmatrix}\begin{bmatrix}\cos y & -\sin y & 0\\ \sin y & \cos y & 0\\ 0&0&1\end{bmatrix}$$

$$=\begin{bmatrix}\cos x\cos y-\sin x\sin y+0 & -\cos x\sin y-\sin x\cos y+0 & 0+0+0\\ \sin x\cos y+\cos x\sin y+0 & -\sin x\sin y+\cos x\cos y+0 & 0+0+0\\ 0+0+0 & 0+0+0 & 0+0+1\end{bmatrix}$$

$$=\begin{bmatrix}\cos(x+y) & -\sin(x+y) & 0\\ \sin(x+y) & \cos(x+y) & 0\\ 0&0&1\end{bmatrix}$$

$\left[\begin{array}{l}\because \cos(A+B)=\cos A\cos B-\sin A\sin B\\ \text{and } \sin(A+B)=\sin A\cos B+\cos A\sin B\end{array}\right]$

$=f(x+y)=$ R.H.S.

Q. 17. Let $A=\begin{bmatrix}2&-1\\3&4\end{bmatrix}$, $B=\begin{bmatrix}5&2\\7&4\end{bmatrix}$, $C=\begin{bmatrix}2&5\\3&8\end{bmatrix}$, find a matrix D such that $CD-AB=O$.★

Sol. Let D be the matrix of order 2×2, $D=\begin{bmatrix}a&b\\c&d\end{bmatrix}$

Given, $\quad CD-AB=0$

$$CD=AB$$

$$\begin{bmatrix}2&5\\3&8\end{bmatrix}\begin{bmatrix}a&b\\c&d\end{bmatrix}=\begin{bmatrix}2&-1\\3&4\end{bmatrix}\begin{bmatrix}5&2\\7&4\end{bmatrix}$$

$$\begin{bmatrix}2a+5c & 2b+5d\\ 3a+8c & 3b+8d\end{bmatrix}=\begin{bmatrix}3&0\\43&22\end{bmatrix}$$

On comparing we get

$$2a+5c=3 \quad \ldots\text{(i)}$$

$$2b+5d=0$$

★ are board exam questions from previous years

$$\Rightarrow \quad b = \frac{-5}{2}d$$

$$3a + 8c = 43 \quad \ldots\text{(ii)}$$

$$3b + 8d = 22 \quad \ldots\text{(iii)}$$

Substitute the value of $b = \frac{-5d}{2}$ in equation (iii),

$$3\left(\frac{-5d}{2}\right) + 8d = 22$$

$$\frac{-15d}{2} + 8d = 22$$

$$-15d + 16d = 44$$

$$d = 44$$

Also, $b = \frac{-5}{2}d$

$$b = \frac{-5}{2} \times 44 = -110$$

From equation (i),

$$a = \frac{3-5c}{2}$$

Substitute in equation (ii)

$$3\left(\frac{3-5c}{2}\right) + 8c = 43$$

$$9 - 15c + 16c = 86$$

$$c = 77$$

$$a = \frac{3-5\times77}{2} = \frac{3-385}{2}$$

$$= \frac{-382}{2} = -191$$

$$D = \begin{bmatrix} -191 & -110 \\ 77 & 44 \end{bmatrix}$$

Q. 18. Find matrix *A* such that★

$$\begin{bmatrix} 2 & -1 \\ 1 & 0 \\ -3 & 4 \end{bmatrix} A = \begin{bmatrix} -1 & -8 \\ 1 & -2 \\ 9 & 22 \end{bmatrix}$$

Sol. We have

$$\begin{bmatrix} 2 & -1 \\ 1 & 0 \\ -3 & 4 \end{bmatrix}_{3\times2} A = \begin{bmatrix} -1 & -8 \\ 1 & -2 \\ 9 & 22 \end{bmatrix}_{3\times2}$$

$\therefore$ *A* is of order 2×2.

Let $A = \begin{bmatrix} a & b \\ c & d \end{bmatrix}$

Then, $\begin{bmatrix} 2 & -1 \\ 1 & 0 \\ -3 & 4 \end{bmatrix}\begin{bmatrix} a & b \\ c & d \end{bmatrix} = \begin{bmatrix} -1 & -8 \\ 1 & -2 \\ 9 & 22 \end{bmatrix}$

$$\begin{bmatrix} 2a-c & 2b-d \\ a & b \\ -3a+4c & -3b+4d \end{bmatrix} = \begin{bmatrix} -1 & -8 \\ 1 & -2 \\ 9 & 22 \end{bmatrix}$$

By equality of matrices, on comparing, we get

$$2a - c = -1$$

$$2b - d = -8$$

$$a = 1$$

$$b = -2$$

and $-3a + 4c = 9$

$$-3b + 4d = 22$$

On solving, the equations, we get

$$a = 1, b = -2, c = 3, d = 4$$

Hence, $A = \begin{bmatrix} 1 & -2 \\ 3 & 4 \end{bmatrix}$

Q. 19. To promote the making of toilets for women, an organisation tried to generate awareness through (i) house cells, (ii) letters and (iii) announcements. The cost for each mode per attempt is given below :

(i) ₹ 50 (ii) ₹ 20 (iii) ₹ 40

The number of attempts made in three villages, *X*, *Y* and *Z* are given below :

	(i)	(ii)	(iii)
X	400	300	100
Y	300	250	75
Z	500	400	150

Find the total cost incurred by the organisation for the three villages separately, using matrices.★

Sol. The number of attempts made in three villages *X*, *Y* and *Z* can be represented by the 3×3 matrix.

$$X = \begin{bmatrix} 400 & 300 & 100 \\ 300 & 250 & 75 \\ 500 & 400 & 150 \end{bmatrix}$$

and the cost for each mode per attempt can be represented by the 3×1 matrix.

$$Y = \begin{bmatrix} 50 \\ 20 \\ 40 \end{bmatrix}$$

$\therefore$ By matrix multiplication the cost incurred by the organisation for the three villages.

$$XY = \begin{bmatrix} 400 & 300 & 100 \\ 300 & 250 & 75 \\ 500 & 400 & 150 \end{bmatrix}\begin{bmatrix} 50 \\ 20 \\ 40 \end{bmatrix}$$

★ are board exam questions from previous years

$$XY = \begin{bmatrix} 30,000 \\ 23,000 \\ 39,000 \end{bmatrix}$$

Hence the total cost incurred by the organisation for the three villages separately as ₹ 30,000, ₹ 23,000 and ₹ 39,000.

Q. 20. Three schools *A*, *B* and *C* organised a mela for collecting funds for helping the rehabilitation of flood victims. They sold hand made fans, mats and plates from recycled material as a cost of ₹ 25, ₹ 100 and ₹ 50 each. The number of articles sold are given below :

School / Article	*A*	*B*	*C*
Hand-fans	**40**	**25**	**35**
Mats	**50**	**40**	**50**
Plates	**20**	**30**	**40**

Find the fund collected by each school separately by selling the above articles. Also find the total funds collected for the purpose.*

Sol. The number of articles sold by each school can be represented by the 3 × 3 matrix.

$$X = \begin{bmatrix} 40 & 25 & 35 \\ 50 & 40 & 50 \\ 20 & 30 & 40 \end{bmatrix}$$

and the cost of each article can be represented by the 1 × 3 matrix.

$$Y = [25 \ 100 \ 50]$$

⇒ Funds collected by each school separately is given by the matrix multiplication.

$$YX = [25 \ 100 \ 50] \begin{bmatrix} 40 & 25 & 35 \\ 50 & 40 & 50 \\ 20 & 30 & 40 \end{bmatrix}$$

$$YX = [7000 \ 6125 \ 7875]$$

Hence the funds collected by schools *A*, *B* and *C* are ₹ 7,000, ₹ 6,125 and ₹ 7,875 respectively.

The total funds collected for flood victims

$$= ₹\,(7,000 + 6,125 + 7,875)$$
$$= ₹\,21,000$$

Chapter 4. Determinants

Q. 1. If *a*, *b*, *c* are in A.P., then evaluate the determinant $\begin{vmatrix} x+2 & x+3 & x+2a \\ x+3 & x+4 & x+2b \\ x+4 & x+5 & x+2c \end{vmatrix}$

Sol. Let $\Delta = \begin{vmatrix} x+2 & x+3 & x+2a \\ x+3 & x+4 & x+2b \\ x+4 & x+5 & x+2c \end{vmatrix}$

* are board exam questions from previous years

Since *a*, *b*, *c* are in A.P.

So $2b = a + c$

$$\Delta = \begin{vmatrix} x+2 & x+3 & x+2a \\ x+3 & x+4 & x+a+c \\ x+4 & x+5 & x+2c \end{vmatrix}$$

Applying $R_1 \to R_1 - R_2$ and $R_3 \to R_3 - R_2$, we have

$$\Delta = \begin{vmatrix} -1 & -1 & a-c \\ x+3 & x+4 & x+a+c \\ 1 & 1 & c-a \end{vmatrix}$$

Applying $R_1 \to R_1 + R_3$, we have

$$\Delta = \begin{vmatrix} 0 & 0 & 0 \\ x+3 & x+4 & x+a+c \\ 1 & 1 & c-a \end{vmatrix}$$

Since, all the elements of first row are zero.

Hence $\Delta = 0.$

Q. 2. Prove that :

$$\begin{vmatrix} a^2+1 & ab & ac \\ ab & b^2+1 & bc \\ ca & cb & c^2+1 \end{vmatrix} = 1 + a^2 + b^2 + c^2.\star$$

[NCERT]

Sol. Consider,

$$\text{L.H.S.} = \begin{vmatrix} a^2+1 & ab & ac \\ ab & b^2+1 & bc \\ ca & cb & c^2+1 \end{vmatrix}$$

Taking common *a*, *b* and *c* from R_1, R_2 and R_3 respectively, we get

$$= abc \begin{vmatrix} a+1/a & b & c \\ a & b+1/b & c \\ a & b & c+1/c \end{vmatrix}$$

Applying $R_2 \to R_2 - R_1$ and $R_3 \to R_3 - R_1$, we get

$$= abc \begin{vmatrix} a+1/a & b & c \\ -1/a & 1/b & 0 \\ -1/a & 0 & 1/c \end{vmatrix}$$

Multiply and divide C_1 by *a*, C_2 by *b* and C_3 by *c* and taking common $\frac{1}{a}, \frac{1}{b}$ and $\frac{1}{c}$ from C_1, C_2 and C_3 respectively, we get

$$= abc \times \frac{1}{abc} \begin{vmatrix} a^2+1 & b^2 & c^2 \\ -1 & 1 & 0 \\ -1 & 0 & 1 \end{vmatrix}$$

$$= \begin{vmatrix} a^2+1 & b^2 & c^2 \\ -1 & 1 & 0 \\ -1 & 0 & 1 \end{vmatrix}$$

Expanding along R_3, we get

$$= -1 \times (b^2 \times 0 - 1 \times c^2) - 0((a^2+1) \times 0 + 1 \times c^2) + 1(a^2+1+b^2)$$

$$= c^2 + a^2 + 1 + b^2$$

$$= 1 + a^2 + b^2 + c^2$$

= R.H.S.

Q. 3. If A is a skew-symmetric matrix of order 3, then prove that $\det A = 0$.★

Sol. Given, A is a skew-symmetric matrix of order 3.

So, $\quad A = -A^T$

Now, $\quad |A| = |-A^T|$

$$|A| = (-1)^3 |A^T| \quad \left[\begin{array}{l} \because |kA| = k^n |A| \\ \text{Where } n \text{ is of order } A \end{array}\right]$$

$$|A| = -|A| \quad [\because |AT| = |A|]$$

$$|A| = -|A|$$

$$|A| + |A| = 0$$

$\therefore \quad |A| = 0$

i.e., $\quad \det A = 0.$

Q. 4. Show that :

$$\begin{vmatrix} 1+a & 1 & 1 \\ 1 & 1+b & 1 \\ 1 & 1 & 1+c \end{vmatrix} = abc\left(1+\frac{1}{a}+\frac{1}{b}+\frac{1}{c}\right) = abc + bc + ca + ab.$$

[NCERT]

Sol. Consider,

$$\text{L.H.S.} = \begin{vmatrix} 1+a & 1 & 1 \\ 1 & 1+b & 1 \\ 1 & 1 & 1+c \end{vmatrix}$$

Taking out factors a, b, c common from R_1, R_2 and R_3, we get

$$= abc \begin{vmatrix} \frac{1}{a}+1 & \frac{1}{a} & \frac{1}{a} \\ \frac{1}{b} & 1+\frac{1}{b} & \frac{1}{b} \\ \frac{1}{c} & \frac{1}{c} & \frac{1}{c}+1 \end{vmatrix}$$

Applying $R_1 \to (R_1 + R_2 + R_3)$, we have

$$= abc \begin{vmatrix} 1+\frac{1}{a}+\frac{1}{b}+\frac{1}{c} & 1+\frac{1}{a}+\frac{1}{b}+\frac{1}{c} & 1+\frac{1}{a}+\frac{1}{b}+\frac{1}{c} \\ \frac{1}{b} & 1+\frac{1}{b} & \frac{1}{b} \\ \frac{1}{c} & \frac{1}{c} & \frac{1}{c}+1 \end{vmatrix}$$

$$= (abc)\left(1+\frac{1}{a}+\frac{1}{b}+\frac{1}{c}\right) \begin{vmatrix} 1 & 1 & 1 \\ \frac{1}{b} & 1+\frac{1}{b} & \frac{1}{b} \\ \frac{1}{c} & \frac{1}{c} & \frac{1}{c}+1 \end{vmatrix}$$

Now applying $C_2 \to C_2 - C_1$, $C_3 \to C_3 - C_1$ we get

$$= abc\left(1+\frac{1}{a}+\frac{1}{b}+\frac{1}{c}\right) \begin{vmatrix} 1 & 0 & 0 \\ \frac{1}{b} & 1 & 0 \\ \frac{1}{c} & 0 & 1 \end{vmatrix}$$

Expanding along row R_1

$$= abc\left(1+\frac{1}{a}+\frac{1}{b}+\frac{1}{c}\right) \left\{1\begin{vmatrix} 1 & 0 \\ 0 & 1 \end{vmatrix} - 0\begin{vmatrix} 1/b & 0 \\ 1/c & 1 \end{vmatrix} - 0\begin{vmatrix} 1/b & 1 \\ 1/c & 0 \end{vmatrix}\right\}$$

$$= abc\left(1+\frac{1}{a}+\frac{1}{b}+\frac{1}{c}\right) \{1(1-0)\}$$

$$= abc\left(1+\frac{1}{a}+\frac{1}{b}+\frac{1}{c}\right)$$

$$= abc + bc + ca + ab$$

= R.H.S.

Q. 5. Using properties of determinants, prove that :★

$$\begin{vmatrix} x & x+y & x+2y \\ x+2y & x & x+y \\ x+y & x+2y & x \end{vmatrix} = 9y^2(x+y).$$

Sol. Consider, $\begin{vmatrix} x & x+y & x+2y \\ x+2y & x & x+y \\ x+y & x+2y & x \end{vmatrix}$

Applying $C_1 \to C_1 + C_2 + C_3$, we get

$$= \begin{vmatrix} 3x+3y & x+y & x+2y \\ 3x+3y & x & x+y \\ 3x+3y & x+2y & x \end{vmatrix}$$

$$= 3(x+y)\begin{vmatrix} 1 & x+y & x+2y \\ 1 & x & x+y \\ 1 & x+2y & x \end{vmatrix}$$

Applying $R_1 \to R_1 - R_2$ and $R_3 \to R_3 - R_2$, we get

$$= 3(x+y)\begin{vmatrix} 0 & y & y \\ 1 & x & x+y \\ 0 & 2y & -y \end{vmatrix}$$

$$= 3(x+y)[-(-y^2 - 2y^2)]$$

Expanding along C_1

$$= 3(x+y) \cdot 3y^2$$

$$= 9y^2(x+y).$$

Q. 6. If the value of the determinant $\begin{vmatrix} a & 1 & 1 \\ 1 & b & 1 \\ 1 & 1 & c \end{vmatrix}$ is positive, then prove that $abc > -8$.

★ are board exam questions from previous years

Sol. Let $\Delta = \begin{vmatrix} a & 1 & 1 \\ 1 & b & 1 \\ 1 & 1 & c \end{vmatrix}$

On applying $C_1 \to C_1 - C_2$ and $C_2 \to C_2 - C_3$

$$= \begin{vmatrix} a-1 & 0 & 1 \\ 1-b & b-1 & 1 \\ 0 & 1-c & c \end{vmatrix}$$

On expanding along R_1

$$= (a-1)\begin{vmatrix} b-1 & 1 \\ 1-c & c \end{vmatrix} - 0\begin{vmatrix} 1-b & 1 \\ 0 & c \end{vmatrix} + 1\begin{vmatrix} 1-b & b-1 \\ 0 & 1-c \end{vmatrix}$$

$$= (a-1)\{(b-1)c-(1-c)\} - 0 + 1\{(1-b)(1-c)-0\}$$

$$= (a-1)(bc-c-1+c)+(1-c-b+bc)$$

$$= (a-1)(bc-1)+(1-c-b+bc)$$

$$= abc-a-bc+1+1-c-b+bc$$

$$= abc-a-b-c+2$$

Since $\Delta > 0$ (given)

$\therefore\ abc-a-b-c+2>0$

$\Rightarrow\ abc+2>a+b+c$

$abc+2>3(abc)^{1/3}$ $\left[\because AM > GM \Rightarrow \dfrac{a+b+c}{3} > (abc)^{1/3}\right]$

Let $(abc)^{1/3} = x$, then

$x^3+2>3x$

$\Rightarrow\ x^3-3x+2>0$

$\Rightarrow\ (x-1)^2(x+2)>0$

$\Rightarrow\ (x+2)>0$

$\Rightarrow\ x>-2$

$\Rightarrow\ (abc)^{1/3}>-2$

$\Rightarrow\ abc>-8.$

Q. 7. If $\begin{vmatrix} a & a^2 & 1+a^3 \\ b & b^2 & 1+b^3 \\ c & c^2 & 1+c^3 \end{vmatrix} = 0$ and the vectors $\vec{A}(1, a, a^2)$, $\vec{B}(1, b, b^2)$, $\vec{C}(1, c, c^2)$ are non-coplanar, then prove that the product $abc = -1$.

Sol. Let $\Delta = \begin{vmatrix} a & a^2 & 1+a^3 \\ b & b^2 & 1+b^3 \\ c & c^2 & 1+c^3 \end{vmatrix} = 0$

$$\Rightarrow \begin{vmatrix} a & a^2 & 1 \\ b & b^2 & 1 \\ c & c^2 & 1 \end{vmatrix} + \begin{vmatrix} a & a^2 & a^3 \\ b & b^2 & b^3 \\ c & c^2 & c^3 \end{vmatrix} = 0$$

$$\Rightarrow \begin{vmatrix} a & a^2 & 1 \\ b & b^2 & 1 \\ c & c^2 & 1 \end{vmatrix} + abc\begin{vmatrix} 1 & a & a^2 \\ 1 & b & b^2 \\ 1 & c & c^2 \end{vmatrix} = 0$$

$$\Rightarrow \begin{vmatrix} 1 & a & a^2 \\ 1 & b & b^2 \\ 1 & c & c^2 \end{vmatrix} + abc\begin{vmatrix} 1 & a & a^2 \\ 1 & b & b^2 \\ 1 & c & c^2 \end{vmatrix} = 0$$

$$\Rightarrow \begin{vmatrix} 1 & a & a^2 \\ 1 & b & b^2 \\ 1 & c & c^2 \end{vmatrix}(1+abc) = 0$$

Since $(1, a, a^2)$, $(1, b, b^2)$ and $(1, c, c^2)$ are non-coplanar.

$$\Rightarrow \begin{vmatrix} 1 & a & a^2 \\ 1 & b & b^2 \\ 1 & c & c^2 \end{vmatrix} \neq 0$$

$\therefore\ abc = -1.$

Q. 8. If $x = -9$ is a root of the equation

$$\begin{vmatrix} x & 3 & 7 \\ 2 & x & 2 \\ 7 & 6 & x \end{vmatrix} = 0$$

then find the other two roots of the above equation.

Sol. Let $\Delta = \begin{vmatrix} x & 3 & 7 \\ 2 & x & 2 \\ 7 & 6 & x \end{vmatrix} = 0$

On expanding along R_1

$$x(x^2-12)-3(2x-14)+7(12-7x)=0$$

$$\Rightarrow\ x^3-67x+126=0$$

Given, $x=-9$ is one root of this equation. Let α and β be the other two roots of the above equation.

$\therefore\ (x+9)(x-\alpha)(x-\beta) = x^3-67x+126$

$$\therefore\ (x-\alpha)(x-\beta) = \frac{x^3-67x+126}{x+9}$$

$$= \frac{(x+9)(x^2-9x+14)}{(x+9)}$$

$\Rightarrow\ (x-\alpha)(x-\beta) = x^2-9x+14$

$\therefore\ (x-\alpha)(x-\beta) = (x-2)(x-7)$

$\Rightarrow\ \alpha = 2$ and $\beta = 7$

$\Rightarrow$ Other two roots of the above equation are 2 and 7.

Q. 9. If $\begin{vmatrix} a+b & b+c & c+a \\ b+c & c+a & a+b \\ c+a & a+b & b+c \end{vmatrix} = k\begin{vmatrix} a & b & c \\ b & c & a \\ c & a & b \end{vmatrix}$, then find the value of k.

Sol. Given that

$$\begin{vmatrix} a+b & b+c & c+a \\ b+c & c+a & a+b \\ c+a & a+b & b+c \end{vmatrix} = k\begin{vmatrix} a & b & c \\ b & c & a \\ c & a & b \end{vmatrix}$$

Let us consider

$$\text{L.H.S.} = \begin{vmatrix} a+b & b+c & c+a \\ b+c & c+a & a+b \\ c+a & a+b & b+c \end{vmatrix}$$

On applying $C_1 \to C_1 + C_2 + C_3$

$$= \begin{vmatrix} 2(a+b+c) & b+c & c+a \\ 2(a+b+c) & c+a & a+b \\ 2(a+b+c) & a+b & b+c \end{vmatrix}$$

$$= 2\begin{vmatrix} a+b+c & b+c & c+a \\ a+b+c & c+a & a+b \\ a+b+c & a+b & b+c \end{vmatrix}$$

On applying $C_1 \to C_1 - C_2$

$$= 2\begin{vmatrix} a & b+c & c+a \\ b & c+a & a+b \\ c & a+b & b+c \end{vmatrix}$$

On applying $C_3 \to C_3 - C_1$

$$= 2\begin{vmatrix} a & b+c & c \\ b & c+a & a \\ c & a+b & b \end{vmatrix}$$

On applying $C_2 \to C_2 - C_3$

$$= 2\begin{vmatrix} a & b & c \\ b & c & a \\ c & a & b \end{vmatrix}$$

But $\begin{vmatrix} a+b & b+c & c+a \\ b+c & c+a & a+b \\ c+a & a+b & b+c \end{vmatrix} = k\begin{vmatrix} a & b & c \\ b & c & a \\ c & a & b \end{vmatrix}$

On comparing both sides of the above equation, we get

$$k = 2.$$

Q. 10. Prove that a, b and c are in G.P. if

$$\begin{vmatrix} a & b & a\alpha+b \\ b & c & b\alpha+c \\ a\alpha+b & b\alpha+c & 0 \end{vmatrix} = 0.$$

Sol. Given, $\begin{vmatrix} a & b & a\alpha+b \\ b & c & b\alpha+c \\ a\alpha+b & b\alpha+c & 0 \end{vmatrix} = 0$

On applying $R_3 \to R_3 - \alpha R_1 - R_2$

$$\Rightarrow \begin{vmatrix} a & b & a\alpha+b \\ b & c & b\alpha+c \\ 0 & 0 & -[\alpha(a\alpha+b)+(b\alpha+c)] \end{vmatrix} = 0$$

On expanding along R_3

$$\Rightarrow \quad [\alpha(a\alpha+b)+(b\alpha+c)]\begin{vmatrix} a & b \\ b & c \end{vmatrix} = 0$$

$$\Rightarrow \quad -[\alpha(a\alpha+b)+(b\alpha+c)](ac-b^2) = 0$$

Either $\quad \alpha(a\alpha+b)+c(b\alpha+c) = 0 \quad$ ...(i)

or $\quad ac - b^2 = 0 \quad$ ...(ii)

From equation (ii),

$$b^2 = ac$$

a, b, c are in G.P.

Q. 11. If

$$f(x) = \begin{vmatrix} 1 & x & x+1 \\ 2x & x(x-1) & x(x+1) \\ 3x(x-1) & x(x-1)(x-2) & x(x-1)(x+1) \end{vmatrix}$$

then evaluate $f(11)$.

Sol. Given,

$$f(x) = \begin{vmatrix} 1 & x & x+1 \\ 2x & x(x-1) & x(x+1) \\ 3x(x-1) & x(x-1)(x-2) & x(x-1)(x+1) \end{vmatrix}$$

On taking x common from R_2 and $x(x-1)$ common from R_3.

$$f(x) = x^2(x-1)\begin{vmatrix} 1 & x & (x+1) \\ 2 & (x-1) & (x+1) \\ 3 & (x-2) & (x+1) \end{vmatrix}$$

On taking $(x+1)$ common from column C_3,

$$\Rightarrow \quad f(x) = x^2(x-1)(x+1)\begin{vmatrix} 1 & x & 1 \\ 2 & x-1 & 1 \\ 3 & x-2 & 1 \end{vmatrix}$$

On applying $R_2 \to R_2 - R_1$ and $R_3 \to R_3 - R_1$

$$\Rightarrow \quad f(x) = x^2(x-1)(x+1)\begin{vmatrix} 1 & x & 1 \\ 1 & -1 & 0 \\ 2 & -2 & 0 \end{vmatrix}$$

$$\Rightarrow \quad f(x) = x^2(x-1)(x+1)(-2+2) = 0$$

$$\Rightarrow \quad f(x) = 0 \text{ for all } x$$

$$\Rightarrow \quad f(11) = 0.$$

Q. 12. By using properties of determinants, show that :

(i) $\begin{vmatrix} a-b-c & 2a & 2a \\ 2b & b-c-a & 2b \\ 2c & 2c & c-a-b \end{vmatrix} = (a+b+c)^3$

[NCERT]

(ii) $\begin{vmatrix} x+y+2z & x & y \\ z & y+z+2x & y \\ z & x & z+x+2y \end{vmatrix} = 2(x+y+z)^3.$

[NCERT]

Sol. (i) Let $\Delta = \begin{vmatrix} a-b-c & 2a & 2a \\ 2b & b-c-a & 2b \\ 2c & 2c & c-a-b \end{vmatrix}$

Applying $R_1 \to R_1 + R_2 + R_3$, we have

$$\Delta = \begin{vmatrix} a+b+c & a+b+c & a+b+c \\ 2b & b-c-a & 2b \\ 2c & 2c & c-a-b \end{vmatrix}$$

Taking $(a+b+c)$ common from row R_1

$$= (a+b+c)\begin{vmatrix} 1 & 1 & 1 \\ 2b & b-c-a & 2b \\ 2c & 2c & c-a-b \end{vmatrix}$$

Applying $C_2 \to C_2 - C_1$, $C_3 \to C_3 - C_1$, we have

$$\Delta = (a+b+c)$$

$$\begin{vmatrix} 1 & 0 & 0 \\ 2b & -(a+b+c) & 0 \\ 2c & 0 & -(a+b+c) \end{vmatrix}$$

$$= (a+b+c)^3\begin{vmatrix} 1 & 0 & 0 \\ 2b & -1 & 0 \\ 2c & 0 & -1 \end{vmatrix}$$

Expanding along C_3, we have

$$\Delta = (a+b+c)^3(-1)(-1)$$
$$= (a+b+c)^3.$$

(ii) Let $\Delta = \begin{vmatrix} x+y+2z & x & y \\ z & y+z+2x & y \\ z & x & z+x+2y \end{vmatrix}$

Applying $C_1 \to C_1 + C_2 + C_3$, we have

$$\Delta = \begin{vmatrix} 2(x+y+z) & x & y \\ 2(x+y+z) & y+z+2x & y \\ 2(x+y+z) & x & z+x+2y \end{vmatrix}$$

$$= 2(x+y+z)\begin{vmatrix} 1 & x & y \\ 1 & y+z+2x & y \\ 1 & x & z+x+2y \end{vmatrix}$$

Applying $R_2 \to R_2 - R_1$ and $R_3 \to R_3 - R_1$, we have

$$\Delta = 2(x+y+z)\begin{vmatrix} 1 & x & y \\ 0 & x+y+z & 0 \\ 0 & 0 & x+y+z \end{vmatrix}$$

Taking $(x+y+z)$ common from R_2 and R_3, we have

$$\Delta = 2(x+y+z)^3\begin{vmatrix} 1 & x & y \\ 0 & 1 & 0 \\ 0 & 0 & 1 \end{vmatrix}$$

Expanding along R_3, we have

$$\Delta = 2(x+y+z)^3\left[1\begin{vmatrix} 1 & x \\ 0 & 1 \end{vmatrix}\right]$$

$$= 2(x+y+z)^3(1)(1\times 1 - 0\times x)$$
$$= 2(x+y+z)^3.$$

Q. 13. Using matrix method, solve $x - 2y = 4$ and $-3x + 5y = -7$.

Sol. Given,

$$x - 2y = 4$$
$$-3x + 5y = -7$$

The above system of equations can be represented in the matrix form as

$$\begin{bmatrix} 1 & -2 \\ -3 & 5 \end{bmatrix}\begin{bmatrix} x \\ y \end{bmatrix} = \begin{bmatrix} 4 \\ -7 \end{bmatrix}$$

or $AX = B$,

where $A = \begin{bmatrix} 1 & -2 \\ -3 & 5 \end{bmatrix}$, $X = \begin{bmatrix} x \\ y \end{bmatrix}$

and $B = \begin{bmatrix} 4 \\ -7 \end{bmatrix}$

Now, $|A| = \begin{vmatrix} 1 & -2 \\ -3 & 5 \end{vmatrix} = 5 - 6 = -1 \neq 0$

$\therefore A^{-1}$ exists.

Hence, the system has unique solution given by

$$X = A^{-1}B$$

Now $\operatorname{adj} A = \begin{bmatrix} 5 & 2 \\ 3 & 1 \end{bmatrix}$

So $A^{-1} = \frac{1}{|A|}\operatorname{adj} A = \frac{1}{-1}\begin{bmatrix} 5 & 2 \\ 3 & 1 \end{bmatrix}$

$$= \begin{bmatrix} -5 & -2 \\ -3 & -1 \end{bmatrix}$$

Now, $X = A^{-1}B$

$$\begin{bmatrix} x \\ y \end{bmatrix} = \begin{bmatrix} -5 & -2 \\ -3 & -1 \end{bmatrix}\begin{bmatrix} 4 \\ -7 \end{bmatrix}$$

$$= \begin{bmatrix} -20+14 \\ -12+7 \end{bmatrix}$$

$\therefore$ $\begin{bmatrix} x \\ y \end{bmatrix} = \begin{bmatrix} -6 \\ -5 \end{bmatrix}$

$\therefore$ $x = -6$ and $y = -5$.

Q. 14. Let $A = \begin{bmatrix} 5 & 2 \\ 3 & -1 \end{bmatrix}$ and $B = \begin{bmatrix} -13 \\ 1 \end{bmatrix}$, find a matrix X such that $AX = B$.

Sol. Given, $A = \begin{bmatrix} 5 & 2 \\ 3 & -1 \end{bmatrix}$ and $B = \begin{bmatrix} -13 \\ 1 \end{bmatrix}$

$$AX = B$$

Multiply by A^{-1} on both sides

$$A^{-1}AX = A^{-1}B$$

$$IX = A^{-1}B \qquad [\because A^{-1}A = I]$$

$\Rightarrow \quad X = A^{-1}B \quad \ldots(i)$

Now, $A = \begin{bmatrix} 5 & 2 \\ 3 & -1 \end{bmatrix}$

$|A| = \begin{vmatrix} 5 & 2 \\ 3 & -1 \end{vmatrix} = -5-6 = -11 \neq 0$

So A is a non-singular matrix.

As a result A^{-1} exists.

Now cofactors of A are

$C_{11} = -1, C_{12} = -3$

$C_{21} = -2, C_{22} = 5$

Then, $\text{adj}(A) = \begin{bmatrix} C_{11} & C_{12} \\ C_{21} & C_{22} \end{bmatrix}^T$

$= \begin{bmatrix} -1 & -3 \\ -2 & 5 \end{bmatrix}^T = \begin{bmatrix} -1 & -2 \\ -3 & 5 \end{bmatrix}$

$A^{-1} = \frac{1}{|A|} \text{adj} A$

$= \frac{1}{-11}\begin{bmatrix} -1 & -2 \\ -3 & 5 \end{bmatrix}$

Now by equation (i)

$X = \frac{1}{-11}\begin{bmatrix} -1 & -2 \\ -3 & 5 \end{bmatrix}\begin{bmatrix} -13 \\ 1 \end{bmatrix}$

$= \frac{1}{-11}\begin{bmatrix} 13 & -2 \\ -39 & +5 \end{bmatrix}$

$\begin{bmatrix} x \\ y \end{bmatrix} = \frac{-1}{11}\begin{bmatrix} 11 \\ -34 \end{bmatrix} = \begin{bmatrix} -1 \\ 34/11 \end{bmatrix}$

$\therefore \quad x = -1$ and $y = 34/11$

[comparing corresponding elements]

$\Rightarrow \quad X = \begin{bmatrix} -1 \\ 34/11 \end{bmatrix}.$

Q. 15. Given $A = \begin{bmatrix} 2 & -3 \\ -4 & 7 \end{bmatrix}$, compute A^{-1} and show that $2A^{-1} = 9I - A$.★

Sol. Given, $A = \begin{bmatrix} 2 & -3 \\ -4 & 7 \end{bmatrix}$

$|A| = 14 - 12 = 2$

$A^{-1} = \frac{1}{2}\begin{bmatrix} 7 & 3 \\ 4 & 2 \end{bmatrix}$

$\text{L.H.S.} = 2A^{-1} = 2\left(\frac{1}{2}\right)\begin{bmatrix} 7 & 3 \\ 4 & 2 \end{bmatrix}$

$\Rightarrow \quad \text{L.H.S.} = \begin{bmatrix} 7 & 3 \\ 4 & 2 \end{bmatrix} \quad \ldots(i)$

$\text{R.H.S.} = 9I - A = 9\begin{bmatrix} 1 & 0 \\ 0 & 1 \end{bmatrix} - \begin{bmatrix} 2 & -3 \\ -4 & 7 \end{bmatrix}$

$= \begin{bmatrix} 9 & 0 \\ 0 & 9 \end{bmatrix} - \begin{bmatrix} 2 & -3 \\ -4 & 7 \end{bmatrix}$

$\Rightarrow \quad \text{R.H.S.} = \begin{bmatrix} 7 & 3 \\ 4 & 2 \end{bmatrix} \quad \ldots(ii)$

From equations (i) and (ii)

L.H.S. = R.H.S.

Q. 16. Test the consistency of the following system of equations $5x + 7y = 2$ and $10x + 14y = 20$. Also find the values of x and y.

Sol. Given, system of equations are

$5x + 7y = 2 \quad \ldots(i)$

and $10x + 14y = 20 \quad \ldots(ii)$

The above system of equations can be represented in the matrix form as

$AX = B$

Where $A = \begin{bmatrix} 5 & 7 \\ 10 & 14 \end{bmatrix}, X = \begin{bmatrix} x \\ y \end{bmatrix}$

and $B = \begin{bmatrix} 2 \\ 20 \end{bmatrix}$

Now, $|A| = \begin{vmatrix} 5 & 7 \\ 10 & 14 \end{vmatrix} = 5 \times 14 - 7 \times 10$

$= 70 - 70 = 0$

$\therefore$ A is a singular matrix.

Its solution is inconsistent.

Cofactors of matrix A,

$C_{11} = 14, C_{12} = -10$

$C_{21} = -7, C_{22} = 5$

$\therefore \quad \text{adj}(A) = \begin{bmatrix} 14 & -10 \\ -7 & 5 \end{bmatrix}^T = \begin{bmatrix} 14 & -7 \\ -10 & 5 \end{bmatrix}$

$\therefore \quad (\text{adj}(A))B = \begin{bmatrix} 14 & -7 \\ 10 & 5 \end{bmatrix}\begin{bmatrix} 2 \\ 20 \end{bmatrix}$

$= \begin{bmatrix} 14 \times 2 - 7 \times 20 \\ 10 \times 2 + 5 \times 20 \end{bmatrix}$

$= \begin{bmatrix} -112 \\ 120 \end{bmatrix}$

$\Rightarrow \quad (\text{adj}(A))B \neq 0$

$\therefore$ Thus, equations are inconsistent and have no solutions.

Q. 17. By using properties of determinants, show that:

$$\begin{vmatrix} -a^2 & ab & ac \\ ba & -b^2 & bc \\ ca & cb & -c^2 \end{vmatrix} = 4a^2b^2c^2.$$ **[NCERT]**

★ are board exam questions from previous years

Sol. Let $\Delta = \begin{vmatrix} -a^2 & ab & ac \\ ba & -b^2 & bc \\ ca & cb & -c^2 \end{vmatrix}$

Taking out common factors a, b, c from R_1, R_2 and R_3 respectively

$$= abc \begin{vmatrix} -a & b & c \\ a & -b & c \\ a & b & -c \end{vmatrix}$$

Taking out common factors a, b, c from C_1, C_2 and C_3 respectively

$$= a^2b^2c^2 \begin{vmatrix} -1 & 1 & 1 \\ 1 & -1 & 1 \\ 1 & 1 & -1 \end{vmatrix}$$

Applying $R_2 \to R_2 + R_1$ and $R_3 \to R_3 + R_1$, we have

$$\Delta = a^2b^2c^2 \begin{vmatrix} -1 & 1 & 1 \\ 0 & 0 & 2 \\ 0 & 2 & 0 \end{vmatrix}$$

$$= a^2b^2c^2 (-1) \begin{vmatrix} 0 & 2 \\ 2 & 0 \end{vmatrix}$$

$$= a^2b^2c^2 (-1)(0-4)$$

$$= 4a^2b^2c^2.$$

Q. 18. If the system of linear equations

$$x + y + z = 2$$

$$2x + y - z = 3$$

$$3x + 2y + kz = 4$$

has unique solution, then prove that $k \neq 0$.

Sol. We know that system of linear equations

$$a_1x + b_1y + c_1 = A \quad \ldots\text{(i)}$$

$$a_2x + b_2y + c_2 = B \quad \ldots\text{(ii)}$$

and $$a_3x + b_3y + c_3 = C \quad \ldots\text{(iii)}$$

has unique solution, if

$$\begin{vmatrix} a_1 & b_1 & c_1 \\ a_2 & b_2 & c_2 \\ a_3 & b_3 & c_3 \end{vmatrix} \neq 0$$

Given system of linear equations

$$x + y + z = 2$$

$$2x + y - z = 3$$

$$3x + 2y + kz = 4$$

$\therefore$ The system has unique solution, if

$$\begin{vmatrix} 1 & 1 & 1 \\ 2 & 1 & -1 \\ 3 & 2 & k \end{vmatrix} \neq 0$$

$$\Rightarrow \quad 1(k+2) - 1(2k+3) + 1(4-3) \neq 0$$

$$\Rightarrow \quad k + 2 - 2k - 3 + 1 \neq 0$$

$$\Rightarrow \quad -k \neq 0$$

$$\Rightarrow \quad k \neq 0.$$

Q. 19. If $A = \begin{bmatrix} 1 & 5 \\ 6 & 7 \end{bmatrix}$, find adj A.

Sol. Let C_{ij} be the cofactor of a_{ij} in $A = [a_{ij}]$. Then,

$$C_{11} = \text{cofactor of } a_{11} = (-1)^{1+1}\, 7 = 7$$

$$C_{12} = \text{cofactor of } a_{12} = (-1)^{1+2}\, 6 = -6$$

$$C_{21} = \text{cofactor of } a_{13} = (-1)^{2+1}\, 5 = -5$$

$$C_{22} = \text{cofactor of } a_{22} = (-1)^{2+2}\, 1 = 1$$

So adj A is

$$\text{adj } A = \begin{bmatrix} 7 & -6 \\ -5 & 1 \end{bmatrix}^T$$

$$\text{adj } A = \begin{bmatrix} 7 & -5 \\ -6 & 1 \end{bmatrix}$$

Q. 20. Find the inverse of matrix

$$A = \begin{bmatrix} \sin\theta & \cos\theta \\ -\cos\theta & \sin\theta \end{bmatrix}$$

Sol. Given, $A = \begin{bmatrix} \sin\theta & \cos\theta \\ -\cos\theta & \sin\theta \end{bmatrix}$

$$\therefore \quad |A| = \begin{vmatrix} \sin\theta & \cos\theta \\ -\cos\theta & \sin\theta \end{vmatrix}$$

$$= \sin^2\theta + \cos^2\theta = 1$$

$$|A| \neq 0$$

Since $|A|$ is not equal to zero, so A is invertible. Let C_{ij} be the cofactor of a_{ij} in $A = [a_{ij}]$. Then

$$C_{11} = \sin\theta,\ C_{12} = \cos\theta,\ C_{21} = -\cos\theta,\ C_{22} = \sin\theta$$

$$\therefore \quad \text{adj } A = \begin{vmatrix} \sin\theta & -\cos\theta \\ \cos\theta & \sin\theta \end{vmatrix}$$

$$\therefore \quad A^{-1} = \frac{1}{|A|} \text{adj } A$$

$$= \frac{1}{1}\begin{bmatrix} \sin\theta & -\cos\theta \\ \cos\theta & \sin\theta \end{bmatrix}$$

$$= \begin{bmatrix} \sin\theta & -\cos\theta \\ \cos\theta & \sin\theta \end{bmatrix}.$$

Q. 21. Using properties of determinants, prove that :

$$\begin{vmatrix} \mathbf{3a} & \mathbf{-a+b} & \mathbf{-a+c} \\ \mathbf{-b+a} & \mathbf{3b} & \mathbf{-b+c} \\ \mathbf{-c+a} & \mathbf{-c+b} & \mathbf{3c} \end{vmatrix} = \mathbf{3(a+b+c)(ab+bc+ca)}.$$

[NCERT]

Sol. Let $\Delta = \begin{vmatrix} 3a & -a+b & -a+c \\ -b+a & 3b & -b+c \\ -c+a & -c+b & 3c \end{vmatrix}$

Applying $C_1 \to C_1 + C_2 + C_3$, we have

$$\Delta = \begin{vmatrix} a+b+c & -a+b & -a+c \\ a+b+c & 3b & -b+c \\ a+b+c & -c+b & 3c \end{vmatrix}$$

$$= (a+b+c)\begin{vmatrix} 1 & -a+b & -a+c \\ 1 & 3b & -b+c \\ 1 & -c+b & 3c \end{vmatrix}$$

Applying $R_2 \to R_2 - R_1$ and $R_3 \to R_3 - R_1$, we have

$$= (a+b+c)\begin{vmatrix} 1 & -a+b & -a+c \\ 0 & 2b+a & -b+a \\ 0 & -c+a & 2c+a \end{vmatrix}$$

Expanding along C_1, we have

$= (a+b+c)\,[(2b+a)(2c+a) - (-c+a)(-b+a)]$

$= (a+b+c)\,[4bc + 2ab + 2ac + a^2 - bc + ac + ab - a^2]$

$= (a+b+c)\,(3ab + 3ac + 3bc)$

$= 3(a+b+c)\,(ab+bc+ca).$

Q. 22. Using properties of determinants, show that ΔABC is isosceles if*

$$\begin{vmatrix} \mathbf{1} & \mathbf{1} & \mathbf{1} \\ \mathbf{1+\cos A} & \mathbf{1+\cos B} & \mathbf{1+\cos C} \\ \mathbf{\cos^2 A+\cos A} & \mathbf{\cos^2 B+\cos B} & \mathbf{\cos^2 C+\cos C} \end{vmatrix} = \mathbf{0}$$

Sol. We have,

$$\begin{vmatrix} 1 & 1 & 1 \\ 1+\cos A & 1+\cos B & 1+\cos C \\ \cos^2 A+\cos A & \cos^2 B+\cos B & \cos^2 C+\cos C \end{vmatrix} = 0$$

Applying $C_2 \to C_2 - C_1$ and $C_3 \to C_3 - C_1$

$$\begin{vmatrix} 1 & 0 & 0 \\ 1+\cos A & \cos B-\cos A & \cos C-\cos A \\ \cos^2 A+\cos A & (\cos B-\cos A)(1+\cos A+\cos B) & (\cos C-\cos A)(1+\cos A+\cos C) \end{vmatrix} = 0$$

Taking $(\cos B - \cos A)$ and $(\cos C - \cos A)$ as common from C_2 and C_3 respectively.

$$(\cos B - \cos A)\,(\cos C - \cos A)\begin{vmatrix} 1 & 0 & 1 \\ 1+\cos A & 1 & 1 \\ \cos^2 A+\cos A & 1+\cos A+\cos B & 1+\cos A+\cos C \end{vmatrix} = 0$$

* are board exam questions from previous years

Expanding along R_1

$(\cos B - \cos A)\,(\cos C - \cos A)$

$[1 + \cos A + \cos C - 1 - \cos A - \cos B] = 0$

$(\cos B - \cos A)\,(\cos C - \cos A)\,(\cos C - \cos B) = 0$

Either $\cos B = \cos A$ or $\cos C = \cos A$ or $\cos C = \cos B$

i.e., either $BC = AC$ or $BC = AB$ or $AC = AB$

Hence, ΔABC is isosceles.

Q. 23. Using properties of determinants, prove that :*

$$\begin{bmatrix} \mathbf{a^2+2a} & \mathbf{2a+1} & \mathbf{1} \\ \mathbf{2a+1} & \mathbf{a+2} & \mathbf{1} \\ \mathbf{3} & \mathbf{3} & \mathbf{1} \end{bmatrix} = \mathbf{(a-1)^3.}$$

Sol. Let $\Delta = \begin{bmatrix} a^2+2a & 2a+1 & 1 \\ 2a+1 & a+2 & 1 \\ 3 & 3 & 1 \end{bmatrix} = \text{L.H.S.}$

Applying, $R_1 \to R_1 - R_2$, $R_2 \to R_2 - R_3$

$$\Delta = \begin{bmatrix} a^2-2a & a-1 & 1 \\ 2a-2 & a+1 & 1 \\ 3 & 3 & 1 \end{bmatrix}$$

$$= (a-1)^2\begin{bmatrix} a+1 & 1 & 0 \\ 2 & 1 & 0 \\ 3 & 3 & 1 \end{bmatrix}$$

[Taking $(a-1)$ as common from R_1 and R_2]

Now, expanding along C_3

$D = (a-1)^2\,\{(a+1)\cdot 1 - (2)1\}$

$= (a-1)^2\,(a-1) = (a-1)^3$

$=$ R.H.S.

Q. 24. Using properties of determinants, prove that :*

$$\begin{vmatrix} \mathbf{1} & \mathbf{1} & \mathbf{1+3x} \\ \mathbf{1+3x} & \mathbf{1} & \mathbf{1} \\ \mathbf{1} & \mathbf{1+3x} & \mathbf{1} \end{vmatrix} = \mathbf{9(3xyz+xy+yz+zx).}$$

Sol. $\text{L.H.S.} = \begin{vmatrix} 1 & 1 & 1+3x \\ 1+3y & 1 & 1 \\ 1 & 1+3z & 1 \end{vmatrix}$

Taking x, y and z common from R_1, R_2 and R_3 respectively, we get

$$= xyz \begin{vmatrix} \frac{1}{x} & \frac{1}{x} & \frac{1}{x}+3 \\ \frac{1}{y}+3 & \frac{1}{y} & \frac{1}{y} \\ \frac{1}{z} & \frac{1}{z}+3 & \frac{1}{z} \end{vmatrix}$$

On applying $R_1 \to R_1 + R_2 + R_3$, we get

$$= xyz \begin{vmatrix} \frac{1}{x}+\frac{1}{y}+\frac{1}{z}+3 & \frac{1}{x}+\frac{1}{y}+\frac{1}{z}+3 & \frac{1}{x}+\frac{1}{y}+\frac{1}{z}+3 \\ \frac{1}{y}+3 & \frac{1}{y} & \frac{1}{y} \\ \frac{1}{z} & \frac{1}{z}+3 & \frac{1}{y} \end{vmatrix}$$

Taking common $\left(\frac{1}{x}+\frac{1}{y}+\frac{1}{z}+3\right)$ from R_1, we get

$$= xyz\left(\frac{1}{x}+\frac{1}{y}+\frac{1}{z}+3\right) \begin{vmatrix} 1 & 1 & 1 \\ \frac{1}{y}+3 & \frac{1}{y} & \frac{1}{y} \\ \frac{1}{z} & \frac{1}{z}+3 & \frac{1}{z} \end{vmatrix}$$

On applying $C_2 \to C_2 - C_1$ and $C_3 \to C_3 - C_1$, we get

$$= xyz\left(\frac{1}{x}+\frac{1}{y}+\frac{1}{z}+3\right) \begin{vmatrix} 1 & 0 & 0 \\ \frac{1}{y}+3 & -3 & -3 \\ \frac{1}{z} & 3 & 0 \end{vmatrix}$$

On expanding along R_1, we get

$$= xyz\left(\frac{1}{x}+\frac{1}{y}+\frac{1}{z}+3\right) 1(0+9)$$

$$= xyz\left(\frac{yz+xz+xy+3xyz}{xyz}\right)(9)$$

$$= 9(3xyz+xy+yz+zx)$$

= R.H.S.

Chapter 5. Continuity and Differentiability

Q. 1. Find the relationship between *a* and *b* so that the function '*f*' defined by :*

$$f(x) = \begin{cases} ax+1, \text{ if } x \le 3 \\ bx+3, \text{ if } x > 3 \end{cases} \text{ is continuous at } x = 3$$

Sol. Since $f(x)$ is continuous at $x = 3$, we have :

$$\lim_{x\to 3^-} f(x) = \lim_{x\to 3^+} f(x) = f(3) \quad \ldots(i)$$

Now,
$$\lim_{x\to 3^-} f(x) = \lim_{x\to 3^-} (ax+1)$$

$$\lim_{h\to 0} [a(3-h)+1] = 3a+1$$

Similarly,

$$\lim_{x\to 3^+} (bx+3) = \lim_{h\to 0} [b(3+h)+3]$$
$$= 3b+3$$

From (i), we can equate :

$$3a+1 = 3b+3$$

$$\Rightarrow \quad 3(a-b) = 2$$

or
$$a - b = \frac{2}{3}.$$

Q. 2. Find the value of *a*, if the function *f*(*x*) is defined by :

$$f(x) = \begin{cases} 2x-1, & x < 2 \\ a, & x = 2 \\ x+1, & x > 2 \end{cases} \textbf{ is continuous at } x = 2.\star$$

Sol. Since $f(x)$ is continuous at $x = 2$, we have

$$\lim_{x\to 2^-} f(x) = \lim_{x\to 2^+} f(x) = f(2) \quad \ldots(i)$$

Now
$$\lim_{x\to 2^-} f(x) = \lim_{x\to 2^-} (2x-1)$$
$$= \lim_{h\to 0} [2(2-h)-1] = 3$$

Also, we are given

$$f(2) = a$$

Therefore, from (i), we can write :

$$f(2) = 3 \text{ or } a = 3.$$

Q. 3. Is the absolute value function continuous for all values of $x \in R$?

Sol. Consider the function $f(x) = |x|$ for all $x \in R$.

Then,
$$f(x) = \begin{cases} x, \text{ if } x \ge 0 \\ -x, \text{ if } x < 0 \end{cases}$$

Then, f is continuous at every $x \neq 0$, as there is no 'abrupt change' in value.

At $x = 0$, $\quad f(0) = 0$

$$\lim_{x\to 0^-} f(x) = \lim_{h\to 0} (-h) = 0$$

and
$$\lim_{x\to 0^+} f(x) = \lim_{h\to 0} h = 0$$

Thus, f is also continuous at $x = 0$.

Q. 4. Is the function $f(x) = \frac{1}{x}$ continuous for all values of $x \in R$?

Sol. Consider the function for $f(x) = \frac{1}{x}$, $x \neq 0$

* are board exam questions from previous years

As the point $x = 0$ is excluded, it is clear that

$$\lim_{x\to a^-} f(x) = \lim_{x\to a^+} f(x) = \frac{1}{a} = f(a) \text{ for } a \neq 0$$

Thus, f is continuous on $R - \{0\}$.

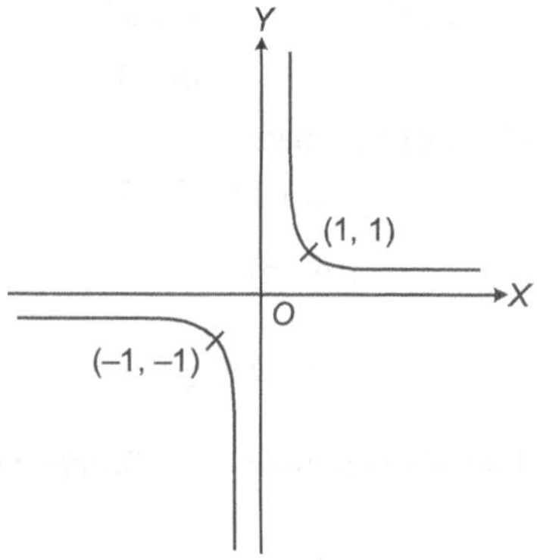

However, the function is not defined at $x = 0$ and is hence not continuous at this point.

Q. 5. Find all points of discontinuity of the given function :

$$f(x) = \frac{x^2 + 2x - 1}{x^2 - 1}$$

Sol. $f(x)$ is a rational function which is continuous at all real numbers except where the denominator is 0. At these points, the function is not defined.

So, the points of discontinuity are where $x^2 - 1 = 0$

Factorising, $(x-1)(x+1) = 0$

$\Rightarrow \qquad x = -1, 1$

Therefore, $f(x)$ is discontinuous at $x = 1, -1$.

Q. 6. Test the function f for continuity in $[0, \infty]$.

$$f(x) = \begin{cases} 1 + x^2, & \text{if } 0 \le x \le 1 \\ 2 - x, & \text{if } x > 1 \end{cases}$$

Sol. In the interval $[0, 1)$, the function $f(x) = 1 + x^2$ is a polynomial and hence continuous.

In the interval $(1, \infty)$, it is again a polynomial and hence continuous.

Now, testing for continuity at $x = 1$.

Note that :

$$\lim_{x\to 1^-} f(x) = \lim_{x\to 1} (1 + x^2) = 2$$

$$\lim_{x\to 1^+} f(x) = \lim_{x\to 1} 2 - x = 1$$

Since, $\quad \lim_{x\to 1^-} f(x) = 2 \neq 1 = \lim_{x\to 1^+} f(x)$

the function is not continuous at $x = 1$.

The function is continuous in (0, 1) and $(1, \infty)$ and discontinuous at $x = 1$.

Q. 7. What is the value of k for which the function

$$f(x) = \begin{cases} \dfrac{\sin 2x}{5x}, & x \neq 0 \\ k, & x = 0 \end{cases} \quad \text{is continuous at } x = 0 \text{ ?}$$

Sol. Since $f(x)$ is continuous at $x = 0$, therefore

$$\lim_{x\to 0^-} f(x) = \lim_{x\to 0^+} f(x) = f(0)$$

$$\Rightarrow \quad \lim_{h\to 0} f(0-h) = \lim_{h\to 0} f(0+h) = k$$

$$\Rightarrow \quad \lim_{h\to 0} \frac{\sin(-2h)}{-5h} = \lim_{h\to 0} \frac{\sin 2h}{5h} = k$$

$$\Rightarrow \quad \lim_{h\to 0} \frac{\sin 2h}{2h} \times \frac{2}{5} = k$$

$$\Rightarrow \quad k = \frac{2}{5}$$

Q. 8. Find the value of 'a' for which the function f defined as :

$$f(x) = \begin{cases} a \sin \dfrac{\pi}{2}(x+1), & x \le 0 \\ \dfrac{\tan x - \sin x}{x^3}, & x > 0 \end{cases} \quad \text{is continuous at } x = 0.\star$$

Sol. Since $f(x)$ is continuous at $x = 0$, we have

$$\lim_{x\to 0^-} a \sin \frac{\pi}{2}(x+1) = \lim_{h\to 0} a \sin \frac{\pi}{2}(-h+1)$$

$$= \lim_{h\to 0} \frac{a \sin \frac{\pi}{2}(-h+1)}{\frac{\pi(-h+1)}{2}} \times \frac{\pi(-h+1)}{2}$$

$$\left[\text{Multiplying and Dividing by } \frac{\pi(-h+1)}{2}\right]$$

$$= \frac{a\pi}{2} \qquad \left[\because \lim_{x\to 0} \frac{\sin x}{x} = 1\right]$$

$$\lim_{x\to 0^+} \frac{\tan x - \sin x}{x^3} = \lim_{h\to 0} \frac{\tan h - \sin h}{h^3}$$

$$= \lim_{h\to 0} \frac{\left(\frac{\sin h}{\cos h}\right) - \sin h}{h^3}$$

$$= \lim_{h\to 0} \frac{\sin h - \sin h \cos h}{h^3 . \cos h}$$

$$= \lim_{h\to 0} \left(\frac{\sin h}{h} \cdot \frac{1 - \cos h}{h^2 . \cos h}\right)$$

$$= \lim_{h\to 0} \frac{\sin h}{h} \cdot \lim_{h\to 0} \frac{1 - \cos h}{h^2} \times \lim_{h\to 0} \frac{1}{\cos h}$$

$\star$ are board exam questions from previous years

$$= 1 \times \lim_{h \to 0} \frac{1 - \cos h}{h^2} \times 1$$

$$= \lim_{h \to 0} \frac{\left(2 \sin^2 \frac{h}{2}\right)}{\left(h^2 \times \frac{4}{4}\right)}$$

$$= \frac{2}{4} \lim_{h \to 0} \left[\frac{\sin \frac{h}{2}}{\frac{h}{2}}\right]^2$$

$$= \frac{1}{2} \times 1$$

$$\therefore \quad \frac{a\pi}{2} = \frac{1}{2}$$

or $$a = \frac{1}{\pi}$$

Q. 9. Find all points of discontinuity of the function

$$f(x) = \begin{cases} -x^2 & x < -1 \\ -1 & x = -1 \\ \dfrac{x}{2-x} & -1 < x < 0 \\ 0 & x = 0 \\ \dfrac{5x^2}{(3x-1)} & x > 0 \end{cases}$$

Sol. Test each interval and its each points in turn for continuity.

$(-\infty, -1)$: The function is a polynomial hence continuous.

$x = -1$:

$$\lim_{x \to -1^-} f(x) = \lim_{x \to -1} -x^2 = -(-1)^2 = -1$$

$$\lim_{x \to -1^+} f(x) = \lim_{x \to -1} \frac{x}{2-x} = -\frac{1}{3}$$

Since the left-hand limit is not equal to the right-hand limit, $\lim_{x \to -1} f(x)$ does not exist.

Thus, the function is not continuous at $x = -1$.

$(-1, 0)$: The function is $f(x) = \dfrac{x}{2-x}$, a rational function not defined at $x = 2$. Since 2 is not in this interval, the function is continuous.

$x = 0$: $$\lim_{x \to 0^-} f(x) = \lim_{x \to 0} \frac{x}{2-x} = 0$$

$$\lim_{x \to 0^+} f(x) = \lim_{x \to 0} \frac{5x^2}{(3x-1)} = 0$$

Also $$f(0) = 0$$

So, $$\lim_{x \to 0^+} f(x) = \lim_{x \to 0^-} f(x) = f(0)$$

Thus, the function is continuous at $x = 0$.

$(0, \infty)$: The function is a rational function in this interval.

The denominator is 0 when $x = \dfrac{1}{3}$.

Since $\dfrac{1}{3} \in (0, \infty)$, this is a point of discontinuity of the function.

So, the function is discontinuous at $x = -1, \dfrac{1}{3}$ and is continuous at all other points of the real line.

Q. 10. Examine the continuity of the greatest integer function.

Sol. Consider the function $f : R \to R$ defined by

$$f(x) = [x], x \in R$$

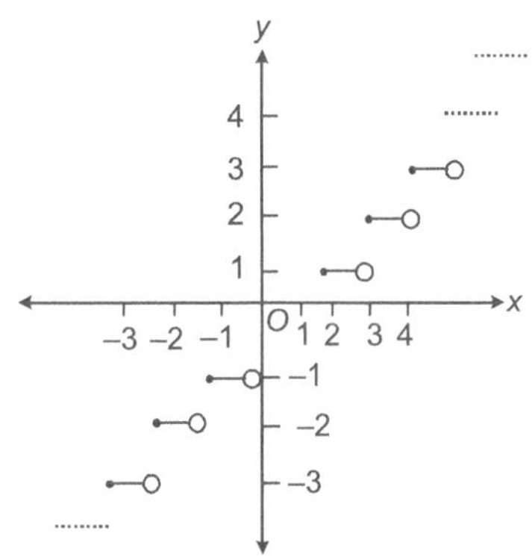

(Graph of a greatest integer function)

$f(x) = [x]$ where x denotes the greatest integer less than or equal to x.

For this function, if n is any integer, then

$$\lim_{x \to n^-} f(x) = n - 1, \quad \lim_{x \to n^+} f(x) = n$$

Thus for any integer n, f is not continuous at $x = n$.

However, if 'a' is not an integer, then there is an integer n such that $n - 1 < a < n$ so that

$$\lim_{x \to a^-} f(x) = \lim_{x \to a^+} f(x) = n - 1 = f(a)$$

Thus, f is continuous at 'a' if and only if a is not an integer.

Q. 11. Show that $f(x)=\begin{cases}12x-13, & x\le 3\\ 2x^2+5, & x>3\end{cases}$ **is differentiable at $x=3$. Also, find $f'(3)$.**

Sol. $\text{L.H.D.} = \lim_{x\to 3^-}\dfrac{f(x)-f(3)}{x-3}$

$$= \lim_{h\to 0}\frac{f(3-h)-f(3)}{(3-h)-3}$$

$$= \lim_{h\to 0}\frac{[12(3-h)-13]-[12(3)-13]}{-h}$$

$$= \lim_{h\to 0}\frac{36-12h-13-36+13}{-h}$$

$$= \lim_{h\to 0}\frac{-12h}{-h} = 12$$

$$\text{R.H.D.} = \lim_{x\to 3^+}\frac{f(x)-f(3)}{x-3}$$

$$= \lim_{h\to 0}\frac{f(3+h)-f(3)}{(3+h)-3}$$

$$= \lim_{h\to 0}\frac{[2(3+h)^2+5]-[12(3)-13]}{h}$$

$$= \lim_{h\to 0}\frac{2(9+h^2+6h)+5-36+13}{h}$$

$$= \lim_{h\to 0}\frac{18+2h^2+12h+18-36}{h}$$

$$= \lim_{h\to 0}\frac{2h(h+6)}{h} = 12$$

Since L.H.D. = R.H.D., $f(x)$ is differentiable at $x=3$ and $f'(3)=12$.

Q. 12. If $f(9)=9$ and $f'(9)=4$, then what is the value of $\lim_{x\to 9}\dfrac{\sqrt{f(x)}-3}{\sqrt{x}-3}$.

Sol. We have, $\lim_{x\to 9}\dfrac{\sqrt{f(x)}-3}{\sqrt{x}-3}$

$$= \lim_{x\to 9}\frac{\sqrt{f(x)}-3}{\sqrt{x}-3}\times\frac{\sqrt{f(x)}+3}{\sqrt{x}+3}\times\frac{\sqrt{x}+3}{\sqrt{f(x)}+3}$$

$$= \lim_{x\to 9}\frac{f(x)-9}{x-9}\times\frac{\sqrt{x}+3}{\sqrt{f(x)}+3}$$

$$= f'(9)\times\frac{\sqrt{9}+3}{\sqrt{f(9)}+3}$$

$$= f'(9)\times\frac{3+3}{3+3}$$

$$= f'(9) = 4.$$

Q. 13. Find the value of c in Rolle's theorem for the function $f(x)=x^3-3x$ in $[-\sqrt{3}, 0]$. ★

Sol. We know that the polynomial function $f(x) = x^3 - 3x$ is everywhere continuous and differentiable.

So, $f(x)$ is continuous on $[-\sqrt{3}, 0]$.

Also $f(x)$ is differentiable on $(-\sqrt{3}, 0)$

Now $\quad f(-\sqrt{3}) = (-\sqrt{3})^3 - 3(-\sqrt{3})$

$$= -3\sqrt{3}+3\sqrt{3} = 0$$

and $\quad f(0) = (0)^3 - 3(0) = 0$

$\therefore \quad f(-\sqrt{3}) = f(0)$

Thus, all three conditions of Rolle's theorem are satisfied.

Now, there must exist $c \in (-\sqrt{3}, 0)$ such that $f'(c)=0$

Now, $\quad f'(x) = 3x^2-3$

Then, $\quad f'(c) = 0$

$\Rightarrow \quad 3c^2-3 = 0$

$3(c^2-1) = 0$

$c = \pm 1$

But $\quad c \ne 1$ as $1 \notin (-\sqrt{3}, 0)$

$\therefore \quad c = -1 \in (-\sqrt{3}, 0)$

Thus required value of c is -1.

Q. 14. If $y = 2\cos(\log x) + 3\sin(\log x)$, prove that:

$$x^2\frac{d^2y}{dx^2}+x\frac{dy}{dx}+y=0.\ \bigstar$$

Sol. Given, $\quad y = 2\cos(\log x) + 3\sin(\log x)$

On differentiating both sides w.r.t. x, we get

$$\frac{dy}{dx} = -2\sin(\log x)\cdot\frac{1}{x} + 3\cos(\log x)\cdot\frac{1}{x}$$

$$x\frac{dy}{dx} = -2\sin(\log x) + 3\cos(\log x)$$

Again differentiating both sides w.r.t. x, we get

$$x\frac{d^2y}{dx^2}+\frac{dy}{dx} = -2\cos(\log x)\cdot\frac{1}{x} - 3\sin(\log x)\cdot\frac{1}{x}$$

$$x^2\frac{d^2y}{dx^2}+x\cdot\frac{dy}{dx} = -(2\cos(\log x)+3\sin(\log x))$$

$$x^2\frac{d^2y}{dx^2}+x\frac{dy}{dx} = -y$$

$$\Rightarrow \quad x^2\frac{d^2y}{dx^2}+x\frac{dy}{dx}+y = 0$$

Q. 15. If $(\cos x)^y = (\cos y)^x$, find $\dfrac{dy}{dx}$. ★

Sol. Taking log on both sides, we get

$$y\log(\cos x) = x\log(\cos y)$$

$[\because \log x^n = n\log x]$

Differentiating both sides of the above equation w.r.t. x, we get

$$y\frac{d}{dx}[\log(\cos x)] + \log(\cos x)\frac{d}{dx}(y)$$

$$= x\frac{d}{dx}[\log(\cos y)] + \log(\cos y)\frac{d}{dx}(x)$$

★ are board exam questions from previous years

$$\Rightarrow y \cdot \frac{1}{\cos x} \cdot \frac{d}{dx}(\cos x) + \log(\cos x)\frac{dy}{dx}$$

$$= x \cdot \frac{1}{\cos y} \cdot \frac{d}{dx}(\cos y) + \log(\cos y) \cdot 1$$

$$\Rightarrow y \cdot \frac{1}{\cos x} \cdot (-\sin x) + \log(\cos x) \cdot \frac{dy}{dx}$$

$$= x \cdot \frac{1}{\cos y} \cdot (-\sin y) \cdot \frac{dy}{dx} + \log(\cos y)$$

$$\Rightarrow -y \cdot \tan x + \log(\cos x) \cdot \frac{dy}{dx}$$

$$= -x \cdot \tan y \cdot \frac{dy}{dx} + \log(\cos y)$$

$$\Rightarrow [x.\tan y + \log(\cos x)] \cdot \frac{dy}{dx} = y.\tan x + \log(\cos y)$$

$$\Rightarrow \quad \frac{dy}{dx} = \frac{y \cdot \tan x + \log(\cos y)}{x \cdot \tan y + \log(\cos x)}$$

Q. 16. For the curve $\sqrt{x} + \sqrt{y} = 1$, find the value of $\frac{dy}{dx}$ at $\left(\frac{1}{4}, \frac{1}{4}\right)$.

Sol. Given curve is

$$\sqrt{x} + \sqrt{y} = 1$$

Differentiating with respect to x, we get

$$\frac{1}{2\sqrt{x}} + \frac{1}{2\sqrt{y}}\frac{dy}{dx} = 0$$

$$\Rightarrow \quad \frac{1}{2\sqrt{y}}\frac{dy}{dx} = -\frac{1}{2\sqrt{x}}$$

$$\Rightarrow \quad \frac{dy}{dx} = -\frac{\sqrt{y}}{\sqrt{x}}$$

$$\Rightarrow \quad \left(\frac{dy}{dx}\right)_{\left(\frac{1}{4}, \frac{1}{4}\right)} = -\frac{\sqrt{\frac{1}{4}}}{\sqrt{\frac{1}{4}}}$$

$$= -1$$

Q. 17. If $y = \cos^{-1}\left(\frac{2x}{1+x^2}\right)$ then find $\frac{dy}{dx}$.

Sol. Given,

$$y = \cos^{-1}\left(\frac{2x}{1+x^2}\right)$$

$$\frac{dy}{dx} = -\frac{1}{\sqrt{1 - \frac{4x^2}{(1+x^2)^2}}} \frac{d}{dx}\left(\frac{2x}{1+x^2}\right)$$

$$= -\frac{(1+x^2)}{\sqrt{(1-x^2)^2}} \cdot \frac{2(1+x^2) - 4x^2}{(1+x^2)^2}$$

$$\Rightarrow \quad \frac{dy}{dx} = -\frac{2(1+x^2)}{(1-x^2)} \cdot \frac{(1-x^2)}{(1+x^2)^2}$$

$$= -\frac{2}{1+x^2}$$

Q. 18. Differentiate $\tan^{-1}\left[\frac{\sqrt{1+x^2} - 1}{x}\right]$ w.r.t. x.★

Sol. Let $\quad y = \tan^{-1}\left[\frac{\sqrt{1+x^2} - 1}{x}\right]$

Put $\quad x = \tan\theta \Rightarrow \theta = \tan^{-1} x$

$$\therefore \quad y = \tan^{-1}\left[\frac{\sqrt{1+\tan^2\theta} - 1}{\tan\theta}\right]$$

$$\Rightarrow \quad y = \tan^{-1}\left[\frac{\sec\theta - 1}{\tan\theta}\right]$$

$$[\because 1 + \tan^2\theta = \sec^2\theta]$$

$$\Rightarrow \quad y = \tan^{-1}\left[\frac{\frac{1}{\cos\theta} - 1}{\frac{\sin\theta}{\cos\theta}}\right]$$

$$= \tan^{-1}\left[\frac{1 - \cos\theta}{\sin\theta}\right]$$

$$\Rightarrow \quad y = \tan^{-1}\left[\frac{2\sin^2\frac{\theta}{2}}{2\sin\frac{\theta}{2}\cos\frac{\theta}{2}}\right]$$

$$= \tan^{-1}\left[\frac{\sin\frac{\theta}{2}}{\cos\frac{\theta}{2}}\right] = \tan^{-1}\left[\tan\frac{\theta}{2}\right]$$

$$\Rightarrow \quad y = \frac{\theta}{2} = \frac{1}{2}\tan^{-1}$$

$$\Rightarrow \quad \frac{dy}{dx} = \frac{1}{2} \cdot \frac{1}{1+x^2}$$

$$= \frac{1}{2(1+x^2)}$$

Q. 19. If $x = \sqrt{a^{\sin^{-1} t}}$, $y = \sqrt{a^{\cos^{-1} t}}$, show that $\frac{dy}{dx} = \frac{-y}{x}$.★

Sol. Given $x = \sqrt{a^{\sin^{-1} t}}$

Differentiating w.r.t. x, we get

$$\frac{dx}{dt} = \frac{1}{2}(a^{\sin^{-1} t})^{-\frac{1}{2}} \frac{d}{dt}(a^{\sin^{-1} t})$$

★ are board exam questions from previous years

$$\left[\because \frac{d}{dx}(x^n) = nx^{n-1}\right]$$

$$= \frac{1}{2}(a^{\sin^{-1}t})^{-\frac{1}{2}}.a^{\sin^{-1}t}.\log a.\frac{d}{dt}(\sin^{-1}t)$$

$$\left[\because \frac{d}{dx}(a^x) = a^x.\log a\right]$$

$$= \frac{1}{2}(a^{\sin^{-1}t})^{-\frac{1}{2}}.(a^{\sin^{-1}t})^1.\log a.\frac{1}{\sqrt{1-t^2}}$$

$$= \frac{1}{2}(a^{\sin^{-1}t})^{\frac{1}{2}}.\log a.\frac{1}{\sqrt{1-t^2}}$$

$$\therefore \quad \frac{dx}{dt} = \frac{\left(\frac{1}{2}\right)\sqrt{a^{\sin^{-1}t}}.\log a}{\sqrt{1-t^2}} \quad \text{...(i)}$$

Also, $y = \sqrt{a^{\cos^{-1}t}}$

Differentiating w.r.t. t, we get

$$\frac{dy}{dt} = \frac{1}{2}(a^{\cos^{-1}t})^{-\frac{1}{2}}.\frac{d}{dt}(a^{\cos^{-1}t})$$

$$\left[\because \frac{d}{dx}(x^n) = nx^{n-1}\right]$$

$$= \frac{1}{2}(a^{\cos^{-1}t})^{-\frac{1}{2}}.a^{\cos^{-1}t}.\log a.\frac{d}{dt}(\cos^{-1}t)$$

$$\left[\because \frac{d}{dx}(a^x) = a^x.\log a\right]$$

$$= \frac{1}{2}(a^{\cos^{-1}t})^{-\frac{1}{2}}.(a^{\cos^{-1}t})^1.\log a.\frac{-1}{\sqrt{1-t^2}}$$

$$= \frac{1}{2}(a^{\cos^{-1}t})^{\frac{1}{2}}.\log a.\frac{-1}{\sqrt{1-t^2}}$$

$$\therefore \quad \frac{dy}{dt} = \frac{\left(-\frac{1}{2}\right)\sqrt{a^{\cos^{-1}t}}.\log a}{\sqrt{1-t^2}} \quad \text{...(ii)}$$

Dividing equation (ii) by equation (i), we get :

$$\frac{dy}{dx} = \frac{\left(\frac{dy}{dt}\right)}{\left(\frac{dx}{dt}\right)} = \frac{\left[\frac{\left(-\frac{1}{2}\right)\sqrt{a^{\cos^{-1}t}}.\log a}{\sqrt{1-t^2}}\right]}{\left[\frac{\left(\frac{1}{2}\right)\sqrt{a^{\sin^{-1}t}}.\log a}{\sqrt{1-t^2}}\right]}$$

$$\frac{dy}{dx} = -\frac{\sqrt{a^{\cos^{-1}t}}}{\sqrt{a^{\sin^{-1}t}}} = -\frac{y}{x}$$

[substituting the given values of x and y]

Q. 20. If $y = \dfrac{\log (x-2)^2}{\log e^x}$ for $x \neq 0, 2$ then find $y'(3)$.

Sol. We have,

$$y = \frac{\log (x-2)^2}{\log e^x}$$

$$= \frac{2\log (x-2)}{x \log e} = \frac{2}{x}\log (x-2)$$

$$[\because \log e = 1]$$

$$y'(x) = \frac{2}{x}\cdot\frac{1}{x-2} - \frac{2}{x^2}\log (x-2)$$

$$y'(3) = \frac{2}{3}\cdot\frac{1}{1} - \frac{2}{9}\log 1 = \frac{2}{3} \qquad [\because \log 1 = 0]$$

Q. 21. If $y = (x)^{x^2}$, then find $\dfrac{dy}{dx}$.

Sol. We have, $y = (x)^{x^2}$

Taking logarithm on both sides, we get

$$\log y = x^2 \log x$$

Differentiating both sides with respect to x, we get

$$\frac{1}{y}\frac{dy}{dx} = x^2\cdot\frac{1}{x} + \log x\cdot 2x = x + 2x\log x$$

$$\Rightarrow \quad \frac{dy}{dx} = (x)^{x^2}.x(2\log x + 1)$$

$$= x^{(x^2+1)}(2\log x + 1)$$

Q. 22. If $x = \tan\left[\dfrac{1}{a}\log y\right]$, show that

$$(1+x^2)\frac{d^2y}{dx^2} + (2x-a)\frac{dy}{dx} = 0.\star$$

Sol. Given, $x = \tan\left[\dfrac{1}{a}\log y\right]$

$$\Rightarrow \quad \tan^{-1}x = \frac{1}{a}\log y$$

$$\Rightarrow \quad \log y = a\tan^{-1}x$$

Differentiating both sides of the above equation w.r.t. x, we get

$$\frac{1}{y}\cdot\frac{dy}{dx} = a\cdot\frac{1}{1+x^2}$$

$$\left[\because \frac{d}{dx}(\tan^{-1}x) = \frac{1}{1+x^2}\right]$$

$$\Rightarrow \quad (1+x^2)\frac{dy}{dx} = a.y$$

★ are board exam questions from previous years

Differentiating both sides of the above equation again w.r.t. x, we get

$$(1+x^2)\frac{d}{dx}\left(\frac{dy}{dx}\right)+\frac{dy}{dx}\cdot\frac{d}{dx}(1+x^2) = \frac{d}{dx}(ay)$$

$$\Rightarrow \quad (1+x^2)\frac{d^2y}{dx^2}+\frac{dy}{dx}.(2x) = a.\frac{dy}{dx}$$

$$\Rightarrow \quad (1+x^2)\frac{d^2y}{dx^2}+(2x-a)\frac{dy}{dx} = 0$$

which is the desired expression.

Q. 23. If $x = \log t + \sin t, y = e^t + \cos t$, then find $\frac{dy}{dx}$.

Sol. We have

$$x = \log t + \sin t, y = e^t + \cos t$$

$$\Rightarrow \quad \frac{dx}{dt} = \frac{1}{t} + \cos t$$

and $$\frac{dy}{dt} = e^t - \sin t$$

$$\therefore \quad \frac{dy}{dx} = \frac{dy}{dt}\times\frac{dt}{dx} = \frac{t(e^t - \sin t)}{1+t\cos t}$$

Q. 24. If $x = a(\theta + \sin\theta)$ and $y = a(1 - \cos\theta)$, find $\frac{d^2y}{dx^2}$.★

Sol. We have,

$$x = a(\theta + \sin\theta)$$

and $$y = a(1-\cos\theta)$$

$$\Rightarrow \quad \frac{dx}{d\theta} = a(1+\cos\theta)$$

and $$\frac{dy}{d\theta} = a\sin\theta$$

$$\Rightarrow \quad \frac{dy}{dx} = \frac{a\sin\theta}{a(1+\cos\theta)}$$

$$= \frac{2\sin\frac{\theta}{2}\cos\frac{\theta}{2}}{2\cos^2\frac{\theta}{2}} = \tan\frac{\theta}{2}$$

Now, $$\frac{d^2y}{dx^2} = \frac{d}{dx}\left(\tan\frac{\theta}{2}\right)$$

$$= \frac{d}{d\theta}\left(\tan\frac{\theta}{2}\right)\cdot\frac{d\theta}{dx}$$

$$\left[\because \frac{d}{dx}[f(\theta)] = \frac{d}{d\theta}f(\theta).\frac{d\theta}{dx}\right]$$

$$= \sec^2\frac{\theta}{2}\cdot\frac{d}{d\theta}\left[\left(\frac{\theta}{2}\right)\right]\cdot\frac{d\theta}{dx}$$

$$= \frac{1}{2}\cdot\sec^2\frac{\theta}{2}\times\frac{1}{a(1+\cos\theta)}$$

$$\left[\because \frac{dx}{d\theta} = a(1+\cos\theta) \therefore \frac{dx}{d\theta} = \frac{1}{a(1+\cos\theta)}\right]$$

$$= \frac{1}{2a}\cdot\sec^2\frac{\theta}{2}\cdot\frac{1}{2\cos^2\frac{\theta}{2}}$$

$$= \frac{1}{4a}\cdot\sec^4\frac{\theta}{2}$$

Q. 25. Find the second derivative of $\cos^{-1}x$.

Sol. Let $\quad y = \cos^{-1}x$

$$\Rightarrow \quad y_1 = \frac{-1}{\sqrt{1-x^2}} = -(1-x^2)^{-1/2}$$

$$\Rightarrow \quad y_2 = -1\times\frac{-1}{2}(1-x^2)^{-3/2}\times -2x$$

$$= \frac{-x}{(1-x^2)^{3/2}}$$

Q. 26. If $y = \sin x + e^x$ then find $\frac{d^2x}{dy^2}$.

Sol. Given, $\quad y = \sin x + e^x$

$$\frac{dy}{dx} = \cos x + e^x$$

$$\frac{dx}{dy} = \frac{1}{\cos x + e^x}$$

$$\frac{d^2x}{dy^2} = (\cos x + e^x)^{-2}(-\sin x + e^x)\frac{dx}{dy}$$

$$= \frac{-\sin x + e^x}{(\cos x + e^x)^2}\cdot(\cos e^x + e^x)^{-1}$$

$$= \frac{e^x - \sin x}{(\cos x + e^x)^3}$$

Q. 27. Differentiate $\tan^{-1}\left(\frac{1+\cos x}{\sin x}\right)$ with respect to x.★

Sol. Let $$y = \tan^{-1}\left(\frac{1+\cos x}{\sin x}\right)$$

$$= \tan^{-1}\left(\frac{2\cos^2\frac{x}{2}}{2\sin\frac{x}{2}\cos\frac{x}{2}}\right)$$

$$= \tan^{-1}\left(\cot\frac{x}{2}\right)$$

$$= \tan^{-1}\left[\tan\left(\frac{\pi}{2}-\frac{x}{2}\right)\right]$$

★ are board exam questions from previous years

$$= \frac{\pi}{2} - \frac{x}{2}$$

Differentiate w.r.t. x,

$$\frac{dy}{dx} = \frac{d}{dx}\left(\frac{\pi}{2} - \frac{x}{2}\right) = -\frac{1}{2}$$

Q. 28. Differentiate $\left(\frac{1}{x}\right)^x$ with respect to x.

Sol. Let $\quad y = \left(\frac{1}{x}\right)^x$

Taking log on both sides, we get

$$\Rightarrow \quad \log y = x \log\left(\frac{1}{x}\right)$$

$$\Rightarrow \quad \log y = x(\log 1 - \log x)$$

$$\Rightarrow \quad \log y = -x \log x$$

Differentiating both sides with respect to x, we get

$$\frac{1}{y}\frac{dy}{dx} = -\left(x.\frac{1}{x} + \log x.1\right)$$

$$\Rightarrow \quad \frac{1}{y}\frac{dy}{dx} = -(1 + \log x)$$

$$\Rightarrow \quad \frac{dy}{dx} = -y(1 + \log x)$$

$$= -\left(\frac{1}{x}\right)^x (1 + \log x)$$

Q. 29. If $\log(x^2 + y^2) = 2\tan^{-1}\left(\frac{y}{x}\right)$, show that

$$\frac{dy}{dx} = -\frac{(x+y)}{x-y}$$

Sol. We have :

$$\log(x^2 + y^2) = 2\tan^{-1}\left(\frac{y}{x}\right)$$

Differentiating both sides w.r.t. x, we get

$$\frac{1}{x^2+y^2}\cdot\frac{d}{dx}(x^2+y^2) = \frac{2}{1+\left(\frac{y}{x}\right)^2}\frac{d}{dx}\left(\frac{y}{x}\right)$$

$$= \frac{2}{1+\frac{y^2}{x^2}}\left[\frac{x\frac{dy}{dx} - y.1}{x^2}\right]$$

$$= \frac{2x^2}{x^2+y^2}\left(\frac{x\frac{dy}{dx} - y}{x^2}\right)$$

$$= \frac{1}{x^2+y^2}\left(-2y + 2x\frac{dy}{dx}\right)$$

$$\Rightarrow \left(x + y\cdot\frac{dy}{dx}\right)\frac{2}{x^2+y^2} = \frac{2}{x^2+y^2}\left(x\frac{dy}{dx} - y\right)$$

$$\Rightarrow \quad x + y\cdot\frac{dy}{dx} = x\frac{dy}{dx} - y$$

$$\Rightarrow \quad \frac{dy}{dx}(x-y) = -x-y$$

$$\Rightarrow \quad \frac{dy}{dx} = -\frac{(x+y)}{x-y}$$

Q. 30. Find the derivative of

$\tan^{-1}\left(\frac{\sqrt{1+x^2}+\sqrt{1-x^2}}{\sqrt{1+x^2}-\sqrt{1-x^2}}\right)$ with respect to x.★

Sol. Let $\quad y = \tan^{-1}\left(\frac{\sqrt{1+x^2}+\sqrt{1-x^2}}{\sqrt{1+x^2}-\sqrt{1-x^2}}\right)$

Put $\quad x^2 = \cos\theta$

$$\Rightarrow \quad \theta = \cos^{-1} x^2$$

$$\therefore \quad y = \tan^{-1}\left(\frac{\sqrt{1+\cos\theta}+\sqrt{1-\cos\theta}}{\sqrt{1+\cos\theta}-\sqrt{1-\cos\theta}}\right)$$

$$= \tan^{-1}\left[\frac{\sqrt{2\cos^2\frac{\theta}{2}}+\sqrt{2\sin^2\frac{\theta}{2}}}{\sqrt{2\cos^2\frac{\theta}{2}}-\sqrt{2\sin^2\frac{\theta}{2}}}\right]$$

$$= \tan^{-1}\left(\frac{\sqrt{2}\left(\cos\frac{\theta}{2}+\sin\frac{\theta}{2}\right)}{\sqrt{2}\left(\cos\frac{\theta}{2}-\sin\frac{\theta}{2}\right)}\right)$$

$$= \tan^{-1}\left(\frac{\cos\frac{\theta}{2}+\sin\frac{\theta}{2}}{\cos\frac{\theta}{2}-\sin\frac{\theta}{2}}\right)$$

$$= \tan^{-1}\left(\frac{\frac{\cos\frac{\theta}{2}}{\cos\frac{\theta}{2}}+\frac{\sin\frac{\theta}{2}}{\cos\frac{\theta}{2}}}{\frac{\cos\frac{\theta}{2}}{\cos\frac{\theta}{2}}-\frac{\sin\frac{\theta}{2}}{\cos\frac{\theta}{2}}}\right)$$

$$= \tan^{-1}\left(\frac{1+\tan\frac{\theta}{2}}{1-\tan\frac{\theta}{2}}\right)$$

★ are board exam questions from previous years

$$= \tan^{-1}\left[\tan\left(\frac{\pi}{4}+\frac{\theta}{2}\right)\right]$$

$$= \frac{\pi}{4}+\frac{\theta}{2} = \frac{\pi}{4}+\frac{1}{2}(\cos^{-1}x^2)$$

$$\therefore \quad \frac{dy}{dx} = \frac{d}{dx}\left(\frac{\pi}{4}\right)+\frac{1}{2}\frac{d}{dx}(\cos^{-1}x^2)$$

$$= 0+\frac{1}{2}\left(-\frac{1}{\sqrt{1-x^4}}\right)\times\frac{d}{dx}(x^2)$$

$$= -\frac{1}{2\sqrt{1-x^4}}\times 2x$$

$$= -\frac{x}{\sqrt{1-x^4}}$$

Q. 31. If

$$\boldsymbol{y = (\log_{\cos x}\sin x)(\log_{\sin x}\cos x)^{-1}+\sin^{-1}\left(\frac{2x}{1+x^2}\right),}$$

find $\frac{dy}{dx}$ at $x = \frac{\pi}{4}$.

Sol. We have,

$$y = (\log_{\cos x}\sin x)(\log_{\sin x}\cos x)^{-1}+\sin^{-1}\left(\frac{2x}{1+x^2}\right)$$

$$= \frac{\log\sin x}{\log\cos x}\left(\frac{\log\cos x}{\log\sin x}\right)^{-1}+\sin^{-1}\left(\frac{2x}{1+x^2}\right)$$

$$= \frac{\log\sin x}{\log\cos x}\times\frac{\log\sin x}{\log\cos x}+\sin^{-1}\left(\frac{2x}{1+x^2}\right)$$

$$= \left(\frac{\log\sin x}{\log\cos x}\right)^2+\sin^{-1}\left(\frac{2x}{1+x^2}\right)$$

$$\Rightarrow \quad y = u+v$$

Where $u = \left(\dfrac{\log\sin x}{\log\cos x}\right)^2$

and $v = \sin^{-1}\left(\dfrac{2x}{1+x^2}\right)$

$$\therefore \quad \frac{dy}{dx} = \frac{du}{dx}+\frac{dv}{dx} \qquad \ldots(i)$$

Now, $u = \left(\dfrac{\log\sin x}{\log\cos x}\right)^2$

$$\therefore \quad \frac{du}{dx} = 2\left(\frac{\log\sin x}{\log\cos x}\right)\times\frac{d}{dx}\left(\frac{\log\sin x}{\log\cos x}\right)$$

$$= 2\left(\frac{\log\sin x}{\log\cos x}\right)\times\left[\frac{\log\cos x\times\frac{\cos x}{\sin x}-\log\sin x\times\left(-\frac{\sin x}{\cos x}\right)}{(\log\cos x)^2}\right]$$

$$= 2\frac{\log\sin x}{(\log\cos x)^3}[\cot x\log\cos x+\tan x\log\sin x]$$

$$\therefore \quad \left.\frac{du}{dx}\right|_{x=\frac{\pi}{4}} = \frac{2\log\left(\frac{1}{\sqrt{2}}\right)}{\left[\log\left(\frac{1}{\sqrt{2}}\right)\right]^3}\left[2\log\left(\frac{1}{\sqrt{2}}\right)\right]$$

$$= \frac{4}{\log\left(\frac{1}{\sqrt{2}}\right)} = \frac{-4}{\log\sqrt{2}}$$

$$= \frac{-8}{\log 2}$$

Also $v = \sin^{-1}\left(\dfrac{2x}{1+x^2}\right)$

Put $x = \tan\theta$

$\Rightarrow \quad \theta = \tan^{-1}x$

$$\therefore \quad v = \sin^{-1}\left(\frac{2\tan\theta}{1+\tan^2\theta}\right)$$

$$= \sin^{-1}(\sin 2\theta)$$

$$= 2\theta = 2\tan^{-1}x$$

$$\therefore \quad \frac{dv}{dx} = \frac{2}{1+x^2}$$

$$\Rightarrow \quad \left.\frac{dv}{dx}\right|_{x=\frac{\pi}{4}} = \frac{32}{16+\pi^2}$$

$$\therefore \quad \left.\frac{dy}{dx}\right|_{x=\frac{\pi}{4}} = \frac{-8}{\log 2}+\frac{32}{16+\pi^2}$$

Q. 32. Differentiate $(\sin x)^{\cos x}+(\cos x)^{\sin x}$ with respect to x.

Sol. Let $y = (\sin x)^{\cos x}+(\cos x)^{\sin x} = U+V,$

Where $U = (\sin x)^{\cos x}$ and $V = (\cos x)^{\sin x}$

$$\therefore \quad \frac{dy}{dx} = \frac{dU}{dx}+\frac{dV}{dx} \qquad \ldots(i)$$

Now, $U = (\sin x)^{\cos x}$

Taking log on both sides, we get

$$\log U = \cos x.\log(\sin x)$$

Differentiating both sides w.r.t. x, we get

$$\frac{1}{U}\frac{dU}{dx} = \cos x.\frac{1}{\sin x}\cos x+\log(\sin x)(-\sin x)$$

$$\Rightarrow \quad \frac{dU}{dx} = (\sin x)^{\cos x}\left[\frac{\cos^2 x}{\sin x} - \sin x \log(\sin x)\right] \quad \text{...(ii)}$$

Also, $V = (\cos x)^{\sin x}$

Taking log on both sides, we get

$$\log V = \sin x.\log(\cos x)$$

Differentiating both sides w.r.t. x, we get

$$\frac{1}{V}\frac{dV}{dx} = \frac{\sin x}{\cos x}(-\sin x) + \log(\cos x).\cos x$$

$$\Rightarrow \quad \frac{dV}{dx} = (\cos x)^{\sin x}$$

$$\times\left[\frac{-\sin^2 x}{\cos x} + \cos x.\log(\cos x)\right] \quad \text{...(iii)}$$

From (i), (ii) and (iii), we have

$$\frac{dy}{dx} = (\sin x)^{\cos x}\left[\frac{\cos^2 x}{\sin x} - \sin x.\log(\sin x)\right]$$

$$+ (\cos x)^{\sin x}\left[\frac{-\sin^2 x}{\cos x} + \cos x.\log(\cos x)\right]$$

Q. 33. If $x = e^t(\cos t + \sin t)$; $y = e^t(\cos t - \sin t)$, find y_1, y_2 and y_3 at $t = 0$.

Sol. Given, $x = e^t(\cos t + \sin t)$

$$y = e^t(\cos t - \sin t)$$

$$\therefore \quad y_1 = \frac{dy}{dx} = \frac{dy/dt}{dx/dt}$$

$$= \frac{e^t(\cos t - \sin t) + e^t(-\sin t - \cos t)}{e^t(\cos t + \sin t) + e^t(-\sin t + \cos t)}$$

$$= \frac{-2e^t \sin t}{2e^t \cos t} = -\tan t$$

$$\therefore \quad (y_1)_{t=0} = -\tan 0 = 0$$

$$y_2 = \frac{d^2y}{dx^2} = -\sec^2 t \cdot \frac{dt}{dx}$$

$$= \frac{-\sec^2 t}{2e^t \cos t}$$

$$= \frac{-1}{2}e^{-t}\sec^3 t$$

$$(y_2)_{t=0} = \frac{-1}{2}$$

$$y_3 = -\frac{1}{2}[-e^{-t}\sec^3 t + e^{-t}3\sec^2 t \sec t \tan t] \cdot \frac{dt}{dx}$$

$$= \frac{\frac{-1}{2}[-e^{-t}\sec^3 t + e^{-t}.3\sec^3 t \tan t]}{2e^t \cos t}$$

$$\therefore \quad (y_3)_{t=0} = \frac{1}{4}$$

Q. 34. If $y = x + \tan x$, prove that

$$\cos^2 x\frac{d^2y}{dx^2} - 2y + 2x = 0.$$

Sol. We have, $y = x + \tan x$

Differentiating w.r.t. x, we get

$$\frac{dy}{dx} = 1 + \sec^2 x$$

Differentiating once again, we get

$$\frac{d^2y}{dx^2} = \frac{d}{dx}(1 + \sec^2 x)$$

$$= 2\sec x.\sec x \tan x$$

$$= \frac{2\tan x}{\cos^2 x}$$

or $\quad \cos^2 x\dfrac{d^2y}{dx^2} = 2\tan x$

$$\Rightarrow \quad \cos^2 x\frac{d^2y}{dx^2} - 2y + 2x = 0 \quad [\because \tan x = y - x]$$

Q. 35. If $y = \cot^{-1}\left(\dfrac{\sqrt{1+\sin x} + \sqrt{1-\sin x}}{\sqrt{1+\sin x} - \sqrt{1-\sin x}}\right)$, find dy/dx.★

Sol. We have

$$y = \cot^{-1}\left(\frac{\sqrt{1+\sin x} + \sqrt{1-\sin x}}{\sqrt{1+\sin x} - \sqrt{1-\sin x}}\right)$$

$$= \cot^{-1}\left(\frac{\sqrt{1+\cos\left(\frac{\pi}{2}-x\right)} + \sqrt{1-\cos\left(\frac{\pi}{2}-x\right)}}{\sqrt{1+\cos\left(\frac{\pi}{2}-x\right)} - \sqrt{1-\cos\left(\frac{\pi}{2}-x\right)}}\right)$$

$$= \cot^{-1}\left(\frac{\sqrt{2\cos^2\left(\frac{\pi}{4}-\frac{x}{2}\right)} + \sqrt{2\sin^2\left(\frac{\pi}{4}-\frac{x}{2}\right)}}{\sqrt{2\cos^2\left(\frac{\pi}{4}-\frac{x}{2}\right)} - \sqrt{2\sin^2\left(\frac{\pi}{4}-\frac{x}{2}\right)}}\right)$$

$$= \cot^{-1}\left(\frac{\sqrt{2}\cos\left(\frac{\pi}{4}-\frac{x}{2}\right) + \sqrt{2}\sin\left(\frac{\pi}{4}-\frac{x}{2}\right)}{\sqrt{2}\cos\left(\frac{\pi}{4}-\frac{x}{2}\right) - \sqrt{2}\sin\left(\frac{\pi}{4}-\frac{x}{2}\right)}\right)$$

★ are board exam questions from previous years

$$= \cot^{-1}\left(\frac{\cos\left(\frac{\pi}{4}-\frac{x}{2}\right)+\sin\left(\frac{\pi}{4}-\frac{x}{2}\right)}{\cos\left(\frac{\pi}{4}-\frac{x}{2}\right)-\sin\left(\frac{\pi}{4}-\frac{x}{2}\right)}\right)$$

$$= \cot^{-1}\left(\frac{1+\tan\left(\frac{\pi}{4}-\frac{x}{2}\right)}{1-\tan\left(\frac{\pi}{4}-\frac{x}{2}\right)}\right)$$

[Dividing the Numerator and the Denominator by $\cos\left(\frac{\pi}{4}-\frac{x}{2}\right)$]

$$= \cot^{-1}\left[\tan\left\{\frac{\pi}{4}+\left(\frac{\pi}{4}-\frac{x}{2}\right)\right\}\right]$$

$$= \cot^{-1}\left\{\tan\left(\frac{\pi}{2}-\frac{x}{2}\right)\right\}$$

$$= \cot^{-1}\left(\cot\frac{x}{2}\right)$$

$$= \frac{x}{2}$$

$$\therefore \quad \frac{dy}{dx} = \frac{d}{dx}\left(\frac{x}{2}\right) = \frac{1}{2}$$

Q. 36. If $y = Pe^{ax} + Qe^{bx}$, show that :

$$\frac{d^2y}{dx^2} - (a+b)\frac{dy}{dx} + aby = 0. \star$$

Sol. Given,

$$y = Pe^{ax} + Qe^{bx} \quad \ldots(i)$$

Differentiating w.r.t. x, we get

$$\frac{dy}{dx} = P.e^{ax}\,a + Q.e^{bx}.b \quad \ldots(ii)$$

Again differentiating, we get

$$\frac{d^2y}{dx^2} = aPe^{ax}.a + bQ.e^{bx}.b$$
$$= a^2Pe^{ax} + b^2Qe^{bx}$$

$$\therefore \quad \text{L.H.S.} = \frac{d^2y}{dx^2} - (a+b)\frac{dy}{dx} + aby$$
$$= a^2Pe^{ax} + b^2Qe^{bx} - (a+b)(aPe^{ax} + bQe^{bx}) + ab(Pe^{ax} + Qe^{bx})$$
$$= e^{ax}[a^2P - (a+b)aP + abP] + e^{bx}[b^2Q - (a+b)bQ + abQ]$$
$$= e^{ax}.0 + e^{bx}.0$$
$$= 0 = \text{R.H.S.}$$

Q. 37. If $x = a\cos\theta + b\sin\theta$, $y = a\sin\theta - b\cos\theta$ show that :

$$y^2\frac{d^2y}{dx^2} - x\frac{dy}{dx} + y = 0. \star$$

Sol. We have,

$$x = a\cos\theta + b\sin\theta \quad \ldots(i)$$
$$y = a\sin\theta - b\cos\theta \quad \ldots(ii)$$

On squaring and adding (i) and (ii), we get

$$x^2 + y^2 = (a\cos\theta + b\sin\theta)^2 + (a\sin\theta - b\cos\theta)^2$$
$$= a^2\cos^2\theta + b^2\sin^2\theta + 2ab\cos\theta\sin\theta + a^2\sin^2\theta + b^2\cos^2\theta - 2ab\cos\theta\sin\theta$$
$$= a^2(\cos^2\theta + \sin^2\theta) + b^2(\sin^2\theta + \cos^2\theta)$$

$$\Rightarrow x^2 + y^2 = a^2 + b^2 \quad \ldots(iii)$$

Now, differentiating both sides of equation (iii) w.r.t. x, we get

$$2x + 2y\frac{dy}{dx} = 0$$

$$\Rightarrow \quad 2y\frac{dy}{dx} = -2x$$

$$\Rightarrow \quad \frac{dy}{dx} = -\frac{x}{y} \quad \ldots(iv)$$

Again, differentiating both sides of (iv) w.r.t. x, we get

$$\frac{d^2y}{dx^2} = -\left(\frac{y\times 1 - x\times\frac{dy}{dx}}{y^2}\right)$$

$$\frac{d^2y}{dx^2} = -\left[\frac{y - x\left(-\frac{x}{y}\right)}{y^2}\right] \quad \text{[from (iv)]}$$

$$\frac{d^2y}{dx^2} = -\left[\frac{y^2+x^2}{y^3}\right] \quad \ldots(v)$$

Now, we have

$$\text{L.H.S.} = y^2\frac{d^2y}{dx^2} - x\frac{dy}{dx} + y$$
$$= y^2\left(\frac{-y^2-x^2}{y^3}\right) - x\left(-\frac{x}{y}\right) + y$$
$$= \frac{-y^2-x^2}{y} + \frac{x^2}{y} + y$$
$$= \frac{-y^2-x^2+x^2+y^2}{y}$$
$$= 0 = \text{R.H.S.}$$

Q. 38. If $x = a\sin 2t(1+\cos 2t)$ and $y = b\cos 2t(1-\cos 2t)$, find the values of $\frac{dy}{dx}$ at $t = \frac{\pi}{4}$ and $t = \frac{\pi}{3}$. $\star$

Sol. We have,

$$x = a\sin 2t(1+\cos 2t)$$

$$\Rightarrow \quad \frac{dx}{dt} = 2a\cos 2t(1+\cos 2t) - 2a\sin^2 2t$$

and $\quad y = b\cos 2t(1-\cos 2t)$

$$\Rightarrow \quad \frac{dy}{dt} = 2b\cos 2t\sin 2t - 2b\sin 2t(1-\cos 2t)$$

$$\therefore \quad \frac{dy}{dx} = \frac{dy}{dt}\times\frac{dt}{dx}$$

★ are board exam questions from previous years

$$= \frac{2b \cos 2t \sin 2t - 2b \sin 2t(1-\cos 2t)}{2a \cos 2t(1+\cos 2t) - 2a \sin^2 2t}$$

$$= \frac{b[\cos 2t \sin 2t - \sin 2t(1-\cos 2t)]}{a[\cos 2t(1+\cos 2t) - \sin^2 2t]}$$

Since we know that at $t = \frac{\pi}{4}$, $\sin 2t = 1$ and $\cos 2t = 0$.

So, $$\left(\frac{dy}{dx}\right)_{t=\frac{\pi}{4}} = \frac{b[0.1-1(1-0)]}{a[0(1+0)-1^2]} = \frac{b}{a}$$

Also, we know that at $t = \frac{\pi}{3}$, $\sin 2t = \frac{\sqrt{3}}{2}$ and

$\cos 2t = -\frac{1}{2}$

So, $$\left(\frac{dy}{dx}\right)_{t=\frac{\pi}{3}} = \frac{b\left[\left(-\frac{1}{2}\right)\left(\frac{\sqrt{3}}{2}\right) - \left(\frac{\sqrt{3}}{2}\right)\left(1+\frac{1}{2}\right)\right]}{a\left[\left(-\frac{1}{2}\right)\left(1-\frac{1}{2}\right) - \left(\frac{\sqrt{3}}{2}\right)^2\right]}$$

$$= \frac{b\left[-\frac{\sqrt{3}}{4} - \frac{3\sqrt{3}}{4}\right]}{a\left[-\frac{1}{4} - \frac{3}{4}\right]}$$

$$= \frac{b[-\sqrt{3}]}{a[-1]} = \sqrt{3}\left(\frac{b}{a}\right)$$

Q. 39. If $e^y(x+1) = 1$, then show that

$$\frac{d^2 y}{dx^2} = \left(\frac{dy}{dx}\right)^2. \star$$

Sol. We have,

$$e^y(x+1) = 1$$

$$e^y = \frac{1}{x+1}$$

$$\log e^y = \log_e\left(\frac{1}{x+1}\right)$$

[Taking log of both sides]

$$y = -\log(x+1)$$

On differentiate both sides w.r.t. x

$$\frac{dy}{dx} = -\frac{1}{x+1} \quad \ldots(i)$$

Again differentiate w.r.t. x

$$\frac{d^2y}{dx^2} = \frac{1}{(x+1)^2}$$

$$\frac{d^2y}{dx^2} = \left(-\frac{1}{(x+1)}\right)^2$$

$$\frac{d^2y}{dx^2} = \left(\frac{dy}{dx}\right)^2 \quad \text{[using (i)]}$$

Q. 40. If $x^m y^n = (x+y)^{m+n}$, prove that $\frac{d^2 y}{dx^2} = 0$.★

Sol. Given,

$$x^m y^n = (x+y)^{m+n}$$

Taking log on both sides

$$\log x^m y^n = \log(x+y)^{m+n}$$

$$m \log x + n \log y = (m+n)\log(x+y)$$

Differentiate both sides w.r.t. x, we get

$$\frac{m}{x} + \frac{n}{y}\cdot\frac{dy}{dx} = \frac{m+n}{(x+y)}\cdot\left[1+\frac{dy}{dx}\right]$$

$$\frac{m}{x} + \frac{n}{y}\cdot\frac{dy}{dx} = \frac{m+n}{x+y} + \frac{m+n}{x+y}\frac{dy}{dx}$$

$$\frac{dy}{dx}\left(\frac{n}{y} - \frac{m+n}{x+y}\right) = \frac{m+n}{x+y} - \frac{m}{x}$$

$$\frac{dy}{dx}\left(\frac{nx+ny-my-ny}{y(x+y)}\right) = \frac{mx+nx-mx-my}{x(x+y)}$$

$$\frac{dy}{dx} = \frac{y}{x}\left(\frac{nx-my}{nx-my}\right) = \frac{y}{x}$$

Again differentiating w.r.t. x, we get

$$\frac{d^2y}{dx^2} = \frac{x\frac{dy}{dx} - y}{x^2}$$

$$= \frac{x \times \frac{y}{x} - y}{x^2} = \frac{y-y}{x^2} = 0.$$

Chapter 6. Applications of Derivatives

Q. 1. Using differentials find the approximate value of tan 46°, if it is being given that 1° = 0·01745 radians.

Sol. Let $y = f(x) = \tan x$,

$$x = 45° = \left(\frac{\pi}{4}\right) \text{ and } x + \Delta x = 46°$$

Then, $\Delta x = 1° = 0{\cdot}01745$ radians

For $x = \frac{\pi}{4}$, we have

$$y = f(\pi/4) = \tan \pi/4 = 1$$

Let $dx = \Delta x = 0{\cdot}01745$

Now, $y = \tan x$

$$\Rightarrow \quad \frac{dy}{dx} = \sec^2 x$$

Now, $$\left(\frac{dy}{dx}\right)_{\text{at } x=\pi/4} = \sec^2 \pi/4 = 2$$

$$\therefore \quad \Delta y = \frac{dy}{dx}\Delta x$$

★ are board exam questions from previous years

$\Rightarrow \quad \Delta y = 2(0{\cdot}01745) = 0{\cdot}03490$

$\Rightarrow \quad \Delta y = 0{\cdot}03490$

Hence, $\tan 46° = y + \Delta y = 1 + 0{\cdot}03490 = 1{\cdot}03490.$

Q. 2. Use differentials to approximate $(25)^{1/3}$.

Sol. Let $\quad y = (25)^{1/3}$

Let $\quad x = 27$ and $x + \Delta x = 25$

$\Rightarrow \quad \Delta x = -2$

For $x = 27$,

$$y = (27)^{1/3} = 3$$

Now $\quad y = x^{1/3}$

Differentiate y with respect to x, we get

$$\frac{dy}{dx} = \frac{1}{3x^{2/3}} \quad \ldots(i)$$

$$\left.\frac{dy}{dx}\right|_{x=27} = \frac{1}{3(27)^{2/3}}$$

$$= \frac{1}{27}$$

$$\Delta y = \frac{dy}{dx}\Delta x$$

$$= \frac{1}{27}(-2) = \frac{-2}{27}$$

or $\quad \Delta y = \dfrac{-2}{27}$

Hence $\quad (25)^{1/3} = y + \Delta y$

$$= 3 - \frac{2}{27} = \frac{81-2}{27}$$

$$= \frac{79}{27} = 2{\cdot}926.$$

Q. 3. Find the approximate value of $f(3{\cdot}02)$, where $f(x) = 3x^2 + 5x + 3$. **[NCERT]**

Sol. Let $\quad x = 3$ and $\Delta x = 0{\cdot}02.$

Then $\quad f(3{\cdot}02) = f(x + \Delta x)$

$$= 3(x + \Delta x)^2 + 5(x + \Delta x) + 3$$

Note that $\Delta y = f(x + \Delta x) - f(x)$. Therefore

$$f(x + \Delta x) = f(x) + \Delta y$$

$$\approx f(x) + f'(x)\,\Delta x \quad (\text{as } dx = \Delta x)$$

or $\quad f(3{\cdot}02) = (3x^2 + 5x + 3) + (6x + 5)\,\Delta x$

$$= [3(3)^2 + 5(3) + 3] + [6(3) + 5]\,[0{\cdot}02]$$

$$(\text{as } x = 3, \Delta x = 0{\cdot}02)$$

$$= (27 + 15 + 3) + (18 + 5)\,(0{\cdot}02)$$

$$= 45 + 0{\cdot}46 = 45{\cdot}46$$

Hence, approximate value of $f(3{\cdot}02)$ is $45{\cdot}46$.

Q. 4. Prove that the tangents to the curve $y = x^2 - 5x + 6$ at the points (2, 0) and (3, 0) are at right angles.

Sol. Given the equation of the curve is

$$y = x^2 - 5x + 6$$

Differentiating w.r.t. x, we get

$$\frac{dy}{dx} = 2x - 5$$

Now, $\quad m_1 =$ Slope of the tangent at (2, 0)

$$= \left(\frac{dy}{dx}\right)_{\text{at }(2,\,0)} = 2 \times 2 - 5 = -1$$

$m_2 =$ Slope of the tangent at (3, 0)

$$= \left(\frac{dy}{dx}\right)_{\text{at }(3,\,0)} = 2 \times 3 - 5 = 1$$

Clearly, product of the slopes $= m_1 \times m_2 = -1$.

Thus, the tangents to the given curve at (2, 0) and (3, 0) are at right angles.

Q. 5. Find the point at which the function f given by $f(x) = (x - 2)^4 (x + 1)^3$ has :

(i) local maxima (ii) local minima (iii) points of inflection.

Sol. Given function is

$$f(x) = (x - 2)^4 (x + 1)^3$$

Differentiate $f(x)$ with respect to x, we get

$$f'(x) = 4(x - 2)^3 (x + 1)^3 + 3(x - 2)^4 (x + 1)^2$$

$$= (x - 2)^3 (x + 1)^2 [4(x + 1) + 3(x - 2)]$$

$$\Rightarrow \quad f'(x) = (x - 2)^3 (x + 1)^2 (7x - 2)$$

$$\Rightarrow \quad f'(x) = (x - 2)^2 (x + 1)^2 (x - 2)(7x - 2)$$

For maximum and minimum value of x, put $f'(x) = 0$

Therefore $\quad x = 2, -1, \dfrac{2}{7}$

Since $(x - 2)^2 (x + 1)^2$ is always positive, so, sign of $f'(x)$ depends upon the sign of $(x - 2)(7x - 2)$.

The changes in signs of $f'(x)$ as x increases through $\dfrac{2}{7}$ and 2 are shown on number line.

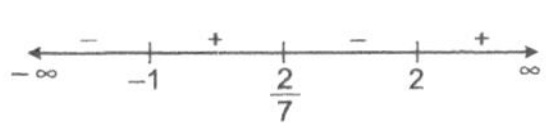

(i) Clearly, $f'(x)$ changes its sign from positive to negative as increases through $\dfrac{2}{7}$.

So, $x = \dfrac{2}{7}$ is a point of local maximum.

(ii) We observe that $f'(x)$ changes its sign from negative to positive as x increases through 2.

So, $x = 2$ is a point of local minimum.

(iii) There is no change in the sign of $f'(x)$ as increases through -1.

So, $x = -1$ is a point of inflection.

Q. 6. Find the maximum and minimum values of $f(x) = -|x-1| + 5$ for all $x \in R$.

Sol. We have, $f(x) = -|x-1| + 5$ for all $x \in R$

Clearly, $|x-1| \geq 0$ for all $x \in R$

Taking minus sign both side, we get

$$-|x-1| \leq 0 \text{ for all } x \in R$$

$$-|x-1| + 5 \leq 5 \text{ for all } x \in R$$

$\Rightarrow$ $f(x) \leq 5$ for all $x \in R$ [by given function]

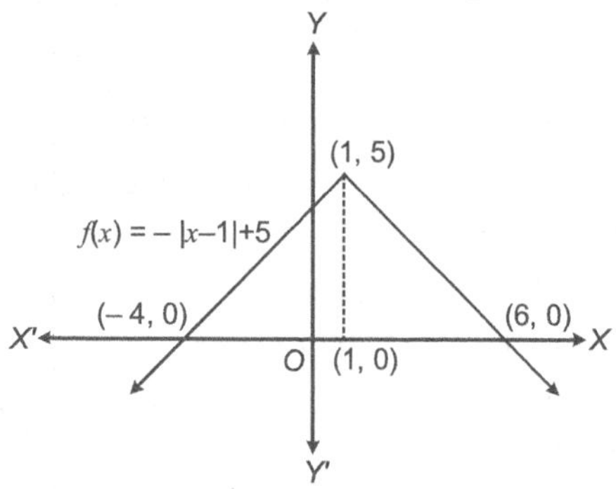

So, 5 is the maximum value of $f(x)$.

Now, $f(x) = 5$

$\Rightarrow$ $-|x-1| + 5 = 5$

$\Rightarrow$ $|x-1| = 0 \Rightarrow x = 1$

Thus, $f(x)$ attains the maximum value 5 at $x = 1$.

Since $f(x)$ can be made as small as possible. Therefore the minimum value of $f(x)$ does not exist.

Q. 7. Find the value(s) of x for which $y = [x(x-2)]^2$ is an increasing function.★

Sol. Given

$$y = [x(x-2)]^2$$

$\Rightarrow$ $y = x^2(x-2)^2$

$$= f(x) \text{ (Let)}$$

Differentiating w.r.t. x, we get

$$f'(x) = \frac{dy}{dx}$$

$$= 2x(x-2)^2 + 2x^2(x-2)$$

$$= 2x(x-2)(x-2+x)$$

$$= 4x(x-1)(x-2)$$

For y to be an increasing function, $\frac{dy}{dx} > 0$

$\Rightarrow x(x-1)(x-2) > 0$

$\Rightarrow$ $x = 0, 1, 2$

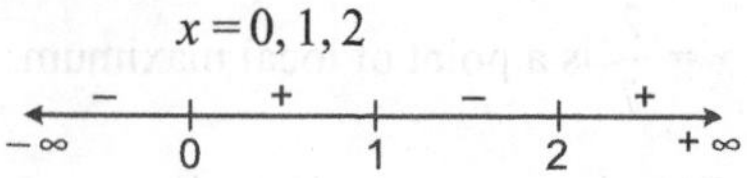

★ are board exam questions from previous years

Interval	Test Value	Sign of $f'(x)$ $f'(x) = 4x(x-1)(x-2)$	Nature of function $f(x) = y$
$(-\infty, 0)$	$x = -1$	$(-)(-)(-) = - < 0$	Strictly decrearsing
$(0, 1)$	$x = 0.5$	$(+)(-)(-) = + > 0$	Strictly increasing
$(1, 2)$	$x = 1.5$	$(+)(+)(-) = - < 0$	Strictly decreasing
$(2, \infty)$	$x = 3$	$(+)(+)(+) = + > 0$	Strictly increasing

$\therefore$ y is an increasing function in $[0, 1] \cup [2, \infty)$.

Q. 8. Find the intervals in which the function $f(x) = 3x^4 - 4x^3 - 12x^2 + 5$ is :★

(i) strictly increasing

(ii) strictly decreasing

Sol. Here $f(x) = 3x^4 - 4x^3 - 12x^2 + 5$

$\Rightarrow$ $f'(x) = 12x^3 - 12x^2 - 24x$

$$= 12x(x^2 - x - 2)$$

$$= 12x[x^2 - 2x + x - 2]$$

$$= 12x[x(x-2) + (x-2)]$$

$$= 12x(x+1)(x-2)$$

The critical values for f are -1, 0 and 2.

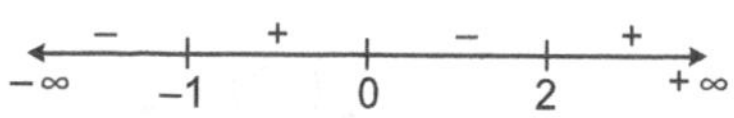

Interval	Test Value	Sign of $f'(x)$ $= 12x(x+1)(x-2)$	Nature of function $f(x)$
$(-\infty, -1)$	$x = -1.5$	$(-)(-)(-) = - < 0$	Strictly decreasing
$(-1, 0)$	$x = -0.5$	$(-)(+)(-) = + > 0$	Strictly increasing
$(0, 2)$	$x = 1$	$(+)(+)(-) = - < 0$	Strictly decreasing
$(2, \infty)$	$x = 3$	$(+)(+)(+) = + > 0$	Strictly increasing

$\therefore$ f is strictly increasing in $(-1, 0) \cup (2, \infty)$ and strictly decreasing in $(-\infty, -1) \cup (0, 2)$.

Q. 9. Show that the function $f(x) = x^3 - 3x^2 + 6x - 100$ is increasing on R.★

Sol. We have, $f(x) = x^3 - 3x^2 + 6x - 100$

then, $f'(x) = 3x^2 - 6x + 6$

$$= 3(x^2 - 2x + 1) + 3$$

$$= 3(x-1)^2 + 3 > 0 \text{ for all } x \in R$$

Hence, the function $f(x)$ is increasing on R.

Q. 10. Show that the function $f(x) = 4x^3 - 18x^2 + 27x - 7$ is always increasing on R.★

Sol. Given, $f(x) = 4x^3 - 18x^2 + 27x - 7$

Differentiating $f(x)$ w.r.t. x, we get

$$f'(x) = 12x^2 - 36x + 27$$
$$= 3(4x^2 - 12x + 9)$$
$$= 3(2x - 3)^2 \text{ for any } x \in R$$
$$3 > 0 \text{ and } (2x - 3)^2 \geq 0$$
$$\therefore \quad f'(x) \geq 0$$

$\Rightarrow$ The function is always increasing on R.

Q. 11. Find the intervals in which the function $f(x) = \frac{x^4}{4} - x^3 - 5x^2 + 24x + 12$ is (i) strictly increasing, (ii) strictly decreasing.★

Sol. Given function is

$$f(x) = \frac{x^4}{4} - x^3 - 5x^2 + 24x + 12$$

$$\Rightarrow \quad f'(x) = \frac{4x^3}{4} - 3x^2 - 10x + 24$$

For critical points, put $f'(x) = 0$

$$\therefore \quad x^3 - 3x^2 - 10x + 24 = 0$$
$$(x - 2)(x^2 - x - 12) = 0$$
$$(x - 2)(x - 4)(x + 3) = 0$$
$$\Rightarrow \quad x = 2, 4, -3$$

Therefore, we have the intervals $(-\infty, -3)$, $(-3, 2)$, $(2, 4)$ and $(4, \infty)$.

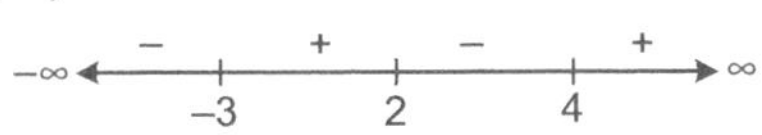

Since $f'(x) > 0$ in $(-3, 2) \cup (4, \infty)$.

$\therefore f(x)$ is increasing in interval $(-3, 2) \cup (4, \infty)$.

And $\quad f'(x) < 0$ in $(-\infty, -3) \cup (2, 4)$

$\therefore f(x)$ is decreasing in $(-\infty, -3) \cup (2, 4)$.

Q. 12. At what points will the tangent to the curve $f(x) = 2x^3 - 15x^2 + 36x - 21$ be parallel to X-axis ? Also, find the equations of tangents to the curve at those points.

Sol. Given equation of the curve is

$$f(x) = 2x^3 - 15x^2 + 36x - 21$$

Differentiate $f(x)$ with respect to x, we get

$$f'(x) = 6x^2 - 30x + 36 = 6(x^2 - 5x + 6)$$
$$= 6(x - 2)(x - 3)$$

For tangent parallel to X-axis,

$$f'(x) = 0$$
$$6(x - 2)(x - 3) = 0$$
$$\Rightarrow \quad x = 2, x = 3$$

At $x = 2, f(x) = 2.2^3 - 15.2^2 + 36.2 - 21 = 7$

At $x = 3, f(x) = 2.2^3 - 15.3^2 + 36.3 - 21 = 6$

Therefore, required points are (2, 7) and (3, 6).

Hence, the equation of the tangent at (2, 7) is

$$y - y_1 = m(x - x_1)$$
$$y - 7 = 0(x - 2)$$
$$y = 7$$

and the equation of the tangent at (3, 6) is

$$y - y_1 = m(x - x_1)$$
$$y - 6 = 0(x - 3)$$
$$y = 6.$$

Q. 13. Find the point of the curve $y = x^3$ at which the slope of the tangent is equal to the y-coordinate of the point. [NCERT]

Sol. Given, the equation of the curve is

$$y = x^3 \quad \text{...(i)}$$

Differentiate y with respect to x, we get

$$\frac{dy}{dx} = 3x^2 \quad \text{...(ii)}$$

Given, the slope of the tangent is equal to y-coordinate of the point.

Let the required point on the curve be $P(x_1, y_1)$.

Since (x_1, y_1) lies on $y = x^3$.

$$\Rightarrow \quad y_1 = x_1^3 \quad \text{...(iii)}$$

$$\left(\frac{dy}{dx}\right)_{(x_1, y_1)} = 3x_1^2$$

$$\therefore \quad 3x_1^2 = y_1 \quad \text{[Given]}$$
$$\Rightarrow \quad 3x_1^2 = x_1^3 \quad \text{[from (iii)]}$$
$$\Rightarrow \quad x_1^2(x_1 - 3) = 0$$
$$\text{or} \quad x_1 = 0, x_1 = 3$$

When $x_1 = 0, y_1 = 0$ and when $x_1 = 3, y_1 = 27$.

Hence, required points are (0, 0) and (3, 27).

Q. 14. Find the value of p for which the curves $x^2 = 9p(9 - y)$ and $x^2 = p(y + 1)$ cut each other at right angles.★

Sol. Given, $\quad x^2 = 9p(9 - y) \quad$ ...(i)

and $\quad x^2 = p(y + 1) \quad$ ...(ii)

From (i) and (ii), we get

$$9p(9 - y) = p(y + 1)$$
$$81p - 9py = py + p$$
$$10py = 80p$$
$$y = 8$$
$$\therefore \quad x^2 = 9p$$

Now, differentiating (i) and (ii) w.r.t. x, we get

$$2x = -9p\frac{dy}{dx}$$

★ are board exam questions from previous years

$$\Rightarrow \quad \frac{dy}{dx} = -\frac{2x}{9p}$$

and $$2x = p\frac{dy}{dx}$$

$$\Rightarrow \quad \frac{dy}{dx} = \frac{2x}{p}$$

Since curves cut each other at right angles.

$$\therefore \quad \left(\frac{2x}{p}\right)\left(-\frac{2x}{9p}\right) = -1 \qquad [\because m_1m_2 = -1]$$

$$\Rightarrow \quad \frac{-4x^2}{9p^2} = -1$$

$$\Rightarrow \quad \frac{4}{9p^2}(9p) = 1 \qquad (\because x^2 = 9p)$$

$$\therefore \quad p = 4.$$

Q. 15. Find the equations of the tangent and normal to the curve $\frac{x^2}{a^2} - \frac{y^2}{b^2} = 1$ at the point $(\sqrt{2}a, b)$.★

Sol. The given curve is

$$\frac{x^2}{a^2} - \frac{y^2}{b^2} = 1$$

Differentiating w.r.t. x, we get

$$\frac{2x}{a^2} - \frac{2y}{b^2}\cdot\frac{dy}{dx} = 0$$

$$\Rightarrow \quad \frac{dy}{dx} = \frac{\frac{x}{a^2}}{\frac{y}{b^2}} = \frac{b^2}{a^2}\cdot\frac{x}{y}$$

At $P(\sqrt{2}a, b)$

Slope of the tangent,

$$m_1 = \frac{b^2}{a^2}\cdot\frac{\sqrt{2}a}{b} = \sqrt{2}.\frac{b}{a}$$

and slope of the normal,

$$m_2 = -\frac{a}{b\sqrt{2}}$$

$\therefore$ Equation of the tangent at P is

$$y - b = \sqrt{2}\frac{b}{a}(x - \sqrt{2}a)$$

$$\Rightarrow \quad y = \sqrt{2}\frac{b}{a}x - 2b + b$$

$$\Rightarrow \quad y = \sqrt{2}\frac{b}{a}x - b$$

$$\Rightarrow \sqrt{2}bx - ay - ab = 0$$

Also equation of the normal at P is

$$y - b = \frac{-a}{\sqrt{2}b}(x - \sqrt{2}a)$$

$$\Rightarrow \quad y = \frac{-ax}{\sqrt{2}.b} + b + \frac{a^2}{b}$$

$$\Rightarrow \sqrt{2}by - \sqrt{2}b^2 = -ax + \sqrt{2}a^2$$

$$\Rightarrow \quad ax + \sqrt{2}by - \sqrt{2}(a^2 + b^2) = 0.$$

Q. 16★. Find the equation of the tangent and normal to the curve $x = a\sin^3\theta$ and $y = a\cos^3\theta$ at $\theta = \frac{\pi}{4}$.

Sol. The given curve is

$$x = a\sin^3\theta;\ y = a\cos^3\theta \qquad ...(i)$$

At, $\theta = \frac{\pi}{4}$

$$x = a\sin^3\frac{\pi}{4} = a\left(\frac{1}{\sqrt{2}}\right)^3 = \frac{a}{2\sqrt{2}}$$

and $$y = a\cos^3\frac{\pi}{4} = a\left(\frac{1}{\sqrt{2}}\right)^3 = \frac{a}{2\sqrt{2}}$$

$\Rightarrow P\left(\frac{a}{2\sqrt{2}}, \frac{a}{2\sqrt{2}}\right)$ is a point on (i) corresponding to

$$\theta = \frac{\pi}{4}$$

Differentiating (i) w.r.t. θ, we get

$$\frac{dx}{d\theta} = 3a\sin^2\theta\cos\theta$$

and $$\frac{dy}{d\theta} = -3a\cos^2\theta\sin\theta$$

$$\therefore \quad \frac{dy}{dx} = \frac{dy/d\theta}{dx/d\theta} = \frac{-3a\cos^2\theta\sin\theta}{3a\sin^2\theta\cos\theta} = -\cot\theta$$

$\therefore$ Slope of the tangent at $\theta = \frac{\pi}{4}$ is -1

and slope of the normal at $\theta = \frac{\pi}{4}$ is 1

Hence equation of the tangent at P is

$$y - \frac{a}{2\sqrt{2}} = -1.\left(x - \frac{a}{2\sqrt{2}}\right)$$

★ are board exam questions from previous years

$$\therefore \quad x + y = \frac{a}{\sqrt{2}}$$

and equation of the normal at P is

$$y - \frac{a}{2\sqrt{2}} = 1.\left(x - \frac{a}{2\sqrt{2}}\right)$$

$$\therefore \quad y = x.$$

Q. 17. The equation of tangent at (2, 3) on the curve $y^2 = ax^3 + b$ is $y = 4x - 5$. Find the value of a and b.★

Sol. Since the point (2, 3) lies on the curve

$$y^2 = ax^3 + b \quad \ldots(i)$$

$$\Rightarrow \quad (3)^2 = a(2)^3 + b$$

$$9 = 8a + b$$

On differentiating (i) w.r.t. x,

$$2y\frac{dy}{dx} = 3ax^2$$

$$\Rightarrow \quad \frac{dy}{dx} = \frac{3ax^2}{2y}$$

$$\left(\frac{dy}{dx}\right)_{(2,\,3)} = \frac{3a(2)^2}{2(3)} = 2a$$

The required equation of the tangent at (2, 3) is

$$(y - 3) = 2a(x - 2)$$

$$y - 3 = 2ax - 4a$$

$$y = 2ax + 3 - 4a$$

Comparing the equation with $y = 4x - 5$, we get

$$4 = 2a \Rightarrow a = 2$$

From equation (i), we get

$$9 = 8(2) + b$$

$$\Rightarrow \quad b = -7.$$

Thus, $a = 2, b = -7$.

Q. 18. Show that the equation of normal at any point on the curve $x = 3\cos t - \cos^3 t$ and $y = 3\sin t - \sin^3 t$ is $4(y\cos^3 t - x\sin^3 t) = 3\sin 4t$.★

Sol. Given,

$$x = 3\cos t - \cos^3 t$$

$$\Rightarrow \quad \frac{dx}{dt} = -3\sin t + 3\cos^2 t \sin t$$

and

$$y = 3\sin t - \sin^3 t$$

$$\Rightarrow \quad \frac{dy}{dt} = 3\cos t - 3\sin^2 t\cos t$$

Slope of the tangent,

$$\frac{dy}{dx} = \frac{\frac{dy}{dt}}{\frac{dx}{dt}} = \frac{3\cos t - 3\sin^2 t\cos t}{-3\sin t + 3\cos^2 t\sin t}$$

$$= \frac{3\cos t[\cos^2 t]}{-3\sin t[\sin^2 t]}$$

$$\frac{dy}{dx} = \frac{-\cos^3 t}{\sin^3 t}$$

$$\therefore \text{ Slope of the normal} = \frac{\sin^3 t}{\cos^3 t}$$

The equation of the normal is given by

$$\frac{y - (3\sin t - \sin^3 t)}{x - (3\cos t - \cos^3 t)} = \frac{\sin^3 t}{\cos^3 t}$$

$$\Rightarrow y\cos^3 t - 3\sin t\cos^3 t + \sin^3 t\cos^3 t$$

$$= x\sin^3 t - 3\cos t\sin^3 t + \sin^3 t\cos^3 t$$

$$\Rightarrow \quad y\cos^3 t - x\sin^3 t = 3(\sin t\cos^3 t - \cos t\sin^3 t)$$

$$\Rightarrow \quad y\cos^3 t - x\sin^3 t = 3\sin t\cos t(\cos^2 t - \sin^2 t)$$

$$\Rightarrow \quad y\cos^3 t - x\sin^3 t = \frac{3}{2}\sin 2t\cos 2t = \frac{3}{4}\sin 4t$$

$$\Rightarrow 4(y\cos^3 t - x\sin^3 t) = 3\sin 4t.$$

Q. 19. An open tank with a square base and vertical sides is to be constructed from a metal sheet so as to hold a given quantity of water. Show that the cost of the material will be least when depth of the tank is half of its width.

Sol. Let, dimensions of tank are as : length $= x$, width $= x$ and height $= y$ units respectively.

Let volume of the tank be V.

Therefore, $\quad V = x^2y \quad \ldots(i)$

The cost of the material will be least if the total surface area is least. Let S denote the total surface area. Then

$$S = x^2 + 4xy \quad \ldots(ii)$$

We have to minimize S subject to the condition that the volume is constant.

Now, $\quad S = x^2 + 4xy$

$$\Rightarrow \quad S = x^2 + 4\frac{V}{x} \quad \text{[from (i)]}$$

Differentiate S with respect to x, we get

$$\frac{dS}{dx} = 2x - \frac{4V}{x^2}$$

Again, differentiate $\frac{dS}{dx}$ with respect to x, we get

$$\frac{d^2S}{dx^2} = 2 + \frac{8V}{x^3}$$

For maximum or minimum values of S, we must have

$$\frac{dS}{dx} = 0$$

★ are board exam questions from previous years

$\therefore \quad 2x - \frac{4V}{x^2} = 0$

$\Rightarrow \quad 2x^3 = 4V$

$\Rightarrow \quad 2x^3 = 4x^2y$ [from (i)]

$\Rightarrow \quad x = 2y$

Clearly, $\frac{d^2S}{dx^2} = 2 + \frac{8V}{x^3} > 0$ for all x

Hence, S is minimum when $x = 2y$ *i.e.*, the depth (height) of the tank is half of its width.

Q. 20. If the radius of a sphere is measured as 9 cm with an error of 0·03 cm, then find the approximate error in calculating its volume. [NCERT]

Sol. Let r be the radius of the sphere and Δr be the error in measuring the radius. Then $r = 9$ cm and $\Delta r = 0.03$ cm. Now, the volume V of the sphere is given by

$$V = \frac{4}{3}\pi r^3$$

or $\frac{dV}{dr} = 4\pi r^2$

$$\therefore \quad dV = \left(\frac{dV}{dr}\right)\Delta r = (4\pi r^2)\,\Delta r$$
$$= 4\pi(9)^2(0{\cdot}03)$$
$$= 9{\cdot}72\pi \text{ cm}^3$$

Thus, the approximate error in calculating the volume is 9.72π cm^3.

Q. 21. The internal box with a square base and vertical sides is to contain 1024 cm^3. The material for the top and bottom costs ₹ 5 per cm^2 and the material for the sides costs ₹ 2·50 per cm^2. Find the least cost of the box.★

Sol. Let the length, breadth and height of the metal box be x cm, x cm and y cm respectively. It is given that

$$x^2y = 1024$$

$\Rightarrow \quad y = \frac{1024}{x^2}$ ...(i)

Let c be the total cost in rupees of material used to construct the box.

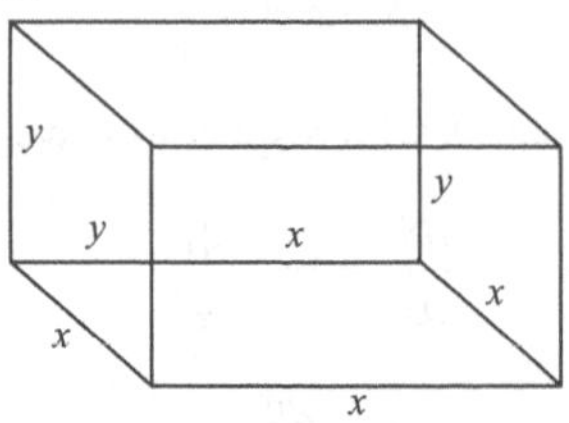

Then, $c = 5x^2 + 5x^2 + \frac{5}{2} \times 4xy$

$$= 10x^2 + 10xy$$
$$= 10x^2 + 10x \times \frac{1024}{x^2}$$
$$= 10x^2 + \frac{10240}{x}$$

★ are board exam questions from previous years

$\Rightarrow \quad \frac{dc}{dx} = 20x - \frac{10240}{x^2}$

and $\frac{d^2c}{dx^2} = 20 + \frac{20480}{x^3}$

For maxima or minima

Put $\frac{dc}{dx} = 0$

$\Rightarrow \quad 20x - \frac{10240}{x^2} = 0$

$\Rightarrow \quad x^3 = 512$

$\Rightarrow \quad x^3 = 8^3$

$\Rightarrow \quad x = 8$

$$\left(\frac{d^2c}{dx^2}\right)_{x=8} = 20 + \frac{20480}{8^3} > 0$$

Thus, the cost of the box is least when $x = 8$, putting $x = 8$ in equation (i) we obtain $y = 16$, so the dimensions of the box are $8 \times 8 \times 16$.

Putting, $x = 8$ and $y = 16$ in $c = 10x^2 + 10xy$

$$c = 10 \times 64 + 10 \times 8 \times 16 = 1920$$

Hence, the least cost of the box is ₹ 1920.

Q. 22. In a kite festival, a kite is at a height of 120 cm and 130 m string is out. If the kite is moving horizontally at the rate of 5·2 m/sec, find the rate at which the string is being pulled out at that instant.

Sol. Let at any instant,

Length of string $= y$ m

and horizontal distance of the kite $= x$ m

then $\frac{dx}{dt} = 5{\cdot}2$ m/sec (given)

Then, $y^2 = x^2 + (120)^2$

Differentiate above equation w.r.t. t

$\Rightarrow \quad 2y \cdot \frac{dy}{dt} = 2x \cdot \frac{dx}{dt}$

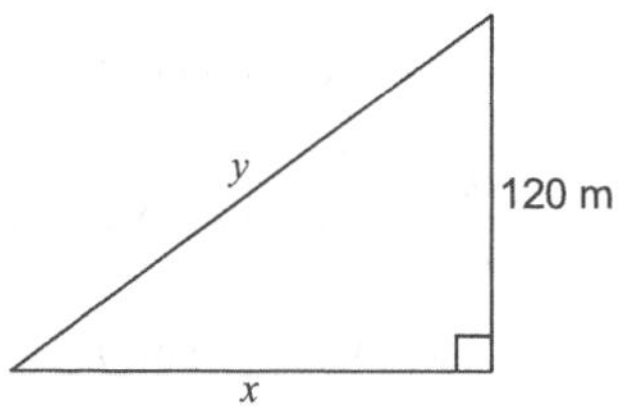

$\Rightarrow \quad \frac{dy}{dt} = \frac{x}{y} \cdot \frac{dx}{dt} = \frac{x}{y}(5{\cdot}2)$ ...(i)

When $y = 130$ m, then

$$x = \sqrt{(130)^2 - (120)^2}$$
$$= \sqrt{2500}$$

$= 50 \text{ m}$

Hence the rate at which the string is being pulled out at the instant when string's length is 130 m.

$$= \frac{50}{130} \times (5\cdot2)$$

$$= \frac{5 \times 52}{130} = 2 \text{ m/sec}$$

Q. 23. An expensive square piece of golden colour board of side 24 centimetres, is to be made into a box by cutting a square from each corner and folding the flaps to form a box. What should be the side of the square piece to be cut from each corner of the board to hold maximum volume and minimize the wastage ?

Sol. Let length of a side of the square cut out $= x$ cm

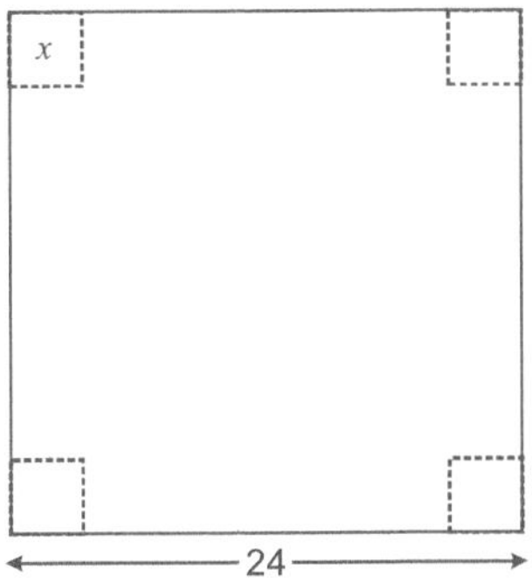

Volume, $V = (24 - 2x)^2 . x$

$$\frac{dV}{dx} = 2(24-2x)(-2)x + (24-2x)^2$$

$$= (24-2x)(-4x+24-2x)$$

$$= (24-2x)(24-6x)$$

$$= 12(12-x)(4-x)$$

Put $\quad \frac{dV}{dx} = 0$ *i.e.*, $(12-x)(4-x) = 0$

Since, $\quad x \neq 12,$

$\therefore \quad x = 4$

Hence, the volume will be maximum when length of a side of the square cut out = 4 cm.

Chapter 7. Integrals

Q. 1. Evaluate $\int \frac{e^x}{\sqrt{5 - 4e^x - e^{2x}}} dx$

Sol. We have, $\int \frac{e^x}{\sqrt{5-4e^x-e^{2x}}} dx$

Put $\quad e^x = t$

Differentiate both sides w.r.t. x

$$e^x \, dx = dt$$

$$\int \frac{dt}{\sqrt{5-4t-t^2}} = \int \frac{dt}{\sqrt{4+5-(t^2+4t+4)}}$$

[Add and subtract 4]

$$= \int \frac{dt}{\sqrt{(3)^2 - (t+2)^2}}$$

$$\left[\because \int \frac{dx}{\sqrt{a^2-x^2}} = \sin^{-1}\frac{x}{a} + c\right]$$

$$= \sin^{-1}\frac{t+2}{3} + c$$

$$= \sin^{-1}\left(\frac{e^x+2}{3}\right) + c$$

Q. 2. Evaluate $\int x \sin^{-1} x \, dx$

Sol. Let $I = \int x \sin^{-1} x \, dx = \int \underset{\text{I}}{\sin^{-1} x} \, \underset{\text{II}}{x} \, dx$ (ILATE)

$$= \sin^{-1} x \int x\,dx - \int\left[\frac{d}{dx}(\sin^{-1} x).\int x\,dx\right] dx$$

$$= \sin^{-1} x \cdot \frac{x^2}{2} - \int \frac{1}{\sqrt{1-x^2}} \cdot \frac{x^2}{2} dx$$

$$= \frac{x^2 \sin^{-1} x}{2} + \frac{1}{2}\int \frac{-x^2}{\sqrt{1-x^2}} dx$$

$$= \frac{x^2 \sin^{-1} x}{2} + \frac{1}{2}\int \frac{1-x^2-1}{\sqrt{1-x^2}} dx$$

[Add and subtract 1]

$$= \frac{x^2 \sin^{-1} x}{2} + \frac{1}{2}\int \frac{1-x^2}{\sqrt{1-x^2}} dx - \frac{1}{2}\int \frac{dx}{\sqrt{1-x^2}}$$

$$= \frac{x^2 \sin^{-1} x}{2} + \frac{1}{2}\int \sqrt{1-x^2}\, dx - \frac{1}{2}\int \frac{dx}{\sqrt{1-x^2}}$$

We know that

$$\int \sqrt{a^2-x^2}\, dx = \frac{x}{2}\sqrt{a^2-x^2} + \frac{a^2}{2}\sin^{-1}\frac{x}{a} + c$$

and $\quad \int \frac{dx}{\sqrt{a^2-x^2}} = \sin^{-1}\frac{x}{a} + c$

$$\therefore \quad I = \frac{x^2 \sin^{-1} x}{2} + \frac{1}{2}\left(\frac{x}{2}\sqrt{1-x^2} + \frac{1}{2}\sin^{-1}\frac{x}{1}\right) - \frac{1}{2}\sin^{-1}\frac{x}{1} + c$$

$$= \frac{x^2 \sin^{-1} x}{2} + \frac{x\sqrt{1-x^2}}{4} + \frac{1}{4}\sin^{-1} x - \frac{1}{2}\sin^{-1} x + c$$

$$= \frac{x^2 \sin^{-1} x}{2} + \frac{x\sqrt{1-x^2}}{4} + \frac{\sin^{-1} x - 2\sin^{-1} x}{4} + c$$

$$= \frac{x^2 \sin^{-1} x}{2} + \frac{x\sqrt{1-x^2}}{4} - \frac{1}{4}\sin^{-1} x + c$$

$$= \sin^{-1} x\left(\frac{x^2}{2} - \frac{1}{4}\right) + \frac{x\sqrt{1-x^2}}{4} + c$$

$$= \frac{2x^2-1}{4}\sin^{-1} x + \frac{x\sqrt{1-x^2}}{4} + c$$

$$= \frac{(2x^2-1)\sin^{-1} x}{4} + \frac{x\sqrt{1-x^2}}{4} + c$$

Q. 3. Find $\int_0^{\pi} \frac{x \tan x}{\sec x \operatorname{cosec} x} dx$

Sol. Let $I = \int_0^{\pi} \frac{x \tan x}{\sec x \operatorname{cosec} x} dx$ ...(i)

$$I = \int_0^{\pi} \frac{(\pi - x) \tan (\pi - x)}{\sec (\pi - x) \operatorname{cosec} (\pi - x)} dx$$

$$\left[\because \int_0^a f(x)\, dx = \int_0^a f(a-x)\, dx\right]$$

$$I = \int_0^{\pi} \frac{\pi(-\tan x)}{(-\sec x) \operatorname{cosec} x} dx - \int_0^{\pi} \frac{x(-\tan x)}{(-\sec x) \operatorname{cosec} x} dx$$

$$I = \int_0^{\pi} \frac{\pi \tan x}{\sec x \operatorname{cosec} x} dx - \int_0^{\pi} \frac{x \tan x}{\sec x \operatorname{cosec} x} dx \quad \text{...(ii)}$$

Add equations (i) and (ii),

$$2I = \int_0^{\pi} \frac{x \tan x}{\sec x \operatorname{cosec} x} dx + \pi \int_0^{\pi} \frac{\tan x}{\sec x \operatorname{cosec} x} dx - \int_0^{\pi} \frac{x \tan x}{\sec x \operatorname{cosec} x} dx$$

$$= \pi \int_0^{\pi} \frac{\frac{\sin x}{\cos x}}{\frac{1}{\cos x} \cdot \frac{1}{\sin x}} dx = \pi \int_0^{\pi} \sin^2 x \, dx$$

$$= \pi \int_0^{\pi} \frac{1 - \cos 2x}{2} dx = \frac{\pi}{2} \int_0^{\pi} (1 - \cos 2x)\, dx$$

$$= \frac{\pi}{2} \left(x - \frac{\sin 2x}{2} \right)_0^{\pi}$$

$$= \frac{\pi}{2} \left[\left(\pi - \frac{\sin 2\pi}{2} \right) - \left(0 - \frac{\sin 0}{2} \right) \right]$$

$$= \frac{\pi}{2} [(\pi - 0) - (0 - 0)]$$

$$2I = \frac{\pi}{2} (\pi)$$

$$I = \frac{\pi^2}{4}$$

Q. 4. Evaluate $\int_0^{\pi} \frac{x}{1 + \sin x} dx$

Sol. Let $I = \int_0^{\pi} \frac{x}{1 + \sin x} dx$

$$I = \int_0^{\pi} \frac{(\pi - x)}{1 + \sin (\pi - x)} dx$$

$$\left[\because \int_0^a f(x)\, dx = \int_0^a f(a-x)\, dx\right]$$

$$= \int_0^{\pi} \frac{\pi}{1 + \sin x} dx - \int_0^{\pi} \frac{x}{1 + \sin x} dx$$

$$I = \pi \int_0^{\pi} \frac{dx}{1 + \sin x} - I$$

$$2I = \pi \int_0^{\pi} \frac{1}{1 + \sin x} \times \frac{1 - \sin x}{1 - \sin x} dx$$

[By rationalising the denominator]

$$= \pi \int_0^{\pi} \frac{1 - \sin x}{1 - \sin^2 x} dx$$

$$= \pi \int_0^{\pi} \frac{1 - \sin x}{\cos^2 x} dx$$

$$= \pi \left[\int_0^{\pi} \frac{1}{\cos^2 x} dx - \int_0^{\pi} \frac{\sin x}{\cos^2 x} dx \right]$$

$$= \pi \left[\int_0^{\pi} \sec^2 x \, dx - \int_0^{\pi} \sec x \tan x \, dx \right]$$

$$= \pi \left[(\tan x)_0^{\pi} - (\sec x)_0^{\pi} \right]$$

$$= \pi[(\tan \pi - \tan 0) - (\sec \pi - \sec 0)]$$

$$I = \frac{\pi}{2} [(0 - 0) - (-1 - 1)]$$

$$\therefore \quad I = \frac{\pi}{2} [0 - (-2)]$$

$$I = \frac{\pi}{2} \times 2$$

$$\therefore \quad I = \pi$$

Q. 5. Find $\int \frac{1 - x^2}{x(1 - 2x)} dx$

Sol. Let $I = \int \frac{1 - x^2}{x(1 - 2x)} dx$

$$= \int \frac{1 - x^2}{x - 2x^2} dx$$

$\because$ deg (num) = deg (den)

$\therefore$

$$\begin{array}{r|l} & \quad 1/2 \\ -2x + x^2 & -x^2 + 1 \\ & -x^2 + \frac{1}{2}x \\ & + \quad - \\ \hline & -\frac{1}{2}x + 1 \end{array}$$

$$I = \int \left(\frac{1}{2} + \frac{-\frac{1}{2}x + 1}{x(1 - 2x)} \right) dx$$

$$I = \frac{1}{2}x + \int \frac{-\frac{1}{2}x + 1}{x(1 - 2x)} dx \quad \text{...(i)}$$

Let $I_1 = \dfrac{-\frac{1}{2}x+1}{x(1-2x)} = \dfrac{A}{x}+\dfrac{B}{1-2x}$

$-\frac{1}{2}x+1 = A(1-2x)+Bx$ ...(ii)

If $x = 0$

then from equation (ii)

$1 = A$

If $x = \frac{1}{2}$

then from equation (ii)

$$-\frac{1}{4}+1 = \frac{B}{2}$$

$$\frac{-1+4}{4} = \frac{B}{2}$$

$$\frac{3}{4} = \frac{B}{2}$$

$$\therefore \quad B = \frac{3}{2}$$

$$\therefore \quad I_1 = \int \frac{1}{x}dx + \frac{3}{2}\int \frac{dx}{1-2x}$$

$$= \log|x| + \frac{3}{2(-2)}\int \frac{-2}{1-2x}dx$$

$$= \log|x| - \frac{3}{4}\log|1-2x|$$

$$\left[\because \int \frac{f'(x)}{f(x)}dx = \log|f(x)|+c\right]$$

$$\therefore \quad I = \frac{1}{2}x + \log|x| - \frac{3}{4}\log|1-2x|+c$$

Q. 6. Find $\int \frac{x+2}{\sqrt{(x-2)(x-3)}}dx$

Sol. Let $I = \int \frac{x+2}{\sqrt{(x-2)(x-3)}}dx$

$$I = \int \frac{x+2}{\sqrt{x^2-5x+6}}dx$$

$$= \frac{1}{2}\int \frac{2x+4}{\sqrt{x^2-5x+6}}dx = \frac{1}{2}\int \frac{2x+4-5+5}{\sqrt{x^2-5x+6}}dx$$

$$= \frac{1}{2}\int \frac{2x-5}{\sqrt{x^2-5x+6}}dx + \frac{9}{2}\int \frac{dx}{\sqrt{x^2-5x+\frac{25}{4}+6-\frac{25}{4}}}$$

$$= \frac{1}{2}\int (x^2-5x+6)^{-1/2}.(2x-5)\,dx + \frac{9}{2}\int \frac{dx}{\sqrt{\left(x-\frac{5}{2}\right)^2-\frac{1}{4}}}$$

$$= \frac{1}{2}\left[\frac{(x^2-5x+6)^{-1/2+1}}{-\frac{1}{2}+1}\right] + \frac{9}{2}\int \frac{dx}{\sqrt{\left(x-\frac{5}{2}\right)^2-\left(\frac{1}{2}\right)^2}}$$

$$I = \sqrt{x^2-5x+6} + \frac{9}{2}\log\left|x-\frac{5}{2}+\sqrt{x^2-5x+6}\right|+c$$

Q. 7. Evaluate $\int (\sqrt{\tan x}+\sqrt{\cot x})\,dx$

Sol. Let $I = \int (\sqrt{\tan x}+\sqrt{\cot x})\,dx$

$$= \int \left(\sqrt{\tan x}+\frac{1}{\sqrt{\tan x}}\right)dx$$

$$= \int \frac{\tan x+1}{\sqrt{\tan x}}dx$$

Put $\sqrt{\tan x} = t$

$\tan x = t^2$

Differentiate both sides w.r.t. x

$\sec^2 x\,dx = 2t\,dt$

$$dx = \frac{2t}{\sec^2 x}dt$$

$$= \frac{2t}{1+\tan^2 x}dt$$

$$dx = \frac{2t}{1+t^4}dt$$

$$\therefore \quad I = \int \frac{t^2+1}{t}\cdot\frac{2t}{1+t^4}dt$$

$$I = 2\int \frac{t^2+1}{t^4+1}dt$$

$$I = 2\int \frac{t^2\left(1+\frac{1}{t^2}\right)}{t^2\left(t^2+\frac{1}{t^2}\right)}dt$$

Let $t-\frac{1}{t} = y$

Squaring both sides and then differentiating both sides

$$t^2+\frac{1}{t^2}-2 = y^2,$$

$$t^2+\frac{1}{t^2} = y^2+2,\ 1+\frac{1}{t^2} = \frac{dy}{dt}$$

$$\Rightarrow \left(1+\frac{1}{t^2}\right)dt = dy$$

$$\therefore \quad I = 2\int \frac{dy}{y^2+2} = 2\int \frac{dy}{y^2+(\sqrt{2})^2}$$

$$\therefore \quad I = 2\cdot\frac{1}{\sqrt{2}}\tan^{-1}\frac{y}{\sqrt{2}}+c$$

$$= \frac{2}{\sqrt{2}}\tan^{-1}\frac{t-\frac{1}{t}}{\sqrt{2}}+c$$

$$= \sqrt{2}\tan^{-1}\frac{t^2-1}{\sqrt{2}t}+c$$

$$\therefore \quad I = \sqrt{2}\tan^{-1}\left(\frac{\tan x-1}{\sqrt{2}\sqrt{\tan x}}\right)+c$$

$$I = \sqrt{2}\tan^{-1}\left(\frac{\tan x-1}{\sqrt{2\tan x}}\right)+c$$

Q. 8. Find $\int \frac{2x}{(x^2+1)(x^2+3)}dx$ **[NCERT]**

Sol. We have, $\int \frac{2x}{(x^2+1)(x^2+3)}dx$

Put $x^2 = t$

Differentiate both sides w.r.t. x

$$2x\,dx = dt$$

$$\therefore \quad I = \int \frac{dt}{(t+1)(t+3)}$$

$$\frac{1}{(t+1)(t+3)} = \frac{A}{t+1}+\frac{B}{t+3}$$

$$1 = A(t+3)+B(t+1)$$

Put $t = -1$

then $1 = A(-1+3)$

$$1 = 2A$$

$$A = \frac{1}{2}$$

If $t = -3$

then $1 = B(-3+1)$

$$1 = -2B$$

$$B = -\frac{1}{2}$$

$$\therefore \quad I = \int \frac{\frac{1}{2}}{t+1}dt + \int \frac{-\frac{1}{2}}{t+3}dt$$

$$I = \frac{1}{2}\log|t+1| - \frac{1}{2}\log|t+3|+c$$

$$= \frac{1}{2}\left[\log\left|\frac{t+1}{t+3}\right|\right]+c$$

$$I = \frac{1}{2}\log\left|\frac{x^2+1}{x^2+3}\right|+c$$

Q. 9. Using properties of integration evaluate :

$$\int_0^{\pi/4} \log(1+\tan x)\,dx$$

Sol. Let $I = \int_0^{\pi/4}\log(1+\tan x)\,dx$

$$= \int_0^{\pi/4}\log\left[1+\tan\left(\frac{\pi}{4}-x\right)\right]dx$$

$$\left[\because \int_0^a f(x)\,dx = \int_0^a f(a-x)\,dx\right]$$

$$\therefore \quad I = \int_0^{\pi/4}\log\left[1+\frac{\tan\frac{\pi}{4}-\tan x}{1+\tan\frac{\pi}{4}\tan x}\right]dx$$

$$= \int_0^{\pi/4}\log\left(1+\frac{1-\tan x}{1+\tan x}\right)dx$$

$$= \int_0^{\pi/4}\log\left|\frac{1+\tan x+1-\tan x}{1+\tan x}\right|dx$$

$$= \int_0^{\pi/4}\log\left|\frac{2}{1+\tan x}\right|dx$$

$$= \int_0^{\pi/4}\log 2\,dx - \int_0^{\pi/4}\log|1+\tan x|\,dx$$

$$= \log 2\,[x]_0^{\pi/4} - I$$

$$I = \log 2\left[\frac{\pi}{4}-0\right]-I$$

$$2I = \frac{\pi}{4}\log 2$$

$$I = \frac{\pi}{8}\log 2$$

Q.10. Evaluate : $\int_0^{\pi/4}\frac{\sin x+\cos x}{9+16\sin 2x}dx$

Sol. Let $I = \int_0^{\pi/4}\frac{\sin x+\cos x}{9+16\sin 2x}dx$

$$= \int_0^{\pi/4}\frac{\sin x+\cos x}{9+16-16+16\sin 2x}dx$$

$$= \int_0^{\pi/4}\frac{\sin x+\cos x}{25-16(1-\sin 2x)}dx$$

$$= \int_0^{\pi/4}\frac{\sin x+\cos x}{25-16(\sin^2 x+\cos^2 x-2\sin x\cos x)}dx$$

$$I = \int_0^{\pi/4}\frac{\sin x+\cos x}{25-16(\sin x-\cos x)^2}dx$$

Put $\sin x - \cos x = t$

Differentiate both sides w.r.t. x

$$\cos x+\sin x = \frac{dt}{dx}$$

$$(\cos x+\sin x)\,dx = dt$$

If $x = 0$, $t = \sin 0 - \cos 0$

$$t = 0-1 = -1$$

If $x = \frac{\pi}{4}$, $t = \sin\frac{\pi}{4} - \cos\frac{\pi}{4}$

$$= \frac{1}{\sqrt{2}} - \frac{1}{\sqrt{2}}$$

$t = 0$

$$\therefore \quad I = \int_{-1}^{0} \frac{dt}{25 - 16t^2} = \frac{1}{16}\int_{-1}^{0} \frac{dt}{\frac{25}{16} - t^2}$$

$$= \frac{1}{16}\int_{-1}^{0} \frac{dt}{\left(\frac{5}{4}\right)^2 - t^2}$$

$$= \frac{1}{16} \times \frac{1}{2 \times \frac{5}{4}} \log \left| \frac{\frac{5}{4} + t}{\frac{5}{4} - t} \right|_{-1}^{0}$$

$$\left[\because \int \frac{dx}{a^2 - x^2} = \frac{1}{2a} \log \left| \frac{a + x}{a - x} \right| + c\right]$$

$$I = \frac{1}{40}\left\{\log \left| \frac{\frac{5}{4} + 0}{\frac{5}{4} - 0} \right| - \log \left| \frac{\frac{5}{4} - 1}{\frac{5}{4} + 1} \right|\right\}$$

$$= \frac{1}{40}\left\{\log \frac{5}{4} \times \frac{4}{5} - \log \frac{\frac{1}{4}}{\frac{9}{4}}\right\}$$

$$= \frac{1}{40}\left[\log 1 - \log \frac{1}{4} \times \frac{4}{9}\right]$$

$$= \frac{1}{40}(\log 1 - \log 9)$$

$$= \frac{1}{40}(-\log 9) = -\frac{1}{40}\log 3^2$$

$$= -\frac{1}{40} \cdot 2 \log 3$$

$$I = -\frac{1}{20}\log 3$$

Q. 11. Evaluate : $\displaystyle\int \frac{2}{(1 - x)(1 + x^2)} dx$

Sol. Let $\dfrac{1}{(1-x)(1+x^2)} = \dfrac{A}{1-x} + \dfrac{Bx + C}{1 + x^2}$

$$1 = A(1 + x^2) + (Bx + C)(1 - x)$$

Put $\quad x = 1$

$\Rightarrow \quad 1 = 2A$

$\Rightarrow \quad A = \dfrac{1}{2}$

Put $\quad x = 0$

$\Rightarrow \quad 1 = A + C$

$\Rightarrow \quad 1 = \dfrac{1}{2} + C$

$\Rightarrow \quad C = \dfrac{1}{2}$

Put $\quad x = 2$

$\Rightarrow \quad 1 = 5A + (2B + C)(1 - 2)$

$\Rightarrow \quad 1 = 5A - 2B - C$

$\Rightarrow \quad 1 = \dfrac{5}{2} - 2B - \dfrac{1}{2}$

$\Rightarrow \quad 1 - \dfrac{5}{2} + \dfrac{1}{2} = -2B$

$\Rightarrow \quad \dfrac{2 - 5 + 1}{2} = -2B$

$\Rightarrow \quad \dfrac{-2}{2} = -2B$

$\Rightarrow \quad 1 = 2B$

$\Rightarrow \quad B = \dfrac{1}{2}$

$$\therefore \quad I = 2\int \frac{dx}{(1-x)(1+x^2)}$$

$$= 2\left[\int \frac{\frac{1}{2}}{1-x} dx + \int \frac{\frac{1}{2}x + \frac{1}{2}}{1 + x^2}\right]$$

$$= 2 \times \frac{1}{2}\int \frac{dx}{1-x} + \frac{1}{2} \times 2\int \frac{x+1}{1+x^2} dx$$

$$= \int \frac{dx}{1-x} + \int \frac{x}{1+x^2} dx + \int \frac{dx}{1+x^2}$$

$$I = -\log|1 - x| + \frac{1}{2}\log|1 + x^2| + \tan^{-1} x + c$$

Q. 12. Evaluate : $\displaystyle\int_{-1}^{2} |x^3 - x|\, dx$

Sol. Clearly, $f(x) = x^3 - x = x(x - 1)(x + 1)$

The sign of $f(x)$ for different values of x are given in figure

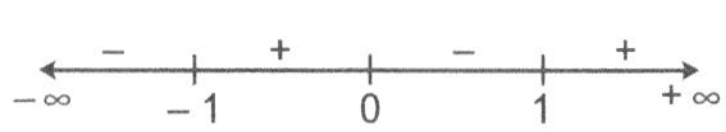

We observe that

$$f(x) > 0 \text{ for all } x \in (-1, 0) \cup (1, 2)$$
$$f(x) < 0 \text{ for all } x \in (0, 1)$$

$$\therefore \int_{-1}^{2} |x^3 - x|\, dx$$

$$= \int_{-1}^{0} (x^3 - x)\, dx + \int_{0}^{1} -(x^3 - x)\, dx + \int_{1}^{2} (x^3 - x)\, dx$$

$$= \left[\frac{x^4}{4} - \frac{x^2}{2}\right]_{-1}^{0} + \left[-\frac{x^4}{4} + \frac{x^2}{2}\right]_{0}^{1} + \left[\frac{x^4}{4} - \frac{x^2}{2}\right]_{1}^{2}$$

$$= \left\{(0 - 0) - \left(\frac{1}{4} - \frac{1}{2}\right)\right\} + \left\{\left(-\frac{1}{4} + \frac{1}{2}\right) - 0\right\}$$

$$+\left\{\left(\frac{16}{4}-\frac{4}{2}\right)-\left(\frac{1}{4}-\frac{1}{2}\right)\right\}$$

$$=-\frac{1}{4}+\frac{1}{2}-\frac{1}{4}+\frac{1}{2}+4-2-\frac{1}{4}+\frac{1}{2}$$

$$=-\frac{1}{4}+\frac{1}{2}-\frac{1}{4}+\frac{1}{2}+2-\frac{1}{4}+\frac{1}{2}$$

$$=\frac{-1+2-1+2+8-1+2}{4}=\frac{11}{4}$$

Q. 13. Evaluate : $\int \frac{dx}{x(x^5+3)}$

Sol. Let $I=\int \frac{dx}{x(x^5+3)}$

Put $x^5=t$

Differentiate both sides w.r.t. x.

$$5x^4=\frac{dt}{dx}$$

$$dx=\frac{dt}{5x^4}$$

$$\therefore \quad I=\frac{1}{5}\int \frac{dt}{x^5(t+3)}=\frac{1}{5}\int \frac{dt}{t(t+3)}$$

$$\frac{1}{t(t+3)}=\frac{A}{t}+\frac{B}{t+3}$$

$$\Rightarrow \quad 1=A(t+3)+Bt$$

If $t=0$, $1=3A$

$$A=\frac{1}{3},$$

If $t=-3$

$$1=-3B$$

$$B=-\frac{1}{3}$$

$$\therefore \quad I=\frac{1}{5}\left[\int \frac{\frac{1}{3}}{t}\,dt+\int \frac{-\frac{1}{3}}{t+3}\,dt\right]$$

$$=\frac{1}{15}\log t-\frac{1}{15}\log|t+3|+c$$

$$=\frac{1}{15}\log\left|\frac{t}{t+3}\right|+c$$

$$I=\frac{1}{15}\log\left|\frac{x^5}{x^5+3}\right|+c$$

Q. 14. Find $\int \frac{e^{2x}-e^{-2x}}{e^{2x}+e^{-2x}}dx$ **[NCERT]**

Sol. Let $I=\int \frac{e^{2x}-e^{-2x}}{e^{2x}+e^{-2x}}dx$

Multiply and divide by 2

$$I=\frac{1}{2}\int \frac{2e^{2x}-2e^{-2x}}{e^{2x}+e^{-2x}}dx$$

$$=\frac{1}{2}\log|e^{2x}+e^{-2x}|+c$$

$$\left[\because \int \frac{f'(x)}{f(x)}dx=\log|f(x)|+c\right]$$

Q. 15. Evaluate $\int \frac{x^3 \sin(\tan^{-1}x^4)}{1+x^8}dx$ **[NCERT]**

Sol. Let $I=\int \frac{x^3\sin(\tan^{-1}x^4)}{1+x^8}dx$

Put $\tan^{-1}x^4=t$

Differentiate w.r.t. x

$$\frac{1}{1+x^8}\cdot 4x^3=\frac{dt}{dx}$$

$$\frac{x^3}{1+x^8}dx=\frac{dt}{4}$$

$$I=\frac{1}{4}\int \sin t\,dt$$

$$=\frac{1}{4}(-\cos t)+c$$

$$I=-\frac{1}{4}\cos(\tan^{-1}x^4)+c$$

Q. 16. Find $\int \cos 2x \cos 4x \cos 6x\,dx$ **[NCERT]**

Sol. Let $I=\frac{1}{2}\int(2\cos 2x\cos 4x)\cos 6x\,dx$

$$\Rightarrow I=\frac{1}{2}\int(\cos 6x+\cos 2x)\cos 6x\,dx$$

$$[\because 2\cos A\cos B=\cos(A+B)+\cos(A-B)]$$

$$=\frac{1}{2}\int \cos 6x\cos 6x\,dx+\frac{1}{2}\int \cos 2x\cos 6x\,dx$$

$$=\frac{1}{4}\int 2\cos 6x\cos 6x\,dx+\frac{1}{4}\int 2\cos 2x\cos 6x\,dx$$

$$=\frac{1}{4}\int(\cos 12x+\cos 0)\,dx+\frac{1}{4}\int \cos 8x+\cos 4x\,dx$$

$$=\frac{1}{4}\left[\int(\cos 12x+1+\cos 8x+\cos 4x)\,dx\right]$$

$$=\frac{1}{4}\left[\frac{\sin 12x}{12}+x+\frac{\sin 8x}{8}+\frac{\sin 4x}{4}\right]+c$$

$$=\frac{1}{4}\left[\frac{\sin 12x}{12}+\frac{\sin 8x}{8}+\frac{\sin 4x}{4}+x\right]+c$$

Q. 17. Find $\int \tan^3 2x\sec 2x\,dx$ **[NCERT]**

Sol. Let $I=\int \tan^2 2x\tan 2x\sec 2x\,dx$

$$=\int(\sec^2 2x-1)\tan 2x\sec 2x\,dx$$

Put $\sec 2x=t$

Differentiate both sides w.r.t. x

$$2\sec 2x\tan 2x=\frac{dt}{dx}$$

$\therefore \quad \sec 2x \tan 2x\, dx = \frac{dt}{2}$

$$I = \frac{1}{2}\int (t^2 - 1)\, dt$$

$$\Rightarrow \quad I = \frac{1}{2}\left[\frac{t^3}{3} - t\right] + c$$

$$\Rightarrow \quad I = \frac{1}{2}\left[\frac{\sec^3 2x}{3} - \sec 2x\right] + c$$

$$\Rightarrow \quad I = \frac{\sec^3 2x}{6} - \frac{\sec 2x}{2} + c$$

Q. 18. Evaluate : $\int \frac{\sin^6 x + \cos^6 x}{\sin^2 x . \cos^2 x} dx.$★

Sol. Let $\quad I = \int \frac{\sin^6 x + \cos^6 x}{\sin^2 x \cos^2 x} dx$

Here, Numerator = $\sin^6 x + \cos^6 x$

$= (\sin^2 x + \cos^2 x)(\sin^4 x + \cos^4 x - \sin^2 x \cos^2 x)$

$[\because a^3 + b^3 = (a + b)(a^2 + b^2 - ab)]$

$= 1.[(\sin^2 x + \cos^2 x)^2 - 2 \sin^2 x \cos^2 x - \sin^2 x \cos^2 x]$

$= 1 - 3 \sin^2 x \cos^2 x$

$$\therefore \quad I = \int \frac{1 - 3\sin^2 x \cos^2 x}{\sin^2 x \cos^2 x} dx$$

$$= \int \frac{1}{\sin^2 x \cos^2 x} dx - 3\int 1\, dx + c$$

Multiplying and dividing denominator by $\cos^2 x$ in first Integral

$$= \int \frac{\sec^4 x}{\tan^2 x} dx - 3x + c$$

$$= \int \frac{\sec^2 x . \sec^2 x}{\tan^2 x} dx - 3x + c$$

$$= \int \frac{(1 + \tan^2 x)\sec^2 x}{\tan^2 x} dx - 3x + c$$

Put $\tan x = t$

$\Rightarrow \quad \sec^2 x\, dx = dt$

$$\Rightarrow \quad I = \int \frac{(t^2 + 1)}{t^2} dt - 3x + c$$

$$= \int \left[1 + \frac{1}{t^2}\right] dt - 3x + c$$

$$= t - \frac{1}{t} - 3x + c$$

$= \tan x - \cot x - 3x + c.$

Q. 19. Evaluate : $\int (x - 3)\sqrt{x^2 + 3x - 18}\, dx.$★

Sol. Let $\quad I = \int (x-3)\sqrt{x^2 + 3x - 18}\, dx$

$$= \int \left[\frac{1}{2}(2x+3) - \frac{9}{2}\right]\sqrt{x^2 + 3x - 18}\, dx$$

$$= \frac{1}{2}\int (2x+3)\sqrt{x^2 + 3x - 18}\, dx - \frac{9}{2}\int \sqrt{x^2 + 3x - 18}\, dx$$

Putting, $x^2 + 3x - 18 = t$ in the first integral.

$\Rightarrow \quad (2x + 3)\, dx = dt$

$$\therefore \quad I = \frac{1}{2}\int \sqrt{t}\, dt - \frac{9}{2}\int \sqrt{\left(x + \frac{3}{2}\right)^2 - \left(\frac{9}{2}\right)^2}\, dx$$

$$= \frac{1}{2}\frac{t^{3/2}}{3/2} - \frac{9}{2}\left[\frac{1}{2}\left(x + \frac{3}{2}\right)\sqrt{\left(x + \frac{3}{2}\right)^2 - \left(\frac{9}{2}\right)^2} - \frac{(9/2)^2}{2}\log\left|\left(x + \frac{3}{2}\right) + \sqrt{\left(x + \frac{3}{2}\right)^2 - \left(\frac{9}{2}\right)^2}\right|\right] + c$$

$$\left[\because \int \sqrt{x^2 - a^2}\, dx = \frac{1}{2}x\sqrt{x^2 - a^2} - \frac{a^2}{2}\log|x + \sqrt{x^2 - a^2}| + c\right]$$

$$= \frac{1}{3}(x^2 + 3x - 18)^{3/2} - \frac{9}{4}\left[\left(x + \frac{3}{2}\right)\sqrt{x^2 + 3x - 18} - \frac{81}{4}\log\left|x + \frac{3}{2} + \sqrt{x^2 + 3x - 18}\right|\right] + c$$

Q. 20. Evaluate : $\int_0^{\pi} \frac{4x \sin x}{1 + \cos^2 x} dx.$★

Sol. Let $\quad I = \int_0^{\pi} \frac{4x \sin x}{1 + \cos^2 x} \quad \ldots(i)$

$$\Rightarrow \quad I = \int_0^{\pi} \frac{4(\pi - x)\sin(\pi - x)}{1 + \cos^2(\pi - x)} dx$$

$$\left[\because \int_0^a f(x)dx = \int_0^a f(a - x)dx\right]$$

$$\therefore \quad I = \int_0^{\pi} \frac{4(\pi - x)\sin x}{1 + \cos^2 x} dx \quad \ldots(ii)$$

Adding equations (i) and (ii), we get

$$2I = 4\pi\int_0^{\pi} \frac{\sin x}{1 + \cos^2 x} dx$$

$$\Rightarrow \quad I = 2\pi\int_0^{\pi} \frac{\sin \pi}{1 + \cos^2 x} dx$$

Put $\quad \cos x = t$

$\Rightarrow \quad \sin x\, dx = -\, dt$

If $\quad x = 0$

$\Rightarrow \quad t = 1$

If $\quad x = \pi$

$\Rightarrow \quad t = -1$

$$\therefore \quad I = 2\pi\int_1^{-1} \frac{-\, dt}{1 + t^2}$$

$$= 2\pi\int_{-1}^{1} \frac{dt}{t^2 + 1}$$

$$\left[\because \int_a^b f(x)\, dx = -\int_b^a f(x)\, dx\right]$$

★ are board exam questions from previous years

$$= 2\pi\left[\tan^{-1} t\right]_{-1}^{1}$$

$$= 2\pi[\tan^{-1} 1 - \tan^{-1}(-1)]$$

$$= 2\pi\left[\frac{\pi}{4} - \left(-\frac{\pi}{4}\right)\right]$$

$$= 2\pi\left[\frac{2\pi}{4}\right] = \pi^2.$$

Q. 21. Evaluate : $\int \frac{x+2}{\sqrt{x^2+5x+6}}\, dx.$ ★

Sol. Let $x+2 = A\frac{d}{dx}(x^2+5x+6)+B$

$\Rightarrow \quad x+2 = A(2x+5)+B \quad \ldots$(i)

$\Rightarrow \quad x+2 = 2Ax+5A+B$

On equating the coefficients of x and constant term on both sides, we get

$2A = 1$

$\Rightarrow \quad A = \frac{1}{2}$

and $\quad 5A+B = 2$

$\Rightarrow \quad \frac{5}{2}+B = 2$

$\Rightarrow \quad B = 2-\frac{5}{2} = -\frac{1}{2}$

∴ Equation (i) becomes,

$$x+2 = \frac{1}{2}(2x+5)-\frac{1}{2}$$

$$\therefore \int \frac{x+2}{\sqrt{x^2+5x+6}}\,dx = \int \frac{\frac{1}{2}(2x+5)-\frac{1}{2}}{\sqrt{x^2+5x+6}}\,dx$$

$$= \frac{1}{2}\int \frac{2x+5}{\sqrt{x^2+5x+6}}\,dx - \frac{1}{2}\int \frac{dx}{\sqrt{x^2+5x+6}}$$

Putting, $x^2+5x+6 = t$

$\Rightarrow \quad (2x+5)\,dx = dt$

$$= \frac{1}{2}\int t^{-\frac{1}{2}}\,dt - \frac{1}{2}\int \frac{dx}{\sqrt{\left(x+\frac{5}{2}\right)^2-\left(\frac{1}{2}\right)^2}}+c$$

$$= \frac{1}{2}\frac{t^{\frac{1}{2}}}{\frac{1}{2}} - \frac{1}{2}\log\left|\left(x+\frac{5}{2}\right)+\sqrt{\left(x+\frac{5}{2}\right)^2-\left(\frac{1}{2}\right)^2}\right|+c$$

$$\left[\because \int \frac{dx}{\sqrt{x^2-a^2}} = \log|x+\sqrt{x^2-a^2}|\right]$$

$$= \sqrt{x^2+5x+6}-\frac{1}{2}\log\left|\left(x+\frac{5}{2}\right)+\sqrt{x^2+5x+6}\right|+c$$

★ are board exam questions from previous years

Q. 22. Evaluate $\int \frac{1}{\cos(x-a)\cos(x-b)}\,dx$ **[NCERT]**

Sol. Let $\quad I = \int \frac{1}{\cos(x-a)\cos(x-b)}\,dx$

Divide and multiply by $\sin(b-a)$

$$I = \frac{1}{\sin(b-a)}\int \frac{\sin(b-a)}{\cos(x-a)\cos(x-b)}\,dx$$

$$= \frac{1}{\sin(b-a)}\int \frac{\sin\{(x-a)-(x-b)\}}{\cos(x-a)\cos(x-b)}\,dx$$

(By adding and subtracting x)

$$= \frac{1}{\sin(b-a)}\int \frac{\sin(x-a)\cos(x-b)-\cos(x-a)\sin(x-b)}{\cos(x-a)\cos(x-b)}\,dx$$

$$= \frac{1}{\sin(b-a)}\int \frac{\sin(x-a)\cos(x-b)}{\cos(x-a)\cos(x-b)}\,dx - \frac{1}{\sin(b-a)}\int \frac{\cos(x-a)\sin(x-b)}{\cos(x-a)\cos(x-b)}\,dx$$

$$= \frac{1}{\sin(b-a)}\int \tan(x-a)\,dx - \frac{1}{\sin(b-a)}\int \tan(x-b)\,dx$$

$$= \frac{1}{\sin(b-a)}\{-\log|\cos(x-a)|+\log|\cos(x-b)|\}$$

$$\left[\because \int \tan x\,dx = -\log\cos x + c\right]$$

$$I = \frac{1}{\sin(b-a)}\log\left|\frac{\cos(x-b)}{\cos(x-a)}\right|+c$$

Q. 23. Find $\int \frac{1}{\sec x+\operatorname{cosec} x}\,dx$ **[NCERT]**

Sol. Given, $\int \frac{1}{\sec x+\operatorname{cosec} x}\,dx$

$$= \int \frac{1}{\frac{1}{\cos x}+\frac{1}{\sin x}}\,dx = \int \frac{1}{\frac{\sin x+\cos x}{\sin x\cos x}}\,dx$$

$$= \int \frac{\sin x\cos x}{\sin x+\cos x}\,dx$$

$$= \frac{1}{2}\int \frac{2\sin x\cos x}{\sin x+\cos x}\,dx \quad [\because \text{Divide and multiply by } 2]$$

$$= \frac{1}{2}\int \frac{\sin 2x}{\sin x+\cos x}\,dx \quad [\because \sin 2A = 2\sin A\cos A]$$

$$= \frac{1}{2}\int \frac{(1+\sin 2x)-1}{\sin x+\cos x}\,dx \quad [\text{add and subtract } 1]$$

$$= \frac{1}{2}\int \frac{(\sin^2 x+\cos^2 x+2\sin x\cos x)-1}{\sin x+\cos x}\,dx$$

$$[\because \sin^2 x+\cos^2 x=1]$$

$$= \frac{1}{2}\int \frac{(\sin x+\cos x)^2-1}{\sin x+\cos x}\,dx$$

$$= \frac{1}{2}\int \frac{(\sin x+\cos x)^2}{\sin x+\cos x}\,dx-\frac{1}{2}\int\frac{1}{\sin x+\cos x}\,dx$$

$$= \frac{1}{2}\int(\sin x+\cos x)\,dx$$

$$-\frac{1}{2}\int \frac{1}{\sqrt{2}\left(\frac{1}{\sqrt{2}}\sin x+\frac{1}{\sqrt{2}}\cos x\right)}\,dx$$

$$= \frac{1}{2}\int(\sin x+\cos x)\,dx-\frac{1}{2\sqrt{2}}\int\frac{1}{\cos\left(x-\frac{\pi}{4}\right)}\,dx$$

$$= \frac{1}{2}[-\cos x+\sin x]-\frac{1}{2\sqrt{2}}\int \sec\left(x-\frac{\pi}{4}\right)dx$$

$$= \frac{-1}{2}\cos x+\frac{1}{2}\sin x$$

$$-\frac{1}{2\sqrt{2}}\log\left|\sec\left(x-\frac{\pi}{4}\right)+\tan\left(x-\frac{\pi}{4}\right)\right|+c$$

Q. 24. Evaluate : $\int (3-2x).\sqrt{2+x-x^2}\,dx$.★

Sol. Let $\quad I=\int (3-2x).\sqrt{2+x-x^2}\,dx$

Let $\quad 3-2x=\lambda(2+x-x^2)+\mu$

$\Rightarrow \quad 3-2x=\lambda(1-2x)+\mu \qquad$ ...(i)

$\Rightarrow \quad 3-2x=(-2\lambda)x+\lambda+\mu$

Equating the coefficients of like terms, we get

$$-2\lambda=-2$$

$\Rightarrow \quad \lambda=1$

and $\quad \lambda+\mu=3$

$$\mu=3-1=2$$

$\therefore$ Equation (i) becomes

$$3-2x=(1-2x)+2$$

$$I=\int\{(1-2x)+2\}\sqrt{2+x-x^2}\,dx$$

$$I=\int(1-2x)\sqrt{2+x-x^2}\,dx+2\int\sqrt{2+x-x^2}\,dx$$

$$I=I_1+2I_2 \qquad \text{...(ii)}$$

Now, $\quad I_1=\int(1-2x)\sqrt{2+x-x^2}\,dx$

Let $\quad 2+x-x^2=t$

$$(1-2x)\,dx=dt$$

$\therefore \quad I_1=\int\sqrt{t}\,dt$

$$I_1=\frac{t^{1/2+1}}{\frac{1}{2}+1}+C_1$$

$$I_1=\frac{2}{3}t^{3/2}+C_1$$

$$I_1=\frac{2}{3}(2+x-x^2)^{3/2}+C_1 \qquad \text{...(iii)}$$

and $\quad I_2=\int\sqrt{2+x-x^2}\,dx$

$$I_2=\int\sqrt{2+\frac{1}{4}-\left(x^2-x+\frac{1}{4}\right)}\,dx$$

$$I_2=\int\sqrt{\left(\frac{3}{2}\right)^2-\left(x-\frac{1}{2}\right)^2}\,dx$$

$$I_2=\frac{1}{2}\left(x-\frac{1}{2}\right)\sqrt{\left(\frac{3}{2}\right)^2-\left(x-\frac{1}{2}\right)^2}+\frac{9/4}{2}\sin^{-1}\left(\frac{x-\frac{1}{2}}{\frac{3}{2}}\right)+C_2$$

$$I_2=\frac{1}{2}\left(x-\frac{1}{2}\right)\sqrt{2+x-x^2}+\frac{9}{8}\sin^{-1}\left(\frac{2x-1}{3}\right)+C_2 \quad \text{...(iv)}$$

From equation (ii), (iii) and (iv), we get

$$I=\frac{2}{3}(2+x-x^2)^{3/2}+\left(x-\frac{1}{2}\right)\sqrt{2+x-x^2}+\frac{9}{4}\sin^{-1}\left(\frac{2x-1}{3}\right)+C$$

where $\quad C=C_1+C_2$.

Q. 25. Find $\int \frac{2x}{(x^2+1)(x^2+2)^2}\,dx$.★

Sol. Let $\quad I=\int\frac{2x}{(x^2+1)(x^2+2)^2}\,dx$

Put $\quad x^2=t$

$$2x\,dx=dt$$

$$I=\int\frac{dt}{(t+1)(t+2)^2}$$

Let $\quad \frac{1}{(t+1)(t+2)^2}=\frac{A}{t+1}+\frac{B}{t+2}+\frac{C}{(t+2)^2} \qquad$ ...(A)

★ are board exam questions from previous years

$$\frac{1}{(t+1)(t+2)^2} = \frac{A(t+2)^2 + B(t+1)(t+2) + C(t+1)}{(t+1)(t+2)^2}$$

$$1 = A(t+2)^2 + B(t+1)(t+2) + C(t+1)$$

Equating coefficients, of t^2, t and constant terms on both sides, we get

$$A + B = 0 \qquad \ldots(i)$$

$$4A + 3B + C = 0 \qquad \ldots(ii)$$

$$4A + 2B + C = 1 \qquad \ldots(iii)$$

Subtracting equation (iii) from (ii), we get

$$B = -1$$

Substitute in equation (i)

$\Rightarrow \qquad A = 1$

Substitute the values of A and B in (ii), we get

$$4 - 3 + C = 0$$

$$C = -1$$

From equation (A)

$$\frac{1}{(t+1)(t+2)^2} = \frac{1}{t+1} + \left(\frac{-1}{t+2}\right) + \left(\frac{-1}{(t+2)^2}\right)$$

$$\int \frac{1}{(t+1)(t+2)^2}\,dt = \int \frac{1}{t+1}\,dt - \int \frac{1}{t+2}\,dt - \int \frac{1}{(t+2)^2}\,dt$$

$$= \log(t+1) - \log(t+2) + \frac{1}{t+2} + c$$

Where c is constant of integration.

$$\therefore \int \frac{2x\,dx}{(x^2+1)(x^2+2)^2} = \log\left|\frac{x^2+1}{x^2+2}\right| + \frac{1}{x^2+2} + c$$

Q. 26. Find : $\displaystyle\int_0^{\pi/4} \frac{dx}{\cos^3 x\sqrt{2\sin 2x}}$ ★

Sol. Let $\displaystyle I = \int_0^{\pi/4} \frac{dx}{\cos^3 x\sqrt{2\sin 2x}}$

Multiplying and dividing denominator by $\cos x$

$$\int_0^{\pi/4} \frac{dx}{\cos^4 x \dfrac{\sqrt{2\sin 2x}}{\cos x}} = \int_0^{\pi/4} \frac{\sec^4 x\,dx}{\sqrt{\dfrac{2\sin 2x}{\cos^2 x}}}$$

$$I = \int_0^{\pi/4} \frac{(1+\tan^2 x)\sec^2 x}{2\sqrt{\dfrac{\sin x\cos x}{\cos^2 x}}}\,dx$$

$[\because \sin 2x = 2\sin x\cos x]$

$$I = \int_0^{\pi/4} \frac{(1+\tan^2 x)\sec^2 x}{2\sqrt{\tan x}}\,dx$$

Put $\qquad \tan x = t$

$\Rightarrow \quad \sec^2 x\,dx = dt$, we get

$$\therefore \quad I = \int_0^1 \frac{(1+t^2)\,dt}{2\sqrt{t}}$$

$$I = \frac{1}{2}\int_0^1 \left(\frac{1}{\sqrt{t}} + t^{3/2}\right)dt \qquad \left(\begin{array}{l}\because \text{ when } x = 0, t = 0 \\ \text{and } x = \dfrac{\pi}{4}, t = 1\end{array}\right)$$

$$I = \frac{1}{2}\left[2\sqrt{t} + \frac{2}{5}t^{5/2}\right]_0^1$$

$$I = \left[\sqrt{t} + \frac{1}{5}t^{5/2}\right]_0^1$$

$$I = \left[\sqrt{1} + \frac{1}{5}1^{5/2}\right] - (0)$$

$$\therefore \quad I = \frac{6}{5}$$

Q. 27. Evaluate : $\displaystyle\int \frac{3x-1}{(x-1)(x-2)(x-3)}\,dx$ **[NCERT]**

Sol. $\displaystyle\frac{3x-1}{(x-1)(x-2)(x-3)} = \frac{A}{x-1} + \frac{B}{x-2} + \frac{C}{x-3}$

$$3x - 1 = A(x-2)(x-3) + B(x-1)(x-3) + C(x-1)(x-2)$$

If $x = 1$, then

$$2 = A(-1)(-2)$$

$$2A = 2$$

$$A = 1$$

If $x = 2$, then

$$5 = B(1)(-1)$$

$$B = -5$$

If $x = 3$, then

$$8 = C(2)(1)$$

$$C = 4$$

$$\therefore \int \frac{3x-1}{(x-1)(x-2)(x-3)}\,dx = \int \frac{dx}{x-1} + \int \frac{-5}{x-2}\,dx + \int \frac{4}{x-3}\,dx$$

$$\int \frac{3x-1}{(x-1)(x-2)(x-3)}\,dx = \log|x-1| - 5\log|x-2| + 4\log|x-3| + c$$

Q. 28. Find : $\displaystyle\int \frac{(3\sin\theta - 2)\cos\theta}{5 - \cos^2\theta - 4\sin\theta}\,d\theta$.★

Sol. Let $\displaystyle I = \int \frac{(3\sin\theta - 2)\cos\theta}{5 - \cos^2\theta - 4\sin\theta}.d\theta$

$$\Rightarrow \quad I = \int \frac{(3\sin\theta - 2)\cos\theta}{4 + 1 - \cos^2\theta - 4\sin\theta}.d\theta$$

★ are board exam questions from previous years

$$I = \int \frac{(3\sin\theta - 2)\cos\theta . d\theta}{4 + \sin^2\theta - 4\sin\theta}$$

$[\because 1 - \cos^2\theta = \sin^2\theta]$

$$I = \int \frac{(3\sin\theta - 2)\cos\theta . d\theta}{(\sin\theta - 2)^2}$$

Put $\sin\theta = t$

$\Rightarrow \cos\theta . d\theta = dt$

$\therefore \quad I = \int \frac{(3t-2)}{(t-2)^2} . dt$

Consider,

$$\frac{3t-2}{(t-2)^2} = \frac{A}{(t-2)} + \frac{B}{(t-2)^2}$$

$$3t - 2 = A(t-2) + B$$

On comparing, we get

$A = 3$ and $B = 4$

$$I = \int \left(\frac{3}{t-2} + \frac{4}{(t-2)^2}\right) dt$$

$$I = 3\log|t-2| - \frac{4}{(t-2)} + c$$

$$I = 3\log|\sin\theta - 2| - \frac{4}{(\sin\theta - 2)} + c$$

or $$I = 3\log|2 - \sin\theta| + \frac{4}{(2 - \sin\theta)} + c$$

Q. 29. Find : $\int \frac{\sqrt{x}}{\sqrt{a^3 - x^3}} dx.$★

Sol. Let $I = \int \frac{\sqrt{x}}{\sqrt{a^3 - x^3}} dx$

Put $x^{3/2} = t$

$$\frac{3}{2}\sqrt{x}.dx = dt$$

$$\Rightarrow \quad \sqrt{x}.dx = \frac{2}{3}.dt$$

Putting the values in I, we get

$$I = \frac{2}{3}\int \frac{1}{\sqrt{(a^{3/2})^2 - t^2}} dt$$

$$I = \frac{2}{3}\sin^{-1}\left(\frac{t}{a^{3/2}}\right) + c$$

or $$I = \frac{2}{3}\sin^{-1}\left(\frac{x}{a}\right)^{3/2} + c$$

Q. 30. Evaluate : $\int_0^{3/2} |x\cos\pi x|\, dx.$★

Sol. We have,

$$0 < x < \frac{3}{2} \Rightarrow 0 < \pi x < \frac{3\pi}{2}$$

Now, $0 < x < \frac{1}{2}$

$$0 < \pi x < \frac{\pi}{2}$$

$\Rightarrow \cos\pi x > 0$

$\Rightarrow x\cos\pi x > 0$

$\Rightarrow |x\cos\pi x| = x\cos\pi x$ for $0 < \pi x < \frac{\pi}{2}$

and $\frac{1}{2} < x < \frac{3}{2}$

$\Rightarrow \frac{\pi}{2} < \pi x < \frac{3\pi}{2}$

$\Rightarrow \cos\pi x < 0$

$\Rightarrow x\cos\pi x < 0$

$\Rightarrow |x\cos\pi x| = -x\cos\pi x$ for $\pi < \pi x < \frac{3\pi}{2}$

$$\therefore \int_0^{3/2} |x\cos\pi x|\, dx$$

$$= \int_0^{1/2} |x\cos\pi x|\, dx + \int_{1/2}^{3/2} |x\cos\pi x|\, dx$$

$$= \int_0^{1/2} (x\cos\pi x)\, dx + \int_{1/2}^{3/2} (-x\cos\pi x)\, dx$$

$$= \int_0^{1/2} (x\cos\pi x)\, dx - \int_{1/2}^{3/2} (x\cos\pi x)\, dx$$

$$= \left[\left(\left|\frac{x\sin\pi x}{\pi}\right|_0^{1/2} - \int_0^{1/2} \frac{\sin\pi x}{\pi} dx\right) - \left(\left|\frac{x\sin\pi x}{\pi}\right|_{1/2}^{3/2} - \int_{1/2}^{3/2} \frac{\sin\pi x}{\pi} dx\right)\right]$$

$$= \left[\left\{\frac{1}{\pi}\left(\frac{1}{2}\sin\frac{\pi}{2} - 0\right) + \left|\frac{\cos\pi x}{\pi^2}\right|_0^{1/2}\right\} - \left\{\frac{1}{\pi}\left(\frac{3}{2}\sin\frac{3\pi}{2} - \frac{1}{2}\sin\frac{\pi}{2}\right) + \left|\frac{\cos\pi x}{\pi^2}\right|_{1/2}^{3/2}\right\}\right]$$

$$= \left\{\frac{1}{2\pi} + \frac{1}{\pi^2}\left(\cos\frac{\pi}{2} - \cos 0\right) - \frac{1}{\pi}\left(\frac{3}{2}(-1) - \frac{1}{2}(1)\right) + \frac{1}{\pi^2}\left(\cos\frac{3\pi}{2} - \cos\frac{\pi}{2}\right)\right\}$$

$$= \frac{1}{2\pi} + \frac{1}{\pi^2}(0-1) - \frac{1}{\pi}\left(-\frac{3}{2} - \frac{1}{2}\right) + \frac{1}{\pi^2}(0-0)$$

$$= \frac{1}{2\pi} - \frac{1}{\pi^2} + \frac{2}{\pi}$$

$$= \frac{1}{2\pi} + \frac{2}{\pi} - \frac{1}{\pi^2}$$

★ are board exam questions from previous years

$$= \frac{\pi + 4\pi - 2}{2\pi^2}$$

$$= \frac{5\pi - 2}{2\pi^2}$$

Q. 31. Evaluate $\int_0^{\pi/2} \sqrt{\sin \phi} \cos^5 \phi \, d\phi$ [NCERT]

Sol. Let $I = \int_0^{\pi/2} \sqrt{\sin \phi} \cos^5 \phi \, d\phi$

$$= \int_0^{\pi/2} \sqrt{\sin \phi} \cos^4 \phi \cos \phi \, d\phi$$

$$= \int_0^{\pi/2} \sqrt{\sin \phi} (\cos^2 \phi)^2 \cos \phi \, d\phi$$

$$I = \int_0^{\pi/2} \sqrt{\sin \phi} (1 - \sin^2 \phi)^2 \cos \phi \, d\phi$$

Put $\sin \phi = t$

$\Rightarrow$ $\cos \phi \, d\phi = dt$

When $\phi = 0, t = \sin 0 = 0$

When $\phi = \frac{\pi}{2}, t = \sin \frac{\pi}{2} \Rightarrow t = 1$

$\therefore$ $I = \int_0^1 \sqrt{t}(1 - t^2)^2 \, dt$

$$= \int_0^1 \sqrt{t}(1 + t^4 - 2t^2) \, dt$$

$$= \int_0^1 (t^{1/2} + t^{9/2} - 2t^{5/2}) \, dt$$

$$I = \left[\frac{t^{3/2}}{3/2} + \frac{t^{11/2}}{11/2} - 2\frac{t^{7/2}}{7/2}\right]_0^1$$

$$= \left[\frac{2}{3}t^{3/2} + \frac{2}{11}t^{11/2} - 2 \times \frac{2}{7}t^{7/2}\right]_0^1$$

$$= \left(\frac{2}{3} + \frac{2}{11} - \frac{4}{7}\right) - 0$$

$$= \frac{154 + 42 - 132}{231}$$

$$= \frac{196 - 132}{231} = \frac{64}{231}$$

Q. 32. Find $\int_0^1 \sin^{-1}\left(\frac{2x}{1+x^2}\right) dx$ [NCERT]

Sol. Put $x = \tan \theta$

$\Rightarrow$ $\theta = \tan^{-1} x$

When $x = 1, \theta = \tan^{-1} 1 = \frac{\pi}{4}$

When $x = 0, \theta = \tan^{-1} 0 = 0$

Also, $\frac{d\theta}{dx} = \frac{1}{1+x^2}$

$\Rightarrow$ $dx = (1 + x^2) \, d\theta$

$\Rightarrow$ $dx = (1 + \tan^2 \theta) \, d\theta = \sec^2 \theta \, d\theta$

$\therefore$ $I = \int_0^{\pi/4} \sin^{-1}\left(\frac{2 \tan \theta}{1 + \tan^2 \theta}\right) . \sec^2 \theta \, d\theta$

$$= \int_0^{\pi/4} \sin^{-1} (\sin 2\theta) \sec^2 \theta \, d\theta$$

$$= \int_0^{\pi/4} 2\theta \sec^2 \theta \, d\theta = 2\int_0^{\pi/4} \underset{\text{I}}{\theta} \underset{\text{II}}{\sec^2 \theta} \, d\theta$$

$$= 2\left[\theta \int_0^{\pi/4} \sec^2 \theta \, d\theta - \int_0^{\pi/4} 1. \tan \theta \, d\theta\right]$$

$$= 2[\theta \tan \theta + \log |\cos \theta|]_0^{\pi/4}$$

$$= 2\left\{\frac{\pi}{4} \tan \frac{\pi}{4} + \log \left|\cos \frac{\pi}{4}\right|\right\} - \{0 + \log 1\}$$

$$= 2\left[\frac{\pi}{4} + \log \frac{1}{\sqrt{2}}\right]$$

$$= \frac{\pi}{2} + 2 \log 2^{-1/2} = \frac{\pi}{2} - 2 \times \frac{1}{2} \log 2$$

$$= \frac{\pi}{2} - \log 2$$

Q. 33. Evaluate :

$$\int_0^{\pi/2} \frac{\sin^{3/2} x}{\sin^{3/2} x + \cos^{3/2}} dx \quad \textbf{[NCERT]}$$

Sol. Let $I = \int_0^{\pi/2} \frac{\sin^{3/2} dx}{\sin^{3/2} dx + \cos^{3/2} x} dx$ ...(i)

We know that

$$\int_0^a f(x) \, dx = \int_0^a f(a - x) \, dx$$

$\therefore$ $I = \int_0^{\pi/2} \frac{\sin^{3/2}\left(\frac{\pi}{2} - x\right)}{\sin^{3/2}\left(\frac{\pi}{2} - x\right) + \cos^{3/2}\left(\frac{\pi}{2} - x\right)} dx$

$$I = \int_0^{\pi/2} \frac{\cos^{3/2} x}{\cos^{3/2} x + \sin^{3/2} x} dx \quad ...(ii)$$

Adding equations (i) and (ii)

$$2I = \int_0^{\pi/2} \frac{\sin^{3/2} x + \cos^{3/2} x}{\sin^{3/2} x + \cos^{3/2} x} dx$$

$$2I = \int_0^{\pi/2} 1 dx$$

$$= [x]_0^{\pi/2}$$

$$2I = \frac{\pi}{2}$$

$$I = \frac{\pi}{4}$$

Q. 34. Evaluate $\int_{-5}^{5} |x+2|\, dx$ [NCERT]

Sol. We know that

$$\int_a^b f(x)\,dx = \int_a^c f(x)\,dx + \int_c^b f(x)\,dx$$

$$\therefore \quad I = \int_{-5}^{-2} -(x+2)\,dx + \int_{-2}^{5} (x+2)\,dx$$

$$= \left[-\frac{x^2}{2} - 2x\right]_{-5}^{-2} + \left[\frac{x^2}{2} + 2x\right]_{-2}^{5}$$

$$= \left\{\left(-\frac{4}{2}+4\right) - \left(\frac{-25}{2}+10\right)\right\} + \left\{\left(\frac{25}{2}+10\right) - \left(\frac{4}{2}-4\right)\right\}$$

$$= \frac{4}{2} + \frac{5}{2} + \frac{45}{2} + \frac{4}{2} = \frac{58}{2}$$

$$\therefore \quad I = 29$$

Q. 35. Find $\int_{-\pi/2}^{\pi/2} \sin^7 x$ [NCERT]

Sol. We know that

$$\int_{-a}^{a} f(x)\,dx = \begin{cases} 0, \text{ if } f(x) \text{ is odd} \\ 2\int_0^a f(x)\,dx, \text{ if } f(x) \text{ is even} \end{cases}$$

Here $f(x) = \sin^7 x$

$f(-x) = \sin^7(-x) = -\sin^7 x$

$\therefore f(x)$ is odd function.

$$\therefore \int_{-\pi/2}^{\pi/2} \sin^7 x\,dx = 0$$

Q. 36. $\int_0^{2\pi} \cos^5 x\,dx$ [NCERT]

Sol. Since

$$\int_0^{2a} f(x)\,dx = \begin{cases} 0, \text{ if } f(x) \text{ is odd} \\ 2\int_0^a f(x)\,dx, \text{ if } f(x) \text{ is even} \end{cases}$$

We have, $\int_0^{2\pi} \cos^5 x\,dx$

Here $f(x) = \cos^5 x$

Then, $f(2\pi - x) = \cos^5(2\pi - x)$

$= \cos^5 x$

$\therefore$ $f(x) =$ is an even function.

$$\therefore \int_0^{2\pi} \cos^5 x\,dx = 2\int_0^{\pi} \cos^5 x\,dx$$

Now, $f(\pi - x) = \cos^5(\pi - x)$

$= -\cos^5 x = -f(x)$

$\because \cos^5 x$ is odd

$$\therefore \int_0^{2\pi} \cos^5 x\,dx = 2 \times 0 = 0.$$

Q. 37. Find : $\int \frac{e^x\,dx}{(e^x-1)^2\,(e^x+2)}$.★

Sol. Let $I = \int \frac{e^x\,dx}{(e^x-1)^2\,(e^x+2)}$

★ are board exam questions from previous years

Putting, $e^x = t$ and $e^x\,dx = dt$, we get

$$I = \int \frac{dt}{(t-1)^2(t+2)}$$

Let, $\frac{1}{(t-1)^2(t+2)} = \frac{A}{t-1} + \frac{B}{(t-1)^2} + \frac{C}{t+2}$

$\Rightarrow$ $1 = A(t-1)(t+2) + B(t+2) + C(t-1)^2$

Putting, $t = 1$

$\Rightarrow$ $1 = 3B \Rightarrow B = 1/3$

Putting, $t = -2$

$\Rightarrow$ $1 = 9C \Rightarrow C = 1/9$

Putting, $t = 0$

$\Rightarrow$ $1 = -2A + 2B + C$

$\Rightarrow$ $1 = -2A + \frac{2}{3} + \frac{1}{9}$

$\Rightarrow$ $A = -1/9$

$$\therefore \quad I = \frac{-1}{9}\int \frac{1}{(t-1)}\,dt + \frac{1}{3}\int \frac{1}{(t-1)^2}\,dt + \frac{1}{9}\int \frac{1}{(t+2)}\,dt$$

$$I = \frac{-1}{9}\log|t-1| - \frac{1}{3\cdot(t-1)} + \frac{1}{9}\log|t+2| + c$$

$$\therefore \quad I = \frac{1}{9}\log\left|\frac{e^x+2}{e^x-1}\right| - \frac{1}{3(e^x-1)} + c.$$

Q. 38. Find : $\int \frac{(3\sin x - 2)\cos x}{13 - \cos^2 x - 7\sin x}\,dx$.★

Sol. Let $I = \int \frac{(3\sin x - 2)\cos x}{13 - \cos^2 x - 7\sin x}\,dx$

$$= \int \frac{(3\sin x - 2)\cos x}{\sin^2 x - 7\sin x + 12}\,dx$$

$[\because \cos^2 x = 1 - \sin^2 x]$

Put $\sin x = t \Rightarrow \cos x\,dx = dt$

$$I = \int \frac{(3t-2)}{t^2 - 7t + 12}\,dt = \int \frac{3t-2}{(t-3)(t-4)}\,dt$$

Let, $\frac{3t-2}{(t-3)(t-4)} = \frac{A}{t-3} + \frac{B}{t-4}$ ...(A)

$$\frac{3t-2}{(t-3)(t-4)} = \frac{A(t-4) + B(t-3)}{(t-3)(t-4)}$$

$$3t - 2 = A(t-4) + B(t-3)$$

Equating co-efficient of t and constant term on both sides, we get

$A + B = 3$ ...(i)

$-4A - 3B = -2$ ...(ii)

Solving the above two equations, we get

$A = -7, B = 10$

From equation A

$$\frac{3t-2}{(t-3)(t-4)} = \frac{-7}{t-3} + \frac{10}{t-4}$$

$$\int \frac{3t-2}{(t-3)(t-4)}\,dt = -7\int \frac{1}{t-3}\,dt + 10\int \frac{1}{t-4}\,dt$$

$$I = -7 \log|t-3| + 10 \log|t-4| + c$$

Where c is constant of integration.

$\therefore \quad I = 10 \log|\sin x - 4| - 7 \log|\sin x - 3| + c.$

Q. 39. Find $\int \frac{2 \cos x}{(1-\sin x)(1+\sin^2 x)} dx$.★

Sol. Let $\quad I = \int \frac{2\cos x}{(1-\sin x)(1+\sin^2 x)} dx$

Put $\quad \sin x = t$

$\Rightarrow \quad \cos x \, dx = dt$

$$\therefore \quad I = \int \frac{2dt}{(1-t)(1+t^2)}$$

Let $\frac{2}{(1-t)(1+t^2)} = \frac{A}{1-t} + \frac{Bt+C}{1+t^2}$

$$2 = A(1+t^2) + (Bt+C)(1-t)$$

$$\Rightarrow \quad 2 = (A-B)t^2 + (B-C)t + A + C$$

Put $t = 1$,

We get, $\quad 2 = 2A$

$\Rightarrow \quad A = 1$

Comparing coefficients of t^2 and t on both sides,

$A - B = 0$

$\Rightarrow \quad B = A$

$\Rightarrow \quad B = 1$

Also, $\quad B - C = 0$

$\Rightarrow \quad B = C = 1$

$$\therefore \quad I = \int \left(\frac{1}{1-t} + \frac{t+1}{t^2+1}\right) dt$$

$$= \frac{\log(1-t)}{-1} + \frac{1}{2}\int \frac{2t}{t^2+1}.dt + \int \frac{1}{t^2+1}.dt$$

$$= -\log(1-t) + \frac{1}{2}\log(t^2+1) + \tan^{-1} t + c$$

$$= -\log(1-\sin x) + \frac{1}{2}\log(\sin^2 x + 1) + \tan^{-1}(\sin x) + c$$

Q. 40. Evaluate : $\int_0^{\pi} \frac{x \tan x}{\sec x + \tan x} dx$.★

Sol. Let $I = \int_0^{\pi} \frac{x \tan x}{\sec x + \tan x} dx. \quad \ldots(i)$

Now $\quad I = \int_0^{\pi} \frac{(\pi - x)\tan(\pi - x)}{\sec(\pi - x) + \tan(\pi - x)} dx$

$$I = \int_0^{\pi} \frac{(\pi - x)\tan x}{\sec x + \tan x} dx \quad \ldots(ii)$$

Adding (i) and (ii), we get

$$2I = \int_0^{\pi} \frac{x \tan x}{\sec x + \tan x} dx + \int_0^{\pi} \frac{(\pi - x)\tan x}{\sec x + \tan x}$$

$$2I = \pi \int_0^{\pi} \frac{\tan x}{\sec x + \tan x} dx$$

$$I = \frac{\pi}{2}\int_0^{\pi} \frac{\tan x}{\sec x + \tan x} \times \frac{\sec x - \tan x}{\sec x - \tan x} dx$$

$$I = \frac{\pi}{2}\int_0^{\pi} \frac{(\sec x \tan x - \tan^2 x)}{\sec^2 x - \tan^2 x} dx$$

$$I = \frac{\pi}{2}\int_0^{\pi} (\sec x \tan x - \sec^2 x + 1)\, dx$$

$$I = \frac{\pi}{2}[\sec x - \tan x + x]_0^{\pi}$$

$$I = \frac{\pi}{2}[(\sec \pi - \tan \pi + \pi) - (\sec 0 - \tan 0 + 0)]$$

$$I = \frac{\pi}{2}[(-1 - 0 + \pi) - (1 - 0)]$$

$$\therefore \quad I = \frac{\pi(\pi - 2)}{2}.$$

Chapter 9. Differential Equations

Q. 1. Solve the following differential equation :★

$$(1+x)^2 \frac{dy}{dx} + y = \tan^{-1} x.$$

Sol. Given differential equation is,

$$(1+x)^2 \frac{dy}{dx} + y = \tan^{-1} x \quad \ldots(i)$$

Divide by $(1 + x^2)$ in above equation, we get

$$\frac{dy}{dx} + \frac{1}{1+x^2} y = \frac{\tan^{-1} x}{1+x^2}$$

The above equation is of the form $\frac{dy}{dx} + Py = Q$

Where, $\quad P = \frac{1}{1+x^2}$ and $Q = \frac{\tan^{-1} x}{1+x^2}$

$\therefore \quad$ I.F. $= e^{\int P\,dx}$

$$= e^{\int \frac{1}{1+x^2} dx}$$

$$= e^{\tan^{-1} x}$$

Solution of the equation is given by,

$$y.\text{I.F} = \int Q\,\text{I.F.}\,dx + c$$

$$ye^{\tan^{-1} x} = \int \frac{e^{\tan^{-1} x} \tan^{-1} x}{1+x^2} dx + c \quad \ldots(ii)$$

Put $\quad \tan^{-1} x = t$

Differentiate both sides w.r.t. x

$$\frac{1}{1+x^2} = \frac{dt}{dx}$$

$$\Rightarrow \quad \frac{dx}{1+x^2} = dt$$

★ are board exam questions from previous years

$\therefore$ Equation (ii) becomes

$$ye^{\tan^{-1}x} = \int e^t.t\,dt + c$$

$$ye^{\tan^{-1}x} = te^t - \int 1.e^t.\,dt + c$$
$$= t\,e^t - e^t + c$$
$$= \tan^{-1}x\; e^{\tan^{-1}x} - e^{\tan^{-1}x} + c$$

$$\Rightarrow \quad ye^{\tan^{-1}x} = e^{\tan^{-1}x}(\tan^{-1}x - 1) + c$$

$$\Rightarrow \quad y = \tan^{-1}x - 1 + c\,e^{-\tan^{-1}x}$$

Q. 2. Find the particular solution, satisfying given condition, for the following differential equation :

$$\frac{dy}{dx} - \frac{y}{x} + \text{cosec}\,\frac{y}{x} = 0;\ y = 0 \text{ when } x = 1.$$

Sol. Given differential equation is

$$\frac{dy}{dx} - \frac{y}{x} + \text{cosec}\,\frac{y}{x} = 0 \qquad \text{...(i)}$$

Put $\frac{y}{x} = t \Rightarrow y = xt$

Differentiate both sides w.r.t. x

$$\frac{dy}{dx} = x\frac{dt}{dx} + t$$

Put in equation (i)

$$x\frac{dt}{dx} + t - t + \text{cosec}\,t = 0$$

$$\Rightarrow \quad x\frac{dt}{dx} = -\text{cosec}\,t$$

$$\Rightarrow \quad x\frac{dt}{dx} = -\frac{1}{\sin t}$$

$$\Rightarrow \quad \sin t\,dt = -\frac{dx}{x}$$

On integrating both sides, we get

$$-\cos t = -\log x + c$$

$$\Rightarrow \quad \log x - \cos t = c$$

$$\log x - \cos\left(\frac{y}{x}\right) = c, \qquad \text{...(ii)} \left[\because t = \frac{y}{x}\right]$$

$$\log x - \cos\left(\frac{y}{x}\right) = -1$$

When $y = 0$ and $x = 1$

$$\log 1 - \cos 0 = c$$

$$\therefore \quad c = -1$$

Put the value of c in equation (ii)

$$\Rightarrow \quad \log x + 1 = \cos\left(\frac{y}{x}\right)$$

$$\Rightarrow \quad \log x + \log e = \cos\left(\frac{y}{x}\right)$$

$$\Rightarrow \quad \log|xe| = \cos\left(\frac{y}{x}\right)$$

Q. 3. Form the differential equation representing the family of curves given by $(x-a)^2 + 2y^2 = a^2$, where a is an arbitrary constant.

Sol. Given curve is

$$(x-a)^2 + 2y^2 = a^2$$

$$\Rightarrow \quad x^2 + a^2 - 2ax + 2y^2 = a^2$$

$$\Rightarrow \quad x^2 - 2ax + 2y^2 = 0 \qquad \text{...(i)}$$

Differentiate both sides w.r.t. x

$$2x - 2a + 4y\frac{dy}{dx} = 0$$

$$\Rightarrow \quad x - a + 2y\frac{dy}{dx} = 0$$

$$\Rightarrow \quad a = x + 2y\frac{dy}{dx}$$

Put the value of a in equation (i)

$$x^2 - 2x\left(x + 2y\frac{dy}{dx}\right) + 2y^2 = 0$$

$$\Rightarrow \quad x^2 - 2x^2 - 4xy\frac{dy}{dx} + 2y^2 = 0$$

$$\Rightarrow \quad -x^2 - 4xy\frac{dy}{dx} + 2y^2 = 0$$

$$\Rightarrow \quad 2y^2 - x^2 = 4xy\frac{dy}{dx}$$

Q. 4. Solve the following differential equation :★

$$\cos^2 x\frac{dy}{dx} + y = \tan x$$

Sol. Given differential equation is

$$\cos^2 x\frac{dy}{dx} + y = \tan x$$

$$\Rightarrow \quad \frac{dy}{dx} + \frac{1}{\cos^2 x}y = \frac{\tan x}{\cos^2 x}$$

$$\frac{dy}{dx} + (\sec^2 x)y = \tan x \sec^2 x \qquad \text{...(i)}$$

Here, $P = \sec^2 x$, $Q = \tan x \sec^2 x$

$$\text{I.F.} = e^{\int P\,dx} = e^{\int \sec^2 x\,dx} = e^{\tan x}$$

Solution of above equation will be given by

$$y.\,\text{I.F.} = \int Q\,\text{I.F.}\,dx + c$$

$$y.e^{\tan x} = \int \tan x \sec^2 x\, e^{\tan x}\,dx + c \qquad \text{...(ii)}$$

★ are board exam questions from previous years

Let $I = \int e^{\tan x} \tan x \sec^2 x\, dx$

Put $\tan x = t$

Differentiate both sides w.r.t. x

$$\sec^2 x = \frac{dt}{dx}$$

$$\Rightarrow \quad \sec^2 x\, dx = dt$$

$$\therefore \quad \int e^t . t\, dt = t\int e^t\, dt - \int \left(\frac{d}{dt}(t) \cdot \int e^t\, dt\right) dt$$

$$= t\, e^t - \int e^t\, dt$$

$$= te^t - e^t$$

$$= \tan x\, e^{\tan x} - e^{\tan x}$$

From equation (ii)

$$ye^{\tan x} = e^{\tan x}(\tan x - 1) + c$$

$$y = \tan x - 1 + ce^{-\tan x}.$$

Q. 5. Solve the differential equation

$$\boldsymbol{x \log x \frac{dy}{dx} + y = 2 \log x.}$$

Sol. Given differential equation is,

$$x \log x \frac{dy}{dx} + y = 2 \log x$$

$$\Rightarrow \quad \frac{dy}{dx} + \frac{1}{x \log x} y = \frac{2}{x} \qquad \text{...(i)}$$

Here $P = \dfrac{1}{x\log x}$, $Q = \dfrac{2}{x}$.

$$\text{I.F.} = e^{\int P\, dx} = e^{\int \frac{1}{x \log x} dx}$$

$$= e^{\int \frac{1/x}{\log x} dx}$$

$$= e^{\log(\log x)} = \log x$$

Solution of given equation will be given by,

$$y.\text{I.F.} = \int Q\, \text{I.F.}\, dx + c$$

$$\Rightarrow \quad y \log x = 2\int \log x \cdot \frac{1}{x} dx + c$$

$$\Rightarrow \quad y \log x = \frac{2(\log x)^2}{2} + c$$

$$\left[\because \int [f(x)]^n \cdot f'(x)\, dx = \frac{[f(x)]^{n+1}}{n+1} + c\right]$$

$$\Rightarrow \quad y = \log x + \frac{c}{\log x}$$

Q. 6. Form the differential equation of the family of circles touching the *X*-axis at origin.★

Sol. Given circle touches the *X*-axis at origin.

$\therefore$ Centre will lie on *Y*-axis.

Equation of the family of circles touching *X*-axis at origin is

$$(x-0)^2 + (y-a)^2 = a^2$$

$$\Rightarrow \quad x^2 + y^2 + a^2 - 2ay = a^2$$

$$\Rightarrow \quad x^2 + y^2 - 2ay = 0 \qquad \text{...(i)}$$

Differentiate w.r.t. x, we get

$$2x + 2y\frac{dy}{dx} - 2a\frac{dy}{dx} = 0$$

$$\Rightarrow \quad 2a\frac{dy}{dx} = 2x + 2y\frac{dy}{dx}$$

$$\Rightarrow \quad a = \frac{x + y\frac{dy}{dx}}{\frac{dy}{dx}}$$

Put the value of a in equation (i)

$$x^2 + y^2 - 2 \times \left(\frac{x + y\frac{dy}{dx}}{\frac{dy}{dx}}\right) y = 0$$

$$\Rightarrow \quad x^2\frac{dy}{dx} + y^2\frac{dy}{dx} - 2xy - 2y^2\frac{dy}{dx} = 0$$

$$\Rightarrow \quad x^2\frac{dy}{dx} - y^2\frac{dy}{dx} - 2xy = 0$$

$$\Rightarrow \quad (x^2 - y^2)\frac{dy}{dx} = 2xy$$

which is the required differential equation.

Q. 7. Find the particular solution of the differential equation satisfying the given condition :

$$\boldsymbol{x^2dy + (xy + y^2)\, dx = 0; \; y = 1 \text{ when } x = 1}$$

Sol. Given differential equation is

$$x^2dy + (xy + y^2)\, dx = 0 \qquad \text{...(i)}$$

$$\Rightarrow \quad x^2dy = -(xy + y^2)\, dx$$

$$\Rightarrow \quad x^2\frac{dy}{dx} = -(xy + y^2)$$

$$\Rightarrow \quad \frac{dy}{dx} = -\frac{xy}{x^2} - \frac{y^2}{x^2}$$

$$\Rightarrow \quad \frac{dy}{dx} = -\frac{y}{x} - \left(\frac{y}{x}\right)^2 \qquad \text{...(ii)}$$

Put $\quad \dfrac{y}{x} = t \qquad$ ...(iii)

$$\Rightarrow \quad y = xt$$

Differentiate w.r.t. x

$$\frac{dy}{dx} = x\frac{dt}{dx} + t \qquad \text{...(iv)}$$

From equations (ii), (iii) and (iv)

$$x\frac{dt}{dx} + t = -t - t^2$$

$$\Rightarrow \quad x\frac{dt}{dx} = -t - t^2 - t$$

★ are board exam questions from previous years

$$\Rightarrow \quad x\frac{dt}{dx} = -2t - t^2$$

$$\Rightarrow \quad x\frac{dt}{dx} = -t(2+t)$$

$$\Rightarrow \quad \frac{dt}{t(2+t)} = -\frac{dx}{x}$$

On integrating both sides, we get

$$\int \frac{dt}{t(2+t)} = -\int \frac{dx}{x} \quad ...(v)$$

Now, $\frac{1}{t(2+t)} = \frac{A}{t} + \frac{B}{2+t}$

$\Rightarrow \quad 1 = A(2+t) + Bt$

If $\quad t = -2$

$\Rightarrow \quad 1 = -2B + 0$

$\Rightarrow \quad B = \frac{-1}{2}$

If $\quad t = 0$

$\Rightarrow \quad 1 = 2A$

$\Rightarrow \quad A = \frac{1}{2}$

$$\therefore \quad \int \frac{dt}{t(2+t)} = \frac{1}{2}\int \frac{dt}{t} - \frac{1}{2}\int \frac{dt}{2+t}$$

$$= \frac{1}{2}\log t - \frac{1}{2}\log|2+t| + c \quad ...(vi)$$

Equation (v) becomes

$$\frac{1}{2}\log t - \frac{1}{2}\log|2+t| = -\log x + c$$

$$\Rightarrow \quad [\log t - \log|2+t|] = -2\log x + 2c$$

$$\Rightarrow \quad \log t - \log|2+t| = -\log x^2 + 2c$$

$$\Rightarrow \quad \log\left|\frac{t}{2+t}\right| + \log x^2 = 2c$$

$$\Rightarrow \quad \log\left|\frac{t}{2+t} \times x^2\right| = 2c$$

$$\Rightarrow \quad \log\left|\frac{\frac{y}{x}}{2+\frac{y}{x}} \times x^2\right| = 2c \quad \left[\because t = \frac{y}{x}\right]$$

$$\Rightarrow \quad \log\left|\frac{yx^2}{2x+y}\right| = 2c$$

$$\Rightarrow \quad \frac{yx^2}{2x+y} = e^{2c}$$

$$\Rightarrow \quad \frac{yx^2}{2x+y} = A, \text{ where } A = e^{2c}$$

When $x = 1, y = 1$

$$\frac{1}{3} = A$$

Hence $\quad \frac{x^2y}{2x+y} = \frac{1}{3}$

$\Rightarrow \quad 3x^2y = 2x + y$

is required particular solution.

Q. 8. Find the general solution of the differential equation.★

$$x\log x \cdot \frac{dy}{dx} + y = \frac{2}{x}\log x$$

Sol. Given differential equation is

$$x\log x \cdot \frac{dy}{dx} + y = \frac{2}{x}\log x \quad ...(i)$$

Divide by $x\log x$ on both sides

$$\frac{dy}{dx} + \frac{1}{x\log x}y = \frac{2}{x^2} \quad ...(ii)$$

Here, $P = \frac{1}{x\log x}$, $Q = \frac{2}{x^2}$,

$$\text{I.F.} = e^{\int \frac{1}{x\log x}dx}$$

$$= e^{\log(\log x)}$$

$$\text{I.F.} = \log x$$

Multiply equation (ii) by I.F.

$$\log x\frac{dy}{dx} + \log x \cdot \frac{1}{x\log x}y = \frac{2}{x^2}\log x$$

$$\log x\frac{dy}{dx} + \frac{1}{x}y = \frac{2}{x^2}\log x$$

$$\frac{d}{dx}(y\log x) = \frac{2}{x^2}\log x$$

Integrating both sides

$$y\log x = \int \frac{2}{x^2}\log x\,dx + c$$

$$= \int 2\log x \cdot x^{-2}\,dx + c$$

$$= 2\left[\log x \cdot \frac{x^{-1}}{-1} - \int \frac{1}{x}\cdot\frac{x^{-1}}{-1}dx\right] + c$$

$$= 2\left[-\frac{\log x}{x} + \int x^{-2}dx\right] + c$$

$$y\log x = 2\left[-\frac{\log x}{x} + \frac{x^{-1}}{-1}\right] + c$$

$$y\log x = 2\left[-\frac{\log x}{x} - \frac{1}{x}\right] + c$$

$$y\log x = \frac{2}{x}[-\log x - 1] + c$$

$$y\log x = \frac{-2}{x}(\log x + 1) + c$$

is the required equation.

★ are board exam questions from previous years

Q. 9. Find the particular solution of the differential equation satisfying the given condition.★

$$\frac{dy}{dx} = y \tan x, \text{ given that } y = 1 \text{ when } x = 0.$$

Sol. Given differential equation is

$$\frac{dy}{dx} = y \tan x \qquad ...(i)$$

$$\frac{dy}{y} = \tan x \, dx$$

On integrating both sides, we get,

$$\log y = \log \sec x + \log c$$
$$\log y = \log c \sec x$$
$$y = c \sec x \qquad ...(ii)$$

When $y = 1$ and $x = 0$

Equation (ii) becomes

$$1 = c \sec 0$$
$$c = 1$$

Put it in equation (ii)

$y = \sec x$ is required solution.

Q. 10. Solve the following differential equation :★

$$(x^2+1)\frac{dy}{dx} + 2xy = \sqrt{x^2+4}$$

Sol. Given differential equation is,

$$(x^2+1)\frac{dy}{dx} + 2xy = \sqrt{x^2+4} \qquad ...(i)$$

Divide by (x^2+1)

$$\frac{dy}{dx} + \frac{2x}{x^2+1}y = \frac{\sqrt{x^2+4}}{x^2+1} \qquad ...(ii)$$

Here, $P = \frac{2x}{x^2+1}$, $Q = \frac{\sqrt{x^2+4}}{x^2+1}$

$$\text{I.F.} = e^{\int P\,dx} = e^{\int \frac{2x}{x^2+1}dx}$$
$$\text{I.F.} = e^{\log|x^2+1|}$$
$$\therefore \quad \text{I.F.} = (x^2+1)$$

Multiply equation (ii) by I.F.

$$(x^2+1)\frac{dy}{dx} + 2xy = \sqrt{x^2+4}$$

$$\frac{d}{dx}[y(x^2+1)] = \sqrt{x^2+2^2}$$

Integrating both sides

$$y(x^2+1) = \frac{1}{2}x\sqrt{x^2+4} + \frac{1}{2}(2)^2 \log|x+\sqrt{x^2+4}| + C$$

$$y(x^2+1) = \frac{x}{2}\sqrt{x^2+4} + 2\log|x+\sqrt{x^2+4}| + C$$

is the required solution of given differential equation.

★ are board exam questions from previous years

Q. 11. Solve the following differential equation :★

$$(x^3+x^2+x+1)\frac{dy}{dx} = 2x^2+x$$

Sol. Given differential equation is,

$$(x^3+x^2+x+1)\frac{dy}{dx} = 2x^2+x$$

$$dy = \frac{(2x^2+x)}{x^3+x^2+x+1}dx$$

On integrating both sides, we get

$$\int dy = \int \frac{2x^2+x}{x^3+x^2+x+1}dx \qquad ...(i)$$

Also
$$\frac{2x^2+x}{x^3+x^2+x+1} = \frac{2x^2+x}{x^2(x+1)+1(x+1)}$$
$$= \frac{2x^2+x}{(x+1)(x^2+1)}$$
$$= \frac{A}{x+1} + \frac{Bx+C}{x^2+1}$$

$$\Rightarrow \quad 2x^2+x = A(x^2+1) + (Bx+C)(x+1)$$

When $x = -1$

$$2 - 1 = 2A$$
$$A = \frac{1}{2}$$

When $x = 0$

$$0 = A + C$$
$$0 = \frac{1}{2} + C$$
$$C = \frac{-1}{2}$$

When $x = 1$

$$3 = 2A + 2B + 2C$$
$$3 = 2 \times \frac{1}{2} + 2B + 2\left(-\frac{1}{2}\right)$$
$$3 = 1 + 2B - 1$$
$$2B = 3 \Rightarrow B = \frac{3}{2}$$

From equation (i)

$$y = \frac{1}{2}\int \frac{dx}{x+1} + \int \frac{\frac{3}{2}x - \frac{1}{2}}{x^2+1}dx + c'$$

$$= \frac{1}{2}\log|x+1| + \frac{3}{2}\int \frac{x}{x^2+1}dx - \frac{1}{2}\int \frac{dx}{x^2+1} + c'$$

$$= \frac{1}{2}\log|x+1| + \frac{3}{2} \times \frac{1}{2}\int \frac{2x}{x^2+1}dx - \frac{1}{2}\tan^{-1}x + c'$$

$$y = \frac{1}{2}\log|x+1| + \frac{3}{4}\log|x^2+1| - \frac{1}{2}\tan^{-1}x + c'$$

is the required solution.

Q. 12. If $y = \text{cosec}^{-1}x, x > 1$, then show that★

$$x(x^2-1)\frac{d^2y}{dx^2}+(2x^2-1)\frac{dy}{dx}=0$$

Sol. Given, $y = \text{cosec}^{-1} x$

Differentiating both sides w.r.t. x, we have

$$\frac{dy}{dx} = -\frac{1}{x\sqrt{x^2-1}}$$

$$\Rightarrow \quad x\sqrt{x^2-1}\frac{dy}{dx} = -1$$

Again differentiating both sides w.r.t. x, we have

$$x\sqrt{x^2-1}\frac{d^2y}{dx^2}+\frac{dy}{dx}\left[\frac{d}{dx}(x\sqrt{x^2-1})\right]=0$$

$$\Rightarrow \quad x\sqrt{x^2-1}\frac{d^2y}{dx^2}+\frac{dy}{dx}\left[x\frac{1}{2\sqrt{x^2-1}}\cdot 2x+\sqrt{x^2-1}\right]=0$$

$$\Rightarrow \quad x\sqrt{x^2-1}\frac{d^2y}{dx^2}+\frac{dy}{dx}\left(\frac{x^2}{\sqrt{x^2-1}}+\sqrt{x^2-1}\right)=0$$

$$\Rightarrow \quad x\sqrt{x^2-1}\frac{d^2y}{dx^2}+\frac{dy}{dx}\left(\frac{x^2+x^2-1}{\sqrt{x^2-1}}\right)=0$$

$$\Rightarrow \quad x\sqrt{x^2-1}\frac{d^2y}{dx^2}+\frac{2x^2-1}{\sqrt{x^2-1}}\frac{dy}{dx}=0$$

$$\Rightarrow \quad x(x^2-1)\frac{d^2y}{dx^2}+(2x^2-1)\frac{dy}{dx}=0$$

Q. 13. Show that the following differential equation is homogeneous and then solve it :★

$$y\,dx + x\log\left(\frac{y}{x}\right)dy - 2x\,dy = 0$$

Sol. Given equation is

$$y\,dx + x\log\left(\frac{y}{x}\right)dy - 2x\,dy = 0$$

Divide by x :

$$\Rightarrow \quad \frac{y}{x}dx+\log\left(\frac{y}{x}\right)dy-2dy=0$$

$$\Rightarrow \quad \frac{y}{x}+\log\left(\frac{y}{x}\right)\frac{dy}{dx}-2\frac{dy}{dx}=0$$

$$\Rightarrow \quad \left(\log\frac{y}{x}-2\right)\frac{dy}{dx}=-\frac{y}{x}$$

$$\Rightarrow \quad \left(2-\log\frac{y}{x}\right)\frac{dy}{dx}=\frac{y}{x}$$

$$\Rightarrow \quad \frac{dy}{dx}=\frac{y/x}{2-\log y/x} \quad ...(i)$$

$\therefore$ It is a homogeneous differential equation of the form $f\left(\frac{y}{x}\right)$

To solve it put $\frac{y}{x}=t$

$$\Rightarrow \quad y = xt$$

Differentiate both sides w.r.t. x, we get

$$\frac{dy}{dx}=x\frac{dt}{dx}+t$$

Put in equation (i)

$$\Rightarrow \quad x\frac{dt}{dx}+t=\frac{t}{2-\log t}$$

$$\Rightarrow \quad x\frac{dt}{dx}=\frac{t}{2-\log t}-\frac{t}{1}$$

$$\Rightarrow \quad x\frac{dt}{dx}=\frac{t-2t+t\log t}{2-\log t}$$

$$\Rightarrow \quad x\frac{dt}{dx}=\frac{t\log t-t}{2-\log t}$$

$$\frac{2-\log t}{t(\log t-1)}dt=\frac{dx}{x}$$

$$\frac{1-(\log t-1)}{t(\log t-1)}dt=\frac{dx}{x}$$

$$\left(\frac{1}{t(\log t-1)}-\frac{1}{t}\right)dt=\frac{dx}{x}$$

$$\left(\frac{1/t}{(\log t-1)}-\frac{1}{t}\right)dt=\frac{dx}{x}$$

On integrating both sides, we get

$$\int\frac{1/t}{\log t-1}dt-\int\frac{1}{t}dt=\int\frac{dx}{x}+c$$

$$\Rightarrow \quad \log|\log t-1|-\log|t|=\log|x|+c$$

$$\Rightarrow \log|\log t-1|-\log|t|-\log|x|=c$$

$$\Rightarrow \quad \log\left|\frac{\log t-1}{tx}\right|=c$$

$$\Rightarrow \quad \left|\frac{\log t-1}{tx}\right|=e^c$$

$$\Rightarrow \quad \frac{\log t-1}{tx}=\pm e^c=A$$

$$\Rightarrow \quad \frac{\log\frac{y}{x}-1}{\frac{y}{x}\times x}=A \quad \left[\because t=\frac{y}{x}\right]$$

$$\Rightarrow \quad \log\left(\frac{y}{x}\right)-1=Ay$$

which is the required general solution.

★ are board exam questions from previous years

Q. 14. Solve the following differential equation :★

$$(x^2-1)\frac{dy}{dx}+2xy=\frac{1}{x^2-1} \quad : |x| \neq 1$$

Sol. Given differential equation is

$$(x^2-1)\frac{dy}{dx}+2xy = \frac{1}{x^2-1}$$

Divide by (x^2-1)

$$\frac{dy}{dx}+\frac{2x}{x^2-1}y = \frac{1}{(x^2-1)^2} \quad ...(i)$$

Here, $P=\frac{2x}{x^2-1}$, $Q=\frac{1}{(x^2-1)^2}$

$$\text{I.F.} = e^{\int P\,dx}$$

$$= e^{\int \frac{2x}{x^2-1}dx} = e^{\log(x^2-1)}$$

$$= x^2-1$$

Multiply equation (i) by I.F.

$$(x^2-1)\frac{dy}{dx}+2xy = \frac{1}{(x^2-1)}$$

$$\frac{d}{dx}[y(x^2-1)] = \frac{1}{x^2-1}$$

Integrating both sides

$$y(x^2-1) = \int \frac{1}{x^2-1}dx + c$$

$$y(x^2-1) = \frac{1}{2}\log\left|\frac{x-1}{x+1}\right| + c$$

$$y = \frac{1}{2(x^2-1)}\log\left|\frac{x-1}{x+1}\right| + \frac{c}{x^2-1}$$

which is the required solution.

Q. 15. Solve the following differential equation :★

$$(1+y^2)(1+\log x)\,dx + x\,dy = 0$$

Sol. Given differential equation is

$$(1+y^2)(1+\log x)\,dx + x\,dy = 0$$

$$(1+y^2)(1+\log x)\,dx = -x\,dy$$

$$\frac{(1+\log x)}{x}dx = -\frac{dy}{1+y^2}$$

On integrating both sides

$$\int(1+\log x)\frac{1}{x}dx = -\int\frac{dy}{1+y^2}+c$$

$$\frac{(1+\log x)^2}{2} = -\tan^{-1}y + c$$

$$\left[\because \int [f(x)]^n \cdot f'(x) = \frac{[f(x)]^{n+1}}{n+1}\right]$$

$$\frac{1}{2}(1+\log x)^2 + \tan^{-1}y = c$$

which is required solution.

Q. 16. Form the differential equation of the family of parabolas having vertex at the origin and axis along positive *Y*-axis.★

Sol. Equation of parabola whose vertex is at origin and axis along positive *Y*-axis is

$$x^2 = 4ay \quad ...(i)$$

where a is an arbitrary constant.

Differentiate equation (i) w.r.t x

$$2x = 4a\frac{dy}{dx}$$

$$\therefore \quad 4a = \frac{2x}{\frac{dy}{dx}}$$

Put in equation (i)

$$x^2 = \frac{2x}{\frac{dy}{dx}}y$$

$$\Rightarrow \quad y = \frac{x^2}{2x}\frac{dy}{dx}$$

$$\Rightarrow \quad 2y = x\frac{dy}{dx}$$

which is required differential equation.

Q. 17. Solve the following differential equation :★

$$e^x \tan y\,dx + (1-e^x)\sec^2 y\,dy = 0$$

Sol. Given differential equation is,

$$e^x \tan y\,dx + (1-e^x)\sec^2 y\,dy = 0$$

$$(1-e^x)\sec^2 y\,dy = -e^x \tan y\,dx$$

$$\frac{\sec^2 y}{\tan y}dy = -\frac{e^x}{1-e^x}dx$$

Integrating both sides

$$\log|\tan y| = \log|1-e^x| + c$$

$$\left[\because \int\frac{f'(x)}{f(x)}dx = \log|f(x)| + c\right]$$

$$\Rightarrow \log|\tan y| - \log|1-e^x| = c$$

$$\Rightarrow \quad \log\left|\frac{\tan y}{1-e^x}\right| = c$$

$$\Rightarrow \quad \frac{\tan y}{1-e^x} = e^c$$

$$\Rightarrow \quad \tan y = e^c(1-e^x)$$

which is the required solution

Q. 18. Solve the following differential equation :

$$x\,dy - (y+2x^2)\,dx = 0.★$$

Sol. Given differential equation

$$x\,dy - (y+2x^2)\,dx = 0$$

$$x\,dy = (y+2x^2)\,dx$$

★ are board exam questions from previous years

$$\Rightarrow \quad \frac{dy}{dx} = \frac{(y+2x^2)}{x}$$

$$\Rightarrow \quad \frac{dy}{dx} = \frac{y}{x}+2x$$

$$\Rightarrow \quad \frac{dy}{dx}-\frac{1}{x}y = 2x \quad \text{...(i)}$$

Here, $P=-\frac{1}{x}$, $Q=2x$

$$\text{I.F.} = e^{\int P\,dx} = e^{-\int \frac{1}{x}dx} = e^{-\log x} = e^{\log x^{-1}} = x^{-1}$$

Multiply equation (i) by I.F.

$$x^{-1}\frac{dy}{dx}-\frac{1}{x^2}y = 2$$

$$\frac{d}{dx}\left(\frac{y}{x}\right) = 2$$

Integrating both sides :

$$\frac{y}{x} = \int 2\,dx + c$$

$$\frac{y}{x} = 2x + c$$

$$y = 2x^2 + cx$$

which is required solution.

Q. 19. Solve the following differential equation :★

$$(y+3x^2)\frac{dx}{dy} = x$$

Sol. Given differential equation is

$$(y+3x^2)\frac{dx}{dy} = x$$

$$x\frac{dy}{dx}-y = 3x^2$$

$$\frac{dy}{dx}-\frac{y}{x} = 3x \quad \text{...(i)}$$

Here $P=-\frac{1}{x}$, $Q=3x$

$$\text{I.F.} = e^{\int P\,dx} = e^{-\int\frac{1}{x}dx} = e^{-\log x}$$

$$= e^{\log x^{-1}} = x^{-1} = \frac{1}{x}$$

Multiply equation (i) by I.F.

$$\frac{1}{x}\frac{dy}{dx}-\frac{1}{x^2}y = 3$$

$$\frac{d}{dx}\left(y\cdot\frac{1}{x}\right) = 3$$

Integrating both sides

$$\frac{y}{x} = 3x + c$$

which is required solution.

★ are board exam questions from previous years

Q. 20. Form the differential equation of the family of ellipses having foci on Y-axis and centre at origin.

[NCERT]

Sol. The equation of family of ellipses having foci on Y-axis and centre at origin is :

$$\frac{x^2}{b^2}+\frac{y^2}{a^2} = 1 \quad \text{...(i)}$$

where, b and a are semi-major and semi-minor axis respectively.

Differentiate w.r.t. x

$$\frac{2x}{b^2}+\frac{2yy'}{a^2} = 0$$

$$\Rightarrow \quad \frac{yy'}{a^2} = -\frac{x}{b^2}$$

$$\Rightarrow \quad \frac{yy'}{x} = -\frac{a^2}{b^2} \quad \text{...(ii)}$$

Again differentiating w.r.t. x :

$$\frac{x\frac{d}{dx}(yy')-yy'\cdot 1}{x^2} = 0$$

$$\Rightarrow \quad x\{yy''+y'y'\}-yy' = 0$$

$$\Rightarrow \quad x(yy''+y'^2)-yy' = 0$$

which is the required differential equation.

Q. 21. Find the particular solution of the following differential equation :★

$$xy\frac{dy}{dx} = (x+2)(y+2);\ y=-1 \text{ when } x=1$$

Sol. Given differential equation is

$$xy\frac{dy}{dx} = (x+2)(y+2)$$

$$\frac{y}{y+2}dy = \frac{(x+2)}{x}dx$$

On integrating both sides, we have

$$\int\frac{y}{y+2}dy = \int\frac{x+2}{x}dx \quad \text{...(i)}$$

Put $y+2=t \Rightarrow y=t-2$

Differentiate w.r.t. y

$$1 = \frac{dt}{dy} \quad \therefore \ dy = dt$$

Equation (i) becomes :

$$\int\frac{t-2}{t}dt = \int 1\,dx + 2\int\frac{dx}{x}$$

$$\int\left(1-\frac{2}{t}\right)dt = x + 2\log x + c$$

$$t-2\log t = x+2\log x+c$$

$$y+2-2\log(y+2) = x+2\log x+c \quad \text{...(ii)}$$

When $x=1$ and $y=-1$

$$-1+2-2\log 1 = 1+2\log 1+c$$

$$1 = 1+c$$

$$\therefore \quad c = 0$$

Equation (ii) becomes :

$$y+2-2\log(y+2)=x+2\log x$$

which is the particular solution of the given differential equation.

Q. 22. Find the general solution of the differential equation $\dfrac{dy}{dx}=\dfrac{1-\cos x}{1+\cos x}$ **[NCERT]**

Sol. Given differential equation is,

$$\frac{dy}{dx} = \frac{1-\cos x}{1+\cos x}$$

$$dy = \left(\frac{1-\cos x}{1+\cos x}\right)dx$$

$$dy = \left(\frac{2\sin^2\frac{x}{2}}{2\cos^2\frac{x}{2}}\right)dx$$

$$dy = \tan^2\frac{x}{2}\,dx$$

$$dy = \left(\sec^2\frac{x}{2}-1\right)dx$$

Integrating both sides :

$$y = \int\left(\sec^2\frac{x}{2}-1\right)dx + c$$

$$y = \frac{\tan\frac{x}{2}}{\frac{1}{2}} - x + c$$

$$y = 2\tan\frac{x}{2} - x + c$$

which is the required general solution.

Q. 23. Find the general solution of the differential equation $\dfrac{dy}{dx}+y=1,\ (y\neq 1)$ **[NCERT]**

Sol. Given differential equation is,

$$\frac{dy}{dx}+y = 1$$

$$\Rightarrow \quad \frac{dy}{dx} = 1-y$$

$$\Rightarrow \quad \frac{dy}{1-y} = dx$$

Integrating both sides

$$-\log(1-y) = x+c$$

$$\Rightarrow \quad \log(1-y) = -x-c$$

$$\Rightarrow \quad 1-y = \pm e^{-x-c} = e^{-x}.e^{-c}$$

$$\Rightarrow \quad 1-y = Ae^{-x}, \quad \text{where } A=e^{-c}$$

$$\Rightarrow \quad y = 1-Ae^{-x}$$

which is the required general solution.

Q. 24. Find the general solution of the differential equation $\sec^2 x\tan y\,dx+\sec^2 y\tan x\,dy=0$. [NCERT]

Sol. Given differential equation is

$$\sec^2 x\tan y\,dx+\sec^2 y\tan x\,dy=0$$

$$\Rightarrow \quad \sec^2 y\tan x\,dy = -\sec^2 x\tan y\,dx$$

$$\frac{\sec^2 y}{\tan y}dy = -\frac{\sec^2 x}{\tan x}dx$$

Integrating both sides

$$\int\frac{\sec^2 y}{\tan y}dy = -\int\frac{\sec^2 x}{\tan x}+c$$

$$\log|\tan y| = -\log|\tan x|+c$$

$$\log|\tan y|+\log|\tan x| = c$$

$$\log|\tan x\tan y| = c$$

$$|\tan x\tan y| = e^c$$

$$\tan x\tan y = \pm e^c$$

$$\tan x\tan y = A \quad \text{where } A=\pm e^c$$

which is the required general solution of the given differential equation.

Q. 25. For the differential equation

$$xy\frac{dy}{dx}=(x+2)(y+2)$$

find the solution curve passing through the point (1, – 1).

Sol. Given differential equation is

$$xy\frac{dy}{dx} = (x+2)(y+2)$$

$$\frac{y}{y+2}dy = \frac{x+2}{x}dx$$

Integrating both sides

$$\int\frac{y}{y+2}dy = \int\frac{x+2}{x}dx+c$$

$$\int\left(1-\frac{2}{y+2}\right)dy = \int\left(1+\frac{2}{x}\right)dx+c$$

$$y-2\log|y+2| = x+2\log x+c \quad \text{...(i)}$$

Point (1, – 1) lies on this curve

$$-1-2\log 1 = 1+2\log 1+c$$

$$c = -2$$

Put the value of c in equation (i)

$$y-2\log|y+2| = x+2\log x-2$$

which is the required solution.

Q. 26. Find the particular solution of the following differential equation :★

$$\frac{dy}{dx} = 1 + x^2 + y^2 + x^2y^2,$$

given that $y = 1$ when $x = 0$.

Sol. Given differential equation is

$$\frac{dy}{dx} = 1 + x^2 + y^2 + x^2y^2$$

$$= (1 + x^2) + y^2(1 + x^2)$$

$$\frac{dy}{dx} = (1 + x^2)(1 + y^2)$$

$$\frac{dy}{1+y^2} = (1 + x^2)\,dx$$

On integrating both sides, we have

$$\int \frac{dy}{1+y^2} = \int (1 + x^2)\,dx$$

$$\tan^{-1} y = x + \frac{x^3}{3} + c \quad ...(i)$$

Put $y = 1$ and $x = 0$ in equation (i)

$$\tan^{-1} 1 = 0 + 0 + c$$

$$c = \frac{\pi}{4}$$

Equation (i) becomes :

$$\tan^{-1} y = x + \frac{x^3}{3} + \frac{\pi}{4}$$

is the required particular solution of given equation.

Q. 27. Find the particular solution of

$$\frac{dy}{dx} + 2y \tan x = \sin x$$

satisfying condition $x = \frac{\pi}{3}$ when $y = 0$. [NCERT]

Sol. Given differential equation is,

$$\frac{dy}{dx} + 2y \tan x = \sin x \quad ...(i)$$

Here $P = 2 \tan x$, $Q = \sin x$

$$\text{I.F.} = e^{\int P\,dx} = e^{2\int \tan x\,dx}$$

$$= e^{2(-\log \cos x)}$$

$$= e^{\log (\cos x)^{-2}} = \frac{1}{\cos^2 x}$$

Multiplying equation (i) by I.F.

$$\frac{1}{\cos^2 x}\frac{dy}{dx} + 2y\frac{\tan x}{\cos^2 x} = \frac{\sin x}{\cos^2 x}$$

$$\frac{d}{dx}(y \sec^2 x) = \sec x \tan x$$

Integrating both sides

$$y \sec^2 x = \int \sec x \tan x\,dx + c$$

$$y \sec^2 x = \sec x + c \quad ...(ii)$$

When $x = \frac{\pi}{3}$, $y = 0$

$$0 = \sec\frac{\pi}{3} + c$$

$$\therefore \quad c = -2$$

Put in equation (ii)

$$y \sec^2 x = \sec x - 2$$

which is the required solution.

Q. 28. Find the particular solution of the differential equation $(1 - y^2)(1 + \log x)\,dx + 2xy\,dy = 0$, given that $y = 0$ when $x = 1$.★

Sol. The given differential equation is

$$(1 - y^2)(1 + \log x)\,dx + 2xy\,dy = 0$$

$$\frac{(1 + \log x)}{x}\,dx = \frac{-2y}{(1 - y^2)}\,dy$$

On integrating both sides, we have

$$\int \frac{1 + \log x}{x}\,dx = \int \frac{-2y}{(1 - y^2)}\,dy$$

In first integral,

put $\quad 1 + \log x = t$

$$\frac{1}{x}\,dx = dt$$

Also in second integral,

put $\quad 1 - y^2 = u$

$$-2y\,dy = du$$

$$\therefore \quad \int t.dt = \int \frac{1}{u}\,du$$

$$\Rightarrow \quad \frac{t^2}{2} - \log|u| = c$$

$$\text{or} \quad \frac{1}{2}(1 + \log x)^2 - \log|1 - y^2| = c$$

It is given that $y = 0$ when $x = 1$

$$\text{So,} \quad \frac{1}{2}(1 + \log 1)^2 - \log|1 - 0^2| = c$$

$$\Rightarrow \quad c = \frac{1}{2}$$

$$\therefore \quad \frac{(1 + \log x)^2}{2} - \log|1 - y^2| = \frac{1}{2}$$

$$\text{or} \quad (1 + \log x)^2 - 2\log|1 - y^2| = 1$$

It is the required particular solution.

Q. 29. Solve the differential equation :★

$$y + x\frac{dy}{dx} = x - y\frac{dy}{dx}$$

★ are board exam questions from previous years

Sol. We have,

$$y+x\frac{dy}{dx}=x-y\frac{dy}{dx}$$

$$x\frac{dy}{dx}+y\frac{dy}{dx}=x-y$$

$$\frac{dy}{dx}=\frac{x-y}{x+y} \quad \text{...(i)}$$

Which is a homogeneous differential equation.

Putting $y=Vx \Rightarrow \frac{dy}{dx}=V+x\frac{dV}{dx}$ in (i), we get

$$V+x\frac{dV}{dx}=\frac{x-Vx}{x+Vx}$$

$$V+x\frac{dV}{dx}=\frac{1-V}{1+V}$$

$$x\frac{dV}{dx}=\frac{1-V}{1+V}-V$$

$$\Rightarrow \quad x\frac{dV}{dx}=\frac{1-V-V-V^2}{1+V}$$

$$\Rightarrow \quad x\frac{dV}{dx}=\frac{1-2V-V^2}{1+V}$$

$$\Rightarrow \quad \frac{1+V}{1-2V-V^2}dV=\frac{dx}{x},\ x\neq 0 \quad \text{...(ii)}$$

Putting $t=1-2V-V^2$

$$\Rightarrow \quad dt=(-2-2V)\,dV$$

$$\Rightarrow \quad -\frac{1}{2}dt=(1+V)\,dV$$

Now, equation (ii) becomes,

$$-\frac{1}{2t}dt=\frac{dx}{x}$$

On integrating above equation, we get

$$-\frac{1}{2}\int\frac{1}{t}\,dt=\int\frac{1}{x}dx$$

$$-\frac{1}{2}\log|t|=\log|x|+\log c$$

$$-\frac{1}{2}|1-2V-V^2|=\log\{c|x|\}$$

$$\log\{c|x|\}+\log|1-2V-V^2|^{1/2}=0$$

$$c|x|\left(1-\frac{2y}{x}-\frac{y^2}{x^2}\right)^{1/2}=0$$

$$c(x^2-2xy-y^2)^{1/2}=0.$$

★ are board exam questions from previous years

Q. 30. Find the general solution of the differential equation $\frac{dy}{dx}-y=\sin x$.★

Sol. We have,

$$\frac{dy}{dx}-y=\sin x$$

Which is a linear differential equation of the form

$$\frac{dy}{dx}+Py=Q$$

Where, $P=-1$ and $Q=\sin x$

Now, $\text{I.F.}=e^{\int P\,dx}=e^{\int -1\,dx}=e^{-x}$

So, the required solution is

$$ye^{-x}=\int e^{-x}\sin x\,dx+c_1 \quad \text{...(i)}$$

Let $$I=\int e^{-x}\sin x\,dx \quad \text{...(ii)}$$

$$I=\sin x\int e^{-x}\,dx-\int\left(\frac{d}{dx}(\sin x)\int e^{-x}\,dx\right)dx+c_2$$

$$I=-\sin x\,e^{-x}+\int\cos x e^{-x}\,dx+c_2$$

$$I=-\sin x\,e^{-x}$$
$$+\cos x\int e^{-x}\,dx-\int\left(\frac{d}{dx}(\cos x)\int e^{-x}\,dx\right)dx$$

$$I=-\sin x\,e^{-x}-\cos x\cdot e^{-x}-\int\sin x\cdot e^{-x}\,dx+c_2$$

$$I=-\sin x\cdot e^{-x}-\cos x\cdot e^{-x}-I+c_2 \quad \text{[using (ii)]}$$

$$2I=-e^{-x}(\sin x+\cos x)+c_2$$

$$I=\frac{-1}{2}e^{-x}(\sin x+\cos x)+c_2$$

By equation (i),

$$ye^{-x}=\frac{-1}{2}e^{-x}(\sin x+\cos x)+c_1+c_2$$

$$2y=-(\sin x+\cos x)+2ce^{x} \quad [\because c_1+c_2=c]$$

$\therefore$ $2y=2ce^x-\sin x-\cos x$ is the required solution.

Q. 31. Show that the family of curves for which $\frac{dy}{dx}=\frac{x^2+y^2}{2xy}$, is given by $x^2-y^2=cx$.★

Sol. Given family of curves

$$\frac{dy}{dx}=\frac{x^2+y^2}{2xy} \quad \text{...(i)}$$

Put $$y=vx$$

$$\frac{dy}{dx}=v+x\frac{dv}{dx}$$

From equation (i)

$$v+x\frac{dv}{dx}=\frac{x^2+(vx)^2}{2x(vx)}=\frac{x^2(1+v^2)}{x^2(2v)}$$

$$x\frac{dv}{dx}=\frac{1+v^2}{2v}-v$$

$$= \frac{1+v^2-2v^2}{2v} = \frac{1-v^2}{2v}$$

$$\frac{2v}{1-v^2}\,dv = \frac{dx}{x}$$

Integrating both sides

$$\int \frac{2v}{1-v^2}\,dv = \int \frac{dx}{x}$$

$$-\log|1-v^2| + \log c = \log x$$

Where log c is constant of integration.

$$\log\left|\frac{c}{1-\frac{y^2}{x^2}}\right| = \log x$$

$$\log\left|\frac{cx^2}{x^2-y^2}\right| = \log x$$

$$\frac{cx^2}{x^2-y^2} = x$$

$$cx = x^2 - y^2.$$

Chapter 10. Vectors

Q. 1. Show that the sum of three vectors determined by the medians of a triangle directed from the vertices is zero.

Sol. Let AD, BE and CF be the medians of a ΔABC.

Using the triangle law of addition, in triangle ABC, we have,

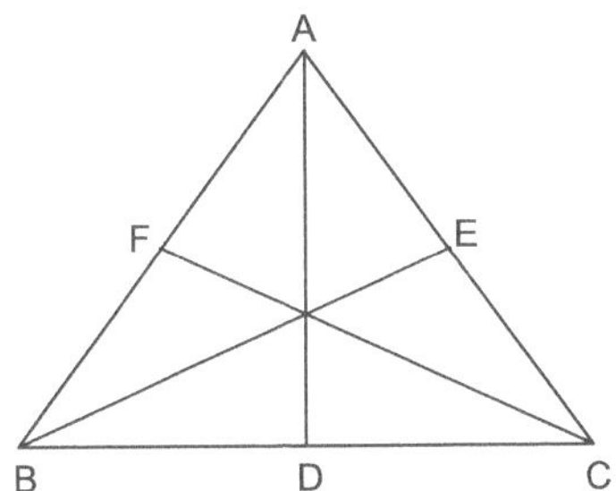

$$\vec{AD} = \vec{AB} + \vec{BD}$$

$$= \vec{AB} + \frac{1}{2}\vec{BC}$$

(since, D is the mid-point of $\vec{BC}$, so $\vec{BD} = \frac{1}{2}\vec{BC}$)

Similarly, $\vec{BE} = \vec{BC} + \frac{1}{2}\vec{CA}$

and $\vec{CF} = \vec{CA} + \frac{1}{2}\vec{AB}$

$$\therefore \quad \vec{AD} + \vec{BE} + \vec{CF} = \vec{AB} + \frac{1}{2}\vec{BC} + \vec{BC} + \frac{1}{2}\vec{CA} + \vec{CA} + \frac{1}{2}\vec{AB}$$

$$= \frac{3}{2}[\vec{AB} + \vec{BC} + \vec{CA}]$$

But $\vec{AC} = \vec{AB} + \vec{BC}$

Hence, $\vec{AD} + \vec{BE} + \vec{CF} = \frac{3}{2}[\vec{AC} + \vec{CA}] = 0$

Q. 2. If $\vec{a}, \vec{b}$ are the position vectors of A and B respectively, show that the position vectors of point C on the line AB is such that $\vec{AC} = 3\vec{AB}$ is $3\vec{b} - 2\vec{a}$.

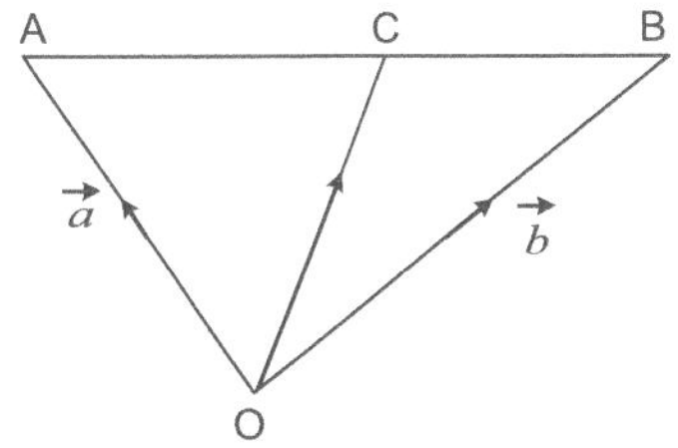

Sol. Let $\vec{v}$ be the position vector of the point C.

Now, $\vec{AC} = 3\vec{AB}$. (Given)

$$\Rightarrow \quad (\vec{v} - \vec{a}) = 3(\vec{b} - \vec{a})$$

$$\Rightarrow \quad (\vec{v} - \vec{a}) = 3\vec{b} - 3\vec{a}$$

$$\Rightarrow \quad \vec{v} = 3\vec{b} - 3\vec{a} + \vec{a}$$

$$= 3\vec{b} - 2\vec{a}$$

Q. 3. If $\vec{a} = 2\hat{i} + 4\hat{j} - 5\hat{k}$ and $\vec{b} = \hat{i} + 2\hat{j} + 3\hat{k}$, find the unit vector in the direction of $\vec{a} + \vec{b}$.

Sol. We have,

$$\vec{a} + \vec{b} = (2\hat{i} + 4\hat{j} - 5\hat{k}) + (\hat{i} + 2\hat{j} + 3\hat{k})$$

$$= 3\hat{i} + 6\hat{j} - 2\hat{k}$$

$$\Rightarrow \quad |\vec{a} + \vec{b}| = \sqrt{(3)^2 + (6)^2 + (-2)^2}$$

$$= \sqrt{(9 + 36 + 4)}$$

$$= \sqrt{49} = 7$$

Hence, the unit vector in the direction of $\vec{a} + \vec{b}$

$$= \frac{\vec{a} + \vec{b}}{|\vec{a} + \vec{b}|}$$

$$= \frac{1}{7}(3\hat{i} + 6\hat{j} - 2\hat{k})$$

$$= \frac{3}{7}\hat{i} + \frac{6}{7}\hat{j} - \frac{2}{7}\hat{k}$$

Q. 4. If *ABCD* is a quadrilateral and *E* and *F* are the mid-points of *AC* and *BD*, prove that

$$\vec{AB} + \vec{AD} + \vec{CB} + \vec{CD} = 4\vec{EF}.$$

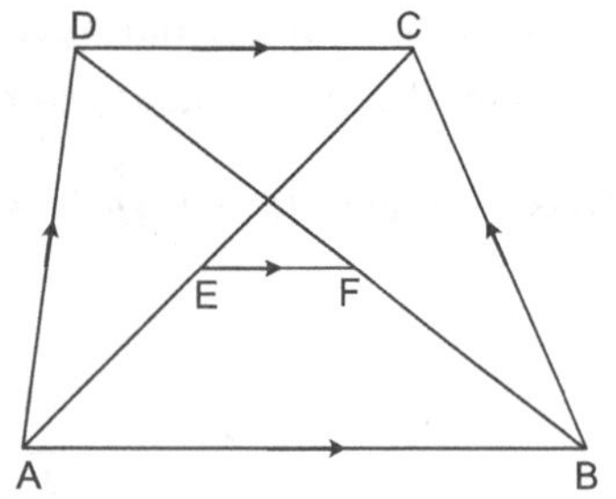

Sol. Let $\vec{a}, \vec{b}, \vec{c}$ and $\vec{d}$ be the position vectors of the vertices of the quadrilateral *ABCD* relative to some chosen point *O* as the origin.

Since, *E* and *F* are the mid-points of the sides *AC* and *BD* respectively, so the position vectors of *E* and *F* are

$$\frac{\vec{a}+\vec{c}}{2} \text{ and } \frac{\vec{b}+\vec{d}}{2}$$

We have, $\vec{AB} + \vec{AD} + \vec{CB} + \vec{CD}$

$$= (\vec{b}-\vec{a})+(\vec{d}-\vec{a})+(\vec{b}-\vec{c})+(\vec{d}-\vec{c})$$

$$= 2(\vec{b}+\vec{d}-\vec{c}-\vec{a})$$

$$= 4\left(\frac{\vec{b}+\vec{d}}{2}\right) - 4\left(\frac{\vec{c}+\vec{a}}{2}\right)$$

= 4 [Position vector of *F* – Position vector of *E*]

$$= 4\vec{EF}$$

Q. 5. If *P* is the mid-point of the side *BC* of a triangle *ABC*, prove that $\vec{AB}^2 + \vec{AC}^2 = 2(\vec{AP}^2 + \vec{BP}^2)$.

Sol. We have

$$\vec{AB}^2 + \vec{AC}^2 = |\vec{AB}|^2 + |\vec{AC}|^2$$

[The square of the vector is equal to the square of its modulus]

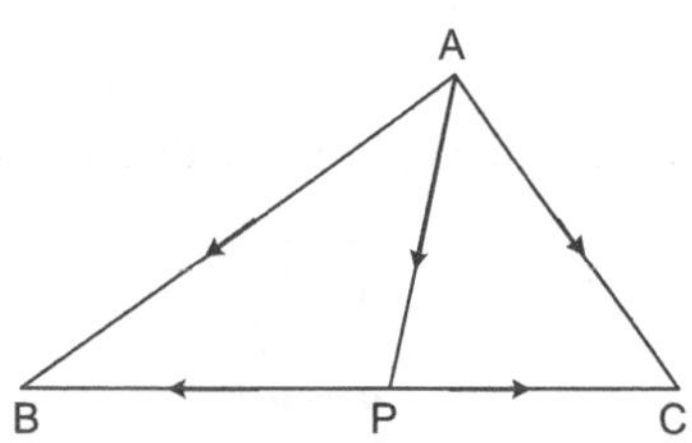

$$= (\vec{AP}+\vec{PB})^2 + (\vec{AP}+\vec{PB})^2$$

$$= (\vec{AP}+\vec{PB})^2 + (\vec{AP}-\vec{PB})^2$$

$$= 2(\vec{AP}^2 + \vec{PB}^2)$$

★ are board exam questions from previous years

Q. 6. Find the ratio in which the point $7\hat{i}-\hat{k}$ divides the join of $-2\hat{i}+3\hat{j}+5\hat{k}$ and $\hat{i}+2\hat{j}+3\hat{k}$.

Sol. Suppose the point $7\hat{i}-\hat{k}$ is *C*, $-2\hat{i}+3\hat{j}+5\hat{k}$ is *A* and $\hat{i}+2\hat{j}+3\hat{k}$ is *B*.

Let *C* divides *AB* in the ratio $1 : m$.

Then ,

$$7\hat{i}-\hat{k} = \frac{(-2\hat{i}+3\hat{j}+5\hat{k})m+(\hat{i}+2\hat{j}+3\hat{k})}{1+m}$$

Equating the coefficients of $\hat{i}, \hat{j}$ and $\hat{k}$, we get

$$m = -\frac{1}{2}$$

Thus, *C* is the point of external division and divides the join of *A* and *B* in the ratio 2 : 3.

Q. 7. Find a vector of magnitude 5 units and parallel to the resultant of $\vec{a} = 2\hat{i} + 3\hat{j} - \hat{k}$ and $\vec{b} = \hat{i} - 2\hat{j} + \hat{k}$.★

Sol. Given that $\vec{a} = 2\hat{i}+3\hat{j}-\hat{k}$

and $\vec{b} = \hat{i}-2\hat{j}+\hat{k}$

Now, resultant of above vectors

$$= \vec{a}+\vec{b}$$

$$= (2\hat{i}+3\hat{j}-\hat{k}) + (\hat{i}-2\hat{j}+\hat{k})$$

$$= 3\hat{i}+\hat{j}$$

Let $\vec{a}+\vec{b} = \vec{c}$

$\therefore$ $\vec{c} = 3\vec{i}+\vec{j}$

Now, we find unit vector in the direction of $\vec{c}$

$$= \frac{\vec{c}}{|\vec{c}|}$$

$$= \frac{3\hat{i}+\hat{j}}{\sqrt{(3)^2+(1)^2}} = \frac{3\hat{i}+\hat{j}}{\sqrt{10}}$$

$$= \frac{3}{\sqrt{10}}\hat{i} + \frac{1}{\sqrt{10}}\hat{j}.$$

Finally, vector of magnitude 5 units and parallel to resultant of $\vec{a}$ and $\vec{b}$ is given by

$$5\left(\frac{3}{\sqrt{10}}\hat{i}+\frac{1}{\sqrt{10}}\hat{j}\right) = \frac{15}{\sqrt{10}}\hat{i}+\frac{5}{\sqrt{10}}\hat{j}$$

Q. 8. Let $\vec{a} = \hat{i}+\hat{j}+\hat{k}$, $\vec{b} = 4\hat{i}-2\hat{j}+3\hat{k}$ and $\vec{c} = \hat{i}-2\hat{j}+\hat{k}$. Find a vector of magnitude 6 units which is parallel to the vector $2\vec{a}-\vec{b}+3\vec{c}$.★

Sol. Given that $\vec{a} = \hat{i}+\hat{j}+\hat{k}$,

$$\vec{b} = 4\hat{i}-2\hat{j}+3\hat{k}$$

and $$\vec{c} = \hat{i}-2\hat{j}+\hat{k}$$

First we find the vector $2\vec{a}-\vec{b}+3\vec{c}$

$$\therefore\ 2\vec{a}-\vec{b}+3\vec{c} = 2(\hat{i}+\hat{j}+\hat{k})-(4\hat{i}-2\hat{j}+3\hat{k}) + 3(\hat{i}-2\hat{j}+\hat{k})$$

$$= 2\hat{i}+2\hat{j}+2\hat{k} - 4\hat{i}+2\hat{j}-3\hat{k} + 3\hat{i}-6\hat{j}+3\hat{k}$$

$$\therefore\ 2\vec{a}-\vec{b}+3\vec{c} = \hat{i}-2\hat{j}+2\hat{k}$$

Now, we find a unit vector in the direction of vector $2\vec{a}-\vec{b}+3\vec{c}$ which is equal to

$$\frac{\hat{i}-2\hat{j}+2\hat{k}}{\sqrt{(1)^2+(-2)^2+(2)^2}} = \frac{\hat{i}-2\hat{j}+2\hat{k}}{\sqrt{9}}$$

$$= \frac{\hat{i}-2\hat{j}+2\hat{k}}{3}$$

$\therefore$ Vector of magnitude 6 units parallel to the vector $2\vec{a}-\vec{b}+3\vec{c}$ is

$$= 6\left(\frac{1}{3}\hat{i}-\frac{2}{3}\hat{j}+\frac{2}{3}\hat{k}\right)$$

$$= 2\hat{i}-4\hat{j}+4\hat{k}$$

Q. 9. Find the position vector of a point *R* which divides the line joining two points *P* and *Q* whose position vectors are $2\vec{a}+\vec{b}$ and $\vec{a}-3\vec{b}$ respectively, externally in the ratio 1 : 2. Also show that *P* is the mid-point of line segment *PQ*.★

Sol. Given that $\vec{OP}$ = position vector of $P = 2\vec{a}+\vec{b}$

$\vec{OQ}$ = position vector of $Q = \vec{a}-3\vec{b}$

Let $\vec{OR}$ = position vector of point *R*, which divides *PQ* externally in the ratio 1 : 2.

We know that position vector of point *R*, which divides line *PQ* externally in the ratio 1 : 2 is given by

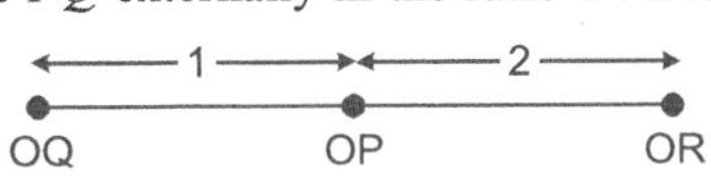

★ are board exam questions from previous years

$$\vec{OR} = \frac{m(\vec{OQ})-n(\vec{OP})}{m-n} \quad ...(i)$$

Given, $m=1, n=2$

Putting above values in equation (i), we get

$$\vec{OR} = \frac{1(\vec{a}-3\vec{b})-2(2\vec{a}+\vec{b})}{1-2}$$

$$= \frac{\vec{a}-3\vec{b}-4\vec{a}-2\vec{b}}{-1}$$

$$= \frac{-3\vec{a}-5\vec{b}}{-1} = 3\vec{a}+5\vec{b}$$

Hence, $\vec{OR} = 3\vec{a}+5\vec{b}$

Also, we have to show that *P* is the mid-point of *RQ*.

$\therefore$ We have to show that

$$\vec{OP} = \frac{\vec{OR}+\vec{OQ}}{2}$$

We have $\vec{OR} = 3\vec{a}+5\vec{b}$

$\vec{OQ} = \vec{a}-3\vec{b}$

$$\frac{\vec{OR}+\vec{OQ}}{2} = \frac{(3\vec{a}+5\vec{b})+(\vec{a}-3\vec{b})}{2}$$

$$= \frac{4\vec{a}+2\vec{b}}{2}$$

$$= \frac{2(2\vec{a}+\vec{b})}{2}$$

$$= 2\vec{a}+\vec{b} = \vec{OP}$$

Hence,

$$\frac{\vec{OR}+\vec{OQ}}{2} = \vec{OP}$$

$\therefore$ *P* is mid-point of line segment *RQ*.

Q. 10. If $\vec{a}, \vec{b}, \vec{c}$ are three vectors such that $|\vec{a}| = 5$, $|\vec{b}| = 12$, $|\vec{c}| = 13$ and $\vec{a}+\vec{b}+\vec{c} = 0$, then find the value of $\vec{a}.\vec{b}+\vec{b}.\vec{c}+\vec{c}.\vec{a}$.★

Sol. Given that $\vec{a}+\vec{b}+\vec{c} = 0$

Multiplying both sides by $(\vec{a}+\vec{b}+\vec{c})$, we get

$$(\vec{a}+\vec{b}+\vec{c})\cdot(\vec{a}+\vec{b}+\vec{c}) = 0.(\vec{a}+\vec{b}+\vec{c})$$

$$\Rightarrow \vec{a}.\vec{a}+\vec{a}.\vec{b}+\vec{a}.\vec{c} + \vec{b}.\vec{a}+\vec{b}.\vec{b}+\vec{b}.\vec{c} + \vec{c}.\vec{a}+\vec{c}.\vec{b}+\vec{c}.\vec{c} = 0$$

$$\Rightarrow |\vec{a}|^2+\vec{a}.\vec{b}+\vec{a}.\vec{c} + \vec{b}.\vec{a}+|\vec{b}|^2+\vec{b}.\vec{c} + \vec{c}.\vec{a}+\vec{c}.\vec{b}+|\vec{c}|^2 = 0$$

$$\Rightarrow |\vec{a}|^2 + |\vec{b}|^2 + |\vec{c}|^2 + 2(\vec{a}.\vec{b} + \vec{b}.\vec{c} + \vec{c}.\vec{a}) = 0$$

$$\Rightarrow (5)^2 + (12)^2 + (13)^2 + 2(\vec{a}.\vec{b} + \vec{b}.\vec{c} + \vec{c}.\vec{a}) = 0$$

$$\Rightarrow 25 + 144 + 169 + 2(\vec{a}.\vec{b} + \vec{b}.\vec{c} + \vec{c}.\vec{a}) = 0$$

$$\Rightarrow 2(\vec{a}.\vec{b} + \vec{b}.\vec{c} + \vec{c}.\vec{a}) = -338$$

Hence, $\vec{a}.\vec{b} + \vec{b}.\vec{c} + \vec{c}.\vec{a} = -169.$

Q. 11. Find a unit vector perpendicular to each of the vectors $\vec{a} + \vec{b}$ and $\vec{a} - \vec{b}$, where $\vec{a} = 3\hat{i} + 2\hat{j} + 2\hat{k}$ and $\vec{b} = \hat{i} + 2\hat{j} - 2\hat{k}$.★

Sol. Given that

$$\vec{a} = 3\hat{i} + 2\hat{j} + 2\hat{k}$$

and $$\vec{b} = \hat{i} + 2\hat{j} - 2\hat{k}$$

Now, $$\vec{a} - \vec{b} = (3\hat{i} + 2\hat{j} + 2\hat{k}) - (\hat{i} + 2\hat{j} - 2\hat{k}) = 2\hat{i} + 4\hat{j}$$

$$\vec{a} + \vec{b} = (3\hat{i} + 2\hat{j} + 2\hat{k}) + (\hat{i} + 2\hat{j} - 2\hat{k}) = 4\hat{i} + 4\hat{j}$$

Let $$\vec{a} + \vec{b} = \vec{c}$$

and $$\vec{a} - \vec{b} = \vec{d}$$

Therefore, we have

$$\vec{c} = 4\hat{i} + 4\hat{j}$$

and $$\vec{d} = 2\hat{i} + 4\hat{k}$$

We know that a unit vector which is perpendicular to both vectors $\vec{c}$ and $\vec{d}$ is given by

$$\hat{n} = \frac{\vec{c} \times \vec{d}}{|\vec{c} \times \vec{d}|} \quad \text{...(i)}$$

$$\therefore \quad |\vec{c} \times \vec{d}| = \begin{vmatrix} \hat{i} & \hat{j} & \hat{k} \\ 4 & 4 & 0 \\ 2 & 0 & 4 \end{vmatrix}$$

$$= \hat{i}(16-0) - \hat{j}(16-0) + \hat{k}(0-8)$$

$$\therefore \quad \vec{c} \times \vec{d} = 16\hat{i} - 16\hat{j} - 8\hat{k} \quad \text{...(ii)}$$

and $$|\vec{c} \times \vec{d}| = \sqrt{(16)^2 + (-16)^2 + (-8)^2}$$

★ are board exam questions from previous years
★★ are frequently asked board exam questions

$$= \sqrt{256 + 256 + 64}$$

$$= \sqrt{576}$$

$$= 24 \quad \text{...(iii)}$$

Hence, the required unit vector

$$= \frac{16\hat{i} - 16\hat{j} - 8\hat{k}}{24}$$

[By using equations (i), (ii) and (iii)]

$$= \frac{8(2\hat{i} - 2\hat{j} - \hat{k})}{24}$$

$$= \frac{2}{3}\hat{i} - \frac{2}{3}\hat{j} - \frac{1}{3}\hat{k}$$

Q. 12. If $\vec{a}, \vec{b}, \vec{c}$ are three vectors such that $|\vec{a}| = 3, |\vec{b}| = 4, |\vec{c}| = 5$ and each one of these is perpendicular to the sum of other two, find $|\vec{a} + \vec{b} + \vec{c}|$.★★

Sol. Given that $|\vec{a}| = 3, |\vec{b}| = 4, |\vec{c}| = 5$ ...(i)

Also, given that each of the vectors $\vec{a}, \vec{b}, \vec{c}$ is perpendicular to sum of the other two vectors

i.e., $$\vec{a} \perp (\vec{b} + \vec{c})$$

$$\Rightarrow \vec{a}.(\vec{b}.\vec{c}) = 0$$

[∵ when two vectors are perpendicular, their dot product is zero]

$$\Rightarrow \vec{a}.\vec{b} + \vec{a}.\vec{c} = 0 \quad \text{...(ii)}$$

$$\vec{b} \perp (\vec{c} + \vec{a})$$

$$\Rightarrow \vec{b}.(\vec{c}.\vec{a}) = 0$$

$$\Rightarrow \vec{b}.\vec{c} + \vec{b}.\vec{a} = 0 \quad \text{...(iii)}$$

and $$\vec{c} \perp (\vec{a} + \vec{b})$$

$$\Rightarrow \vec{c}.(\vec{a} + \vec{b}) = 0$$

$$\Rightarrow \vec{c}.\vec{a} + \vec{c}.\vec{b} = 0 \quad \text{...(iv)}$$

Now, adding equations (ii), (iii) and (iv), we get

$$2(\vec{a}.\vec{b} + \vec{b}.\vec{c} + \vec{c}.\vec{a}) = 0$$

$$\Rightarrow \vec{a}.\vec{b} + \vec{b}.\vec{c} + \vec{c}.\vec{a} = 0 \quad \text{...(v)}$$

Now, consider $(\vec{a} + \vec{b} + \vec{c}).(\vec{a} + \vec{b} + \vec{c})$

$$= \vec{a}.\vec{a} + \vec{a}.\vec{b} + \vec{a}.\vec{c} + \vec{b}.\vec{a} + \vec{b}.\vec{b} + \vec{b}.\vec{c} + \vec{c}.\vec{a} + \vec{c}.\vec{b} + \vec{c}.\vec{c}$$

We know that

$$\vec{x}\cdot\vec{x} = |\vec{x}|^2,$$

$$\therefore \quad |\vec{a}+\vec{b}+\vec{c}|^2 = |\vec{a}|^2 + |\vec{b}|^2 + |\vec{c}|^2 + 2(\vec{a}\cdot\vec{b}+\vec{b}\cdot\vec{c}+\vec{c}\cdot\vec{a})$$

$$(\because \vec{a}^2 = |\vec{a}|^2)$$

From equations (i) and (v), we get

$$|\vec{a}+\vec{b}+\vec{c}|^2 = (3)^2+(4)^2+(5)^2+2(0)$$

$$\therefore \quad |\vec{a}+\vec{b}+\vec{c}|^2 = 9+16+25 = 50$$

Hence, $|\vec{a}+\vec{b}+\vec{c}| = \sqrt{50}$

$$= \sqrt{25\times 2} = 5\sqrt{2}$$

Q. 13. If $\vec{a}, \vec{b}$ and $\vec{c}$ are three mutually perpendicular vectors of equal magnitude, show that $\vec{a}+\vec{b}+\vec{c}$ is equally inclined to $\vec{a}, \vec{b}$ and $\vec{c}$. Also find the angle.★

Sol. Given that $\vec{a}, \vec{b}$ and $\vec{c}$ are three mutually perpendicular vectors of equal magnitude.

i.e., $\quad |\vec{a}| = |\vec{b}| = |\vec{c}| = x$ (say) ...(i)

and $\quad \vec{a}\cdot\vec{b} = 0$ ...(ii)

$\quad \vec{b}\cdot\vec{c} = 0$ ...(iii)

and $\quad \vec{c}\cdot\vec{a} = 0$ ...(iv)

To show $\vec{a}+\vec{b}+\vec{c}$ is equally inclined to $\vec{a}, \vec{b}$ and $\vec{c}$ *i.e.*, $\vec{a}+\vec{b}+\vec{c}$ makes equal angle with $\vec{a}, \vec{b}$ and $\vec{c}$.

Now, we know that angle between two vectors $\vec{x}$ and $\vec{y}$ is given by

$$\cos\theta = \frac{\vec{x}\cdot\vec{y}}{|\vec{x}||\vec{y}|}$$

Using above result, we find the angle between $\vec{a}+\vec{b}+\vec{c}$ and $\vec{a}, \vec{b}$ and $\vec{c}$. Let θ_1 be the angle between $\vec{a}+\vec{b}+\vec{c}$ and $\vec{a}$

$$\therefore \quad \cos\theta_1 = \frac{(\vec{a}+\vec{b}+\vec{c})\cdot\vec{a}}{|\vec{a}+\vec{b}+\vec{c}|\,|\vec{a}|}$$

$$= \frac{\vec{a}\cdot\vec{a}+\vec{b}\cdot\vec{a}+\vec{c}\cdot\vec{a}}{|\vec{a}+\vec{b}+\vec{c}|\,|\vec{a}|}$$

$$= \frac{|\vec{a}|^2+0+0}{|\vec{a}+\vec{b}+\vec{c}|\,|\vec{a}|}$$

[From equations (ii) and (iv)]

$$\therefore \quad \cos\theta_1 = \frac{|\vec{a}|}{|\vec{a}+\vec{b}+\vec{c}|}$$

$$= \frac{x}{|\vec{a}+\vec{b}+\vec{c}|} \quad ...(v)$$

Now, let θ_2 be the angle between $\vec{a}+\vec{b}+\vec{c}$ and $\vec{b}$

$$\therefore \quad \cos\theta_2 = \frac{(\vec{a}+\vec{b}+\vec{c})\cdot\vec{b}}{|\vec{a}+\vec{b}+\vec{c}|\,|\vec{b}|}$$

$$= \frac{\vec{a}\cdot\vec{b}+\vec{b}\cdot\vec{b}+\vec{c}\cdot\vec{b}}{|\vec{a}+\vec{b}+\vec{c}|\,|\vec{b}|}$$

$$= \frac{0+|\vec{b}|^2+0}{|\vec{a}+\vec{b}+\vec{c}|\,|\vec{b}|}$$

$$= \frac{|\vec{b}|}{|\vec{a}+\vec{b}+\vec{c}|}$$

[From equations (ii) and (iii)]

$$\therefore \quad \cos\theta_2 = \frac{x}{|\vec{a}+\vec{b}+\vec{c}|} \quad ...(vi)$$

Again, let θ_3 be angle between $\vec{a}+\vec{b}+\vec{c}$ and $\vec{c}$

$$\therefore \quad \cos\theta_3 = \frac{(\vec{a}+\vec{b}+\vec{c})\cdot\vec{c}}{|\vec{a}+\vec{b}+\vec{c}|\,|\vec{c}|}$$

$$= \frac{\vec{a}\cdot\vec{c}+\vec{b}\cdot\vec{c}+\vec{c}\cdot\vec{c}}{|\vec{a}+\vec{b}+\vec{c}|\,|\vec{c}|}$$

$$= \frac{0+0+|\vec{c}|^2}{|\vec{a}+\vec{b}+\vec{c}|\,|\vec{c}|}$$

[From equations (iii) and (iv)]

$$= \frac{|\vec{c}|}{|\vec{a}+\vec{b}+\vec{c}|}$$

$$= \frac{x}{|\vec{a}+\vec{b}+\vec{c}|}$$

★ are board exam questions from previous years

$$\therefore \quad \cos\theta_3 = \frac{x}{|\vec{a}+\vec{b}+\vec{c}|} \quad \text{...(vii)}$$

Now, from equations (v), (vi) and (vii), we get

$$\cos\theta_1 = \cos\theta_2 = \cos\theta_3$$

$$\Rightarrow \quad \theta_1 = \theta_2 = \theta_3$$

Hence, $\vec{a}+\vec{b}+\vec{c}$ is equally inclined to $\vec{a}$, $\vec{b}$ and $\vec{c}$.

Also, angle between the vectors is

$$\cos^{-1}\left(\frac{x}{|\vec{a}+\vec{b}+\vec{c}|}\right).$$

Q. 14. Using vectors, find the area of triangle with vertices $A(2, 3, 5)$, $B(3, 5, 8)$ and $C(2, 7, 8)$.★

Sol. Let $\vec{OA} = 2\hat{i}+3\hat{j}+5\hat{k}$

$\vec{OB} = 3\hat{i}+5\hat{j}+8\hat{k}$

and $\vec{OC} = 2\hat{i}+7\hat{j}+8\hat{k}$

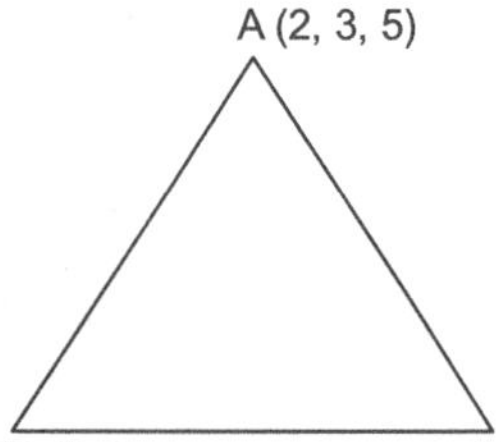

Now, $\vec{AB} = \vec{OB}-\vec{OA} = \left(3\hat{i}+5\hat{j}+8\hat{k}\right) - \left(2\hat{i}+3\hat{j}+5\hat{k}\right)$

$\vec{AB} = \hat{i}+2\hat{j}+3\hat{k}$

and $\vec{AC} = \vec{OC}-\vec{OA}$

$= \left(2\hat{i}+7\hat{j}+8\hat{k}\right)-\left(2\hat{i}+3\hat{j}+5\hat{k}\right)$

$= 4\hat{j}+3\hat{k}$

We know that area of triangle ABC

$$= \frac{1}{2}|\vec{AB}\times\vec{AC}|$$

Now, $\vec{AB}\times\vec{AC} = \begin{vmatrix} \hat{i} & \hat{j} & \hat{k} \\ 1 & 2 & 3 \\ 0 & 4 & 3 \end{vmatrix}$

$= \hat{i}(6-12)-\hat{j}(3-0)+\hat{k}(4-0)$

$\vec{AB}\times\vec{AC} = -6\hat{i}-3\hat{j}+4\hat{k}$

$\therefore |\vec{AB}\times\vec{BC}| = \sqrt{(-6)^2+(-3)^2+(4)^2}$

$= \sqrt{36+9+16} = \sqrt{61}$

$\therefore$ Area of triangle ABC

$= \frac{1}{2}\sqrt{61}$ sq. units.

Q. 15. Show that the points A, B, C with position vectors $2\hat{i}-\hat{j}+\hat{k}$, $\hat{i}-3\hat{j}-5\hat{k}$ and $3\hat{i}-4\hat{j}-4\hat{k}$ respectively, are the vertices of a right-angled triangle. Hence find the area of the triangle.★

Sol. Given, the position vectors of the points A, B and C are $2\hat{i}-\hat{j}+\hat{k}$, $\hat{i}-3\hat{j}-5\hat{k}$ and $3\hat{i}-4\hat{j}-4\hat{k}$ respectively.

Then, $\vec{OA} = 2\hat{i}-\hat{j}+\hat{k}$,

$\vec{OB} = \hat{i}-3\hat{j}-5\hat{k}$

and $\vec{OC} = 3\hat{i}-4\hat{j}-4\hat{k}$

Now, $\vec{AB} = \vec{OB}-\vec{OA}$

$= \hat{i}-3\hat{j}-5\hat{k}-(2\hat{i}-\hat{j}+\hat{k})$

$= -\hat{i}-2\hat{j}-6\hat{k}$

$\vec{BC} = \vec{OC}-\vec{OB}$

$= 3\hat{i}-4\hat{j}-4\hat{k}-(\hat{i}-3\hat{j}-5\hat{k})$

$= 2\hat{i}-\hat{j}+\hat{k}$

and $\vec{CA} = \vec{OA}-\vec{OC}$

$= 2\hat{i}-\hat{j}+\hat{k}-(3\hat{i}-4\hat{j}-4\hat{k})$

$= -\hat{i}+3\hat{j}+5\hat{k}$

$\vec{AB}+\vec{BC}+\vec{CA}$

$= (-\hat{i}-2\hat{j}-6\hat{k})+(2\hat{i}-\hat{j}+\hat{k}) + (-\hat{i}+3\hat{j}+5\hat{k})$

$= 0$

Thus, A, B and C forms a triangle.

Also, $\vec{BC}\cdot\vec{CA} = (2\hat{i}-\hat{j}+\hat{k})(-\hat{i}+3\hat{j}+5\hat{k})$

$= -2-3+5 = 0$

$\therefore \quad \vec{BC} \perp \vec{CA}$

$\Rightarrow \quad \angle C = 90°$

Hence, A, B, C are the vertices of a right angled triangle.

Now, Area of $\Delta ABC = \frac{1}{2}|\vec{BC}\times\vec{CA}|$

$\therefore \quad \vec{BC}\times\vec{CA} = \begin{vmatrix} \hat{i} & \hat{j} & \hat{k} \\ 2 & -1 & 1 \\ -1 & 3 & 5 \end{vmatrix}$

★ are board exam questions from previous years

$= -8\hat{i} - 11\hat{j} + 5\hat{k}$

$|\vec{BC} \times \vec{CA}| = \sqrt{(-8)^2 + (-11)^2 + (5)^2}$

$= \sqrt{64+121+25} = \sqrt{210}$

Hence, Area of $\Delta ABC = \frac{1}{2}\sqrt{210}$ sq. units.

Q. 16. The scalar product of vector $\hat{i} + \hat{j} + \hat{k}$ with the unit vector along the sum of vectors $2\hat{i} + 4\hat{j} - 5\hat{k}$ and $\lambda\hat{i} + 2\hat{j} + 3\hat{k}$ is equal to one. Find the value of λ.★★

Sol. Let $\vec{a} = \hat{i} + \hat{j} + \hat{k}$,

$\vec{b} = 2\hat{i} + 4\hat{j} - 5\hat{k}$

and $\vec{c} = \lambda\hat{i} + 2\hat{j} + 3\hat{k}$

Now, given that dot product of $\vec{a}$ with unit vector along vector $\vec{b} + \vec{c}$ is one.

Now, $\vec{b} + \vec{c} = (2\hat{i} + 4\hat{j} - 5\hat{k}) + (\lambda\hat{i} + 2\hat{j} + 3\hat{k})$

$\vec{b} + \vec{c} = (2+\lambda)\hat{i} + 6\hat{j} - 2\hat{k}$

Unit vector along $\vec{b} + \vec{c}$

$= \frac{\vec{b} + \vec{c}}{|\vec{b} + \vec{c}|}$

$= \frac{(2+\lambda)\hat{i} + 6\hat{j} - 2\hat{k}}{\sqrt{(2+\lambda)^2 + (6)^2 + (-2)^2}}$

$= \frac{(2+\lambda)\hat{i} + 6\hat{j} - 2\hat{k}}{\sqrt{4+\lambda^2+4\lambda+36+4}}$

$= \frac{(2+\lambda)\hat{i} + 6\hat{j} - 2\hat{k}}{\sqrt{\lambda^2+4\lambda+44}}$

Now, as per given condition, we get

$(\hat{i} + \hat{j} + \hat{k}).\frac{(2+\lambda)\hat{i} + 6\hat{j} - 2\hat{k}}{\sqrt{\lambda^2+4\lambda+44}} = 1$

$\Rightarrow \frac{(2+\lambda)+6-2}{\sqrt{\lambda^2+4\lambda+44}} = 1$

$\Rightarrow \lambda + 6 = \sqrt{\lambda^2+4\lambda+44}$

Squaring both sides, we get

$\lambda^2 + 4\lambda + 44 = (\lambda+6)^2$

$\Rightarrow \lambda^2 + 4\lambda + 44 = \lambda^2 + 36 + 12\lambda$

$\Rightarrow 4\lambda + 44 = 36 + 12\lambda$

$\Rightarrow 8\lambda = 8$

$\Rightarrow \lambda = 1$

Q. 17. If $\vec{a} \times \vec{b} = \vec{c} \times \vec{d}$ and $\vec{a} \times \vec{c} = \vec{b} \times \vec{d}$, show that $\vec{a} - \vec{d}$ is parallel to $\vec{b} - \vec{c}$, where $\vec{a} \neq \vec{d}$ and $\vec{b} \neq \vec{c}$.★

Sol. Given that $\vec{a} \times \vec{b} = \vec{c} \times \vec{d}$ ...(i)

and $\vec{a} \times \vec{c} = \vec{b} \times \vec{d}$ ...(ii)

Subtracting equation (ii) from equation (i), we get

$(\vec{a} \times \vec{b}) - (\vec{a} \times \vec{c}) = (\vec{c} \times \vec{d}) - (\vec{b} \times \vec{d})$

$\Rightarrow (\vec{a} \times \vec{b}) - (\vec{a} \times \vec{c}) + (\vec{b} \times \vec{d}) - (\vec{c} \times \vec{d}) = 0$

$\Rightarrow \vec{a} \times (\vec{b} - \vec{c}) + (\vec{b} - \vec{c}) \times d = 0$

$\Rightarrow \vec{a} \times (\vec{b} - \vec{c}) - \vec{d} \times (\vec{b} - \vec{c}) = 0$

$(\because \vec{a} \times \vec{b} = -\vec{b} \times \vec{a})$

$\Rightarrow (\vec{a} - \vec{d}) \times (\vec{b} - \vec{c}) = 0$

$\therefore$ We have shown that cross product of vectors $\vec{a} - \vec{d}$ and $\vec{b} - \vec{c}$ is a zero vector, so $\vec{a} - \vec{d}$ is parallel to $\vec{b} - \vec{c}$.

Q. 18. Three vectors $\vec{a}, \vec{b}, \vec{c}$ satisfy the condition $\vec{a} + \vec{b} + \vec{c} = 0$.★

Find the value of $\vec{a} \times \vec{b} + \vec{b} \times \vec{c} + \vec{c} \times \vec{a}$, if $|\vec{a}| = 1$, $|\vec{b}| = 4$ and $|\vec{c}| = 2$.

Sol. Given that for three vectors $\vec{a}, \vec{b}, \vec{c}$

$\vec{a} + \vec{b} + \vec{c} = 0$ ...(i)

Taking dot product with vector $\vec{a} + \vec{b} + \vec{c}$ on both sides of equation (i), we get

$(\vec{a} + \vec{b} + \vec{c}) \cdot (\vec{a} + \vec{b} + \vec{c})$

$= 0.(\vec{a} + \vec{b} + \vec{c})$

$\Rightarrow \vec{a}.\vec{a} + \vec{a}.\vec{b} + \vec{a}.\vec{c} + \vec{b}.\vec{a} + \vec{b}.\vec{b} + \vec{b}.\vec{c} + \vec{c}.\vec{a} + \vec{c}.\vec{b} + \vec{c}.\vec{c} = 0$

$\Rightarrow |a|^2 + |b|^2 + |c|^2 + 2(\vec{a}.\vec{b} + \vec{b}.\vec{c} + \vec{c}.\vec{a}) = 0$

★ are board exam questions from previous years
★★ are frequently asked board exam questions

$\Rightarrow (1)^2+(4)^2+(2)^2+2(\vec{a}.\vec{b}+\vec{b}.\vec{c}+\vec{c}.\vec{a})=0$

[∵ given that $|\vec{a}|=1, |\vec{b}|=4, |\vec{c}|=2$]

$\Rightarrow \quad 1+16+4+2(\vec{a}.\vec{b}+\vec{b}.\vec{c}+\vec{c}.\vec{a}) = 0$

or $\quad 2(\vec{a}.\vec{b}+\vec{b}.\vec{c}+\vec{c}.\vec{a}) = -21$

Hence, $\quad \vec{a}.\vec{b}+\vec{b}.\vec{c}+\vec{c}.\vec{a} = -\dfrac{21}{2}$

Q.19. If $\vec{a} = \hat{i}+\hat{j}+\hat{k}$ and $\vec{b} = \hat{j}-\hat{k}$, find a vector $\vec{c}$ such that $\vec{a}\times\vec{c} = \vec{b}$ and $\vec{a}.\vec{c} = 3$ ★

Sol. Given that $\quad \vec{a} = \hat{i}+\hat{j}+\hat{k}$

and $\quad \vec{b} = \hat{j}-\hat{k}$

Let $\quad \vec{c} = c_1\hat{i}+c_2\hat{j}+c_3\hat{k}$

Also, given that $\quad \vec{a}\times\vec{c} = \vec{b}$

$$\Rightarrow \begin{vmatrix} \hat{i} & \hat{j} & \hat{k} \\ 1 & 1 & 1 \\ c_1 & c_2 & c_3 \end{vmatrix} = \hat{j}-\hat{k}$$

$\Rightarrow \hat{i}(c_3-c_2)-\hat{j}(c_3-c_1)+\hat{k}(c_2-c_1) = \hat{j}-\hat{k}$

Comparing coefficients of $\hat{i}, \hat{j}$ and $\hat{k}$ on both sides, we get

$c_3-c_2 = 0 \quad \Rightarrow \quad c_2 = c_3 \quad$ ...(i)

$c_3-c_1 = -1 \quad \Rightarrow \quad c_1-c_3 = 1 \quad$ ...(ii)

and $\quad c_2-c_1 = -1 \quad \Rightarrow \quad c_1-c_2 = 1 \quad$ ...(iii)

Again, given that $\quad \vec{a}.\vec{c} = 3$

$\therefore (\hat{i}+\hat{j}+\hat{k}).(c_1\hat{i}+c_2\hat{j}+c_3\hat{k}) = 3$

$\Rightarrow \quad c_1+c_2+c_3 = 3 \quad$ ...(iv)

Putting $c_2 = c_3$ from equation (i) in equation (iv), we get

$c_1+2c_3 = 3 \quad$ ...(v)

Subtracting equation (ii) from equatoin (v), we get

$c_1+2c_3 = 3$

$c_1-c_3 = 1$

$- \quad + \quad -$

$3c_3 = 2$

$\Rightarrow \quad c_3 = \dfrac{2}{3}$

$\because \quad c_2 = c_3$

$\therefore \quad c_3 = \dfrac{2}{3}$

★ are board exam questions from previous years

Putting $c_2 = c_3 = \dfrac{2}{3}$ in equation (iv), we get

$c_1+\dfrac{2}{3}+\dfrac{2}{3} = 3$

$\Rightarrow \quad c_1+\dfrac{4}{3} = 3$

$\Rightarrow \quad c_1 = 3-\dfrac{4}{3}$

$= \dfrac{9-4}{3} = \dfrac{5}{3}$

$\therefore \quad c_1 = \dfrac{5}{3}$

Hence, the required vector is

$\vec{c} = c_1\hat{i}+c_2\hat{j}+c_3\hat{k}$

$= \dfrac{5}{3}\hat{i}+\dfrac{2}{3}\hat{j}+\dfrac{2}{3}\hat{k}.$

Q. 20. Find the angle of inclination of any two diagonals of a cube.

Sol. The angle between the diagonals of a cube is the same for any size of the cube. Therefore, let us consider a unit cube, where its each edge is of unit length. Hence, let us represent its co-terminal edges *OA*, *OB* and *OC* by $\hat{i}, \hat{j}$ and $\hat{k}$, the usual unit vectors. Consider any two diagonals, say, OF and *GC*. Let the angle between *OF* and *GC* be θ. Now, the position vector of F is $\hat{i}+\hat{j}+\hat{k}$ and the position vector of *G* is $\hat{i}+\hat{j}$ and the position vector of *C* is $\hat{k}$.

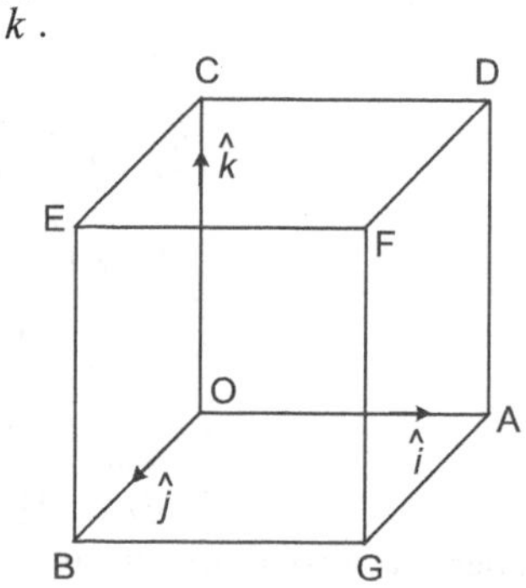

So, $\quad \vec{OF} = \hat{i}+\hat{j}+\hat{k}$

and $\quad \vec{GC} = \hat{k}-(\hat{i}+\hat{j})$

$= -\hat{i}-\hat{j}+\hat{k}.$

Hence, $\vec{OF}.\vec{GC} = (\hat{i}+\hat{j}+\hat{k}).(-\hat{i}-\hat{j}+\hat{k})$

$\therefore \quad |\vec{OF}||\vec{GC}|\cos\theta = -1-1+1$

$\Rightarrow \left(\sqrt{1+1+1}\right)\left(\sqrt{(-1)^2+(-1)^2+1^2}\right)\cos\theta = 1$

$\Rightarrow \quad (\sqrt{3})(\sqrt{3})\cos\theta = -1$

$$\Rightarrow \qquad \cos\theta = -\frac{1}{3}$$

$$\Rightarrow \qquad \theta = \cos^{-1}\left(-\frac{1}{3}\right)$$

Q. 21. Show that the points with position vectors $2\hat{i}+\hat{j}-\hat{k}, 3\hat{i}-2\hat{j}+\hat{k}$ and $\hat{i}+4\hat{j}-3\hat{k}$ are collinear.

Sol. Let
$$\vec{a} = 2\hat{i}+\hat{j}-\hat{k}$$
$$\vec{b} = 3\hat{i}-2\hat{j}+\hat{k}$$
$$\vec{c} = \vec{i}+4\hat{j}-3\hat{k}$$

We know that if three points with position vectors $\vec{a}, \vec{b}$ and $\vec{c}$ are collinear, then

$$\vec{a}\times\vec{b}+\vec{b}\times\vec{c}+\vec{c}\times\vec{a} = 0$$

Now,
$$\vec{a}\times\vec{b} = \begin{vmatrix} \hat{i} & \hat{j} & \hat{k} \\ 2 & 1 & -1 \\ 3 & -2 & 1 \end{vmatrix}$$
$$= \hat{i}(1-2)-\hat{j}(2+3)+\hat{k}(-4-3)$$
$$= -\hat{i}-5\hat{j}-7\hat{k}$$

$$\vec{b}\times\vec{c} = \begin{vmatrix} \hat{i} & \hat{j} & \hat{k} \\ 3 & -2 & 1 \\ 1 & 4 & -3 \end{vmatrix}$$
$$= \hat{i}(6-4)-\hat{j}(-9-1)+\hat{k}(12+2)$$
$$= 2\hat{i}+10\hat{j}+14\hat{k}$$

$$\vec{c}\times\vec{a} = \begin{vmatrix} \hat{i} & \hat{j} & \hat{k} \\ 1 & 4 & -3 \\ 2 & 1 & -1 \end{vmatrix}$$
$$= \hat{i}(-4+3)-\hat{j}(-1+6)+\hat{k}(1-8)$$
$$= -\hat{i}-5\hat{j}-7\hat{k}$$

$$\therefore\ \vec{a}\times\vec{b}+\vec{b}\times\vec{c}+\vec{c}\times\vec{a} = (-\hat{i}-5\hat{j}-7\hat{k})$$
$$+ (2\hat{i}+10\hat{j}+14\hat{k}) + (-\hat{i}-5\hat{j}-7\hat{k}) = 0$$

Hence, the given points are collinear.

Q. 22. If $\hat{a}$ and $\hat{b}$ are unit vectors and θ is the angle between them, then show that

$$\sin\frac{\theta}{2} = \frac{|\hat{a}-\hat{b}|}{2}$$

Sol.
$$(\hat{a}-\hat{b})^2 = |\hat{a}|^2+|\hat{b}|^2-2\hat{a}.\hat{b}$$
$$= 1+1-2.1.1\cos\theta$$
$$= 2-2\cos\theta$$
$$= 2(1-\cos\theta)$$
$$= 2\left(2\sin^2\frac{\theta}{2}\right)$$
$$= 4\sin^2\frac{\theta}{2}$$

$$\Rightarrow \qquad |\hat{a}-\hat{b}|^2 = 4\sin^2\frac{\theta}{2}$$

[Since, square of a vector = square of its modulus]

$$\Rightarrow \qquad |\hat{a}-\hat{b}| = 2\sin\frac{\theta}{2}$$

Hence,
$$\sin\frac{\theta}{2} = \frac{|\hat{a}-\hat{b}|}{2}$$

Q. 23. A parallelogram, whose diagonals are equal, is a rectangle. Prove this using vectors.

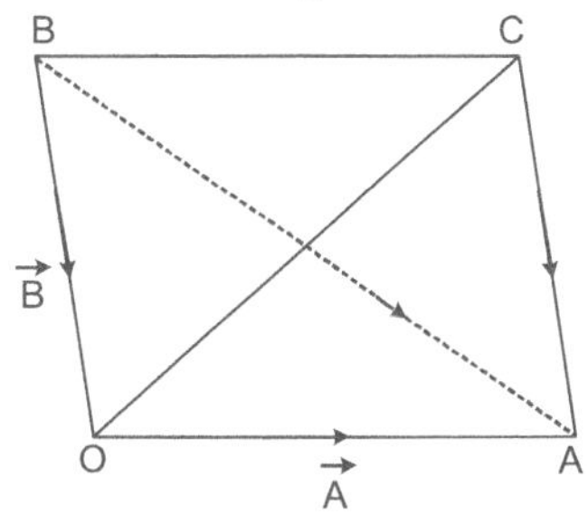

Sol. Let *OACB* be the parallelogram with $\vec{OA} = \vec{a}, \vec{OB} = \vec{b}$

Now,
$$\vec{OC} = \vec{a}+\vec{b}$$

(Using triangle law of vector addition)

and
$$\vec{BA} = \vec{a}-\vec{b}$$
$$\vec{OC} = \vec{BA}$$

(since the diagonals of a parallelogram are equal)

$$\Rightarrow \qquad OC^2 = BA^2$$
$$\Rightarrow \qquad \vec{OC}.\vec{OC} = \vec{BA}.\vec{BA}$$
$$\Rightarrow \qquad (\vec{a}+\vec{b}).(\vec{a}+\vec{b}) = (\vec{a}-\vec{b}).(\vec{a}-\vec{b})$$
$$\Rightarrow \vec{a}.\vec{a}+\vec{b}.\vec{a}+\vec{a}.\vec{b}+\vec{b}.\vec{b}$$
$$= \vec{a}.\vec{a}-\vec{b}.\vec{a}-\vec{a}.\vec{b}+\vec{b}.\vec{b}$$

But
$$\vec{a}.\vec{b} = \vec{b}.\vec{a}$$

Therefore,
$$4\,\vec{a}.\vec{b} = 0$$
$$\Rightarrow \qquad \vec{a}.\vec{b} = 0$$

$\Rightarrow$ *OA* is perpendicular to *OB*.

Therefore, the parallelogram is a rectangle.

Q. 24. If θ is the angle between two vectors $\hat{i}-2\hat{j}+3\hat{k}$ and $3\hat{i}-2\hat{j}+\hat{k}$, find sin θ.★

Sol. Let $\vec{a}=\hat{i}-2\hat{j}+3\hat{k}$

$\vec{b}=3\hat{i}-2\hat{j}+\hat{k}$

We know $\vec{a}.\vec{b}=|\vec{a}||\vec{b}|\cos\theta$

$$\Rightarrow \cos\theta=\frac{\vec{a}.\vec{b}}{|\vec{a}||\vec{b}|}$$

$$=\frac{(\hat{i}-2\hat{j}+3\hat{k}).(3\hat{i}-2\hat{j}+\hat{k})}{|\hat{i}-2\hat{j}+3\hat{k}||3\hat{i}-2\hat{j}+\hat{k}|}$$

$$=\frac{1(3)+(-2)(-2)+3(1)}{\sqrt{1+4+9}\sqrt{9+4+1}}$$

$$=\frac{3+4+3}{14}=\frac{10}{14}=\frac{5}{7}$$

Now, $\sin\theta=\sqrt{1-\cos^2\theta}$

$$=\sqrt{1-\left(\frac{5}{7}\right)^2}=\sqrt{1-\frac{25}{49}}=\sqrt{\frac{49-25}{49}}$$

$$\Rightarrow \sin\theta=\sqrt{\frac{24}{49}}=\frac{2\sqrt{6}}{7}.$$

Q. 25. Show that the four points *A*, *B*, *C* and *D* with position vectors $4\hat{i}+5\hat{j}+\hat{k}, -\hat{j}-\hat{k}, 3\hat{i}+9\hat{j}+4\hat{k}$ and $4(-\hat{i}+\hat{j}+\hat{k})$ respectively are coplanar.★

Sol. Given, the position vectors are

$$A(4\hat{i}+5\hat{j}+\hat{k}),\ B(-\hat{j}-\hat{k}),\ C(3\hat{i}+9\hat{j}+4\hat{k})$$

and $D[4(-\hat{i}+\hat{j}+\hat{k})]$

These points will be coplanar if

$$[\vec{AB}\ \vec{AC}\ \vec{AD}]=0$$

Now, $\vec{AB}=(-\hat{j}-\hat{k})-(4\hat{i}+5\hat{j}+\hat{k})$

$=-4\hat{i}-6\hat{j}-2\hat{k}$

$\vec{AC}=(3\hat{i}+9\hat{j}+4\hat{k})-(4\hat{i}+5\hat{j}+\hat{k})$

$=-\hat{i}+4\hat{j}+3\hat{k}$

$\vec{AD}=(-4\hat{i}+4\hat{j}+4\hat{k})-(4\hat{i}+5\hat{j}+\hat{k})$

$=-8\hat{i}-\hat{j}+3\hat{k}$

$$\therefore [\vec{AB}\ \vec{AC}\ \vec{AD}]=\begin{vmatrix}-4 & -6 & -2\\ -1 & 4 & 3\\ -8 & -1 & 3\end{vmatrix}$$

$=-4(12+3)+6(-3+24)-2(1+32)$

$=-60+126-66$

$=-126+126=0$

⇒ The given points *A*, *B*, *C* and *D* are coplanar.

Q. 26. Find the value of *x* such that the points *A*(3, 2, 1), *B*(4, *x*, 5), *C*(4, 2, – 2) and *D*(6, 5, – 1) are coplanar.★

Sol. Given, the points *A*(3, 2, 1), *B*(4, *x*, 5), *C*(4, 2, – 2) and *D*(6, 5, – 1).

$\therefore\ \vec{AB}=4\hat{i}+x\hat{j}+5\hat{k}-(3\hat{i}+2\hat{j}+\hat{k})$

$=\hat{i}+(x-2)\hat{j}+4\hat{k}$

$\vec{BC}=4\hat{i}+2\hat{j}-2\hat{k}-(4\hat{i}+x\hat{j}+5\hat{k})$

$=(2-x)\hat{j}-7\hat{k}$

$\vec{CD}=6\hat{i}+5\hat{j}-\hat{k}-(4\hat{i}+2\hat{j}-2\hat{k})$

$=2\hat{i}+3\hat{j}+\hat{k}$

Since the points are coplanar, so $[\vec{AB}\,\vec{BC}\,\vec{CD}]=0$

or $$\begin{vmatrix}1 & x-2 & 4\\ 0 & 2-x & -7\\ 2 & 3 & 1\end{vmatrix}=0$$

$1(2-x+21)-(x-2)(0+14)+4(0-(4-2x))=0$

$\Rightarrow 1(23-x)-(x-2)(14)+4(2x-4)=0$

$\Rightarrow 23-x-14x+28+8x-16=0$

$\Rightarrow -7x+35=0$

$\Rightarrow x=5.$

Q. 27. Find the value of λ, if four points with position vectors $3\hat{i}+6\hat{j}+9\hat{k}$, $\hat{i}+2\hat{j}+3\hat{k}$, $2\hat{i}+3\hat{j}+\hat{k}$ and $4\hat{i}+6\hat{j}+\lambda\hat{k}$ are coplanar.★

Sol. Let the four points be *A*, *B*, *C* and *D* with position vectors $3\hat{i}+6\hat{j}+9\hat{k}$, $\hat{i}+2\hat{j}+3\hat{k}$, $2\hat{i}+3\hat{j}+\hat{k}$ and $4\hat{i}+6\hat{j}+\lambda\hat{k}$ respectively.

Then, $\vec{OA}=3\hat{i}+6\hat{j}+9\hat{k}$

$\vec{OB}=\hat{i}+2\hat{j}+3\hat{k}$

$\vec{OC}=2\hat{i}+3\hat{j}+\hat{k}$

$\vec{OD}=4\hat{i}+6\hat{j}+\lambda\hat{k}$

Then, $\vec{AB}=\vec{OB}-\vec{OA}$

$=\hat{i}+2\hat{j}+3\hat{k}-(3\hat{i}+6\hat{j}+9\hat{k})$

$=-2\hat{i}-4\hat{j}-6\hat{k}$

$\vec{BC}=\vec{OC}-\vec{OB}$

★ are board exam questions from previous years

$$= 2\hat{i}+3\hat{j}+\hat{k}-(\hat{i}+2\hat{j}+3\hat{k})$$

$$= \hat{i}+\hat{j}-2\hat{k}$$

and $\quad \vec{CD} = \vec{OD}-\vec{OC}$

$$= 4\hat{i}+6\hat{j}+\lambda\hat{k}-(2\hat{i}+3\hat{j}+\hat{k})$$

$$= 2\hat{i}+3\hat{j}+(\lambda-1)\hat{k}$$

Since the points are coplanar, then $[\vec{AB}, \vec{BC}, \vec{CD}] = 0$

i.e., $$\begin{vmatrix} -2 & -4 & -6 \\ 1 & 1 & -2 \\ 2 & 3 & \lambda-1 \end{vmatrix} = 0$$

$$-2(\lambda-1+6)+4(\lambda-1+4)-6(3-2)=0$$

$$-2(\lambda+5)+4(\lambda+3)-6(1)=0$$

$$-2\lambda-10+4\lambda+12-6=0$$

$$2\lambda=4$$

$\therefore \quad \lambda = 2.$

Q. 28. Prove that, for any three vectors $\vec{a}, \vec{b}, \vec{c}$

$$[\vec{a}+\vec{b} \quad \vec{b}+\vec{c} \quad \vec{c}+\vec{a}] = 2[\vec{a} \quad \vec{b} \quad \vec{c}]$$ ★

Sol. L.H.S. $= [\vec{a}+\vec{b} \quad \vec{b}+\vec{c} \quad \vec{c}+\vec{a}]$

$$= (\vec{a}+\vec{b}).[(\vec{b}+\vec{c})\times(\vec{c}+\vec{a})]$$

$$= (\vec{a}+\vec{b}).[\vec{b}\times\vec{c}+\vec{b}\times\vec{a}+\vec{c}\times\vec{c}+\vec{c}\times\vec{a}]$$

$$= (\vec{a}+\vec{b}).[\vec{b}\times\vec{c}-\vec{a}\times\vec{b}+\vec{0}+\vec{c}\times\vec{a}]$$

$$(\because \vec{c}\times\vec{c}=0)$$

$$= \vec{a}.(\vec{b}\times\vec{c})-\vec{a}.(\vec{a}\times\vec{b})+\vec{a}.(\vec{c}\times\vec{a}) + \vec{b}.(\vec{b}\times\vec{c})-\vec{b}.(\vec{a}\times\vec{b})+\vec{b}.(\vec{c}\times\vec{a})$$

$$= \vec{a}.(\vec{b}\times\vec{c})-0+0+0-0+\vec{b}.(\vec{c}\times\vec{a})$$

$$= [\vec{a} \quad \vec{b} \quad \vec{c}]+[\vec{b} \quad \vec{c} \quad \vec{a}]$$

$$\left[\begin{array}{l} \because \quad \vec{a}.(\vec{b}\times\vec{c}) = [\vec{a}\ \vec{b}\ \vec{c}] \\ \text{and } \vec{b}.(\vec{c}\times\vec{a}) = [\vec{b}\ \vec{c}\ \vec{a}] \end{array}\right]$$

$$= [\vec{a} \quad \vec{b} \quad \vec{c}]+[\vec{a} \quad \vec{b} \quad \vec{c}]$$

$$= 2[\vec{a} \quad \vec{b} \quad \vec{c}] = \text{R.H.S.}$$

Q. 29. The two adjacent sides of a parallelogram are $2\hat{i}-4\hat{j}-5\hat{k}$ and $2\hat{i}+2\hat{j}+3\hat{k}$. Find the two unit vectors parallel to its diagonals. Using the diagonal vectors, find the area of the parallelogram. ★

Sol. Let *ABCD* be a parallelogram such that

$$\vec{AB} = \vec{a} = 2\hat{i}-4\hat{j}-5\hat{k}$$

and $$\vec{BC} = \vec{b} = 2\hat{i}+2\hat{j}+3\hat{k}$$

Then, $\quad \vec{AB}+\vec{BC} = \vec{AC}$

$$\vec{AC} = \vec{a}+\vec{b} = 4\hat{i}-2\hat{j}-2\hat{k}$$

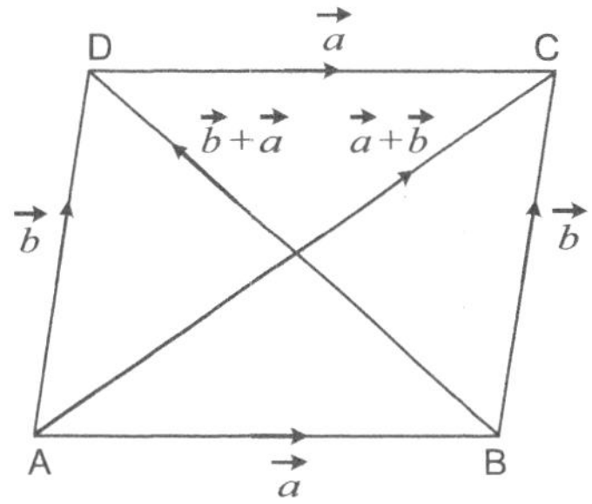

and $\quad \vec{AB}+\vec{BD} = \vec{AD}$

$$\vec{BD} = \vec{AD}-\vec{AB}$$

$$\vec{BD} = \vec{b}-\vec{a} = 0\hat{i}+6\hat{j}+8\hat{k}$$

Now, $\quad \vec{AC} = 4\hat{i}-2\hat{j}-2\hat{k}$

$$|\vec{AC}| = \sqrt{(4)^2+(-2)^2+(-2)^2}$$

$$= \sqrt{24} = 2\sqrt{6}$$

and $\quad |\vec{BD}| = 0\hat{i}+6\hat{j}+8\hat{k}$

$$|\vec{BD}| = \sqrt{(0)^2+(6)^2+(8)^2}$$

$$= \sqrt{100} = 10$$

Unit vector along $\vec{AC}$

$$= \frac{\vec{AC}}{|\vec{AC}|} = \frac{1}{2\sqrt{6}}(4\hat{i}-2\hat{j}-2\hat{k})$$

$$= \frac{1}{\sqrt{6}}(2\hat{i}-\hat{j}-\hat{k})$$

Unit vector along $\vec{BD}$

$$= \frac{\vec{BD}}{|\vec{BD}|} = \frac{1}{10}(6\hat{j}+8\hat{k}) = \frac{1}{5}(3\hat{j}+4\hat{k})$$

Now, area of parallelogram

$$= \frac{1}{2}|\vec{AC}\times\vec{BD}| \qquad \ldots(i)$$

$$\Rightarrow \quad \vec{AC}\times\vec{BD} = \begin{vmatrix} \hat{i} & \hat{j} & \hat{k} \\ 4 & -2 & -2 \\ 0 & 6 & 8 \end{vmatrix}$$

★ are board exam questions from previous years

$$= -4\hat{i} - 32\hat{j} + 24\hat{k}$$

and $|\vec{AC} \times \vec{BD}| = \sqrt{(-4)^2 + (-32)^2 + (24)^2}$

$$= \sqrt{1616}$$

$$= 2\sqrt{404}$$

Area of parallelogram

$$= \frac{1}{2}|\vec{AC} \times \vec{BD}|$$

$$= \sqrt{404}$$

$$= 2\sqrt{101} \text{ sq. units}$$

Q. 30. Show that the vectors $\vec{a}, \vec{b}$ and $\vec{c}$ are coplanar if $\vec{a} + \vec{b}, \vec{b} + \vec{c}$ and $\vec{c} + \vec{a}$ are coplanar.★

Sol. As $\vec{a} + \vec{b}, \vec{b} + \vec{c}, \vec{c} + \vec{a}$ are coplanar.

So, $[\vec{a} + \vec{b} \;\; \vec{b} + \vec{c} \;\; \vec{c} + \vec{a}] = 0$ ...(i)

Now, if $\vec{a}, \vec{b}$ and $\vec{c}$ are coplanar vectors then $[\vec{a} \; \vec{b} \; \vec{c}] = 0$

Consider $[\vec{a} + \vec{b} \;\; \vec{b} + \vec{c} \;\; \vec{c} + \vec{a}]$

$$= [(\vec{a} + \vec{b}) \times (\vec{b} + \vec{c})].(\vec{c} + \vec{a})$$

$$= [\vec{a} \times \vec{b} + \vec{a} \times \vec{c} + \vec{b} \times \vec{b} + \vec{b} \times \vec{c}].(\vec{c} + \vec{a})$$

$$= (\vec{a} \times \vec{b}).\vec{c} + (\vec{a} \times \vec{b}).\vec{a} + (\vec{a} \times \vec{c}).\vec{c} + (\vec{a} \times \vec{c}).\vec{a}$$

$$+ (\vec{b} \times \vec{c}).\vec{c} + (\vec{b} \times \vec{c}).\vec{a}$$

$$= (\vec{a} \times \vec{b}).\vec{c} + 0 + 0 + 0 + 0 + (\vec{b} \times \vec{c}).\vec{a}$$

$$= [\vec{a} \; \vec{b} \; \vec{c}] + [\vec{b} \; \vec{c} \; \vec{a}]$$

$$= [\vec{a} \; \vec{b} \; \vec{c}] + [\vec{a} \; \vec{b} \; \vec{c}]$$

$$= 2[\vec{a} \; \vec{b} \; \vec{c}]$$

From equation (i),

$$2[\vec{a} \; \vec{b} \; \vec{c}] = [\vec{a} + \vec{b} \;\; \vec{b} + \vec{c} \;\; \vec{c} + \vec{a}] = 0$$

$$\Rightarrow \quad [\vec{a} \; \vec{b} \; \vec{c}] = 0$$

$\therefore$ $\vec{a}, \vec{b}$ and $\vec{c}$ are coplanar if $\vec{a} + \vec{b}, \vec{b} + \vec{c}$ and $\vec{c} + \vec{a}$ are coplanar.

Chapter 11. Three-Dimensional Geometry

Q. 1. Prove that if the direction cosines of a line *PQ* are *l*, *m*, *n* then the direction cosines of *QP* are

$-l, -m, -n$.

Sol. Let *PQ* make angles α, β, γ with the positive direction of the coordinate axes.

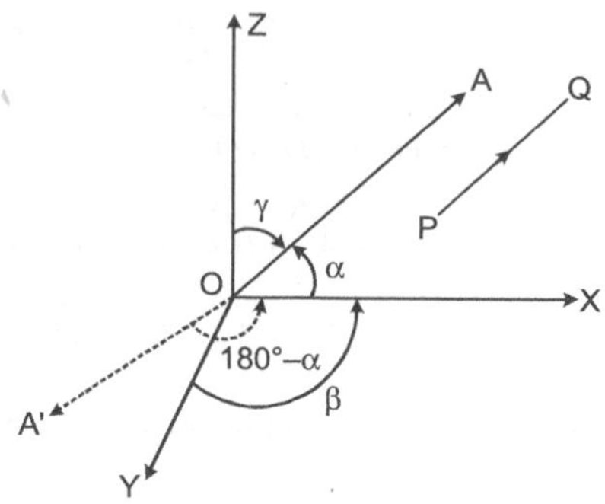

Then $\cos\alpha = l$, $\cos\beta = m$, $\cos\gamma = n$. Now through *O*, draw a line *OA* parallel to and having the same sense as that of *PQ*. Then the angles which *OA* makes with the coordinate axes will be α, β, γ.

Now, the angles which *QP* makes with the coordinate axes will be the angles which *AO* produced (in the figure it is *OA'*) makes with the axes.

Therefore, the direction cosines of *QP* are $\cos(180° - \alpha)$, $\cos(180° - \beta)$, $\cos(180° - \gamma)$ or $-\cos\alpha$, $-\cos\beta$ and $-\cos\gamma$

i.e., $\quad -l, -m, -n.$

Q. 2. Find the direction cosines of a line, which makes equal angles with the coordinate axes.

Sol. It is given that

$$\alpha = \beta = \gamma$$

so that $\quad l = m = n = \cos\alpha$

Putting this value of *l*, *m*, *n* in $l^2 + m^2 + n^2 = 1$, we get

$$3l^2 = 1$$

or $\quad l^2 = 1/3$

or $\quad l = \pm\frac{1}{\sqrt{3}}$

There are four lines which make equal angles with the axes.

Their direction cosines are

$$\frac{1}{\sqrt{3}}, \frac{1}{\sqrt{3}}, \frac{1}{\sqrt{3}}; -\frac{1}{\sqrt{3}}, \frac{1}{\sqrt{3}}, \frac{1}{\sqrt{3}}; \frac{1}{\sqrt{3}}, \frac{-1}{\sqrt{3}}, \frac{1}{\sqrt{3}};$$

$$\frac{1}{\sqrt{3}}, \frac{1}{\sqrt{3}}, \frac{-1}{\sqrt{3}}$$

Now, if it is assumed (as is generally the case) that the line makes equal acute angles with the positive directions of the axes, then its direction cosines are

$$\frac{1}{\sqrt{3}}, \frac{1}{\sqrt{3}}, \frac{1}{\sqrt{3}}$$

Q. 3. The direction cosines of a line are proportional to 2, – 3, 6. Find the actual values of the direction cosines.

Sol. If *l*, *m*, *n* are the direction cosines of the line then

$$l : m : n = 2 : -3 : 6$$

Hence $\quad l = \frac{a}{\sqrt{a^2 + b^2 + c^2}}, \; m = \frac{b}{\sqrt{a^2 + b^2 + c^2}},$

★ are board exam questions from previous years

$$n = \frac{c}{\sqrt{a^2+b^2+c^2}}$$

$$l = \frac{2}{\sqrt{2^2+(-3)^2+6^2}},\ m = \frac{-3}{\sqrt{2^2+(-3)^2+6^2}},$$

$$n = \frac{6}{\sqrt{2^2+(-3)^2+6^2}}$$

i.e., $\quad l = \frac{2}{7},\ m = \frac{-3}{7},\ n = \frac{6}{7}$

Q. 4. If $P(1, 2, 3)$ and $Q(-1, 2, 1)$ are two points in space, find the direction cosines of PQ.

Sol. Let l, m, n be the direction cosines of the line PQ.

Then, $\quad l : m : n = (x_2 - x_1) : (y_2 - y_1) : (z_2 - z_1)$

$l : m : n = (-1-1) : (2-2) : (1-3)$

or $\quad l : m : n = -2 : 0 : -2$

This shows that $-2, 0, -2$ are the direction ratios of the line PQ.

Hence $l = \frac{a}{\sqrt{a^2+b^2+c^2}},\ m = \frac{b}{\sqrt{a^2+b^2+c^2}},$

$$n = \frac{c}{\sqrt{a^2+b^2+c^2}}$$

$$l = \frac{-2}{\sqrt{(-2)^2+(-2)^2}};\ m = \frac{0}{\sqrt{(-2)^2+(-2)^2}};$$

$$n = \frac{-2}{\sqrt{(-2)^2+(-2)^2}}$$

or $\quad l = \frac{-1}{\sqrt{2}},\ m = 0,\ n = \frac{-1}{\sqrt{2}}$

Q. 5. Find the direction cosines of the line, which is perpendicular to the lines whose direction cosines are proportional to $1, -1, 2$ and $2, 1, -1$.

Sol. Let l, m, n be the direction cosines of the required line.

Then $\quad l(1) + m(-1) + n(2) = 0$

and $\quad l(2) + m(1) + n(-1) = 0$

$\Rightarrow \quad l - m + 2n = 0 \quad \ldots$(i)

and $\quad 2l + m - n = 0 \quad \ldots$(ii)

On solving (i) and (ii) by cross multiplication, we get

$$\frac{l}{1-2} = \frac{m}{4+1} = \frac{n}{1+2}$$

$$\Rightarrow \quad \frac{l}{-1} = \frac{m}{5} = \frac{n}{3} = \frac{\sqrt{l^2+m^2+n^2}}{\sqrt{(-1)^2+(5)^2+(3)^2}}$$

$$\Rightarrow \quad \frac{l}{-1} = \frac{m}{5} = \frac{n}{3} = \frac{1}{\sqrt{35}}$$

$[\because l^2 + m^2 + n^2 = 1]$

$$\therefore \quad l = \frac{-1}{\sqrt{35}},\ m = \frac{5}{\sqrt{35}} \text{ and } n = \frac{3}{\sqrt{35}}$$

Q. 6. A line passes through the points $A(6, -7, -1)$ and $B(2, -3, 1)$. Find direction ratios and direction cosines of the line so directed that the angle α is acute.

Sol. The direction ratios of a vector passing through $P(x_1, y_1, z_1)$ and $Q(x_2, y_2, z_2)$ are

$$(x_2 - x_1),\ (y_2 - y_1) \text{ and } (z_2 - z_1)$$

Therefore, direction ratios of AB are

$$2-6,\ -3+7,\ 1+1$$

i.e., $-4, 4, 2$ or $-2, 2, 1$.

We know that if a, b, c are direction ratios of a vector then its direction cosines are

$$\pm\frac{a}{\sqrt{a^2+b^2+c^2}},\ \pm\frac{b}{\sqrt{a^2+b^2+c^2}},\ \pm\frac{c}{\sqrt{a^2+b^2+c^2}}$$

Therefore, direction cosines of AB are

$$\pm\frac{-2}{\sqrt{(-2)^2+(2)^2+(1)^2}},\ \pm\frac{2}{\sqrt{(-2)^2+(2)^2+(1)^2}},$$

$$\pm\frac{1}{\sqrt{(-2)^2+(2)^2+(1)^2}}$$

i.e., $\quad \pm\frac{-2}{3},\ \pm\frac{2}{3},\ \pm\frac{1}{3}$

Since α is acute, $\cos\alpha > 0$.

Therefore, the direction cosines of AB are $\frac{2}{3}, \frac{-2}{3}, \frac{-1}{3}$

(Taking negative sign)

Q. 7. Find the vector equation of the line which is parallel to the vector $2\hat{i} - \hat{j} + 3\hat{k}$ and which passes through the point $(5, -2, 4)$. Also, find its Cartesian equation.

Sol. The vector equation of the line passing through the point $(5, -2, 4)$ and parallel to the vector $2\hat{i} - \hat{j} + 3\hat{k}$ is given by

$$\vec{r} = (5\hat{i} - 2\hat{j} + 4\hat{k}) + \lambda(2\hat{i} - \hat{j} + 3\hat{k}) \quad \ldots\text{(i)}$$

Let $\quad \vec{r} = x\hat{i} + y\hat{j} + z\hat{k}$

Then,

$$(x\hat{i} + y\hat{j} + z\hat{k}) = (5\hat{i} - 2\hat{j} + 4\hat{k}) + \lambda(2\hat{i} - \hat{j} + 3\hat{k})$$

$$(x\hat{i} + y\hat{j} + z\hat{k}) = (5+2\lambda)\hat{i} + (-2-\lambda)\hat{j} + (4+3\lambda)\hat{k}$$

$\Rightarrow \quad x = 5 + 2\lambda$

$y = -2 - \lambda$

$z = 4 + 3\lambda$

$$\Rightarrow \quad \frac{x-5}{2} = \frac{y+2}{-1} = \frac{z-4}{-3} = \lambda$$

$$i.e., \quad \frac{x-5}{2} = \frac{y+2}{-1} = \frac{z-4}{-3}$$

is the Cartesian equation of the required line.

Q. 8. Find the vector equation of the line passing through the point $A(1, 2, -1)$ and parallel to the line $5x - 25 = 14 - 7y = 35z$.★

Sol. Given line is

$$5x-25 = 14-7y = 35z$$

$$\Rightarrow \quad 5(x-5) = -7(y-2) = 35z$$

$$\Rightarrow \quad \frac{x-1}{3} = \frac{y-2}{-1/7} = \frac{z-0}{1/35}$$

$$\Rightarrow \quad \frac{x-5}{7} = \frac{y-2}{-5} = \frac{z-0}{1}$$

Direction ratios of this line are 7, – 5, 1.

∴ Vector equation of the line which passes through the point $A(1, 2, -1)$ and whose direction ratios are proportional to 7, – 5, 1 is

$$\vec{r} = \hat{i}+2\hat{j}-\hat{k}+\lambda(7\hat{i}-5\hat{j}+\hat{k}).$$

Q. 9. Find the value of λ, so that following lines are perpendicular to each other.

$$\frac{x+5}{5\lambda+2} = \frac{2-y}{5} = \frac{1-z}{-1} \text{ and } \frac{x}{1} = \frac{2y+1}{4\lambda} = \frac{1-z}{-3}.★$$

Sol. Given equation of lines are

$$\frac{x+5}{5\lambda+2} = \frac{2-y}{5} = \frac{1-z}{-1}$$

$$\text{and} \quad \frac{x}{1} = \frac{2y+1}{4\lambda} = \frac{1-z}{-3}$$

Above equations can be written as

$$\frac{x+5}{5\lambda+2} = \frac{y-2}{-5} = \frac{z-1}{1} \quad \ldots(i)$$

$$\text{and} \quad \frac{x-0}{1} = \frac{2\left(y+\frac{1}{2}\right)}{4\lambda} = \frac{z-1}{3}$$

$$\text{or} \quad \frac{x-0}{1} = \frac{y+\frac{1}{2}}{2\lambda} = \frac{z-1}{3} \quad \ldots(ii)$$

Comparing equations (i) and (ii) with one point form of line $\frac{x-x_1}{a} = \frac{y-y_1}{b} = \frac{z-z_1}{c}$, we get

$$a_1 = 5\lambda+2,\ b_1 = -5,\ c_1 = 1$$

$$\text{and} \quad a_2 = 1,\ b_2 = 2\lambda,\ c_2 = 3$$

Since, two lines are perpendicular,

$$\therefore \quad a_1a_2 + b_1b_2 + c_1c_2 = 0$$

$$\Rightarrow \quad 1(5\lambda+2)+2\lambda(-5)+3(1) = 0$$

$$\Rightarrow \quad 5\lambda+2-10\lambda+3 = 0$$

$$\Rightarrow \quad -5\lambda+5 = 0$$

$$\Rightarrow \quad 5\lambda = 5$$

$$\Rightarrow \quad \lambda = 1$$

★ are board exam questions from previous years

Q. 10. Find the angle between pair of lines given by $\vec{r} = 2\hat{i}-5\hat{j}+\hat{k}+\lambda(3\hat{i}+2\hat{j}+6\hat{k})$ and $\vec{r} = 7\hat{i}-6\hat{j}-6\hat{k}+\mu(\hat{i}+2\hat{j}+2\hat{k})$.★

Sol. The given equation of lines are

$$\vec{r} = 2\hat{i}-5\hat{j}+\hat{k}+\lambda(3\hat{i}+2\hat{j}+6\hat{k}) \quad \ldots(i)$$

$$\text{and} \quad \vec{r} = 7\hat{i}-6\hat{j}-6\hat{k}+\mu(\hat{i}+2\hat{j}+2\hat{k}) \quad \ldots(ii)$$

Comparing equations (i) and (ii) with vector form of equation of line which is $\vec{r} = \vec{a}+\lambda\vec{b}$, we get

$$\vec{a_1} = 2\hat{i}-5\hat{j}+\hat{k}$$

$$\vec{b_1} = 3\hat{i}+2\hat{j}+6\hat{k}$$

$$\vec{a_2} = 7\hat{i}-6\hat{j}-6\hat{k}$$

$$\vec{b_2} = \hat{i}+2\hat{j}+2\hat{k}$$

Now, we know that angle between two lines is given by

$$\cos\theta = \frac{|\vec{b_1}.\vec{b_2}|}{|\vec{b_1}|.|\vec{b_2}|}$$

$$\Rightarrow \quad \cos\theta = \left|\frac{(3\hat{i}+2\hat{j}+6\hat{k}).(\hat{i}+2\hat{j}+2\hat{k})}{\sqrt{(3)^2+(2)^2+(6)^2}.\sqrt{(1)^2+(2)^2+(2)^2}}\right|$$

$$\Rightarrow \quad \cos\theta = \left|\frac{3+4+12}{\sqrt{49}\times\sqrt{9}}\right|$$

$$\therefore \quad \cos\theta = \left|\frac{19}{7\times3}\right|$$

$$\text{or} \quad \cos\theta = \frac{19}{21}$$

Hence, angle between two lines is

$$\theta = \cos^{-1}\left(\frac{19}{21}\right).$$

Q. 11. Find the angle between the following pair of lines

$$\frac{-x+2}{2} = \frac{y-1}{7} = \frac{z+3}{-3}$$

$$\text{and} \quad \frac{x+2}{-1} = \frac{2y-8}{4} = \frac{z-5}{4}.$$

Also check whether the lines are parallel or perpendicular.★

Sol. Given equations of two lines are

$$\frac{x-2}{-2} = \frac{y-1}{7} = \frac{z+3}{-3}$$

$$\text{and} \quad \frac{x+2}{-1} = \frac{2y-8}{4} = \frac{z-5}{4}$$

Above equations can be written as

$$\frac{x-2}{2} = \frac{y-1}{7} = \frac{z+3}{-3} \quad \ldots(i)$$

and $$\frac{x+2}{-1} = \frac{y-4}{2} = \frac{z-5}{4} \quad \ldots(ii)$$

Comparing equation (i) and equation (ii) with $\frac{x-x_1}{a} = \frac{y-y_1}{b} = \frac{z-z_1}{c}$, we get

$$a_1 = 2, b_1 = 7, c_1 = -3$$

and $$a_2 = -1, b_2 = 2, c_2 = 4$$

Now, we know that angle between two lines is given by

$$\cos\theta = \frac{a_1a_2 + b_1b_2 + c_1c_2}{\sqrt{a_1^2+b_1^2+c_1^2}\sqrt{a_2^2+b_2^2+c_2^2}}$$

$$\therefore \cos\theta = \frac{(2)(-1)+(7)(2)+(-3)(4)}{\sqrt{(2)^2+(7)^2+(-3)^2}.\sqrt{(-1)^2+(2)^2+(4)^2}}$$

$$\Rightarrow \cos\theta = \frac{-2+14-12}{\sqrt{62}\times\sqrt{21}} = \frac{0}{\sqrt{62}\times\sqrt{21}} = 0$$

$$\Rightarrow \cos\theta = \cos\frac{\pi}{2} \quad \left(\because \cos\frac{\pi}{2} = 0\right)$$

$$\Rightarrow \theta = \frac{\pi}{2}$$

Hence, angle between them is $\frac{\pi}{2}$.

Since, angle between the lines is $\frac{\pi}{2}$, therefore, the above pair of lines are perpendicular to each other.

Q. 12. By computing the shortest distance, determine whether the following lines intersects or not $\frac{x-1}{2} = \frac{y+1}{3} = z$ and $\frac{x+1}{5} = \frac{y-2}{1} = \frac{z-2}{0}$.

Sol. The given equations of the two lines are

$$\frac{x-1}{2} = \frac{y+1}{3} = \frac{z-0}{1} \quad \ldots(i)$$

and $$\frac{x+1}{5} = \frac{y-2}{1} = \frac{z-2}{0} \quad \ldots(ii)$$

The vector equations of the above two lines are

$$\vec{r} = (\hat{i} - \hat{j} + 0\hat{k}) + \lambda(2\hat{i} + 3\hat{j} + \hat{k})$$

$$\vec{r} = (-\hat{i} + 2\hat{j} + 2\hat{k}) + \mu(5\hat{i} + \hat{j})$$

Comparing these equations with $\vec{r} = \vec{a_1} + \lambda\vec{b_1}$ and $\vec{r} = \vec{a_2} + \mu\vec{b_2}$, we have

$$\vec{a_1} = \hat{i} - \hat{j}$$

$$\vec{a_2} = -\hat{i} + 2\hat{j} + 2\hat{k}$$

$$\vec{b_1} = 2\hat{i} + 3\hat{j} + \hat{k}$$

$$\vec{b_2} = 5\hat{i} + \hat{j}$$

Now, $$|\vec{a_2} - \vec{a_1}| = -2\hat{i} + 3\hat{j} + 2\hat{k}$$

and $$\vec{b_1} \times \vec{b_2} = \begin{vmatrix} \hat{i} & \hat{j} & \hat{k} \\ 2 & 3 & 1 \\ 5 & 1 & 0 \end{vmatrix}$$

$$= \hat{i}(0-1) - \hat{j}(0-5) + \hat{k}(2-15)$$

$$= -\hat{i} + 5\hat{j} - 13\hat{k}$$

Also, $$\vec{b_1} \times \vec{b_2} = \sqrt{(-1)^2 + (5)^2 + (-13)^2}$$

$$= \sqrt{1+25+169} = \sqrt{195}$$

$$(\vec{a_2} - \vec{a_1}).(\vec{b_1} \times \vec{b_2}) = (-2\hat{i}+3\hat{j}+2\hat{k}).(-\hat{i}+5\hat{j}-13\hat{k})$$

$$= 2 + 15 - 26 = -9$$

Therefore, the shortest distance between the lines

$$= \left|\frac{-9}{\sqrt{195}}\right|$$

$$= \frac{9}{\sqrt{195}} \geq 0$$

Since, $d \neq 0$

Hence, the given lines do not intersect.

Q. 13. Check whether the lines

$$l_1 : \vec{r} = \hat{i} - \hat{j} + t(2\hat{i} + \hat{k})$$

and $$l_2 : \vec{r} = 2\hat{i} - \hat{j} + u(\hat{i} + \hat{j} - \hat{k})$$

are skew lines or not. Hence, find the shortest distance between them.

Sol. Clearly they are not parallel, since the l_1 is parallel to $2\hat{i} + \hat{k}$ and line l_2 is parallel to $\hat{i} + \hat{j} - \hat{k}$.

If they intersect, then for some value of t and u.

$$2\hat{i} - \hat{j} + u(\hat{i} + \hat{j} - \hat{k}) = \hat{i} - \hat{j} + t(2\hat{i} + \hat{k})$$

Equating the coefficients, we get

$$2 + u = 1 + 2t, \quad \ldots(i)$$

$$\Rightarrow \quad -1 + u = -1 \quad \ldots(ii)$$

and $$-u = t \quad \ldots(iii)$$

From (ii), we get $u = 0$

From (iii), we get $t = 0$

Putting these values in (i), we get 2 = 1, which is not possible.

Thus, these are indeed skew lines.

The shortest distance between them is

$$\left|\frac{[(2\hat{i}+\hat{k})\times(\hat{i}+\hat{j}-\hat{k})].(2\hat{i}-\hat{j}-\hat{i}+\hat{j})}{|(2\hat{i}+\hat{k})\times(\hat{i}+\hat{j}-\hat{k})|}\right| = \frac{1}{\sqrt{14}} \text{ unit}$$

Q. 14. Find the points on the line $\frac{x+2}{3}=\frac{y+1}{2}=\frac{z-3}{2}$ at a distance of 5 units from the point $P(1, 3, 3)$.★

Sol. Given equation of line is

$$\frac{x+2}{3}=\frac{y+1}{2}=\frac{z-3}{2}$$

We find any random point on the line as follows :

$$\frac{x+2}{3}=\frac{y+1}{2}=\frac{z-3}{2}=\lambda \text{ (say)}$$

$$\Rightarrow \quad \frac{x+2}{3}=\lambda, \frac{y+1}{2}=\lambda, \frac{z-3}{2}=\lambda$$

$$\Rightarrow \quad x=3\lambda-2, y=2\lambda-1, z=2\lambda+3$$

$\therefore$ We have the point $Q(3\lambda-2, 2\lambda-1, 2\lambda+3)$ ...(i)

Now, given that distance between point $P(1, 3, 3)$ and $Q(3\lambda-2, 2\lambda-1, 2\lambda+3)$ is 5 units.

i.e., $\quad PQ=5$

$$\Rightarrow \sqrt{(3\lambda-2-1)^2+(2\lambda-1-3)^2+(2\lambda+3-3)^2}=5$$

Squaring both sides, we get

$$(3\lambda-3)^2+(2\lambda-4)^2+(2\lambda)^2=25$$

$$\Rightarrow \quad 9\lambda^2+9-18\lambda+4\lambda^2+16-16\lambda+4\lambda^2=25$$

$$\Rightarrow \quad 17\lambda^2-34\lambda=0$$

$$\Rightarrow \quad 17\lambda(\lambda-2)=0$$

$\Rightarrow$ Either $\quad 17\lambda=0$ or $\lambda-2=0$

$$\Rightarrow \quad \lambda=0 \text{ or } 2$$

Putting $\lambda = 0$ and $\lambda = 2$ in equation (i), we get the required points as $(-2, -1, 3)$ and $(4, 3, 7)$.

Q. 15. Find the shortest distance between lines

$$\vec{r} = 6\hat{i}+2\hat{j}+2\hat{k}+\lambda(\hat{i}-2\hat{j}+2\hat{k})$$

and $$\vec{r} = -4\hat{i}-\hat{k}+\mu(3\hat{i}-2\hat{j}-2\hat{k})^{\star}$$

Sol. Given equations of lines are

$$\vec{r} = 6\hat{i}+2\hat{j}+2\hat{k}+\lambda(\hat{i}-2\hat{j}+2\hat{k}) \quad \text{...(i)}$$

and $$\vec{r} = -4\hat{i}-\hat{k}+\mu(3\hat{i}-2\hat{j}-2\hat{k}) \quad \text{...(ii)}$$

We know that vector form of equation of line is

$$\vec{r} = \vec{a}+\lambda\vec{b} \quad \text{...(iii)}$$

Comparing equations (i) and (ii) with equation (iii), we get

$$\vec{a_1} = 6\hat{i}+2\hat{j}+2\hat{k}, \ \vec{b_1} = \hat{i}-2\hat{j}+2\hat{k}$$

$$\vec{a_2} = -4\hat{i}-\hat{k}, \ \vec{b_2} = 3\hat{i}-2\hat{j}-2\hat{k}$$

We know that the shortest distance between two lines is given by

$$d=\left|\frac{(\vec{b_1}\times\vec{b_2}).(\vec{a_2}-\vec{a_1})}{|\vec{b_1}\times\vec{b_2}|}\right| \quad \text{...(v)}$$

$$\text{Now } \vec{b_1}\times\vec{b_2} = \begin{vmatrix} \hat{i} & \hat{j} & \hat{k} \\ 1 & -2 & 2 \\ 3 & -2 & -2 \end{vmatrix}$$

$$= \hat{i}(4+4)-\hat{j}(-2-6)+\hat{k}(-2+6)$$

$$\Rightarrow \quad \vec{b_1}\times\vec{b_2} = 8\hat{i}+8\hat{j}+4\hat{k} \quad \text{...(vi)}$$

$$\text{and } |\vec{b_1}\times\vec{b_2}| = \sqrt{(8)^2+(8)^2+(4)^2}$$

$$= \sqrt{64+64+16}$$

$$= \sqrt{144}=12 \quad \text{...(vii)}$$

$$\text{Also, } \quad \vec{a_2}-\vec{a_1} = (-4\hat{i}-\hat{k})-(6\hat{i}+2\hat{j}+2\hat{k})$$

$$\Rightarrow \quad \vec{a_2}-\vec{a_1} = -10\hat{i}-2\hat{j}-3\hat{k}$$

So from equations (iv), (v) and (vi) and (vii), we get

$$d=\left|\frac{(8\hat{i}+8\hat{j}+4\hat{k}).(-10\hat{i}-2\hat{j}-3\hat{k})}{12}\right|$$

$$\Rightarrow \quad d=\left|\frac{-80-16-12}{12}\right|$$

$$=\left|\frac{-108}{12}\right|=\frac{108}{12}=9$$

$\therefore$ Shortest distance = 9 units.

Q. 16. Find the shortest distance between the lines $l_1: \frac{x-1}{1}=\frac{y-2}{-1}=\frac{z-1}{1}$ and $l_2: \frac{x-2}{2}=\frac{y+1}{1}=\frac{z+1}{2}$.★

Sol. Given equation of lines are

$$\frac{x-1}{1}=\frac{y-2}{-1}=\frac{z-1}{1} \quad \text{...(i)}$$

and $$\frac{x-2}{2}=\frac{y+1}{1}=\frac{z+1}{2} \quad \text{...(ii)}$$

Here, $\quad x_1=1, y_1=2, z_1=1$

$\quad a_1=1, b_1=-1, c_1=1$

and $\quad x_2=2, y_2=-1, z_2=-1$

$\quad a_2=2, b_2=1, c_2=2$

★ are board exam questions from previous years

Now, we know that the shortest distance between two lines is given by

$$d = \frac{\begin{vmatrix} x_2 - x_1 & y_2 - y_1 & z_2 - z_1 \\ a_1 & b_1 & c_1 \\ a_2 & b_2 & c_2 \end{vmatrix}}{\sqrt{(b_1c_2 - b_2c_1)^2 + (c_1a_2 - c_2a_1)^2 + (a_1b_2 - a_2b_1)^2}}$$

$$\therefore \quad d = \frac{\begin{vmatrix} 2-1 & -1-2 & -1-1 \\ 1 & -1 & 1 \\ 2 & 1 & 2 \end{vmatrix}}{\sqrt{(-2-1)^2 + (2-2)^2 + (1+2)^2}}$$

$$\Rightarrow \quad d = \frac{\begin{vmatrix} 1 & -3 & -2 \\ 1 & -1 & 1 \\ 2 & 1 & 2 \end{vmatrix}}{\sqrt{9+0+9}}$$

$$= \left| \frac{1(-2-1) + 3(2-2) - 2(1+2)}{\sqrt{18}} \right|$$

$$= \left| \frac{-3+0-6}{\sqrt{18}} \right| = \frac{9}{3\sqrt{2}} = \frac{3}{\sqrt{2}}$$

Hence, the required shortest distance is $\frac{3}{\sqrt{2}}$ units.

Q. 17. Find the shortest distance between lines

$$\vec{r} = (1+\lambda)\hat{i} + (2-\lambda)\hat{j} + (\lambda+1)\hat{k}$$

and $\vec{r} = (2\hat{i} - \hat{j} - \hat{k}) + \mu(2\hat{i} + \hat{j} + 2\hat{k})$.★

Sol. Given equation of lines are

$$\vec{r} = (1+\lambda)\hat{i} + (2-\lambda)\hat{j} + (\lambda+1)\hat{k} \quad \text{...(i)}$$

and $\vec{r} = (2\hat{i} - \hat{j} - \hat{k}) + \mu(2\hat{i} + \hat{j} + 2\hat{k})$ ...(ii)

Equation (i) can be written as

$$\vec{r} = \hat{i} + \lambda\hat{i} + 2\hat{j} - \lambda\hat{j} + \lambda\hat{k} + \hat{k}$$

or $\vec{r} = (\hat{i} + 2\hat{j} + \hat{k}) + \lambda(\hat{i} - \hat{j} + \hat{k})$ ...(iii)

Comparing equations (ii) and (iii) with vector form of line $\tau = a + \lambda b$, we get

$$\vec{a}_1 = 2\hat{i} - \hat{j} - \hat{k}, \; b_1 = 2\hat{i} + \hat{j} + 2\hat{k}$$

and $\vec{a}_2 = \hat{i} + 2\hat{j} + \hat{k}, \; b_2 = \hat{i} - \hat{j} + \hat{k}$

Now, we know that shortest distance between two lines is given by

$$d = \left| \frac{(\vec{b}_1 \times \vec{b}_2).(\vec{a}_2 - \vec{a}_1)}{|\vec{b}_1 \times \vec{b}_2|} \right| \quad \text{...(iv)}$$

Now, $\vec{b}_1 \times \vec{b}_2 = \begin{vmatrix} \hat{i} & \hat{j} & \hat{k} \\ 2 & 1 & 2 \\ 1 & -1 & 1 \end{vmatrix}$

$$= \hat{i}(1+2) - \hat{j}(2-2) + \hat{k}(-2-1)$$

$$\Rightarrow \quad \vec{b}_1 \times \vec{b}_2 = 3\hat{i} - 3\hat{k} \quad \text{...(v)}$$

and $|\vec{b}_1 \times \vec{b}_2| = \sqrt{(3)^2 + (-3)^2} = \sqrt{9+9}$

$$= 3\sqrt{2} \quad \text{...(vi)}$$

Also, $\vec{a}_2 - \vec{a}_1 = (\hat{i} + 2\hat{j} + \hat{k}) - (2\hat{i} - \hat{j} - \hat{k})$

$$= -\hat{i} + 3\hat{j} + 2\hat{k} \quad \text{...(vii)}$$

From equations (iv), (v), (vi) and (vii), we get

$$d = \left| \frac{(3\hat{i} - 3\hat{k}).(-\hat{i} + 3\hat{j} + 2\hat{k})}{3\sqrt{2}} \right|$$

$$= \left| \frac{-3+0-6}{3\sqrt{2}} \right| = \left| \frac{-9}{3\sqrt{2}} \right|$$

$$= \frac{9}{3\sqrt{2}} = \frac{3}{\sqrt{2}}$$

$$\therefore \quad d = \frac{3\sqrt{2}}{2} \text{ units.}$$

Q. 18. Find the vector and Cartesian equation of line passing through point (1, 2, – 4) and perpendicular to two lines $\frac{x-8}{3} = \frac{y+19}{-16} = \frac{z-10}{7}$ and $\frac{x-15}{3} = \frac{y-29}{8} = \frac{z-5}{-5}$.

Sol. Let the required equation of line passing through (1, 2, – 4) is

$$\frac{x-1}{a} = \frac{y-2}{b} = \frac{z+4}{c} \quad \text{...(i)}$$

Given that line (i) is perpendicular to lines

$$\frac{x-8}{3} = \frac{y+19}{-16} = \frac{z-10}{7} \quad \text{...(ii)}$$

and $\frac{x-15}{3} = \frac{y-29}{8} = \frac{z-5}{-5}$ ...(iii)

We know that when two lines are perpendicular then we have $a_1a_2 + b_1b_2 + c_1c_2 = 0$, where a_1, b_1, c_1 and a_2, b_2, c_2 are the direction ratios of two lines. Using this property first in equations (i) and (ii) and then in equations (i) and (iii), we get

$$3a - 16b + 7c = 0 \quad \text{...(iv)}$$

$$(\because a_1a_2 + b_1b_2 + c_1c_2 = 0)$$

★ are board exam questions from previous years

and $\quad 3a+8b-5c=0 \quad$ ...(v)

Subtracting equation (v) from equation (iv), we get

$$3a-16b=-7c$$
$$3a+8b=5c$$
$$- \quad - \quad -$$
$$-24b=-12c$$

or $\quad b=\frac{c}{2}$

Putting $b=\frac{c}{2}$ in equation (iv), we get

$$3a-16\left(\frac{c}{2}\right)+7c=0$$

$\Rightarrow \quad 3a-8c+7c=0$

$\Rightarrow \quad 3a-c=0$

$\Rightarrow \quad a=\frac{c}{3}$

Putting $a=\frac{c}{3}$ and $b=\frac{c}{2}$ in equation (i), we get the required equation of line as

$$\frac{x-1}{\left(\frac{c}{3}\right)}=\frac{y-2}{\left(\frac{c}{2}\right)}=\frac{z+4}{c}$$

$$\Rightarrow \quad \frac{x-1}{2c}=\frac{y-2}{3c}=\frac{z+4}{6c}$$

$$\frac{x-1}{2}=\frac{y-2}{3}=\frac{z+4}{6}$$

Also, the vector equation of line is

$$r=(\hat{i}+2\hat{j}-4\hat{k})+\lambda(2\hat{i}+3\hat{j}+6\hat{k})$$

Q. 19. Find the length and foot of perpendicular drawn from the point (2, – 1, 5) to line

$$\frac{\boldsymbol{x-11}}{\mathbf{10}}=\frac{\boldsymbol{y+2}}{\mathbf{-4}}=\frac{\boldsymbol{z+8}}{\mathbf{-11}}.\star$$

Sol. Let AB be the line whose equation is

$$\frac{x-11}{10}=\frac{y+2}{-4}=\frac{z+8}{-11}$$

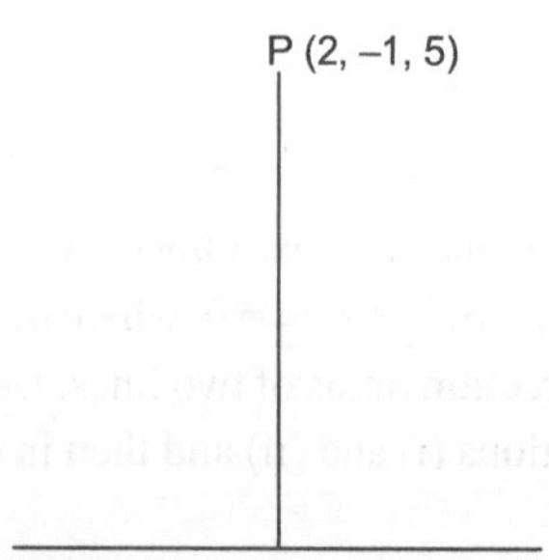

Let Q be the foot of perpendicular and PQ be the length of perpendicular.

★ are board exam questions from previous years

Any point Q on the given line is given by

$$\frac{x-11}{10}=\frac{y+2}{-4}=\frac{z+8}{-11}=\lambda \text{ (say)}$$

$\Rightarrow \quad x=10\lambda+11, y=-4\lambda-2, z=-11\lambda-8$

$\therefore$ Coordinates of Q are $(10\lambda+11,-4\lambda-2,-11\lambda-8)$

Now, direction ratios of the line PQ are

$$(10\lambda+11-2,-4\lambda-2+1,-11\lambda-8-5)$$

$\therefore$ Direction ratios of $PQ=(10\lambda+9,-4\lambda-1,-11\lambda-13)$

$\because \quad PQ \perp AB$

$\therefore \quad a_1a_2+b_1b_2+c_1c_2=0$

Where $a_1=10\lambda+9, b_1=-4\lambda-1, c_1=-11\lambda-13$ (Direction ratios of PQ)

and $a_2=10, b_2=-4, c_2=-11$ (Direction ratios of AB)

$\therefore$ We get

$$10(10\lambda+9)-4(-4\lambda-1)-11(-11\lambda-13)=0$$

$\Rightarrow \quad 100\lambda+90+16\lambda+4+121\lambda+143=0$

$\Rightarrow \quad 237\lambda+237=0$

$\Rightarrow \quad \lambda=-1$

Putting $\lambda=-1$ in coordinates of point Q_1 we get

Foot of perpendicular

$$=Q(-10+11,4-2,11-8)$$
$$=Q(1,2,3)$$

Also, length of perpendicular PQ is given by

$$PQ=\sqrt{(2-1)^2+(-1-2)^2+(5-3)^2}$$
$$=\sqrt{1+9+4}$$
$$=\sqrt{14} \text{ units}$$

Q. 20. Find the perpendicular distance of point (1, 0, 0) from the line $\frac{x-1}{2}=\frac{y+1}{-3}=\frac{z+10}{8}$. Also, find the coordinates of foot of perpendicular and equation of perpendicular.★

Sol. Let AB be the line whose equation is given as

$$\frac{x-1}{2}=\frac{y+1}{-3}=\frac{z+10}{8}$$

Let P be the foot of perpendiculat drawn from point $O(1, 0, 0)$ on line AB.

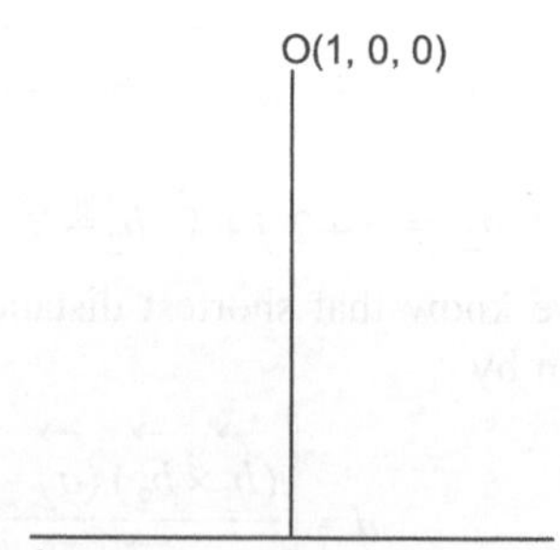

Let $\dfrac{x-1}{2} = \dfrac{y+1}{-3} = \dfrac{z+10}{8} = \lambda$

$\Rightarrow \dfrac{x-1}{2} = \lambda, \dfrac{y+1}{-3} = \lambda, \dfrac{z+10}{8} = \lambda$

$\Rightarrow x = 2\lambda + 1, y = -3\lambda - 1, z = 8\lambda - 10$

$\therefore$ Coordinates of P are

$$(2\lambda + 1, -3\lambda - 1, 8\lambda - 10)$$

$\Rightarrow$ Direction ratios of line OP

$$= (2\lambda + 1 - 1, -3\lambda - 1 - 0, 8\lambda - 10 - 0)$$
$$= (2\lambda, -3\lambda - 1, 8\lambda - 10)$$

Since, $OP \perp AB$

$\therefore \quad a_1a_2 + b_1b_2 + c_1c_2 = 0$

Where a_1, b_1, c_1 are direction ratios of line OP and a_2, b_2, c_2 are direction ratios of line AB.

$\therefore$ We get $a_1 = 2\lambda, b_1 = -3\lambda - 1, c_1 = 8\lambda - 10$

and $a_2 = 2, b_2 = -3, c_2 = 8$

$\therefore$ From equation (i)

$$2(2\lambda) - 3(-3\lambda - 1) + 8(8\lambda - 10) = 0$$

$\Rightarrow 4\lambda + 9\lambda + 3 + 64\lambda - 80 = 0$

$\Rightarrow 77\lambda - 77 = 0$

$\Rightarrow 77\lambda = 77$

$\Rightarrow \lambda = 1$

Coordinates of point P

$$= (2\lambda + 1, -3\lambda - 1, 8\lambda - 10)$$
$$= (3, -4, -2)$$

Now, since P is the foot of perpendicular, so coordinates of foot of perpendicular = $(3, -4, -2)$.

Now, we find the perpendicular distance of the point O (1, 0, 0) and the line which is equal to OP.

$\therefore$ Using distance formula,

$$OP = \sqrt{(3-1)^2 + (-4-0)^2 + (-2-0)^2}$$
$$= \sqrt{4+16+4} = \sqrt{24} = 2\sqrt{6} \text{ units}$$

$\therefore \quad OP = 2\sqrt{6}$ units .

Finally, equation of perpendicular OP, where $O(1, 0, 0)$ and $P(3, -4, -2)$ is given by

$$\frac{x-1}{3-1} = \frac{y-0}{-4-0} = \frac{z-0}{-2-0}$$

$$\Rightarrow \frac{x-1}{2} = \frac{y}{-4} = \frac{z}{-2}$$

which is the required equation of perpendicular.

Q. 21. Find the vector and Cartesian equation of the plane containing the two lines

$$\vec{r} = 2\hat{i} + \hat{j} - 3\hat{k} + \lambda(\hat{i} + 2\hat{j} + 5\hat{k})$$

and $$\vec{r} = 3\hat{i} + 3\hat{j} - 7\hat{k} + \mu(3\hat{i} - 2\hat{j} + 5\hat{k})$$

Sol. Here $\vec{a} = 2\hat{i} + \hat{j} - 3\hat{k}$

$$\vec{a}' = 3\hat{i} + 3\hat{j} - 7\hat{k}$$
$$\vec{b} = \hat{i} + 2\hat{j} + 5\hat{k}$$
$$\vec{b}' = 3\hat{i} - 2\hat{j} + 5\hat{k}$$

Vector equation of plane is

$$[\vec{r} - (2\hat{i} + \hat{j} - 3\hat{k})].(\hat{i} + 2\hat{j} + 5\hat{k}) \times (3\hat{i} - 2\hat{j} + 5\hat{k})] = 0$$

$$\begin{vmatrix} \hat{i} & \hat{j} & \hat{k} \\ 1 & 2 & 5 \\ 3 & -2 & 5 \end{vmatrix} = 20\hat{i} + 10\hat{j} - 8\hat{k}$$

$$[\vec{r} - (2\hat{i} + \hat{j} - 3\hat{k})].(20\hat{i} + 10\hat{j} - 8\hat{k}) = 0$$

$$\vec{r}.(20\hat{i} + 10\hat{j} - 8\hat{k}) - (40 + 10 + 24) = 0$$

$\Rightarrow \vec{r}.(20\hat{i} + 10\hat{j} - 8\hat{k}) = 74$

or $\vec{r}.(10\hat{i} + 5\hat{j} - 4\hat{k}) = 37 \quad$ ...(i)

is the required vector equation.

Let $\vec{r} = x\hat{i} + y\hat{j} + z\hat{k}$

Using this in equation (i), we get

$10x + 5y - 4z = 37$ as the equation of plane in Cartesian form.

Q. 22. Find the ratio in which the join of $A(2, 1, 5)$ and $B(3, 4, 3)$ is divided by the plane $2x + 2y - 2z = 1$. Also, find the coordinates of the point of division.

Sol. Suppose the plane $2x + 2y - 2z = 1$ divides the line joining the points $A(2, 1, 5)$ and $B(3, 4, 3)$ at a point C in the ratio $\lambda : 1$.

Then, the coordinates of C are

$$\left(\frac{3\lambda+2}{\lambda+1}, \frac{4\lambda+1}{\lambda+1}, \frac{3\lambda+5}{\lambda+1}\right) \quad ...(i)$$

Since, the point C lies on the plane $2x + 2y - 2z = 1$, therefore, coordinates of C must satisfy the equation of the plane.

That is,

$$2\left(\frac{3\lambda+2}{\lambda+1}\right) + 2\left(\frac{4\lambda+1}{\lambda+1}\right) - 2\left(\frac{3\lambda+5}{\lambda+1}\right) = 1$$

$\Rightarrow 6\lambda + 4 + 8\lambda + 2 - 6\lambda - 10 = \lambda + 1$

$\Rightarrow 8\lambda - 4 = \lambda + 1$

$$\Rightarrow \quad 8\lambda - \lambda = 1 + 4$$

$$\Rightarrow \quad 7\lambda = 5$$

$$\Rightarrow \quad \lambda = \frac{5}{7}$$

So, the required ratio is $\frac{5}{7}$: 1 or 5 : 7.

Putting $\lambda = \frac{5}{7}$ in (i), the coordinates of the point of division C are $\left(\frac{29}{12}, \frac{9}{4}, \frac{25}{6}\right)$.

Q. 23. Find the direction cosines of the perpendicular from the origin to the plane $\vec{r}.(6\hat{i} - 3\hat{j} - 2\hat{k}) + 1 = 0$.

Sol. The given equation of the plane is

$$\vec{r}.(6\hat{i} - 3\hat{j} - 2\hat{k}) + 1 = 0$$

$$\Rightarrow \quad \vec{r}.(6\hat{i} - 3\hat{j} - 2\hat{k}) = -1$$

$$\Rightarrow \quad \vec{r}.\vec{n} = -1 \qquad \ldots\text{(i)}$$

where $\quad \vec{n} = 6\hat{i} - 3\hat{j} - 2\hat{k}$

Now $\quad |\vec{n}| = \sqrt{(6)^2 + (-3)^2 + (-2)^2}$

$$= \sqrt{36 + 9 + 4}$$

$$= \sqrt{49} = 7$$

Therefore, from (i), we have

$$\vec{r}.\vec{n} = -1$$

$$\Rightarrow \quad \vec{r}\cdot\frac{\vec{n}}{|\vec{n}|} = \frac{-1}{|\vec{n}|}$$

$$\Rightarrow \quad \vec{r}.\hat{n} = p$$

where $\quad \hat{n} = \dfrac{\vec{n}}{|\vec{n}|}$

$$= \frac{6}{7}\hat{i} - \frac{3}{7}\hat{j} - \frac{2}{7}\hat{k}$$

and $\quad p = \dfrac{-1}{7}$

Therefore, direction cosines of $\hat{n}$ are $\frac{6}{7}, \frac{-3}{7}, \frac{-2}{7}$.

Q. 24. Find the distance of the point (– 1, – 5, – 10) from the point of intersection of the line $\frac{x-2}{3} = \frac{y+1}{4} = \frac{z-2}{12}$ and the plane $x - y + z = 5$.

Sol. Equation of the given line is

$$\frac{x-2}{3} = \frac{y+1}{4} = \frac{z-2}{12} = k \text{ (say)}$$

$$\Rightarrow \quad x = 3k + 2$$

$$\Rightarrow \quad y = 4k - 1$$

$$\Rightarrow \quad z = 12k + 2$$

The coordinates of any point P on the line are

$$(3k + 2, 4k - 1, 12k + 2) \qquad \ldots\text{(i)}$$

Now, P will be the point of intersection of the line and the plane only when P lies on the plane also.

The equation of the plane is

$$x - y + z = 5$$

As P lies on it therefore, we have

$$(3k + 2) - (4k - 1) + (12k + 2) = 5$$

$$\Rightarrow \quad 11k + 5 = 5$$

$$\Rightarrow \quad 11k = 0$$

$$\Rightarrow \quad k = 0$$

Putting the value of k in (i), we get the coordinates of the point of intersection of the line and plane as (2, – 1, 2).

Now, distance of the point (– 1, – 5, – 10) from the point P is given by

$$d = \sqrt{(2+1)^2 + (-1+5)^2 + (2+10)^2}$$

$$= \sqrt{(3)^2 + (4)^2 + (12)^2}$$

$$= \sqrt{9 + 16 + 144}$$

$$= \sqrt{169} = 13$$

Therefore, the required distance is 13 units.

Q. 25. Find the angle between the planes $\vec{r}.(2\hat{i} + 3\hat{j} - 6\hat{k}) = 3$ and $\vec{r}.(\hat{i} - 2\hat{j} + 2\hat{k}) = 5$.

Sol. We know that the angle between the planes

$\vec{r}.\vec{n}_1 = d_1$ and $\vec{r}.\vec{n}_2 = d_2$ is given by

$$\cos\theta = \frac{\vec{n}_1.\vec{n}_2}{|\vec{n}_1||\vec{n}_2|}$$

Here $\quad \vec{n}_1 = 2\hat{i} + 3\hat{j} - 6\hat{k}$

and $\quad \vec{n}_2 = \hat{i} - 2\hat{j} + 2\hat{k}$

$$\text{So, } \cos\theta = \frac{(2\hat{i} + 3\hat{j} - 6\hat{k}).(\hat{i} - 2\hat{j} + 2\hat{k})}{\sqrt{(2)^2 + (3)^2 + (-6)^2}\sqrt{(1)^2 + (-2)^2 + (2)^2}}$$

$$\Rightarrow \quad \cos\theta = \frac{2(1) + 3(-2) - 6(2)}{\sqrt{4 + 9 + 36}\sqrt{1 + 4 + 4}}$$

$$\Rightarrow \quad \cos\theta = \frac{2 - 6 - 12}{\sqrt{49}\sqrt{9}}$$

$$\Rightarrow \quad \cos\theta = \frac{-16}{21}$$

This is the cosine of the obtuse angle between the given planes.

So, the acute angle between the given planes is $\cos^{-1}\left(\frac{-16}{21}\right)$.

Q. 26. Find the angle between the planes $2x - y + z = 6$ and $x + y + 2z = 7$.

Sol. We know that the angles between the planes $a_1x + b_1y + c_1z + d_1 = 0$ and $a_2x + b_2y + c_2z + d_2 = 0$ is given by

$$\cos\theta = \frac{a_1a_2 + b_1b_2 + c_1c_2}{\sqrt{a_1^2 + b_1^2 + c_1^2}\sqrt{a_2^2 + b_2^2 + c_2^2}}$$

Therefore, the angle between $2x - y + z = 6$ and $x + y + 2z = 7$ is given by

$$\cos\theta = \frac{(2)(1) + (-1)(1) + (1)(2)}{\sqrt{(2)^2 + (-1)^2 + (1)^2}\sqrt{(1)^2 + (1)^2 + (2)^2}}$$

$$\cos\theta = \frac{2 - 1 + 2}{\sqrt{4+1+1}\sqrt{1+1+4}}$$

$$\Rightarrow \quad \cos\theta = \frac{3}{\sqrt{6}\sqrt{6}}$$

$$\Rightarrow \quad \cos\theta = \frac{3}{6}$$

$$\Rightarrow \quad \cos\theta = \frac{1}{2}$$

$$\Rightarrow \quad \theta = 60°$$

Hence, the angle between the given planes is 60°.

Q. 27. Find the value of λ for which the points with position vector $\hat{i} - \hat{j} + 3\hat{k}$ and $3\hat{i} + \lambda\hat{j} + 3\hat{k}$ are equidistant from the plane $\vec{r}.(5\hat{i} + 2\hat{j} - 7\hat{k}) + 9 = 0$.

Sol. Distance of point $\hat{i} - \hat{j} + 3\hat{k}$ and $\vec{r}.(5\hat{i} + 2\hat{j} - 7\hat{k}) + 9 = 0$ is

$$d_1 = \left|\frac{(\hat{i} - \hat{j} + 3\hat{k}).(5\hat{i} + 2\hat{j} - 7\hat{k}) + 9}{\sqrt{5^2 + 2^2 + 7^2}}\right|$$

$$d_1 = \left|\frac{5 - 2 - 21 + 9}{\sqrt{78}}\right|$$

$$d_1 = \frac{9}{\sqrt{78}}$$

Distance of $\vec{b'} = 3\hat{i} + \lambda\hat{j} + 3\hat{k}$ from the plane, $\vec{r}.(5\hat{i} + 2\hat{j} - 7\hat{k}) + 9 = 0$ is

$$d_2 = \left|\frac{(3\hat{i} + \lambda\hat{j} + 3\hat{k}).(5\hat{i} + 2\hat{j} - 7\hat{k}) + 9}{\sqrt{78}}\right|$$

$$= \left|\frac{15 + 2\lambda - 21 + 9}{\sqrt{78}}\right|$$

$$= \left|\frac{3 + 2\lambda}{\sqrt{78}}\right| = \frac{3 + 2\lambda}{\sqrt{78}}$$

$$\because \quad d_1 = d_2 \qquad \text{(given)}$$

$$\therefore \quad \frac{9}{\sqrt{78}} = \frac{3 + 2\lambda}{\sqrt{78}}$$

$$\Rightarrow \quad 6 = 2\lambda$$

$$\Rightarrow \quad \lambda = 3$$

Q. 28. Find the equation of the plane passing through the points (– 1, 2, 0), (2, 2, – 1) and parallel to the line

$$\frac{x - 1}{1} = \frac{2y + 1}{2} = \frac{z + 1}{-1}. \star$$

Sol. The equation of a plane passing through (–1, 2, 0) is

$$a(x + 1) + b(y - 2) + c(z - 0) = 0 \qquad \text{...(i)}$$

It also passes through (2, 2, –1).

$$\therefore \quad a(2 + 1) + b(2 - 2) + c(-1 - 0) = 0$$

$$3a + 0b - c = 0 \qquad \text{...(ii)}$$

The given line is

$$\frac{x - 1}{1} = \frac{2y + 1}{2} = \frac{z + 1}{-1}$$

i.e.,

$$\frac{x - 1}{1} = \frac{y + 1/2}{1} = \frac{z + 1}{-1}$$

∴ d.r.'s of line are 1, 1, – 1.

The plane (i) is parallel to the given line

$$a + b - c = 0 \qquad \text{...(iii)}$$

Solving (ii) and (iii) by cross multiplication, we get

$$\frac{a}{1} = \frac{b}{2} = \frac{c}{3}$$

i.e., direction ratios of normal to the plane are 1, 2, 3.

$$\therefore \quad 1(x + 1) + 2(y - 2) + 3(z - 0) = 0$$

i.e., $x + 2y + 3z = 3$.

which is the required equation of plane.

Q. 29. Find the vector equation of the plane passing through the intersection of the planes $\vec{r}.(2\hat{i} - 7\hat{j} + 4\hat{k}) = 3$ and $\vec{r}.(3\hat{i} - 5\hat{j} + 4\hat{k}) + 11 = 0$ and passing through the point (– 2, 1, 3).

Sol. The equations of the given planes are

$$\vec{r}\cdot\vec{n_1} = d_1 \text{ and } \vec{r}\cdot\vec{n_2} = d_2$$

where $\vec{n_1} = 2\hat{i} - 7\hat{j} + 4\hat{k}$

★ are board exam questions from previous years

$$\vec{n_2} = 3\hat{i} - 5\hat{j} + 4\hat{k}$$

$$d_1 = 3 \text{ and } d_2 = -11$$

Therefore, the equation of the plane passing through the intersection of these two planes is

$$\vec{r}(\vec{n_1} + \lambda\vec{n_2}) = d_1 + \lambda d_2 \quad \text{...(i)}$$

$$\Rightarrow \vec{r}[\hat{i}(2+3\lambda) - \hat{j}(7+5\lambda) + \hat{k}(4+4\lambda)] = 3 - 11\lambda \quad \text{...(ii)}$$

If equation (ii) passes through (– 2, 1, 3), then the vector $(-2\hat{i} + \hat{j} + 3\hat{k})$ should satisfy it.

$$\therefore (-2\hat{i} + \hat{j} + 3\hat{k}).[\hat{i}(2+3\lambda) - \hat{j}(7+5\lambda) + \hat{k}(4+4\lambda)] = 3 - 11\lambda$$

$$\Rightarrow -2(2+3\lambda) + 1(-7-5\lambda) + 3(4+4\lambda) = 3 - 11\lambda$$

$$\Rightarrow -4 - 6\lambda - 7 - 5\lambda + 12 + 12\lambda = 3 - 11\lambda$$

$$\Rightarrow 1 + \lambda = 3 - 11\lambda$$

$$\Rightarrow 12\lambda = 2$$

$$\Rightarrow \lambda = \frac{1}{6}$$

Substituting the value of λ in equation (ii), we get

$$\vec{r}\cdot\left[\hat{i}\left(2+3\cdot\frac{1}{6}\right) - \hat{j}\left(7+5\cdot\frac{1}{6}\right) + \hat{k}\left(4+4\cdot\frac{1}{6}\right)\right] = 3 - 11\cdot\frac{1}{6}$$

$$\Rightarrow \vec{r}\cdot\left[\frac{5}{2}\hat{i} - \frac{47}{6}\hat{j} + \frac{14}{3}\hat{k}\right] = \frac{7}{6}$$

$$\Rightarrow \vec{r}.[15\hat{i} - 47\hat{j} + 28\hat{k}] = 7$$

Hence $\vec{r}.[15\hat{i} - 47\hat{j} + 28\hat{k}] = 7$ is the required equation of the plane.

Q. 30. Write the equation of the plane whose intercepts on the coordinate axes are – 4, 2 and 3.

Sol. We know that the equation of a plane whose intercepts on the coordinate axes are *a*, *b* and *c* respectively is

$$\frac{x}{a} + \frac{y}{b} + \frac{z}{c} = 1$$

Here, $a = -4, b = 2$ and $c = 3$

So, the equation of the required plane is

$$\frac{x}{-4} + \frac{y}{2} + \frac{z}{3} = 1$$

$$\Rightarrow \frac{-3x + 6y + 4z}{12} = 1$$

$$\Rightarrow -3x + 6y + 4z = 12$$

Therefore, $3x - 6y - 4z + 12 = 0$.

This is the equation of the required plane.

Q. 31. Find out that the following lines are coplanar or not

$$\vec{r} = 2\hat{i} + \hat{j} - 3\hat{k} + \lambda(\hat{i} + 2\hat{j} + 5\hat{k})$$

and $$\vec{r} = 3\hat{i} + 3\hat{j} - 7\hat{k} + \mu(3\hat{i} - 2\hat{j} + 5\hat{k})$$

Sol. Here

$$\vec{a} = 2\hat{i} + \hat{j} - 3\hat{k},$$

$$\vec{a'} = 3\hat{i} + 3\hat{j} - 7\hat{k}$$

$$\vec{b} = \hat{i} + 2\hat{j} + 5\hat{k}$$

$$\vec{b'} = 3\hat{i} - 2\hat{j} + 5\hat{k}$$

Lines are coplanar if

$$(\vec{a'} - \vec{a}).(\vec{b} \times \vec{b'}) = 0 \quad \text{...(A)}$$

Now, $$\vec{a'} - \vec{a} = \hat{i} + 2\hat{j} - 4\hat{k} \quad \text{...(i)}$$

and $$\vec{b} \times \vec{b'} = \begin{vmatrix} \hat{i} & \hat{j} & \hat{k} \\ 1 & 2 & 5 \\ 3 & -2 & 5 \end{vmatrix} \quad \text{...(ii)}$$

Using (i) and (ii) in (A), we get

$$(\hat{i} + 2\hat{j} - 4\hat{k}).(20\hat{i} + 10\hat{j} - 8\hat{k})$$

$$= 20 + 20 - 32$$

$$= 8 \neq 0$$

So, the lines are not coplanar.

Q. 32. Find the equation of the plane passing through the point (– 1, 2, 1) and perpendicular to the line joining the points (– 3, 1, 2) and (2, 3, 4). Find also the perpendicular distance from origin to the plane.

Sol. The required plane passes through the point (– 1, 2, 1) having position vector $\vec{a} = -\hat{i} + 2\hat{j} + \hat{k}$ and is perpendicular to the line joining the points $A(-3, 1, 2)$ and $B(2, 3, 4)$.

Therefore, a vector normal to the plane is given by

$$\vec{n} = \vec{AB}$$

= Position vector of *B* – Position vector of *A*

$$= (2\hat{i} + 3\hat{j} + 4\hat{k}) - (-3\hat{i} + \hat{j} + 2\hat{k})$$

$$= 5\hat{i} + 2\hat{j} + 2\hat{k}$$

We know that the vector equation of a plane passing through a point having position vector $\vec{a}$ and normal to vector $\vec{n}$ is given by

$$(\vec{r} - \vec{a}).\vec{n} = 0$$

$$\Rightarrow \vec{r}.\vec{n} = \vec{a}.\vec{n}$$

Therefore, the equation of the required plane is

$$\vec{r}.(5\hat{i}+2\hat{j}+2\hat{k}) = (-\hat{i}+2\hat{j}+\hat{k}).(5\hat{i}+2\hat{j}+2\hat{k})$$

$$\Rightarrow \vec{r}.(5\hat{i}+2\hat{j}+2\hat{k}) = (-1)(5)+(2)(2)+(1)(2) = -5+4+2=1$$

$$\Rightarrow \vec{r}.(5\hat{i}+2\hat{j}+2\hat{k}) = 1 \quad \text{...(i)}$$

To find the distance of this plane from the origin, we reduce its equation to normal form.

We have

$$\vec{n} = 5\hat{i}+2\hat{j}+2\hat{k}$$

$$\Rightarrow |\vec{n}| = \sqrt{(5)^2+(2)^2+(2)^2} = \sqrt{25+4+4}$$

$$\Rightarrow |\vec{n}| = \sqrt{33}$$

Dividing (i) throughout by

$$\Rightarrow |\vec{n}| = \sqrt{33}$$

We get $\vec{r}.\left(\frac{5}{\sqrt{33}}\hat{i}+\frac{2}{\sqrt{33}}\hat{j}+\frac{2}{\sqrt{33}}\hat{k}\right) = \frac{1}{\sqrt{33}}$

So, the perpendicular distance from the origin to the plane is $\frac{1}{\sqrt{33}}$ units.

Q. 33. Find the equation of plane (s) passing through the intersection of planes $x+3y+6=0$ and $3x-y-4z=0$ whose perpendicular distance from origin is unity.★

Sol. Let the required equation of plane is

$$(x+3y+6)+\lambda(3x-y-4z)=0 \quad \text{...(i)}$$

Above equation can be written as

$$x+3y+6+3\lambda x-\lambda y-4\lambda z=0$$

$$\Rightarrow x(1+3\lambda)+y(3-\lambda)-4\lambda z+6=0 \quad \text{...(ii)}$$

Which is the general form of equation of plane.

Now, given that perpendicular distance of plane (i) from origin, *i.e.*, (0, 0, 0) is unity, *i.e.*, one.

∵ Distance of point (x_1, y_1, z_1) from a plane $ax+by+cz+d=0$ is given as

$$d=\left|\frac{ax_1+by_1+cz_1+d}{\sqrt{a^2+b^2+c^2}}\right|$$

Here $a=1+3\lambda, b=3-\lambda, c=-4\lambda$

$(x_1, y_1, z_1)=(0,0,0)$

∴ We have

$$\left|\frac{(1+3\lambda)(0)+(3-\lambda)(0)-4\lambda(0)+6}{\sqrt{(1+3\lambda)^2+(3-\lambda)^2+(-4\lambda)^2}}\right|=1$$

$$\Rightarrow \left|\frac{6}{\sqrt{1+9\lambda^2+6\lambda+9+\lambda^2-6\lambda+16\lambda^2}}\right|=1$$

$$\Rightarrow \frac{6}{\sqrt{26\lambda^2+10}}=1$$

$$\Rightarrow 6=\sqrt{26\lambda^2+10}$$

Squaring both sides, we get

$$36=26\lambda^2+10$$

$$\Rightarrow 26\lambda^2=26$$

$$\Rightarrow \lambda^2=1$$

$$\Rightarrow \lambda=\pm 1$$

Now putting $\lambda = 1$ in equation (i), we get

$$x+3y+6+3x-y-4z=0$$

$$\Rightarrow 4x+2y-4z+6=0$$

$$\Rightarrow 2x+y-2z+3=0 \quad \text{...(iii)}$$

And putting $\lambda = -1$ in equation (i), we get

$$x+3y+6-3x+y+4z=0$$

$$\Rightarrow -2x+4y+4z+6=0$$

$$\Rightarrow x-2y-2z-3=0 \quad \text{...(iv)}$$

Equations (iii) and (iv) are the required equations of the plane.

Q. 34. Find the Cartesian equation of the plane passing through points $A(0, 0, 0)$ and $B(3, -1, 2)$, parallel to the line $\frac{x-4}{1}=\frac{y+3}{-4}=\frac{z+1}{7}$.★

Sol. Equation of plane passing through the point $A(0, 0, 0)$ is

$$a(x-0)+b(y-0)+c(z-0)=0$$

$$\Rightarrow ax+by+cz=0 \quad \text{...(i)}$$

Since, the plane (i) passes through the point $B(3, -1, 2)$.

Put $x=3, y=-1, z=2$ in equation (i), we get

$$3a-b+2c=0 \quad \text{...(ii)}$$

Also, the plane (i) is parallel to the line

$$\frac{x-4}{1}=\frac{y+3}{-4}=\frac{z+1}{7}$$

If the plane is parallel to the line, then normal to the plane is perpendicular to the line.

$$\therefore a_1a_2+b_1b_2+c_1c_2=0$$

$$\therefore a(1)+b(-4)+c(7)=0$$

$$\Rightarrow a-4b+7c=0 \quad \text{...(iii)}$$

Now, multiplying equation (iii) by 3 and subtracting it from equation (ii), we get

$$\begin{array}{r} 3a-b+2c=0 \\ 3a-12b+21c=0 \\ -\quad+\quad- \\ \hline 11b-19c=0 \end{array}$$

$$\Rightarrow b=\frac{19}{11}c$$

Putting $b=\frac{19c}{11}$ in equation (ii), we get

$$3a-\frac{19c}{11}+2c=0$$

★ are board exam questions from previous years

$$\Rightarrow \quad 3a + \frac{-19c + 22c}{11} = 0$$

$$\Rightarrow \quad 3a + \frac{3c}{11} = 0$$

$$\Rightarrow \quad 3a = -\frac{3c}{11}$$

$$\Rightarrow \quad a = -\frac{c}{11}$$

Now, putting $a = -\frac{c}{11}$ and $b = \frac{19}{11}c$ in equation (i), we get the required equation of plane as

$$-\frac{c}{11}x + \frac{19c}{11}y + cz = 0$$

$$\Rightarrow \quad -\frac{x}{11} + \frac{19y}{11} + z = 0$$

$$\Rightarrow \quad -x + 19y + 11z = 0$$

$$\Rightarrow \quad x - 19y - 11z = 0$$

Q. 35. Find the equation of plane that contains the point (1, – 1, 2) and is perpendicular to each of planes $2x + 3y - 2z = 5$ and $x + 2y - 3z = 8$. ★

Sol. Equation of plane passing through point $(1, -1, 2)$ is given by

$$a(x-1) + b(y+1) + c(z-2) = 0 \quad \ldots\text{(i)}$$

Now, given that plane (i) is perpendicular to planes

$$2x + 3y - 2z = 5 \quad \ldots\text{(ii)}$$

and $$x + 2y - 3z = 8 \quad \ldots\text{(iii)}$$

We know that when two planes $a_1x + b_1y + c_1z = d_1$ and $a_2x + b_2y + c_2z = d_2$ are perpendicular then

$$a_1a_2 + b_1b_2 + c_1c_2 = 0$$

$\therefore$ we get $\quad 2a + 3b - 2c = 0$ [from eq. (i) and (ii)]

and $\quad a + 2b - 3c = 0$ [from eq. (i) and (iii)]

or we may write

$$2a + 3b = 2c \quad \ldots\text{(iv)}$$

$$a + 2b = 3c \quad \ldots\text{(v)}$$

Multiplying equation (v) by 2 and subtracting it from equation (iv), we get

$$2a + 3b = 2c$$
$$2a + 4b = 6c$$
$$- \quad\quad -$$
$$-b = -4c$$

$$\Rightarrow \quad b = 4c$$

Putting $b = 4c$ in equation (v), we get

$$a + 8c = 3c$$

$$\Rightarrow \quad a = -5c$$

Now, putting $a = -5c$ and $b = 4c$ in equation (i), we get the required equation of plane as

$$-5c(x-1) + 4c(y+1) + c(z-2) = 0$$

$$\Rightarrow \quad -5(x-1) + 4(y+1) + (z-2) = 0$$

(divide both sides by 'c')

$$\Rightarrow \quad -5x + 5 + 4y + 4 + z - 2 = 0$$

$$\Rightarrow \quad 5x - 4y - z - 7 = 0.$$

Chapter 12. Linear Programming

Q. 1. Find graphically, the maximum value of $Z = 2x + 5y$, subject to constraints given below :★

$$2x + 4y \leq 8$$
$$3x + y \leq 6$$
$$x + y \leq 4$$
$$x \geq 0, y \geq 0$$

Sol. We first convert the inequalities into equations to obtain lines

$$2x + 4y = 8 \quad \ldots\text{(i)}$$
$$3x + y = 6 \quad \ldots\text{(ii)}$$
$$x + y = 4 \quad \ldots\text{(iii)}$$
$$x = 0$$
and $$y = 0$$

We need to maximize the objective function

$$Z = 2x + 5y$$

These lines are drawn and the feasible, region of the L.P.P. is the shaded region.

The point of intersection of equations (i) and (ii) is $B(1{\cdot}6, 1{\cdot}2)$.

The coordinates of the corner points of the feasible region are $O(0, 0)$, $A(0, 2)$, $B(1{\cdot}6, 1{\cdot}2)$ and $C(2, 0)$.

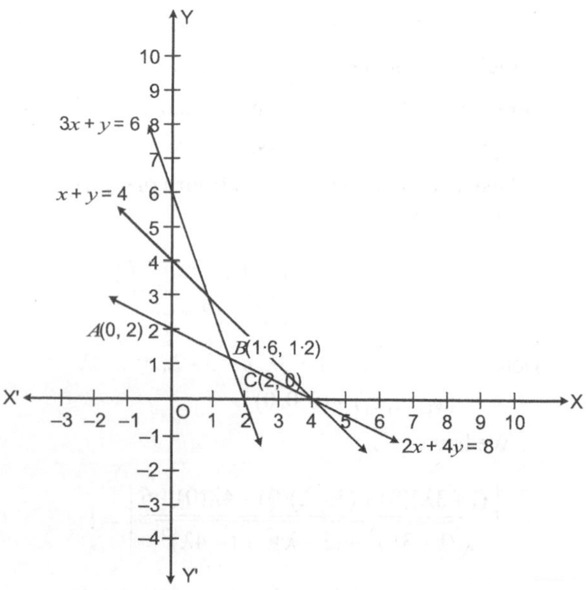

★ are board exam questions from previous years

The value of the objective function at these points are given in the following table :

Corner Points	Objective function $Z = 2x + 5y$
$O(0, 0)$	$2 \times 0 + 5 \times 0 = 0$
$A(0, 2)$	$2 \times 0 + 5 \times 2 = 10$ ←Maximum
$B(1{\cdot}6, 1{\cdot}2)$	$2 \times 1{\cdot}6 + 5 \times 1{\cdot}2 = 9{\cdot}2$
$C(2, 0)$	$2 \times 2 + 5 \times 0 = 4$

Out of these values of Z, the maximum value of Z is 10 which is attained at the point (0, 2). Thus the maximum value of Z is 10.

Q. 2. Solve the following L.P.P. graphically :★

Maximise $Z = 4x + y$

Subject to following constraints

$$x + y \leq 50,$$
$$3x + y \leq 90,$$
$$x \geq 10$$
$$x, y \geq 0$$

Sol. $x + y = 50$ $\quad$ $3x + y = 90$

x	0	50
y	50	0

x	0	30
y	90	0

$x = 10$ is a line which is parallel to Y-axis.

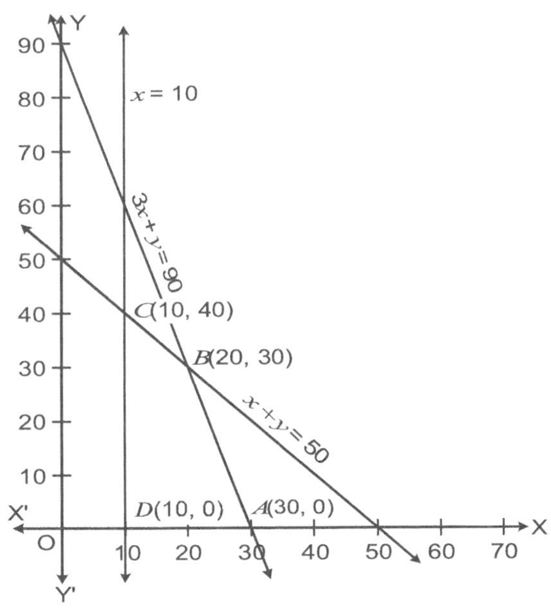

Corner point	$Z = 4x + y$
$A(30, 0)$	$Z = 120 + 0 = 120$ ← Maximum
$B(20, 30)$	$Z = 80 + 30 = 110$
$C(10, 40)$	$Z = 40 + 40 = 80$
$D(10, 0)$	$Z = 40 + 0 = 40$

∴ Maximum value is 120 at (30, 0).

Q. 3. Solve the following L.P.P. graphically :★

Maximise $Z = 20x + 10y$

Subject to the following constraints :

$$x + 2y \leq 28,$$
$$3x + y \leq 24,$$
$$x \geq 2$$
$$x, y \geq 0$$

Sol. $x + 2y = 28$ $\quad$ $3x + y = 24$

x	0	28
y	14	0

x	0	8
y	24	0

$x = 2$ is a line which is perpendicular to X-axis.

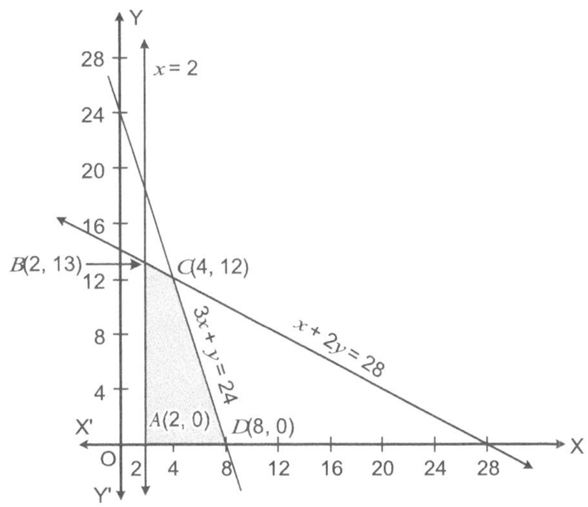

Corner point	$Z = 20x + 10y$
$A(2, 0)$	$Z = 40 + 0 = 40$
$B(2, 13)$	$Z = 40 + 130 = 170$
$C(4, 12)$	$Z = 80 + 120 = 200$ ← Maximum
$D(8, 0)$	$Z = 160 + 0 = 160$

∴ Maximum value is 200 at (4, 12).

Q. 4. Solve the following linear programming problem graphically :★

Maximise $Z = 7x + 10y$

Subject to the constraints

$$4x + 6y \leq 240$$
$$6x + 3y \leq 240$$
$$x \geq 10$$
$$x \geq 0, y \geq 0$$

Sol. We have, maximize

$$Z = 7x + 10y$$

Subject to the constraints :

$$4x + 6y \leq 240$$
$$6x + 3y \leq 240$$
$$x \geq 10$$
$$x \geq 0, y \geq 0$$

Converting the given inequalities into equations, we obtain the following equations,

$$4x + 6y = 240$$
$$6x + 3y = 240$$
$$x = 10$$

★ are board exam questions from previous years

Then, $4x+6y=240$

x	0	60
y	40	0

$6x+3y=240$

x	0	40
y	80	0

and $x=10$ is a line parallel to Y-axis.

Plotting these points on the graph, we get the shaded feasible region, *i.e.*, *DEFGD*.

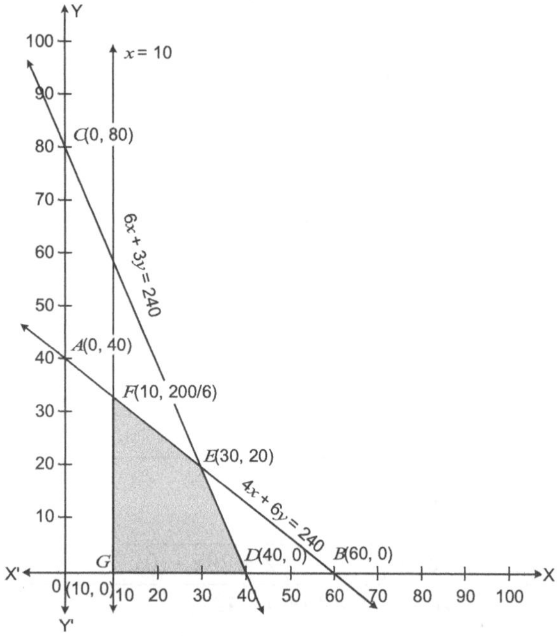

Corner point	$Z = 7x + 10y$
$D(40, 0)$	$7(40)+10(0)=280$
$E(30, 20)$	$7(30)+10(20)=410$
$F(10, 200/6)$	$7(10)+10(200/6)=403.33$
$G(10, 0)$	$7(10)+10(0)=70$

Clearly the maximum value of Z is 410 at (30, 20).

Q. 5. Solve the following linear programming problem graphically :★

Maximise $\mathbf{Z=34x+45y}$

under the following constraints

$$x+y\leq 300$$
$$2x+3y\leq 70$$
$$x\geq 0, y\geq 0$$

Sol. We have, maximise $Z=34x+45y$

Subject to the constraints

$$x+y\leq 300$$
$$2x+3y\leq 70$$
$$x\geq 0, y\geq 0$$

Converting the given inequalities into equations, we obtain the following equations

$$x+y=300$$
$$2x+3y=70$$

Then, $x+y=300$

x	0	300
y	300	0

and $2x+3y=70$

x	0	35
y	70/3	0

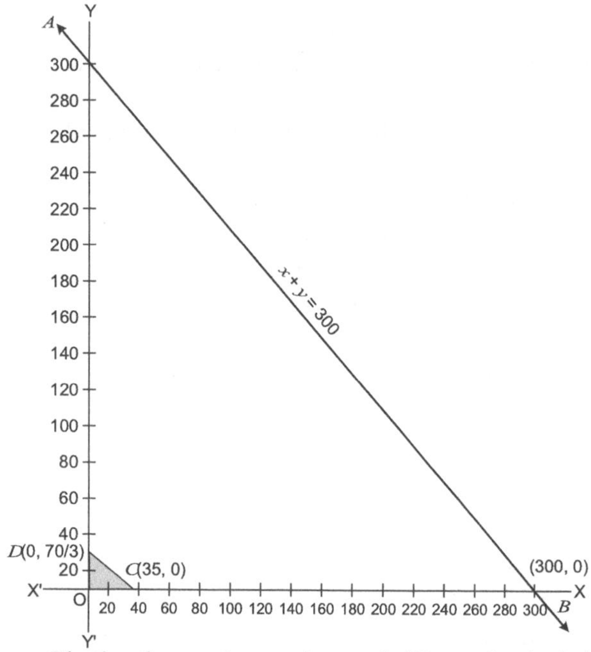

Plotting these points on the graph. We get the shaded feasible region *i.e.*, *OCDO*.

Corner point	$Z=34x+45y$
$O(0, 0)$	$34(0)+45(0)=0$
$C(35, 0)$	$34(35)+45(0)=1190$
$D(0, 70/3)$	$34(0)+45(70/3)=1050$

Clearly, the maximum value of Z is 1190 at (35, 0).

Q. 6. Maximize $\mathbf{Z=x+2y}$

subject to the constraints

$$x+2y\geq 100,\quad 2x-y\leq 0$$
$$2x+y\leq 200,\quad x, y\geq 0$$

Solve the above LPP graphically.★

Sol. Given, Maximise $Z=x+2y$

Subject to the constraints

$x+2y\geq 100$ $\quad$ $2x-y\leq 0$

$2x+y\leq 200$ $\quad$ $x, y\geq 0$

Converting the inequations into equations we obtain the lines $x+2y=100$, $2x-y=0$, $2x+y=200$.

Then, $x+2y=100$

x	0	100
y	50	0

$2x-y=0$

x	10	20
y	20	40

★ are board exam questions from previous years

and $\qquad 2x + y = 200$

x	0	100
y	200	0

Plotting these points on the graph, we get the shaded feasible region, *i.e.*, *ADECA*.

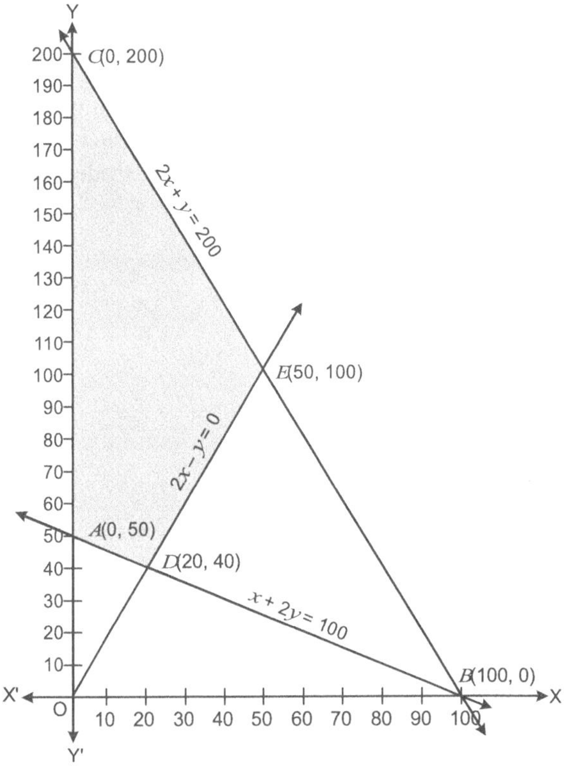

Corner points	$Z = x + 2y$
$A(0, 50)$	$(0)+2(50)=100$
$D(20, 40)$	$20+2(40)=100$
$E(50, 100)$	$50+2(100)=250$
$C(0, 200)$	$0+2(200)=400$ maximum

Clearly, the maximum value of Z is 400 at (0, 200).

Q. 7. There are two types of fertilisers '*A*' and '*B*'. '*A*' consists of 12% nitrogen and 5% phosphoric acid whereas '*B*' consists of 4% nitrogen and 5% phosphoric acid. After testing the soil conditions, farmer finds that he needs at least 12 kg of nitrogen and 12 kg of phosphoric acid for his crops. If '*A*' costs ₹ 10 per kg and '*B*' cost ₹ 8 per kg, then graphically determine how much of each type of fertiliser should be used so that nutrient requirements are met at a minimum cost.★

Sol. Let the quantity of fertiliser *A* and *B* be *x* and *y* respectively.

To minimize : $\qquad Z = ₹\,(10x + 8y)$

Subject to the constraints :

$$\frac{12}{100}x + \frac{4}{100}y \geq 12$$

or $\qquad 12x + 4y \geq 1200$

and $\qquad \dfrac{5}{100}x + \dfrac{5}{100}y \geq 12$

or $\qquad 5x + 5y \geq 1200$

and $\qquad x \geq 0, y \geq 0$

i.e., $\qquad 3x + y \geq 300,$

$\qquad x + y \geq 240,$

$\qquad x \geq 0, y \geq 0$

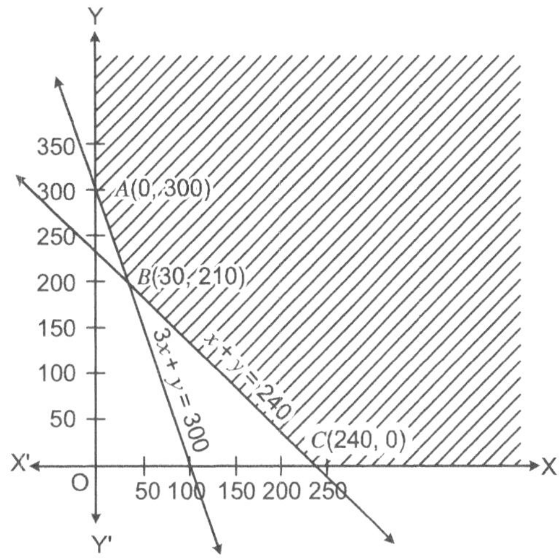

Corner point	$Z = 10x + 8y$
$A(0, 300)$	$Z = 10\times0+8\times300 \;= ₹\,2400$
$B(30, 210)$	$Z = 10\times30+8\times210 = ₹\,1980$
$C(240, 0)$	$Z = 10\times240+8\times0 \;= ₹\,2400$

Clearly, Z is minimum for $x = 30$ and $y = 210$ and the minimum value of Z is ₹ 1980.

Hence, the quantity of fertiliser *A* needed is 30 kg and of fertiliser *B* is 210 kg.

Q. 8. Solve the following L.P.P. graphically :★

Minimise $\qquad Z = 5x + 10y$

Subject to $\qquad x + 2y \leq 120$

Constraints $\qquad x + y \geq 60$

$\qquad x - 2y \geq 0$

and $\qquad x, y \geq 0$

Sol. Minimize $Z = 5x + 10y$

$x + 2y = 120 \qquad x + y = 60 \qquad x - 2y = 0$

x	0	120
y	60	0

x	0	60
y	60	0

x	20	60
y	10	30

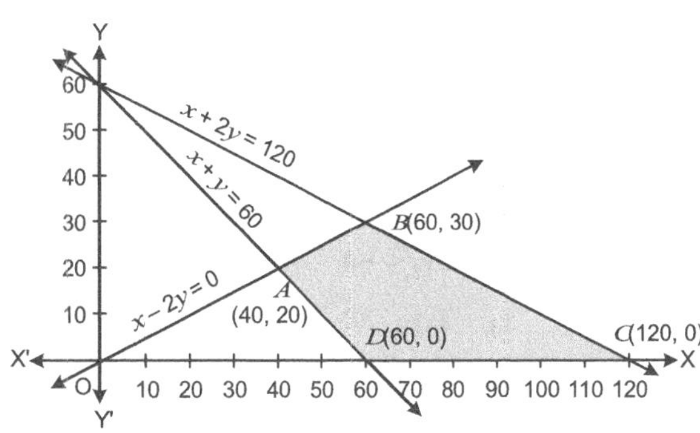

★ are board exam questions from previous years

The shaded region $ABCDA$ represented by the given constraints is the feasible region. Corner points of the common shaded region are $A(40, 20)$, $B(60, 30)$, $C(120, 0)$ and $D(60, 0)$. Value of Z at each corner point is given as :

Corner Point	$Z = 5x + 10y$
$A(40, 20)$	$200 + 200 = 400$
$B(60, 30)$	$300 + 300 = 600$
$C(120, 0)$	$600 + 0 = 600$
$D(60, 0)$	$300 + 0 = 300 \leftarrow$ Minimum

Hence, minimum value of Z is 300 at (60, 0).

Q. 9. Solve the following LPP graphically. Maximize $Z = 15x + 20y$ subject to the constraints :

$$2x + 3y \geq 60$$
$$x + 4y \leq 20$$
$$x \geq 0,$$
$$y \geq 0$$

Sol. We convert the inequations into equations and draw the lines as follows :

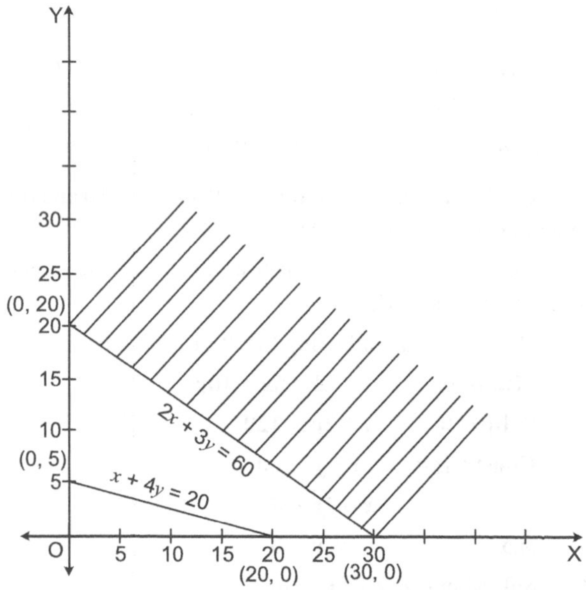

In the figure, the shaded portions in the graph shows the feasible regions for each constraint but there is no feasible region which is common to all the constraints. Thus, irrespective of the objective function to be maximized, the conditions of the constraints are such that there is no solution to the problem.

Q. 10. Make a graphical representation of the set of constraints in the following LPP :

Maximize $z = 4x_1 + 3x_2$

Subject to : $x_1 + x_2 \leq 50$

$$x_1 + 2x_2 \leq 80$$
$$2x_1 + x_2 \geq 20$$
$$x_1, x_2 \geq 0$$

Find the extreme point of the feasible region. Hence, find the maximum value of the objective function.

Sol. Considering the inequations as equations, we draw the graph. The three constraints of the problem are indicated by straight lines (i), (ii) and (iii) respectively.

$$x_1 + x_2 = 50 \quad ...(i)$$
$$x_1 + 2x_2 = 80 \quad ...(ii)$$
$$2x_1 + x_2 = 20 \quad ...(iii)$$

These three constraints together with the non-negativety constraints $x_1, x_2 \geq 0$, of the decision variable enclose the feasible region as shown by the shaded region $ABCDEA$ of the graph.

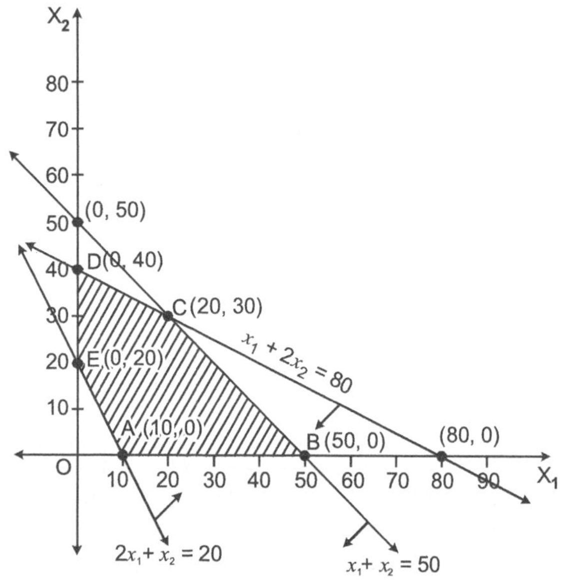

The extreme points of the region, as seen from the graph are : $A(10, 0)$, $B(50, 0)$, $C(20, 30)$, $D(0, 40)$, $E(0, 20)$ and the values of the objective function at these points are : 40, 200, 170, 120 and 60 respectively.

Maximum value of the objective function is 200 at (50, 0).

Hence, the optimal solution is $x_1 = 50, x_2 = 0$.

Q. 11. A small firm manufactures necklaces and bracelets. The total number of necklaces and bracelets that it can handle per day is at most 24. If takes one hour to make a bracelet and half an hour to make a necklace. The maximum number of hours available per day is 16. If the profit on a necklace is ₹ 100 and that on a bracelet is ₹ 300. Formulate on L.P.P. for finding how many of each should be produced daily to maximize the profit ? It is being given that at least one of each must be produced.★

Sol. Let the manufacturer produces x pieces of necklaces and y pieces of bracelets.

Since total number of necklaces and bracelets that can be handle per day are 24 so $x + y \leq 24$ ...(i)

To make bracelet, one needs one hour and half an hour is need to make necklace and maximum time available is 16 hours

so, $$\frac{1}{2}x + y \leq 16 \quad ...(ii)$$

Now let Z be the profit and we have to maximize it so our LPP will be

$$\text{Maximize } Z = 100x + 300y$$

Subject to constraints

$$x + y \leq 24$$

$$\frac{1}{2}x + y \leq 16$$

and $x \geq 1; y \geq 1.$

Q. 12. Two tailors A and B earn ₹ 300 and ₹ 400 per day respectively. A can stitch 6 shirts and 4 pairs of trousers while B can stitch 10 shirts and 4 pairs of trousers per day. To find how many days should each of them work and if it is desired to produce at least 60 shorts and 32 pairs of trousers at a minimum labour cost, formulate this as an LPP.★

Sol. Let x and y be the number of days for which the tailors A and B work respectively.

Total cost per day = ₹ $(300x + 400y)$

Let Z denote the total cost in rupees, then,

$$Z = 300x + 400y$$

Since in one day 6 shirts are stitched by tailor A and 10 shirts are stitched by tailor B and it is desired to produced atleast 60 shirts.

$$\therefore \quad 6x + 10y \geq 60$$

It is given that 4 pairs of trousers are stitched by each tailor A and B per day to produce at least 32 pairs of trousers.

$$\therefore \quad 4x + 4y \geq 32$$

Finally, the no. of shirts and pair of trousers cannot be negative.

$$\therefore \quad x \geq 0, y \geq 0$$

Thus, the mathematical formulation of the given LPP is as follows :

Minimize $Z = 300x + 400y$

Subject to : $6x + 10y \geq 60$

$4x + 4y \geq 32$

$x \geq 0, y \geq 0.$

Chapter 13. Probability

Q. 1. Ram can find a solution of 60% of the problems given in mathematics examination paper and Shyam can find out 80% of it. What is the probability that atleast one of them will solve the problem that is selected at random?

Sol. Let us consider the events :

A = Ram finds the solution

B = Shyam finds the solution

Since, A and B are independent events.

So $P(A) = \frac{60}{100} = \frac{3}{5}$ and $P(B) = \frac{80}{100} = \frac{4}{5}$

$\therefore$ Required Probability

$$= P(A \cup B)$$

$$= P(A) + P(B) - P(A \cap B)$$

$$= P(A) + P(B) - P(A)\, P(B)$$

$$[\because P(A \cap B) = P(A)\, P(B)]$$

$$= \frac{3}{5} + \frac{4}{5} - \frac{3}{5} \times \frac{4}{5}$$

$$= \frac{7}{5} - \frac{12}{25} = \frac{23}{25}$$

Q. 2. A box has 2 apples out of which one is red and other is green. What is the probability that both of them are bad, if it is known that (i) the red apple is bad, (ii) the green apple is bad.

Sol. Let B_i and G_i denote the apples to be bad and good respectively where $i \in \{R, G\}$. R stands for red apple and G for green apple.

Then sample space is $S = \{G_R G_G, B_R G_G, B_R B_G, G_R B_G\}$

Let us consider the events

A = both apples are bad

B = red apple is bad

C = green apple is bad

Then, $A = \{B_R B_G\}, B = \{B_R G_G, B_R B_G\}$ and $C = \{B_R B_G, G_R B_G\}$

and $A \cap B = \{B_R B_G\}, A \cap C = \{B_R B_G\}$

$\therefore$ (i) Required probability

$$= P(A/B)$$

$$= \frac{P(A \cap B)}{P(B)} = \frac{1/4}{2/4} = \frac{1}{2}$$

(ii) Required probability

$$= P(A/C)$$

$$= \frac{P(A \cap C)}{P(C)} = \frac{1/4}{2/4} = \frac{1}{2}$$

Q. 3. A bag contains 5 green and 12 blue marbles. A second bag contains 7 green and 9 blue marbles. One marble is picked at random from first bag and mixed up with the marbles in the second bag. Then, a marble is randomly picked from it. Find the probability that the picked marble is green.

Sol. Let us consider the events :

X = Marble picked from first bag is green

Y = Marble picked from first bag is blue

and A = Marble picked from second bag is green

Now, $P(X) = \frac{5}{17}, \; P(Y) = \frac{12}{17}$

Also, probability of picking up a green marble from second bag if X has already occured

$$P(A/X) = \frac{8}{17}$$

Similarly, $P(A/Y) = \frac{7}{17}$

★ are board exam questions from previous years

∴ By the law of total probability, we get

$$P(A) = P(X)P(A/X) + P(Y)P(A/Y)$$

$$= \frac{5}{17} \times \frac{8}{17} + \frac{12}{17} \times \frac{7}{17}$$

$$= \frac{40}{289} + \frac{84}{289} = \frac{124}{289}$$

Q. 4. Given that the events A and B are such that $P(A) = \frac{1}{2}$, $P(A \cup B) = \frac{3}{5}$ and $P(B) = p$. Find p if they are:

(i) mutually exclusive events, (ii) independent events. **[NCERT]**

Sol. Given, $P(A) = \frac{1}{2}$ and $P(B) = p$, $P(A \cup B) = \frac{3}{5}$

(i) When A & B are mutually exclusive,

Then, $A \cap B = \phi$

∴ $P(A \cap B) = 0$

It is know that

$$P(A \cup B) = P(A) + P(B) - P(A \cap B)$$

$$\frac{3}{5} = \frac{1}{2} + p - 0$$

$$\Rightarrow \quad p = \frac{3}{5} - \frac{1}{2} = \frac{6-5}{10} = \frac{1}{10}$$

(ii) When A and B are independent,

Then, $P(A \cap B) = P(A) \cdot P(B)$

i.e., $P(A \cap B) = \frac{1}{2} \times p$

$$P(A \cup B) = P(A) + P(B) - P(A \cap B)$$

$$\frac{3}{5} = \frac{1}{2} + p - \frac{p}{2}$$

$$\Rightarrow \quad \frac{p}{2} = \frac{1}{10}$$

$$\Rightarrow \quad p = \frac{1}{5}$$

Q. 5. Two students are trying to solve a problem and the probability of solving the problem independently by the first and second student are $\frac{1}{4}$ and $\frac{1}{5}$ respectively. If both try to solve the problem independently, find the probability that (i) the problem is solved, (ii) exactly one of them solves the problem.

Sol. Let A and B be the events that problem is solved by first and second student respectively.

Probability of solving the problem by A

$$P(A) = \frac{1}{4}$$

Probability of solving the problem by B

$$P(B) = \frac{1}{5}$$

Since the problem is being solved independently

$$\therefore \quad P(A \cap B) = P(A) \cdot P(B) = \frac{1}{4} \times \frac{1}{5} = \frac{1}{20}$$

$$P(\overline{A}) = 1 - P(A) = 1 - \frac{1}{4} = \frac{3}{4}$$

$$P(\overline{B}) = 1 - P(B) = 1 - \frac{1}{5} = \frac{4}{5}$$

(i) Probability that the problem is solved $= P(A \cup B)$

$$= P(A) + P(B) - P(A \cap B)$$

$$= \frac{1}{4} + \frac{1}{5} - \frac{1}{20}$$

$$= \frac{5+4}{20} - \frac{1}{20}$$

$$= \frac{8}{20}$$

(ii) Probability that exactly one of them solves the problem is given by

$$P(A).P(\overline{B}) + P(B).P(\overline{A})$$

$$= \frac{1}{4} \times \frac{4}{5} + \frac{1}{5} \times \frac{3}{4} = \frac{4+3}{20} = \frac{7}{20}$$

Q. 6. Evaluate $P(A \cup B)$, if $2P(A) = P(B) = \frac{5}{13}$ and $P(A/B) = \frac{2}{5}$. **[NCERT]**

Sol. Given, $2P(A) = \frac{5}{13}$

$$\Rightarrow \quad P(A) = \frac{5}{26} \text{ and } P(B) = \frac{5}{13}$$

Also, $P(A/B) = \frac{2}{5}$

We know that

$$P(A/B) = \frac{P(A \cap B)}{P(B)} = \frac{2}{5}$$

$$\Rightarrow \quad P(A \cap B) = \frac{2}{5} \times P(B) = \frac{2}{5} \times \frac{5}{13}$$

$$\Rightarrow \quad P(A \cap B) = \frac{2}{13}$$

Also we know that

$$P(A \cup B) = P(A) + P(B) - P(A \cap B)$$

$$= \frac{5}{26} + \frac{5}{13} - \frac{2}{13} = \frac{5}{26} + \frac{3}{13}$$

$$= \frac{11}{26}$$

Q. 7. An instructor has a question bank consisting of 300 easy True/False questions, 200 difficult True/False questions, 500 easy MCQ's and 400 difficult MCQ's. If a

question is selected at random from the question bank, what is the probability that it will be an easy question given that it is a MCQ ? [NCERT]

Sol. The given data can be tabulated as :

	True/False	MCQ's	Total
Easy	300	500	800
Difficult	200	400	600
Total	500	900	1400

Let us denote E = easy questions

M = MCQ's

D = Difficult questions

T = True/False questions

Total number of questions = 1400

Total number of MCQ's = 900

$$\therefore \quad P(\text{selecting a MCQ}) = P(M) = \frac{900}{1400} = \frac{9}{14}$$

$$P(\text{selecting easy MCQ's}) = \frac{500}{1400} = \frac{5}{14}$$

Now, $P(E/M)$ represents the probability that a randomly selected question will be an easy question, given that it is a MCQ.

$$\therefore \quad P(E/M) = \frac{P(E \cap M)}{P(M)} = \frac{5/14}{9/14} = \frac{5}{9}$$

Q. 8. A black and a red die are rolled together. Find the conditional probability of obtaining the sum 8, given that the red die resulted in a number less than 4.★

Sol. Let A be event that the sum of observations is 8.

$$\therefore \quad A = \{(2, 6), (3, 5), (5, 3), (4, 4), (6, 2)\}$$

$$\Rightarrow \quad n(A) = 5$$

$$\Rightarrow \quad P(A) = \frac{5}{36}$$

Let B be event that the observation on red die is less than 4.

$\therefore B = \{(1, 1), (2, 1), (3, 1), (4, 1), (5, 1), (6, 1), (1, 2), (2, 2), (3, 2), (4, 2), (5, 2), (6, 2), (1, 3), (2, 3), (3, 3), (4, 3), (5, 3), (6, 3)\}$

$$\Rightarrow \quad n(B) = 18$$

$$\Rightarrow \quad P(B) = \frac{18}{36} = \frac{1}{2}$$

Clearly $A \cap B = \{(5, 3), (6, 2)\}$

$$\therefore \quad n(A \cap B) = 2$$

$$\therefore \quad P(A \cap B) = \frac{2}{36} = \frac{1}{18}$$

$$P(A/B) = \frac{P(A \cap B)}{P(B)} = \frac{\frac{1}{18}}{\frac{1}{2}}$$

$$= \frac{2}{18} = \frac{1}{9}$$

Q. 9. An electronic assembly consists of two sub-systems, say A and B. From previous testing procedures, the following probabilities are assumed to be known :

$$P(A \text{ fails}) = 0{\cdot}2$$

$$P(B \text{ fails alone}) = 0{\cdot}15$$

$$P(A \text{ and } B \text{ fail}) = 0{\cdot}15$$

Evaluate the following probabilities :

(i) $P(A)$ fails given that B has failed, (ii) $P(A)$ fails alone. [NCERT]

Sol. Let the event in which A fails and B fails be denoted by E_A and E_B respectively.

Then, $P(E_A) = 0{\cdot}2$

$$P(E_A \text{ and } E_B) = 0{\cdot}15$$

$$P(B \text{ fails alone}) = P(E_B) - P(E_A \text{ and } E_B)$$

$$\therefore \quad 0{\cdot}15 = P(E_B) - 0{\cdot}15$$

$$\therefore \quad P(E_B) = 0{\cdot}15 + 0{\cdot}15 = 0{\cdot}3.$$

(i)
$$P(E_A/E_B) = \frac{P(E_A \cap E_B)}{P(E_B)}$$

$$= \frac{0{\cdot}15}{0{\cdot}3} = 0{\cdot}5.$$

(ii)
$$P(A \text{ fails alone}) = P(E_A) - P(E_A \text{ and } E_B)$$

$$= 0{\cdot}2 - 0{\cdot}15$$

$$= 0{\cdot}05.$$

Q. 10. A coin is tossed thrice. Let the event E be 'the first throw result in a head' and the event F be 'the last throw results in a tail'. Find whether the events E & F are independent.

Sol. When a coin is tossed three times, the sample space is given by

$$S = \{HHH, HHT, HTH, THH, HTT, TTH, THT, TTT\}$$

Now, E = event that the first throw results in a head

$$\therefore \quad E = \{HHH, HHT, HTH, HTT\}$$

and F = event that the last throw results in a tail

$$\therefore \quad F = \{HHT, THT, HTT, TTT\}$$

So, $E \cap F = \{HHT, HTT\}$

Clearly, $n(E) = 4$, $n(F) = 4$, $n(E \cap F) = 2$ and $n(S) = 8$

$$\therefore \quad P(E) = \frac{n(E)}{n(S)} = \frac{4}{8} = \frac{1}{2}$$

$$P(F) = \frac{n(F)}{n(S)} = \frac{4}{8} = \frac{1}{2}$$

★ are board exam questions from previous years

and $P(E\cap F)=\dfrac{n(E\cap F)}{n(S)}=\dfrac{2}{8}=\dfrac{1}{4}$

Now, $P(E\cap F)=P(E)\times P(F)=\dfrac{1}{2}\times\dfrac{1}{2}=\dfrac{1}{4}$

Hence 'E' and 'F' are independent events.

Q. 11. Prove that if E and F are independent events, then the events E and F are also independent.★

Sol. Since E and F are independent events :

$\therefore\quad P(E\cap F)=P(E)\,.\,P(F)\quad\ldots$(i)

Now $P(E\cap F')=P(E)-P(E\cap F)$

$=P(E)-P(E)\,P(F)$ [From (i)]

$=P(E)\,[1-P(F)]$

$=P(E)\,P(F')$

$\therefore$ E and F' are also independent.

Q. 12. A die, whose faces are marked 1, 2, 3 in red and 4, 5, 6 in green is tossed. Let A be the event "number obtained is even" and B be the event "number obtained is red". Find if A and B are independent events.★

Sol. Since A is the event of number obtained is even.

$\therefore\quad A=\{2,4,6\}$

and B is the event of number obtained is red.

$\therefore\quad B=\{1,2,3\}$

$\therefore\quad A\cap B=\{2\}$

So, $P(A)=\dfrac{3}{6}=\dfrac{1}{2};$

$P(B)=\dfrac{3}{6}=\dfrac{1}{2};$

$P(A\cap B)=\dfrac{1}{6}$

Now, $P(A\cap B)\neq P(A)\cdot P(B)$

$\dfrac{1}{6}\neq\dfrac{1}{4}$

Hence, the events A and B are not independent events.

Q. 13. A problem in mathematics is given to three students whose chances of solving it correctly are $\dfrac{1}{2},\dfrac{1}{3}$ and $\dfrac{1}{4}$ respectively. What is the probability that only one of them solves it correctly ?

Sol. Let A, B, C be the given students and let E_1, E_2, E_3 be the events that the problem is solved by A, B, C respectively. Then, $\bar{E}_1$, $\bar{E}_2$, $\bar{E}_3$ are the events that the given problem is not solved by A, B, C respectively. Then,

$$P(E_1)=\frac{1}{2},\ P(E_2)=\frac{1}{3}\text{ and }P(E_3)=\frac{1}{4}$$

$$P(\bar{E}_1)=\left(1-\frac{1}{2}\right)=\frac{1}{2};\ P(\bar{E}_2)=\left(1-\frac{1}{3}\right)=\frac{2}{3}$$

and $P(\bar{E}_3)=\left(1-\dfrac{1}{4}\right)=\dfrac{3}{4}$

P(exactly one of them solves the problem)

$=P[(E_1\cap\bar{E}_2\cap\bar{E}_3)\text{ or }(\bar{E}_1\cap E_2\cap\bar{E}_3)\text{ or }(\bar{E}_1\cap\bar{E}_2\cap E_3)]$

$=P(E_1\cap\bar{E}_2\cap\bar{E}_3)+P(\bar{E}_1\cap E_2\cap\bar{E}_3)$

$+\,P(\bar{E}_1\cap\bar{E}_2\cap E_3)$

Now since $(E_1,\bar{E}_2,\bar{E}_3)$; $(\bar{E}_1,E_2,\bar{E}_3)$; $(\bar{E}_1,\bar{E}_2,E_3)$ are independent events respectively.

$\therefore$ P(only one of A, B, C solve the problem correctly)

$=[P(E_1)\times P(\bar{E}_2)\times P(\bar{E}_3)]$

$+\,[P(\bar{E}_1)\times P(E_2)\times P(\bar{E}_3)]$

$+\,[P(\bar{E}_1)\times P(\bar{E}_2)\times P(E_3)]$

$$=\left(\frac{1}{2}\times\frac{2}{3}\times\frac{3}{4}\right)+\left(\frac{1}{2}\times\frac{1}{3}\times\frac{3}{4}\right)+\left(\frac{1}{2}\times\frac{2}{3}\times\frac{1}{4}\right)$$

$$=\frac{6+3+2}{24}=\frac{11}{24}$$

Hence, the required probability is $\dfrac{11}{24}$.

Q. 14. Three groups of children contain 3 girls and 1 boy; 2 girls and 2 boy and 1 girl and 3 boys. One child is selected at random from each group. Find the chance that the three children selected comprise 1 girl and 2 boys.

Sol. Let G_1, G_2, G_3 be the events of selecting a girl from the first, second and third group respectively. Let B_1, B_2, B_3 be the events of selecting a boy from the first, second and third group respectively.

Then $P(G_1)=\dfrac{3}{4};\ P(G_2)=\dfrac{2}{4}=\dfrac{1}{2};\ P(G_3)=\dfrac{1}{4}$

$P(B_1)=\dfrac{1}{4};\ P(B_2)=\dfrac{2}{4}=\dfrac{1}{2};\ P(B_3)=\dfrac{3}{4}$

Now, P(selecting 1 girl and 2 boys)

$=P[(G_1B_2B_3)\text{ or }(B_1G_2B_3)\text{ or }(B_1B_2G_3)]$

$=P(G_1B_2B_3)+P(B_1G_2B_3)+P(B_1B_2G_3)$

$=[P(G_1)\times P(B_2)\times P(B_3)]+[P(B_1)\times P(G_2)\times P(B_3)]$

$+\,[P(B_1)\times P(B_2)\times P(G_3)]$

[$\because$ G_1, G_2, G_3 and B_1, B_2, B_3 are independent events respectively]

$$=\left(\frac{3}{4}\times\frac{1}{2}\times\frac{3}{4}\right)+\left(\frac{1}{4}\times\frac{1}{2}\times\frac{3}{4}\right)+\left(\frac{1}{4}\times\frac{1}{2}\times\frac{1}{4}\right)$$

$$=\frac{9}{32}+\frac{3}{32}+\frac{1}{32}=\frac{13}{32}$$

Hence, the chances of selecting 1 girl and 2 boys are $\dfrac{13}{32}$.

★ are board exam questions from previous years

Q. 15. Five gems are drawn from a pack which contains 7 blue gems and 4 black gems. What is the probability that :

(i) all will be blue ? (ii) 3 will be blue and 2 black

Sol. Total number of gems in the pack = 7 + 4 = 11

Hence the number of elementary events = $^{11}C_5$

(or number of ways in which 5 gems can be drawn out of 11)

(i) Total number of blue gems = 7

5 gems can be drawn out of 7 in $^7C_5$ ways.

$$\therefore \quad \text{Required probability} = \frac{^7C_5}{^{11}C_5}$$

$$= \frac{7!}{2!5!} \times \frac{5!6!}{11!}$$

$$= \frac{6 \times 5 \times 4 \times 3}{11 \times 10 \times 9 \times 8}$$

$$= \frac{1}{22}$$

(ii) Three blue gems out of 7 blue and 2 black gems out of 4 black gems can be drawn in $^7C_3 \times {^4C_2}$ ways.

$$\therefore \quad \text{Required probability} = \frac{^7C_3 \times {^4C_2}}{^{11}C_5}$$

$$= \frac{7!}{4!3!} \times \frac{4!}{2!2!} \times \frac{5!6!}{11!}$$

$$= \frac{5 \times 4 \times 6 \times 5 \times 4 \times 3}{11 \times 10 \times 9 \times 8 \times 2 \times 1}$$

$$= \frac{5}{11}$$

Q. 16. A bag contains 4 balls. Two balls are drawn at random (without replacement) and are found to be white. What is the probability that all balls in the bag are white ?*

Sol. There are three possible cases. Thy are :

A_1 : The bag has 2 white balls and 2 balls and 2 balls of other colour

A_2 : The bag has 3 white balls and 1 ball of other colour

A_3 : The bag has 4 white balls

Now, $P(A_1) = P(A_2) = P(A_2) = P(A_3) = \frac{1}{3}$

Let E be the event of selecting 2 white balls.

$$\therefore \quad P\left(\frac{E}{A_1}\right) = \frac{^2C_2}{^4C_2} = \frac{1}{6}$$

$$P\left(\frac{E}{A_2}\right) = \frac{^3C_2}{^4C_2} = \frac{1}{2}$$

$$P\left(\frac{E}{A_3}\right) = \frac{4C_2}{^4C_2} = 1$$

Applying Bayes theorem,

$$\therefore \quad P\left(\frac{A_3}{E}\right) = \frac{P(A_3)P\left(\frac{P}{A_3}\right)}{P(A_1)\,P\left(\frac{E}{A_1}\right) + P(A_2)P\left(\frac{E}{A_2}\right) + P(A_3)P\left(\frac{E}{A_3}\right)}$$

$$= \frac{\frac{1}{3} \times 1}{\frac{1}{3}\left(\frac{1}{6} + \frac{1}{2} + 1\right)} = \frac{3}{5}$$

Q. 17. In a multiple choice test with five possible answers, the probability of a student guessing the correct answer by actually knowing it is *p* and the probability of his guessing the correct answer without knowing it is $\frac{1}{5}$. If the student has answered the questions correctly, then what is the probability that he actually knew the answer ?

Sol. Let A denote the event that the student has answered the questions correctly.

Let B denote the event that the student actually knows the answer, and $\bar{B}$, the event that he does not know the answer.

We have to find $P(B/A)$

Now, $\quad P(A/B) = 1,\ P(A/\bar{B}) = \frac{1}{5},$

$$P(B) = p,\ P(\bar{B}) = 1 - p$$

So Bayes' theorem

$$P\left(\frac{B}{A}\right) = \frac{P(A/B)P(B)}{P(A/B)P(B) + P(A/\bar{B})P(\bar{B})}$$

$$= \frac{1.p}{1.p + \frac{1}{5}(1-p)}$$

$$= \frac{p}{p + \frac{1-p}{5}}$$

$$= \frac{5p}{4p+1}$$

Q. 18. The contents of bags I, II and III are as follows :

Bag I : 1 white, 2 black and 3 red balls

Bag II : 2 white, 1 black and 1 red ball

Bag III : 4 white, 5 black and 3 red balls

One bag is chosen at random and three balls are drawn from it. Two happen to be the red and one white. What is the probability that they came from bag III ?

★ are board exam questions from previous years

Sol. Let B_i : Bag i is chosen, $i = 1, 2, 3$.

Therefore, the probability of choosing bag B_i is $P(B_i)$

$$= \frac{1}{3}, \forall_i = 1, 2, 3$$

Let A : Two of the balls drawn are red and one is white

Then clearly,

$$P(A/B_1) = \frac{{}^1C_1 \times {}^3C_2}{{}^6C_3}$$

$$= \frac{3\,(3 \times 2 \times 1)}{6 \times 5 \times 4} = \frac{3}{20}$$

$$P(A/B_2) = \frac{{}^2C_1 \times 0}{{}^4C_3}$$

[∵ Two red balls cannot be chosen from the one available]

$$= 0$$

$$P(A/B_3) = \frac{{}^4C_1 \times {}^3C_2}{{}^{12}C_3}$$

$$= \frac{4 \times 3\,(3 \times 2 \times 1)}{12 \times 11 \times 10} = \frac{3}{55}$$

By Bayes' theorem,

$$P(B_3/A) = \frac{P(A/B_3)P(B_3)}{\sum_{i=1}^{3} P(A/B_i)\,P(B_i)}$$

$$= \frac{\frac{3}{55} \times \frac{1}{3}}{\frac{3}{20} \times \frac{1}{3} + 0 + \frac{3}{55} \times \frac{1}{3}}$$

$$= \frac{\frac{3}{55}}{\frac{(33+12)}{220}} = \frac{12}{45} = \frac{4}{15}$$

Therefore, the required probability is $\frac{4}{15}$.

Q. 19. A man is known to speak the truth 3 out of 4 times. He throws a die and reports that it is a six. Find the probability that it is actually a six.

Sol. Let E be the event that the man reports that six occurs in the throwing of the die. Let S_1 be the event that six occurs and S_2 be the event that six does not occur.

Then, $P(S_1)$ = probability that six occurs $= \frac{1}{6}$

$P(S_2)$ = probability that six does not occur

$$= 1 - \frac{1}{6} = \frac{5}{6}$$

$P(E/S_1)$ = probability that the man reports that six occurs when six has actually occurred = probability that the man speaks the truth $= \frac{3}{4}$

$P(E/S_2)$ = Probability that the man reports that six occurs when six has not actually occurred = probability that the man does not speak the truth

$$= 1 - \frac{3}{4} = \frac{1}{4}$$

Hence, by Bayes' theorem,

$P(S_1/E)$ = probability that the report of the man that six has occurred is actually a six

$$= \frac{P(S_1)\,P(E/S_1)}{P(S_1)\,P(E/S_1) + P(S_2)\,P(E/S_2)}$$

$$= \frac{\frac{1}{6} \times \frac{3}{4}}{\frac{1}{6} \times \frac{3}{4} + \frac{5}{6} \times \frac{1}{4}}$$

$$= \frac{\frac{1}{8}}{\frac{3+5}{24}}$$

$$= \frac{1}{8} \times \frac{24}{8} = \frac{3}{8}$$

Hence, the required probability is $\frac{3}{8}$.

Q. 20. In a competitive examination, an examinee either guesses or copies or knows the answer to a multiple choice question with four options. The probability that he makes a guess is $\frac{1}{3}$ and the probability that he copies the answer is $\frac{1}{6}$. The probability that the answer is correct, given that he copied it is $\frac{1}{8}$ and the guesses it is $\frac{1}{4}$ and when he knows the answer to it is 1.

Find the probability that he knows the answer to the question, given that he answered it correctly.

Sol. Let B_1, B_2 and B_3 be the events that the examinee guess, copies and knows the answer, respectively.

$$\therefore \quad P(B_1) = \frac{1}{3},\ P(B_2) = \frac{1}{6},$$

$$P(B_3) = 1 - [P(B_1) + P(B_2)]$$

$$= 1 - \frac{1}{3} - \frac{1}{6} = \frac{1}{2}$$

Let A be the event that the answer is correct.

Then,

$$P(A/B_1) = \frac{1}{4},\ P(A/B_2) = \frac{1}{8},\ P(A/B_3) = 1$$

Using Bayes' theorem

$$P(B_3/A) = \frac{P(B_3)P(A/B_3)}{P(B_1)P(A/B_1) + P(B_2)P(A/B_2) + P(B_3)P(A/B_3)}$$

$$= \frac{\frac{1}{2} \times 1}{\frac{1}{3} \times \frac{1}{4} + \frac{1}{6} \times \frac{1}{8} + \frac{1}{2} \times 1}$$

$$= \frac{\frac{1}{2}}{\frac{1}{12} + \frac{1}{48} + \frac{1}{2}} = \frac{\frac{1}{2}}{\frac{4+1+24}{48}}$$

$$= \frac{24}{29}$$

Q. 21. In a school, 8% of the boys and 2% of the girls have an I.Q. of more than 120. In the school, 60% of the students are boys. A student with an I.Q. more than 120 is selected. Find the probability that the student selected is a boy.

Sol. Let E = event that the student has an I.Q. of more than 120.

Let B and G denote the events of choosing a boy and a girl respectively.

Therefore, $P(B) = P(\text{choosing a boy}) = \frac{60}{100} = 0{\cdot}6$

and $P(G) = P(\text{choosing a girl}) = \frac{40}{100} = 0{\cdot}4$ ...(i)

We have to find P (student selected is a boy given that his I.Q. is more than 120)

i.e., we have to find $P(B / E)$.

Given : $P\left(\frac{E}{B}\right) = \frac{8}{100} = 0{\cdot}08$

and $P(E / G) = \frac{2}{100} = 0{\cdot}02$ ...(ii)

By Bayes' theorem,

$$P(B / E) = \frac{P(B).P(E / B)}{P(B).P(E / B) + P(G).P(E / G)}$$

$$= \frac{0{\cdot}6 \times 0{\cdot}08}{(0{\cdot}6 \times 0{\cdot}08) + (0{\cdot}4 \times 0{\cdot}02)} \quad \text{[using (i) and (ii)]}$$

$$= \frac{0{\cdot}048}{0{\cdot}048 + 0{\cdot}008}$$

$$= \frac{0{\cdot}048}{0{\cdot}056} = \frac{6}{7}$$

Q. 22. If a machine is set correctly, it produces 10% of defective items. If it is set incorrectly, it produces 10% good items. Chances for setting to be correct and incorrect are in the ratio 7 : 3. After a setting is made, the first item produced is found to be good item. What is the chance that the machine was set correctly ?

Sol. Let E_1 and E_2 be the events that the set-up is correct and incorrect respectively.

Let E be the event that the items produces are good ones.

Given that

$$\frac{\text{Setting is correct}}{\text{Setting is incorrect}} = \frac{7}{3}$$

$\Rightarrow$ $P(E_1) = \frac{7}{10}, \; P(E_2) = \frac{3}{10}$ ...(i)

10% defective items means 90% good items.

Therefore, $P(E/E_1) = \frac{90}{100} = \frac{9}{10}$

and $P(E/E_2) = \frac{10}{100} = \frac{1}{10}$ ...(ii)

Therefore by Bayes' theorem,

$$P(E_1/E) = \frac{P(E_1).P(E / E_1)}{P(E_1).P(E / E_2) + P(E_2).P(E / E_2)}$$

$$= \frac{\frac{7}{10} \times \frac{9}{10}}{\frac{7}{10} \times \frac{9}{10} + \frac{3}{10} \times \frac{1}{10}} \quad \text{[using (i) and (ii)]}$$

$$= \frac{\frac{63}{100}}{\frac{63+3}{100}} = \frac{63}{66} = \frac{21}{22}$$

Q. 23. Three urns contain 6 red, 4 black; 4 red, 6 black; 5 red, 5 black balls respectively. One of the urns is selected at random and a ball is drawn from it. If the ball drawn is red, find the probability that it was drawn from the first urn.★

Sol. Let E_1, E_2, E_3 be the events of choosing the first, second and third urn respectively.

Let A be the event of drawing a red ball.

Since an urn is selected at random from 3 urns, therefore,

$$P(E_1) = P(E_2) = P(E_3) = \frac{1}{3}$$

P (drawing a red ball when 1st urn is choosen)

$= P(A/E_1) = \frac{6}{10}$

Similarly, $P(A/E_2) = \frac{4}{10}$ and $P(A/E_3) = \frac{5}{10}$

$P(\text{red ball is drawn from the first urn}) = P(E_1/A)$

$$= \frac{P(E_1){\cdot}P(A / E_1)}{P(E_1){\cdot}P(A / E_1) + P(E_2){\cdot}P(A / E_2) + P(E_3){\cdot}P(A / E_3)}$$

$$= \frac{\frac{1}{3} \times \frac{6}{10}}{\frac{1}{3} \times \frac{6}{10} + \frac{1}{3} \times \frac{4}{10} + \frac{1}{3} \times \frac{5}{10}}$$

$$= \frac{6}{15} = \frac{2}{5}$$

Q. 24. A letter is known to have come either from TATANAGAR or CALCUTTA. On the envelope just two consecutive letters *TA* are visible. What is the probability that the letter has come from CALCUTTA?

★ are board exam questions from previous years

Sol. Let E_1 be the event that the letter is from CALCUTTA and E_2 be the event that the letter is from TATANAGAR.

Let A denote the event that two consecutive letters visible on the envelope are TA.

Since, the letters have come either from CALCUTTA or TATANAGAR,

$$\therefore \quad P(E_1) = \frac{1}{2} = P(E_2)$$

If E_1 has occurred, then it means that the letters is from CALCUTTA.

There are seven consecutive letters namely CA, AL, LC, CU UT, TT, TA out of which one case is favourable.

$$\Rightarrow \quad P(A/E_1) = \frac{1}{7}$$

If E_2 has occurred, then the letter came from TATANAGAR.

There are eight consecutive letters namely TA, AT, TA, AN, NA, AG, GA, AR out of which two cases are favourable.

$$\therefore \quad P(A/E_2) = \frac{2}{8}$$

Using Bayes' theorem

$$P(E_1/A) = \frac{P(E_1).P(A/E_1)}{P(E_1).P(A/E_1)+P(E_2).P(A/E_2)}$$

$$= \frac{\frac{1}{2}\times\frac{1}{7}}{\left(\frac{1}{2}\times\frac{1}{7}\right)+\left(\frac{1}{2}\times\frac{2}{8}\right)}$$

$$= \frac{4}{11}$$

Q. 25. Often it is taken that a truthful person commands more respect in the society. A man is known to speak the truth 4 out of 5 times. He throws a die and reports that it is a six. Find the probability that it is actually a six.★

Sol. Let E_1, E_2 and A be the events defined as follows :

E_1 = Six occurs

E_2 = Six does not occur

and A = the man reports that it is a six

We have, $P(E_1) = \frac{1}{6}$

$$P(E_2) = \frac{5}{6}$$

Now $P(A/E_1)$ = Probability that the man reports that there is a six on the die given that six has occured on the die = 4/5 (probability that the man speaks truth) and $P(A/E_2)$ = Probability that the man reports that there is a six on the die given that six has not occurred on the die (probability that the man does not speak truth).

★ are board exam questions from previous years

$$= 1-\frac{4}{5} = \frac{1}{5}$$

By Bayes' theorem, we have

$$P(E_1/A) = \frac{P(E_1)\,P(A/E_1)}{P(E_1)\,P(A/E_1)+P(E_2)\,P(A/E_2)}$$

$$= \frac{\frac{1}{6}\times\frac{4}{5}}{\frac{1}{6}\times\frac{4}{5}+\frac{5}{6}\times\frac{1}{5}}$$

$$= \frac{4}{4+5} = \frac{4}{9}$$

Q. 26. Suppose a girl throws a die. If she gets 1 and 2 she tosses a coin three times and notes the number of tails. If she gets 3, 4, 5 and 6, she tosses a coin once and notes whether a 'head' or 'tail', is obtained. If obtained exactly one 'tail'. What is the probability that she threw 3, 4, 5 and 6 with the die ?★

Sol. Let E_1 be the event that girl gets 1 or 2 on the roll and E_2 be the event that girls gets 3, 4, 5 or 6 on the roll of a die.

$$\therefore \quad P(E_1) = \frac{2}{6} = \frac{1}{3}$$

$$P(E_2) = \frac{4}{6} = \frac{2}{3}$$

Let A be event that she gets exactly one tail.

If she tossed coin 3 times and gets exactly one tail then possible outcomes are HTH, HHT, THH.

$$\therefore \quad P(A/E_1) = \frac{3}{8}$$

If she tossed coin only once and exactly one tail shows, then

$$P(A/E_2) = \frac{1}{2}$$

$$\therefore \quad P(E_2/A) = \frac{P(E_2).P(A/E_2)}{P(E_1)\cdot P(A/E_1)+P(E_2)\cdot P(A/E_2)}$$

$$= \frac{\frac{1}{2}\times\frac{2}{3}}{\frac{1}{2}\times\frac{2}{3}+\frac{3}{8}\times\frac{1}{3}} = \frac{\frac{1}{3}}{\frac{1}{8}+\frac{1}{3}} = \frac{1/3}{11/24}$$

$$= \frac{8}{11}$$

Q. 27. A city's data shows 5 men out of 100 and 25 women out of 1000 are colour blind. If a colourblind person is chosen at random, then find the probability of this person being a male (assuming males and females are in equal proportion in the city).

Sol. Let

M = Event that a person chosen at random is a male

F = Event that a person chosen at random is a female

C = Event that a person chosen at random is colour blind

Therefore, required probability $= P(M/C)$ given that

$P(M) = P(F) = \frac{1}{2}$ (equal proportion)

Now, $\quad P(C/M) = \frac{5}{100}$ and $P(C/F) = \frac{25}{1000}$

Hence, by Bayes' theorem

$$P(M/C) = \frac{P(C/M)\,P(M)}{P(C/M)\,P(M) + P(C/F)\,P(F)}$$

$$= \frac{\frac{5}{100} \times \frac{1}{2}}{\frac{5}{100} \times \frac{1}{2} + \frac{25}{1000} \times \frac{1}{2}}$$

$$= \frac{\frac{5}{200}}{\frac{5}{200} + \frac{25}{2000}}$$

$$= \frac{2}{3}$$

Q. 28. Suppose that 90% of the people have diabetes. What is the probability that at most 6 of a random sample of 10 people are diabetic ?

Sol. A person can be either diabetic or non-diabetic. It is given that 90% of the people are diabetic.

$$\therefore \quad p = P(\text{diabetic}) = \frac{90}{100} = \frac{9}{10}$$

$$q = P(\text{non-diabetic}) = 1 - \frac{9}{10} = \frac{1}{10}$$

Using binomial distribution, the probability that more than 6 people are diabetic is given by

$$\sum_{r=7}^{10} {}^{10}C_r\, p^r q^{10-r} = \sum_{r=7}^{10} {}^{10}C_r \left(\frac{9}{10}\right)^r \left(\frac{1}{10}\right)^{10-r}$$

Therefore, the probability that at most 6 people are diabetic

$$= 1 - P(\text{more than 6 people are diabetic})$$

$$= 1 - \sum_{r=7}^{10} {}^{10}C_r (0{\cdot}9)^r (0{\cdot}1)^{10-r}$$

Q. 29. In a treasure hunt, a participant has to collect 10 hints. The probability that he will collect each hint is 5/6. What is the probability that he will miss collecting less than 2 hints ?

Sol. Let p and q respectively be the probabilities that the participant will collect and don't collect the hints.

$$\therefore \quad p = \frac{5}{6} \text{ and } q = 1 - p = 1 - \frac{5}{6} = \frac{1}{6}$$

Let X be the random variable that represents the number of times the player will not collect the hints.

$\therefore$ By binomial distribution, we obtain

$$P(X = x) = {}^nC_x\, p^{n-x} q^x$$

$P(\text{player not collecting less than 2 hints}) = P(X < 2)$

$$= P(X=0) + P(X=1)$$

$$= {}^{10}C_0\,(q)^0\,(p)^{10} + {}^{10}C_1\,(q)\,(p)^9$$

$$= \left(\frac{5}{6}\right)^{10} + 10 \times \frac{1}{6} \times \left(\frac{5}{6}\right)^9$$

$$= \left[\frac{5}{6}\right]^9 \left[\frac{5}{6} + \frac{10}{6}\right]$$

$$= \left[\frac{5}{6}\right]^9 \left[\frac{5}{2}\right]$$

$$= \frac{5^{10}}{6^9 \times 2}$$

$$= \frac{5^{10}}{2 \times 6^9}$$

Q. 30. How many times must a man toss a fair coin so that the probability of having at least one tail is more than 90% ? [NCERT]

Sol. Let the man toss the coin n times. The n tosses are n Bernouli trails.

Probability (p) of getting a tail at the toss of a coin is $\frac{1}{2}$.

$$\therefore \quad P(X = x) = {}^nC_x\, p^x q^{n-x}$$

$$= {}^nC_x \left(\frac{1}{2}\right)^x \left(\frac{1}{2}\right)^{n-x} = {}^nC_x \cdot \left(\frac{1}{2}\right)^n$$

It is given that,

$$P(\text{getting at least one tail}) > \frac{90}{100}$$

$$P(X \geq 1) > 0{\cdot}9$$

$$\therefore \quad 1 - P(X=0) > 0{\cdot}9$$

$$1 - {}^nC_0 \left(\frac{1}{2}\right)^n > 0{\cdot}9$$

$$\Rightarrow \quad {}^nC_0 \cdot \frac{1}{2^n} < 0{\cdot}1$$

$$\Rightarrow \quad \frac{1}{2^n} < 0{\cdot}1$$

$$\Rightarrow \quad 2^n > \frac{1}{0{\cdot}1}$$

$$2^n > 10$$

The minimum value of n that satisfies the given inequality is 4.

Thus, the man should toss the coin 4 or more than 4 times.

Q. 31. If on an average, out of 10 satellites launched, one fails then what is the probability that out of 5 satellites, at least 4 are launched safely ?

Sol. $P(\text{launching a satellite safely}) = \frac{9}{10}$

[∵ only one fails]

P(satellite is not launched safely/it fails)

$$= 1 - \frac{9}{10} = \frac{1}{10}$$

Let X be the random variable showing the number of satellites that are launched safely.

P(at least 4 are successfully launched)

$= P$(4 successfully launched) or P(5 successfully launched)

$$= P(X=4) + P(X=5)$$

$$= {}^5C_4\left(\frac{9}{10}\right)^4\left(\frac{1}{10}\right)^1 + {}^5C_5\left(\frac{9}{10}\right)^5\left(\frac{1}{10}\right)^0$$

$$= \frac{1}{2}\left(\frac{9}{10}\right)^4 + \left(\frac{9}{10}\right)^5 = \left(\frac{9}{10}\right)^4\left[\frac{1}{2} + \frac{9}{10}\right]$$

$$= \left(\frac{9}{10}\right)^4\left[\frac{7}{5}\right]$$

Q. 32. An unbiased die is thrown again and again until three fours are obtained. Find the probability of obtaining 3rd four in the sixth throw of the die.

Sol. Let 'p' be the probability of obtaining a four in a single throw of the die. Then,

$$p = \frac{1}{6} \text{ and } q = 1 - \frac{1}{6} = \frac{5}{6}$$

Obtaining 3rd four in the sixth throw of the die means that in the first five throws, there are 2 fours and the 3rd four is obtained in 6th throw. Therefore,

Required probability

$= P$[getting 2 fours in first five throws]

$-$ P[getting four in 6th throw]

$$= (5C_2p^2q^{5-2})(p) = {}^5C_2\left(\frac{1}{6}\right)^2\left(\frac{5}{6}\right)^3 \times \frac{1}{6}$$

$$= 10 \times \left(\frac{1}{6}\right)^3 \times \left(\frac{5}{6}\right)^3$$

$$= \frac{625}{23328}$$

Q. 33. 5 cards are drawn successively with replacement from a well-shuffled deck of 52 cards. What is the probability that :

(i) all 5 cards drawn are spades

(ii) at least 3 cards drawn are spades

(iii) at most 2 cards drawn are spades

Sol. Let E be the event of getting a spade in a single draw.

Then, $P(E) = \frac{13}{52} = \frac{1}{4}$

$$\therefore \quad p = \frac{1}{4},\ q = 1 - p = \frac{3}{4},\ n = 5$$

Let X be the random variable denoting the number of spades

(i) P (all 5 cards drawn are spades)

$$= P(X=5) = P(5) = {}^5C_5p^5q^0$$

$$= {}^5C_5\left(\frac{1}{4}\right)^5\left(\frac{3}{4}\right)^0$$

$$= \left(\frac{1}{4}\right)^5 = \frac{1}{1024}$$

(ii) P (at least 3 cards drawn are spades) $= P(X > 3)$

$$= P(X=3) + P(X=4) + P(X=5)$$

$$= P(3) + P(4) + P(5)$$

$$= {}^5C_3\left(\frac{1}{4}\right)^3\left(\frac{3}{4}\right)^2 + {}^5C_4\left(\frac{1}{4}\right)^4\left(\frac{3}{4}\right) + {}^5C_5\left(\frac{1}{4}\right)^5\left(\frac{3}{4}\right)^0$$

$$= \frac{5\times4}{1\times2}\times\frac{1}{64}\times\frac{9}{16} + 5\times\frac{1}{256}\times\frac{3}{4} + 1\times\frac{1}{1024}$$

$$= \frac{45}{512} + \frac{15}{1024} + \frac{1}{1024} = \frac{106}{1024} = \frac{53}{512}$$

(ii) P (at most 2 cards drawn are spades) $= P(X \le 2)$

$$= P(X=0) + P(X=1) + P(X=2)$$

$$= P(0) + P(1) + P(2)$$

$$= {}^5C_0\left(\frac{1}{4}\right)^0\left(\frac{3}{4}\right)^5 + {}^5C_1\left(\frac{1}{4}\right)^1\left(\frac{3}{4}\right)^4 + {}^5C_2\left(\frac{1}{4}\right)^2\left(\frac{3}{4}\right)^3$$

$$= \left(\frac{3}{4}\right)^3\left[\frac{9}{16} + 5\times\frac{1}{4}\times\frac{3}{4} + \frac{5\times4}{1\times2}\times\frac{1}{16}\right]$$

$$= \frac{27}{64}\left[\frac{9}{16} + \frac{15}{16} + \frac{10}{16}\right]$$

$$= \frac{27}{64}\times\frac{34}{16} = \frac{918}{1024} = \frac{459}{512}$$

Q. 34. The random variable X can take only the values 0, 1, 2, 3. Given that $P(X=0) = P(X=1) = p$ and $P(X=2) = P(X=3)$ such that $\Sigma p_ix_i^2 = 2\Sigma p_ix_i$, find the value of p.★

★ are board exam questions from previous years

Sol. Given, $P(X=0) = P(X=1) = p$ and $P(X=2) = P(X=3)$

Let $P(X=2) = P(X=3) = k$

X	0	1	2	3
$P(X)$	p	p	k	k

Also given that

$$\Sigma p_i x_i^2 = 2\Sigma p_i x_i$$

$$\Rightarrow \quad 0 + p + 4k + 9k = 2(0 + p + 2k + 3k)$$

$$\Rightarrow \quad p + 13k = 2(p + 5k)$$

$$\Rightarrow \quad 3k = p \qquad \ldots(i)$$

Also we know that

$$\Sigma p_i = 1$$

$$\Rightarrow \quad p + p + k + k = 1$$

$$2p + 2k = 1$$

$$6k + 2k = 1 \qquad [\text{Using (i)}]$$

$$8k = 1$$

or $\quad k = \frac{1}{8}$

From equation (i) $\quad p = \frac{3}{8}$.

Q. 35. There are 4 cards numbered 1, 3, 5 and 7, one number on one card. Two cards are drawn at random without replacement. Let X denote the sum of the numbers on the two drawn cards. Find the mean and variance of X.★

Sol. Given, X denote the sum of the numbers on the two drawn cards.

Then, X can take values 4, 6, 8, 10, 12.

and sample space (S) = {(1, 3), (1, 5), (1, 7), (3, 1), (3, 5), (3, 7), (5, 1), (5, 3), (5, 7), (7, 1), (7, 3), (7, 5)}

So the probability distribution of X is given below:

X	4	6	8	10	12
$P(X)$	2/12	2/12	4/12	2/12	2/12

Computation of Mean and Variance

x_i	$P(X = x_i) = p_i$	$p_i x_i$	$p_i x_i^2$
4	2/12	8/12	32/12
6	2/12	12/12	72/12
8	4/12	32/12	256/12
10	2/12	20/12	200/12
12	2/12	24/12	288/12
		$\Sigma p_i x_i = 96/12$	$\Sigma p_i x_i^2 = 848/12$

We have, $\quad \Sigma p_i x_i = \frac{96}{12} = 8$

$\therefore \quad \overline{X} = \text{Mean} = \Sigma p_i x_i = 8$

★ are board exam questions from previous years

Now, $\quad \text{Var}(X) = \Sigma p_i x_i^2 - (\Sigma p_i x_i)^2$

$$= \frac{848}{12} - (8)^2$$

$$= \frac{848}{12} - 64$$

$$= \frac{80}{12} = \frac{20}{3}$$

Hence, Mean = 8 and Variance = $\frac{20}{3}$

Q. 36. In a game, a man wins ₹ 5 for getting a number greater than 4 and loses ₹ 1 otherwise, when a fair die is thrown. The man decided to throw a die thrice but to quit as and when he gets a number greater than 4. Find the expected value of price amount he wins/loses.★

Sol. Let 'n' denote the number of throws required to get a number greater than 4 and X denote the amount won/lost. The man may get a number greater than 4 in the very first throw of the die or in second throw or in the third throw.

Thus, we have the following probability distribution for X.

Number of throws (n):	1	2	3	3
Amount won/lost (X):	5	4	3	– 3
Probability $[P(X)]$:	$\frac{2}{6}$	$\frac{4}{6}\times\frac{2}{6}$	$\frac{4}{6}\times\frac{4}{6}\times\frac{2}{6}$	$\frac{4}{6}\times\frac{4}{6}\times\frac{4}{6}$

$$E(X) = \left(5\times\frac{2}{6}\right) + \left(4\times\frac{4}{6}\times\frac{2}{6}\right) + \left(3\times\frac{4}{6}\times\frac{4}{6}\times\frac{2}{6}\right) + (-3)\times\left(\frac{4}{6}\times\frac{4}{6}\times\frac{4}{6}\right)$$

$$= \frac{19}{9}$$

Expected amount he could win is ₹ $\frac{19}{9}$

Q. 37. If the sum of mean and variance of a binomial distribution for 5 tails is 3·2, find the distribution.

Sol. We know that, for binomial distribution

Mean = np and variance = npq

Given $\quad n = 5$ and Mean + Variance = 3·2

$\therefore \quad np + npq = 3{\cdot}2, n = 5$

$\Rightarrow \quad 5p + 5pq = 3{\cdot}2$

$$p + p(1-p) = \frac{3{\cdot}2}{5}$$

$$p + p - p^2 = 0{\cdot}64$$

$$\Rightarrow \quad p^2 - 2p + 0{\cdot}64 = 0$$

$$\Rightarrow \quad p = \frac{-(-2) \pm \sqrt{4 - 4(1)(0{\cdot}64)}}{2(1)}$$

$$= \frac{2 \pm \sqrt{4 - 2{\cdot}56}}{2} = \frac{2 \pm \sqrt{1{\cdot}44}}{2}$$

$$p = \frac{2 \pm 1{\cdot}2}{2} = \frac{3{\cdot}2}{2} \text{ or } \frac{0{\cdot}8}{2}$$

p cannot exceed '1'.

$$\therefore \quad p \neq \frac{3{\cdot}2}{2} \text{ and } p = \frac{0{\cdot}8}{2} = 0{\cdot}4 = \frac{4}{10} = \frac{2}{5}$$

Thus, $n = 5, p = 0{\cdot}4$ and $q = (1 - p) = (1 - 0{\cdot}4) = 0{\cdot}6$.

Let X denote the binomial variate then, the required distribution is

$$P(X = r) = {}^nC_r \,.\, p^r \,.\, q^{(n-r)} = {}^5C_r\,(0{\cdot}4)^r\,(0{\cdot}6)^{5-r}$$

where $r = 0, 1, 2, 3, 4, 5.$

Q. 38. For 6 trials of experiment, let X be binomial variate which satisfies the relation $9P(X = 4) = P(X = 2)$. Find the probability of success.★

Sol. Let p denote the probability of getting success and q be the probability of failure.

Since, $P(X = r) = {}^nC_r\, p^r\, q^{n-r}$

$\therefore$ $P(X = 4) = {}^6C_4\, p^4\, q^{6-4}$

and $P(X = 2) = {}^6C_2\, p^2\, q^{6-2}$

we have $9P(X = 4) = P(X = 2)$

$\Rightarrow$ $9\,{}^6C_4\, p^4 q^2 = {}^6C_2\, p^2 q^4$

$\Rightarrow$ $9p^2 = q^2 \qquad [\because {}^6C_4 = {}^6C_2]$

$\Rightarrow$ $9p^2 = (1 - p)^2 \qquad [\because p + q = 1]$

$\Rightarrow$ $9p^2 = 1^2 + p^2 - 2p$

$\Rightarrow$ $9p^2 - p^2 + 2p - 1 = 0$

$\Rightarrow$ $8p^2 + 2p - 1 = 0$

$\Rightarrow$ $8p^2 + 4p - 2p - 1 = 0$

$\Rightarrow$ $(4p - 1)(2p + 1) = 0$

$\therefore$ $p = \frac{1}{4}, \frac{-1}{2}$

Since probability cannot be –ve.

$\Rightarrow$ $p = \frac{1}{4}$

Hence probability of success is $\frac{1}{4}$.

Q. 39. Calculate $P(r)$ for $r = 1, 2, 3, 4, 5$, taking $n = 5$ and $p = \frac{1}{6}$ with the help of the recurrence formula of binomial distribution.

Sol. Given, $p = \frac{1}{6}$, $q = 1 - p = 1 - \frac{1}{6} = \frac{5}{6}$, $n = 5$

The recurrence formula is :

$$P(r+1) = \frac{n-r}{r+1}\,\frac{p}{q}\,P(r)$$

i.e., $$P(r+1) = \frac{5-r}{r+1}\,\frac{1}{5}\,P(r)$$

Now, $P(0) = P(X = 0)$

$$= {}^nC_0\, p^0\, q^n$$

$$= {}^5C_0 \left(\frac{1}{6}\right)^0 \left(\frac{5}{6}\right)^5 = \left(\frac{5}{6}\right)^5$$

$= 0{\cdot}40$ (approx.)

$$P(1) = \frac{5-0}{0+1}\,\frac{1}{5}\,P(0)$$

$\Rightarrow$ $P(1) = 5 \times \frac{1}{5} P(0) = 0{\cdot}40$

$$P(2) = \frac{5-1}{1+1} \times \frac{1}{5}\,P(1)$$

$\Rightarrow$ $P(2) = \frac{4}{2 \times 5}\,P(1) = \frac{2}{5} \times 0{\cdot}40$

$= 0{\cdot}4 \times 0{\cdot}4 = 0{\cdot}16$

$$P(3) = \frac{5-2}{2+1} \times \frac{1}{5}\,P(2)$$

$P(3) = \frac{1}{5} \times 0{\cdot}16 = 0{\cdot}2 \times 0{\cdot}16$

$= 0{\cdot}032$

$$P(4) = \frac{5-3}{3+1} \times \frac{1}{5}\,P(4)$$

$P(4) = \frac{1}{10} \times 0{\cdot}032 = 0{\cdot}0032$

$$P(5) = \frac{5-4}{4+1} \times \frac{1}{5}\,P(4)$$

$= \frac{1}{5} \cdot \frac{1}{5} P(4)$

$= \frac{1}{25} \times 0{\cdot}0032$

$= 0{\cdot}04 \times 0{\cdot}0032$

$= 0{\cdot}000128$

★ are board exam questions from previous years

Assertion and Reasoning Based Questions

| Set 4 |

The following questions consist of two statements, one labelled as **Assertion (A)** and the other labelled as **Reason (R).** You are to examine these two statements carefully and decide if the Assertion (A) and the Reason (R) are individually true and if so, whether the Reason (R) is the correct explanation for the given Assertion (A). Select your answer to these items using the codes given then select the correct option.

(A) Both (A) and (R) are individually true and (R) is the correct explanation of (A)
(B) Both (A) and (R) are individually true but (R) is not the correct explanation of (A).
(C) (A) is true but (R) is false.
(D) (A) is false but (R) is true.

Chapter 1. Relations and Functions

1. **Assertion (A) :** If $f(x) = \text{In}\, x^2$ and $g(x) = 2\,\text{In}\, x$, then $f(x) = g(x)$.
   Reason (R) : For $x < 0$, $g(x)$ is not defined.
2. **Assertion (R) :** The function $f(x) = |x|$ is not one-one.
   Reason (R) : The negative real number are not the images of any real numbers.
3. **Assertion (A) :** If $f : R \to R$ and $g : R \to R$ be two mappings such that $f(x) = \sin x$ and $g(x) = x^2$, then $fog \neq gof$.
   Reason (R) : $(fog)x = f(x)\, g(x) = (gof)x$.
4. **Assertion (A) :** A function $y = f(x)$ is defined by $x^2 - \text{arc}\cos y = \pi$, then domain of $f(x)$ is R.
   Reason (R) : $\cos^{-1} y \in [0, \pi]$.
5. **Assertion (A) :** If $f(x)$ is odd function and $g(x)$ is even function, then $f(x) + g(x)$ is neither even nor odd.
   Reason (R) : Odd function is symmetrical in opposite quadrants and even function is symmetrical about the y-axis.
6. **Assertion (A) :** Every even function $y = f(x)$ is not one-one, $\forall\, x \in D_f$.
   Reason (R) : Even function is symmetrical about the y-axis.
7. **Assertion (A) :** A function $f : R \to R$ be defined by $f(x) = x - [x]$ (where $[x]$ is greatest integer $\leq x$) for all $x \in R$. f is not invertiable.
   Reason (R) : $f(x)$ is periodic function.
8. **Assertion (A) :** The function $f(x) = x^2 - x + 1$, $x \geq \frac{1}{2}$ and $g(x) = \frac{1}{2} + \sqrt{\left(x - \frac{3}{4}\right)}$, then the number of solutions of the equation $f(x) = g(x)$ is two.
   Reason (R) : $f(x)$ and $g(x)$ are mutually inversion.
9. **Assertion (A) :** $f(x) = \sin x + \cos ax$ is a periodic function.
   Reason (R) : a is rational number.
10. **Assertion (A) :** The least period of the function, $f(x) = \cos(\cos x) + \cos(\sin x) + \sin 4x$ is π.
    Reason (R) : $\because f(x + \pi) = f(x)$.
11. **Assertion (A) :** If $f(x + y) + f(x - y) = 2f(x) \cdot f(y)$ $\forall\, x, y \in R$ and $f(0) \neq 0$, then $f(x)$ is an even function.
    Reason (R) : If $f(-x) = f(x)$, then $f(x)$ is an even function.
12. **Assertion (A) :** The equation $x^4 = (\lambda x - 1)^2$ has atmost two real solutions (is $\lambda > 0$).
    Reason (R) : Curves $f(x) = x^4$ and $g(x) = (\lambda x - 1)^2$ has atmost two points.
13. **Assertion (A) :** The domains of $f(x) = \sqrt{\cos(\sin x)}$ and $g(x) = \sqrt{\sin(\cos x)}$ are same.
    Reason (R) : $-1 \leq \cos(\sin x) \leq 1$ and $-1 \leq \sin(\cos x) \leq 1$
14. **Assertion (A) :** If $f(x) = x^5 - 16x + 2$, then $f(x) = 0$ has only one root in the interval $[-1, 1]$.
    Reason (R) : $f(-1)$ and $f(1)$ are of opposite sign.
15. **Assertion (A) :** The domain of the function $f(x) = \sin^{-1} x + \cos^{-1} x + \tan^{-1} x$ is $[-1, 1]$.
    Reason (R) : $\sin^{-1} x$ and $\cos^{-1} x$ is defined in $|x| \leq 1$ and $\tan^{-1} x$ defined for all x.
16. **Assertion (A) :** The period of $f(x) = \sin 3x \cos [3x] - \cos 3x \sin [3x]$ is $\frac{1}{3}$ where [ ] denotes the greatest integer function $\leq x$.
    Reason (R) : The period of $\{x\}$ is 1, where $\{x\}$ denotes the fractional part function of x.

Answers :

1. (D) $\because \text{In}\, x^2 = 2\,\text{In}\,|x| \neq 2\,\text{In}\, x$ and $g(x)$ is defined only when $x > 0$

2. (C) The function $f(x) = |x|$ is many one

$\therefore \quad f(-x) = f(x)$.

i.e., not one-one and the image of $-x$ is x.

3. (C) $\because \quad (fog)\,x = f(g(x)) = f(x^2) = \sin x^2$

and $\quad (gof)x = g(f(x)) = g(\sin x) = \sin^2 x$

$\Rightarrow \quad fog \neq gof$

4. (D) $\because x^2 - \text{arc}\cos y = \pi$

$\Rightarrow \cos^{-1} y = (x^2 - \pi)$

$\because 0 \le \cos^{-1} y \le \pi$

$\Rightarrow 0 \le x^2 - \pi \le \pi$

$\Rightarrow \pi \le x^2 \le 2\pi$

$\therefore x \in [-\sqrt{2\pi}, -\sqrt{\pi}] \cup [\sqrt{\pi} - \sqrt{2\pi}]$

5. (B) $\because f(x)$ is odd

$\Rightarrow \quad f(-x) = -f(x)$

and $g(x)$ is even $\Rightarrow g(-x) = g(x)$

Let $\quad F(x) = f(x) + g(x)$

$\therefore \quad F(-x) = f(-x) + g(-x) = -f(x) + g(x)$

$\ne \pm F(x)$

$\therefore F(x)$ is neither even nor odd.

6. (A) Since every even function is symmetrical about the y-axis.

$\therefore$ any line parallel to x-axis cuts the graph at more than one point, then $f(x)$ is not one-one $\forall\, x \in D_f$.

7. (A) $\because f(x)$ is periodic function

$\Rightarrow f(x)$ is many one.

Hence, $f(x)$ is not invertible.

8. (D) Let $\quad y = x^2 - x + 1$

$\Rightarrow \quad x^2 - x + 1 - y = 0$

$$\therefore \quad x = \frac{1 \pm \sqrt{1 - 4 \cdot 1 \cdot (1 - y)}}{2}$$

$$= \frac{1}{2} \pm \sqrt{\left(y - \frac{3}{4}\right)}$$

$$= \frac{1}{2} + \sqrt{\left(y - \frac{3}{4}\right)}, \left(\because x \ge \frac{1}{2}\right)$$

$= g(y)$

$\therefore \quad y = g^{-1}(x) \Rightarrow f(x) = g^{-1}(x)$

Hence, $f(x)$ and $g(x)$ are mutually inversion.

$\Rightarrow$ The graph of the original and inversion functions can intersect only on the straight line $y = x$

$\therefore \quad x = f(x)$

$\Rightarrow \quad x = x^2 - x + 1$

$\Rightarrow \quad x^2 - 2x + 1 = 0$

$\Rightarrow \quad (x - 1)^2 = 0$

$\therefore \quad x = 1$

9. (A) $\because$ Period of $\sin x$ is $\frac{2\pi}{1}$ and period of $\cos ax = \frac{2\pi}{a}$

Hence, period of $f(x) = \text{LCM of } \left\{\frac{2\pi}{1}, \frac{2\pi}{a}\right\}$

$$= \frac{\text{LCM of } \{2\pi, 2\pi\}}{\text{HCF of } \{1, a\}} = \frac{2\pi}{k}$$

when k is HCF of 1 and a.

$\Rightarrow \quad \frac{1}{k} = \text{integer} = q \text{ (say) } (\ne 0)$

and $\quad \frac{a}{k} = \text{integer} = p \text{ (say)}$

$\therefore \quad \frac{a/k}{1/k} = \frac{p}{q}$

$\Rightarrow \quad a = \frac{p}{q}$

$\Rightarrow a$ is rational.

10. (D) Let $f(x)$ be periodic with period $\lambda, \lambda \ne 0, \lambda > 0$

$\therefore \quad f(x + \lambda) = f(x)$

$\Rightarrow \cos(\cos(x + \lambda)) + \cos(\sin(x + \lambda)) + \sin(4(x + \lambda)$

$= \cos(\cos x) + \cos(\sin x) + \sin 4x$

Put $x = 0$,

$\cos(\cos \lambda) + \cos(\sin \lambda) + \sin(4\lambda)$

$= \cos(1) + \cos(0) + \sin(0)$

$= \cos(\sin \pi/2) + \cos(\cos \pi/2) + \sin(2\pi)$

$\Rightarrow \lambda = \pi/2$

11. (B) Given $f(x + y) + f(x - y) = 2f(x) f(y)$ ...(i)

Replacing x by y and y by x in equation (i), then

$f(y + x) + f(y - x) = 2f(y) f(x)$ ...(ii)

$\therefore$ From equations (i) and (ii), we get

$f(y - x) = f(x - y)$

Putting $y = 2x$, then $f(x) = f(-x)$

Hence, $f(x)$ is an even function.

12. (D) $\because \quad x^4 = (\lambda x - 1)^2$

$x^2 = \pm(\lambda x - 1)$

$\Rightarrow \quad x^2 = \lambda x - 1$

$\therefore \quad x^2 - \lambda x + 1 = 0$

$$\Rightarrow \quad x = \frac{\lambda \pm \sqrt{(\lambda^2 - 4)}}{2}$$

$\therefore \quad \lambda^2 - 4 \ge 0$

$\Rightarrow \lambda \in (-\infty, -2] \cup [2, \infty)$

but $\lambda > 0 \therefore \lambda \in [2, \infty)$

and $\quad x^2 = -(\lambda x - 1)$

$x^2 + \lambda x - 1 = 0$

$$x = \frac{-\lambda \pm \sqrt{(\lambda^2 + 4)}}{2}$$

$\lambda > 0$

$\Rightarrow$ Infinite solutions.

cut at two points.

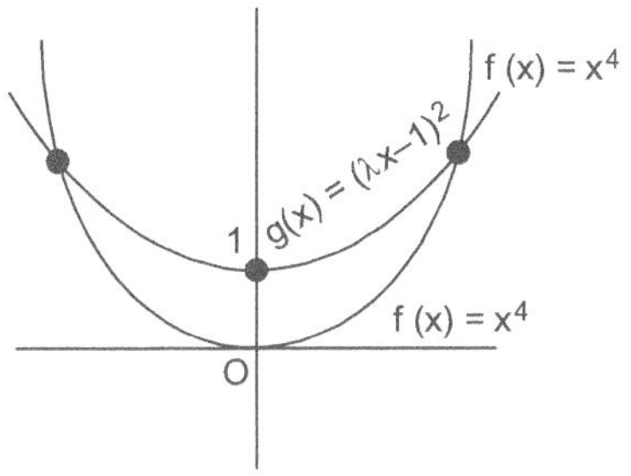

13. (D) $\cos(\sin x) \geq 0$

$$\Rightarrow 2n\pi - \frac{\pi}{2} \leq \sin(x) \leq 2n\pi + \frac{\pi}{2}, n \in I$$

but $\quad -1 \leq \sin x \leq 1$

$\Rightarrow \quad -1 \leq \sin x \leq 1$

$\Rightarrow \quad x \in \text{R}$

$\therefore \quad D_f = \text{R}$

Also, $\sin(\cos x) \geq 0$

$\Rightarrow \quad 2n\pi \leq \cos x \leq 2n\pi + \pi, n \in \text{I}$

but $\quad -1 \leq \cos x \leq 1$

$\Rightarrow \quad 0 \leq \cos x \leq 1$

$$\Rightarrow \quad x \in \left[2p\pi - \frac{\pi}{2}, 2p\pi + \frac{\pi}{2}\right], p \in \text{I}$$

$$\therefore \quad D_g = \left[2p\pi - \frac{\pi}{2}, 2p\pi + \frac{\pi}{2}\right], p \in \text{I}$$

$\Rightarrow \quad D_f \neq D_g$

14. (B) $\because\ f(-1) = -1 + 16 + 2 = 17$

and $f(1) = 1 - 16 + 2 = -13$

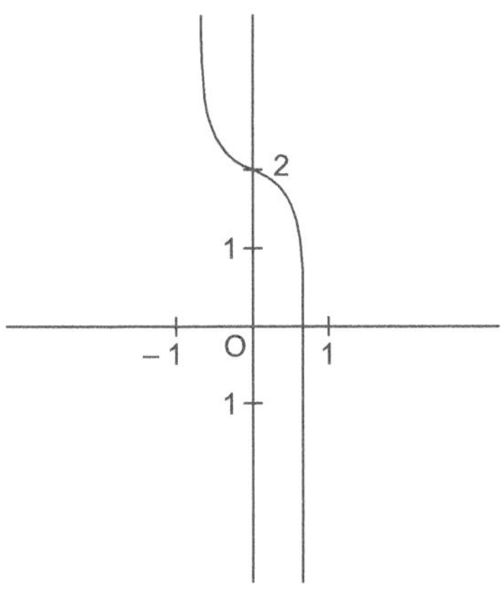

15. (A) $\because \sin^{-1} x$ is defined in $[-1, 1]$

$\cos^{-1} x$ is defined in $[-1, 1]$

and $\tan^{-1} x$ is defined in R.

Hence, $f(x)$ is defined in $[-1, 1]$.

16. (A) $\because \quad f(x) = \sin 3x \cos[3x] - \cos 3x \sin[3x]$

$= \sin(3x - [3x])$

$= \sin[\{3x\}]$

$\because \quad \{x\} = x - \{x\}$ is periodic with period 1

$\therefore$ Period of $\{3x\}$ is $\frac{1}{3}$.

Chapter 2. Inverse Trigonometric Functions

1. **Assertion (A) :** $\sin^{-1}(\sin 3) = 3$

   Reason (R) : For principal values $\sin^{-1}(\sin x) = x$

2. **Assertion (A) :** If $x < 0$,

$$\tan^{-1} x + \tan^{-1}\left(\frac{1}{x}\right) = \frac{\pi}{2}$$

   Reason (R) : $\tan^{-1} x + \cot^{-1} x = \frac{\pi}{2}, \forall x \in R$

3. **Assertion (A) :** $\tan^{-1}\left(\frac{3}{4}\right) + \tan^{-1}\left(\frac{1}{7}\right) = \frac{\pi}{4}$

   Reason (R) : For $x > 0, y > 0$

$$\tan^{-1}\left(\frac{x}{y}\right) + \tan^{-1}\left(\frac{x-y}{y+x}\right) = \frac{\pi}{4}$$

4. **Assertion (A) :**

   If $x = \text{cosec}(\tan^{-1}(\cos(\cot^{-1}(\sec(\sin^{-1} a)))))$ and $y = \sec(\cot^{-1}(\sin(\tan^{-1}(\text{cosec}(\cos^{-1} a)))))$, where $a \in [0, 1]$, then $x = y$

   Reason (R) :

$$\sin^{-1}\left(\frac{1}{x}\right) = \text{cosec}^{-1} x, \forall x \in (-\infty, -1] \cup [1, \infty)$$

$$\cos^{-1}\left(\frac{1}{x}\right) = \sec^{-1} x, \forall x \in (-\infty, -1] \cup [1, \infty)$$

$$\tan^{-1}\left(\frac{1}{x}\right) = \begin{cases} \cot^{-1} x, & x > 0 \\ -\pi + \cot^{-1} x, & x < 0 \end{cases}$$

5. **Assertion (A) :** The solution of system of equations

$$\cos^{-1} x + (\sin^{-1} y)^2 = \frac{p\pi^2}{4}$$

   and $(\cos^{-1} x)(\sin^{-1} y)^2 = \frac{\pi^4}{16}$ is

   $x = \cos\frac{\pi^2}{4}$ and $y = \pm 1, \forall p \in I.$

   Reason (R) : AM $\geq$ GM

6. **Assertion (A) :** If $p > q > 0$ and $pr < -1 < qr$, then

$$\tan^{-1}\left(\frac{p-q}{1+pq}\right) + \tan^{-1}\left(\frac{q-r}{1+qr}\right) + \tan^{-1}\left(\frac{r-p}{1+rp}\right) = \pi$$

   Reason (R) : $\tan^{-1} x - \tan^{-1} y = \tan^{-1}\left(\frac{x-y}{1+xy}\right)$ for all x, y

7. Assertion (A) : $\sin^{-1} 2x + \sin^{-1} 3x = \frac{\pi}{3}$

$$\Rightarrow \qquad x = \sqrt{\left(\frac{3}{76}\right)} \text{ only}$$

Reason (R) : Sum of two negative angles cannot be positive.

8. Assertion (A) : If $\sum_{i=1}^{2n} \sin^{-1} x_i = n\pi, n \in N$

Then, $\sum_{i=1}^{n} x_i = \sum_{i=1}^{n} x_i^2 = \sum_{i=1}^{n} x_i^3$

Reason (R) : $-\frac{\pi}{2} \le \sin^{-1} x \le \frac{\pi}{2}, \forall x \in [-1, 1]$

Answers :

1. (D) $\because 3 \approx 171°$ (lies in II quadrant)

$\therefore \quad \sin^{-1} \sin 3 = 3 - \pi \neq 3$

But $\sin^{-1} \sin x = x$ for principal values.

2. (D) For $x < 0, \tan^{-1}\left(\frac{1}{x}\right) = -\pi + \cot^{-1} x$

$$\therefore \tan^{-1} x + \tan^{-1}\left(\frac{1}{x}\right) = \tan^{-1} - \pi + \cot^{-1} x$$

$$= \frac{\pi}{2} - \pi$$

$$= -\frac{\pi}{2} \neq \frac{\pi}{2}$$

and $\tan^{-1} x + \cot^{-1} x = \frac{\pi}{2}, \forall x \in R$

3. (A) $\because \tan^{-1}\left(\frac{3}{4}\right) + \tan^{-1}\left(\frac{1}{7}\right) = \tan^{-1}\left(\frac{\frac{3}{4}+\frac{1}{7}}{1-\frac{3}{4}\cdot\frac{1}{7}}\right)$

$$= \tan^{-1}(1) = \frac{\pi}{4}$$

Also, $\frac{4-3}{4+3} = \frac{1}{7} \qquad (\because 4 > 0, 3 > 0)$

$\therefore$ (R) is the correct reason of (A).

4. (B) $\because x = \text{cosec}(\tan^{-1}(\cos(\cot^{-1}(\sec(\sin^{-1} a)))))$

$$= \text{cosec}\left(\frac{\pi}{2} - \cot^{-1}(\cos(\cot^{-1}(\sec(\sin^{-1} a))))\right)$$

$$= \sec(\cot^{-1}(\cos(\cot^{-1}(\sec(\sin^{-1} a)))))$$

$$\sec\left(\cot^{-1}\left(\cos\left(\frac{\pi}{2} - \tan^{-1}(\sec(\sin^{-1} a\right)\right)\right)$$

$$= \sec(\cot^{-1}(\sin(\tan^{-1}(\sec(\sin^{-1} a)))))$$

$$= \sec\cot^{-1}\left(\sin\left(\tan^{-1}\left(\sec\left(\frac{\pi}{2} - \cos^{-1} a\right)\right)\right)\right)$$

$$= \sec(\cot^{-1}(\sin(\tan^{-1}(\text{cosec}(\cos^{-1} a)))))$$

$$= y$$

5. (A) $\because$ AM $\ge$ GM

$$\therefore \quad \frac{\cos^{-1} x + (\sin^{-1} y)^2}{2} \ge \sqrt{(\cos^{-1} x)(\sin^{-1} y)^2}$$

$$\Rightarrow \qquad \frac{p\pi^2}{8} \ge \frac{\pi^2}{4}$$

$$\Rightarrow \qquad p \ge 2$$

Thus, we conclude that the only value of p that satisfies all conditions is $p = 2$.

Then, $\cos^{-1} x = (\sin^{-1} y)^2$

$$\Rightarrow \quad (\cos^{-1} x)^2 = \frac{\pi^4}{16}$$

$$\Rightarrow \quad \cos^{-1} x = \pm\frac{\pi^2}{4}$$

$$\Rightarrow \quad x = \cos\left(\pm\frac{\pi^2}{4}\right)$$

$$\therefore \quad x = \cos\left(\frac{\pi^2}{4}\right)$$

Also, $(\sin^{-1} y)^4 = \frac{\pi^4}{16}$

$$\Rightarrow \quad \sin^{-1} y = \pm\frac{\pi}{2}$$

$$\therefore \quad y = \sin\left(\pm\frac{\pi}{2}\right) = \pm 1$$

6. (D) Since, $p, q > 0$, therefore, $pq > 0$.

So, $\tan^{-1}\left(\frac{p-q}{1+pq}\right) = \tan^{-1} p - \tan^{-1} q \qquad$ ...(i)

Since, $qr > -1$

$\therefore \quad \tan^{-1}\left(\frac{q-r}{1+qr}\right) = \tan^{-1} q - \tan^{-1} r \qquad$ ...(ii)

Since, $pr < -1$ and $r < 0$

$\therefore \quad \tan^{-1}\left(\frac{r-p}{1+rp}\right) = \pi + \tan^{-1} r - \tan^{-1} p \qquad$ ...(iii)

On adding equations (i), (ii), and (iii), we get

$$\tan^{-1}\left(\frac{p-q}{1+pq}\right) + \tan^{-1}\left(\frac{q-r}{1+qr}\right) + \tan^{-1}\left(\frac{r-p}{1+rp}\right) = \pi$$

7. (A) $\because \quad \sin^{-1} 2x + \sin^{-1} 3x = \frac{\pi}{3}$

$$\Rightarrow \frac{\pi}{2} - \cos^{-1} 2x + \frac{\pi}{2} - \cos^{-1} 3x = \frac{\pi}{3}$$

$$\Rightarrow \cos^{-1} 2x + \cos^{-1} 3x = \frac{2\pi}{3}$$

$$\Rightarrow \cos^{-1}\left\{6x^2 - \sqrt{(1-(2x)^2}\sqrt{1-(3x)^2}\right\} = \frac{2\pi}{3}$$

$$\Rightarrow 6x^2 - \sqrt{(1-13x^2+36x^4)} = -\frac{1}{2}$$

$$\Rightarrow \left(6x^2+\frac{1}{2}\right)^2 = 1 - 13x^2 + 36x^4$$

$$\Rightarrow 19x^2 = \frac{3}{4}$$

$$\Rightarrow x = \pm\sqrt{\frac{3}{76}}$$

But sum of two –ve number cannot be $\pi/3$.

$\therefore \quad x = \sqrt{\frac{3}{76}}$ is only solution.

8. (A) Since, maximum value of $\sin^{-1} x_i$ is $\frac{\pi}{2}$

$\therefore \sum_{i=1}^{2n} \sin^{-1} x_i = n\pi$ is possible, if

$$x_1 = x_2 = x_3 = ... = x_{2n} = 1$$

$\therefore \sum_{i=1}^{n} x_i = 1 + 1 + 1 + ...$ upto n times $= n$

$\therefore \sum_{i=1}^{n} x_i^2 = 1^2 + 1^2 + 1^2 + 1^2 + ...$ upto n times $= n$

and $\sum_{i=1}^{n} x_i^3 = 1^3 + 1^3 + 1^3 + ...$ upto n times $= n$

Hence, $\sum_{i=1}^{n} x_i = \sum_{i=1}^{n} x_i^2 = \sum_{i=1}^{n} x_i^3 = n$

Chapter 3. Matrices

1. **Assertion (A) :** If $A = \begin{pmatrix} 3 & -3 & 4 \\ 2 & -3 & 4 \\ 0 & -1 & 1 \end{pmatrix}$, then adj (adj A) = A.

   Reason (R) : | adj (adj A) | = $|A|^{(n-1)^2}$, A be n rowed non-singular matrix

2. **Assertion (A) :** If $A = \begin{pmatrix} a & 0 & 0 \\ 0 & b & 0 \\ 0 & 0 & c \end{pmatrix}$, then

$$A^{-1} = \begin{pmatrix} \frac{1}{a} & 0 & 0 \\ 0 & \frac{1}{b} & 0 \\ 0 & 0 & \frac{1}{c} \end{pmatrix}$$

   Reason (R) : The inverse of a diagonal matrix is a diagonal matrix.

3. **Assertion (A) :** The rank of a unit matrix of order $n \times n$ is n.

   Reason (R) : The rank of a non-singular matrix of order $n \times n$ is not n.

4. **Assertion (A) :** The matrix $\begin{bmatrix} a & 0 & 0 & 0 \\ 0 & b & 0 & 0 \\ 0 & 0 & c & 0 \end{bmatrix}$ is a diagonal matrix.

   Reason (R) : $A = [a_{ij}]$ is a square matrix such that $a_{ij} = 0 \ \forall \ i \neq j$, then A is called diagonal matrix.

5. **Assertion (A) :** The inverse of the matrix $A = \begin{pmatrix} 1 & 1 & 1 \\ 1 & 2 & 3 \\ 1 & 4 & 7 \end{pmatrix}$ does not exist.

   Reason (R) : The matrix A is singular.

6. **Assertion (A) :** If A is a matrix of order $n \times n$, then det $(kA) = k^n$ det (A) or $|kA| = k^n|A|$.

   Reason (R) : If B is a matrix obtained from A by multiplying any row or column by a scalar k, then det $B = k$ det A or $|B| = k|A|$.

7. **Assertion (A) :** The matrix

$$A = \frac{1}{3}\begin{pmatrix} 1 & -2 & 2 \\ -2 & 1 & 2 \\ -2 & -2 & -1 \end{pmatrix}$$

   is an orthogonal matrix.

   Reason (R) : If A and B are orthogonal, then AB is also orthogonal.

8. **Assertion (A) :** If A is a skew-symmetric matrix of order 3 × 3, then det $(A) = 0$ or $|A| = 0$.

   Reason (R) : If A is a square matrix, then det (A) = det (A') = det $(-A')$.

9. **Assertion (A) :** The inverse of $A = \begin{pmatrix} 3 & 4 \\ 3 & 5 \end{pmatrix}$ does not exist.

   Reason (R) : The matrix A is non-singular.

10. **Assertion (A) :** If a matrix of order 2 × 2, commutes with every matrix of order 2 × 2, then it is scalar matrix.

    Reason (R) : A scalar matrix of order 2 × 2 commutes with every 2 × 2 matrix.

Answers :

1. (B) $\because$ $\text{adj}(\text{adj} A) = |A|^{n-2} A$

Here, $n = 3$

$\therefore$ $(\text{adj})(\text{adj} A) = |A| A$ ...(i)

Now, $|A| = \begin{vmatrix} 3 & -3 & 4 \\ 2 & -3 & 4 \\ 0 & -1 & 1 \end{vmatrix}$

$= 3(-3+4) + 3(2) + 4(-2) = 1$

From equation (i),

$\text{adj}(\text{adj}) A = A$

2. (A) $\because$ $A = \begin{pmatrix} a & 0 & 0 \\ 0 & b & 0 \\ 0 & 0 & c \end{pmatrix}$

$\therefore$ $|A| = abc$

and $\text{adj} A = \begin{pmatrix} bc & 0 & 0 \\ 0 & ca & 0 \\ 0 & 0 & ab \end{pmatrix}$

$\therefore$ $A^{-1} = \dfrac{\text{adj} A}{|A|} = \begin{pmatrix} \frac{1}{a} & 0 & 0 \\ 0 & \frac{1}{b} & 0 \\ 0 & 0 & \frac{1}{c} \end{pmatrix}$

3. (C) In a unit matrix of order $n \times n$,

Number of non-zero rows $= n$.

$\therefore$ Rank is n.

The rank of non-singular matrix of order $n \times n$ is n.

$\because$ For non-singular, $|A| \neq 0$.

4. (D) $A = \begin{bmatrix} a & 0 & 0 & 0 \\ 0 & b & 0 & 0 \\ 0 & 0 & c & 0 \end{bmatrix}$ is not a diagonal matrix.

Since, A is not a square matrix.

5. (A) $\because$ $|A| = \begin{vmatrix} 1 & 1 & 1 \\ 1 & 2 & 3 \\ 1 & 4 & 7 \end{vmatrix} = 0$ [$\because A$ is singular]

$\therefore$ A^{-1} does not exist.

6. (A) If $A = \begin{pmatrix} a_{11} & a_{12} & a_{13} & \cdots & a_{1n} \\ a_{21} & a_{22} & a_{23} & \cdots & a_{2n} \\ \vdots & \vdots & \vdots & & \vdots \\ a_{m1} & a_{m2} & a_{m3} & \cdots & a_{mn} \end{pmatrix}$

$\therefore$ $kA = \begin{pmatrix} ka_{11} & ka_{12} & ka_{13} & \cdots & ka_{1n} \\ ka_{21} & ka_{22} & ka_{23} & \cdots & ka_{2n} \\ \vdots & \vdots & \vdots & & \vdots \\ ka_{m1} & ka_{m2} & ka_{m3} & \cdots & ka_{mn} \end{pmatrix}$

$\therefore$ $|kA| = k^n |A|$

7. (B) $\because AA' = \frac{1}{3}\begin{pmatrix} 1 & -2 & 2 \\ -2 & 1 & 2 \\ -2 & -2 & -1 \end{pmatrix} \cdot \frac{1}{3}\begin{pmatrix} 1 & -2 & -2 \\ -2 & 1 & -2 \\ 2 & 2 & -1 \end{pmatrix}$

$= \frac{1}{9}\begin{pmatrix} 9 & 0 & 0 \\ 0 & 9 & 0 \\ 0 & 0 & 9 \end{pmatrix} = \begin{pmatrix} 1 & 0 & 0 \\ 0 & 1 & 0 \\ 0 & 0 & 1 \end{pmatrix} = I$

$\therefore$ A is orthogonal.

Also, if A and B are orthogonal, then AB is orthogonal.

8. (C) $\because$ $A = \begin{pmatrix} 0 & -c & b \\ c & 0 & a \\ -b & -a & 0 \end{pmatrix}$

$A = -A'$ [$\because A$ is skew-symmetric]

$\therefore$ $\det(A) = \det(-A')$

$= -\det(A')$

$= -\det A$

$\therefore$ $\det A = 0$

$\because$ $\det A' = \det(-A')$ is not true.

$\therefore$ $\det(-A') = (-1)^3 \det(A') = -\det A'$.

9. (D) $\because |A| = \begin{vmatrix} 3 & 4 \\ 3 & 5 \end{vmatrix} = 15 - 12 = 3 \neq 0$

$\therefore$ A is non-singular matrix.

$\therefore$ A^{-1} is exist.

10. (B) Let $A = \begin{pmatrix} a & b \\ c & d \end{pmatrix}, B = \begin{pmatrix} x & y \\ z & u \end{pmatrix}$

$\Rightarrow$ $AB = BA$

$\Rightarrow$ $\begin{pmatrix} ax+bz & ay+bu \\ cx+dz & cy+du \end{pmatrix} = \begin{pmatrix} ax+cy & bx+dy \\ az+cu & bz+du \end{pmatrix}$

On comparing, then

$ax + bz = ax + cy$

$\Rightarrow$ $bz = cy$

$\Rightarrow$ $\dfrac{z}{c} = \dfrac{y}{b} = \lambda$ (say)

$\therefore$ $y = b\lambda, z = c\lambda$ ...(i)

$ay + bu = bx + dy$

$\Rightarrow$ $ab\lambda + bu = bx + bd\lambda$ [From (i)]

$\Rightarrow$ $a\lambda + u = x + d\lambda = k$ (let)

For $\lambda = 0, y = 0, z = 0, u = k, x = k,$

Then, $B = \begin{pmatrix} k & 0 \\ 0 & k \end{pmatrix}$ = scalar matrix.

Also, if $A = \begin{pmatrix} a & b \\ c & d \end{pmatrix}$ and $B = \begin{pmatrix} k & 0 \\ 0 & k \end{pmatrix}$,

Then, $AB = BA = \begin{pmatrix} ak & bk \\ ck & dk \end{pmatrix} = kA.$

Chapter 4. Determinants

1. **Assertion (A) :**

$$\begin{vmatrix} a^2+x^2 & ab-cx & ac+bx \\ ab+cx & b^2+x^2 & bc-ax \\ ac-bx & bc+ax & c^2+x^2 \end{vmatrix} = \begin{vmatrix} x & c & -b \\ -c & x & a \\ b & -a & x \end{vmatrix}^2$$

Reason (R) : $\Delta^c = \Delta^{n-1}$ where n is order of determinant, and Δ^c is the determinant of cofactors of Δ.

2. **Assertion (A) :** The system of equations possess a non trivial solution over the set of rationals $x + ky + 3z = 0$, $3x + ky - 2z = 0$, $2x + 3y - 4z = 0$, then the value of k is 31/2.

Reason (R) : For non trivial solution $\Delta = 0$.

3. **Assertion (A) :**

$$\begin{vmatrix} \cos(\theta+\alpha) & \cos(\theta+\beta) & \cos(\theta+\gamma) \\ \sin(\theta+\alpha) & \sin(\theta+\beta) & \sin(\theta+\gamma) \\ \sin(\beta-\gamma) & \sin(\gamma-\alpha) & \sin(\alpha-\beta) \end{vmatrix}$$

is independent of θ.

Reason (R) : If $f(\theta) = c$, then $f(\theta)$ is independent of θ.

4. **Assertion (A) :** If $\Delta(x) = \begin{vmatrix} f_1(x) & f_2(x) \\ g_1(x) & g_2(x) \end{vmatrix}$,

then $\Delta'(x) \neq \begin{vmatrix} f_1'(x) & f_2'(x) \\ g_1'(x) & g_2'(x) \end{vmatrix}$

Reason (R) : $\frac{d}{dx}\{f(x)g(x)\} \neq \frac{d}{dx}f(x)\frac{d}{dx}g(x).$

5. **Assertion (A) :** If $\Delta(x) = \begin{vmatrix} f(x) & g(x) \\ a & b \end{vmatrix}$, then

$$\int \Delta(x)\,dx = \begin{vmatrix} \int f(x)\,dx & \int g(x)\,dx \\ a & b \end{vmatrix}$$

Reason (R) : $\int \lambda f(x)\,dx = \lambda \int f(x)\,dx$

6. **Assertion (A) :** If a, b, c are distinct and x, y, z are not all zero and $ax + by + cz = 0$, $bx + cy + ax = 0$, $cx + ay + bz = 0$, then $a + b + c = 0$.

Reason (R) : $a^2 + b^2 + c^2 > ab + bc + ca$, if a, b, c are distinct.

Answers :

1. (A) Let $\Delta = \begin{vmatrix} x & c & -b \\ -c & x & a \\ b & -a & x \end{vmatrix}$

$$\therefore \quad \Delta^c = \begin{vmatrix} a^2+x^2 & cx+ab & ac-bx \\ ab-cx & b^2+x^2 & ax+bc \\ ac+bx & bc-ax & c^2+x^2 \end{vmatrix}^T$$

$$= \begin{vmatrix} a^2+x^2 & ab-cx & ac+bx \\ cx+ab & b^2+x^2 & bc-ax \\ ac-bx & ax+bc & c^2+x^2 \end{vmatrix} = \Delta^2$$

$$= \begin{vmatrix} x & c & -b \\ -c & x & a \\ b & -a & x \end{vmatrix}^2$$

2. (D) $\because \quad \Delta = \begin{vmatrix} 1 & k & 3 \\ 3 & k & -2 \\ 2 & 3 & -4 \end{vmatrix}$

Applying $R_2 \to R_2 - 3R_1$ and $R_3 \to R_3 - 2R_1$, then

$$\Delta = \begin{vmatrix} 1 & k & 3 \\ 0 & -2k & -11 \\ 0 & 3-2k & -10 \end{vmatrix}$$

$= 20k + 33 - 22k = 0$

$\therefore \quad k = 33/2$

3. (B) Let $f(\theta) = \begin{vmatrix} \cos(\theta+\alpha) & \cos(\theta+\beta) & \cos(\theta+\gamma) \\ \sin(\theta+\alpha) & \sin(\theta+\beta) & \sin(\theta+\gamma) \\ \sin(\beta-\gamma) & \sin(\gamma-\alpha) & \sin(\alpha-\beta) \end{vmatrix}$

$$\therefore f'(\theta) = \begin{vmatrix} -\sin(\theta+\alpha) & -\sin(\theta+\beta) & -\sin(\theta+\gamma) \\ \sin(\theta+\alpha) & \sin(\theta+\beta) & \sin(\theta+\gamma) \\ \sin(\beta-\gamma) & \sin(\gamma-\alpha) & \sin(\alpha-\beta) \end{vmatrix}$$

$$+ \begin{vmatrix} \cos(\theta+\alpha) & \cos(\theta+\beta) & \cos(\theta+\gamma) \\ \cos(\theta+\alpha) & \cos(\theta+\beta) & \cos(\theta+\gamma) \\ \sin(\beta-\gamma) & \sin(\gamma-\alpha) & \sin(\alpha-\beta) \end{vmatrix}$$

$$+ \begin{vmatrix} \cos(\theta+\alpha) & \cos(\theta+\beta) & \cos(\theta+\gamma) \\ \sin(\theta+\alpha) & \sin(\theta+\beta) & \sin(\theta+\gamma) \\ 0 & 0 & 0 \end{vmatrix}$$

$= 0 + 0 + 0 = 0$

$f'(\theta) = 0$

$\Rightarrow \therefore \quad f(\theta) = c$

4. (A) $\because \quad \frac{d}{dx}f(x)g(x) = f(x)\frac{d}{dx}g(x) + g(x)\frac{d}{dx}f(x)$

$\Rightarrow \frac{d}{dx}f(x)g(x) \neq \frac{d}{dx}f(x)\frac{d}{dx}g(x)$

$$\Rightarrow \quad \Delta(x) = \begin{vmatrix} f_1(x) & f_2(x) \\ g_1(x) & g_2(x) \end{vmatrix}$$

$$= f_1(x)\, g_2(x) - f_2(x)\, g_1(x)$$

$$\therefore \frac{d}{dx}\{\Delta(x)\} = \{f_1(x)\, g_2'(x) + g_2'(x) f_1(x)\}$$

$$- \{f_2(x)\, g_1'(x) + g_1(x) f_2'(x)\}$$

$$\Delta'(x) = \begin{vmatrix} f_1'(x) & f_2'(x) \\ g_1(x) & g_2(x) \end{vmatrix} + \begin{vmatrix} f_1(x) & f_2(x) \\ g_1'(x) & g_2'(x) \end{vmatrix}$$

$$\neq \begin{vmatrix} f'(x) & f_2'(x) \\ g_1'(x) & g_2'(x) \end{vmatrix}$$

5. (A) $\because \quad \Delta x = \begin{vmatrix} f(x) & g(x) \\ a & b \end{vmatrix}$

$$= bf(x) - ag(x)$$

$$\therefore \int \Delta(x)\, dx = \int \{bf(x) - ag(x)\}\, dx$$

$$= b\int f(x)\, dx - a\int g(x)\, dx$$

$$= \begin{vmatrix} \int f(x)\, dx & \int g(x)\, dx \\ a & b \end{vmatrix}$$

6. (A) $\because \Delta = \begin{vmatrix} a & b & c \\ b & c & a \\ c & a & b \end{vmatrix}$

$$= -(a^3 + b^3 + c^3 - 3abc)$$

$$= -(a + b + c)(a^2 + b^2 + c^2 - ab - bc - ca) = 0$$

$[\because a + b + c = 0]$ (non-trival)

and $a^2 + b^2 + c^2 - ab - bc - ca$

$$= \frac{1}{2}\{(a-b)^2 + (b-c)^2 + (c-a^2)\} > 0.$$

Chapter 5. Continuity and Differentiability

1. **Assertion (A) :** The function $f(x) = |x|$ is discontinuous at $x = 0$.

   Reason (R) : The function $f(x) = |x|$ is non-differentiable at $x = 0$

2. **Assertion (A) :** $f(x) = \dfrac{1}{\{x\}}$ is discontinuous for integral values of x, where { } denotes the fractional part function.

   Reason (R) : For integral values of x, $f(x)$ is not defined.

3. **Assertion (A) :** The function $f(x)$ in the figure is differentiable at $x = a$.

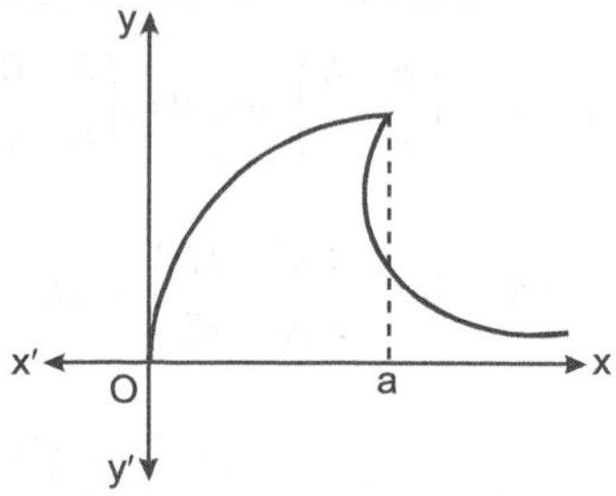

   Reason (R) : The function $f(x)$ is continuous at $x = a$

4. **Assertion (A) :** $f(x) = \sqrt{(x-2)} + \sqrt{(2-x)}$ is continuous at $x = 2$

   Reason (R) : $f(x)$ is a point function.

5. **Assertion (A) :** The function

$$f(x) = \begin{cases} [x] + \sqrt{x - [x]}, & x \geq 0 \\ \sin x, & x < 0 \end{cases},$$

   where [ ] denotes the greatest integer function, is continuous everywhere.

   Reason (R) : $f(x)$ is a periodic function

6. **Assertion (A) :** The function $f(x) = \{x\}$, where { } denotes the fractional part function, is discontinuous at $x = 1$.

   Reason (R) : $\lim_{x \to 1^-} f(x) \neq \lim_{x \to 1^+} f(x)$

7. **Assertion (A) :** The function

$$f(x) = \frac{(27 - 2x)^{1/3} - 3}{9 - 3(243 + 5x)^{1/5}}$$

   is continuous eveywhere is $f(0) = 2$.

   Reason (R) : For continuous function

$$f(0) = \lim_{x \to 0} f(x)$$

8. **Assertion (A) :** If $f(x)$ is continuous, then $|f(|x|)|$ is also continuous.

   Reason (R) : If $|f(x)| \leq |x| \ \forall x \in R$, then $|f(x)|$ is continuous at $x = 0$.

Answers :

1. (D) It is clear from the figure $f(x)$ is continuous at $x = 0$ but non-differentiable at $x = 0$.

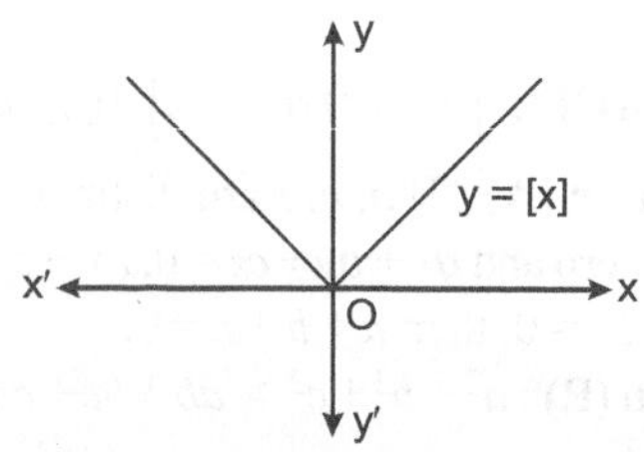

2. (A) If $x \in I$, then $\{x\} = 0$

$\because \quad 0 \leq \{x\} < 1$

or $\quad \{x\} = x - \{x\} = 0$

$\Rightarrow \quad x = [x] \ \ (\text{for } x \in I)$

3. (D) At $x = a$, two tangents can be drawn. Hence, $f(x)$ is non-differentiable at $x = a$ but continuous at $x = a$.

4. (D) $\sqrt{x-2}$ is defined for $x \geq 2$

and $\sqrt{(2-x)}$ is defined for $x \leq 2$

$\therefore f(x)$ is defined only when at $x = 2$

$\therefore \quad f(2) = 0$

Hence, $f(x)$ is discontinuous at $x = 2$

And $f(x)$ is a point function.

[$\because$ domain and range consist of one value only]

5. (C) Hence, $f(x)$ is continuous everywhere but non-periodic function.

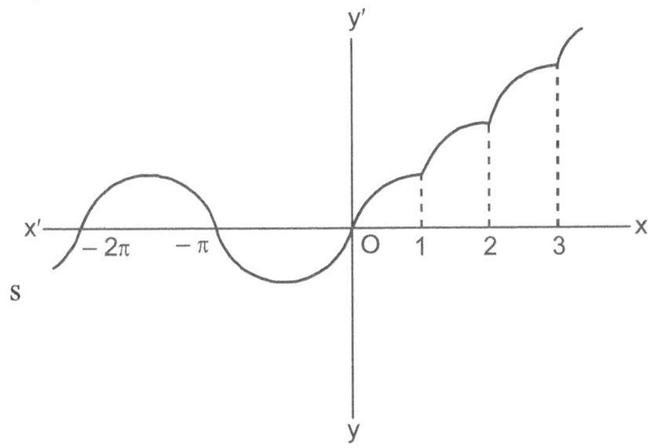

6. (A) It is clear from figure

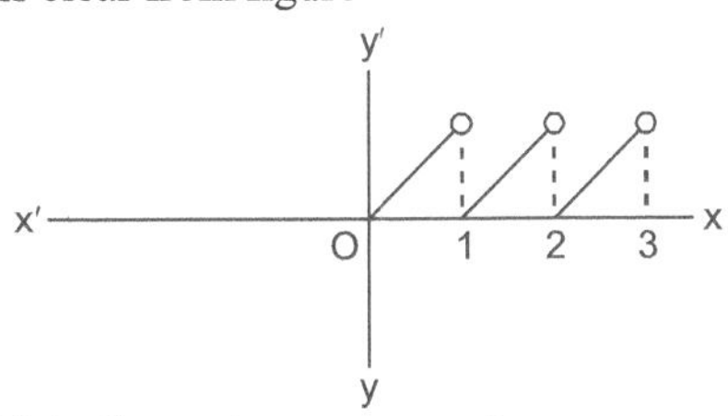

$f(x)$ is discontinuous at $x = 1$

$$\because \lim_{x \to 1^-} f(x) \neq \lim_{x \to 1^+} f(x)$$

7. (A) $f(0) = \text{RHL} = \lim_{x \to 0^+} f(x) = \lim_{h \to 0} f(h)$

$$= \lim_{h \to 0} \frac{(27-2h)^{1/3} - 3}{9 - 3(243+5h)^{1/5}}$$

$$= \frac{1}{3} \lim_{h \to 0} \frac{\dfrac{(27-2h)^{1/3} - (27)^{1/3}}{(27-2h) - 27} \times (-2h)}{\dfrac{(243)^{1/5} - (243+5h)^{1/5}}{243 - (243+5h)} \times (-15h)}$$

$$= \frac{1}{3} \cdot \frac{2}{5} \cdot \frac{\frac{1}{3}(27)^{-2/3}}{\frac{1}{5}(243)^{-4/5}}$$

$$= \frac{2}{15} \cdot \frac{5}{3} \cdot \frac{3^{-2}}{3^{-4}} = \frac{2}{9} \cdot 9 = 2$$

8.

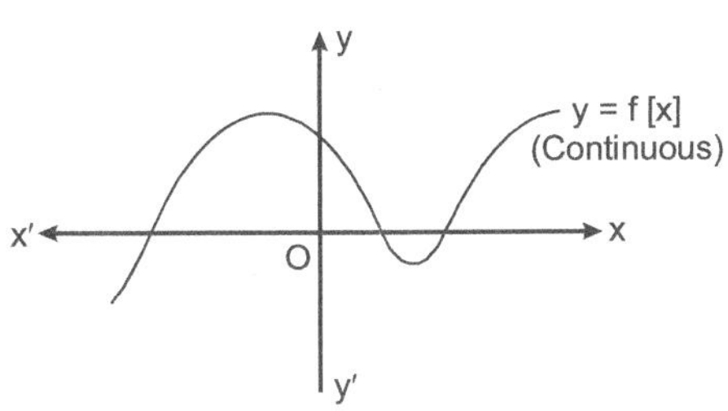

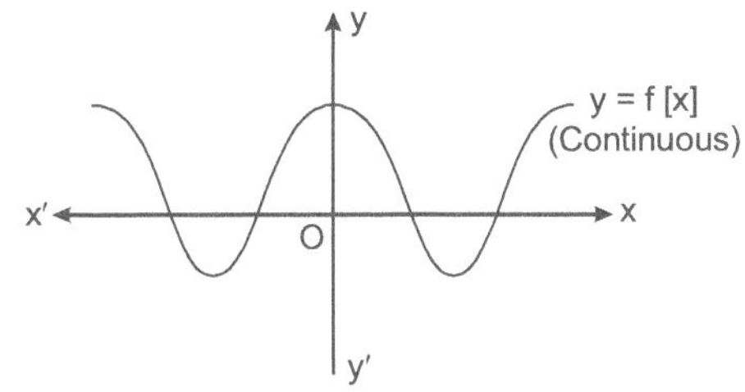

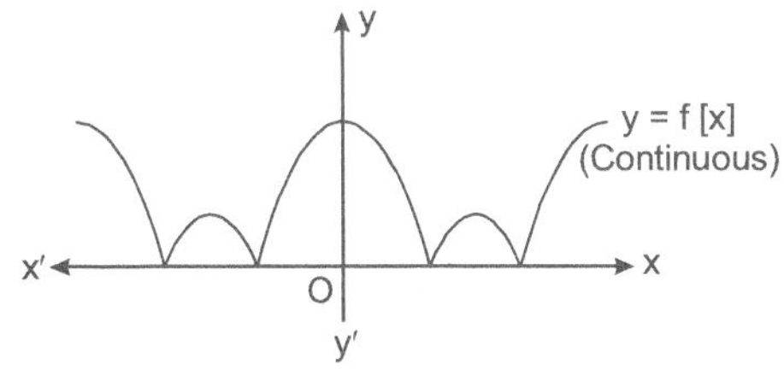

Also, $\because |f(x)| \leq |x|$

$\Rightarrow \quad |f(0)| \leq |0|$

$\Rightarrow \quad |f(0)| \leq 0$

$\therefore \quad |f(0)| = 0$

$\Rightarrow \quad f(0) = 0$

$\because \quad |f(x)| \leq |x|$

$$\lim_{x \to 0} |f(x)| \leq \lim_{x \to 0} |x|$$

$$\Rightarrow \quad \lim_{x \to 0} |f(x)| \leq 0$$

$$\text{But} \quad \lim_{x \to 0} |f(x)| \geq 0$$

$$\therefore \quad \lim_{x \to 0} |f(x)| = 0\, |f(0)|$$

$$\text{RHL} = \text{V.F.}$$

Hence, $|f(x)|$ is continuous at $x = 0$.

Chapter 6. Applications of Derivatives

1. **Assertion (A) :** The points of contact of the vertical tangents to $-x = 2 - 3 \sin \theta$, $y = 3 + 2 \cos \theta$ are $(-1, 3)$ and $(5, 3)$.

   Reason (R) : For vertical tangent, $\dfrac{dx}{d\theta} = 0$.

2. **Assertion (A) :** If $y^2 = 3 + 2x - x^2$ then, at $(3, 0)$ and $(-1, 0)$ tangent is perpendicular to x-axis.

   Reason (R) : At $(3, 0)$ and $(-1, 0)$, $\dfrac{dy}{dx} = \infty$.

3. **Assertion (A) :** The points on the curve $y^2 = x + \sin x$ at which the tangent is parallel to x-axis lie on a straight line.

   Reason (R) : Tangent is parallel to x-axis, then

   $$\frac{dy}{dx} = 0 \text{ or } \frac{dy}{dx} = \infty.$$

4. **Assertion (A) :** Equation of tangents to the curve $f(x) = x^2$ at the point where slope of tangent is equal to functional value of the curve is $4x - y - 4 = 0, y = 0$.

   Reason (R) : $f'(x) = f(x)$.

5. Assertion (A) : If S be the area of a circle having radius x and A is the area of an equilateral triangle having side πx at any instant, then $\frac{dA}{dt} > \frac{dS}{dt}$

Reason (R) : $A > S$

6. Assertion (A) : The ordinate of a point describing the circle $x^2 + y^2 = 25$ decreases at the rate of 1.5 cm/s. The rate of change of the abscissa of the point when ordinate equals 4 cm is 2 cm/s.

Reason (R) : $x\,dx + y\,dy = 0$.

7. Assertion (A) : The possible percentage error in computing the parallel resistance R of three resistances R_1, R_2, R_3 from the formula

$$\frac{1}{R} = \frac{1}{R_1} + \frac{1}{R_2} + \frac{1}{R_3},$$

if R_1, R_2, R_3 are each in error by plus 1.2% is 1.2%.

Reason (R) : $\frac{\Delta R}{R^2} = \frac{\Delta R_1}{R_1^2} + \frac{\Delta R_2}{R_2^2} + \frac{\Delta R_3}{R_3^2}$.

8. Assertion (A) : Two cyclists move on two separate roads angled at 60° to each other at the rate of 5 m/s and 10 m/s respectively. The rate which they are separating from each other is $5\sqrt{3}$ m/s.

Reason (R) : Using sine rule.

Answers :

1. (A) For vertical tangent $\frac{dx}{d\theta} = 0$

$\therefore -3\cos\theta = 0 \Rightarrow \cos\theta = 0 \Rightarrow \theta = \pi/2, 3\pi/2$

At $\theta = \pi/2$; $x = 2 - 3 = -1$, $y = 3 + 0 = 3$ *i.e.*, $(-1, 3)$

and At $\theta = 3\pi/2$; $x = 2 + 3 = 5$ and $y = 3 + 0 = 3$

i.e., (5, 3)

2. (A) $2y\frac{dy}{dx} = (2 - 2x)$

$$\Rightarrow \frac{dy}{dx} = \left(\frac{1-x}{y}\right)$$

$$\left.\frac{dy}{dx}\right|_{(3,0)} = \infty \text{ and } \left.\frac{dy}{dx}\right|_{(-1,0)} = \infty$$

i.e., tangents makes an angle $\frac{\pi}{2}$ with x-axis.

3. (D) $\because \quad y^2 = x + \sin x \qquad$ ...(i)

$\therefore \quad 2y\frac{dy}{dx} = 1 + \cos x = 0 \qquad \left[\because \frac{dy}{dx} = 0\right]$

$\therefore \quad \cos x = -1,$

then $\sin x = 0$

From equation (i), $y^2 = x$ (Parabola).

4. (A) Given $f'(x) = f(x)$

$\Rightarrow 2x = x^2$

$\Rightarrow x = 0, 2$

at $x = 0, y = 0$

and At $x = 2, y = 4$

So we have to find equation of tangents at (0, 0) and (2, 4).

At (0, 0),

$f'(0) = 0$ and at (2, 4), $f'(2) = 4$

$\therefore$ Tangents are $y - 0 = 0(x - 0)$ and $y - 4 = 4(x - 2)$

i.e., $\quad y = 0$ and $4x - y - 4 = 0$.

5. (B) $\because S = \pi x^2$

$$\Rightarrow \frac{dS}{dt} = 2\pi x\frac{dx}{dt}$$

and $\quad A = \frac{\sqrt{3}}{4}\pi^2 x^2$

$$\Rightarrow \frac{dA}{dt} = 2\pi^2\frac{\sqrt{3}}{4}x\frac{dx}{dt}$$

then $$\frac{dS}{dA} = \frac{\frac{dS}{dt}}{\frac{dA}{dt}} = \frac{2\pi x\frac{dx}{dt}}{\pi^2\frac{\sqrt{3}x}{2}\frac{dx}{dt}} = \frac{4}{\pi\sqrt{3}} < 1$$

$$\therefore \quad \frac{dS}{dt} < \frac{dA}{dt} \text{ or } \frac{dA}{dt} > \frac{dS}{dt}$$

But $$\frac{A}{S} = \frac{\frac{\sqrt{3}}{4}\pi^2x^2}{\pi x^2} = \frac{\sqrt{3}}{4}\pi > 1$$

$\therefore \quad A > S$

6. (A) $\because \quad x^2 + y^2 = 25$

$\Rightarrow \quad 2x\,dx + 2y\,dy = 0$

or $$\frac{dy}{dx} = -\frac{x}{y}$$

$$\Rightarrow \frac{\frac{dy}{dt}}{\frac{dx}{dt}} = -\frac{x}{y}$$

$$\Rightarrow \frac{-1.5}{\frac{dx}{dt}} = -\frac{3}{4} \qquad \left[\because x = \sqrt{(25 - y)^2}\right]$$

or $$\frac{dx}{dt} = \frac{1.5 \times 4}{3} = 2 \text{ cm/s}$$

7. (A) $\because \quad \frac{1}{R} = \frac{1}{R_1} + \frac{1}{R_2} + \frac{1}{R_3}$

$$\Rightarrow -\frac{1}{R^2}\Delta R = -\frac{1}{R_1^2}\Delta R_1 - \frac{1}{R_2^2}\Delta R_2 - \frac{1}{R_3^2}\Delta R_3$$

$$\Rightarrow \frac{1}{R}\left(\frac{\Delta R}{R}\right) = \frac{1}{R_1}\left(\frac{\Delta R_1}{R_1}\right) + \frac{1}{R_2}\left(\frac{\Delta R_2}{R_2}\right) + \frac{1}{R_3}\left(\frac{\Delta R_3}{R_3}\right)$$

$$\Rightarrow \frac{1}{R}\left(\frac{\Delta R}{R}\times 100\right) = \frac{1}{R_1}\left(\frac{\Delta R_1}{R_1}\times 100\right) + \frac{1}{R_2}\left(\frac{\Delta R_2}{R_2}\times 100\right) + \frac{1}{R_3}\left(\frac{\Delta R_3}{R_3}\times 100\right)$$

$$\Rightarrow \frac{1}{R}\left(\frac{\Delta R}{R}\times 100\right) = (1.2)\left(\frac{1}{R_1}+\frac{1}{R_2}+\frac{1}{R_3}\right)$$

$$= (1.2)\left(\frac{1}{R}\right)$$

$$\therefore \quad \frac{\Delta R}{R}\times 100 = 1.2\%$$

8. (C) $\because \quad \cos 60° = \dfrac{x^2+(2x)^2-y^2}{2\cdot x\cdot 2x}$

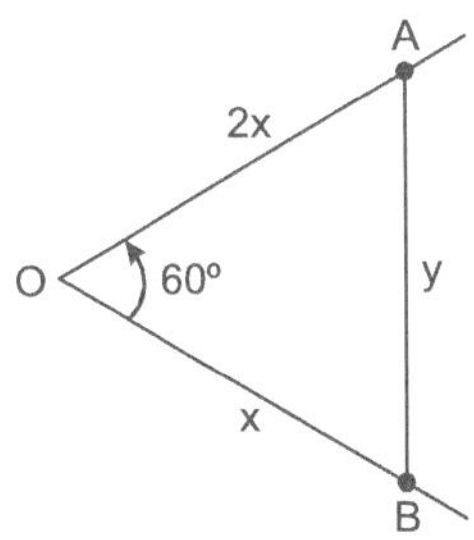

$$\Rightarrow \quad 2x^2 = 5x^2 - y^2$$

$$\Rightarrow \quad y = x\sqrt{3}$$

$$\therefore \quad \frac{dy}{dt} = \sqrt{3}\,\frac{dx}{dt} = 5\sqrt{3} \text{ m/s.}$$

Chapter 7. Integrals

1. **Assertion (A):** $\int e^x(\tan x + \sec^2 x)dx = e^x\cdot \tan x + c$

   Reason (R): $\int e^x\{f(x)+f'(x)\}\,dx = e^x f(x) + c$

2. **Assertion (A) :** If $I_n = \int \cot^n x\,dx$, then $5(I_6 + I_4) = -\cot^5 x$

   Reason (R) : If $I_n = \int \cot^n x dx$

   then $I_n = -\dfrac{\cot^{n-1} x}{n} - I_{n-2}$, where $n \geq 2$

3. **Assertion (A) :** $\int \dfrac{dx}{\sqrt{1-x^2)^2}} = \sin^{-1} x^2 + c$

   Reason (R) : $\int \dfrac{dx}{\sqrt{1-x^2}} = \sin^{-1} x + c$

4. **Assertion (A) :** $\int_{-1}^{1} |x|\,dx$ can be found while $\int |x|\,dx$ can not be found.

   Reason (R) : $|x|$ is non-differentiable at $x = 0$.

5. **Assertion (A) :**

   $$\int \frac{(3-2x)}{\sqrt{(4+2x-x^2)}}dx = 2\sqrt{(4+2x-x^2)} + \sin^{-1}\left(\frac{x-1}{\sqrt{5}}\right) + c$$

   Reason (R) :

   $$\int \frac{dx}{\sqrt{(a^2-x^2)}} = \frac{x}{2}\sqrt{a^2-x^2} + \frac{a^2}{2}\sin^{-1}\left(\frac{x}{a}\right)$$

6. **Assertion (A) :** $\int \tan 5x \tan 3x \tan 2x\,dx$

   $$= \frac{\text{In}\,|\sec 5x|}{5} - \frac{\text{In}\,|\sec 3x|}{3} - \frac{\text{In}\,|\sec 2x|}{2} + c$$

   Reason (R) :

   $\tan 5x - \tan 3x - \tan 2x = \tan 5x \tan 3x \tan 2x$

7. **Assertion (A) :** If $y(x-y)^2 = x$, then

   $$\int \frac{dx}{(x-3y)} = \frac{1}{2}\,\text{In}\,\{(x-y)^2 - 1\}$$

   Reason (R) : $\int \dfrac{dx}{(x-3y)} = \text{In}\,(x-3y) + c$

8. **Assertion (A) :** $\int e^{x^2}dx = e^{x^2} + c$

   Reason (R) : $\int e^x dx = e^x + c$

9. **Assertion (A) :** If $\int \dfrac{1}{f(x)}dx = 2\,\text{In}\,|f(x)| + c$, then

   $f(x) = \dfrac{x}{2}$

   Reason (R) : When $f(x) = \dfrac{x}{2}$ then

   $$\int \frac{1}{f(x)}dx = \int \frac{2}{x}dx = 2\,\text{In}\,|x| + c$$

10. **Assertion (A) :** The function $F(x) = \int \sin^2 x\,dx$ satisfies $F(x+\pi) = F(x)\ \forall\ x \in R$

    Reason (R) : $\sin^2(x+\pi) = \sin^2 x$

11. **Assertion (A) :** $F(x) = \int_0^x \text{In}\left(\dfrac{1-t}{1+t}\right)dt$ is an even function.

    Reason (R) : In $\left(\dfrac{1-t}{1+t}\right)$ is an odd function.

12. **Assertion (A) :** $\int_0^1 e^{-x}\cos^2 x\,dx < \int_0^1 e^{-x^2}\cos^2 x\,dx$

    Reason (R) : $\int_a^b f(x)\,dx < \int_a^b g(x)\,dx\ \forall f(x) \geq g(x)$

13. Assertion (A) : $\int_0^{\pi/2} \frac{\sin x}{x} dx < \frac{\pi}{2}$

Reason (R) : $\lim_{x \to 0} \frac{\sin x}{x} = 1$

14. Assertion (A) :

$$\lim_{n\to\infty}\left(\frac{1}{n+1}+\frac{1}{n+2}+\frac{1}{n+3}+...+\frac{1}{2n}\right) = \text{In } 3.$$

Reason (R) : $\lim_{n\to\infty} \sum_{r=1}^{n} \frac{1}{n} f\left(\frac{r}{n}\right) = \int_0^1 f(x)\, dx,$

symbols have their usual meaning.

15. Assertion (A) : $\int_{\pi/6}^{\pi/3} \frac{(\sin x)^{\sqrt{2}}}{(\sin x)^{\sqrt{2}} + (\cos x)^{\sqrt{2}}} dx = \frac{\pi}{12}$

Reason (R) : $\int_{\pi/6}^{\pi/3} \frac{f(x)\, dx}{f(x) + f\left(\frac{\pi}{2} - x\right)} dx = \frac{\pi}{12}$

16. Assertion (A) : If

$$\int_0^1 e^{\sin x} dx = \lambda, \text{ then } \int_0^{200} e^{\sin x} dx = 200\lambda$$

Reason (R) : $\int_0^{na} f(x)\, dx = n\int_0^a f(x)\, dx,\ n \in I$ and $f(a+x) = f(x)$

17. Assertion (A) : $\int_1^6 \{x+5\}^2 dx = 41,$

where { } denotes the fractional part function.

Reason (R) : $\{x+5\}$ is a periodic function.

18. Assertion (A) : $\int_0^{2\pi} \sin^3 x\, dx = 0$

Reason (R) : $\sin^3 x$ is an odd function.

19. Assertion (A) : $\int_{-\pi/2}^{\pi/2} |\sin x|\, dx = 2$

Reason (R) : $\int_a^b f(x)\, dx = \int_a^c f(x)\, dx + \int_c^b f(x)\, dx,$

where $c \in (a, b)$

20. Assertion (A) : $\int_{\pi/2}^{3\pi/2} [2\sin x]\, dx = 0,$

where [ ] denotes the greatest integer function.

Reason (R) : 2 sin x decreasing function in $\left(\frac{\pi}{2}, \frac{3\pi}{2}\right)$.

21. Assertion (A) : $\int_0^t \cos x\, dx = \sin t$

Reason (R) : cos x is continuous in any closed interval $[0, t]$.

22. Assertion (A) : $\int_0^{1/4} \frac{dx}{1+(\cot 2\pi x)^{\sqrt{2}}} = \frac{1}{4}$

Reason (R) : If $a + b = \pi/2$, then

$$\int_a^b \frac{f(\sin x)}{f(\sin x)+f(\cos x)} dx = \frac{(b-a)}{2}$$

23. Assertion (A) : $\int_{-1}^{1} \frac{|x|}{x} dx = 2$

Reason (R) : $\frac{|x|}{x}$ is undefined at $x = 0$

24. Assertion (A) : $\int_0^2 f(x)\, dx = \frac{4(\sqrt{2}-1)}{3},$

where $f(x) = \begin{cases} x^2, & \text{for } 0 \le x < 1 \\ \sqrt{x}, & \text{for } 1 \le x \le 2 \end{cases}$

Reason (R) : $f(x)$ is continuous in $[0, 2]$

25. Assertion (A) : $\int_0^{\pi/2} (\sin^6 x + \cos^6 x)\, dx$ lies in the interval $\left(\frac{\pi}{8}, \frac{\pi}{2}\right)$.

Reason (R) : $\sin^6 x + \cos^6 x$ is periodic with period $\pi/2$.

26. Assertion (A) : If $f(x) = x - x^2 + 1$ and $g(x) = \max \{f(t) : 0 \le t \le x\}$,

then $\int_0^1 g(x)\, dx = \frac{29}{24}$

Reason (R) : $f(x)$ is increasing in $\left(0, \frac{1}{2}\right)$ and decreasing in $\left(\frac{1}{2}, 1\right)$.

Answers :

1. (A) $\because \int e^x \{f(x) + f'f(x)\}\, dx = \int e^x f(x)\, dx$

$$+ \int e^x f'(x)\, dx$$

$$= f(x)\cdot e^x - \int f'(x)\, e^x dx + \int e^x f'(x)\, dx$$

$$= e^x f(x)$$

For $f(x) = \tan x$

$$\int e^x (\tan x + \sec^2 x)\, dx = e^x \cdot \tan x + c$$

2. (C) $\because I_n = \int \cot^n x\, dx = \int \cot^{n-2} x \cdot (\text{cosec}^2 x - 1)\, dx$

$$= \int \cot^{n-2} x\, \text{cosec}^2 x\, dx - I_{n-2}$$

$$= -\frac{\cot^{n-1} x}{(n-1)} - I_{n-2}$$

For $n = 6$

$$5(I_6 + I_4) = -\cot^5 x$$

3. (D) The formula $\int \frac{dx}{\sqrt{1-x^{2n}}} = \sin^{-1} x^n$ is valid only for $n = 1$.

4. (B) $\because$ $\int_{-1}^{1} |x|\, dx = \int_{-1}^{0} |x|\, dx + \int_{0}^{1} |x|\, dx$

$$= \int_{-1}^{0} (-x) dx + \int_{0}^{1} (x)\, dx$$

$$= -\left\{\frac{x^2}{2}\right\}_{-1}^{0} + \left\{\frac{x^2}{2}\right\}_{0}^{1}$$

$$= -\left(0 - \frac{1}{2}\right) + \left(\frac{1}{2} - 0\right) = 1$$

and $\int |x|\, dx$ cannot be found, since condition on x is not given.

Also, $|x|$ is non-differentiable at $x = 0$.

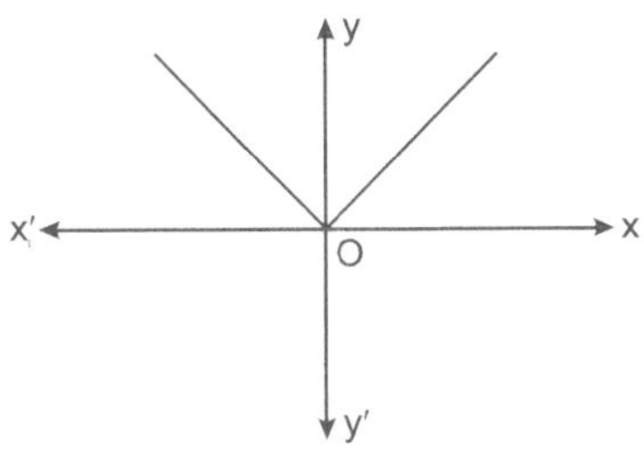

5. (C) Let $I = \int \frac{(3-2x)}{\sqrt{(4+2x-x^2)}} dx$

$$= \int \frac{(2-2x)}{\sqrt{(4+2x-x^2)}} dx + \int \frac{dx}{(4+2x-x^2)}$$

$$= 2\sqrt{(4+2x-x^2)} + \int \frac{dx}{\sqrt{5-(x-1)^2}}$$

$$= 2\sqrt{(4+2x-x^2)} + \sin^{-1}\left(\frac{x-1}{\sqrt{5}}\right) + c$$

and $\int \sqrt{(a^2-x^2)}\, dx$

$$= \frac{x}{2}\sqrt{(a^2-x^2)} + \frac{a^2}{2} \sin^{-1}\left(\frac{x}{2}\right) + c$$

6. (A) $\because$ $5x = 3x + 2x$

$$\therefore \quad \tan 5x = \frac{\tan 3x + \tan 2x}{1 - \tan 3x \tan 2x}$$

$\therefore$ $\tan 5x - \tan 3x - \tan 2x = \tan 5x \tan 3x \tan 2x$

7. (C) Let $P = \int \frac{dx}{(x-3y)} = \frac{1}{2}$ In $\{(x-y)^2 - 1\}$

$$\therefore \quad P = \int \frac{dx}{(x-3y)}$$

$$\Rightarrow \quad \frac{dP}{dx} = \frac{1}{(x-3y)} \quad ...(i)$$

Also, $P = \frac{1}{2}$ In $\{(x-y)^2 - 1\}$

$$\therefore \quad \frac{dP}{dx} = \frac{2(x-y)\left(1-\frac{dy}{dx}\right)}{2\{(x-y)^2-1\}} = \frac{(x-y)\left(1-\frac{dy}{dx}\right)}{(x-y)^2-1} \quad ...(ii)$$

Given, $y(x-y)^2 = x$

In $y + 2$ In $(x - y) =$ In x

$$\Rightarrow \quad \frac{1}{y}\frac{dy}{dx} + \frac{2}{(x-y)}\left(1 - \frac{dy}{dx}\right) = \frac{1}{x}$$

$$\Rightarrow \quad \frac{dy}{dx}\left(\frac{1}{y} - \frac{2}{x-y}\right) = \frac{1}{x} - \frac{2}{x-y} = \frac{x-y-2x}{x(x-y)}$$

$$\Rightarrow \quad \frac{dy}{dx}\left(\frac{x-3y}{y(x-y)}\right) = -\frac{(x+y)}{x(x-y)}$$

or $$\frac{dy}{dx} = -\frac{y(x+y)}{x(x-3y)}$$

Now, from equation (ii),

$$\frac{dP}{dx} = \frac{(x-y)\left\{1 + \frac{y(x+y)}{x(x-2y)}\right\}}{(x-y)^2 - 1}$$

$$= \frac{(x-y)\left\{\frac{x^2-2xy+y^2}{x(x-3y)}\right\}}{\left(\frac{x}{y}-1\right)}$$

$$= \frac{y(x-y)^2}{x(x-3y)} = \frac{1}{x-3y} \quad ...(iii)$$

$\therefore$ It is true from (i),

Hence, $\int \frac{dy}{x-3y} = \frac{1}{2}$ In $\{(x-y)^2 - 1\}$

$\because$ y is variable

$\therefore$ $\int \frac{dx}{x-2y} \neq$ In $(x - 3y)$

8.(D) $\because$ $\int e^{x^2} dx$ cannot expressed in terms of elementary function, then integral is known as inexpressible or that is "cannot be found".

9. (A) $\because$ $\int \frac{1}{f(x)} dx = 2$ In $|f(x)| + c$

Differentiating both sides w.r.t. x, then

$$\because \quad \frac{1}{f(x)} = \frac{2}{f(x)} f'(x)$$

or $$f'(x) = \frac{1}{2}$$

$$\therefore \quad f(x) = \frac{x}{2} + c$$

If $f(0) = 0$, then $f(x) = \frac{x}{2}$

10. (A) $\because$ $F(x+\pi) = \int \sin^2(x+\pi)\,dx$

$= \int \sin^2 x\,dx$ $[\because \sin^2(x+\pi) = \sin^2 x]$

$= F(x)$

11. (A) $\because$ $F(x) = \int_0^x \text{In}\left(\frac{1-t}{1+t}\right)dt$

$\therefore F'(x) = \text{In}\left(\frac{1-x}{1+x}\right)$

$\Rightarrow F'(-x) = \text{In}\left(\frac{1+x}{1-x}\right) = -\text{In}\left(\frac{1-x}{1+x}\right) = -F'(x)$

$\Rightarrow F'(x)$ is an odd function.

Hence, $F'(x)$ is an odd function.

12. (C) $\because 0 < x < 1$

Then $x > x^2$

$\Rightarrow -x < -x^2$

$\Rightarrow e^x < e^{-x^2}$

$\Rightarrow e^{-x}\cos^2 x < e^{-x^2}\cos^2 x$

Then, $\int_0^1 e^{-x}\cos^2 x\,dx < \int_0^1 e^{-x^2}\cos^2 x\,dx$

and if $f(x) \geq g(x)$

Then, $\int_a^b f(x)\,dx \geq \int_a^b g(x)\,dx$

13. (B) For $x > 0$

$\sin x < x$

$\Rightarrow \frac{\sin x}{x} < 1$

$\Rightarrow \int_0^{\pi/2} \frac{\sin x}{x}dx < \int_0^{\pi/2} 1\cdot dx$

$\Rightarrow \int_0^{\pi/2} \frac{\sin x}{x}dx < \frac{\pi}{2}$

and $\lim_{x\to 0}\frac{\sin x}{x} = 1$

14. (D) $\because \lim_{n\to\infty}\left(\frac{1}{n+1}+\frac{1}{n+2}+\frac{1}{n+3}+...+\frac{1}{2n}\right)$

$= \lim_{n\to\infty}\sum_{r=1}^{n}\frac{1}{(n+r)}$

$= \lim_{n\to\infty}\sum_{r=1}^{n}\frac{1}{n}\cdot\frac{1}{\left(1+\frac{r}{n}\right)}$

$= \int_0^1 \frac{dx}{1+x} = [\text{In}(1+x)]_0^1$

$= \text{In } 2 - \text{In } 1$

$= \text{In } 2 - 0$

$= \text{In } 2$

15. (A) Let $I = \int_{\pi/6}^{\pi/3}\frac{f(x)\,dx}{f(x)+f\left(\frac{\pi}{2}-x\right)}$ ...(i)

$= \int_{\pi/6}^{\pi/3}\frac{f\left(\frac{\pi}{3}+\frac{\pi}{6}-x\right)dx}{f\left(\frac{\pi}{3}+\frac{\pi}{6}-x\right)+f\left[\frac{\pi}{2}-\left(\frac{\pi}{3}+\frac{\pi}{6}-x\right)\right]}$

(by property)

$= \int_{\pi/6}^{\pi/3}\frac{f\left(\frac{\pi}{2}-x\right)dx}{f\left(\frac{\pi}{2}-x\right)+f(x)}$ ...(ii)

Adding equations (i) and (ii), then

$2I = \int_{\pi/6}^{\pi/3} 1.\,dx = \frac{\pi}{3}-\frac{\pi}{6} = \frac{\pi}{6}$

$\therefore \quad I = \frac{\pi}{12}$

Also for $f(x) = (\sin x)^{\sqrt{2}}$, $I = \frac{\pi}{12}$

16. (D) $\because$ Period of $e^{\sin x}$ is 2π

$\therefore \int_0^{200} e^{\sin x}dx \neq 200\lambda$

17. (D) $\because$ $\int_1^6 \{x+5\}^2 dx = \int_0^5 \{x+6\}^2 dx$ (by property)

$= \int_0^5 \{x\}^2\,dx$

$= 5\int_0^1 \{x\}^2 dx$

[$\because$ { } is periodic with period 1]

$= 5\int_0^1 x^2 dx$

$= \frac{5}{3}$

18. (B) $\because I = \int_0^{2\pi}\sin^3 x\,dx = \int_0^{2\pi}(1-\cos^2 x)\sin x\,dx$

Put $\cos x = t$

$\Rightarrow \sin x\,dx = -dt$

Then, $I = \int_1^1 (1-t)^2(-dt) = 0$

19. (B) $\because$ $|\sin x|$ is an even function

$\therefore \int_{-\pi/2}^{\pi/2}|\sin x|\,dx = 2\int_0^{\pi/2}|\sin x|\,dx$

$= 2\int_0^{\pi/2}\sin x\,dx$

$= -2(\cos x)_0^{\pi/2} = -2(0-1) = 2$

20. (D) $\because \int_{\pi/2}^{3\pi/2} [2\sin x]\, dx$

$$= \int_{\pi/2}^{5\pi/6} [2\sin x]\, dx + \int_{5\pi/6}^{\pi} [2\sin x]\, dx + \int_{\pi}^{7\pi/6} [2\sin x]\, dx + \int_{7\pi/6}^{3\pi/2} [2\sin x]\, dx$$

$$= \int_{\pi/2}^{5\pi/6} 1.\, dx + 0 - \int_{\pi}^{7\pi/6} 1.\, dx - 2\int_{7\pi/6}^{3\pi/2} 1.\, dx$$

$$= \frac{\pi}{3} - \frac{\pi}{6} - \frac{2\pi}{3}$$

$$= -\frac{\pi}{2}$$

$\left[\because 2\sin x \text{ is decreasing function is } \left(\frac{\pi}{2}, \frac{3\pi}{2}\right)\right]$

21. (A) $\because$ $\cos x$ is continuous in $[0, t]$

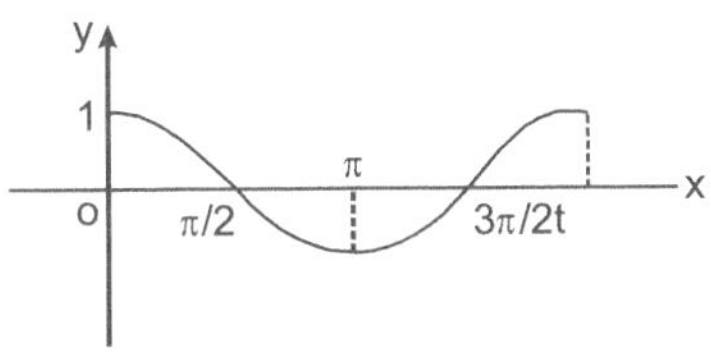

$\therefore \int_0^t \cos x\, dx = [\sin x]_0^t = \sin t - \sin 0 = \sin t$

22. (D) Let $I = \int_0^{1/4} \frac{dx}{1+(\cot 2\pi x)^{\sqrt{2}}}$

$$= \frac{1}{2\pi}\int_0^{\pi/2} \frac{dx}{1+(\cot x)^{\sqrt{2}}}$$

$$= \frac{1}{2\pi}\int_0^{\pi/2} \frac{(\sin x)^{\sqrt{2}}}{(\sin x)^{\sqrt{2}}+(\cos x)^{\sqrt{2}}}\, dx$$

$$= \frac{1}{2\pi}\left(\frac{\frac{\pi}{2}-0}{2}\right) = \frac{1}{8}$$

and let $I_1 = \int_a^b \frac{f(\sin x)}{f(\sin x)+f(\cos x)}\, dx$ ...(i)

$$I_1 = \int_a^b \frac{f(\sin(a+b-x))}{f(\sin(a+b-x))+f(\cos(a+b-x))}\, dx$$

$$= \int_a^b \frac{f(\cos x)}{f(\cos x)+f(\sin x)}\, dx \quad \text{...(ii)}$$

Adding equations (i) and (ii), then

$$2I_1 = \int_a^b dx = b-a$$

$$\therefore \quad I_1 = \frac{(b-a)}{2}$$

23. (D) The function $\frac{|x|}{x}$ is undefined at $x = 0$

$$\therefore \quad \int_{-1}^{1} \frac{|x|}{x}\, dx = \int_{-1}^{0} \frac{|x|}{x}\, dx + \int_0^1 \frac{|x|}{x}\, dx$$

$$= \int_{-1}^{0} \frac{-x}{x}\, dx + \int_0^1 \frac{x}{x}\, dx$$

$$= \int_{-1}^{0} dx + \int_0^1 dx$$

$$= -(0-(-1)) + (1-0)$$

$$= 0$$

24. (D) $\because$ $f(x)$ is continuous in $[0, 2]$

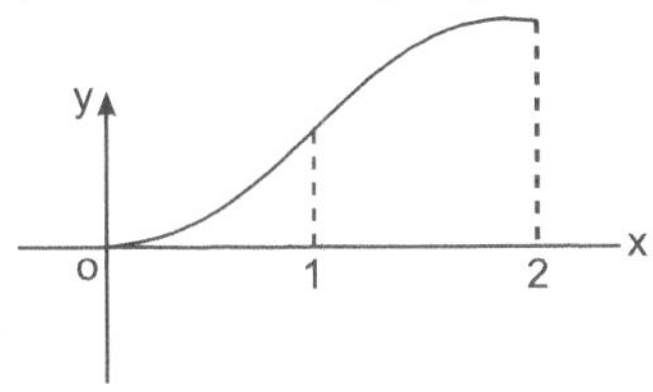

$$\therefore \quad \int_0^2 f(x)\, dx = \int_0^1 f(x)\, dx + \int_1^2 f(x)\, dx$$

$$= \int_0^1 x^2 dx + \int_1^2 \sqrt{x}\, dx$$

$$= \frac{1}{3} + \frac{2}{3}(2^{3/2}-1)$$

$$= \frac{1}{3} + \frac{4\sqrt{2}}{3} - \frac{2}{3}$$

$$= \left(\frac{4\sqrt{2}-1}{3}\right)$$

25. (B) $\because$ $\sin^6 x + \cos^6 x$

$= (\sin^2 x)^3 + (\cos^2 x)^3$

$= (\sin^2 x + \cos^2 x)^3 - 3\sin^2 x \cos^2 x (\sin^2 x + \cos^2 x)$

$= 1 - 3\sin^2 x \cos^2 x$

$= 1 - \frac{3}{4}\sin^2 2x$ $\left[\because \text{period } \frac{\pi}{2}\right]$

$\therefore$ Least and greatest value of $\sin^6 x + \cos^6 x$ are

$\frac{1}{4}$ and 1

Hence,

$$\left(\frac{\pi}{2}-0\right)\times\frac{1}{4} < \int_0^{\pi/2} (\sin^6 x + \cos^6 x)dx < \left(\frac{\pi}{2}-0\right)\times 1$$

$$\Rightarrow \quad \frac{\pi}{8} < \int_0^{\pi/2} (\sin^6 x + \cos^6 x)\, dx < \frac{\pi}{2}$$

26. (A) $\because$ $f(x) = x - x^2 + 1$

$\therefore$ $f'(x) = 1 - 2x$

$f'(x) > 0 \Rightarrow 1 - 2x > 0 \Rightarrow x < \frac{1}{2}$

and $\quad f'(x) < 0 \Rightarrow 1-2x<0 \Rightarrow x > \frac{1}{2}$

$\Rightarrow f(x)$ is increasing in $\left(0, \frac{1}{2}\right)$ and decreasing in $\left(\frac{1}{2}, 1\right)$

Now, $\quad g(x) = \max \{f(t); 0 \le t \le x\}$

$$= \begin{cases} x - x^2 + 1, & 0 \le x \le \frac{1}{2} \\ \frac{1}{2} - \frac{1}{4} + 1, & \frac{1}{2} \le x \le 1 \end{cases}$$

$$= \begin{cases} x - x^2 + 1, & 0 \le x \le \frac{1}{2} \\ \frac{5}{4}, & \frac{1}{2} \le x \le 1 \end{cases}$$

$$\therefore \int_0^1 g(x)\,dx = \int_0^{1/2} (x - x^2 + 1)\,dx + \int_{1/2}^1 \frac{5}{4}\,dx$$

$$= \left[\frac{x^2}{2} - \frac{x^3}{3} + x\right]_0^{1/2} + \frac{5}{4}[x]_{1/2}^1$$

$$= \left(\frac{1}{8} - \frac{1}{24} + \frac{1}{2}\right) + \frac{5}{4}\left(1 - \frac{1}{2}\right)$$

$$= \frac{1}{8} - \frac{1}{24} + \frac{1}{2} + \frac{5}{8}$$

$$= \frac{3 - 1 + 12 + 15}{24} = \frac{29}{24}$$

Chapter 8. Applications of the Integrals

1. **Assertion (A) :** Area enclosed by the curve $y = e^{x^3}$ between the lines $x = a, x = b$ and x-axis is $\int_a^b e^{x^3}\,dx$.

   Reason (R) : e^{x^3} is an increasing function.

2. **Assertion (A) :** The area of the function $y = \sin^2 x$ from 0 to π will be more than that of curve $y = \sin x$ from 0 to π.

   Reason (R) : $t^2 < t$, if $0 < t < 1$.

3. **Assertion (A) :** The area bounded by the curves $y = x^2 + 2x - 3$ and the line $y = \lambda x + 1$ is least, if $\lambda = 2$.

   Reason (R) : The area bounded by the curves $y = x^2 + 2x - 3$ and $y = \lambda x + 1$ is $\{(\lambda - 2)^2 + 16)^{3/2}\}$ sq. unit

4. **Assertion (A) :** Let f be a real value function satisfying $f\left(\frac{x}{y}\right) = f(x) - f(y)$ and $\lim_{x \to 0} \frac{f(1+x)}{x} = 3$.

   Then the area bounded by the curve $y = f(x)$, the y-axis and the line $y = 3$ is $3e$ sq unit.

   Reason (R) : The function $f(x)$ is concave down.

5. **Assertion (A) :** If $f(x) = (x-1)(x-2)(x-3)$, then area enclosed by $|f(x)|$ between the lines $x = 2.2, 2.8$ and x-axis is equal to

   $$\int_{2.2}^{2.8} (x-1)(x-2)(x-3)\,dx.$$

   Reason (R) :

   $(x-1)(x-2)(x-3) \le 0 \ \forall \ x \in [2.2, 2.8]$

Answers :

1. (B) $\because \quad y = e^{x^3}$

$\therefore \quad \frac{dy}{dx} = e^{x^3} \cdot 3x^2 > 0$

$\Rightarrow y$ is an increasing function.

and area enclosed by the curve $y = e^{x^3}$ between the lines $x = a, x = b$ and x-axis is $\int_a^b e^{x^3}\,dx$.

2. (D) For $0 < t < 1$

$t^2 < 1$

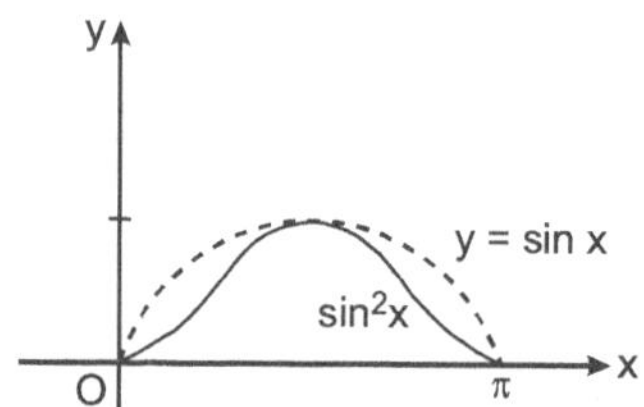

$\therefore \quad \sin^2 x < \sin x$

$\Rightarrow \int_0^\pi \sin^2 x\,dx < \int_0^\pi \sin x\,dx$

3. (C) The given curves are

$$y = x^2 + 2x - 3 \quad ...(i)$$

and $\quad y = \lambda x + 1 \quad ...(ii)$

Solving equations (i) and (ii), we get

$$x^2 + (2 - \lambda)x - 4 = 0$$

α, β are the roots of the quadratic equation, then

$$\alpha + \beta = \lambda - 2, \alpha\beta = -4$$

Hence, required area

$$S(\lambda) = \left|\int_\alpha^\beta [(\lambda x + 1) - (x^2 + 2x - 3)]\,dx\right|$$

$$= \left|\left\{4x + (\lambda - 2)\frac{x^2}{2} - \frac{x^3}{3}\right\}_\alpha^\beta\right|$$

$$= \left|4(\beta - \alpha) + \frac{(\lambda - 2)}{2}(\beta^2 - \alpha^2) - \frac{1}{3}(\beta^2 - \alpha^3)\right|$$

$$= \sqrt{(\beta + \alpha)^2 - 4\beta\alpha}$$

$$\left|\left\{4 + \frac{(\lambda - 2)}{2}(\beta + \alpha) - \frac{1}{3}\{(\alpha + \beta)^2 - \alpha\beta\}\right\}\right|$$

$= \frac{1}{6}\{(\lambda-2)^2+16\}^{3/2}$

For least value of $S(\lambda)$,

$\lambda - 2 = 0$

$\therefore \quad \lambda = 2$

4. (B) Given, $f\left(\frac{x}{y}\right) = f(x) - f(y)$ ...(i)

Putting $x = y$, then

$\Rightarrow \quad f(1) = 0$

$$\therefore \quad f'(x) = \lim_{h \to 0} \frac{f(x+h)-f(x)}{h}$$

$$= \lim_{h \to 0} \frac{f\left(1+\frac{h}{x}\right)}{h} \quad \text{[From (i)]}$$

$$= \lim_{h \to 0} \frac{f\left(1+\frac{h}{x}\right)}{x \cdot \frac{h}{x}}$$

$$= \frac{3}{x} \quad \left[\because \lim_{x \to 0} \frac{f(1+x)}{x} = 3\right]$$

$\therefore \quad f(x) = 3 \text{ In } x + c$

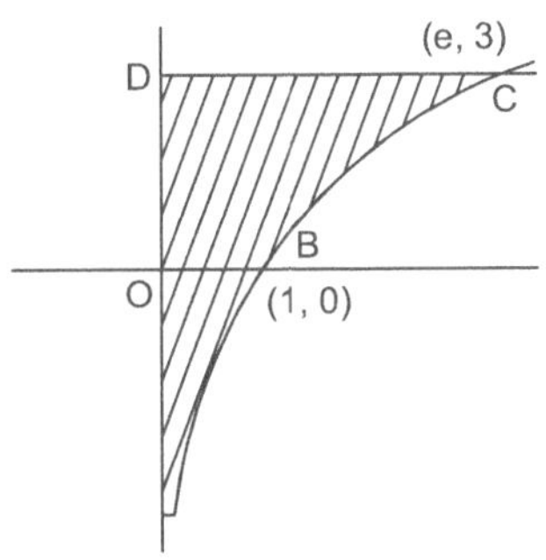

Putting $x = 1$, then

$f(1) = 0 + c = 0$

$\Rightarrow f(x) = 3 \text{ In } x = y$ (say)

$\therefore \quad x = e^{y/3}$

$$\therefore \text{ Required area } = \int_{-\infty}^{3} x\, dy$$

$$= \int_{-\infty}^{3} e^{y/3} dy$$

$$= 3\,\{e^{y/3}\}_{-\infty}^{3} = 3\,(e-0)\,3e \text{ sq. unit.}$$

$$\therefore \quad f''(x) = -\frac{3}{x^2} < 0$$

$\Rightarrow f(x)$ is concave down.

5. (D) It is clear from the figure for $x \in [2.2, 2.8]$

$\Rightarrow (x-1)(x-2)(x-3) \le 0$

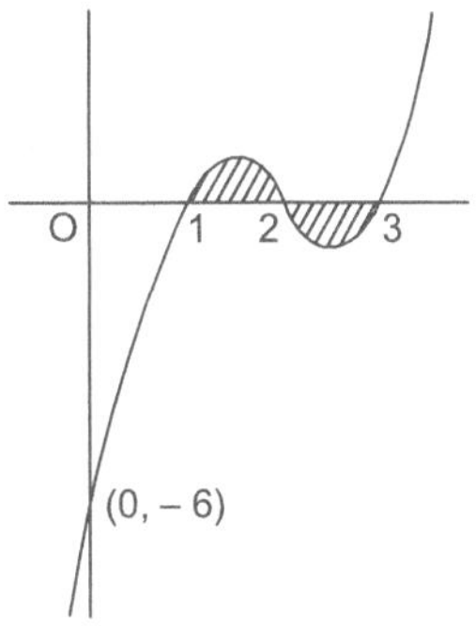

$$\text{Required area } = \left|\int_{2.2}^{2.8} f(x)\, dx\right|$$

$$= \left|\int_{2.2}^{2.8} (x-1)(x-2)(x-3)\, dx\right|$$

Chapter 9. Differential Equations

1. **Assertion (A) :** The elimination of four arbitrary constants in $y = (c_1 + c_2 + c_3ec_4)\, x$ results into a differential equation of the first order $x\frac{dy}{dx} = y$.

   Reason (R) : Elimination of n arbitrary constants requires, in general, a differential equation of the nth order.

2. **Assertion (A) :** $y = a \sin x + b \cos x$ is a general solution of $y'' + y = 0$.

   Reason (R) : $y = a \sin x + b \cos x$ is a trigonometric function.

3. **Assertion (A) :** Order of the differential equation whose solution is $y = c_1e^{x+c_2} + c_3e^{x+c_4}$ is 4.

   Reason (R) : Order of the differential equation is equal to the number of independent arbitrary constant mentioned in the solution of differential equation.

4. **Assertion (A) :** The degree of the differential equation $\frac{d^2y}{dx^2} + \frac{dy}{dx} = \text{In}\left(\frac{d^2y}{dx^2}\right)$ is 2.

   Reason (R) : The degree of a differential equation which can be written as polynomial in the derivatives is the degree of the derivative of the higher order occuring in it.

5. **Assertion (A) :** The differential equation of all circles in a plane must be of order 3.

   Reason (R) : If three points are non-collinear, then only one circle always passes through these points.

6. **Assertion (A) :** The differential equation of the form $y\,f(xy)\,dx + x\,\phi(xy)\,dy = 0$ can be converted to homogeneous forms by substitution $xy = v$

   Reason (R) : All differential equation of first order and first degree become homogeneous, if we put $y = vx$

Answers :

1. **(A)** Let $c_1 + c_2 + c_3 e c_4 = A$ (constant)

Then, $y = Ax$

$$\Rightarrow \frac{dy}{dx} = A$$

or $$y = x\frac{dy}{dx}$$

or $$x\frac{dy}{dx} = y$$

2. **(B)** $\because$ $y = a\sin + b\sin x$ ...(i)

$\therefore$ $y' = a\cos x - b\sin x$

$\Rightarrow$ $y'' = -a\sin x - b\cos x = -y$ [From (i)]

or $y'' + y = 0$

3. **(D)** $\because$ $y = (c_1 e^{c_2} + c_3 e^{c_4})\, ex = ce^x$ (say)

$$\therefore \frac{dy}{dx} = ce^x = y$$

$\therefore$ Order is 1.

4. **(D)** $\because$ The given equation cannot be written as a polynomial in all the differentials.

$\therefore$ Degree of the equation is not defined.

5. **(A)** Let $x^2 + y^2 + 2gx + 2fy + c = 0$.

Here three constants.

$\therefore$ Order = 3

$\because$ Circle passes through three non-collinear points, then we get three constants g, t, *c*.

6. **(D)** $xy = v$

$$\therefore x\frac{dy}{dx} + y = \frac{dv}{dx}$$

Then, the given equation reduces to

$$\frac{v}{x}f(v) + x\phi(v)\left[\frac{1}{x}\left(\frac{dv}{dx} - y\right)\right] = 0$$

$$\Rightarrow \frac{v}{x}f(v) + \phi(v)\frac{dv}{dy} - y\phi(v) = 0$$

$$\Rightarrow \left\{\frac{v(f(v) - \phi(v))}{x}\right\} + \phi(v)\frac{dv}{dx} = 0$$

$$\Rightarrow \frac{dx}{x} + \frac{\phi(v)dv}{v(f(v)) - \phi(v)} = 0$$

which is variable separable form.

Chapter 10. Vectors

1. **Assertion (A) :** $|\vec{a}| = |\vec{b}|$ does not implies that $\vec{a} = \vec{b}$

Reason (R) : If $\vec{a} = \vec{b}$, then $\vec{a}\cdot\vec{b} = |\vec{a}|^2 = |\vec{b}|^2$

2. **Assertion (A) :** If $\vec{a}, \vec{b}, \vec{c}$ are unit vectors such that $\vec{a} + \vec{b} + \vec{c} = 0$, then $\vec{a}\cdot\vec{b} + \vec{b}\cdot\vec{c} + \vec{c}\cdot\vec{a} = -3/2$.

Reason (R) : $(\vec{x} + \vec{y})^2 = |\vec{x}|^2 + |\vec{y}|^2 + 2(\vec{x}\cdot\vec{y})$

3. **Assertion (A) :** Three points with position vectors $\vec{a}, \vec{b}, \vec{c}$ are collinear if $\vec{a}\times\vec{b} + \vec{b}\times\vec{c} + \vec{c}\times\vec{a} = 0$

Reason (R) : Three points *A*, *B*, *C* are collinear iff $\vec{AB} = t\,\vec{BC}$, where *t* is scalar.

4. **Assertion (A) :** If a force $\vec{F}$ passes through $Q(\vec{b})$, then moment of force $\vec{F}$ about $P\ (\vec{a})$ is $\vec{F}\times\vec{r}$ when $\vec{r} = \vec{PQ}$.

Reason (R): Moment is a vector.

5. **Assertion (A) :** If $\vec{a}, \vec{b}, \vec{c}$ are coplanar, then $\vec{a}\times\vec{b}$, $\vec{b}\times\vec{c}$ and $\vec{c}\times\vec{a}$ are also coplanar.

Reason (R) : $[\vec{a}\times\vec{b}\ \ \vec{b}\times\vec{c}\ \ \vec{c}\times\vec{a}] = 2[\vec{a}\ \ \vec{b}\ \ \vec{c}]^2$

6. **Assertion (A) :** The scalar product of a force $\vec{F}$ and displacement $\vec{r}$ is equal to the work done.

Reason (R) : Work done is not a scalar.

7. **Assertion (A) :** In a ΔABC, $\vec{AB} + \vec{BC} + \vec{CA} = 0$

Reason (R) : If $\vec{AB} = \vec{a}$, $\vec{BC} = \vec{b}$, then $\vec{AC} = \vec{a} + \vec{b}$ (Triangle law of addition).

8. **Assertion (A) :** For $a = -\frac{1}{\sqrt{3}}$ the volume of the parallelopiped formed by vectors $\hat{i} + a\hat{j}$, $a\hat{i} + \hat{j} + \hat{k}$ and $\hat{j} + a\hat{k}$ is maximum.

Reason (R) : The volume of the parallelopiped having three coterminous edges and

9. **Assertion (A) :** If $\vec{a}$ is perpendicular to $\vec{b}$ and $\vec{c}$, then $\vec{a}\times(\vec{b}\times\vec{c}) = 0$.

Reason (R) : $\vec{b}$ is perpendicular $\vec{c}$ then $\vec{b}\times\vec{c} = 0$.

10. **Assertion (A) :** If $|\vec{a}| = 2$, $|\vec{b}| = 3$, $|2\vec{a} - \vec{b}| = 5$, then $|2|\,2\vec{a} + \vec{b}| = 5$

Reason (R) : $|\vec{p} - \vec{q}| = |\vec{p} + \vec{q}|$

11. **Assertion (A) :** If in a ΔABC,

$\vec{BC} = \frac{\vec{e}}{|\vec{e}|} - \frac{\vec{f}}{|\vec{f}|}$ and $\vec{AC} = \frac{2\vec{e}}{|\vec{e}|}$; $|\vec{e}| \neq |\vec{f}|$, then the value of $\cos 2A + \cos 2B + \cos 2C$ is 1.

Reason (R) : If in ΔABC, $\angle C = 90°$, then $\cos 2A + \cos 2B + \cos 2C = -1$

12. Assertion (A) : If $\vec{a} \times \vec{b} = \vec{c} \times \vec{d}$ and $\vec{a} \times \vec{c} = \vec{b} \times \vec{d}$, then $\vec{a} - \vec{d}$ is perpendicular to $\vec{b} - \vec{c}$.

Reason (R) : If $\vec{p}$ is perpendicular to $\vec{q}$, then $\vec{p} \cdot \vec{q} = 0$

13. Assertion (A) : If $\vec{r} \cdot \vec{a} = 0$, $\vec{r} \cdot \vec{b} = 0$, $\vec{r} \cdot \vec{c} = 0$ for some non-zero vector $\vec{r}$, then $\vec{a}, \vec{b}, \vec{c}$ are coplanar, vectors.

Reason (R) : If $\vec{a}, \vec{b}, \vec{c}$ are coplanar, then $\vec{a} + \vec{b} + \vec{c} = 0$

14. Assertion (A) : If $\vec{a}$ and $\vec{b}$ are reciprocal vectors, then $\vec{a} \cdot \vec{b} = 1$.

Reason (R) : If $\vec{a}$ and $\vec{b}$ are reciprocal, then $\vec{a} = \lambda \vec{b}, \lambda \in R^+$ and $|\vec{a}|\,|\vec{b}| = 1$

15. Assertion (A) : For any two vectors $\vec{a}$ and $\vec{b}$,

$$(\vec{a} \times \hat{i}) \cdot (\vec{b} \times \hat{i}) + (\vec{a} \times \hat{j}) \cdot (\vec{b} \times \hat{j}) + (\vec{a} \times \hat{k}) \cdot (\vec{b} \times \hat{k}) = 2\,\vec{a} \cdot \vec{b}$$

Reason (R) :

$$(\vec{a} \cdot \hat{i})(\vec{b} \cdot \hat{i}) + (\vec{a} \cdot \hat{j})(\vec{b} \cdot \hat{j})$$

Answers :

1. (B) If $\vec{a} = \vec{b}$, then $|\vec{a}| = |\vec{b}|$

$\therefore \quad \vec{a} \cdot \vec{b} = \vec{a} \cdot \vec{a} = (\vec{a})^2 = |\vec{a}|^2$

and $\quad \vec{a} \cdot \vec{b} = \vec{b} \cdot \vec{b} = (\vec{b})^3 = |\vec{b}|^2$

$\therefore \quad \vec{a} \cdot \vec{b} = |\vec{a}|^2 = |\vec{b}|^2$

Suppose, if $\vec{a} = a_1 \hat{i} + a_2 \hat{j}$

Then, $\quad |\vec{a}| = |a_1 \hat{i} + a_2 \hat{j}|$

$= |a_1 \hat{i} - a_2 \hat{j}|$

$= |-a_1 \hat{i} + a_2 \hat{j}|$

$= |-a_1 \hat{i} - a_2 \hat{j}|$

Hence, $|\vec{a}| = |\vec{b}|$

does not implies that $\vec{a} = \vec{b}$

2. (A) $\because |\vec{a}| = 1, |\vec{b}| = 1, |\vec{c}| = 1$

$\therefore (\vec{a} + \vec{b} + \vec{c})^2 = |\vec{a}|^2 + |\vec{b}|^2 + |\vec{c}|^2 + 2(\vec{a} \cdot \vec{b} + \vec{b} \cdot \vec{c} + \vec{c} \cdot \vec{a})$

$\Rightarrow \quad 0 = 1 + 1 + 1 + 2(\vec{a} \cdot \vec{b} + \vec{b} \cdot \vec{c} + \vec{c} \cdot \vec{a})$

$\therefore \quad \vec{a} \cdot \vec{b} + \vec{b} \cdot \vec{c} + \vec{c} \cdot \vec{a} = -3/2$

3. (A) $\because \vec{OA} = \vec{a}, \vec{OB} = \vec{b}, \vec{OC} = \vec{c}$

$\therefore \quad \vec{AB} = \vec{OB} - \vec{OA} = (\vec{b} - \vec{a})$

and $\quad \vec{BC} = \vec{OC} - \vec{OB} = (\vec{c} - \vec{b})$

$\vec{AB}$ is parallel to $\vec{BC}$.

$\therefore \quad \vec{AB} \times \vec{BC} = 0$

$\Rightarrow \quad (\vec{b} - \vec{a}) \times (\vec{c} - \vec{b}) = 0$

$\Rightarrow \quad \vec{b} \times \vec{c} - \vec{b} \times \vec{b} - \vec{a} \times \vec{c} + \vec{a} \times \vec{b} = 0$

$\Rightarrow \quad \vec{b} \times \vec{c} - 0 + \vec{c} \times \vec{a} + \vec{a} \times \vec{b} = 0$

Hence, $\quad \vec{a} \times \vec{b} + \vec{b} \times \vec{c} + \vec{c} \times \vec{a} = 0$

4. (D) Then, moment of force $\vec{F}$ about point $P(\vec{a})$ is $\vec{r} \times \vec{f}$ when $\vec{r} = \vec{PQ}$

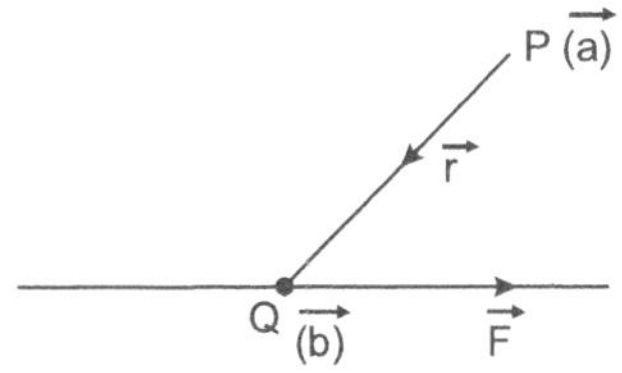

5. (C) $\because [\vec{a} \times \vec{b} \;\; \vec{b} \times \vec{c} \;\; \vec{c} \times \vec{a}]$

$= (\vec{a} \times \vec{b}) \cdot \{(\vec{b} \times \vec{c}) \times (\vec{c} \times \vec{a})\}$

$= (\vec{a} \times \vec{b}) \cdot \{(\vec{b} \cdot (\vec{c} \times \vec{a}))\,\vec{c}(-\vec{c} \cdot (\vec{c} \times \vec{a}))\,\vec{b}\}$

$= (\vec{a} \times \vec{b}) \cdot \{[\vec{b}\,\vec{c}\,\vec{a}]\,\vec{c} - 0\}$

$= (\vec{a} \times \vec{b} \cdot \vec{c})[\vec{b}\,\vec{c}\,\vec{a}] = [\vec{a}\,\vec{b}\,\vec{c}]^2$

$\because \vec{a}, \vec{b}, \vec{c}$ are coplanar.

$\therefore \quad [\vec{a}\,\vec{b}\,\vec{c}] = 0$

and then

$[\vec{a} \times \vec{b} \;\; \vec{b} \times \vec{c} \;\; \vec{c} \times \vec{a}] = 0$

Hence, $\vec{a} \times \vec{b}$, $\vec{b} \times \vec{c}$ and $\vec{c} \times \vec{a}$ are also coplanar.

6. (C) $\because$ Work done, $W = \vec{F} \cdot \vec{r}$

$\therefore$ Work done is a Scalar quantity.

7. (A) $\because \quad \vec{AB} + \vec{BC} + \vec{CA} = \vec{AC} + \vec{CA}$

$= \vec{AC} - \vec{AC}$

$= 0$

8. (D)

$$V = \begin{vmatrix} 1 & a & 0 \\ a & 1 & 1 \\ 0 & 1 & a \end{vmatrix} = a - 1 - a^3$$

$\therefore \quad \frac{dV}{da} = 1 - 3a^2 = 0$

$\therefore \quad a = \pm \frac{1}{\sqrt{3}}$

$\Rightarrow \quad \frac{d^2V}{da^2} = -6a$

$\left.\frac{d^2V}{da^2}\right|_{a=\frac{1}{\sqrt{3}}} = -\frac{6}{\sqrt{3}}(\text{-ve})$

$\therefore$ V is maximum at $a = \frac{1}{\sqrt{3}}$

9. (C) $\because \quad \vec{a} \times (\vec{b} \times \vec{c}) = (\vec{a} \cdot \vec{c})\vec{b} - (\vec{a} \cdot \vec{b})\vec{c}$

$= 0 - 0 \quad [\because \vec{a} \perp \vec{b} \text{ and } \vec{a} \perp \vec{c}]$

$= 0 \quad [\because \vec{a} \cdot \vec{b} = 0, \vec{a} \cdot \vec{c} = 0]$

10. (C) $\because \quad |2\vec{a} - \vec{b}| = 5$

$\Rightarrow \quad |2\vec{a} - \vec{b}|^2 = 5^2$

$\Rightarrow \quad 4a^2 + b^2 - 4\vec{a} \cdot \vec{b} = 25$

$\Rightarrow \quad 16 + 9 - 4\vec{a} \cdot \vec{b} = 25$

$\therefore \quad \vec{a} \cdot \vec{b} = 0$

$\therefore \quad |2\vec{a} + \vec{b}| = \sqrt{|2\vec{a} + \vec{b}|^2}$

$= \sqrt{[4a^2 + b^2 + 4(\vec{a} \cdot \vec{b})]}$

$= \sqrt{(16 + 9 + 0)}$

$= 5$

$\therefore \quad |\vec{p} - \vec{q}| = |\vec{p} + \vec{q}|$

is possible only when $\vec{p} \perp \vec{q}$.

11. (B) $\because \quad \vec{BA} = \vec{BC} - \vec{AC}$

$= \left(\frac{\vec{e}}{|\vec{e}|} + \frac{\vec{f}}{|\vec{f}|}\right) - \left(\frac{2\vec{e}}{|\vec{e}|}\right)$

$= -\left(\frac{\vec{e}}{|\vec{e}|} + \frac{\vec{f}}{|\vec{f}|}\right)$

Now, $\vec{BA} \cdot \vec{BC} = -\left(\frac{\vec{e}}{|\vec{e}|} + \frac{\vec{f}}{|\vec{f}|}\right)\left(\frac{\vec{e}}{|\vec{e}|} - \frac{\vec{f}}{|\vec{f}|}\right)$

$= -\left(\frac{e^2}{e^2} - \frac{f^2}{f^2}\right)$

$= -(1 - 1) = 0$

$\Rightarrow \quad \angle B = 90°$

$\therefore \quad \cos 2A + \cos 2C = 2\cos(A + C)\cos(A - C)$

$= 2\cos(180° - B)\cos(A - C)$

$= 2\cos 90° \cos(A - C)$

$= 0$

$\therefore \cos 2A + \cos 2B + \cos 2C = -1$

Also, if $\angle C = 90°$.

Then, $\cos 2A + \cos 2B$

$= 2\cos(A + B)\cos(A - B)$

$= 2\cos(180° - C)\cos(A - B)$

$= 2\cos 90° (A - B)$

$= 0$

$\therefore \cos 2A + \cos 2B + \cos 2C = -1$

12. (C) We have, $\vec{a} \times \vec{b} = \vec{c} \times \vec{d}$ ...(i)

and $\vec{a} \times \vec{c} = \vec{b} \times \vec{d}$ ...(ii)

$\because (\vec{a} - \vec{d}) \times (\vec{b} - \vec{c}) = \vec{a} \times \vec{b} - \vec{a} \times \vec{c} - \vec{d} \times \vec{b} + \vec{d} \times \vec{c}$

$= \vec{c} \times \vec{d} - \vec{b} \times \vec{d} + \vec{b} \times \vec{d} - \vec{c} \times \vec{d}$

$= 0$ [From (i) and (ii)]

$\therefore \vec{a} - \vec{d}$ and $\vec{b} - \vec{c}$ are parallel.

13. (B) We have, $\left.\begin{array}{l} \vec{r} \cdot \vec{a} = 0 \Rightarrow \vec{r} \perp \vec{a} \\ \vec{r} \cdot \vec{b} = 0 \Rightarrow \vec{r} \perp \vec{b} \\ \vec{r} \cdot \vec{c} = 0 \Rightarrow \vec{r} \perp \vec{c} \end{array}\right\}$

$\Rightarrow \vec{a}, \vec{b}, \vec{c}$ are coplanar vectors

Also, $\because \quad \vec{a} + \vec{b} + \vec{c} = 0$

$\therefore \quad \vec{a} \times \vec{b} + \vec{b} \times \vec{c} + \vec{c} \times \vec{a} = 0$

Also $\vec{a} \cdot (\vec{a} \times \vec{b}) + \vec{a} \cdot (\vec{b} \times \vec{c}) + \vec{a} \cdot (\vec{c} \times \vec{a}) = 0$

$0 + [\vec{a}\,\vec{b}\,\vec{c}] + 0 = 0$

$\therefore \quad [\vec{a}\,\vec{b}\,\vec{c}] = 0$

Hence, $\vec{a}, \vec{b}, \vec{c}$ are coplanar.

14. (A) If $\vec{a}$ and $\vec{b}$ are reciprocal, then

$\vec{a} = \lambda\vec{b},\ \lambda \in R^+$ and $|\vec{a}||\vec{b}| = 1$

$\Rightarrow \quad |\vec{a}| = |\lambda||\vec{b}|$

$\therefore \quad |\lambda| = \frac{|\vec{a}|}{|\vec{b}|} = \frac{1}{|\vec{b}|^2}$

$\because \quad \lambda \in R^+$

$$\therefore \quad \lambda = \frac{1}{|\vec{b}|^2}$$

$$\therefore \quad \vec{a} = \frac{\vec{b}}{|\vec{b}|^2}$$

$$\Rightarrow \quad \vec{a}\cdot\vec{b} = \frac{\vec{b}}{|\vec{b}|^2}\cdot\vec{b} = \frac{|\vec{b}|^2}{|\vec{b}|^2} = 1$$

15. (A) $\because \ (\vec{a}\times\hat{i})\cdot(\vec{b}\times\hat{i}) = \begin{vmatrix} \vec{a}\cdot\vec{b} & \vec{a}\cdot\hat{i} \\ \vec{b}\cdot\hat{i} & \hat{i}\cdot\hat{i} \end{vmatrix}$

$$= (\vec{a}\cdot\vec{b}) - (\vec{a}\cdot\hat{i})(\vec{b}\cdot\hat{i})$$

Similarly,

$$(\vec{a}\times\hat{j})\cdot(\vec{b}\times\hat{j}) = (\vec{a}\cdot\vec{b}) - (\vec{a}\cdot\hat{j})(\vec{b}\cdot\hat{j})$$

and $(\vec{a}\times\hat{k})\cdot(\vec{b}\times\hat{k}) = (\vec{a}\cdot\vec{b}) - (\vec{a}\cdot\hat{k})(\vec{b}\cdot\hat{k})$

Adding, we get

$$(\vec{a}\times\hat{i})\cdot(\vec{b}\times\hat{i}) + (\vec{a}\times\hat{j})\cdot(\vec{b}\times\hat{j}) + (\vec{a}\times\hat{k})\cdot(\vec{b}\times\hat{k})$$

$$= 3\vec{a}\cdot\vec{b} - \{(\vec{a}\cdot\hat{i})(\vec{b}\cdot\hat{i}) + (\vec{a}\cdot\hat{j})(\vec{b}\cdot\hat{j}) + (\vec{a}\cdot\hat{k})\cdot(\vec{b}\cdot\hat{k})$$

Now, let $\vec{a} = a_1\hat{i} + a_2\hat{j} + a_3\hat{k}$

and $\vec{b} = b_1\hat{i} + b_2\hat{j} + b_3\hat{k}$

$$\therefore (\vec{a}\cdot\hat{i})(\vec{b}\cdot\hat{j}) + (\vec{a}\cdot\hat{j})(\vec{b}\cdot\hat{j}) + (\vec{a}\cdot\hat{k})(\vec{b}\cdot\hat{k})$$

$$= a_1b_1 + a_2b_2 + a_3b_3$$

$$= \vec{a}\cdot\vec{b}$$

Hence,

$$(\vec{a}\times\hat{i})\cdot(\vec{b}\times\hat{i}) + (\vec{a}\times\hat{j})(\vec{b}\times\hat{j}) + (\vec{a}\times\hat{k})(\vec{b}\times\hat{k})$$

$$= 2\vec{a}\,\vec{b}$$

Chapter 11. Three-Dimensional Geometry

1. **Assertion (A) :** The points $A(2, 9, 12)$, $B(1, 8, 8)$, $C(-2, 11, 8)$ and $D(-1, 12, 12)$ are the vertices of a rhombus.

   Reason (R) : $AB = BC = CD = DA$ and $AC \neq BD$.

2. **Assertion (A) :** The direction ratios of the line joining origin and point (x, y, z) must be x, y, z.

   Reason (R) : If $P(x, y, z)$ is a point in space and $OP = r$, then direction cosines of OP are $\frac{x}{r}, \frac{y}{r}, \frac{z}{r}$.

3. **Assertion (A) :** The equation of the plane through the intersection of the planes $x + y + z = 6$ and $2x + 3y + 4z + 5 = 0$ and the point $(4, 4, 4)$ is $29x + 23y + 17z = 276$

   Reason (R) : Equation of the plane through the line of intersection of the planes $P_1 = 0$ and $P_2 = 0$ is $P_1 + \lambda P_2 = 0$, $\lambda \neq 0$

4. **Assertion (A) :** The equation

   $$2x^2 + 6y^2 + 4z^2 + 18yz + 2zx + xy = 0$$

   represents a pair of perpendicular planes.

   Reason (R) : A pair of planes given by $ax^2 + by^2 + cz^2 + 2fyz + 2hxy = 0$ are perpendicular, if $a + b + c = 0$

5. **Assertion (A) :** The points $(2, 1, 5)$ and $(3, 4, 3)$ lie on opposite side of the plane $2x + 2y - 2z - 1 = 0$

   Reason (R) : The algebric perpendicular distance from the given points to the line have opposite sign.

6. **Assertion (A) :** If centroid and circumcentre of a triangle are known its orthocentre can be found.

   Reason (R) : Centroid, orthocentre and circumcentre of a triangle are collinear.

7. **Assertion (A) :** The shortest distance between the skew lines

   $$\frac{x+3}{-4} = \frac{y-6}{3} = \frac{z}{2} \text{ and } \frac{x+2}{-4} = \frac{y}{1} = \frac{z-7}{1} \text{ is } 9$$

   Reason (R) : Two lines are skew lines if there exists no plane passing through them.

8. **Assertion (A) :** Equation of the polar to the sphere $x^2 + y^2 + z^2 = 1$ with respect to the point $(1, 2, 3)$ is $x + 2y + 3z = 1$.

   Reason (R) : The point $(1, 2, 3)$ lies outside the sphere $x^2 + y^2 + z^2 = 1$.

9. **Assertion (A) :** Two spheres of radii r_1 and r_2 cut orthogonally, then radius of the common circle is

   $$\frac{r_1r_2}{\sqrt{(r_1^2 + r_2^2)}}$$

   Reason (R) : If two spheres

   $$x^2 + y^2 + z^2 + 2ux + 2vy + 2wz + d = 0$$

   and $x^2 + y^2 + z^2 + 2u'x + 2v'y + 2w'z + d' = 0$

   orthogonally, then

   $$2uu' + 2vv' + 2ww' = d + d'$$

10. **Assertion (A) :** The spheres $x^2 + y^2 + z^2 + 2ax + c = 0$ and $x^2 + y^2 + z^2 + 2by + c = 0$ touch if

    $$\frac{1}{a^2} + \frac{1}{b^2} + \frac{1}{c}$$

    Reason (R) : Two spheres with centre C_1 and C_2 and radii r_1, r_2 touch each other if $|r_1 \pm r_2| = |C_1C_2|$.

11. **Assertion (A) :** The spheres $x^2 + y^2 + z^2 = 64$ and $x^2 + y^2 + z^2 - 12x + 4y - 6z + 48 = 0$ touch externally.

    Reason (R) : The distance between centres of spheres is equal to the sum of their radii.

12. Assertion (A) : A homogeneous equation of second degree in x, y and z represents a pair of plane passing through origin.

Reason (R) : A homogeneous equation of second degree in x, y and z can be factorised into homogeneous linear factors.

Answers :

1. (D) $\because \quad AB = \sqrt{(2-1)^2+(9-8)^2+(12-8)^2}$

$= \sqrt{18} = 3\sqrt{2}$

$BC = \sqrt{(1+2)^2+(8-11)^2+(8-8)^2}$

$= \sqrt{18} = 3\sqrt{2}$

$CD = \sqrt{(-2+1)^2+(11-12)^2+(8-12)^2}$

$= \sqrt{18} = 3\sqrt{2}$

$DA = \sqrt{(-1-2)^2+(12-9)^2+(12-12)^2}$

$= \sqrt{18} = 3\sqrt{2}$

$AC = \sqrt{(2+2)^2+(9-11)^2+(12-8)^2}$

$= \sqrt{36} = 6$

and $BD = \sqrt{(1+1)^2+(8-12)^2+(8-12)^2}$

$= \sqrt{36} = 6$

$\therefore AB = BC = CD = DA$ and $AC = BD$.

Then, it is a square not a rhombus.

2. (A) $\because$ Direction ratios of OP are $x-0$, $y-0$, $z-0$ *i.e.*, x, y, z.

$\therefore$ Direction cosines of OP are $\frac{x}{r}, \frac{y}{r}, \frac{z}{r}$

3. (A) $\because$ Equation of the plane through the line of intersection of the planes $x+y+z=6$ and $2x+3y+4z+5=0$ is

$$(x+y+z-6)+\lambda(2x+3y+4z+5)=0 \quad \text{...(i)}$$

Since, it passes through (4, 4, 4), then

$(4+4+4-6)+\lambda(8+12+16+5) = 0$

$\Rightarrow \quad 6+41\lambda = 0$

$\therefore \quad \lambda = -\frac{6}{41}$

Then from equation (i),

$41(x+y+z-6)-6(2x+3y+4z+5)=0$

$\Rightarrow \quad 29x+23y+17z = 276$

4. (D) Here, $a=2$, $b=-6$, $c=4$, $f=9$, $g=1$ and $h=\frac{1}{2}$

$\therefore abc+2fgh-af^2-bg^2-ch^2$

$= (2)(-6)(4)+2(9)(1)\left(\frac{1}{2}\right) - (2)(81)+6(1)^2-4\left(\frac{1}{2}\right)^2$

$= -48+9-162+6-1$

$= -196 \neq 0$

$\therefore$ Given equation cannot represent a pair of planes.

5. (A) Algebraic distance of (2, 1, 5) from the plane $2x+2y-2z-1=0$ is

$$\frac{2(2)+2(1)-2(5)-1}{\sqrt{(2)^2+(2)^2+(-2)^2}} = -\frac{5}{2\sqrt{3}} \quad \text{...(i)}$$

and algebic distance of (3, 4, 3) from the plane $2x+2y-2z-1=0$ is

$$\frac{2(3)+2(4)-2(3)-1}{\sqrt{\{(2)^2+(2)^2+(-2)^2\}}} = \frac{7}{2\sqrt{3}} \quad \text{...(ii)}$$

Since equations (i) and (ii) are of opposite sign. Hence, points (2, 1, 5) and (3, 4, 3) lie on opposite sides of the plane $2x+2y-2z-1=0$

6. (B) $\therefore$ Orthocentre, nine point centre, centroid and circumcentre are collinear and centroid divides orthocentre and circumcentre in the ratio 2 : 1 (internally).

$\therefore \quad \alpha = \frac{x+2\gamma}{2+1}$

$\Rightarrow \quad x = 3\alpha-2\gamma$

and $\quad \beta = \frac{y+2\delta}{2+1}$

$\Rightarrow \quad y = 3\beta-2\delta$

$\therefore$ Orthocentre is $(3\alpha-2\gamma, 3\beta-2\delta)$.

7. (B) Let l, m, n be the D.C.'s of the line of the common perpendicular (or SD) to the two given lines, Then, we have

$-4l+3m+2n = 0$

and $\quad -4l+1m+1n = 0$

Solving these, we get

$$\frac{l}{3-2} = \frac{m}{-8+4} = \frac{n}{-4+12}$$

or $\frac{l}{1} = \frac{m}{-4} = \frac{n}{8} = \frac{\sqrt{(l^2+m^2+n^2)}}{\sqrt{(1)^2+(-4)^2+(8)^2}} = \frac{1}{9}$

$\therefore$ D.C.'s of SD are $\frac{1}{9}, \frac{4}{-9}, \frac{8}{9}$.

Also, $A(-3, 6, 0)$ is a point on first line and $B(-2, 0, 7)$ is a point on second line, then

$$SD = \left|(-2+3)\frac{1}{9}+(0-6)\left(-\frac{4}{9}\right)+(7-0)\left(\frac{8}{9}\right)\right|$$

$= 9$

And two lines are said to be skew lines or non intersecting lines if they do not lie in the same plane.

8. (B) Equation of the polar to the sphere $x^2 + y^2 + z^2 = 1$ with respect to the point (1, 2, 3) is $x \cdot 1 + y \cdot 2 + z \cdot 3 = 1$ *i.e.*, $x + 2y + 3z = 1$

Let $f(x, y, z) = x^2 + y^2 + z^2 - 1$

$\therefore$ $f(1, 2, 3) = 1^2 + 2^2 + 3^2 - 1$

$= 13 > 0$

$\therefore$ Point (1, 2, 3) lies outside the sphere $x^2 + y^2 + z^2 = 1$; for polar point may be inside or outside of sphere.

9. (A) Let the equation of the common circle be

$$x^2 + y^2 = a^2, z = 0 \quad ...(i)$$

Its radius is evidently a and we are to evaluate it. Now, let the equations of the two given spheres through this circle be

$$(x^2 + y^2 - a^2) + 2\lambda z + z^2 = 0 \quad ...(ii)$$

and $$(x^2 + y^2 - a^2) + 2\mu z + z^2 = 0 \quad ...(iii)$$

(Here an extra term z^2 has been introduced in each equation, so that it may represent a sphere)

From equation (ii), the radius of the sphere

$$= \sqrt{(-\lambda)^2 - (-a)^2} = \sqrt{(\lambda^2 + a^2)} = r_1 \text{ (given)}$$

and similarly from equation (iii) the radius of the sphere

$$= \sqrt{(\mu^2 + a^2)} = r_2 \text{ (given)}$$

Also, as the spheres (ii) and (iii) cut each other orthogonally, so we have

$$2\lambda\mu = -a^2 - a^2$$

or $$\lambda^2\mu^2 = a^4$$

or $$(r_1^2 - a^2)(r_2^2 - a^2) = a^4$$

$\Rightarrow$ $$r_1^2 r_2^2 - a^2(r_1^2 + r_2^2) = 0$$

or $$a = \frac{r_1 r_2}{\sqrt{(r_1^2 + r_2^2)}}$$

10. (A) Here, $C_1 \equiv (-a, 0, 0)$, $r_1 = \sqrt{(a^2 - c)}$

and $C_2 \equiv (0, -b, 0)$, $r_2 = \sqrt{(b^2 - c)}$

$\therefore$ $|C_1C_2| = \sqrt{(a^2 + b^2)}$

For touch

$$|C_1C_2| = r_1 \pm r_2$$

$\Rightarrow$ $$\sqrt{(a^2 + b^2)} = \sqrt{(a^2 - c)} \pm \sqrt{(b^2 - c)}$$

Squaring both sides, then

$$a^2 + b^2 = a^2 - c + b^2 - c \pm 2\sqrt{(a^2 - c)}\sqrt{(b^2 - c)}$$

$\Rightarrow$ $$c = \pm\sqrt{(a^2 - c)(b^2 - c)}$$

Again, squaring both sides, then

$$c^2 = (a^2 - c)(b^2 - c)$$

$\Rightarrow$ $$c^2 = a^2b^2 - c(a^2 + b^2) + c^2$$

or $$\frac{1}{a^2} + \frac{1}{b^2} = \frac{1}{c}.$$

11. (D) $\because$ $C_1 \equiv (0, 0, 0), r_1 = 8$

and $C_2 \equiv (6, -2, 3)$,

$$r_2 = \sqrt{(36 + 4 + 9 - 48)} = 1$$

$\because$ $$|C_1C_2| = \sqrt{36 + 4 + 9} = \sqrt{49} = 7$$

$$= r_1 - r_2$$

Hence, spheres touch internally.

12.(A) A homogeneous equation of second degree in x, y and z.

$$ax^2 + by^2 + cz^2 + 2fyz + 2gzx + 2hxy = 0 \quad ...(i)$$

Let the equations of the planes represented by equation (i) be $l_1x + m_1y + n_1z = 0$ and $l_2x + m_2y + n_2z) = 0$.

Then, equations do not contain the constant terms as otherwise their product will not be homogeneous.

So, we have

$$ax^2 + by^2 + cz^2 + 2fyz + 2gzx + 2hxy$$
$$\equiv (l_1x + m_1y + n_1z)(l_2x + m_2y + n_2z) = 0$$

$\because$ equation (i) does not contain any constant term, then, these planes will obviously pass through the origin.

And $l_1x + m_1y + n_1z = 0$

and $l_2x + m_2y + n_2z = 0$

are homogeneous linear factors.

Chapter 13. Probability

1. **Assertion (A) :** If A is any event and $P(B) = 1$, then A and B are independent.

   Reason (R) : If $P(A \cap B) = P(A) \cdot P(B)$, then A and B are independent.

2. **Assertion (A) :** If A, B, C are three events such that $P(A) = \frac{2}{3}, P(B) = \frac{1}{4}$ and $P(C) = \frac{1}{6}$, then A, B, C are mutually exclusive events.

   Reason (R) : If $P(A \cup B \cup C) = P(A) + P(B) + P(C)$, then A, B, C are mutually exclusive events.

3. **Assertion (A) :** The probability that A and B can solve a problem is $\frac{1}{3}$ and $\frac{1}{4}$ respectively, then the probability that problem will be solved is $\frac{7}{12}$.

   Reason (R) : Above mentioned events are independent events.

4. **Assertion (A) :** If A and B be mutually exclusive events in a sample space such that $P(A) = 0.3$ and $P(B) = 0.6$, then $P(\bar{A} \cap \bar{B}) = 0.28$

Reason (R) : If A and B are mutually exclusive events, then $P(A \cap B) = 0$.

5. **Assertion (A) :** Out of 5 tickets consecutively numbered, three are drawn at random, the chance that the numbers on then are in AP is $\frac{2}{5}$.

Reason (R) : Out of $(2n + 1)$ tickets consecutively numbered three are drawn at random, the chance that the numbers on them are in AP is $\frac{(4n-2)}{(4n^2-1)}$.

6. **Assertion (A) :** If A and B are two independent events such that $P(A) = \frac{1}{2}$ and $P(B) = \frac{1}{5}$, then $P(A/B) = \frac{1}{2}$

Reason (R) : $P(A/B) = P(A)$

7. **Assertion (A) :** If A and B are two events such that $P(A) = \frac{1}{2}$ and $P(B) = \frac{2}{3}$, then $\frac{1}{6} \leq P(A \cap B) \leq \frac{1}{2}$

Reason (R) : $P(A \cup B) \leq \max\ \{P(A), (B)\}$ and $P(A \cap B) \geq \min\ \{P(A), P(B)\}$

8. **Assertion (A) :** A number is chosen at random from the numbers 1, 2, 3, ..., $6n + 3$. Let A and B be defined as follows

A : Number is divisible by 2

B : Number is divisible by 3

Then, A and B are independent.

Reason (R) : If events A and B are independent, then $P(A \cap B) = P(A) \cdot P(B)$

9. **Assertion (A) :** If a leap year is selected at random, the chance it will contain 53 Sundays is $\frac{2}{7}$.

Reason (R) : A leap year has 366 days.

10. **Assertion (A) :** A and B are two candidates seeking admission in IIT. The probability that A is selected is 0.5 and the probability that both A and B are selected is at most 0.3. Then, the probability of B getting selected is 0.9.

Reason (R) : If E_1 and E_2 are the events of A and B selected respectively, then

$$P(E_1 \cap E_2) = P(E_1) \cdot P(E_2)$$

11. **Assertion (A) :** If A and B be two events in a sample space such that $P(A) = 0.3$, $P(B) = 0.3$, then $P(A \cap \bar{B})$ can not be found.

Reason (R) : $P(A \cap \bar{B}) = P(A) - P(A \cap B)$

12. **Assertion (A) :** If $P(A/B) \geq P(A)$, then $P(B/A) \geq P(B)$

Reason (R) : $P(A/B) = \frac{P(A \cap B)}{P(B)}$

13. **Assertion (A) :** If the probability of an event A is 0.4 and that of B is 0.3, then the probability of neither A nor B occuring depends upon the fact that A and B, are mutually exclusive or not.

Reason (R) : Two events are mutually exclusive, if they do not occur simultaneously.

14. **Assertion (A) :** Ankit and Rahul are weak students in Mathematics and their chances of solving a problem correctly are $\frac{1}{8}$ and $\frac{1}{12}$ respectively. They are given a question and they obtain the same answer. If the probability of a common mistake is $\frac{1}{1001}$, then the probability that the answer was correct is $\frac{13}{14}$.

Reason (R) : If E_1 and E_2 are mutually exclusive and exhaustive events with non-zero probabilities of a random experiment and E is any other event of the same experiment, then

$$P\left(\frac{E_1}{E}\right) = \frac{P\left(\frac{E}{E_1}\right) \cdot P(E_1)}{P\left(\frac{E}{E_1}\right) \cdot P(E_1) + P\left(\frac{E}{E_2}\right) \cdot P(E_2)}$$

Answers :

1. (A) If A and B are independent, then

$$\begin{aligned} P(A \cap B) &= P(A) \cdot P(B) \\ &= P(A) \qquad [\because P(B) = 1] \end{aligned}$$

and

$$\begin{aligned} P(A \cup B) &= P(A) + P(B) - P(A \cap B) \\ &= P(A) + P(B) - P(A) \\ &= P(B) \\ &= 1 \text{ which is true} \end{aligned}$$

2. (D) For mutually exclusive events

$$\begin{aligned} P(A \cup B \cup C) &= P(A) + P(B) + P(C) \\ &= \frac{2}{3} + \frac{1}{4} + \frac{1}{6} \\ &= \frac{13}{12} > 1 \end{aligned}$$

Which is not possible

3. (D) Required probability

$= 1 -$ Probability that problem will not be solved

$= 1 - P(\bar{A} \cap \bar{B})$

$= 1 - P(\bar{A})\, P(\bar{B})$

$= 1 - [1 - P(A)]\,[1 - P(B)]$

$= 1 - \left(1 - \frac{1}{3}\right)\left(1 - \frac{1}{4}\right)$

$= 1 - \frac{2}{3} \times \frac{3}{4} = 1 - \frac{1}{2} = \frac{1}{2}$

4. (D) $\because$ A and B are mutually exclusive

$$\therefore \quad P(A \cup B) = P(A) + P(B) - P(A \cap B)$$
$$= P(A) + P(B) - 0$$
$$= 0.3 + 0.6$$
$$= 0.9$$
$$\therefore \quad P(\bar{A} \cap \bar{B}) = P(\overline{A \cup B}) = 1 - P(A \cup B)$$
$$= 1 - 0.9 = 0.1$$

5. (C) Total ways $= {}^{2n+1}C_3 = \frac{(2n+1)2n\,(2n-1)}{1 \cdot 2 \cdot 3}$

$$= \frac{n\,(4n^2 - 1)}{3}$$

Let the three numbers a, b, c are drawn where $a < b < c$ and given a, b, c are in AP.

$$\therefore \quad b = \frac{a+c}{2} \text{ or } 2b = a + c \qquad \text{...(i)}$$

It is clear that from equation (i) a and c are both odd or both even.

Out of $(2n + 1)$ tickets consecutively numbers either $(n + 1)$ of them will be odd and n of them will be even.

$$\therefore \quad \text{Favourable ways} = {}^{n+1}C_2 + {}^{n}C_2$$
$$= \frac{(n+1)n}{1 \cdot 2} + \frac{n(n-1)}{1 \cdot 2}$$
$$= \frac{n}{2}\,(n + 1 + n - 1)$$
$$= n^2$$

$$\therefore \text{ Required probability} = \frac{n^2}{\frac{n(4n^2-1)}{3}} = \frac{3n}{(4n^2-1)}$$

6. (A) $\because$ A and B are independent

$$\therefore \quad P(A \cap B) = P(A) \cdot P(B)$$
$$\Rightarrow \quad \frac{P(A \cap B)}{P(B)} = P(A)$$
$$\Rightarrow \quad P(A/B) = P(A) = \frac{1}{2}$$

7. (C) We have

$$P(A \cup B) \geq \max\{P(A), P(B)\} = \frac{2}{3}$$

or $\quad P(A \cup B) \geq \frac{2}{3}$

$\Rightarrow P(A \cap B) = P(A) + P(B) - P(A \cup B) \geq P(A) + P(B) - 1$

$$= \frac{1}{2} + \frac{2}{3} - 1$$
$$= \frac{1}{6}$$

or $\quad P(A \cap B) \geq \frac{1}{6}$ ...(i)

and $P(A \cap B) \leq \min\{P(A), P(B)\} = \frac{1}{2}$

$\therefore \quad P(A \cap B) \leq \frac{1}{2}$ ...(ii)

From equations (i) and (ii), we get

$$\frac{1}{6} \leq P(A \cap B) \leq \frac{1}{2}$$

8. (D) We have

$$A = \{2, 4, 6, 8, \ldots, 6n, 6n+2\}$$
$$B = \{3, 6, 9, 12, \ldots, 6n, 6n+3\}$$

and $\quad A \cap B = \{6, 12, 18, \ldots, 6n\}$

Here, $n(A) = 3n + 1$, $n(B) = 2n + 1$ and $n(A \cap B) = n$

$$\therefore P(A) = \frac{n(A)}{6n+3} = \frac{3n+1}{6n+3},\ P(B) = \frac{n(B)}{6n+3} = \frac{1}{3}$$

and $\quad P(A \cap B) = \frac{n(A \cap B)}{6n+3} = \frac{n}{(6n+3)}$

Since, $\quad P(A \cap B) \neq P(A)\,P(B)$

$\therefore$ A and B are not independent.

9. (D) A leap year has 367 days

i.e., 52 complete weaks and 3 days more.

The three days will be three consecutive days of a week.

The probability that 53 Sundays $= \frac{3}{7}$

10. (D) Given, $\quad P(E_1 \cap E_2) \leq 0.3$

$$\Rightarrow \quad P(E_1) \cdot P(E_2) \leq 0.3$$
$$\Rightarrow \quad (0.5)\,P(E_2) \leq (0.3)$$
$$\therefore \quad P(E_2) \leq \frac{(0.3)}{(0.5)}$$
$$\Rightarrow \quad P(E_2) \leq 0.6$$
$$P(E_2) \neq 0.9$$

11. (A) $\because P(A \cap \bar{B}) = P(A) = P(A \cap B)$

$$\Rightarrow \quad P(A \cap \bar{B}) = 0.3 - P(A \cap B)$$

$\therefore P(A \cap \bar{B})$ cannot be found.

12. (A) $\because \quad P(A/B) \geq P(A)$

$$\Rightarrow \quad \frac{P(A \cap B)}{P(B)} \geq P(A)$$

$$\text{or} \quad \frac{P(A \cap B)}{P(A)} \geq P(B)$$

$$\text{or} \quad \frac{P(B \cap A)}{P(A)} \geq P(B)$$

$$\therefore \quad P(B/A) \geq P(B)$$

13. (B) Two or more events, associated with the same space are called mutually exclusive events if they cannot occur simultaneously.

If events A and B are mutually exclusive, then

$$A \cap B = \phi \text{ i.e., } P(A \cap B) = 0$$

$$\begin{aligned} \text{Now, } P(\overline{A} \cap \overline{B}) &= P(\overline{A \cup B}) = 1 - P(A \cup B) \\ &= 1 - \{P(A) + P(B) - P(A \cap B)\} \\ &= 1 - P(A) - P(B) + P(A \cap B) \\ &= 1 - 0.4 - 0.3 + 0 \\ &= 0.3 \end{aligned}$$

and if A and B are independent, then

$$P(A \cap B) = P(A)\, P(B)$$

$$\begin{aligned} \therefore \quad P(\overline{A} \cap \overline{B}) &= P(\overline{A}) \cdot P(\overline{B}) \\ &= [1 - P(A)]\,[1 - P(B)] \\ &= (1 - 0.4)\,(1 - 0.3) \\ &= (0.6)\,(0.7) \\ &= 0.42 \end{aligned}$$

14. (A) Let E_1 be the events of both getting the correct answer and E_2 the event of both getting wrong answers. Let E be the event of both obtaining the same answer.

$$\therefore \quad P(E_1) = \frac{1}{8} \times \frac{1}{12} = \frac{1}{96}$$

$$\text{and} \quad P(E_2) = \left(1 - \frac{1}{8}\right)\left(1 - \frac{1}{12}\right) = \frac{7}{8} \times \frac{11}{12} = \frac{77}{96}$$

$$\text{and } P\left(\frac{E_1}{E_2}\right) = 1,$$

$$P\left(\frac{E}{E_2}\right) = \frac{1}{1001}$$

$$\therefore \quad P\left(\frac{E_1}{E}\right) = \frac{P\left(\frac{E}{E_1}\right) \cdot P(E_1)}{P\left(\frac{E}{E_1}\right) \cdot P(E_1) + P\left(\frac{E}{E_2}\right) \cdot P(E_2)}$$

$$= \frac{1 \times \frac{1}{96}}{1 \times \frac{1}{96} + \frac{1}{1001} \times \frac{77}{96}}$$

$$= \frac{13}{14}.$$

●●

Case Based Questions | Set 5 |

Chapter 1. Relations and Functions

Q.1. A retailer purchases two types of tea priced at ₹ 120 and ₹ 160 per kg. He wants to sell the mixtures of these two types of tea at a price of ₹ 150 per kg. If he bought total of 200 kg of tea.

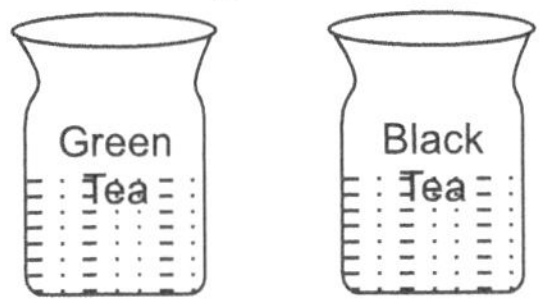

Give the answer of the followng questions:

(i) If $f(x)$ is function for the total purchase price of green tea + black tea where x denotes the quantity (in kg) for Green tea, formulate the function $f(x)$:

(a) $3200 + 4x$ (b) $32,000 - 40x$

(c) $30,000 + 4x$ (d) $60,000 - 2x$

Solution: (b)

Green tea $= x$

Black tea $= 200 - x$

$f(x) = 120x + (200 - x)\,160$

$120x + 32000 - 160x$

$f(x) = ₹\ 32000 - 40x$

(ii) How much is the quantity of green tea to be purchased to earn the same revenue?

(a) 60 kg (b) 40 kg

(c) 50 kg (d) 100 kg

Solution: (c)

Total revenue $150 \times 200 = 30,000$

$32,000 - 40x = 30,000$

$2000 = 40x$

$x = 50$

(iii) Find the value of $f(400) - f(200)$

(a) 8000 (b) -8000

(c) -5000 (d) 5000

Solution: (b)

$f(400) = 32000 - 40 \times 400$

$= 32000 \times 16000$

$= ₹\ 16000$

$f(200) = 32000 - 8000$

$= ₹\ 24000$

$f(400) - f(200) = 16000 - 24000 = -8000$

(iv) If $f(x) = x^2 - \frac{1}{x}$ and $g(x) = 2x + 5$; find fog

(a) $\frac{5x^3 + 60x^2 + 150x}{2x+5} + 124$ (b) $\frac{5x^2 + 210 + 124x}{2x+5}$

(c) $\frac{x^2 - 210}{x+5}$ (d) $\frac{210 + 124x}{2x+5}$

Solution: (a)

$$\text{fog} = f[g(x)] = f(2x+5) = (2x+5)^2 - \frac{1}{2x+5}$$

$$= \frac{(2x+5)^3 - 1}{2x+5}$$

$$= \frac{8x^3 + 125 + 60x^2 + 150x - 1}{2x+5}$$

$$= \frac{8x^3 + 60x^2 + 150x + 124}{2x+5}$$

Ans.

(v) In above function find gof?

(a) $x^2 - \frac{2}{x} + 5$ (b) $x^2 - \frac{2}{x} - 5$

(c) $2x^2 - \frac{2}{x} + 5$ (d) $2x^2 + \frac{2}{x} + 5$

Solution: (c)

$$\text{gof} = g[f(x)]$$

$$g\left(x^2 - \frac{1}{x}\right)$$

$$= 2\left[x^2 - \frac{1}{x}\right] + 5$$

$$= 2x^2 - \frac{2}{x} + 5$$

Q.2. A vegetable seller deals in oranges. When the price of orange is ₹ 30 per unit, its demand and supply are 600 & 900 units respectively.

When the price becomes ₹ 20 per unit, demand & supply changes to 1000 & 700 units respetively.

Give the answer of the following questions:

(i) If demand function is in the linear form and denoted by $p = a + bx$, where x in the no of units demanded and p is price percent; the demand function is:

(a) $p = 45 - \frac{1}{40}x_d$ **(b)** $p = 15 - \frac{1}{40}x_s$

(c) $40\,p = 15 + x_d$ **(d)** $p = 40 - \frac{x_d}{15}$

Solution: (a)

$$30 = a + b\,600$$
$$20 = a + b\,1000$$
$$- \quad - \quad -$$
$$10 = -400\,b$$
$$b = -\frac{10}{400} = -\frac{1}{40}$$
$$30 = a - \frac{1}{40} \times 600$$
$$a = 45$$

demand f^n is $\Rightarrow$ $\boxed{p = 45 - \frac{1}{40}x_d}$

(ii) If supply function is also in linear form and denoted by $p = q + rx_s$ where p is price per unit and x_s is units supplied, the supply function is:

(a) $p = 15 - \frac{1}{40}x_s$ **(b)** $p = -15 + \frac{1}{20}x_s$

(c) $p = \frac{x_s}{15} - 20$ **(d)** $p = x_s + \frac{15}{20}$

Solution: (b)

$$30 = a + 900\,x_s$$
$$20 = a + 700\,x_s$$
$$- \quad - \quad -$$
$$10 = 200\,xs$$
$$x_s = \frac{10}{200} = \frac{1}{20}$$
$$30 = a + 900 \times \frac{1}{20}$$
$$30 = a + 45$$
$$a = -15$$
$$\boxed{p = -15 + \frac{1}{20}x_s}$$

(iii) Find the price p, when demand is equal to supply.

(a) 30 **(b)** 50

(c) 25 **(d)** 40

Solution: (c)

$$p = 45 - \frac{1}{40}x_d$$
$$40p = 1800 - x_d$$
$$p = -15 + \frac{1}{20}x_s$$
$$20p = -300 + x_s$$
$$\because \quad x_d = x_s$$
$$1800 - 40p = 20p + 300$$
$$\Rightarrow \quad 1500 = 60p$$
$$\Rightarrow \quad \boxed{p = 25}$$

(iv) Find the quantity when demand is equal to supply.

(a) 800 **(b)** 400

(c) 500 **(d)** 300

Solution: (a)

$$x_d = x_s = 1800 - 40p$$
$$= 1800 - 40 \times 25$$
$$= 800 \text{ units}$$

(v) Find the number of units demanded if price rises to 40 per unit.

(a) 600 **(b)** 800

(c) 900 **(d)** 200

Solution: (d)

$$x_d = 1800 - 40p$$
$$x_d = 1800 - 40 \times 40$$
$$x_d = 200 \text{ units}$$

Chapter 2. Inverse Trigonometric Functions

Q.1. Two men on either side of temple of 30m height abserve its top at the angle of elevation α and β respectively. The distance between the two men is $40\sqrt{3}\ m$ and distance between men A and the temple is $30\sqrt{3}\ m$.

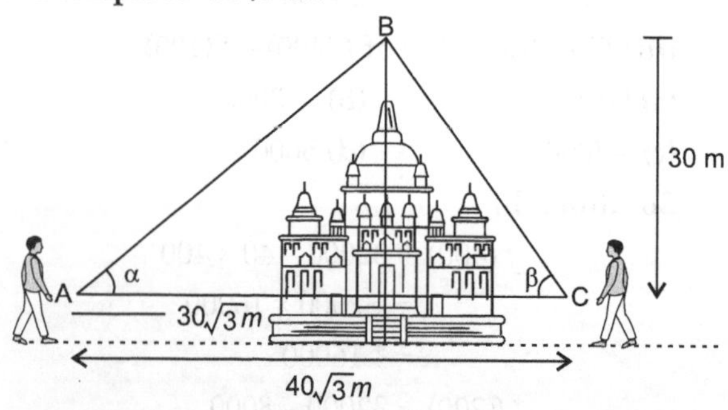

Based on above information answer the following questions:

(i) Find ∠CAB = α =

(a) $\sin^{-1}\frac{1}{2}$ **(b)** $\sin^{-1}\frac{2}{\sqrt{3}}$

(c) $\sin^{-1}\frac{\sqrt{3}}{2}$ **(d)** $\sin^{-1} 2.$

Solution: (a)

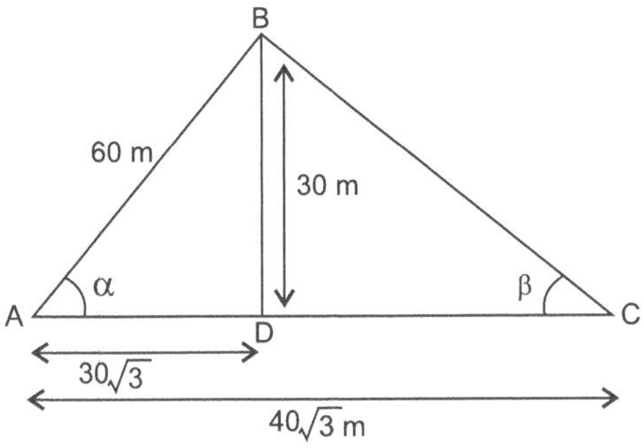

$$\sin\alpha = \frac{BD}{AB}$$

$$\sin\alpha = \frac{30}{60} = \frac{1}{2}$$

$$AB = \sqrt{(30\sqrt{3})^2 + 30^2}$$

$$= \sqrt{2700+900}$$

$$= \sqrt{3600} = 60.$$

$$\boxed{\alpha = \sin^{-1}\frac{1}{2}}$$

(ii) ∠CAB = α = ?

(a) $\cos^{-1}\frac{1}{5}$ **(b)** $\cos^{-1}\frac{2}{5}$

(c) $\cos^{-1}\frac{\sqrt{3}}{2}$ **(d)** $\cos^{-1}\frac{4}{5}$

Solution: (c)

$$\cos\alpha = \frac{AD}{AB} = \frac{30\sqrt{3}}{60} = \frac{\sqrt{3}}{2}$$

$$\alpha = \cos^{-1}\left(\frac{\sqrt{3}}{2}\right)$$

(iii) ∠BCA = β = ?

(a) $\tan^{-1}\frac{1}{2}$ **(b)** $\tan^{-1} 2$

(c) $\tan^{-1}\frac{1}{\sqrt{3}}$ **(d)** $\tan^{-1}\sqrt{3}$

Solution: (d)

$$\tan\beta = \frac{BD}{DC} = \frac{30}{10\sqrt{3}} = \frac{3}{\sqrt{3}} = \sqrt{3}$$

$$\boxed{\beta = \tan^{-1}\sqrt{3}}$$

(iv) ∠ABC = ?

(a) $\frac{\pi}{4}$ **(b)** $\frac{\pi}{6}$

(c) $\frac{\pi}{2}$ **(d)** $\frac{\pi}{3}$

Solution: (c)

$$BC = \sqrt{30^2 + (10\sqrt{3})^2}$$

$$= \sqrt{900+300} = \sqrt{1200} = 20\sqrt{3}\,m$$

$$\because\ AB^2 + BC^2 = \sqrt{60^2 + 1200} = \sqrt{3600+1200}$$

$$= \sqrt{4800} = 40\sqrt{3} = 4C$$

∵ By convex of pythagoras theory

$$\angle ABC = \frac{\pi}{2}$$

(v) Domain and range of $\cos^{-1} x$?

(a) (– 1, 1), (0 π) **(b)** [– 1, 1], (0, π)

(c) [0, π], [– 1, 1] **(d)** (– 1, 1), $\left[-\frac{\pi}{2}, \frac{\pi}{2}\right]$

Solution: (c)

Q.2. The government of India is planning to fix a hoarding board at the face of a building on the road of a market for awareness of covid-19. Ram, Rahim and Raja are three persons who are working on this project. A is considred to be a person veweing the hoarding board 20 m away from the building, standing at the edge of a path way nearby. Ram, Rahim and Raja is suggested to the firm to place the hoarding board at three different locations namely C, D & E. C is the length 10m from the ground level. For the viewer A the angle of elevation of D is double the angle of elevation of C. The angle of elevation of E is the triple the angle of elevation of C. Look at the figure and based on these information answer the following:

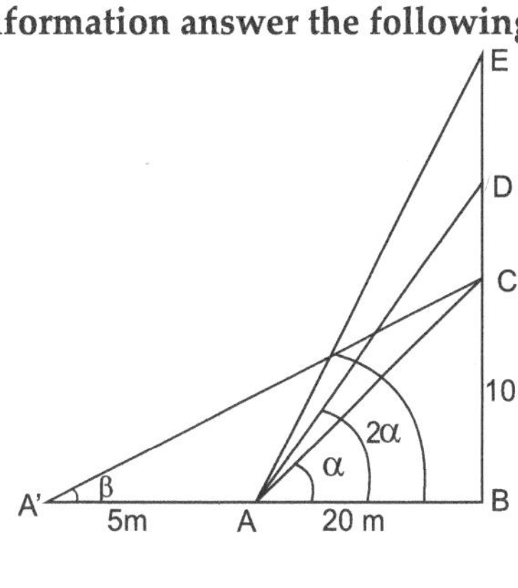

(i) Measure of ∠CAB = ?

(a) $\tan^{-1}\left(\frac{1}{3}\right)$ **(b)** $\tan^{-1}\left(\frac{1}{2}\right)$

(c) $\tan^{-1}(1)$ **(d)** $\tan^{-1} 3$

Solution: (b)

$$\tan\alpha = \frac{10}{20} = \frac{1}{2}$$

$$\boxed{\alpha = \tan^{-1}\frac{1}{2}}$$

(ii) ∠DAB = ?

(a) $\tan^{-1}\frac{3}{4}$ **(b)** $\tan^{-1} 3$

(c) $\tan^{-1}\frac{4}{3}$ **(d)** $\tan^{-1}(4)$

Solution: (c)

$$\tan 2\alpha = \frac{2\tan\alpha}{1-\tan^2\alpha}$$

$$= \frac{2\times\frac{1}{2}}{1-\frac{1}{4}} = \frac{1}{\frac{3}{4}} = \frac{4}{3}$$

$$2\alpha = \tan^{-1}\left(\frac{4}{3}\right)$$

(iii) Measure of ∠EAB = ?

(a) $\tan^{-1} 11$ **(b)** $\tan^{-1} 3$

(c) $\tan^{-1}\frac{2}{11}$ **(d)** $\tan^{-1}\frac{11}{2}$

Solution: (d)

$$\tan 3\alpha = \frac{3\tan\alpha - \tan^3\alpha}{1-3\tan^2\alpha}$$

$$= \frac{3\times\frac{1}{2}-\left(\frac{1}{2}\right)^3}{1-3\left(\frac{1}{2}\right)^2} = \frac{\frac{3}{2}-\frac{1}{8}}{1-\frac{3}{4}} = \frac{\frac{12-1}{8}}{\frac{1}{4}}$$

$$\tan 3\alpha = \frac{11}{8}\times 4 = \frac{11}{2}$$

(iv) A′ is auother viewer standing on the same line of observation across the road. If width of the road in 5m, then the difference between ∠CAB & ∠CA′B is

(a) $\tan^{-1}\frac{1}{2}$ **(b)** $\tan^{-1}\frac{1}{8}$

(c) $\tan^{-1}\frac{2}{5}$ **(d)** $\tan^{-1}\frac{1}{12}$

(iv) Solution: (d)

$$\angle CAB - \angle CA'B = \alpha - \beta \qquad (\alpha > \beta)$$

$$\tan\beta = \frac{10}{25} = \frac{2}{5}$$

$$\tan(\alpha-\beta) = \frac{\tan\alpha - \tan\beta}{1+\tan\alpha+\tan\beta}$$

$$= \frac{\frac{1}{2}-\frac{2}{5}}{1+\frac{1}{2}\times\frac{2}{5}} = \frac{\frac{5-4}{10}}{\frac{10+2}{10}} = \frac{1}{12}$$

$$\alpha-\beta = \tan^{-1}\left(\frac{1}{12}\right)$$

(v) Range of $\tan^{-1} x$ = ?

(a) $\left(\frac{-\pi}{2}, \frac{\pi}{2}\right)$ **(b)** $(0, \pi)$

(c) $[0, \pi]$ **(d)** $\left[\frac{-\pi}{2}, \frac{\pi}{2}\right]$

Solution: (a)

Chapter 3. Matrices

Q. 1. A firm produces three products P_1, P_2 and P_3 requiring the mix-up of three materials M_1, M_2 and M_3. Per unit requirement of each product for each material (in units) is represented by matrix A as follows :

$$A = \begin{matrix} \\ P_1 \\ P_2 \\ P_3 \end{matrix}\begin{matrix} M_1 & M_2 & M_3 \\ \end{matrix}\begin{bmatrix} 2 & 3 & 1 \\ 4 & 2 & 5 \\ 2 & 4 & 2 \end{bmatrix}$$

The per unit cost of material M_1, M_2 and M_3 is ₹ 5, ₹ 10 and ₹ 5 respectively.

Attempt the following questions :

(i) If C represents the matrix showing cost of material, the matrix C will be expressed as :

(a) $C = \begin{bmatrix} 10 \\ 5 \\ 5 \end{bmatrix}$ **(b)** $C = [5\ \ 10\ \ 15]$

(c) $C = \begin{bmatrix} 5 \\ 10 \\ 5 \end{bmatrix}$ **(d)** $C = \begin{bmatrix} 15 \\ 10 \\ 20 \end{bmatrix}$

Solution. (c) $C = \begin{matrix} M_1 \\ M_2 \\ M_3 \end{matrix}\begin{bmatrix} 5 \\ 10 \\ 5 \end{bmatrix}$ **Ans.**

(ii) Expressed Matrix B in the order of 3 × 1 if the firm produces 100 units of each product :

(a) $\begin{bmatrix} 100 \\ 100 \\ 100 \end{bmatrix}$ **(b)** [100 100 100]

(c) $\begin{bmatrix} 100 \\ 200 \\ 300 \end{bmatrix}$ **(d)** $\begin{bmatrix} 10 \\ 20 \\ 30 \end{bmatrix}$

Solution. (a) $B = \begin{matrix} P_1 \\ P_2 \\ P_3 \end{matrix}\begin{bmatrix} 100 \\ 100 \\ 100 \end{bmatrix}$ **Ans.**

(iii) Find the total requirement of each material for 100 units of each product :

(a) $\begin{bmatrix} 900 \\ 800 \\ 800 \end{bmatrix}$ **(b)** $\begin{bmatrix} 90 \\ 80 \\ 80 \end{bmatrix}$

(c) $\begin{bmatrix} 190 \\ 200 \\ 80 \end{bmatrix}$ **(d)** $\begin{bmatrix} 800 \\ 900 \\ 800 \end{bmatrix}$

Solution. (d) The total requirement of each material is given by A′B.

$$A' = \begin{bmatrix} 2 & 4 & 2 \\ 3 & 2 & 4 \\ 1 & 5 & 2 \end{bmatrix}$$

$$A'B = \begin{bmatrix} 2 & 4 & 2 \\ 3 & 2 & 4 \\ 1 & 5 & 2 \end{bmatrix}\begin{bmatrix} 100 \\ 100 \\ 100 \end{bmatrix} = \begin{bmatrix} 800 \\ 900 \\ 800 \end{bmatrix}$$ **Ans.**

(iv) Find per unit cost of production of each product as per the unit cost of material given above :

(a) $\begin{bmatrix} 45 \\ 65 \\ 60 \end{bmatrix}$ **(b)** $\begin{bmatrix} 65 \\ 60 \\ 45 \end{bmatrix}$

(c) $\begin{bmatrix} 60 \\ 65 \\ 40 \end{bmatrix}$ **(d)** $\begin{bmatrix} 50 \\ 60 \\ 70 \end{bmatrix}$

Solution. (a) The per unit cost production of each product is given by

$$AC = \begin{bmatrix} 2 & 3 & 1 \\ 4 & 2 & 5 \\ 2 & 4 & 2 \end{bmatrix}\begin{bmatrix} 5 \\ 10 \\ 5 \end{bmatrix} = \begin{bmatrix} 45 \\ 65 \\ 60 \end{bmatrix}$$ **Ans.**

(v) Find the total cost of production if the firm produces 200 units of each product :

(a) 28,000 **(b)** 30,000
(c) 34,000 **(d)** 32,000

Solution. (c) Total cost of production of 200 units

$$= [200\ 200\ 200]\begin{bmatrix} 45 \\ 65 \\ 60 \end{bmatrix}$$

= 9,000 + 13,000 + 12,000 = ₹ 34,000 **Ans.**

Q. 2. A concert is organised to earn the revenue of ₹ 1,80,000 which can be distributed among the needy people during the period of Covid-19. The concert hall has 4,000 seats which are divided into two sections A and B. The cost of a ticket in Section A is ₹ 50 and that of Section B is ₹ 40. All the seats are occupied.

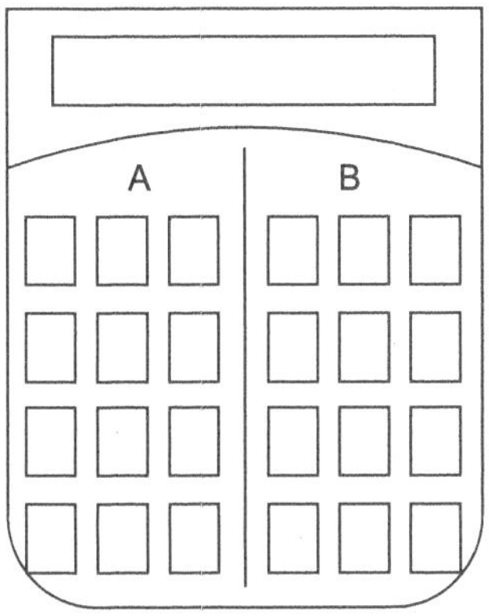

Attempt the following questions :

(i) If x and y are number of seats of Section A and B respectively, then write the equation related to number of seats :

(a) $x - y = 4000$ **(b)** $x + y = 400$
(c) $x + y = 4000$ **(d)** $2x + y = 4000$

Solution. (b) Let x and y be the no. of seats in section A and section B respectively.

$\therefore \quad x + y = 4000$ **Ans.**

(ii) What will be relation between x and y for the total revenue ?

(a) $5x + 4y = 18000$ **(b)** $50x + 40y = 18000$
(c) $5x + 4y = 1800$ **(d)** $x + y = 180$

Solution. (a) $50x + 40y = 1,80,000$
$5x + 4y = 18,000$ **Ans.**

(iii) If equations are to be solved through the matrix method in the form of AX = B, then what will be matrix A and matrix B :

(a) $A = \begin{bmatrix} 1 & 1 \\ 5 & 4 \end{bmatrix}, B = \begin{bmatrix} 4000 \\ 18000 \end{bmatrix}$

(b) $A = \begin{bmatrix} 5 & 4 \\ 1 & 1 \end{bmatrix}, B = \begin{bmatrix} 4000 \\ 18000 \end{bmatrix}$

(c) $A = \begin{bmatrix} 1 & 5 \\ 1 & 4 \end{bmatrix}, B = \begin{bmatrix} 18000 \\ 4000 \end{bmatrix}$

(d) $A = \begin{bmatrix} 1 & 4 \\ 1 & 5 \end{bmatrix}, B = \begin{bmatrix} 4000 \\ 18000 \end{bmatrix}$

Solution. (a) Writing these equation in matrix form AX = B

where $\begin{bmatrix} 1 & 1 \\ 5 & 4 \end{bmatrix}\begin{bmatrix} x \\ y \end{bmatrix} = \begin{bmatrix} 4000 \\ 18000 \end{bmatrix}$

So, $A = \begin{bmatrix} 1 & 1 \\ 5 & 4 \end{bmatrix}, X = \begin{bmatrix} x \\ y \end{bmatrix}, B = \begin{bmatrix} 4000 \\ 18000 \end{bmatrix}$ **Ans.**

(iv) The value of X will be equal to $A^{-1}B$, then A^{-1}:

(a) $\begin{bmatrix} -4 & 1 \\ -5 & 1 \end{bmatrix}$ (b) $\begin{bmatrix} -4 & 1 \\ 5 & -1 \end{bmatrix}$

(c) $\begin{bmatrix} 4 & -5 \\ -1 & 1 \end{bmatrix}$ (d) $\begin{bmatrix} -4 & 5 \\ -1 & -1 \end{bmatrix}$

Solution. (b)

$$X = A^{-1}B$$

$$\because \quad A^{-1} = \frac{\text{adj } A}{|A|}$$

$$|A| = \begin{bmatrix} 1 & 1 \\ 5 & 4 \end{bmatrix} = 4 - 6 = -1$$

$$\therefore \quad \text{adj } A = \begin{bmatrix} 4 & -1 \\ -5 & 1 \end{bmatrix}$$

$$\text{So,} \quad A^{-1} = \frac{1}{-1}\begin{bmatrix} 4 & -1 \\ -5 & 1 \end{bmatrix} = \begin{bmatrix} -4 & 1 \\ 5 & -1 \end{bmatrix}$$

Ans..

(v) How many seats are there in Section A and Section B of Concert Hall ?

(a) 2000, 4000 (b) 4000, 2000

(c) 2000, 2000 (d) 6000, 4000

Solution. (c)

$$X = -1\begin{bmatrix} 4 & -1 \\ -5 & 1 \end{bmatrix}\begin{bmatrix} 4000 \\ 18000 \end{bmatrix}$$

$$= \begin{bmatrix} -4 & 1 \\ 5 & -1 \end{bmatrix}\begin{bmatrix} 400 \\ 18000 \end{bmatrix}$$

$$= \begin{bmatrix} 16000 + 18000 \\ 20000 - 18000 \end{bmatrix} = \begin{bmatrix} 2000 \\ 2000 \end{bmatrix}$$

$\therefore$ x = 2000 seats in Sec. A

y = 2000 seats in Sec. B. **Ans.**

Chapter 6. Applications of Derivatives

Q. 1. An architect designs a building for a multi-national company. The floor consists of a rectangular region with semicircular ends having a perimeter of 200 m as show below :

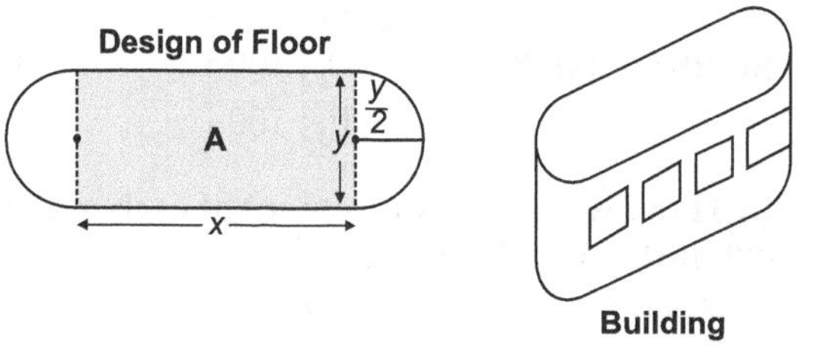

Based on the above information answer the following :

(i) If x and y represents the length and breadth of the rectangular region, then the relation between the variables is :

(a) $x + \pi y = 100$ (b) $2x + \pi y = 200$

(c) $\pi x + y = 50$ (d) $x + y = 100$

Solution. (b) Perimeter= 200 m

$$\Rightarrow \quad 2x + 2\pi\left(\frac{y}{2}\right) = 200$$

$$\Rightarrow \quad 2x + \pi y = 200$$

Hence (b) is correct option. **Ans.**

(ii) The area of the rectangular region A expressed as a function of x is :

(a) $\frac{2}{\pi}(100x - x^2)$ (b) $\frac{1}{\pi}(100x - x^2)$

(c) $\frac{x}{\pi}(100 - x)$ (d) $\pi y^2 + \frac{2}{\pi}(100x - x^2)$

Solution. (a) Area of rectangular region

$$= L \times B$$

$$\Rightarrow \quad A = xy = x\left(\frac{200 - 2x}{\pi}\right) \quad \text{[From (i)]}$$

$$\therefore \quad A = \frac{2}{\pi}(100x - x^2)$$

Hence (a) is correct option. **Ans.**

(iii) The maximum value of area A is :

(a) $\frac{\pi}{3200}$ m^2 (b) $\frac{3200}{\pi}$ m^2

(c) $\frac{5000}{\pi}$ m^2 (d) $\frac{1000}{\pi}$ m^2

Solution. (c)

$$\frac{dA}{dx} = \frac{2}{\pi}(100 - 2x)$$

Put $\frac{dA}{dx} = 0 \Rightarrow x = 50$

$$\frac{d^2A}{dx^2} = \frac{2}{\pi}(-2) < 0$$

$\therefore$ A is maximum when $x = 50$

$$\therefore \text{Maximum area} = \frac{2}{\pi}(5000 - 2500) = \frac{5000}{\pi} \text{ m}^2$$

Hence (c) is correct option. **Ans.**

(iv) The CEO of the multi-national company is interested in maximizing the area of the whole floor including the semi-circular ends. For this to happen the value of x should be :

(a) 0 m **(b)** 30 m

(c) 50 m **(d)** 80 m

Solution. (a) Area of whole floor

$$= xy + \pi\left(\frac{y}{2}\right)^2$$

$$= \frac{(200 - \pi y)y}{2} + \frac{\pi y^2}{4}$$

$$= 100y - \frac{\pi y^2}{2} + \frac{\pi y^2}{4}$$

$$A_1 = 100y - \frac{\pi y^2}{4}$$

$$\Rightarrow \frac{dA_1}{dy} = 100 - \frac{2\pi y}{4}$$

$$\Rightarrow \frac{dA_1}{dy} = 0$$

$$\Rightarrow 200 = \pi y$$

$$\Rightarrow y = \frac{200}{\pi}$$

$$\Rightarrow \frac{d^2A_1}{dy^2} = -\frac{\pi}{2} < 0$$

$\therefore$ A_1 is maximum when

$$y = \frac{200}{\pi} \text{ and } x = \left[\frac{200 - \pi\left(\frac{200}{\pi}\right)}{2}\right] = 0$$

$$\Rightarrow x = 0$$

Hence (a) is correct option. **Ans.**

(v) The extra area generated if the area of the whole floor is maximized is :

(a) $\frac{3000}{\pi}$ m^2 **(b)** $\frac{5000}{\pi}$ m^2

(c) $\frac{7000}{\pi}$ m^2

(d) No change both areas are equal.

Solution. (d) When the whole area is maximized, there will be no further increase in area.

Hence (d) is correct option.

Q. 2. A new township is to be constructed along the curve $y = x^3 - 11x + 5$. Two attraction point *i.e.*, Amusement park and shopping mall is to be made on the this curve. At a point (1, – 5) on the curve amusement park is to be constructed and on another point shopping mall is to be constructed at which tangent has the equation $y = x - 11$.

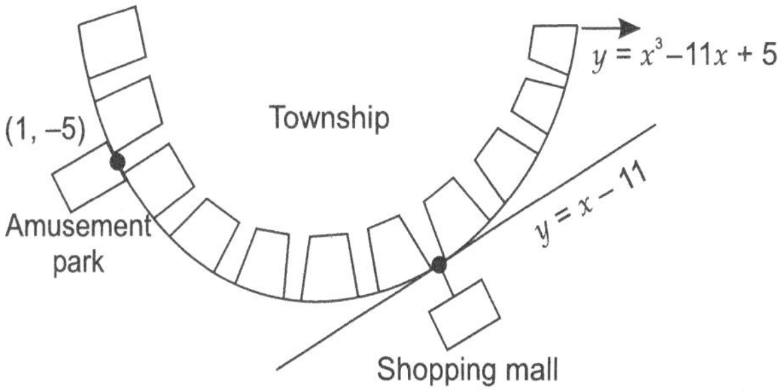

Give the answers of the following questions :

(i) Find the slope of the normal to the curve at (1, – 5) on which amusement park is to be constructed ?

(a) $\frac{1}{8}$ **(b)** 8

(c) – 8 **(d)** $\frac{-1}{8}$

Solution. (a) Given equation of curve is

$$y = x^3 - 11x + 5$$

$$\frac{dy}{dx} = 3x^2 - 11 \text{ at } (1, -5)$$

$$\frac{dy}{dx} = 3(1)^2 - 11 = -8$$

Slope of normal to the curve is $\frac{1}{8}$. **Ans.**

(ii) Find the equation of normal to the curve along which amusement park is to be constructed ?

(a) $8y - x - 41 = 0$ **(b)** $8y + x = 41$

(c) $8y + x + 41 = 0$ **(d)** $8y - x + 41 = 0$

Solution. (d) Equation of normal is

$$y - y_1 = \frac{-1}{\frac{dy}{dx}}(x - x_1)$$

$$y+5 = \frac{1}{8}(x-1)$$

$$8y+40 = x-1$$

$$8y-x+41 = 0. \qquad \textbf{Ans.}$$

(iii) Find the point of x where curve along which township is parallel to X-axis ?

(a) $\pm\sqrt{\frac{11}{3}}$ (b) $\pm\frac{11}{3}$

(c) ± 4 (d) ± 2

Solution. (a) For curve parallel to X-axis slope *i.e.,*

$$\frac{dy}{dx} = 0$$

$$3x^2 - 11 = 0$$

$$x^2 = \frac{11}{3}$$

$$x = \pm\sqrt{\frac{11}{3}} \qquad \textbf{Ans.}$$

(iv) Find the points on the curve at which tangent $y = x - 11$ is given ?

(a) (– 2, 19) (b) (2, – 9)

(c) (– 2, 9) (d) (2, – 19)

Solution. (b) Slope of tangent = Slope of curve

$$3x^2 - 11 = 1$$

$$3x^2 = 12$$

$$x^2 = 4$$

$$x = \pm 2$$

If $x = 2, y = -9$, if $x = -2, y = 19$

But (– 2, 19) does not lie on the tangent. **Ans.**

(v) What is the slope of tangent at the curve ?

(a) 3 (b) 1

(c) – 1 (d) $\frac{1}{2}$

Solution. (b) Slope of tangent is

$$y = x - 11$$

$$\frac{dy}{dx} = 1. \qquad \textbf{Ans.}$$

Chapter 7. Integrals

Q. 1. Consider the following equations of curve $x^2 = y$ and $y = x$. On the basis of above informations, answer the following questions :

(i) The point of intersection of both the curve is :

(a) (0, 0), (2, 2) (b) (0, 0), (1, 1)

(c) (0, 0), (–1, –1) (d) (0, 0), (–2, –2)

Solution. (b) $x^2 = y$ & $y = x$

$$x^2 = x$$

$$x^2 - x = 0$$

$$x(x-1) = 0$$

$$x = 0, x = 1$$

$$y = 0, y = 1$$

Intersecting points are (0, 0), (1, 1)

(ii) The graph of the above given curve :

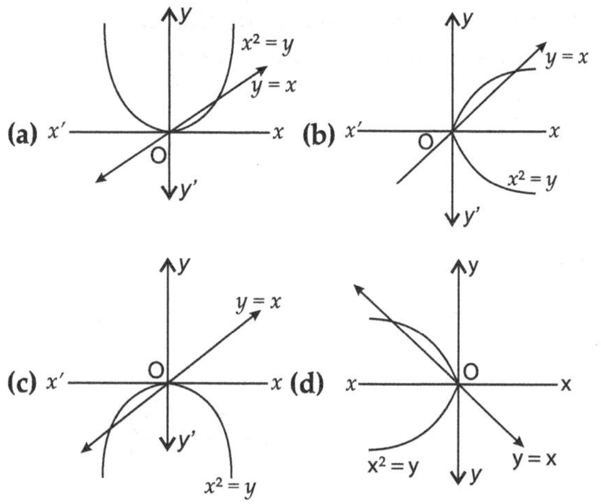

Solution. (a)

(iii) The value of integral $\int_0^1 x\,dx$ is:

(a) $\frac{1}{4}$ (b) $\frac{1}{3}$

(c) $\frac{1}{2}$ (d) 1

Solution. (c)

$$\int_0^1 x\,dx = \left.\frac{x^2}{2}\right|_0^1 = \frac{1}{2} - 0 = \frac{1}{2}$$

(iv) The value of integral $\int_0^1 x^2\,dx$ is :

(a) $\frac{1}{4}$ (b) $\frac{1}{3}$

(c) $\frac{1}{2}$ (d) 1

Solution. (b)

$$\int_0^1 x^2\,dx = \left.\frac{x^3}{3}\right|_0^1 = \frac{1}{3} - 0 = \frac{1}{3}$$

(v) The value of area bounded by the curves $x^2 = y$ & $y = x$ is :

(a) $\frac{1}{8}$ (b) $\frac{1}{4}$

(c) $\frac{1}{3}$ (d) $\frac{1}{2}$

Solution. (a)

$$\int_0^1 x dx - \int_0^1 x^2 dx$$

$$\frac{1}{2} - \frac{1}{3} = \frac{1}{6}$$

Chapter 8. Applications of the Integrals

Q. 1. A school is to be constructed for the rural area children on the wasted land which is lying under the area covered by the lines $3x - 2y + 1 = 0$, $2x + 3y - 21 = 0$ and $x - 5y + 9 = 0$.

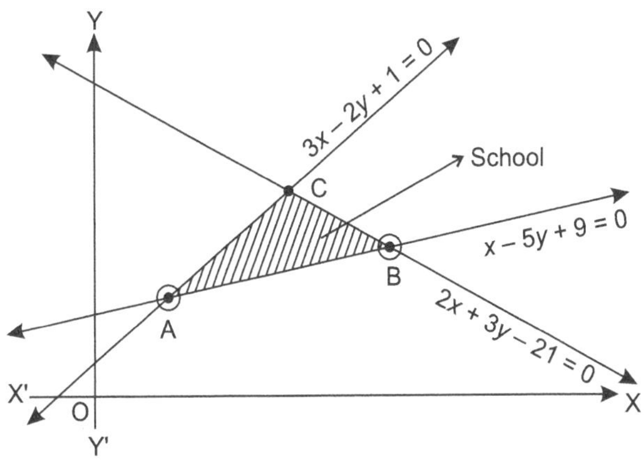

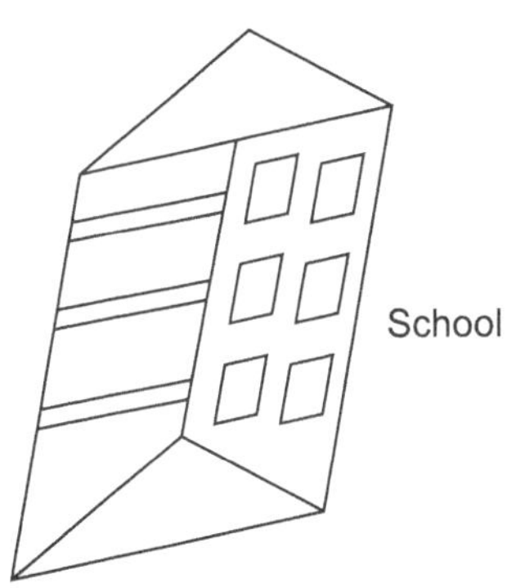

Answer the following questions based on above case study :

(i) What is coordinates of point A ?

(a) (1, 2) (b) (2, 1)

(c) (2, 2) (d) (1, 1)

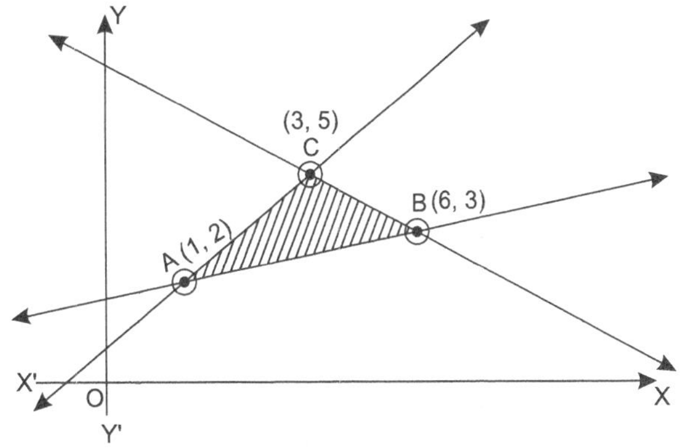

Equation of AB :

$$x - 5y + 9 = 0 \quad ...(i)$$

$$5y = x + 9$$

$$y = \frac{x+9}{5}$$

Equation of CB :

$$2x + 3y - 21 = 0 \quad ...(ii)$$

$$3y = 21 - 2x$$

$$y = 7 - \frac{2}{3}x$$

Equation of AC :

$$3x - 2y + 1 = 0 \quad ...(iii)$$

$$2y = 3x + 1$$

$$y = \frac{3}{2}x + \frac{1}{2}$$

Solution. (a) Solving (i) and (iii) we get co-ordinates of A

$$A = (1, 2).$$

(ii) Find the coordinates of B :

(a) (6, 3) **(b)** (1, 6)

(c) (1, 3) **(d)** (2, 5)

Solution. (a) On solving (i) and (ii), we get coordinates of B.

So, B = (6, 3)..

(iii) Write the coordinates of point C :

(a) (6, 3) **(b)** (3, 5)

(c) (1, 2) **(d)** (0, 1)

Solution. (b) On solving (ii) and (iii), we get coordinates of C.

So, C = (3, 5)

(iv) Write the limits of equation of AC which is to be integrated :

(a) $\int_1^6 f(x)dx$ **(b)** $\int_0^3 f(x)dx$

(c) $\int_3^6 f(x)dx$ **(d)** $\int_1^3 f(x)dx$

Solution. (d) Equation of AC is in terms of x is

$$y = \frac{3}{2}x + \frac{1}{2}$$

It will be integrated from 1 to 3 $\int_1^3 f(x)dx$.

(v) What is the total area coverd by the school ?

(a) $\frac{2}{15}$ sq unit **(b)** $\frac{13}{2}$ sq unit

(c) 15 sq unit **(d)** 10 sq unit

Solution. (b) Area will be:

$$\int_1^3 eq^n \text{ of AC}\, dx + \int_3^6 eq^n \text{ of CB}\, dx - \int_1^6 eq^n \text{ of AB}\, dx$$

$$= \int_1^3 \left(\frac{3}{2}x + \frac{1}{2}\right) dx + \int_3^6 \left(7 - \frac{2}{3}x\right) dx - \int_1^6 \frac{x+9}{5} dx$$

$$= \left[\frac{3}{2}\cdot\frac{x^2}{2} + \frac{1}{2}x\right]_1^3 + \left[7x - \frac{2}{3}\frac{x^2}{2}\right]_3^6 - \left[\frac{1}{5}\left(\frac{x^2}{2} + 9x\right)\right]_1^6$$

$$= \left[\frac{3}{4}(3)^2 + \frac{1}{2}(3) - \frac{3}{4}(1)^2 - \frac{1}{2}\right] + \left[\left(42 - \frac{36}{3}\right) - (21 - 3)\right] - \frac{1}{5}\left[\left(\frac{36}{2} + 54\right) - \left(\frac{1}{2} + 9\right)\right]$$

$$= \frac{13}{2} \text{ sq units.}$$

Q.2. Two roads are to be constructed along the curves $y^2 = 2x$ and $x^2 = 2y$ which are intersected with each other. The common portion is to be covered by the green grass and planting colour flower plants as shown below :

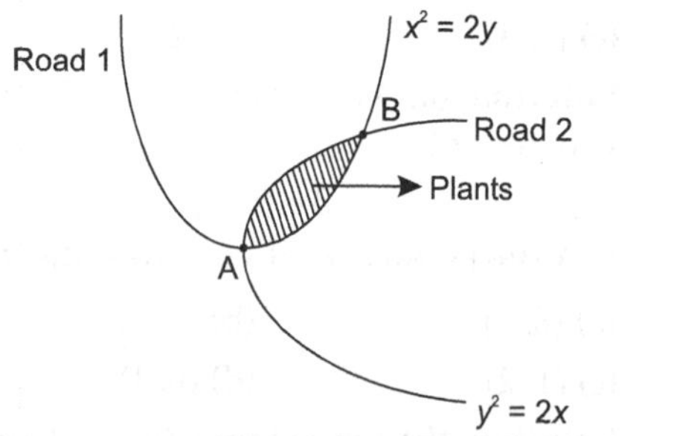

Give the answers of the following questions :

(i) Find the points at which both the curves intersect with each other.

(a) (0, 2) **(b)** (2, – 1)

(c) (– 1, – 1) **(d)** (2, 2)

Solution. (a) Given curves are

$$y^2 = 2x \quad \text{...(i)}$$

$$\Rightarrow \quad y = \sqrt{2x}$$

$$x^2 = 2y \quad \text{...(ii)}$$

$$y = \frac{x^2}{2}$$

Putting the value of y in (i),

$$\left(\frac{x^2}{2}\right)^2 = 2x$$

$$\Rightarrow \quad \frac{x^4}{4} = 2x$$

$$\Rightarrow \quad x^4 = 8x$$

$$\Rightarrow \quad x^4 - 8x = 0$$

$$\Rightarrow \quad x(x^3 - 8) = 0$$

Thus, $x = 0, x = 2$. **Ans.**

(ii) Express $x^2 = 2y$ in the form which is to be integrated.

(a) $\sqrt{2}y$ **(b)** $\sqrt{2y}$

(c) $\frac{x^2}{2}$ **(d)** $2x^2$

Solution. (c) $x^2 = 2y$

$$y = \frac{x^2}{2}.$$

(iii) Write the function which is to be integrated.

(a) $\int \sqrt{2x}\, dx$ **(b)** $\int \frac{x^2}{2} dx$

(c) $\int (\sqrt{2} + 3x)\, dx$ **(d)** $\int \left(\sqrt{2x} - \frac{x^2}{2}\right) dx$

Solution. (d) The function which is to be integrated $\int \left(\sqrt{2x} - \frac{x^2}{2}\right) dx$.

(iv) Write the limit on X-axis between which the above function will be integrated.

(a) 0 to 4 **(b)** 0 to 2

(c) – 2 to 2 **(d)** – 2 to 0

Solution. (b)

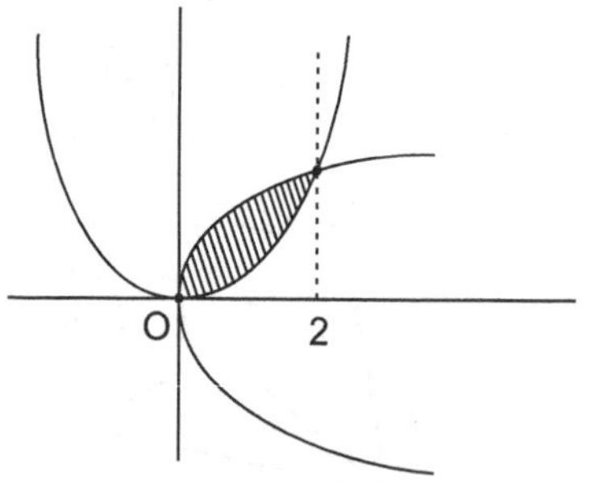

Given function will be integrated w.r.t. x from 0 to 2.

(v) Find the area in which plantation need to be done.

(a) 8 sq. unit **(b)** $\frac{8}{3}$ sq. unit

(c) $\frac{4}{3}$ sq. unit **(d)** 4 sq. unit

Solution. (c) Area $= \int_0^2 \left(\sqrt{2x} - \frac{x^2}{2} \right) dx$

$$\sqrt{2} \cdot \frac{x^{3/2}}{3/2} - \frac{1}{2} \cdot \frac{x^3}{3} \Big|_0^2 = \frac{2\sqrt{2}}{3} x^{3/2} - \frac{1}{6} x^3 \Big|_0^2$$

$$= \left[\frac{2\sqrt{2}}{3} \cdot (2)^{3/2} - \frac{1}{6} (2)^3 \right] - 0$$

$$= \frac{2\sqrt{2}}{3} \cdot 2\sqrt{2} - \frac{4}{3}$$

$$= \frac{8}{3} - \frac{4}{3} = \frac{4}{3} \text{ sq. units.}$$ **Ans.**

Chapter 9. Differential Equations

Q.1. Consider the given equation $\frac{dy}{dx} + py = Q$.

The above equation is known as linear differential equation

Here, IF = $e^{\int p dx}$ and solution is given by

y. IF = $\int$ QIF $dx + c$

Consider the given equation

$(1 + \sin x) \frac{dy}{dx} + y \cos x + x = 0$

On the above information, answer the following questions:

(i) The value of P and Q are:

(a) $\frac{\cos x}{1+\sin x}, \frac{-x}{1+\sin x}$ **(b)** $\frac{-x}{1+\sin x}, \frac{\cos x}{1+\sin x}$

(c) $\frac{-\cos x}{1+\sin x}, \frac{x}{1+\sin x}$ **(d)** $\frac{x}{1+\sin x}, \frac{-\cos x}{1+\sin x}$

Solution: (a)

$$(1 + \sin x)\frac{dy}{dx} + \cos x\, y = -x$$

$$\frac{dy}{dx} + \frac{\cos x}{1+\sin x} y = \frac{-x}{1+\sin x}$$

$$p = \frac{\cos x}{1+\sin x}, Q = \frac{-x}{1+\sin x}$$

(ii) The value of integrating factor (I.F.) is

(a) $\cos x$ **(b)** $1 + \sin x$

(c) $\frac{1}{1+\sin x}$ **(d)** $1 - \sin x$

Solution: (b)

$$\text{If} = e^{\int P dx} = e^{\int \frac{\cos x}{1+\sin x} dx}$$

$$e^{\int \log(1+\sin x)} = 1 + \sin x$$

(iii) Solution of given equation is:

(a) $y\,(1 = \sin x) = \frac{x^2}{2} + C$

(b) $y\,(1 + \sin x) = x^2 + C$

(c) $y\,(1 + \sin x) = \frac{-x^2}{2} + C$

(d) None of these

Solution: (c)

Given equation is

$$y.\text{ IF} = \int \text{Q.IF}\, dx + C$$

$$y\,(1 + \sin x) = \int \frac{-x}{1+\sin x}(1+\sin x) dx$$

$$y\,(1 + \sin x) = \frac{-x^2}{2} + C$$

(iv) If $y(0) = 1$ then y =

(a) $\frac{2-x^2}{2(1+\sin x)}$ **(b)** $\frac{1+x^2}{2(1+\sin x)}$

(c) $\frac{1-x^2}{4(1+\sin x)}$ **(d)** $\frac{1+x^2}{4(1+\sin x)}$

Solution: (a)

Put $x = 0, y = 1$

$$1(1 + \sin 0) = \frac{0}{2} + C$$

$$1 = C$$

$$y\,(1 + \sin x) = \frac{-x^2}{2} + 1$$

$$y = \frac{2-x^2}{2(1+\sin x)}$$

(v) Value of $y\left(\frac{\pi}{2}\right)$ =

(a) $\frac{4+\pi^2}{4}$ **(b)** $\frac{8-\pi^2}{16}$

(c) $\frac{4+\pi^2}{2}$ **(d)** $\frac{4+\pi^2}{16}$

Solution: (b)

Put $y = \frac{\pi}{2}$ in (iv)

$$y = \frac{2 - \frac{\pi^2}{4}}{2\left(1 + \sin\frac{\pi}{2}\right)} = \frac{8-\pi^2}{2(1+1)4}$$

$$= \frac{8-\pi^2}{16}$$

Q.2. It is given that at any point (x, y) on the curve, the product of the slope of its tangent and y – coordinate of the point is equal to the x – coordinate of the point.

(i) What is the slope of tangent?

(a) $\frac{dx}{dy}$ (b) $\frac{dy}{dx}$

(c) $\frac{-dx}{dy}$ (d) $\frac{-dy}{dx}$

Solution: (b)

$$\text{Slope of tangent} = \frac{dy}{dx}$$

(ii) Write the differential equation.

(a) $y\frac{dy}{dx} = x$ (b) $x\frac{dy}{dx} = y$

(c) $y\frac{dx}{dy} = 1$ (d) $x\frac{dx}{dy} = 1$

Solution: (a)

Product of slope of tangent $y = x$

$$\frac{dy}{dx}.y = x$$

(iii) Equation of curve in general form is:

(a) $y^2 = x^2 + C$ (b) $y = x + C$

(c) $y^2 = x^2 + 2C$ (d) $y^2 = x + C$

(iii) Solution: (c)

$$y\frac{dy}{dx} = x$$

$$y\frac{dy}{dx} = x\,dx$$

Integraling both sides

$$\int y dy = \int x dx$$

$$\frac{y^2}{2} = \frac{x^2}{2} + C$$

$$\boxed{y^2 = x^2 + 2C}$$

(iv) If the above curve passes through the point (0, –2), the value of C is:

(a) 4 (b) 2

(c) 1 (d) – 1

Solution: (b)

Passes through point (0, – 2)

$$(-2)^2 = (0)^2 + 2C$$

$$4 = 2C$$

$$C = 2$$

(v) The equation of curve is:

(a) $y^2 = x^2 + 2$ (b) $y = x + 2$

(c) $y^2 - x^2 = 4$ (d) $y^2 = x + 4$

Solution: (c)

$$y^2 = x^2 + 4$$

$$y^2 - x^2 = 4$$

Chapter 10. Vectors

Q. 1. Raja was flying a kite from a point A in the direction of $\vec{a} = 5\hat{i} + \hat{j} + 4\hat{k}$. His friend Sohan was watching him from the point B which is horizontal to A with the position vector $\vec{b} = 2\hat{i} + 6\hat{j} + 3\hat{k}$. The point B is just below the kite.

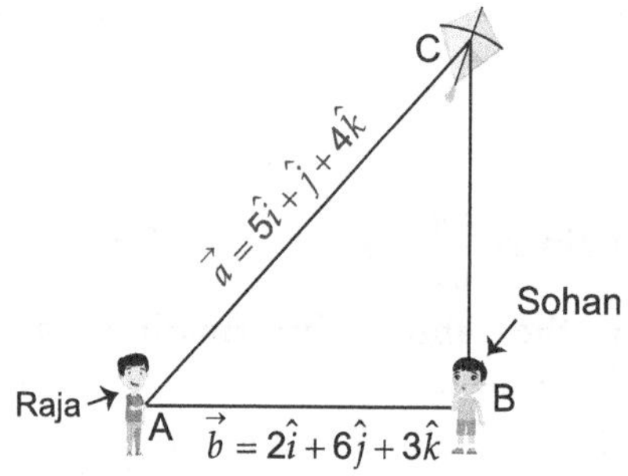

Answer the following questions :

(i) Find the distance between AB ?

(a) 4 (b) 5

(c) 6 (d) 7

Solution. (d) Distance between

$$AB = |\vec{AB}|$$

$$= \sqrt{2^2 + 6^2 + 3^2} = \sqrt{4 + 36 + 9} = 7 \text{ units.}$$ **Ans.**

(ii) Find the length of the string ?

(a) $\sqrt{77}$ units (b) $\sqrt{42}$ units

(c) $\sqrt{11}$ units (d) 10 units

Solution. (b) Length of string is

$$|\vec{AC}| = \sqrt{5^2 + 1^2 + 4^2} = \sqrt{25 + 1 + 16} = \sqrt{42} \text{ unit.}$$ **Ans.**

(iii) Find the projection of $\vec{a}$ on $\vec{b}$?

(a) 1 (b) 2

(c) 4 (d) 3

Solution. **(c)** Projection of $\vec{a}$ on $\vec{b}$ = $\dfrac{\vec{a}\cdot\vec{b}}{|\vec{b}|}$

$$\vec{a}\cdot\vec{b} = (5\hat{i}+\hat{j}+4\hat{k}).(2\hat{i}+6\hat{j}+3\hat{k})$$

$$= 10+6+12 = 28$$

$$\text{Projection} = \frac{28}{7} = 4 \qquad \textbf{Ans.}$$

(iv) Find a unit vector in the direction of $\vec{a}$ and $\vec{b}$?

(a) $\dfrac{\hat{i}+\hat{j}+\hat{k}}{3}$ (b) $\dfrac{\hat{i}+\hat{j}+\hat{k}}{\sqrt{3}}$

(c) $\hat{i}-\hat{j}+\hat{k}$ (d) $\dfrac{\hat{i}-\hat{j}+\hat{k}}{\sqrt{3}}$

Solution. **(b)** Unit vector in the direction of $\vec{a}+\vec{b}$ is

$$\frac{\vec{a}+\vec{b}}{|\vec{a}+\vec{b}|} = \frac{7\hat{i}+7\hat{j}+7\hat{k}}{\sqrt{7^2+7^2+7^2}} = \frac{7(\hat{i}+\hat{j}+\hat{k})}{\sqrt{147}}$$

$$= \frac{7(\hat{i}+\hat{j}+\hat{k})}{7\sqrt{3}} \cdot = \frac{\hat{i}+\hat{j}+\hat{k}}{\sqrt{3}} \qquad \textbf{Ans.}$$

(v) Find the angle between $\vec{a}$ and $\vec{b}$:

(a) $\cos^{-1}\left(\dfrac{4}{\sqrt{42}}\right)$ (b) $\cos^{-1}\dfrac{4}{77}$

(c) $\cos^{-1}\dfrac{4}{11}$ (d) $\cos^{-1}\dfrac{4}{25}$

Solution. **(a)** Angle between $\vec{a}$ and $\vec{b}$

$$\cos\theta = \frac{\vec{a}\cdot\vec{b}}{|\vec{a}||\vec{b}|} = \frac{28}{7\times\sqrt{42}}$$

$$\theta = \cos^{-1}\left(\frac{4}{\sqrt{42}}\right) \qquad \textbf{Ans.}$$

Answer the following questions :

Q. 2. Two friends Raja and Rahul are standing on a point O from where they started to move in different directions with the position vector of $2\hat{i}-\hat{j}+\hat{k}$, $\hat{i}-3\hat{j}-5\hat{k}$ respectively. Raja reaches to his office at point A and Rahul reaches his client office at point B which are opposite to each other. They decided to meet at a coffee shop situated on the way of AB coffee shop divided the road AB in the ratio of 2 : 1.

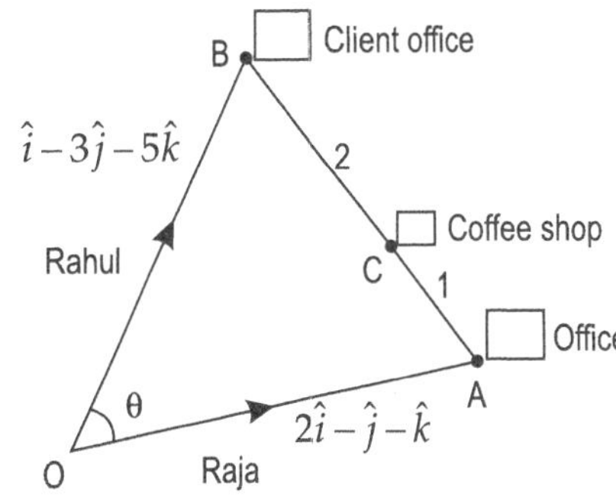

Give answers of the following questions (any 4) :

(i) How much distance is covered by Raja to reach his office 'A' ?

(a) 6 units (b) $\sqrt{6}$ units

(c) $\sqrt{10}$ units (d) $\sqrt{5}$ units

Solution. **(b)**

$$\vec{OA} = 2\hat{i}-\hat{j}+\hat{k}$$

$$|\vec{OA}| = \sqrt{2^2+(-1)^2+1^2}$$

$$= \sqrt{4+1+1} = \sqrt{6} \text{ units} \qquad \textbf{Ans.}$$

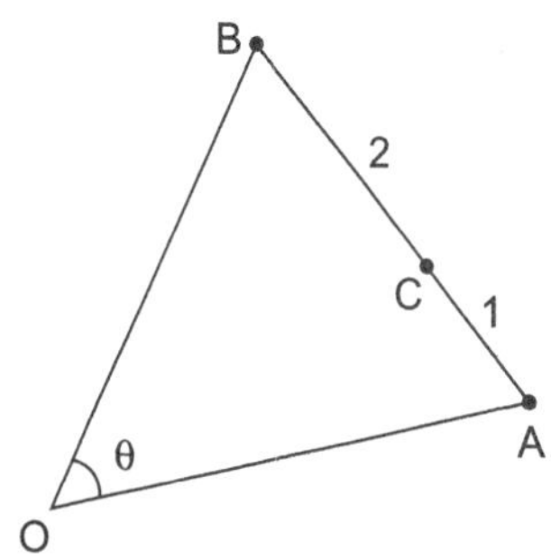

(ii) What is the distance between Rahul's starting point and terminating point ?

(a) $\sqrt{35}$ (b) 6 units

(c) $\sqrt{6}$ (d) $\sqrt{40}$

Solution. **(a)**

$$\vec{OB} = \hat{i}-3\hat{j}-5\hat{k}$$

Distance covered by Rahul is

$$|\vec{OB}| = \sqrt{1^2+(-3)^2+(-5)^2}$$

$$= \sqrt{1+9+25} = \sqrt{35} \text{ unit.} \qquad \textbf{Ans.}$$

(iii) What is the distance between Raja's office and Rahul's client office ?

(a) $\sqrt{6}$ (b) $\sqrt{35}$

(c) 6 (d) $\sqrt{41}$

Solution. (d) Distance between Raja office and Rahul's client office will be $|\overrightarrow{AB}|$.

$$\overrightarrow{AB} = \overrightarrow{OB} - \overrightarrow{OA}$$
$$= (\hat{i} - 3\hat{j} - 5\hat{k}) - (2\hat{i} - \hat{j} + \hat{k})$$
$$= -\hat{i} - 2\hat{j} - 6\hat{k}$$
$$|\overrightarrow{AB}| = \sqrt{(-1)^2 + (-2) + (-6)^2}$$
$$= \sqrt{1+4+36} = \sqrt{41} \text{ unit.}$$ **Ans.**

(iv) What are the coordinates of the point of coffee shop ?

(a) $\left(\frac{5}{3}, \frac{-5}{3}, -1\right)$ (b) $\left(\frac{-5}{3}, \frac{5}{3}, -1\right)$

(c) $\left(\frac{5}{3}, \frac{5}{3}, 1\right)$ (d) $\left(-1, \frac{5}{3}, \frac{-5}{3}\right)$

Solution. (a)

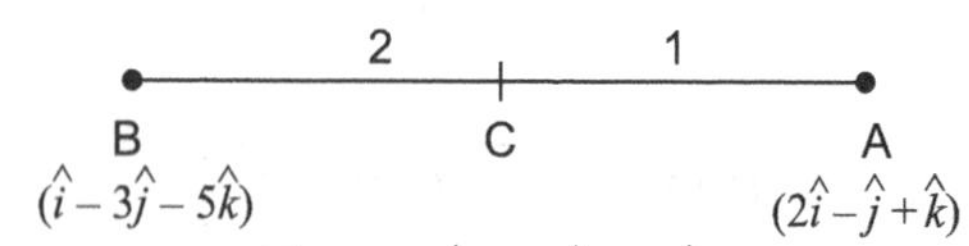

Let $\overrightarrow{OC} = x\hat{i} + y\hat{j} + z\hat{k}$

Which divides BA in the ratio 2 : 1. By section formula coordinates of C will be :

$$x = \frac{2(2)+1}{3}, y = \frac{2(-1)+1(-3)}{2+1},$$
$$z = \frac{2(1)+1(-5)}{2+1}$$
$$x = \frac{4+1}{3} = \frac{5}{3}, y = \frac{-2-3}{3} = \frac{-5}{3},$$
$$z = \frac{2-5}{3} = \frac{-3}{3} = -1$$

Coordinates of $C\left(\frac{5}{3}, \frac{-5}{3}, -1\right)$ **Ans.**

(v) If θ is the angle between $\overrightarrow{OA}$ and $\overrightarrow{OB}$, what is the value of θ ?

(a) 0° (b) 60°

(c) 30° (d) 90°

Solution. (d)

$$\cos\theta = \frac{a_1a_2 + b_1b_2 + c_1c_2}{\sqrt{a_1^2+b_1^2+c_1^2}\sqrt{a_2^2+b_2^2+c_2^2}}$$
$$\cos\theta = \frac{2(1)+(-1)(-3)+(1)(-5)}{\sqrt{6}\sqrt{35}}$$
$$\cos\theta = \frac{2+3-5}{\sqrt{6}\sqrt{35}} = 0$$
$$\theta = 90°$$ **Ans.**

Chapter 11. Three-Dimensional Geometry

Q.1. A Badminton Hall is to be constructed in the form of a cube in a sports complex in the 3-D as given below :

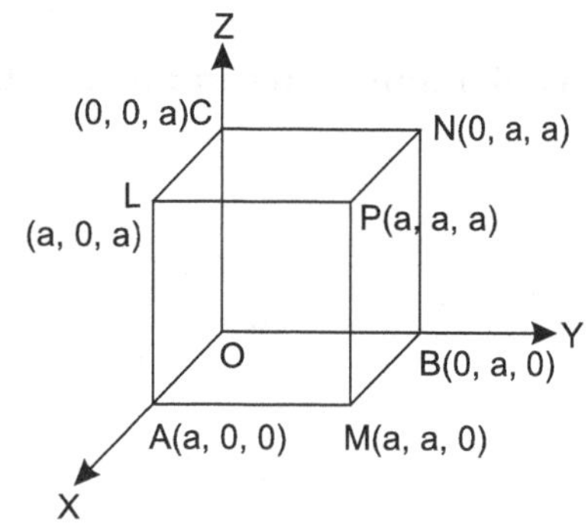

Length of side of cube be a units four diagonals are OP, MC, AN and BL.

From the above information given answer of the following questions :

(i) Find the direction cosines of diagonal AN :

(a) $\left(\frac{-1}{\sqrt{3}}, \frac{1}{\sqrt{3}}, \frac{1}{\sqrt{3}}\right)$ (b) $\left(\frac{1}{\sqrt{3}}, \frac{1}{\sqrt{3}}, \frac{1}{\sqrt{3}}\right)$

(c) $\left(\frac{-1}{\sqrt{3}}, \frac{-1}{\sqrt{3}}, \frac{-1}{\sqrt{3}}\right)$ (d) $\left(\frac{-1}{\sqrt{3}}, \frac{-1}{\sqrt{3}}, \frac{1}{\sqrt{3}}\right)$

Directions ratios of diagonal AN $= \overrightarrow{ON} - \overrightarrow{OA}$

$$\overrightarrow{ON} = 0\hat{i} + a\hat{j} + a\hat{k}$$
$$\overrightarrow{OA} = a\hat{i} + 0\hat{j} + 0\hat{k}$$
$$\overrightarrow{AN} = (x_2 - x_1)\hat{i} + (y_2 - y_1)\hat{j} + (z_2 - z_1)\hat{k}$$
$$= (0-a)\hat{i} + (a-0)\hat{j} + (a-0)\hat{k}$$
$$= -a\hat{i} + a\hat{j} + a\hat{k}$$

Solution. (a) d.r.'s = $\langle -a, a, a\rangle$

$$\text{d.c.'s} = \frac{-a}{\sqrt{a^2+a^2+a^2}}, \frac{a}{\sqrt{a^2+a^2+a^2}}, \frac{a}{\sqrt{a^2+a^2+a^2}}$$
$$= \left(\frac{-1}{\sqrt{3}}, \frac{1}{\sqrt{3}}, \frac{1}{\sqrt{3}}\right)$$

(ii) Write the direction cosines of diagonal MC :

(a) $\left(\frac{1}{\sqrt{3}}, \frac{1}{\sqrt{3}}, \frac{1}{\sqrt{3}}\right)$ (b) $\left(\frac{-1}{\sqrt{3}}, \frac{-1}{\sqrt{3}}, \frac{1}{\sqrt{3}}\right)$

(c) $\left(\frac{-1}{\sqrt{3}}, \frac{-1}{\sqrt{3}}, \frac{-1}{\sqrt{3}}\right)$ (d) $\left(\frac{1}{\sqrt{3}}, \frac{-1}{\sqrt{3}}, \frac{-1}{\sqrt{3}}\right)$

Solution. (b) Coordinates of diagonal MC

$$M = (a, a, 0), C = (0, 0, a)$$

$$\text{d.r.'s of MC} = (0-a)\hat{i} + (a-0)\hat{j} + (a-0)\hat{k}$$

$$= -a\hat{i} + a\hat{j} + a\hat{k}$$

$$\text{d.c.'s of MC} = \frac{-a}{\sqrt{(-a)^2+(-a)^2+a^2}}, \frac{-a}{\sqrt{(-a)^2+(-a)^2+a^2}}, \frac{a}{\sqrt{(-a)^2+(-a)^2+a^2}}$$

$$= \frac{-1}{\sqrt{3}}, \frac{-1}{\sqrt{3}}, \frac{1}{\sqrt{3}}.$$

(iii) Find the angle between the diagonals MC and AN :

(a) $\cos^{-1}\left(\frac{-1}{3}\right)$ (b) $\cos^{-1}\frac{1}{3}$

(c) $-\cos^{-1}\frac{1}{3}$ (d) $\cos\frac{1}{6}$

Solution. (b) Let θ be angle between AN and MC.

$$\cos\theta = |\, l_1 l_2 + m_1 m_2 + n_1 n_2 \,|$$

$$= \frac{-1}{\sqrt{3}} \times \left(\frac{-1}{\sqrt{3}}\right) + \frac{1}{\sqrt{3}}\left(\frac{-1}{\sqrt{3}}\right) + \frac{1}{\sqrt{3}} \times \frac{1}{\sqrt{3}}$$

$$\cos\theta = \left|\frac{1}{3} - \frac{1}{3} + \frac{1}{3}\right| = \frac{1}{3}$$

$$\theta = \cos^{-1}\left(\frac{1}{3}\right).$$

(iv) Write the Cartesian equation of the side of the room BN :

(a) $\frac{x}{2} = \frac{y-a}{a} = \frac{z}{a}$ (b) $x = y - a = z$

(c) $\frac{x}{0} = \frac{y-a}{0} = \frac{z}{a}$ (d) $x = y - a = \frac{z}{a}$

Solution. (c) Cartesian equation of side BN where B(0, a, 0), N(0, a, a) is

$$\frac{x-x_1}{x_2-x_1} = \frac{y-y_1}{y_2-y_1} = \frac{z-z_1}{z_2-z_1}$$

$$\frac{x-0}{0-0} = \frac{y-a}{a-a} = \frac{z-0}{a-0}$$

$$\frac{x}{0} = \frac{y-a}{0} = \frac{z}{a}.$$

(v) Write the direction ratios of the side of the room of Z-axis :

(a) (0, a, 0) (b) (a, a, a)

(c) (a, 0, 0) (d) (0, 0, a)

Solution. (d) direction ratios of Z-axis = $\langle 0, 0, a \rangle$.

Q.2. One light pole is inserted in the middle of the pond which is full of water. The pond is in the shape of plane whose equation is $\vec{r}.(2\hat{i} - \hat{j} + \hat{k}) + 3 = 0$ and pole is inserted in such a way that the length which is inside the water is exactly equal to the length outside the water. The light is fixed at the top of the pole whose position vector is $\hat{i} + 3\hat{j} + 4\hat{k}$.

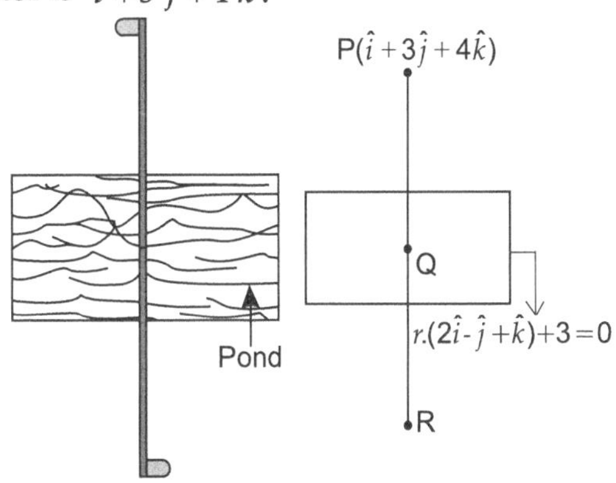

Attempt the following questions on the basis of above case study:

(i) What are the direction ratios of the light pole ?

(a) (2, – 1, – 1) (b) (– 2, 1, 1)

(c) (2, – 1, 1) (d) (2, 1, 1)

Solution. (c) Given equation of plane

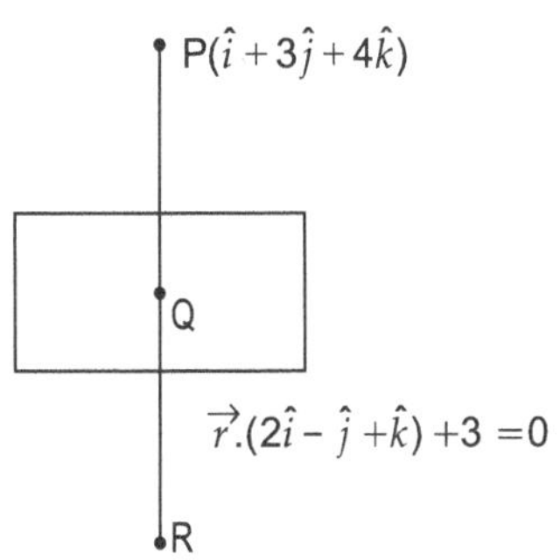

$$\vec{r}.(2\hat{i}-\hat{j}+\hat{k})+3=0$$

Cartesian form is $2x - y + z + 3 = 0$

d.r.'s of plane is equal to the line PQ.

dr's of line = (2, – 1, 1).

(ii) Find the foot of the pole on the water :

(a) (1, 4, – 3) **(b)** (1, – 4, 3)

(c) (1, 4, 3) **(d)** (– 1, 4, 3)

Solution. (d) Equation of line PQ passing through (1, 3, 4) with d.r.'s (2, – 1, 1) is

$$\frac{x-1}{2} = \frac{y-3}{-1} = \frac{z-4}{1} = \lambda \text{ (say)}$$

$$x = 2\lambda + 1, y = -\lambda + 3, z = \lambda + 4$$

Let Q($2\lambda + 1, -\lambda + 3, \lambda + 4$) be the coordinates of foot of perpendicular for same value of λ.

Point Q lie on the plane.

∴ Satisfy its equation

$$2(2\lambda + 1) - (-\lambda + 3) + (\lambda + 4) + 3 = 0$$

$$4\lambda + 2 + \lambda - 3 + \lambda + 4 + 3 = 0$$

$$6\lambda + 6 = 0$$

$$\Rightarrow \lambda = -1$$

Coordinates of Q = (– 1, 4, 3).

(iii) If R be the point of image which is end point of light pole, find the coordinates of R :

(a) (– 3, 5, 2) **(b)** (3, 5, 2)

(c) (3, – 5, 2) **(d)** (3, 5, – 2)

Solution. (a) R be the point of image.

∴ Q will be the mid-point of PR.

Let R(x_1, y_1, z_1)

$$\therefore \frac{1+x_1}{2} = -1, \frac{3+y_1}{2} = 4, \frac{4+z_1}{2} = 3$$

$$1 + x_1 = -2, 3 + y_1 = 8, z_1 = 6 - 4$$

$$x_1 = -3, y_1 = 5, z_1 = 2$$

Point of image (– 3, 5, 2).

(iv) Find the total length of the pole PR :

(a) $2\sqrt{7}$ **(b)** $2\sqrt{6}$

(c) $3\sqrt{6}$ **(d)** $2\sqrt{8}$

Solution. (b) Length of pole is PR.

Coordinates of P and R : P(1, 3, 4), R(– 3, 5, 2)

$$PR = \sqrt{(-3-1)^2 + (5-3)^2 + (2-4)^2}$$

$$= \sqrt{16+4+4} = \sqrt{24} = 2\sqrt{6} \text{ units.}$$

(v) What is Cartesian equation of the pole PR ?

(a) $\frac{x-1}{2} = \frac{y-3}{-1} = \frac{z-4}{1}$

(b) $\frac{x+1}{-2} = \frac{y-3}{1} = \frac{z-4}{-1}$

(c) $\frac{x+1}{-2} = \frac{y-3}{-1} = \frac{z-4}{1}$

(d) $\frac{x+1}{3} = \frac{y+3}{2} = \frac{z+4}{1}$

Solution. (a) Equation of PR passing through (1, 3, 4) with d.r.'s (2, – 1, 1) is

$$\frac{x-1}{2} = \frac{y-3}{-1} = \frac{z-4}{1}.$$ **Ans.**

Chapter 12. Linear Programming

Q.1. A furniture dealer deals in only two items—tables and chairs. He took a loan of ` 50,000 from the bank to invest in this business. He took a room on rent for the storage of furniture which has a storage space of at most 60 pieces. A table costs ₹ 2500 while a chair costs ₹ 500. He estimates that from the sale of one table, he can make a profit of ₹ 250 and that from the sale of one chair a profit of ₹ 75. Find the number of tables and chairs he should buy from the available money so as to maximize his total profit, assuming that he can sell all the items which he buys.

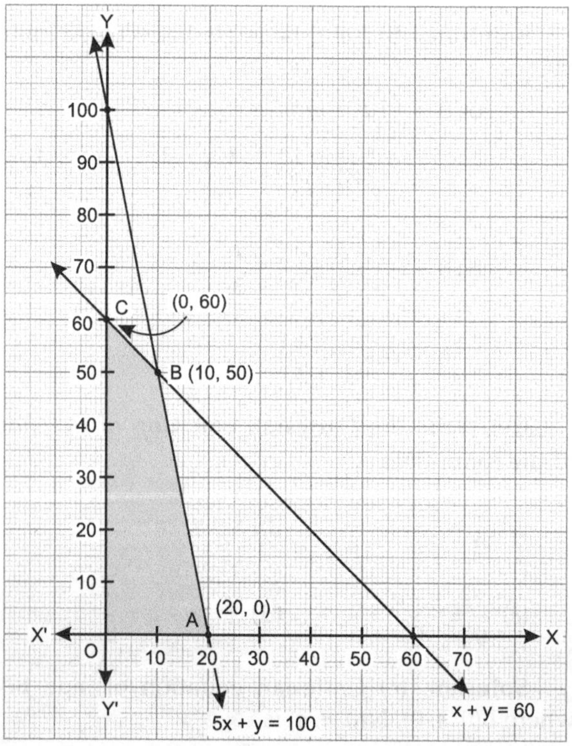

Fig. : Graphical representation of optimization problem

Based on the above information answer the following :

(i) Which constraint shows the linear objective function using decision variables x and y ?

(a) Z = 250x + 75y **(b)** Z = 250x – 75y

(c) Z = 275x – 50y **(d)** Z = 75x – 250y

Solution. (a) The linear objective function using decision variables x and y is :

$$Z = 250x + 75y$$

(ii) Which equation shows that "he can store only 60 pieces of chairs and tables" ?

(a) $x + y \geq 60$ **(b)** $x - y \geq 60$

(c) $x - y = 60$ **(d)** $x + y \leq 60$

Solution. (d) Following equation shows that "he can store only 60 pieces of chairs and tables :

i.e., $x + y \leq 60$

(iii) Choose correct investment constraint by furniture dealer :

(a) $2500x + 500y \leq 50000$

(b) $2500x - 500y \leq 50000$

(c) $500x - 2500y \leq 50000$

(d) $500x + 2500y \leq 50000$

Solution. (a) Correct investment constraint by furniture dealer is :

$$₹\,2500x + ₹\,500y \leq ₹\,50000$$

(iv) In which case total profit would be ₹ 6250 ?

(a) 10 tables and 50 chairs

(b) 50 tables and 10 chairs

(c) 5 tables and 55 chairs

(d) 20 tables and 40 chairs

Solution.(a) He may choose to buy 10 tables and 50 chairs, as he can store only 60 pieces. Total profit in this case would be ₹ (10 × 250 + 50 × 75), *i.e.,* ₹ 6250.

10 tables and 50 chairs

(v) Due to this storage space maximum of 60 pieces, his investment is limited to a maximum of :

(a) ₹ 55000 **(b)** ₹ 50000

(c) ₹ 58000 **(d)** ₹ 45000

Solution. (b) Due to his storage space maximum of 60 pieces, his investment is limited to a maximum of ₹ 50000.

Q.2. A cooperative society of farmers has 50 hectare of land to grow two crops X and Y. The estimation of income from crops per hectare is 10,500 and 9,000 respectively. To control weeds, a runny herbicide has used for both crops at rates of 20 liters and 10 liters per hectare. Additionally, no more than 800 liters of herbicide should be used for protecting fish and natural world using a fishpond which collect drainage from this ground.

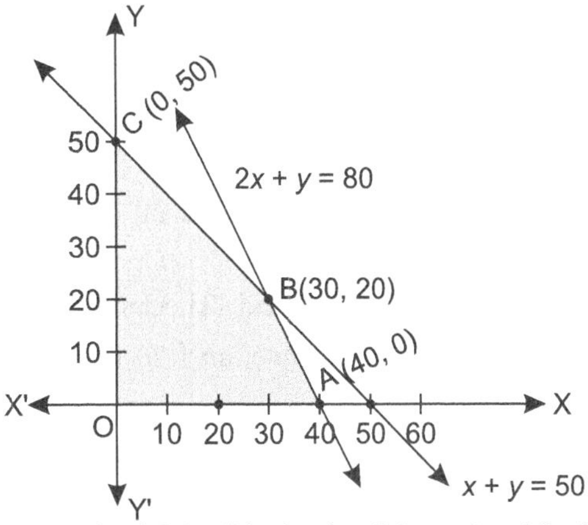

Fig. : Graph of OABC is the feasible region (shaded)

Corner Point	Z = 10500x + 9000y
$O(0, 0)$	0
$A(40, 0)$	420000
$B(30, 20)$	495000
$C(0, 50)$	450000

Based on the above information answer the following questions :

(i) Which constraint shows the total income per hectare on both crops ?

(a) Z = 9000x + 10500y

(b) Z = 10500x + 9000y

(c) Z = 8000x + 10500y

(d) Z = 10500x – 9000y

Solution. (b) Constraint, which shows the total income per hectare on both crops is

$$Z = 10500x + 9000y$$ **Ans.**

(ii) Which equation shows the constraint related to land ?

(a) $x + y \geq 50$ **(b)** $x - y \geq 50$

(c) $x + y \leq 50$ **(d)** $x - y \leq 50$

Solution. (c) This equation is of ground :

i.e., $x + y \leq 50$. **Ans.**

(iii) What is the characteristics of the feasible region OABC ?

(a) Unbounded **(b)** Bounded

(c) Both (a) and (b) **(d)** None of these

Solution. (b) Bounded is the characteristics of the feasible section (covered). **Ans.**

(iv) The maximum income would be got by people is :

(a) ₹ 4,95,000 **(b)** ₹ 4,90,000

(c) ₹ 4,85,000 **(d)** ₹ 4,80,000

Solution. (a) The people will get the maximum income of ₹ 4,95,000. **Ans.**

(v) How much ground should be billed to each crops so as to maximize the total income of the people ?

(a) 30 hectare for crop x and 20 hectare from crop y

(b) 20 hectare for crop x and 30 hectare from crop y

(c) 20 hectare for crop x and 40 hectare from crop y

(d) 40 hectare for crop x and 20 hectare from crop y

Solution. (a) Ground should be allocated to each crop so as to maximize the total income of the people.

i.e., 30 hectare for crop x and 20 hectare from crop y. **Ans.**

Chapter 13. Probability

Q.1. In an office three employees Vinay, Sonia and Iqbal process incoming copies of a certain form. Vinay process 50% of the forms. Sonia processes 20% and Iqbal the remaining 30% of the forms. Vinay has an error rate of 0.06, Sonia has an error rate of 0.04 and Iqbal has an error rate of 0.03.

$$P(V) = 50\% = \frac{1}{2},$$

$$P(S) = 20\% = \frac{1}{5},$$

$$P(I) = 30\% = \frac{3}{10}$$

Let E be the event that error occured.

$$P(E/V) = \frac{6}{100}$$

$$P(E/S) = \frac{4}{100}$$

$$P(E/I) = \frac{30}{100}$$

Based on the above information answer the following :

(i) The conditional probability that an error is committed in processing given that Sonia processed the form is :

(a) 0.0210 (b) 0.04

(c) 0.47 (d) 0.06

Solution. $P(E/S) = \frac{4}{100} = 0.04$

Hence (b) is correct option.

(ii) The probability that Sonia processed the form and comitted and error is :

(a) 0.005 (b) 0.006

(c) 0.008 (d) 0.68

Solution. (c) $P(E \cap S) = P(E/S) \cdot P(S)$

$$= \frac{4}{100} \times \frac{20}{100} = 0.008$$ **Ans**

Hence (c) is correct option.

(iii) The total probability of committing an error in processing the form is :

(a) 0 (b) 0.047

(c) 0.234 (d) 1

Solution. (b) $P(E) = P(V)\,P(E/V) + P(S)\,P(E/S) + P(I)\,P(E/I)$

$$= \frac{50}{100} \times \frac{6}{100} + \frac{20}{100} \times \frac{4}{100} + \frac{30}{100} \times \frac{3}{100} = 0{\cdot}047.$$ **Ans**

Hence (b) is correct option

(iv) The manager of the company wants to do a quality check. During inspection he selects a form at random from the days output of processed forms. If the form selected at random has an error, the probability that the form is not processed by Vinay is :

(a) 1 (b) 30/47

(c) 20/47 (d) 17/47

Solution. (d)

$$P(V/E) = \frac{P(V)\,P(E/V)}{P(E)}$$

$$= \frac{\frac{50}{100} \times \frac{6}{100}}{\frac{47}{1000}} = \frac{30}{47}$$

$\therefore$ P(Not processed by Vinay)

$$= 1 - \frac{30}{47}$$

$$= \frac{17}{47}$$ **Ans.**

Hence (d) is correct option.

(v) Let A be the event of committing an error in processing the form and let E_1, E_2 and E_3 be the events that Vinay, Sonia and Iqbal processed the form. The value of $\sum_{i=1}^{3} P(E_i / A)$ is :

(a) 0 **(b)** 0.03

(c) 0.06 **(d)** 1

Solution. (d)

$$\sum_{i=1}^{3} P(E_i/A) = 1$$

[∵ Sum probabilities of all the possible event is 1]

Q. 2. Sanjay, Ajay and Vijay are three best friends. After completing their MBA from IIM Lucknow. They apply for the job in the same company for the post of manager in finance department. chances of selection of Sanjay, Ajay and Vijay are in the ratio of 1 : 2 : 4. The probabilities that Sanjay can increase the profits of the company by his efforts is 0.8 whereas probabilities for the same task of Ajay and Vijay are 0.5 and 0.3 respectively.

From the above information attempt the following questions :

(i) If E_1, E_2 and E_3 are the events of selecting Sanjay, Ajay and Vijay, then find $P(E_1)$, $P(E_2)$ and $P(E_3)$:

(a) $\frac{1}{7}, \frac{2}{7}, \frac{4}{7}$ **(b)** $\frac{2}{7}, \frac{4}{7}, \frac{1}{7}$

(c) $\frac{4}{7}, \frac{2}{7}, \frac{1}{7}$ **(d)** $\frac{4}{7}, \frac{1}{7}, \frac{2}{7}$

Solution. (a)

E_1 be event of selecting Sanjay

E_2 be event of selecting Ajay

E_3 be event of selecting Vijay.

$$P(E_1) = \frac{1}{7}$$

$$P(E_2) = \frac{2}{7}$$

$$P(E_3) = \frac{4}{7}$$ **Ans.**

(ii) A be event of increase the profit then find if it is due to Ajay :

(a) $\frac{3}{5}$ **(b)** $\frac{2}{3}$

(c) $\frac{1}{2}$ **(d)** $\frac{2}{5}$

Solution. (c)

A be the event of increasing profits.

$$P(A/E_1) = 0{\cdot}8$$
$$P(A/E_2) = 0{\cdot}5$$
$$P(A/E_3) = 0{\cdot}3$$

$P(E_2/A)$

$$= \frac{P(E_2)\,P(A/E_1)}{P(E_1)\,P(A/E_1) + P(E_2)\,P(A/E_2) + P(E_3)\,P(A/E_3)}$$

$$= \frac{\frac{2}{7} \times 0{\cdot}5}{\frac{1}{7} \times 0{\cdot}8 \times \frac{2}{7} \times 0{\cdot}5 + \frac{4}{7} \times 0{\cdot}3}$$ **Ans.**

$$= \frac{\frac{1}{7}}{\frac{0{\cdot}8}{7} + \frac{1}{7} + \frac{1{\cdot}2}{7}} = \frac{1}{2}$$ **Ans.**

(iii) Find $P(E_1/A)$:

(a) 1 **(b)** $\frac{1}{5}$

(c) $\frac{3}{5}$ **(d)** $\frac{2}{5}$

Solution. (d)

$P(E_1/A)$

$$= \frac{P(E_1)\,P(A/E_1)}{P(E_1)\,P(A/E_1) + P(E_2)\,P(A/E_2) + P(E_3)\,P(A/E_3)}$$

$$= \frac{\frac{1}{7} \times 0{\cdot}8}{\frac{1}{7} \times 0{\cdot}8 + \frac{2}{7} \times 0{\cdot}5 + \frac{4}{7} \times 0{\cdot}3} = \frac{\frac{0{\cdot}8}{7}}{\frac{0{\cdot}8 + 1 + 1{\cdot}2}{7}} = \frac{0{\cdot}8}{20} = \frac{2}{5}.$$

Ans.

(iv) If E_1, E_2 and E_3 be events of not increase the profits, then find the probability that it is due to the appointment of Vijay ?

(a) $\frac{5}{10}$ **(b)** $\frac{7}{10}$

(c) $\frac{6}{10}$ **(d)** $\frac{9}{10}$

Solution. (d)

P be event of not increase in profits.

$$P(P/E_1) = 1 - 0{\cdot}8 = 0{\cdot}2 = \frac{2}{10}$$

$$P(P/E_2) = 1 - 0{\cdot}5 = 0{\cdot}5 = \frac{5}{10}$$

$$P(P/E_3) = 1 - 0{\cdot}3 = 0{\cdot}7 = \frac{7}{10}$$

$P(E_3/P)$

$$= \frac{P(E_3)\,P(P/E_3)}{P(E_1)\,P(P/E_1) + P(E_2)\,P(P/E_2) + P(E_3)\,P(P/E_3)}$$

$$= \frac{\frac{4}{7} \times \frac{7}{10}}{\frac{1}{7} \times \frac{2}{10} + \frac{2}{7} \times \frac{5}{10} + \frac{4}{7} \times \frac{7}{10}} = \frac{\frac{28}{70}}{\frac{2}{70} + \frac{10}{70} + \frac{28}{70}}$$

$$= \frac{28}{40} = \frac{7}{10}.$$

Ans

(v) Find the probability that profit does not increase due to the appointment of Sanjay ?

(a) $\frac{7}{20}$ **(b)** 1

(c) $\frac{3}{20}$ **(d)** $\frac{1}{20}$

Solution. (d) $P(E_1/P)$

$$= \frac{P(E_1)\,P(P/E_1)}{P(E_1)\,P(P/E_1) + P(E_2)\,P(P/E_2) + P(E_3)\,P(P/E_3)}$$

$$= \frac{\frac{1}{7} \times \frac{2}{10}}{\frac{1}{7} \times \frac{2}{10} + \frac{2}{7} \times \frac{5}{10} + \frac{4}{7} \times \frac{7}{10}} = \frac{\frac{2}{70}}{\frac{2}{70} + \frac{10}{70} + \frac{28}{70}}$$

$$= \frac{2}{40} = \frac{1}{20}.$$

Ans.

Long Questions

| Set 6 |

Chapter 1. Relations and Functions

Q. 1. Find f^{-1} if it exists for $f: A \to B$ where
(i) $A = \{0, -1, -3, 2\}$; $B = \{-9, -3, 0, 6\}$ and $f(x) = 3x$
(ii) $A = \{1, 3, 5, 7, 9\}$; $B = \{0, 1, 9, 25, 49, 81\}$ and $f(x) = x^2$

Sol. (i) $f: A \to B$, such that

$$f(x) = 3x$$

$$\Rightarrow \quad f(0) = 0$$

$$f(-1) = -3$$

$$f(-3) = -9$$

$$f(2) = 6$$

Therefore,

$$\text{domain } (f) = \{0, -1, -3, 2\} = A$$

and $\quad \text{range } (f) = \{0, -3, -9, 6\} = B$

f is one-one and onto.

$\Rightarrow f^{-1}$ exists.

$$f^{-1}(0) = 0$$

$$f^{-1}(-3) = -1$$

$$f^{-1}(-9) = -3$$

$$f^{-1}(6) = 2$$

$$f^{-1}(x) = \frac{x}{3}$$

In other words,

$$f^{-1}(x) = \frac{x}{3}$$

Thus, $f^{-1}: B \to A$ such that

$$f^{-1}(x) = \frac{x}{3}$$

or $f^{-1} = \{(0, 0), (-3, -1), (-9, -3), (6, 2)\}$

Now, $fof^{-1}: A \to A$ such that

$$(fof^{-1})(x) = f(f^{-1}(x))$$

$$= f\left(\frac{x}{3}\right)$$

$$= 3\left(\frac{x}{3}\right) = x$$

so it is the identity function.

(ii) We cannot define f^{-1} in this case, since f is not onto.

$$f(x) = x^2$$

$$\Rightarrow \quad f(1) = 1$$

$$f(3) = 9$$

$$f(5) = 25$$

$$f(7) = 49$$

$$f(9) = 81$$

$\Rightarrow f = \{(1, 1), (3, 9), (5, 25), (7, 49), (9, 81)\}$

But $0 \in B$ does not have a pre-image in A.

Thus, f is not surjective.

$\Rightarrow f^{-1}$ does not exist.

Q. 2. If $f: Q \to Q$ and $g: Q \to Q$ are two functions defined by $f(x) = 2x$ and $g(x) = x + 2$, show that f and g are bijective and verify that $(gof)^{-1} = f^{-1}og^{-1}$.

Sol. Given $f: Q \to Q$ such that

$$f(x) = 2x$$

Let x and y be two elements of Q (domain), then

$$f(x) = f(y)$$

$$\Rightarrow \quad 2x = 2y$$

$$\Rightarrow \quad x = y$$

$$\therefore \quad f(x) = f(y)$$

$$\Rightarrow \quad x = y \text{ for all } x, y \in Q \text{ (domain)}$$

Therefore, f is injective.

Let y be an arbitrary element of Q (codomain).

$$\therefore \quad f(x) = y \Rightarrow 2x = y$$

$$\Rightarrow \quad x = \frac{y}{2}$$

Therefore, every element y in codomain has a pre-image x in the domain given by $x = \frac{y}{2}$

Therefore, f is surjective.

$\Rightarrow f$ is bijective.

Now, given $g: Q \to Q$ such that

$$g(x) = x + 2$$

Let x and y be two elements of Q (domain)

$$\therefore \quad g(x) = g(y)$$

$$\Rightarrow \quad x + 2 = y + 2$$

$$\Rightarrow \quad x = y$$

$$\therefore \quad g(x) = g(y)$$

$$\Rightarrow \quad x = y$$

for all $x, y \in Q$ (domain)

Therefore, g is injective.

Let y be an arbitrary element of Q (codomain)

$$\therefore \quad g(x) = y$$

$\Rightarrow \quad x+2=y$

$\Rightarrow \quad x=y-2$

Therefore, every element y in the codomain has a pre-image x in the domain given by $x=y-2$

Therefore, g is surjective.

$\Rightarrow g$ is bijective.

Inverse of f: Let $x \in Q$ (domain) and $y \in Q$ (co-domain) such that

$$f(x)=y$$

$$\Rightarrow \quad 2x=y$$

$$\Rightarrow \quad x=\frac{y}{2}$$

$$\Rightarrow \quad f^{-1}(y)=\frac{y}{2}$$

$\therefore f^{-1}: Q \rightarrow Q$ is defined by $f^{-1}(x)=\frac{x}{2}$ for all $x \in Q$.

Inverse of g : Let $x \in Q$ (domain) and $z \in Q$ (co-domain) such that

$$g(x)=z$$

$$\Rightarrow \quad x+2=z$$

$$\Rightarrow \quad x=z-2$$

$$\Rightarrow \quad g^{-1}(z)=z-2$$

$\therefore g^{-1}: Q \rightarrow Q$ is defined by $g^{-1}(x)=x-2$ for all $x \in Q$.

Now, $(gof)(x)=g(f(x))$

$=g(2x)$

$=2x+2$

Let $(gof)(x)=y$

$\Rightarrow \quad 2x+2=y$

$\Rightarrow \quad 2x=y-2$

$$\Rightarrow \quad x=\frac{y-2}{2}$$

$$\Rightarrow \quad (gof)^{-1}(y)=\frac{y-2}{2}$$

$\therefore (gof)^{-1}: Q \rightarrow Q$

such that $(gof)^{-1}(x)=\dfrac{x-2}{2}$ ...(i)

Now, $(f^{-1}og^{-1})(x)=f^{-1}[g^{-1}(x)]$

$=f^{-1}[x-2]$

$\therefore \quad (f^{-1}o\,g^{-1})(x)=\dfrac{x-2}{2}$ ...(ii)

By (i) and (ii),

$$(gof)^{-1}=f^{-1}og^{-1}.$$

Q. 3. Show that the function $f: R \rightarrow R$ defined by $f(x) = \dfrac{x}{x^2+1}, \forall x \in R$ is neither one-one nor onto. Also, if $g: R \rightarrow R$ is defined as $g(x) = 2x - 1$, find $fog\,(x)$.★

Sol. Given, $f(x)=\dfrac{x}{x^2+1}, \forall x \in R$

For one-one :

Let $f(x)=f(y)$

$$\frac{x}{x^2+1}=\frac{y}{y^2+1}$$

$$xy^2+x=yx^2+y$$

$$xy(y-x)=y-x$$

$$xy=1$$

$$x=\frac{1}{y}$$

Since $x \neq y$, therefore, $f(x)$ is not one-one

For onto :

$$f(x)=y$$

$$\frac{x}{x^2+1}=y$$

$$x=yx^2+y$$

$$x^2y+y-x=0$$

x cannot be expressed in terms of y

$\Rightarrow f(x)$ is not onto

As $\quad g(x)=2x-1$

$\therefore \quad fog\,(x)=f[g(x)]=f(2x-1)$

$$=\frac{2x-1}{(2x-1)^2+1}=\frac{2x-1}{4x^2-4x+2}$$

Q. 4. Consider $f: R-\left\{-\frac{4}{3}\right\} \rightarrow R-\left\{\frac{4}{3}\right\}$ given by $f(x)=\dfrac{4x+3}{3x+4}$. Show that f is bijective. Fine the inverse of f and hence find $f^{-1}(0)$ and x such that $f^{-1}(x)=2$.★

Sol. For one-one :

Let $x, y \in R-\left\{-\frac{4}{3}\right\}$

and $\quad f(x)=f(y)$

$$\Rightarrow \quad \frac{4x+3}{3x+4}=\frac{4y+3}{3y+4}$$

$\Rightarrow \quad (4x+3)(3y+4)=(3x+4)(4y+3)$

$\Rightarrow \quad 12xy+16x+9y+12=12xy+9x+16y+12$

$\Rightarrow \quad 16x+9y-9x-16y=0$

$\Rightarrow \quad 7x-7y=0$

$\Rightarrow \quad 7(x-y)=0$

$\Rightarrow \quad x-y=0$

$\Rightarrow \quad x=y$

Hence, f is one-one function.

For onto :

Let $y \in R-\left\{\frac{4}{3}\right\}$

Then, $\quad f(x)=y$

$$\Rightarrow \quad \frac{4x+3}{3x+4}=y$$

★ are board exam questions from previous years

$\Rightarrow \quad 4x+3 = 3xy+4y$

$\Rightarrow \quad (4-3y)x = 4y-3$

$\Rightarrow \quad x = \dfrac{4y-3}{4-3y}$ ...(i)

As $y \in R - \left\{\dfrac{4}{3}\right\}$, $x = \dfrac{4y-3}{4-3y} \in R$

Also $\dfrac{4y-3}{4-3y} \neq -\dfrac{4}{3}$

Since if, $\dfrac{4y-3}{4-3y} = -\dfrac{4}{3}$

$\Rightarrow \quad 12y-9 = -16+12y$

$\Rightarrow \quad 9 = 16$ which is not possible.

Thus, $x = \dfrac{4y-3}{4-3y} \in R - \left\{-\dfrac{4}{3}\right\}$ such that

$$f(x) = f\left(\frac{4y-3}{4-3y}\right)$$

$$f(x) = \frac{4\left(\frac{4y-3}{4-3y}\right)+3}{3\left(\frac{4y-3}{4-3y}\right)+4}$$

$$= \frac{16y-12+12-9y}{12y-9+16-12y} = \frac{7y}{7}$$

$$f(x) = y$$

Hence every element $y \in R - \left\{\dfrac{4}{3}\right\}$ has its pre image $x \in R - \left\{-\dfrac{4}{3}\right\}$.

Hence f is onto.

$\Rightarrow f$ is one-one and onto, so f is invertible.

Now, $f(x) = y$

$f^{1}(y) = x$

$f^{1}(y) = \dfrac{4y-3}{4-3y}$ [From eq. (i)]

$f^{1}(0) = \dfrac{4\times 0-3}{4-3\times 0}$

$= \dfrac{-3}{4}$

Also given, $f^{-1}(x) = 2$

$\dfrac{4x-3}{4-3x} = 2$

$4x-3 = 8-6x$

$10x = 11$

$\Rightarrow \quad x = \dfrac{11}{10}$.

Q. 5. Consider $f : R_+ \to [-5, \infty)$, given by $f(x) = 9x^2 + 6x - 5$. Show that f is invertible with $f^{-1}(y) = \left(\dfrac{\sqrt{y+6}-1}{3}\right)$.★

Hence find :

(i) $f^{-1}(10)$ (ii) y if $f^{-1}(y) = \dfrac{4}{3}$,

Where R_+ is the set of all non-negative real numbers.

Sol. For one-one :

Let $x_1, x_2 \in R_+$

and $f(x_1) = f(x_2)$

$\Rightarrow \quad 9x_1^2+6x_1-5 = 9x_2^2+6x_2^2-5$

$9(x_1^2-x_2^2)+6(x_1-x_2) = 0$

$(x_1-x_2)(9(x_1+x_2)+6) = 0$

Since $9(x_1+x_2)+6 > 0$ $[\because x_1, x_2 \in R_+]$

$\Rightarrow \quad x_1 - x_2 = 0$

or $x_1 = x_2$

$\therefore$ Function is one-one

For onto :

Let $y \in [-5, \infty)$ such that $f(x) = y$

$\Rightarrow \quad 9x^2+6x-5 = y$

$(3x)^2+2.(3x).1+1^2-1^2-5 = y$

$(3x+1)^2-6 = y$

$(3x+1)^2 = y+6$

$3x+1 = \sqrt{y+6}$

$x = \dfrac{-1+\sqrt{y+6}}{3} \in R_+$

$\therefore$ Function is onto

Since f is both one-one and onto

$\therefore$ Function is invertible.

Now, $f(x) = y$

$\Rightarrow \quad x = f^{-1}(y) = \dfrac{-1+\sqrt{y+6}}{3}$

(i) $f^{-1}(10) = \dfrac{-1+\sqrt{16}}{3}$

$= \dfrac{-1+4}{3} = 1$

(ii) $f^{-1}(y) = \dfrac{4}{3}$

$\dfrac{4}{3} = \dfrac{-1+\sqrt{y+6}}{3}$

$4 = -1+\sqrt{y+6}$

★ are board exam questions from previous years

$$5 = \sqrt{y+6}$$

Squaring on both sides

$$25 = y+6$$

or $\quad y = 19.$

Q. 6. Consider $f : R^+ \rightarrow [-9, \infty)$ given by $f(x) = 5x^2 + 6x - 9$. Prove that f is invertible with

$$f^{-1}(y) = \left(\frac{\sqrt{54+5y}-3}{5}\right). \star$$

Sol. To prove f is invertible, we have to prove that f is one-one and onto.

For one-one :

Let $x_1, x_2 \in R^+$ then

$$f(x_1) = f(x_2)$$

$\Rightarrow \quad 5x_1^2 + 6x_1 - 9 = 5x_2^2 + 6x_2 - 9$

$\Rightarrow \quad 5(x_1^2 - x_2^2) + 6(x_1 - x_2) = 0$

$\Rightarrow \quad (x_1 - x_2)(5x_1 + 5x_2 + 6) = 0$

$\Rightarrow \quad x_1 - x_2 = 0$ as $5x_1 + 5x_2 + 6 \neq 0$

$\Rightarrow \quad x_1 = x_2$

i.e., f is one-one function.

For onto :

Let $\quad f(x) = y$

$\Rightarrow \quad y = 5x^2 + 6x - 9$

$$5x^2 + 6x - (9+y) = 0$$

$$x = \frac{-6 \pm \sqrt{36 + 4 \times 5 \times (9+y)}}{10}$$

$$= \frac{-6 \pm \sqrt{216+20y}}{10}$$

$$= \frac{\pm\sqrt{54+5y}-3}{5}$$

$$x = \frac{\sqrt{54+5y}-3}{5} \quad [\because x \in R^+]$$

Clearly $\forall\, y \in [-9, \infty)$, the value of $x \in R^+$

$\Rightarrow f$ is onto function.

Hence, f is one-one and onto function.

$\Rightarrow f$ is invertible function with

$$f^{-1}(y) = \frac{\sqrt{54+5y}-3}{5}$$

Q. 7. Let N denote the set of all natural numbers and R be the relation on $N \times N$ defined by $(a, b)\, R\, (c, d) \Leftrightarrow ad(b+c) = bc(a+d)$. Check whether R is an equivalence relation on $N \times N$.

Sol. We observe the following properties :

Reflexivity : Let (a, b) be an arbitrary element of $N \times N$.

Then, $\quad (a, b) \in N \times N$

$\Rightarrow \quad a, b \in N$

$\Rightarrow \quad ab\,(b+a) = ba(a+b)$

[By commutativity of addition and multiplication on N]

$\Rightarrow \quad (a, b)\, R\, (a, b)$

Thus, $(a, b)\, R\, (a, b)$ for all $(a, b) \in N \times N$.

So, R is reflexive on $N \times N$.

Symmetric : Let $(a, b), (c, d) \in N \times N$ be such that $(a, b)\, R\, (c, d)$

Then, $(a, b)\, R\, (c, d)$

$\Rightarrow \quad ad(b+c) = bc(a+d)$

$\Rightarrow \quad cb(d+a) = da(c+b)$

[By commutativity of addition and multiplication on N]

$\Rightarrow \quad (c, d)\, R\, (a, b)$

Thus, if $(a, b)\, R\, (c, d)$

$\Rightarrow (c, d)\, R\, (a, b)$ for all $(a, b), (c, d) \in N \times N$.

So, R is symmetric on $N \times N$.

Transitivity : Let $(a, b), (c, d), (e, f) \in N \times N$ such that $(a, b)\, R\, (c, d)$ and $(c, d)\, R\, (e, f)$

Then, $(a, b)\, R\, (c, d)$

$\Rightarrow \quad ad(b+c) = bc(a+d)$

$\Rightarrow \quad \dfrac{b+c}{bc} = \dfrac{a+d}{ad}$

$\Rightarrow \quad \dfrac{1}{b} + \dfrac{1}{c} = \dfrac{1}{a} + \dfrac{1}{d} \quad \ldots(i)$

and $(c, d)\, R\, (e, f)$

$\Rightarrow \quad cf(d+e) = de(c+f)$

$\Rightarrow \quad \dfrac{d+e}{de} = \dfrac{c+f}{cf}$

$\Rightarrow \quad \dfrac{1}{d} + \dfrac{1}{e} = \dfrac{1}{c} + \dfrac{1}{f} \quad \ldots(ii)$

Adding (i) and (ii), we get

$$\left(\frac{1}{b}+\frac{1}{c}\right)+\left(\frac{1}{d}+\frac{1}{e}\right) = \left(\frac{1}{a}+\frac{1}{d}\right)+\left(\frac{1}{c}+\frac{1}{f}\right)$$

$\Rightarrow \quad \dfrac{1}{b} + \dfrac{1}{e} = \dfrac{1}{a} + \dfrac{1}{f}$

$\Rightarrow \quad \dfrac{b+e}{be} = \dfrac{a+f}{af}$

$\Rightarrow \quad af(b+e) = be(a+f)$

★ are board exam questions from previous years

$\Rightarrow (a, b)\, R\, (e, f)$

Thus, if $(a, b)\, R\, (c, d)$ and $(c, d)\, R\, (e, f)$

$\Rightarrow (a, b)\, R\, (e, f)$ for all $(a, b), (c, d), (e, f) \in N \times N$

So, R is transitive on $N \times N$.

Hence, R being reflexive, symmetric and transitive, is an equivalence relation on $N \times N$.

Q. 8. Show that the relation R defined on the set A of all triangles in a plane as $R = \{(T_1, T_2) : T_1$ is congruent to $T_2\}$ is an equivalence relation.

Sol. Given,

$$R = \{(T_1, T_2) : T_1 \text{ is congruent to } T_2\}$$

Reflexivity : Let $(T, T) \in R \;\forall\; T \in A$

Since every triangle is congruent to itself

i.e., R is reflexive.

Symmetric : Let $(T_1, T_2) \in R$

then, T_1 is congruent to T_2

$\Rightarrow T_2$ is congruent to T_1

$\Rightarrow \qquad (T_2, T_1) \in R$

So, R is symmetric.

Transitivity : Let $T_1, T_2, T_3 \in A$ such that $(T_1, T_2) \in R$ and $(T_2, T_3) \in R$. Then,

$\Rightarrow T_1$ is congruent to T_2 and T_2 is congruent to T_3

$\Rightarrow T_1$ is congruent to T_3

$\Rightarrow \qquad (T_1, T_3) \in R$

So, R is transitive.

Hence, R is an equivalene relation on set A.

Q. 9. Show that the relation R defined on the set A of all polygons as

$R = \{(P_1, P_2) : P_1$ and P_2 have same number of sides$\}$

is an equivalence relation. What is the set of all elements in A related to the right angle triangle T with sides 3, 4 and 5 ?

Sol. Given, $R = \{(P_1, P_2) : P_1$ and P_2 have same number of sides$\}$

Reflexivity : Let P be any polygon in A.

Then, P and P have same number of sides.

$\Rightarrow \qquad (P, P) \in R$

Thus, $\qquad (P, P) \in R \;\forall\; P \in A$

So, R is reflexive.

Symmetric : Let P_1 and P_2 be two polygons in A such that

$$(P_1, P_2) \in R$$

Then, $\qquad (P_1, P_2) \in R$

Now, P_1 and P_2 have same number of sides.

$\Rightarrow P_2$ and P_1 have same number of sides

$$(P_2, P_1) \in R$$

So, R is symmetric on A.

Transitivity : Let P_1, P_2 and P_3 be three polygons in A such that $(P_1, P_2) \in R$ and $(P_2, P_3) \in R$.

Then, $(P_1, P_2) \in R \Rightarrow P_1$ and P_2 have same number of sides.

and $(P_2, P_3) \in R \Rightarrow P_2$ and P_3 have same number of sides.

$\therefore$ P_1 and P_3 have same number of sides.

$\Rightarrow \qquad (P_1, P_3) \in R$

Thus, if $\qquad (P_1, P_2) \in R$ and $(P_2, P_3) \in R$

$\Rightarrow \qquad (P_1, P_3) \in R$

So, R is transitive.

Hence, R is an equivalence relation on A.

Let P be a polygon in A such that $(P, T) \in R$. Then, polygon P and triangle T have same number of sides.

Thus, P is any triangle in A with sides 3, 4 and 5.

Q. 10. If S is the set of all real numbers which satisfy the function $f(x) = \dfrac{2x+1}{2x^3 + 3x^2 + x} > 0$ then find all the values from the set S.

Sol. Since, $\qquad f(x) = \dfrac{2x+1}{2x^3 + 3x^2 + x} > 0$

$$\Rightarrow \qquad f(x) = \frac{2x+1}{x(2x^2 + 3x + 1)} > 0$$

$$\Rightarrow \qquad f(x) = \frac{2x+1}{x(2x+1)(x+1)} > 0$$

$$\Rightarrow \qquad f(x) = \frac{1}{x(x+1)} > 0, x \neq -\frac{1}{2}$$

[$\because$ $(2x + 1)$ cancels out in the numerator and denominator]

By using the sign rule (or otherwise) we can now find out the region on the number line where $\dfrac{1}{x(1+x)} > 0$ will be true. The region is $-\infty$ to -1 and 0 to ∞, where neither -1 nor 0 are included in the solution set. Hence, we write the solution as :

$$f(x) \in (-\infty, -1) \cup (0, \infty)$$

Q. 11. Show that the relation R on the set A of points in a plane is given by

$R = \{(P, Q)$: Distance of the point P from the origin = Distance of point Q from origin$\}$ is an equivalence relation.

Further, show that the set of all points related to a point $P \neq (0, 0)$ is the circle passing through P with origin as centre. [HOTS]

Sol. If O is the origin, then

$$R = \{(P, Q) : OP = OQ\}$$

Reflexivity : For any point $P \in A$

$$OP = OP$$

$\Rightarrow$ $(P, P) \in R$

i.e., R is reflexive.

Symmetric : Let $P, Q \in A$ such that

$(P, Q) \in R$

$\Rightarrow$ $OP = OQ$

$\Rightarrow$ $OQ = OP$

$\Rightarrow$ $(Q, P) \in R$

i.e., R is symmetric.

Transitivity : Let $P, Q, S \in A$ such that

$(P, Q) \in R$ and $(Q, S) \in R$

$\Rightarrow$ $OP = OQ$ and $OQ = OS$

$\Rightarrow$ $OP = OS$

$\Rightarrow$ $(P, S) \in R$

i.e., R is transitive.

Now we have, R is reflexive, symmetric and transitive.

Therefore, R is an equivalence relation.

Let $P, Q, R....$ be points in the set A such that

$(P, Q), (P, R)... \in R$

$\Rightarrow$ $OP = OQ; OP = OR;...$ [where O is origin]

$\Rightarrow$ $OP = OQ = OR = ...$

i.e., All points $P, Q, R... \in A$, which are related to P are equidistant from origin 'O'.

Hence, set of all points of A related to P is the circle passing through P having origin as centre.

Q. 12. Let $A = \{x \in Z : 0 \le x \le 12\}$. Show that $R = \{(a, b) : a, b \in A, \ | a - b | \text{ is divisible by } 4\}$ is an equivalence relation. Find the set of all elements related to 1. Also write the equivalence class [2].★

Sol. Given, $R = \{(a, b) : a, b \in A, | a - b | \text{ is divisible by } 4\}$

Reflexivity : For any $a \in A$

$| a - a | = 0$, which is divisible by 4

$\Rightarrow (a, a) \in R$

So, R is reflexive.

Symmetric : Let $(a, b) \in R$

$\Rightarrow |a - b|$ is divisible by 4

$\Rightarrow |b - a|$ is divisible by 4

$\Rightarrow$ $(b, a) \in R$

So, R is symmetric.

Transitivity : Let $(a, b) \in R$ and $(b, c) \in R$

$\Rightarrow |a - b|$ is divisible by 4

$|a - b| = 4k$

$a - b = \pm 4k$ ...(i)

★ are board exam questions from previous years

Also $| b - c |$ is divisible by 4

$| b - c | = 4m$

$b - c = \pm 4m$ ...(ii)

Adding equation (i) and (ii)

$a - b + b - c = \pm 4(k + m)$

$a - c = \pm 4 (k + m)$

$\Rightarrow$ $(a, c) \in R$

So, R is transitive.

$\Rightarrow R$ is reflexive, symmetric and transitive

$\therefore$ R is an equivalence relation.

Let x be an element of R such that $(x, 1) \in R$

Then $|x - 1|$ is divisible by 4

$x - 1 = 0, 4, 8, 12, ...$

$\Rightarrow$ $x = 1, 5, 9$ $[\because x \le 12]$

$\therefore$ Set of all elements of A which are related to 1 are $\{1, 5, 9\}$.

Equivalence class of 2 *i.e.*

$[2] = \{(a, 2) : a \in A, | a - 2 | \text{ is divisible by } 4\}$

$\Rightarrow$ $| a - 2 | = 4k$ [k is whole number, $k \le 3$]

$\Rightarrow$ $a = 2, 6, 10$

Therefore, equivalence class [2] is $\{2, 6, 10\}$.

Q. 13. Let $f : N \to N$ be a function defined as $f(x) = 9x^2 + 6x - 5$. Show that $f : N \to S$ is the range of f, is invertible. Find the inverse of f and hence find $f^{-1}(43)$ and $f^{-1}(163)$.★

Sol. Given, $f(x) = 9x^2 + 6x - 5$

Let $x_1, x_2 \in N$ and $f(x_1) = f(x_2)$

$\Rightarrow$ $9x_1^2 + 6x_1 - 5 = 9x_2^2 + 6x_2 - 5$

$\Rightarrow$ $9(x_1^2 - x_2^2) + 6(x_1 - x_2) = 0$

$\Rightarrow$ $(x_1 - x_2) [9(x_1 + x_2) + 6] = 0$

$\Rightarrow$ $x_1 - x_2 = 0$

as $9x_1 + 9x_2 + 6 \ne 0, x_1, x_2 \in N$

$\Rightarrow$ $x_1 = x_2$

$\therefore f$ is one-one function.

Let $y = 9x^2 + 6x - 5$

$\Rightarrow$ $y = (3x + 1)^2 - 1 - 5 = (3x + 1)^2 - 6$

$\Rightarrow$ $(3x + 1)^2 = y + 6$

$\Rightarrow$ $3x + 1 = \sqrt{y + 6}$

$\Rightarrow$ $x = \dfrac{\sqrt{y + 6} - 1}{3}$ as $x \in N$

$\Rightarrow$ $\sqrt{y + 6} - 1 > 0$

$\Rightarrow$ $y + 6 > 1$

$\Rightarrow$ $y > -5$ and $y \in N$

So, the function is invertible if the range of the function $f(x)$ is $\{1, 2, 3, \ldots\}$.

$f: N \to S$ is onto as codomain = Range

Hence, f is invertible.

Therefore, the inverse of the function $f(x)$ is $f^{-1}(y)$, *i.e.*, x.

Now, $$f^{-1}(y) = \frac{\sqrt{y+6}-1}{3}$$

$$f^{-1}(43) = \frac{\sqrt{43+6}-1}{3} = 2$$

$$f^{-1}(163) = \frac{\sqrt{163+6}-1}{3} = 4.$$

Chapter 2. Inverse Trigonometric Functions

Q. 1. If $\cos^{-1}\frac{x}{a} + \cos^{-1}\frac{y}{b} = \alpha$, prove that :

$$\frac{x^2}{a^2} - \frac{2xy}{ab}\cos\alpha + \frac{y^2}{b^2} = \sin^2\alpha.$$

Sol. Given, $$\cos^{-1}\frac{x}{a} + \cos^{-1}\frac{y}{b} = \alpha$$

$$\Rightarrow \cos^{-1}\left[\frac{xy}{ab} - \sqrt{\left(1-\frac{x^2}{a^2}\right)\left(1-\frac{y^2}{b^2}\right)}\right] = \alpha$$

$$\Rightarrow \cos\alpha = \frac{xy}{ab} - \sqrt{\left(1-\frac{x^2}{a^2}\right)\left(1-\frac{y^2}{b^2}\right)}$$

$$\Rightarrow \sqrt{\left(1-\frac{x^2}{a^2}\right)\left(1-\frac{y^2}{b^2}\right)} = \frac{xy}{ab} - \cos\alpha$$

$$\Rightarrow \left(1-\frac{x^2}{a^2}\right)\left(1-\frac{y^2}{b^2}\right) = \frac{x^2y^2}{a^2b^2} + \cos^2\alpha - 2\cos\alpha\frac{xy}{ab}$$

$$\Rightarrow 1 - \frac{y^2}{b^2} - \frac{x^2}{a^2} + \frac{x^2y^2}{a^2b^2} = \frac{x^2y^2}{a^2b^2} + \cos^2\alpha - 2\cos\alpha\frac{xy}{ab}$$

$$\Rightarrow 1 - \cos^2\alpha = \frac{x^2}{a^2} + \frac{y^2}{b^2} - 2\cos\alpha\frac{xy}{ab}$$

$$\Rightarrow \frac{x^2}{a^2} + \frac{y^2}{b^2} - \frac{2xy}{ab}\cos\alpha = \sin^2\alpha$$

Q. 2. Solve the following :

$\sin(2\cos^{-1}(\cot(2\tan^{-1}x))) = 0.$

Sol. We have, $\sin(2\cos^{-1}(\cot(2\tan^{-1}x))) = 0$

$$\Rightarrow \sin\left(2\cos^{-1}\left(\cot\left(\tan^{-1}\frac{2x}{1-x^2}\right)\right)\right) = 0$$

$$\Rightarrow \sin\left(2\cos^{-1}\left(\cot\left(\cot^{-1}\left(\frac{1-x^2}{2x}\right)\right)\right)\right) = 0$$

$$\Rightarrow \sin\left(2\cos^{-1}\left(\frac{1-x^2}{2x}\right)\right) = 0$$

$$\Rightarrow \sin\left[\sin^{-1}\left[2\left\{\frac{1-x^2}{x}\right\}\sqrt{1-\left\{\frac{1-x^2}{2x}\right\}^2}\right]\right] = 0$$

$$\left[\because 2\cos^{-1}x = \sin^{-1}(2x\sqrt{1-x^2})\right]$$

$$\Rightarrow 2\left\{\frac{1-x^2}{x}\right\}\sqrt{1-\left\{\frac{1-x^2}{2x}\right\}^2} = 0$$

$$\Rightarrow \frac{1-x^2}{2} = 0 \quad \text{or} \quad \sqrt{1-\left\{\frac{1-x^2}{2x}\right\}^2} = 0$$

$$\Rightarrow 1-x^2 = 0 \quad \text{or} \quad \left\{\frac{1-x^2}{2x}\right\}^2 = 1$$

$$\Rightarrow x = \pm 1 \text{ or } (1-x^2)^2 = 4x^2$$

$$\Rightarrow x = \pm 1 \text{ or } (1-x^2)^2 - 2(x) = 0$$

$$\Rightarrow x = \pm 1 \text{ or } [(1-x^2)-(2x)]$$

$$[(1-x^2)+(2x)] = 0$$

$$\Rightarrow x = \pm 1 \text{ or } [(1-x^2)-(2x)] = 0$$

$$\text{or } [(1-x^2)+(2x)] = 0$$

$$\Rightarrow x = \pm 1 \text{ or } x^2-2x-1 = 0$$

$$\text{or } x^2+2x-1 = 0$$

$$\Rightarrow x = \pm 1 \text{ or } x = -1 \pm \sqrt{2},\ x = 1 \pm \sqrt{2}$$

Q. 3. Express in the simplest form :

$$\tan^{-1}\left(\frac{\cos x - \sin x}{\cos x + \sin x}\right)$$

Sol. Given, $\tan^{-1}\left(\frac{\cos x - \sin x}{\cos x + \sin x}\right)$

$$= \tan^{-1}\left(\frac{\cos x - \cos\left(\frac{\pi}{2} - x\right)}{\cos x + \cos\left(\frac{\pi}{2} - x\right)}\right)$$

$$\left[\because \cos\left(\frac{\pi}{2} - \theta\right) = \sin\theta\right]$$

$$= \tan^{-1}\left[\frac{-2\sin\dfrac{\left(x+\dfrac{\pi}{2}-x\right)}{2}\sin\dfrac{\left(x-\dfrac{\pi}{2}+x\right)}{2}}{2\cos\dfrac{\left(x+\dfrac{\pi}{2}-x\right)}{2}\cos\dfrac{\left(x-\dfrac{\pi}{2}+x\right)}{2}}\right]$$

$$= \tan^{-1}\left[\frac{-\sin\left(\dfrac{\pi}{4}\right)\sin\left(\dfrac{2x}{2}-\dfrac{\pi}{4}\right)}{\cos\left(\dfrac{\pi}{4}\right)\cos\left(\dfrac{2x}{2}-\dfrac{\pi}{4}\right)}\right]$$

$$= \tan^{-1}\left[\frac{-\sin\dfrac{\pi}{4}\sin\left(x-\dfrac{\pi}{4}\right)}{\cos\dfrac{\pi}{4}\cos\left(x-\dfrac{\pi}{4}\right)}\right]$$

$$= \tan^{-1}\left(-\tan\frac{\pi}{4}\tan\left(x-\frac{\pi}{4}\right)\right)$$

$$= \tan^{-1}\left(\tan\left(\frac{\pi}{4}-x\right)\right)$$

$$= \frac{\pi}{4}-x$$

Q. 4. Prove that :

$$\mathbf{4\tan^{-1}\frac{1}{5}-\tan^{-1}\frac{1}{70}+\tan^{-1}\frac{1}{99}=\frac{\pi}{4}.}$$

Sol. L.H.S.

$$= 4\tan^{-1}\frac{1}{5}-\tan^{-1}\frac{1}{70}+\tan^{-1}\frac{1}{99}$$

$$= 2\left(2\tan^{-1}\frac{1}{5}\right)-\tan^{-1}\frac{1}{70}+\tan^{-1}\frac{1}{99}$$

$$= 2\tan^{-1}\left[\frac{2\times\dfrac{1}{5}}{1-\left(\dfrac{1}{5}\right)^2}\right]-\tan^{-1}\frac{1}{70}+\tan^{-1}\frac{1}{99}$$

$$= 2\tan^{-1}\frac{5}{12}-\left(\tan^{-1}\frac{1}{70}-\tan^{-1}\frac{1}{99}\right)$$

$$= \tan^{-1}\left(\frac{2\times\dfrac{5}{12}}{1-\left(\dfrac{5}{12}\right)^2}\right)-\tan^{-1}\left(\frac{\dfrac{1}{70}-\dfrac{1}{99}}{1+\dfrac{1}{70}\times\dfrac{1}{99}}\right)$$

$$= \tan^{-1}\left(\frac{120}{119}\right)-\tan^{-1}\left(\frac{29}{6931}\right)$$

$$= \tan^{-1}\left(\frac{120}{119}\right)-\tan^{-1}\left(\frac{1}{239}\right)$$

$$= \tan^{-1}\left(\frac{\dfrac{120}{119}-\dfrac{1}{239}}{1+\dfrac{120}{119}\times\dfrac{1}{239}}\right)$$

$$= \tan^{-1}1 = \frac{\pi}{4}$$

= R.H.S.

Q. 5. Solve for x :

$$\mathbf{\sin^{-1}\frac{3x}{5}+\sin^{-1}\frac{4x}{5}=\sin^{-1}x.}$$

Sol. We have, $\sin^{-1}\dfrac{3x}{5}+\sin^{-1}\dfrac{4x}{5}=\sin^{-1}x$

$$\Rightarrow \sin^{-1}\left[\frac{3x}{5}\sqrt{1-\frac{16x^2}{25}}+\frac{4x}{5}\sqrt{1-\frac{9x^2}{25}}\right]=\sin^{-1}x$$

$$\Rightarrow \frac{3x}{5}\sqrt{1-\frac{16x^2}{25}}+\frac{4x}{5}\sqrt{1-\frac{9x^2}{25}}=x$$

$$\Rightarrow \frac{3x}{5}\frac{\sqrt{25-16x^2}}{5}+\frac{4x}{5}\frac{\sqrt{25-9x^2}}{5}=x$$

$$\Rightarrow 3x\sqrt{25-16x^2}+4x\sqrt{25-9x^2}=25x$$

$$\Rightarrow x\,[3\sqrt{25-16x^2}+4\sqrt{25-9x^2}-25]=0$$

$$\Rightarrow x=0 \text{ or } 3\sqrt{25-16x^2}+4\sqrt{25-9x^2}-25=0$$

$$\Rightarrow 4\sqrt{25-9x^2}=25-3\sqrt{25-16x^2}$$

$$\Rightarrow 16(25-9x^2)=625+9(25-16x^2)-150\sqrt{25-16x^2}$$

$$\Rightarrow 150\sqrt{25-16x^2}=450$$

$$\Rightarrow 25-16x^2=9$$

$$\Rightarrow 16=16x^2 \Rightarrow x^2=1$$

$$\Rightarrow x=\pm 1$$

Thus, $x = 0, 1 - 1$

Q. 6. Write in the simplest form : $\tan^{-1}\sqrt{\dfrac{a-x}{a+x}}$

Sol. Let $x = a\cos\theta$

$$\therefore \quad \tan^{-1}\sqrt{\frac{a-x}{a+x}}=\tan^{-1}\sqrt{\frac{a-a\cos\theta}{a+a\cos\theta}}$$

$$= \tan^{-1}\sqrt{\frac{1-\cos\theta}{1+\cos\theta}}$$

$$= \tan^{-1}\sqrt{\frac{2\sin^2\theta/2}{2\cos^2\theta/2}}$$

$$= \tan^{-1}\sqrt{\tan^2\theta/2}$$

$$= \tan^{-1}(\tan\theta/2)$$

$$= \frac{\theta}{2}$$

$\because \quad x = a\cos\theta$

$\Rightarrow \quad \cos\theta = \frac{x}{a}$

$\Rightarrow \quad \theta = \cos^{-1}\left(\frac{x}{a}\right)$

$\therefore \quad \tan^{-1}\sqrt{\frac{a-x}{a+x}} = \frac{\theta}{2}$

$$= \frac{1}{2}\cos^{-1}\frac{x}{a}$$

Q. 7. Show that :

$$\tan^{-1}\left(\frac{1}{\sqrt{3}}\tan\frac{x}{2}\right) = \frac{1}{2}\cos^{-1}\left(\frac{1+2\cos x}{2+\cos x}\right)$$

Sol. We know, $2\tan^{-1}y = \cos^{-1}\left(\frac{1-y^2}{1+y^2}\right)$ ...(i)

Substituting $y = \frac{1}{\sqrt{3}}\tan\frac{x}{2}$, we get

$$2\tan^{-1}\left(\frac{1}{\sqrt{3}}\tan\frac{x}{2}\right) = \cos^{-1}\left\{\frac{1-\left(\frac{1}{\sqrt{3}}\tan\frac{x}{2}\right)^2}{1+\left(\frac{1}{\sqrt{3}}\tan\frac{x}{2}\right)^2}\right\}$$

$$= \cos^{-1}\left\{\frac{3-\tan^2\frac{x}{2}}{3+\tan^2\frac{x}{2}}\right\} = \cos^{-1}\left\{\frac{3-\frac{1-\cos x}{1+\cos x}}{3+\frac{1-\cos x}{1+\cos x}}\right\}$$

$$= \cos^{-1}\left\{\frac{3+3\cos x-1+\cos x}{3+3\cos x+1-\cos x}\right\}$$

$$= \cos^{-1}\left\{\frac{2+4\cos x}{4+2\cos x}\right\}$$

$$= \cos^{-1}\left\{\frac{1+2\cos x}{2+\cos x}\right\} = \text{R.H.S.}$$

Hence, $\tan^{-1}\left(\frac{1}{\sqrt{3}}\tan\frac{x}{2}\right) = \frac{1}{2}\cos^{-1}\left\{\frac{1+2\cos x}{2+\cos x}\right\}$

Chapter 3. Matrices

Q. 1. If $A = \begin{bmatrix} 0 & -\tan\alpha/2 \\ \tan\alpha/2 & 0 \end{bmatrix}$ and I is a 2×2 unit matrix then prove that

$$I + A = (I - A)\begin{bmatrix} \cos\alpha & -\sin\alpha \\ \sin\alpha & \cos\alpha \end{bmatrix}$$ **[NCERT]**

Sol. L.H.S. $= I + A$

$$= \begin{bmatrix} 1 & 0 \\ 0 & 1 \end{bmatrix} + \begin{bmatrix} 0 & -\tan\alpha/2 \\ \tan\alpha/2 & 0 \end{bmatrix}$$

$$= \begin{bmatrix} 1 & -\tan\alpha/2 \\ \tan\alpha/2 & 1 \end{bmatrix}$$

R.H.S. $= (I-A)\begin{bmatrix} \cos\alpha & -\sin\alpha \\ \sin\alpha & \cos\alpha \end{bmatrix}$

$$= \left(\begin{bmatrix} 1 & 0 \\ 0 & 1 \end{bmatrix} - \begin{bmatrix} 0 & -\tan\alpha/2 \\ \tan\alpha/2 & 0 \end{bmatrix}\right)\begin{bmatrix} \cos\alpha & -\sin\alpha \\ \sin\alpha & \cos\alpha \end{bmatrix}$$

$$= \begin{bmatrix} 1 & \tan\alpha/2 \\ -\tan\alpha/2 & 1 \end{bmatrix}\begin{bmatrix} \cos\alpha & -\sin\alpha \\ \sin\alpha & \cos\alpha \end{bmatrix}$$

$$= \begin{bmatrix} \cos\alpha + \tan\alpha/2\sin\alpha & -\sin\alpha + \tan\alpha/2\cos\alpha \\ -\tan\alpha/2\cos\alpha + \sin\alpha & \tan\alpha/2\sin\alpha + \cos\alpha \end{bmatrix}$$

$$= \begin{bmatrix} \cos\alpha + \frac{\sin\alpha/2}{\cos\alpha/2}\sin\alpha & -\sin\alpha + \frac{\sin\alpha/2}{\cos\alpha/2}\cos\alpha \\ -\frac{\sin\alpha/2}{\cos\alpha/2}\cos\alpha + \sin\alpha & \frac{\sin\alpha/2}{\cos\alpha/2}\sin\alpha + \cos\alpha \end{bmatrix}$$

$$= \begin{bmatrix} \frac{\cos\alpha\cos\alpha/2 + \sin\alpha/2\sin\alpha}{\cos\alpha/2} & \frac{-\cos\alpha/2\sin\alpha + \sin\alpha/2\cos\alpha}{\cos\alpha/2} \\ \frac{-\sin\alpha/2\cos\alpha + \sin\alpha\cos\alpha/2}{\cos\alpha/2} & \frac{\sin\alpha\sin\alpha/2 + \cos\alpha\cos\alpha/2}{\cos\alpha/2} \end{bmatrix}$$

$$= \frac{1}{\cos\alpha/2}\begin{bmatrix} \cos(\alpha-\alpha/2) & \sin(\alpha/2-\alpha) \\ \sin(\alpha-\alpha/2) & \cos(\alpha-\alpha/2) \end{bmatrix}$$

$$= \frac{1}{\cos\alpha/2}\begin{bmatrix} \cos\alpha/2 & -\sin\alpha/2 \\ \sin\alpha/2 & \cos\alpha/2 \end{bmatrix}$$

$$= \begin{bmatrix} 1 & -\tan\alpha/2 \\ \tan\alpha/2 & 1 \end{bmatrix}$$

= L.H.S.

Q. 2. If $A_\alpha = \begin{bmatrix} \cos\alpha & \sin\alpha \\ -\sin\alpha & \cos\alpha \end{bmatrix}$, prove that :

(i) $A_\alpha \cdot A_\beta = A_{\alpha+\beta}$

(ii) $A_\alpha \cdot A_{(-\alpha)} = I$.

Sol. Given, $A_\alpha = \begin{bmatrix} \cos\alpha & \sin\alpha \\ -\sin\alpha & \cos\alpha \end{bmatrix}$

and $\quad A_\beta = \begin{bmatrix} \cos\beta & \sin\beta \\ -\sin\beta & \cos\beta \end{bmatrix}$

$$A_{(-\alpha)} = \begin{bmatrix} \cos(-\alpha) & \sin(-\alpha) \\ -\sin(-\alpha) & \cos(-\alpha) \end{bmatrix}$$

$$= \begin{bmatrix} \cos\alpha & -\sin\alpha \\ \sin\alpha & \cos\alpha \end{bmatrix}$$

(i) $A_\alpha A_\beta = \begin{bmatrix} \cos\alpha & \sin\alpha \\ -\sin\alpha & \cos\alpha \end{bmatrix}\begin{bmatrix} \cos\beta & \sin\beta \\ -\sin\beta & \cos\beta \end{bmatrix}$

$$= \begin{bmatrix} \cos\alpha\cos\beta - \sin\alpha\sin\beta & \cos\alpha\sin\beta + \sin\alpha\cos\beta \\ -\sin\alpha\cos\beta - \cos\alpha\sin\beta & -\sin\alpha\sin\beta + \cos\alpha\cos\beta \end{bmatrix}$$

$$= \begin{bmatrix} \cos(\alpha+\beta) & \sin(\alpha+\beta) \\ -\sin(\alpha+\beta) & \cos(\alpha+\beta) \end{bmatrix} = A_{\alpha+\beta}$$

(ii) $A_\alpha A_{(-\alpha)} = \begin{bmatrix} \cos\alpha & \sin\alpha \\ -\sin\alpha & \cos\alpha \end{bmatrix}\begin{bmatrix} \cos\alpha & -\sin\alpha \\ \sin\alpha & \cos\alpha \end{bmatrix}$

$$= \begin{bmatrix} \cos^2\alpha + \sin^2\alpha & -\cos\alpha\sin\alpha + \sin\alpha\cos\alpha \\ -\sin\alpha\cos\alpha + \sin\alpha\cos\alpha & \sin^2\alpha + \cos^2\alpha \end{bmatrix}$$

$$= \begin{bmatrix} 1 & 0 \\ 0 & 1 \end{bmatrix} = I.$$

Q. 3. If $A = \begin{bmatrix} 3 & -5 \\ -4 & 2 \end{bmatrix}$, then show that $A^2 - 5A - 14I = 0$. Hence find A^{-1}.

Sol. Given, $A = \begin{bmatrix} 3 & -5 \\ -4 & 2 \end{bmatrix}$

$$\Rightarrow \quad A^2 = \begin{bmatrix} 3 & -5 \\ -4 & 2 \end{bmatrix}\begin{bmatrix} 3 & -5 \\ -4 & 2 \end{bmatrix}$$

$$= \begin{bmatrix} 29 & -25 \\ -20 & 24 \end{bmatrix}$$

$A^2 - 5A - 14I$

$$= \begin{bmatrix} 29 & -25 \\ -20 & 24 \end{bmatrix} - 5\begin{bmatrix} 3 & -5 \\ -4 & 2 \end{bmatrix} - 14\begin{bmatrix} 1 & 0 \\ 0 & 1 \end{bmatrix}$$

[where I is an identity matrix]

$$= \begin{bmatrix} 29 & -25 \\ -20 & 24 \end{bmatrix} + \begin{bmatrix} -15 & 25 \\ 20 & -10 \end{bmatrix} + \begin{bmatrix} -14 & 0 \\ 0 & -14 \end{bmatrix}$$

$$= \begin{bmatrix} 29-15-14 & -25+25+0 \\ -20+20+0 & 24-10-14 \end{bmatrix}$$

$$= \begin{bmatrix} 0 & 0 \\ 0 & 0 \end{bmatrix} = 0$$

$$\Rightarrow \quad A^2 - 5A - 14I = 0$$

On pre-multiplying by A^{-1}, we get

$$A^{-1} \cdot A^2 - 5A^{-1}A - 14A^{-1}I = 0$$

$$A - 5I - 14A^{-1} = 0$$

$$\Rightarrow \quad A^{-1} = \frac{1}{14}(A - 5I)$$

$$= \frac{1}{14}\left\{\begin{bmatrix} 3 & -5 \\ -4 & 2 \end{bmatrix} - 5\begin{bmatrix} 1 & 0 \\ 0 & 1 \end{bmatrix}\right\}$$

$$= \frac{1}{14}\begin{bmatrix} -2 & -5 \\ -4 & -3 \end{bmatrix}$$

Q. 4. Using elementary row operations, find the inverse of the matrix

$$A = \begin{bmatrix} 3 & -1 & -2 \\ 2 & 0 & -1 \\ 3 & -5 & 0 \end{bmatrix}$$

Sol. We know that,

$$A = IA$$

or $$\begin{bmatrix} 3 & -1 & -2 \\ 2 & 0 & -1 \\ 3 & -5 & 0 \end{bmatrix} = \begin{bmatrix} 1 & 0 & 0 \\ 0 & 1 & 0 \\ 0 & 0 & 1 \end{bmatrix} A$$

On applying $R_1 \to R_1 - R_2$

$$\begin{bmatrix} 1 & -1 & -1 \\ 2 & 0 & -1 \\ 3 & -5 & 0 \end{bmatrix} = \begin{bmatrix} 1 & -1 & 0 \\ 0 & 1 & 0 \\ 0 & 0 & 1 \end{bmatrix} A$$

On applying $R_2 \to R_2 + (-2)R_1$ and $R_3 \to R_3 + (-3)R_1$

$$\Rightarrow \begin{bmatrix} 1 & -1 & -1 \\ 0 & 2 & 1 \\ 0 & -2 & 3 \end{bmatrix} = \begin{bmatrix} 1 & -1 & 0 \\ -2 & 3 & 0 \\ -3 & 3 & 1 \end{bmatrix} A$$

On applying $R_2 \to \frac{1}{2}R_2$

$$\Rightarrow \begin{bmatrix} 1 & -1 & -1 \\ 0 & 1 & 1/2 \\ 0 & -2 & 3 \end{bmatrix} = \begin{bmatrix} 1 & -1 & 0 \\ -1 & 3/2 & 0 \\ -3 & 3 & 1 \end{bmatrix} A$$

On applying $R_1 \to R_1 + R_2$ and $R_3 \to R_3 + 2R_2$

$$\Rightarrow \begin{bmatrix} 1 & 0 & -1/2 \\ 0 & 1 & 1/2 \\ 0 & 0 & 4 \end{bmatrix} = \begin{bmatrix} 0 & 1/2 & 0 \\ -1 & 3/2 & 0 \\ -5 & 6 & 1 \end{bmatrix} A$$

On applying $R_3 \to \frac{1}{4}R_3$

$$\begin{bmatrix} 1 & 0 & -1/2 \\ 0 & 1 & 1/2 \\ 0 & 0 & 1 \end{bmatrix} = \begin{bmatrix} 0 & 1/2 & 0 \\ -1 & 3/2 & 0 \\ -5/4 & 3/2 & 1/4 \end{bmatrix} A$$

On applying $R_1 \to R_1 + \left(\frac{1}{2}\right)R_3$ and $R_2 \to R_2 + \left(\frac{-1}{2}\right)R_3$

$$\begin{bmatrix} 1 & 0 & 0 \\ 0 & 1 & 0 \\ 0 & 0 & 1 \end{bmatrix} = \begin{bmatrix} -5/8 & 5/4 & 1/8 \\ -3/8 & 3/4 & -1/8 \\ -5/4 & 3/2 & 1/4 \end{bmatrix} A$$

$$\therefore \qquad A^{-1} = \begin{bmatrix} -5/8 & 5/4 & 1/8 \\ -3/8 & 3/4 & -1/8 \\ -5/4 & 3/2 & 1/4 \end{bmatrix}$$

Q. 5. Express the following matrix as the sum of a symmetric and skew-symmetric matrix and verify your result :★

$$\begin{bmatrix} \mathbf{3} & \mathbf{-2} & \mathbf{-4} \\ \mathbf{3} & \mathbf{-2} & \mathbf{-5} \\ \mathbf{-1} & \mathbf{1} & \mathbf{2} \end{bmatrix}$$

Sol. Let, $A = \begin{bmatrix} 3 & -2 & -4 \\ 3 & -2 & -5 \\ -1 & 1 & 2 \end{bmatrix}$

$$\Rightarrow \qquad A^T = \begin{bmatrix} 3 & 3 & -1 \\ -2 & -2 & 1 \\ -4 & -5 & 2 \end{bmatrix}$$

So $A + A^T = \begin{bmatrix} 3 & -2 & -4 \\ 3 & -2 & -5 \\ -1 & 1 & 2 \end{bmatrix} + \begin{bmatrix} 3 & 3 & -1 \\ -2 & -2 & 1 \\ -4 & -5 & 2 \end{bmatrix}$

$$= \begin{bmatrix} 6 & 1 & -5 \\ 1 & -4 & -4 \\ -5 & -4 & 4 \end{bmatrix}$$

and $A - A^T = \begin{bmatrix} 3 & -2 & -4 \\ 3 & -2 & -5 \\ -1 & 1 & 2 \end{bmatrix} - \begin{bmatrix} 3 & 3 & -1 \\ -2 & -2 & 1 \\ -4 & -5 & 2 \end{bmatrix}$

$$= \begin{bmatrix} 0 & -5 & -3 \\ 5 & 0 & -6 \\ 3 & 6 & 0 \end{bmatrix}$$

Let $P = \dfrac{A + A^T}{2} = \begin{bmatrix} 3 & 1/2 & -5/2 \\ 1/2 & -2 & -2 \\ -5/2 & -2 & 2 \end{bmatrix}$

$$Q = \frac{A - A^T}{2} = \begin{bmatrix} 0 & -5/2 & -3/2 \\ 5/2 & 0 & -3 \\ 3/2 & 3 & 0 \end{bmatrix}$$

then $P^T = \begin{bmatrix} 3 & 1/2 & -5/2 \\ 1/2 & -2 & -2 \\ -5/2 & -2 & 2 \end{bmatrix}^T$

$$= \begin{bmatrix} 3 & 1/2 & -5/2 \\ 1/2 & -2 & -2 \\ -5/2 & -2 & 2 \end{bmatrix} = P$$

and $Q^T = \begin{bmatrix} 0 & -5/2 & -3/2 \\ 5/2 & 0 & -3 \\ 3/2 & 3 & 0 \end{bmatrix}^T$

$$= \begin{bmatrix} 0 & 5/2 & 3/2 \\ -5/2 & 0 & 3 \\ -3/2 & -3 & 0 \end{bmatrix}$$

$$= -\begin{bmatrix} 0 & -5/2 & -3/2 \\ 5/2 & 0 & -3 \\ 3/2 & 3 & 0 \end{bmatrix}$$

$$= -Q$$

Thus P is a symmetric and Q is a skew-symmetric matrix.

Also $P + Q = \begin{bmatrix} 3 & 1/2 & -5/2 \\ 1/2 & -2 & -2 \\ -5/2 & -2 & 2 \end{bmatrix} + \begin{bmatrix} 0 & -5/2 & -3/2 \\ 5/2 & 0 & -3 \\ 3/2 & 3 & 0 \end{bmatrix}$

$$= \begin{bmatrix} 3 & -2 & -4 \\ 3 & -2 & -5 \\ -1 & 1 & 2 \end{bmatrix} = A$$

Thus, we have expressed matrix A as the sum of a symmetric and skew-symmetric matrix.

Q. 6. If $A = \begin{bmatrix} 1 & 1 & 1 \\ 1 & 1 & 1 \\ 1 & 1 & 1 \end{bmatrix}$, prove that :

$$A^n = \begin{bmatrix} 3^{n-1} & 3^{n-1} & 3^{n-1} \\ 3^{n-1} & 3^{n-1} & 3^{n-1} \\ 3^{n-1} & 3^{n-1} & 3^{n-1} \end{bmatrix}, \; n \in N$$

by using mathematical induction. [NCERT]

Sol. Here, $A = \begin{bmatrix} 1 & 1 & 1 \\ 1 & 1 & 1 \\ 1 & 1 & 1 \end{bmatrix}$

Let $P(n) : A^n = \begin{bmatrix} 3^{n-1} & 3^{n-1} & 3^{n-1} \\ 3^{n-1} & 3^{n-1} & 3^{n-1} \\ 3^{n-1} & 3^{n-1} & 3^{n-1} \end{bmatrix}$

Putting $n = 1$, we get

$$P(1) : A^1 = \begin{bmatrix} 3^{1-1} & 3^{1-1} & 3^{1-1} \\ 3^{1-1} & 3^{1-1} & 3^{1-1} \\ 3^{1-1} & 3^{1-1} & 3^{1-1} \end{bmatrix}$$

★ are board exam questions from previous years

$$= \begin{bmatrix} 3^0 & 3^0 & 3^0 \\ 3^0 & 3^0 & 3^0 \\ 3^0 & 3^0 & 3^0 \end{bmatrix}$$

$$= \begin{bmatrix} 1 & 1 & 1 \\ 1 & 1 & 1 \\ 1 & 1 & 1 \end{bmatrix} \quad \ldots(i)$$

which is true for $n = 1$.

Let the result be true for $n = K$.

Then $$P(K) : A^K = \begin{bmatrix} 3^{K-1} & 3^{K-1} & 3^{K-1} \\ 3^{K-1} & 3^{K-1} & 3^{K-1} \\ 3^{K-1} & 3^{K-1} & 3^{K-1} \end{bmatrix} \quad \ldots(ii)$$

Putting $n = K + 1$, we get

$$P(K+1) : A^{K+1} = \begin{bmatrix} 3^K & 3^K & 3^K \\ 3^K & 3^K & 3^K \\ 3^K & 3^K & 3^K \end{bmatrix}$$

Now, $\quad$ L.H.S $= A^{K+1} = A^K \cdot A^1$

$$= \begin{bmatrix} 3^{K-1} & 3^{K-1} & 3^{K-1} \\ 3^{K-1} & 3^{K-1} & 3^{K-1} \\ 3^{K-1} & 3^{K-1} & 3^{K-1} \end{bmatrix} \begin{bmatrix} 1 & 1 & 1 \\ 1 & 1 & 1 \\ 1 & 1 & 1 \end{bmatrix}$$

[using eqs. (i) and (ii)]

$$= \begin{bmatrix} 3^{K-1}+3^{K-1}+3^{K-1} & 3^{K-1}+3^{K-1}+3^{K-1} & 3^{K-1}+3^{K-1}+3^{K-1} \\ 3^{K-1}+3^{K-1}+3^{K-1} & 3^{K-1}+3^{K-1}+3^{K-1} & 3^{K-1}+3^{K-1}+3^{K-1} \\ 3^{K-1}+3^{K-1}+3^{K-1} & 3^{K-1}+3^{K-1}+3^{K-1} & 3^{K-1}+3^{K-1}+3^{K-1} \end{bmatrix}$$

$$= \begin{bmatrix} 3 \times 3^{K-1} & 3 \times 3^{K-1} & 3 \times 3^{K-1} \\ 3 \times 3^{K-1} & 3 \times 3^{K-1} & 3 \times 3^{K-1} \\ 3 \times 3^{K-1} & 3 \times 3^{K-1} & 3 \times 3^{K-1} \end{bmatrix}$$

$$= \begin{bmatrix} 3^K & 3^K & 3^K \\ 3^K & 3^K & 3^K \\ 3^K & 3^K & 3^K \end{bmatrix} = \text{R.H.S}$$

Therefore, the result is true for $n = K + 1$, whenever it is true for $n = K$. So, by principle of mathematical induction, it is true for all $n \in N$.

Q. 7. Determine the product

$$\begin{bmatrix} -4 & 4 & 4 \\ -7 & 1 & 3 \\ 5 & -3 & -1 \end{bmatrix} \begin{bmatrix} 1 & -1 & 1 \\ 1 & -2 & -2 \\ 2 & 1 & 3 \end{bmatrix}$$

and use it to solve the system of equations :★

$$x - y + z = 4; x - 2y - 2z = 9; 2x + y + 3z = 1.$$

Sol. Given, $$\begin{bmatrix} -4 & 4 & 4 \\ -7 & 1 & 3 \\ 5 & -3 & -1 \end{bmatrix} \begin{bmatrix} 1 & -1 & 1 \\ 1 & -2 & -2 \\ 2 & 1 & 3 \end{bmatrix}$$

$$= \begin{bmatrix} -4+4+8 & 4-8+4 & -4-8+12 \\ -7+1+6 & 7-2+3 & -7-2+9 \\ 5-3-2 & -5+6-1 & 5+6-3 \end{bmatrix}$$

$$= \begin{bmatrix} 8 & 0 & 0 \\ 0 & 8 & 0 \\ 0 & 0 & 8 \end{bmatrix} = 8I$$

[where I is identity matrix]

Let $$A = \begin{bmatrix} -4 & 4 & 4 \\ -7 & 1 & 3 \\ 5 & -3 & -1 \end{bmatrix}$$

and $$B = \begin{bmatrix} 1 & -1 & 1 \\ 1 & -2 & -2 \\ 2 & 1 & 3 \end{bmatrix}$$

$\Rightarrow \quad AB = 8I$

$ABB^{-1} = 8\,IB^{-1}$

$A = 8\,B^{-1}$

$$B^{-1} = \frac{A}{8} \quad \ldots(i)$$

Given equations are

$$x - y + z = 4$$
$$x - 2y - 2z = 9$$
$$2x + y + 3z = 1$$

$$\Rightarrow \begin{bmatrix} 1 & -1 & 1 \\ 1 & -2 & -2 \\ 2 & 1 & 3 \end{bmatrix} \begin{bmatrix} x \\ y \\ z \end{bmatrix} = \begin{bmatrix} 4 \\ 9 \\ 1 \end{bmatrix}$$

$\Rightarrow \quad BX = C$

where $$X = \begin{bmatrix} x \\ y \\ z \end{bmatrix} \text{ and } C = \begin{bmatrix} 4 \\ 9 \\ 1 \end{bmatrix}$$

$\Rightarrow \quad X = B^{-1}C$

$$\Rightarrow \quad X = \frac{A}{8}C \quad \text{[Using (i)]}$$

$$= \frac{1}{8}\begin{bmatrix} -4 & 4 & 4 \\ -7 & 1 & 3 \\ 5 & -3 & -1 \end{bmatrix} \begin{bmatrix} 4 \\ 9 \\ 1 \end{bmatrix}$$

$$= \frac{1}{8}\begin{bmatrix} -16+36+4 \\ -28+9+3 \\ 20-27-1 \end{bmatrix} = \frac{1}{8}\begin{bmatrix} 24 \\ -16 \\ -8 \end{bmatrix}$$

$$\begin{bmatrix} x \\ y \\ z \end{bmatrix} = \begin{bmatrix} 3 \\ -2 \\ -1 \end{bmatrix}$$

On comparing corresponding elements, we get

$\Rightarrow \quad x = 3, y = -2$ and $z = -1$.

★ are board exam questions from previous years

Chapter 4. Determinants

Q. 1. Prove that :

$$\begin{vmatrix} ka & k^2+a^2 & 1 \\ kb & k^2+b^2 & 1 \\ kc & k^2+c^2 & 1 \end{vmatrix} = k(a-b)(c-a)(b-c).$$

Sol. Let $\Delta = \begin{vmatrix} ka & k^2+a^2 & 1 \\ kb & k^2+b^2 & 1 \\ kc & k^2+c^2 & 1 \end{vmatrix}$

$$= k\begin{vmatrix} a & k^2+a^2 & 1 \\ b & k^2+b^2 & 1 \\ c & k^2+c^2 & 1 \end{vmatrix}$$

$$= k\begin{vmatrix} a & k^2 & 1 \\ b & k^2 & 1 \\ c & k^2 & 1 \end{vmatrix} + k\begin{vmatrix} a & a^2 & 1 \\ b & b^2 & 1 \\ c & c^2 & 1 \end{vmatrix}$$

$$= k^3\begin{vmatrix} a & 1 & 1 \\ b & 1 & 1 \\ c & 1 & 1 \end{vmatrix} + k\begin{vmatrix} a & a^2 & 1 \\ b & b^2 & 1 \\ c & c^2 & 1 \end{vmatrix}$$

Let $\Delta = \Delta_1 + \Delta_2$ ...(i)

where $\Delta_1 = k^3\begin{vmatrix} a & 1 & 1 \\ b & 1 & 1 \\ c & 1 & 1 \end{vmatrix}$ and $\Delta_2 = k\begin{vmatrix} a & a^2 & 1 \\ b & b^2 & 1 \\ c & c^2 & 1 \end{vmatrix}$

Since two columns of Δ_1 are equal, so

$\Delta_1 = 0$

$$\Delta_2 = k\begin{vmatrix} a & a^2 & 1 \\ b & b^2 & 1 \\ c & c^2 & 1 \end{vmatrix}$$

On applying $R_2 \to R_2 - R_1$ and $R_3 \to R_3 - R_1$

$$= k\begin{vmatrix} a & a^2 & 1 \\ b-a & b^2-a^2 & 0 \\ c-a & c^2-a^2 & 0 \end{vmatrix}$$

On expanding along C_3

$$= k.1\begin{vmatrix} b-a & b^2-a^2 \\ c-a & c^2-a^2 \end{vmatrix}$$

$$= k(b-a)(c-a)\begin{vmatrix} 1 & b+a \\ 1 & c+a \end{vmatrix}$$

$$= k(b-a)(c-a)(c+a-b-a)$$

$$= k(b-a)(c-a)(c-b)$$

$\Rightarrow \quad \Delta_2 = k(a-b)(b-c)(c-a)$

By using equation (i)

$$\Delta = \Delta_1 + \Delta_2$$

$$\Delta = k(a-b)(b-c)(c-a).$$

Q. 2. If 1, ω and ω² are the cube roots of unity, then what values of x satisfies,

$$\begin{vmatrix} x+\omega^2 & \omega & 1 \\ \omega & \omega^2 & 1+x \\ 1 & x+\omega & \omega^2 \end{vmatrix} = 0.$$

Sol. Since 1, ω and ω^2 are the cube roots of unity then

$$1+\omega+\omega^2 = 0 \text{ and } \omega^3 = 1$$

Given, $\begin{vmatrix} x+\omega^2 & \omega & 1 \\ \omega & \omega^2 & 1+x \\ 1 & x+\omega & \omega^2 \end{vmatrix} = 0$

On applying $C_1 \to C_1 + C_2 + C_3$

$$\begin{vmatrix} x+\omega^2+\omega+1 & \omega & 1 \\ \omega+\omega^2+1+x & \omega^2 & 1+x \\ 1+x+\omega+\omega^2 & x+\omega & \omega^2 \end{vmatrix} = 0$$

$$\Rightarrow \quad \begin{vmatrix} x & \omega & 1 \\ x & \omega^2 & 1+x \\ x & x+\omega & \omega^2 \end{vmatrix} = 0$$

$$\Rightarrow \quad x\begin{vmatrix} 1 & \omega & 1 \\ 1 & \omega^2 & 1+x \\ 1 & x+\omega & \omega^2 \end{vmatrix} = 0$$

On applying $R_2 \to R_2 - R_1$ and $R_3 \to R_3 - R_1$

$$\Rightarrow \quad x\begin{vmatrix} 1 & \omega & 1 \\ 0 & \omega^2-\omega & x \\ 0 & x & \omega^2-1 \end{vmatrix} = 0$$

On expanding along C_1

$$\Rightarrow \quad x \times 1\begin{vmatrix} \omega^2-\omega & x \\ x & \omega^2-1 \end{vmatrix} = 0$$

$$\Rightarrow \quad x[(\omega^2-\omega)(\omega^2-1)-x^2] = 0$$

$$\Rightarrow \quad x[\omega^4-\omega^2-\omega^3+\omega-x^2] = 0$$

$$\Rightarrow \quad x[\omega-\omega^2-\omega^3+\omega-x^2] = 0 \qquad [\because \omega^3 = 1]$$

$$\Rightarrow \quad x[-(1+\omega+\omega^2)+3\omega-x^2] = 0$$

$$\Rightarrow \quad x(3\omega-x^2) = 0$$

$$\Rightarrow \quad x = 0$$

or $\quad 3\omega - x^2 = 0$

$$x^2 = 3\omega.$$

Q. 3. Prove that the following three lines are concurrent

$$\boldsymbol{ax+by+c=0}$$

$$\boldsymbol{bx+cy+a=0}$$

$$\boldsymbol{cx+ay+b=0}$$

Only when $\boldsymbol{a^3+b^3+c^3 = 3abc.}$

Sol. We know that lines

$$a_1x+b_1y+c_1 = 0$$

$$a_2x+b_2y+c_2 = 0$$

and $a_3x + b_3y + c_3 = 0$

are concurrent, if $\begin{vmatrix} a_1 & b_1 & c_1 \\ a_2 & b_2 & c_2 \\ a_3 & b_3 & c_3 \end{vmatrix} = 0$

Given lines are

$$ax + by + c = 0 \quad \text{...(i)}$$
$$bx + cy + a = 0 \quad \text{...(ii)}$$
$$cx + ay + b = 0 \quad \text{...(iii)}$$

For the above lines to be concurrent,

$$\begin{vmatrix} a & b & c \\ b & c & a \\ c & a & b \end{vmatrix} = 0$$

On applying $R_1 \to R_1 + R_2 + R_3$

$$\begin{vmatrix} a+b+c & b+c+a & c+a+b \\ b & c & a \\ c & a & b \end{vmatrix} = 0$$

$$\Rightarrow \quad (a+b+c)\begin{vmatrix} 1 & 1 & 1 \\ b & c & a \\ c & a & b \end{vmatrix} = 0$$

On applying $C_2 \to C_2 - C_1$ and $C_3 \to C_3 - C_1$

$$\Rightarrow \quad (a+b+c)\begin{vmatrix} 1 & 0 & 0 \\ b & c-b & a-b \\ c & a-c & b-c \end{vmatrix} = 0$$

$$\Rightarrow \quad (a+b+c)\left(1\begin{vmatrix} c-b & a-b \\ a-c & b-c \end{vmatrix}\right) = 0$$

$$\Rightarrow \quad (a+b+c)\,[(c-b)(b-c) - (a-c)(a-b)] = 0$$

$$\Rightarrow (a+b+c)\,[(bc - c^2 - b^2 + bc) - (a^2 - ab - ac + bc)] = 0$$

$$\Rightarrow \quad (a+b+c)(-a^2 - b^2 - c^2 + bc + ca + ab) = 0$$

$$\Rightarrow \quad (a+b+c)(a^2 + b^2 + c^2 - ab - bc - ca) = 0$$

So, either $a + b + c = 0$

or $\quad (a^2 + b^2 + c^2 - bc - ca - ab) = 0 \quad$ ...(iv)

$$a^2 + b^2 + c^2 = ab + bc + ca$$

Also $(a + b + c)(a^2 + b^2 + c^2 - ab - bc - ca)$

$$= a^3 + b^3 + c^3 - 3abc = 0$$

$\therefore \quad a^3 + b^3 + c^3 = 3abc.$

Q. 4. If $f(x) = \begin{vmatrix} x^n & \sin x & \cos x \\ n! & \sin(n\pi/2) & \cos(n\pi/2) \\ a & a^2 & a^3 \end{vmatrix}$ **then what will be the value of** $\dfrac{d^n}{dx^n}f(x)$ **at** $x = 0$ **for** $n = (2m+1)$**?**

Sol. We know that, if

$$f(x) = \begin{vmatrix} g(x) & p(x) & q(x) \\ a & b & c \\ d & e & f \end{vmatrix}$$

where $g(x)$, $p(x)$ and $q(x)$ are differentiable function with respect to x, then

$$f'(x) = \begin{vmatrix} g'(x) & p'(x) & q'(x) \\ a & b & c \\ d & e & f \end{vmatrix}$$

Given, $\quad f(x) = \begin{vmatrix} x^n & \sin x & \cos x \\ n! & \sin\left(\dfrac{n\pi}{2}\right) & \cos\left(\dfrac{n\pi}{2}\right) \\ a & a^2 & a^3 \end{vmatrix}$

$$\frac{d^n}{dx^n}f(x) = \begin{vmatrix} \dfrac{d^n}{dx^n}(x^n) & \dfrac{d^n}{dx^n}(\sin x) & \dfrac{d^n}{dx^n}(\cos x) \\ n! & \sin\left(\dfrac{n\pi}{2}\right) & \cos\left(\dfrac{n\pi}{2}\right) \\ a & a^2 & a^3 \end{vmatrix}$$

$$= \begin{vmatrix} n! & \sin\left(\dfrac{n\pi}{2}+x\right) & \cos\left(\dfrac{n\pi}{2}+x\right) \\ n! & \sin\left(\dfrac{n\pi}{2}\right) & \cos\left(\dfrac{n\pi}{2}\right) \\ a & a^2 & a^3 \end{vmatrix}$$

At $x = 0$

$$\frac{d^n}{dx^n}f(x) = \begin{vmatrix} n! & \sin\left(\dfrac{n\pi}{2}\right) & \cos\left(\dfrac{n\pi}{2}\right) \\ n! & \sin\left(\dfrac{n\pi}{2}\right) & \cos\left(\dfrac{n\pi}{2}\right) \\ a & a^2 & a^3 \end{vmatrix}$$

Now $\quad R_1 = R_2$

$$\therefore \quad \frac{d^n}{dx^n}f(x) = 0.$$

Q. 5. If $f(x) = \begin{vmatrix} x^3 & x^4 & 3x^2 \\ 1 & -6 & 4 \\ p & p^2 & p^3 \end{vmatrix}$**, where** p **is a constant, then prove that** $\dfrac{d^3 f(x)}{dx^3}$ **is proportional to** x**.**

Sol. Given, $\quad f(x) = \begin{vmatrix} x^3 & x^4 & 3x^2 \\ 1 & -6 & 4 \\ p & p^2 & p^3 \end{vmatrix} \quad$ ...(i)

where p is a constant.

We know that

$$\text{If } f(x) = \begin{vmatrix} g(x) & p(x) & q(x) \\ h(x) & t(x) & r(x) \\ u(x) & v(x) & w(x) \end{vmatrix}$$

where each element of $f(x)$ is differentiable function of x, then

$$f'(x)=\begin{vmatrix} g'(x) & p'(x) & q'(x) \\ h(x) & t(x) & r(x) \\ u(x) & v(x) & w(x)\end{vmatrix}+\begin{vmatrix} g(x) & p(x) & q(x) \\ h'(x) & t'(x) & r'(x) \\ u(x) & v(x) & w(x)\end{vmatrix}$$

$$+\begin{vmatrix} g(x) & p(x) & q(x) \\ h(x) & t(x) & r(x) \\ u'(x) & v'(x) & w'(x)\end{vmatrix}$$

On differentiating equation (i) both sides with respect to x, we get

$$f'(x)=\begin{vmatrix} 3x^2 & 4x^3 & 6x \\ 1 & -6 & 4 \\ p & p^2 & p^3\end{vmatrix}+\begin{vmatrix} x^3 & x^4 & 3x^2 \\ 0 & 0 & 0 \\ p & p^2 & p^3\end{vmatrix}$$

$$+\begin{vmatrix} x^3 & x^4 & 3x^2 \\ 1 & -6 & 4 \\ 0 & 0 & 0\end{vmatrix}$$

$$=\begin{vmatrix} 3x^2 & 4x^3 & 6x \\ 1 & -6 & 4 \\ p & p^2 & p^3\end{vmatrix}+0+0$$

Similarly $f''(x)=\begin{vmatrix} 6x & 12x^2 & 6 \\ 1 & -6 & 4 \\ p & p^2 & p^3\end{vmatrix}+0+0$

and $f'''(x)=\begin{vmatrix} 6 & 24x & 0 \\ 1 & -6 & 4 \\ p & p^2 & p^3\end{vmatrix}+0+0$

Since $f'''(x)$ contains only x.

$\Rightarrow f''(x)$ is proportional to x.

Q. 6. Find the equations of lines passing through (2, – 1), (4, 0) and (7, 1), (1, 3) using determinant method. Find the point of intersection (if it exists) using matrix method.

Sol. The equation of the line passing through (2, – 1) and (4, 0) is

$$\begin{vmatrix} x & y & 1 \\ 2 & -1 & 1 \\ 4 & 0 & 1\end{vmatrix}=0$$

$\Rightarrow \quad x(-1-0)-y(2-4)+1(0+4)=0$

$\Rightarrow \quad -x+2y+4=0$

$\Rightarrow \quad x-2y=4$

The equation of the line passing through (7, 1) and (1, 3) is

$$\begin{vmatrix} x & y & 1 \\ 7 & 1 & 1 \\ 1 & 3 & 1\end{vmatrix}=0$$

$\Rightarrow \quad x(1-3)-y(7-1)+1(21-1)=0$

$\Rightarrow \quad -2x-6y+20=0$

$\Rightarrow \quad x+3y=10$

$\therefore$ The equations of lines are

$$x-2y=4 \quad \ldots\text{(i)}$$

$$x+3y=10 \quad \ldots\text{(ii)}$$

The above system of equations can be written in the matrix form as

$$\begin{bmatrix} 1 & -2 \\ 1 & 3\end{bmatrix}\begin{bmatrix} x \\ y\end{bmatrix}=\begin{bmatrix} 4 \\ 10\end{bmatrix}$$

or $\quad AX=B,$

where $\quad A=\begin{bmatrix} 1 & -2 \\ 1 & 3\end{bmatrix}, x=\begin{bmatrix} x \\ y\end{bmatrix}$

and $\quad B=\begin{bmatrix} 4 \\ 10\end{bmatrix}$

Now $\quad |A|=\begin{vmatrix} 1 & -2 \\ 1 & 3\end{vmatrix}$

$$=3+2=5\neq 0$$

$\therefore A^{-1}$ can be calculated.

It means that the above system of equations has a unique solution.

$$\text{adj } A=\begin{bmatrix} 3 & -1 \\ 2 & 1\end{bmatrix}'=\begin{bmatrix} 3 & 2 \\ -1 & 1\end{bmatrix}$$

Now, $\quad X=A^{-1}B$

$$=\frac{1}{|A|}\text{adj } A.\,B$$

$$=\frac{1}{5}\begin{bmatrix} 3 & 2 \\ -1 & 1\end{bmatrix}\begin{bmatrix} 4 \\ 10\end{bmatrix}$$

$$=\frac{1}{5}\begin{bmatrix} 12+20 \\ -4+10\end{bmatrix}=\frac{1}{5}\begin{bmatrix} 32 \\ 6\end{bmatrix}$$

$\therefore \quad X=\begin{bmatrix} 32/5 \\ 6/5\end{bmatrix}$

$$\begin{bmatrix} x \\ y\end{bmatrix}=\begin{bmatrix} 32/5 \\ 6/5\end{bmatrix}$$

$\therefore \quad x=\frac{32}{5}$ and $y=\frac{6}{5}$

$\Rightarrow$ Point of intersection of given lines is $\left(\frac{32}{5},\frac{6}{5}\right)$.

Q. 7. If $A = \begin{bmatrix} 2 & -3 & 5 \\ 3 & 2 & -4 \\ 1 & 1 & -2 \end{bmatrix}$, find A^{-1}. Use it to solve the system of equations :★

$$\mathbf{2x - 3y + 5z = 11}$$
$$\mathbf{3x + 2y - 4z = -5}$$
$$\mathbf{x + y - 2z = -3}$$

Sol. The given system of equations are,

$$2x - 3y + 5z = 11$$
$$3x + 2y - 4z = -5$$
$$x + 2y - 2z = -3$$

$$|A| = \begin{bmatrix} 2 & -3 & 5 \\ 3 & 2 & -4 \\ 1 & 1 & -2 \end{bmatrix}$$

$$= 2(-4+4) + 3(-6+4) + 5(3-2)$$

$$|A| = 0 - 6 + 5 = -1 \neq 0$$

$\therefore$ A^{-1} can be calculated

Co-factors of matrix A,

$$C_{11} = (-1)^{1+1}\begin{vmatrix} 2 & -4 \\ 1 & -2 \end{vmatrix} = 0$$

$$C_{12} = (-1)^{1+2}\begin{vmatrix} 3 & -4 \\ 1 & -2 \end{vmatrix} = 2$$

$$C_{13} = (-1)^{1+3}\begin{vmatrix} 3 & 2 \\ 1 & 1 \end{vmatrix} = 1$$

$$C_{21} = (-1)^{2+1}\begin{vmatrix} -3 & 5 \\ 1 & -2 \end{vmatrix} = -1$$

$$C_{22} = (-1)^{2+2}\begin{vmatrix} 2 & 5 \\ 1 & -2 \end{vmatrix} = -9$$

$$C_{23} = (-1)^{2+3}\begin{vmatrix} 2 & -3 \\ 1 & 1 \end{vmatrix} = -5$$

$$C_{31} = (-1)^{3+1}\begin{vmatrix} -3 & 5 \\ 2 & -4 \end{vmatrix} = 2$$

$$C_{32} = (-1)^{3+2}\begin{vmatrix} 2 & 5 \\ 3 & -4 \end{vmatrix} = 23$$

$$C_{33} = (-1)^{3+3}\begin{vmatrix} 2 & -3 \\ 3 & 2 \end{vmatrix} = 13$$

$$\therefore \text{ Matrix of cofactors} = \begin{bmatrix} 0 & 2 & 1 \\ -1 & -9 & -5 \\ 2 & 23 & 13 \end{bmatrix}$$

$$\text{Adj. } A = \begin{bmatrix} 0 & -1 & 2 \\ 2 & -9 & 23 \\ 1 & -5 & 13 \end{bmatrix}$$

★ are board exam questions from previous years

Now,
$$A^{-1} = \frac{\text{Adj } A}{|A|} = \frac{1}{(-1)}\begin{bmatrix} 0 & -1 & 2 \\ 2 & -9 & 23 \\ 1 & -5 & 13 \end{bmatrix}$$

$$= \begin{bmatrix} 0 & 1 & -2 \\ -2 & 9 & -23 \\ -1 & 5 & -13 \end{bmatrix}$$

$$X = A{-}1\,B$$

$$= \begin{bmatrix} 0 & 1 & -2 \\ -2 & 9 & -23 \\ -1 & 5 & -13 \end{bmatrix}\begin{bmatrix} 11 \\ -5 \\ -3 \end{bmatrix}$$

$$\begin{bmatrix} x \\ y \\ z \end{bmatrix} = \begin{bmatrix} 0-5+6 \\ -22-45+69 \\ -11-25+39 \end{bmatrix} = \begin{bmatrix} 1 \\ 2 \\ 3 \end{bmatrix}$$

$$x = 1, y = 2, z = 3.$$

Q. 8. Using matrix method, solve the following system of linear equations :

$$\mathbf{2x + y + z = 1}$$
$$\mathbf{x - 2y - 3z = 1}$$
$$\mathbf{3x + 2y + 4z = 5}$$

Sol. The given system of equations are

$$2x + y + z = 1$$
$$x - 2y - 3z = 1$$
$$3x + 2y + 4z = 5$$

The given system of equations can be represented as

$$\begin{bmatrix} 2 & 1 & 1 \\ 1 & -2 & -3 \\ 3 & 2 & 4 \end{bmatrix}\begin{bmatrix} x \\ y \\ z \end{bmatrix} = \begin{bmatrix} 1 \\ 1 \\ 5 \end{bmatrix}$$

or $\quad AX = B$

where $A = \begin{bmatrix} 2 & 1 & 1 \\ 1 & -2 & -3 \\ 3 & 2 & 4 \end{bmatrix}$, $X = \begin{bmatrix} x \\ y \\ z \end{bmatrix}$, $B = \begin{bmatrix} 1 \\ 1 \\ 5 \end{bmatrix}$

$$|A| = \begin{vmatrix} 2 & 1 & 1 \\ 1 & -2 & -3 \\ 3 & 2 & 4 \end{vmatrix}$$

$$= 2(-8+6) - 1(4+9) + 1(2+6)$$
$$= -4 - 13 + 8 = -9 \neq 0$$

Thus A is a non-singular matrix.

$\therefore$ A^{-1} can be calculated.

Cofactors of matrix A

$$C_{11} = \begin{vmatrix} -2 & -3 \\ 2 & 4 \end{vmatrix} = -2$$

$$C_{12} = -\begin{vmatrix} 1 & -3 \\ 3 & 4 \end{vmatrix} = -13$$

$$C_{13}=\begin{vmatrix}1 & -2\\ 3 & 2\end{vmatrix}=8$$

$$C_{21}=-\begin{vmatrix}1 & 1\\ 2 & 4\end{vmatrix}=-2$$

$$C_{22}=\begin{vmatrix}2 & 1\\ 3 & 4\end{vmatrix}=5$$

$$C_{23}=-\begin{vmatrix}2 & 1\\ 3 & 2\end{vmatrix}=-1$$

$$C_{31}=\begin{vmatrix}1 & 1\\ -2 & -3\end{vmatrix}=-1$$

$$C_{32}=-\begin{vmatrix}2 & 1\\ 1 & -3\end{vmatrix}=7$$

$$C_{33}=\begin{vmatrix}2 & 1\\ 1 & -2\end{vmatrix}=-5$$

$\therefore$ Matrix of cofactors $=\begin{bmatrix}-2 & -13 & 8\\ -2 & 5 & -1\\ -1 & 7 & -5\end{bmatrix}$

$$\therefore \text{ adj } A=\begin{bmatrix}-2 & -13 & 8\\ -2 & 5 & -1\\ -1 & 7 & -5\end{bmatrix}^T=\begin{bmatrix}-2 & -2 & -1\\ -13 & 5 & 7\\ 8 & -1 & -5\end{bmatrix}$$

$$A^{-1}=\frac{1}{|A|}\text{ adj } A=\frac{1}{-9}\begin{bmatrix}-2 & -2 & -1\\ -13 & 5 & 7\\ 8 & -1 & -5\end{bmatrix}$$

$$X=A^{-1}B$$

$$=\frac{1}{-9}\begin{bmatrix}-2 & -2 & -1\\ -13 & 5 & 7\\ 8 & -1 & -5\end{bmatrix}\begin{bmatrix}1\\ 1\\ 5\end{bmatrix}$$

$$=\frac{1}{9}\begin{bmatrix}-2-2-5\\ -13+5+35\\ 8-1-25\end{bmatrix}=\frac{-1}{9}\begin{bmatrix}-9\\ 27\\ -18\end{bmatrix}$$

$$\begin{bmatrix}x\\ y\\ z\end{bmatrix}=\begin{bmatrix}1\\ -3\\ 2\end{bmatrix}$$

On comparing corresponding elements

$x=1, y=-3$ and $z=2$.

Q. 9. Let $A=\begin{bmatrix}1 & -2 & 1\\ -2 & 3 & 1\\ 1 & 1 & 5\end{bmatrix}$, verify that $[\text{adj } A]^{-1}=\text{adj }(A^{-1})$. **[NCERT]**

Sol. Given, $A=\begin{bmatrix}1 & -2 & 1\\ -2 & 3 & 1\\ 1 & 1 & 5\end{bmatrix}$

$$\Rightarrow \quad |A|=\begin{vmatrix}1 & -2 & 1\\ -2 & 3 & 1\\ 1 & 1 & 5\end{vmatrix}$$

$$=1(15-1)-(-2)(-10-1)+1(-2-3)$$

$$=14-22-5$$

$$=-13\neq 0$$

Cofactors of $|A|$ are

$A_{11}=15-1=14,$

$A_{12}=-(-10-1)=11,$

$A_{13}=-2-3=-5,$

$A_{21}=-(-10-1)=11,$

$A_{22}=5-1=4$

$A_{23}=-(1+2)=-3,$

$A_{31}=(-2-3)=-5,$

$A_{32}=-(1+2)=-3$

$A_{33}=3-4=-1$

$$\text{Now, adj }(A)=\begin{bmatrix}14 & 11 & -5\\ 11 & 4 & -3\\ -5 & -3 & -1\end{bmatrix}^T$$

$$=\begin{bmatrix}14 & 11 & -5\\ 11 & 4 & -3\\ -5 & -3 & -1\end{bmatrix}$$

$$\therefore \quad A^{-1}=\frac{1}{|A|}\text{ adj }(A)=-\frac{1}{13}\begin{bmatrix}14 & 11 & -5\\ 11 & 4 & -3\\ -5 & -3 & -1\end{bmatrix}$$

$$=\begin{bmatrix}-\frac{14}{13} & -\frac{11}{13} & \frac{5}{13}\\ -\frac{11}{13} & -\frac{4}{13} & \frac{3}{13}\\ \frac{5}{13} & \frac{3}{13} & \frac{1}{13}\end{bmatrix}$$

$$\text{Here adj }(A)=\begin{bmatrix}14 & 11 & -5\\ 11 & 4 & -3\\ -5 & -3 & -1\end{bmatrix}=B\text{ (Let)}$$

$$\therefore \quad |B|=|\text{adj }(A)|=\begin{vmatrix}14 & 11 & -5\\ 11 & 4 & -3\\ -5 & -3 & -1\end{vmatrix}$$

$$=14(-4-9)-11(-11-15)-5(-33+20)$$

$$=-182+286+65=169\neq 0$$

Cofactors of $|B|$ are

$B_{11}=(-4-9)=-13,$

$B_{12}=-(-11-15)=26,$

$B_{13}=(-33+20)=-13,$

$B_{21}=-(-11-15)=26,$

$B_{22}=-14-25=-39,$

$B_{23}=-(-42+55)=-13,$

$B_{31}=(-33+20)=-13,$

$B_{32}=-(-42+55)=-13,$

$B_{33}=56-121=-65,$

Now, $\text{adj}(B)=\text{adj}(\text{adj}A)=\begin{bmatrix}-13 & 26 & -13\\ 26 & -39 & -13\\ -13 & -13 & -65\end{bmatrix}^T$

$$=\begin{bmatrix}-13 & 26 & -13\\ 26 & -39 & -13\\ -13 & -13 & -65\end{bmatrix}$$

Now, $B^{-1}=(\text{adj}A)^{-1}=\dfrac{1}{|\text{adj}(A)|}\{\text{adj}(\text{adj}A)\}$

$$\Rightarrow\quad (\text{adj}A)^{-1}=\frac{1}{169}\begin{bmatrix}-13 & 26 & -13\\ 26 & -39 & -13\\ -13 & -13 & -65\end{bmatrix}$$

$$=\frac{1}{13}\begin{bmatrix}-1 & 2 & -1\\ 2 & -3 & -1\\ -1 & -1 & -5\end{bmatrix}\qquad \text{...(i)}$$

Cofactors of $|A^{-1}|$ are

$$A_{11}=\frac{-13}{169},\ A_{12}=\frac{26}{169},\ A_{13}=\frac{-13}{169},$$

$$A_{21}=\frac{26}{169},\ A_{22}=\frac{-39}{169},\ A_{23}=\frac{-13}{169}$$

$$A_{31}=-\frac{13}{169},\ A_{32}=-\frac{13}{169},\ A_{33}=-\frac{65}{169}$$

Now, $\text{adj}(A^{-1})=\begin{bmatrix}-\frac{13}{169} & \frac{26}{169} & -\frac{13}{169}\\ \frac{26}{169} & -\frac{39}{169} & -\frac{13}{169}\\ -\frac{13}{169} & -\frac{13}{169} & -\frac{65}{169}\end{bmatrix}^T$

$$=\begin{bmatrix}-\frac{13}{169} & \frac{26}{169} & -\frac{13}{169}\\ \frac{26}{169} & -\frac{39}{169} & -\frac{13}{169}\\ -\frac{13}{169} & -\frac{13}{169} & -\frac{65}{169}\end{bmatrix}$$

$$=\frac{1}{13}\begin{bmatrix}-1 & 2 & -1\\ 2 & -3 & -1\\ -1 & -1 & -5\end{bmatrix}\qquad \text{...(ii)}$$

From equations (i) and (ii), we get

$\text{adj}(A^{-1})=(\text{adj}A)^{-1}.$

Q. 10. Using the property of determinants and without expanding, prove that :★

$$\begin{vmatrix}b+c & q+r & y+z\\ c+a & r+p & z+x\\ a+b & p+q & x+y\end{vmatrix}=2\begin{vmatrix}a & p & x\\ b & q & y\\ c & r & z\end{vmatrix}$$

Sol. Let $\Delta=\begin{vmatrix}b+c & q+r & y+z\\ c+a & r+p & z+x\\ a+b & p+q & x+y\end{vmatrix}$

$$=\begin{vmatrix}b+c & q+r & y+z\\ c+a & r+p & z+x\\ a & p & x\end{vmatrix}+\begin{vmatrix}b+c & q+r & y+z\\ c+a & r+p & z+x\\ b & q & y\end{vmatrix}$$

$$=\Delta_1+\Delta_2\ (\text{say})\qquad \text{...(i)}$$

Now, $\Delta_1=\begin{vmatrix}b+c & q+r & y+z\\ c+a & r+p & z+x\\ a & p & x\end{vmatrix}$

Applying $R_2\to R_2-R_3$, we have

$$\Delta_1=\begin{vmatrix}b+c & q+r & y+z\\ c & r & z\\ a & p & x\end{vmatrix}$$

Applying $R_1\to R_1-R_2$, we have

$$\Delta_1=\begin{vmatrix}b & q & y\\ c & r & z\\ a & p & x\end{vmatrix}$$

Applying $R_1\leftrightarrow R_3$ and $R_2\leftrightarrow R_3$, we have

$$\Delta_1=(-1)^2\begin{vmatrix}a & p & x\\ b & q & y\\ c & r & z\end{vmatrix}=\begin{vmatrix}a & p & x\\ b & q & y\\ c & r & z\end{vmatrix}\qquad \text{...(ii)}$$

$$\Delta_2=\begin{vmatrix}b+c & q+r & y+z\\ c+a & r+p & z+x\\ b & q & y\end{vmatrix}$$

Applying $R_1\to R_1-R_3$, we have

$$\Delta_2=\begin{vmatrix}c & r & z\\ c+a & r+p & z+x\\ b & q & y\end{vmatrix}$$

★ are board exam questions from previous years

Applying $R_2 \to R_2 - R_1$, we have

$$\Delta_2 = \begin{vmatrix} c & r & z \\ a & p & x \\ b & q & y \end{vmatrix}$$

Applying $R_1 \leftrightarrow R_2$ and $R_2 \leftrightarrow R_3$, we have

$$\Delta_2 = (-1)^2 \begin{vmatrix} a & p & x \\ b & q & y \\ c & r & z \end{vmatrix} = \begin{vmatrix} a & p & x \\ b & q & y \\ c & r & z \end{vmatrix} \quad \ldots\text{(iii)}$$

By equations (i), (ii) and (iii), we have

$$\Delta = 2\begin{vmatrix} a & p & x \\ b & q & y \\ c & r & z \end{vmatrix}.$$

Q. 11. If a, b and c are real numbers and

$$\Delta = \begin{vmatrix} b+c & c+a & a+b \\ c+a & a+b & b+c \\ a+b & b+c & c+a \end{vmatrix} = 0$$

Show that either $a+b+c=0$ or $a=b=c$.

[NCERT]

Sol. Let $\Delta = \begin{vmatrix} b+c & c+a & a+b \\ c+a & a+b & b+c \\ a+b & b+c & c+a \end{vmatrix}$

Applying $C_1 \to C_1 + C_2 + C_3$

$$\Delta = \begin{vmatrix} 2(a+b+c) & c+a & a+b \\ 2(a+b+c) & a+b & b+c \\ 2(a+b+c) & b+c & c+a \end{vmatrix}$$

Taking common $2(a+b+c)$ from column C_1

$$= 2(a+b+c)\begin{vmatrix} 1 & c+a & a+b \\ 1 & a+b & b+c \\ 1 & b+c & c+a \end{vmatrix}$$

Applying $R_2 \to R_2 - R_1$ and $R_3 \to R_3 - R_1$

$$= 2(a+b+c)\begin{vmatrix} 1 & c+a & a+b \\ 0 & b-c & c-a \\ 0 & b-a & c-b \end{vmatrix}$$

Expanding along C_1, we get

$$= 2(a+b+c)\,[(b-c)(c-b)-(b-a)(c-a)]$$

$$= 2(a+b+c)(bc-b^2-c^2+bc-bc+ab+ac-a^2)$$

$$\Delta = 2(a+b+c)(ab+bc+ca-a^2-b^2-c^2)$$

It is given that

$$\Delta = 0$$

$$\therefore (a+b+c)\times 2(ab+bc+ca-a^2-b^2-c^2) = 0$$

Either $a+b+c = 0$

or $2[ab+bc+ca-a^2-b^2-c^2] = 0$

$\Rightarrow$ $2a^2+2b^2+2c^2-2ab-2bc-2ca = 0$

$\Rightarrow$ $(a^2+b^2-2ab)+(b^2+c^2-2bc)+(c^2+a^2-2ca) = 0$

$\Rightarrow$ $(a-b)^2+(b-c)^2+(c-a)^2 = 0$

$\Rightarrow$ $(a-b)^2 = (b-c)^2 = (c-a)^2 = 0$

$\Rightarrow$ $(a-b) = (b-c) = (c-a) = 0$

$\Rightarrow$ $a = b = c$

Hence, if $\Delta = 0$, then either $a+b+c=0$

or $a=b=c$.

Q. 12. If $A = \begin{pmatrix} 2 & 3 & 10 \\ 4 & -6 & 5 \\ 6 & 9 & -20 \end{pmatrix}$, find A^{-1}. Using A^{-1} solve the system of equations :★

$$\frac{2}{x}+\frac{3}{y}+\frac{10}{z} = 2;\ \frac{4}{x}-\frac{6}{y}+\frac{5}{z} = 5;\ \frac{6}{x}+\frac{9}{y}-\frac{20}{z} = -4$$

Sol. Given, $A = \begin{bmatrix} 2 & 3 & 10 \\ 4 & -6 & 5 \\ 6 & 9 & -20 \end{bmatrix}$

$$|A| = 2(120-45)-3(-80-30)+10(36+36)$$

$$= 2\times 75 - 3\times(-110)+10\times 72$$

$$= 150+330+720 = 1200$$

$$A_{11} = 75, A_{21} = 150, A_{31} = 75$$

$$A_{12} = 110, A_{22} = -100, A_{32} = 30$$

$$A_{13} = 72, A_{23} = 0, A_{33} = -24$$

$$\text{Cofactor of } A = \begin{bmatrix} 75 & 110 & 72 \\ 150 & -100 & 0 \\ 75 & 30 & -24 \end{bmatrix}$$

$$\text{adj } A = \begin{bmatrix} 75 & 150 & 75 \\ 110 & -100 & 30 \\ 72 & 0 & -24 \end{bmatrix}$$

$$A^{-1} = \frac{\text{adj } A}{|A|}$$

$$= \frac{1}{1200}\begin{bmatrix} 75 & 150 & 75 \\ 110 & -100 & 30 \\ 72 & 0 & -24 \end{bmatrix}$$

Given, system of equations is

$$\frac{2}{x}+\frac{3}{y}+\frac{10}{z} = 2$$

★ are board exam questions from previous years

$$\frac{4}{x}-\frac{6}{y}+\frac{5}{z}=5$$

$$\frac{6}{x}+\frac{9}{y}-\frac{20}{z}=-4$$

Let $\frac{1}{x}=u, \frac{1}{y}=v, \frac{1}{z}=w$

Therefore, the given linear equations becomes

$$2u+3v+10w=2$$
$$4u-6v+5w=5$$
$$6u+9v-20w=-4$$

$$A=\begin{bmatrix}2 & 3 & 10\\4 & -6 & 5\\6 & 9 & -20\end{bmatrix}, X=\begin{bmatrix}u\\v\\w\end{bmatrix}, B=\begin{bmatrix}2\\5\\-4\end{bmatrix}$$

$$X=A^{-1}B$$

$$\begin{bmatrix}u\\v\\w\end{bmatrix}=\begin{bmatrix}2 & 3 & 10\\4 & -6 & 5\\6 & 9 & -20\end{bmatrix}^{-1}\begin{bmatrix}2\\5\\-4\end{bmatrix}$$

$$=\frac{1}{1200}\begin{bmatrix}75 & 150 & 75\\110 & -100 & 30\\72 & 0 & -24\end{bmatrix}\begin{bmatrix}2\\5\\-4\end{bmatrix}$$

$$=\frac{1}{1200}\begin{bmatrix}150+750-300\\220-500-120\\144+0+96\end{bmatrix}$$

$$=\frac{1}{1200}\begin{bmatrix}600\\-400\\240\end{bmatrix}=\begin{bmatrix}1/2\\-1/3\\1/5\end{bmatrix}$$

$u=1/2 \Rightarrow x=2$

$v=-1/3 \Rightarrow y=-3$

and $w=1/5 \Rightarrow z=5$

$\Rightarrow x=2, y=-3$ and $z=5$.

Q. 13. The monthly incomes of Aryan and Babban are in the ratio 3 : 4 and their monthly expenditures are in the ratio 5 : 7. If each saves ₹ 15,000 per month, find their monthly incomes using matrix method. ★

Sol. Let the monthly incomes of Aryan and Babban be $3x$ and $4x$ respectively.

Suppose their monthly expenditures are $5y$ and $7y$ respectively.

Since each saves ₹ 15,000 per month.

Monthly saving of Aryan : $3x-5y=15,000$

Monthly saving of Babban : $4x-7y=15,000$

The above system of equations can be written in the matrix form as follows :

$$\begin{bmatrix}3 & -5\\4 & -7\end{bmatrix}\begin{bmatrix}x\\y\end{bmatrix}=\begin{bmatrix}15000\\15000\end{bmatrix}$$

$$AX=B$$

Where $A=\begin{bmatrix}3 & -5\\4 & -7\end{bmatrix}, X=\begin{bmatrix}x\\y\end{bmatrix}$ and $B=\begin{bmatrix}15000\\15000\end{bmatrix}$

Now, $|A|=\begin{vmatrix}3 & -5\\4 & -7\end{vmatrix}=-21-(-20)=-1\neq 0$

$$\text{Adj } A=\begin{bmatrix}-7 & -4\\5 & 3\end{bmatrix}^T=\begin{bmatrix}-7 & 5\\-4 & 3\end{bmatrix}$$

So, $A^{-1}=\frac{1}{|A|}\text{ adj } A=-1\begin{bmatrix}-7 & 5\\-4 & 3\end{bmatrix}=\begin{bmatrix}7 & -5\\4 & -3\end{bmatrix}$

$\therefore \quad X=A^{-1}B$

$$\Rightarrow \begin{bmatrix}x\\y\end{bmatrix}=\begin{bmatrix}7 & -5\\4 & -3\end{bmatrix}\begin{bmatrix}15000\\15000\end{bmatrix}$$

$$\Rightarrow \begin{bmatrix}x\\y\end{bmatrix}=\begin{bmatrix}105000-75000\\60000-45000\end{bmatrix}$$

$$\Rightarrow \begin{bmatrix}x\\y\end{bmatrix}=\begin{bmatrix}30000\\15000\end{bmatrix}$$

$\Rightarrow x=30,000$ and $y=15,000$

Therefore, monthly income of Aryan = 3 × 30,000 = ₹ 90,000

Monthly income of Babban = 4 × 30,000 = ₹ 1,20,000

Q. 14. A trust invested some money in two type of bonds. The first bond pays 10% interest and second bond pays 12% interest. The trust received ₹ 2,800 at interest. However, if trust had interchanged money in bonds, they would have got ₹ 100 less as interest. Using matrix method, find the amount invested by the trust.★

Sol. Let the amount invested by the trust in first and second bond be x and y respectively.

$$\text{Interest from first bond}=\frac{10\times x\times 1}{100}=\frac{10x}{100}$$

$$\text{Interest from second bond}=\frac{12\times y\times 1}{100}=\frac{12y}{100}$$

Interest received by trust = ₹ 2,800

According to the question,

$$\frac{10x}{100}+\frac{12y}{100}=2{,}800$$

$$10x+12y=2{,}80{,}000 \quad \text{...(i)}$$

and $$\frac{12x}{100}+\frac{10y}{100}=2{,}700$$

$$12x+10y=2{,}70{,}000 \quad \text{...(ii)}$$

This system of equations can be written in matrix form as follows :

$$\begin{bmatrix}10 & 12\\12 & 10\end{bmatrix}\begin{bmatrix}x\\y\end{bmatrix}=\begin{bmatrix}2{,}80{,}000\\2{,}70{,}000\end{bmatrix}$$

★ are board exam questions from previous years

or $AX = B$

where $A = \begin{bmatrix} 10 & 12 \\ 12 & 10 \end{bmatrix}, X = \begin{bmatrix} x \\ y \end{bmatrix}$

and $B = \begin{bmatrix} 2,80,000 \\ 2,70,000 \end{bmatrix}$

Now, $|A| = \begin{vmatrix} 10 & 12 \\ 12 & 10 \end{vmatrix}$

$= 100 - 144 = -44 \neq 0$

So, A^{-1} exists and the solution of the given system of equations is given by

$$X = A^{-1}B$$

Let c_{ij} be the cofactor of a_{ij} in $A = [a_{ij}]$. Then,

$$c_{11} = 10, c_{12} = -12, c_{21} = -12, c_{22} = 10$$

$\therefore$ $\text{adj } A = \begin{bmatrix} 10 & -12 \\ -12 & 10 \end{bmatrix}^T = \begin{bmatrix} 10 & -12 \\ -12 & 10 \end{bmatrix}$

So, $A^{-1} = \dfrac{1}{|A|}(\text{adj } A) = -\dfrac{1}{44}\begin{bmatrix} 10 & -12 \\ -12 & 10 \end{bmatrix}$

Hence, the solution is given by

$$X = A^{-1}B$$

$$= -\frac{1}{44}\begin{bmatrix} 10 & -12 \\ -12 & 10 \end{bmatrix}\begin{bmatrix} 2,80,000 \\ 2,70,000 \end{bmatrix}$$

$$\begin{bmatrix} x \\ y \end{bmatrix} = -\frac{1}{44}\begin{bmatrix} 28,00,000 - 32,40,000 \\ -33,60,000 + 27,00,000 \end{bmatrix}$$

$$\begin{bmatrix} x \\ y \end{bmatrix} = -\frac{1}{44}\begin{bmatrix} -4,40,000 \\ -6,60,000 \end{bmatrix} = \begin{bmatrix} 10,000 \\ 15,000 \end{bmatrix}$$

$\Rightarrow$ $x = 10,000$ and $y = 15,000$

$\Rightarrow$ $A = x + y = 10,000 + 15,000$

$= ₹\,25,000$

Hence, the amount invested by the trust is ₹ 25,000.

Q. 15. Two societies want to award their members on the values of honesty, helpfulness and character. The society 'Bharat' wants to award ₹ *x* each, ₹ *y* each and ₹ *z* each for three respective values to 1, 2 and 3 members respectively with a total award money of ₹ 2,000. Society 'Prem' wants to award ₹ 2,500 to award its 4, 1 and 3 members on the respective values. If the total amount of award for one prize on each value is ₹ 900, using matrices, find the award money for each value.★

Sol. Consider that award money for honesty, helpfulness and character are *x*, *y* and *z* respectively.

So $x + 2y + 3z = 2000$

$4x + y + 3z = 2500$

$x + y + z = 900$

★ are board exam questions from previous years

The given system of equations can be represented in the form of matrix as

$$\begin{bmatrix} 1 & 2 & 3 \\ 4 & 1 & 3 \\ 1 & 1 & 1 \end{bmatrix}\begin{bmatrix} x \\ y \\ z \end{bmatrix} = \begin{bmatrix} 2000 \\ 2500 \\ 900 \end{bmatrix}$$

or $AX = B$ ...(i)

$$|A| = \begin{vmatrix} 1 & 2 & 3 \\ 4 & 1 & 3 \\ 1 & 1 & 1 \end{vmatrix}$$

$= 1(1-3) - 2(4-3) + 3(4-1)$

$= -2 - 2 + 9$

$= 5 \neq 0$

$\therefore$ A^{-1} can be calculated.

Cofactors of matrix *A*

$C_{11} = \begin{vmatrix} 1 & 3 \\ 1 & 1 \end{vmatrix} = -2$

$C_{12} = -\begin{vmatrix} 4 & 3 \\ 1 & 1 \end{vmatrix} = -1$

$C_{13} = \begin{vmatrix} 4 & 1 \\ 1 & 1 \end{vmatrix} = 3$

$C_{21} = -\begin{vmatrix} 2 & 3 \\ 1 & 1 \end{vmatrix} = 1$

$C_{22} = \begin{vmatrix} 1 & 3 \\ 1 & 1 \end{vmatrix} = -2$

$C_{23} = -\begin{vmatrix} 1 & 2 \\ 1 & 1 \end{vmatrix} = 1$

$C_{31} = \begin{vmatrix} 2 & 3 \\ 1 & 3 \end{vmatrix} = 3$

$C_{32} = -\begin{vmatrix} 1 & 3 \\ 4 & 3 \end{vmatrix} = 9$

$C_{33} = \begin{vmatrix} 1 & 2 \\ 4 & 1 \end{vmatrix} = -7$

$\therefore$ Matrix of cofactors of $A = \begin{bmatrix} -2 & -1 & 3 \\ 1 & -2 & 1 \\ 3 & 9 & -7 \end{bmatrix}$

$\Rightarrow \text{adj } A = \begin{bmatrix} -2 & -1 & 3 \\ 1 & -2 & 1 \\ 3 & 9 & -7 \end{bmatrix}^T = \begin{bmatrix} -2 & 1 & 3 \\ -1 & -2 & 9 \\ 3 & 1 & -7 \end{bmatrix}$

By equation (i),

$$X = A^{-1}B$$

$$= \frac{1}{|A|}(\text{adj } A)\,B$$

$$= \frac{1}{5}\begin{bmatrix} -2 & 1 & 3 \\ -1 & -2 & 9 \\ 3 & 1 & -7 \end{bmatrix}\begin{bmatrix} 2000 \\ 2500 \\ 900 \end{bmatrix}$$

$$= \frac{1}{5}\begin{bmatrix} -4000+2500+2700 \\ -2000-5000+8100 \\ 6000+2500-6300 \end{bmatrix}$$

$$\begin{bmatrix} x \\ y \\ z \end{bmatrix} = \frac{1}{5}\begin{bmatrix} 1200 \\ 1100 \\ 2200 \end{bmatrix} = \begin{bmatrix} 240 \\ 220 \\ 440 \end{bmatrix}$$

$\Rightarrow \qquad x = 240, y = 220$ and $z = 440$

i.e., ₹ 240 for honesty, ₹ 220 for helpfulness and ₹ 440 for character.

Q. 16. To promote "Swachh Bharat Abhiyan" three NGO's decided to award their members on the basis of their contribution, approach and promotion of this program. NGO 'Help India' wants to award ₹ 3,000 to award its 1, 1 and 1 members on the respective values. NGO 'Karma' wants to award ₹ 6,000 to 4, 1 and 3 members while NGO 'Dharma' wants to award ₹ 10,000 to 2, 3 and 4 members in the respective values. If ₹ *x*, ₹ *y* and ₹ *z* are respected award money for contribution, approach and promotion, find the award money for each value.

Sol. Let contribution, approach and promotion awards money are ₹ x, ₹ y and ₹ z respectively.

So the system of equations :

$$x + y + z = 3000 \qquad \ldots(i)$$

$$4x + y + 3z = 6000 \qquad \ldots(ii)$$

$$2x + 3y + 4z = 10000 \qquad \ldots(iii)$$

Now the system of equations can be written in matrix form as

$$\begin{bmatrix} 1 & 1 & 1 \\ 4 & 1 & 3 \\ 2 & 3 & 4 \end{bmatrix}\begin{bmatrix} x \\ y \\ z \end{bmatrix} = \begin{bmatrix} 3000 \\ 6000 \\ 10000 \end{bmatrix}$$

or $\qquad AX = B \qquad \ldots(iv)$

where $A = \begin{bmatrix} 1 & 1 & 1 \\ 4 & 1 & 3 \\ 2 & 3 & 4 \end{bmatrix}$, $X = \begin{bmatrix} x \\ y \\ z \end{bmatrix}$ and $B = \begin{bmatrix} 3000 \\ 6000 \\ 10000 \end{bmatrix}$

Now, $\qquad |A| = \begin{vmatrix} 1 & 1 & 1 \\ 4 & 1 & 3 \\ 2 & 3 & 4 \end{vmatrix}$

$$= 1(4-9) - 1(16-6) + 1(12-2)$$

$$= -5 - 10 + 10$$

$$= -5 \neq 0$$

$\therefore$ A is a non-singular matrix.

$\Rightarrow$ A^{-1} can be calculated.

Now, cofactors of A are

$$C_{11} = (-1)^{1+1}\begin{vmatrix} 1 & 3 \\ 3 & 4 \end{vmatrix} = -5$$

$$C_{12} = (-1)^{1+2}\begin{vmatrix} 4 & 3 \\ 2 & 4 \end{vmatrix} = -10$$

$$C_{13} = (-1)^{1+3}\begin{vmatrix} 4 & 1 \\ 2 & 3 \end{vmatrix} = 10$$

$$C_{21} = (-1)^{2+1}\begin{vmatrix} 1 & 1 \\ 3 & 4 \end{vmatrix} = -1$$

$$C_{22} = (-1)^{2+2}\begin{vmatrix} 1 & 1 \\ 2 & 4 \end{vmatrix} = 2$$

$$C_{23} = (-1)^{2+3}\begin{vmatrix} 1 & 1 \\ 2 & 3 \end{vmatrix} = -1$$

$$C_{31} = (-1)^{3+1}\begin{vmatrix} 1 & 1 \\ 1 & 3 \end{vmatrix} = 2$$

$$C_{32} = (-1)^{3+2}\begin{vmatrix} 1 & 1 \\ 4 & 3 \end{vmatrix} = 1$$

$$C_{33} = (-1)^{3+3}\begin{vmatrix} 1 & 1 \\ 4 & 1 \end{vmatrix} = -3$$

$\therefore$ Matrix of cofactors of $A = \begin{bmatrix} -5 & -10 & 10 \\ -1 & 2 & -1 \\ 2 & 1 & -3 \end{bmatrix}$

$$\text{adj } A = \begin{bmatrix} -5 & -10 & 10 \\ -1 & 2 & -1 \\ 2 & 1 & -3 \end{bmatrix}^T = \begin{bmatrix} -5 & -1 & 2 \\ -10 & 2 & 1 \\ 10 & -1 & -3 \end{bmatrix}$$

$$A^{-1} = \frac{1}{|A|} \text{adj } A$$

$$= \frac{-1}{5}\begin{bmatrix} -5 & -1 & 2 \\ -10 & 2 & 1 \\ 10 & -1 & -3 \end{bmatrix}$$

Now, by equation (iv)

$$X = A^{-1}B$$

$$= \frac{-1}{5}\begin{bmatrix} -5 & -1 & 2 \\ -10 & 2 & 1 \\ 10 & -1 & -3 \end{bmatrix}\begin{bmatrix} 3000 \\ 6000 \\ 10000 \end{bmatrix}$$

$$= \frac{-1}{5}\begin{bmatrix} -15000-6000+20000 \\ -30000+12000+10000 \\ 30000-6000-30000 \end{bmatrix}$$

$$= \frac{-1}{5}\begin{bmatrix} -1000 \\ -8000 \\ -6000 \end{bmatrix}$$

$$\begin{bmatrix} x \\ y \\ z \end{bmatrix} = \begin{bmatrix} 200 \\ 1600 \\ 1200 \end{bmatrix}$$

Hence award moneys are ₹ 200, ₹ 1,600 and ₹ 1,200 for contribution, approach and promotion.

Q. 17. In year 2017, 1 packet of tea and 3 packets of sugar together cost ₹ 90. Next year the rate of tea increased by 20% and that of sugar by 15%. So the same amount of tea and sugar cost ₹ 106·5. Find the rates of sugar and tea per packet in year 2017 and next year using matrix method.

Sol. Let the price of tea packet is ₹ x and the price of sugar packet is ₹ y in year 2017.

So the system of equations can be written as

$$x + 3y = 90 \qquad \ldots(i)$$

$$\frac{120x}{100} + \frac{115(3y)}{100} = 106{\cdot}5 \qquad \ldots(ii)$$

$$\Rightarrow \quad 1{\cdot}2x + 3{\cdot}45y = 106{\cdot}5$$

Now, the system of equations can be written in matrix form as :

$$\begin{bmatrix} 1 & 3 \\ 1{\cdot}2 & 3{\cdot}45 \end{bmatrix}\begin{bmatrix} x \\ y \end{bmatrix} = \begin{bmatrix} 90 \\ 106{\cdot}5 \end{bmatrix}$$

or $\qquad AX = B \qquad \ldots(iii)$

where $A = \begin{bmatrix} 1 & 3 \\ 1{\cdot}2 & 3{\cdot}45 \end{bmatrix}$, $X = \begin{bmatrix} x \\ y \end{bmatrix}$ and $B = \begin{bmatrix} 90 \\ 106{\cdot}5 \end{bmatrix}$

Now $\qquad |A| = \begin{vmatrix} 1 & 3 \\ 1{\cdot}2 & 3{\cdot}45 \end{vmatrix}$

$$= 1 \times 3{\cdot}45 - 1{\cdot}2 \times 3$$

$$= 3{\cdot}45 - 3{\cdot}6 = -0{\cdot}15 \neq 0$$

$\therefore$ A^{-1} can be calculated.

Now, cofactors of A are

$$C_{11} = (-1)^{1+1}\, 3{\cdot}45 = 3{\cdot}45$$

$$C_{12} = (-1)^{1+2}\, 1{\cdot}2 = -1{\cdot}2$$

$$C_{21} = (-1)^{2+1}\, 3 = -3$$

$$C_{22} = (-1)^{2+2}\, 1 = 1$$

$\therefore$ Matrix of cofactors of $A = \begin{bmatrix} 3{\cdot}45 & -1{\cdot}2 \\ -3 & 1 \end{bmatrix}$

$\therefore \qquad \text{adj } A = \begin{bmatrix} 3{\cdot}45 & -1{\cdot}2 \\ -3 & 1 \end{bmatrix}^T = \begin{bmatrix} 3{\cdot}45 & -3 \\ -1{\cdot}2 & 1 \end{bmatrix}$

Now, we know that

$$A^{-1} = \frac{1}{|A|}\text{ adj } A$$

$$= \frac{1}{-0{\cdot}15}\begin{bmatrix} 3{\cdot}45 & -3 \\ -1{\cdot}2 & 1 \end{bmatrix}$$

Now by equation (iii)

$$X = A^{-1}B$$

$$= \frac{-1}{0{\cdot}15}\begin{bmatrix} 3{\cdot}45 & -3 \\ -1{\cdot}2 & 1 \end{bmatrix}\begin{bmatrix} 90 \\ 106{\cdot}5 \end{bmatrix}$$

$$= \frac{-1}{0{\cdot}15}\begin{bmatrix} 3{\cdot}45 \times 90 - 3 \times 106{\cdot}5 \\ -1{\cdot}2 \times 90 + 1 \times 106{\cdot}5 \end{bmatrix}$$

$$= \frac{-1}{0{\cdot}15}\begin{bmatrix} 310{\cdot}5 - 319{\cdot}5 \\ -108 + 106{\cdot}5 \end{bmatrix}$$

$$= \frac{-1}{0{\cdot}15}\begin{bmatrix} -9 \\ -1{\cdot}5 \end{bmatrix}$$

$$\begin{bmatrix} x \\ y \end{bmatrix} = \begin{bmatrix} 60 \\ 10 \end{bmatrix}$$

Hence $\qquad x = ₹\, 60, y = ₹\, 10$

So price of one tea pocket is ₹ 60 and price of one sugar packet is ₹ 10.

Q. 18. To promote "Beti Bachao Beti Padhao" program an NGO awards those who take any of the following steps additionally. From the table given below, form a set of simultaneous equations and solve using matrix method to find the amount given exactly to each step.

S. No.	Step	No. of people in Delhi	No. of people in Bangalore	No. of people in Chennai
1.	Save Girl Child	2	1	2
2.	Girls Education	1	2	3
3.	Adult literacy	3	2	1
	Amount received	₹ 7,000	₹ 5,000	₹ 5,000

Sol. Let the award money to save girl child is ₹ x, for girls education is ₹ y and for adult literacy is ₹ z.

So system of equations

$$2x + y + 3z = 7000 \qquad \ldots(i)$$

$$x + 2y + 2z = 5000 \qquad \ldots(ii)$$

$$2x + 3y + z = 5000 \qquad \ldots(iii)$$

The given system of equations can be written in matrix form as

$$\begin{bmatrix} 2 & 1 & 3 \\ 1 & 2 & 2 \\ 2 & 3 & 1 \end{bmatrix}\begin{bmatrix} x \\ y \\ z \end{bmatrix} = \begin{bmatrix} 7000 \\ 5000 \\ 5000 \end{bmatrix}$$

or $AX = B$

$$X = A^{-1}B \quad \ldots\text{(iv)}$$

where $A = \begin{bmatrix} 2 & 1 & 3 \\ 1 & 2 & 2 \\ 2 & 3 & 1 \end{bmatrix}$, $X = \begin{bmatrix} x \\ y \\ z \end{bmatrix}$ and $B = \begin{bmatrix} 7000 \\ 5000 \\ 5000 \end{bmatrix}$

Now, $|A| = \begin{vmatrix} 2 & 1 & 3 \\ 1 & 2 & 2 \\ 2 & 3 & 1 \end{vmatrix}$

$= 2(2-6) - 1(1-4) + 3(3-4)$

$= 2 \times (-4) - 1(-3) + 3(-1)$

$= -8 \neq 0.$

$\therefore A^{-1}$ is a non-singular matrix.

$\therefore A^{-1}$ can be calculated.

Cofactors of A

$$C_{11} = (-1)^{1+1} \begin{vmatrix} 2 & 2 \\ 3 & 1 \end{vmatrix} = -4$$

$$C_{12} = (-1)^{1+2} \begin{vmatrix} 1 & 2 \\ 2 & 1 \end{vmatrix} = 3$$

$$C_{13} = (-1)^{1+3} \begin{vmatrix} 1 & 2 \\ 2 & 3 \end{vmatrix} = -1$$

$$C_{21} = (-1)^{2+1} \begin{vmatrix} 1 & 3 \\ 3 & 1 \end{vmatrix} = 8$$

$$C_{22} = (-1)^{2+2} \begin{vmatrix} 2 & 3 \\ 2 & 1 \end{vmatrix} = -4$$

$$C_{23} = (-1)^{2+3} \begin{vmatrix} 2 & 1 \\ 2 & 3 \end{vmatrix} = -4$$

$$C_{31} = (-1)^{3+1} \begin{vmatrix} 1 & 3 \\ 2 & 2 \end{vmatrix} = -4$$

$$C_{32} = (-1)^{3+2} \begin{vmatrix} 2 & 3 \\ 1 & 2 \end{vmatrix} = -1$$

$$C_{33} = (-1)^{3+1} \begin{vmatrix} 2 & 1 \\ 1 & 2 \end{vmatrix} = 3$$

$$\therefore \text{Matrix of cofactors} = \begin{bmatrix} -4 & 3 & -1 \\ 8 & -4 & -4 \\ -4 & -1 & 3 \end{bmatrix}$$

$$\text{adj } A = \begin{bmatrix} -4 & 3 & -1 \\ 8 & -4 & -4 \\ -4 & -1 & 3 \end{bmatrix}^T = \begin{bmatrix} -4 & 8 & -4 \\ 3 & -4 & -1 \\ -1 & -4 & 3 \end{bmatrix}$$

Now by equation (iv)

$$X = A^{-1}B$$

$$= \frac{1}{|A|} (\text{adj } A)\, B$$

$$= \frac{1}{-8} \begin{bmatrix} -4 & 8 & -4 \\ 3 & -4 & -1 \\ -1 & -4 & 3 \end{bmatrix} \begin{bmatrix} 7000 \\ 5000 \\ 5000 \end{bmatrix}$$

$$= \frac{-1}{8} \begin{bmatrix} -4 \times 7000 + 8 \times 5000 - 4 \times 5000 \\ 3 \times 7000 - 4 \times 5000 - 1 \times 5000 \\ -1 \times 7000 - 4 \times 5000 + 3 \times 5000 \end{bmatrix}$$

$$= \frac{-1}{8} \begin{bmatrix} -28000 + 40000 - 20000 \\ 21000 - 20000 - 5000 \\ -7000 - 20000 + 15000 \end{bmatrix}$$

$$= \frac{-1}{8} \begin{bmatrix} -8000 \\ -4000 \\ -12000 \end{bmatrix}$$

$$\begin{bmatrix} x \\ y \\ z \end{bmatrix} = \begin{bmatrix} 1000 \\ 500 \\ 1500 \end{bmatrix}$$

$\Rightarrow \quad x = ₹\,1000, y = ₹\,500$ and $z = ₹\,1500$

Q. 19. In a legislative assembly election, a political group hired a public relations firm to promote its candidate in 3 ways namely posters, pamphlets and public addressing system.

The cost per contract

$$A = \begin{bmatrix} \mathbf{200} \\ \mathbf{300} \\ \mathbf{100} \end{bmatrix} \begin{matrix} \textbf{Posters} \\ \textbf{Pamphlets} \\ \textbf{Public addressing system} \end{matrix}$$

Number of contracts of each type made by a city is given as

$$\boldsymbol{B = [200 \;\; 500 \;\; 1000]}$$

Find the amount spent by the city on these ways.

Sol. Given, the cost per contract

$$A = \begin{bmatrix} 200 \\ 300 \\ 100 \end{bmatrix} \begin{matrix} \text{Posters} \\ \text{Pamphlets} \\ \text{Public addressing system} \end{matrix}$$

Number of contracts $B = [200 \quad 500 \quad 1000]$

Total amount spend on each way

$$= AB$$

$$= \begin{bmatrix} 200 \\ 300 \\ 100 \end{bmatrix} [200 \;\; 500 \;\; 1000]$$

Since number of columns of A = Number of rows of $B = 1$

$\therefore$ Cost $= 200 \times 200 + 300 \times 500 + 100 \times 1000$

$= 40000 + 150000 + 100000$

$= 290000$

Hence, amount spent by city is ₹ 2,90,000.

∴ Cost for postors, pamphlets and public addressing system are ₹ 40000, ₹ 150000 and ₹ 100000 respectively.

Q. 20. A total amount of ₹ 7,000 is deposited in three different savings bank accounts with annual interest rate of 5%, 8% and 8½% respectively. The total annual interest from these three accounts is ₹ 550. Equal amounts have been deposited in the 5% and 8% savings accounts. Find the amount deposited in each of the three accounts, with the help of matrices.★

Sol. Let the amount deposited in three accounts be ₹ x, ₹ y and ₹ z.

Then according to the given condition,

$$x + y + z = 7000 \quad \ldots(i)$$

$$\frac{5x}{100} + \frac{8y}{100} + \frac{17z}{200} = 550$$

$$\Rightarrow \quad 5x + 8y + 8{\cdot}5z = 55000 \quad \ldots(ii)$$

and $$x = y \text{ or } x - y = 0 \quad \ldots(iii)$$

The system of equations can be written in matrix form as

$$AX = B \quad \ldots(iv)$$

where $A = \begin{bmatrix} 1 & 1 & 1 \\ 5 & 8 & 8{\cdot}5 \\ 1 & -1 & 0 \end{bmatrix}$, $X = \begin{bmatrix} x \\ y \\ z \end{bmatrix}$ and $B = \begin{bmatrix} 7000 \\ 55000 \\ 0 \end{bmatrix}$

Now, $$|A| = \begin{vmatrix} 1 & 1 & 1 \\ 5 & 8 & 8{\cdot}5 \\ 1 & -1 & 0 \end{vmatrix}$$

$$= 1(0 + 8{\cdot}5) - 1(0 - 8{\cdot}5) + 1(-5 - 8)$$

$$= 8{\cdot}5 + 8{\cdot}5 - 13$$

$$= 4 \neq 0$$

∴ A is a non-singular matrix.

∴ A^{-1} can be calculated.

Now cofactors of matrix A are

$$C_{11} = \begin{vmatrix} 8 & 8{\cdot}5 \\ -1 & 0 \end{vmatrix} = 8{\cdot}5$$

$$C_{12} = -\begin{vmatrix} 5 & 8{\cdot}5 \\ 1 & 0 \end{vmatrix} = 8{\cdot}5$$

$$C_{13} = \begin{vmatrix} 5 & 8 \\ 1 & -1 \end{vmatrix} = -13$$

$$C_{21} = -\begin{vmatrix} 1 & 1 \\ -1 & 0 \end{vmatrix} = -1$$

$$C_{22} = \begin{vmatrix} 1 & 1 \\ 1 & 0 \end{vmatrix} = -1$$

$$C_{23} = -\begin{vmatrix} 1 & 1 \\ 1 & -1 \end{vmatrix} = 2$$

$$C_{31} = \begin{vmatrix} 1 & 1 \\ 8 & 8{\cdot}5 \end{vmatrix} = 0{\cdot}5$$

$$C_{32} = -\begin{vmatrix} 1 & 1 \\ 5 & 8{\cdot}5 \end{vmatrix} = -3{\cdot}5$$

$$C_{33} = \begin{vmatrix} 1 & 1 \\ 5 & 8 \end{vmatrix} = 3$$

$$\therefore \quad \text{adj}(A) = \begin{bmatrix} 8{\cdot}5 & 8{\cdot}5 & -13 \\ -1 & -1 & 2 \\ 0{\cdot}5 & -3{\cdot}5 & 3 \end{bmatrix}^T$$

$$= \begin{bmatrix} 8{\cdot}5 & -1 & 0{\cdot}5 \\ 8{\cdot}5 & -1 & -3{\cdot}5 \\ -13 & 2 & 3 \end{bmatrix}$$

$$\therefore \quad A^{-1} = \frac{1}{|A|}\text{adj}\, A = \frac{1}{4}\begin{bmatrix} 8{\cdot}5 & -1 & 0{\cdot}5 \\ 8{\cdot}5 & -1 & -3{\cdot}5 \\ -13 & 2 & 3 \end{bmatrix}$$

Now, equation (iv) can be written as

$$X = A^{-1}B$$

$$\begin{bmatrix} x \\ y \\ z \end{bmatrix} = \frac{1}{4}\begin{bmatrix} 8{\cdot}5 & -1 & 0{\cdot}5 \\ 8{\cdot}5 & -1 & -3{\cdot}5 \\ -13 & 2 & 3 \end{bmatrix}\begin{bmatrix} 7000 \\ 55000 \\ 0 \end{bmatrix}$$

$$= \frac{1}{4}\begin{bmatrix} 59500 - 55000 + 0 \\ 59500 - 55000 + 0 \\ -91000 + 110000 + 0 \end{bmatrix}$$

$$= \frac{1}{4}\begin{bmatrix} 4500 \\ 4500 \\ 19000 \end{bmatrix} = \begin{bmatrix} 1125 \\ 1125 \\ 4750 \end{bmatrix}$$

On comparing the corresponding elements, we get

$x =$ ₹ 1125, $y =$ ₹ 1125 and $z =$ ₹ 4750.

Q. 21. The management committee of a residential colony decided to award some of its members (say x) for honesty, some (say y) for cooperating (helping others) and some others (say z) for supervising the workers to keep the colony neat and clean. The sum of all the awardees is 12. Three times the sum of awardees for cooperation and supervision added to two times the number of awardees for honesty is 33. If the sum of the number of awardees of honesty and supervision is twice the number of awardess for helping others, using matrix method, find the number of awardees of each category.★

Sol. Let number of members for honesty, helping others and supervising the workers be x, y and z.

★ are board exam questions from previous years

Then according to given conditions,

$$x+y+z=12 \quad \ldots(i)$$

$$2x+3y+3z=33 \quad \ldots(ii)$$

$$x-2y+z=0 \quad \ldots(iii)$$

The system of equations can be written in form of matrix as

$$AX=B$$

$$\begin{bmatrix}1 & 1 & 1\\2 & 3 & 3\\1 & -2 & 1\end{bmatrix}\begin{bmatrix}x\\y\\z\end{bmatrix}=\begin{bmatrix}12\\33\\0\end{bmatrix}$$

where $A=\begin{bmatrix}1 & 1 & 1\\2 & 3 & 3\\1 & -2 & 1\end{bmatrix}$, $X=\begin{bmatrix}x\\y\\z\end{bmatrix}$ and $B=\begin{bmatrix}12\\33\\0\end{bmatrix}$

Now, $|A|=\begin{vmatrix}1 & 1 & 1\\2 & 3 & 3\\1 & -2 & 1\end{vmatrix}$

$$=1(3+6)-1(2-3)+1(-4-3)$$

$$=9+1-7$$

$$=3\neq 0$$

$\therefore$ A is a non-singular matrix.

$\therefore$ A^{-1} can be calculated.

Cofactors of matrix A

$$C_{11}=(-1)^{1+1}\begin{vmatrix}3 & 3\\-2 & 1\end{vmatrix}=9$$

$$C_{12}=(-1)^{1+2}\begin{vmatrix}2 & 3\\1 & 1\end{vmatrix}=1$$

$$C_{13}=(-1)^{1+3}\begin{vmatrix}2 & 3\\1 & -2\end{vmatrix}=-7$$

$$C_{21}=(-1)^{2+1}\begin{vmatrix}1 & 1\\-2 & 1\end{vmatrix}=-3$$

$$C_{22}=(-1)^{2+2}\begin{vmatrix}1 & 1\\1 & 1\end{vmatrix}=0$$

$$C_{23}=(-1)^{2+3}\begin{vmatrix}1 & 1\\1 & -2\end{vmatrix}=3$$

$$C_{31}=(-1)^{3+1}\begin{vmatrix}1 & 1\\3 & 3\end{vmatrix}=0$$

$$C_{32}=(-1)^{3+2}\begin{vmatrix}1 & 1\\2 & 3\end{vmatrix}=-1$$

$$C_{33}=(-1)^{3+3}\begin{vmatrix}1 & 1\\2 & 3\end{vmatrix}=1$$

$\therefore$ Matrix of cofactors of $A=\begin{bmatrix}9 & 1 & -7\\-3 & 0 & 3\\0 & -1 & 1\end{bmatrix}$

$$\therefore \text{adj } A=\begin{bmatrix}9 & 1 & -7\\-3 & 0 & 3\\0 & -1 & 1\end{bmatrix}^T=\begin{bmatrix}9 & -3 & 0\\1 & 0 & -1\\-7 & 3 & 1\end{bmatrix}$$

$$A^{-1}=\frac{1}{|A|}\text{adj } A$$

$$=\frac{1}{3}\begin{bmatrix}9 & -3 & 0\\1 & 0 & -1\\-7 & 3 & 1\end{bmatrix}$$

By equation (iv)

$$X=A^{-1}B$$

$$=\frac{1}{3}\begin{bmatrix}9 & -3 & 0\\1 & 0 & -1\\-7 & 3 & 1\end{bmatrix}\begin{bmatrix}12\\33\\0\end{bmatrix}$$

$$=\frac{1}{3}\begin{bmatrix}9\times12-3\times33+0\times0\\1\times12+0\times33-1\times0\\-7\times12+3\times33+1\times0\end{bmatrix}$$

$$=\frac{1}{3}\begin{bmatrix}108-99+0\\12+0-0\\-84+99+0\end{bmatrix}=\frac{1}{3}\begin{bmatrix}9\\12\\15\end{bmatrix}$$

$$\begin{bmatrix}x\\y\\z\end{bmatrix}=\begin{bmatrix}3\\4\\5\end{bmatrix}$$

So $\quad x=3, y=4$ and $z=5$

Thus number of awards for honesty, helping others and supervision are respectively 3, 4 and 5.

Chapter 5. Continuity and Differentiability

Q. 1. In each of the following, verify the hypothesis of Rolle's theorem and find a point in the interval where the derivative vanishes.

(i) $f(x)=x^2-4$ on the interval $[-2, 2]$

(ii) $g(x)=x^2-3x+5$ on the interval $[1, 2]$

(iii) $h(x)=e^{1-x^2}$ on $[1, 1]$

Sol. (i) The given function f is a polynomial and hence is continuous as well as differentiable everywhere.

In particular, f is continuous on $[-2, 2]$ and differentiable on $(-2, 2)$.

Also, $\quad f(-2)=(-2)^2-4=4-4=2^2-4=f(2)$.

Therefore, f satisfies all the hypothesis of Rolle's theorem.

Now, to find a point in $(-2, 2)$, we equate the first derivative of f to zero.

This implies, $f'(x) = 2x = 0 \Rightarrow x = 0$

Thus, the point $c = 0$ lies in $(-2, 2)$ and $f'(0) = 0$.

(ii) Again, g is polynomial and satisfies the conditions of Rolle's theorem.

$$g'(x) = 2x - 3 = 0 \text{ gives } x = 3/2$$

As 3/2 lies in the given interval, we have found the required point is (3/2, 1/2).

(iii) The function $(1 - x^2)$ as well as the exponential function e^{1-x^2} are continuous and differentiable on R.

Hence, the composite function h is also continuous and differentiable on R. Therefore, it satisfies all the conditions of Rolle's theorem.

Now, $\frac{d}{dx}(h(x)) = e^{1-x^2} \times \frac{d}{dx}(1-x^2) = e^{1-x^2} \times (-2x)$

Equating this to 0, we get

$$-2xe^{1-x^2} = 0$$

As the exponential never vanishes, we must have $x = 0$

Thus, $h'(x)$ vanishes at $x = 0$, which belongs to the given interval as required, hence the point where the derivative vanishes is $(0, e)$.

Q. 2. (i) Find $\frac{dy}{dx}$, when $x = a(t + \sin t)$ and $y = a(1 - \cos t)$.

(ii) If $y = \sin^{-1}\left(\frac{2x}{1+x^2}\right) + \sec^{-1}\left(\frac{1+x^2}{1-x^2}\right)$, prove that $\frac{dy}{dx} = \frac{4}{1+x^2}$

Sol. (i) We have :

$$x = a(t + \sin t)$$

$$y = a(1 - \cos t)$$

$$\therefore \quad \frac{dx}{dt} = a(1 + \cos t)$$

$$\text{and} \quad \frac{dy}{dt} = a \sin t$$

$$\therefore \quad \frac{dy}{dx} = \frac{dy}{dt} \times \frac{dt}{dx}$$

$$= \frac{a \sin t}{a(1 + \cos t)}$$

$$= \left(\frac{2 \sin \frac{t}{2} \cos \frac{t}{2}}{2 \cos^2 \frac{t}{2}}\right)$$

$$\left[\because \sin t = 2 \sin \frac{t}{2} \cos \frac{t}{2} \text{ and } (1 + \cos t) = 2 \cos^2 \frac{t}{2}\right]$$

$$= \tan \frac{t}{2}$$

(ii) We have

$$y = \sin^{-1}\left(\frac{2x}{1+x^2}\right) + \sec^{-1}\left(\frac{1+x^2}{1-x^2}\right)$$

$$= \sin^{-1}\left(\frac{2x}{1+x^2}\right) + \cos^{-1}\left(\frac{1-x^2}{1+x^2}\right)$$

Put $\quad x = \tan \theta \Rightarrow \theta = \tan^{-1} x$

$$\therefore \quad y = \sin^{-1}\left(\frac{2 \tan \theta}{1 + \tan^2 \theta}\right) + \cos^{-1}\left(\frac{1 - \tan^2 \theta}{1 + \tan^2 \theta}\right)$$

$$= \sin^{-1}(\sin 2\theta) + \cos^{-1}(\cos 2\theta)$$

$$= 2\theta + 2\theta = 4\theta = 4 \tan^{-1} x$$

$$\therefore \quad \frac{dy}{dx} = \frac{d}{dx}(4 \tan^{-1} x) = \frac{4}{1+x^2}$$

Q. 3. Verify Rolle's theorem for $f(x) = |x|$ in $[-1, 1]$.

Sol. We have

$$f(x) = x = \begin{cases} -x, & -1 \le x \le 0 \\ x, & 0 \le x \le 1 \end{cases}$$

$f(x)$ being a polynomial function is continuous as well as derivable for all $x < 0$ and for all $x > 0$, except possibly at $x = 0$.

So, consider the point $x = 0$.

We have $\quad \lim_{x \to 0^-} f(x) = \lim_{x \to 0} (-x) = 0$

and $\quad \lim_{x \to 0^+} f(x) = \lim_{x \to 0} x = 0$

Therefore, $\quad \lim_{x \to 0^-} f(x) = \lim_{x \to 0^+} f(x) = f(0)$

Thus, $f(x)$ is continuous at $x = 0$.

Hence, $f(x)$ is continuous in the interval $[-1, 1]$.

Now, L.H.D. (at $x = 0$)

$$= \lim_{x \to 0^-} \frac{f(x) - f(0)}{x - 0}$$

$$= \lim_{x \to 0} \frac{-x - 0}{x}$$

$$[\because f(x) = -x \text{ for } x < 0 \text{ and } f(0) = 0]$$

$$= \lim_{x\to 0} \frac{-x}{x} = \lim_{x\to 0} -1 = -1$$

and R.H.D. (at $x = 0$),

$$= \lim_{x\to 0^+} \frac{f(x)-f(0)}{x-0}$$

$$= \lim_{x\to 0} \frac{x-0}{x-0} = \lim_{x\to 0} \frac{x}{x} = 1$$

Therefore, L.H.D. $\neq$ R.H.D.

$\Rightarrow f(x)$ is not derivable at $x = 0 \in (-1, 1)$.

Hence, Rolle's theorem is not applicable to $f(x) = |x|$ in the interval $[-1, 1]$.

Q. 4. Using Lagrange's Mean Value Theorem, find a point on the curve $y = \sqrt{x-2}$ defined in the interval [2, 3], where the tangent is parallel to the chord joining the end points of the given interval on the curve.

Sol. Let $f(x) = \sqrt{x-2}$

Since for each $x \in [2, 3]$, the function $f(x)$ attains a unique definite value, so $f(x)$ is continuous in [2, 3].

Also, $f'(x) = \dfrac{1}{2\sqrt{x-2}}$ exist for all $x \in (2, 3)$

So, $f(x)$ is differentiable on (2, 3).

Consequently, there must exist one $c \in (2, 3)$ such that

$$f'(c) = \frac{f(3)-f(2)}{3-2}$$

Now, $f(x) = \sqrt{x-2}$

$$\Rightarrow \quad f'(x) = \frac{1}{2\sqrt{x-2}}$$

Also, $f(3) = 1, f(2) = 0$

$$\Rightarrow \quad f'(c) = \frac{f(3)-f(2)}{3-2}$$

$$\Rightarrow \quad \frac{1}{2\sqrt{c-2}} = \frac{1-0}{3-2}$$

$$\Rightarrow \quad \frac{1}{2\sqrt{c-2}} = 1$$

$$\Rightarrow \quad \frac{1}{\sqrt{c-2}} = 2$$

$$\Rightarrow \quad 1 = 4(c-2)$$

$$\Rightarrow \quad \frac{1}{4} = c-2$$

$$\Rightarrow \quad c = 2 + \frac{1}{4}$$

$$\Rightarrow \quad c = \frac{9}{4}$$

Thus, $c = \dfrac{9}{4} \in (2, 3)$

Now, $f(c) = \sqrt{\dfrac{9}{4} - 2} = \dfrac{1}{2}$

Thus, $(c, f(c))$ *i.e.*, (9/4, 1/2) is a point on the curve $y = \sqrt{x-2}$ such that the tangent at it is parallel to the chord joining the end points of the curve.

Q. 5. If $y = (x+\sqrt{x^2-1})^m$, prove that :

$$(x^2-1)\left(\frac{dy}{dx}\right)^2 = m^2y^2.$$

Sol. Given, $y = (x+\sqrt{x^2-1})^m$

Let $t = x + \sqrt{x^2-1}$

$\Rightarrow \quad y = t^m$

Differentiating w.r.t. x, we get

$$\frac{dy}{dt} = mt^{m-1} \qquad \text{...(i)}$$

$\because \quad t = x + \sqrt{x^2-1}$

Differentiating w.r.t. x, we get

$$\frac{dt}{dx} = 1 + \frac{1}{2} \times \frac{2x}{\sqrt{x^2-1}}$$

$$= 1 + \frac{x}{\sqrt{x^2-1}}$$

$$= \frac{\sqrt{x^2-1}+x}{\sqrt{x^2-1}}$$

$$\therefore \quad \frac{dy}{dx} = \frac{dy}{dt} \times \frac{dt}{dx}$$

$$= mt^{m-1} \times \frac{\sqrt{x^2-1}+x}{\sqrt{x^2-1}}$$

$$= m.(x+\sqrt{x^2-1})^{m-1} . \frac{(x+\sqrt{x^2-1})}{\sqrt{x^2-1}}$$

$$= m\frac{(x+\sqrt{x^2-1})^m}{\sqrt{x^2-1}} = \frac{my}{\sqrt{x^2-1}}$$

Squaring both sides, we get

$$\left(\frac{dy}{dx}\right)^2 = \frac{m^2y^2}{x^2-1}$$

$$\Rightarrow \quad (x^2-1)\cdot\left(\frac{dy}{dx}\right)^2 = m^2y^2.$$

Q. 6. It is given that Rolle's theorem holds good for the function $f(x) = x^3 + ax^2 + bx, x \in [1, 2]$ at the point $x = 4/3$. Find the values of a and b.

Sol. Given,

$$f(x) = x^3 + ax^2 + bx$$

Now, $f(x)$ is a polynomial.

So, it is continuous on R *i.e.*, on [1, 2].

$$\therefore \quad f'(x) = 3x^2 + 2ax + b$$

which exists in the interval (1, 2).

$\therefore f(x)$ is derivable on (1, 2).

$$\therefore \quad f(1) = (1)^3 + a(1)^2 + b(1) = 1 + a + b$$

and

$$f(2) = (2)^3 + a(2)^2 + b(2) = 8 + 4a + 2b$$

It is given that Rolle's theorem holds for $f(x)$ on [1, 2]

then

$$f(1) = f(2)$$

$$1 + a + b = 8 + 4a + 2b$$

$$3a + b + 7 = 0 \quad \text{...(i)}$$

Since, Rolle's theorem holds at point $x = \frac{4}{3}$

$$\Rightarrow \quad f'(c) = 0$$

$$3\left(\frac{4}{3}\right)^2 + 2a\left(\frac{4}{3}\right) + b = 0$$

$$\frac{16}{3} + \frac{8a}{3} + b = 0$$

$$16 + 8a + 3b = 0 \quad \text{...(ii)}$$

Solving (i) and (ii), we get

$$a = -5 \text{ and } b = 8.$$

Q. 7. Find the values of a and b such that the following function $f(x)$ is a continuous function.

$$\begin{cases} 5, & x \le 2 \\ ax + b, & 2 < x < 10 \\ 21, & x \ge 10 \end{cases}$$

Sol. Given the function is continuous. For the continuity at $x = 2$ and $x = 10$.

Continuity at $x = 2$:

$$\text{L.H.L.} = \lim_{x\to 2^-} f(x) = \lim_{x\to 2} 5 = 5$$

$$\text{R.H.L.} = \lim_{x\to 2^+} f(x) = \lim_{x\to 2^+} (ax+b)$$

$$= \lim_{h\to 0} [a(2+h)+b] \quad [\because x = 2+h]$$

$$= 2a + b$$

Since, $f(x)$ is continuous at $x = 2$.

So, L.H.L. = R.H.L. = $f(2)$

$$\therefore \quad 2a + b = 5 \quad \text{...(i)}$$

Continuity at $x = 10$

$$\text{L.H.L.} = \lim_{x\to 10^-} f(x) = \lim_{x\to 10^-} (ax+b)$$

$$= \lim_{h\to 0} [a(10-h)+b] \quad [\because x = 10-h]$$

$$= 10a + b$$

$$\text{R.H.L.} = \lim_{x\to 10^+} f(x) = \lim_{x\to 10^+} 21 = 21$$

Since, $f(x)$ is continuous at $x = 10$.

So, L.H.L. = R.H.L. = $f(10)$

$$10a + b = 21 \quad \text{...(ii)}$$

Solving (i) and (ii), we get

$$a = 2 \text{ and } b = 1.$$

Chapter 6. Applications of Derivatives

Q. 1. Find the equation of all lines of slope – 1 that are tangents to the curve $y = \frac{1}{x-1}$, $x \neq 1$.

Sol. Let (x_1, y_1) be the point of contact of a line of slope – 1 which touch the curve, $y = \frac{1}{x-1}$.

On differentiating with respect to x, we get

$$\frac{dy}{dx} = -\frac{1}{(x-1)^2}$$

Value of $\frac{dy}{dx}$ at point (x_1, y_1) is

$$\left(\frac{dy}{dx}\right)_{(x_1, y_1)} = -1$$

$$\Rightarrow \quad -\frac{1}{(x_1-1)^2} = -1$$

$\Rightarrow \quad (x_1-1)^2=1$

$\Rightarrow \quad x_1-1=\pm 1$

$\Rightarrow \quad x_1=0, x_1=2$

Since (x_1, y_1) lies on the curve $y=\dfrac{1}{x-1}$

Therefore, $\quad y_1=\dfrac{1}{x_1-1}$

Now, $\quad x_1=0 \Rightarrow y_1=\dfrac{1}{-1}=-1$

and $\quad x_1=2 \Rightarrow y_1=\dfrac{1}{2-1}=1$

Thus, the coordinates of the points of contact are $(0,-1)$ and $(2, 1)$.

The equations of the tangent at $(0,-1)$ and $(2, 1)$ are respectively.

$$y+1=-1(x-0) \text{ and } (y-1)=-1(x-2)$$

$\Rightarrow \quad x+y+1=0$ and $x+y-3=0$.

Q. 2. Find the equation of the tangent line to the curve $x=1-\cos\theta, y=\theta-\sin\theta$ at $\theta=\dfrac{\pi}{4}$.

Sol. Given, the curve

$$x=1-\cos\theta, y=\theta-\sin\theta \quad \ldots(i)$$

The value of x and y at $\theta=\dfrac{\pi}{4}$

Therefore, $\quad x=1-\cos\dfrac{\pi}{4}=1-\dfrac{1}{\sqrt{2}}$

and $\quad y=\dfrac{\pi}{4}-\sin\dfrac{\pi}{4}=\dfrac{\pi}{4}-\dfrac{1}{\sqrt{2}}$

So coordinate of the point of contact are

$$\left(1-\frac{1}{\sqrt{2}}, \frac{\pi}{4}-\frac{1}{\sqrt{2}}\right)$$

Differentiate x and y with respect to θ, we get

$$\frac{dx}{d\theta}=\sin\theta \text{ and } \frac{dy}{d\theta}=1-\cos\theta \quad \ldots(ii)$$

Now, $\quad \dfrac{dy}{dx}=\dfrac{\dfrac{dy}{d\theta}}{\dfrac{dx}{d\theta}}=\dfrac{1-\cos\theta}{\sin\theta}$

[from (i) and (ii)]

The value of $\dfrac{dy}{dx}$ at $\theta=\dfrac{\pi}{4}$, we have

$$\frac{dy}{dx}=\frac{1-\cos\frac{\pi}{4}}{\sin\frac{\pi}{4}}=\frac{1-\frac{1}{\sqrt{2}}}{\frac{1}{\sqrt{2}}}=\sqrt{2}-1$$

So, the equation of the tangent line at $\theta=\dfrac{\pi}{4}$ is

$$y-\left(\frac{\pi}{4}-\frac{1}{\sqrt{2}}\right)=(\sqrt{2}-1)\left\{x-\left(1-\frac{1}{\sqrt{2}}\right)\right\}$$

$\Rightarrow \quad (\sqrt{2}-1)x-y=2(\sqrt{2}-1)-\dfrac{\pi}{4}$

Q. 3. Find the equations of the normal lines to the curve $y=4x^3-3x+5$ which are parallel to the line $9y+x+3=0$. **[HOTS]**

Sol. Given $\quad y=4x^3-3x+5$

Differentiate w.r.t. x, we get

$$\frac{dy}{dx}=12x^2-3$$

$\therefore$ Slope of normal is $-\dfrac{1}{12x^2-3}$.

Also equation of the normal line to the curve $y=4x^3-3x+5$ is parallel to the line $9y+x+3=0$.

$\Rightarrow \quad 9y=-(x+3)$

$\Rightarrow \quad y=-\dfrac{x}{9}-\dfrac{1}{3}$

Slope of above line $=-\dfrac{1}{9}$

Hence $\quad \dfrac{-1}{12x^2-3}=-\dfrac{1}{9}$

$$12x^3-3=9$$

$$12x^2=12$$

$$x^2=1$$

$$x=\pm 1$$

At $x=1$; $\quad y=4-3+5=6$

At $x=-1$; $\quad y=4(-1)-3(-1)+5=4$

Equation of normal at $(1, 6)$ is

$$y-6=-\frac{1}{9}(x-1)$$

$\Rightarrow \quad x+9y=55$

Equation of normal at $(-1, 4)$ is

$$y-4=-\frac{1}{9}(x+1)$$

$\Rightarrow \quad x+9y=35.$

Q. 4. Find the intervals in which the function $f(x)=x^3+\dfrac{1}{x^3}, x\neq 0$ is (i) increasing and (ii) decreasing.

Sol. The given function is

$$f(x)=x^3+\frac{1}{x^3}, \ x\neq 0$$

Differentiate $f(x)$ with respect to x, we get

$$\therefore \quad f'(x) = 3x^2 - \frac{3}{x^4} = \frac{3}{x^4}(x^6 - 1)$$

$$= \frac{3}{x^4}(x^2-1)(x^4+x^2+1)$$

(i) For $f(x)$ to be increasing, we must have

$$\therefore \quad \frac{3}{x^4}(x^2-1)(x^4+x^2+1) > 0$$

$$\Rightarrow \quad x^2 - 1 > 0$$

$$\Rightarrow \quad (x-1)(x+1) > 0$$

Either $\quad x > 1$ or $x < -1$

Hence, $f(x)$ is increasing in $\{x : x > 1\} \cup \{x : x < -1\}$.

(ii) For $f(x)$ to be decreasing, we must have

$$f'(x) < 0$$

$$\therefore \quad \frac{3}{x^4}(x^2-1)(x^4+x^2+1) < 0$$

$$\Rightarrow \quad x^2 - 1 < 0$$

$\Rightarrow$ Hence $f(x)$ is decreasing on $(-1, 0) \cup (0, 1)$. (as $x \neq 0$)

Q. 5. Find the intervals in which the function f given by

$$\mathbf{f(x) = \sin x + \cos x, 0 \leq x \leq 2\pi}$$

is strictly increasing or strictly decreasing. [NCERT]

Sol. Given,

$$f(x) = \sin x + \cos x, 0 \leq x \leq 2\pi$$

Differentiate $f(x)$ with respect to x, we get

$$\therefore \quad f'(x) = \cos x - \sin x$$

$$= \sqrt{2}\left(\frac{1}{\sqrt{2}}\cos x - \frac{1}{\sqrt{2}}\sin x\right)$$

$$= \sqrt{2}\left(\sin\frac{\pi}{4}\cos x - \cos\frac{\pi}{4}\sin x\right)$$

$$= \sqrt{2}\sin\left(\frac{\pi}{4} - x\right)$$

$$= -\sqrt{2}\sin\left(x - \frac{\pi}{4}\right)$$

For $f(x)$ to be increasing it must be $f'(x) > 0$

Therefore, $\quad -\sqrt{2}\sin\left(x - \frac{\pi}{4}\right) > 0$

$$\Rightarrow \quad \sin\left(x - \frac{\pi}{4}\right) < 0$$

$$\therefore \quad \pi < x - \frac{\pi}{4} < 2\pi$$

$$\Rightarrow \quad \frac{5\pi}{4} < x < \frac{9\pi}{4}$$

$$\Rightarrow \quad \frac{5\pi}{4} < x < 2\pi \text{ or } 2\pi < x < \frac{9\pi}{4}$$

$$\Rightarrow \quad \frac{5\pi}{4} < x < 2\pi \text{ or } 0 < x < \frac{\pi}{4}$$

Hence, $f(x)$ is increasing function on

$$0 < x < \frac{\pi}{4} \text{ or } \frac{5\pi}{4} < x < 2\pi$$

i.e., $\quad \left(0, \frac{\pi}{4}\right) \cup \left(\frac{5\pi}{4}, 2\pi\right)$

For $f(x)$ to be decreasing function, we must have

$$f'(x) < 0$$

$$\Rightarrow \quad \sqrt{2}\sin\left(\frac{\pi}{4} - x\right) < 0$$

$$\Rightarrow \quad -\sqrt{2}\sin\left(x - \frac{\pi}{4}\right) < 0$$

$$\Rightarrow \quad \sin\left(x - \frac{\pi}{4}\right) < 0$$

$$\Rightarrow \quad 0 < x - \frac{\pi}{4} < \pi$$

$$\Rightarrow \quad \frac{\pi}{4} < x < \frac{5\pi}{4}$$

$$\Rightarrow \quad x \in \left(\frac{\pi}{4}, \frac{5\pi}{4}\right)$$

Hence $f(x)$ is decreasing on $\left(\frac{\pi}{4}, \frac{5\pi}{4}\right)$.

Q. 6. Find intervals in which the function given by $f(x) = \sin 3x, 0 \leq x \leq \frac{\pi}{2}$ is (i) increasing (ii) decreasing. [NCERT]

Sol. Given function is

$$f(x) = \sin 3x$$

Differentiating $f(x)$ w.r.t. x, we get

$$f'(x) = 3\cos 3x$$

For increasing or decreasing, it must be $f'(x) = 0$ we get

$$\cos 3x = 0$$

$$\therefore \quad 3x = \frac{\pi}{2}, \frac{3\pi}{2}$$

So $\quad x = \frac{\pi}{6}$ and $\frac{\pi}{2}$

But it is given that

$$0 \leq x \leq \frac{\pi}{2} \text{ implies } 0 \leq 3x \leq \frac{3\pi}{2}$$

The point $\frac{\pi}{6}$ divides the interval $0 \le x \le \frac{\pi}{2}$ into two disjoint intervals $0 \le x < \frac{\pi}{6}$ and $\frac{\pi}{6} < x \le \frac{\pi}{2}$.

When $\quad 0 \le x < \frac{\pi}{6}$

$$f'(x) = 3 \cos 3x$$

Also $\quad 0 \le 3x < \frac{3\pi}{6}$

$$0 \le 3x < \frac{\pi}{2}$$

$$\Rightarrow \quad f'(x) > 0 \quad \left(\text{As } \cos\theta > 0 \text{ for } < \theta < \frac{\pi}{2}\right)$$

When $\quad \frac{\pi}{6} < x \le \frac{\pi}{2}$

$$f'(x) = 3 \cos 3x$$

Also $\quad \frac{3\pi}{6} < 3x \le \frac{3\pi}{2}$

$$\frac{\pi}{2} < 3x \le \frac{3\pi}{2}$$

$$\Rightarrow \quad f'(x) < 0$$

As $\quad \cos\theta < 0$ for $\frac{\pi}{2} < \theta \le \frac{3\pi}{2}$

Therefore, $f(x)$ is strictly increasing in $0 \le x < \frac{\pi}{6}$ and strictly decreasing in $\frac{\pi}{6} < x \le \frac{\pi}{2}$.

Also, the given function is continuous at $x = 0$ and $x = \frac{\pi}{6}$.

Therefore, we know that $f(x)$ is increasing on $0 \le x \le \frac{\pi}{6}$ and decreasing on $\frac{\pi}{6} \le x \le \frac{\pi}{2}$.

Q.7. Find the intervals in which $f(x) = (x + 1)^3 (x - 3)^3$ is increasing or decreasing. **[NCERT]**

Sol. Given function is

$$f(x) = (x+1)^3 (x-3)^3$$

Differentiating $f(x)$ with respect to x, we get

$$\Rightarrow \quad f'(x) = \left\{3(x+1)^2 \frac{d}{dx}(x+1)\right\}(x-3)^3 + (x+1)^3.\left\{3(x-3)^2 \frac{d}{dx}(x-3)\right\}$$

$$\Rightarrow \quad f'(x) = 3(x+1)^2 (x-3)^3 + 3(x+1)^3 (x-3)^2$$

$$\Rightarrow \quad f'(x) = 3(x+1)^2 (x-3)^2 [x+1+x-3]$$

$$\Rightarrow \quad f'(x) = 6(x+1)^2 (x-3)^2 (x-1)$$

For increasing value of $f(x)$ we must have

$$f'(x) > 0$$

Therefore, $6(x+1)^2 (x-3)^2 (x-1) > 0$

$$\Rightarrow \quad x - 1 > 0 \text{ and } x \ne -1, 3$$

$$\Rightarrow \quad x > 1 \text{ and } x \ne -1, 3$$

$$\Rightarrow \quad x \in (1, 3) \cup (3, \infty)$$

So, $f(x)$ is increasing in $(1, 3) \cup (3, \infty)$.

For $f(x)$ to be decreasing, we must have

$$f'(x) < 0$$

$$\Rightarrow \quad 6(x+1)^2 (x-3)^2 (x-1) < 0$$

$$\Rightarrow \quad x - 1 < 0 \text{ and } x \ne -1, 3$$

$$\Rightarrow \quad x < 1 \text{ and } x \ne -1, 3$$

$$\Rightarrow \quad x \in (-\infty, -1) \cup (-1, 1)$$

So, $f(x)$ is decreasing in $(-\infty, 1) \cup (-1, 1)$.

Q. 8. The combined resistance R of two resistor R_1 and R_2 ($R_1, R_2 > 0$) is given by $\frac{1}{R} = \frac{1}{R_1} + \frac{1}{R_2}$.

If $R_1 + R_2 = C$ (a constant), show that the maximum resistance R is obtained by choosing $R_1 = R_2$.

Sol. Given, $\frac{1}{R} = \frac{1}{R_1} + \frac{1}{R_2}$

and $\quad R_1 + R_2 = C$

$$\Rightarrow \quad \frac{1}{R} = \frac{R_1 + R_2}{R_1 R_2} = \frac{C}{R_1 R_2} = \frac{C}{R_1(C - R_1)}$$

$$\Rightarrow \quad R = \frac{R_1 C - R_1^2}{C} = R_1 - \frac{R_1^2}{C}$$

On differentiating R with respect to R_1, we get

$$\frac{dR}{dR_1} = 1 - \frac{2R_1}{C} \quad \text{...(i)}$$

Again differentiating $\frac{dR}{dR_1}$ with respect to R_1, we get

$$\frac{d^2 R}{dR_1^2} = -\frac{2}{C}$$

For maximum or minimum value, we must have $\frac{dR}{dR_1} = 0$

$$1 - \frac{2R_1}{C} = 0 \quad \text{[from (i)]}$$

$$\Rightarrow \quad R_1 = \frac{C}{2}$$

Now, $\frac{d^2R}{dR_1^2} = -\frac{2}{C} < 0$ for all values of R_1

Thus, R is maximum when $R_1 = \frac{C}{2}$.

Putting $R_1 = \frac{C}{2}$ in $R_1 + R_2 = C$, we get

$$R_2 = C - \frac{C}{2} = \frac{C}{2}$$

Hence, R is maximum when $R_1 = R_2 = \frac{C}{2}$.

Q. 9. Show that the rectangle of maximum area that can be inscribed in a circle is a square.

Sol. Let $ABCD$ be a rectangle of length = x units and breadth = y units.

Let O be the centre of the circle, in which rectangle $ABCD$ is inscribed and r be its radius. Let

$$\angle CAB = \theta$$

In ΔABC, $\cos\theta = \frac{x}{2r}$

$\therefore$ $x = 2r\cos\theta$ ...(i)

and $\sin\theta = \frac{y}{2r}$

$\therefore$ $y = 2r\sin\theta$ ...(ii)

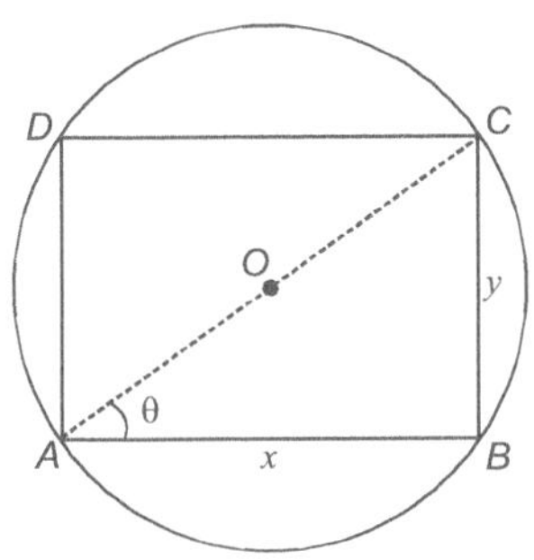

Now, area of rectangle $ABCD$

$= A = x.y$

$= 2r\cos\theta . 2r\sin\theta$ [from (i) and (ii)]

$= 2r^2\sin 2\theta$

Differentiate A with respect to θ, we get

$$\frac{dA}{d\theta} = 4r^2\cos 2\theta$$

For maxima value it must be $\frac{dA}{d\theta} = 0$

Therefore, $4r^2\cos 2\theta = 0$

$\Rightarrow$ $\cos 2\theta = 0 \Rightarrow \cos 2\theta = \cos\frac{\pi}{2}$

$\Rightarrow$ $2\theta = \frac{\pi}{2} \Rightarrow \theta = \frac{\pi}{4}$

Again differentiate $\frac{dA}{d\theta}$ with respect to θ, we get

$$\frac{d^2A}{d\theta^2} = -8r^2\sin 2\theta$$

Now value of $\frac{d^2A}{d\theta^2}$ at $\theta = \frac{\pi}{4}$ is

$$\left.\frac{d^2A}{d\theta^2}\right|_{\theta=\frac{\pi}{4}} = -8r^2\sin 2\left(\frac{\pi}{4}\right)$$

$$= -8r^2\sin\frac{\pi}{2}$$

$$= -8r^2 < 0$$

$\therefore$ A is maximum at $\theta = \frac{\pi}{4}$

and $x = 2r \cdot \frac{1}{\sqrt{2}}$ and $y = 2r \cdot \frac{1}{\sqrt{2}}$

$\therefore$ $x = y$

$\Rightarrow ABCD$ is a square.

Hence area of rectangle $ABCD$ is maximum when rectangle is a square.

Q. 10. Find the shortest distance between the line $y - x = 1$ and the curve $x = y^2$. **[HOTS]**

Sol. Let $y = t$, then

$x = t^2$ [by given curve]

Therefore, $P(t^2, t)$ be any point on the curve $x = y^2$

The perpendicular distance between the curve and given line is

$$S = \left|\frac{t - t^2 - 1}{\sqrt{2}}\right|$$

$\Rightarrow$ $S = \left|\frac{t^2 - t + 1}{\sqrt{2}}\right|$

$\Rightarrow$ $S = \frac{t^2 - t + 1}{\sqrt{2}}$

Differentiating S w.r.t. t, we get

$$\frac{dS}{dt} = \frac{2t - 1}{\sqrt{2}}$$

Again, differentiating $\frac{dS}{dt}$ w.r.t t, we get

$$\frac{d^2S}{dt^2} = \frac{2}{\sqrt{2}} = \sqrt{2}$$

For maximum or minimum value it must be $\frac{dS}{dt} = 0$, we get

$$2t - 1 = 0$$

$$\Rightarrow \quad t = \frac{1}{2}$$

For $t = \frac{1}{2}$, we have

$$\frac{d^2S}{dt^2} > 0$$

So, S is minimum when $t = \frac{1}{2}$

Then minimum value of S is given by

$$S = \frac{\left(\frac{1}{2}\right)^2 - \frac{1}{2} + 1}{\sqrt{2}} = \frac{3\sqrt{2}}{8}.$$

Q. 11. A jet of an enemy is flying along the curve $x^2 + 4$. A soldier is placed at the point (5, 3). What is the nearest distance between the soldier and the jet?

[HOTS]

Sol. Let the point (x, y) be the nearest point on the curve $y = x^2 + 4$.

Also, let D be the distance of the point (x, y) and the point (5, 3).

$$D = \sqrt{(x-5)^2 + (y-3)^2}$$

$$D^2 = (x-5)^2 + (y-3)^2$$

Let $\quad D^2 = A$

$$A = (x-5)^2 + (y-3)^2$$

$$A = (x-5)^2 + (x^2+1)^2 \qquad [\because y = x^2 + 4]$$

Differentiate A w.r.t. x, we get

$$\frac{dA}{dx} = 2(x-5) + 2(x^2+1) \times 2x$$

For maximum and minimum value of A,

Put $\quad \frac{dA}{dx} = 0$

$$\Rightarrow \quad 2(x-5) + 4x^3 + 4x = 0$$

$$\Rightarrow \quad 2(2x^3 + 2x + x - 5) = 0$$

$$\Rightarrow \quad (2x^3 + 3x - 5) = 0$$

$$\Rightarrow \quad (x-1)(2x^2 + 2x + 5) = 0$$

$\Rightarrow x = -1$, no solution for $2x^2 + 2x + 5 = 0$

Now, differentiate $\frac{dA}{dx}$ w.r.t. x, we get

$$\frac{d^2A}{dx^2} = 2 + \frac{d}{dx}[4x(x^2+1)]$$

$$= 2 + 4\frac{d}{dx}(x^3 + x)$$

$$\left.\frac{d^2A}{dx^2}\right|_{x=1} = 2 + 4(3x^2+1) = 2 + 4[3(1)^2 + 1]$$

$$= 2 + 4(4) = 18 > 0$$

Hence $A = D^2$ has minimum value for $x = 1$

Now from equation (i)

$$A = (1-5)^2 + (1+1)^2$$

$$= 16 + 4 = 20$$

$$D = \sqrt{A} = \sqrt{20} = 4{\cdot}47$$

Thus distance between jet and soldier is 4·47 unit.

Q. 12. In an activity organised in the school, Rohan was given the task to put the slogan 'Satyamev Jayate' on a trapezium shaped card sheet. If length of three sides of a trapezium other than base are equal 20 cm, then find the area of the trapezium when it is maximum.

Sol. Let $ABCD$ be a trapezium such that DC is parallel to AB and $AD = 20$ cm $= DC = BC$.

Now draw perpendiculars DP and CQ from D and C, respectively on AB.

$$\because \quad \Delta APD \approx \Delta BQC$$

Therefore, $\quad PA = QB = x$ cm

In ΔAPD, we have

$$AD^2 = AP^2 + PD^2$$

[Pythagoras theorem]

$$\Rightarrow \quad PD^2 = AD^2 - AP^2$$

$$\Rightarrow \quad PD = \sqrt{AD^2 - AP^2}$$

$$PD = \sqrt{400 - x^2} \text{ cm}$$

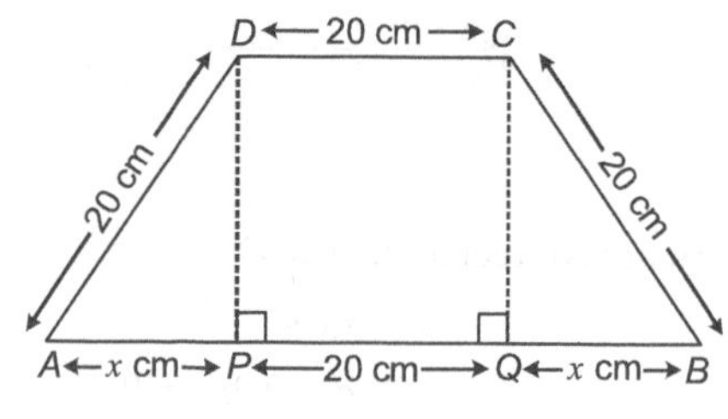

Similarly, in ΔBQC,

$$QC = \sqrt{400 - x^2} \text{ cm}$$

If S denotes the area of the trapezium $ABCD$, then

$$S = \frac{1}{2}(\text{Sum of parallel sides}) \times \text{Height}$$

$$\Rightarrow \quad S = \frac{1}{2}(AB + DC) \times PD$$

$$\Rightarrow \quad S = \frac{1}{2}[(2x + 20) + 20] \times \sqrt{400 - x^2}$$

$$\Rightarrow \quad S = (x+20) \times \sqrt{400-x^2} \quad \ldots(i)$$

On differentiating above w.r.t. 'x', we get

$$\frac{dS}{dx} = 1\sqrt{400-x^2} + (x+20)\left[\frac{-2x}{2\sqrt{400-x^2}}\right]$$

$$\frac{dS}{dx} = \frac{(400-x^2)-x^2-20x}{\sqrt{400-x^2}}$$

$$= \frac{400-2x^2-20x}{\sqrt{400-x^2}}$$

Again differentiating w.r.t. 'x', we get

$$\frac{d^2S}{dx^2} = \frac{\sqrt{400-x^2}(-4x-20)-(400-2x^2-20x)\left(\frac{-2x}{2\sqrt{400-x^2}}\right)}{(400-x^2)}$$

$$= \frac{(400-x^2)(-4x-20)+(400x-2x^3-20x^2)}{(400-x^2)\sqrt{400-x^2}}$$

$$= \frac{(-1600x-8000+4x^3+20x^2)+(400x-2x^3-20x^2)}{(400-x^2)\sqrt{400-x^2}} \quad \ldots(ii)$$

For maxima or minima, we have

$$\frac{dS}{dx} = 0$$

$$\Rightarrow \quad \frac{400-2x^2-20x}{\sqrt{400-x^2}} = 0$$

$$\Rightarrow \quad 400-2x^2-20x = 0$$

$$\Rightarrow \quad 2x^2+20x-400 = 0$$

$$\Rightarrow \quad 2(x^2+10x-200) = 0$$

$$\Rightarrow \quad x^2+10x-200 = 0$$

$$\Rightarrow \quad x^2+20x-10x-200 = 0$$

$$\Rightarrow \quad x(x+20)-10(x+20) = 0$$

$$\Rightarrow \quad (x+20)(x-10) = 0$$

$$\Rightarrow \quad x = -20, x = 10$$

$$\Rightarrow \quad x = 10,$$

since x represents distance it cannot be negative.

Putting $x = 10$ in equation (ii), we get

$$\left.\frac{d^2S}{dx^2}\right]_{\text{at } x = 10}$$

$$= \frac{[-1600(10)-8000+4(10)^3+20(10)^2] + [400(10)-2(10)^3-20(10^2)]}{(400-100)\sqrt{400-100}}$$

$$= \frac{-16000-8000+4\cdot(1000)+20(100) + 400(10)-2(1000)-20(100)}{(300)\sqrt{300}}$$

$$= \frac{-28000+10000}{300\sqrt{300}}$$

$$= \frac{-18000}{300\sqrt{300}}$$

$$= \frac{-60}{\sqrt{300}} < 0$$

Hence, the area of trapezium is maximum when $x = 10$ and the area is given by

$$S = (10+20)\sqrt{400-100}$$

$$= 30(\sqrt{300})$$

$$= 30 \times 10\sqrt{3}$$

$$= 300\sqrt{3} \text{ cm}^2$$

Q. 13. Show that the height of the cylinder of maximum volume that can be inscribed in a sphere of radius a is $\frac{2a}{\sqrt{3}}$.★

Sol. Let radius and height of a cylinder be r, h respectively.

Let radius of sphere $= a$, $OL = x$ and $\angle OAL = \theta$

In ΔOLA, $\quad OA^2 = OL^2 + AL^2$

$$\Rightarrow \quad AL = \sqrt{a^2-x^2}$$

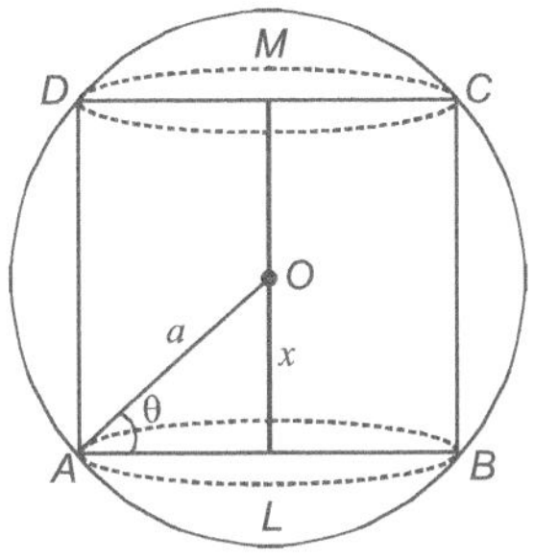

Let V be the volume of the cylinder.

Then, $\quad V = \pi(AL)^2 \times LM$

$$\Rightarrow \quad V = \pi(AL)^2 \times 2(OL)$$

$$\Rightarrow \quad V = \pi(a^2-x^2) \times 2x$$

$$\Rightarrow \quad V = 2\pi(a^2x-x^3)$$

On differentiating V with respect to x, we get

$$\frac{dV}{dx} = 2\pi(a^2-3x^2)$$

Again, differentiating $\frac{dV}{dx}$ with respect to x, we get

$$\frac{d^2V}{dx^2} = -12\pi x$$

★ are board exam questions from previous years

For maximum or minimum values of V, we must have

$$\frac{dV}{dx}=0$$

$$\Rightarrow \quad 2\pi(a^2-3x^2)=0$$

$$\Rightarrow \quad x=\frac{a}{\sqrt{3}}$$

$$\therefore \quad \left(\frac{d^2V}{dx^2}\right)_{x=a/\sqrt{3}}=-12\pi\times\frac{a}{\sqrt{3}}<0$$

Hence, V is maximum when $x=\frac{a}{\sqrt{3}}$ and hence height

$$h=2x=\frac{2a}{\sqrt{3}}\cdot$$

Q. 14. Show that the semi-vertical angle of the right circular cone of maximum volume and of given slant height is $\tan^{-1}\sqrt{2}$.

Sol. According to question, let α be the semi-vertical angle of cone OAB of given slant height say l.

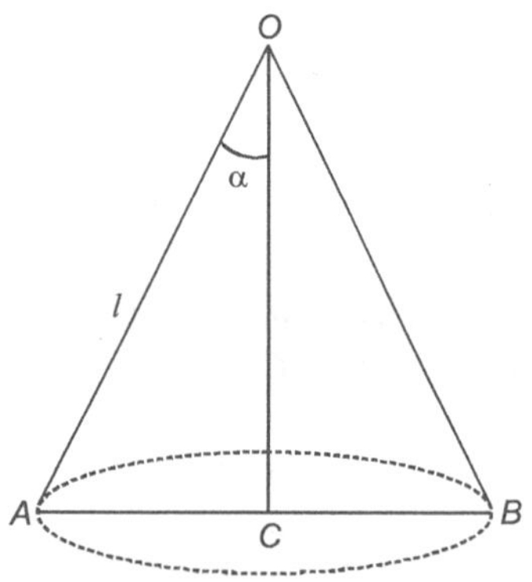

In ΔOAC, $\cos\alpha=\frac{OC}{OA}=\frac{OC}{l}$

$$\Rightarrow \quad OC=l\cos\alpha \quad \ldots(i)$$

and $\sin\alpha=\frac{AC}{OA}=\frac{AC}{l}$

$$\Rightarrow \quad AC=l\sin\alpha \quad \ldots(ii)$$

Let V be the volume of the cone.

Then $V=\frac{1}{3}\pi(AC)^2(OC)$

$$\Rightarrow \quad V=\frac{1}{3}\pi(l\sin\alpha)^2(l\cos\alpha)$$

[from (i), (ii)]

$$\Rightarrow \quad V=\frac{1}{3}\pi l^3(\sin\alpha)^2\cos\alpha$$

Differentiate V with respect to α, we get

$$\therefore \quad \frac{dV}{d\alpha}=\frac{1}{3}\pi l^3[-\sin^3\alpha+2\sin\alpha\cos^2\alpha]$$

$$=\frac{1}{3}\pi l^3\sin\alpha[-\sin^2\alpha+2\cos^2\alpha] \quad \ldots(i)$$

For maximum volume of V we have $\frac{dV}{d\alpha}=0$

Therefore, $\frac{1}{3}\pi l^3\sin\alpha[-\sin^2\alpha+2\cos^2\alpha]=0$

$$\Rightarrow \quad 2\cos^2\alpha=\sin^2\alpha$$

$$\Rightarrow \quad \tan^2\alpha=2$$

$$\Rightarrow \quad \tan\alpha=\sqrt{2}$$

$$\therefore \quad \cos\alpha=\frac{1}{\sqrt{1+\tan^2\alpha}}=\frac{1}{\sqrt{3}}$$

Again differentiate (i) with respect to α, we get

$$=\frac{\pi}{3}l^3[-3\sin^2\alpha\cos\alpha+2\sin\alpha(-2\cos\alpha)\sin\alpha+2\cos^3\alpha]$$

$$=\frac{\pi}{3}l^3[-3\sin^2\alpha\cos\alpha-4\sin^2\alpha\cos\alpha+2\cos^3\alpha]$$

$$=\frac{\pi}{3}l^3[-3(1-\cos^2\alpha)\cos\alpha-4\cos\alpha(1-\cos^2\alpha)+2\cos^3\alpha]$$

$$=\frac{\pi}{3}l^3[-3\cos\alpha+3\cos^3\alpha-4\cos\alpha+4\cos^3\alpha+2\cos^3\alpha]$$

$$=\frac{\pi}{3}l^3[9\cos^3\alpha-7\cos\alpha]$$

$$=\frac{\pi}{3}l^3\cos^3\alpha(9-7\sec^2\alpha)$$

$$=\frac{\pi}{3}l^3\cos^3\alpha[9-7(1+\tan^2\alpha)]$$

$$=\frac{\pi}{3}l^3\cos^3\alpha[2-7\tan^2\alpha]$$

$$\therefore \quad \left.\frac{d^2V}{d\alpha^2}\right|_{\tan\alpha=\sqrt{2}}=\frac{\pi}{3}l^3\left(\frac{1}{\sqrt{3}}\right)^3[2-14]$$

$$=-12\times\frac{\pi}{3}l^3\left(\frac{1}{\sqrt{3}}\right)^3<0$$

$\therefore$ V is maximum when $\tan\alpha=\sqrt{2}$ or $\alpha=\tan^{-1}\sqrt{2}$

Hence when semi-vertical angle of the cone is $\tan^{-1}\sqrt{2}$ then volume of right circular cone is maximum.

Q. 15. Show that the volume of the greatest right circular cylinder that can be inscribed in a right circular cone of height h and semi-vertical angle α is $\frac{4}{27}\pi h^3\tan^2\alpha$.

Also show that height of the cylinder is $\frac{h}{3}$ ·[NCERT]

Sol. Let r = radius of the cone, h = height of the cone and R = radius of the cylinder, H = height of the cylinder.

Let α be the semi-vertical angle of the cone.

Now, $\Delta AEF \sim \Delta ADC$

Then by property of similarity,

$$\frac{AE}{AD} = \frac{EF}{DC}$$

or $$\frac{h-H}{h} = \frac{R}{r}$$

Also, in ΔADC

$$\tan\alpha = \frac{r}{h}$$

or $$r = h\tan\alpha$$

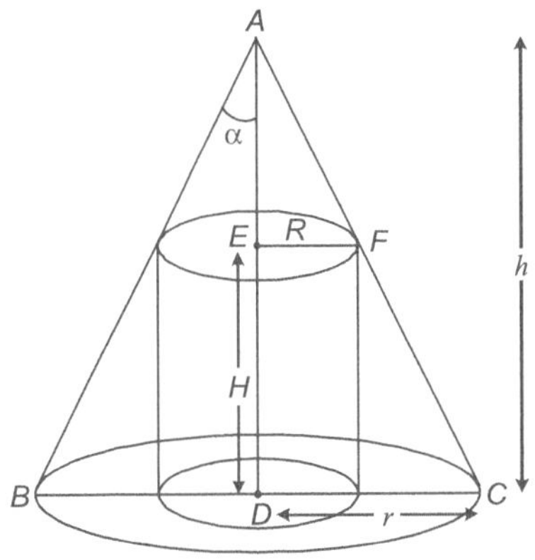

In ΔAEF, $\tan\alpha = \dfrac{R}{h-H}$

or $$R = (h-H)\tan\alpha$$

Let V be the volume of the cylinder.

$$\therefore \quad V = \pi(h-H)^2\tan^2\alpha \times H$$

$$= \pi\tan^2\alpha\,(h^2H + H^3 - 2hH^2)$$

Differentiate V with respect to H, we get

$$\therefore \quad \frac{dV}{dH} = \pi\tan^2\alpha\,(h^2 + 3H^2 - 4hH) \quad \ldots(i)$$

Again differentiate equation (i) with respect to H, we get

$$\frac{d^2V}{dH^2} = \pi\tan^2\alpha\,(-4h + 6H)$$

Now, for maximum value it must be $\dfrac{dV}{dH} = 0$

$\Rightarrow 3H^2 - 4hH + h^2 = 0$

$$\therefore \quad H = \frac{4h \pm \sqrt{16h^2 - 12h^2}}{6} = \frac{4h \pm 2h}{6}$$

$$= h \text{ and } \frac{h}{3}$$

But $H \neq h$ so $H = \dfrac{h}{3}$

Therefore $\left.\dfrac{d^2V}{dH^2}\right|_{H=\frac{h}{3}} = \pi\tan^2\alpha\,[-4h + 2h] < 0$

$\therefore$ V is maximum at $H = \dfrac{h}{3}$

$$\text{Maximum volume} = \pi H(h-H)^2\tan^2\alpha$$

$$= \pi\frac{h}{3}\left(h - \frac{h}{3}\right)^2\tan^2\alpha$$

$$= \pi\cdot\frac{h}{3}\left(\frac{4h^2}{9}\right)\tan^2\alpha$$

$$= \frac{4}{27}\pi h^3\tan^2\alpha$$

Hence cylinder has maximum volume when height of the cylinder is $\dfrac{h}{3}$.

Q. 16. Show that the altitude of the right circular cone of maximum volume that can be described in a sphere of radius r is $\dfrac{4r}{3}$. Also show that the maximum volume of the cone is $\dfrac{8}{27}$ of the volume of the sphere.★

Sol. From the figure $OA = OC = r$ (radius of the sphere)

From right angled ΔOBC,

$$BC = r\sin\theta,$$
$$OB = r\cos\theta$$

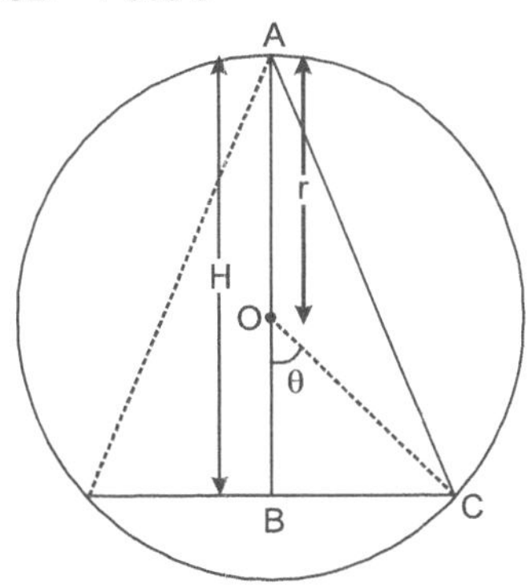

V = Volume of the inscribed cone

$$= \frac{1}{3}\pi R^2 H = \frac{1}{3}\pi(BC)^2.AB$$

$$= \frac{1}{3}\pi r^2\sin^2\theta(r + r\cos\theta) \quad \ldots(i)$$

$[\because AB = AO + OB]$

$$\Rightarrow \quad V = \frac{\pi}{3}r^3(\sin^2\theta + \sin^2\theta\cos\theta)$$

Differentiating w.r.t. θ, we get

$$\Rightarrow \frac{dV}{d\theta} = \frac{\pi r^3}{3}[2\sin\theta\cos\theta + 2\sin\theta\cos^2\theta - \sin^3\theta]$$

For maxima or minima,

★ are board exam questions from previous years

$$\frac{dV}{d\theta} = 0$$

$\Rightarrow \quad \sin\theta\,[2\cos\theta + 2\cos^2\theta - \sin^2\theta] = 0$

$\Rightarrow \quad 2\cos\theta + 2\cos^2\theta - (1 - \cos^2\theta) = 0$

$[\because \sin\theta \neq 0]$

$\Rightarrow \quad 3\cos^2\theta + 2\cos\theta - 1 = 0$

$\Rightarrow \quad 3\cos^2\theta + 3\cos\theta - \cos\theta - 1 = 0$

$\Rightarrow \quad 3\cos\theta(\cos\theta + 1) - 1(\cos\theta + 1) = 0$

$\Rightarrow \quad (3\cos\theta - 1)(\cos\theta + 1) = 0$

$\Rightarrow \quad 3\cos\theta - 1 = 0 \qquad [\because \cos\theta \neq 0]$

$\Rightarrow \quad \cos\theta = \frac{1}{3}$

$\therefore \quad H = r + r\cos\theta$

$$= r + r\cdot\frac{1}{3} = \frac{4r}{3}$$

Also V changes sign from +ve to –ve for this value of θ.

$\Rightarrow V$ is maximum.

$\therefore$ Maximum volume of the inscribed cone,

$$V = \frac{\pi}{3}r^3(1 - \cos^2\theta)(1 + \cos\theta)$$

[From (i)]

$$= \frac{\pi}{3}r^3\left(1 - \frac{1}{9}\right)\left(1 + \frac{1}{3}\right)$$

$$= \frac{\pi}{3}r^3\,\frac{32}{27} = \frac{8}{27}\left(\frac{4}{3}\pi r^3\right)$$

$$= \frac{8}{27} \text{ Volume of the sphere.}$$

Q. 17. Prove that the semi-vertical angle of the right circular cone of given volume and least curved surface area is $\cot^{-1}\sqrt{2}$.★

Sol. Let r, h, l, V and S be the base radius, height, slant height, volume (given) and curved surface area of the cone respectively. Then

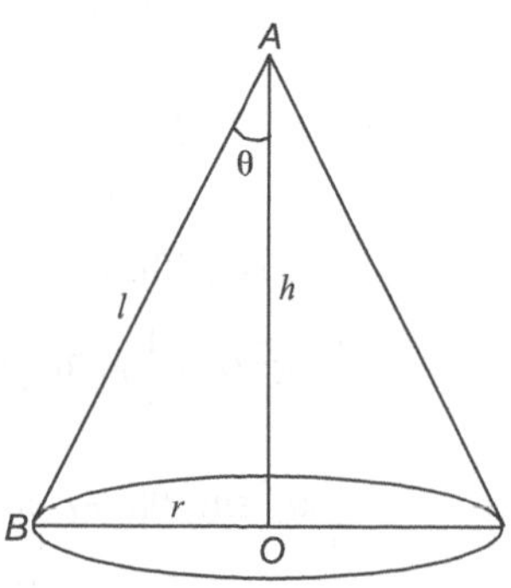

$\therefore$ Slant height $l^2 = r^2 + h^2 \qquad \ldots(i)$

$$V = \frac{1}{3}\pi r^2 h \text{ (given)}$$

$\Rightarrow \quad r^2 = \frac{3V}{\pi h} \qquad \ldots(ii)$

and $\quad S = \pi r l = \pi r\sqrt{r^2 + h^2} \quad (\because l^2 = r^2 + h^2)$

$\Rightarrow \quad S^2 = \pi^2 r^2(r^2 + h^2)$

$$= \pi^2.\frac{3V}{\pi h}\left(\frac{3V}{\pi h} + h^2\right) \quad \text{[using (ii)]}$$

$\therefore \quad S^2 = 3\pi V\left(\frac{3V}{\pi h^2} + h\right)$

For S to be least, S^2 is also least.

$\therefore \quad \frac{dS^2}{dh} = 3\pi V.\left(\frac{-6V}{\pi h^3} + 1\right)$

and $\quad \frac{d^2S^2}{dh^2} = 3\pi V\left(\frac{-6V}{\pi}\right)\cdot\frac{-3}{h^4}$

$$= \frac{54V^2}{h^4} \qquad \ldots(iii)$$

For max. or min. S (and so S^2),

$$\frac{dS^2}{dh} = 0$$

$\Rightarrow \quad 6V = \pi h^3$

$\Rightarrow \quad h = \left(\frac{6V}{\pi}\right)^{1/3} \qquad \ldots(iv)$

$\therefore$ From (3),

$$\frac{d^2S^2}{dh^2} = \frac{54V^2}{h^4} > 0 \text{ (+ve)}$$

$\Rightarrow S^2$ and therefore S is least.

In right angled ΔAOB.

$$\cot\theta = \frac{h}{r}$$

$$= \frac{h}{\sqrt{3V/\pi h}} = \sqrt{\frac{\pi}{3V}}\,h^{3/2}$$

[from (ii)]

$\Rightarrow \quad \cot\theta = \sqrt{\frac{\pi}{3V}}\cdot\sqrt{\frac{6V}{\pi}} = \sqrt{2} \qquad \text{[from (iv)]}$

$\therefore$ The semi-vertical angle,

$$\theta = \cot^{-1}(\sqrt{2})$$

Q.18. If the sum of the lengths of the hypotenuse and a side of a right angled triangle is given. Show that the area of the triangle is maximum when the angle between them is $\frac{\pi}{3}$.

Sol. Let in ΔABC,

Base $BC = x$ and hypotenuse $AB = y$

Given, $\quad x + y = k$, where k is a constant

Let α be the angle between the base and hypotenuse and also S be the area of the triangle.

Then $\quad$ area of the triangle $= \frac{1}{2} \times \text{base} \times \text{height}$

★ are board exam questions from previous years

$$\therefore \quad S = \frac{1}{2} BC \times AC$$

$$= \frac{1}{2} x\sqrt{y^2 - x^2}$$

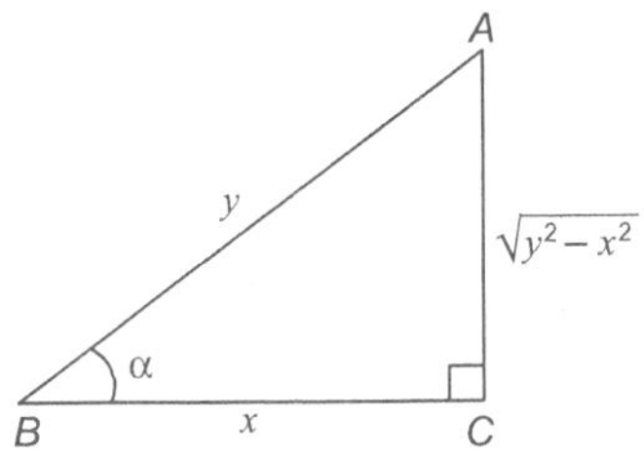

Squaring both sides, we get

$$S^2 = \frac{x^2}{4}(y^2 - x^2)$$

$$= \frac{x^2}{4}[(k-x)^2 - x^2]$$

$$\Rightarrow \quad S^2 = \frac{x^2}{4}[k^2 - 2kx] = \frac{k^2x^2 - 2kx^3}{4} \quad \text{...(i)}$$

Differentiating equation (i) w.r.t. x, we get

$$2S\frac{dS}{dx} = \frac{2k^2x - 6kx^2}{4}$$

$$\text{or} \quad S\frac{dS}{dx} = \frac{k^2x - 3kx^2}{4} \quad \text{...(ii)}$$

For maximum or minimum it must be $\frac{dS}{dx} = 0$

$$\therefore \quad \frac{k^2x - 3kx^2}{4} = 0$$

$$\Rightarrow \quad x = \frac{k}{3}$$

Again, differentiating equation (ii) w.r.t. x, we get

$$\left(\frac{dS}{dx}\right)^2 + S\frac{d^2S}{dx^2} = \frac{k^2 - 6kx}{4}$$

When $x = \frac{k}{3}$, $\frac{dS}{dx} = 0$. Put $\frac{dS}{dx} = 0$ and $x = \frac{k}{3}$, in above equation, we get

$$\frac{d^2S}{dx^2} = \frac{-k^2}{4S} < 0$$

Thus, S is maximum when $x = \frac{k}{3}$

$$\Rightarrow \quad y = k - \frac{k}{3} = \frac{2k}{3}$$

In ΔABC, $\cos\alpha = \frac{x}{y}$

$$\Rightarrow \quad \cos\alpha = \frac{k/3}{2k/3} = \frac{1}{2}$$

Hence, $\alpha = \frac{\pi}{3}$.

Q. 19. Prove that the height of the cylinder of maximum volume that can be inscribed in a sphere of radius R is $\frac{2R}{\sqrt{3}}$. Also find the maximum volume.★

Sol. From the figure,

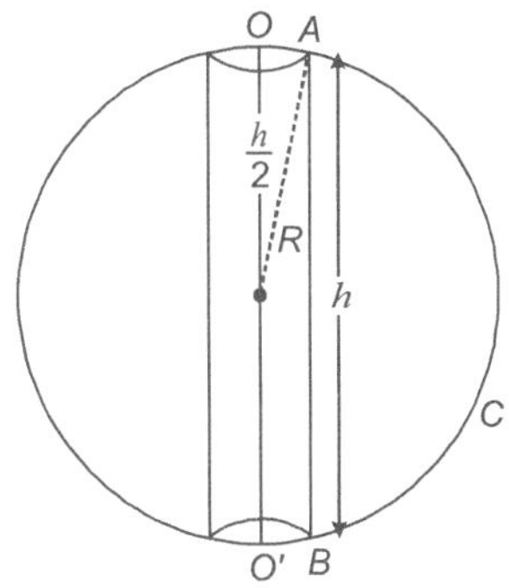

$$\left(\frac{h}{2}\right)^2 + r^2 = R^2$$

$$\Rightarrow \quad r^2 = R^2 - \frac{h^2}{4} \quad \text{...(i)}$$

Now, V = Volume of the cylinder inscribed in a sphere

$$\therefore \quad V = \pi r^2 h$$

$$= \pi h\left(R^2 - \frac{h^2}{4}\right) \quad \text{[using (i)]}$$

$$\therefore \quad V = \pi\left(R^2 h - \frac{h^3}{4}\right)$$

Differentiating w.r.t. h, we get

$$\Rightarrow \quad \frac{dV}{dh} = \pi\left(R^2 - \frac{3h^2}{4}\right) \quad \text{...(ii)}$$

$$\text{and} \quad \frac{d^2V}{dh^2} = \pi\left(0 - \frac{3}{4}.2h\right) \quad \text{...(iii)}$$

For maxima or minima

$$\frac{dV}{dh} = 0$$

From (ii),

$$\Rightarrow \quad R^2 - \frac{3}{4}h^2 = 0$$

$$\Rightarrow \quad h^2 = \frac{4}{3}R^2$$

$$\Rightarrow \quad h = \frac{2R}{\sqrt{3}}$$

For the value of h, from (iii),

$$\frac{d^2V}{dh^2} = -\frac{3}{2}\pi\frac{2R}{\sqrt{3}} = -\sqrt{3}\pi R < 0 \text{ (–ve)}$$

$\Rightarrow V$ is maximum.

Also maximum value of V

$$= \pi.\frac{2R}{\sqrt{3}}\left(R^2 - \frac{1}{4}.\frac{4}{3}R^2\right)$$

★ are board exam questions from previous years

$$= \pi.\frac{2R}{\sqrt{3}}.\frac{2}{3}R^2 = \frac{4\pi}{3\sqrt{3}}R^3$$

$$= \frac{4\sqrt{3}}{9}\pi R^3 \text{ cu. unit.}$$

Q. 20. Of all the closed right circular cylindrical cans of volume 128 π cm^3, find the dimensions of the can which has minimum surface area.★

Sol. Let r cm be the radius and h cm be the height of the closed cylindrical cans of given volume = 128π cm^3.

Then Volume, $V = \pi r^2 h = 128\pi$

$$\Rightarrow \quad h = \frac{128}{r^2} \quad \ldots(i)$$

Also surface area,

$$S = 2\pi rh + 2\pi r^2$$

$$= 2\pi\left(r \cdot \frac{128}{r^2} + r^2\right)$$

$$\Rightarrow \quad S = 2\pi\left(r^2 + \frac{128}{r}\right)$$

Differentiating w.r.t. r, we get

$$\therefore \quad \frac{dS}{dr} = 2\pi\left(2r - \frac{128}{r^2}\right)$$

and $$\frac{d^2S}{dr^2} = 2\pi\left(2 + \frac{128 \times 2}{r^3}\right)$$

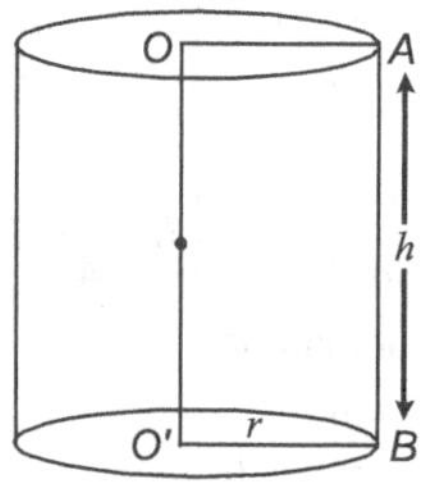

For maxima or minima,

$$\frac{dS}{dr} = 0$$

$$\Rightarrow 2\pi\left(2r - \frac{128}{r^2}\right) = 0$$

$$\Rightarrow \quad 2r^3 = 128$$

$$\Rightarrow \quad r^3 = 64$$

$$\Rightarrow \quad r = 4 \text{ cm}$$

and $$\frac{d^2S}{dr^2} = 2\pi\left(2 + \frac{256}{64}\right) = 12\pi > 0$$

$\Rightarrow S$ is minimum.

The dimensions of such a can are

$$r = 4 \text{ cm}$$

and $$h = \frac{128}{r^2} = \frac{128}{4^2} = \frac{128}{16} = 8 \text{ cm.}$$

Q. 21. Prove that the least perimeter of an isosceles triangle in which a circle of radius r can be inscribed is $6\sqrt{3}r$.★

Sol. Let ABC be an isosceles triangle with $AB = AC$ and a circle with centre O and radius r, touching sides AB, BC, CA at D, F, E respectively.

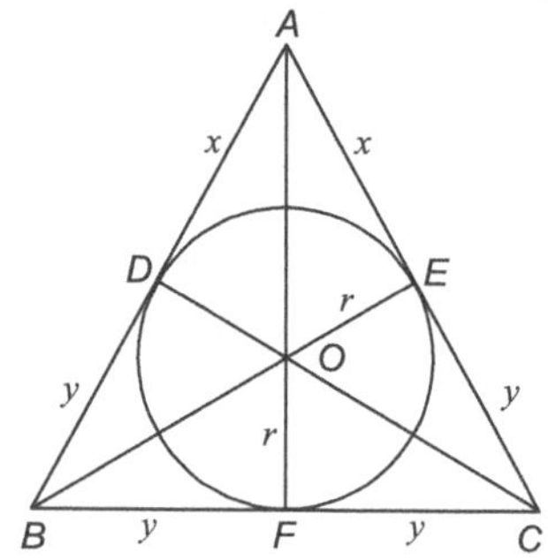

In ΔABC,

Let $AD = AE = x$, $BD = BF = y$ and $CF = CE = y$

($\because$ Tangents drawn from an external point are equal)

Now, ar (ΔABC) = ar (ΔAOB) + ar (ΔAOC) + ar (ΔBOC)

$$\Rightarrow \frac{1}{2} \times 2y(r + \sqrt{r^2 + x^2})$$

$$= \frac{1}{2}\{2yr + (x+y)r + (x+y)r\}$$

$$\Rightarrow \quad y(r + \sqrt{r^2 + x^2}) = \frac{1}{2}\{2yr + 2(x+y)r\}$$

$$\Rightarrow \quad y(r + \sqrt{r^2 + x^2}) = r(y + x + y)$$

$$\Rightarrow \quad yr + y(\sqrt{r^2 + x^2}) = 2yr + rx$$

$$\Rightarrow \quad y(\sqrt{r^2 + x^2}) = rx + yr$$

Squaring both sides, we get

$$y^2(r^2 + x^2) = r^2x^2 + y^2r^2 + 2r^2xy$$

$$y^2r^2 + y^2x^2 = r^2x^2 + y^2r^2 + 2r^2xy$$

$$y^2x = r^2x + 2r^2y$$

$$x = \frac{2r^2y}{y^2 - r^2}$$

Now, P(Perimeter of ΔABC) = $2x + 4y$

$$P = \frac{4r^2y}{y^2 - r^2} + 4y$$

Differentiate above equation w.r.t. to y, we get

$$\frac{dP}{dy} = \frac{4r^2(y^2 - r^2) - 4r^2y(2y)}{(y^2 - r^2)^2} + 4$$

★ are board exam questions from previous years

$$= \frac{4r^2[y^2 - r^2 - 2y^2]}{(y^2 - r^2)^2} + 4$$

$$\frac{dP}{dy} = \frac{-4r^2(r^2 + y^2)}{(y^2 - r^2)^2} + 4$$

For maxima and minima at P, put

$$\frac{dP}{dy} = 0$$

$$\Rightarrow \quad \frac{-4r^2(r^2 + y^2)}{(y^2 - r^2)^2} + 4 = 0$$

$$\Rightarrow \quad r^2(r^2 + y^2) = (y^2 - r^2)^2$$

$$\Rightarrow \quad r^4 + r^2y^2 = y^4 + r^4 - 2y^2r^2$$

$$\Rightarrow \quad 3y^2r^2 = y^4$$

$$\Rightarrow \quad y^2 = 3r^2$$

$$\Rightarrow \quad y = \sqrt{3}r$$

Now, again differentiate w.r.t. y, we get

$$\frac{d^2P}{dy^2} = \frac{-4r^2(2y)(y^2 - r^2)^2 + 4r^2(r^2 + y^2)2(y^2 - r^2) \times 2y}{(y^2 - r^2)^4}$$

$$= \frac{4r^2(y^2 - r^2)[-2y(y^2 - r^2) + 4y(r^2 + y^2)]}{(y^2 - r^2)^4}$$

$$= \frac{4r^2y(y^2 - r^2)[-2y^2 + 2r^2 + 4r^2 + 4y^2]}{(y^2 - r^2)^4}$$

$$\frac{d^2P}{dy^2} = \frac{4r^2y[2y^2 + 6r^2]}{(y^2 - r^2)^3}$$

$$\left.\frac{d^2P}{dy^2}\right|_{y = \sqrt{3}r} = \frac{6\sqrt{3}}{r} > 0$$

Henec perimeter P of ΔABC is least for $y = \sqrt{3}r$ and least perimeter is

$$P = 4y + \frac{4r^2y}{y^2 - r^2}$$

$$= 4\sqrt{3}r + \frac{4r^2\sqrt{3}r}{2r^2} = 6\sqrt{3}r$$

Q. 22. *AB* is the diameter of a circle and *C* is any point on the circle. Show that the area of triangle *ABC* is maximum, when it is an isosceles triangle.★

Sol. Let r be the radius of the circle then

$AB = 2r$ [$\because AB$ is diameter]

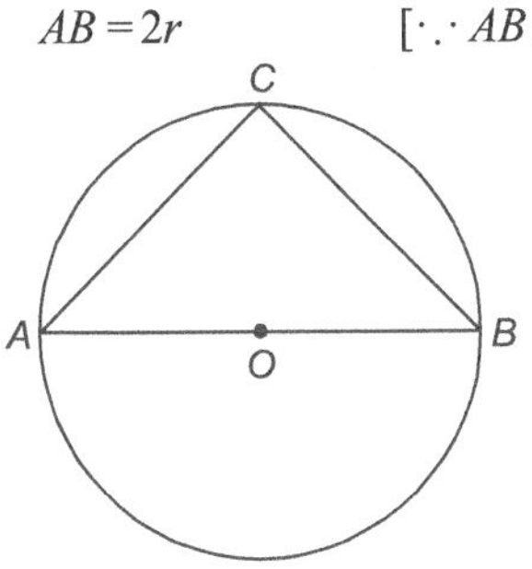

Let $BC = x$ units

We know that angle subtended by diameter in a circle is right angle.

$\therefore \quad \angle C = 90°$

Then, $AC = \sqrt{(AB)^2 - (BC)^2}$

$$AC = \sqrt{(2r)^2 - (x)^2} = \sqrt{4r^2 - x^2} \quad \ldots(i)$$

Now, area of $\Delta ABC(A) = \frac{1}{2}(AC)(BC)$

$$A = \frac{1}{2}\sqrt{4r^2 - x^2}\,(x)$$

$$\Rightarrow \quad \frac{dA}{dx} = \frac{1}{2}\left\{\sqrt{4r^2 - x^2} + x\frac{1}{2\sqrt{4r^2 - x^2}} \cdot \frac{d}{dx}(4r^2 - x^2)\right\}$$

$$\frac{dA}{dx} = \frac{1}{2}\left\{\sqrt{4r^2 - x^2} - \frac{x^2}{\sqrt{4r^2 - x^2}}\right\}$$

$$= \frac{1}{2}\left\{\frac{4r^2 - x^2 - x^2}{\sqrt{4r^2 - x^2}}\right\}$$

$$\frac{dA}{dx} = \frac{1}{2}\left\{\frac{4r^2 - 2x^2}{\sqrt{4r^2 - x^2}}\right\}$$

The critical numbers of x are given by $\frac{dA}{dx} = 0$

$$\Rightarrow \quad \frac{1}{2}\left\{\frac{4r^2 - 2x^2}{\sqrt{4r^2 - x^2}}\right\} = 0$$

$$\Rightarrow \quad 4r^2 - 2x^2 = 0$$

$$\Rightarrow \quad 4r^2 = 2x^2$$

$$x = \sqrt{2}r$$

Now, $\frac{dA}{dx} = \frac{1}{2}\left\{\frac{4r^2 - 2x^2}{\sqrt{4r^2 - x^2}}\right\}$

$$\frac{d^2A}{dx^2} = \frac{1}{2}\left\{(-4x)\frac{1}{\sqrt{4r^2 - x^2}} + (4r^2 - 2x^2)\left(\frac{-1}{2}\right)(4r^2 - x^2)^{-3/2}\frac{d}{dx}(4r^2 - x^2)\right\}$$

$$\frac{d^2A}{dx^2} = \frac{1}{2}\left\{\frac{-4x}{\sqrt{4r^2 - x^2}} + \frac{x(4r^2 - 2x^2)}{(4r^2 - x^2)^{3/2}}\right\}$$

$$\Rightarrow \quad \left(\frac{d^2A}{dx^2}\right)_{x = \sqrt{2}r} = \frac{1}{2}\left\{\frac{-4(\sqrt{2}r)}{\sqrt{4r^2 - 2r^2}} + \frac{\sqrt{2}r(4r^2 - 4r^2)}{(4r^2 - 2r^2)^{3/2}}\right\}$$

$$= \frac{-2\sqrt{2}r}{\sqrt{2}r} = -2 < 0$$

Thus, A is maximum when $x = \sqrt{2}r$

Putting $x = \sqrt{2}r$ in (i)

$$AC = \sqrt{4r^2 - (\sqrt{2}r)^2}$$

$$\therefore \quad AC = \sqrt{2}r$$

★ are board exam questions from previous years

$\therefore \quad BC = AC = \sqrt{2}r$

Hence, A is maximum when the triangle is isosceles.

Q. 23. Show that the surface area of a closed cuboid with square base and given volume is minimum, when it is a cube.★

Sol. Let the length and breadth of the cuboid of square base be x and height be y.

Then, volume of the cuboid $(V) = x^2y$

$$y = \frac{V}{x^2} \quad \text{...(i)}$$

Now, surface area of cuboid $(S) = 2(x^2 + yx + yx)$

$$S = 2(x^2 + 2xy)$$

$$S = 2\left(x^2 + 2x\left(\frac{V}{x^2}\right)\right) \quad \text{[using (i)]}$$

$$S = 2\left(x^2 + \frac{2V}{x}\right)$$

$$\therefore \quad \frac{dS}{dx} = 2\left(2x - \frac{2V}{x^2}\right)$$

For maxima and minima, we have.

$$\frac{dS}{dx} = 0 \Rightarrow 2\left(2x - \frac{2V}{x^2}\right) = 0$$

$$\Rightarrow \quad 4\left(x - \frac{V}{x^2}\right) = 0$$

$$x = \frac{V}{x^2} \Rightarrow x = \sqrt[3]{V}$$

Now, $$\frac{dS}{dx} = 2\left(2x - \frac{2V}{x^2}\right) = 4\left(x - \frac{V}{x^2}\right)$$

$$\frac{d^2S}{dx^2} = 4\left(1 + \frac{2V}{x^3}\right)$$

$$\Rightarrow \quad \left[\frac{d^2S}{dx^2}\right]_{x=\sqrt[3]{V}} = 4\left(1 + \frac{2V}{V}\right) 12 > 0$$

Thus, S is minimum when $x = \sqrt[3]{V}$

Putting $V = x^3$ in (i), we get,

$$y = \frac{V}{x^2} = \frac{x^3}{x^2} = x$$

$$\Rightarrow \quad y = x$$

Hence, it is a cube since the length, breadth and height of a cube are equal.

Q. 24. Water is running into a conical vessel, 15 cm deep and 5 cm in radius, at the rate of 0·1 cm³/sec. When the water is 6 cm deep, find at what rate is

(i) the water level rising ?

(ii) the water-surface area increasing ?

(iii) the wetted surface of the vessel increasing ?

Sol. In a conical vessel,

let $\quad VO' = h, O'A' = r$ and $VA' = l$

Let α be the semi-vertical angle of the cone.

Then, in ΔAOV and $\Delta A'O'V$

$$\tan\alpha = \frac{OA}{VO} = \frac{5}{15} = \frac{1}{3} \quad \text{...(i)}$$

and $$\tan\alpha = \frac{O'A'}{VO'} = \frac{r}{h} \quad \text{...(ii)}$$

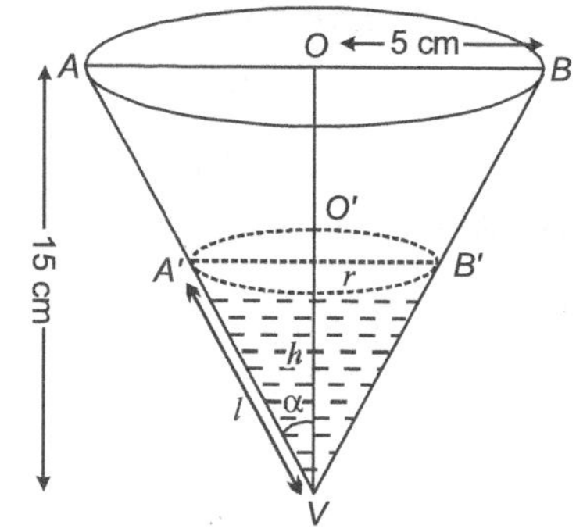

From equations (i) and (ii), we get

$$\frac{1}{3} = \frac{r}{h}$$

$$\Rightarrow \quad 3r = h.$$

(i) Let V be the volume of water in the cone.

We have, $$V = \frac{1}{3}\pi r^2 h$$

$$= \frac{1}{3}\pi\left(\frac{h}{3}\right)^2 h = \frac{\pi}{27}h^3 \quad \left[\because r = \frac{h}{3}\right]$$

Differentiate V with respect to t, we get

$$\frac{dV}{dt} = \frac{3\pi}{27}h^2\frac{dh}{dt}$$

$$\Rightarrow \quad 0{\cdot}1 = \frac{3\pi}{27}h^2\frac{dh}{dt}$$

$$\Rightarrow \quad \frac{dh}{dt} = \frac{2{\cdot}7}{3\pi h^2}$$

$$\Rightarrow \quad \left(\frac{dh}{dt}\right)_{h=6} = \frac{2.7}{3\pi.6^2} = \frac{1}{40\pi} \text{ cm/sec}$$

Thus, the water level is rising at the rate of $\frac{1}{40\pi}$ cm/sec.

(ii) Let A be the water surface area at any time t.

We know that, $A = \pi r^2$

★ are board exam questions from previous years

Therefore, $A = \pi \frac{h^2}{9}$

Differentiate A with respect to t, we get

$$\frac{dA}{dt} = \frac{2\pi h}{9}\frac{dh}{dt}$$

When, $h = 6$, $\frac{dh}{dt} = \frac{1}{40\pi}$, we have

$$\frac{dA}{dt} = \frac{2\pi \times 6}{9} \times \frac{1}{40\pi} = \frac{1}{30} \text{ cm}^2\text{/sec}$$

Thus, the water surface area is increasing at the rate of $\frac{1}{30}$ cm²/sec.

(iii) Let S be the wetted surface area of the vessel at any time t.

Then, $S = \pi r l$

We have, $l^2 = VA'^2 = VO'^2 + O'A'^2$

$\Rightarrow$ $l^2 = h^2 + r^2$

$\Rightarrow$ $l^2 = h^2 + \frac{h^2}{9}$

$\Rightarrow$ $l = \frac{\sqrt{10}h}{3}$

$\therefore$ $S = \pi r l$

$\Rightarrow$ $S = \pi\left(\frac{h}{3}\right)\left(\frac{\sqrt{10}h}{3}\right)$

$\Rightarrow$ $S = \frac{\pi}{9}\sqrt{10}h^2$

Differentiate S with respect to t, we get

$$\Rightarrow \quad \frac{dS}{dt} = \frac{2\pi\sqrt{10}h}{9}\frac{dh}{dt} \quad \text{...(i)}$$

Substitute $h = 6$, $\frac{dh}{dt} = \frac{1}{40\pi}$ in (i), we get

$$\frac{dS}{dt} = \frac{2\pi\sqrt{10}}{9} \times 6 \times \frac{1}{40\pi}$$

$$= \frac{\sqrt{10}}{30} \text{ cm}^2\text{/sec.}$$

Thus, the wetted surface area of the vessel is increasing at the rate of $\frac{\sqrt{10}}{30}$ cm²/sec.

Q. 25. Find the maximum area of the greatest isosceles triangle that can be inscribed in the ellipse $\frac{x^2}{a^2} + \frac{y^2}{b^2} = 1$ with its vertex at one end of major axis.

[NCERT]

Sol. Given equation of the ellipse is

$$\frac{x^2}{a^2} + \frac{y^2}{b^2} = 1$$

Let ABC be an isosceles triangle having the vertex at $A(a, 0)$, $B(a\cos\theta, b\sin\theta)$ and $C(a\cos\theta, -b\sin\theta)$.

$\therefore$ Area of $\Delta ABC = \frac{1}{2} \times BC \times AM$

$$= \frac{1}{2}(2b\sin\theta)(a - a\cos\theta)$$

$$= \frac{1}{2}(2ab\sin\theta)(1 - \cos\theta)$$

$\Rightarrow$ $A = ab(\sin\theta - \sin\theta\cos\theta)$

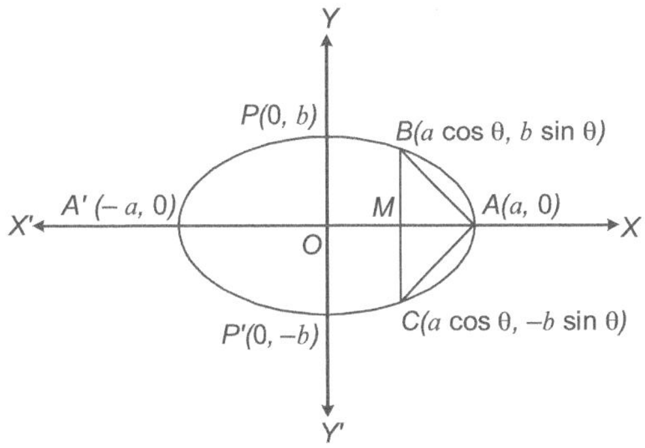

Differentiate A with respect to θ, we get

$$\frac{dA}{d\theta} = ab(\cos\theta - \cos^2\theta + \sin^2\theta)$$

$$= ab(\cos\theta - \cos 2\theta) \quad \text{...(i)}$$

For maximum or minimum value of A it must be

$$\frac{dA}{d\theta} = 0$$

$\therefore$ $ab(\cos\theta - \cos 2\theta) = 0$

$\Rightarrow$ $\cos\theta = \cos 2\theta$

$\Rightarrow$ $\theta = \frac{2\pi}{3}$

Again, differentiate (i) with respect to θ, we get

$$\frac{d^2A}{d\theta^2} = ab(-\sin\theta + 2\sin 2\theta)$$

$$\left.\frac{d^2A}{d\theta^2}\right|_{\theta = \frac{2\pi}{3}} = ab\left(-\sin\frac{2}{3}\pi + 2\sin\frac{4}{3}\pi\right)$$

$$= ab\left(\frac{-\sqrt{3}}{2} - \frac{2\sqrt{3}}{2}\right) < 0$$

Hence A has maximum area when $\theta = \frac{2\pi}{3}$. The maximum area A is given by

$$A = ab\left(\sin\frac{2\pi}{3} - \sin\frac{2\pi}{3}\cos\frac{2\pi}{3}\right)$$

$$= ab\left(\frac{\sqrt{3}}{2} + \frac{\sqrt{3}}{2} \times \frac{1}{2}\right) = \frac{3\sqrt{3}}{4}ab$$

Q. 26. A manufacturer can sell x items at a price of ₹ $\left(5-\frac{x}{100}\right)$ each. The cost price of x items is ₹ $\left(\frac{x}{5}+500\right)$. Find the number of items he should sell to earn maximum profit.

Sol. Let $S(x)$ be the selling price and $C(x)$ be the cost price of x items.

Given, manufacturer sell x items at a price of rupees $\left(5-\frac{x}{100}\right)$ each.

Then selling price of x items

$$S(x) = x\left(\frac{5-x}{100}\right)$$

$$= 5x - \frac{x^2}{100}$$

Also cost of x items $C(x) = \frac{x}{5} + 500$

$$\therefore \quad \text{Profit } P(x) = \left(5x - \frac{x^2}{100}\right) - \left(\frac{x}{5} + 500\right)$$

$$= \frac{24x}{5} - \frac{x^2}{100} - 500$$

Differentiating $P(x)$ w.r.t. x, we get

$$P'(x) = \frac{24}{5} - \frac{2x}{100}$$

For maximum/minimum it must be $P'(x) = 0$

$$\therefore \quad \frac{24}{5} - \frac{x}{50} = 0$$

$$\Rightarrow \quad x = \frac{24}{5} \times 50 = 240$$

Again, differentiating $P'(x)$ w.r.t. x, we get

$$P''(x) = -\frac{2}{100} < 0$$

Therefore, $P(x)$ is maximum for $x = 240$.

Hence, the manufacturer should sell 240 items to earn maximum profit.

Q. 27. A window is in the form of a rectangle surmounted by a semicircular opening. The total perimeter of the window is 10 m. Find the dimensions of the window to admit maximum light through the whole opening.★

Sol. Let $ABCD$ be a window of rectangular form surmounted by a semicircle with diameter AB. Given, Perimeter of the window $(P) = 10$ m.

Let the length and breadth of the rectangle be $2x$ and $2y$ respectively.

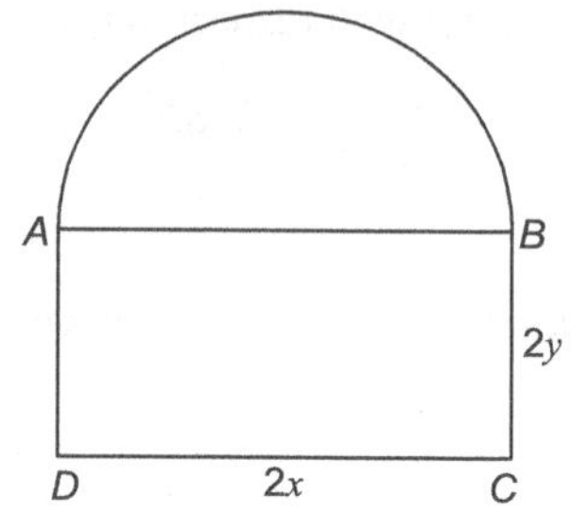

Since, $\quad P = 10$ m

i.e., $\quad 2x + 4y + \pi x = 10$

$\Rightarrow \quad 4y = 10 - 2x - \pi x$

$\Rightarrow \quad 4xy = 10x - 2x^2 - \pi x^2 \quad \ldots$(i)

Now, area of the window (A)

$$= (2x)(2y) + \frac{1}{2}\pi x^2$$

$$A = 4xy + \frac{1}{2}\pi x^2$$

$$A = 10x - 2x^2 - \pi x^2 + \frac{1}{2}\pi x^2$$

[using (i)]

$$A = 10x - 2x^2 - \frac{1}{2}\pi x^2$$

On differentiating w.r.t. x.

$$\frac{dA}{dx} = 10 - 4x - \pi x$$

The critical numbers of x are given by $\frac{dA}{dx} = 0$

$\Rightarrow \quad 10x - 4x - \pi x = 0$

$\Rightarrow \quad -x(4+\pi) = -10$

$$x = \frac{10}{4+\pi}$$

Now, $\quad \frac{dA}{dx} = 10 - 4x - \pi x$

$$\frac{d^2A}{dx^2} = -4 - \pi = -(4+\pi) < 0$$

Thus, A is maximum when $x = \frac{10}{4+\pi}$ m

Now, length of the window $= 2x = \frac{20}{4+\pi}$ m and

Width of the window

$$= 2y = \frac{10 - 2x - \pi x}{2} \quad \text{[using (i)]}$$

$$= 5 - \frac{10}{4+\pi} - \frac{\pi}{2} \cdot \frac{10}{4+\pi}$$

★ are board exam questions from previous years

$$= \frac{10(4+\pi)-20-10\pi}{2(4+\pi)}$$

$$= \frac{40+10\pi-20-10\pi}{2(4+\pi)}$$

$$= \frac{10}{4+\pi} \text{ m}$$

Also, radius of the semi-circle is $\left(\frac{10}{4+\pi}\right)$ m

Hence, the dimensions of the rectangular part of the window are $\frac{20}{4+\pi}$ m and $\frac{10}{4+\pi}$ m.

Q. 28. A window has the shape of a rectangle surmounted by an equilateral triangle. If the perimeter of the window is 12 cm, find the dimensions of the rectangle that will produce the largest area of the window.

Sol. Let x and y be the length and breadth respectively of the rectangular window and x be the side of equilateral triangle.

Let A be the area of window, then

$$A = xy + \frac{\sqrt{3}}{4}x^2 \quad \text{...(i)}$$

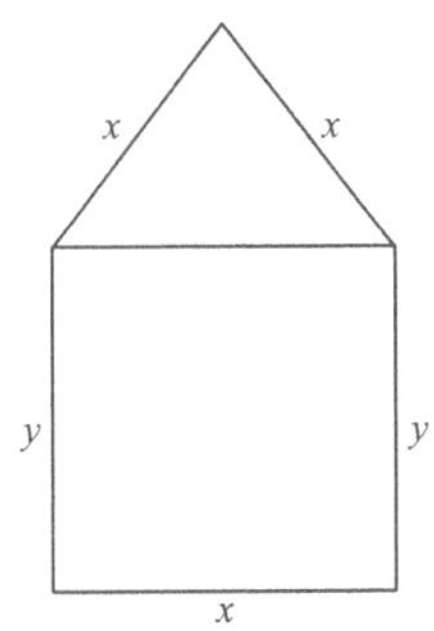

Also $\quad x+2y+2x = 12$

$\Rightarrow \quad 3x+2y = 12$

$\Rightarrow \quad y = \frac{12-3x}{2}$

$\therefore \quad A = x.\frac{(12-3x)}{2} + \frac{\sqrt{3}}{4}x^2 \quad$ [from (i)]

$$A = 6x - \frac{3x^2}{2} + \frac{\sqrt{3}}{4}x^2$$

On differentiating with respect to x, we get

$\Rightarrow \quad A' = 6 - 3x + \frac{\sqrt{3}}{2}x$

Now, for maximum or minimum value of A,

Put $\quad A' = 0$

$$6 - 3x + \frac{\sqrt{3}}{2}x = 0$$

$\Rightarrow \quad x = \frac{12}{6-\sqrt{3}}$

Again, differentiate A' with respect to x, we get

$$A'' = -3 + \frac{\sqrt{3}}{2} < 0 \text{ (for any value of } x)$$

$$A''\big]_{x = \frac{12}{6-\sqrt{3}}} < 0$$

i.e., A is maximum if $x = \frac{12}{6-\sqrt{3}}$ and $y = \frac{12-3\left(\frac{12}{6-\sqrt{3}}\right)}{2}$

i.e., For largest area of window, dimensions of rectangle are

$$x = \frac{12}{6-\sqrt{3}} \text{ and } y = \frac{18-6\sqrt{3}}{6-\sqrt{3}}.$$

Q. 29. The volume of spherical balloon being inflated changes at a constant rate. If initially its radius is 3 units and after 3 seconds it is 6 units, find the radius of the balloon after *t* seconds. [NCERT]

Sol. Let the radius of the balloon of volume V at time t seconds be r, then

$$\frac{d}{dt}(V) = \text{constant}$$

$\Rightarrow \quad \frac{d}{dt}\left(\frac{4}{3}\pi r^3\right) = \text{constant}$

$\Rightarrow \quad \frac{4}{3}\pi.3r^2 \cdot \frac{dr}{dt} = k \text{ (say)}$

$\Rightarrow \quad 4\pi r^2 \frac{dr}{dt} = k$

$\Rightarrow \quad 4\pi r^2 dr = k\, dt \quad$ ...(i)

Integrating both sides

$$\frac{4\pi r^3}{3} = kt + c \quad \text{...(ii)}$$

When $t = 0, r = 3$

$$c = 4 \times \frac{27\pi}{3}$$

$$c = 36\pi \quad \text{...(iii)}$$

Put in equation (ii)

$$\frac{4\pi r^3}{3} = kt + 36\pi \quad \text{...(iv)}$$

Also when $t = 3, r = 6$

$\Rightarrow \quad \frac{4\pi(216)}{3} = 3k + 36\pi$

$\Rightarrow \quad 3k = 288\pi - 36\pi$

$$\Rightarrow \quad 3k = 252\pi$$

$$\Rightarrow \quad k = 84\pi$$

Equation (iv) becomes

$$\frac{4}{3}\pi r^3 = 84\pi t + 36\pi$$

$$\Rightarrow \quad \frac{4}{3}\pi r^3 = 12\pi(7t+3)$$

$$\Rightarrow \quad r^3 = \frac{12\pi \times 3(7t+3)}{4\pi}$$

$$\Rightarrow \quad r^3 = 9(7t+3)$$

$$\Rightarrow \quad r = \{9(7t+3)\}^{1/3}$$

Q. 30. A point on the hypotenuse of a right triangle is at distances a and b from the sides of the triangle. Show that the minimum length of the hypotenuse is $(a^{2/3} + b^{2/3})^{3/2}$. **[NCERT]**

Sol. Let ABC be a right angled triangle, Point P on AB as

$$PL = a \text{ and } PM = b$$

Let $\quad \angle BAC = \theta$

In ΔAPL, $AP = a \operatorname{cosec}\theta$ and in ΔBMP, $BP = b\sec\theta$ ...(i)

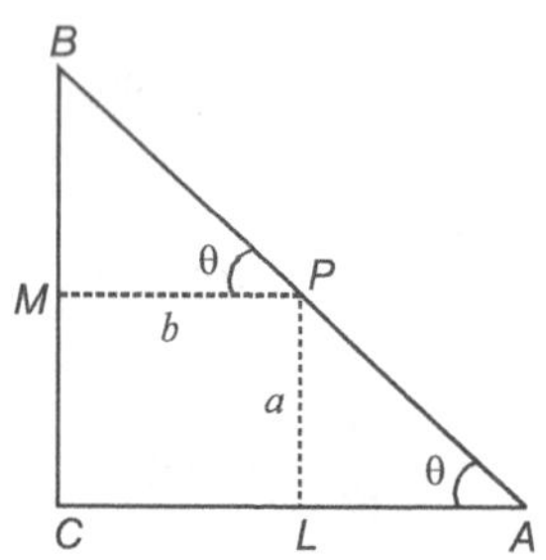

Let l be the length of the hypotenuse AB. Then

$$l = AP + BP$$

$$\Rightarrow \quad l = a\operatorname{cosec}\theta + b\sec\theta \quad \text{[from (i)]}$$

Differentiating l w.r.t. θ, we get

$$\frac{dl}{d\theta} = -a\operatorname{cosec}\theta\cot\theta + b\sec\theta\tan\theta$$

Again, differentiate $\frac{dl}{d\theta}$ w.r.t. θ, we get

$$\frac{d^2l}{d\theta^2} = a\operatorname{cosec}^3\theta + a\operatorname{cosec}\theta\cot^2\theta + b\sec^3\theta + b\sec\theta\tan^2\theta$$

$$= a\operatorname{cosec}\theta(\operatorname{cosec}^2\theta + \cot^2\theta) + b\sec\theta(\sec^2\theta + \tan^2\theta)$$

$$= a\operatorname{cosec}\theta(1 + 2\cot^2\theta) + b\sec\theta(1 + 2\tan^2\theta)$$

For maximum or minimum it must have $\frac{dl}{d\theta} = 0$, we get

$$-a\operatorname{cosec}\theta\cot\theta + b\sec\theta\tan\theta = 0$$

$$\Rightarrow \quad -\frac{a\cos\theta}{\sin^2\theta} + \frac{b\sin\theta}{\cos^2\theta} = 0$$

$$\Rightarrow \quad \tan^3\theta = \frac{a}{b}$$

$$\Rightarrow \quad \tan\theta = \left(\frac{a}{b}\right)^{1/3}$$

We can write θ as $\sin\theta = \dfrac{a^{1/3}}{\sqrt{a^{2/3} + b^{2/3}}}$

and $\quad \cos\theta = \dfrac{b^{1/3}}{\sqrt{a^{2/3} + b^{2/3}}}$

Since θ is acute angle, we have $\dfrac{d^2l}{d\theta^2} > 0$

For $\quad \tan\theta = \left(\dfrac{a}{b}\right)^{1/3}$

Thus, l is minimum when $\tan\theta = \left(\dfrac{a}{b}\right)^{1/3}$

The minimum value of l is given by

$$l = a\operatorname{cosec}\theta + b\sec\theta$$

$$= a\sqrt{1+\cot^2\theta} + b\sqrt{1+\tan^2\theta}$$

$$\Rightarrow \quad l = a\sqrt{1+\left(\frac{b}{a}\right)^{2/3}} + b\sqrt{1+\left(\frac{a}{b}\right)^{2/3}}$$

$$\Rightarrow \quad l = (a^{2/3} + b^{2/3})^{3/2}.$$

Chapter 7. Integrals

Q. 1. Evaluate : $\displaystyle\int_{\pi/6}^{\pi/3} \frac{dx}{1+\sqrt{\cot x}}$. ★

Sol. Let $\displaystyle I = \int_{\pi/6}^{\pi/3} \frac{dx}{1+\sqrt{\cot x}}$

$$= \int_{\pi/6}^{\pi/3} \frac{dx}{1+\dfrac{\sqrt{\cos x}}{\sqrt{\sin x}}}$$

$$= \int_{\pi/6}^{\pi/3} \frac{\sqrt{\sin x}}{\sqrt{\sin x}+\sqrt{\cos x}}\,dx \quad \text{...(i)}$$

$$I = \int_{\pi/6}^{\pi/3} \frac{\sqrt{\sin(\pi/2 - x)}}{\sqrt{\sin(\pi/2-x)} + \cos(\pi/2 - x)}\,dx$$

$$\left[\because \int_a^b f(x)\,dx = \int_a^b (a+b-x)\,dx \text{ and } a+b = \frac{\pi}{6}+\frac{\pi}{3} = \frac{\pi}{2}\right]$$

★ are board exam questions from previous years

$$\Rightarrow \quad I = \int_{\pi/6}^{\pi/3} \frac{\sqrt{\cos x}}{\sqrt{\cos x} + \sqrt{\sin x}}\, dx \qquad \ldots\text{(ii)}$$

Adding (i) and (ii), we get

$$\therefore \quad I + I = \int_{\pi/6}^{\pi/3} \left[\frac{\sqrt{\sin x} + \sqrt{\cos x}}{\sqrt{\sin x} + \sqrt{\cos x}}\, dx\right]$$

$$\Rightarrow \quad 2I = \int_{\pi/6}^{\pi/3} 1.dx$$

$$2I = [x]_{\pi/6}^{\pi/3} = \frac{\pi}{3} - \frac{\pi}{6} = \frac{\pi}{6}$$

$$\therefore \quad I = \frac{\pi}{12}$$

Q. 2. Evaluate : ★

$$\int \frac{1}{\sin^4 x + \sin^2 x \cos^2 x + \cos^4 x}\, dx.$$

Sol. Let $I = \int \frac{1}{\sin^4 x + \sin^2 x \cos^2 x + \cos^4 x}\, dx$

Dividing by $\cos^4 x$ in Nr. and Dr., we get

$$= \int \frac{\sec^4 x\, dx}{\tan^4 x + \tan^2 x + 1}$$

$$= \int \frac{\sec^2 x \cdot \sec^2 x}{\tan^4 x + \tan^2 x + 1}\, dx$$

$$= \int \frac{(1 + \tan^2 x)\sec^2 x\, dx}{\tan^4 x + \tan^2 x + 1}$$

$$[\because \sec^2 x = 1 + \tan^2 x]$$

Put $\tan x = t$

$\Rightarrow \quad \sec^2 x\, dx = dt$

$$\therefore \quad I = \int \frac{1 + t^2}{t^4 + t^2 + 1}\, dt$$

$$= \int \frac{1 + \frac{1}{t^2}}{t^2 + 1 + \frac{1}{t^2}}\, dt = \int \frac{1 + \frac{1}{t^2}}{\left(t - \frac{1}{t}\right)^2 + 3}\, dt$$

Again put

$$t - \frac{1}{t} = y$$

$$\Rightarrow \quad \left(1 + \frac{1}{t^2}\right) dt = dy$$

Thus, $$I = \int \frac{dy}{y^2 + (\sqrt{3})^2}$$

$$= \frac{1}{\sqrt{3}} \tan^{-1}\left(\frac{y}{\sqrt{3}}\right) + C$$

$$\left(\because \int \frac{dx}{x^2 + a^2} = \frac{1}{a} \tan^{-1} \frac{x}{a} + C\right)$$

$$= \frac{1}{\sqrt{3}} \tan^{-1}\left(t - \frac{1}{t}\right) / \sqrt{3} + C$$

$$= \frac{1}{\sqrt{3}} \tan^{-1}\left(\frac{\tan x - \cot x}{\sqrt{3}}\right) + C$$

Q. 3. Evaluate : $\int_0^{\pi/2} \frac{x \sin x \cos x}{\sin^4 x + \cos^4 x}\, dx.$★

Sol. Let $$I = \int_0^{\pi/2} \frac{x \sin x \cos x}{\sin^4 x + \cos^4 x}\, dx \qquad \ldots\text{(i)}$$

$$\therefore \quad I = \int_0^{\pi/2} \frac{\left(\frac{\pi}{2} - x\right) \sin\left(\frac{\pi}{2} - x\right) \cos\left(\frac{\pi}{2} - x\right)}{\sin^4\left(\frac{\pi}{2} - x\right) + \cos^4\left(\frac{\pi}{2} - x\right)}\, dx$$

$$\left[\because \int_0^a f(x)\, dx = \int_0^a f(a - x)\, dx\right]$$

$$= \int_0^{\pi/2} \frac{\left(\frac{\pi}{2} - x\right) \cos x \sin x}{\cos^4 x + \sin^4 x}\, dx$$

$$\left[\because \sin\left(\frac{\pi}{2} - x\right) = \cos x \text{ and } \cos\left(\frac{\pi}{2} - x\right) = \sin x\right]$$

$$\therefore \quad I = \int_0^{\pi/2} \frac{\left(\frac{\pi}{2} - x\right) \sin x \cos x}{\sin^4 x + \cos^4 x}\, dx \qquad \ldots\text{(ii)}$$

Adding (i) and (ii), we get

$$2I = \int_0^{\pi/2} \frac{\frac{\pi}{2} \sin x \cos x}{\sin^4 x + \cos^4 x}$$

Dividing Nr and Dr by $\cos^4 x$, we get

$$I = \frac{\pi}{4} \int_0^{\pi/2} \frac{\tan x \sec^2 x}{\tan^4 x + 1}\, dx$$

Put $\tan^2 x = t$

$\Rightarrow \quad 2 \tan x \sec^2 x\, dx = dt$

Also, $x = 0$

$\Rightarrow \quad t = 0$

and $x = \frac{\pi}{4}$

$\Rightarrow \quad t = \infty$

$$\therefore \quad I = \frac{\pi}{8} \int_0^{\infty} \frac{dt}{t^2 + 1} = \frac{\pi}{8} \left[\tan^{-1} t\right]_0^{\infty}$$

$$= \frac{\pi}{8} [\tan^{-1} \infty - \tan^{-1} 0]$$

$$= \frac{\pi}{8}\left(\frac{\pi}{2} - 0\right) = \frac{\pi^2}{16}$$

★ are board exam questions from previous years

Q. 4. Evaluate :★

$$\int_0^{\pi/4} \frac{\sin x+\cos x}{16+9\sin 2x} \, . \, dx$$

Sol. Let $I = \int_0^{\pi/4} \frac{\sin x+\cos x}{16+9\sin 2x} \, . \, dx$

$$= \int_0^{\pi/4} \frac{\sin x+\cos x}{16+9[1-(\sin x-\cos x)^2]} \, . \, dx$$

$$= \int_0^{\pi/4} \frac{\sin x+\cos x}{25-9(\sin x-\cos x)^2]} \, . \, dx$$

Put $\sin x - \cos x = t$

When $x = 0, t = -1$

When $x = \frac{\pi}{4}, t = 0$

$$I = \int_{-1}^{0} \frac{dt}{25-9t^2} = \frac{1}{9}\int_{-1}^{0} \frac{dt}{\frac{25}{9}-t^2}$$

$$= \frac{1}{9}\int_{-1}^{0} \frac{dt}{\left(\frac{5}{3}\right)^2-t^2}$$

$$= \frac{1}{9}\left[\frac{1}{2\times\frac{5}{3}}\log\left|\frac{\frac{5}{3}+t}{\frac{5}{3}-t}\right|\right]_{-1}^{0}$$

$$= \frac{1}{9}\left[\frac{1}{\frac{10}{3}}\left(\log\left|\frac{\frac{5}{3}}{\frac{5}{3}}\right| - \log\left|\frac{\frac{2}{3}}{\frac{8}{3}}\right|\right)\right]$$

$$= \frac{1}{30}\left[\log 1 - \log\frac{1}{4}\right]$$

$$= \frac{1}{30}[\log 1 - (\log 1 - \log 4)]$$

$$= \frac{1}{30}\log 4 = \frac{1}{30}\log (2)^2 = \frac{2}{30}\log 2$$

$$= \frac{1}{15}\log 2.$$

Q. 5. Find $\int \frac{1}{x-x^3} dx$ **[NCERT]**

Sol. $\int \frac{1}{x-x^3} dx = \int \frac{dx}{x(1-x^2)}$

$$= \int \frac{dx}{x(1-x)(1+x)}$$

$$\frac{1}{x(1-x)(1+x)} = \frac{A}{x}+\frac{B}{1-x}+\frac{C}{1+x}$$

$$1 = A(1-x^2)+B(x)(1+x)+C(x)(1-x)$$

Put $x = 1$ then

$$1 = 2B$$

$$B = \frac{1}{2}$$

Put $x = 0$ then

$$1 = A(1)$$

$$A = 1$$

Put $x = -1$ then

$$1 = C(-1)(1+1)$$

$$-2C = 1$$

$$C = -\frac{1}{2}$$

$$I = \int\frac{dx}{x}+\frac{1}{2}\int\frac{dx}{1-x}-\frac{1}{2}\int\frac{dx}{1+x}$$

$$= \log x+\frac{1}{2}\frac{\log|1-x|}{-1}-\frac{1}{2}\log(1+x)+c$$

$$= \log x-\frac{1}{2}\{\log(1-x)+\log(1+x)\}+c$$

$$= \log x-\frac{1}{2}\log\{(1-x)(1+x)\}+c$$

$$I = \log x-\frac{1}{2}\log(1-x^2)+c$$

Q. 6. Evaluate : $\int_0^1 \frac{dx}{\sqrt{1+x}-\sqrt{x}}$ **[NCERT]**

Sol. Rationalising the denominator

$$I = \int_0^1 \frac{\sqrt{1+x}+\sqrt{x}}{(\sqrt{1+x}-\sqrt{x})(\sqrt{1+x}+\sqrt{x})} dx$$

$$= \int_0^1 \frac{\sqrt{1+x}+\sqrt{x}}{(\sqrt{1+x})^2-(\sqrt{x})^2} dx$$

$$I = \int_0^1 \frac{\sqrt{1+x}+\sqrt{x}}{1+x-x} dx$$

$$= \int_0^1 (\sqrt{1+x}+\sqrt{x})\, dx$$

$$= \left[\frac{(1+x)^{1/2+1}}{1/2+1}\right]_0^1 + \left[\frac{x^{1/2+1}}{1/2+1}\right]_0^1$$

$$= \frac{2}{3}\left[(1+x)^{3/2}+(x)^{3/2}\right]_0^1$$

$$I = \frac{2}{3}[(1+1)^{3/2}-(1+0)^{3/2}]+\frac{2}{3}[1-0]$$

$$= \frac{2}{3}(2^{3/2}-1)+\frac{2}{3}$$

$$= \frac{2}{3}\cdot 2^{3/2}-\frac{2}{3}+\frac{2}{3}$$

$$= \frac{2}{3}(8)^{1/2} = \frac{2}{3}\cdot 2\sqrt{2}$$

$$I = \frac{4\sqrt{2}}{3}$$

★ are board exam questions from previous years

Q. 7. Evaluate : $\int (\sqrt{\cot x} + \sqrt{\tan x})\, dx$.★

Sol. Let $I = \int (\sqrt{\cot x} + \sqrt{\tan x})\, dx$

Put $\sqrt{\tan x} = t$

$\Rightarrow \tan x = t^2$

$\Rightarrow \sec^2 x\, dx = 2t\, dt$

$$\Rightarrow dx = \frac{2t\, dt}{1+\tan^2 x} \quad [\because \sec^2 x = 1 + \tan^2 x]$$

$$= \frac{2t}{1+t^4}\, dt$$

$$\therefore \quad I = \int \left(\frac{1}{t} + t\right) \cdot \frac{2t}{1+t^4}\, dt$$

$$= 2\int \frac{t^2+1}{t^4+1}\, dt$$

$$= 2\int \frac{1+\frac{1}{t^2}}{t^2+\frac{1}{t^2}}\, dt$$

$$= 2\int \frac{1+\frac{1}{t^2}}{\left(t-\frac{1}{t}\right)^2+2}\, dt$$

Again put,

$$t - \frac{1}{t} = y$$

$$\Rightarrow \left(1+\frac{1}{t^2}\right) dt = dy$$

$$\Rightarrow \quad I = 2\int \frac{dy}{y^2+(\sqrt{2})^2} + c$$

$$= 2 \cdot \frac{1}{\sqrt{2}} \tan^{-1}\left(\frac{y}{\sqrt{2}}\right) + c$$

$$= \sqrt{2} \tan^{-1}\left(t - \frac{1}{t}\right) / \sqrt{2} + c$$

$$= \sqrt{2} \tan^{-1}\left[\frac{\sqrt{\tan x} - \sqrt{\cot x}}{\sqrt{2}}\right] + c$$

Q. 8. Evaluate : $\int_{\pi/4}^{\pi/2} \cos 2x \log \sin x\, dx$

Sol. Let $I = \int_{\pi/4}^{\pi/2} \underset{\text{II}}{\cos 2x} (\underset{\text{I}}{\log \sin x})\, dx$

$$= \left[\frac{\log \sin x \cdot \sin 2x}{2}\right]_{\pi/4}^{\pi/2} - \int_{\pi/4}^{\pi/2} \frac{1}{\sin x} \cdot \cos x \cdot \frac{\sin 2x}{2}\, dx$$

$$= \left[\frac{1}{2} \log \sin \frac{\pi}{2} \sin \pi - \frac{1}{2} \log \sin \frac{\pi}{4} \sin \frac{\pi}{2}\right] - \int_{\pi/4}^{\pi/2} \frac{1}{2}\frac{\cos x}{\sin x} \cdot 2 \sin x \cos x\, dx$$

$$= 0 - \frac{1}{2} \log \sin \frac{\pi}{4} \cdot \sin \frac{\pi}{2} - \int_{\pi/4}^{\pi/2} \cos^2 x\, dx$$

$$= -\frac{1}{2} \log \frac{1}{\sqrt{2}} - \int_{\pi/4}^{\pi/2} \frac{1+\cos 2x}{2}\, dx$$

$$= -\frac{1}{2} \log 2^{-1/2} - \frac{1}{2}\left[x + \frac{\sin 2x}{2}\right]_{\pi/4}^{\pi/2}$$

$$= \frac{1}{4} \log 2 - \frac{1}{2}\left[\frac{\pi}{2} + \frac{1}{2} \sin \pi - \frac{\pi}{4} - \frac{1}{2} \sin \frac{\pi}{2}\right]$$

$$= \frac{1}{4} \log 2 - \frac{1}{2}\left[\frac{\pi}{2} + 0 - \frac{\pi}{4} - \frac{1}{2}\right]$$

$$= \frac{1}{4} \log 2 - \frac{\pi}{4} + \frac{\pi}{8} + \frac{1}{4}$$

$$= \frac{1}{4} \log 2 + \frac{\pi - 2\pi}{8} + \frac{1}{4}$$

$$= \frac{1}{4} \log 2 - \frac{\pi}{8} + \frac{1}{4}$$

Q. 9. Evaluate :

$$\int_0^{\pi/2} 2 \sin x \cos x \tan^{-1} (\sin x)\, dx$$

Sol. Let $I = \int_0^{\pi/2} 2 \sin x \cos x \tan^{-1} (\sin x)\, dx$

Put $\sin x = t$

$\cos x\, dx = dt$

if $x = 0, \quad t = 0$

if $x = \frac{\pi}{2}, \quad t = 1$

$$\therefore \quad I = 2\int_0^1 \underset{\text{II}}{t} \underset{\text{I}}{\tan^{-1} t}\, dt$$

$$= 2\left[\tan^{-1} t \int_0^1 t\, dt - \int_0^1 \frac{1}{1+t^2} \cdot \frac{t^2}{2}\, dt\right]$$

$$= 2\left[\frac{t^2 \tan^{-1} t}{2}\right]_0^1 - \frac{2}{2}\int_0^1 \frac{t^2}{1+t^2}\, dt$$

$$= 2\left\{\frac{1 \cdot \tan^{-1} 1}{2} - 0\right\} - \frac{2}{2}\int_0^1 \frac{1+t^2-1}{1+t^2}\, dt$$

$$= 2 \times \frac{\pi}{8} - \int_0^1 \frac{1+t^2}{1+t^2}\, dt + \int_0^1 \frac{dt}{1+t^2}$$

$$= \frac{\pi}{4} - \int_0^1 dt + \int_0^1 \frac{dt}{1+t^2}$$

$$= \frac{\pi}{4} - [t]_0^1 + [\tan^{-1} t]_0^1$$

$$= \frac{\pi}{4} - (1-0) + (\tan^{-1} 1 - \tan^{-1} 0)$$

$$= \frac{\pi}{4} - 1 + \frac{\pi}{4} = \frac{2\pi}{4} - 1$$

$$I = \frac{\pi}{2} - 1$$

★ are board exam questions from previous years

Q. 10. Calculate $\int_1^4 (x^2 - x)\, dx$ as a limit of sum.

Sol. $\int_1^4 (x^2 - x)\, dx = \int_a^b f(x)\, dx$

where $a = 1, b = 4, h = \frac{4-1}{n} = \frac{3}{n}$

$f(x) = x^2 - x$, as $h \to 0, n \to \infty$

$$\therefore \int_1^4 (x^2 - x)\, dx = \lim_{h \to 0} h[f(a) + f(a+h) + f(a+2h) + \ldots + f[a + (n-1)h]$$

$$= \lim_{h \to 0} h[1^2 - 1 + (1+h)^2 - (1+h) + (1+2h)^2 - (1+2h) + \ldots + \{1 + (n-1)h\}^2 - \{1 + (n-1)h\}]$$

$$= \lim_{h \to 0} h[1 + h^2 + 2h - 1 - h + 1 + 2^2h^2 + 2.2h - 1 - 2h + \ldots + 1 + (n-1)^2 h^2 + 2(n-1)h]$$

$$= \lim_{h \to 0} h[h^2\{1^2 + 2^2 + 3^2 + \ldots + (n-1)^2\} + 2h(1 + 2 + 3 + \ldots + (n-1)]$$

$$= \lim_{h \to 0} h^2 \left[h \frac{n(n-1)(2n-1)}{6} + \frac{n(n-1)}{2}\right]$$

$$= \lim_{h \to 0} \left[h^3 \frac{n(n-1)(2n-1)}{6} + h^2 \frac{n(n-1)}{2}\right]$$

$$= \lim_{n \to \infty} \left[\frac{27}{6} \cdot \frac{n}{n} \cdot \frac{n-1}{n} \cdot \frac{2n-1}{n} + \frac{9}{n^2} \cdot \frac{n(n-1)}{2}\right]$$

$$= \lim_{n \to \infty} \left[\frac{27}{6}\left(1 - \frac{1}{n}\right)\left(2 - \frac{1}{n}\right) + \frac{9}{2} \cdot \frac{n}{n} \cdot \frac{n-1}{n}\right]$$

$$= \lim_{n \to \infty} \left[\frac{9}{2}\left(1 - \frac{1}{n}\right)\left(2 - \frac{1}{n}\right) + \frac{9}{2}\left(1 - \frac{1}{n}\right)\right]$$

$$= \frac{9}{2}(1)\,(2) + \frac{9}{2}(1)$$

$$= 9 + \frac{9}{2}$$

$$= \frac{18+9}{2}$$

$$\int_1^4 (x^2 - x)\, dx = \frac{27}{2}$$

Q. 11. Evaluate $\int_2^5 (x^2 + 3)\, dx$ as limit of sum.

Sol. $\int_2^5 (x^2 + 3)\, dx$

$a = 2, b = 5, h = \frac{3}{n}$, as $h \to 0, n \to \infty$

$$= \lim_{h \to 0} h[f(2) + f(2+h) + \ldots + f[2 + (n-1)h]\}$$

$$= \lim_{h \to 0} h[2^2 + 3 + 2^2 + h^2 + 4h + 3 + \ldots + 2^2 + (n-1)^2 h^2 + 2 \cdot (n-1)h + 3]$$

$$= \lim_{h \to 0} h\Big[\underbrace{(4 + 4 + \ldots + 4}_{n \text{ times}} + \underbrace{(3 + 3 + \ldots 3)}_{n \text{ times}} + h^2\{(1 + 2^2 + \ldots + (n-1)^2\} + 4h\{1 + 2 + \ldots + (n-1)\}\Big]$$

$$= \lim_{h \to 0} h\left[4n + 3n + h^2 \cdot \frac{n(n-1)(2n-1)}{6} + 4h\frac{n(n-1)}{2}\right]$$

$$= \lim_{h \to 0} \left[7nh + h^3 \frac{n(n-1)(2n-1)}{6} + 2h^2\, n(n-1)\right]$$

$$= \lim_{n \to \infty} \left[7n \cdot \frac{3}{n} + \frac{27}{6} \cdot \frac{n}{n}\left(\frac{n-1}{n}\right)\left(\frac{2n-1}{n}\right) + 18 \cdot \frac{n}{n}\left(\frac{n-1}{n}\right)\right]$$

$$= \lim_{n \to \infty} \left[21 + \frac{27}{6}\left(1 - \frac{1}{n}\right)\left(2 - \frac{1}{n}\right) + 18\left(1 - \frac{1}{n}\right)\right]$$

$$= \left[21 + \frac{27}{6} \times 2 + 18(1)\right]$$

$$= \frac{21}{1} + \frac{27}{3} + \frac{18}{1}$$

$$= 21 + 9 + 18 = 48$$

Q. 12. Evaluate $\int_0^4 (x + e^{2x})\, dx$ as limit of sum. **[NCERT]**

Sol. Here $f(x) = x + e^{2x}$, $a = 0, b = 4, h = \frac{4}{n}$

$$\int_0^4 (x + e^{2x})\, dx = \lim_{h \to 0} h[f(0) + f(0+h) + f(0+2h) + \ldots + f\{0 + (n-1)h\}]$$

$$= \lim_{h \to 0} h[0 + 1 + h + e^{2h} + 2h + e^{2.2h} + \ldots + e^{n-1h} + (n-1)h]$$

$$= \lim_{h \to 0} h[h\{1 + 2 + 3 + \ldots + (n-1)\} + 1 + e^{2h} + e^{4h} + \ldots + e^{(n-1)h}]$$

$$= \lim_{h \to 0} h\left[h\frac{n(n-1)}{2} + 1\left\{\frac{(e^{2h})^n - 1}{e^{2h} - 1}\right\}\right]$$

$$= \lim_{h \to 0} \left[\frac{h^2 n(n-1)}{2} + \frac{2h}{2}\, \frac{e^{2hn} - 1}{e^{2h} - 1}\right]$$

$$= \lim_{n \to \infty} \left[\frac{16}{n^2}\, \frac{n(n-1)}{2} + \frac{1}{2}\, \frac{\dfrac{e^{2 \cdot \frac{4}{n} \cdot n} - 1}{e^{2h} - 1}}{2h}\right]$$

$$= \frac{16}{2}(1 - 0) + \frac{1}{2}\, \frac{(e^8 - 1)}{1}$$

$$= \frac{16}{2} + \frac{e^8 - 1}{2}$$

$$= 8 + \frac{e^8 - 1}{2} = \frac{16 + e^8 - 1}{2}$$

$$= \frac{15 + e^8}{2}$$

Q. 13. $\int_1^3 (x^2+3x+e^x)\,dx$ **as the limit of sum.★**

Sol. We have, $\int_1^3 (x^2+3x+e^x)\,.\,dx$

Here $f(x) = x^2 + 3x + e^x$, $a = 1$, $b = 3$, $nh = b - a = 3 - 1 = 2$

By limit of sum, we have

$$\int_a^b f(x)dx = \lim_{h\to 0} h\,[f(a+h)+f(a+2h)+\ldots\ldots. +f(a+(n-1)h]$$

where $nh = b - a$

$$\therefore \int_1^3 (x^2+3x+e^x)\,.dx$$

$$= \lim_{h\to 0} h\,[(1+3+e)+\{(1+h)^2+3(1+h)+e^{1+h}\}+\{(1+2h)^2+3(1+2h)+e^{1+2h}\}+\ldots.]$$

$$= \lim_{h\to 0} h\,[4+e+(1+h^2+2h+3+3h+e^{1+h})+(1+4h^2+4h+3+6h+e^{1+2h})+\ldots..]$$

$$= \lim_{h\to 0} h\,[4+e+(4+h^2+5h+e^{1+h})+(4+4h^2+10h+e^{1+2h})+\ldots...]$$

$$= \lim_{h\to 0} h\,[4n+e(1+e^h+e^{2h}+\ldots.)+h^2(1^2+2^2+\ldots.)+5h(1+2+\ldots.)\}$$

$$= \lim_{h\to 0} h\left[4n+e\left(\frac{e^{nh}-1}{e^h-1}\right)+h^2\,\frac{n(n-1)(2n-1)}{6}+\frac{5h\,n(n-1)}{2}\right]$$

$$= \lim_{h\to 0}\left[4nh+e.\frac{h}{e^h-1}(e^{nh}-1)+\frac{nh(nh-h)(2nh-h)}{6}+\frac{5nh(nh-h)}{2}\right]$$

$$= 4(2)+e.(e^2-1)+\frac{2(2-0)(4-0)}{6}+\frac{5(2)(2-0)}{2}$$

$$= 8+(e^3-e)+\frac{8}{3}+10$$

$$= \frac{24+8+30}{3}+e^3-e$$

$$= \frac{62}{3}+e(e^2-1).$$

Q. 14. Evaluate $\int_1^3 (e^{2-3x}+x^2+1)\,dx$ **as a limit of sum.★**

Sol. We have, $\int_a^b f(x)\,dx$

$$= \lim_{h\to 0} h\,[f(a)+f(a+h)+f(a+2h)+\ldots +f(a+(n-1)h)]$$

where $h = \frac{b-a}{n}$

Here, $a = 1, b = 3$

and $f(x) = e^{2-3x}+x^2+1$

$\therefore$ $h = \frac{2}{n}$

$\Rightarrow$ $hn = 2$

Now, $f(a) = f(1)$

$$= e^{2-3\times 1}+1^2+1$$

$$f(a+h) = f(1+h) = e^{2-3(1+h)}+(1+h)^2+1$$

$$f(a+2h) = f(1+2h) = e^{2-3(1+2h)}+(1+2h)^2+1$$

$$f(a+(n-1)h) = f(1+(n-1)h) = e^{2-3[1+(n-1)h]}+[1+(n-1)h]^2+1$$

Adding these equations, we get

$$f(a)+f(a+h)+f(a+2h)+\ldots+f(a+(n-1)h)$$

$$= [e^{2-3\times 1}+1^2+1]+[e^{2-3\times(1+h)}+(1+h)^2+1]+[e^{2-3\times(1+2h)}+(1+2h)^2+1]+\ldots+\{e^{2-3[1+(n-1)h]}+[1+(n-1)h]^2+1$$

$$\therefore \int_1^3 (e^{2-3x}+x^2+1)\,dx$$

$$= \lim_{h\to 0} h\,[e^2\cdot(e^{-3}+e^{-3(1+h)}+e^{-3(1+2h)}+\ldots+e^{-3(1+n-1)h)})]+\lim_{h\to 0} h\,[1^2+(1+h)^2+(1+2h)^2+\ldots+\{1+(n-1)h\}]^2+\lim_{h\to 0} h(1+1+\ldots n\text{ times})$$

$$= \lim_{h\to 0} h\left\{e^2\times e^{-3}\,\frac{(1-e^{-3nh})}{1-e^{-3h}}\right\}+\lim_{h\to 0}\{h(1+1+\ldots n\text{ times})+2h^2[1+2^2+3+\ldots+(n-1)]+h^3(1^2+2^2+3^2+\ldots+(n-1)^2)\}+\lim_{h\to 0} h(n)$$

$\left[\because \text{sum to } n \text{ terms of G.P. } S_n = \frac{a(1-r^n)}{1-r}\right]$

$$= \lim_{h\to 0} h\left\{\frac{e^{-1}(1-e^{-6})}{1-e^{-3h}}\right\}+\lim_{h\to 0}\left[nh+2h^2\times\frac{(n-1)n}{2}+h^3\times\frac{(n-1)n(2n-1)}{6}\right]+\lim_{h\to 0} h(n)$$

$$= \lim_{h\to 0} h\left(\frac{e^{-1}(1-e^{-6})}{1-e^{-3h}}\right)+\lim_{h\to 0}\left[2nh+(nh-h)nh+\frac{(nh-h)nh(2nh-h)}{6}\right]$$

★ are board exam questions from previous years

$$= \frac{1}{e}\left(1-\frac{1}{e^6}\right)\times\frac{\lim_{h\to 0} e^{3h}}{3\times\lim_{h\to 0}\left(\frac{e^{3h}-1}{3h}\right)}$$

$$+\lim_{h\to 0}\left[2\times 2+(2-h)\times 2+\frac{(2-h)\times 2(2\times 2-h)}{6}\right]$$

$$= \frac{1}{e}\left(1-\frac{1}{e^6}\right)\times\frac{1}{3\times 1}+\left(4+4+\frac{8}{3}\right)$$

$$= \frac{1}{3e}\left(1-\frac{1}{e^6}\right)+\frac{32}{3}$$

Chapter 8. Applications of the Integrals

Q. 1. Find the area of the region enclosed between the two curves

$$(x-6)^2+y^2=36 \text{ and } x^2+y^2=36$$

Sol. Given curves are :

$$(x-6)^2+(y-0)^2=6^2 \quad \text{...(i)}$$

and $$x^2+y^2=6^2 \quad \text{...(ii)}$$

Curve (i) is a circle whose radius is ± 6 and centre is (6, 0) and curve (ii) is also a circle whose radius is ± 6 and centre is origin (0, 0).

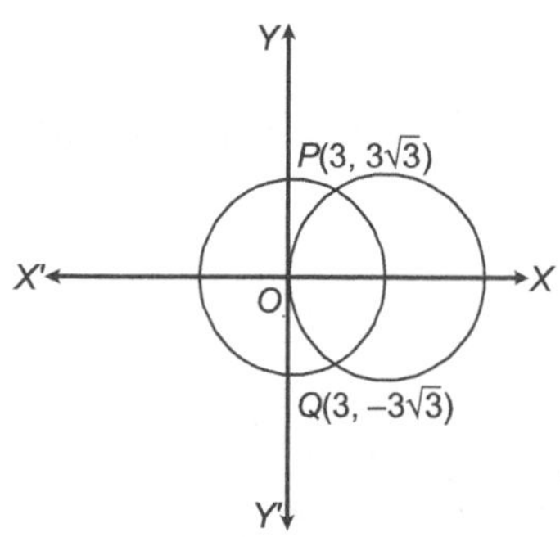

To find points of intersection subtract equation (ii) from equation (i)

$$(x-6)^2-x^2=0$$

$$x^2+36-12x-x^2=0$$

$$36-12x=0$$

$$x=3$$

On putting $x = 2$ in equation (ii) we get,

$$9+y^2=36$$

$$y^2=27$$

$$y=\pm 3\sqrt{3}$$

∴ Area bounded by the curves

$$= 2\int_0^3 \sqrt{36-(x-6)^2}\,dx+2\int_3^6\sqrt{36-x^2}\,dx$$

★ are board exam questions from previous years

$$\because \sqrt{a^2-x^2}=\frac{x}{2}\sqrt{a^2-x^2}+\frac{a^2}{2}\sin^{-1}\frac{x}{a}+c$$

∴ Required area

$$= 2\left[\frac{(x-6)\sqrt{36-(x-6)^2}}{2}+\frac{36}{2}\sin^{-1}\frac{x-6}{6}\right]_0^3$$

$$+2\left[\frac{x}{2}\sqrt{36-x^2}+\frac{36}{2}\sin^{-1}\frac{x}{6}\right]_3^6$$

$$= 2\left[\left\{\frac{-3}{2}\sqrt{36-9}+18\sin^{-1}\left(-\frac{1}{2}\right)\right\}\right.$$

$$\left.-\left\{\frac{(0-6)\times 0}{2}+18\sin^{-1}(-1)\right\}\right]$$

$$+2\left[\left\{\frac{6}{2}\times 0+18\sin^{-1}1\right\}-\left\{\frac{3}{2}\sqrt{27}+18\sin^{-1}\frac{1}{2}\right\}\right]$$

$$= 2\left[\frac{-3\sqrt{27}}{2}+\left(-18\times\frac{\pi}{6}\right)-18\left(-\frac{\pi}{2}\right)\right]$$

$$+2\left[18\times\frac{\pi}{2}-\frac{9\sqrt{3}}{2}-18\times\frac{\pi}{6}\right]$$

$$= -9\sqrt{3}-6\pi+18\pi+18\pi-9\sqrt{3}-6\pi$$

$$= -18\sqrt{3}+24\pi$$

$$= 6(4\pi-3\sqrt{3}) \text{ sq. units.}$$

Q. 2. Using integration, find the area of the region bounded between the two circles.★

$$x^2+y^2=4 \text{ and } (x-2)^2+y^2=4.$$

Sol. Given, $$x^2+y^2=4 \quad \text{...(i)}$$

$$(x-2)^2+y^2=4 \quad \text{...(ii)}$$

Equation (i) is a circle with centre (0, 0) and radius 2 and equation (ii) is a circle of centre (2, 0) and radius 2 units. From equations (i) and (ii)

$$(x-2)^2+4-x^2=4$$

$$x^2+4-4x+4-x^2=4$$

$$-4x+8-4=0$$

$$-4x=-4$$

$$x=1$$

If $x=1$, $$1+y^2=4$$

$$y^2=3$$

$$y=\pm\sqrt{3}$$

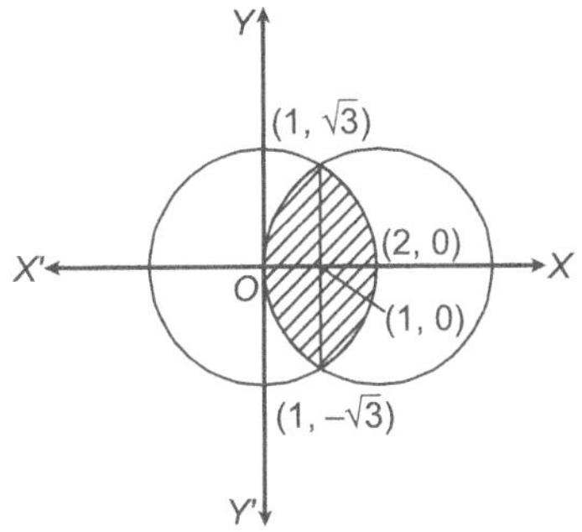

Required area

$$= 2\int_0^1 \text{Area under circle (ii)}\, dx + \int_1^2 \text{Area under circle (i)}\, dx$$

$$= 2\int_0^1 \sqrt{4-(x-2)^2}\, dx + 2\int_1^2 \sqrt{4-x^2}\, dx$$

$$= 2\left[\frac{(x-2)\sqrt{4-(x-2)^2}}{2} + \frac{4}{2}\sin^{-1}\left(\frac{x-2}{2}\right)\right]_0^1 + 2\left[\frac{x\sqrt{4-x^2}}{2} + \frac{4}{2}\sin^{-1}\frac{x}{2}\right]_1^2$$

$$= 2\left[\left\{\frac{-1\sqrt{4-1}}{2} + 2\sin^{-1}\left(\frac{-1}{2}\right)\right\} - \{0 + 2\sin^{-1}(-1)\}\right] + 2\left[\{0 + 2\sin^{-1}(1)\} - \left\{\frac{1\sqrt{3}}{2} + 2\sin^{-1}\frac{1}{2}\right\}\right]$$

$$= 2\left[-\frac{\sqrt{3}}{2} - 2\cdot\frac{\pi}{6} + 2\cdot\frac{\pi}{2}\right] + 2\left[2\times\frac{\pi}{2} - \frac{\sqrt{3}}{2} - 2\cdot\frac{\pi}{6}\right]$$

$$= 2\left[-\frac{\sqrt{3}}{2} - \frac{\pi}{3} + \pi\right] + 2\left[\pi - \frac{\sqrt{3}}{2} - \frac{\pi}{3}\right]$$

$$= -\sqrt{3} - \frac{2\pi}{3} + 2\pi + 2\pi - \sqrt{3} - \frac{2\pi}{3}$$

$$= -2\sqrt{3} + 4\pi - \frac{4\pi}{3}$$

$$= -2\sqrt{3} + \frac{12\pi - 4\pi}{3}$$

$$= \left(\frac{8\pi}{3} - 2\sqrt{3}\right) \text{ sq. units.}$$

Q. 3. Find the area bounded by curves $(x-1)^2 + y^2 = 1$ and $x^2 + y^2 = 1$ which lies to the right of $x = 0$ and to the left of $x = 1$. **[NCERT]**

Sol. $(x-1)^2 + y^2 = 1$ is a circle of centre (1, 0) and radius 1 units and $x^2 + y^2 = 1$ is a circle of centre (0, 0) and radius 1 units.

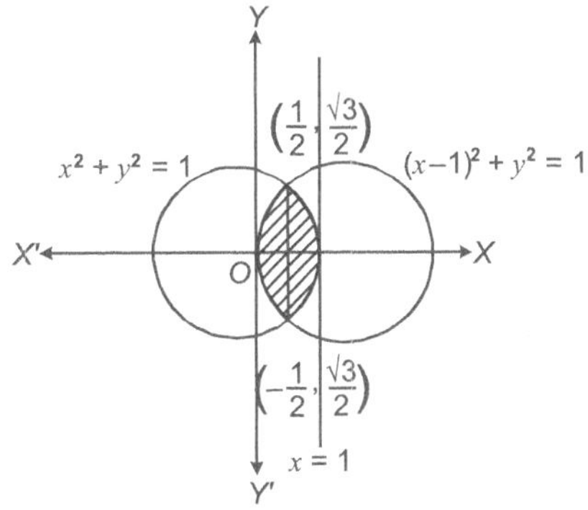

$$x^2 + y^2 = 1 \quad \ldots\text{(i)}$$

$$(x-1)^2 + y^2 = 1 \quad \ldots\text{(ii)}$$

From equations (i) and (ii)

$$(x-1)^2 + 1 - x^2 = 1$$

$$x^2 + 1 - 2x + 1 - x^2 = 1$$

$$-2x + 2 = 1$$

$$-2x = -1$$

$$x = \frac{1}{2}$$

When $x = \frac{1}{2}$ from equation (i)

$$\frac{1}{4} + y^2 = 1$$

$$y^2 = 1 - \frac{1}{4}$$

$$y = \frac{\pm\sqrt{3}}{2}$$

Required area

$$= 2\int_0^{1/2} \sqrt{1-(x-1)^2}\, dx + 2\int_{1/2}^1 \sqrt{1-x^2}\, dx$$

$$= 2\left[\frac{x-1}{2}\sqrt{1-(x-1)^2} + \frac{1}{2}\sin^{-1}\frac{x-1}{1}\right]_0^{1/2} + 2\left[\frac{x}{2}\sqrt{1-x^2} + \frac{1}{2}\sin^{-1}\frac{x}{1}\right]_{1/2}^1$$

$$= 2\left[\left\{\frac{\frac{1}{2}-1}{2}\sqrt{1-\left(\frac{1}{2}-1\right)^2} + \frac{1}{2}\sin^{-1}\left(\frac{1}{2}-1\right)\right\} - \left\{\frac{0-1}{2}\sqrt{1-(0-1)^2} + \frac{1}{2}\sin^{-1}\left(\frac{0-1}{1}\right)\right\}\right]$$

$$+ 2\left[\left\{\frac{1}{2}\times 0 + \frac{1}{2}\sin^{-1}(1)\right\} - \left\{\frac{1/2}{2}\sqrt{1-\left(\frac{1}{2}\right)^2} + \frac{1}{2}\sin^{-1}\left(\frac{1}{2}\right)\right\}\right]$$

$$= 2\left[\left\{-\frac{1}{4}\sqrt{1-\frac{1}{4}} + \frac{1}{2}\sin^{-1}\left(-\frac{1}{2}\right)\right\}\right.$$

$$-\left\{-\frac{1}{2}\times 0+\frac{1}{2}\sin^{-1}(-1)\right\}\Bigg]$$

$$+2\left[\frac{1}{2}\times\frac{\pi}{2}-\left\{\frac{1}{4}\sqrt{1-\frac{1}{4}}+\frac{1}{2}\times\frac{\pi}{6}\right\}\right]$$

$$=2\left[-\frac{1}{4}\times\frac{\sqrt{3}}{2}-\frac{1}{2}\times\frac{\pi}{6}-\left\{-\frac{1}{2}\times\frac{\pi}{2}\right\}\right]$$

$$+2\left[\frac{\pi}{4}-\left\{\frac{1}{4}\times\frac{\sqrt{3}}{2}+\frac{\pi}{12}\right\}\right]$$

$$=2\left(\frac{-\sqrt{3}}{8}-\frac{\pi}{12}\right)+2\times\frac{\pi}{4}+2\times\frac{\pi}{4}-2\times\frac{\sqrt{3}}{8}-2\times\frac{\pi}{12}$$

$$=\frac{-\sqrt{3}}{4}-\frac{\pi}{6}+\frac{\pi}{2}+\frac{\pi}{2}-\frac{\sqrt{3}}{4}-\frac{\pi}{6}$$

$$=\frac{-2\sqrt{3}}{4}-\frac{2\pi}{6}+\frac{2\pi}{2}=\frac{-\sqrt{3}}{2}-\frac{\pi}{3}+\frac{\pi}{1}$$

$$=\frac{-\sqrt{3}}{2}+\frac{3\pi-\pi}{3}$$

$$=\left(\frac{-\sqrt{3}}{2}+\frac{2\pi}{3}\right)\text{ sq. units}$$

Q. 4. Prove that the curve $y^2=4x$ and $x^2=4y$ divide the area of the square bounded by $x=0$, $x=4, y=4$ and $y=0$ into three equal parts.

Sol. The given curves are

$$y^2=4x \quad \text{...(i)}$$

and

$$x^2=4y \quad \text{...(ii)}$$

are right hand and upward parabolas respectively with vertices at (0, 0).

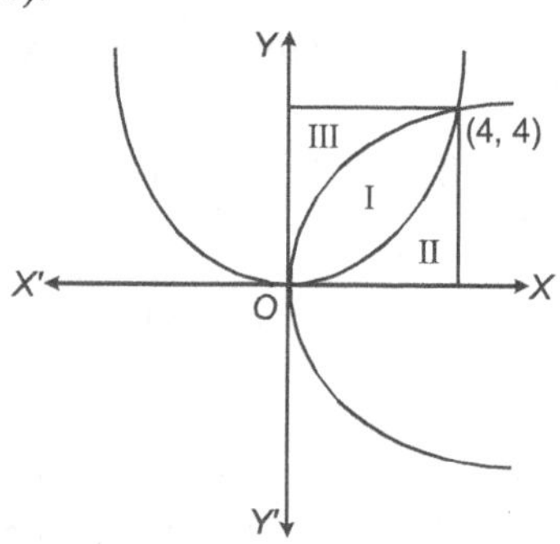

From equation (ii)

$$y=\frac{x^2}{4} \quad \text{...(iii)}$$

Put in equation (i)

$$\left(\frac{x^2}{4}\right)^2=4x$$

$$\frac{x^4}{16}=4x$$

$$x^4-64x=0$$

$$x(x^3-64)=0$$

$$x=0, 4$$

From equation (iii)

If $x=0, y=0$ and if $x=4, y=\frac{16}{4}=4$

Area of part I :

$$=\int_0^4 (\text{Area under } y^2=4x)\,dx$$

$$-\int_0^4 (\text{Area under } x^2=4y)\,dx$$

$$=\int_0^4 \sqrt{4x}\,dx-\int_0^4 \frac{x^2}{4}\,dx$$

$$=2\int_0^4 \sqrt{x}\,dx-\frac{1}{4}\int_0^4 x^2\,dx$$

$$=2\left(\frac{x^{3/2}}{3/2}\right)_0^4-\frac{1}{4}\left(\frac{x^3}{3}\right)_0^4$$

$$=\frac{4}{3}(4)^{3/2}-\frac{1}{12}(64)$$

$$=\frac{32}{3}-\frac{16}{3}=\frac{16}{3}\text{ sq. units} \quad \text{...(A)}$$

Area of part II :

$$\int_0^4 (\text{Area under } x^2=4y)\,dx$$

$$=\int_0^4 \frac{x^2}{4}\,dx=\frac{1}{4}\left(\frac{x^3}{3}\right)_0^4$$

$$=\frac{1}{12}(64)=\frac{16}{3}\text{ sq. units} \quad \text{...(B)}$$

Area of part III :

$$\int_0^4 (\text{Area under } y^2=4x)\,dy$$

$$=\int_0^4 \frac{y^2}{4}\,dy=\frac{1}{4}\left(\frac{y^3}{3}\right)_0^4$$

$$=\frac{1\times 64}{4\times 3}$$

$$=\frac{16}{3}\text{ sq. units} \quad \text{...(C)}$$

Hence, from (A), (B), (C) :

Area of region I = Area of region II = Area of region III

Q.5. Find the area of the figure bounded by the graphs of the function $y=x^2$ and $y=2x-x^2$.

Sol. Given $y=x^2$ is a upward parabola and $y=2x-x^2$ is also a parabola.

$$y = x^2 \quad \ldots(i)$$
$$y = 2x - x^2 \quad \ldots(ii)$$
$$y = -(x^2 - 2x + 1) + 1$$
$$y = -(x-1)^2 + 1$$
$$\Rightarrow \quad (y-1) = -(x-1)^2$$

(i) Equation is a parabola with vertex (0, 0), facing upwards and equation (ii) is also a parabola with vertex (1, 1), facing downwards.

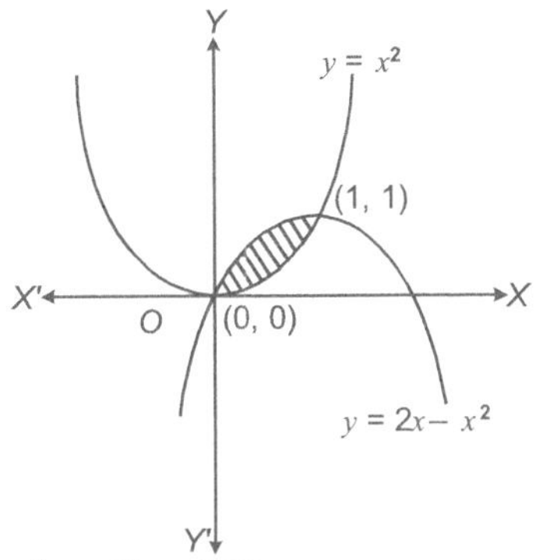

From equations (i) and (ii)

$$x^2 = 2x - x^2$$
$$2x^2 - 2x = 0$$
$$2x(x-1) = 0$$
$$x = 0, 1$$

If $x = 0$, $\quad y = 0$

If $x = 1$, $\quad y = 1$

$$\therefore \text{ Required area} = \int_0^1 (2x - x^2)\, dx - \int_0^1 x^2\, dx$$
$$= \left(2\frac{x^2}{2} - \frac{x^3}{3}\right)_0^1 - \left(\frac{x^3}{3}\right)_0^1$$
$$= \left(1 - \frac{1}{3}\right) - (0) - \left(\frac{1}{3} - 0\right)$$
$$= \frac{2}{3} - \frac{1}{3}$$
$$= \frac{1}{3} \text{ sq. units}$$

Q. 6. Find the area of the region $\{(x, y) : y^2 \leq 6ax$ and $x^2 + y^2 \leq 16a^2\}$ using method of integration.★

Sol. $\quad x^2 + y^2 \leq 16a^2 \quad \ldots(i)$

$$y^2 \leq 6ax$$

is a circle with centre (0, 0) and radius $4a$ and $y^2 \leq 6ax$ is left hand parabola

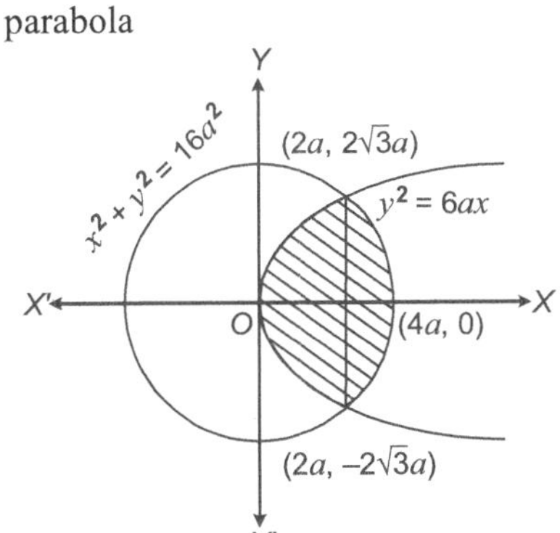

★ are board exam questions from previous years

From equations (i) and (ii)

$$x^2 + 6ax - 16a^2 = 0$$
$$x^2 + 8ax - 2ax - 16a^2 = 0$$
$$x(x + 8a) - 2a(x + 8a) = 0$$
$$(x + 8a)(x - 2a) = 0$$
$$x = 2a, -8a$$

But $x \neq -8a$ as $x > 0$.

$\Rightarrow x = 2a, \quad y^2 = 6a(2a)$

$$y^2 = 12a^2$$
$$y = \pm 2\sqrt{3}a$$

Required area

$$= 2\int_0^{2a} (\text{Area under parabola})\, dx + 2\int_{2a}^{4a} (\text{Area under circle})\, dx$$

$$= 2\left[\int_0^{2a} \sqrt{6ax}\, dx + \int_{2a}^{4a} \sqrt{16a^2 - x^2}\, dx\right]$$

$$= 2\left[\sqrt{6a}\,\frac{x^{3/2}}{3/2}\right]_0^{2a} + 2\left[\frac{x\sqrt{16a^2 - x^2}}{2} + \frac{16a^2}{2}\sin^{-1}\frac{x}{4a}\right]_{2a}^{4a}$$

$$= 2\left[\sqrt{6a}\cdot\frac{2}{3}\cdot 2a\sqrt{2a}\right] + 2\left[0 + \frac{16a^2}{2}\sin^{-1}(1) - \frac{2a\sqrt{12a^2}}{2} - 8a^2\cdot\sin^{-1}\left(\frac{1}{2}\right)\right]$$

$$= \frac{16\sqrt{3}a^2}{3} + 16a^2\cdot\frac{\pi}{2} - 4\sqrt{3}a^2 - 16a^2\cdot\frac{\pi}{6}$$

$$= \frac{16a^2\sqrt{3}}{3} + 8a^2\pi - 4\sqrt{3}a^2 - 8a^2\cdot\frac{\pi}{3}$$

$$= \frac{16a^2\sqrt{3}}{3} + \frac{24a^2\pi - 8a^2\pi}{3} - 4\sqrt{3}a^2$$

$$= \frac{16a^2\sqrt{3}}{3} + \frac{16a^2\pi}{3} - 4\sqrt{3}a^2$$

$$= \frac{16a^2\sqrt{3} - 12a^2\sqrt{3}}{3} + \frac{16a^2\pi}{3}$$

$$= \frac{4a^2\sqrt{3}}{3} + \frac{16a^2\pi}{3}$$

$$= \frac{4a^2}{3}(\pi + \sqrt{3}) \text{ sq. units.}$$

Q. 7. Find the area of region $\{(x, y) : y^2 \leq 4x, 4x^2 + 4y^2 \leq 9\}$ using method of integration. ★

Sol. $y^2 = 4x$ is a right hand parabola and $4x^2 + 4y^2 = 9$ is a circle.

$$y^2 = 4x \quad \ldots(i)$$
$$x^2 + y^2 = \frac{9}{4}$$
$$x^2 + y^2 = \left(\frac{3}{2}\right)^2 \quad \ldots(ii)$$

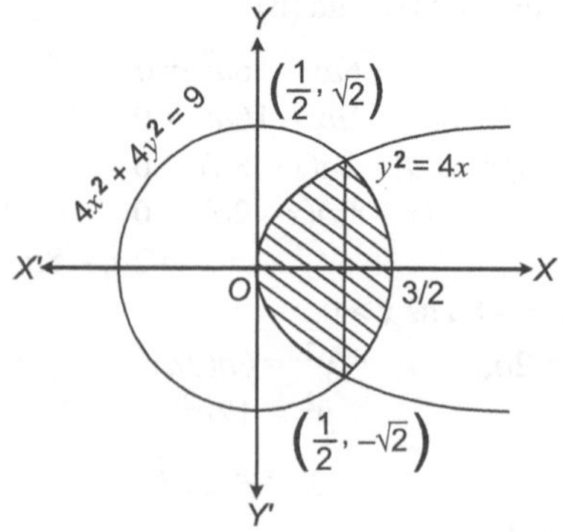

From equations (i) and (ii)

$$x^2 + 4x = \frac{9}{4}$$
$$4x^2 + 16x - 9 = 0$$
$$4x^2 + 18x - 2x - 9 = 0$$
$$2x(2x + 9) - 1(2x + 9) = 0$$
$$(2x + 9)(2x - 1) = 0$$
$$x = -\frac{9}{2}, \frac{1}{2}$$

If $x = \frac{1}{2}$, $\quad y^2 = 4 \times \frac{1}{2}$

$$y = \pm\sqrt{2}$$

Required area $= 2\int_0^{1/2}$ Area under parabola dx

$+2\int_{1/2}^{3/2}$ Area under circle dx

$$= 2\int_0^{1/2} \sqrt{4x}\, dx + 2\int_{1/2}^{3/2} \sqrt{\left(\frac{3}{2}\right)^2 - x^2}\, dx$$

$$= 2\left[2\left|\frac{x^{3/2}}{3/2}\right|_0^{1/2} + \left|\frac{x\sqrt{\frac{9}{4} - x^2}}{2} + \frac{\left(\frac{3}{2}\right)^2 \sin^{-1}\left(\frac{x}{\frac{3}{2}}\right)}{2}\right|_{1/2}^{3/2}\right]$$

$$= \frac{2}{3} \times 4\left[\left(\frac{1}{2}\right)^{3/2} - 0\right] + 2\left[\frac{\frac{3}{2}\sqrt{\frac{9}{4} - \frac{9}{4}}}{2} + \frac{9}{8}\sin^{-1} 1\right.$$
$$\left. - \frac{\frac{1}{2}\sqrt{\frac{9}{4} - \frac{1}{4}}}{2} - \frac{9}{8}\sin^{-1}\frac{1}{2} \times \frac{2}{3}\right]$$

$$= \frac{8}{3} \times \left(\frac{1}{2}\right)^{3/2} + 0 + \frac{9}{8} \times \frac{\pi}{2} \times 2 - \frac{2}{4} \times \sqrt{2} - \frac{9 \times 2}{8}\sin^{-1}\frac{1}{3}$$

$$= \frac{8}{3} \times \frac{1}{2\sqrt{2}} + \frac{9\pi}{16} \times 2 - \frac{1}{2}\sqrt{2} - \frac{9}{4}\sin^{-1}\frac{1}{3}$$

$$= \frac{4}{3} \times \frac{\sqrt{2}}{2} + \frac{9\pi}{8} - \frac{\sqrt{2}}{2} - \frac{9}{4}\sin^{-1}\frac{1}{3}$$

$$= \frac{2\sqrt{2}}{3} - \frac{\sqrt{2}}{2} + \frac{9\pi}{8} - \frac{9}{4}\sin^{-1}\frac{1}{3}$$

$$= \frac{\sqrt{2}}{6} + \frac{9\pi}{8} - \frac{9}{4}\sin^{-1}\frac{1}{3} \text{ sq. units.}$$

Q. 8. Using integration find the area of the region $\{(x, y) : x^2 + y^2 \le 2ax, y^2 \ge ax, x, y \ge 0\}$.★

Sol. We have,

$$\{(x, y) : x^2 + y^2 \le 2ax, y^2 \ge ax, x, y \ge 0\}$$

Consider $\quad x^2 + y^2 = 2ax \quad$ ...(i)

$\quad y^2 = ax \quad$ ...(ii)

If, $\quad x = 0, y = 0$

Solving equation (i) and (ii), we get

$$x^2 + ax = 2ax$$
$$\Rightarrow \quad x^2 - ax = 0$$
$$x(x - a) = 0$$
$$\therefore \quad x = 0, a$$

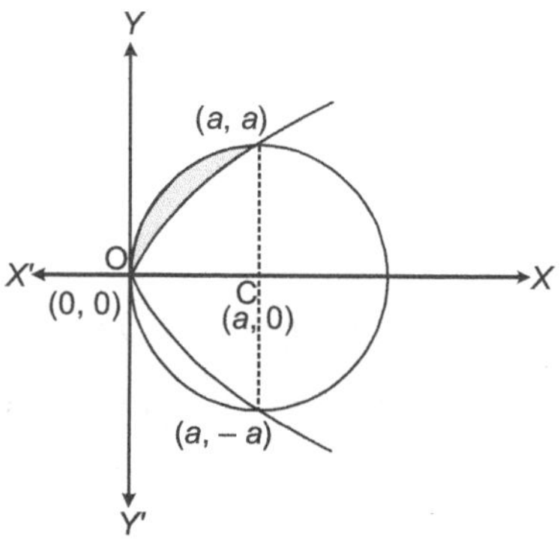

So, points of intersections of equations (i) and (ii) are (0, 0) and $(a, \pm a)$. Also, equation (i) can be written as

$(x - a)^2 + (y - 0)^2 = a^2$ whose centre is at $(a, 0)$ and radius is of 'a' units.

$\therefore$ Required area

$$= \int_0^a y_1\, dx - \int_0^a y_2\, dx$$

$$= \int_0^a \sqrt{a^2 - (x - a)^2}\, dx - \sqrt{a}\int_0^a \sqrt{x}\, dx$$

$$= \left[\frac{x - a}{2}\sqrt{a^2 - (x - a)^2} + \frac{a^2}{2}\sin^{-1}\left(\frac{x - a}{a}\right)\right]_0^a$$
$$- \frac{2}{3}\sqrt{a}\left[x^{3/2}\right]_0^a$$

$$= [0 + 0] - \left[0 + \frac{a^2}{2}\left(-\frac{\pi}{2}\right)\right] - \frac{2}{3}\sqrt{a}\left[a^{3/2} - 0\right]$$

$$= \left(\frac{\pi}{4} - \frac{2}{3}\right)a^2 \text{ sq. units.}$$

★ are board exam questions from previous years

Q. 9.. Find the area of circle $x^2 + y^2 = 16$ exterior to the parabola $y^2 = 6x$.

Sol. Given $x^2 + y^2 = 16$ is a circle of radius 4 units and centre (0, 0) and $y^2 = 6x$ is right hand parabola.

$$x^2 + y^2 = 16 \quad \text{...(i)}$$

$$y^2 = 6x \quad \text{...(ii)}$$

From equations (i) and (ii)

$$x^2 + 6x - 16 = 0$$
$$x^2 + 8x - 2x - 16 = 0$$
$$x(x+8) - 2(x+8) = 0$$
$$(x-2)(x+8) = 0$$
$$x = 2, -8$$

$x \neq -8$ as it gives imaginary values of y.

When $x = 2$ $\quad y = \pm\sqrt{12}$

$y = \pm 2\sqrt{3}$

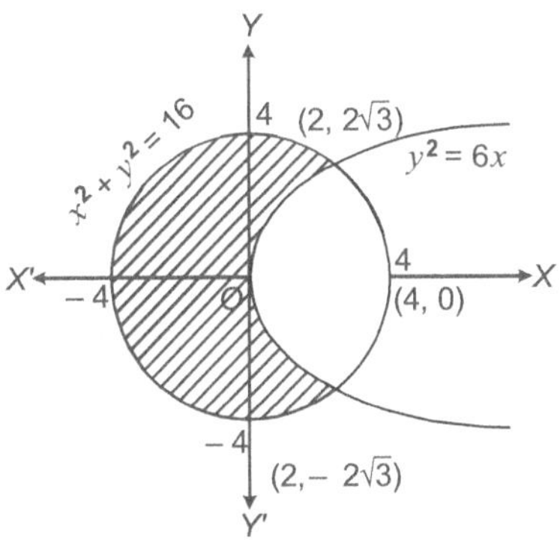

Area of shaded region

$$= 2\int_0^2 \sqrt{6x}\, dx + 2\int_2^4 \sqrt{16 - x^2}\, dx$$

$$= 2\sqrt{6}\left[\frac{x^{3/2}}{3/2}\right]_0^2 + 2\left[\frac{x}{2}\sqrt{16-x^2} + \frac{16}{2}\sin^{-1}\frac{x}{4}\right]_2^4$$

$$= 2\sqrt{6} \times \frac{2}{3}\{2^{3/2} - 0\}$$

$$+ 2\left[\left\{\frac{4}{2}\times 0 + 8\sin^{-1}(1)\right\} - \left\{\frac{2}{2}\sqrt{16-4} + 8\sin^{-1}\frac{1}{2}\right\}\right]$$

$$= \frac{4\sqrt{6}}{3} \times 2\sqrt{2} + 16 \times \frac{\pi}{2} - 2\sqrt{12} - 16 \times \frac{\pi}{6}$$

$$= \frac{8\sqrt{12}}{3} + 8\pi - 2\times 2\sqrt{3} - \frac{8}{3}\pi$$

$$= \frac{16\sqrt{3}}{3} - \frac{4\sqrt{3}}{1} + \frac{24\pi - 8\pi}{3}$$

$$= \frac{4\sqrt{3}}{3} + \frac{16\pi}{3}$$

$$\therefore \text{ Required area} = \pi(4)^2 - \left\{\frac{4\sqrt{3}}{3} + \frac{16\pi}{3}\right\}$$

$$= 16\pi - \frac{4\sqrt{3}}{3} - \frac{16\pi}{3}$$

$$= \frac{48\pi - 16\pi}{3} - \frac{4\sqrt{3}}{3}$$

$$= \frac{32\pi}{3} - \frac{4\sqrt{3}}{3}$$

$$= \frac{4}{3}\{8\pi - \sqrt{3}\} \text{ sq. units}$$

Q. 10. Find the area of the smaller part of the circle $x^2 + y^2 = a^2$ cut off by the line $x = \frac{a}{\sqrt{2}}$. [NCERT]

Sol. Given, $x^2 + y^2 = a^2$ is a circle with centre (0, 0) and radius a units and $x = \frac{a}{\sqrt{2}}$ is a line parallel to Y-axis.

$$x^2 + y^2 = a^2 \quad \text{...(i)}$$

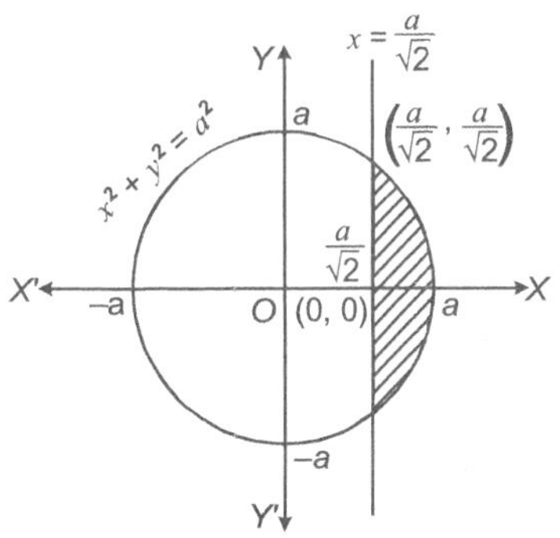

Put $\quad x = \frac{a}{\sqrt{2}}$ in equation (i)

$$\left(\frac{a}{\sqrt{2}}\right)^2 + y^2 = a^2$$

$$\frac{a^2}{2} + y^2 = a^2$$

$$y^2 = a^2 - \frac{a^2}{2}$$

$$y^2 = \frac{a^2}{2}$$

$$y = \pm\frac{a}{\sqrt{2}}$$

$$\text{Required area} = 2\int_{a/\sqrt{2}}^{a} \text{Area under circle } dx$$

$$= 2\int_{a/\sqrt{2}}^{a} \sqrt{a^2 - x^2}\, dx$$

$$= 2\left[\frac{x}{2}\sqrt{a^2 - x^2} + \frac{a^2}{2}\sin^{-1}\frac{x}{a}\right]_{a/\sqrt{2}}^{a}$$

$$= 2\left\{\frac{a}{2}\times 0 + \frac{a^2}{2}\sin^{-1}\left(\frac{a}{a}\right)\right\}$$

$$- \left\{\frac{a/\sqrt{2}}{2}\sqrt{a^2 - \frac{a^2}{2}} + \frac{a^2}{2}\sin^{-1}\left(\frac{a}{\sqrt{2}a}\right)\right\}$$

$$=2\left[0+\frac{a^2}{2}\sin^{-1}(1)-\frac{a}{2\sqrt{2}}\sqrt{\frac{a^2}{2}}-\frac{a^2}{2}\sin^{-1}\frac{1}{\sqrt{2}}\right]$$

$$=2\left[\frac{a^2}{2}\times\frac{\pi}{2}-\frac{a}{2\sqrt{2}}\times\frac{a}{\sqrt{2}}-\frac{a^2}{2}\times\frac{\pi}{4}\right]$$

$$=2\left[\frac{a^2\pi}{4}-\frac{a^2}{4}-\frac{a^2\pi}{8}\right]$$

$$=2\left[\frac{2a^2\pi-a^2\pi}{8}-\frac{a^2}{4}\right]$$

$$=2\times\frac{a^2\pi}{8}-\frac{a^2}{4}\times 2$$

$$=\left(\frac{a^2\pi}{4}-\frac{a^2}{2}\right) \text{ square units.}$$

Q. 11. Using integration, find the area of the triangle formed by positive X-axis and tangent and normal to the circle $x^2+y^2=4$ at $(1, \sqrt{3})$. ★

Sol. The equation of the given circle is $x^2+y^2=4$.

The equation of the normal to the circle at $(1, \sqrt{3})$ is same as the line joining the points $(1, \sqrt{3})$ and $(0, 0)$ which is given by

$$y-0=\frac{\sqrt{3}-0}{1-0}(x-0)$$

$$\left[\because y-y_1=\frac{y_2-y_1}{x_2-x_1}(x-x_1)\right]$$

$$y=\sqrt{3}x \quad \text{...(i)}$$

So, the slope of the normal is $\sqrt{3}$.
We know that,
Slope of normal × Slope of tangent = – 1.

$\therefore$ The slope of tangent $=\frac{-1}{\sqrt{3}}$

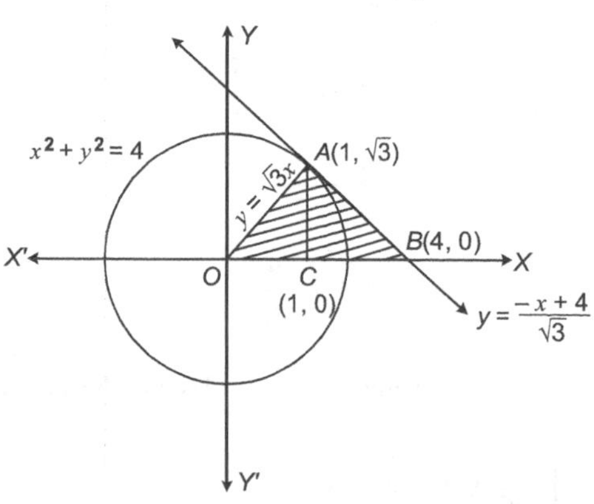

Now, the equation of the tangent to the circle at $(1, \sqrt{3})$ is given by

$$y-\sqrt{3}=\frac{-1}{\sqrt{3}}(x-1) \quad [\because y-y_1=m(x-x_1)]$$

$$\sqrt{3}y-3=-x+1$$

$$y=\frac{-x+4}{\sqrt{3}} \quad \text{...(ii)}$$

★ are board exam questions from previous years

Putting $y=0$ in (ii), we get $x=4$.

Thus, AOB is triangle formed by the tangent, normal and the positive X-axis.

Now, Area of ΔAOB

$$=\text{Area of }\Delta AOC+\text{Area of }\Delta ACB$$

$$=\int_0^1 y\,dx+\int_1^4 y\,dx$$

$$=\int_0^1 \sqrt{3x}\,dx+\int_1^4\left(\frac{-x+4}{\sqrt{3}}\right)dx$$

$$=\sqrt{3}\int_0^1 x\,dx-\frac{1}{\sqrt{3}}\int_1^4 x\,dx+\frac{4}{\sqrt{3}}\int_1^4 1\cdot dx$$

$$=\sqrt{3}\left[\frac{x^2}{2}\right]_0^1-\frac{1}{\sqrt{3}}\left[\frac{x^2}{2}\right]_1^4+\frac{4}{\sqrt{3}}[x]_1^4$$

$$=\sqrt{3}\left(\frac{1}{2}-0\right)-\frac{1}{\sqrt{3}}\left(\frac{16}{2}-\frac{1}{2}\right)+\frac{4}{\sqrt{3}}(4-1)$$

$$=\frac{\sqrt{3}}{2}-\frac{15}{2\sqrt{3}}+\frac{12}{\sqrt{3}}$$

$$=\frac{\sqrt{3}}{2}+\frac{9}{2\sqrt{3}}=\frac{\sqrt{3}}{2}+\frac{3\sqrt{3}}{2}$$

$$=2\sqrt{3}\text{ sq. units}$$

Hence, the area of the triangle so formed is $2\sqrt{3}$ sq. units.

Q. 12. Using integration, find the area of the following region :

$$\{(x, y): |x-1|\le y\le\sqrt{5-x^2}\}$$

Sol. Given curves are,

$$y=\sqrt{5-x^2} \quad \text{...(i)}$$

and

$$|x-1|=y \quad \text{...(ii)}$$

From equation (i),

$$y^2=5-x^2 \quad \text{...(iii)}$$

$$x^2+y^2=(\sqrt{5})^2$$

It is a circle with centre (0, 0) and radius $\sqrt{5}$ units and from equation (ii),

$$y=\begin{cases}-(x-1), & \text{if } x<1\\ (x-1), & \text{if } x\ge 1\end{cases}$$

From equations (ii) and equation (iii)

$$(|x-1|)^2=5-x^2$$
$$x^2+1-2x-5+x^2=0$$
$$2x^2-2x-4=0$$
$$x^2-x-2=0$$
$$x^2-2x+x-2=0$$
$$x(x-2)+1(x-2)=0$$

$$(x-2)(x+1) = 0$$
$$x = 2, -1$$

When $x = 2$, $\quad y = 1$

When $x = -1$ $\quad y = |-2|$

$$y = 2$$

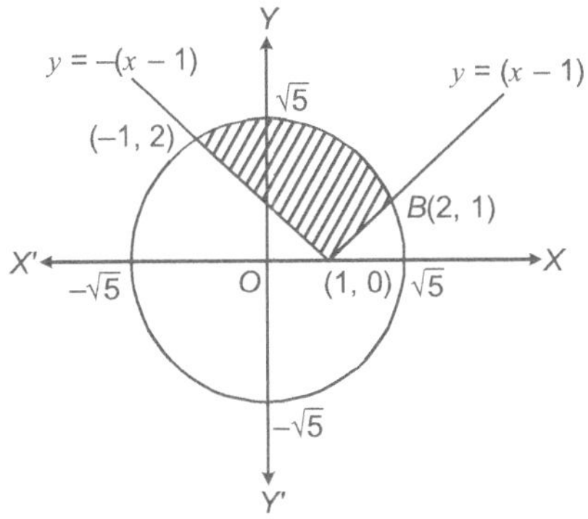

∴ Required area

$$= \int_{-1}^{2} \sqrt{5-x^2}\, dx - \int_{-1}^{1} -(x-1)\, dx - \int_{1}^{2} (x-1)\, dx$$

$$= \int_{-1}^{2} \sqrt{(\sqrt{5})^2 - x^2}\, dx + \int_{-1}^{1} (x-1)\, dx - \int_{1}^{2} (x-1)\, dx$$

$$= \left[\frac{x}{2}\sqrt{5-x^2} + \frac{5}{2}\sin^{-1}\frac{x}{\sqrt{5}}\right]_{-1}^{2} + \left(\frac{x^2}{2} - x\right)_{-1}^{1} - \left(\frac{x^2}{2} - x\right)_{1}^{2}$$

$$= \left\{\left\{\frac{2}{2}\sqrt{5-4} + \frac{5}{2}\sin\frac{2}{\sqrt{5}}\right\} - \left\{\frac{-1}{2}\sqrt{5-1} + \frac{5}{2}\sin^{-1}\left(\frac{-1}{\sqrt{5}}\right)\right\}\right\}$$

$$+ \left[\left\{\frac{1}{2} - 1\right\} - \left\{\frac{1}{2} + 1\right\}\right] - \left[\left\{\frac{4}{2} - 2\right\} - \left\{\frac{1}{2} - 1\right\}\right]$$

$$= 1 + \frac{5}{2}\sin^{-1}\frac{2}{\sqrt{5}} + \frac{1}{2} \times 2 - \frac{5}{2}\left\{-\sin^{-1}\left(\frac{1}{\sqrt{5}}\right)\right\}$$

$$+ \left\{\frac{-1}{2} - \frac{3}{2} + \frac{1}{2}\right\}$$

$$= 1 + \frac{5}{2}\sin^{-1}\frac{2}{\sqrt{5}} + 1 + \frac{5}{2}\sin^{-1}\frac{1}{\sqrt{5}} + \left\{\frac{-1-3+1}{2}\right\}$$

$$= 2 + \frac{5}{2}\sin^{-1}\left(\frac{2}{\sqrt{5}}\right) - \frac{3}{2} + \frac{5}{2}\sin^{-1}\left(\frac{1}{\sqrt{5}}\right)$$

$$= \frac{4-3}{2} + \frac{5}{2}\sin^{-1}\left(\frac{2}{\sqrt{5}}\right) + \frac{5}{2}\sin^{-1}\left(\frac{1}{\sqrt{5}}\right)$$

$$= \frac{1}{2} + \frac{5}{2}\left\{\sin^{-1}\left(\frac{2}{\sqrt{5}}\right) + \sin^{-1}\left(\frac{1}{\sqrt{5}}\right)\right\} \text{ sq. units}$$

Q. 13. Find the area bounded by the circle $x^2 + y^2 = 16$ and the line $\sqrt{3}y = x$ in the first quadrant, using integration.★

Sol.

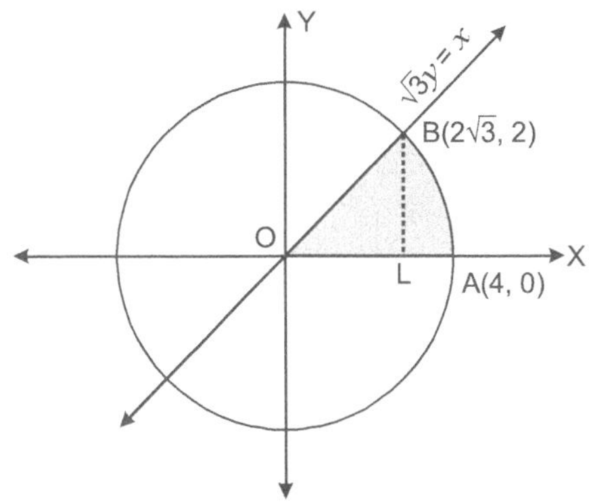

Given, $\quad x = \sqrt{3}y$

and $\quad x^2 + y^2 = 16$

$$\Rightarrow \quad (\sqrt{3}y)^2 + y^2 = 16$$
$$4y^2 = 16$$
$$y^2 = 4$$
$$\Rightarrow \quad y = 2$$
$$\therefore \quad x = \sqrt{3}y = 2\sqrt{3}$$

∴ $B(2\sqrt{3}, 2)$ is the point of intersection in first quadrant.

Required area = area under OBL + area under LBA

$$= \int_{0}^{2\sqrt{3}} \frac{x}{\sqrt{3}}\, dx + \int_{2\sqrt{3}}^{4} \sqrt{16-x^2}\, dx$$

$$= \frac{1}{\sqrt{3}}\left(\frac{x^2}{2}\right)_0^{2\sqrt{3}} + \left[\frac{x}{2}\sqrt{16-x^2} + \frac{16}{2}\sin^{-1}\frac{x}{4}\right]_{2\sqrt{3}}^{4}$$

$$= \frac{1}{2\sqrt{3}}(12-0) + (0 + 8\sin^{-1} 1)$$

$$- \left(\frac{2\sqrt{3}}{2}\sqrt{16-12} + 8\sin^{-1}\frac{2\sqrt{3}}{4}\right)$$

$$= \frac{6}{\sqrt{3}} + 8 \times \frac{\pi}{2} - 2\sqrt{3} - 8\sin^{-1}\frac{\sqrt{3}}{2}$$

$$= \frac{6\sqrt{3}}{3} + 4\pi - 2\sqrt{3} - 8 \times \frac{\pi}{3}$$

$$= 2\sqrt{3} + \frac{4\pi}{3} - 2\sqrt{3}$$

$$= \frac{4\pi}{3} \text{ sq. units.}$$

Q. 14. Find the area of the region bounded by the parabola $y = x^2$ and $y = |x|$.★ [NCERT]

Sol. Given curves are

$$y = x^2 \quad \text{...(i)}$$

and $\quad y = |x| \quad$ ...(ii)

$$\Rightarrow \quad x^2 - |x| = 0$$
$$|x|\{|x| - 1\} = 0$$

★ are board exam questions from previous years

$|x|=0$ or $|x|=1$

$x=0$ $x=-1$ or $+1$

If $x=0$, $y=0$

If $x=\pm 1$, $y=1$

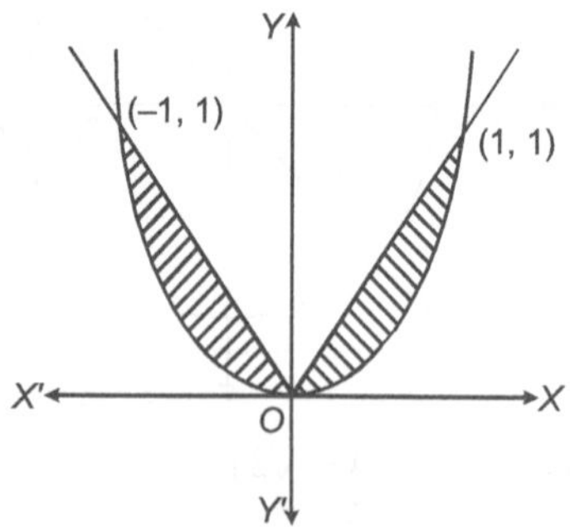

$\therefore$ $y=x^2$ is upward parabola with vertex (0, 0)

$$\text{Required area } = 2\int_0^1 (|x|-x^2)\,dx$$

$$= 2\left[\frac{x^2}{2}-\frac{x^3}{3}\right]_0^1$$

$$= 2\left[\left(\frac{1}{2}-\frac{1}{3}\right)-(0)\right]$$

$$=2\left(\frac{3-2}{6}\right)$$

$$= \frac{1}{3} \text{ sq. units.}$$

Q. 15. Using integration, find the area bounded by the curve $x^2 = 4y$ and the line $x = 4y - 2$.★

Sol. Given, $x^2 = 4y$ is a parabola symmetric to Y-axis and $x = 4y - 2$ is a line

If $y=0, x=-2$,   If $y=1, x=2$

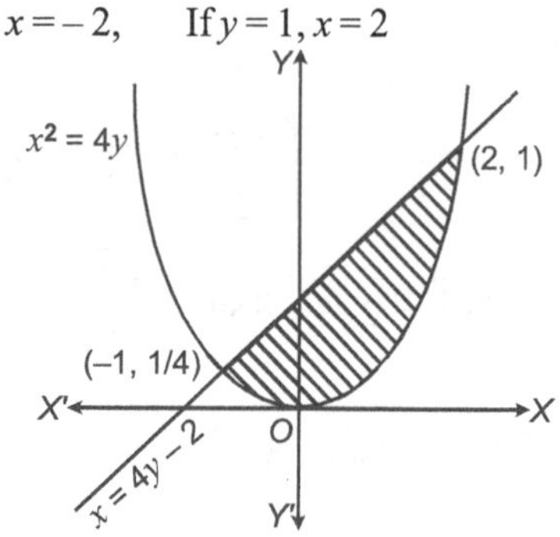

Put $x = 4y-2$ in $x^2 = 4y$

$$(4y-2)^2 = 4y$$

$$16y^2+4-16y-4y = 0$$

$$16y^2-20y+4 = 0$$

$$4y^2-5y+1 = 0$$

$$4y^2-4y-y+1 = 0$$

$$4y(y-1)-1(y-1) = 0$$

$$(y-1)(4y-1) = 0$$

$$y = 1, \frac{1}{4}$$

When $y=1, x=2$

When $y=\frac{1}{4}, x=-1$

Required area

$$= \int_{-1}^{2} \text{Area under line } dx - \int_{-1}^{2} \text{Area under curve } dx$$

$$= \int_{-1}^{2} \frac{x+2}{4}\,dx - \int_{-1}^{2}\frac{x^2}{4}\,dx$$

$$= \frac{1}{4}\left(\frac{x^2}{2}+2x\right)_{-1}^{2} - \frac{1}{4}\left(\frac{x^3}{3}\right)_{-1}^{2}$$

$$= \frac{1}{4}\left[\left(\frac{4}{2}+4\right)-\left(\frac{1}{2}-2\right)\right]-\frac{1}{4}\left[\frac{8}{3}-\left(\frac{-1}{3}\right)\right]$$

$$= \frac{1}{4}\left[6-\left(\frac{1-4}{2}\right)\right]-\frac{1}{4}\left[\frac{8}{3}+\frac{1}{3}\right]$$

$$= \frac{1}{4}\left(\frac{6}{1}+\frac{3}{2}\right)-\frac{1}{4}\times\frac{9}{3}$$

$$= \frac{1}{4}\times\frac{15}{2}-\frac{9}{12}$$

$$= \frac{15}{8}-\frac{9}{12} = \frac{9}{8} \text{ sq. units.}$$

Q. 16. The area between $x = y^2$ and $x = 4$ is divided into two equal parts by the line $x = a$, find the value of a.

[NCERT]

Sol. Given $x = y^2$ is a parabola symmetric to positive X-axis with vertex (0, 0).

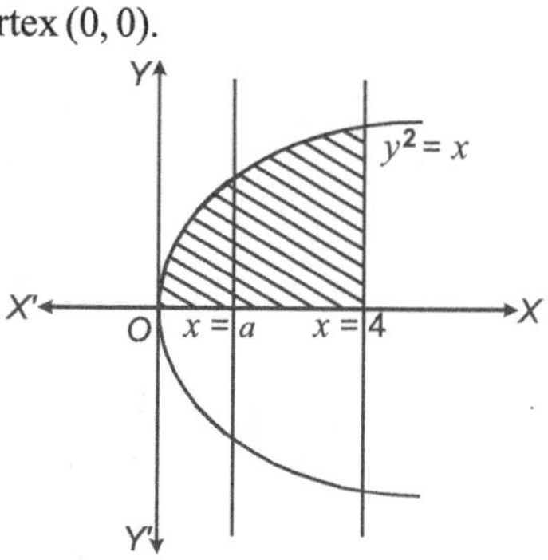

We have given that

$$\int_0^a \sqrt{x}\,dx = \int_a^4 \sqrt{x}\,dx$$

$$\left[\frac{x^{3/2}}{3/2}\right]_0^a = \left[\frac{x^{3/2}}{3/2}\right]_a^4$$

$$\frac{2}{3}(a)^{3/2} = \frac{2}{3}(4^{3/2}-a^{3/2})$$

$$\frac{2}{3}a^{3/2}+\frac{2}{3}a^{3/2} = \frac{2}{3}\times 8$$

★ are board exam questions from previous years

$$\frac{4}{3}a^{3/2} = \frac{8\times 2}{3}$$

$$a\sqrt{a} = \frac{16}{3}\times\frac{3}{4}$$

$$a\sqrt{a} = 4$$

$$a^{3/2} = 4$$

$$a = 4^{2/3}$$

$$= (2^2)^{2/3}$$

$$a = 2^{4/3}$$

Q. 17. Find the area of the region bounded by the curves $y = x^2 + 2$, $y = x$, $x = 0$ and $x = 3$.

Sol. Given, $y = x^2 + 2$ ...(i)

and $y = x$ ...(ii)

From equations (i) and (ii)

$$x = x^2 + 2$$

$$x^2 - x + 2 = 0$$

which shows the value of x do not exist.

$\therefore$ line (ii) and equation (i) do not intersect.

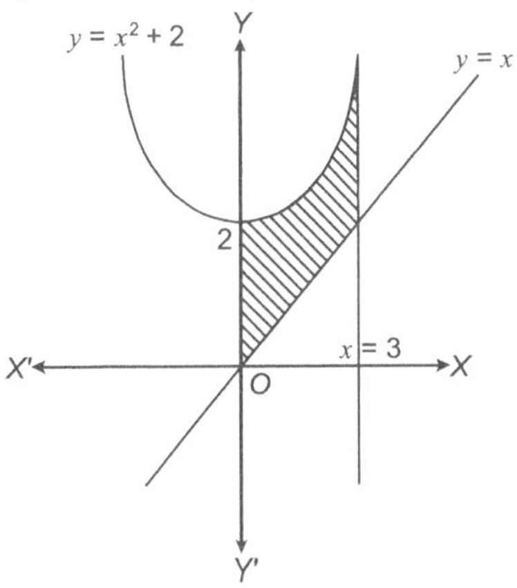

$$\text{Required area} = \int_0^3 (x^2+2)\,dx - \int_0^3 x\,dx$$

$$= \left[\frac{x^3}{3}+2x\right]_0^3 - \left(\frac{x^2}{2}\right)_0^3$$

$$= \left(\frac{27}{3}+6\right) - 0 - \left(\frac{9}{2}-0\right)$$

$$= 15 - \frac{9}{2} = \frac{30-9}{2}$$

$$= \frac{21}{2} \text{ sq. units}$$

Q. 18. Using integration, find the area of the following region :

$$\left\{(x, y) : \frac{x^2}{9}+\frac{y^2}{4} \le 1 \le \frac{x}{3}+\frac{y}{2}\right\}$$

Sol. Given equations are

$$\frac{x^2}{9}+\frac{y^2}{4} = 1 \quad \text{...(i)}$$

$$\frac{x}{3}+\frac{y}{2} = 1 \quad \text{...(ii)}$$

Equation (i) represents an ellipse whose major axis is along X-axis and minor axis is along Y-axis and equation (ii) represents a straight line.

From equation (ii), we have

$$\frac{y}{2} = 1 - \frac{x}{3}$$

Put this value of y in equation (i)

$$\frac{x^2}{9}+\left(1-\frac{x}{3}\right)^2 = 1$$

$$\frac{x^2}{9}+1+\frac{x^2}{9}-\frac{2x}{3} = 1$$

$$\frac{2x^2}{9}-\frac{2x}{3} = 0$$

$$\frac{2x^2-6x}{9} = 0$$

$$2x(x-3) = 0$$

$$x = 0, 3$$

When $x = 0$ $\quad \frac{y}{2} = 1$

$$y = 2$$

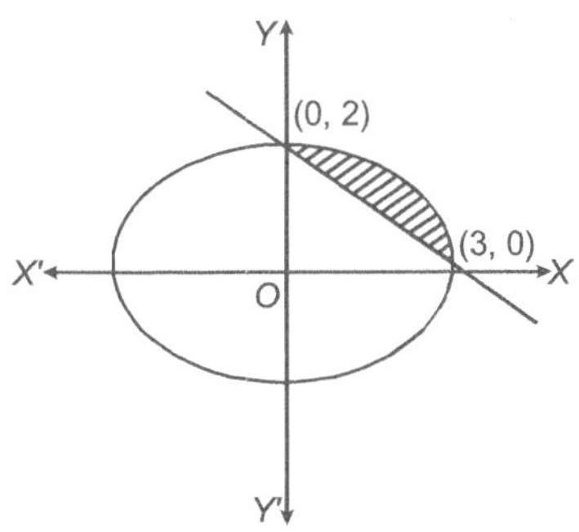

When $x = 3$ $\quad \frac{y}{2} = 1 - \frac{3}{3}$

$$\frac{y}{2} = 0$$

$$y = 0$$

Required area

$$= \int_0^3 \text{Area under ellipse } dx - \int_0^3 \text{Area under line } dx$$

$$= \int_0^3 \frac{2}{3}\sqrt{9-x^2}\,dx - \int_0^3 \frac{2}{3}(3-x)\,dx$$

$$= \frac{2}{3}\left[\frac{x}{2}\sqrt{9-x^2}+\frac{9}{2}\sin^{-1}\frac{x}{3}\right]_0^3 - \frac{2}{3}\left[3x-\frac{x^2}{2}\right]_0^3$$

$$= \frac{2}{3}\left[\left\{\frac{3}{2}\times 0+\frac{9}{2}\sin^{-1}(1)\right\}-\left\{0+\frac{9}{2}\sin^{-1}(0)\right\}\right]$$

$$- \frac{2}{3}\left[\left\{9-\frac{9}{2}\right\}-\{0\}\right]$$

$$= \frac{2}{3}\left(\frac{9}{2}\times\frac{\pi}{2}-0\right)-\frac{2}{3}\left(\frac{9}{2}-0\right)$$

$$= \frac{2}{3}\times\frac{9}{4}\pi-3$$

$$= \frac{3\pi}{2}-\frac{3}{1}$$

$$= \frac{3\pi-6}{2}$$

$$= \frac{3}{2}(\pi-2) \text{ sq. units}$$

Q. 19. Using integration find the area of the region bounded by the triangle whose vertices are (1, 3), (2, 5) and (3, 4).

Sol. Let $A(1, 3)$, $B(2, 5)$ and $C(3, 4)$ are vertices of ΔABC.

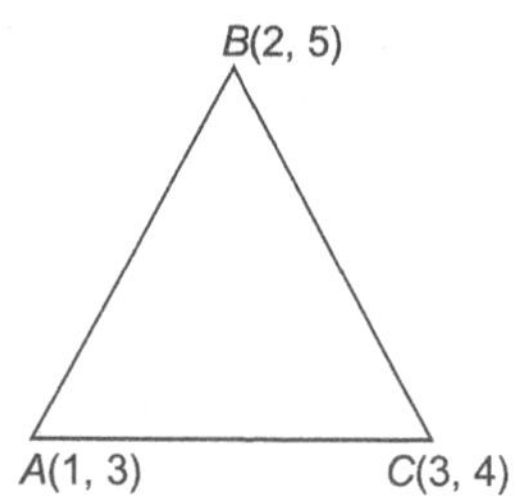

Equation of line AB :

$$y-3 = \frac{5-3}{2-1}(x-1)$$

$$y-3 = 2(x-1)$$

$$2x-2 = y-3$$

$$2x-y = -1$$

$$y = 2x+1 \quad \text{...(i)}$$

Equation of line BC :

$$y-5 = \frac{4-5}{3-2}(x-2)$$

$$y-5 = -1(x-2)$$

$$y-5 = -x+2$$

$$x+y = 7$$

$$y = -x+7 \quad \text{...(ii)}$$

Equation of line CA :

$$y-4 = \frac{3-4}{1-3}(x-3)$$

$$y-4 = \frac{-1}{-2}(x-3)$$

$$2y-8 = x-3$$

$$2y = x-3+8$$

$$2y = x+5$$

$$y = \frac{x+5}{2} \quad \text{...(iii)}$$

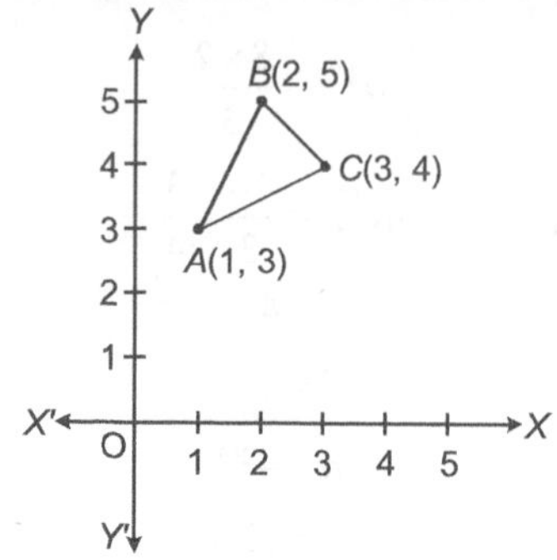

Required area :

$$= \int_1^2 (\text{eq. of } AB)\,dx + \int_2^3 (\text{eq. of } BC)\,dx - \int_1^3 (\text{eq. of } AC)\,dx$$

$$= \int_1^2 (2x+1)\,dx + \int_2^3 (-x+7)\,dx - \int_1^3 \left(\frac{x+5}{2}\right)dx$$

$$= \left(\frac{2x^2}{2}+x\right)_1^2 + \left(-\frac{x^2}{2}+7x\right)_2^3 - \frac{1}{2}\left(\frac{x^2}{2}+5x\right)_1^3$$

$$= [(4+2)-(1+1)]+\left[\left(-\frac{9}{2}+21\right)-\left(-\frac{4}{2}+14\right)\right] - \frac{1}{2}\left[\left(\frac{9}{2}+15\right)-\left(\frac{1}{2}+5\right)\right]$$

$$= 6-2+\left(-\frac{9}{2}+21+2-14\right)-\frac{1}{2}\left(\frac{9}{2}+15-\frac{1}{2}-5\right)$$

$$= 4+\left[\frac{-9+18}{2}\right]-\frac{1}{2}\left[\frac{9+20-1}{2}\right]$$

$$= 4+\frac{9}{2}-\frac{1}{2}\times\frac{28}{2}$$

$$= \frac{8+9-14}{2} = \frac{3}{2} \text{ sq. units.}$$

Q. 20. Using the method of integration, find the area of the region bounded by the lines

$2x + y = 4$, $3x - 2y = 6$ and $x - 3y + 5 = 0$

Sol. The given lines are :

$$2x+y = 4 \quad \text{...(i)}$$

$$\Rightarrow \quad y = 4-2x$$

x	0	2
y	4	0

and

$$3x-2y = 6 \quad \text{...(ii)}$$

$$\Rightarrow \quad 2y = 3x-6$$

$$\Rightarrow \quad y = \frac{3x-6}{2}$$

x	0	2
y	−3	0

and

$$x-3y+5 = 0 \quad \text{...(iii)}$$

$$3y = x+5$$

$$y = \frac{x+5}{3}$$

x	1	– 2
y	2	1

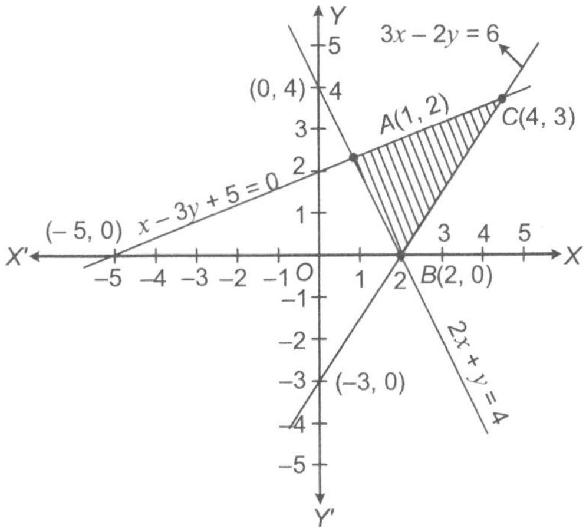

Coordinates of C:

Multiplying equation (iii) by 3 and then subtracting equation (ii) from it, we get

$$\begin{aligned} 3x - 9y &= -15 \\ 3x - 2y &= 6 \\ - \quad + \quad & \quad - \\ \hline -7y &= -21 \\ y &= 3 \end{aligned}$$

Put in equation (iii)

$$\begin{aligned} x - 9 + 5 &= 0 \\ x &= 4 \end{aligned}$$

$\Rightarrow$ Coordinates of C are (4, 3).

Coordinates of A:

Multiplying equation (iii) by 2 and then subtracting equation (i) from it we get,

$$\begin{aligned} 2x - 6y + 10 &= 0 \\ 2x + y - 4 &= 0 \\ - \quad - \quad + & \\ \hline -7y + 14 &= 0 \\ y &= 2 \end{aligned}$$

Put $y = 2$ in equation (iii)

$$\begin{aligned} x - 6 + 5 &= 0 \\ x &= 1 \end{aligned}$$

$\Rightarrow$ Coordinates of A are (1, 2)

Coordinates of B are (2, 0)

Required area

$$= \int_1^4 (\text{Under line iii})\, dx - \int_1^2 (\text{Under line i})\, dx - \int_2^4 (\text{Under line ii})\, dx$$

$$= \int_1^4 \frac{x+5}{3}\, dx - \int_1^2 (4-2x)\, dx - \int_2^4 \left(\frac{3x-6}{2}\right) dx$$

$$= \frac{1}{3}\left(\frac{x^2}{2} + 5x\right)_1^4 - \left(4x - \frac{2x^2}{2}\right)_1^2 - \frac{1}{2}\left(\frac{3x^2}{2} - 6x\right)_2^4$$

$$= \frac{1}{3}\left[\left(\frac{16}{2} + 20\right) - \left(\frac{1}{2} + 5\right)\right] - \left[\left(8 - \frac{8}{2}\right) - \left(4 - \frac{2}{2}\right)\right] - \frac{1}{2}\left[\left(\frac{48}{2} - 24\right) - \left(\frac{12}{2} - 12\right)\right]$$

$$= \frac{1}{3}\left[28 - \frac{11}{2}\right] - [4 - 3] - \frac{1}{2}(0 + 6)$$

$$= \frac{1}{3} \times \frac{45}{2} - 1 - 3$$

$$= \frac{15}{2} - \frac{4}{1}$$

$$= \frac{7}{2} \text{ sq. units.}$$

Q. 21. Using integration, find the area of the region bounded by the triangle whose vertices are (– 1, 2), (1, 5) and (3, 4).★

Sol. Let $A(-1, 2)$; $B(1, 5)$ and $C(3, 4)$.

Equation of AB is

$$y - 5 = \frac{5-2}{1+1}(x-1)$$

$$\Rightarrow \quad y = \frac{3}{2}x + \frac{7}{2} \quad \text{...(i)}$$

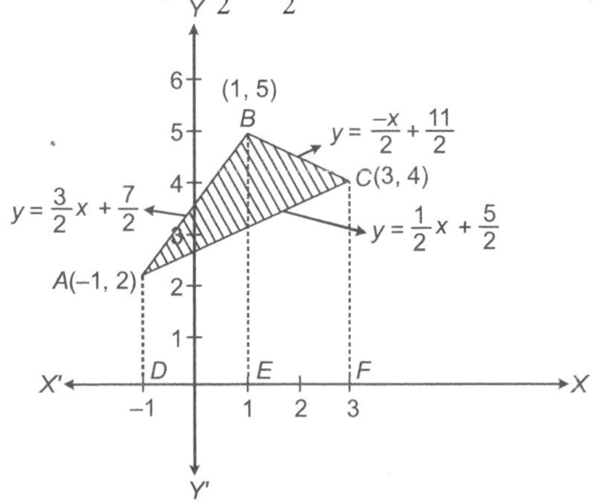

Equation of BC is

$$y - 4 = \frac{4-5}{3-1}(x-3)$$

$$\Rightarrow \quad y = -\frac{1}{2}x + \frac{11}{2} \quad \text{...(ii)}$$

Equation of AC is

$$y - 2 = \frac{4-2}{3+1}(x+1)$$

$$\Rightarrow \quad y = \frac{1}{2}x + \frac{5}{2} \quad \text{...(iii)}$$

★ are board exam questions from previous years

Area of the required triangular region, ABC

$=$ Area of trapezium $ADEB$ + Area of trapezium $BEFC$

$-$ Area of the trapezium $ADFC$

$$= \int_{-1}^{1} y_{AB}\, dx + \int_{1}^{3} y_{BC}\, dx - \int_{-1}^{3} y_{AC}\, dx$$

$$= \int_{-1}^{1} \left(\frac{3}{2}x + \frac{7}{2}\right) dx + \int_{1}^{3} \left(\frac{-1}{2}x + \frac{11}{2}\right) dx - \int_{-1}^{3} \left(\frac{1}{2}x + \frac{5}{2}\right) dx$$

$$= \left[\frac{3}{4}x^2 + \frac{7}{2}x\right]_{-1}^{1} + \left[-\frac{x^2}{4} + \frac{11}{2}x\right]_{1}^{3} - \left[\frac{x^2}{4} + \frac{5}{2}x\right]_{-1}^{3}$$

$$= \left(\frac{3}{4} + \frac{7}{2}\right) - \left(\frac{3}{4} - \frac{7}{2}\right) + \left(-\frac{9}{4} + \frac{33}{2}\right) - \left(\frac{-1}{4} + \frac{11}{2}\right) - \left(\frac{9}{4} + \frac{15}{2}\right) + \left(\frac{1}{4} - \frac{5}{2}\right)$$

$$= \frac{3}{4} + \frac{7}{2} - \frac{3}{4} + \frac{7}{2} - \frac{9}{4} + \frac{33}{2} + \frac{1}{4} - \frac{11}{2} - \frac{9}{4} - \frac{15}{2} + \frac{1}{4} - \frac{5}{2}$$

$= 7 - 4 + 1 = 4$ sq. units.

Q. 22. Using integration, find the area of the triangular region whose sides have the equations

$$y = 2x + 1, y = 3x + 1 \text{ and } x = 4.$$

Sol. Given $\quad y = 2x + 1 \quad$ ...(i)

$y = 3x + 1 \quad$ ...(ii)

and $\quad x = 4 \quad$ ...(iii)

From equations (i) and (ii)

$$3x + 1 = 2x + 1$$

$$x = 0$$

When $x = 0$, $\quad y = 1$

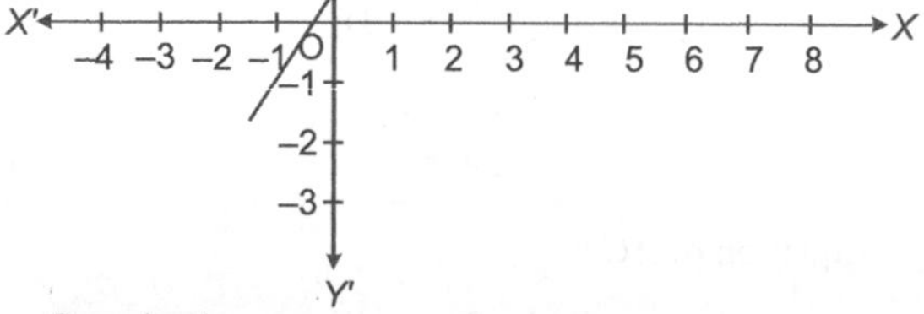

Required area

$$= \int_{0}^{4} (3x+1)\, dx - \int_{0}^{4} (2x+1)\, dx$$

$$= \left[\frac{3x^2}{2} + x\right]_{0}^{4} - \left[\frac{2x^2}{2} + x\right]_{0}^{4}$$

$$= \left[\left\{\frac{48}{2} + 4\right\} - \{0\}\right] - [\{16 + 4\} - \{0\}]$$

$= 28 - 20$

$= 8$ sq. units

Q. 23. Sketch the graph of $y = |x + 3|$ and evaluate :

$$\int_{-6}^{0} |x + 3|\, dx.$$

Sol. Given $\quad y = |x + 3|$

Required area

$$= \int_{-6}^{0} |x+3|\, dx$$

$$= \int_{-6}^{-3} -(x+3)\, dx + \int_{-3}^{0} (x+3)\, dx$$

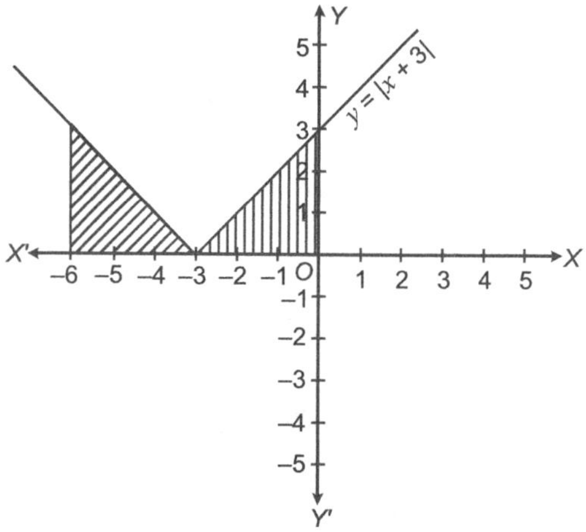

$$= -\left[\frac{x^2}{2} + 3x\right]_{-6}^{-3} + \left[\frac{x^2}{2} + 3x\right]_{-3}^{0}$$

$$= -\left[\left\{\frac{9}{2} - 9\right\} - \left\{\frac{36}{2} - 18\right\}\right] + \left[\{0 + 0\} - \left\{\frac{9}{2} - 9\right\}\right]$$

$$= -\left\{\frac{-9}{2} - 0\right\} + \left\{0 - \left(\frac{-9}{2}\right)\right\}$$

$$= \frac{9}{2} + \frac{9}{2}$$

$$= \frac{18}{2} = 9 \text{ sq. units}$$

Q. 24. Using the method of integration, find the area of the region bounded by the lines :

$$3x - 2y + 1 = 0, 2x + 3y - 21 = 0 \text{ and } x - 5y + 9 = 0.$$

Sol. Given equations are

$3x - 2y + 1 = 0 \quad$ ...(i)

$2x + 3y - 21 = 0 \quad$ ...(ii)

$x - 5y + 9 = 0 \quad$ ...(iii)

From equations (i) and (ii),

Multiplying equation (i) by 2 and equation (ii) by 3 and then subtracting we get

$$6x - 4y + 2 = 0$$
$$6x + 9y - 63 = 0$$
$$- \quad - \quad +$$
$$\overline{-13y + 65 = 0}$$
$$y = 5$$

Put value of y in equation (i), we get

$$3x-2(5)+1=0$$
$$3x-10+1=0$$
$$3x-9=0$$
$$x=3$$

So, equations (i) and (ii) intersect at (3, 5)

Multiplying equation (iii) by 2, we get

$$2x-10y+18=0 \quad \text{... (iv)}$$

From equations (ii) and (iv)

$$2x+3y-21=0$$
$$2x-10y+18=0$$
$$\underline{-\quad+\quad-} \quad \text{[By subtracting]}$$
$$13y-39=0$$
$$y=3$$

Put value of y in equation (iii)

$$x-15+9=0$$
$$x-6=0$$
$$x=6$$

So equations (ii) and (iii) intersect at (6, 3).

Multiplying equation (iii) by 3, we get

$$3x-15y+27=0 \quad \text{...(v)}$$

From equation (i) and (v), we get

$$3x-2y+1=0$$
$$3x-15y+27=0$$
$$\underline{-\quad+\quad-} \quad \text{[By subtracting]}$$
$$13y-26=0$$
$$y=2$$

Put $y=2$ in equation (iii)

$$x-10+9=0$$
$$x-1=0$$
$$x=1$$

So equations (i) and (iii) intersect at (1, 2)

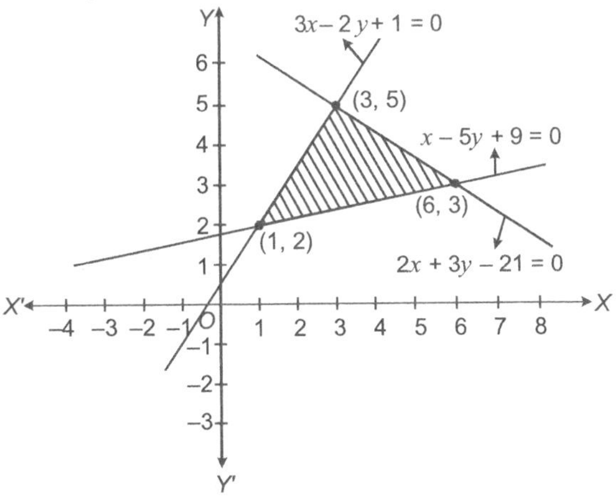

From equation (i), $\quad y=\frac{1}{2}(3x+1)$

From equation (ii), $\quad y=\frac{1}{3}(-2x+21)$

From equation (iii), $\quad y=\frac{1}{5}(x+9)$

★ are board exam questions from previous years

Required area

$$=\int_1^3 \frac{1}{2}(3x+1)\,dx+\int_3^6 \frac{1}{3}(-2x+21)\,dx-\int_1^6 \frac{1}{5}(x+9)\,dx$$

$$=\frac{1}{2}\left[\frac{3x^2}{2}+x\right]_1^3+\frac{1}{3}\left[\frac{-2x^2}{2}+21x\right]_3^6-\frac{1}{5}\left[\frac{x^2}{2}+9x\right]_1^6$$

$$=\frac{1}{2}\left[\left\{\frac{27}{2}+3\right\}-\left\{\frac{3}{2}+1\right\}\right]+\frac{1}{3}\left[\left\{\frac{-72}{2}+126\right\}-\left\{\frac{-18}{2}+63\right\}\right]$$
$$-\frac{1}{5}\left[\left\{\frac{36}{2}+54\right\}-\left\{\frac{1}{2}+9\right\}\right]$$

$$=\frac{1}{2}\left[\frac{33}{2}-\frac{5}{2}\right]+\frac{1}{3}[90-54]-\frac{1}{5}\left[72-\frac{19}{2}\right]$$

$$=\frac{1}{2}\times\frac{28}{2}+\frac{1}{3}\times 36-\frac{1}{5}\left(\frac{125}{2}\right)$$

$$=7+12-\frac{25}{2}$$

$$=\frac{19}{1}-\frac{25}{2}=\frac{38-25}{2}$$

$$=\frac{13}{2}=6.5 \text{ sq. units.}$$

Q. 25. Using integration, find the area of region bounded by the triangle whose vertices are (–2, 1), (0, 4) and (2, 3).★

Sol. The vertices of the ΔABC are $A(-2, 1)$, $B(0, 4)$ and $C(2, 3)$

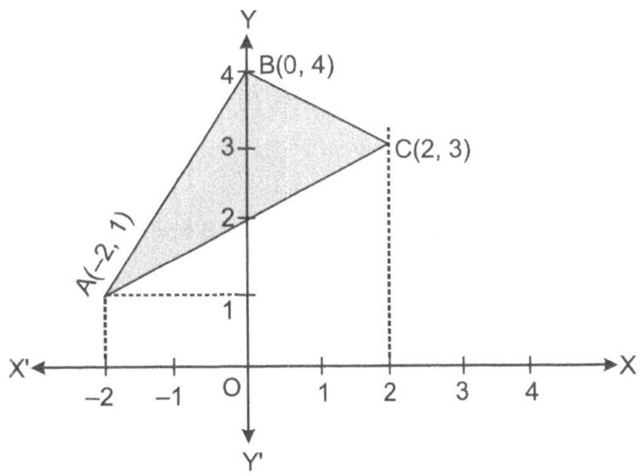

Equation of the side AB is

$$y-y_1=\frac{y_2-y_1}{x_2-x_1}(x_1-x_1)$$

$$y-1=\frac{4-1}{0-(-2)}(x-(-2))$$

$$=\frac{3}{2}(x+2)$$

$$y=\frac{3}{2}x+4$$

Equation of the side BC is

$$y-4=\frac{3-4}{2-0}(x-0)=\frac{-1}{2}x$$

$$y=\frac{-1}{2}x+4$$

Equation of the side AC is

$$y-1=\frac{3-1}{2-(-2)}(x-(-2))$$

$$=\frac{1}{2}(x+2)$$

$$y=\frac{x}{2}+2$$

Required area = Shaded area

$$=\int_{-2}^{0}\left(\frac{3}{2}x+4\right)dx+\int_{0}^{2}\left(\frac{-1}{2}x+4\right)dx-\int_{-2}^{2}\left(\frac{1}{2}x+2\right)dx$$

$$=\left[\frac{3}{2}\frac{x^2}{2}+4x\right]_{-2}^{0}+\left[\frac{-1}{2}\frac{x^2}{2}+4x\right]_{0}^{2}-\left[\frac{1}{2}\frac{x^2}{2}+2x\right]_{-2}^{2}$$

$$=\frac{3}{4}(0-4)+4(0+2)-\frac{1}{4}(4-0)+4(2-0)$$

$$-\frac{1}{4}(4-4)-2(2+2)$$

$=-3+8-1+8-0-8=4$ sq. units.

Q. 26. Using the method of integration find the area of the triangle *ABC*, coordinates of whose vertices are *A*(4, 1), *B*(6, 6) and *C*(8, 4).★

Sol. We have, $A(4, 1)$, $B(6, 6)$ and $(8,4)$

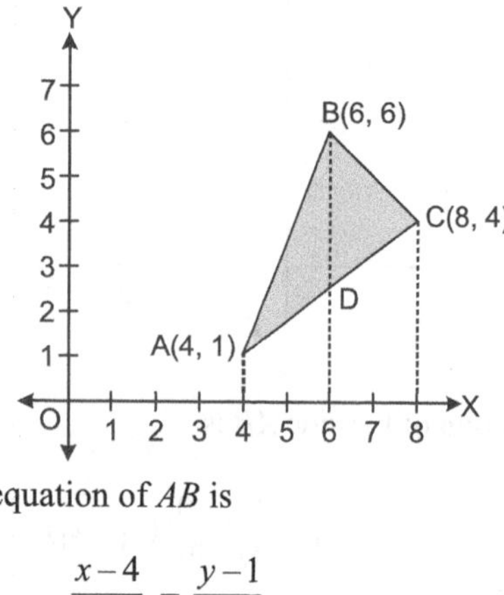

Then, equation of AB is

$$\frac{x-4}{6-4}=\frac{y-1}{6-1}$$

$$\frac{x-4}{2}=\frac{y-1}{5}$$

$$5x-20=2y-2$$

$$5x-2y-18=0 \quad ...(i)$$

Equation of BC is

$$\frac{x-6}{8-6}=\frac{y-6}{4-6}$$

$$\frac{x-6}{2}=\frac{y-6}{-2}$$

$$-2x+12=2y-12$$

$$-2x-2y+24=0$$

$$x+y-12=0 \quad ...(ii)$$

and equation of CA is

$$\frac{x-8}{4-8}=\frac{y-4}{1-4}$$

$$\frac{x-8}{-4}=\frac{y-4}{-3}$$

$$-3x+24=-4y+16$$

$$-3x+4y+8=0 \quad ...(iii)$$

Clearly, Area of ΔABC = Area of ADB + Area of BDC

$$\therefore \text{ Area of } \Delta ADB=\int_{4}^{6}(y_2-y_1)\,dx$$

$$=\int_{4}^{6}\left\{\frac{5x-18}{2}-\left(\frac{3x-8}{4}\right)\right\}dx$$

Similarly, we have

$$\text{Area of } \Delta BDC=\int_{6}^{8}(y_4-y_3)\,dx$$

$$=\int_{6}^{8}\left\{12-x-\left(\frac{3x-8}{4}\right)\right\}dx$$

$$\therefore \text{ Area of } \Delta ABC=\int_{4}^{6}\left\{\frac{5x-18}{2}-\left(\frac{3x-8}{4}\right)\right\}dx$$

$$+\int_{6}^{8}\left\{12-x-\left(\frac{3x-8}{4}\right)\right\}dx$$

$$=\frac{1}{4}\int_{4}^{6}(7x-28)\,dx+\frac{1}{4}\int_{6}^{8}(56-7x)\,dx$$

$$=\frac{1}{4}\left[\left[\frac{7x^2}{2}-28x\right]_4^6+\left[56x-\frac{7x^2}{2}\right]_6^8\right]$$

$$\text{Area of } \Delta ABC=\frac{1}{4}\left[\left(\frac{252}{2}-168-\frac{112}{2}+112\right)\right.$$

$$\left.+\left(448-\frac{448}{2}-336+\frac{252}{2}\right)\right]$$

★ are board exam questions from previous years

$$= \frac{1}{4}[70-56-98+112]$$

$$= \frac{28}{4} = 7 \text{ sq. units.}$$

Chapter 9. Differential Equations

Q. 1. Solve the differential equation :

$$\left(x\cos\frac{y}{x}+y\sin\frac{y}{x}\right)y-\left(y\sin\frac{y}{x}-x\cos\frac{y}{x}\right)x\frac{dy}{dx}=0$$

Sol. Given differential equation is

$$\left(x\cos\frac{y}{x}+y\sin\frac{y}{x}\right)y-\left(y\sin\frac{y}{x}-x\cos\frac{y}{x}\right)x\frac{dy}{dx}=0$$

$$\left(x\cos\frac{y}{x}+y\sin\frac{y}{x}\right)y=\left(y\sin\frac{y}{x}-x\cos\frac{y}{x}\right)x\frac{dy}{dx}$$

$$\frac{dy}{dx} = \frac{\left(x\cos\frac{y}{x}+y\sin\frac{y}{x}\right)y}{\left(y\sin\frac{y}{x}-x\cos\frac{y}{x}\right)x}$$

$$= \frac{x\left(\cos\frac{y}{x}+\frac{y}{x}\sin\frac{y}{x}\right)\frac{y}{x}}{x\left(\frac{y}{x}\sin\frac{y}{x}-\cos\frac{y}{x}\right)\frac{x}{x}}$$

$$\frac{dy}{dx} = \frac{\left(\cos\frac{y}{x}+\frac{y}{x}\sin\frac{y}{x}\right)\frac{y}{x}}{\frac{y}{x}\sin\frac{y}{x}-\cos\frac{y}{x}} \quad \text{...(i)}$$

Put $\frac{y}{x} = t$

$y = xt$

Differentiate w.r.t. x

$$\frac{dy}{dx} = x\frac{dt}{dx}+t$$

Put in equation (i)

$$x\frac{dt}{dx}+t = \frac{t(\cos t+t\sin t)}{t\sin t-\cos t}$$

$$x\frac{dt}{dx} = \frac{t\cos t+t^2\sin t}{t\sin t-\cos t}-t$$

$$= \frac{t\cos t+t^2\sin t-t^2\sin t+t\cos t}{t\sin t-\cos t}$$

$$x\frac{dt}{dx} = \frac{2t\cos t}{t\sin t-\cos t}$$

$$\frac{t\sin t-\cos t}{2t\cos t}dt = \frac{dx}{x}$$

$$\left(\frac{t\sin t}{2t\cos t}-\frac{\cos t}{2t\cos t}\right)dt = \frac{dx}{x}$$

Integrating both sides

$$\int\left(\frac{1}{2}\tan t-\frac{1}{2t}\right)dt = \int\frac{dx}{x}+c$$

$$\Rightarrow \frac{1}{2}(-\log|\cos t|-\log|t|) = \log|x|+c$$

$$\Rightarrow \log|\cos t|+\log|t| = -2\log|x|-2c$$

$$\Rightarrow \log(t\cos t) = -\log x^2-2c$$

$$\Rightarrow \log(t\cos t)+\log x^2 = -2c$$

$$\Rightarrow \log(t\cos t\times x^2) = -2c$$

$$\Rightarrow x^2 t\cos t = e^{-2c}$$

$$\Rightarrow \frac{y}{x}\left(\cos\frac{y}{x}\right)x^2 = A \qquad \left[\because t=\frac{y}{x}\right]$$

(where $A = e^{-2c}$)

$$\Rightarrow xy\cos\frac{y}{x} = A$$

which is required solution.

Q. 2. Solve the following differential equation :

$$x\frac{dy}{dx}+y-x+xy\cot x=0$$

Sol. Given differential equation is

$$x\frac{dy}{dx}+y-x+xy\cot x=0$$

$$x\frac{dy}{dx}+y(1+x\cot x)=x$$

$$\frac{dy}{dx}+\frac{1+x\cot x}{x}y=1 \quad \text{...(i)}$$

Here $P=\frac{1+x\cot x}{x}$, $Q=1$

$$\text{I.F.} = e^{\int P\,dx} = e^{\int\left(\frac{1}{x}+\cot x\right)}$$

$$= e^{\log x+\log\sin x} = e^{\log x\sin x}$$

$$\text{I.F.} = x\sin x$$

Multiply equation (i) by I.F.

$$x\sin x\frac{dy}{dx}+x\sin x\,\frac{1+x\cot x}{x}y = x\sin x$$

$$\frac{d}{dx}(xy\sin x) = x\sin x$$

Integrating both sides :

$$xy\sin x = -x\cos x-\int(-\cos x)\,dx+c$$

$$xy\sin x = -x\cos x+\sin x+c$$

which is the required solution.

Q. 3. Find the particular solution of the following differential equation :

$$x\frac{dy}{dx}-y+x\sin\frac{y}{x}=0,$$

given that when $x=2, y=\pi$.

★ are board exam questions from previous years

Sol. Given differential equation is,

$$x\frac{dy}{dx} - y + x\sin\frac{y}{x} = 0 \qquad ...(i)$$

Divide by x

$$\frac{dy}{dx} - \frac{y}{x} + \sin\frac{y}{x} = 0$$

$$\Rightarrow \quad \frac{dy}{dx} = \frac{y}{x} - \sin\frac{y}{x} \qquad ...(ii)$$

Put $\frac{y}{x} = t \Rightarrow y = xt$

Differentiate w.r.t. x

$$\frac{dy}{dx} = x\frac{dt}{dx} + t$$

Put in equation (ii)

$$x\frac{dt}{dx} + t = t - \sin t$$

$$\Rightarrow \quad x\frac{dt}{dx} = -\sin t$$

$$\Rightarrow \quad \frac{dt}{\sin t} = -\frac{dx}{x}$$

$$\Rightarrow \quad \text{cosec } t\, dt = -\frac{dx}{x}$$

On integrating both sides, we have

$$\log|\text{cosec } t - \cot t| = -\log x + \log c$$

$$\Rightarrow \log\left|\text{cosec}\frac{y}{x} - \cot\frac{y}{x}\right| = -\log x + \log c$$

$$\Rightarrow \log\left|\text{cosec}\frac{y}{x} - \cot\frac{y}{x}\right| = \log\frac{c}{x}$$

$$\Rightarrow \quad \text{cosec}\frac{y}{x} - \cot\frac{y}{x} = \frac{c}{x} \qquad ...(iii)$$

Put $x = 2, y = \pi$

$$\Rightarrow \quad \text{cosec}\frac{\pi}{2} - \cot\frac{\pi}{2} = \frac{c}{2}$$

$$\Rightarrow \quad 1 - 0 = \frac{c}{2}$$

$$\Rightarrow \quad c = 2$$

Put in equation (iii)

$$\text{cosec}\frac{y}{x} - \cot\frac{y}{x} = \frac{2}{x}$$

which is the required solution of given differential equation.

Q. 4. Form the differential equation of the family of circles in the second quadrant and touching the coordinate axes.★

Sol. To form differential equation of the family of circles in the second quadrant and touching the coordinate axes, let us assume $(-a, a)$ be the coordinates of centre of circles.

Equation representing the family of circles is :

$$(x+a)^2 + (y-a)^2 = a^2 \qquad ...(i)$$

★ are board exam questions from previous years

Differentiate both sides w.r.t. x

$$2(x+a) + 2(y-a)\frac{dy}{dx} = 0$$

$$\Rightarrow \quad 2x + 2a + 2y\frac{dy}{dx} - 2a\frac{dy}{dx} = 0$$

$$\Rightarrow \quad x + y\frac{dy}{dx} + a - a\frac{dy}{dx} = 0$$

$$\Rightarrow \quad x + y\frac{dy}{dx} + a\left(1 - \frac{dy}{dx}\right) = 0$$

$$\Rightarrow \quad x + y\frac{dy}{dx} = a\left(\frac{dy}{dx} - 1\right)$$

$$\Rightarrow \quad a = \frac{x + y\frac{dy}{dx}}{\frac{dy}{dx} - 1}$$

$$\Rightarrow \quad a = \frac{x + yy'}{y' - 1} \qquad \text{where } y' = \frac{dy}{dx}$$

Put in equation (i)

$$\Rightarrow \quad \left(x + \frac{x+yy'}{y'-1}\right)^2 + \left(y - \frac{x+yy'}{y'-1}\right)^2 = \left(\frac{x+yy'}{y'-1}\right)^2$$

$$\Rightarrow \left(\frac{xy' - x + x + yy'}{y'-1}\right)^2 + \left(\frac{yy' - y - x - yy'}{y'-1}\right)^2 = \left(\frac{x+yy'}{y'-1}\right)^2$$

$$\Rightarrow \quad \left(\frac{y'(x+y)}{y'-1}\right)^2 + \left(\frac{-y-x}{y'-1}\right)^2 = \left(\frac{x+yy'}{y'-1}\right)^2$$

$$\Rightarrow y'^2(x+y)^2 + (x+y)^2 = (x+yy')^2$$

$$\Rightarrow \quad (x+y)^2(y'^2 + 1) = (x+yy')^2$$

which is the required differential equation.

Q. 5. Find the particular solution of the differential equation :

$$\frac{dx}{dy} + x\cot y = 2y + y^2\cot y, y \neq 0$$

given that $x = 0$, when $y = \frac{\pi}{2}$.

Sol. Given, differential equation is

$$\frac{dx}{dy} + x\cot y = 2y + y^2\cot y \qquad ...(i)$$

Here $P = \cot y$, $Q = 2y + y^2\cot y$

$$\text{I.F.} = e^{\int \cot y\, dy} = e^{\log \sin y} = \sin y$$

Multiply equation (i) by I.F.

$$\sin y\frac{dx}{dy} + x\cos y = 2y\sin y + y^2\cos y$$

$$\frac{d}{dy}(x \sin y) = 2y \sin y + y^2 \cos y$$

Integrating both sides

$$x \sin y = 2\int y \sin y\, dy + \int y^2 \cos y\, dy$$

$$= 2\left\{\sin y \frac{y^2}{2} - \int \cos y \frac{y^2}{2}\, dy\right\}$$

$$+ \int y^2 \cos y\, dy + c$$

$$= y^2 \sin y - \int y^2 \cos y\, dy + \int y^2 \cos y\, dy + c$$

$$x \sin y = y^2 \sin y + c \qquad \text{...(ii)}$$

When $x = 0,\ y = \frac{\pi}{2}$

Equation (ii) becomes

$$0 = \frac{\pi^2}{4} \sin \frac{\pi}{2} + c$$

$$c = -\frac{\pi^2}{4}$$

∴ Equation (ii) will become :

$$x \sin y = y^2 \sin y - \frac{\pi^2}{4}$$

$$x = y^2 - \frac{\pi^2}{4} \operatorname{cosec} y$$

which is the required particular solution.

Q. 6. Show that the differential equation

$$x\frac{dy}{dx}\sin\frac{y}{x} + x - y\sin\frac{y}{x} = 0$$

is homogeneous and find the particular solution when $x = 1$ and $y = \frac{\pi}{2}$.

Sol. Given differential equation is,

$$x\frac{dy}{dx}\sin\frac{y}{x} + x - y\sin\frac{y}{x} = 0 \qquad \text{...(i)}$$

$$x\frac{dy}{dx}\sin\frac{y}{x} = y \sin\frac{y}{x} - x$$

$$\frac{dy}{dx} = \frac{y\sin\frac{y}{x} - x}{x\sin\frac{y}{x}}$$

$$\frac{dy}{dx} = \frac{\frac{y}{x}\sin\frac{y}{x} - 1}{\sin\frac{y}{x}} \qquad \text{...(ii)}$$

which is homogeneous differential equation of the form $f\left(\frac{y}{x}\right)$

Put $\frac{y}{x} = t, \quad y = xt$

Differentiate both sides w.r.t. x

$$\frac{dy}{dx} = x\frac{dt}{dx} + t$$

Put in equation (ii)

$$x\frac{dt}{dx} + t = \frac{t\sin t - 1}{\sin t}$$

$$\Rightarrow \quad x\frac{dt}{dx} = \frac{t\sin t - 1 - t\sin t}{\sin t}$$

$$\Rightarrow \quad x\frac{dt}{dx} = -\frac{1}{\sin t}$$

$$\Rightarrow \quad \sin t\, dt = -\frac{dx}{x}$$

Integrating both sides :

$$-\cos t = -\log x + c$$

$$\Rightarrow \quad -\cos\frac{y}{x} = -\log x + c \qquad \text{...(iii)}$$

$$\left[\because t = \frac{y}{x}\right]$$

Put $x = 1$ and $y = \frac{\pi}{2}$

$$\Rightarrow \quad -\cos\frac{\pi}{2} = -\log 1 + c$$

$$\Rightarrow \quad c = 0$$

Equation (iii) becomes :

$$-\cos\frac{y}{x} = -\log x + 0$$

$$\Rightarrow \quad \cos\frac{y}{x} = \log x$$

which is the required particular solution.

Q. 7. Find the particular solution of the differential equation $2ye^{x/y}\, dx + (y - 2xe^{x/y})\, dy = 0$, given that $x = 0$ when $y = 1$.

Sol. Given differential equation is,

$$2ye^{x/y}\, dx + (y - 2xe^{x/y})\, dy = 0$$

$$(2xe^{x/y} - y)dy = 2ye^{x/y}\, dx$$

$$\frac{dy}{dx} = \frac{2ye^{x/y}}{2xe^{x/y} - y}$$

$$= \frac{2e^{x/y}}{2\frac{x}{y}e^{x/y} - 1}$$

$$\therefore \quad \frac{dx}{dy} = \frac{2\frac{x}{y}e^{x/y} - 1}{2e^{x/y}} \qquad \text{...(i)}$$

Put $\frac{x}{y} = t, \quad x = yt$

Differentiate w.r.t. y

$$\frac{dx}{dy} = y\frac{dt}{dy}+t$$

Put in equation (i)

$$y\frac{dt}{dy}+t = \frac{2t\,e^t-1}{2e^t}$$

$$\Rightarrow \quad y\frac{dt}{dy} = \frac{2te^t-1-2te^t}{2e^t}$$

$$\Rightarrow \quad y\frac{dt}{dy} = -\frac{1}{2e^t}$$

$$\Rightarrow \quad 2e^t\,dt = -\frac{dy}{y}$$

Integrating both sides

$$2e^t = -\log y + c$$

$$\Rightarrow \quad 2e^{x/y} = -\log y + c \quad \text{...(ii)}$$

When $x = 0, y = 1$

$$2 = c$$

Put in equation (ii)

$$2e^{x/y} + \log y = 2$$

which is the required solution of given differential equation.

Q. 8. In each of the following verify that the given function is a solution of the corresponding differential equation. [NCERT]

(i) $y = x^2 + 2x + c \;:\; y' - 2x - 2 = 0$

Sol. Given, $y = x^2 + 2x + c$

Differentiate both sides w.r.t. x

$$\frac{dy}{dx} = 2x + 2$$

or $\quad y' = 2x + 2$

Now, substitute the value of y' in the given differential equation, we have

$$\text{L.H.S.} = y' - 2x - 2$$
$$= 2x + 2 - 2x - 2 = 0 = \text{R.H.S.}$$

Hence, $\quad y = x^2 + 2x + c$

is a solution of the given differential equation.

(ii) $y = \sqrt{1+x^2} \;:\; y' = \dfrac{xy}{1+x^2}$

Sol. Given, $\quad y = \sqrt{1+x^2} \quad$ ...(i)

Differentiate w.r.t. x

$$\frac{dy}{dx} = \frac{1}{2}(1+x^2)^{-1/2}\cdot 2x$$

$$\frac{dy}{dx} = \frac{x}{\sqrt{1+x^2}} \quad \text{...(ii)}$$

Divide equation (ii) by equation (i)

$$\frac{\frac{dy}{dx}}{y} = \frac{\frac{x}{\sqrt{1+x^2}}}{\sqrt{1+x^2}}$$

$$\frac{dy}{dx} = \frac{xy}{1+x^2} \quad \text{or} \quad y' = \frac{xy}{1+x^2}$$

Hence, $\quad y = \sqrt{1+x^2}$

is the solution of the given differential equation.

(iii) $y = x \sin x : xy' = y + x\sqrt{x^2 - y^2}$

$(x \neq 0$ and $x > y$ or $x < y)$

Sol. Given,

$$y = x \sin x \quad \text{....(i)}$$
$$y' = x\cos x + \sin x$$

Multiply both sides by x

$$xy' = x^2\cos x + x\sin x$$
$$= x^2\cos x + y$$
$$= x^2(\sqrt{1-\sin^2 x}) + y$$
$$= x\sqrt{x^2-(x\sin x)^2} + y$$
$$xy' = x\sqrt{x^2-y^2} + y$$

Hence, $\quad y = x\sin x$

is a solution of the given differential equation.

(iv) $xy = \log y + c : y' = \dfrac{y^2}{1-xy}$ **$(xy \neq 1)$**

Sol. Given differential equation is,

$$xy = \log y + c$$

Differentiate w.r.t. x

$$x\frac{dy}{dx} + y = \frac{1}{y}\frac{dy}{dx} + 0$$
$$xy'y + y^2 = y'$$
$$y^2 = y'(1-xy)$$
$$\frac{y^2}{1-xy} = y'$$

Hence, $\quad xy = \log y + c$

is a solution of given differential equation.

(v) $y - \cos y = x : (y\sin y + \cos y + x)y' = y$

Sol. Given,

$$y - \cos y = x \quad \text{...(i)}$$

Differentiate w.r.t. x:

$$\frac{dy}{dx} + \sin y\frac{dy}{dx} = 1$$
$$y' + (\sin y)\,y' = 1$$
$$y'(1+\sin y) = 1 \quad \text{...(ii)}$$

Given differential equation is

$$(y\sin y + \cos y + x)y' = (y\sin y + y)\,y'$$

[from equation (i)]

$$= y(1+\sin y)y'$$
$$= y \times 1 \qquad [\because\ y'(1+\sin y)=1]$$
$$= y$$

$\because \quad (y\sin y+\cos y+x)\,y' = y$

$\therefore \quad y-\cos y = x$

is a solution of the given differential equation.

(vi) $\quad x+y=\tan^{-1}y : y^2y'+y^2+1=0$

Sol. Given

$$x+y = \tan^{-1} y$$

Differentiate w.r.t. x

$$1+y' = \frac{1}{1+y^2}y'$$

$\Rightarrow \quad (1+y')(1+y^2) = y'$

$\Rightarrow 1+y^2y'+y^2+y' = y'$

$\Rightarrow \quad y^2y'+y^2+1 = 0$

Hence, $\quad x+y = \tan^{-1} y$

is the solution of given differential equation.

Q. 9. In each of the following form a differential equation representing the given family of curves by eliminating arbitrary constants *a* and *b*. [NCERT]

(i) $\dfrac{x}{a}+\dfrac{y}{b}=1$

Sol. Given equation is,

$$\frac{x}{a}+\frac{y}{b} = 1 \qquad ...(i)$$

Differentiate w.r.t. x

$$\frac{1}{a}+\frac{1}{b}y' = 0$$

Again, differentiating w.r.t x

$$0+\frac{1}{b}y'' = 0$$

$$y = 0$$

which is the required differential equation.

(ii) $y^2 = a(b^2-x^2)$

Sol. Given equation is,

$$y^2 = a(b^2-x^2) \qquad ...(i)$$

Differentiate w.r.t. x

$$2yy_1 = a(-2x)$$
$$-ax = yy_1 \qquad ...(ii)$$

Again differentiating w.r.t. x

$$-a = yy_2+y_1y_1 \qquad ...(iii)$$

Put the value of a in equation (ii)

$$(yy_2+y_1^2)x = yy_1$$
$$yy_1 = x(y_1^2+yy_2)$$

which is the required differential equation.

(iii) $y = ae^{3x}+be^{-2x}$

Sol. Given equation is,

$$y = ae^{3x}+be^{-2x} \qquad ...(i)$$

Differentiating w.r.t. x

$$y' = 3ae^{3x}-2be^{-2x} \qquad ...(ii)$$

Multiply equation (i) by 3 and subtract from (ii)

$$y'-3y = -5be^{-2x} \qquad ...(iii)$$

Again, differentiating w.r.t. x

$$y''-3y' = -5be^{-2x}(-2)$$
$$y''-3y' = 10be^{-2x} \qquad ...(iv)$$

Multiply equation (iii) by 2 and add to equation (iv)

$$y''-3y'+2y'-6y = 0$$
$$y''-y'-6y = 0$$

which is the required differential equation.

(iv) $y = e^{2x}(a+bx)$

Sol. Given equation is

$$y = e^{2x}(a+bx) \qquad ...(i)$$

Differentiate both sides w.r.t. x

$$y' = e^{2x}\cdot b+(a+bx)\cdot 2\cdot e^{2x}$$
$$y' = be^{2x}+2e^{2x}(a+bx)$$
$$y' = be^{2x}+2y \qquad \text{[from eq. (i)]}$$
$$y'-2y = be^{2x} \qquad ...(ii)$$

Again, differentiating w.r.t. x

$$y''-2y' = b\cdot 2e^{2x}$$

$\Rightarrow \quad y''-2y' = 2be^{2x}$

$\Rightarrow \quad y''-2y' = 2(y'-2y) \qquad$ [from eq. (ii)]

$\Rightarrow \quad y''-2y' = 2(y'-2y)$

$\Rightarrow \quad y''-2y'-2y'+4y = 0$

$\Rightarrow \quad y''-4y'+4y = 0$

which is the required differential equation.

Q. 10. Show that the given differential equation is homogeneous and solve each of them.

(i) $(x^2+xy)\,dy = (x^2+y^2)\,dx$ **[NCERT]**

Sol. The given differential equation is,

$$(x^2+xy)\,dy = (x^2+y^2)\,dx$$

$$\frac{dy}{dx} = \frac{x^2+y^2}{x^2+xy}$$

$$= \frac{x^2\left(1+\dfrac{y^2}{x^2}\right)}{x^2\left(1+\dfrac{y}{x}\right)}$$

$$\Rightarrow \quad \frac{dy}{dx} = \frac{1+\left(\frac{y}{x}\right)^2}{1+\frac{y}{x}} \quad \text{...(i)}$$

$\therefore$ It is homogeneous equation of the form $f\left(\frac{y}{x}\right)$

Put $\frac{y}{x} = t, \Rightarrow y = xt$

Differentiate w.r.t. x

$$\frac{dy}{dx} = x\frac{dt}{dx} + t$$

Put in equation (i)

$$x\frac{dt}{dx} + t = \frac{1+t^2}{1+t}$$

$$\Rightarrow \quad x\frac{dt}{dx} = \frac{1+t^2}{1+t} - t$$

$$\Rightarrow \quad x\frac{dt}{dx} = \frac{1+t^2-t-t^2}{1+t}$$

$$\Rightarrow \quad x\frac{dt}{dx} = \frac{1-t}{1+t}$$

$$\Rightarrow \quad \frac{1+t}{1-t}dt = \frac{dx}{x}$$

Integrating both sides

$$\int\left(-1+\frac{2}{-t+1}\right)dt = \frac{dx}{x}$$

$$\Rightarrow \quad -t-2\log|-t+1| = \log x + c$$

$$\Rightarrow \quad -2\log|1-t| = \log x + t + c$$

$$\Rightarrow \quad -2\log|1-t| - \log x = t + c$$

$$\Rightarrow \quad -(2\log|1-t| + \log x) = t + c$$

$$\Rightarrow \quad \log(1-t)^2 + \log x = -t - c$$

$$\Rightarrow \quad \log x(1-t)^2 = -t - c$$

$$\Rightarrow \quad \log x\left(1-\frac{y}{x}\right)^2 = -\frac{y}{x} - c \quad \left(\because t = \frac{y}{x}\right)$$

$$\Rightarrow \quad x\left(1-\frac{y}{x}\right)^2 = e^{-y/x-c}$$

$$\Rightarrow \quad x\left(1-\frac{y}{x}\right)^2 = e^{-y/x}\cdot e^{-c}$$

$$\Rightarrow \quad \frac{x(x-y)^2}{x^2} = e^{-y/x}\cdot A,$$

where $A = e^{-c}$

$$\Rightarrow \quad (x-y)^2 = xe^{-y/x}A$$

which is the required solution.

(ii) $x^2\frac{dy}{dx} = x^2 - 2y^2 + xy$ **[NCERT]**

Sol. Given differential equation is

$$x^2\frac{dy}{dx} = x^2 - 2y^2 + xy$$

$$\Rightarrow \quad \frac{dy}{dx} = \frac{x^2}{x^2} - 2\frac{y^2}{x^2} + \frac{xy}{x^2}$$

$$\Rightarrow \quad \frac{dy}{dx} = 1 - 2\left(\frac{y}{x}\right)^2 + \frac{y}{x} \quad \text{...(i)}$$

which is homogeneous equation of the form $f\left(\frac{y}{x}\right)$

Put $\frac{y}{x} = t \Rightarrow \quad y = xt$

Differentiate w.r.t. x

$$\frac{dy}{dx} = x\frac{dt}{dx} + t$$

Put in equation (i)

$$x\frac{dt}{dx} + t = 1 - 2t^2 + t$$

$$\Rightarrow \quad x\frac{dt}{dx} = 1 - 2t^2$$

$$\Rightarrow \quad \frac{dt}{1-2t^2} = \frac{dx}{x}$$

Integrating both sides

$$\frac{1}{2}\int\frac{dt}{\left(\frac{1}{\sqrt{2}}\right)^2 - t^2} = \int\frac{dx}{x} + c$$

$$\Rightarrow \frac{1}{2}\cdot\frac{1}{2\times\frac{1}{\sqrt{2}}}\log\left|\frac{\frac{1}{\sqrt{2}}+t}{\frac{1}{\sqrt{2}}-t}\right| = \log x + c$$

$$\Rightarrow \quad \frac{1}{2\sqrt{2}}\log\left|\frac{1+\sqrt{2}t}{1-\sqrt{2}t}\right| = \log x + c$$

$$\Rightarrow \quad \frac{1}{2\sqrt{2}}\log\left|\frac{1+\sqrt{2}\frac{y}{x}}{1-\sqrt{2}\frac{y}{x}}\right| = \log x + c$$

$$\Rightarrow \quad \frac{1}{2\sqrt{2}}\log\left|\frac{x+\sqrt{2}y}{x-\sqrt{2}y}\right| = \log x + c$$

which is the required solution.

(iii) $\left\{x\cos\left(\frac{y}{x}\right)+y\sin\left(\frac{y}{x}\right)\right\}y\,dx = \left\{y\sin\left(\frac{y}{x}\right)-x\cos\left(\frac{y}{x}\right)\right\}x\,dy$ **[NCERT]**

Sol. The given differential equation is,

$$\left\{x\cos\left(\frac{y}{x}\right)+y\sin\left(\frac{y}{x}\right)\right\}y\,dx = \left\{y\sin\left(\frac{y}{x}\right)-x\cos\left(\frac{y}{x}\right)\right\}x\,dy$$

$$\Rightarrow xy\left\{\cos\left(\frac{y}{x}\right)+\frac{y}{x}\sin\left(\frac{y}{x}\right)\right\}dx = x^2\left\{\frac{y}{x}\sin\left(\frac{y}{x}\right)-\cos\left(\frac{y}{x}\right)\right\}dy$$

$$\Rightarrow \frac{dy}{dx} = \frac{xy\left\{\cos\left(\frac{y}{x}\right)+\frac{y}{x}\sin\left(\frac{y}{x}\right)\right\}}{x^2\left\{\frac{y}{x}\sin\left(\frac{y}{x}\right)-\cos\left(\frac{y}{x}\right)\right\}}$$

$$\Rightarrow \frac{dy}{dx} = \frac{y}{x}\left\{\frac{\cos\left(\frac{y}{x}\right)+\frac{y}{x}\sin\left(\frac{y}{x}\right)}{\frac{y}{x}\sin\left(\frac{y}{x}\right)-\cos\left(\frac{y}{x}\right)}\right\} \quad ...(i)$$

which is homogeneous equation of the form $f\left(\frac{y}{x}\right)$

Put $\frac{y}{x}=t \qquad y=xt$

Differentiate w.r.t. x

$$\frac{dy}{dx} = x\frac{dt}{dx}+t$$

Put in equation (i)

$$\Rightarrow x\frac{dt}{dx}+t = \frac{t(\cos t+t\sin t)}{t\sin t-\cos t}$$

$$\Rightarrow x\frac{dt}{dx} = \frac{t\cos t+t^2\sin t-t^2\sin t+t\cos t}{t\sin t-\cos t}$$

$$\Rightarrow x\frac{dt}{dx} = \frac{2t\cos t}{t\sin t-\cos t}$$

$$\Rightarrow \frac{t\sin t-\cos t}{2t\cos t}dt = \frac{dx}{x}$$

$$\Rightarrow \frac{1}{2}\left[\frac{t\sin t}{t\cos t}-\frac{\cos t}{t\cos t}\right] = \frac{dx}{x}$$

$$\Rightarrow \left(\tan t-\frac{1}{t}\right)dt = \frac{2}{x}dx$$

Integrating both sides :

$$-\log|\cos t|-\log t = 2\log x+c$$

$$\Rightarrow 2\log x+\log|\cos t|+\log t = -c$$

$$\Rightarrow \log|x^2(\cos t)\,t| = -c$$

$$\Rightarrow |tx^2\cos t| = e^{-c}$$

$$\Rightarrow \frac{y}{x}x^2\cos\frac{y}{x} = \pm e^{-c}$$

$$\Rightarrow xy\cos\frac{y}{x} = A$$

which is the required solution.

(iv) $y\,dx+x\log\left(\frac{y}{x}\right)dy-2x\,dy=0$ **[NCERT]**

Sol. Given differential equation is,

$$\Rightarrow y\,dx+x\log\left(\frac{y}{x}\right)dy-2x\,dy=0$$

$$\Rightarrow y\,dx = 2x\,dy-x\log\left(\frac{y}{x}\right)dy$$

$$\Rightarrow y\,dx = \left\{2x-x\log\left(\frac{y}{x}\right)\right\}dy$$

$$\Rightarrow \frac{dy}{dx} = \frac{y}{2x-x\log\left(\frac{y}{x}\right)} = \frac{y}{x\left\{2-\log\left(\frac{y}{x}\right)\right\}}$$

$$\Rightarrow \frac{dy}{dx} = \frac{y/x}{2-\log\left(\frac{y}{x}\right)} \quad ...(i)$$

which is homogeneous equation of the form $f\left(\frac{y}{x}\right)$

Put $\frac{y}{x}=t \Rightarrow \qquad y=xt$

Differentiate both sides w.r.t. x

$$\frac{dy}{dx} = x\frac{dt}{dx}+t$$

Put in equation (i)

$$x\frac{dt}{dx}+t = \frac{t}{2-\log t}$$

$$\Rightarrow x\frac{dt}{dx} = \frac{t}{2-\log t}-t$$

$$\Rightarrow x\frac{dt}{dx} = \frac{t-2t+t\log t}{2-\log t}$$

$$\Rightarrow \frac{2-\log t}{t\log t-t} = \frac{dx}{x}$$

$$\Rightarrow \frac{1-(\log t-1)}{t(\log t-1)} = \frac{dx}{x}$$

$$\Rightarrow \frac{1}{t(\log t-1)}-\frac{\log t-1}{t(\log t-1)} = \frac{dx}{x}$$

$$\Rightarrow \frac{1/t}{\log t-1}-\frac{1}{t} = \frac{dx}{x}$$

Integrating both sides

$$\Rightarrow \log|\log t-1|-\log t = \log|x|+c$$

$$\Rightarrow \log|\log t-1|-\log|t|-\log x = c$$

$$\Rightarrow \log\left|\frac{\log t-1}{tx}\right| = c$$

$$\Rightarrow \frac{\log t-1}{tx} = \pm e^c$$

$$\Rightarrow \frac{\log\left(\frac{y}{x}\right)-1}{\frac{y}{x}\times x} = \pm e^c \qquad \left(\because \frac{y}{x}=t\right)$$

$$\Rightarrow \frac{\log\left(\frac{y}{x}\right)-1}{y} = A \text{ where } A=\pm e^c$$

$$\Rightarrow \log\left(\frac{y}{x}\right)-1 = Ay$$

which is the required solution.

Q. 11. Find the particular solution of the differential equation $(\tan^{-1} y - x)\, dy = (1 + y^2)\, dx$, given that $x = 1$, when $y = 0$.★

Sol. Given,

$$(\tan^{-1} y-x)dy = (1+y^2)dx$$

$$\frac{\tan^{-1} y-x}{1+y^2}\, dy = dx \qquad ...(i)$$

Putting $\tan^{-1} y = t$ and $\frac{1}{1+y^2}=\frac{dt}{dy}$ in (i), we get

$$(t-x)\, dt = dx$$

$$\frac{dx}{dt} = t-x$$

$$\frac{dx}{dt}+x = t \qquad ...(ii)$$

Here, $\quad$ I.F. $= e^{\int 1\, dt} = e^t$

Hence, the solution of the differential equation (ii) is given by

$$x(\text{I.F.}) = \int(\text{I.F.})\, t\, dt + C = \int e^t\, t\, dt + C,$$

where C is arbitrary constant

$$xe^t = te^t - \int e^t\, dt + C$$

$$xe^t = te^t - e^t + C$$

$$xe^{\tan^{-1}y} = e^{\tan^{-1}y}(\tan^{-1} y-1)+C \qquad ...(iii)$$

$$[\because t=\tan^{-1} y]$$

★ are board exam questions from previous years

It is given that $x = 1$ when $y = 0$

So, $\quad e^0 = e^0(0-1)+C$

$\Rightarrow \quad C = 2$

Putting $C = 2$ in (iii), we get

$$xe^{\tan^{-1}y} = e^{\tan^{-1}y}(\tan^{-1} y-1)+2$$

$$\therefore \quad x = \tan^{-1}y + 2e^{\tan y} - 1$$

is the particular solution the given differential equation.

Q. 12. Find the particular solution of the differential equation $2xy+y^2-2x^2\frac{dy}{dx}=0$, given $y = 2$ when $x = 1$ **[NCERT]**

Sol. Given differential equation is,

$$2xy+y^2-2x^2\frac{dy}{dx}=0$$

$$\Rightarrow 2x^2\frac{dy}{dx} = 2xy+y^2$$

$$\Rightarrow \frac{dy}{dx} = \frac{2xy}{2x^2}+\frac{y^2}{2x^2}$$

$$\Rightarrow \frac{dy}{dx} = \frac{y}{x}+\frac{1}{2}\left(\frac{y}{x}\right)^2 \qquad ...(i)$$

which is homogeneous equation of the form $f\left(\frac{y}{x}\right)$

Put $\frac{y}{x}=t \qquad \Rightarrow y = xt$

Differentiate w.r.t. x

$$\frac{dy}{dx} = x\frac{dt}{dx}+t$$

Put in equation (i)

$$x\frac{dt}{dx}+t = t+\frac{1}{2}t^2$$

$$\Rightarrow x\frac{dt}{dx} = \frac{t^2}{2}$$

$$\Rightarrow \frac{2dt}{t^2} = \frac{dx}{x}$$

$$\Rightarrow 2t^{-2}\, dt = \frac{dx}{x}$$

Integrating both sides

$$\frac{2t^{-1}}{-1} = \log x + c$$

$$-\frac{2}{t} = \log x + c$$

$$\frac{-2x}{y} = \log x + c \qquad \left[\because t=\frac{y}{x}\right] \quad ...(ii)$$

When $x = 1, y = 2$

$$-\frac{2}{2} = \log 1 + c$$

$$\therefore \quad c = -1$$

Put in equation (ii)

$$-\frac{2x}{y} = \log x - 1$$

$$\frac{2x}{y} = 1 - \log x$$

which is the required solution.

Q. 13. Solve the differential equation $x\frac{dy}{dx} + y = x - \cos x + \sin x$, given that $y = 1$, when $x = \frac{\pi}{2}$.★

Sol. Given differential equation is :

$$x\frac{dy}{dx} + y = x\cos x + \sin x$$

$$\frac{dy}{dx} + \frac{y}{x} = \cos x + \frac{\sin x}{x}$$

which is of the form $\frac{dy}{dx} + Py = Q$

Where $P = \frac{1}{x}, Q = \cos x + \frac{\sin x}{x}$

$$\text{I. F.} = e^{\int P dx} = e^{\int \frac{1}{x} dx} = e^{\log x} = x$$

Required solution is

$$y.\text{I.F.} = \int \text{Q.I.F} + c$$

$$y.x = \int\left(\cos x + \frac{\sin x}{x}\right) x\,dx + c$$

$$= \int x\cos x\,dx + \int \sin x\,dx + c$$

$$= x.\int \cos x\,dx - \int\left[\frac{d}{dx}(x).\int \cos x\,dx\right]dx - \cos x + c$$

$$xy = x\sin x - \int \sin x\,dx - \cos x + c$$

$$xy = x\sin x + \cos x - \cos x + c$$

$$\Rightarrow \quad xy = x\sin x + c \quad \text{...(i)}$$

Given, $y = 1$ when $x = \frac{\pi}{2}$

From eq. (i)

$$1 \times \frac{\pi}{2} = \frac{\pi}{2}\sin\frac{\pi}{2} + c$$

$$\frac{\pi}{2} = \frac{\pi}{2} + c$$

$$\Rightarrow \quad c = 0$$

Substitute the value of $c = 0$ in (i), we get

$$xy = x\sin x$$

$y = \sin x$, which is the required solution.

Q. 14. Find the particular solution of the differential equation $(x - y)\frac{dy}{dx} = (x + 2y)$, given that $y = 0$ when $x = 1$.★

Sol. We have, $(x-y)\frac{dy}{dx} = (x+2y)$

$$\frac{dy}{dx} = \frac{x+2y}{x-y} \quad \text{...(i)}$$

Putting $y = Vx$ and $\frac{dy}{dx} = V + x\frac{dV}{dx}$

$$V + x\frac{dV}{dx} = \frac{x+2Vx}{x-Vx}$$

$$V + x\frac{dV}{dx} = \frac{1+2V}{1-V}$$

$$x\frac{dV}{dx} = \frac{1+2V}{1-V} - V$$

$$x\frac{dV}{dx} = \frac{1+2V-V+V^2}{1-V}$$

$$\frac{1-V}{1+V+V^2}dV = \frac{dx}{x}$$

On integrating both sides, we get

$$\int\frac{1-V}{1+V+V^2}dV = \int\frac{dx}{x}$$

$$\frac{3}{2}\int\frac{1}{1+V+V^2}dV - \frac{1}{2}\int\frac{1+2V}{1+V+V^2}dV = \int\frac{dx}{x}$$

$$\frac{3}{2}\int\frac{dV}{\left(V+\frac{1}{2}\right)^2 + \left(\frac{\sqrt{3}}{2}\right)^2} - \frac{1}{2}\int\frac{1+2V}{1+V+V^2}dV = \int\frac{dx}{x}$$

$$\sqrt{3}\tan^{-1}\left(\frac{2V+1}{\sqrt{3}}\right) - \frac{1}{2}\log|1+V+V^2| = \log|x| + c$$

$$\sqrt{3}\tan^{-1}\left(\frac{2y+x}{\sqrt{3}x}\right) - \frac{1}{2}\log|x^2+xy+y^2| = c \quad \text{...(ii)}$$

Now, given $y = 0$, when $x = 1$.

So, $\sqrt{3}\tan^{-1}\left(\frac{2(0)+1}{\sqrt{3}}\right) - \frac{1}{2}\log|1+0+0| = c$

★ are board exam questions from previous years

$$c = \sqrt{3}\tan^{-1}\left(\frac{1}{\sqrt{3}}\right) = \frac{\sqrt{3}\pi}{6}$$

Putting the value of c in (ii), we get

$$\sqrt{3}\tan^{-1}\left(\frac{2y+x}{\sqrt{3}x}\right) - \frac{1}{2}\log|x^2+xy+y^2| = \frac{\sqrt{3}\pi}{6}$$

Which is the required solution.

Q. 15. Find the general solution of the differential equation

(i) $\frac{dy}{dx} + 3y = e^{-2x}$ **[NCERT]**

Sol. We have,

$$\frac{dy}{dx} + 3y = e^{-2x} \quad ...(i)$$

It is linear equation of the form of

$$\frac{dy}{dx} + Py = Q$$

Here, $P = 3, Q = e^{-2x}$

$$\text{I.F.} = e^{\int P\,dx} = e^{\int 3\,dx} = e^{3x}$$

Multiply equation (i) by I.F.

$$e^{3x}\frac{dy}{dx} + 3ye^{3x} = e^{3x}\cdot e^{-2x}$$

$$\frac{d}{dx}(ye^{3x}) = e^x$$

Integrating both sides

$$ye^{3x} = e^x + c$$

$$y = e^{-2x} + ce^{-3x}$$

which is required solution.

(ii) $\cos^2 x\frac{dy}{dx} + y = \tan x \quad \left(0 \le x < \frac{\pi}{2}\right)$ **[NCERT]**

Sol. Given differential equation is,

$$\cos^2 x\frac{dy}{dx} + y = \tan x$$

$$\frac{dy}{dx} + \frac{1}{\cos^2 x}y = \frac{\tan x}{\cos^2 x}$$

$$\frac{dy}{dx} + (\sec^2 x)y = \tan x\sec^2 x \quad ...(i)$$

Here, $P = \sec^2 x, Q = \tan x\sec^2 x$

$$\text{I.F.} = e^{\int P\,dx} = e^{\int \sec^2 x\,dx} = e^{\tan x}$$

Multiply equation (i) by I.F.

$$e^{\tan x}\frac{dy}{dx} + (\sec^2 x)\,y\,e^{\tan x} = \tan x\sec^2 x\,e^{\tan x}$$

$$\frac{d}{dx}(ye^{\tan x}) = \tan x\sec^2 x\,e^{\tan x}$$

Integrating both sides

$$y\,e^{\tan x} = \int \tan x\sec^2 x\,e^{\tan x}\,dx + C$$

Put $\tan x = t$

$$\sec^2 x\,dx = dt$$

$$\therefore \quad ye^{\tan x} = \int t\,e^t\,dt + C$$

$$= t\,e^t - \int e^t\,dt + C$$

$$ye^{\tan x} = te^t - e^t + C$$

$$ye^{\tan x} = e^t(t-1) + C$$

$$ye^{\tan x} = e^{\tan x}(\tan x - 1) + C \quad (\because t = \tan x)$$

$$y = (\tan x - 1) + ce^{-\tan x}$$

which is the required solution.

(iii) $x\log x\frac{dy}{dx} + y = \frac{2}{x}\log x.$ **[NCERT]**

Sol. Given differential equation is ,

$$x\log x\frac{dy}{dx} + y = \frac{2}{x}\log x \quad ...(i)$$

Divide by $x\log x$

$$\frac{dy}{dx} + \frac{1}{x\log x}y = \frac{2}{x}\cdot\log x\,\frac{1}{x\log x}$$

$$\frac{dy}{dx} + \frac{1}{x\log x}y = \frac{2}{x^2} \quad ...(ii)$$

Here, $P = \frac{1}{x\log x}, \; Q = \frac{2}{x^2}$

$$\text{I.F.} = e^{\int P\,dx} = e^{\int \frac{1/x}{\log x}dx}$$

$$= e^{\log(\log x)} = \log x$$

Multiply equation (ii) by $\log x$

$$\log x\frac{dy}{dx} + \frac{1}{x}y = \frac{2}{x^2}\log x$$

$$\frac{d}{dx}(y\log x) = \frac{2}{x^2}\log x$$

Integrating both sides :

$$y\log x = \int \frac{2}{x^2}\log x\,dx + c$$

$$y\log x = 2\int \log x\cdot x^{-2}\,dx + c$$

$$y\log x = 2\left\{\log x\frac{x^{-1}}{-1} - \int \frac{1}{x}\cdot\frac{x^{-1}}{-1}\,dx\right\} + c$$

$$= 2\left\{-\frac{\log x}{x} + \int \frac{1}{x^2}\,dx\right\} + c$$

$$= 2\left\{-\frac{\log x}{x} + \frac{x^{-1}}{-1}\right\} + c$$

$$y \log x = -\frac{2 \log x}{x} - \frac{2}{x} + c$$

$$xy \log x = -2 \log x - 2 + cx$$

which is the required solution.

(iv) $(x+y)\dfrac{dy}{dx} = 1$ **[NCERT]**

Sol. Given differential equation is

$$(x+y)\frac{dy}{dx} = 1$$

$$(x+y) = \frac{dx}{dy}$$

$$\frac{dx}{dy} - x = y \quad \text{...(i)}$$

which is in the form of

$$\frac{dx}{dy} + Px = Q$$

Here $P = -1, Q = y$

$$\text{I.F.} = e^{\int P\,dy} = e^{\int -dy} = e^{-y}$$

Multiply equation (i) by I.F.

$$e^{-y}\frac{dx}{dy} - e^{-y}x = e^{-y}y$$

$$\frac{d}{dx}(xe^{-y}) = ye^{-y}$$

Integrating both sides :

$$xe^{-y} = \int ye^{-y}\,dy + c$$

$$\Rightarrow \quad xe^{-y} = y(-e^{-y}) - \int(-e^{-y})\,dy + c$$

$$\Rightarrow \quad xe^{-y} = -ye^{-y} + (-e^{-y}) + c$$

$$\Rightarrow \quad xe^{-y} = -e^{-y}(y+1) + c$$

$$x = -(y+1) + ce^{y}$$

which is the required general solution.

(v) $(x+3y^2)\dfrac{dy}{dx} = y, (y > 0)$ **[NCERT]**

Sol. Given differential equation is, $(x+3y^2)\dfrac{dy}{dx} = y$

$$x + 3y^2 = y\frac{dx}{dy}$$

$$y\frac{dx}{dy} - x = 3y^2$$

$$\frac{dx}{dy} - \frac{1}{y}x = 3y \quad \text{...(i)}$$

which is in the form of $\dfrac{dx}{dy} + Px = Q$

Here $P = -\dfrac{1}{y}$, $Q = 3y$

$$\text{I.F.} = e^{\int P\,dy} = e^{\int -\frac{1}{y}dy} = e^{-\log y} = e^{\log y^{-1}}$$

$$\text{I.F.} = y^{-1} = \frac{1}{y}$$

Multiply equation (i) by $\dfrac{1}{y}$

$$\frac{1}{y}\frac{dx}{dy} - \frac{1}{y^2}x = 3$$

$$\frac{d}{dy}\left(\frac{x}{y}\right) = 3$$

Integrating both sides

$$\frac{x}{y} = 3y + c$$

$$x = 3y^2 + cy$$

which is the required general solution.

Chapter 10. Vectors

Q. 1. Find a vector of magnitude $\sqrt{51}$ which makes equal angles with the unit vectors $\vec{a} = \frac{1}{3}(\hat{i} - 2\hat{j} + 2\hat{k})$, $\vec{b} = \frac{1}{5}(-4\hat{i} - 3\hat{k})$ and $\vec{c} = \hat{j}$

Sol. Let the required vector be

$$\vec{r} = \alpha\hat{i} + \beta\hat{j} + \gamma\hat{k}$$

Where $\alpha^2 + \beta^2 + \gamma^2 = 51$ (given)

$$\Rightarrow \quad |\vec{r}| = \sqrt{51} \quad \text{...(i)}$$

Let θ be the angle which $\vec{a}, \vec{b}$ and $\vec{c}$ make with $\vec{r}$

Then, $$\cos\theta = \frac{\vec{r}\cdot\vec{a}}{|\vec{r}||\vec{a}|} = \frac{\vec{r}\cdot\vec{b}}{|\vec{r}||\vec{b}|} = \frac{\vec{r}\cdot\vec{c}}{|\vec{r}||\vec{c}|}$$

$$= \frac{\vec{r}\cdot\vec{a}}{\sqrt{51}\times 1} = \frac{\vec{r}\cdot\vec{b}}{\sqrt{51}\times 1} = \frac{\vec{r}\cdot\vec{c}}{\sqrt{51}\times 1}$$

[Since $|\vec{r}| = \sqrt{51}$ and $\vec{a}, \vec{b}$ and $\vec{c}$ are unit vectors]

$$\Rightarrow \quad \vec{r}\cdot\vec{a} = \vec{r}\cdot\vec{b} = \vec{r}\cdot\vec{c}$$

$$\Rightarrow (\alpha\hat{i} + \beta\hat{j} + \gamma\hat{k}) \cdot \left(\frac{\hat{i}}{3} - \frac{2\hat{j}}{3} + \frac{2\hat{k}}{3}\right)$$

$$= (\alpha\hat{i} + \beta\hat{j} + \gamma\hat{k}) \left(\frac{-4\hat{i}}{5} - \frac{3\hat{k}}{5}\right)$$

$= (\alpha\hat{i}+\beta\hat{j}+\gamma\hat{k}).\hat{j}$

$\Rightarrow \quad \frac{1}{3}(\alpha-2\beta+2\gamma) = \frac{1}{5}(-4\alpha-3\gamma) = \beta$

$\Rightarrow \quad 5\alpha-10\beta+10\gamma = -12\alpha-9\gamma = 15\beta$

$\Rightarrow \quad 5\alpha-25\beta+10\gamma = 0$

and $\quad 12\alpha+15\beta+9\gamma = 0$

$\Rightarrow \quad \alpha-5\beta+2\gamma = 0 \quad \text{...(i)}$

$\Rightarrow \quad 4\alpha+5\beta+3\gamma = 0 \quad \text{...(ii)}$

Hence, $\quad \frac{\alpha}{-15-10} = \frac{\beta}{8-3} = \frac{\gamma}{5+20}$

$\Rightarrow \quad \frac{\alpha}{-25} = \frac{\beta}{5} = \frac{\gamma}{25} = \lambda$, (say)

$\Rightarrow \quad \frac{\alpha}{5} = \frac{\beta}{-1} = \frac{\gamma}{-5} = \lambda$

then $\quad \alpha = 5\lambda, \beta = -\lambda, \gamma = -5\lambda$

Putting these values of α, β and γ in (i), we get,

$25\lambda^2+\lambda^2+25\lambda^2 = 51$

$\Rightarrow \quad 51\lambda^2 = 51$

$\Rightarrow \quad \lambda^2 = 1$

$\Rightarrow \quad \lambda = \pm 1$

Hence, the required vectors are

$$\vec{r} = \pm(5\hat{i}-\hat{j}-5\hat{k})$$

Q. 2. Find the angles of a triangle *ABC* whose vertices are *A*(1, 2, 3), *B* (2, 4, 8) and *C*(3, 6, 7).

Sol. Let *O* be the origin.

Then, $\quad \vec{OA} = \hat{i}+2\hat{j}+3\hat{k}$,

$\vec{OB} = 2\hat{i}+4\hat{j}+8\hat{k}$

and $\quad \vec{OC} = 3\hat{i}+6\hat{j}+7\hat{k}$

$\therefore \quad \vec{AB} = \vec{OB} - \vec{OA}$

$= (2\hat{i}+4\hat{j}+8\hat{k}) - (\hat{i}+2\hat{j}+3\hat{k})$

$= \hat{i}+2\hat{j}+5\hat{k}$,

$|\vec{AB}| = \sqrt{30}$,

$\vec{BC} = \vec{OC} - \vec{OB}$

$= (3\hat{i}+6\hat{j}+7\hat{k}) - (2\hat{i}+4\hat{j}+8\hat{k})$

$= \hat{i}+2\hat{j}-\hat{k}$,

$|\vec{BC}| = \sqrt{6}$

$\vec{CA} = \vec{OA} - \vec{OC}$

$= (\hat{i}+2\hat{j}+3\hat{k}) - (3\hat{i}+6\hat{j}+7\hat{k})$

$= -2\hat{i}-4\hat{j}-4\hat{k}$,

$|\vec{CA}| = \sqrt{36} = 6$

Now $\quad \vec{AB}+\vec{BC} = 2\hat{i}+4\hat{j}+4\hat{k}$

Also, $\quad \vec{AB}+\vec{BC}+\vec{CA} = \vec{0}$

Which show that *A*, *B* and *C* are the vertices of a triangle.

Now, $\vec{AB}\cdot\vec{BC} = (\hat{i}+2\hat{j}+5\hat{k}).(\hat{i}+2\hat{j}-\hat{k})$

$= 1+4-5 = 0$

$\Rightarrow AB \perp BC$ and therefore, $B = 90°$

Now, $\angle A$ is the angle between $\vec{AB}$ and $\vec{AC}$

and $\vec{AB}\cdot\vec{AC} = (\hat{i}+2\hat{j}+5\hat{k})\cdot(2\hat{i}+4\hat{j}+4\hat{k})$

$= 2+8+20 = 30$

$$\cos A = \frac{\vec{AB}\cdot\vec{AC}}{|\vec{AB}||\vec{AC}|} = \frac{30}{\sqrt{30}\times 6} = \frac{\sqrt{30}}{6}$$

But $\angle C$ is the angle between $\vec{CA}$ and $\vec{CB}$

$$\text{Hence, } \cos C = \frac{\vec{CA}.\vec{CB}}{|\vec{CA}||\vec{CB}|}$$

$$= \frac{(-2\hat{i}-4\hat{j}-4\hat{k}).(-\hat{i}-2\hat{j}+\hat{k})}{6\sqrt{6}}$$

$$= \frac{2+8-4}{6\sqrt{6}} = \frac{6}{6\sqrt{6}}$$

$$= \frac{1}{\sqrt{6}} = \frac{\sqrt{6}}{6}$$

Hence, *ABC* is a right-angled triangle, right-angled at *B*. The other two angles are

$$\cos^{-1}\frac{\sqrt{30}}{6} \text{ and } \cos^{-1}\frac{\sqrt{6}}{6}$$

Q. 3. If the diagonals of a parallelogram are determined by the vector $\vec{a}$ and $\vec{b}$, show that its area is $\frac{1}{2}|\vec{a}\times\vec{b}|$.

Sol. Let $ABCD$ be a parallelogram. Its diagonals AC and BD intersect at O.

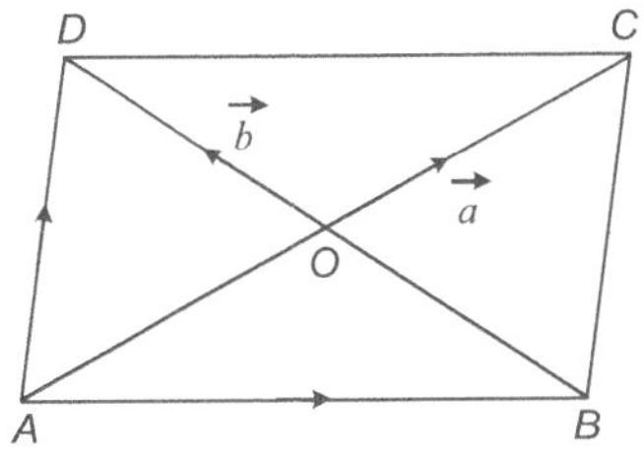

Let $\vec{AC} = \vec{a}$

and $\vec{BD} = \vec{b}$

Since the diagonals of a parallelogram bisect each other, O is the mid-point of AC and BD.

$$\therefore \quad \vec{OD} = \frac{1}{2}\vec{BD} = \frac{\vec{b}}{2},$$

$$\vec{OC} = \frac{1}{2}\vec{AC} = \frac{\vec{a}}{2},$$

$$\vec{OB} = -\frac{1}{2}\vec{BD} = -\frac{1}{2}\vec{b}$$

and $$\vec{OA} = -\frac{1}{2}\vec{AC} = -\frac{1}{2}\vec{a}$$

$$\therefore \quad \vec{AB} = \vec{OB} - \vec{OA}$$

$$= \left(-\frac{1}{2}\vec{b}\right) - \left(-\frac{1}{2}\vec{a}\right)$$

$$= \frac{1}{2}(\vec{a} - \vec{b})$$

and $$\vec{AD} = \vec{OD} - \vec{OA} = \frac{\vec{b}}{2} - \left(-\frac{\vec{a}}{2}\right)$$

$$= \frac{1}{2}(\vec{a} + \vec{b})$$

Therefore, the area of parallelogram $ABCD$

$$= |\vec{AB} \times \vec{AD}|$$

$$= |\frac{1}{2}(\vec{a} - \vec{b}) \times \frac{1}{2}(\vec{a} + \vec{b})|$$

$$= |\frac{1}{4}(\vec{a} \times \vec{a} + \vec{a} \times \vec{b} - \vec{b} \times \vec{a} - \vec{b} \times \vec{b})|$$

$$= \frac{1}{4}|0 + \vec{a} \times \vec{b} + \vec{a} \times \vec{b} - 0|$$

$$= \frac{1}{4}|2(\vec{a} \times \vec{b})|$$

$$= \frac{1}{2}|\vec{a} \times \vec{b}|$$

Chapter 11. Three-Dimensional Geometry

Q. 1. Find the direction cosines of the two lines which are connected by the relations.

$$\mathbf{l - 5m + 3n = 0 \text{ and } 7l^2 + 5m^2 - 3n^2 = 0.}$$

Sol. Given lines are,

$$l - 5m + 3n = 0 \quad \text{...(i)}$$

$$7l^2 + 5m^2 - 3n^2 = 0 \quad \text{...(ii)}$$

Putting $l = (5m - 3n)$ from (i) into (ii), we get

$$7(5m - 3n)^2 + 5m^2 - 3n^2 = 0$$

$$175m^2 + 63n^2 - 210mn + 5m^2 - 3n^2 = 0$$

$$\Rightarrow \quad 180m^2 + 60n^2 - 210\,mn = 0$$

or $$6m^2 - 7mn + 2n^2 = 0$$

or $$6\left(\frac{m}{n}\right)^2 - 7\left(\frac{m}{n}\right) + 2 = 0$$

or $$6p^2 - 7p + 2 = 0$$

where $$p = \frac{m}{n}$$

or $$(3p - 2)(2p - 1) = 0$$

$$\therefore \quad p = \frac{2}{3} \text{ or } p = \frac{1}{2}$$

So, $m/n = 2/3$ or $m/n = 1/2$ (since $p = m/n$)

Now $$\frac{m}{n} = \frac{2}{3}$$

$$\Rightarrow \quad \frac{m}{2} = \frac{n}{3}$$

$$\Rightarrow \quad \frac{5m - 3n}{5 \times 2 - 3 \times 3} = \frac{l}{1} \quad [\because 5m - 3n = l]$$

Thus $$\frac{l}{1} = \frac{m}{2} = \frac{n}{3} = \frac{\sqrt{l^2 + m^2 + n^2}}{1^2 + 2^2 + 3^2} = \frac{1}{\sqrt{14}}$$

$$\therefore \quad l = \frac{1}{\sqrt{14}},\ m = \frac{2}{\sqrt{14}},\ n = \frac{3}{\sqrt{14}}$$

Again $$\frac{m}{n} = \frac{1}{2} \Rightarrow \frac{m}{1} = \frac{n}{2}$$

$$\Rightarrow \quad \frac{5m - 3n}{5 \times 1 - 3 \times 2} = \frac{l}{-1}$$

Therefore $$\frac{l}{-1} = \frac{m}{1} = \frac{n}{2}$$

$$= \frac{\sqrt{l^2 + m^2 + n^2}}{\sqrt{(-1^2) + 1^2 + 2^2}}$$

$$= \frac{1}{\sqrt{6}}$$

So $\quad l = \frac{-1}{\sqrt{6}},\ m = \frac{1}{\sqrt{6}}$ and $n = \frac{2}{\sqrt{6}}$

Hence, the direction cosines of the lines are

$$\left(\frac{1}{\sqrt{14}}, \frac{2}{\sqrt{14}}, \frac{3}{\sqrt{14}}\right) \text{ and } \left(\frac{-1}{\sqrt{6}}, \frac{1}{\sqrt{6}}, \frac{2}{\sqrt{6}}\right)$$

Q. 2. Show that the straight lines whose direction cosines are given by the equations $al + bm + cn = 0$ and $ul^2 + vm^2 + wn^2 = 0$ are perpendicular. If $a^2(v + w) + b^2(u + w) + c^2(v + u) = 0$ are parallel, then show that

$$\frac{a^2}{u} + \frac{b^2}{v} + \frac{c^2}{w} = 0.$$

Sol. Given, $\quad al + bm + cn = 0 \quad$ ...(i)

and $\quad ul^2 + vm^2 + wn^2 = 0 \quad$ ...(ii)

Putting $l = -\frac{(bm + cn)}{a}$ from (i) into (ii), we get

$$\frac{u[-(bm + cn)]^2}{a^2} + vm^2 + wn^2 = 0$$

$$\frac{u(b^2m^2 + c^2n^2 + 2bcmn)}{a^2} + vm^2 + wn^2 = 0$$

$$(ub^2 + va^2)\, m^2 + (uc^2 + wa^2)\, n^2 + 2u\, b\, c\, mn = 0$$

or $(b^2u + a^2v)\left(\frac{m}{n}\right)^2 + 2ubc\left(\frac{m}{n}\right)$

$$+ (c^2u + a^2w) = 0$$

[dividing throughout by n^2] ...(iii)

Let $\frac{m_1}{n_1}$ and $\frac{m_2}{n_2}$ be the roots of (iii).

Then, $\quad \frac{m_1}{n_1} \cdot \frac{m_2}{n_2} = \frac{c^2u + a^2w}{b^2u + a^2v}$

$$\therefore \frac{m_1m_2}{c^2u + a^2w} = \frac{n_1n_2}{b^2u + a^2v} = \frac{l_1l_2}{b^2w + c^2v} = k$$

[by symmetry]

So, $l_1l_2 + m_1m_2 + n_1n_2$

$$= k(b^2w + c^2v + c^2u + a^2w + b^2u + a^2v)$$

For perpendicularity,

we have $\quad l_1l_2 + m_1m_2 + n_1n_2 = 0$

i.e., $\quad a^2(v + w) + b^2(w^2 + u) + c^2(u + v) = 0$

For the given lines to be parallel, the direction cosines must be equal and so the roots of (iii) must be equal.

Thus, we must have

$$4u^2b^2c^2 = 4(b^2u + a^2v) \times (c^2u + a^2w) = 0$$

or $$\frac{a^2}{u} + \frac{b^2}{v} + \frac{c^2}{w} = 0.$$

Q. 3. Find the equations of the two lines through the origin, which intersect the line $\frac{x-3}{2} = \frac{y-3}{1} = \frac{z}{1}$ at angle of $\frac{\pi}{3}$.

Sol. The equation of the given line is

$$\frac{x-3}{2} = \frac{y-3}{1} = \frac{z-0}{1} = \lambda \text{ (say)} \quad \text{...(i)}$$

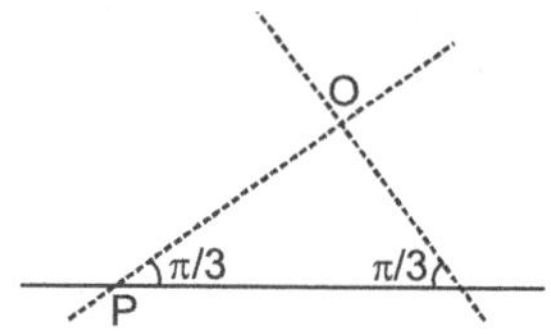

Any point on the line (i) is

$$P(2\lambda + 3, \lambda + 3, \lambda) \quad \text{...(ii)}$$

Now, direction ratios of OP are

$$2\lambda + 3 - 0, \lambda + 3 - 0, \lambda - 0$$

i.e., $2\lambda + 3, \lambda + 3, \lambda$

Direction ratios of line (i) are 2, 1, 1

Since, OP makes an angle of $\frac{\pi}{3}$ with (i), we get

$$\cos\frac{\pi}{3} = \frac{|2(2\lambda+3) + 1(\lambda+3) + 1(\lambda)|}{\sqrt{(2\lambda+3)^2 + (\lambda+3)^2 + \lambda^2}\sqrt{(2)^2 + (1)^2 + (1)^2}}$$

$$\Rightarrow \frac{1}{2} = \frac{4\lambda + 6 + \lambda + 3 + \lambda}{\sqrt{4\lambda^2 + 12\lambda + 9 + \lambda^2 + 9 + 6\lambda + \lambda^2}\sqrt{6}}$$

$$\Rightarrow \frac{1}{2} = \frac{6\lambda + 9}{\sqrt{6\lambda^2 + 18\lambda + 18}\sqrt{6}}$$

$$\Rightarrow \frac{1}{2} = \frac{3(2\lambda + 3)}{\sqrt{6}\sqrt{6}\sqrt{\lambda^2 + 3\lambda + 3}}$$

$$\Rightarrow \frac{1}{2} = \frac{3(2\lambda + 3)}{6\sqrt{\lambda^2 + 3\lambda + 3}}$$

$$\Rightarrow \frac{1}{2} = \frac{(2\lambda + 3)}{2\sqrt{\lambda^2 + 3\lambda + 3}}$$

$$\Rightarrow \sqrt{\lambda^2 + 3\lambda + 3} = (2\lambda + 3)$$

$$\Rightarrow \lambda^2 + 3\lambda + 3 = (2\lambda + 3)^2$$

$$\Rightarrow \lambda^2 + 3\lambda + 3 = 4\lambda^2 + 12\lambda + 9$$

$$\Rightarrow 3\lambda^2 + 9\lambda + 6 = 0$$

$$\Rightarrow \lambda^2 + 3\lambda + 2 = 0$$

$$\Rightarrow (\lambda + 1)(\lambda + 2) = 0$$

$$\Rightarrow \lambda = -1 \text{ or } \lambda = -2$$

Substituting these values of λ, the two possible positions of P are (1, 2, – 1) and (– 1, 1, – 2).

Hence, the required lines are

$$\frac{x-0}{1-0}=\frac{y-0}{2-0}=\frac{z-0}{-1-0}$$

and $$\frac{x-0}{-1-0}=\frac{y-0}{1-0}=\frac{z-0}{-2-0}$$

that is $$\frac{x}{1}=\frac{y}{2}=\frac{z}{-1}$$

and $$\frac{x}{-1}=\frac{y}{1}=\frac{z}{-2}$$

Therefore $\frac{x}{1}=\frac{y}{2}=\frac{z}{-1}$ and $\frac{x}{-1}=\frac{y}{1}=\frac{z}{-2}$ are the two lines through origin, which intersects the given line at angle of $\frac{\pi}{3}$.

Q. 4. Prove that the lines whose direction cosines are given by $al+bm+cn=0$ and $fmn+gnl+hlm=0$ are

(i) perpendicular if $\frac{f}{a}+\frac{g}{b}+\frac{h}{c}=0$ and

(ii) parallel if $a^2f^2+b^2g^2+c^2h^2-2(bcgh+cahf+abfg)=0$.

Sol. Given that **[HOTS]**

$$al+bm+cn=0 \quad \ldots(i)$$

and $$fmn+gnl+hlm=0 \quad \ldots(ii)$$

From (i), $$al+bm=-cn$$

$$\Rightarrow \quad n=-\left(\frac{al+bm}{c}\right) \quad \ldots(iii)$$

Substituting this value of n in (ii), we get

$$(fm+gl)\left(-\frac{al+bm}{c}\right)+hlm=0$$

$$\Rightarrow \quad -(fm+gl)(al+bm)+hlmc=0$$

$$\Rightarrow \quad -aflm-agl^2-bfm^2-bglm+chlm=0$$

$$\Rightarrow \quad agl^2+(af+bg-ch)lm+bfm^2=0 \quad \ldots(iv)$$

We note that both l and m cannot be zero.

For, if $l=m=0$

Then from (iii), we get $n=0$

which is wrong as $l^2+m^2+n^2=1$.

So, without any loss of generality we may take $m\neq 0$.

Dividing both sides of (iv) by m^2, we get

$$ag\left(\frac{l}{m}\right)^2+(af+bg-ch)\left(\frac{l}{m}\right)+bf=0 \quad \ldots(v)$$

The above equation is quadratic in $\left(\frac{l}{m}\right)$ giving two values of $\left(\frac{l}{m}\right)$.

If (l_1, m_1, n_1) and (l_2, m_2, n_2) are the direction cosines of the two given line (i) and (ii), then

$\frac{l_1}{m_1}, \frac{l_2}{m_2}$ are roots of equation (v)

Therefore product of root $=\frac{bf}{ag}$

$$\Rightarrow \quad \left(\frac{l_1}{m_1}\right)\left(\frac{l_2}{m_2}\right)=\frac{bf}{ag}$$

$$\Rightarrow \quad \frac{l_1l_2}{m_1m_2}=\frac{f/a}{g/b}$$

$$\Rightarrow \quad \frac{l_1l_2}{f/a}=\frac{m_1m_2}{g/b}$$

$$\Rightarrow \quad \frac{l_1l_2}{f/a}=\frac{m_1m_2}{g/b}=\frac{n_1n_2}{h/c}$$

[By symmetry of result]

Let $$\frac{l_1l_2}{f/a}=\frac{m_1m_2}{g/b}=\frac{n_1n_2}{h/c}=\lambda$$

where λ is a non-zero real number.

$$l_1l_2=\lambda\cdot\frac{f}{a},\ m_1m_2=\lambda\cdot\frac{g}{b},\ n_1n_2=\lambda\cdot\frac{h}{c}$$

Now, these lines whose direction cosines are given by (i) and (ii) are at right angles.

If $$l_1l_2+m_1m_2+n_1n_2=0$$

i.e., if $$\lambda\frac{f}{a}+\lambda\frac{g}{b}+\lambda\frac{h}{c}=0$$

i.e., if $$\frac{f}{a}+\frac{g}{b}+\frac{h}{c}=0$$

Also, the two lines will be parallel if,

$$l_1=l_2, m_1=m_2, n_1=n_2$$

i.e., if $$\frac{l_1}{m_1}=\frac{l_2}{m_2}$$

i.e., if equation (v) has equal roots.

i.e., if discriminant of (v) is zero.

i.e., if $(af+bg-ch)^2-4agbf=0$

i.e., if $a^2f^2+b^2g^2+c^2h^2+2abfg-2bgch-2afch+4agbf=0$

i.e., if $a^2f^2+b^2g^2+c^2h^2-2(bcgh+cahf+abfg)=0$

Q. 5. The edges of a rectangular parallelepiped are a, b, c. Show that the angles between the four diagonals are given by

$$\cos^{-1}\left(\frac{a^2\pm b^2\pm c^2}{a^2+b^2+c^2}\right)$$

Sol. Taking O, a corner of the rectangular parallelepiped as the origin and three edges OA, OB, OC through it as the axes.

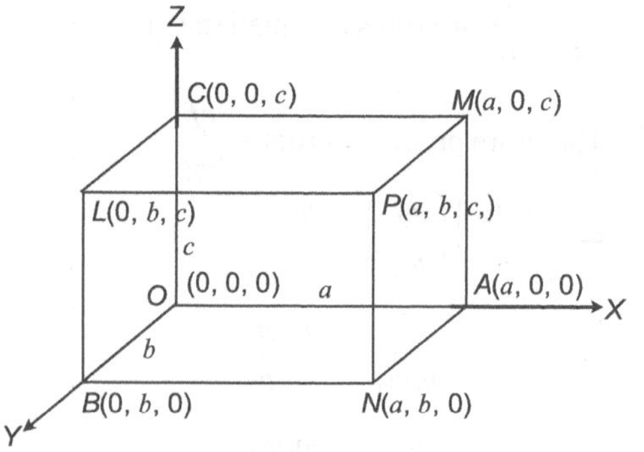

The coordinates of different corners are shown in the figure.

The four diagonals are AL, BM, CN and OP.

The direction ratios of AL are $0-a, b-0, c-0$.

i.e., $-a, b, c$

The direction ratios of BM are $a-0, 0-b, c-0$

i.e., $a, -b, c$

The direction ratios of CN are $a-0, b-0, 0-c$

i.e., $a, b, -c$

and the direction ratios of OP are $a-0, b-0, c-0$

i.e., a, b, c

Let α be the angle between the diagonals OP and AL.

Then $$\cos\alpha = \frac{a(-a)+b(b)+c(c)}{\sqrt{a^2+b^2+c^2}\sqrt{a^2+b^2+c^2}}$$

$$\Rightarrow \quad \cos\alpha = \frac{-a^2+b^2+c^2}{a^2+b^2+c^2}$$

$$\Rightarrow \quad \alpha = \cos^{-1}\left(\frac{-a^2+b^2+c^2}{a^2+b^2+c^2}\right) \quad \ldots(i)$$

Angle between OP and BM is

$$\cos\beta = \frac{a(a)+b(-b)+c(c)}{\sqrt{a^2+b^2+c^2}\sqrt{a^2+b^2+c^2}}$$

$$\Rightarrow \quad \cos\beta = \frac{a^2-b^2+c^2}{a^2+b^2+c^2}$$

$$\Rightarrow \quad \beta = \cos^{-1}\left(\frac{a^2-b^2+c^2}{a^2+b^2+c^2}\right)$$

Angle between OP and CN is

$$\cos\gamma = \frac{a(a)+b(b)+c(-c)}{\sqrt{a^2+b^2+c^2}\sqrt{a^2+b^2+c^2}}$$

$$\Rightarrow \quad \cos\gamma = \frac{a^2+b^2-c^2}{a^2+b^2+c^2}$$

$$\Rightarrow \quad \gamma = \cos^{-1}\left(\frac{a^2+b^2-c^2}{a^2+b^2+c^2}\right) \quad \ldots(iii)$$

Angle between AL and BM is

$$\cos\delta = \frac{-a(a)+b(-b)+c(c)}{\sqrt{a^2+b^2+c^2}\sqrt{a^2+b^2+c^2}}$$

$$\Rightarrow \quad \cos\delta = \frac{-a^2-b^2+c^2}{a^2+b^2+c^2}$$

$$\Rightarrow \quad \cos\delta = \frac{a^2+b^2-c^2}{a^2+b^2+c^2}$$

[when acute angle is taken]

$$\Rightarrow \quad \delta = \cos^{-1}\left(\frac{a^2+b^2-c^2}{a^2+b^2+c^2}\right)$$

This angle is same as the angle between OP and CN.

Similarly, we can show that all other angles between other two diagonals are repeated and we get only three different angles given by (i), (ii) and (iii). Hence, the angles between four diagonals are given by

$$\cos^{-1}\left(\frac{a^2 \pm b^2 \pm c^2}{a^2+b^2+c^2}\right)$$

Q. 6. Find the shortest distance and the vector equation of the line of shortest distance between the following pair of lines

$$\vec{r} = (1-\lambda)\hat{i}+(\lambda-2)\hat{j}+(3-2\lambda)\hat{k}$$

and $$\vec{r} = (\mu+1)\hat{i}+(2\mu-1)\hat{j}-(2\mu+1)\hat{k}$$

Sol. The vector equations of the two given lines are

$$l_1 : \vec{r} = (\hat{i}-2\hat{j}+3\hat{k})+\lambda(-\hat{i}+\hat{j}-2\hat{k}) \quad \ldots(i)$$

and $$l_2 : \vec{r} = (\hat{i}-\hat{j}-\hat{k})+\mu(\hat{i}+2\hat{j}-2\hat{k}) \quad \ldots(ii)$$

Let PQ be the shortest distance between the two lines l_1 and l_2. Then, the line of shortest distance is along PQ.

Let the position vector of P and Q be

$$(1-\lambda)\hat{i}+(\lambda-2)\hat{j}+(3-2\lambda)\hat{k} \quad \ldots(iii)$$

$$(\mu+1)\hat{i}+(2\mu-1)\hat{j}-(2\mu+1)\hat{k} \quad \ldots(iv)$$

respectively.

Then, $PQ = (\mu+\lambda)\hat{i}+(2\mu-\lambda+1)\hat{j}+(-2\mu+2\lambda-4)\hat{k}$

Since $\vec{PQ}$ is perpendicular to the line l_1.

Therefore, $\vec{PQ}$ is perpendicular to $(-\hat{i}+\hat{j}-2\hat{k})$

$\Rightarrow \quad \overrightarrow{PQ}\cdot(-\hat{i}+\hat{j}-2\hat{k})=0$

$\Rightarrow \quad -\mu-\lambda+2\mu-\lambda+1+4\mu-4\lambda+8=0$

$5\mu-6\lambda+9=0 \quad \ldots(v)$

Next, since $\overrightarrow{PQ}$ is perpendicular to line l_2, therefore

$(\mu+\lambda)+2(2\mu-\lambda+1)-2(-2\mu+2\lambda-4)=0$

$\Rightarrow \quad \mu+\lambda+4\mu-2\lambda+2+4\mu-4\lambda+8=0$

$\Rightarrow \quad 9\mu-5\lambda+10=0 \quad \ldots(vi)$

Now solving (v) and (vi), we get

$$\lambda=\frac{31}{29} \text{ and } \mu=\frac{-15}{29}$$

As $\overrightarrow{PQ}=\left(\frac{31}{29}-\frac{15}{29}\right)\hat{i}+\left(\frac{-30}{29}-\frac{31}{29}+1\right)\hat{j}+\left(\frac{30}{29}+\frac{62}{29}-4\right)\hat{k}$

$$=\frac{16}{29}\hat{i}-\frac{32}{29}\hat{j}-\frac{24}{29}\hat{k}$$

$$=\frac{8}{29}(2\hat{i}-4\hat{j}-3\hat{k})$$

$\Rightarrow \quad |\overrightarrow{PQ}|=\frac{8}{29}\sqrt{4+16+9}=\frac{8\sqrt{29}}{29}$

Thus, the shortest distance between the two lines l_1 and l_2 is $\frac{8\sqrt{29}}{29}$ units.

Since, the line PQ passes through P and Q, the vector equation of PQ is

$$\vec{r}=\left(1-\frac{31}{29}\right)\hat{i}+\left(\frac{31}{29}-2\right)\hat{j}+\left(3-\frac{62}{29}\right)\hat{k}+\lambda_1\frac{8}{29}(2\hat{i}-4\hat{j}-3\hat{k})$$

or $\vec{r}=\frac{-2}{29}\hat{i}-\frac{27}{29}\hat{j}+\frac{25}{29}\hat{k}+\frac{8}{29}\lambda_1(2\hat{i}-4\hat{j}-3\hat{k})$

Q. 7. Find the equation of the line of shortest distance between the lines $\frac{x-1}{2}=\frac{y+1}{-1}=\frac{z-3}{4}$ and $\frac{x-1}{3}=\frac{y+6}{4}=\frac{z+1}{2}$. Also, find the shortest distance.

Sol. Let $\frac{x-1}{2}=\frac{y+1}{-1}=\frac{z-3}{4}=\lambda \quad \ldots(i)$

Therefore, a general point on the line (i) is

$$P(2\lambda+1,-\lambda-1,4\lambda+3)$$

Let $\frac{x-1}{3}=\frac{y+6}{4}=\frac{z+1}{2}=\mu \quad \ldots(ii)$

$\Rightarrow \quad x=3\mu+1, y=4\mu-6, z=2\mu-1$

Therefore, a general point on the line (ii) is

$$Q(3\mu+1,4\mu-6,2\mu-1)$$

Let PQ be the shortest distance between the two given lines. Then, the line of shortest distance passes through P and Q.

Direction ratios of PQ are

$$(3\mu-2\lambda, 4\mu+\lambda-5, 2\mu-4\lambda-4)$$

Since, PQ is perpendicular to the line (i),

$2(3\mu-2\lambda)-1(4\mu+\lambda-5)+4(2\mu-4\lambda-4)=0$

$\Rightarrow \quad 10\mu-21\lambda-11=0 \quad \ldots(iii)$

Also, PQ is perpendicular to the line (ii).

$\therefore \quad 3(3\mu-2\lambda)+4(4\mu+\lambda-5)+2(2\mu-4\lambda-4)=0$

$\Rightarrow \quad 29\mu-10\lambda-28=0 \quad \ldots(iv)$

Solving equations (iii) and (iv), we get

$$\frac{\lambda}{-280+319}=\frac{\mu}{110-588}=\frac{1}{-609+100}$$

$\Rightarrow \quad \frac{\lambda}{39}=\frac{\mu}{-478}=\frac{1}{-509}$

$\Rightarrow \quad \lambda=\frac{-39}{509}$

and $\quad \mu=\frac{478}{509}$

So, the required points, where the line of shortest distance meets the given lines are

$$P\left(\frac{431}{509},\frac{-470}{509},\frac{1371}{509}\right) \text{ and } Q\left(\frac{1943}{509},\frac{-186}{509},\frac{447}{509}\right)$$

Therefore, shortest distance $= PQ$

$$=\sqrt{\left(\frac{1943}{509}-\frac{431}{509}\right)^2+\left(\frac{-186}{509}+\frac{470}{509}\right)^2+\left(\frac{447}{509}-\frac{1371}{509}\right)^2}$$

$$=\sqrt{\left(\frac{1512}{509}\right)^2+\left(\frac{284}{509}\right)^2+\left(\frac{-924}{509}\right)^2}$$

$$=\frac{1}{509}\sqrt{2286144+80656+853776}$$

$$=\frac{1}{509}\sqrt{3220576}$$

$$=\frac{4}{509}\sqrt{201286} \text{ units}$$

Also, the line of shortest distance is

$$\frac{x-\frac{431}{509}}{\frac{1512}{509}}=\frac{y+\frac{470}{509}}{\frac{284}{509}}=\frac{z-\frac{1371}{509}}{\frac{-924}{509}}$$

i.e., $\frac{509x-431}{1512}=\frac{509y+470}{284}=\frac{509z-1371}{-924}$

Q. 8. Find the image of the point (1, 6, 3) on the line $\frac{x}{1}=\frac{y-1}{2}=\frac{z-2}{3}$. Also, write the equation of the line joining the given points and its image and find the length of segment joining given point and its image.★

Sol. Here, T is the image of point $P(1, 6, 3)$. Q is the foot of perpendicular PQ on the line AB.

First, we find Q.

Equation of line AB is given by

$$\frac{x}{1}=\frac{y-1}{2}=\frac{z-2}{3} \quad \ldots(i)$$

Let $\frac{x}{1}=\frac{y-1}{2}=\frac{z-2}{3}=\lambda$ (say)

$\Rightarrow \quad x=\lambda, y-1=2\lambda, z-2=3\lambda$

$\Rightarrow \quad x=\lambda, y=2\lambda+1, z=3\lambda+2$

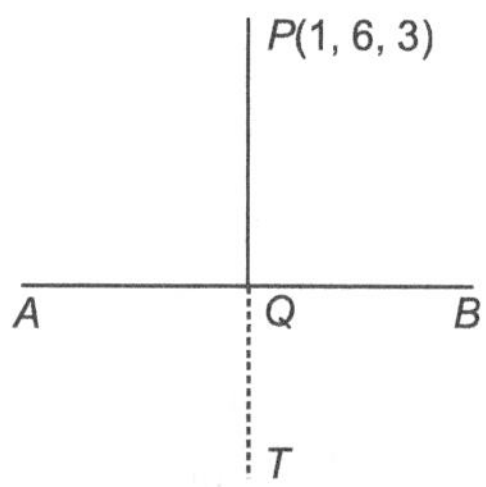

Let coordinates of

$$Q=(\lambda, 2\lambda+1, 3\lambda+2) \quad \ldots(ii)$$

Now, direction ratios of line

$$PQ=(\lambda-1, 2\lambda+1-6, 3\lambda+2-3)$$

$\therefore$ Direction ratios of $PQ=(\lambda-1, 2\lambda-5, 3\lambda-1)$

Now, line $\quad PQ \perp AB$

$\therefore \quad a_1a_2+b_1b_2+c_1c_2=0$

Where $\quad a_1=\lambda-1, b_1=2\lambda-5, c_1=3\lambda-1$

and $\quad a_2=1, b_2=2, c_2=3$

$\therefore \quad 1(\lambda-1)+2(2\lambda-5)+3(3\lambda-1)=0$

$\Rightarrow \quad \lambda-1+4\lambda-10+9\lambda-3=0$

$\Rightarrow \quad 14\lambda-14=0$

$\Rightarrow \quad \lambda=1$

Putting $\lambda=1$ in equation (ii), we get

$$Q(1, 2+1, 3+2)=(1, 3, 5)$$

Now, as discussed earlier Q is the mid-point of PT.

Let coordinates of $T=(x, y, z)$

$\therefore$ Using by mid-point formula,

$Q=$ Mid-point of $P(1, 6, 3)$ and $T(x, y, z)$

$$=\left(\frac{x+1}{2}, \frac{y+6}{2}, \frac{z+3}{2}\right)$$

But $\quad Q=(1, 3, 5)$

$$\therefore \left(\frac{x+1}{2}, \frac{y+6}{2}, \frac{z+3}{2}\right)=(1, 3, 5)$$

$\Rightarrow \quad \frac{x+1}{2}=1, \frac{y+6}{2}=3, \frac{z+3}{2}=5$

$\Rightarrow \quad x=2-1, y=6-6, z=10-3$

$\Rightarrow \quad x=1, y=0, z=7$

$\therefore$ Coordinates of $T=(x, y, z)=(1, 0, 7)$

Hence, coordinates of image of point

$$P(1, 6, 3)=T(1, 0, 7).$$

Q. 9. Write the vector equations of the following lines and hence find the distance between them :★

$$\frac{x-1}{2}=\frac{y-2}{3}=\frac{z+4}{6}, \frac{x-3}{4}=\frac{y-3}{6}=\frac{z+5}{12}$$

Sol. Given equation of lines are

$$\frac{x-1}{2}=\frac{y-2}{3}=\frac{z+4}{6}$$

and $$\frac{x-3}{4}=\frac{y-3}{6}=\frac{z+5}{12}$$

Now, the vector equation of lines are

$$r=(\hat{i}+2\hat{j}-4\hat{k})+\lambda(2\hat{i}+3\hat{j}+6\hat{k}) \quad \ldots(i)$$

[$\because$ Vector form of equation of line is $\tau=a+\lambda b$]

and $r=(3\hat{i}+3\hat{j}-5\hat{k})+\mu(4\hat{i}+6\hat{j}+12\hat{k}) \quad \ldots(ii)$

where $\quad a=x_1\hat{i}+y_1\hat{j}+z_1\hat{k}$

and $\quad b=a\hat{i}+b\hat{j}+c\hat{k}$

Here $\quad a_1=\hat{i}+2\hat{j}-4\hat{k}, b_1=2\hat{i}+3\hat{j}+6\hat{k}$

$\quad a_2=3\hat{i}+3\hat{j}-5\hat{k}, b_2=4\hat{i}+6\hat{j}+12\hat{k}$

Now, $b_1 \times b_2=\begin{vmatrix} \hat{i} & \hat{j} & \hat{k} \\ 2 & 3 & 6 \\ 4 & 6 & 12 \end{vmatrix}$

$=\hat{i}(36-36)-\hat{j}(24-24)+\hat{k}(12-12)$

$=0\hat{i}-0\hat{j}+0\hat{k}=0$

$\because \quad b_1 \times b_2=0$

$\Rightarrow$ Vector b_1 is parallel to b_2

[$\because a \times b=0$, then $a \parallel b$]

Therefore the two lines are parallel.

$\therefore \quad b=(2\hat{i}+3\hat{j}+6\hat{k}) \quad \ldots(iii)$

[Since, direction ratios of given lines are proportional]

Since, the two lines are parallel, we use the formula for shortest distance between two parallel lines.

★ are board exam questions from previous years

We know that

$$d = \left| \frac{b \times (a_2 - a_1)}{|b|} \right| \quad \ldots(iv)$$

∴ From equations (iii) and (iv), we get

$$d = \left| \frac{(2\hat{i} + 3\hat{j} + 6\hat{k}) \times (2\hat{i} + \hat{j} - \hat{k})}{\sqrt{(2)^2 + (3)^2 + (6)^2}} \right| \quad \ldots(v)$$

Now $(2\hat{i} + 3\hat{j} + 6\hat{k}) \times (2\hat{i} + \hat{j} - \hat{k})$

$$= \begin{vmatrix} \hat{i} & \hat{j} & \hat{k} \\ 2 & 3 & 6 \\ 2 & 1 & -1 \end{vmatrix}$$

$$= \hat{i}(-3-6) - \hat{j}(-2-12) + \hat{k}(2-6)$$

$$= -9\hat{i} + 14\hat{j} - 4\hat{k}$$

From equation (v), we get

$$d = \left| \frac{|-9\hat{i} + 14\hat{j} - 4\hat{k}|}{\sqrt{49}} \right|$$

$$= \frac{\sqrt{(-9)^2 + (14)^2 + (-4)^2}}{7}$$

$$\therefore \quad d = \frac{\sqrt{81 + 196 + 16}}{7} = \frac{\sqrt{293}}{7} \text{ units}$$

Q. 10. The points $A(4, 5, 10)$, $B(2, 3, 4)$ and $C(1, 2, -1)$ are three vertices of parallelogram $ABCD$. Find the vector equations of sides AB and BC and also find coordinates of point D.★

Sol. First, we find vector equation of AB where $A(4, 5, 10)$ and $B(2, 3, 4)$.

We know that two points vector formula of line is given by

$$\vec{r} = \vec{a} + \lambda(\vec{b} - \vec{a}) \quad \ldots(i)$$

Where a and b are the position vector of points through which the line is passing through.

Here $\vec{a} = \vec{OA} = 4\hat{i} + 5\hat{j} + 10\hat{k}$

$\vec{b} = \vec{OB} = 2\hat{i} + 3\hat{j} + 4\hat{k}$

∴ Using equation (i), required equation of line AB is

$$\vec{r} = (4\hat{i} + 5\hat{j} + 10\hat{k}) + \lambda[(2\hat{i} + 3\hat{j} + 4\hat{k}) - (4\hat{i} + 5\hat{j} + 10\hat{k})]$$

$$\vec{r} = (4\hat{i} + 5\hat{j} + 10\hat{k}) + \lambda(-2\hat{i} - 2\hat{j} - 6\hat{k})$$

Similarly, vector equation of line BC where $B(2, 3, 4)$ and $C(1, 2, -1)$ is

$$\vec{r} = (2\hat{i} + 3\hat{j} + 4\hat{k}) + \lambda[(\hat{i} + 2\hat{j} - \hat{k}) - (2\hat{i} + 3\hat{j} + 4\hat{k})$$

$$\vec{r} = (2\hat{i} + 3\hat{j} + 4\hat{k}) + \lambda(-\hat{i} - \hat{j} - 5\hat{k})$$

Now let the coordinates of D be (x, y, z).

Mid-point of diagonal BD = Mid-point of diagonal AC

(∵ diagonals of a parallelogram bisect each other)

$$\therefore \left(\frac{x+2}{2}, \frac{y+3}{2}, \frac{z+4}{2}\right) = \left(\frac{4+1}{2}, \frac{5+2}{2}, \frac{10-1}{2}\right)$$

Comparing corresponding coordinates

$$\Rightarrow \quad \frac{x+2}{2} = \frac{5}{2}, \frac{y+3}{2} = \frac{7}{2}, \frac{z+4}{2} = \frac{9}{2}$$

$$\Rightarrow \quad x = 3, y = 4, z = 5$$

∴ Coordinates of point $D(x, y, z) = (3, 4, 5)$.

Q. 11. Find the vector and cartesian equations of a line passing through (1, 2, – 4) and perpendicular to the two lines

$$\frac{x-8}{3} = \frac{y+19}{-16} = \frac{z-10}{7} \text{ and } \frac{x-15}{3} = \frac{y-29}{8} = \frac{z-5}{-5}.★$$

Sol. Cartesian equation of the line passing through $(1, 2-4)$ is

$$\frac{x-1}{a} = \frac{y-2}{b} = \frac{z+4}{c} \quad \ldots(i)$$

Given lines are

$$\frac{x-8}{3} = \frac{y+19}{-16} = \frac{z-10}{7} \quad \ldots(ii)$$

$$\text{and} \quad \frac{x-15}{3} = \frac{y-29}{8} = \frac{z-5}{-5} \quad \ldots(iii)$$

Let $\vec{b_1}, \vec{b_2}, \vec{b_3}$ are parallel vectors of (i), (ii) and (iii) respectively.

$$\therefore \quad \vec{b_1} = a\hat{i} + b\hat{j} + c\hat{k}$$

$$\vec{b_2} = 3\hat{i} - 16\hat{j} + 7\hat{k}$$

$$\vec{b_3} = 3\hat{i} + 8\hat{j} - 5\hat{k}$$

Given that (i) is perpendicular to both (ii) and (iii)

$$\therefore \quad \vec{b_1} \cdot \vec{b_2} = 0$$

$$\Rightarrow 3a - 16b + 7c = 0 \quad \ldots(iv)$$

$$\text{and} \quad \vec{b_1} \cdot \vec{b_3} = 0$$

$$\Rightarrow \quad 3a + 8b - 5c = 0 \quad \ldots(v)$$

★ are board exam questions from previous years

From equations (iv) and (v)

$$\frac{a}{80-56} = \frac{b}{21+15} = \frac{c}{24+48}$$

$$\frac{a}{24} = \frac{b}{36} = \frac{c}{72}$$

$$\Rightarrow \quad \frac{a}{2} = \frac{b}{3} = \frac{c}{6} = \lambda \text{ (say)}$$

$$\Rightarrow \quad a = 2\lambda, b = 3\lambda, c = 6\lambda$$

Putting in equation (i)

$$\frac{x-1}{2\lambda} = \frac{y-2}{3\lambda} = \frac{z+4}{6\lambda}$$

$$\Rightarrow \quad \frac{x-1}{2} = \frac{y-2}{3} = \frac{z+4}{6}$$

Which is the required cartesian equation of line and vector equation of this line is

$$\vec{r} = (\hat{i}+2\hat{j}-4\hat{k})+\lambda(2\hat{i}+3\hat{j}+6\hat{k})$$

Q. 12. A variable plane which remains at a constant distance 3*p* from the origin cuts the coordinate axes at *A*, *B*, *C*. Show that the locus of the centroid of triangle *ABC* is $x^{-2}+y^{-2}+z^{-2}=p^{-2}$.★

Sol. Let the intercepts made by the plane with coordinate axes at *A*, *B*, *C* be *a*, *b*, *c* respectively.

Then, the equation of variable plane is

$$\frac{x}{a}+\frac{y}{b}+\frac{z}{c} = 1 \quad \text{...(i)}$$

Clearly, the coordinates of *A*, *B*, *C* are (*a*, 0, 0), (0, *b*, 0) and (0, 0, *c*).

Since, the plane is at a distance of 3*p* from the origin,

$$\therefore \quad 3p = \frac{1}{\sqrt{\frac{1}{a^2}+\frac{1}{b^2}+\frac{1}{c^2}}}$$

[∵ distance of a point (x_1, y_1, z_1) from a plane $ax+by+cz+d=0$ is $\frac{|ax_1+by_1+cz_1+d|}{\sqrt{a^2+b^2+c^2}}$]

$$\Rightarrow \quad \frac{1}{3p} = \sqrt{\frac{1}{a^2}+\frac{1}{b^2}+\frac{1}{c^2}}$$

$$\Rightarrow \quad \frac{1}{9p^2} = \frac{1}{a^2}+\frac{1}{b^2}+\frac{1}{c^2} \quad \text{...(ii)}$$

Let (*u*, *v*, *w*) be the coordinates of the centroid of triangle *ABC*.

Then, $u=\frac{a}{3}$, $v=\frac{b}{3}$ and $w=\frac{c}{3}$

$$\Rightarrow a = 3u, b = 3v, c = 3w$$

Putting the values of *a*, *b* and *c* in equation (ii), we get

$$\Rightarrow \quad \frac{1}{9p^2} = \frac{1}{9u^2}+\frac{1}{9v^2}+\frac{1}{9w^2}$$

$$\Rightarrow \quad \frac{1}{p^2} = \frac{1}{u^2}+\frac{1}{v^2}+\frac{1}{w^2}$$

For locus, we replace *u* by *x*, *v* by *y* and *w* by *z*.

Therefore, the locus of the centroid of the triangle *ABC* is

$$\frac{1}{p^2} = \frac{1}{x^2}+\frac{1}{y^2}+\frac{1}{z^2}$$

i.e., $x^{-2}+y^{-2}+z^{-2}=p^{-2}$.

Q. 13. Show that the origin lies in the acute angle between the planes $x+2y+2z=9$ and $4x-3y+12z+13=0$.

Sol. To find the equation of the plane bisecting the angle containing the origin, we make the constant terms in the given equation of the same sign. Making constant terms of the same sign in the equations of two planes we obtain,

$$-x-2y-2z+9 = 0 \quad \text{...(i)}$$

and $$4x-3y+12z+13 = 0 \quad \text{...(ii)}$$

The equation of the plane bisecting the angle containing the origin is

$$\frac{-x-2y-2z+9}{\sqrt{(-1)^2+(-2)^2+(-2)^2}} = \frac{4x-3y+12z+13}{\sqrt{(4)^2+(-3)^2+(12)^2}}$$

$$\Rightarrow \quad \frac{-x-2y-2z+9}{\sqrt{9}} = \frac{4x-3y+12z+13}{\sqrt{169}}$$

$$\Rightarrow \quad \frac{-x-2y-2z+9}{3} = \frac{4x-3y+12z+13}{13}$$

$$\Rightarrow 13(-x-2y-2z+9) = 3(4x-3y+12z+13)$$

$$\Rightarrow -13x-26y-26z+117 = 12x-9y+36z+39$$

$$\Rightarrow \quad 25x+17y+62z-78 = 0 \quad \text{...(iii)}$$

Let θ be the angle between (i) and (iii), then

$$\cos\theta = \frac{|(25)(-1)+(17)(-2)+(62)(-2)|}{\sqrt{(-1)^2+(-2)^2+(-2)^2}\sqrt{(25)^2+(17)^2+(62)^2}}$$

$$\cos\theta = \frac{|-25-34-124|}{\sqrt{1+4+4}\sqrt{625+289+3844}}$$

$$\cos\theta = \frac{183}{3\sqrt{4758}}$$

$$\cos\theta = \frac{61}{\sqrt{4758}}$$

★ are board exam questions from previous years

Therefore, $\sin\theta = \sqrt{1-\cos^2\theta}$

$$= \sqrt{1-\frac{(61)^2}{4758}}$$

$$= \sqrt{1-\frac{3721}{4758}}$$

$$= \sqrt{\frac{1037}{4758}}$$

Therefore, $\tan\theta = \frac{\sin\theta}{\cos\theta}$

$$= \frac{\sqrt{1037}}{61}$$

$$= 0.528 < 1$$

So, $25x+17y+62z-78=0$ is also the bisector of the acute angles between the planes (i) and (ii). Hence, the origin lies in the acute angle between the given planes.

Q. 14. Find the image of the point (3, – 2, 1) in the plane $3x-y+4z=2$.

Sol.

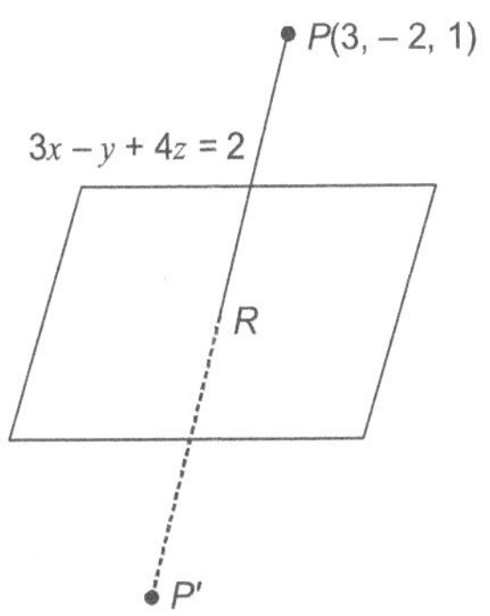

Let P' be the image of the point $P(3, -2, 1)$ in the plane $3x-y+4z=2$.

Then, PP' is normal to the plane.

Therefore, direction ratios of PP' are 3, – 1, 4.

Since, PP' passes through the point $P(3, -2, 1)$ and has direction ratios 3, – 1, 4.

Therefore, equation of PP' is

$$\frac{x-3}{3} = \frac{y+2}{-1} = \frac{z-1}{4}$$

Let, $\frac{x-3}{3} = \frac{y+2}{-1} = \frac{z-1}{4} = k$ (say)

$\Rightarrow$ $x = 3k+3$

$y = -k-2$

$z = 4k+1$

Let the coordinates of P', for some values of k, be

$$P'(3k+3, -k-2, 4k+1) \quad ...(i)$$

Let R be the mid-point of PP'

Therefore, coordinates of R are

$$\left(\frac{3k+3+3}{2}, \frac{-k-2-2}{2}, \frac{4k+1+1}{2}\right)$$

i.e., $\left(\frac{3k+6}{2}, \frac{-k-4}{2}, 2k+1\right)$

Since, the point R lies on the plane $3x-y+4z=2$

$\therefore$ $3(3k+6)+(k+4)+8(2k+1) = 4$

$\Rightarrow$ $9k+18+k+4+16k+8 = 4$

$\Rightarrow$ $26k+26 = 0$

$\Rightarrow$ $26k = -26$

$k = -1$

Putting the value of k in (i), we get the coordinates of P' as $(0, -1, -3)$.

Thus, the image of the point (3, – 2, 1) in the plane $3x-y+4z=2$ is $(0, -1, -3)$.

Q. 15. Find the equation of the plane through the line of intersection of $\vec{r}.(2\hat{i}-3\hat{j}+4\hat{k})=1$ and $\vec{r}.(\hat{i}-\hat{j})+4=0$ and perpendicular to the plane $\vec{r}.(2\hat{i}-\hat{j}+\hat{k})+8=0$. Hence find whether the plane thus obtained contains the line $x-1=2y-4=3z-12$.★

Sol. The equation of the plane passing through the line of intersection of the planes is :

$$r\left[(2\hat{i}-3\hat{j}+4\hat{k})+\lambda(\hat{i}-\hat{j})\right]+4\lambda-1=0$$

$$\vec{r}.\left[(2+\lambda)\hat{i}-(3+\lambda)\hat{j}+4\hat{k}\right]=1-4\lambda$$

Taking $\vec{r} = x\hat{i}+y\hat{j}+z\hat{k}$, we get

$$(2+\lambda)x-(3+\lambda)y+4z = 1-4\lambda \quad ...(i)$$

Cartesian equation of this plane

$$\vec{r}.(2\hat{i}-\hat{j}+\hat{k})+8=0 \text{ is :}$$

$$2x-y+z+8 = 0 \quad ...(ii)$$

$\because$ Planes (i) and (ii) are perpendicular

$\therefore$ $(2+\lambda)2+(3+\lambda)+4 = 0$

$\Rightarrow$ $4+2\lambda+3+\lambda+4 = 0$

$\Rightarrow$ $11+3\lambda = 0$

or $\lambda = \frac{-11}{3}$

From equation (i)

$$\left(2-\frac{11}{3}\right)x-\left(3-\frac{11}{3}\right)y+4z = 1-4\times\frac{-11}{3}$$

$$\frac{-5x}{3}+\frac{2}{3}y+4z = \frac{47}{3}$$

$\Rightarrow$ $-5x+2y+12z = 47$

★ are board exam questions from previous years

Required vector equation of this plane is :

$$\vec{r}.(-5\hat{i}+2\hat{j}+12\hat{k}) = 47 \quad ...(iii)$$

Now, equation of the given line is,

$$x-1 = 2y-4 = 3z-12$$

$$\frac{x-1}{1} = 2(y-2) = 3(z-4)$$

$$\frac{x-1}{1} = \frac{y-2}{1/2} = \frac{z-4}{1/3}$$

$$\Rightarrow \frac{x-1}{6} = \frac{y-2}{3} = \frac{z-4}{2}$$

Vector equation of this line is :

$$\vec{r} = (\hat{i}+2\hat{j}+4\hat{k})+\lambda(6\hat{i}+3\hat{j}+2\hat{k}) \quad ...(iv)$$

Obviously plane (iii) contains the line (iv) since the point $\hat{i}+2\hat{j}+4\hat{k}$ satisfy the equation of plane (iii)

[as $(\hat{i}+2\hat{j}+4\hat{k}).(-5\hat{i}+2\hat{j}+12\hat{k}) = -5+4+48 = 47$]

and vector $-5\hat{i}+2\hat{j}+12\hat{k}$ is perpendicular to $6\hat{i}+3\hat{j}+2\hat{k}$ as $(6\hat{i}+3\hat{j}+2\hat{k}).(-5\hat{i}+2\hat{j}+12\hat{k}) = -30+6+24=0$]

$\therefore$ Plane (iii) contains the line (iv)

Q. 16. Find the equation of plane which contains the line of intersection of planes $\vec{r}\cdot(\hat{i}+2\hat{j}+3\hat{k})-4=0$, $\vec{r}\cdot(2\hat{i}+\hat{j}-\hat{k})+5=0$ and which is perpendicular to plane $(5\hat{i}+3\hat{j}-6\hat{k})+8=0$.★

Sol. Given that the required plane contains the line of intersection of planes whose equations are

$$\vec{r}\cdot(\hat{i}+2\hat{j}+3\hat{k})-4 = 0 \quad ...(i)$$

and $$\vec{r}\cdot(2\hat{i}+\hat{j}-\hat{k})+5 = 0 \quad ...(ii)$$

Equations (i) and (ii) can be written in Cartesian form as

$$(x\hat{i}+y\hat{j}+z\hat{k})\cdot(\hat{i}+2\hat{j}+3\hat{k})-4 = 0$$

and $$(x\hat{i}+y\hat{j}+z\hat{k})\cdot(2\hat{i}+\hat{j}-\hat{k})+5 = 0$$

$$\Rightarrow x+2y+3z-4 = 0$$

and $$2x+y-z+5 = 0$$

So, let the required equation of plane is

$$(x+2y+3z-4)+\lambda(2x+y-z+5) = 0 \quad ...(iii)$$

$$\Rightarrow x+2y+3z-4+2\lambda x+\lambda y-\lambda z+5\lambda = 0$$

$$\Rightarrow x(1+2\lambda)+y(2+\lambda)+z(3-\lambda)+(-4+5\lambda) = 0 \quad ...(iv)$$

Now, given that plane (iv) is perpendicular to the plane $\vec{r}\cdot(5\hat{i}+3\hat{j}-6\hat{k})+8=0$ which in Cartesian form can be written as

$$5x+3y-6z+8 = 0$$

$\therefore$ We have $a_1a_2+b_1b_2+c_1c_2 = 0$

Where $a_1 = 1+2\lambda$, $b_1 = 2+\lambda$, $c_1 = 3-\lambda$

And $a_2 = 5$, $b_2 = 3$, $c_2 = -6$

$$\therefore 5(1+2\lambda)+3(2+\lambda)-6(3-\lambda) = 0$$

$$\Rightarrow 5+10\lambda+6+3\lambda-18+6\lambda = 0$$

$$\Rightarrow 19\lambda-7 = 0$$

$$\Rightarrow \lambda = \frac{7}{19}$$

Putting $\lambda = \frac{7}{19}$ in equation (iii) we get the required equation of plane as

$$(x+2y+3z-4)+\frac{7}{19}(2x+y-z+5) = 0$$

$$\Rightarrow 19x+38y+57z-76+14x+7y-7z+35 = 0$$

$$\Rightarrow 33x+45y+50z-41 = 0$$

In vector form, the required equation is

$$\vec{r}.(33\hat{i}+45\hat{j}+50\hat{k}) = 41$$

Q. 17. Find the distance of the point (– 1, – 5, – 10) from the point of intersection of the line $r = (2\hat{i}-\hat{j}+2\hat{k}) + \lambda(3\hat{i}+4\hat{j}+2\hat{k})$ and the plane $\vec{r}\cdot(\hat{i}-\hat{j}+\hat{k}) = 5$.★

Sol. Given equation of line and plane are

$$\vec{r} = (2\hat{i}-\hat{j}+2\hat{k})+\lambda(3\hat{i}+4\hat{j}+2\hat{k})$$

and $\vec{r}\cdot(\hat{i}-\hat{j}+\hat{k}) = 5$

$$\Rightarrow x\hat{i}+y\hat{j}+z\hat{k} = (2+3\lambda)\hat{i}+(-1+4\lambda)\hat{j} + (2+2\lambda)\hat{k}$$

and $(x\hat{i}+y\hat{j}+z\hat{k})\cdot(\hat{i}-\hat{j}+\hat{k}) = 5$

(put $r = x\hat{i}+y\hat{j}+z\hat{k}$)

Above equations in Cartesian form can be written as

$$\frac{x-2}{3} = \frac{y+1}{4} = \frac{z-2}{2} \quad ...(i)$$

and $$x-y+z = 5 \quad ...(ii)$$

First, we solve equations (i) and (ii) and find their point of intersection. Let the point of intersection be Q.

★ are board exam questions from previous years

Let $\frac{x-2}{3} = \frac{y+1}{4} = \frac{z-2}{2} = \lambda$ (say)

$\Rightarrow \quad \frac{x-2}{3} = \lambda, \frac{y+1}{4} = \lambda, \frac{z-2}{2} = \lambda$

$\Rightarrow \quad x = 3\lambda + 2, y = 4\lambda - 1, z = 2\lambda + 2$

$\therefore$ any point Q on the given line is

$$Q(3\lambda+2, 4\lambda-1, 2\lambda+2)$$

$\because$ Plane also passes through Q, so coordinates of Q satisfies equation (ii).

$$(3\lambda+2)-(4\lambda-1)+(2\lambda+2)=5$$

$\Rightarrow \quad 3\lambda+2-4\lambda+1+2\lambda+2=5$

$\Rightarrow \quad \lambda = 0$

Putting $\lambda = 0$ in $Q(3\lambda + 2, 4\lambda - 1, 2\lambda + 2)$ we get the point of intersection as $Q(2, -1, 2)$.

Now, using distance formula, the required distance

$$PQ = \sqrt{(2+1)^2 + (-1+5)^2 + (2+10)^2}$$

$$= \sqrt{9+16+144}$$

$$= \sqrt{144+25}$$

$$= \sqrt{169} = 13.$$

Hence, the required distance = 13 units.

Q. 18. Find the equation of plane passing through the line of intersection of planes $\vec{r}\cdot(\hat{i}+\hat{j}+\hat{k}) = 1$ and $\vec{r}\cdot(2\hat{i}+3\hat{j}-\hat{k})+4=0$ and parallel to X-axis.★

Sol. Given equation of planes are

$$\vec{r}\cdot(\hat{i}+\hat{j}+\hat{k}) = 1$$

and $\quad \vec{r}\cdot(2\hat{i}+3\hat{j}-\hat{k})+4 = 0$

Put $\vec{r} = x\hat{i}+y\hat{j}+z\hat{k}$ then above equations can be written in Cartesian form as

$$x+y+z-1=0 \quad \text{...(i)}$$

and $\quad 2x+3y-z+4=0 \quad$ ...(ii)

Let the required equation of plane passing through the line of intersection of planes (i) and (ii) is

$$(x+y+z-1)+\lambda(2x+3y-z+4)=0$$

$\Rightarrow \quad x+y+z-1+2\lambda x+3\lambda y-\lambda z+4\lambda=0$

$\Rightarrow \quad x(1+2\lambda)+y(1+3\lambda)+z(1-\lambda)+(-1+4\lambda)=0$ ...(iii)

$\therefore$ Direction ratios of the above plane are $1 + 2\lambda$, $1+3\lambda$, $1-\lambda$.

Also, direction ratios of the X-axis are (1, 0, 0).

Now, since given that above plane (iii) is parallel to the X-axis.

$\therefore$ We have $a_1a_2 + b_1b_2 + c_1c_2 = 0$

where $a_1 = 1+2\lambda, b_1 = 1+3\lambda, c_1 = 1-\lambda$

and $\quad a_2 = 1, b_2 = 0, c_2 = 0$

$\therefore \quad 1(1+2\lambda)+0(1+3\lambda)+0(1-\lambda)=0$

$\Rightarrow \quad 1+2\lambda = 0$

$\Rightarrow \quad 2\lambda = -1$

$\Rightarrow \quad \lambda = -\frac{1}{2}$

Putting $\lambda = -\frac{1}{2}$ in equation (iii), we get the required equation of plane as

$$x\left(1-\frac{2\times 1}{2}\right)+y\left(1-\frac{3}{2}\right)+z\left(1+\frac{1}{2}\right)+\left(-1-\frac{4}{2}\right)=0$$

$\Rightarrow \quad -\frac{y}{2}+\frac{3z}{2}-\frac{6}{2}=0$

$\Rightarrow \quad y-3z+6=0$

Q. 19. Find the equation of plane passing through the line of intersection of planes $2x + y - z = 3$ and $5x - 3y + 4z + 9 = 0$ and parallel to line

$$\frac{x-1}{2} = \frac{y-3}{4} = \frac{z-5}{5}.\ ★$$

Sol. Given equation of planes are

$$2x+y-z-3=0 \quad \text{...(i)}$$

and $\quad 5x-3y+4z+9=0 \quad$ ...(ii)

Let the required equation of plane which passes through the line of intersection of planes (i) and (ii) is

$$(2x+y-z-3)+\lambda(5x-3y+4z+9)=0 \quad \text{...(iii)}$$

$\Rightarrow x(2+5\lambda)+y(1-3\lambda)+z(-1+4\lambda)+(-3+9\lambda)=0$ ...(iv)

Here, direction ratios of plane are $2 + 5\lambda$, $1 - 3\lambda$, $-1 + 4\lambda$. Given that the plane (i) is parallel to the line whose equation as

$$\frac{x-1}{2} = \frac{y-3}{4} = \frac{z-5}{5}$$

Direction ratios of the line are 2, 4, 5.

Since, the plane is parallel to the line.

$\therefore$ We have $a_1a_2 + b_1b_2 + c_1c_2 = 0$

where $\quad a_1 = 2+5\lambda, b_1 = 1-3\lambda, c_1 = -1+4\lambda$

and $\quad a_2 = 2, b_2 = 4, c_2 = 5$

$\Rightarrow \quad 2(2+5\lambda)+4(1-3\lambda)+5(-1+4\lambda)=0$

$\Rightarrow \quad 4+10\lambda+4-12\lambda-5+20\lambda=0$

$\Rightarrow \quad 18\lambda+3=0$

or $\quad \lambda = -\frac{3}{18} = -\frac{1}{6}$

★ are board exam questions from previous years

$$\therefore \qquad \lambda = -\frac{1}{6}$$

Putting $\lambda = -\frac{1}{6}$ in equation (iii), we get the required equation of plane is

$$(2x+y-z-3)-\frac{1}{6}(5x-3y+4z+9)=0$$

$$\Rightarrow \quad 12x+6y-6z-18-5x+3y-4z-9=0$$

$$\Rightarrow \quad 7x+9y-10z-27=0$$

Q. 20. Find the equation of plane passing through the point (– 1, 3, 2) and perpendicular to each of the planes $x+2y+3z=5$ and $3x+3y+z=5$.★

Sol. Let the required equation of plane passing through (– 1, 3, 2) is

$$a(x+1)+b(y-3)+c(z-2)=0 \qquad \ldots(i)$$

Given that plane (i) is perpendicular to the planes whose equations are

$$x+2y+3z=5 \qquad \ldots(ii)$$

and
$$3x+3y+z=5 \qquad \ldots(iii)$$

We have $\quad a_1a_2+b_1b_2+c_1c_2=0$

Using the above result first in equation (i), (ii) and then in eq. (i) and (iii), we get

$$a+2b+3c=0 \qquad \ldots(iv)$$

and
$$3a+3b+c=0 \qquad \ldots(v)$$

Multiplying (iv) by '3' and subtracting it from equation (v), we get

$$\begin{array}{r} 3a+3b+\ \ c=0 \\ 3a+6b+9c=0 \\ -\quad -\quad - \\ \hline -3b-8c=0 \end{array}$$

$$\Rightarrow \qquad -3b=8c$$

or
$$b=-\frac{8c}{3}$$

Putting $b=-\frac{8c}{3}$ in equation (iv), we get

$$a+2\left(-\frac{8c}{3}\right)+3c=0$$

$$\Rightarrow \qquad a-\frac{16c}{3}+3c=0$$

or
$$a=\frac{16c}{3}-3c=\frac{16c-9c}{3}=\frac{7c}{3}$$

$$\therefore \qquad a=\frac{7c}{3}$$

Finally, putting $a=\frac{7c}{3}$ and $b=-\frac{8c}{3}$ in equation (i), we get the required equation of plane as

$$\frac{7c}{3}(x+1)-\frac{8c}{3}(y-3)+c(z-2)=0$$

Divide both sides by 'c', we get

$$\frac{7}{3}(x+1)-\frac{8}{3}(y-3)+(z-2)=0$$

$$\Rightarrow \qquad 7x+7-8y+24+3z-6=0$$

or
$$7x-8y+3z+25=0.$$

Q. 21. Find the vector equation of plane passing through the points $A(2,2,-1)$, $B(3,4,2)$ and $C(7,0,6)$. Also find the Cartesian equation of plane.★

Sol. First we check whether the points are collinear or not.

Given points are $A(2,2,-1)$, $B(3,4,2)$ and $C(7,0,6)$.

$$\therefore \qquad AB=\sqrt{(3-2)^2+(4-2)^2+(2+1)^2}$$
$$=\sqrt{1+4+9}$$
$$=\sqrt{14}$$

$$BC=\sqrt{(7-3)^2+(0-4)^2+(6-2)^2}$$
$$=\sqrt{16+16+16}=\sqrt{48}=4\sqrt{3}$$

$$\therefore \qquad CA=\sqrt{(2-7)^2+(2-0)^2+(-1-6)^2}$$
$$=\sqrt{25+4+49}$$
$$=\sqrt{78}$$

$$\because \qquad AB+BC\neq CA$$

so points A, B, C are not collinear.

Now, the equation of plane passing through three non-collinear points (x_1,y_1,z_1), (x_2,y_2,z_2) and (x_3,y_3,z_3) is given by

$$\begin{vmatrix} x-x_1 & y-y_1 & z-z_1 \\ x_2-x_1 & y_2-y_1 & z_2-z_1 \\ x_3-x_1 & y_3-y_1 & z_3-z_1 \end{vmatrix}=0$$

$$\therefore \text{ We get } \begin{vmatrix} x-2 & y-2 & z+1 \\ 3-2 & 4-2 & 2+1 \\ 7-2 & 0-2 & 6+1 \end{vmatrix}=0$$

where $\quad (x_1,y_1,z_1)=(2,2,-1),\ (x_2,y_2,z_2)=(3,4,2)$

and $\quad (x_3,y_3,z_3)=(7,0,6)$

$$\Rightarrow \qquad \begin{vmatrix} x-2 & y-2 & z+1 \\ 1 & 2 & 3 \\ 5 & -2 & 7 \end{vmatrix}=0$$

Expanding along R_1

$$\Rightarrow (x-2)(14+6)-(y-2)(7-15)+(z+1)(-2-10)=0$$

$$\Rightarrow \quad (x-2)\,20-(y-2)(-8)+(z+1)(-12)=0$$

$$\Rightarrow \quad 20x-40+8y-16-12z-12=0$$

$$\Rightarrow \quad 20x+8y-12z-68=0$$

★ are board exam questions from previous years

Dividing both sides by '4', we get

$5x+2y-3z=17$ is the required Cartesian equation of plane.

Also, we have to find the vector equation of plane.

We know that vector form of Cartesian equation $ax+by+cz=d$ of plane is given by

$$\vec{r}\cdot(a\hat{i}+b\hat{j}+c\hat{k})=d$$

∴ Required vector equation of plane is

$$\vec{r}\cdot(5\hat{i}+2\hat{j}-3\hat{k})=17$$

Q. 22. Find the equation of plane passing through the point (1, 2, 1) and perpendicular to line joining points (1, 4, 2) and (2, 3, 5). Also, find the coordinates of foot of the perpendicular and the perpendicular distance of the point (4, 0, 3) from the above found plane.★

Sol. First we find equation of plane passing through point $R(1, 2, 1)$ and is perpendicular to line PQ where $P(1, 4, 2)$ and $Q(2, 3, 5)$.

Direction ratios of the line PQ are

$$=(2-1, 3-4, 5-2)=(1,-1,3)$$

Let the required equation of plane which passes through point $R(1, 2, 1)$ is

$$a(x-1)+b(y-2)+c(z-1)=0 \quad \text{...(i)}$$

Plane (i) is perpendicular to line PQ.

Since, line is perpendicular to the plane, then direction ratios of normal to the plane is proportional to the direction ratios of a line, *i.e.*, $a \propto 1, b \propto -1, c \propto 3$.

$$\therefore \quad 1(x-1)-1(y-2)+3(z-1)=0$$

$$\text{or} \quad x-1-y+2+3z-3=0$$

$$\text{or} \quad x-y+3z-2=0 \quad \text{...(ii)}$$

Is the required equation of plane.

Next, we have to find the foot of perpendicular and perpendicular distance of the point (4, 0, 3) from above plane.

So, let $B(x_1, y_1, z_1)$ be the foot of perpendicular on above plane (ii). So it must satisfy equation (ii).

$$\therefore \quad x_1-y_1+3z_1-2=0 \quad \text{...(iii)}$$

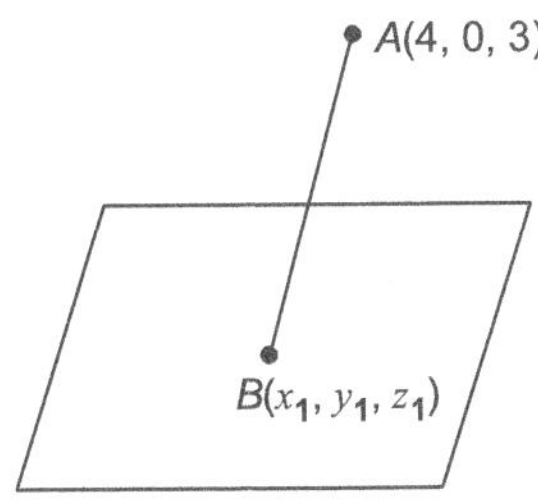

Also, direction ratios of line AB normal to above plane (i) are given by

$$x_1-4, y_1-0, z_1-3$$

∵ Direction ratios of line AB and plane (iii) are proportional.

$$\therefore \quad \frac{x_1-4}{1}=\frac{y_1-0}{-1}=\frac{z_1-3}{3}$$

$$\text{Let} \quad \frac{x_1-4}{1}=\frac{y_1}{-1}=\frac{z_1-3}{3}=\lambda \quad \text{(say)}$$

$$\Rightarrow \quad x_1=\lambda+4, y_1=-\lambda, z_1=3\lambda+3 \quad \text{...(iv)}$$

Putting above values of x_1, y_1 and z_1 in eq. (iii), we get

$$\lambda+4+\lambda+9\lambda+9-2=0$$

$$\Rightarrow \quad 11\lambda+11=0$$

$$\Rightarrow \quad 11\lambda=-11$$

$$\Rightarrow \quad \lambda=-1$$

Putting $\lambda=-1$ in equation (iv), we get the required foot of perpendicular as

$$B(x_1, y_1, z_1)=B(\lambda+4, -\lambda, 3\lambda+3)=B(3, 1, 0)$$

Also, perpendicular distance AB where $A(4, 0, 3)$ and $B(3, 1, 0)$ is given by using distance formula *i.e.*,

$$AB=\sqrt{(3-4)^2+(1-0)^2+(0-3)^2}$$

$$=\sqrt{1+1+9}$$

$$=\sqrt{11} \text{ units}$$

Q. 23. Find the equation of plane determined by the points $A(3,-1,2)$, $B(5,2,4)$ and $C(-1,-1,6)$. Also, find the distance of point (6, 5, 9) from plane.★★

Sol. Given points are $A(3,-1,2)$, $B(5,2,4)$ and $C(-1,-1,6)$.

First we check whether the points are collinear or not. Using distance formula, we have

$$AB=\sqrt{(5-3)^2+(2+1)^2+(4-2)^2}$$

$$=\sqrt{4+9+4}$$

$$=\sqrt{17}$$

$$BC=\sqrt{(-1-5)^2+(-1-2)^2+(6-4)^2}$$

$$=\sqrt{36+9+4}$$

$$=\sqrt{49}$$

$$CA=\sqrt{(3+1)^2+(-1+1)^2+(2-6)^2}$$

$$=\sqrt{16+0+16}$$

$$=\sqrt{32}$$

$$\because \quad AB+BC \neq CA$$

∴ Given points are non-collinear.

★ are board exam questions from previous years
★★ are frequently asked board exam questions

Now, we know that equation of plane passing through three non-collinear points (x_1, y_1, z_1), (x_2, y_2, z_2) and (x_3, y_3, z_3) is given by

$$\begin{vmatrix} x-x_1 & y-y_1 & z-z_1 \\ x_2-x_1 & y_2-y_1 & z_2-z_1 \\ x_3-x_1 & y_3-y_1 & z_3-z_1 \end{vmatrix} = 0 \quad ...(i)$$

We have

$$(x_1, y_1, z_1) = A(3, -1, 2), (x_2, y_2, z_2) = B(5, 2, 4)$$

and $\quad (x_3, y_3, z_3) = C(-1, -1, 6)$

Putting above values in equation (i), we get the required equation as

$$\begin{vmatrix} x-3 & y+1 & z-2 \\ 5-3 & 2+1 & 4-2 \\ -1-3 & -1+1 & 6-2 \end{vmatrix} = 0$$

$$\Rightarrow \begin{vmatrix} x-3 & y+1 & z-2 \\ 2 & 3 & 2 \\ -4 & 0 & 4 \end{vmatrix} = 0$$

$$\Rightarrow (x-3)(12-0) - (y+1)(8+8) + (z-2)(0+12) = 0$$

$$\Rightarrow 12x - 36 - 16y - 16 + 12z - 24 = 0$$

$$\Rightarrow 12x - 16y + 12z - 76 = 0$$

$$\Rightarrow 3x - 4y + 3z - 19 = 0$$

Next, we find distance of point $P(6, 5, 9)$ to the above plane (ii) by using the formula

$$d = \frac{|Ax_1 + By_1 + Cz_1 + D|}{\sqrt{A^2 + B^2 + C^2}}$$

Between plane $Ax + By + Cz + D = 0$ and point (x_1, y_1, z_1).

Here, $A = 3$, $B = -4$, $C = 3$, $D = -19$, $x_1 = 6$, $y_1 = 5$, $z_1 = 9$

$$\therefore \quad d = \left|\frac{18 - 20 + 27 - 19}{\sqrt{9+16+9}}\right| = \left|\frac{6}{\sqrt{34}}\right| = \frac{6}{\sqrt{34}}$$

$$\therefore \quad \text{Distance} = \frac{6\sqrt{34}}{34} \text{ units} = \frac{3\sqrt{34}}{17} \text{ units.}$$

Q. 24. Find the distance of point $(-2, 3, -4)$ from the line $\frac{x+2}{3} = \frac{2y+3}{4} = \frac{3z+4}{5}$ measured parallel to the plane $4x + 12y - 3z + 1 = 0$.★★

Sol. Let $P(-2, 3, -4)$ be the given point and the given line is

$$\frac{x+2}{3} = \frac{2y+3}{4} = \frac{3z+4}{5}$$

or $$\frac{x+2}{3} = \frac{2\left(y+\frac{3}{2}\right)}{4} = \frac{3\left(z+\frac{4}{3}\right)}{5}$$

or $$\frac{x+2}{3} = \frac{y+\frac{3}{2}}{2} = \frac{z+\frac{4}{3}}{\left(\frac{5}{3}\right)} \quad ...(i)$$

Now, any random point T on the given line (i) is calculated as

$$\frac{x+2}{3} = \frac{y+\frac{3}{2}}{2} = \frac{z+\frac{4}{3}}{\left(\frac{5}{3}\right)} = \lambda \text{ (say)}$$

$$\Rightarrow \quad \frac{x+2}{3} = \lambda, \frac{y+\frac{3}{2}}{2} = \lambda, \frac{z+\frac{4}{3}}{\frac{5}{3}} = \lambda$$

$$\Rightarrow \quad x = 3\lambda - 2, y = \frac{4\lambda - 3}{2}, z = \frac{5\lambda - 4}{3}$$

$\therefore$ Coordinates of T are

$$\left(3\lambda - 2, \frac{4\lambda - 3}{2}, \frac{5\lambda - 4}{3}\right)$$

Now, direction ratios of line PT are

$$\left(3\lambda - 2 + 2, \frac{4\lambda - 3}{2} - 3, \frac{5\lambda - 4}{3} + 4\right) = \left(3\lambda, \frac{4\lambda - 9}{2}, \frac{5\lambda + 8}{3}\right)$$

Since, the line PT is parallel to the plane $4x + 12 - 3z + 1 = 0$.

$\therefore$ We have $a_1a_2 + b_1b_2 + c_1c_2 = 0$

where $\quad a_1 = 3\lambda, b_1 = \frac{4\lambda - 9}{2}, c_1 = \frac{5\lambda + 8}{3}$ and

$a_2 = 4, b_2 = 12, c_2 = -3$

[Direction ratios of plane]

$\therefore$ We get $4(3\lambda) + 12\left(\frac{4\lambda - 9}{2}\right) - 3\left(\frac{5\lambda + 8}{3}\right) = 0$

$$\Rightarrow \quad 12\lambda + 24\lambda - 54 - 5\lambda - 8 = 0$$

$$\Rightarrow \quad 31\lambda - 62 = 0$$

$$\Rightarrow \quad 31\lambda = 62$$

$$\Rightarrow \quad \lambda = 2$$

$\therefore$ Coordinates of T are

$$\left(6 - 2, \frac{8-3}{2}, \frac{10-4}{3}\right) = \left(4, \frac{5}{2}, 2\right)$$

$\therefore$ The required distance = distance between the points $P(-2, 3, -4)$ and $T\left(4, \frac{5}{2}, 2\right)$

★★ are frequently asked board exam questions

$\therefore$ Using distance formula, we have

$$d = \sqrt{(4+2)^2 + \left(\frac{5}{2} - 3\right)^2 + (2+4)^2}$$

$$= \sqrt{36 + \frac{1}{4} + 36}$$

$$= \sqrt{\frac{144+1+144}{4}}$$

$$= \sqrt{\frac{289}{4}} = \frac{17}{2} \text{ units}$$

Q. 25. From the point $P(1, 2, 4)$, a perpendicular is drawn on the plane $2x + y - 2z + 3 = 0$. Find the equation, the length and the coordinates of foot of perpendicular.★

Sol. Let PT be the perpendicular drawn from the point $P(1, 2, 4)$ to the plane whose equation is given by

$$2x + y - 2z + 3 = 0 \qquad \text{...(i)}$$

We have to find equation, length and coordinates of foot of perpendicular.

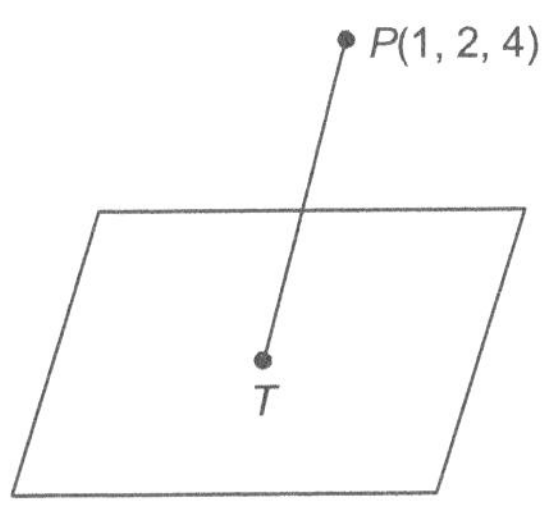

From eq. (i), direction ratios of plane are 2, 1, – 2. Since the line PT is normal to the plane, so direction ratios of line normal to plane are 2, 1, – 2.

$\therefore$ Equation of line PT where $P(1, 2, 4)$ is given as

$$\frac{x-1}{2} = \frac{y-2}{1} = \frac{z-4}{-2} \qquad \text{...(ii)}$$

Equation of line passing through one point is

Let $$\frac{x-1}{2} = \frac{y-2}{1} = \frac{z-4}{-2} = \lambda \text{ (say)}$$

$$\Rightarrow \frac{x-1}{2} = \lambda, \frac{y-2}{1} = \lambda, \frac{z-4}{-2} = \lambda$$

$$\Rightarrow x = 2\lambda + 1, y = \lambda + 2, z = -2\lambda + 4$$

$\therefore$ Coordinates of any random point are

$$T(2\lambda + 1, \lambda + 2, -2\lambda + 4) \qquad \text{...(iii)}$$

From the figure, we see that T lies on the given plane. So, we put $x = 2\lambda + 1, y = \lambda + 2, z = -2\lambda + 4$ in eq. (i),

$$2(2\lambda + 1) + (\lambda + 2) - 2(-2\lambda + 4) + 3 = 0$$

$$\Rightarrow 4\lambda + 2 + \lambda + 2 + 4\lambda - 8 + 3 = 0$$

$$\Rightarrow 9\lambda - 1 = 0$$

$$\Rightarrow \lambda = \frac{1}{9}$$

★ are board exam questions from previous years

Putting value of λ in equation (iii), we get the foot of perpendicular,

$$\left(\frac{2}{9} + 1, \frac{1}{9} + 2, -\frac{2}{9} + 4\right) = T\left(\frac{11}{9}, \frac{19}{9}, \frac{34}{9}\right)$$

Also, length of perpendicular PT = distance between points P and T

$$= \sqrt{\left(1 - \frac{11}{9}\right)^2 + \left(2 - \frac{19}{9}\right)^2 + \left(4 - \frac{34}{9}\right)^2}$$

$$\left[\because (x_1, y_1, z_1) = (1, 2, 4) \text{ and } (x_2, y_2, z_2) = \left(\frac{11}{9}, \frac{19}{9}, \frac{34}{9}\right)\right]$$

$$= \sqrt{\left(-\frac{2}{9}\right)^2 + \left(-\frac{1}{9}\right)^2 + \left(\frac{2}{9}\right)^2}$$

$$= \sqrt{\frac{4}{81} + \frac{1}{81} + \frac{4}{81}}$$

$$= \sqrt{\frac{9}{81}} = \sqrt{\frac{1}{9}} = \frac{1}{3}$$

Finally, the equation of perpendicular PT where $P(1, 2, 4)$ and $T\left(\frac{11}{9}, \frac{19}{9}, \frac{34}{9}\right)$ is given as

$$\frac{x-1}{\frac{11}{9} - 1} = \frac{y-2}{\frac{19}{9} - 2} = \frac{z-4}{\frac{34}{9} - 4}$$

$$\Rightarrow \frac{x-1}{\left(\frac{2}{9}\right)} = \frac{y-2}{\left(\frac{1}{9}\right)} = \frac{z-4}{\left(-\frac{2}{9}\right)}$$

$$\Rightarrow \frac{x-1}{2} = \frac{y-2}{1} = \frac{z-4}{-2}$$

Q. 26. Find the equation of plane passing through points (3, 4, 1) and (0, 1, 0) and parallel to line $\frac{x+3}{2} = \frac{y-3}{7} = \frac{z-2}{5}$.★

Sol. Equation of plane passing through the point (3, 4, 1) is given as

$$a(x-3) + b(y-4) + c(z-1) = 0 \qquad \text{...(i)}$$

[using one point form]

Since, given plane (i) is also passing through point (0, 1, 0).

So, this point also satisfies equation of plane (i)

$$\therefore \quad a(0-3) + b(1-4) + c(0-1) = 0$$

$$\text{or} \quad -3a - 3b - c = 0$$

$$\text{or} \quad 3a + 3b + c = 0 \qquad \text{...(ii)}$$

Also, given that plane (i) is parallel to the line

$$\frac{x+3}{2} = \frac{y-3}{7} = \frac{z-2}{5}$$

$\therefore \quad a_1a_2 + b_1b_2 + c_1c_2 = 0$ ...(iii)

[$\because$ line is parallel to the plane, therefore, normal to the plane is perpendicular to the line]

where $\quad a_1 = a, b_1 = b, c_1 = c$

and $\quad a_2 = 2, b_2 = 7, c_2 = 5$

$\therefore$ We get

$$2a + 7b + 5c = 0 \quad \text{...(iii)}$$

Multiplying equation (ii) by 2 and equation (iii) by 3 then subtracting, we get

$$6a + 6b + 2c = 0$$
$$6a + 21b + 15c = 0$$
$$- \quad - \quad -$$
$$-15b - 13c = 0$$

$\Rightarrow \quad -15b = 13c$

or $\quad b = \dfrac{-13}{15}c$

Putting $b = \dfrac{-13c}{15}$ in equation (ii), we get

$$3a + 3\left(-\frac{13c}{15}\right) + c = 0$$

$\Rightarrow \quad 3a - \dfrac{13}{5}c + c = 0$

$\Rightarrow \quad 3a - \dfrac{8c}{5} = 0$

$\Rightarrow \quad 3a = \dfrac{8c}{5}$

$\Rightarrow \quad a = \dfrac{8c}{15}$

Putting $a = \dfrac{8c}{15}$ and $b = -\dfrac{13c}{15}$ in eq. (i), we get the required equation of plane as

$$\frac{8c}{15}(x-3) - \frac{13c}{15}(y-4) + c(z-1) = 0$$

Divide both sides by 'c', we get

$$\frac{8}{15}(x-3) - \frac{13}{15}(y-4) + z - 1 = 0$$

$\Rightarrow \quad 8x - 24 - 13y + 52 + 15z - 15 = 0$

$\Rightarrow \quad 8x - 13y + 15z + 13 = 0$

Q. 27. Find the distance of the point $P(3, 4, 4)$ from the point, where the line joining the points $A(3, -4, -5)$ and $B(2, -3, 1)$ intersects the plane $2x + y + z = 7$.★

Sol. Equation of the line joining the points $A(3, -4, -5)$ and $B(2, -3, 1)$ is

$$\Rightarrow \quad \frac{x-3}{2-3} = \frac{y+4}{-3+4} = \frac{z+5}{1+5}$$

$$\Rightarrow \quad \frac{x-3}{-1} = \frac{y+4}{1} = \frac{z+5}{6} = k \text{ (say)}$$

Coordinates of any point on the line are

$$Q(3-k, k-4, 6k-5)$$

The given line intersects the plane $2x + y + z = 7$

So, $\quad 2(3-k) + k - 4 + 6k - 5 = 7$

$\Rightarrow \quad 6 - 2k + k - 4 + 6k - 5 = 7$

$\Rightarrow \quad 5k = 10$

$\Rightarrow \quad k = 2$

$\therefore$ Coordinates of a point where line intersect plane are

$$(3-k, k-4, 6k-5) = (3-2, 2-4, 12-5) = (1, -2, 7)$$

Now, the distance between the point $P(3, 4, 4)$ and $Q(1, -2, 7)$ is given by

$$PQ = \sqrt{(3-1)^2 + (4+2)^2 + (4-7)^2}$$

$$PQ = \sqrt{(2)^2 + (6)^2 + (-3)^2}$$

$$PQ = \sqrt{4 + 36 + 9} = \sqrt{49}$$

$$PQ = 7 \text{ units.}$$

Q. 28. Find the distance between the point $(-1, -5, -10)$ and the point of intersection of the line $\dfrac{x-2}{3} = \dfrac{y+1}{4} = \dfrac{z-2}{12}$ and the plane $x - y + z = 5$.★

Sol. Let $\quad \dfrac{x-2}{3} = \dfrac{y+1}{4} = \dfrac{z-2}{12} = k$

$\Rightarrow \quad x = 3k + 2,$

$\quad y = 4k - 1,$

$\quad z = 12k + 2$

Coordinates of any point on the line are

$$(3k + 2, 4k - 1, 12k + 2)$$

The point of intersection of the line $\dfrac{x-2}{3} = \dfrac{y+1}{4} = \dfrac{z-2}{12}$ and the plane $x - y + z = 5$ will also be in the form $(3k + 2, 4k - 1, 12k + 2)$ and it will satisfy the equation of plane.

Now, putting, $\quad x = 3k + 2,$

$\quad y = 4k - 1$

and $\quad z = 12k + 2$ in equation of plane

we get $\quad 3k + 2 - (4k - 1) + 12k + 2 = 5$

$\Rightarrow \quad 11k + 5 = 5$

$\Rightarrow \quad 11k = 0$

$\Rightarrow \quad k = 0$

$\therefore \quad x = 2, y = -1, z = 2.$

★ are board exam questions from previous years

Hence, the point of intersection of $\frac{x-2}{3}=\frac{y+1}{4}=\frac{z-2}{12}$ and the plane $x-y+z=5$ is $(2,-1,2)$.

$\therefore$ Distance between the points $(-1,-5,-10)$ and $(2,-1,2)$.

Using distance formula

$$=\sqrt{(2+1)^2+(-1+5)^2+(2+10)^2}$$

$$=\sqrt{3^2+4^2+12^2}$$

$$=\sqrt{169}=13 \text{ units}$$

Hence, the distance between the point $(-1,-5,-10)$ and the point of intersection of the line $\frac{x-2}{3}=\frac{y+1}{4}=\frac{z-2}{12}$ and the plane $x-y+z=5$ is 13 units.

Q. 29. If the lines $\frac{x-1}{2}=\frac{y+1}{3}=\frac{z-1}{4}$ and $\frac{x-3}{1}=\frac{y-k}{2}=\frac{z}{1}$ intersect each other then find the value of k and hence find the equation of the plane containing these lines.★

Sol. The coordinates of any point on first line are

$$\frac{x-1}{2}=\frac{y+1}{3}=\frac{z-1}{4}=\lambda$$

i.e., $x=2\lambda+1$

$y=3\lambda-1,$

$z=4\lambda+1$

i.e., $(2\lambda+1, 3\lambda-1, 4\lambda+1)$

and the coordinates of any point on second line are

$$\frac{x-3}{1}=\frac{y-k}{2}=\frac{z}{1}=\mu$$

i.e., $x=\mu+3,$

$y=2\mu+k,$

$z=\mu$

i.e., $(\mu+3, 2\mu+k, \mu)$

if these two lines intersect each other, then

$$2\lambda+1=\mu+3,\ 3\lambda-1=2\mu+k,\ 4\lambda+1=\mu$$

i.e., $2\lambda-\mu=2,\ 3\lambda-2\mu=k+1,\ 4\lambda-\mu=-1$

Solving, $2\lambda-\mu=2$ and $4\lambda-\mu=-1$, we get

$$\lambda=\frac{-3}{2} \text{ and } \mu=-5$$

Now, substituting the values of λ and μ in $3\lambda-2\mu=k+1$, we get

$$k=\frac{9}{2}$$

Now, we have $\vec{b_1}=2\hat{i}+3\hat{j}+4\hat{k}$ and $\vec{b_2}=\hat{i}+2\hat{j}+\hat{k}$.

So, the required plane contains both lines and it passes through a point $\vec{a}(1,-1,1)$ and perpendicular vector $\vec{n}$, given by $\vec{n}=\vec{b_1}\times\vec{b_2}$

$$\vec{n}=\begin{vmatrix}\hat{i} & \hat{j} & \hat{k}\\ 2 & 3 & 4\\ 1 & 2 & 1\end{vmatrix}$$

$$=\hat{i}(3-8)-\hat{j}(2-4)+\hat{k}(4-3)$$

$$=-5\hat{i}+2\hat{j}+\hat{k}$$

$\therefore$ The equation of plane passing through $\vec{a}$ and perpendicular to $\vec{n}$ is given by

$$(\vec{r}-\vec{a})\cdot\vec{n}=0$$

$$[\vec{r}-(\hat{i}-\hat{j}+\hat{k})].(-5\hat{i}+2\hat{j}+\hat{k})=0$$

$$\Rightarrow \vec{r}(-5\hat{i}+2\hat{j}+\hat{k})=(\hat{i}-\hat{j}+\hat{k})\cdot(5\hat{i}+2\hat{j}+\hat{k})$$

$$\Rightarrow \vec{r}\cdot(-5\hat{i}+2\hat{j}+\hat{k})=-6$$

Writing $\vec{r}=x\hat{i}+y\hat{j}+z\hat{k}$

or $5x-2y-z-6=0.$

Q. 30. Find the coordinates of the point P where the line through $A(3,-4,-5)$ and $B(2,-3,1)$ crosses the plane passing through three points $L(2,2,1)$, $M(3,0,1)$ and $N(4,-1,0)$. Also, find the ratio in which P divides the line segment AB.★

Sol. The equation of the plane passing through three given points can be given by

$$\begin{vmatrix}x-2 & y-2 & z-1\\ 3-2 & 0-2 & 1-1\\ 4-2 & -1-2 & 0-1\end{vmatrix}=0$$

$$\Rightarrow \begin{vmatrix}x-2 & y-2 & z-1\\ 1 & -2 & 0\\ 2 & -3 & -1\end{vmatrix}=0$$

Solving the above determinant, we get

$$\Rightarrow (x-2)(2-0)-(y-2)(-1-0)+(z-1)(-3+4)=0$$

$$\Rightarrow (2x-4)+(y-2)+(z-1)=0$$

$$\Rightarrow 2x+y+z-7=0$$

Therefore, the equation of the plane is

$$2x+y+z-7=0.$$

Now, the equation of the line passing through two given points is

$$\frac{x-3}{2-3}=\frac{y+4}{-3+4}=\frac{z+5}{1+5}=\lambda$$

★ are board exam questions from previous years

$$\Rightarrow \quad \frac{x-3}{-1}=\frac{y+4}{1}=\frac{z+5}{6}=\lambda$$

$$\Rightarrow \quad x=(-\lambda+3), y=(\lambda-4), z=(6\lambda-5)$$

At the point of intersection, the above point satisfy the equation of the plane $2x+y+z-7=0$.

Putting the values of x, y and z in the equation of the plane, we get the value of λ as

$$2(-\lambda+3)+(\lambda-4)+(6\lambda-5)-7=0$$

$$\Rightarrow \quad -2\lambda+6+\lambda-4+6\lambda-5-7=0$$

$$\Rightarrow \quad 5\lambda=10$$

$$\Rightarrow \quad \lambda=2$$

Thus, the point of intersection is $P(1,-2,7)$.

Now, let P divide the line AB in the ratio $m:n$.

By the section formula, we have

$$1=\frac{2m+3n}{m+n}$$

$$\Rightarrow \quad m+2n=0$$

$$\Rightarrow \quad m=-2n$$

$$\Rightarrow \quad \frac{m}{n}=\frac{-2}{1}$$

Hence, P divides the line segment AB externally in the ratio 2 : 1.

Chapter 12. Linear Programming

Q. 1. Solve the following LPP by the graphical method.

Maximize $\quad Z=3x+4y$

Subject to the constraints

$$2x+2y\leq 80$$

$$2x+4y\leq 120$$

$$x, y\geq 0$$

Sol. Converting the given inequations, we obtain the following equations.

$$2x+2y=80$$

$$2x+4y=120$$

$$x=0 \text{ and } y=0$$

Region represented by $2x+2y\leq 80$

The line $2x+2y=80$ meets the coordinate axis at A (40, 0) and B (0, 40) respectively. Join these points to obtain the line. Clearly (0, 0) satisfies the inequation $2x+2y\leq 80$. So the region containing the origin represents the solution set of the inequation $2x+2y\leq 80$.

Region represented by $2x+4y\leq 120$

The line $2x+4y=120$ meets the coordinate axis at $C(60, 0)$ and D (0, 30) respectively. Join these points to obtain the line $2x+4y=120$. Clearly (0, 0) satisfies the inequation $2x+4y\leq 120$. So the region containing the origin represents the solution set of this inequation.

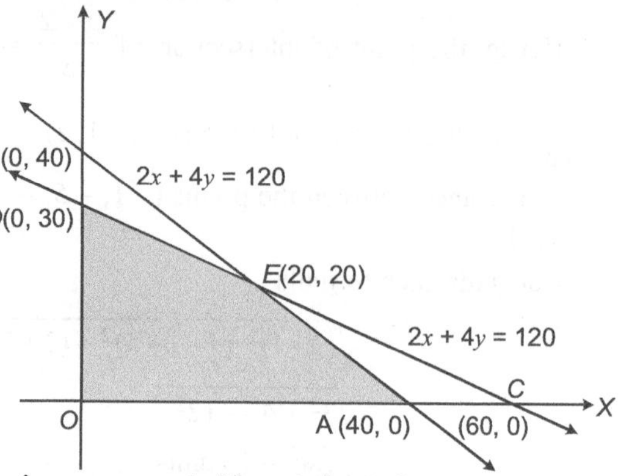

Region represented by $x\geq 0, y\geq 0$

Since every point in the first quadrant satisfies these inequations so that first quadrant is the region represented by the inequations.

$$x\geq 0, \text{ and } y\geq 0$$

The shaded region $OAED$ in the figure represents the common region of the above inequations. This region is the feasible region of the given LPP.

The coordinates of the vertices of the shaded region are O (0, 0), A (40, 0), E (20, 20) and D (0, 30). These points may be obtained by solving the equations of the corresponding intersecting lines, simultaneously. The value of the objective function at these points are given in the following table.

Corner point (x, y)	Value of the objective function $Z=3x+4y$
$O(0,0)$	$Z=0$
$A(40,0)$	$Z=120$
$E(20,20)$	$Z=140$
$D(0,30)$	$Z=120$

The maximum value of Z is thus given by the coordinates of E. Hence $x=20, y=20$ is the optimal solution of the given LPP. The optimal value of Z is 140.

Q. 2. Solve the following LPP by graphical method

Minimize $\quad Z=2x+3y$

Subject to the constraints

$$x+y\leq 4$$

$$3x+y\geq 4$$

$$x+5y\geq 4$$

$$x\leq 3$$

$$y\leq 3$$

$$x, y\geq 0$$

Sol. Converting the given inequations into equations we obtain the following equations

$$x+y=4$$

$$3x+y=4$$

$$x+5y=4$$

$$x = 3$$
$$y = 3$$
$$x = 0$$
and $$y = 0$$

Consider, $x + y = 4$

$$\frac{x}{4} + \frac{y}{4} = 1$$

This line cuts X-axis at A (4, 0) and Y-axis at B (0, 4). Therefore, AB is the line $x + y = 4$. Clearly (0, 0) satisfies $x + y \le 4$. Therefore, closed half plane containing (0, 0) is the graph of $x + y \le 4$.

Consider, $3x + y = 4$

$$\Rightarrow \quad \frac{x}{(4/3)} + \frac{y}{4} = 1$$

This line cuts X-axis at $C\left(\frac{4}{3}, 0\right)$ and B (0, 4). Therefore CB is the line $3x + y = 4$. Clearly (0, 0) does not satisfy $3x + y \ge 4$. Therefore, closed half plane not containing (0, 0) is the graph of $3x + y \ge 4$

Consider, $x + 5y = 4$

$$\Rightarrow \quad \frac{x}{4} + \frac{y}{(4/5)} = 1$$

This line cuts X-axis at A (4, 0) and F (0, 0.8). Therefore, AF is the line $x + 5y = 4$. Clearly (0, 0) doesn't satisfy $x + 5y \ge 4$.

Therefore, closed half plane not containing (0, 0) is the graph of $x + 5y \ge 4$.

Also $x = 3$ is the line parallel to Y-axis, shown by GH. Clearly (0, 0) satisfies $x \le 3$. So, closed half plane containing (0, 0) is the graph of $x \le 3$

and $y = 3$ is the line parallel to X-axis, shown by JK.

Clearly (0, 0) satisfies $y \le 3$. So, closed half plane containing (0, 0) is the graph of $y \le 3$. Thus, the shaded region $PQRST$ is the feasible region.

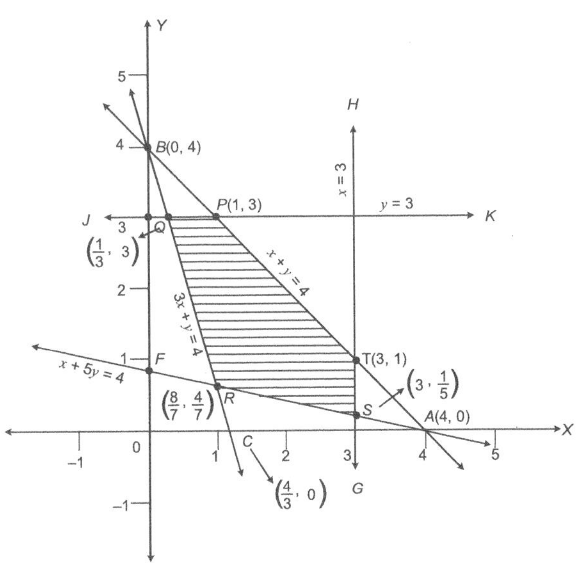

P lies on the intersection of AB and JK. Solving $x + y = 4$ and $y = 3$, we get, P (1, 3). Q lies on the intersection of CB and JK.

Solving $3x + y = 4$ and $y = 3$, we get, $Q\left(\frac{1}{3}, 3\right)$. R lies on the intersection of CB and AF.

Solving $3x + y = 4$ and $x + 5y = 4$, we get, $R\left(\frac{8}{7}, \frac{4}{7}\right)$. S lies on the intersection of AF and GH.

Solving $x + 5y = 4$ and $x = 3$, we get, $S\left(3, \frac{1}{5}\right)$. T lies on the intersection of AB and GH.

Solving $x + y = 4$ and $x = 3$, we get, T (3, 1).

At P (1, 3), we have $Z = 11$

At $Q\left(\frac{1}{3}, 3\right)$, we have

$$Z = \frac{29}{3} = 9.66$$

At $R\left(\frac{8}{7}, \frac{4}{7}\right)$, we have

$$Z = \frac{28}{7} = 4$$

At $S\left(3, \frac{1}{5}\right)$, we have

$$Z = \frac{33}{5} = 6.6$$

At T (3, 1), we have

$$Z = 9$$

So, the minimum value of $Z = 2x + 3y$ is 4 at $R\left(\frac{8}{7}, \frac{4}{7}\right)$.

Q. 3. Find the maximum and minimum value of $Z = 5x + 2y$, subject to the constraints.

$$2x + 3y \ge 6$$
$$x - 2y \le 2$$
$$3x + 2y \le 12$$
$$-3x + 2y \le 3$$
$$x \ge 0$$
$$y \ge 0$$

Sol. Converting the given inequation into equations, we obtain the following equations,

$$2x + 3y = 6$$
$$x - 2y = 2$$
$$3x + 2y = 12$$
$$-3x + 2y = 3$$
$$x = 0$$
and $$y = 0$$

Consider, $2x + 3y = 6$

$$\Rightarrow \quad \frac{x}{3} + \frac{y}{2} = 1$$

This line meets X-axis at A (3, 0), Y-axis at B (0, 2). Then AB is the line $2x + 3y = 6$. Also (0, 0) does not satisfy $2x + 3y \geq 6$. So, closed half plane not containing (0, 0) is the graph of $2x + 3y \geq 6$.

Again, $x - 2y = 2$

$$\Rightarrow \quad \frac{x}{2} + \frac{y}{(-1)} = 1$$

This line meets X-axis at C (2, 0), Y-axis at D (0, –1). Therefore, CD is the line $x - 2y = 2$. Also (0, 0) satisfies $x - 2y \leq 2$. So, closed half plane containing (0, 0) is the graph of $x - 2y \leq 2$.

Consider, $3x + 2y = 12$

$$\Rightarrow \quad \frac{x}{4} + \frac{y}{6} = 1$$

This line meets X-axis at E (4, 0), and F (0, 6). Therefore, EF is the line $3x + 2y = 12$. Clearly (0, 0) satisfies $3x + 2y \leq 12$. So, closed half plane containing (0, 0) is the graph of $3x + 2y \leq 12$.

Consider, $-3x + 2y = 3$

$$\Rightarrow \quad \frac{x}{(-1)} + \frac{y}{(3/2)} = 1$$

This line meets X-axis at G (–1, 0) and Y-axis at $H\left(0, \frac{3}{2}\right)$. Therefore, GH is the line $-3x + 2y = 3$. Clearly (0, 0) satisfies $-3x + 2y \leq 3$. Therefore, closed half plane containing (0, 0) is the graph of $-3x + 2y \leq 3$.

$x = 0$ is the Y-axis and $y = 0$ is the X-axis.

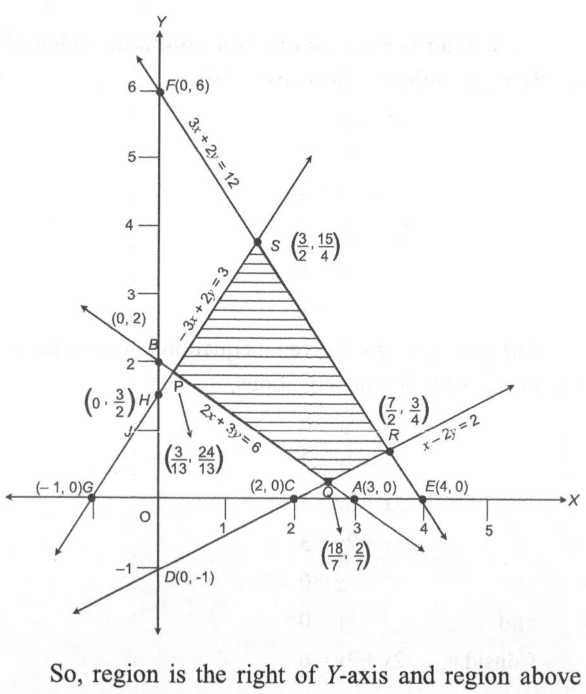

So, region is the right of Y-axis and region above X-axis are the graphs of $x \geq 0$ and $y \geq 0$ respectively the shaded region $PQRS$ is the feasible region.

The coordinates of the corner points (vertices) of the shaded feasible region are $P\left(\frac{3}{13}, \frac{24}{13}\right)$, $Q\left(\frac{18}{7}, \frac{2}{7}\right)$, $R\left(\frac{7}{2}, \frac{3}{4}\right)$ and $S\left(\frac{3}{2}, \frac{15}{4}\right)$.

At $P\left(\frac{3}{13}, \frac{24}{13}\right)$, we have

$$Z = 5\left(\frac{3}{13}\right) + 2\left(\frac{24}{13}\right)$$

$$= \frac{63}{13} = 4.84$$

At $Q\left(\frac{18}{7}, \frac{2}{7}\right)$, we have

$$Z = 5\left(\frac{18}{7}\right) + 2\left(\frac{2}{7}\right)$$

$$= \frac{94}{7} = 13.42$$

At $R\left(\frac{7}{2}, \frac{3}{4}\right)$, we have

$$Z = 5\left(\frac{7}{2}\right) + 2\left(\frac{3}{4}\right) = 19$$

At $S\left(\frac{3}{2}, \frac{15}{4}\right)$, we have

$$Z = 5\left(\frac{3}{2}\right) + 2\left(\frac{15}{4}\right) = 15$$

So, (i) Maximum value of Z is 19 at $R\left(\frac{7}{2}, \frac{3}{4}\right)$.

(ii) Minimum value is Z is 4.84 at $P\left(\frac{3}{13}, \frac{24}{13}\right)$.

Q. 4. Find the optimal solution to the following linear programming problem using graphical method.

Minimize $Z = 5x + 8y$

Subject to

$$x \leq 400$$
$$y \geq 200$$
$$x + y = 500$$
$$x, y \geq 0$$

[HOTS]

Sol. We first convert the inequations into equations to obtain the lines

$x = 400$, $y = 200$, $x + y = 500$, $x = 0$ and $y = 0$.

The line $x = 400$ is parallel to Y-axis and the line $y = 200$ is parallel to X-axis

The line $x + y = 500$ meets the coordinate axes at A (500, 0) and B (0, 500)

Join these points to obtain the line $x + y = 500$. Also $x = 0$ is the Y-axis and $y = 0$ is the X-axis.

For the inequality $x \le 400$ the feasible combination will lie on and to the left side of the line (*i.e.*, towards the origin).

For the inequality $y \ge 200$ the feasible combination will lie on and to the right side of the line (*i.e.* away from the origin).

And for the line $x + y = 500$, all feasible combinations will lie on the line only. These points lie on the line $x + y = 500$ restricted by the feasible region given by the inequalities $x \le 400$ and $y \ge 200$ as shown in the figure. Here B and E are the only feasible points.

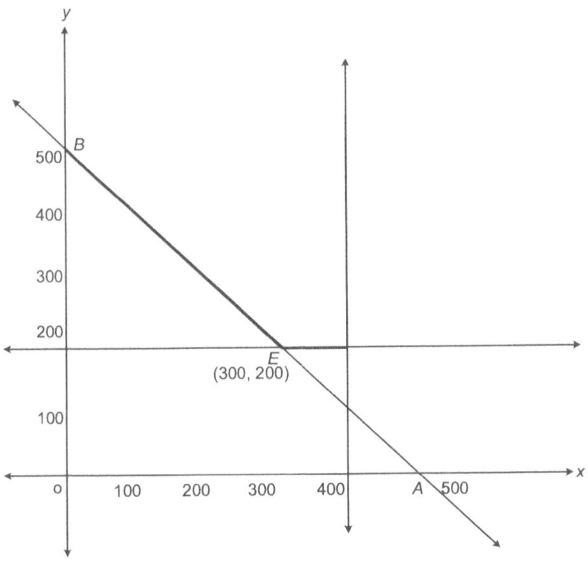

The coordinates of B are (0, 500).

The point E lies at the intersection of the lines $x + y = 500$ and $y = 200$.

Therefore the coordinates of point E are (300, 200).

The value of the objective function at these feasible points are as follows.

At B (0, 500); $Z = 5(0) + 8(500)$
$= 4{,}000$

At E (300, 200); $Z = 5(300) + 8(200)$
$= 1500 + 1600$
$= 3{,}100$

Hence, optimal solution is at $x = 300$ and $y = 200$ resulting in minimum value of Z equal to 3,100.

Q. 5. Solve graphically, the following linear programming problem to minimize the cost.

$$Z = 3x + 2y$$

Subject to the following constraints

$$5x + y \ge 10$$
$$x + y \ge 6$$
$$x + 4y \ge 12$$
$$x, y \ge 0$$

Sol. To solve the LPP graphically, we first convert the inequations into equations.

Consider $5x + y = 10$

This line meets the coordinate axis at A (2, 0) and B (0, 10). Therefore, AB is the line $5x + y = 10$. As O (0, 0) does not satisfy $5x + y \ge 10$, therefore, we shade the region not containing (0, 0).

Consider $x + y = 6$

This line meets the coordinate axis at C (6, 0) and D (0, 6). Therefore, CD is the line $x + y = 6$. Clearly, (0, 0) does not satisfy $x + y \ge 6$. Therefore, we shade the region not containing the origin.

Consider $x + 4y = 12$

This line meets the coordinate axis at E (12, 0) and F (0, 3). Therefore, EF is the line $x + 4y = 12$. Clearly, (0, 0) does not satisfy $x + 4y \ge 12$. Therefore, we shade the region not containing this origin.

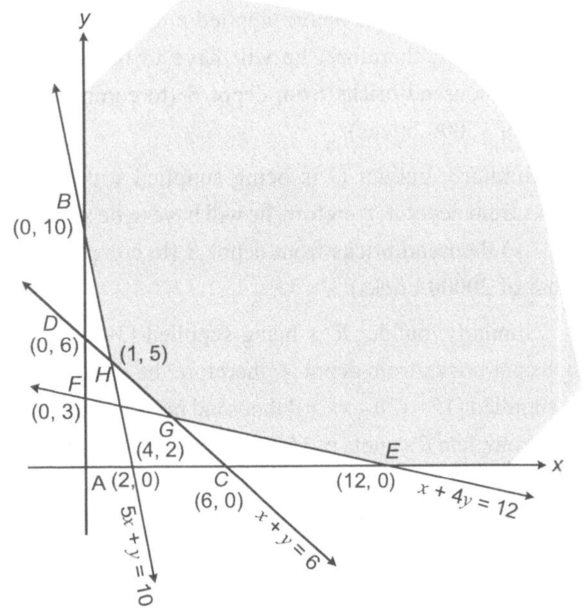

The shaded region $BHGE$ represents the common region of the above inequations. The region is the feasible region of the given LPP.

The coordinates of the vertices (corner points) of the shaded feasible region are B (0, 10), H (1, 5), G (4, 2) and E (12, 0). These points have been obtained by solving the equations of the corresponding intersecting lines simultaneously.

The value of the objective function at the points are given as follows

At B (0, 10), $C = 3(0) + 2(10) = 20$

At H (1, 5), $C = 3(1) + 2(5) = 13$

At G (4, 2), $C = 3(4) + 2(2) = 16$

At E (12, 0), $C = 3(12) + 2(0) = 36$

Clearly C is minimum at H (1, 5). Therefore, minimum cost is 13 when $x = 1$ and $y = 5$.

Q. 6. A brick manufacturer has two depots, *A* and *B*, which stocks of 30,000 and 20,000 bricks respectively. He receives orders from three builders *P*, *Q* and *R* for 15,000, 20,000 and 15,000 bricks respectively. The costs for transporting 1000 bricks to the builders from the depots are given below :

From / To	*P*	*Q*	*R*
A	40	20	30
B	20	60	40

How should the manufacturer fulfil the orders so as to keep the cost of transportation minimum?

Sol. Let the depot *A* transport x thousand bricks to builder *P* and y thousand bricks to builder *Q*, then it will have to supply $(30 - x - y)$ thousand bricks to *R*.

Now, builder *P* is being supplied x thousand bricks from depot *A*, therefore, he will have to be supplied $(15 - x)$ thousand bricks from depot *B* (to complete *P*'s quota of 15000 bricks).

Similarly builder *Q* is being supplied y thousand bricks from depot *A*, therefore, he will have to be supplied $(20 - y)$ thousand bricks from depot *B* (to complete *Q*'s quota of 20000 bricks).

Similarly, builder *R* is being supplied $(30 - x - y)$ thousand bricks from depot *A*, therefore, he will have to be supplied $[15 - (30 - x - y)]$ thousand bricks from depot *B* (to complete *R*'s quota of 15000 bricks).

The above discussion may be shown schematically as :

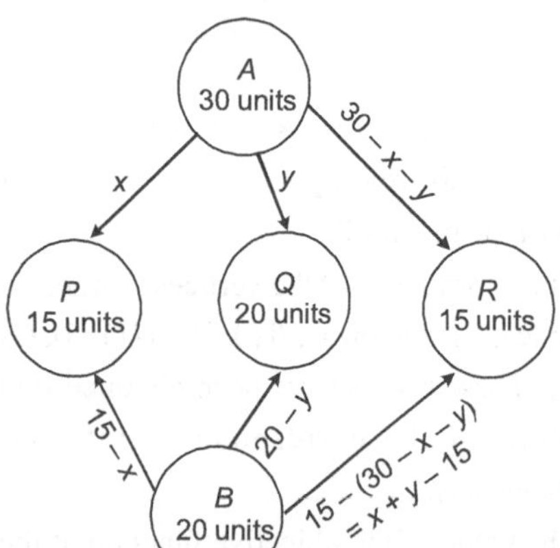

Then the above LPP can be stated mathematically as follows :

$$\text{Minimize} \quad Z = 40x + 20y + 30(30 - x - y)$$
$$+ 20(15 - x) + 60(20 - y)$$
$$+ 40[15 - (30 - x - y)]$$
$$\Rightarrow \quad Z = 40x + 20y + 900 - 30x - 30y$$
$$+ 300 - 20x + 1200 - 60y$$
$$- 600 + 40x + 40y$$
$$\text{Minimize} \quad Z = 30x - 30y + 1800$$

[in units or 1000 bricks]

The constraints would be that all the quantities of bricks supplied from depots *A* and *B* to builders *P*, Q and *R* are non-negative.

Therefore, the constraints are :

$$x \geq 0$$
$$y \geq 0$$
$$30 - x - y \geq 0$$
$$\text{or} \quad x + y \leq 30$$
$$15 - x \geq 0$$
$$\text{or} \quad x \leq 15$$
$$20 - y \geq 0$$
$$\text{or} \quad y \leq 20$$
$$x + y - 15 \geq 0$$
$$\text{or} \quad x + y \geq 15$$

To solve the LPP graphically, we first convert inequations into equations and then draw the corresponding lines.

The feasible region of the LPP is shaded in the figure. The coordinates of the corner points of the feasible region are A_2 (15, 0), *P* (15, 15), *Q* (10, 20), B_3 (0, 20) and B_2 (0, 15). These points have been obtained by solving the corresponding intersecting lines simultaneously.

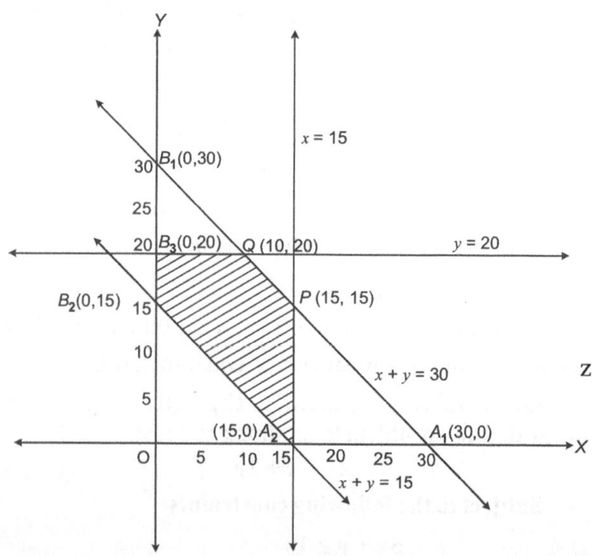

The values of the objective function at the corner points of the feasible region are given in the following table :

Point (x, y)	Value of the Objective Function: $Z = 30x - 30y + 1800$
$A_2 (15, 0)$	$Z = (30 \times 15) - (30 \times 0) + 1800 = 2250$
$P(15, 15)$	$Z = (30 \times 15) - (30 \times 15) + 1800 = 1800$
$Q(10, 20)$	$Z = (30 \times 10) - (30 \times 20) + 1800 = 1500$
$B_3 (0, 20)$	$Z = (30 \times 0) - (30 \times 20) + 1800 = 1200$
$B_2 (0, 15)$	$Z = (30 \times 0) - (30 \times 15) + 1800 = 1350$

We can see from the table above that Z is minimum at $x = 0, y = 20$ and the minimum value of Z is 1200.

Thus, the manufacturer should supply 0, 20 and 10 thousand bricks to builders P, Q and R from depot A and 15, 0 and 5 thousand bricks to builders P, Q and R from depot B, respectively. In this case, the minimum transportation cost will be ₹1200.

Q. 7. A dietician mixes together two kinds of food to prepare a special diet in such a way that the mixture contains at least 6 units of vitamin A, 7 units of vitamin B, 11 units of vitamin C and 9 units of vitamin D. The vitamin contents of one kg of food X and food Y are given below :

	Vitamin A	Vitamin B	Vitamin C	Vitamin D
Food X	1	1	1	2
Food Y	2	1	3	1

One kg of food X costs ₹ 5 whereas one kg of food Y costs ₹ 8. Find the least cost of the mixture which will produce the special diet.

Sol. Let x kg of food X be mixed with y kg of food Y.

$$\left.\begin{array}{l} x \geq 0 \\ y \geq 0 \end{array}\right\} \quad \text{...(i)}$$

The conditions about vitamin contents in the food give us the following :

$$x + 2y \geq 6 \quad \text{...(ii)}$$
$$x + y \geq 7 \quad \text{...(iii)}$$
$$x + 3y \geq 11 \quad \text{...(iv)}$$
$$2x + y \geq 9 \quad \text{...(v)}$$

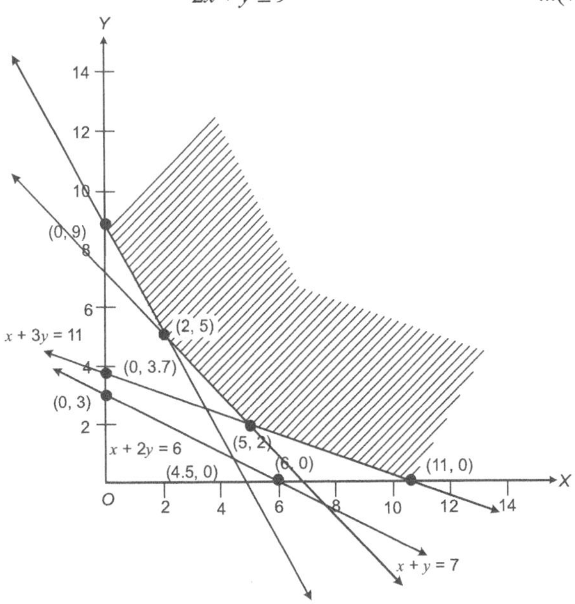

The cost of the mixture is given by :

$$Z = 5x + 8y \quad \text{...(vi)}$$

Our objective is to minimize (vi) subject to the constraints (i) to (v). The shaded region in the figure is the feasible region. The region satisfying the inequations (i) to (v) has vertices on its boundaries at the points (0, 9), (2, 5), (5, 2) and (11, 0).

At the point (0, 9) :

$$Z = 0 + 72 = 72$$

At the point (2, 5) :

$$Z = 10 + 40 = 50$$

At the point (5, 2) :

$$Z = 25 + 16 = 41 \quad \text{Least}$$

At the point (11, 0) :

$$Z = 55 + 0 = 55$$

Thus, the least cost of the mixture is ₹ 41.

Q. 8. A manufacturing unit makes a profit of ₹ 600 on each 21" TV set it produces and ₹ 400 profit on each 14" TV set. To produce a 21" TV the manufacturing unit requires :

1 hour on machine X

1 hour on machine Y and

4 hours on machine Z.

To produce a 14" TV, the manufacturing unit requires:

2 hours on X

1 hour on Y and

1 hour on Z.

In a given day, machine X, Y and Z can work a maximum of 16, 9 and 24 hours respectively.

How many 21" TV sets and how many 14" TV sets should be produced per day by the manufacturing unit to maximize its profit ?

Sol. Let x be the number of 21" TV sets and y be the number of 14" TV sets produced (and sold).

Then, the problem can be formulated as :

Maximize $\quad P = 600x + 400y$

Subject to the constraints

$x + 2y \leq 16$ (machine x constraint)

$x + y \leq 9$ (machine y constraint)

$4x + y \leq 24$ (machine z constraint)

$x \geq 0$,

$y \geq 0$ (since, number of TV sets cannot be negative)

Consider $\quad x + 2y = 16$

$$\Rightarrow \quad \frac{x}{16} + \frac{y}{8} = 1$$

This line cuts the X-axis at A (16, 0) and Y-axis at B (0, 8). Therefore, AB is the line $x + 2y = 16$. Clearly,

(0, 0) satisfies $x + 2y \leq 16$. Therefore, closed half plane containing (0, 0) is the graph at $x + 2y \leq 16$.

Consider $\quad x+y=9$

$$\Rightarrow \quad \frac{x}{9}+\frac{y}{9}=1$$

This line cuts the X-axis at C (9, 0) and Y-axis at D (0, 9). Therefore, CD is the line $x+y=9$. Clearly, (0, 0) satisfies $x+y \leq 9$. Therefore, closed half plane containing (0, 0) is the graph of $x+y \leq 9$.

Next consider

$$4x+y=24$$

$$\Rightarrow \quad \frac{x}{6}+\frac{y}{24}=1$$

This line cuts the X-axis at E (6, 0) and Y-axis at F (0, 24). Clearly, (0, 0) satisfies $4x + y \leq 24$. Therefore, closed half plane containing (0, 0) is the graph of $4x+y \leq 24$.

Also, $x = 0$ is the Y-axis and $y = 0$ is the X-axis. So, the region to the right of Y-axis and the region above X-axis are the graph of $x \leq 0$ and $y \leq 0$ respectively. The shaded region $OBHGE$ is the feasible region.

The coordinates of the corner points of the shaded feasible region are :

O (0, 0), B (0, 8), H (2, 7), G (5, 4) and E (6, 0).

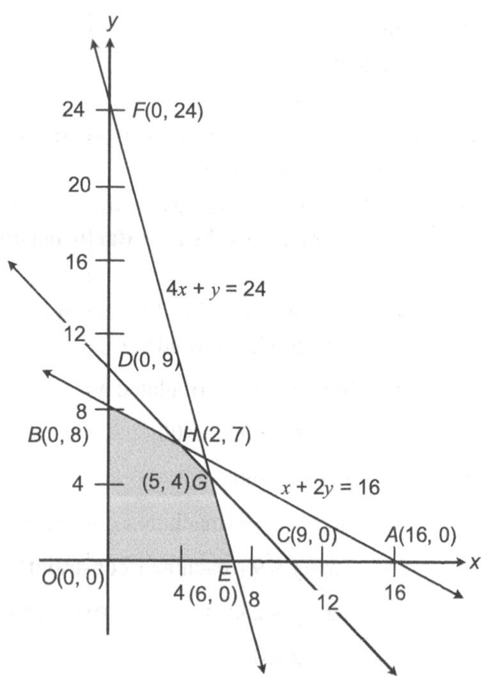

Now, at O (0, 0), we have

$$P=600\,(0)+400\,(0)=0$$

At E (6, 0), we have

$$P=600\,(6)+400\,(0)=3600$$

At G (5, 4), we have

$$P=600\,(5)+400\,(4)=4600$$

At H (2, 7), we have

$$P=600\,(2)+400\,(7)=4000$$

At B (0, 8), we have

$$P=600\,(0)+400\,(8)=3200$$

Thus, the maximum value of P is 4,600 at G (5, 4). Hence, for maximum profit, *i.e.* ₹ 4,600 a day, five 21" TV sets and four 14" TV sets should be produced.

Q. 9. The animal feed company must produce at least 200 kg of mixture consisting of ingredients X_1 and X_2. The ingredient X_1 costs ₹ 3 per kg and X_2 costs ₹ 5 per kg. Not more than 80 kg of X_1 can be used and at least 60 kg of X_2 must be used. The problem is to determine the quantities X_1 and X_2 so that the cost of the mixture may come to be minimum.

(i) Give a mathematical formulation of the above said linear programming problem.

(ii) Indicate the feasible region on a graph paper.

(iii) Use graphical method to solve the problem.

Sol. Let x_1 denote the weight in kilograms of the ingredient X_1, and let x_2 denote the weight in kilogram of the ingredient X_2.

Then, the given problem can be formulated mathematically as an LPP as :

Minimize $\quad Z=3x_1+5x_2$

Subject to

$$x_1+x_2 \geq 200$$
$$x_1 \leq 80$$
$$x_2 \geq 60$$
$$x_1, x_2 \geq 0$$

The feasible region of the above LPP is represented by the shaded region in the figure.

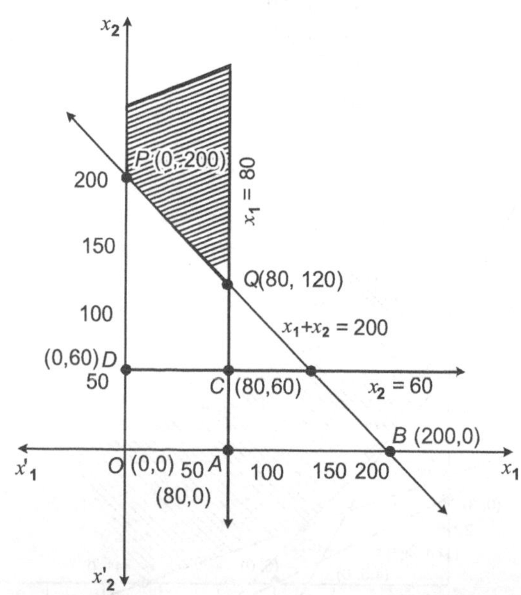

The coordinates of the corner feasible region are P (0, 200) and Q (80, 120). The objective function is

$$Z=3x_1+5x_2$$

Value of Z at

$$P(0, 200) = 5 \times 200$$
$$= 1{,}000$$

Value of Z at $Q(80, 120)$

$$Q(80, 120) = 3 \times 80 + 5 \times 120$$
$$= 240 + 600$$
$$= 840$$

The cost will be minimum at Q (80, 120) and the minimum value of $Z = 840$.

Q. 10. A manufacturer makes two types of calculators. Deluxe sells for \$12 and standard sells for \$10. It costs \$9 to produce a deluxe and \$8 to produce a standard calculator. The manufacturer can produce from 200 to 300 deluxe calculators and from 100 to 250 standard calculators in one week, but not more than 500 total calculators.

How many calculators of each type should the manufacturer produce per week to maximize the profits? Solve this problem using corner point method.

Sol. Let x be the number of deluxe calculators manufactured per week, and y be the number of standard calculators manufactured per week.

The profit per deluxe calculator is \$3 and the profit per standard calculator is \$2. Thus the problem can be formulated as :

Maximize $Z = 3x + 2y$

Subject to the constraints

$$200 \le x \le 300$$
$$100 \le y \le 250$$
$$x + y \le 500$$
$$x, y \ge 0$$

The graph can be drawn as :

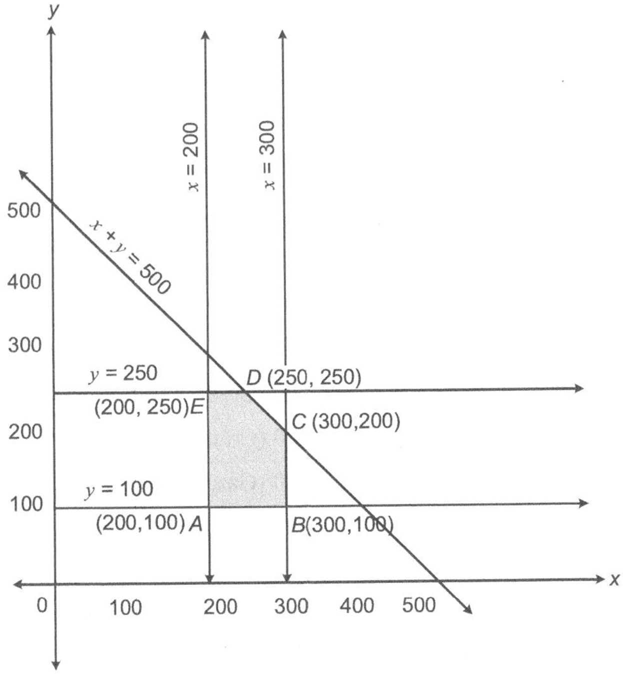

We first convert the inequations into equations and draw the corresponding lines. The feasible region of the LPP is shaded in the figure. The coordinates of the corner points of the feasible region $ABCDE$ are A (200, 100), B(300, 100), C(300, 200), D(250, 250) and E(200, 250).

These points have been obtained by solving the corresponding intersecting lines simultaneously.

The value of the objective function at these points are as follows.

At A (200, 100), we have

$$Z = 3(200) + 2(100)$$
$$= 800$$

At B (300, 100), we have

$$Z = 3(300) + 2(100) = 1100$$

At C (300, 200), we have

$$Z = 3(300) + 2(200) = 1300$$

At D (250, 250), we have

$$Z = 3(250) + 2(250) = 1250$$

At E (200, 250), we have

$$Z = 3(200) + 2(250) = 1100$$

Clearly, Z is maximum at $x = 300, y = 200$. The optimal value of Z is 1300.

Hence for maximum profit (\$1300 per week) 300 deluxe calculators and 200 standard calculators should be produced by the manufacturing unit.

Q. 11. A company has two factories located at P and Q and has sales depots situated at A, B and C. The weekly requirement of the sales depots at A, B, C is respectively 5, 5 and 4 units, while the production capacity of the factories at P and Q respectively is 8 and 6 units. The cost of transportation from factories to the sales depots (per unit) are given below.

From \ To	Cost (in ₹)		
	A	*B*	*C*
P	16	10	15
Q	10	12	10

How many units should be transported from each factory to each depot in order that the transportation cost is minimum ?

Sol. The given data can be represented schematically as :

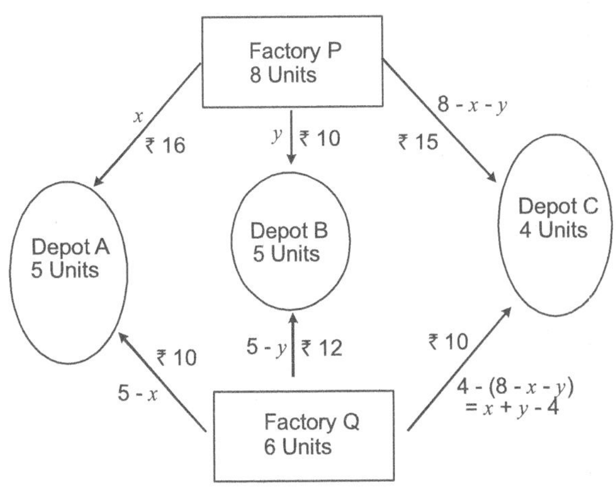

We can see from the problem that the total weekly production (of P and Q) = 8 + 6 = 14 units, and the total weekly requirement at depots $A, B, C = 5 + 5 + 4 = 14$ units.

So, there is no mismatch between supply and demand. Let factory P supply x units per week to depot A and y units to depot B, which means it would supply $(8 - x - y)$ units to depot C.

$$0 \le x \le 5$$
$$0 \le y \le 5$$
$$0 \le 8 - x - y \le 4$$

This gives the constraints as :

$$x + y \le 8$$
$$x + y \ge 4$$
$$0 \le x \le 5$$
$$0 \le y \le 5$$

As depot A' s requirement is 5 units and it receives x units from factory P, it must then receive $(5 - x)$ units from factory Q.

Similarly, depot B receives $(5 - y)$ units from factory Q and depot C receives $[4 - (8 - x - y)]$ units or $(x + y - 4)$ units from factory Q.

Thus total transportation cost (in rupees) or the objective function

$$Z = 16x + 10y + 15(8 - x - y) + 10(5 - x) + 12(5 - y) + 10(x + y - 4)$$

or $$Z = x - 7y + 190$$

Hence, the given problem can be formulated as an LPP as :

Minimize $Z = x - 7y + 190$

Subject to
$$x + y \ge 4$$
$$x + y \le 8$$
$$0 \le x \le 5$$
$$0 \le y \le 5$$

The feasible region corresponding to these constraints is shown in the graph below.

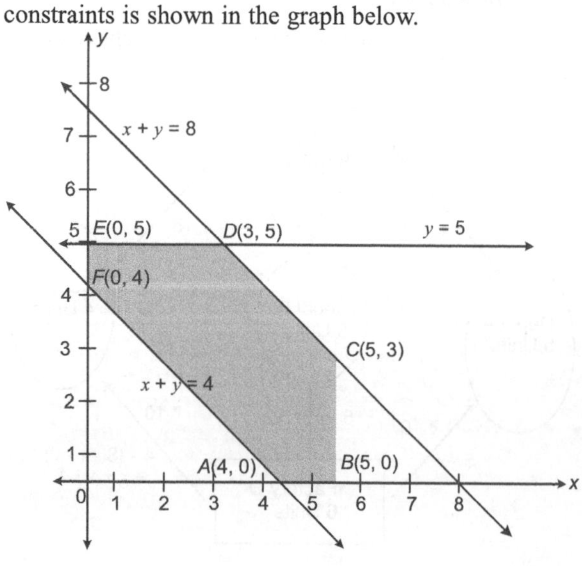

The coordinates of the vertices of shaded feasible region $ABCDEF$ are $A(4, 0)$, $B(5, 0)$, $C(5, 3)$, $D(3, 5)$, $E(0, 5)$ and $F(0, 4)$.

We calculate the values of Z at the six corner points.

At $A(4, 0)$, $Z = 4 - 7(0) + 190 = 194$

At $B(5, 0)$, $Z = 5 - 7(0) + 190 = 195$

At $C(5, 3)$, $Z = 5 - 7(3) + 190 = 174$

At $D(3, 5)$, $Z = 3 - 7(5) + 190 = 158$

At $E(0, 5)$, $Z = 0 - 7(5) + 190 = 155$

At $F(0, 4)$, $Z = 0 - 7(4) + 190 = 162$

We see that Z is minimum at point $E(0, 5)$. The minimum cost is ₹ 155 at $x = 0$ and $y = 5$.

Thus, for minimum transportation cost factory P should supply 0, 5, 3 units to depot A, B, C respectively and factory Q should supply 5, 0, 1 units to depot A, B, C respectively.

Q. 12. The manager of an oil refinery must decide on the optimal mix of two possible blending processes, one using certain quantity of units of Crude *A* and the other using certain other quantity of units of Crude *B*, which form the input to produce two different qualities of gasoline, namely : gasoline *X* and gasoline *Y*.

	Input (Units)		Output (Units)	
Process	**Crude *A***	**Crude *B***	**Gasoline *X***	**Gasoline *Y***
1	5	3	5	8
2	4	5	5	4

The maximum quantity available of crude *A* and *B* are 200 units and 150 units respectively. Market requirements show that at least 100 units of gasoline *X* and 80 units of gasoline *Y* must be produced. The profit seen from production process 1 and production process 2 is ₹ 300 and ₹ 400 respectively.

Formulate the problem as LPP and solve it by graphical method.

Sol. Let x and y denote the number of production runs of process 1 and process 2 respectively.

Then the appropriate of the problem is :

Maximize $Z = 300x + 400y$

Subject to the constraints

$$5x + 4y \le 200 \text{ (Crude } A \text{ constraint)}$$
$$3x + 5y \le 150 \text{ (Crude } B \text{ constraint)}$$
$$5x + 4y \ge 100 \text{ (Gasoline } X \text{ constraint)}$$
$$8x + 4y \ge 80 \text{ (Gasoline } Y \text{ constraint)}$$
$$x, y \ge 0$$

The feasible region of the above LPP is represented by the shaded region in the figure.

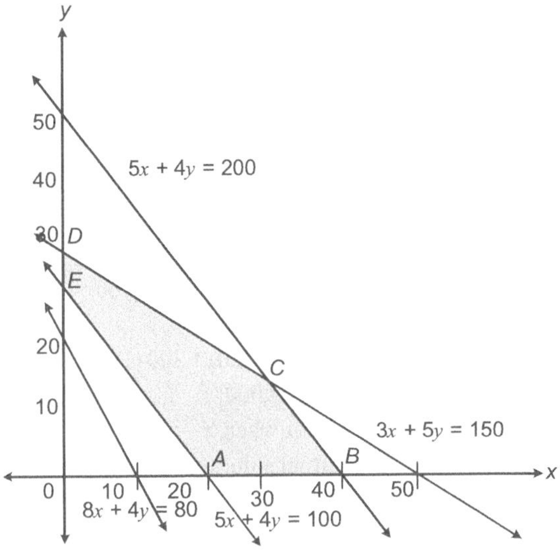

The coordinates of the corner feasible region $ABCDE$ are : A (20, 0), B (40, 0), $C\left(\frac{400}{13}, \frac{150}{13}\right)$, D (0, 30) and E (0, 25). These points are obtained by solving the corresponding lines simultaneously. It remains only to evaluate the value if the objectives functions of each corner point

At A (20, 0), $\quad Z = 300(20) + 400(0) = 6000$

At B (40, 0), $\quad Z = 300(40) + 400(0) = 12000$

At $C\left(\frac{400}{13}, \frac{150}{13}\right)$, $\quad Z = 300\left(\frac{400}{13}\right) + 400\left(\frac{150}{13}\right)$

$$= \frac{180000}{13}$$

At D (0, 30), $\quad Z = 300(0) + 400(30) = 12000$

At E (0, 25), $\quad Z = 300(0) + 400(25) = 10000$

Since maximum value of Z occurs at the corner point $C\left(\frac{400}{13}, \frac{150}{13}\right)$. Therefore, optimum solution is $x = \frac{400}{13}$, $y = \frac{150}{13}$ and maximum value of $Z = ₹\ \frac{180000}{13}$.

Q. 13. A firm plans to purchase at least 200 quintals of scrap containing high quality metal *X* and low quality metal *Y*. It decided that the scrap to be purchased must contain at least 100 quintals of *X*-metal and no more than 35 quintals of *Y*-metal. The firm can purchase the scrap from two supplies (*A* and *B*) in unlimited quantities. The percentage of *X* and *Y* metals in terms of weights in the scrap supplied by *A* and *B* is given below :

Metal	Supplier *A*	Supplier *B*
X	25%	75%
Y	10%	20%

The prices of *A*'s scrap is ₹ 200 per quintals and that of *B*'s Scrap is ₹ 400 per quintal.

The firm wants to determine the quantities that it should buy from the two suppliers so that the total cost is minimized. Solve using the graphical method.

Sol. Let the firm purchase x quintal of scrap from supplier A and y quintal of scrap from supplier B.

Then the LP model of the given problem can be expressed as :

Minimize $\quad Z = 200x + 400y$

Subject to the constraints

$$x + y \geq 200$$

$$\frac{25}{100}x + \frac{75}{100}y \geq 100$$

or $\quad x + 3y \geq 400$

$$\frac{10}{100}x + \frac{20}{100}y \leq 35$$

or $\quad x + 2y \leq 350$

$$x, y \geq 0$$

To solve, we first convert inequations into equations and then draw the corresponding lines on the graph.

The feasible region of the LPP is shaded in the figure. The coordinates of the corner points of the feasible region PQR are : P (100, 100), Q (250, 50) and R (50, 150).

The value of the objective function at each of the corner points is as follows :

At P (100, 100), $\quad Z(P) = 200(100) + 400(100) = 60000$

At Q (250, 50), $\quad Z(Q) = 200(250) + 400(50) = 70000$

At R (50, 150), $\quad Z(R) = 200(50) + 400(150) = 70000$

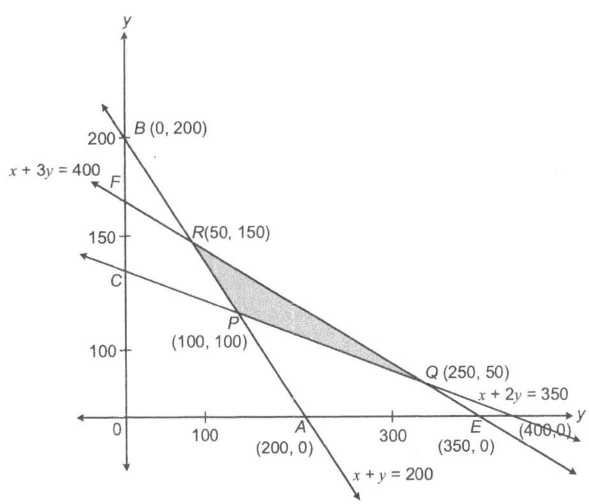

Since Z has the minimum value at the corner point P (100, 100), therefore, the optimum solution is :

$$x = 100$$

$$y = 100 \text{ and}$$

Minimize $\quad Z = ₹\ 60000$

That is, the firm should purchase 100 quintals of scrap from each of the suppliers A and B to minimize its cost.

Q. 14. An aeroplane can carry a maximum of 200 passengers. A profit of ₹ 400 is made on each first class ticket while a profit of ₹ 300 is made on each economy class ticket. The airline reserves at least 20 seats for first class. However, at least 4 times as many passengers prefer to travel by economy class than by the first class.

Determine how many of each type of tickets must be sold in order to maximize the profit for the airline. What is the maximum profit ?

Sol. Let x and y be the number of ticket sold by the airline of the first class and the economy class respectively.

Profit from each first class ticket = ₹ 400

Therefore, profit from x first class tickets = ₹ $400x$

Profit from each economy class ticket = ₹ 300

Therefore, profit from y economy class tickets = ₹ $300y$.

If P denotes the total profit (in ₹), then

$$P = 400x + 300y$$

As the aeroplane can carry a maximum of 200 passengers, therefore

$$x + y \leq 200$$

Also, since at least 20 first class tickets are reserved, therefore,

$$x \geq 20$$

Again, the number of passengers in the economy class is at least 4 times the number of passengers in first class. Therefore,

$$y \geq 4x$$

Since, the number of ticket cannot be negative, therefore

$$x \geq 0$$
$$y \geq 0.$$

Therefore, the mathematical formulation of given LPP is as follows :

Maximize $P = 400x + 300y$

Subject to the constraints

$$x + y \leq 200$$
$$x \geq 20$$
$$y \geq 4x$$
$$x, y \geq 0$$

The set of all feasible solutions of the above LPP is represented by the feasible region shaded in the figure.

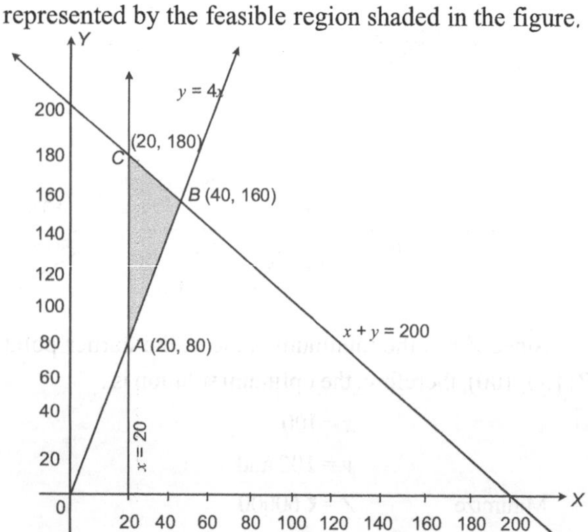

★ are board exam questions from previous years

The coordinates of the corner points of the feasible region ABC are A (20, 80), B (40, 160) and C (20, 180). These points have been obtained by solving the corresponding intersecting lines simultaneously.

The value of the objective function P at the corner points of the feasible region ABC are as follows :

At A (20, 80), $P = 400(20) + 300(80) = 32{,}000$

At B (40, 160), $P = 400(40) + 300(160) = 64{,}000$

At C (20, 180), $P = 400(20) + 300(180) = 62{,}000$

Clearly, P is maximum when $x = 40$, $y = 160$. Thus $x = 40$, $y = 160$ is the optimal solution of the given LPP.

Hence, the airline should sell 40 first class tickets and 160 economy class tickets to maximize the profit that can be obtained is ₹ 64,000.

Q. 15. A manufacturer produces two products *A* and *B*. Both the products are processed on two different machines. The available capacity of first machine is 12 hours and that of second machine is 9 hours per day. Each unit of product *A* requires 3 hours on both machines and each unit of product *B* requires 2 hours on first machine and 1 hour on second machine. Each unit of product *A* is sold at ₹ 7 profit and *B* at a profit of ₹ 4. Find the production level per day for maximum profit graphically.★

Sol. Let the numbers of unit of products A and B to be produced be x and y, respectively.

Product	Machine	
	I(h)	II(h)
A	3	3
B	2	1

Total profit : $Z = 7x + 4y$

We have to maximize $Z = 7x + 4y$.

Subject to constraints :

$$3x + 2y \leq 12$$
$$3x + y \leq 9$$
$$\Rightarrow \quad x \geq 0 \text{ and } y \geq 0$$

The given information can be graphically expressed as follows :

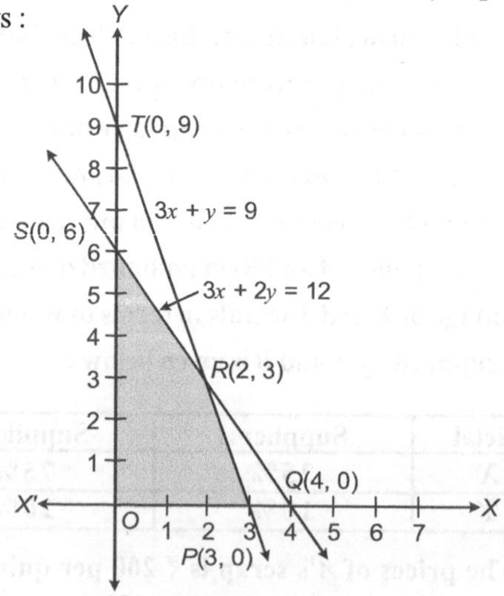

Values of $Z = 7x + 4y$ at the corner points are as follows :

Corner point	$Z = 7x + 4y$
(0, 6)	24
(2, 3)	26 ← **Maximum**
(3, 0)	21

Therefore, the manufacturer has to produce 2 units of product A and 3 units of product B for the maximum profit of ₹ 26.

Q. 16. The company manufactures three kinds of calculators A, B and C in its factorises I and II. The company has got an order for manufacturing at least 6400 calculators of kind A, 4000 of kind B and 4800 of kind C. The daily output of factory I is 50 calculators of kind A, 50 calculators of kind B and 30 calculators of kind C. The daily output of factory II is of 40 calculators of kind A, 20 of kind B and 40 of kind C. The cost per day to run factory I is ₹ 12,000 and of factory II is ₹ 15,000. How many days do the two factories have to be in operation to produce the order with the minimum cost ? Formulate this problem as an LPP and solve it graphically.★

Sol. Let the factories I and II work for x and y number of days respectively.

Thus, the given linear programming problem is

$$\text{Minimize } Z = ₹\,(12000x + 15000y)$$

Subject to the constraints

$$50x + 40y \geq 6400$$
$$50x + 20y \geq 4000$$
$$30x + 40y \geq 4800$$
$$x \geq 0$$

and $$y \geq 0$$

i.e., $$5x + 4y \geq 640$$
$$5x + 2y \geq 400$$
$$3x + 4y \geq 480$$
$$x \geq 0, y \geq 0$$

To solve this L.P.P.

Let us consider the equations

$$L_1 : 5x + 4y = 640 \quad \ldots(1)$$
$$L_2 : 5x + 2y = 400 \quad \ldots(2)$$
$$L_3 : 3x + 4y = 480 \quad \ldots(3)$$

The point of intersection of L_1 and L_2 is $D(32, 120)$ and that of L_1 and L_3 is $C(80, 60)$.

The shaded region is the solution region of the given L.P.P.

Corner Points	Values of the **objective function** $\mathbf{Z = 12000x + 15000y}$
$A(0, 200)$	$12000 \times 0 + 15000 \times 200 = 30,00,000$
$B(160, 0)$	$12000 \times 160 + 15000 \times 0 = 19,20,000$
$C(80, 60)$	$12000 \times 80 + 15000 \times 60 = 18,60,000$
$D(32, 120)$	$12000 \times 32 + 15000 \times 120 = 21,84,000$

Out of these values of Z, the minimum value of Z is 18,60,000 at $x = 80$ and $y = 60$.

Since the feasible region is unbounded so we draw the graph of inequality $12000x + 15000y < 186000$ *i.e.,* $4x + 5y < 620$.

★ are board exam questions from previous years

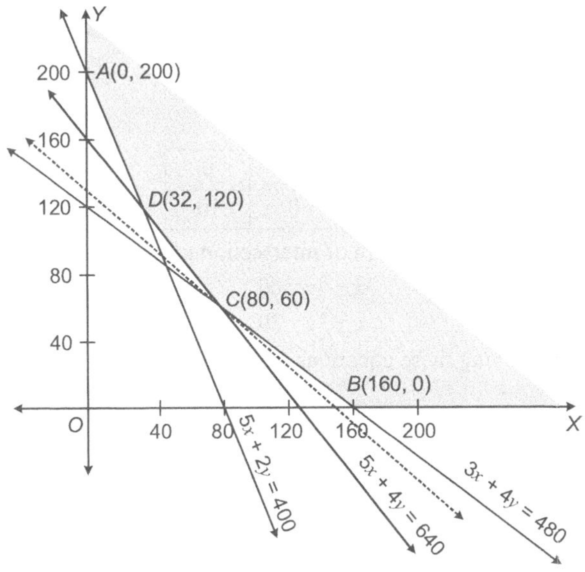

$L : 4x + 5y = 620$

x	0	155
y	124	0

We observe that open half one represented by L have no point common with feasible region.

$$Z = 12000 \times 80 + 15000 \times 60$$
$$= ₹\,18,60,000$$

Hence, the factories I and II work for 80 and 60 number of days respectively.

Q. 17. A manufacturing company makes two types of teaching aids A and B of Mathematics for class XII. Each type of A requires 9 labour hours of fabricating and 1 labour hour for finishing. Each type of B requires 12 labour hours for fabricating and 3 labour hours for finishing. For fabricating and finishing, the maximum labour hours available per week are 180 and 30 respectively. The company makes a profit of ₹ 80 on each piece of type A and ₹ 120 on each piece of type B. How many pieces of type A and type B should be manufactured per week to get a maximum profit ? Make it as an LPP and solve graphically. What is the maximum profit per week ?★

Sol. Let x and y be the number of teaching aids of type A and B respectively. Then the LPP is

Maximize: $$Z = 80x + 120y$$

Subject to constraints :

$$9x + 12y \leq 180$$
$$x + 3y \leq 30$$

and $$x \geq 0, y \geq 0$$

First we draw the lines AB and CD whose equations are

$$9x + 12y = 180$$
$$\Rightarrow \quad 3x + 4y = 60 \quad \ldots(1)$$

x	20	0
y	0	15

and $x+3y=30$ ...(2)

x	30	0
y	0	10

Let P be the point of intersection of the lines

$$3x+4y=60$$

and $$x+3y=30$$

Solving these equations we get point $P(12, 6)$.

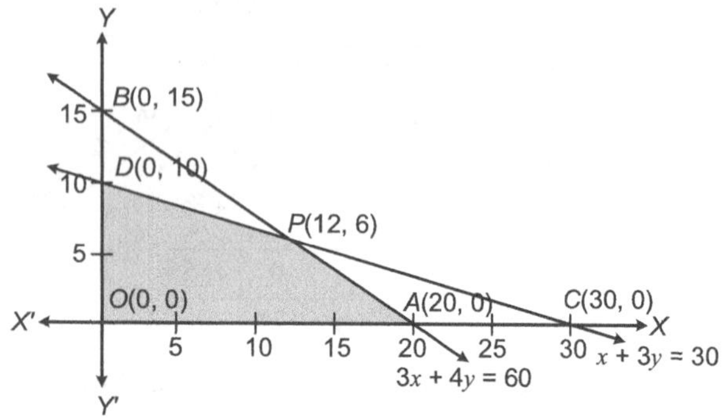

The feasible region is $OAPDO$ which is shaded in the figure.

The vertices of the feasible region are $O(0, 0)$, $A(20, 0)$, $P(12, 6)$ and $D(0, 10)$.

The value of objective function $Z = 80x + 120y$ as follows :

Corner Points	Maximize $Z = 80x+120y$
At **$O(0, 0)$**	0
At **$A(20, 0)$**	1600
At **$P(12, 6)$**	1680 **maximum**
At **$D(0, 10)$**	1200

∴ The profit is maximum of $P(12, 6)$ *i.e.*, when the teaching aids of types A and B are 12 and 6 respectively.

Also maximum profit = ₹ 1680 per week.

Q. 18. A dealer in rural area wishes to purchase a number of sewing machines. He has only ₹ 5,760 to invest and has space for at most 20 items for storage. An electronic sewing machine cost him ₹ 360 and a manually operated sewing machine ₹ 240. He can sell an electronic sewing machine at a profit of ₹ 22 and a manually operated sewing machine at a profit of ₹ 18. Assuming that he can sell all the items that he can buy, how should he invest his money in order to maximize his profit ? Make it as a LPP and solve it graphically.★

Sol. Let the dealer buy x electronic and y manually operated sewing machines. The LPP is

Maximize $Z = 22x + 18y$

Subject to constraints :

$$x+y \le 20$$

$$360x+240y \le 5,760$$

$$\Rightarrow \quad 3x+2y \le 48$$

and $x \ge 0, y \ge 0$

First we draw the lines AB and CD whose equations are

$$x+y=20$$

	A	B
x	20	0
y	0	20

and $3x + 2y = 48$

	C	D
x	16	0
y	0	24

∴ The feasible region is $OCPAO$ which is shaded in the figure.

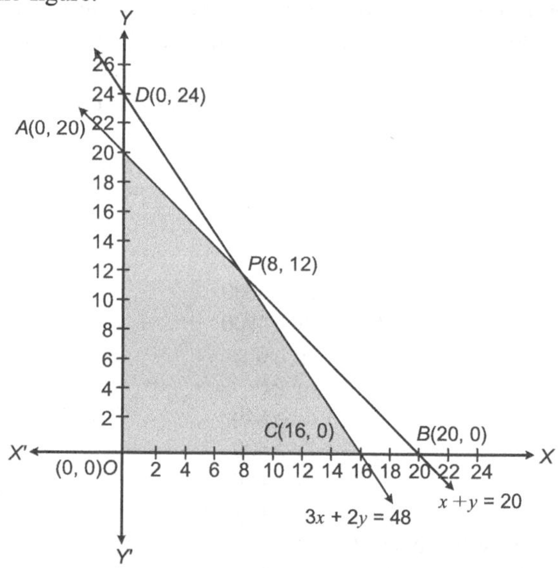

The vertices of the feasible region are $O(0, 0)$, $C(16, 0)$, $A(0, 20)$.

P is the point of intersection of the lines :

$$x+y=20 \text{ and } 3x+2y=48$$

Solving these equations, we get point $P(8, 12)$.

The value of objective function Z = s22x + 18y at these vertices are as follows :

Corner points	$Z = 22x+18y$
At **$O(0,0)$**	$Z=0$
At **$C(16,0)$**	$Z=352$
At **$P(8,12)$**	$Z=392$ **maximum**
At **$A(0,20)$**	$Z=360$

∴ The maximum profit is ₹ 392 when 8 electronic and 12 manually operated machines are purchased.

Q. 19. The cooperative society of farmers has 50 hectares of land to grow two crops A and B. The profits from crops A and B per hectare are estimated as ₹ 10,500 and ₹ 9,000 respectively. To control weeds, a

★ are board exam questions from previous years

liquid herbicide has to be used for crops A and B at the rate of 20 litres and 10 litres per hectare, respectively. Further not more than 800 litres of herbicide should be used in order to protect fish and wildlife using a pond which collects drainage from this land. Keeping in mind that the protection of fish and other wildlife is more important than earning profit, how much land should be allocated to each crop so as to maximize the total profit ? Form an LPP from the above and solve it graphically. ★

Sol. Let x hectares and y hectares be alloted to grow crops A and B respectively. Then the L.P.P. is maximize :

$$Z = 10{,}500x + 9{,}000y \text{ subject to constraints}$$

$$x + y \leq 50$$

$$20x + 10y \leq 800$$

$$\Rightarrow \quad 2x + y \leq 80$$

$$\text{and} \quad x \geq 0, y \geq 0$$

First we draw the line AB and CD whose equations are

$$x + y = 50 \quad \ldots\text{(i)}$$

	A	B
x	0	50
y	50	0

and

$$2x + y = 80 \quad \ldots\text{(ii)}$$

	C	D
x	0	40
y	80	0

$\therefore$ The feasible region is $ODPAO$ which is shaded in the figure.

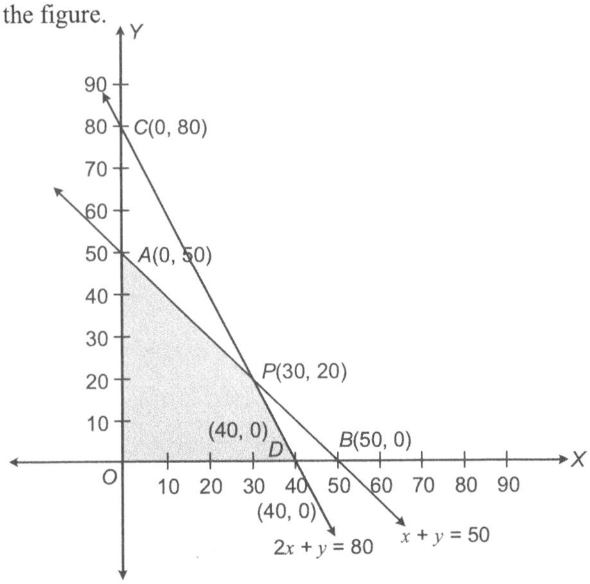

P is the point of intersection of the lines

$$x + y = 50$$
$$2x + y = 80$$

Solving these equations we get point $P(30, 20)$.

The vertices of the feasible region are O (0, 0), D(40, 0), P(30, 20) and A(0, 50). The value of objective function $Z = 10{,}500x + 9{,}000y$ at these vertices are as follows :

Corner points	$Z = 10{,}500x + 9{,}000y$
At **O(0,0)**	$Z = 0$
At **D(40,0)**	$Z = 4{,}20{,}000$
At **P(30,20)**	$Z = 4{,}95{,}000$ **maximum**
At **A(0,50)**	$Z = 4{,}50{,}000$

$\therefore$ The maximum profit is 4,95,000 at point $P(30, 20)$.

Q. 20. A factory manufactures two types of screws A and B, each type requiring the use of two machines, an automatic and a hand-operated. It takes 4 minutes on the automatic and 6 minutes on the hand-operated machines to manufacture a packet of screws 'A' while it takes 6 minutes on the automatic and 3 minutes on the hand-operated machine to manufacture a packet of screws 'B'. Each machine is available for at most 4 hours on any day. The manufacturer can sell a packet of screws 'A' at a profit of 70 paise and screws 'B' at a profit of ₹ 1. Assuming that he can sell all the screws he manufactures, how many packets of each type should the factory owner produce in a day in order to maximize his profit ? Formulate the above LPP and solve it graphically and find the maximum profit.★

Sol. Let no. of packets of screw manufacture each day A be x and that of screw B be y.

Therefore, $x \geq 0, y \geq 0$

Item	Number	Machine A	Machine B	Profit
Screw A	x	4 minutes	6 minutes	₹ 0.7
Screw B	y	6 minutes	3 minutes	₹ 1
Max. time available		4 hrs. = 240 min.	4 hrs. = 240 min.	

The constraints are :

$$4x + 6y \leq 240 \text{ or } 2x + 3y \leq 120$$

$$6x + 3y \leq 240 \text{ or } 2x + y \leq 80$$

$$Z = 0.7x + y$$

So our LPP will be

Max. $\quad Z = 0.7x + y$

such that $\quad 2x + 3y \leq 120,$

$\quad 2x + y \leq 80,$

and $\quad x, y \geq 0.$

$2x + 3y \leq 120$

x	0	60
y	40	0

$2x + y \leq 80$

x	0	40
y	80	0

★ are board exam questions from previous years

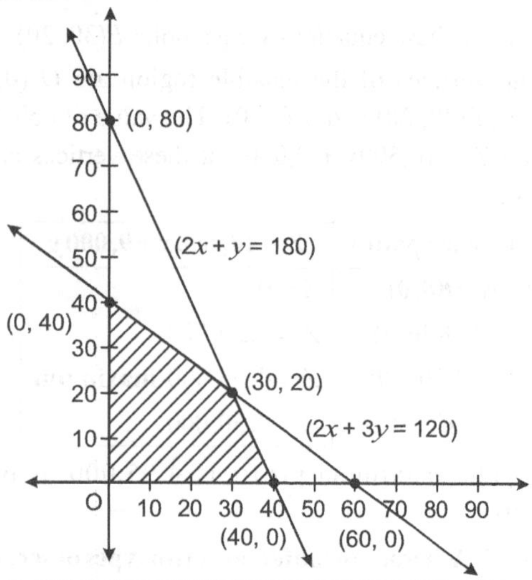

Corner points	Value of z
(0,40)	40
(30,20)	41 **Maximum**
(40,0)	28

Hence, profit will be maximum if company produces 30 packets of screw A and 20 packets of screw B and maximum profit = ₹ 41.

Chapter 13. Probability

Q. 1. A fair dice is rolled consider events $E = \{1, 3, 5\}, F = \{2, 3\}$ and $G = \{2, 3, 4, 5\}$ find

(i) $P(E/F)$ and $P(F/E)$

(ii) $P(E/G)$ and $P(G/E)$

(iii) $P[(E \cup F)/G]$ and $P[(E \cap F)/G]$ [NCERT]

Sol. When a fair dice is rolled, the sample spaces will be $S = \{1, 2, 3, 4, 5, 6\}$

It is given that $E = \{1, 3, 5\}, F = \{2, 3\}$ and $G = \{2, 3, 4, 5\}$.

$$\therefore \quad P(E) = \frac{3}{6} = \frac{1}{2}$$

$$P(F) = \frac{2}{6} = \frac{1}{3}$$

$$P(G) = \frac{4}{6} = \frac{2}{3}$$

(i) $$(E \cap F) = \{3\}$$

$$\therefore \quad P(E \cap F) = \frac{1}{6}$$

$$P(E/F) = \frac{P(E \cap F)}{P(F)}$$

$$= \frac{\frac{1}{6}}{\frac{1}{3}} = \frac{1}{2}$$

$$P(F/E) = \frac{P(F \cap E)}{P(E)}$$

$$= \frac{\frac{1}{6}}{\frac{1}{2}} = \frac{2}{6} = \frac{1}{3}$$

(ii) $$E \cap G = \{3, 5\}$$

$$P(E \cap G) = \frac{2}{6} = \frac{1}{3}$$

$$P(E/G) = \frac{P(E \cap G)}{P(G)}$$

$$= \frac{\frac{1}{3}}{\frac{2}{3}} = \frac{1}{2}$$

$$P(G/E) = \frac{P(G \cap E)}{P(E)}$$

$$= \frac{\frac{1}{3}}{\frac{1}{2}} = \frac{2}{3}$$

(iii) $$E \cup F = \{1, 2, 3, 5\}$$

$$(E \cup F) \cap G = \{1, 2, 3, 5\} \cap \{2, 3, 4, 5\}$$

$$= \{2, 3, 5\}$$

$$E \cap F = \{3\}$$

$$(E \cap F) \cap (G) = \{3\} \cap \{2, 3, 4, 5\}$$

$$= \{3\}$$

$$\text{P}((E \cup F) \cap G) = \frac{3}{6} = \frac{1}{2}$$

$$\therefore \quad P((E \cup F)/G) = \frac{P[(E \cup F) \cap G]}{P(G)}$$

$$= \frac{\frac{1}{2}}{\frac{2}{3}} = \frac{1}{2} \times \frac{3}{2} = \frac{3}{4}$$

$$P(E \cap F) = \frac{1}{6}$$

$$P((E \cap F) \cap G) = \frac{1}{6}$$

$$P((E \cap F)/G) = \frac{P[(E \cap F) \cap G]}{P(G)}$$

$$= \frac{\frac{1}{6}}{\frac{2}{3}} = \frac{1}{6} \times \frac{3}{2} = \frac{1}{4}$$

Q. 2. Two balls are drawn at random with replacement from a box containing 10 black and 8 red balls. Find the probability that[NCERT]

(i) both are red

(ii) first ball is black and second is red

(iii) One of them is black and other is red.

Sol. Total number of balls $= 10$ black $+ 8$ red

$= 18$

Number of black balls $= 10$

Number of red balls $= 8$

(i) Probability of getting a red ball in first draw

$$= \frac{8}{18} = \frac{4}{9}$$

The ball is replaced after the first draw.

$\therefore$ P (getting red ball in 2nd draw)

$$= \frac{8}{18} = \frac{4}{9}.$$

Hence P (getting both red balls)

$$= \frac{4}{9} \times \frac{4}{9}$$

$$= \frac{16}{81}$$

(ii) P (getting first ball black)

$$= \frac{10}{18} = \frac{5}{9}$$

The ball is replaced after first draw

P (getting second ball as red)

$$= \frac{8}{18} = \frac{4}{9}$$

Hence, P (getting first ball as black and second ball as red)

$$= \frac{5}{9} \times \frac{4}{9} = \frac{20}{81}.$$

(iii) P {getting first ball as red}

$$= \frac{4}{9}$$

The ball is replaced after the first draw.

P (getting second ball as black)

$$= \frac{10}{18} = \frac{5}{9}$$

$\therefore$ P (getting first ball as black and second ball as red)

$$= \frac{4}{9} \times \frac{5}{9} = \frac{20}{81}$$

P (One of them is black and other is red)

= P (first ball black and second ball red) + P (first ball red and second ball black)

$$= \frac{20}{81} + \frac{20}{81}$$

$$= \frac{40}{81}.$$

Q. 3. One card is drawn at random from a well shuffled deck of 52 cards. In which of the following cards are the events E and F are independent ?

(i) E : 'the card drawn is a spade'

F : 'the card drawn is an ace'.

(ii) E : 'the card drawn is black'

F : 'the card drawn is a king'.

(iii) E : 'the card drawn is king or queen'

F : 'the card drawn is a queen or jack'. [NCERT]

Sol. (i) In a deck of 52 cards, 13 are spades and 4 cards are aces.

$$\therefore \quad P(E) = \frac{13}{52} = \frac{1}{4} = P(\text{the card drawn is a spade})$$

$$P(F) = \frac{4}{52} = \frac{1}{13} = P(\text{the card drawn is an ace})$$

In a deck of cards, only 1 card is an ace of spades.

$\therefore$ P(the card drawn is spade and ace)

$$= \frac{1}{52} = P(E \cap F)$$

$$P(E) \cdot P(F) = \frac{1}{4} \times \frac{1}{13} = \frac{1}{52}$$

Clearly, $P(E \cap F) = P(E) \cdot P(F)$.

Hence E and F are independent event.

(ii) In a deck of 52 cards, 26 cards are black and 4 cards are king.

$\therefore$ P (the card drawn is black)

$= P(E)$

$$= \frac{26}{52} = \frac{1}{2}$$

P (the card drawn is a king)

$= P(F)$

$$= \frac{4}{52} = \frac{1}{13}$$

In a deck of 52 cards 2 cards are black as well as kings

$\therefore$ $P(E \cap F) = P$ (the card drawn is a black king)

$$= \frac{2}{52} = \frac{1}{26}$$

$$P(E) \times P(F) = \frac{1}{2} \times \frac{1}{13}$$

$$= \frac{1}{26}$$

Clearly, $P(E \cap F) = P(E) \times P(F)$

Hence, E and F are independent events.

(iii) In a deck of 52 cards, 4 cards are kings, 4 cards are queens and 4 cards are jacks.

P (the card drawn is a king or queen)

$$= \frac{8}{52} = \frac{2}{13}$$
$$= P(E)$$

P (the card drawn is a queen or jack)

$$= \frac{8}{52} = \frac{2}{13}$$
$$= P(F)$$

There are 4 cards which are king or queen and queen or jack.

$\therefore$ P (the card drawn is a king / queen or queen / jack)

$$= \frac{4}{52} = \frac{1}{13}$$

i.e., $$P(E \cap F) = \frac{1}{13}$$

$$P(E) \times P(F) = \frac{2}{13} \times \frac{2}{13}$$
$$= \frac{4}{169}$$

Clearly, $P(E \cap F) \neq P(E) \cdot P(F)$

Hence, E and F are not independent events.

Q. 4. A box contains 10 batteries out of which just three are defective. If a random sample of 5 batteries is drawn, find the probabilities that the sample contains :

(i) Exactly one defective battery

(ii) Exactly two defective batteries.

(iii) No defective batteries.

Sol. Out of 10 batteries 5 can be chosen in $^{10}C_5$ ways. So, total number of elementary events = $^{10}C_5$.

(i) There are 3 defective and 7 non-defective batteries. The number of ways of selecting one defective battery out of 3 and 4 non-defective batteries out of 7 is $^3C_1 \times {}^7C_4$.

$\therefore$ Favourable number of elementary events

$$= {}^3C_1 \times {}^7C_4$$

So, required probability

$$= \frac{^3C_1 \times {}^7C_4}{^{10}C_5}$$
$$= \frac{5}{12}$$

(ii) The number of ways of selecting 2 defective batteries out of 3 and 3 non-defective batteries out of 7 is $^3C_2 \times {}^7C_3$.

$\therefore$ Favourable number of elementary events

$$= {}^3C_2 \times {}^7C_3$$

So, the required probability

$$= \frac{^3C_2 \times {}^7C_3}{^{10}C_5}$$
$$= \frac{5}{12}.$$

(iii) No defective battery means all non-defective batteries. The number of way of selecting all 5 non-defective batteries out of 7 is $^7C_5$.

$\therefore$ Favourable number of elementary events

$$= {}^7C_5.$$

So, the required probability is

$$\frac{^7C_5}{^{10}C_5} = \frac{1}{12}$$

Q. 5. In a girls hostel, 60% of the student read Hindi newspaper, 40% read English newspaper and 20% read both Hindi and English newspaper. A student is selected at random.

(i) Find the probability that she reads neither Hindi nor English newspaper.

(ii) If she reads Hindi newspaper, find the probability that she reads English newspaper.

(iii) If she reads English newspaper, find the probability that she reads Hindi newspaper.

Sol. Let H denote the event that students who read Hindi newspaper and E denote the students who read English newspaper

It is given that,

$$P(H) = 60\%$$
$$= \frac{60}{100} = \frac{3}{5}$$
$$P(E) = 40\%$$
$$= \frac{40}{100} = \frac{2}{5}$$
$$P(H \cap E) = 20\%$$
$$= \frac{20}{100} = \frac{1}{5}$$

(i) Probability that a student reads neither Hindi nor English newspaper is

$$P(H \cup E)' = 1 - P(H \cup E)$$
$$= 1 - [P(H) + P(E) - P(H \cap E)]$$

$$= 1 - \left\{\frac{3}{5} + \frac{2}{5} - \frac{1}{5}\right\}$$

$$= 1 - \frac{4}{5} = \frac{1}{5}.$$

(ii) Probability that a randomly chosen student reads English newspaper if she reads a Hindi newspaper is

$$P(E/H) = \frac{P(E \cap H)}{P(H)}$$

$$= \frac{\frac{1}{5}}{\frac{3}{5}} = \frac{1}{3}$$

(iii) Probability that a randomly chosen student reads Hindi newspaper if she reads a English newspaper is

$$P(H/E) = \frac{P(H \cap E)}{P(E)}$$

$$= \frac{\frac{1}{5}}{\frac{2}{5}} = \frac{1}{2}.$$

Q. 6. An insurance company insured 2000 scooter drivers, 4000 car driver and 6000 truck drivers. The probability of accidents are 0.01, 0.03 and 0.15 respectively. One of the insured person meets with an accident. What is the probability that he is a scooter driver?

Sol. Let A, B and C be the respective events that the driver is a scooter driver, a car driver and a truck driver.

Let E be the event that the person meets with an accident.

	Number of scooter driver	= 2000
	Number of car driver	= 4000
and	Number of truck driver	= 6000
∴	Total number of drivers	= 12000

$P(A) = P$ (driver is a scooter driver)

$$= \frac{2000}{12000} = \frac{1}{6}$$

$P(B) = P$ (driver is a car driver)

$$= \frac{4000}{12000} = \frac{1}{3}$$

$P(C) = P$ (driver is a truck driver)

$$= \frac{6000}{12000} = \frac{1}{2}$$

Now, to find that the probability that a driver meets with an accident

P (scooter driver met with an accident)

$$= P(E/A)$$

$$= 0.01 = \frac{1}{100}$$

P (car driver met with an accident)

$$= P(E/B)$$

$$= 0.03 = \frac{3}{100}$$

P (truck driver met with an accident)

$$= P(E/C)$$

$$= 0.15 = \frac{15}{100}$$

The probability that the driver is a scooter driver, who met with the accident is given by $P(A/E)$

By using Bayes' theorem, we obtain

$$P(A/E) = \frac{P(A) \cdot P(E/A)}{P(A) \cdot P(E/A) + P(B) \cdot P(E/B) + P(C) \cdot P(E/C)}$$

$$P(A/E) = \frac{\frac{1}{6} \cdot \frac{1}{100}}{\frac{1}{6} \cdot \frac{1}{100} + \frac{1}{3} \cdot \frac{3}{100} + \frac{1}{2} \cdot \frac{15}{100}}$$

$$= \frac{\frac{1}{6} \cdot \frac{1}{100}}{\frac{1}{100}\left[\frac{1}{6} + 1 + \frac{15}{2}\right]}$$

$$= \frac{1}{6} \times \frac{6}{52}$$

$$= \frac{1}{52}$$

Hence, the required probability is $\frac{1}{52}$.

Q. 7. Find the probability distribution of

(i) Number of heads in two tosses of a coin.

(ii) Number of tails in the simultaneous tosses of three coins.

(iii) Number of heads in four tosses of a coin.

[NCERT]

Sol. (i) When one coin is tossed twice, the sample space is

$$S = \{HH, HT, TH, TT\}$$

Let X represent the number of heads

$\therefore X(HH) = 2; X(HT) = 1; X(TH) = 1; X(TT) = 0$

$\therefore X$ can take the value of 0, 1 or 2.

It is known that

$$P(HH) = P(HT) = P(TH) = P(TT) = \frac{1}{4}$$

$$\therefore \quad P(X = 0) = P(TT)$$

$$= \frac{1}{4}$$

$$P(X = 1) = P(HT) + P(TH)$$

$$= \frac{1}{4} + \frac{1}{4} = \frac{2}{4}$$

$$= \frac{1}{2}$$

$$P(X=2) = P(HH)$$

$$= \frac{1}{4}$$

Thus, the required probability distribution is as follows.

X	0	1	2
$P(X)$	$\frac{1}{4}$	$\frac{1}{2}$	$\frac{1}{4}$

(ii) When three coins are tossed simultaneously, the sample space is {*HHH, HHT, HTH, HTT, THH, THT, TTH, TTT*}.

Let X represent number of tails

Then X can take value of 0, 1, 2 or 3.

$$P(X=0) = P(HHH)$$

$$= \frac{1}{8}$$

$$P(X=1) = P(HHT) + P(HTH) + P(THH)$$

$$= \frac{1}{8} + \frac{1}{8} + \frac{1}{8} = \frac{3}{8}$$

$$P(X=2) = P(HTT) + P(THT) + P(TTH)$$

$$= \frac{1}{8} + \frac{1}{8} + \frac{1}{8} = \frac{3}{8}$$

$$P(X=3) = P(TTT)$$

$$= \frac{1}{8}.$$

Thus, the probability distribution is as follows :

X	0	1	2	3
$P(X)$	$\frac{1}{8}$	$\frac{3}{8}$	$\frac{3}{8}$	$\frac{1}{8}$

(iii) When a coin is tossed four times, the sample space is

$$S = \{HHHH, HHHT, HHTH, HHTT, HTHT, HTHH, HTTH, HTTT, THHH, THHT, THTH, THTT, TTHH, TTHT, TTTH, TTTT\}$$

Let X be the random variable, which represents the number of heads. It can be seen that X can take value of 0, 1, 2, 3, or 4.

$$P(X=0) = P(TTTT)$$

$$= \frac{1}{16}$$

$$P(X=1) = P(TTTH) + P(TTHT) + P(THTT) + P(HTTT)$$

$$= \frac{1}{16} + \frac{1}{16} + \frac{1}{16} + \frac{1}{16}$$

$$= \frac{4}{16} = \frac{1}{4}$$

$$P(X=2) = P(TTHH) + P(HHTT) + P(THHT) + P(THTH) + P(HTHT) + P(HTTH)$$

$$= \frac{1}{16} + \frac{1}{16} + \frac{1}{16} + \frac{1}{16} + \frac{1}{16} + \frac{1}{16}$$

$$= \frac{6}{16} = \frac{3}{8}$$

$$P(X=3) = P(HHHT) + P(HHTH) + P(HTHH) + P(THHH)$$

$$= \frac{1}{16} + \frac{1}{16} + \frac{1}{16} + \frac{1}{16}$$

$$= \frac{4}{16} = \frac{1}{4}$$

$$P(X=4) = P(HHHH)$$

$$= \frac{1}{16}$$

Thus, the probability distribution is as follows :

X	0	1	2	3	4
$P(X)$	$\frac{1}{16}$	$\frac{1}{4}$	$\frac{3}{8}$	$\frac{1}{4}$	$\frac{1}{16}$

Q. 8. Three critics review a film. For the three critics, the odds in favour of the film are (4 : 1), (2 : 3) and (3 : 2) respectively. Find the probability that the majority is in favour of the film.

Sol. Let A, B, C denote the events that the film be favoured by the first, second and third critic respectively. Then,

$$P(A) = \frac{4}{5}; P(B) = \frac{2}{5}; P(C) = \frac{3}{5}$$

$$P(\bar{A}) = 1 - P(A)$$

$$= 1 - \frac{4}{5} = \frac{1}{5}$$

$$P(\bar{B}) = 1 - P(B)$$

$$= 1 - \frac{2}{5} = \frac{3}{5}$$

$$P(\bar{C}) = 1 - P(C)$$

$$= 1 - \frac{3}{5} = \frac{2}{5}$$

Required probability

$= P$ (1 critic favour the film) or P (2 critics favour the film)

$= P[\{A \text{ and } B \text{ and not } C\}$ or $\{A \text{ and } C \text{ and not } B\}$ or $\{B \text{ and } C \text{ and not } A\}] + P[A \text{ and } B \text{ and } C]$.

$$= P(A \cap B \cap \overline{C}) + P(A \cap \overline{B} \cap C) + P(\overline{A} \cap B \cap C) + P(A \cap B \cap C)$$

$$= \{P(A) \times P(B) \times P(\overline{C})\} + \{P(A) \times P(\overline{B}) \times P(C)\} + \{P(\overline{A}) \times P(B) \times P(C)\} + \{P(A) \times P(B) \times P(C)\}$$

$$= \left\{\frac{4}{5} \times \frac{2}{5} \times \frac{2}{5}\right\} + \left\{\frac{4}{5} \times \frac{3}{5} \times \frac{3}{5}\right\} + \left\{\frac{1}{5} \times \frac{2}{5} \times \frac{3}{5}\right\} + \left\{\frac{4}{5} \times \frac{2}{5} \times \frac{3}{5}\right\}$$

$$= \frac{16}{125} + \frac{36}{125} + \frac{6}{125} + \frac{24}{125}$$

$$= \frac{82}{125}$$

Hence, the required probability is $\frac{82}{125}$.

Q. 9. An urn contains 25 balls of which 10 balls bear a mark 'X' and the remaining 15 bear a mark 'Y'. A ball is drawn at random from the urn, its mark is noted down and it is replaced. If 6 balls are drawn in this way, find the probability that [NCERT]

(i) all will bear 'X' mark.

(ii) not more than 2 will bear 'Y' mark.

(iii) at least one ball will bear 'Y' mark

(iv) the number of balls with 'X' mark and 'Y' mark will be equal.

Sol. Total number of ball in the urn = 25

Balls bearing mark 'X' = 10

Balls bearing mark 'Y' = 15

$p = P$ (ball bearing mark 'X')

$$= \frac{10}{25} = \frac{2}{5}$$

$q = P$ (ball bearing mark 'Y')

$$= \frac{15}{25} = \frac{3}{5}$$

Six balls are drawn with replacement. Therefore, the number of trails are Bernoulli trails

Let Z be the random variable that represents the number of balls with 'Y' mark on them in the trails.

Clearly, Z has a binomial distribution with $n = 6$ and $p = \frac{2}{5}$.

$$\therefore \quad P(Z = z) = {}^nC_z p^{n-z} \cdot q^z$$

(i) P (all will bear 'X' mark)

$$= P(Z = 0)$$

$$= {}^6C_0 \left(\frac{2}{5}\right)^6 = \left(\frac{2}{5}\right)^6$$

(ii) P (not more than 2 bear 'Y' mark)

$$= P(Z \leq 2)$$

$$= P(Z = 0) + P(Z = 1) + P(Z = 2)$$

$$= {}^6C_0 \left(\frac{2}{5}\right)^6 + {}^6C_1 \left(\frac{2}{5}\right)^5 \left(\frac{3}{5}\right)^1 + {}^6C_2 \left(\frac{2}{5}\right)^4 \left(\frac{3}{5}\right)^2$$

$$= \left(\frac{2}{5}\right)^4 \left[\left(\frac{2}{5}\right)^2 + 6\left(\frac{2}{5}\right)\left(\frac{3}{5}\right) + 15\left(\frac{3}{5}\right)^2\right]$$

$$= \left(\frac{2}{5}\right)^4 \left[\frac{4}{25} + \frac{36}{25} + \frac{135}{25}\right]$$

$$= \left(\frac{2}{5}\right)^4 \left[\frac{175}{25}\right]$$

$$= 7 \cdot \left(\frac{2}{5}\right)^4.$$

(iii) P (at least one ball bears 'Y' mark)

$$= P(Z \geq 1)$$

$$= 1 - P(Z < 1)$$

$$= 1 - P(Z = 0)$$

$$= 1 - \left[{}^6C_0 \left(\frac{2}{5}\right)^6\right]$$

$$= 1 - \left(\frac{2}{5}\right)^6$$

(iv) P (equal number of ball with 'X' mark and 'Y' mark)

$$= P(Z = 3)$$

$$= {}^6C_3 \left(\frac{2}{5}\right)^3 \left(\frac{3}{5}\right)^3$$

$$= \frac{20 \times 8 \times 27}{15625}$$

$$= \frac{864}{3125}$$

Q. 10. Coloured gifts are placed in three shelves as shown in the following table :

Shelf	Colour of the Gift		
	Gold	Silver	Red
I	1	2	3
II	2	4	1
III	4	5	3

A shelf is selected at random and then two gifts are randomly taken from the selected shelf. They happen to be Gold and red. What is the probability that they are taken from shelf I ?

Sol. Let E_1, E_2, E_3 be the following events :

E_1 = Shelf I is selected

E_2 = Shelf II is selected

E_3 = Shelf III is selected

Let A be the event of getting a gold and red gift.

$$P(E_1) = P(E_2) = P(E_3) = \frac{1}{3}$$

$$P(A/E_1) = \frac{{}^1C_1 \times {}^3C_1}{{}^6C_2}$$

$$= \frac{1\times 3}{\frac{6\times 5}{2\times 1}}$$

$$= \frac{3}{1}\times\frac{2}{30} = \frac{1}{5}$$

$$P(A/E_2) = \frac{{}^2C_1 \times {}^1C_1}{{}^7C_2}$$

$$= \frac{2\times 1}{\frac{7\times 6}{2\times 1}}$$

$$= \frac{2}{1}\times\frac{2}{42} = \frac{2}{21}$$

$$P(A/E_3) = \frac{{}^4C_1 \times {}^3C_1}{{}^{12}C_2} = \frac{4\times 3}{\frac{12\times 11}{2\times 1}}$$

$$= \frac{12}{1}\times\frac{2}{132} = \frac{2}{11}$$

Now using Bayes' theorem,

$$P(E_1/A) = \frac{P(E_1)\cdot P(A/E_1)}{P(E_1)\cdot P(A/E_1) + P(E_2) \cdot P(A/E_2) + P(E_3)\cdot P(A/E_3)}$$

$$= \frac{\frac{1}{3}\times\frac{1}{5}}{\left(\frac{1}{3}\times\frac{1}{5}\right)+\left(\frac{1}{3}\times\frac{2}{21}\right)+\left(\frac{1}{3}\times\frac{2}{11}\right)}$$

$$= \frac{\frac{1}{3\times 5}}{\frac{1}{3}\left[\frac{1}{5}+\frac{2}{21}+\frac{2}{11}\right]}$$

$$= \frac{1}{5}\times\frac{1}{\left[\frac{1}{5}+\frac{2}{21}+\frac{2}{11}\right]}$$

$$= \frac{231}{551}$$

Q. 11. A card from a pack of 52 cards is lost. From the remaining cards of the pack, two cards are drawn at random and are found to be both hearts. Find the probability of the lost card being a heart.

Sol. Let A be the event that the lost card is a heart, B be the event that the card is not a heart.

$$P(A) = \frac{13}{52} = \frac{1}{4},$$

$$P(B) = \frac{39}{52} = \frac{3}{4}$$

Let 'C' be the event that two cards drawn from the remaining pack are hearts.

$$P(C/A) = \frac{{}^{12}C_2}{{}^{51}C_2}$$

(Since if one lost card is of diamond then we have 12 remaining diamonds out of 51 cards now.)

$$= \frac{\frac{12\times 11}{2\times 1}}{\frac{51\times 50}{2\times 1}}$$

$$= \frac{12\times 11}{51\times 50} = \frac{132}{2550}$$

$$P(C/B) = \frac{{}^{13}C_2}{{}^{51}C_2}$$

$$= \frac{\frac{13\times 12}{2\times 1}}{\frac{51\times 50}{2\times 1}} = \frac{13\times 12}{51\times 50}$$

$$= \frac{156}{2550}$$

By Bayes' theorem,

$\therefore$ P (that the lost card being a heart)

$$= P(A/C)$$

$$= \frac{P(A)\cdot P(C/A)}{P(A)\cdot P(C/A) + P(B)\cdot P(C/B)}$$

$$= \frac{\frac{1}{4}\times\frac{132}{2550}}{\left(\frac{1}{4}\times\frac{132}{2550}\right)+\left(\frac{3}{4}\times\frac{156}{2550}\right)}$$

$$= \frac{132}{132+468}$$

$$= \frac{132}{600} = \frac{11}{50}$$

Q. 12. Given three identical bags contain 2 balls each. In bag I, both balls are red, in bag II, both are blue and in bag III, there is one red and one blue balls. A person chooses a bag at random and takes out a ball. If a ball is of red colour, what is the probability that the other ball in the bag is also of red colour?

Sol. Let A be the event that bag I is chosen,

B be the event that bag II is chosen,

C be the event that bag III is chosen,

Let 'E' be the event that ball chosen is of red colour.

We have,

$$P(A) = P(B) = P(C) = \frac{1}{3}$$

$P(E/A)$ = Probability of drawing a red ball from

bag I $= \frac{2}{2} = 1$

$P(E/B)$ = Probability of drawing a red ball from

bag II $= 0$

$P(E/C)$ = Probability of drawing a red ball from

bag III $= \frac{1}{2}$

$\therefore$ Probability that the other ball in the bag is of red colour = Probability that red ball is drawn from the bag I

$= P(A/E)$

$$P(A/E) = \frac{P(A)\cdot P(E/A)}{P(A)\cdot P(E/A) + P(B)\cdot P(E/B) + P(C)\cdot P(E/C)}$$

$$= \frac{\frac{1}{3}\times 1}{\frac{1}{3}\times 1 + \frac{1}{3}\times 0 + \frac{1}{3}\times\frac{1}{2}}$$

$$= \frac{1}{1+\frac{1}{2}} = \frac{2}{3}$$

Q. 13. Find the mean, variance and standard deviation of the number of fours in two tosses of a die.

Sol. In a single toss, we have

$$P\text{ (getting a four)} = \frac{1}{6}$$

$$P\text{ (getting a non-four)} = \left(1-\frac{1}{6}\right) = \frac{5}{6}$$

Let X denote the number of fours in two tosses.

Then, clearly

$$X = \{0, 1, 2\}$$

$P(X=0) = P$ [(non-four in the first draw) and (non-four in the second draw)]

$= P$ [(non-four in the first draw) $\times P$ (non-four in the second draw)]

$$= \left(\frac{5}{6}\times\frac{5}{6}\right) = \frac{25}{36}$$

$P(X=1) = P$ [(four in the first draw) and (non-four in the second draw) or (non-four in the first draw) and (four in the second draw)]

$= P$ [(four in the first draw and non-four in the second draw)] + P [(non-four in the first draw and four in the second draw)]

$$= \left(\frac{1}{6}\times\frac{5}{6}\right)+\left(\frac{5}{6}\times\frac{1}{6}\right)$$

$$= \left[\frac{5}{36}+\frac{5}{36}\right]$$

$$= \frac{10}{36} = \frac{5}{18}$$

$P(X=2) = P$ (four in the first draw) and (four in the second draw)]

$= P$ [(four in the first draw) $\times P$ (four in the second draw)]

$$= \frac{1}{6}\times\frac{1}{6} = \frac{1}{36}$$

Hence, the probability distribution is given by

$X = x_i$	0	1	2
p_i	$\frac{25}{36}$	$\frac{5}{18}$	$\frac{1}{36}$

$\therefore$ Mean,

$$\mu = \Sigma x_i p_i$$

$$= 0\times\frac{25}{36} + 1\times\frac{5}{18} + 2\times\frac{1}{36}$$

$$= \frac{10}{36} + \frac{2}{36}$$

$$= \frac{12}{36} = \frac{6}{18} = \frac{1}{3}$$

Variance,

$$\sigma^2 = \Sigma x_i^2 p_i - \mu^2$$

$$= \left[\left(0 \times \frac{25}{36}\right) + \left(1 \times \frac{5}{18}\right) + \left(4 \times \frac{1}{36}\right) - \frac{1}{9}\right]$$

$$= \frac{5}{18} + \frac{1}{9} - \frac{1}{9} = \frac{5}{18}$$

Standard deviation,

$$\sigma = \sqrt{\frac{5}{18}} = \frac{1}{3} \cdot \sqrt{\frac{5}{2}}$$

Q. 14. Consider the experiment of tossing a coin. If the coin shows tail, toss it again but if it shows head, then throw a die. Find the conditional probability of the event that 'the die shows a number greater than 4' given that 'there is atleast one head'.

Sol. The outcomes of the experiment can be represented in the following diagrammatic manner called the 'tree diagram'.

The sample space of the experiment may be described as

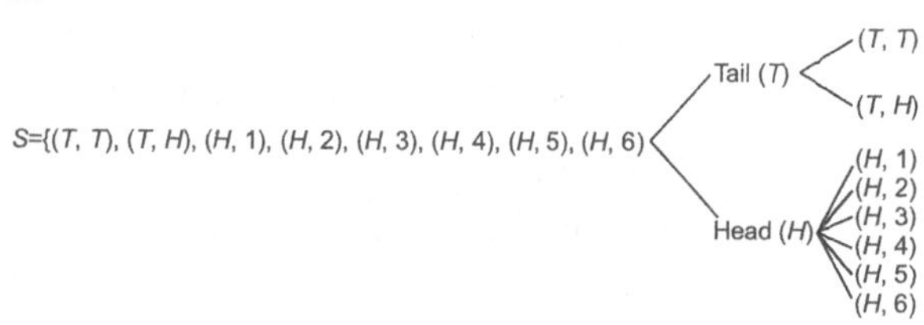

where (T, T) denote that both the tosses result into tail and (H, i) denote the first toss result into a head and the number 'i' appeared on the die for

$$i = 1, 2, 3, 4, 5, 6$$

Thus, the probability assigned to the 8 elementary events (T, T), (T, H), $(H, 1)$, $(H, 2)$, $(H, 3)$, $(H, 4)$, $(H, 5)$, $(H, 6)$, are $\frac{1}{4}, \frac{1}{4}, \frac{1}{12}, \frac{1}{12}, \frac{1}{12}, \frac{1}{12}, \frac{1}{12}, \frac{1}{12}$ respectively which is clear from the figure below.

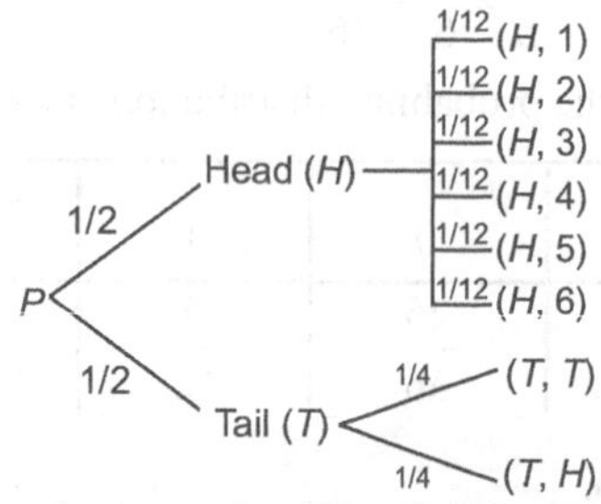

Let 'A' be the event that 'there is atleast one head' and 'B' be the event that 'the die shows a number greater than 4'.

Then

$$A = \{(T, H), (H, 1), (H, 2), (H, 3), (H, 4), (H, 5)\\ (H, 6)\}$$

$$B = \{(H, 5), (H, 6)\}$$

and $A \cap B = \{(H, 5), (H, 6)\}$

$$P(A) = P[(T, H)] + P[(H, 1)] + P[(H, 2)] + P[(H, 3)] + P[(H, 4)] + P[(H, 5)] + P[(H, 6)]$$

$$= \frac{1}{4} + \frac{1}{12} + \frac{1}{12} + \frac{1}{12} + \frac{1}{12} + \frac{1}{12} + \frac{1}{12}$$

$$= \frac{1}{4} + \frac{6}{12}$$

$$= \frac{1}{4} + \frac{1}{2} = \frac{3}{4}$$

and $\quad P(A \cap B) = P[\{(H, 5)\}] + P[\{(H, 6)\}]$

$$= \frac{1}{12} + \frac{1}{12}$$

$$= \frac{1}{6}$$

Hence $\quad P(B/A) = \frac{P(A \cap B)}{P(A)}$

$$= \frac{\frac{1}{6}}{\frac{3}{4}} = \frac{1}{6} \times \frac{4}{3}$$

$$= \frac{2}{9}$$

Q. 15. Jagan, Magan and Gagan take part in a game. If Jagan wins the game 4 times out of 5; Magan wins the game 3 out of 4 times and Gagan wins it 2 out of 3 times, What is the probability that the game is won by at least 2 persons ?

Sol. Let A, B and C be the events that Jagan wins the game, Magan wins the game and Gagan wins the game respectively.

then $\quad P(A) = \frac{4}{5}; P(B) = \frac{3}{4}$ and $P(C) = \frac{2}{3}$.

$$P(\overline{A}) = 1 - \frac{4}{5} = \frac{1}{5}$$

$$P(\overline{B}) = 1 - \frac{3}{4} = \frac{1}{4}$$

and $\quad P(\overline{C}) = 1 - \frac{2}{3} = \frac{1}{3}$

Case I : Jagan, Magan and Gagan all win the game.

In this case, $P(A \text{ and } B \text{ and } C)$

$$= P(A) \cdot P(B) \cdot P(C)$$

[$\because$ A, B and C are independent]

$$=\left[\frac{4}{5}\times\frac{3}{4}\times\frac{2}{3}\right]$$

$$=\frac{2}{5}$$

Case II : Jagan and Magan win but not Gagan.

In this case, P (A and B but not C)

$$=P(A\cap B\cap C)$$

$$=P(A)\cdot P(B)\cdot P(\overline{C})$$

($\because$ A, B and $\overline{C}$ are independent)

$$=\left[\frac{4}{5}\times\frac{3}{4}\times\frac{1}{3}\right]$$

$$=\frac{1}{5}$$

Case III : Jagan and Gagan win but not Magan

In this case , P (A and C and not B)

$$=P(A\cap \overline{B}\cap C)$$

$$=P(A)\cdot P(\overline{B})_3 P(C)$$

($\because$ A, $\overline{B}$ and C are independent)

$$=\left[\frac{4}{5}\times\frac{2}{3}\times\frac{1}{4}\right]$$

$$=\frac{2}{15}$$

Case IV : Magan and Gagan both win but not Jagan.

In this case, P (B & C and not A)

$$=P(B\cap C\cap \overline{A})$$

$$=P(B)\cdot P(C)\cdot P(\overline{A})$$

($\because$ B, C and $\overline{A}$ are independent)

$$=\left[\frac{3}{4}\times\frac{2}{3}\times\frac{1}{5}\right]$$

$$=\frac{1}{10}$$

Clearly, all these are mutually exclusive.

Hence, required probability $=\left(\frac{2}{5}+\frac{1}{5}+\frac{2}{15}+\frac{1}{10}\right)=\frac{5}{6}$

Q. 16. Five bad oranges are accidentally mixed with 20 good ones. If four oranges are drawn one by one successively with replacement, then find the probability distribution of number of bad oranges drawn. Hence, find themean and varianceof thedistribution.★

Sol. Let X denote the number of bad oranges in a draw of 4 oranges from a group of 20 good oranges and 5 bad oranges.

★ are board exam questions from previous years

Since there are 5 bad oranges in the group, therefore X can take values 0, 1, 2, 3, 4.

Now,

$P(X=0)$ = Probability of getting no bad oranges

= Probability of getting 4 good oranges

$$\therefore\ P(X=0)=\left(\frac{20}{25}\right)^4\cdot {}^4C_0$$

$P(X=1)=P$ (getting 1 bad orange)

$=P$ (getting 3 good oranges)

$$\therefore\ P(X=1)=\frac{5}{25}\times\left(\frac{20}{25}\right)^3\cdot {}^4C_1$$

$P(X=2)=P$ (getting 2 bad oranges)

$=P$ (getting 2 good oranges)

$$=\left(\frac{5}{25}\right)^2\times\left(\frac{20}{25}\right)^2\cdot {}^4C_2$$

$P(X=3)=P$ (getting 3 bad oranges)

$=P$ (getting 1 good orange)

$$=\left(\frac{5}{25}\right)^3\times\left(\frac{20}{25}\right)\cdot {}^4C_3$$

$P(X=4)=P$ (getting 4 bad oranges)

$$=\left(\frac{5}{25}\right)^4\cdot {}^4C_4$$

Computation of mean and variance

x_i	0	1	2	3	4
p_i	$\frac{256}{625}$	$\frac{256}{625}$	$\frac{96}{625}$	$\frac{16}{625}$	$\frac{1}{625}$
x_ip_i	0	$\frac{256}{625}$	$\frac{192}{625}$	$\frac{48}{625}$	$\frac{4}{625}$
x_i^2	0	1	4	9	16
$p_ix_i^2$	0	$\frac{256}{625}$	$\frac{384}{625}$	$\frac{144}{625}$	$\frac{16}{625}$

We have,

$$\Sigma p_i x_i=\frac{500}{625}=\frac{4}{5}$$

and $$\Sigma p_ix_i^2=\frac{800}{625}=\frac{32}{25}$$

$$\therefore\quad \overline{X}=\text{Mean}=\Sigma p_ix_i=\frac{4}{5}$$

and $\text{var}(X)=\Sigma p_ix_i^2-(\Sigma p_ix_i)^2$

$$=\frac{32}{25}-\frac{16}{25}=\frac{16}{25}$$

Hence, mean $=\frac{4}{5}$ and variance $=\frac{16}{25}$.

Q. 17. An urn contains 3 white and 6 red balls. Four balls are drawn one by one with replacement from the urn. Find the probability distribution of the number of red balls drawn. Also find mean and variance of the distribution.★

Sol. Let X denote the total number of red balls when four balls are drawn one by one with replacement.

P (getting a red ball in one draw)

$$= \frac{6}{9} = \frac{2}{3}$$

P (getting a white ball in one draw)

$$= \frac{1}{3}$$

X	0	1	2	3	4
$P(X)$	$\left(\frac{1}{3}\right)^4$	$\frac{2}{3}\left(\frac{1}{3}\right)^3 \times {}^4C_1$	$\left(\frac{2}{3}\right)^2 \times \left(\frac{1}{3}\right)^2 \times {}^4C_2$	$\left(\frac{2}{3}\right)^3 \times \left(\frac{1}{3}\right) \times {}^4C_3$	$\left(\frac{2}{3}\right)^4$
	$\frac{1}{81}$	$\frac{8}{81}$	$\frac{24}{81}$	$\frac{32}{81}$	$\frac{16}{81}$

Using the formula for mean, we have

$$\overline{X} = \Sigma P_i X_i$$

$$= \left(0 \times \frac{1}{81}\right) + \left(1 \times \frac{8}{81}\right) + \left(2 \times \frac{24}{81}\right) + \left(3 \times \frac{32}{81}\right) + \left(4 \times \frac{16}{81}\right)$$

$$= \frac{1}{81}[0 + 8 + 48 + 96 + 64]$$

$$= \frac{216}{81} = \frac{8}{3}$$

$$\therefore \text{ Mean } \overline{X} = \frac{8}{3}$$

Using the formula for variance, we have

$$\text{var}(X) = \Sigma P_i X_i^2 - (\Sigma P_i X_i)^2$$

$$= \left\{\left(0 \times \frac{1}{81}\right) + \left(1 \times \frac{8}{81}\right) + \left(4 \times \frac{24}{81}\right) + \left(9 \times \frac{32}{81}\right) + \left(16 \times \frac{16}{81}\right)\right\} - \left(\frac{8}{3}\right)^2$$

$$= \frac{1}{81}\{8 + 96 + 288 + 256\} - \frac{64}{9}$$

$$\text{var}(X) = \frac{648}{81} - \frac{64}{9}$$

$$= \frac{8}{9}$$

Hence, for the obtained distribution mean is $\frac{8}{3}$ and variance is $\frac{8}{9}$.

Q. 18. If A and B are two independent events such that $P(\overline{A} \cap B) = \frac{2}{15}$ and $P(A \cap \overline{B}) = \frac{1}{6}$, then find $P(A)$ and $P(B)$.★

Sol. Let $P(A) = x$ and $P(B) = y$

$$P(\overline{A}) = 1 - P(A) = 1 - x, \; P(\overline{B}) = 1 - y$$

We have,

$$P(\overline{A} \cap B) = \frac{2}{15}$$

and $$P(A \cap \overline{B}) = \frac{1}{6}$$

Now $$P(\overline{A} \cap B) = \frac{2}{15}$$

$$P(\overline{A})P(B) = \frac{2}{15}$$

$$(1 - x)y = \frac{2}{15} \quad \text{...(i)}$$

Since, A and B are independent events so are $\overline{A}$ and B as well as A and $\overline{B}$

Given, $$P(A \cap \overline{B}) = \frac{1}{6}$$

$$P(A)P(\overline{B}) = \frac{1}{6}$$

$$x(1 - y) = \frac{1}{6}$$

$$x = \frac{1}{6 - 6y} \quad \text{...(ii)}$$

Putting the value of x in equation (i), we get

$$\left(1 - \frac{1}{6 - 6y}\right) y = \frac{2}{15}$$

$$-90y^2 + 87y = 12$$

or $$30y^2 - 29y + 4 = 0$$

Solving the quadratic equation, we get

$$y = \frac{4}{5}, \; y = \frac{1}{6}$$

For $y = \frac{4}{5}$; using (ii), we get

$$x = \frac{1}{6 - 6 \times \frac{4}{5}} = \frac{5}{30 - 24} = \frac{5}{6}$$

For $y = \frac{1}{6}$ using (ii), we get

$$x = \frac{1}{6 - 6 \times \frac{1}{6}} = \frac{1}{5}$$

$$\therefore \quad P(A) = \frac{5}{6}, \; P(B) = \frac{4}{5}$$

★ are board exam questions from previous years

or $$P(A) = \frac{1}{5}, P(B) = \frac{1}{6}$$

Q. 19. From a lot of 15 bulbs which include 5 defectives, a sample of 4 bulbs is drawn one by one with replacement. Find the probability distribution of number of defective bulbs. Hence find the mean of the distribution.★

Sol. Let D be the event of drawing a defective bulb and X denote the variable showing the number of defective bulbs in 4 draws. Then

$$P(D) = \frac{5}{15} = \frac{1}{3}$$

$$\Rightarrow \quad P(\bar{D}) = 1 - \frac{1}{3} = \frac{2}{3}$$

The drawn bulb is replaced.

Hence X takes values 0, 1, 2, 3 and 4.

$\therefore$ $P(X=0) = P$(Getting no defective bulb)

$$= {}^4C_0\left(\frac{1}{3}\right)^0\left(\frac{2}{3}\right)^4 = \left(\frac{2}{3}\right)^4 = \frac{16}{81}$$

$$P(X=1) = {}^4C_1\left(\frac{1}{3}\right)\left(\frac{2}{3}\right)^3 = \frac{32}{81}$$

$$P(X=2) = {}^4C_2\left(\frac{1}{3}\right)^2\left(\frac{2}{3}\right)^2 = \frac{24}{81}$$

$$P(X=3) = {}^4C_3\left(\frac{1}{3}\right)^3\left(\frac{2}{3}\right) = \frac{8}{81}$$

$$P(X=4) = {}^4C_4\left(\frac{1}{3}\right)^4 = \frac{1}{81}$$

$\therefore$ The probability distribution is

X	0	1	2	3	4
$P(X=x)$	$\frac{16}{81}$	$\frac{32}{81}$	$\frac{24}{81}$	$\frac{8}{81}$	$\frac{1}{81}$

Mean of the distribution $= \Sigma[P_iX_i]$

$$= \frac{16}{81}\times 0 + \frac{32}{81}\times 1 + \frac{24}{81}\times 2 + \frac{8}{81}\times 3 + \frac{1}{81}\times 4$$

$$= \frac{1}{81}(0+32+48+24+4)$$

$$= \frac{108}{81} = \frac{4}{3}$$

Q. 20. In a factory which manufactures bolts, machine *A*, *B* and *C* manufacture respectively 30%, 50% and 20% of the bolts. Of their outputs 3, 4 and 1 percent respectively are defective bolts. A bolt is drawn at random from the product and is found to be defective. Find the probability that this is not manufactured by machine *B*.★

Sol. Let E_1, E_2, E_3 and A be the events defined as below :

E_1 = the bolt is manufactured by machine A

E_2 = the bolt is manufactured by machine B

E_3 = the bolt is manufactured by machine C

A = the bolt is defective

then $P(E_1)$ = Probability that the bolt drawn is manufactured by machine $A = \frac{30}{100}$

$P(E_2)$ = Probability that the bolt drawn is manufactured by machine $B = \frac{50}{100}$

$P(E_3)$ = Probability that the bolt drawn is manufactured by machine $C = \frac{20}{100}$

$P(A/E_1)$ = Probability that the bolt drawn is defective given that it is manufactured by machine A.

$$P(A/E_1) = \frac{3}{100}$$

Similarly, we have

$$P(A/E_2) = \frac{4}{100}$$

and $$P(A/E_3) = \frac{1}{100}$$

Now, using Bayes' theorem

$P(E_2/A)$ = Probability that the bolt is manufactured by machine B given that the bolt drawn is defective.

$$= \frac{P(E_2)P(A/E_2)}{P(E_1)P(A/E_1) + P(E_2)P(A/E_2) + P(E_3)P(A/E_3)}$$

$$= \frac{\frac{50}{100}\times\frac{4}{100}}{\frac{30}{100}\times\frac{3}{100} + \frac{50}{100}\times\frac{4}{100} + \frac{20}{100}\times\frac{1}{100}}$$

$$= \frac{200}{90+200+20}$$

$$= \frac{200}{310} = \frac{20}{31}$$

Hence, the probability that this is not manufactured by Machine $B = 1 - P(E_2/A)$

$$= 1 - \frac{20}{31} = \frac{11}{31}$$

★ are board exam questions from previous years

Practice Exercise

Chapter 1. Relations and Functions

1. Is $y = f(x) = \frac{x^2}{1+x^2}$ the one-one and onto function provided $f: R \to R$?
2. Prove that the relation R on Q (the set of rational numbers) defined by $(a, b) \in R \to 1 + ab > 0$ is not an equivalence relation.
3. Let $f = \{(1, 3), (2, 6), (4, 12)\}$ and $g = \{(3, 7), (6, 8), (12, 9), (10, 13)$. Find gof.
4. Verify $ho(gof) = (hog)of$ for the following three mappings :
   $f: N \to Z_0$ (the set of non-zero integers), $g: Z_0 \to Q$ and $h: Q \to R$, given by
   $$f(x) = 2x, g(x) = \frac{1}{x}, h(x) = e^x.$$
5. If f, g, h are functions from R to R such that $f(x) = x^2 - 1$, $g(x) = \sqrt{x^2+1}$, $h(x) = \begin{cases} 0, \text{ if } x \le 0 \\ x, \text{ if } x \ge 0 \end{cases}$ then find the composite function $ho(fog)$ and determine whether the function fog is invertible and whether function h is the identity function.
6. Let R be the relation 'is perpendicular to' the set of all straight lines in the plane. Is R reflexive, symmetric or transitive ?
7. If the function $f: R \to R$ be given by $f(x) = x^2 + 2$ and $g: R \to R$ be given by $g(x) = \frac{x}{x-1}$, find fog and gof.
8. If $f: Q \to Q$ and $g: Q \to Q$ are two functions defined by $f(x) = 2x$ and $g(x) = x + 2$, show that f and g are bijective and verify that $(gof)^{-1} = f^{-1}og^{-1}$.
9. Show that the function $f: R \to \{x \in R: -1 < x < 1\}$ defined by $f(x) = \frac{x}{1+|x|}$, $x \in R$ is one-one and onto function.
10. Prove that the relation R in the set $A = \{1, 2, 3, \ldots, 12\}$ given by $R = \{(a, b) : |a - b| \text{ is divisible by } 3\}$, is an equivalence relation. Find all elements related to the element 1.
11. Show that the function f in $A = R - \left\{\frac{2}{3}\right\}$ defined as $f(x) = \frac{4x+3}{6x-4}$ is one-one and onto. Hence, find f^{-1}.★
12. Let $A = \{-1, 0, 1, 2\}, B = \{-4, -2, 0, 2\}$ and $f, g: A \to B$ be functions defined by
    $$f(x) = x^2 - x, x \in A \text{ and } g(x) = 2\left|x - \frac{1}{2}\right| - 1, x \in A.$$
    Are f and g equal ? Justify your answer.

★ are board exam questions from previous years

Answers

1. The function is neither one-one nor onto.
2. R is reflexive; R is symmetric; R is not transitive.
3. $\{(1, 7), (2, 8), (4, 9)\}$
5. $ho(fog) = x^2$; fog is not invertible; h is not the identity function.
6. R is neither reflexive nor transitive, R is symmetric.
7. $(fog)(x) = \frac{x^2}{(x-1)^2} + 2$; $(gof)(x) = \frac{x^2+2}{x^2+1}$
10. 1, 4, 7, 10.
11. $f^{-1}(x) = \frac{4x+3}{6x-4}$
12. f and g are equal.

Chapter 2. Inverse Trigonometric Functions

1. Find the value of the expression
   $\sin[\cot^{-1}\{\cos(\tan^{-1})\}]$ **[NCERT]**
2. Find (with reasons) which of the following functions are not defined :
   (i) $\operatorname{cosec}^{-1}\left(\frac{1}{2}\right)$ (ii) $\tan^{-1}(1)$
   (iii) $\cos^{-1}\left(-\frac{3}{2}\right)$ (iv) $\sin^{-1}\left(\frac{1\cdot7}{1\cdot8}\right)$
   (v) $\cot^{-1}(7)$ (vi) $\sec^{-1}\left(\frac{4}{5}\right)$
3. Find the value of : $\tan^{-1}\left(\frac{1}{2}\right) + \tan^{-1}\left(\frac{1}{3}\right)$.
4. Evaluate : $4\left[2\tan^{-1}\left(\frac{1}{3}\right) + \tan^{-1}\left(\frac{1}{7}\right)\right]$.
5. Prove that : $\tan^{-1}1 + \tan^{-1}2 + \tan^{-1}3 = \pi$
   $$= 2\left[\tan^{-1}\frac{1}{2} + \tan^{-1}\frac{1}{3} + \tan^{-1}1\right].$$
6. Prove that : $\left[\tan^{-1}\left(\frac{1}{4}\right) + \tan^{-1}\left(\frac{2}{9}\right)\right] = \cos^{-1}\left(\frac{2}{\sqrt{5}}\right)$.
7. Show that : $\left[\sin^{-1}\left(\frac{\sqrt{3}}{2}\right) + 2\tan^{-1}\left(\frac{1}{\sqrt{3}}\right)\right] = \frac{2\pi}{3}$.
8. Prove that : $\left[\sin^{-1}\left(\frac{1}{\sqrt{10}}\right) + \sin^{-1}\left(\frac{3}{\sqrt{10}}\right)\right] = \frac{\pi}{2}$.

9. Prove that : $\left[\cos^{-1}\left(\frac{3}{5}\right)+\cos^{-1}\left(\frac{4}{5}\right)\right]=\frac{\pi}{2}$.

10. Prove that : $\sin(\cos^{-1}x)=\cos(\sin^{-1}x)$.

11. Prove that :

$$\sin^{-1}(\cos(\sin^{-1}x))+\cos^{-1}(\sin(\cos^{-1}x))=\frac{\pi}{2}.$$

12. Prove that : $4(\cot^{-1}3+\operatorname{cosec}^{-1}\sqrt{5})=\pi$.

13. Prove that : $\sin^{-1}\frac{4}{5}+\cos^{-1}\frac{2}{\sqrt{5}}=\cot^{-1}\frac{2}{11}$.

14. Find the simplest value of :

$$2\tan^{-1}\left(\frac{1+x}{1-x}\right)+\sin^{-1}\left(\frac{1-x^2}{1+x^2}\right).$$

15. Find the value of x, if

$$\sin^{-1}\left(\frac{2a}{1+a^2}\right)-\cos^{-1}\left(\frac{1-b^2}{1+b^2}\right)=2\tan^{-1}x.$$

16. Solve for x : $\sin^{-1}x+\sin^{-1}(1-x)=\cos^{-1}x$.

17. Solve for x : $\sin^{-1}\frac{5}{x}+\sin^{-1}\frac{12}{x}=\frac{\pi}{2}$.

18. If $\tan^{-1}x+\tan^{-1}y+\tan^{-1}z=\pi$, prove that : $x+y+z=xyz$.

19. If $\tan^{-1}x+\tan^{-1}y+\tan^{-1}z=\frac{\pi}{2}$, prove that : $xy+yz+zx=1$.

20. If $\cos^{-1}x+\cos^{-1}y+\cos^{-1}z=\pi$, prove that : $x^2+y^2+z^2+2xyz=1$.

21. If $\sec^{-1}x=\operatorname{cosec}^{-1}y$, show that : $\frac{1}{x^2}+\frac{1}{y^2}=1$.

22. Simplify : $\cos(2\cos^{-1}x+\sin^{-1}x)$ when $x=\frac{1}{5}$, where

$$0\le\cos^{-1}x\le\pi \text{ and } -\frac{\pi}{2}\le\sin^{-1}x\le\frac{\pi}{2}.$$

23. Solve : $\sin^{-1}\left(\frac{2a}{1+a^2}\right)-\sin^{-1}\left(\frac{2b}{1+b^2}\right)=\cos^{-1}\frac{1-x^2}{1+x^2}$.

24. Solve :

$$3\sin^{-1}\left(\frac{2a}{1+a^2}\right)-5\cos^{-1}\left(\frac{1-a^2}{1+a^2}\right)+4\tan^{-1}\left(\frac{2a}{1+a^2}\right)=\frac{2\pi}{3}.$$

25. If two angles of a triangle are $\tan^{-1}2$ and $\tan^{-1}3$, then find the third angle.

26. If a, b and c are the lengths opposite to angles A, B and C respectively of a triangle and angle C is a right angle, then prove that :

$$\tan^{-1}\left(\frac{a}{b+c}\right)+\tan^{-1}\left(\frac{b}{c+a}\right)=\frac{\pi}{4}.$$

★ are board exam questions from previous years

Answers

1. $\sqrt{\frac{2}{3}}$

2. (i) Not defined (ii) Defined (iii) Not defined (iv) Defined (v) Defined (vi) Not defined

3. $\frac{\pi}{4}$ 4. π

14. π 15. $\frac{a+b}{1-ab}$

16. $x=0$ or $\frac{1}{2}$ 17. $x=13$

22. $-\frac{2\sqrt{6}}{5}$ 23. $\frac{a-b}{1+ab}$

24. $\frac{1}{\sqrt{3}}$ 25. $\tan^{-1}1$

Chapter 3. Matrices

1. If a matrix has 24 elements, then what are the possible orders it can have ? What if it has 13 elements ? **[NCERT]**

2. Write the value of $a-2b$ if

$$\begin{bmatrix} a+4 & 3b \\ 8 & -6 \end{bmatrix}=\begin{bmatrix} 2a+2 & b+2 \\ 8 & a-8b \end{bmatrix}$$

3. Construct a matrix of order 2×2 whose elements are given by $a_{ij}=\frac{(i-2j)^2}{2}$. **[NCERT]**

4. Simplify★ : $\cos\theta\begin{bmatrix} \cos\theta & \sin\theta \\ -\sin\theta & \cos\theta \end{bmatrix}+\sin\theta\begin{bmatrix} \sin\theta & -\cos\theta \\ \cos\theta & \sin\theta \end{bmatrix}$

5. If $2\begin{bmatrix} 3 & 4 \\ 5 & x \end{bmatrix}+\begin{bmatrix} 1 & y \\ 0 & 1 \end{bmatrix}=\begin{bmatrix} 7 & 0 \\ 10 & 5 \end{bmatrix}$, then find $(x-y)$.★

6. If $A=\begin{bmatrix} \cos\alpha & -\sin\alpha \\ \sin\alpha & \cos\alpha \end{bmatrix}$ and $A+A'=I$, then find the value of α. **[NCERT]**

7. If $\begin{bmatrix} 1 & -\sin\theta \\ \sin\theta & 1 \end{bmatrix}\begin{bmatrix} 1 & -\sin\theta \\ \sin\theta & 1 \end{bmatrix}^{-1}=\begin{bmatrix} a & -b \\ b & a \end{bmatrix}$, then find the values of a and b. **[HOTS]**

8. For the matrix $A=\begin{bmatrix} 1 & 5 \\ 6 & 7 \end{bmatrix}$, verify that $A+A'$ is a symmetric matrix. **[NCERT]**

9. Find x from the matrix equation

$$\begin{bmatrix} 1 & 2 \\ 7 & 5 \end{bmatrix}\begin{bmatrix} x \\ 2 \end{bmatrix} = \begin{bmatrix} 5 \\ 17 \end{bmatrix}$$

10. If $A^T = \begin{bmatrix} 3 & 4 \\ -4 & 12 \\ 0 & 4 \end{bmatrix}$ and $B = \begin{bmatrix} -7 & 12 & 1 \\ 1 & 2 & 23 \end{bmatrix}$ then find $A^T - B^T$.

11. If $A' = \begin{bmatrix} 1 & 2 \\ 3 & 1 \end{bmatrix}$ and $B = \begin{bmatrix} -5 & 1 \\ 0 & 1 \end{bmatrix}$, then find $(2A+B)'$.

12. If $A = \begin{bmatrix} 5 & 4 \\ 0 & -2 \\ 3 & 6 \end{bmatrix}$ and $B = \begin{bmatrix} 1 & 2 \\ 3 & -4 \\ 6 & -5 \end{bmatrix}$ then find the matrix X, such that $3A + 2X = 4B$.

13. Find a, b, c, d if

$$\begin{bmatrix} a+b & 4 \\ 3 & c+d \end{bmatrix} + \begin{bmatrix} 6 & a \\ 2d & -1 \end{bmatrix} = \begin{bmatrix} 3b & 3a \\ 3d & 3c \end{bmatrix}$$

14. If A is 3×4 matrix and A^TB and BA^T are defined, then what is the order of matrix B ?

15. Find X and Y if,

$$2X + 5Y = \begin{bmatrix} 2 & 7 \\ 0 & 1 \end{bmatrix} \text{ and } 3X + 2Y = \begin{bmatrix} 7 & -1 \\ 1 & 0 \end{bmatrix}$$

16. If $A = \begin{bmatrix} 1 & 2 \\ -1 & 2 \end{bmatrix}$, $B = \begin{bmatrix} 0 & 4 \\ 5 & 2 \end{bmatrix}$ and $C = \begin{bmatrix} 1 & 2 \\ 3 & -5 \end{bmatrix}$, then show that matrices A, B and C satisfies the property of associativity with respect to multiplication *i.e.*, $A(BC) = (AB)C$.

17. Solve the matrix equation for x :

$$[x \ \ 1]\begin{bmatrix} 2 & 4 \\ -1 & -2 \end{bmatrix} = 0.$$

18. Prove that the square matrix $\begin{bmatrix} 1 & 2 \\ -7 & 0 \end{bmatrix}$ can be expressed as a sum of symmetric and skew symmetric matrices.

19. Prove that square matrix $\begin{bmatrix} 1 & -5 & 2 \\ 7 & 0 & 1 \\ -1 & 2 & 5 \end{bmatrix}$ can be expressed as a sum of symmetric and a skew symmetric matrices.

20. If A and B are symmetric matrices, then prove that $AB - BA$ is a skew symmetric matrix. **[NCERT]**

21. Using elementary transformation, find the inverse of following matrix :★

$$\begin{bmatrix} -1 & 1 & 2 \\ 1 & 2 & 3 \\ 3 & 1 & 1 \end{bmatrix}$$

22. If $A = \begin{bmatrix} x & 2 \\ 2 & -1 \end{bmatrix}$ and $A^{-1} = \begin{bmatrix} 1/5 & 2/5 \\ 2/5 & -1/5 \end{bmatrix}$, then find the value of x.

23. If $A = \begin{bmatrix} 7 & -5 \\ 1 & 2 \end{bmatrix}$, then show that $(A - A')$ is a skew-symmetric matrix, where A' is the transpose of matrix A.

24. If $A = \begin{bmatrix} 1 & 0 & 2 \\ 3 & 2 & 1 \\ 1 & 1 & 3 \end{bmatrix}$, $B = \begin{bmatrix} 0 & -2 & 3 \\ 3 & 1 & 2 \\ 3 & 1 & 5 \end{bmatrix}$ and $C = \begin{bmatrix} 2 \\ -3 \\ 3 \end{bmatrix}$, then prove that :

$$(A+B)C = AC + BC.$$

25. If $\begin{bmatrix} 1 & -\tan\theta \\ \tan\theta & 1 \end{bmatrix}\begin{bmatrix} 1 & \tan\theta \\ -\tan\theta & 1 \end{bmatrix}^{-1} = \begin{bmatrix} a & -b \\ b & a \end{bmatrix}$, then what will be the values of a and b ?

26. If $A = \begin{bmatrix} 0 & -1 & 2 \\ 1 & 0 & 3 \\ -2 & -3 & 0 \end{bmatrix}$, then show that $A + 2A^T$ is equal to A^T.

27. Find the inverse of the matrix using elementary transformations :

$$\begin{bmatrix} 2 & -3 & 3 \\ 2 & 2 & 3 \\ 3 & -2 & 2 \end{bmatrix}$$

28. Two schools A and B decided to award prizes to their students for their regularity (x), honesty (y) and scoring in examination (z). School A decided to award a total of ₹ 11000 for the three values to 5, 4 and 3 students respectively, while school B decided to award ₹ 10,700 for the three values to 4, 3 and 5 students respectively. If all the three prizes together amount to ₹ 2700, then

(i) Represent the above situation by a matrix equation and form linear equation using matrix multiplication method.

(ii) Is it possible to solve the system of equations so obtained using matrices ? If yes then find the amount of all award prizes.

29. An amount of ₹ 800 crores is spent by the government in three schemes. Scheme A is for saving girl child from the cruel parents who don't want girl child and get the abortion before her birth. Scheme B is for saving poor people those die without having food. Scheme C is for good health of senior citizen. Now twice the amount spent on scheme C together with amount spent on scheme A is 1000 crores. Three times the amount spent on scheme A together with amount spent on twice of B and scheme C is ₹ 1600 crores. Find the amount spent on each scheme using matrices.

★ are board exam questions from previous years

30. If $A = \begin{bmatrix} \cos\theta & i\sin\theta \\ i\sin\theta & \cos\theta \end{bmatrix}$, then prove by principle of mathematical induction that

$$A^n = \begin{bmatrix} \cos n\theta & i\sin n\theta \\ i\sin n\theta & \cos n\theta \end{bmatrix} \text{ for all } n \in N.$$

31. Prove that every square matrix can be uniquely expressed as the sum of symmetric matrix and a skew-symmetric matrix.

Answers

1. 1 × 24, 24 × 1, 2 × 12, 12 × 2, 3 × 8, 8 × 3, 4 × 6 and 6 × 4

The possible orders of a matrix having 13 elements are 1 × 13 and 13 × 1.

2. 0 **3.** $\begin{bmatrix} \frac{1}{2} & \frac{9}{2} \\ 0 & 2 \end{bmatrix}$

4. $\begin{bmatrix} 1 & 0 \\ 0 & 1 \end{bmatrix}$ **5.** 0

6. $\pi/3$ **7.** $a = 1, b = 0$

9. $x = 1$ **10.** $\begin{bmatrix} 10 & 3 \\ -16 & 10 \\ -1 & -19 \end{bmatrix}$

11. $\begin{bmatrix} -3 & 4 \\ 7 & 3 \end{bmatrix}$ **12.** $X = \begin{bmatrix} -11/2 & -2 \\ 6 & -5 \\ 15/2 & -19 \end{bmatrix}$

13. $a = 2, b = 4, c = 1, d = 3$

14. 3 × 4

15. $\begin{bmatrix} 31/11 & 19/11 \\ -5/11 & 2/11 \end{bmatrix}, \begin{bmatrix} -8/11 & 19/11 \\ -2/11 & 3/11 \end{bmatrix}$

17. $x = \frac{1}{2}$

21. $A^{-1} = \begin{bmatrix} 1 & -1 & 1 \\ -8 & 7 & -5 \\ 5 & -4 & 3 \end{bmatrix}$

22. 1 **25.** $a = \cos 2\theta$ and $b = \sin 2\theta$

27. $A^{-1} = \begin{bmatrix} -2/5 & 0 & 3/5 \\ -1/5 & 1/5 & 0 \\ 2/5 & 1/2 & -2/5 \end{bmatrix}$

28. (i)
$$5x + 4y + 3z = 11{,}000$$
$$4x + 3y + 5z = 10{,}700$$
$$x + y + z = 2700$$

(ii) Yes, $x = ₹\,1000, y = ₹\,900$ and $z = ₹\,800$.

29.
$$x + y + z = 800$$
$$x + 2z = 1000$$
$$3x + 2y + z = 1600$$

$x = ₹\,\frac{1000}{3}$ crore, $y = ₹\,\frac{400}{3}$ crore, $z = ₹\,\frac{1000}{3}$ crore.

Chapter 4. Determinants

1. Find the value of x if $\begin{vmatrix} x & 2 \\ 1 & 5 \end{vmatrix} = \begin{vmatrix} 1 & 3 \\ 2 & 4 \end{vmatrix}$.

2. Let A be a square matrix of order 3 and $|A| = 4$, find the value of $|2A|$.

3. Let A be a non-singular square matrix of order 3 × 3 Then, find $|\text{adj } A|$.

4. If a singular matrix A is given by $\begin{bmatrix} 1 & 7 \\ 2 & x \end{bmatrix}$, then find the value of x.

5. If $A = \begin{bmatrix} 1 & 3 \\ 2 & x \end{bmatrix}$, for what value of x, A^{-1} exist ?

6. Evaluate the determinants :

(i) $\begin{bmatrix} 1 & 0 & 1 \\ 2 & 3 & 5 \\ 2 & 0 & -7 \end{bmatrix}$ (ii) $\begin{bmatrix} 2 & -3 & 5 \\ -3 & 2 & 5 \\ 0 & 2 & 3 \end{bmatrix}$

(iii) $\begin{bmatrix} 2 & 2 & 9 \\ -2 & 5 & 7 \\ 2 & -2 & 5 \end{bmatrix}$ (iv) $\begin{bmatrix} 0 & 5 & -7 \\ -5 & 0 & 2 \\ 7 & -2 & 0 \end{bmatrix}$

(v) $\begin{bmatrix} 0 & 2 & 3 \\ 2 & 0 & 5 \\ 3 & 5 & 0 \end{bmatrix}$

7. Evaluate the determinants : **[NCERT]**

(i) $\begin{vmatrix} x^2 - x + 1 & x - 1 \\ x + 1 & x + 1 \end{vmatrix}$ (ii) $\begin{vmatrix} 2 & -1 & -2 \\ 0 & 2 & -1 \\ 3 & -5 & 0 \end{vmatrix}$

(iii) $\begin{vmatrix} 3 & -1 & -2 \\ 0 & 0 & -1 \\ 3 & -5 & 0 \end{vmatrix}$ (iv) $\begin{vmatrix} 1 & 1 & -2 \\ 2 & 1 & -3 \\ 5 & 4 & -9 \end{vmatrix}$

8. Find the value of x, if :

(i) $\begin{vmatrix} 2 & 4 \\ 2 & 1 \end{vmatrix} = \begin{vmatrix} 2x & 4 \\ 6 & x \end{vmatrix}$ (ii) $\begin{vmatrix} x & 2 \\ 18 & x \end{vmatrix} = \begin{vmatrix} 6 & 2 \\ 18 & 6 \end{vmatrix}$

9. Find area of triangle with vertices at the point given in each of the following :
(i) (1, 0), (6, 0), (4, 3)
(ii) (– 2, – 3) (3, 2), (– 1, – 8) **[NCERT]**

10. If $A = \begin{bmatrix} 1 & 2 \\ 4 & 2 \end{bmatrix}$, then show that $|2A| = 4|A|$. **[NCERT]**

11. Find x, if $\begin{bmatrix} 1 & 2 & x \\ 1 & 1 & 1 \\ 2 & 1 & -1 \end{bmatrix}$ is singular.

12. Show without expanding at any stage : **[NCERT]**

$$\begin{vmatrix} 1 & bc & a(b+c) \\ 1 & ca & b(c+a) \\ 1 & ab & c(a+b) \end{vmatrix} = 0.$$

13. If A_{ij} is the cofactor of the element a_{ij} of the determinant $\begin{vmatrix} 2 & -3 & 5 \\ 6 & 0 & 4 \\ 1 & 5 & -7 \end{vmatrix}$, then write the value of $a_{32} \cdot A_{32}$.★

14. If $D = \begin{vmatrix} 1 & 2 & 3 \\ 7 & -2 & 1 \\ -1 & 5 & 1 \end{vmatrix}$, what will be the value of $(-1)D$?

Using properties of determinants, prove the following (15 to 17) :

15. $\begin{vmatrix} 1 & 1 & 1 \\ a & b & c \\ a^3 & b^3 & c^3 \end{vmatrix} = (a-b)(b-c)(c-a)(a+b+c)$.★

16. $\begin{vmatrix} a & b & c \\ a-b & b-c & c-a \\ b+c & c+a & a+b \end{vmatrix} = a^3+b^3+c^3-3abc$.

17. $\begin{vmatrix} b^2c^2 & bc & b+c \\ c^2a^2 & ca & c+a \\ a^2b^2 & ab & a+b \end{vmatrix} = 0$ **[HOTS]**

18. Using properties of determinants, prove that

$$\begin{vmatrix} x-3 & x-4 & x-\alpha \\ x-2 & x-3 & x-\beta \\ x-1 & x-2 & x-\gamma \end{vmatrix} = 0$$

where α, β and γ are in A.P.★

19. Without expanding the determinant prove that $(a-b)$ is a factor of the determinant $\begin{vmatrix} 1 & a & a^2 \\ 1 & b & b^2 \\ 1 & c & c^2 \end{vmatrix}$.★

★ are board exam questions from previous years

20. Prove that : $\begin{vmatrix} 1 & x & x^3 \\ 1 & y & y^3 \\ 1 & z & z^3 \end{vmatrix} = (x-y)(y-z)(z-x)(x+y+z)$ **[NCERT]**

21. Using properties of determinants, prove that★

$$\begin{vmatrix} 3x & -x+y & -x+z \\ x-y & 3y & z-y \\ x-z & y-z & 3z \end{vmatrix} = 3(x+y+z)(xy+yz+zx).$$

22. Using properties of determinants, show that

$$\begin{vmatrix} a & a+b & a+2b \\ a+2b & a & a+b \\ a+b & a+2b & a \end{vmatrix} = 9b^2(a+b).$$

23. If $a \neq b \neq c$, what will be the value of x which satisfies the equation

$$\begin{vmatrix} 0 & x-a & x-b \\ x+a & 0 & x-c \\ x+b & x+c & 0 \end{vmatrix} = 0.$$

24. If $\begin{vmatrix} x & 2 & 3 \\ 2 & 3 & x \\ 3 & x & 2 \end{vmatrix} = \begin{vmatrix} 1 & x & 4 \\ x & 4 & 1 \\ 4 & 1 & x \end{vmatrix} = \begin{vmatrix} 0 & 5 & x \\ 5 & x & 0 \\ x & 0 & 5 \end{vmatrix} = 0$, then find the value of x.

25. If $\begin{vmatrix} a+x & a-x & a-x \\ a-x & a+x & a-x \\ a-x & a-x & a+x \end{vmatrix} = 0$, then find the value of x.

26. For what value of k, the following system of linear equations will have infinite solutions

$$x - y + z = 3$$
$$2x + y - z = 2$$
$$-3x - 2ky + 6z = 3$$

27. Prove that : $\begin{vmatrix} y+z & x & x \\ y & z+x & y \\ z & z & x+y \end{vmatrix} = 4xyz$.

28. Evaluate : $\begin{vmatrix} (a^x+a^{-x})^2 & (a^x-a^{-x})^2 & 1 \\ (b^x+b^{-x})^2 & (b^x-b^{-x})^2 & 1 \\ (c^x+c^{-x})^2 & (c^x-c^{-x})^2 & 1 \end{vmatrix}$

29. If $\Delta = \begin{vmatrix} a & b & ax+b \\ b & c & bx+c \\ ax+b & bx+c & 0 \end{vmatrix} = 0$ then prove that a, b and c are in G.P.

30. Find the value of θ between 0 and $\frac{\pi}{2}$ which satisfies the equation

$$\begin{vmatrix} 1+\sin^2\theta & \cos^2\theta & 4\sin 4\theta \\ \sin^2\theta & 1+\cos^2\theta & 4\sin 4\theta \\ \sin^2\theta & \cos^2\theta & 1+4\sin 4\theta \end{vmatrix} = 0.$$

31. If ω is a cube root of unity then evaluate the value of

$$\begin{vmatrix} a & b\omega^2 & a\omega \\ b\omega & c & b\omega^2 \\ c\omega^2 & a\omega & c \end{vmatrix}$$

32. If $A = \begin{bmatrix} 1 & -1 & 1 \\ 2 & 1 & -3 \\ 1 & 1 & -1 \end{bmatrix}$, then find A^{-1} and hence solve the following system of equations :

$$x + 2y + z = 4$$
$$-x + y + z = 0$$
and $$x - 3y + z = 2$$

33. The sum of three numbers is 6. If we multiply third number by 3 and add second number to it, we get 11. By adding first and third numbers, we get double of the second number. Represent graphically and find the numbers using matrix method.

34. Prove that :

$$\Delta = \begin{vmatrix} a+bx & c+dx & p+qx \\ ax+b & cx+d & px+q \\ u & v & w \end{vmatrix} = (1-x^2)\begin{vmatrix} a & c & p \\ b & d & q \\ u & v & w \end{vmatrix}$$

35. Solve system of linear equations, using matrix method,

(i)
$$x - y + 2z = 7$$
$$3x + 4y - 5z = -5$$
$$2x - y + 3z = 12$$

(ii)
$$x + y - z = 3$$
$$2x + 3y + z = 10$$
$$3x - y - 7z = 1$$

36. If $\begin{vmatrix} 1 & 1 & 1 \\ a & b & c \\ a^3 & b^3 & c^3 \end{vmatrix} = (a-b)(b-c)(c-a)(a+b+c)$

where a, b, c are all different, then prove that determinant

$$\begin{vmatrix} 1 & 1 & 1 \\ (x-a)^2 & (x-b)^2 & (x-c)^2 \\ (x-b)(x-c) & (x-c)(x-a) & (x-a)(x-b) \end{vmatrix}$$

vanish when $x = \frac{1}{3}(a+b+c)$.

Answers

1. 0 **2.** 32

3. $|A|^2$ **4.** 14

5. $x \neq 6$

6. (i) 15 (ii) – 65 (iii) 72 (iv) 0 (v) 0

7. (i) $x^3 - x^2 + 2$ (ii) 5 (iii) – 12 (iv) 0

8. (i) $x = 3$ (ii) $x = \pm 6$

9. (i) $\frac{15}{2}$ square units (ii) 15 square units

11. 4 **13.** 110

14. 76 **23.** $x = 0$

24. $x = -5$ **25.** $x = 0, 3a$

26. $k = 3$ **28.** 0

30. $\frac{7\pi}{24}$ and $\frac{11\pi}{24}$ **31.** 0

32. $\frac{1}{10}\begin{bmatrix} 4 & 2 & 2 \\ -5 & 0 & 5 \\ 1 & -2 & 3 \end{bmatrix}$, $x = \frac{9}{5}$, $y = \frac{2}{5}$, $z = \frac{7}{5}$

33. 1, 2 and 3

35. (i) $x = 2, y = 1$ and $z = 3$

(ii) $x = 3, y = 1$ and $z = 1$

Chapter 5. Continuity and Differentiability

1. Given $f(x) = \begin{cases} 2, & x = 1 \\ \dfrac{x^2 - 4x + 3}{x^2 - 1}, & x \neq 1 \end{cases}$.

Verify $f(x)$ is continuous at $x = 1$.

2. For a function, $f(0) = 0$ and $f(x) = \frac{1}{1 + 2^{1/x}}$, $x \neq 0$.

Verify $f(x)$ is continuous at $x = 0$.

3. A function is given as $f(x) = \begin{cases} x - 1, & x < 0 \\ 1/4, & x = 0 \\ x^2, & x > 0 \end{cases}$. Draw the graph of the function and discuss the continuity of the function at $x = 0$.

4. Differentiate the following :

(i) $y = x^2 \sin x$

(ii) $y = \dfrac{x^2 + 1}{x}$

(iii) $y = \dfrac{\sin x + \cos x}{\cos x}$

(iv) $y = e^{x^2 + 1}$

(v) $y = (x^2 + 1)^{1/2}$

(vi) $y = \ln(3x^2 + 2x)^2$

★ are board exam questions from previous years

(vii) $y = \dfrac{x^2 + 2x - 3}{3x - 4}$

(viii) $y = \dfrac{4}{(x^3 + 4)^8}$

(ix) $y = \tan(2x^2 - 3x + 1)$

(x) $y = \sin(3x)\cos(4x)$

(xi) $y = x^2.\sqrt{3x+4}$

(xii) $y = \dfrac{\sin x}{1 + \cos x}$

(xiii) $y = \sin^3(x^2 - 1)$

(xiv) $y = \cos(xy)$

5. If $x = a \sin 2t\,(1 + \cos 2t)$ and $y = b \cos 2t(1 - \cos 2t)$, find $\dfrac{dy}{dx}$ at $t = \dfrac{\pi}{4}$.★

6. Show that the function $f(x) = |x - 1| + |x + 1|$, for all $x \in R$, is not differentiable at the points $x = -1$ and $x = 1$.★

7. If $f(x) = \sqrt{x^2 + 1}$, $g(x) = \dfrac{x+1}{x^2+1}$ and $h(x) = 2x - 3$, then find $f'\,[h'\{g(x)\}]$.★

8. If $y = \sqrt{\tan x + \sqrt{\tan x + \sqrt{\tan x + \sqrt{\tan x + \ldots \text{ to } \infty}}}}$

then prove that : $\dfrac{dy}{dx} = \dfrac{\sec^2 x}{2y - 1}$

9. If $y = \dfrac{x}{2}\sqrt{a^2 - x^2} + \dfrac{a^2}{2}\sin^{-1}\left(\dfrac{x}{a}\right)$, then show that : $\dfrac{dy}{dx} = \sqrt{a^2 - x^2}$.

10. Verify Rolle's theorem for the function $f(x)$ in the interval $[-3, -2]$; $f(x) = x^2 + 5x + 6$.

Answers

1. Discontinuous
2. Discontinuous
3. Discontinuous
4. (i) $2x \sin x + x^2 \cos x$

(ii) $\dfrac{x^2 - 1}{x^2}$

(iii) $\sec^2 x$

(iv) $2xe^{x^2 + 1}$

(v) $\dfrac{x}{(x^2 + 1)^{1/2}}$

(vi) $\dfrac{12x + 4}{3x^2 + 2x}$

(vii) $\dfrac{3x^2 - 8x + 1}{(3x - 4)^2}$

(viii) $\dfrac{-96x^2}{(x^3 + 4)^9}$

(ix) $\sec^2(2x^2 - 3x + 1) \times (4x - 3)$

(x) $\sin(3x)[-4\sin(4x)] + \cos(4x)[3\cos(3x)]$

(xi) $\dfrac{15x^2 + 16x}{2\sqrt{(3x + 4)}}$

(xii) $\dfrac{1}{1 + \cos x}$

(xiii) $6x \sin^2(x^2 - 1)\cos(x^2 - 1)$

(xiv) $\dfrac{-y\sin(xy)}{1 + x\sin(xy)}$

5. $\dfrac{b}{a}$ 7. $\dfrac{2}{\sqrt{5}}$

Chapter 6. Applications of Derivatives

1. A ladder 5 m long is leaning against a wall. The bottom of the ladder is pulled along the ground, away from the wall, at the rate of 2 cm/s. How fast is its height on the wall decreasing when the foot of the ladder is 4 m away from the wall ? **[NCERT]**
2. A balloon, which always remains spherical, has a variable diameter $\dfrac{3}{2}(2x + 1)$. Find the rate of change of its volume with respect to x. **[NCERT]**
3. The total revenue in rupees received from the sale of x units of a product is given by
$$R(x) = 13x^2 + 26x + 15.$$
Find the marginal revenue when $x = 7$. **[NCERT]**
4. Find the point on the curve $y^2 = 8x$ for the abscissa and ordinate change at the same rate.
5. Water is running into an inverted cone at the rate of π cubic metre per minute. The height of the cone is 10 metres and the radius of its base is 5 m. How fast the water level is rising when the water stands 7·5 m above the base.
6. Function $f(x) = \cos x - 2\alpha x$ is monotonic decreasing, then find α. **[HOTS]**
7. The function $f(x) = \sin x - \dfrac{x}{2}$, defined on $x \in \left[-\dfrac{\pi}{3}, \dfrac{\pi}{3}\right]$ is increasing or decreasing ?
8. Find the intervals for which the function f given by
$$f(x) = 2x^3 - 15x^2 + 36x + 1$$
is increasing or decreasing.
9. Show that $f(x) = \tan^{-1}(\sin x + \cos x)$ is an increasing function on the interval $\left(0, \dfrac{\pi}{4}\right)$.

★ are board exam questions from previous years

10. Find points at which the tangent to the curve $y = x^3 - 3x^2 - 9x + 7$ is parallel to the X-axis.

11. Find the equations of all lines having slope 0 which are tangent to the curve $y = \frac{1}{x^2 - 2x + 3}$ · **[NCERT]**

12. Find the points on the curve $x^2 + y^2 - 2x - 3 = 0$ at which the tangents are parallel to the X-axis.

13. Show that the curves $4x = y^2$ and $4xy = k$ cut at right angles, if $k = \sqrt{512}$.

14. Find the equation of the normals to the curve $y = x^3 + 2x + 6$ which are parallel to the line $x + 14y + 4 = 0$.

15. Use differential to approximate $(3{\cdot}968)^{3/2}$.

16. Use differential to approximate $\sqrt{49{\cdot}5}$

17. Find the approximate value of $f(5{\cdot}001)$, where $f(x) = x^3 - 7x^2 + 15$.

18. Find the approximate change in the volume V of a cube of side x meters caused by increasing the side by 2%. **[NCERT]**

19. If the radius of a sphere is measured as 9 m with an error of 0·03 m, then find the approximate error in calculating its surface area.

20. Find local maximum or local minimum value of the function f given by

$$f(x) = \frac{1}{x^2 + 2}.$$

21. Show that e^x has no local maxima or local minima. **[HOTS]**

22. Show that the right circular cone of least curved surface and given volume has an altitude equal to $\sqrt{2}$ times the radius of the base. **[HOTS]**

23. Prove that the function f do not have maxima or minima where $f(x) = \log x$.

24. Find two positive numbers x and y such that their sum is 35 and the product x^2y^5 is maximum. **[NCERT]**

25. It is given that at $x = 1$, the function $x^4 - 62x^2 + ax + 9$ attains its maximum value on the interval [0, 2]. Find the value of a. **[NCERT]**

26. A water tank has the shape of an inverted right circular cone with its axis vertical and vertex lowermost. Its semi-vertical angle is $\tan^{-1}(0{\cdot}5)$. Water is poured into it at a constant rate of 5 cubic metre per hour. Find the rate at which the level of the water is rising at the instant when the depth of water in the tank is 4 m.

27. Let f be a function defined on $[a, b]$ such that $f'(x) > 0$ for all $x \in (a, b)$. Then prove that f is an increasing function on (a, b).

28. Show that the rectangle of maximum perimeter which can be inscribed in a circle of radius a is a square of side $\sqrt{2}a$.

29. An open tank with a square base and vertical sides is to be constructed from a metal sheet so as to hold a given quantity of water. Show that the cost of the material will be least when depth of the tank is half of its width.

30. An open box with square base is to be made of a given quantity of carboard of area a^2. Show that maximum volume of the box is $\frac{a}{6\sqrt{3}}$ cubic units.

Answers

1. $-\frac{8}{3}$ cm/s. **2.** $\frac{27}{8}\pi(2x+1)^2$.

3. ₹ 208 **4.** $-(2, 4)$

5. 0·64 metre/minute. **6.** $-\alpha > \frac{1}{2}$

7. Increasing **8.** $(-\infty, 2) \cup (3, \infty)$

10. $(3, -20)$ and $(-1, 12)$

11. $y = \frac{1}{2}$ **12.** $(1, \pm 2)$

14. $x + 14y - 254 = 0, x + 14y + 86 = 0$

15. 7·904 **16.** 7·035

17. $-3{\cdot}995$ **18.** $0{\cdot}06x^3\,\text{m}^3$

19. $2{\cdot}16\pi\,\text{m}^3$.

20. Local maximum at $x = 0$, local maximum value is $\frac{1}{2}$

24. 25, 10

25. $a = 120$ **26.** $\frac{35}{88}$ m/hr

Chapter 7. Integrals

Integrate the following :

1. $\int x^2 e^x \, dx$ **2.** $\int \tan^{-1} x \, dx$

3. $\int \frac{dx}{x^2 - 9}$ **4.** $\int \frac{3x}{1 + 2x^4} dx$

5. $\int \frac{1 - \cos x}{1 + \cos x} dx$ **6.** $\int x\sqrt{x + 2} \, dx$

7. $\int \frac{e^{\tan^{-1} x}}{1 + x^2} dx$ **8.** $\int_0^{\pi/2} \frac{\sin x}{1 + \cos^2 x} dx$

9. $\int \frac{dx}{\sin(x - a)\sin(x - b)}$ **10.** $\int e^{\sin^2 x} \sin 2x \, dx$

11. $\int_0^{\pi/2} \frac{\cos^2 x}{\cos^2 x + 4\sin^2 x} dx$ **12.** $\int \frac{1}{x^3 + 1} dx$

13. $\int \frac{\cos\theta}{(2 + \sin\theta)(3 + 4\sin\theta)} d\theta$

14. $\int \frac{x}{1+x+x^2+x^3}\,dx$ **15.** $\int \frac{dx}{5+4\cos x}$

16. $\int \frac{\cos\sqrt{x}}{\sqrt{x}}\,dx$ **17.** $\int \sin x \sin 2x \sin 3x\,dx$

18. $\int \sin^3(2x+1)\,dx$ **19.** $\int \frac{1}{\sqrt{9-25x^2}}\,dx$

20. $\int \frac{4x+1}{\sqrt{2x^2+x-3}}\,dx$ **21.** $\int \frac{dx}{x(x^4-1)}$

22. $\int_0^1 \cos^{-1} x\,dx$ **23.** $\int_0^1 x\log(1+2x)\,dx$

24. $\int_2^e \left(\frac{1}{\log x}-\frac{1}{(\log x)^2}\right)dx$

25. $\int_0^\pi \frac{x\sin x}{1+\cos^2 x}\,dx$

26. $\int_0^{\pi/2} \sin^6 x\,dx$ **27.** $\int_0^{\pi/2} \log|\sin x|\,dx$

28. $\int_0^{\pi/2} \frac{\sin^3 x}{\sin^3 x+\cos^3 x}\,dx$ **29.** $\int_1^2 \frac{5x^2}{x^2+4x+3}\,dx$

30. $\int (\sin^{-1} x)^2\,dx$ **31.** $\int \sqrt{10-4x+4x^2}\,dx$

32. $\int_0^{\pi/2} 2\tan^3 x\,dx$ **33.** $\int_0^{\pi/4} \frac{\sin x+\cos x}{9+16\sin 2x}\,dx$

34. $\int \frac{e^x}{(1+e^x)(2+e^x)}\,dx$ **35.** $\int_0^2 x\sqrt{2-x}\,dx$

36. $\int_0^a \frac{\sqrt{x}}{\sqrt{x}+\sqrt{a-x}}\,dx$ **37.** $\int_0^\pi \log(1+\cos x)\,dx$

38. $\int_{-\pi/4}^{\pi/4} \log(\sin x+\cos x)\,dx$

Answers

1. $x^2e^x-2xe^x+2e^x+c$

2. $x\tan^{-1}x-\frac{1}{2}\log|1+x^2|+c$

3. $\frac{1}{6}\log\left|\frac{x-3}{x+3}\right|+c$

4. $\frac{3}{2\sqrt{2}}\tan^{-1}(\sqrt{2}x^2)+c$

5. $2\tan\frac{x}{2}-x+c$

6. $\frac{2}{5}(x+2)^{5/2}-\frac{4}{3}(x+2)^{3/2}+c$

7. $e^{\tan^{-1}x}+c$

8. $\frac{\pi}{4}$

9. $\frac{1}{\sin(b-a)}\log\left|\frac{\sin(x-b)}{\sin(x-a)}\right|+c$

10. $e^{\sin^2 x}+c$ **11.** $\frac{\pi}{6}$

12. $\frac{1}{3}\log|x+1|-\frac{1}{6}\log(x^2-x+1)+\frac{1}{\sqrt{3}}\tan^{-1}\left(\frac{2x-1}{\sqrt{3}}\right)+c$

13. $\frac{1}{5}\{\log|3+4\sin\theta|-\log|2+\sin\theta|\}+c$

14. $-\frac{1}{2}\log|x+1|+\frac{1}{4}\log|x^2+1|+\frac{1}{2}\tan^{-1}x+c$

15. $\frac{2}{3}\tan^{-1}\left(\frac{\tan x/2}{3}\right)+c$

16. $2\sin\sqrt{x}+c$

17. $-\frac{\cos 4x}{16}+\frac{\cos 6x}{24}-\frac{\cos 2x}{8}+c$

18. $-\frac{3}{8}\cos(2x+1)+\frac{1}{24}\cos(6x+3)+c$

19. $\frac{1}{5}\sin^{-1}\left(\frac{5x}{3}\right)+c$ **20.** $2\sqrt{2x^2+x-3}+c$

21. $\frac{1}{4}\log\left|\frac{x^4-1}{x^4}\right|+c$ **22.** 1

23. $\frac{3}{8}\log 3$ **24.** $e-\frac{2}{\log 2}$

25. $\frac{\pi^2}{4}$ **26.** $\frac{5\pi}{32}$

27. $-\frac{\pi}{2}\log 2$ **28.** $\frac{\pi}{4}$

29. $5-10\log\left(\frac{15}{8}\right)+\frac{25}{2}\log\left(\frac{6}{5}\right)$

30. $x(\sin^{-1}x)^2+2\sqrt{1-x^2}\sin^{-1}x-2x+c$

31. $\frac{2x-1}{4}\sqrt{4x^2-4x+10}+\frac{9}{4}\log\left|\frac{2x-1}{2}+\frac{1}{2}\sqrt{4x^2-4x+10}\right|+c$

32. $1-\log 2$ **33.** $\frac{1}{40}\log 9$

34. $\log\left(\frac{1+e^x}{2+e^x}\right)+c$ **35.** $\frac{16\sqrt{2}}{15}$

36. $\frac{1}{2}a$ **37.** $-\pi\log 2$

38. $\frac{-\pi}{4}\log 2$

Chapter 8. Applications of the Integrals

1. Find the area of region bounded by the curve $x^2 = 4y$ and the straight line $x = 4y - 2$.
2. Draw the rough sketch of the curves $y = (x - 1)^2$ and $y = |x - 1|$. Hence, find the area of the region bounded by these curves.
3. Find the area enclosed between the curves $y^2 = 4ax$ and $x^2 = 4ay$, where $a > 0$.
4. Find the area of the region bounded by the parabola $y^2 = 4ax$ and the chord $y = mx$. **[NCERT]**
5. Sketch the graph of $y = |x + 3|$ and evaluate :

$$\int_{-6}^{0} |x+3|\, dx.$$ **[NCERT]**

6. Find the area of region bounded by the curves $y = x^4, x = 1, x = 5$ and X-axis.
7. Find the area of the region bounded by the curves $|x + y| \le 1, |x - y| \le 1$ and $3x^2 + 3y^2 \ge 1$.
8. Find the area of region bounded by the curves $x^2 + y^2 \le 2ax, y^2 \ge ax, x \ge 0, y \ge 0$.
9. Using integration, find the area of the region bounded by the lines $x + 2y = 2, y - x = 1$ and $2x + y = 7$.
10. Using integration, find the area of the triangle whose vertices are $A(0, 5)$, $B(-1, 1)$ and $C(3, 2)$.
11. Find the area of the region given by $\{(x, y) : x^2 \le y \le |x|\}$.
12. Find the area of the region bounded by the curves $y^2 = 4x$ and $x^2 = 4y$.
13. Find the area of the region bounded by the curves $y^2 = 4x, x = 1, x = 4$ and X-axis in the first quadrant.
14. Draw the rough sketch of the following curve $y = 2\cos x, 0 \le x \le 2\pi$, and find the area.
15. Find the area of the region $\{(x, y) : x^2 + y^2 \le 2ax, y^2 \ge ax, x \ge 0, y \ge 0\}$.
16. Draw a rough sketch of the given region $\{(x, y), 0 \le y \le x^2 + 3; 0 \le y \le 2x + 3, 0 \le x \le 3\}$ and find its area using integration.
17. Sketch the region of the ellipse and find its area, using integration $\dfrac{x^2}{a^2} + \dfrac{y^2}{b^2} = 1, a > b$. **[NCERT]**
18. Find the area bounded by lines $x + 2y = 2, y - x = 1$ and $2x + y = 7$.
19. Using integration find the area of the region bounded by the parabola $y = x^2$ and line $y = x$.
20. Using integration find the area of the region : $\{(x, y) : 9x^2 + y^2 \le 36 \text{ and } 3x + y \ge 6\}$

Answers

1. $\frac{9}{8}$ sq. units
2. $\frac{1}{3}$ sq. units
3. $\frac{16}{3}a^2$ sq. units
4. $\frac{8a^2}{3m^3}$ sq. units
5. 9 sq. units
6. 624.8 sq. units
7. $\left(2 - \frac{\pi}{3}\right)$ sq. units
8. $\left(\frac{\pi a^2}{4} - \frac{2a^2}{3}\right)$ sq. units
9. 6 sq. units
10. 7.5 sq. units
11. $\frac{1}{3}$ sq. units
12. $\frac{16}{3}$ sq. units
13. $\frac{28}{3}$ sq. units
14. 8 sq. units
15. $\left(\frac{\pi}{4} - \frac{2}{3}\right)a^2$ sq. units
16. $\frac{50}{3}$ sq. units
17. πab sq. units
18. 6 sq. units
19. $\frac{1}{6}$ sq. units
20. $3(\pi - 2)$ sq. units

Chapter 9. Differential Equations

1. Form the differential equation satisfied by $(x - a)^2 + (y - b)^2 = r^2$, where a and b are arbitrary constants.
2. Find the differential equation consistent with $y = ae^{3x} + be^{5x} + ce^{7x}$; where a, b and c are arbitrary constants.
3. Form a differential equation of all the circles which pass through the origin and whose centre lies on X-axis.
4. Verify that $y = (x^3 - x)\log(cx)$ is a solution of the differential equation

$$(x^3 - x)\frac{dy}{dx} - (3x^2 - 1)\,y = x^5 - 2x^3 + x.$$

5. Show that $y = (a + b)\,e^{2x}$ is a solution of the differential equation $\dfrac{d^2y}{dx^2} - 4\dfrac{dy}{dx} + 4y = 0$.
6. Show that $y = cx + \dfrac{a}{c}$ is a solution of the differential equation $y = x\dfrac{dy}{dx} + a\dfrac{dy}{dx}$.
7. Show that the differential equation of which $x^2 - y^2 = c(x^2 + y^2)^2$ is a solution is

$$(x^3 - 3xy^2)dx = (y^3 - 3x^2y)\,dy.$$

8. Show that the function $y = be^x + ce^{2x}$ is a solution of the differential equation

$$\frac{d^2y}{dx^2} - 3\frac{dy}{dx} + 2y = 0.$$

9. Show that the function $y = (A + Bx)\,e^{3x}$ is a solution of the equation $\dfrac{d^2y}{dx^2} - 6\dfrac{dy}{dx} + 9y = 0$.

10. Solve : $\sqrt{1+x^2}\,dy+\sqrt{1+y^2}\,dx=0$

11. Solve : $3e^x \tan y\,dx+(1-e^x)\sec^2 y\,dy=0$

12. Solve : $\dfrac{dy}{dx}=\sin^3 x\cos^2 x+xe^x$

13. Solve : $\dfrac{dy}{dx}=x\log x$

14. Find the particular solution of differential equation

$\log\dfrac{dy}{dx}=3x+4y$, given that $y=0$ when $x=0$.

15. Solve : $\sec^2 x\tan y\,dx+\sec^2 y\tan x\,dy=0$

16. Solve : $(e^x+1)\,y\,dy=(y+1)\,e^x\,dx$

17. Solve : $\dfrac{dy}{dx}+\dfrac{\cos x\sin y}{\cos y}=0$

18. Find the particular solution of the differential equation $e^{dy/dx}=x+1$, given that $y=3$ when $x=0$

19. Solve : $x\dfrac{dy}{dx}=y-x\tan\dfrac{y}{x}$

20. Solve : $x^2\dfrac{dy}{dx}=x^2-2y^2+xy$

21. Solve : $(y^2-x^2)\,dy-3xy\,dx=0$

22. Solve : $\left(x\sin\dfrac{y}{x}\right)dy=\left(y\sin\dfrac{y}{x}-x\right)dx$

23. Solve : $x^2\dfrac{dy}{dx}-xy=1+\cos\left(\dfrac{y}{x}\right)$,

$x\neq 0$ and $y=\dfrac{\pi}{2}$ if $x=1$

24. Solve : $\dfrac{dy}{dx}=\dfrac{x+y+1}{2x+2y+1}$

25. Solve : $\dfrac{dy}{dx}=(3x+y+4)^2$

26. Solve : $\dfrac{dy}{dx}-2y\cos x=-2\sin 2x$

27. Solve : $(2x-10y^3)\,dy+y\,dx=0,\ y\neq 0$

28. Solve : $y\sin 2x\,dx-(1+y^2+\cos^2 x)\,dy=0$

29. Solve : $\dfrac{dy}{dx}+ay=e^{mx}$

30. Solve : $\dfrac{dy}{dx}+\dfrac{y}{x}=\cos x+\dfrac{\sin x}{x},\ x>0$

31. Solve : $\dfrac{dy}{dx}+x\sin 2y=x^3\cos^2 y$

32. Solve : $dy=\cos x\,(2-y\,\text{cosec}\,x)\,dx$

given that $y=2$ when $x=\dfrac{\pi}{2}$

33. Solve : $\dfrac{dy}{dx}-2y\tan x=y^2\tan^2 x,\ 0<x<\dfrac{\pi}{2}$

34. Solve : $\dfrac{dy}{dx}+\dfrac{4x}{x^2+1}y=-\dfrac{1}{(x^2+1)^3}$

35. Solve : $\dfrac{dy}{dx}+2y\tan x=\sin x,\ y=0$ when $x=\dfrac{\pi}{3}$

Answers

1. $(1+y_1^2)^3=r^2y_2^2$ where $y_1=\dfrac{dy}{dx}$ and $y_2=\dfrac{d^2y}{dx^2}$

2. $y_3-15y_2+71y_1-105y=0$

where $y_3=\dfrac{d^3y}{dx^3},\ y_2=\dfrac{d^2y}{dx^2},\ y_1=\dfrac{dy}{dx}$

3. $y^2-x^2=2xyy_1$

10. $(y+\sqrt{1+y^2})\,(x+\sqrt{1+x^2})=A$

11. $\tan y=A(1-e^x)^3$

12. $y=-\dfrac{1}{3}\cos^3 x+\dfrac{1}{5}\cos^5 x+(x-1)\,e^x+c$

13. $y=\dfrac{x^2}{2}\left(\log x-\dfrac{1}{2}\right)+c$

14. $4e^{3x}+3e^{-4y}=7$

15. $\tan x\tan y=A$

16. $y=\log|(y+1)(e^x+1)|+c$

17. $\sin y=\pm e^c e^{-\sin x}$ or $\sin y=Ae^{-\sin x}$

18. $y=(x+1)\log|x+1|-x+3$

19. $x\sin\dfrac{y}{x}=A$

20. $\log|x|=\dfrac{1}{2\sqrt{2}}\log\left|\dfrac{x+\sqrt{2}y}{x-\sqrt{2}y}\right|+c$

21. $y^2(4x^2-y^2)^3=A$

22. $\cos\left(\dfrac{y}{x}\right)=\log|x|+c$

23. $\tan\left(\dfrac{y}{2x}\right)=-\dfrac{1}{2x^2}+\dfrac{3}{2}$

24. $6y-3x-\log|3x+3y+2|=A$

25. $3x+y+4=\sqrt{3}\tan[\sqrt{3}(x+A)]$

26. $y=2\sin x+1+ce^{2\sin x}$

27. $xy^2=2y^5+c$

28. $-\cos^2 x=1+\dfrac{1}{3}y^2+\dfrac{A}{y}$

29. $y=\dfrac{e^{mx}}{m+a}+ce^{-ax}$

30. $y=\sin x+\dfrac{c}{x}$

31. $\tan y = \frac{1}{2}(x^2 - 1) + ce^{-x^2}$

32. $y \sin x = -\frac{\cos 2x}{2} + \frac{3}{2}$

33. $y^{-1} \sec^2 x = c - \frac{1}{3} \tan^3 x$

34. $y = \frac{c - \tan^{-1} x}{(x^2 + 1)^2}$

35. $y = \cos x - 2\cos^2 x$

Chapter 10. Vectors

1. Let $\vec{a}$ and $\vec{b}$ be the vectors determined by the adjacent sides of a regular hexagon as shown below in the figure.

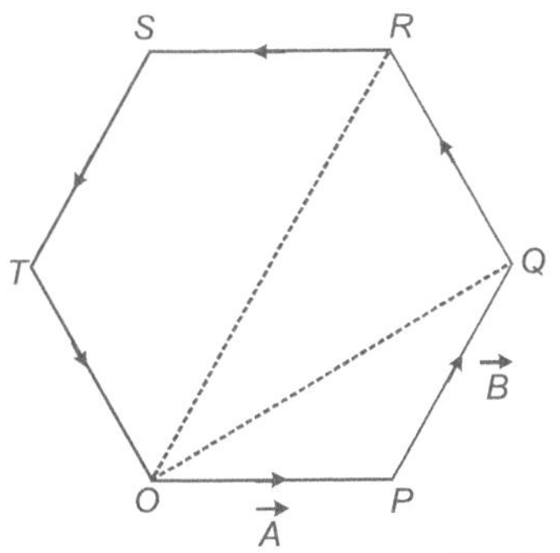

What are the vectors determined by the other sides taken in order ?

2. Let $\vec{a}$ be a vector in the north-east direction with magnitude 5 and $\vec{b}$ be another vector in the north-west direction with magnitude 5. What is the magnitude of $\vec{a} - \vec{b}$?

3. Interpret the equation $|\vec{a} + \vec{b}| = |\vec{a} - \vec{b}|$ geometrically.

4. If $\vec{a} + 3\vec{b} = \vec{c}$ and $\vec{b} - 3\vec{a} = 4\vec{c}$, then show that the directions of $\vec{a}$ and $\vec{b}$ are opposite to each other.

5. Find the component of $\vec{AB}$ with $A(1, 0)$ and $B(5, 0)$ along the line $y = x$.

6. Let $\vec{AB}$ be a vector in space whose magnitude is 10 units. It makes an angle of 60° with the X-axis. Find the component of $\vec{AB}$ along the coordinate axes.

7. Let $\vec{AB}$ be a vector in space whose magnitude is 20 units. It makes an angle of 60° each with the X-axis and the Y-axis. Find the acute angle it makes with Z-axis and also find its components along all the three coordinate axes.

8. Let $\vec{A}, \vec{B}, \vec{C}$ be non-coplanar vectors. Show that $-2\vec{A} + 3\vec{B} + 5\vec{C}$, $\vec{A} + 2\vec{B} + 3\vec{C}$, $7\vec{A} - \vec{C}$ are collinear.

9. Find the terminal point B of the vector $\vec{AB}$, whose initial point A is the point (4, 2) and the components along OX and OY are $5\hat{i}$ and $2\hat{j}$.

10. If the components of a vector are 5, 0 and 3 along OX, OY and OZ respectively and its terminal point is (1, – 1, 1), find its initial point.

11. Find the magnitude of the vector $-4\hat{i} + \hat{j} + 2\hat{k}$.

12. Suppose A, B, C, D are vertices of a parallelogram and A is the point (2, 3), B is (1, 4) and C is (0, – 2). Find the coordinates of D.

13. Find the ratio in which the point $7\hat{i} - \hat{k}$ divides the join of $-2\hat{i} + 3\hat{j} + 5\hat{k}$ and $\hat{i} + 2\hat{j} + 3\hat{k}$.

14. Prove that the internal bisector of angle A of a triangle ABC divides the side BC in the ratio $AB : AC$.

15. Show that the sum of three vectors determined by the medians of a triangle directed from the vertices is zero.

16. Show that the vectors $-2\hat{i} - \hat{j} + \hat{k}$, $\hat{i} - 3\hat{j} - 5\hat{k}$ and $3\hat{i} - 4\hat{j} - 4\hat{k}$ are vertices of right angled triangle.

17. Find the component of $\vec{A} = \hat{i} - 2\hat{j} + \hat{k}$ along $\vec{B} = 4\hat{i} - 4\hat{j} + 7\hat{k}$.

18. Find the scalar product of $\hat{i} + \hat{j} + \hat{k}$ and $\hat{i} + \hat{j} - 3\hat{k}$.

19. Show that $\vec{A} = 2\hat{i} + \hat{j} + \hat{k}$, $\vec{B} = \hat{i} + \hat{j} - 3\hat{k}$ and $\vec{C} = -4\hat{i} + 7\hat{j} + \hat{k}$ are mutually perpendicular.

20. Prove that the vectors $5\hat{i} + 2\hat{j} - \hat{k}$ and $\frac{7}{5}\hat{i} - \hat{j} + 5\hat{k}$ are orthogonal.

21. Determine the unit vector perpendicular to the plane of $\vec{A} = 2\hat{i} - 6\hat{j} - 3\hat{k}$ and $\vec{B} = 4\hat{i} + 3\hat{j} - \hat{k}$.

22. Find the area of parallelogram determined by vectors $(\hat{i}+2\hat{j}+3\hat{k})$ and $(3\hat{i}-2\hat{j}+\hat{k})$.

23. Find the area of triangle with vertices (1, 1, 2), (2, 3, 5) and (1, 5, 5).

24. A cube has four diagonals, connecting opposite vertices. What is the angle between an adjacent pair of diagonals ?

Answers

1. $\vec{QR}=\vec{b}-\vec{a}$

$\vec{RS}=\vec{a}-\vec{b}$

2. $5\sqrt{2}$ along QP

3. rectangle.

5. $2\sqrt{2}$

6. 5 along the Z-axis and $5\sqrt{3}$ along the Y-axis.

7. $45°, 10, 10\sqrt{2}$

9. $(9,4)$

10. $(-4,-1,-2)$

11. $\sqrt{21}$

12. $(1,-3)$

13. Externally in the ratio 2 : 3.

17. $\frac{19}{81}(4\hat{i}-4\hat{j}+7\hat{k})$,

18. -1

21. $\frac{3}{7}\hat{i}-\frac{2}{7}\hat{j}+\frac{6}{7}\hat{k}$

22. $8\sqrt{3}$ square units.

23. $\frac{\sqrt{61}}{2}$ square units

24. $\cos^{-1}\left(\frac{1}{3}\right)$

Chapter 11. Three-Dimensional Geometry

1. Find the distance between the points (2, 4, – 1) and (1, 3, 4) and then find the direction cosines of the lines joining the points.

2. Find the cartesian equation of the line which is parallel to the vector $2\hat{i}-\hat{j}+3\hat{k}$ and which passes through the point (5, – 2, 4).

3. Find the equation of a line passing through (1, – 1, 0) and parallel to the line $\frac{x-2}{3}=\frac{2y+1}{2}=\frac{5-z}{1}$

4. Find the equation of the line that passes through the origin and is parallel to $\frac{x-5}{3}=\frac{y+1}{4}=\frac{z}{1}$.

5. Find the value of k such that the lines $\frac{x-1}{3}=\frac{14-7y}{2k}=\frac{3-z}{2}$ and $\frac{7x-7}{3k}=\frac{5-y}{1}=\frac{z-6}{5}$ are perpendicular to each other.

6. Find the value of k such that lines are perpendicular to each other.

$$\frac{x-2}{3}=\frac{-(y+1)}{3}=\frac{z-2}{0} \text{ and } \frac{x-1}{1}=\frac{2y-3}{3}=\frac{(z+5)}{k}$$

7. Find the angle between the lines

$$\frac{x+4}{3}=\frac{y-1}{5}=\frac{z+3}{4} \text{ and } \frac{x+1}{1}=\frac{y-4}{1}=\frac{z-5}{2}$$

8. Find the angle between two lines, one which has direction ratios 2, 2, 1 while the other one is obtained by joining the points (3, 1, 4) and (7, 2, 12).

9. A vector $\vec{r}$ is equally inclined to the coordinates axes and $|\vec{r}|=15\sqrt{3}$. Find $\vec{r}$.

10. Let the vector equation of two lines are

$$\vec{r}=\hat{i}+2\hat{j}+\hat{k}+t(\hat{i}-\hat{j}+\hat{k}),$$

$$\vec{r}=2\hat{i}-\hat{j}-\hat{k}+u(2\hat{i}+\hat{j}+2\hat{k})$$

Find the shortest distance between them.

11. Check whether the lines $l_1: \vec{r}=\hat{i}-\hat{j}+t(2\hat{i}+\hat{k})$ and $l_2: \vec{R}=2\hat{i}-\hat{j}+u(2\hat{i}+\hat{j}-\hat{k})$ are skew lines or not. Hence, find the shortest distance between them.

12. By computing the shortest distance, determine whether the following lines intersect or not. The lines are $\frac{x-1}{2}=\frac{y+1}{3}=z$ and $\frac{x+1}{5}=\frac{y-2}{1}=z-2$.

13. Find out whether the pair of lines intersect or not :

$$\vec{r}=(1-t)\hat{i}+(t-2)\hat{j}+(3-2k)\hat{k}$$

and $\vec{r}=(s+1)\hat{i}+(2s-1)\hat{j}-(2s+1)\hat{k}$

14. Find the Cartesian equation of the plane through the point with position vector $2\hat{i}-\hat{j}+\hat{k}$ and perpendicular to the vector $4\hat{i}+2\hat{j}-3\hat{k}$.

15. Find the vector and Cartesian equation of the plane containing the two lines

$$\vec{r} = 2\hat{i} + \hat{j} - 3\hat{k} + \lambda(\hat{i} + 2\hat{j} + 5\hat{k})$$

and $$\vec{r} = 3\hat{i} + 3\hat{j} - 7\hat{k} + \mu(3\hat{i} - 2\hat{j} + 5\hat{k})$$

16. Find the equation of a plane passing through the points $A(1, 1, 1)$, $B(1, -1, 1)$ and $C(-7, -3, -5)$.
17. Obtain the equation of a plane passing through $(-2, -2, 2)$ and containing the line joining $(1, 1, 1)$ and $(1, -1, 2)$.
18. Find the equation of the plane through the intersection of $\vec{r}\cdot(\hat{i}+\hat{j}+\hat{k}) = 6$ and $\vec{r}\cdot(2\hat{i}+3\hat{j}+4\hat{k})+5 = 0$ and passing through the point $\hat{i}+\hat{j}+\hat{k}$.
19. Find the vector equation of the plane passing through the intersection of the planes $\vec{r}\cdot(2\hat{i}-7\hat{j}+4\hat{k}) = 3$ and $\vec{r}.(3\hat{i}-5\hat{j}+4\hat{k})+11 = 0$ and passing through the point $(-2, 1, 3)$.
20. Find the intercepts cut-off by the plane $x - 3y - 6z + 8 = 0$ with the coordinate axes.
21. Write the equation of the plane whose intercepts on the coordinate axes are -4, 2 and 3.
22. Find the distance between the parallel planes $x + y + z + 4 = 0$ and $x + y + z + 5 = 0$.
23. Find the distance between the parallel planes $2x - y + 2z + 3 = 0$ and $4x - 2y + 4z + 5 = 0$.
24. Show that the planes $2x + 6y + 6z = 7$ and $3x + 4y - 5y = 8$ are at right angles.
25. If the planes $\vec{r}\cdot(2\hat{i}+\hat{j}+\lambda\hat{k}) = 5$ and $\vec{r}\cdot(3\hat{i}+2\hat{j}+2\hat{k}) = 4$ are perpendicular. Find the value of λ.
26. Find the line represented by the intersection of the planes $x + y + z + 1 = 0$ and $4x + y - 2z + 2 = 0$.
27. Find the distance of the point $\hat{i}+2\hat{j}+3\hat{k}$ from the plane $\vec{r}(\hat{i}+\hat{j}+\hat{k}) = 5$.
28. Find the angle between the plane $4x - y - 2z = 6$ and the line $\frac{x-1}{1} = \frac{y-1}{-2} = \frac{z-2}{1}$
29. If the line $\vec{r} = (\hat{i}-2\hat{j}+\hat{k})+\lambda(2\hat{i}+\hat{j}+2\hat{k})$ is parallel to the plane $\vec{r}.(3\hat{i}-2\hat{j}+m\hat{k})$, then find the value of m.

Answers

1. $\sqrt{27}; \frac{-1}{\sqrt{27}}, \frac{-1}{\sqrt{27}}, \frac{5}{\sqrt{27}}$
2. $\frac{x-5}{2} = \frac{y+2}{-1} = \frac{z-4}{-3}$
3. $\frac{x-3}{1} = \frac{y+1}{1} = \frac{z}{-1}$
4. $\frac{x}{3} = \frac{y}{4} = \frac{z}{1}$
5. $\frac{70}{11}$
6. 2
7. $\cos^{-1}\left[\frac{8}{5\sqrt{3}}\right]$
8. $\cos^{-1}\left(\frac{2}{3}\right)$
9. $15(\hat{i}+\hat{j}+\hat{k})$
10. $\frac{3}{\sqrt{2}}$
11. Skew lines; $\frac{1}{\sqrt{14}}$
12. The lines do not intersect.
13. The lines do not intersect.
14. $4x + 2y - 3z - 3 = 0$
15. $\vec{r}\cdot(10\hat{i}+5\hat{j}-4\hat{k}) = 37, 10x+5y-4z = 37$
16. $3x - 4z + 1 = 0$
17. $x - 3y - 6z + 8 = 0$
18. $\vec{r}\cdot(20\hat{i}+23\hat{j}+26\hat{k}) = 69$
19. $\vec{r}\cdot[15\hat{i}-47\hat{j}+28\hat{k}] = 7$
20. $-8, \frac{8}{3}, \frac{4}{3}$ on X, Y, Z axes respectively
21. $3x - 6y - 4x + 12 = 0$
22. $\frac{1}{\sqrt{3}}$
23. $\frac{1}{6}$
25. $\lambda = -2$
26. $-\frac{1}{3}\hat{i} - \frac{2}{3}\hat{j} - \frac{t}{\sqrt{7}}(\hat{i}-2\hat{j}+\hat{k})$
27. $-\frac{1}{\sqrt{3}}$
28. $0°$
29. -2

Chapter 12. Linear Programming

1. Two foods A and B are available. Food A costs ₹ 2 per unit and food B costs ₹ 4 per unit. A unit of food A contains 10 units of carbohydrates, 20 units of fat and 15 units of protein. A unit of food B contains 25 units of carbohydrates, 10 units of fat and 20 units of protein.

A diet is to contain at least 4000 units of carbohydrates, 500 units of fat and 300 units of protein.

Formulate the problem as LPP so as to find the minimum cost for a diet that consists of a mixture of these two foods and also meets the minimum requirements.

2. A factory uses 3 machines *A, B* and *C* to produce 2 toys. The time required to produce the first type of toy is 6 hours, 8 hours and 12 hours in machines *A, B* and *C* respectively. The time required to make the second type of toy is 8 hours, 4 hours and 4 hours in machines *A, B* and *C* repectively. The maximum available time (in hours) for the machines *A, B, C* is : 380, 300 and 404 respectively.

If the profit on the first type of toy is ₹ 5 while that on the second type of toy is ₹ 3, then formulate the problem as LPP so as to find the number of toys of each type that the factory should produce to maximize its profit.

3. A brick manufacturer has two supply depots *A* and *B*, with stocks of 30,000 and 20,000 bricks repectively. He receives orders from three buildings *P, Q* and *R* for 15,000, 20,000 and 15,000 bricks respectively.

The cost (in ₹) of transporting 1000 bricks from the supply depots to the buildings are tabulated below :

From \ To	*P*	*Q*	*R*
A	40	20	30
B	20	60	40

How should the manufacturer fulfil the orders so as to keep the cost of transportation minimum ?

4. A manufacturer produces two types of steel trunks. He has two machines *A* and *B*. The first type of trunk requires 3 hours on machine *A* and 3 hours on machine *B*. The second type of trunk requires 3 hours on machine *A* and 2 hours on machine *B*. Machines *A* and *B* can work atmost for 18 hours and 15 hours per day respectively. He earns a profit of ₹ 30 and ₹ 25 per trunk of I & II type. How many trunks of each type must he make each day to make maximum profit?

5. An aeroplane can carry a maximum of 200 passengers. A profit of ₹ 400 is made on each business class ticket and a profit of ₹ 300 is made on each economy class ticket. The airlines reserves at least 20 seats for the business class on each flight. However at least four times as many passengers prefer to travel by economy class than by the business class per flight. Determine how many tickets of each type must be sold to maximize profit for the airlines for each flight. Form an LPP and solve it graphically.

6. A manufacturer of surfboards make a standard model and a competition model. Each standard board requires 6 labour-hours for fabricating and 1 labour-hour for finishing. The maximum labour-hour available per week in the fabricating and finishing departments are 120 and 30, respectively. If the company makes a profit of ₹ 40 on each standard board and ₹ 75 on each competition board, how many boards of each type should be manufactured each week to maximize the total weekly profit ?

7. Minimize and maximize,

$$Z = 10x + 5y$$

Subject to constraints

$$2x + y \geq 40$$
$$3x + y \leq 150$$
$$2x - y \geq 0$$
$$x, y \geq 0$$

8. A farmer has a supply of chemical fertilizer of type I which contains 10% nitrogen and 6% phosphoric acid and type II fertilizer which contains 5% nitrogen and 10% phosphoric acid. After testing the soil conditions of a field, it is found that at least 14 kg of nitrogen and 14 kg of phosphoric acid is required for a good crop. The fertilizer type I costs ₹ 2·00 per kg and the type II costs ₹ 3·00 per kg. How many kilograms of each fertilizer should be used to meet the requirement and to keep the cost to the minimum?

9. A small firm manufactures gold rings and chains. The total number of rings and chains manufactured per day is atmost 24. It takes 1 hour to make a ring and 30 minutes to make a chain. The maximum number of hours available per day is 16. If the profit on a ring is ₹ 300 and that on chain is ₹ 190, find the number of rings and chains that should be manufactured per day, so as to earn the maximum profit. Make it as a LPP and solve graphically.

10. A manufacturer produces nuts and bolts. It takes 1 hour of work on machine *A* and 3 hours on machine *B* to produce a package of nuts. It takes 3 hours on machine *A* and 1 hour on machine *B* to produce a package of bolts. He earns a profit of ₹ 17.50 per package on nuts and ₹ 7 per package on bolts. How many packages of each should be produced each so as to maximise his profits if he operates his machines for at the most 12 hours. Form the above as a linear programming problem and solve it graphically. **[NCERT]**

11. A factory makes tennis rackets and cricket bats. A tennis racket takes 1.5 hours of machine time and 3 hours of craftman's time in its making while a cricket bat takes 3 hours of machine time and 1 hour of craftman's time. In a day, the factory has the availability of not more than 42 hours of machine time and 24 hours of craftsman's time. If the profit on a racket and on a bat is ₹ 20 and ₹ 10 respectively, find the number of tennis rackets and cricket bats that the factory must manufacture to earn the maximum profit. Make it as a LPP and solve graphically.

12. A company produces soft drinks that has a contract which requires that a minimum of the units of the chemical A and 60 units of the chemical B go into each bottle of the drink. The chemicals are available in prepared mix packets from two different suppliers. Supplier S has a packet of mix of 4 units of A and 2 units of B that costs ₹ 10. The supplier T has a packet of mix of 1 unit of A and 1 unit of B that costs ₹ 4. How many packets of mixes from S and T should the company purchase to honour the contract requirement and yet minimise cost. Make a LPP and solve graphically.

13. A factory makes two types of items A and B, made of plywood. One piece of item A requires 5 minutes for cutting and 10 minutes for assembling. One piece of item B requires 8 minute for cutting and 8 minutes for assembling. There are 3 hours and 20 minutes available for cutting and 4 hours for assembling. The profit on one piece of item A is ₹ 5 and that of item B is ₹ 6. How many pieces of each type should the factory make so as to maximise profit ? Make it as a LPP and solve it graphically.

14. A merchant plans to sell two types of personal computers – a desktop model and a portable model that will cost ₹ 25,000 and ₹ 40,000 respectively. He estimates that the total monthly demand of computers will not exceed 250 units. Determine the number of units of each type of computers which the merchant should stock to get maximum profit if he does not want to invest more than ₹ 70 lakh and his profit on the desktop model is ₹ 4,500 and on the portable model is ₹ 5,000. Make a LPP and solve it graphically.

Answers

1. LPP : Minimize $Z = 2x + 4y$

   Subject to the constraints

$$10x + 25y \geq 4000$$
$$20x + 10y \geq 500$$
$$15x + 2y \geq 300$$
$$x \geq 0$$
$$y \geq 0$$

2. LPP : Minimize $Z = 5x + 3y$

   Subject to the constraints

$$6x + 8y \leq 380$$
$$8x + 4y \leq 300$$
$$12x + 4y \leq 404$$
$$x \geq 0$$
$$y \geq 0$$

3. To minimize the costs, the manufacturer should supply bricks as given below :

Depot	Bricks Shipping to Building			Total Bricks
	P	Q	R	
A	0	20000	10000	30000
B	15000	0	5000	20000

4. (3, 3)

5. (40, 160)

   LPP : Max. $Z = 400x + 300y$

   Subject to the constraints

$$x + y \leq 200$$
$$x \geq 20$$
$$y \geq 4x$$

6. 12 standard surfboards and 6 competition surfboards for a maximum profit of ₹ 930 per week.

7. Maximum $Z = 600$ at (30 , 60)

   Minimum $Z = 200$ at (10 , 20) and (20 , 0)

8. Type I = 100 kg and Type II = 80 kg; Minimum cost = ₹ 440

9. 8 gold rings and 16 chains, max profit = ₹ 5440.

10. 3 packages of nuts and 3 packages of bolts, max. profit ₹ 73·50.

11. 4 tennis rackets and 12 cricket bats, max profit = ₹ 200.

12. 10 packets from supplier S and 40 packets from supplier T, minimum cost = ₹ 260.

13. 8 pieces of A and 20 pieces of B, maximum profit = ₹ 260.

14. 200 desktop models and 50 portable models maximum profit ₹11,50,000.

Chapter 13. Probability

1. In a class, 40% of the students opted Spanish, 30% opted French and 20% opted both. If a student is selected at random, what is the probability that he has opted Spanish or French ?

2. In a simultaneous throw of a pair of dice, find the probability of getting a total more than 5.

3. Two dice are thrown together. What is the probability that the sum of the numbers on the two faces is divisible by 3 or 5 ?

4. If E_1 and E_2 are two independent events such that $P(E_1) = 0{\cdot}35$ and $P(E_1 \cup E_2) = 0{\cdot}60$, find $P(E_2)$.

5. What is the probability of getting two numbers whose product is odd when two dice are thrown simultaneously ?

6. An unbiased die is tossed. Find the probability of getting a multiple of 2.
7. In a lottery, there are 12 prizes and 28 blanks. A lottery is drawn at random. What is the probability of not getting a prize ?
8. Two dice are tossed. What is the probability that the total score is a prime number ?
9. A bag contains 5 green and 8 red balls. Two balls are drawn at random. Find the probability that they are of the same colour.
10. In a box, there are 8 white, 7 black and 6 violet balls. One ball is picked up randomly. What is the probability that it is neither white nor violet ?
11. Two dice are thrown simultaneously. What is the probability of getting two numbers whose product is even ?
12. Two cards are drawn at random one by one without replacement from a pack of 52 playing cards. Find the probability that both the cards are red.
13. A die is thrown 6 times. If getting an even number is a success, find the probability of getting at least 5 success.
14. A box of bulbs is inspected by examining three randomly selected bulbs drawn without replacement. If all the three bulbs are good, the box is approved for export, otherwise, it is rejected. Find the probability that a box containing 15 bulbs out of which 12 are good and 3 are bad will be approved for sale.
15. Bag A contains 2 white and 3 red balls and bag B contains 4 white and 5 red balls. One ball is drawn at random from one of the bags and it is found to be red. Find probability that it was drawn from bag B.
16. In a debate competition, the odds in favour of competitors A, B, C, D are $1:3$, $1:4$, $1:5$ and $1:6$ respectively. Find the probability that one of them wins the competition.
17. A group housing consists of 80 adults 2·5 of them are ladies and 55 gents; 10 of them are rich and the remaining poor; 20 of them are fair complexioned. What is the probability of selecting a fair complexioned rich lady ?
18. A coin is tossed 5 times. What is the probability of getting at least 3 tails ?
19. A die is thrown again and again until three sixes are obtained. Find the probability of obtaining the 3rd six in the 6th throw of the die.
20. A person buys a lottery ticket in 50 lotteries. In each of the lottery, his chance of winning a prize is $\frac{1}{100}$. What is the probability that he will win a prize
    (i) at least once
    (ii) exactly once
    (iii) at least twice ? **[NCERT]**
21. Urn I contains 3 green and 4 blue balls and urn II contains 5 green and 6 blue balls. One ball is drawn at random from one of the urns and is found to be green. Find the probability that it was drawn from urn II.
22. Two cards are drawn from a pack of 52 cards. What is the probability that both are black or both are queen?
23. A card from a pack of 52 cards is lost. From the remaining cards of the pack, two cards are drawn and are found to be both hearts. Find the probability of the lost card being a heart.
24. Six coins are tossed simultaneously. Find the probability of getting
    (i) 3 heads
    (ii) no head
    (iii) at least one head
    (iv) not more than 3 head
25. On his vacation, Amar visited four cities A, B, C and D in a random order. What is the probability that he visits :
    (i) A before B
    (ii) A before B and B before C
    (iii) A first and B last
    (iv) A just before B
26. Bag A contains 4 black and 6 red balls and bag B contains 7 black and 3 red balls. A die is thrown. If 1 or 2 appears on it, then bag A is chosen, otherwise bag B. If two balls are drawn at random (without replacement) from the selected bag, find the probability of one of them being red and another black.
27. Two dice are thrown. Find the probability that the numbers appeared have a sum 8 if it is known that the second die always exhibits 4. **[HOTS]**
28. Suppose that 5% of men and 0·25% of women have grey hair. A grey haired person is selected at random. What is the probability of this person being male ? Assume that there are equal number of males and females.

29. A committee of 4 students is selected at random from a group consisting of 8 boys and 4 girls. Given that there is at least one girl in the committee, calculate the probability that there are exactly 2 girls in the committee.

30. In a bulb factory, machines A, B and C manufactures 60%, 30% and 10% bulbs respectively. 1%, 2% and 3% of the bulbs produced respectively by A, B and C are found to be defective. A bulb is picked up at random from the product and is found to be defective. Find the probability that this bulb was produced by the machine A.

Answers

1. $\frac{1}{2}$ **2.** $\frac{13}{18}$

3. $\frac{19}{36}$ **4.** $\frac{5}{13}$

5. $\frac{1}{4}$ **6.** $\frac{1}{2}$

7. $\frac{7}{10}$ **8.** $\frac{5}{12}$

9. $\frac{19}{39}$ **10.** $\frac{1}{3}$

11. $\frac{3}{4}$ **12.** $\frac{25}{102}$

13. $\frac{7}{64}$ **14.** $\frac{44}{91}$

15. $\frac{25}{52}$ **16.** $\frac{114}{120}$

17. $\frac{5}{512}$ **18.** $\frac{1}{2}$

19. $\frac{625}{23328}$

20. (i) $1-\left(\frac{99}{100}\right)^{50}$ (ii) $\frac{1}{2}\cdot\left(\frac{99}{100}\right)^{49}$

(iii) $1-\left(\frac{149}{100}\right)\left(\frac{99}{100}\right)^{49}$

21. $\frac{38}{68}$

22. $\frac{55}{221}$ **23.** $\frac{11}{50}$

24. (i) $\frac{5}{16}$, (ii) $\frac{1}{64}$, (iii) $\frac{63}{64}$, (iv) $\frac{21}{32}$

25. (i) $\frac{1}{2}$, (ii) $\frac{1}{6}$ (iii) $\frac{1}{12}$ (iv) $\frac{1}{4}$

26. $\frac{22}{45}$ **27.** $\frac{1}{6}$

28. $\frac{20}{21}$ **29.** $\frac{168}{425}$

30. $\frac{2}{5}$

Topper's Answers - 2019

Section - A

1. Find the order and the degree of the differential equation $x^2 \frac{d^2y}{dx^2} = \left\{1 + \left(\frac{dy}{dx}\right)^2\right\}^4$.

Q1.

$$x^2 \frac{d^2y}{dx^2} = \left\{1 + \left(\frac{dy}{dx}\right)^2\right\}^4$$

ORDER = 2.

DEGREE = 1

2. If $f(x) = x + 7$ and $g(x) = x - 7$, $x \in R$, then find $\frac{d}{dx}$ (fog) (x).

Q2.

$f(x) = x+7$; $g(x) = x-7$

$fog(x) = f(g(x))$

$= f(x-7)$

$= (x-7) + 7$

$= x \quad \forall\, x \in R$

$\frac{d}{dx} fog(x) = \frac{d}{dx}(x) = 1$

$\boxed{\frac{d}{dx} fog(x) = 1}$

3. Find the value of $x - y$, if $2\begin{bmatrix} 1 & 3 \\ 0 & x \end{bmatrix} + \begin{bmatrix} y & 0 \\ 1 & 2 \end{bmatrix} = \begin{bmatrix} 5 & 6 \\ 1 & 8 \end{bmatrix}$

Q3

$$2\begin{bmatrix} 1 & 3 \\ 0 & x \end{bmatrix} + \begin{bmatrix} y & 0 \\ 1 & 2 \end{bmatrix} = \begin{bmatrix} 5 & 6 \\ 1 & 8 \end{bmatrix}$$

$$\Rightarrow \begin{bmatrix} 2 & 6 \\ 0 & 2x \end{bmatrix} + \begin{bmatrix} y & 0 \\ 1 & 2 \end{bmatrix} = \begin{bmatrix} 5 & 6 \\ 1 & 8 \end{bmatrix}$$

$$\Rightarrow \begin{bmatrix} 2+y & 6 \\ 1 & 2x+2 \end{bmatrix} = \begin{bmatrix} 5 & 6 \\ 1 & 8 \end{bmatrix}$$

comparing corresponding elements of each matrix,

$2+y = 5 \qquad 2x+2 = 8$

$\boxed{y = 3} \qquad \boxed{x = 3}$

$x - y = 3 - 3$

$\boxed{x - y = 0}$

4. If a line makes angles 90°, 135°, 45° with the x, y and z axes respectively, find its direction cosines.

OR

Find the vector equation of the line which passes through the point (3, 4, 5) and is parallel to the vector $2\hat{i} + 2\hat{j} - 3\hat{k}$

Q4. $\vec{r} = \vec{a} + \lambda\vec{b}$ where $\vec{a}$ = position vector of pt on line, $\vec{b}$ = parallel vector to line

$\vec{r} = (3\hat{i} + 4\hat{j} + 5\hat{k}) + \lambda(2\hat{i} + 2\hat{j} - 3\hat{k})$

Also, $\vec{r} = (3+2\lambda)\hat{i} + (4+2\lambda)\hat{j} + (5-3\lambda)\hat{k}$

Section – B

5. Examine whether the operation * defined on R by $a * b = ab + 1$ is (i) a binary or not. (ii) if a binary operation, is it associative or not ?

Q5. $a * b = ab + 1$

i) For all $(a, b) \in (R \times R)$

$\Rightarrow ab + 1 \in R$ [∵ If $a, b \in R \Rightarrow ab \in R \Rightarrow (ab+1) \in R$] [∵ Multiplication is binary operation]

∴ a R b relates to a unique element in R and hence is a binary operation from R×R to R

ii) For binary operation to be associative $(a * b) * c = a * (b * c)$

LHS $= (a * b) * c$

$= (ab+1) * c$

$= (ab+1)c + 1$

$= abc + c + 1$

RHS $= a * (b * c)$

$= a * (bc+1)$

$= a(bc+1) + 1$

$= abc + a + 1$

LHS ≠ RHS

∴ It is NOT ASSOCIATIVE

6. If $A = \begin{bmatrix} 2 & 0 & 1 \\ 2 & 1 & 3 \\ 1 & -1 & 0 \end{bmatrix}$, then find $(A^2 - 5A)$.

Q6. $A = \begin{bmatrix} 2 & 0 & 1 \\ 2 & 1 & 3 \\ 1 & -1 & 0 \end{bmatrix}$

$A^2 = AA$

$= \begin{bmatrix} 2 & 0 & 1 \\ 2 & 1 & 3 \\ 1 & -1 & 0 \end{bmatrix} \begin{bmatrix} 2 & 0 & 1 \\ 2 & 1 & 3 \\ 1 & -1 & 0 \end{bmatrix}$

$= \begin{bmatrix} 5 & -1 & 2 \\ 9 & -2 & 5 \\ 0 & -1 & -2 \end{bmatrix}$

$A^2 - 5A = A^2 + (-5)A$

$= \begin{bmatrix} 5 & -1 & 2 \\ 9 & -2 & 5 \\ 0 & -1 & -2 \end{bmatrix} + (-5)\begin{bmatrix} 2 & 0 & 1 \\ 2 & 1 & 3 \\ 1 & -1 & 0 \end{bmatrix}$

$= \begin{bmatrix} 5 & -1 & 2 \\ 9 & -2 & 5 \\ 0 & -1 & -2 \end{bmatrix} + \begin{bmatrix} -10 & 0 & -5 \\ -10 & -5 & -15 \\ -5 & 5 & 0 \end{bmatrix}$

$A^2 - 5A = A^2 + (-5A)$

$= \begin{bmatrix} -5 & -1 & -3 \\ -1 & -7 & -10 \\ -5 & 4 & -2 \end{bmatrix}$

7. Find : $\int \sqrt{1 - \sin 2x}\, dx, \frac{\pi}{4} < x < \frac{\pi}{2}$

OR

Find : $\int \sin^{-1}(2x)\, dx$.

Q7. $I = \int \underset{I}{1} \cdot \underset{II}{\sin^{-1}(2x)}\, dx$

$I = \{\sin^{-1}(2x)\}(x) - \int \frac{(x)(2)}{\sqrt{1-4x^2}} dx$ [INTEGRATION by parts]

$I = x \sin^{-1}(2x) - \frac{1}{2}\int \frac{2x}{\sqrt{(\frac{1}{2})^2 - x^2}} dx$

$I = x\sin^{-1}(2x) + \frac{1}{4}\int \frac{(-8x)\,dx}{\sqrt{1-4x^2}} = x\sin^{-1}(2x) + \frac{1}{4} I_1$ where $I_1 = \int \frac{-8x\, dx}{\sqrt{1-4x^2}}$

Now, $I_1 = \int \frac{dt}{\sqrt{t}}$ let $1 - 4x^2 = t$

$-8x\,dx = dt$

$= 2\sqrt{t} + C$

$= 2\sqrt{1-4x^2} + C$

$I = x\sin^{-1}(2x) + \frac{1}{4}(2\sqrt{1-4x^2}) + C$

$I = x\sin^{-1}(2x) + \frac{\sqrt{1-4x^2}}{2} + C$

8. Form the differential equation representing the family fo curves $y = e^{2x}(a + bx)$, where 'a' and 'b' are arbitrary constants.

Q8. $y = e^{2x}(a + bx)$ —(1)

Diff. both sides w.r.t x.

$\frac{dy}{dx} = e^{2x}(b) + (a+bx)(e^{2x})(2)$

$\Rightarrow \frac{dy}{dx} = e^{2x}(b + 2a + 2bx)$ —(2)

Diff. both sides w.r.t. x.

$\frac{d^2y}{dx^2} = e^{2x}(2b) + (b + 2a + 2bx)(e^{2x})(2)$

$= e^{2x}(4b + 4a + 4bx)$

$\frac{d^2y}{dx^2} = 2e^{2x}(2b + 2a + 2bx)$ —(3)

Subtract $2\times(2)$ eqn from eqn (3)

$$\frac{d^2y}{dx^2} - 2\frac{dy}{dx} = e^{2x}(4b+4a+4bx) - e^{2x}(2b+4a+4bx)$$

$$= e^{2x}(2b)$$

$$\Rightarrow \frac{d^2y}{dx^2} - 2\frac{dy}{dx} = 2(e^{2x})(b) \quad \text{—— (4)}$$

In eq^n (3)

$$\frac{d^2y}{dx^2} = e^{2x}(4b+4a+4bx)$$

$$= 2\left(\frac{d^2y}{dx^2} - 2\frac{dy}{dx}\right) + 4e^{2x}(a+bx) \quad [\text{From (4)}]$$

$$\frac{d^2y}{dx^2} = 2\frac{d^2y}{dx^2} - 4\frac{dy}{dx} + 4y \quad [\text{From (1)}]$$

∴ The required differential eq^n is

$$\boxed{\frac{d^2y}{dx^2} - 4\frac{dy}{dx} + 4y = 0} \Rightarrow y'' - 4y' + 4y = 0.$$

where $y'' = \frac{d^2y}{dx^2}$, $y' = \frac{dy}{dx}$

9. A die is thrown 6 times. If "getting an odd number" is a "sucess", what is the probability of (i) 5 successes ? (ii) atmost 5 successes ?

OR

The random variable X has a probability distribution P(X) of the following form, where 'k' is some number.

$$P(X = x) = \begin{cases} k, & \text{if } x = 0 \\ 2k, & \text{if } x = 1 \\ 3k, & \text{if } x = 2 \\ 0, & \text{otherwise} \end{cases}$$

Determine the value of 'k'

Q9

$$\sum_{i=0}^{2} P(X_i) = 1 \quad \left[\because \sum_{i=0}^{n} P(x_i) = 1\right]$$

[Sum of all probabilities = 1] disjoint exhaustive exclusive

$$\Rightarrow k + 2k + 3k = 1$$

$$\Rightarrow 6k = 1$$

$$\boxed{k = \frac{1}{6}}$$

10. A die marked 1, 2, 3 in red and 4, 5, 6 in green is tossed. Let A be event "number is even" and B be the event "number is marked red". Find whether the events A and B are independent or not.

Q10. $P(A) = \frac{3}{6} = \frac{1}{2}$ $\left[\because P(E) = \frac{\text{No. of favourable outcomes}}{\text{Total no. of possible outcomes}}\right]$

$P(B) = \frac{3}{6} = \frac{1}{2}$ [⇒ Sample space S = {1r, 2r, 3r, 4g, 5g, 6g}, A = {2r, 4g, 6g}, B = {1r, 2r, 3r}, A∩B = {2r}]

$P(A \cap B) = P(\text{no. which is even and red}) = \frac{1}{6}$ [∵ No. which is even and red = {2r}]

$P(A) \cdot P(B) = \frac{1}{2} \times \frac{1}{2} = \frac{1}{4}$

$P(A \cap B) \neq P(A) \cdot P(B)$

∴ Events are not NOT INDEPENDENT.

11. If the sum of two unit vectors is a unit vector, prove that the magnitude of their difference is.

OR

If $\vec{a} = 2\hat{i} + 3\hat{j} + \hat{k}$, $\vec{b} = \hat{i} - 2\hat{j} + \hat{k}$ and $\vec{c} = -3\hat{i} + \hat{j} + 2\hat{k}$, find $\left[\vec{a}\,\vec{b}\,\vec{c}\right]$

Q11. $\vec{a} = 2\hat{i} + 3\hat{j} + \hat{k}$, $\vec{b} = \hat{i} - 2\hat{j} + \hat{k}$, $\vec{c} = -3\hat{i} + \hat{j} + 2\hat{k}$

$[\vec{a}\,\vec{b}\,\vec{c}] = \vec{a} \cdot (\vec{b} \times \vec{c})$

$= \begin{vmatrix} a_1 & a_2 & a_3 \\ b_1 & b_2 & b_3 \\ c_1 & c_2 & c_3 \end{vmatrix}$ where $\vec{a} = a_1\hat{i} + a_2\hat{j} + a_3\hat{k} = 2\hat{i} + 3\hat{j} + \hat{k}$, $\vec{b} = b_1\hat{i} + b_2\hat{j} + b_3\hat{k} = \hat{i} - 2\hat{j} + \hat{k}$, $\vec{c} = c_1\hat{i} + c_2\hat{j} + c_3\hat{k} = -3\hat{i} + \hat{j} + 2\hat{k}$

$= \begin{vmatrix} 2 & 3 & 1 \\ 1 & -2 & 1 \\ -3 & 1 & 2 \end{vmatrix}$

$= 2(-4-1) - 3(2+3) + 1(1-6)$

$= -10 - 15 - 5$

∴ $[\vec{a}\,\vec{b}\,\vec{c}] = -30$

12. Find : $\int \frac{\tan^2 x \sec^2 x}{1 - \tan^6 x} dx.$

Q12. $I = \int \frac{\tan^2 x \cdot \sec^2 x}{1 - \tan^6 x} dx$

let $\tan^3 x = t$

$3\tan^2 x \sec^2 x\, dx = dt$

$I = \frac{1}{3}\int \frac{dt}{(1)^2 - t^2}$

$= \frac{1}{3} \cdot \frac{1}{2.(1)} \log\left|\frac{1+t}{1-t}\right| + C$ $\left[\because \int \frac{dx}{a^2 - x^2} = \log\left|\frac{a+x}{a-x}\right| + c\right]$

$I = \frac{1}{6} \log\left|\frac{1+\tan^3 x}{1-\tan^3 x}\right| + c$

Section – C

13. Solve for x : $\tan^{-1}(2x) + \tan^{-1}(3x) = \frac{\pi}{4}$.

Q13. $\tan^{-1}(2x) + \tan^{-1}(3x) = \pi/4$

$\tan^{-1}A + \tan^{-1}B = \tan^{-1}\left(\frac{A+B}{1-AB}\right)$, $AB < 1$

$\Rightarrow \tan^{-1}\left(\frac{5x}{1-6x^2}\right) = \frac{\pi}{4}$ and $(2x)(3x) < 1$

$6x^2 < 1$

$-\frac{1}{\sqrt{6}} < x < \frac{1}{\sqrt{6}}$

Taking tan on both side

$\Rightarrow \frac{5x}{1-6x^2} = 1$

$\Rightarrow 6x^2 + 5x - 1 = 0$

$\Rightarrow 6x^2 + 6x - x - 1 = 0$

$\Rightarrow (6x-1)(x+1) = 0$

$\Rightarrow x = \frac{1}{6}$ or $x = -1$

$\therefore x = -1$ is not a solution $\left[\because -1 \notin \left(-\frac{1}{\sqrt{6}}, \frac{1}{\sqrt{6}}\right)\right]$

$\Rightarrow \boxed{x = \frac{1}{6}}$ is sol^n.

VERIFICATION

For $x = 1/6$

LHS : $\tan^{-1}\left(\frac{1}{3}\right) + \tan^{-1}\left(\frac{1}{2}\right) = \tan^{-1}\left(\frac{5}{5}\right) = \pi/4 =$ RHS.

LHS = RHS

$\Rightarrow x = 1/6$ is a sol^n

For $(x) = -1$

$\tan^{-1}(-2) - \tan^{-1}(-3) = \frac{\pi}{2} - \cot^{-1}(-2) - \frac{\pi}{2} + \cot^{-1}(-3)$

$= \pi - \cot^{-1}(3) - \pi + \cot^{-1}(2)$

$= \tan^{-1}\left(\frac{1}{2}\right) - \tan^{-1}\left(\frac{1}{3}\right) = \tan^{-1}\left(\frac{\frac{1}{2}-\frac{1}{3}}{1+\frac{1}{6}}\right) = \tan^{-1}\left(\frac{1}{7}\right) \neq$ RHS

LHS $\neq$ RHS

$\therefore x = -1$ is NOT sol^n

14. If $\log(x^2 + y^2) = 2\tan^{-1}\left(\frac{y}{x}\right)$, show that $\frac{dy}{dx} = \frac{x+y}{x-y}$

OR

If $x^y - y^x = a^b$, find $\frac{dy}{dx}$.

Q14. $\log(x^2+y^2) = 2\tan^{-1}\frac{y}{x}$

Diff. both sides w.r.t x

$\frac{1}{x^2+y^2}\left(2x + 2y\frac{dy}{dx}\right) = \frac{2}{1+\frac{y^2}{x^2}} \frac{d}{dx}\left(\frac{y}{x}\right)$ $\left[\because \frac{d}{dx}(\tan^{-1}x) = \frac{1}{1+x^2}; \frac{d}{dx}(\log x) = \frac{1}{x}\right]$

$\Rightarrow \frac{x + yy'}{x^2+y^2} = \frac{x^2}{x^2+y^2}\frac{(xy'-y)}{x^2}$ [where $y' = dy/dx$]

$\Rightarrow x + yy' = xy' - y$

$\Rightarrow x + y = y'(x - y)$

$\Rightarrow \quad y' = \frac{x+y}{x-y}$

$\boxed{\frac{dy}{dx} = \frac{x+y}{x-y}}$

Hence proved

15. Find : $\int \frac{3x+5}{x^2+3x-18}\,dx$.

Q15 $I = \int \frac{3x+5}{x^2+3x-18}\,dx$

$3x+5 = a\left[\frac{d}{dx}(x^2+3x-18)\right] + b$

$= a(2x+3) + b$

comparing coefficients on both sides,

$3 = 2a \qquad 3a+b = 5$

$\boxed{a = \frac{3}{2}} \qquad b = 5 - 3\left(\frac{3}{2}\right) = \frac{1}{2} \Rightarrow \boxed{b = \frac{1}{2}}$

$I = \int \frac{\frac{3}{2}(2x+3) + \frac{1}{2}}{x^2+3x-18}\,dx$

$= \frac{3}{2}\int \frac{(2x+3)\,dx}{x^2+3x-18} + \frac{1}{2}\int \frac{dx}{x^2+3x-18}$

$= \frac{3}{2} I_1 + \frac{1}{2} I_2$

$I_1 = \int \frac{2x+3}{x^2+3x-18}\,dx$

Let $x^2+3x-18 = t$

$(2x+3)\,dx = dt$

$I_1 = \int \frac{dt}{t} = \log|t| + c_1$

$I_1 = \log|x^2+3x-18| + c_1$

Ans. $I_2 = \int \frac{dx}{x^2+3x-18} = \int \frac{dx}{x^2+3x+\left(\frac{3}{2}\right)^2-18-\left(\frac{3}{2}\right)^2}$

$= \int \frac{dx}{\left(x+\frac{3}{2}\right)^2-18-\frac{9}{4}} = \int \frac{dx}{\left(x+\frac{3}{2}\right)^2-\frac{81}{4}}$

$I_2 = \int \frac{dx}{\left(x+\frac{3}{2}\right)^2-\left(\frac{9}{2}\right)^2}$

$= \frac{1}{2\cdot\frac{9}{2}} \log\left|\frac{\left(x+\frac{3}{2}\right)-\frac{9}{2}}{\left(x+\frac{3}{2}\right)+\frac{9}{2}}\right| + c_2$

$= \frac{1}{9}\log\left|\frac{x-3}{x+6}\right| + c_2$

$$I = \frac{3}{2}\log|x^2+3x-18| + c_1 + \frac{1}{18}\log\left|\frac{x-3}{x+6}\right| + c_2$$

$$I = \frac{3}{2}\log|x^2+3x-18| + \frac{1}{18}\log\left|\frac{x-3}{x+6}\right| + c \quad \text{where } c = c_1 + c_2 = \text{constt.}$$

16. Prove that $\int_0^a f(x)\,dx = \int_0^a f(a-x)\,dx$, hence evaluate $\int_0^{\pi} \frac{x\sin x}{1+\cos^2 x}dx$.

Q16. T.P: $\int_0^a f(x)\,dx = \int_0^a f(a-x)\,dx$

LHS : $I_1 = \int_0^a f(x)\,dx$

let $x = a - t$ When $x = 0$, $t = a$

$dx = -dt$ $x = a$, $t = 0$

$$I_1 = -\int_a^0 f(a-t)\,dt$$

$$= \int_0^a f(a-t)\,dt \quad \left[\because \int_a^b f(x)\,dx = -\int_b^a f(x)\,dx\right]$$

$$I_1 = \int_0^a f(a-x)\,dx \quad \left[\because \int_a^b f(t)\,dt = \int_a^b f(x)\,dx\right]$$

LHS = RHS

Hence proved.

$$I = \int_0^{\pi} \frac{x\sin x}{1+\cos^2 x}dx \quad \text{---(1)}$$

$$I = \int_0^{\pi} \frac{(\pi-x)\sin(\pi-x)}{1+\cos^2(\pi-x)}dx \quad \left[\because \int_0^a f(x)\,dx = \int_0^a f(a-x)\,dx\right]$$

$$I = \int_0^{\pi} \frac{(\pi-x)\sin x}{1+\cos^2 x}dx \quad \text{---(2)}$$

Adding eqn (1), (2)

$$2I = \int_0^{\pi} \frac{\pi\sin x}{1+\cos^2 x}dx$$

$$\Rightarrow 2I = \pi\int_0^{\pi} \frac{\sin x}{1+\cos^2 x}dx$$

let $\cos x = t$ when $x = 0$, $t = 1$

$-\sin x\,dx = dt$ $x = \pi$, $t = -1$

$$\Rightarrow I = \frac{\pi}{2}\int_1^{-1} \frac{-dt}{1+t^2}$$

$$= -\frac{\pi}{2}\left[\tan^{-1} t\right]_1^{-1} = -\frac{\pi}{2}\left[-\frac{\pi}{4} - \frac{\pi}{4}\right] = \frac{\pi}{2}\cdot\frac{\pi}{2}$$

$$I = \frac{\pi^2}{4}$$

$$\therefore \int_0^{\pi} \frac{x\sin x}{1+\cos^2 x} = \frac{\pi^2}{4}$$

17. If $\hat{i}+\hat{j}+k, 2\hat{i}+5\hat{j}, 3\hat{i}+2\hat{j}-3\hat{k}$ and $\hat{i}-6\hat{j}-\hat{k}$ respectively are the position vectors of points A, B, C and D, then find the angle between the straight lines AB and CD. Find whether $\overrightarrow{AB}$ and $\overrightarrow{CD}$ are collinear or not.

Q17

$\vec{A} = \hat{i} + \hat{j} + \hat{k}$

$\vec{B} = 2\hat{i} + 5\hat{j} + 0\hat{k}$

$\vec{C} = 3\hat{i} + 2\hat{j} - 3\hat{k}$

$\vec{D} = \hat{i} - 6\hat{j} - \hat{k}$

$\overrightarrow{AB}$ = Position vector of $\vec{B}$ – Position vector of $\vec{A}$

$\vec{m} = \hat{i} + 4\hat{j} - \hat{k}$

$\overrightarrow{CD}$ = Position vector of $\vec{D}$ – Position vector of $\vec{C}$

$\vec{n} = -2\hat{i} - 8\hat{j} + 2\hat{k}$

Angle b/w $\overrightarrow{AB}$ and $\overrightarrow{CD}$ = θ

$$\cos\theta = \frac{\vec{m}\cdot\vec{n}}{|\vec{m}|\,|\vec{n}|}$$

$$= \frac{m_1n_1 + m_2n_2 + m_3n_3}{\sqrt{m_1^2+m_2^2+m_3^2}\,\sqrt{n_1^2+n_2^2+n_3^2}}$$

$$= \frac{-2-32-2}{\sqrt{1+16+1}\,\sqrt{4+64+4}}$$

$$= \frac{-36}{\sqrt{18}\,\sqrt{72}} = \frac{-36}{(3\sqrt{2})(6\sqrt{2})} = \frac{-36}{18\times 2}$$

$$= \frac{-36}{36} = -1$$

$\theta = \cos^{-1}(-1) \Rightarrow \boxed{\theta = \pi}$

$\therefore$ Since, $\overrightarrow{AB}$ and $\overrightarrow{CD}$ are antiparallel, they ARE COLLINEAR.

18. Using properties of determinants, prove the following :

$$\begin{vmatrix} a+b+c & -c & -b \\ -c & a+b+c & -a \\ -b & -a & a+b+c \end{vmatrix} = 2(a+b)(b+c)(c+a).$$

Q18.
$$\Delta = \begin{vmatrix} a+b+c & -c & -b \\ -c & a+b+c & -a \\ -b & -a & a+b+c \end{vmatrix}$$

$$\Delta = \begin{vmatrix} a+b & a+b & -(a+b) \\ -c & a+b+c & -a \\ -b & -a & a+b+c \end{vmatrix}$$ $R_1 \to R_1 + R_2$

$$= (a+b) \begin{vmatrix} 1 & 1 & -1 \\ -c & a+b+c & -a \\ -b & -a & a+b+c \end{vmatrix}$$ Taking (a+b) common from R_1

$$= (a+b) \begin{vmatrix} 1 & 1 & -1 \\ -(b+c) & b+c & b+c \\ -b & -a & a+b+c \end{vmatrix}$$ $R_2 \to R_2 + R_3$

$$= (a+b)(b+c) \begin{vmatrix} 1 & 1 & -1 \\ -1 & 1 & 1 \\ -b & -a & a+b+c \end{vmatrix}$$ Taking (b+c) common from R_2

$$= (a+b)(b+c) \begin{vmatrix} 0 & 0 & -1 \\ 0 & 2 & 1 \\ a+c & b+c & a+b+c \end{vmatrix}$$ $C_1 \to C_1 + C_3$, $C_2 \to C_2 + C_3$

$$\Delta = (a+b)(b+c)(c+a) \begin{vmatrix} 0 & 0 & -1 \\ 0 & 2 & 1 \\ 1 & b+c & a+b+c \end{vmatrix}$$ Taking (a+c) common from C_1

Expanding along C_1

$\Delta = (a+b)(b+c)(c+a)(2)$

$\Delta = 2(a+b)(b+c)(c+a)$

Hence proved

19. If $x = \cos t + \log \tan\left(\frac{t}{2}\right)$, $y = \sin t$, then find the values of $\frac{d^2y}{dt^2}$ and $\frac{d^2y}{dx^2}$ at $t = \frac{\pi}{4}$.

Q19. $y = \sin t$ $\qquad x = \cos t + \log(\tan(t/2))$

Diff. w.r.t t $\qquad$ Diff. w.r.t t

$\frac{dy}{dt} = \cos t$ — (1)

$\frac{dx}{dt} = -\sin t + \frac{1}{\tan t/2} \cdot \sec^2 t/2 \cdot \frac{1}{2}$

Diff. w.r.t t

$= -\sin t + \frac{\cos t/2}{2\sin t/2 \cos^2 t/2}$

$\frac{d^2y}{dt^2} = -\sin t$

$= -\sin t + \frac{1}{\sin t}$ $[\because 2\sin A\cos A = \sin 2A]$

$\left.\frac{d^2y}{dt^2}\right]_{t=\pi/4} = -\sin\frac{\pi}{4}$

$\frac{dx}{dt} = -\sin t + \frac{1}{\sin t}$ — (2) $\Rightarrow \frac{dx}{dt} = \frac{\cos^2 t}{\sin t}$

$\left.\frac{d^2y}{dt^2}\right]_{t=\pi/4} = -\frac{1}{\sqrt{2}}$

P.T.O

Q19 Contd...

Divide eqn (1) by (2)

$$\frac{dy/dt}{dx/dt} = \frac{\cos t}{-\sin t + \frac{1}{\sin t}}$$

$$\frac{dy}{dx} = \frac{\cos t . \sin t}{1 - \sin^2 t} = \frac{\sin t}{\cos t} = \tan t$$

Diff. w.r.t x.

$$\frac{d^2y}{dx^2} = \frac{d(\tan t)}{dt} \frac{dt}{dx}$$

$$= \sec^2 t . \frac{1}{dx/dt} = \frac{\sec^2 t}{-\sin t + \frac{1}{\sin t}}$$

$$= \frac{\sec^2 t . \sin t}{1 - \sin^2 t} = \frac{\sin t}{\cos^2 t . \cos^2 t} = \frac{\sin t}{\cos^4 t}$$

$$= \sin t . \sec^4 t$$

$$\left.\frac{d^2y}{dx^2}\right]_{t=\pi/4} = \frac{1}{\sqrt{2}} (\sqrt{2})^4 = \frac{4}{\sqrt{2}} = 2\sqrt{2}$$

$$\therefore \left.\frac{d^2y}{dx^2}\right]_{t=\pi/4} = 2\sqrt{2}$$

20. Show that the relation R on R defined as R = {(*a*, *b*) : *a* ≤ *b*}, is reflexive, and transitive but not symmetric.

OR

Prove that the function f : N → N, defined by $f(x) = x^2 + x + 1$ is one-one but not onto. Find inverse of f : N → S, where S is range of f.

Q20 $R = \{(a, b) : a \le b\}$

REFLEXIVE :

Every element $a \in \mathbb{R}$ is equal to itself

$\Rightarrow a = a$

$\Rightarrow a \le a$ is true

$\therefore (a, a) \in R$ for all $a \in \mathbb{R}$ where $\mathbb{R}$ = set of all Real nos

The relation is REFLEXIVE R = Relation.

TRANSITIVE :

For all $(a, b) \in R$ and $(b, c) \in R$

$a \le b$ and $b \le c$ where, $a, b, c \in \mathbb{R}$

$\Rightarrow a \le c$

$\Rightarrow (a, c) \in R$

$\therefore$ The ~~set is tr~~ relation is TRANSITIVE

$\because$ for $(a, b), (b, c) \in R$, ~~(a, c) ∈ R~~ $(a, c) \in R$

SYMMETRIC : For relation to be symmetric, for all $(a, b) \in R$, (b, a) should also exist in R

~~$a \leq b$~~ $a \leq b$

$b \nleq a$ → This relation is true only $a = b = 1$.

For eg: $\frac{1}{2} \leq 1 \Rightarrow (\frac{1}{2}, 1) \in R$

but $1 \nleq \frac{1}{2}$ ∴ $(1, \frac{1}{2}) \notin R$

∴ Relation is NOT SYMMETRIC

21. Find the equation of tangent to the curve $y = \sqrt{3x - 2}$ which is parallel to the line $4x - 2y + 5 = 0$. Also, write the equation of normal to the curve at the point of contact.

Q21. $y = \sqrt{3x-2}$

$\left.\frac{dy}{dx}\right|_{(x_1,y_1)} = \frac{3}{2\sqrt{3x_1-2}} = \frac{3}{2y_1}$ = slope of tangent = m

Acc. to ques. m = 2 [∵ Slope of $(4x-2y+5=0)$ is $m_1 = 2$ and ∥el line have equal slope]

$\Rightarrow \frac{3}{2y_1} = 2 \Rightarrow \boxed{y_1 = \frac{3}{4}}$

$\Rightarrow \frac{3}{2\sqrt{3x_1-2}} = 2 \Rightarrow 3 = 4\sqrt{3x_1-2}$

$\Rightarrow 9 = 16(3x_1-2)$ squaring both side

$\Rightarrow 3x_1 - 2 = \frac{9}{16} \Rightarrow x_1 = \frac{41}{48} \Rightarrow \boxed{x_1 = \frac{41}{48}}$

$x_1 = \frac{41}{48}$

$\Rightarrow y_1 = \sqrt{3x_1-2} = \sqrt{3 \times \frac{41}{48} - 2} = \sqrt{\frac{41-32}{16}} = \pm\frac{3}{4} = \left|\frac{3}{4}\right| = \sqrt{\frac{9}{16}}$

At $\boxed{\left(\frac{41}{48}, \frac{3}{4}\right)}$ and $\left(\frac{41}{48}, \frac{-3}{4}\right)$ ~~slope of~~ tangent is parallel to given line.

Not possible (But ∵ $y \nless 0 \Rightarrow y \geq 0$)

EQUATION OF TANGENT

For $x_1 = \frac{41}{48}$, $y_1 = \frac{3}{4}$

$y - \frac{3}{4} = 2\left(x - \frac{41}{48}\right)$

$y - \frac{3}{4} = 2x - \frac{41}{24}$

$2x - y - \frac{41}{24} + \frac{18}{24} = 0$

$\Rightarrow 2x - y - \frac{23}{24} = 0$

$\boxed{48x - 24y - 23 = 0}$

For $x_1 = \frac{41}{48}$, ~~$y_1 = -\frac{3}{4}$~~

$y + \frac{3}{4} = 2\left(x - \frac{41}{48}\right)$

$2x - y - \frac{41}{24} - \frac{18}{24} = 0$

~~$48x - 24y - 59 = 0$~~ ✗

↑ NOT POSSIBLE

Slope of normal $= \frac{-1}{\left[\frac{dy}{dx}\right]_{(x_1,y_1)}} = \frac{-1}{2}$

EQUATION OF NORMAL :

At $\left(\frac{41}{48}, \frac{3}{4}\right)$

$y - \frac{3}{4} = \frac{-1}{2}\left(x - \frac{41}{48}\right)$

$\Rightarrow y - \frac{3}{4} = \frac{-x}{2} + \frac{41}{96}$

$\Rightarrow y + \frac{x}{2} - \frac{3}{4} - \frac{41}{96} = 0$

$\Rightarrow y + \frac{x}{2} + \frac{(-72-41)}{96} = 0$

$\Rightarrow$ $96y + 48x - 113 = 0$

~~At~~ At $\left(\frac{41}{48}, \frac{-3}{4}\right)$

$y + \frac{3}{4} = \frac{-1}{2}\left(x - \frac{41}{48}\right)$

$\Rightarrow y + \frac{3}{4} + \frac{x}{2} - \frac{41}{96} = 0$

$\Rightarrow y + \frac{x}{2} + \frac{72-41}{96} = 0$

$\Rightarrow$ $96y + 48x + 31 = 0$

NOT POSSIBLE

22. Solve the differential equation : $x\,dy - y\,dx = \sqrt{x^2 + y^2}\,dx$, given that $y = 0$ when $x = 1$.

OR

Solve the differential equation : $(1 + x^2)\frac{dy}{dx} + 2xy - 4x^2 = 0$, subject to the initial condition $y(0) = 0$.

Q22

$(1+x^2)\frac{dy}{dx} + 2xy = 4x^2$

$\frac{dy}{dx} + \left(\frac{2x}{1+x^2}\right)y = \frac{4x^2}{1+x^2}$

It is linear DE of form $\frac{dy}{dx} + Py = Q$

$P = \frac{2x}{1+x^2}$ $\quad Q = \frac{4x^2}{1+x^2}$

$I.F = e^{\int P\,dx} = e^{\ln|1+x^2|} = 1+x^2$

Sol^n of DE :

$y(1+x^2) = \int \frac{4x^2(1+x^2)}{1+x^2}dx + C$

$= \int 4x^2\,dx + C$

$y(1+x^2) = \frac{4x^3}{3} + c$

$x = 0, y = 0$

$\therefore C = 0$

$\Rightarrow 3y(1+x^2) = 4x^3$

23. Find the value of λ, so that the lines $\frac{1-x}{3}=\frac{7y-14}{\lambda}=\frac{z-3}{2}$ and $\frac{7-7x}{3\lambda}=\frac{y-5}{l}=\frac{6-z}{5}$ are at right angles. Also, find whether the lines are intersecting or not.

Line I: $\frac{1-x}{3}=\frac{7y-14}{\lambda}=\frac{z-3}{2}$

Line II: $\frac{7-7x}{3\lambda}=\frac{y-5}{1}=\frac{6-z}{5}$

Q23.

$\Rightarrow \frac{x-1}{-3}=\frac{y-2}{\lambda/7}=\frac{z-3}{2}=\beta$ $\qquad$ $\frac{x-1}{\frac{-3\lambda}{7}}=\frac{y-5}{1}=\frac{z-6}{-5}=\mu$

Direction ratios are $(-3,\frac{\lambda}{7},2)$ and $(\frac{-3\lambda}{7},1,-5)$ respectively.
(a_1,a_2,a_3) $\qquad$ (b_1,b_2,b_3)

For lines to be ⊥, $\vec{a}.\vec{b}=0$ where ($\vec{a},\vec{b}$ are ||el vectors of lines)

$a_1b_1+a_2b_2+a_3b_3=0$

$\frac{9\lambda}{7}+\frac{\lambda}{7}-10=0$

$\frac{10\lambda}{7}=10$

$\boxed{\lambda=7}$

∴ For λ = 7, lines are ⊥.

Any point of Line I is $(-3\beta+1,\ \beta+2,\ 2\beta+3)$

Line II is $(-3\mu+1,\ \mu+5,\ -5\mu+6)$

For lines to intersect, they should be equal.

$-3\beta+1=-3\mu+1$ $\qquad$ $\mu+5=\beta+2$ $\qquad$ $2\beta+3=-5\mu+6$

$3\mu-3\beta=0$ $\qquad$ $\mu-\beta+3=0$ —(3) $\qquad$ $2\beta+5\mu=3$ — (2)

$\mu=\beta$ —(1)

From (1), (2)

$7\mu=3$

$\boxed{\mu=3/7}$ $\boxed{\beta=\frac{3}{7}}$

In eqn(3)

$\frac{3}{7}-\frac{3}{7}+3\neq 0$

∴ Values donot satisfy eqn(3)

⇒ LINES do NOT intersect.

Section - D

24. Show that the altitude of the right circular cone of maximum volume that can be inscribed in a sphere of radius r is $\frac{4r}{3}$. Also find the maximum volume of cone.

Q24

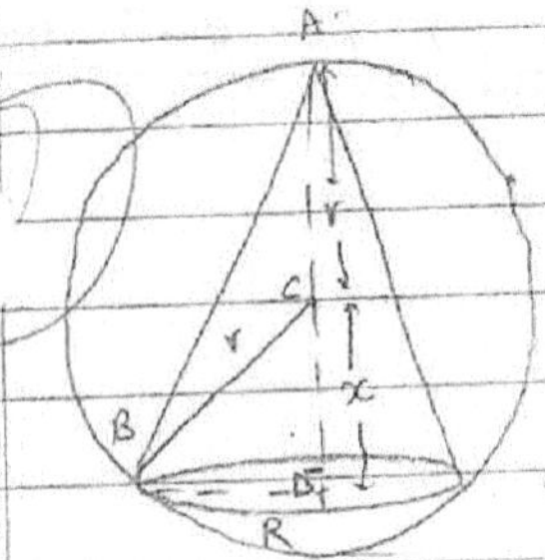

Consider a sphere of radius r.

Then, height of cone = r + x = H

Radius of cone = $\sqrt{r^2 - x^2} = R$

Volume of cone $= V = \frac{1}{3}\pi R^2 H$

$$V = \frac{1}{3}\pi (r^2 - x^2)(r + x)$$

$$= \frac{\pi}{3}(r^3 - rx^2 + xr^2 - x^3).$$

$$\frac{dV}{dx} = \frac{\pi}{3}(-2rx + r^2 - 3x^2)$$

For critical pt $\frac{dV}{dx} = 0$

$$r^2 - 2rx - 3x^2 = 0$$

$$r^2 - 3rx + rx - 3x^2 = 0$$

$$(r + x)(r - 3x) = 0$$

x cannot be −ve $\Rightarrow$ $\boxed{\frac{r}{3} = x}$ is a critical pt

$$\frac{d^2V}{dx^2} = \frac{\pi}{3}(-2r - 6x)$$

$$\left.\frac{d^2V}{dx^2}\right]_{x = r/3} = \frac{\pi}{3}[-2r - 2r] = \frac{-4r\pi}{3} < 0$$

Using Second Derivative test, $\therefore$ $\boxed{x = \frac{r}{3}}$ is a point of maxima.

$$H = r + x$$

$$\boxed{H = \frac{4r}{3}}$$ Hence proved

Max volume of cone $= V_{max} = \frac{\pi}{3}(r + x)(r^2 - x^2)$

$$= \frac{\pi}{3}\left(\frac{4r}{3}\right)\left(r^2 - \frac{r^2}{9}\right)$$

$$= \frac{4\pi r^3 \cdot 8}{81} = \frac{32\pi r^3}{81} \text{ (unit)}^3$$

$$\boxed{V_{max} = \frac{32\pi r^3}{81} \text{ (unit)}^3}$$

25. If $A = \begin{bmatrix} 2 & -3 & 5 \\ 3 & 2 & -4 \\ 1 & 1 & -2 \end{bmatrix}$ then find A^{-1}. Hence solve the following system of equations :

$2x - 3y + 5z = 11,\ 3x + 2y - 4z = -5,\ x + y - 2z = -3.$

OR

Obtain the inverse of the following matrix using elementary operations :

$$A = \begin{bmatrix} -1 & 1 & 2 \\ 1 & 2 & 3 \\ 3 & 1 & 1 \end{bmatrix}$$

Q25. $A = \begin{bmatrix} 2 & -3 & 5 \\ 3 & 2 & -4 \\ 1 & 1 & -2 \end{bmatrix}$

$$\text{adj } A = \begin{bmatrix} 0 & +2 & 1 \\ -1 & -9 & -5 \\ 2 & +23 & 13 \end{bmatrix}^T = \begin{bmatrix} 0 & -1 & 2 \\ 2 & -9 & 23 \\ 1 & -5 & 13 \end{bmatrix}$$

$|A| = 2(0) + 3(-2) + 5(1)$
$= 0 - 6 + 5 = -1$

$$A^{-1} = \frac{1}{|A|} \text{adj } A = -1 \begin{bmatrix} 0 & -1 & 2 \\ 2 & -9 & 23 \\ 1 & -5 & 13 \end{bmatrix}$$

$$A^{-1} = \begin{bmatrix} 0 & 1 & -2 \\ -2 & 9 & -23 \\ -1 & 5 & -13 \end{bmatrix}$$

Given Eq^n :

$2x - 3y + 5z = 11$
$3x + 2y - 4z = -5$
$x + y - 2z = -3$

$AX = B$

Premultiply A^{-1} on both sides

$A^{-1}AX = A^{-1}B$

$IX = A^{-1}B$

$X = A^{-1}B$

$$\begin{bmatrix} x \\ y \\ z \end{bmatrix} = \begin{bmatrix} 0 & 1 & -2 \\ -2 & 9 & -23 \\ -1 & 5 & -13 \end{bmatrix} \begin{bmatrix} 11 \\ -5 \\ -3 \end{bmatrix}$$

$$\begin{bmatrix} x \\ y \\ z \end{bmatrix} = \begin{bmatrix} -5+6 \\ -22-45+69 \\ -11-25+39 \end{bmatrix} = \begin{bmatrix} 1 \\ 2 \\ 3 \end{bmatrix}$$

$x = 1$

$y = 2$

$z = 3$

26. A manufacturer has three machine operators A, B and C. The first operator A produces 1% of defective items, whereas the other two operators B and C produces 5% and 7% defective items respectively. A is on the job for 50% of the time. B on the job 30% of the time and C on the job for 20% of the time. All the items are put into one stockpile and then one item is chosen at random from this and is found to be defective. What is the probability that it was produced by A ?

Q26. A = the Event of choosing a defective item.
E_1 = Item produced by A
E_2 = Item produced by B
E_3 = Item produced by C

$P(E_1) = \frac{50}{100} = \frac{1}{2}$, $P(E_2) = \frac{30}{100} = \frac{3}{10}$, $P(E_3) = \frac{20}{100} = \frac{1}{5}$

$P\left(\frac{A}{E_1}\right) = \frac{1}{100}$, $P\left(\frac{A}{E_2}\right) = \frac{5}{100} = \frac{5}{100}$; $P\left(\frac{A}{E_3}\right) = \frac{7}{100}$.

Using Baye's Thm,

$$P\left(\frac{E_1}{A}\right) = \frac{P(E_1)\,P(A/E_1)}{P(E_1)\,P(A/E_1) + P(E_2)\,P(A/E_2) + P(E_3)\,P(A/E_3)}$$

$$= \frac{\frac{1}{2} \times \frac{1}{100}}{\frac{1}{2} \times \frac{1}{100} + \frac{3}{10} \times \frac{5}{100} + \frac{1}{5} \times \frac{7}{100}}$$

$$= \frac{\frac{1}{200}}{\frac{1}{200} + \frac{3}{200} + \frac{7}{500}} = \frac{\frac{1}{2}}{\frac{1}{2} + \frac{3}{2} + \frac{7}{5}} = \frac{\frac{1}{2}}{2 + \frac{7}{5}} = \frac{\frac{1}{2}}{\frac{17}{5}}$$

$P\left(\frac{E_1}{A}\right) = \frac{5}{34}$ = Probability that defective item was produced by A

27. Find the vector and Cartesian equations of the plane passing through the points (2, 2 – 1), (3, 4, 2) and (7, 0, 6). Also find the vector equation of a plane passing through (4, 3, 1) and parallel to the plane obtained above.

OR

Find the vector equation of the plane that contains the lines $\vec{r} = (\hat{i} + \hat{j}) + \lambda(\hat{i} + 2\hat{j} - \hat{k})$ and the point (– 1, 3, – 4). Also, find the length of the perpendicular drawn from the point (2, 1, 4) to the plane thus obtained.

Q27. Let $\vec{n} = A\hat{i} + B\hat{j} + C\hat{k}$ of the required plane
Eqn of plane is $A(x-2) + B(y-2) + C(z+1) = 0$ —(3)
The plane also passes through (3,4,2) & (7,0,6)

$A + 2B + 3C = 0$ — (1)
$5A - 2B + 7C = 0$ — (2)

Adding (1), (2)
$6A + 10C = 0$
$A = -\frac{5C}{3}$

In eqn(2) $-\frac{25C}{3} + 7C = 2B$

$-\frac{4C}{2 \times 3} = B$

$\Rightarrow \boxed{B = -\frac{2c}{3}}$

Direction ratios of normal to the plane are

$\left(\frac{-5c}{3}, \frac{-2c}{3}, c\right) \equiv$ ~~(−5c, −2c, 3c)~~ $(5, 2, -3)$

$\therefore \vec{n} = 5\hat{i} + 2\hat{j} - 3\hat{k}$

Eq^n of plane: $5(x-2) + 2(y-2) - 3(z+1) = 0$ [From (3)]

$5x + 2y - 3z - 10 - 4 - 3 = 0$

$\boxed{5x + 2y - 3z = 17}$ is cartesian eq^n of plane.

VECTOR EQN:

$\boxed{\vec{r}\cdot\vec{n} = d} \Rightarrow \left[\vec{r} - (2\hat{i} + 2\hat{j} - \hat{k})\right]\cdot\vec{n} = 0$ [$\because (\vec{r} - \vec{a})\cdot\vec{n} = 0$ is eq^n of plane]

$\vec{r}(5\hat{i} + 2\hat{j} - 3\hat{k}) - (2\hat{i} + 2\hat{j} - \hat{k})(5\hat{i} + 2\hat{j} - 3\hat{k}) = 0$

$\Rightarrow \boxed{\vec{r}(5\hat{i} + 2\hat{j} - 3\hat{k}) = 17}$ is vector eq^n of plane.

For plane parallel to above plane, $\vec{n_2} = \vec{n} = 5\hat{i} + 2\hat{j} - 3\hat{k}$

$(\vec{r} - \vec{a_1})\cdot\vec{n} = 0$

here $\vec{a_1} = 4\hat{i} + 3\hat{j} + \hat{k}$

$\left[\vec{r} - (4\hat{i} + 3\hat{j} + \hat{k})\right]\cdot(5\hat{i} + 2\hat{j} - 3\hat{k}) = 0$

$\vec{r}(5\hat{i} + 2\hat{j} - 3\hat{k}) - (20 + 6 - 3) = 0$

$\vec{n_1} = -\frac{5c}{3}\hat{i} - \frac{2c}{3}\hat{j} + c\hat{k} = -\frac{c}{3}(5\hat{i} + 2\hat{j} - 3\hat{k})$

$\therefore$ Eq^n of other plane is

$\boxed{\vec{r}\cdot(5\hat{i} + 2\hat{j} - 3\hat{k}) = 23}$ $\rightarrow$ vector eq^n $\Rightarrow \boxed{\vec{r}\cdot(5\hat{i} + 2\hat{j} - 3\hat{k}) - 23 = 0}$

$5x + 2y - 3z = 23$ $\rightarrow$ cartesian eq^n

28. Using integration, find the area of triangle ABC, whose vertices are A(2, 5), B(4, 7) and C(6, 2).

OR

Find the area of the region lying about x-axis and included between the circle $x^2 + y^2 = 8x$ and inside of the parabola $y^2 = 4x$.

Q28.

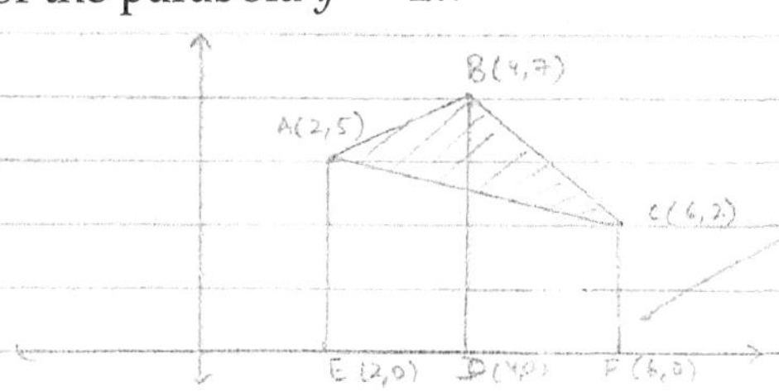

Eq^n of general line:

$y - y_1 = \frac{y_2 - y_1}{x_2 - x_1}(x - x_1)$

Eq^n of line AB:

$(y - 5) = \frac{2}{2}(x - 2)$

$\boxed{y_1 = x + 3}$

Eq^n of line BC:

$(y-2) = \frac{5}{-2}(x-6)$

$\Rightarrow y-2 = -\frac{5}{2}x+15$

$\Rightarrow \boxed{y_2 = -\frac{5x}{2}+17}$

Eq^n of line AC

$(y-5) = \frac{3}{-4}(x-2)$

$y-5 = -\frac{3x}{4}+\frac{3}{2}$

$\boxed{y_3 = -\frac{3x}{4}+\frac{13}{2}}$

Area of shaded region is required area A = ar(ABDE) + ar(BCFD) − ar(ACFE)

$A = \int_2^4 y_1\,dx + \int_4^6 y_2\,dx - \int_2^6 y_3\,dx$

$= \int_2^4 (x+3)\,dx + \int_4^6 \left(-\frac{5x}{2}+17\right)dx - \int_2^6 \left(-\frac{3x}{4}+\frac{13}{2}\right)dx$

$A = \left[\frac{x^2}{2}+3x\right]_2^4 + \left[17x-\frac{5x^2}{4}\right]_4^6 - \left[\frac{13x}{2}-\frac{3x^2}{8}\right]_2^6$

$= [8+12-2-6] + [102-45-68+20] - \left[39-\frac{27}{2}-13+\frac{3}{2}\right]$

$= 12+9-14$

$\boxed{A = 7 \text{ sq. units}}$

29. A manufacturer has employed 5 skilled men and 10 semi-skilled men and makes two models A anb B of an article. The making of one item of model A requires 2 hours work by a skillled man and 2 hours work by a semi-skilled man. One item of model B expected to work more than 8 hours per day. the manufacturer's profit on an item of model A is ₹ 15 and on an item of model B is ₹ 10. How many of items of each model should be made per day in order to maximize daily profit ? Formulate the above LPP and solve it graphically and find the maximum profit.

Q29. let x be no. of A models

y be no. of B models.

Objective $Z = 15x + 10y$ (Maximize)

Subject to constraints:

$x, y \geq 0$

$2x + y \leq 40$ (Skilled man working hrs)

$2x + 3y \leq 80$ (Semi-skilled man working hrs)

$2x+y \leq 40$

$2x+y = 40$

x	0	20
y	40	0

Zero test : TRUE

$2x+3y \leq 80$

$2x+3y = 80$

x	10	40
y	20	0

Zero test : TRUE

GRAPH: On graph paper.

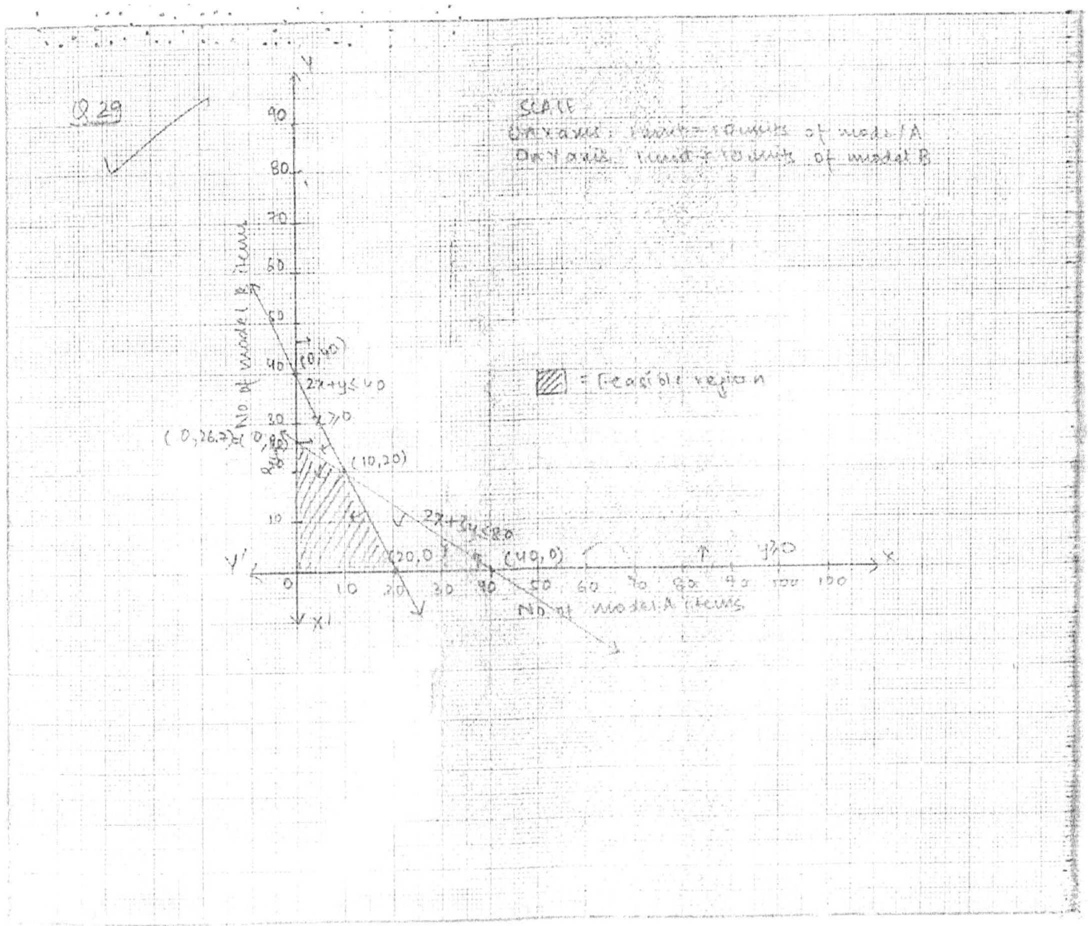

Corner pt	$Z = 15x + 10y$
(0,0)	Z = 0
~~(80/3, 0)~~ (0, 80/3)	Z = ~~400~~ 0 + 800/3 = 266.7 ≈ 267
(10, 20)	Z = 150 + 200 = 350 (MAX)
(20, 0)	Z = 300

∴ No. of model A = 10

No. of model B = 20

Maximum profit = ₹ 350

MATHEMATICS

Outside Delhi [Set-I] 2020

Time allowed : 3 hours **Maximum marks : 80**

General Instructions :

Read the following instructions very carefully and strictly follow them :

(i) This question paper comprises **four** sections : A, B, C and D.
This question paper carries **36** questions. **All** questions are compulsory.

(ii) **Section A** – Question no. **1** to **20** comprises of **20** questions of **one** mark each.

(iii) **Section B** – Question no. **21** to **26** comprises of **6** questions of **two** marks each.

(iv) **Section C** – Question no. **27** to **32** comprises of **6** questions of **four** marks each.

(v) **Section D** – Question no. **33** to **36** comprises of **4** questions of **six** marks each.

(vi) There is no overall choice in the question paper. However, an internal choice has been provided in **3** questions of one mark, **2** questions of two marks, **2** questions of four marks and **2** questions of six marks. Only one of the choices in such questions have to be attempted.

(vii) In addition to this, separate instructions are given with each section and question, wherever necessary.

(viii) Use of calculators is not permitted.

SECTION–A

Question numbers 1 to 10 are multiple choice questions of 1 mark each. Select the correct option :

1. The value of $\sin^{-1}\left(\cos\frac{3\pi}{5}\right)$ is

(a) $\frac{\pi}{10}$ (b) $\frac{3\pi}{5}$

(c) $\frac{-\pi}{10}$ (d) $\frac{-3\pi}{5}$ **[1]**

Sol.

$$\sin^{-1}\left(\cos\frac{3\pi}{5}\right) = \frac{\pi}{2} - \cos^{-1}\left(\cos\frac{3\pi}{5}\right)$$

$$= \frac{\pi}{2} - \frac{3\pi}{5}$$

$$= \frac{5\pi - 6\pi}{10}$$

$$= \frac{-\pi}{10}$$

$\therefore$ Option (c) is correct.

2. If A = [2 – 3 4], B = $\begin{bmatrix}3\\2\\2\end{bmatrix}$, X= [1 2 3] and Y = $\begin{bmatrix}2\\3\\4\end{bmatrix}$, then AB + XY equals.

(a) [28] (b) [24]

(c) 28 (d) 24 **[1]**

Sol. $AB + XY = [2\ \ -3\ \ 4]\begin{bmatrix}3\\2\\2\end{bmatrix} + [1\ 2\ 3]\begin{bmatrix}2\\3\\4\end{bmatrix}$

$$= [8] + [20] = [28]$$

$\therefore$ Option (a) is correct.

3. If $\begin{vmatrix}2 & 3 & 2\\x & x & x\\4 & 9 & 1\end{vmatrix} + 3 = 0$, then the value of x is

(a) 3 (b) 0

(c) –1 (d) 1 **[1]**

Sol. Given : $\begin{vmatrix}2 & 3 & 2\\x & x & x\\4 & 9 & 1\end{vmatrix} + 3 = 0$

$\Rightarrow$ $2(x - 9x) - 3(x - 4x) + 2(9x - 4x) + 3 = 0$

$\Rightarrow$ $-16x + 9x + 10x + 3 = 0$

$\Rightarrow$ $3x = -3$

$\Rightarrow$ $x = -1$

$\therefore$ Option (c) is correct.

4. $\int_0^{\pi/8} \tan^2(2x)\,dx$ is equal to

(a) $\frac{4-\pi}{8}$ (b) $\frac{4+\pi}{8}$

(c) $\frac{4-\pi}{4}$ (d) $\frac{4-\pi}{2}$ **[1]**

Sol. $\int_0^{\pi/8} \tan^2(2x)dx = \int_0^{\pi/8} \left(\sec^2(2x) - 1\right)dx$

$$= \left[\frac{\tan(2x)}{2} - x\right]_0^{\pi/8}$$

$$= \frac{1}{2} - \frac{\pi}{8} - 0$$

$$= \frac{4-\pi}{8}$$

∴ Option (a) is correct.

5. If $\vec{a} \cdot \vec{b} = \frac{1}{2}|\vec{a}||\vec{b}|$, then the angle between $\vec{a}$ and $\vec{b}$ is

(a) 0° **(b) 30°**

(c) 60° **(d) 90°** [1]

Sol. $\vec{a} . \vec{b} = \frac{1}{2}|\vec{a}||\vec{b}| = |\vec{a}||\vec{b}|\cos\theta$

$\Rightarrow$ $\cos\theta = \frac{1}{2}$

$\Rightarrow$ $\theta = 60°$

∴ Option (c) is correct.

6. The two lines $x = ay + b$, $z = cy + d$; and $x = a'y + b'$, $z = c'y + d'$ are perpendicular to each other, if

(a) $\frac{a}{a'} + \frac{c}{c'} = 1$ **(b)** $\frac{a}{a'} + \frac{c}{c'} = -1$

(c) $aa' + cc' = 1$ **(d)** $aa' + cc' = -1$ [1]

Sol. Option (d) is correct.

7. The two planes $x - 2y + 4z = 10$ and $18x + 17y + kz = 50$ are perpendicular, if k is equal to

(a) –4 **(b) 4**

(c) 2 **(d) –2** [1]

Sol. $18 - 34 + 4k = 0$

{using $a_1a_2 + b_1b_2 + c_1c_2 = 0$}

$\Rightarrow$ $4k = 16$

$\Rightarrow$ $k = 4$

∴ Option (b) is correct.

8. In an LPP, if the objective function $z = ax + by$ has the same maximum value on two corner points of the feasible region, then the number of points at which z_{max} occurs is

(a) 0 **(b) 2**

(c) finite **(d) infinite** [1]

Sol. Option (d) is correct.

9. From the set {1, 2, 3, 4, 5}, two numbers a and b ($a \neq b$) are chosen at random. The probability that $\frac{a}{b}$ is an integer is

(a) $\frac{1}{3}$ **(b)** $\frac{1}{4}$

(c) $\frac{1}{2}$ **(d)** $\frac{3}{5}$

Sol. Total possible outcomes of (a, b) = {(1, 2), (1, 3), (1, 4), (1, 5), (2, 1), (2, 3), (2, 4), (2, 5), (3, 1), (3, 2), (3, 4), (3, 5), (4, 1), (4, 2), (4, 3), (4, 5), (5, 1), (5, 2), (5, 3), (5, 4)}

⇒ Number of possible outcomes = 20

Also, favourable outcomes of (a, b) = {(2, 1), (3, 1), (4, 1), (4, 2), (5, 1)}

⇒ Number of possible outcomes = 5

∴ $P\left(\frac{a}{b} \text{ is integer}\right) = \frac{5}{20} = \frac{1}{4}$

∴ Option (b) is correct.

10. A bag contains 3 white, 4 black and 2 red balls. If 2 balls are drawn at random (without replacement), then the probability that both the balls are white is

(a) $\frac{1}{18}$ **(b)** $\frac{1}{36}$

(c) $\frac{1}{12}$ **(d)** $\frac{1}{24}$ [1]

Sol. Required probability = $\frac{^3C_2}{^9C_2} = \frac{1}{12}$

∴ Option (c) is correct.

In Q. numbers 11 to 15, fill in the blanks with correct word/sentence :

11. If $f : R \to R$ be given by $f(x) = (3 - x^3)^{1/3}$, then $fof(x)$ = ________ [1]

Sol. x

Explanation: $(fof)(x) = f(f(x))$

$= f((3 - x^3)^{1/3})$

$= \left[3 - \left[(3-x^3)^{1/3}\right]^3\right]^{1/3}$

$= [3 - 3 + x^3]^{1/3}$

$= x$

12. If $\begin{bmatrix} x+y & 7 \\ 9 & x-y \end{bmatrix} = \begin{bmatrix} 2 & 7 \\ 9 & 4 \end{bmatrix}$, then $x . y$ = ________ [1]

Sol. – 3

Explanation: Comparing elements of both matrices, we get

$x + y = 2$ ...(i)

and $x - y = 4$ ...(ii)

Adding (i) and (ii), we get

$$2x = 6$$

$\Rightarrow$ $x = 3$

From (i), $y = 2 - x$

$= 2 - 3 = -1$

$\therefore$ $x.y = 3(-1) = -3$

13. The number of points of discontinuity of f defined by $f(x) = |x| - |x+1|$ is ______. [1]

Sol. 0

Explanation: Here $f(x) = |x| - |x+1|$

Let $h(x) = |x|$

and $g(x) = |x+1|$

Then $h(x)$ and $g(x)$ are both continuous functions as modulus function and polynomial function are continuous everywhere.

Also $(hog)(x) = h(g(x)) = h(x+1)$

$= |x+1|$

Since composite of two continuous functions is also continuous in its domain

$\therefore (hog)(x) = |x+1|$ is also continuous

Also difference of two continuous functions is continuous everywhere

$\therefore f(x) = |x| - |x+1| = h(x) - (hog)(x)$

is also continuous everywhere.

Hence, the number of points of discontinuity of $f(x) = |x| - |x+1|$ is zero.

14. The slope of the tangent to the curve $y = x^3 - x$ at the point (2, 6) is ______. [1]

OR

The rate of change of the area of a circle with respect to its radius r, when $r = 3$ cm, is ______. [1]

Sol. 11

Explanation: $y = x^3 - x$

$\Rightarrow$ $\frac{dy}{dx} = 3x^2 - 1$

$\therefore$ Slope of tangent at (2, 6) $= 3(2)^2 - 1 = 11$

OR

6π cm^2/cm

Explanation: Area of circle, $A = \pi r^2$

$\therefore$ $\frac{dA}{dr} = 2\pi r$

$$\left[\frac{dA}{dr}\right]_{r=3\text{cm}} = 2 \times \pi \times 3 = 6\pi \text{cm}^2/\text{cm}$$

15. If $\vec{a}$ is a non-zero vector, then $(\vec{a}.\hat{i})\hat{i} + (\vec{a}.\hat{j})\hat{j} + (\vec{a}.\hat{k})\hat{k}$ equals ______. [1]

OR

The projection of the vector $\hat{i} - \hat{j}$ on the vector $i + j$ is______. [1]

Sol. $\vec{a}$

Explanation: Let $\vec{a} = a_1\hat{i} + b_1\hat{j} + c_1\hat{k}$

$\therefore (\vec{a}.\hat{i})\hat{i} + (\vec{a}.\hat{j})\hat{j} + (\vec{a}.\hat{k})\hat{k}$

$= a_1\hat{i} + b_1\hat{j} + c_1\hat{k}$

$= \vec{a}$

OR

0

Explanation: Projection of $\hat{i} - \hat{j}$ on $\hat{i} + \hat{j}$

$$= \frac{(\hat{i}-\hat{j}).(\hat{i}+\hat{j})}{|\hat{i}+\hat{j}|} = 0$$

Q. 16 to 20 are very short answer questions.

16. Find adj A, if A = $\begin{bmatrix} 2 & -1 \\ 4 & 3 \end{bmatrix}$ [1]

Sol. Given : $A = \begin{bmatrix} 2 & -1 \\ 4 & 3 \end{bmatrix}$

For a square matrix of order 2, the adj A can be obtained by interchanging a_{11} and a_{22} and by changing sign of a_{12} and a_{21}.

$\therefore$ $\text{adj A} = \begin{bmatrix} 3 & 1 \\ -4 & 2 \end{bmatrix}$ **Ans.**

17. Find $\int \frac{2^{x+1} - 5^{x-1}}{10^x} dx$ [1]

Sol. $\int \frac{2^{x+1} - 5^{x-1}}{10^x} dx = \int \left(\frac{2.2^x}{10^x}\right) dx - \int \left(\frac{1}{5}.\frac{5^x}{10^x}\right) dx$

$= 2\int \frac{1}{5^x} dx - \frac{1}{5}\int \frac{1}{2^x} dx$

$= 2\int 5^{-x} dx - \frac{1}{5}\int 2^{-x} dx$

$= \frac{-2(5^{-x})}{\log 5} + \frac{1}{5}\frac{(2^{-x})}{\log 2} + C$

$= \frac{-2}{5^x \log 5} + \frac{1}{5.2^x \log 2} + C$ **Ans.**

18. Evaluate $\int_0^{2\pi} |\sin x|\, dx$ **[1]**

Sol. Let $I = \int_0^{2\pi} |\sin x|\, dx$

$$= \int_0^{\pi} \sin x\, dx - \int_{\pi}^{2\pi} \sin x\, dx$$

$$= -[\cos x]_0^{\pi} + [\cos x]_{\pi}^{2\pi}$$

$$= -[\cos \pi - \cos 0] + [\cos 2\pi - \cos \pi]$$

$$= -[-1-1] + [1+1] = 4$$ **Ans.**

19. If $\int_0^a \frac{dx}{1+4x^2} = \frac{\pi}{8}$, then find the value of a. **[1]**

OR

Find $\int \frac{dx}{\sqrt{x}+x}$ **[1]**

Sol. Given,

$$\int_0^a \frac{dx}{1+4x^2} = \frac{\pi}{8}$$

$$\Rightarrow \int_0^a \frac{d}{1^2+(2x)^2} = \frac{\pi}{8}$$

$$\Rightarrow \frac{1}{2}[\tan^{-1} 2x]_0^a = \frac{\pi}{8}$$

$$\Rightarrow \tan^{-1} 2a - 0 = \frac{\pi}{4}$$

$$\Rightarrow \tan^{-1} 2a = \frac{\pi}{4}$$

$$\Rightarrow 2a = \tan\frac{\pi}{4}$$

$$\Rightarrow 2a = 1$$

$$\Rightarrow a = \frac{1}{2}$$ **Ans.**

OR

Let $I = \int \frac{dx}{\sqrt{x}+x} = \int \frac{1}{\sqrt{x}(1+\sqrt{x})} dx$

Let $\sqrt{x} = t$

$$\Rightarrow \frac{1}{2\sqrt{x}} dx = dt$$

$$\Rightarrow \frac{dx}{\sqrt{x}} = 2dt$$

$$\therefore \quad I = 2\int \frac{1}{(1+t)} dt$$

$$= 2 \log |(1+t)| + C$$

$$= 2\log|1+\sqrt{x}| + C$$ **Ans.**

20. Show that the function $y = ax + 2a^2$ is a solution of the differential equation $2\left(\frac{dy}{dx}\right)^2 + x\left(\frac{dy}{dx}\right) - y = 0$ **[1]**

Sol. Given, $y = ax + 2a^2$

$$\Rightarrow \frac{dy}{dx} = a$$

Now, L.H.S. $= 2\left(\frac{dy}{dx}\right)^2 + x\left(\frac{dy}{dx}\right) - y$

$$= 2a^2 + x(a) - (ax + 2a^2)$$

$$= 2a^2 + ax - ax - 2a^2$$

$$\Rightarrow = 0 = \text{R.H.S.}$$

Hence Proved.

SECTION – B

Q. Nos. 21 to 26 carry 2 marks each.

21. Check if the relation R on the set A = {1, 2, 3, 4, 5, 6} defined as R = {(x, y) : y is divisible by x} is (i) symmetric, (ii) transitive. **[2]**

OR

Prove that :

$$\frac{9\pi}{8} - \frac{9}{4}\sin^{-1}\left(\frac{1}{3}\right) = \frac{9}{4}\sin^{-1}\left(\frac{2\sqrt{2}}{3}\right)$$

Sol. Given, A = {1, 2, 3, 4, 5, 6}

R = {(x, y) : y is divisible by x}

(i) $(2, 4) \in$ R {$\because$ 4 is divisible by 2}

But $(4, 2) \notin$ R {$\because$ 2 is not divisible by 4}

$\therefore$ R is not symmetric. **Ans.**

(ii) Let $(a, b) \in$ R & $(b, c) \in$ R

$\Rightarrow$ $b = \lambda a$ and $c = \mu b$

Now, $c = \mu b = \mu(\lambda a) \Rightarrow (a, c) \in$ R

$\Rightarrow$ c is divisible by a

$\therefore$ R is transitive. **Ans.**

OR

L.H.S. $= \frac{9\pi}{8} - \frac{9}{4}\sin^{-1}\frac{1}{3}$

$$= \frac{9}{4}\left(\frac{\pi}{2} - \sin^{-1}\frac{1}{3}\right)$$

$$= \frac{9}{4}\cos^{-1}\frac{1}{3} \qquad \left\{\because \frac{\pi}{2} - \sin^{-1}x = \cos^{-1}x\right\}$$

$$= \frac{9}{4}\sin^{-1}\sqrt{1-\left(\frac{1}{3}\right)^2}$$

$$\left\{\because \cos^{-1}x = \sin^{-1}\sqrt{1-x^2}\right\}$$

$$= \frac{9}{4}\sin^{-1}\sqrt{\frac{8}{9}}$$

$$= \frac{9}{4}\sin^{-1}\frac{2\sqrt{2}}{3} = \text{R.H.S.}$$

Hence Proved.

22. Find the value of $\frac{dy}{dx}$ at $\theta = \frac{\pi}{3}$, if $x = \cos\theta - \cos 2\theta$, $y = \sin\theta - \sin 2\theta$. **[2]**

Sol. Given, $x = \cos\theta - \cos 2\theta$

$\Rightarrow$ $\frac{dx}{d\theta} = -\sin\theta + 2\sin 2\theta$

Also, $y = \sin\theta - \sin 2\theta$

$\Rightarrow$ $\frac{dy}{d\theta} = \cos\theta - 2\cos 2\theta$

Now, $\frac{dy}{dx} = \frac{dy}{d\theta} \times \frac{d\theta}{dx} = \frac{\cos\theta - 2\cos 2\theta}{2\sin 2\theta - \sin\theta}$

$$\Rightarrow \left[\frac{dy}{dx}\right]_{\theta = \frac{\pi}{3}} = \frac{\cos\frac{\pi}{3} - 2\cos\frac{2\pi}{3}}{2\sin\frac{2\pi}{3} - \sin\frac{\pi}{3}} = \sqrt{3}$$

Ans.

23. Show that the function f defined by $f(x) = (x-1)e^x + 1$ is an increasing function for all $x > 0$. **[2]**

Sol. $f(x) = (x-1)e^x + 1, x > 0$

$\therefore$ $f'(x) = (x-1)e^x + e^x(1) + 0$

$= e^x(x - 1 + 1) = xe^x > 0 \ \forall \ x > 0$

$\therefore f'(x)$ is an increasing function for all $x > 0$.

Hence Proved.

24. Find $|\vec{a}|$ and $|\vec{b}|$, if $|\vec{a}| = 2|\vec{b}|$ and $(\vec{a}+\vec{b}).(\vec{a}-\vec{b}) = 12$. **[2]**

OR

Find the unit vector perpendicular to each of the vectors $\vec{a} = 4\hat{i} + 3\hat{j} + \hat{k}$ and $\vec{b} = 2\hat{i} - \hat{j} + 2\hat{k}$.

Sol. Given, $|\vec{a}| = 2|\vec{b}|$ ...(i)

And, $(\vec{a}+\vec{b}).(\vec{a}-\vec{b}) = 12$

$\Rightarrow$ $|\vec{a}|^2 - |\vec{b}|^2 = 12$

$\Rightarrow$ $(2|\vec{b}|)^2 - |\vec{b}|^2 = 12$ [Using (i)]

$\Rightarrow$ $3|\vec{b}|^2 = 12$

$\Rightarrow$ $|\vec{b}|^2 = 4$

$\Rightarrow$ $|\vec{b}| = 2$

From (i), $|\vec{a}| = 2(2) = 4$

$\therefore$ $|\vec{a}| = 4$ and $|\vec{b}| = 2$ **Ans.**

OR

Given, $\vec{a} = 4\hat{i} + 3\hat{j} + \hat{k}$

and $\vec{b} = 2\hat{i} - \hat{j} + 2\hat{k}$

$$\therefore \quad \vec{a} \times \vec{b} = \begin{vmatrix} \hat{i} & \hat{j} & \hat{k} \\ 4 & 3 & 1 \\ 2 & -1 & 2 \end{vmatrix}$$

$$= 7\hat{i} - 6\hat{j} - 10\hat{k}$$

$\therefore$ Unit vector perpendicular to both $\vec{a}$ and $\vec{b}$

$$= \pm\frac{(\vec{a}\times\vec{b})}{|\vec{a}\times\vec{b}|}$$

$$= \pm\frac{(7\hat{i} - 6\hat{j} - 10\hat{k})}{\sqrt{49+36+100}}$$

$$= \pm\frac{(7\hat{i} - 6\hat{j} - 10\hat{k})}{\sqrt{185}}$$ **Ans.**

25. Find the equation of the plane with intercept 3 on the y-axis and parallel to xz-plane. **[2]**

Sol. Equation of any plane parallel to xz-plane is $y = b$ (where b is a constant)

Since its y-intercept is 3,

$\therefore$ Equation of required plane is $y = 3$. **Ans.**

26. Find [P(B/A) + P(A/B)], if P(A) = $\frac{3}{10}$, P(B) = $\frac{2}{5}$ and P(A $\cup$ B) = $\frac{3}{5}$. **[2]**

Sol. We have,

$$P(A\cup B) = P(A) + P(B) - P(A\cap B)$$

$$\Rightarrow \quad \frac{3}{5} = \frac{3}{10}+\frac{2}{5} - P(A\cap B)$$

$$\Rightarrow \quad P(A\cap B) = \frac{3}{10}+\frac{2}{5}-\frac{3}{5}$$

$$\Rightarrow \quad P(A\cap B) = \frac{1}{10}$$

$$\therefore \quad P(B/A) + P(A/B) = \frac{P(A\cap B)}{P(A)}+\frac{P(A\cap B)}{P(B)}$$

$$= \frac{1/10}{3/10}+\frac{1/10}{2/5}$$

$$= \frac{1}{3}+\frac{1}{4} = \frac{7}{12}$$ **Ans.**

SECTION–C

Q. Nos. 27 to 32 carry 4 marks each.

27. Prove that the relation R on Z, defined by R {(x, y) : (x – y) is divisible by 5} is an equivalence relation. **[4]**

Sol. R on Z is defined as :

R = {$(x, y) : (x - y)$ is divisible by 5}

For reflexive : Let $x \in Z$

$\Rightarrow x - x = 0$, which is divisible by 5

$\Rightarrow \quad (x, x) \in R \ \forall x \in Z$

Hence, R is a reflexive relation.

For symmetric : Let $(x, y) \in R$

$\Rightarrow x - y$ is divisible by 5

$\Rightarrow -(x - y)$ is divisible by 5

$\Rightarrow y - x$ is divisible by 5

$\Rightarrow \quad (y, x) \in R$

Hence, R is a symmetric relation.

For transitive : Let $(x, y) \in R$ and $(y, z) \in R$

$\Rightarrow x - y$ and $y - z$ are divisible by 5

$\Rightarrow x - y + y - z$ is divisible by 5

$\Rightarrow x - z$ is divisible by 5

$\Rightarrow (x, z) \in R$

Hence, R is a transitive relation.

Hence, R is an equivalence relation.

Hence Proved.

28. If $y = \sin^{-1}\left(\frac{\sqrt{1+x}+\sqrt{1-x}}{2}\right)$, then show that $\frac{dy}{dx} = \frac{-1}{2\sqrt{1-x^2}}$. **[4]**

OR

Verify the Rolle's Theorem for the function $f(x) = e^x \cos x$ in $\left[-\frac{\pi}{2}, \frac{\pi}{2}\right]$

Sol. Let $x = \cos\theta$

$$\Rightarrow \quad y = \sin^{-1}\left(\frac{\sqrt{1+\cos\theta}+\sqrt{1-\cos\theta}}{2}\right)$$

$$= \sin^{-1}\left(\frac{\sqrt{2}\cos\frac{\theta}{2}+\sqrt{2}\sin\frac{\theta}{2}}{2}\right)$$

$$= \sin^{-1}\left(\frac{1}{\sqrt{2}}\cos\frac{\theta}{2}+\frac{1}{\sqrt{2}}\sin\frac{\theta}{2}\right)$$

$$= \sin^{-1}\left(\sin\frac{\pi}{4}\cos\frac{\theta}{2}+\cos\frac{\theta}{4}\sin\frac{\theta}{2}\right)$$

$$= \sin^{-1}\left[\sin\left(\frac{\theta}{2}+\frac{\pi}{4}\right)\right] = \frac{\theta}{2}+\frac{\pi}{4}$$

$$\therefore \quad y = \frac{1}{2}\cos^{-1}x+\frac{\pi}{4} \qquad [\because x = \cos\theta]$$

$$\Rightarrow \quad \frac{dy}{dx} = -\frac{1}{2\sqrt{1-x^2}}$$ **Hence Proved.**

OR

Given, $f(x) = e^x \cos x$ in $\left[-\frac{\pi}{2}, \frac{\pi}{2}\right]$

We know that e^x and cosine function are continuous everywhere and product of any two continuous function is also continuous in its domain.

Hence, $f(x) = e^x \cos x$ is continuous in $\left[-\frac{\pi}{2}, \frac{\pi}{2}\right]$

Now, $f'(x) = e^x(-\sin x) + \cos x.e^x$

$= e^x (\cos x - \sin x)$

It is defined $\forall x \in \left(-\frac{\pi}{2}, \frac{\pi}{2}\right)$

$\therefore f$ is differentiable in $\left(-\frac{\pi}{2}, \frac{\pi}{2}\right)$

And, $$f\left(-\frac{\pi}{2}\right) = e^{-\pi/2}\cos\left(-\frac{\pi}{2}\right) = 0$$

$$f\left(\frac{\pi}{2}\right) = e^{\pi/2}\cos\left(\frac{\pi}{2}\right) = 0$$

$$\therefore \quad f\left(-\frac{\pi}{2}\right) = f\left(\frac{\pi}{2}\right)$$

∴ All the three postulates of Rolle's theorem are satisfied. Hence there exists a real number,

$c \in \left[-\frac{\pi}{2}, \frac{\pi}{2}\right]$ such that

$$f'(c) = 0$$

$$\Rightarrow \quad e^c(\cos c - \sin c) = 0$$

$$\Rightarrow \quad \cos c - \sin c = 0 \qquad \{\because e^c \neq 0\}$$

$$\Rightarrow \quad \cos c = \sin c$$

$$\Rightarrow \quad \tan c = 1 = \tan\frac{\pi}{4}$$

$$\Rightarrow \quad c = \frac{\pi}{4} \in \left(-\frac{\pi}{2}, \frac{\pi}{2}\right)$$

Hence Verified.

29. Evaluate : $\int_0^{\pi} \frac{x \sin x}{1+\cos^2 x} dx$. [4]

Sol. Let $I = \int_0^{\pi} \frac{x\sin x}{1+\cos^2 x} dx \qquad$ ...(i)

Using $\int_0^{\pi} f(x)dx = \int_0^{a} f(a-x)dx$

$$I = \int_0^{\pi} \frac{(\pi - x)\sin(\pi - x)}{1+\cos^2(\pi - x)} dx$$

$$\Rightarrow \quad I = \int_0^{\pi} \frac{(\pi - x)\sin x}{1+\cos^2 x} dx \qquad \text{...(ii)}$$

On adding equations (i) and (ii), we get

$$\Rightarrow \quad 2I = \pi\int_0^{\pi} \frac{\sin x}{1+\cos^2 x} dx$$

Let $\cos x = t$

$\Rightarrow \quad \sin x\, dx = -dt$

When $x = 0$, $t = 1$

When $x = \pi$, $t = -1$

$$\therefore \quad I = \frac{\pi}{2}\int_1^{-1} \frac{-dt}{1+t^2}$$

$$\Rightarrow \quad I = -\frac{\pi}{2}(\tan^{-1} t)_1^{-1}$$

$$= -\frac{\pi}{2}[\tan^{-1}(-1) - \tan^{-1}(1)]$$

$$= -\frac{\pi}{2}\left(-\frac{\pi}{4} - \frac{\pi}{4}\right)$$

$$= -\frac{\pi}{2}\left(-\frac{\pi}{2}\right)$$

$$\therefore \quad I = \frac{\pi^2}{4} \qquad \textbf{Ans.}$$

30. For the differential equation given below, find a particular solution satisfying the given condition $(x+1)\frac{dy}{dx} = 2e^{-y} + 1$; $y = 0$ when $x = 0$. [4]

Sol. Given,

$$(x+1)\frac{dy}{dx} = \frac{2}{e^y} + 1, \; y = 0, x = 0$$

$$\Rightarrow \quad (x+1)\frac{dy}{dx} = \frac{2+e^y}{e^y}$$

$$\Rightarrow \quad \int \frac{e^y}{e^y+2} dy = \int \frac{dx}{x+1}$$

Let $e^y + 2 = t$

$\Rightarrow \quad e^y dy = dt$

$$\Rightarrow \quad \int \frac{dt}{t} = \int \frac{dx}{x+1}$$

$\Rightarrow \quad \log|t| = \log|x+1| + \log|c|$

$\Rightarrow \quad \log|e^y + 2| = \log|c(x+1)|$

$\Rightarrow \quad e^y + 2 = c(x+1)$

Put $x = 0, y = 0$

$\Rightarrow \quad e^0 + 2 = c(0+1)$

$\Rightarrow \quad 1 + 2 = c$

$\Rightarrow \quad c = 3$

∴ Solution is,

$e^y + 2 = 3(x+1)$

$\Rightarrow \quad e^y = 3x + 1 \qquad$ **Ans.**

31. A manufacturer has three machines I, II and III installed in his factory. Machine I and II are capable of being operated for atmost 12 hours wherease machine III must be operated for atleast 5 hours a day. He produces only two items M and N each requiring the use of all the three machines.

The number of hours required for producing 1 unit of M and N on three machines are given in the following table :

Items	Number of hours required on machines		
	I	II	III
M	1	2	1
N	2	1	1.25

He makes a profit of ₹ 600 and ₹ 400 on one unit of items M and N respectively. How many units of each item should he produce so as to maximize his profit assuming that he can sell all the items that he produced. What will be the maximum profit ? [4]

Sol. Let number of units of item M produced be x and that of item N be y.

$$\text{Profit (in ₹)} = 600x + 400y$$

$\therefore$ LPP is maximized $Z = 600x + 400y$

Subject to constraints :

$$x + 2y \leq 12 \quad \text{...(i)}$$

$$2x + y \leq 12 \quad \text{...(ii)}$$

$$x + 1.25y \geq 5 \quad \text{...(iii)}$$

$$x \geq 0$$

$$y \geq 0$$

To solve this LPP, we draw the lines

$$x + 2y = 12$$

$$2x + y = 12$$

$$x + 1.25y = 5$$

$$x = 0$$

and $y = 0$

$$x + 2y = 12$$

x	0	12
y	6	0

$$2x + y = 12$$

x	0	6
y	12	0

$$x + 1.25y = 5$$

x	0	5
y	4	0

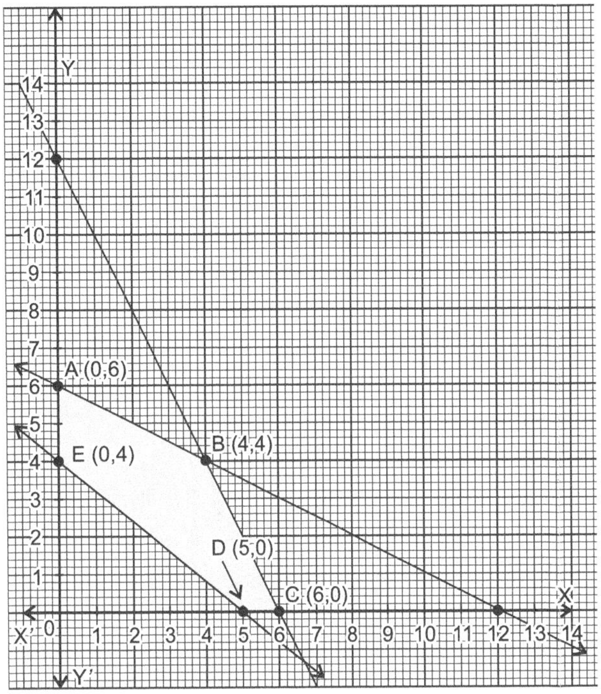

Corner point	$Z = 600x + 400y$
A (0, 6)	Z = 2400
B (4, 4)	Z = 4000 Maximum
C (6, 0)	Z = 3600
D (5, 0)	Z = 3000
E (0, 4)	Z = 1600

$\therefore$ Maximum profit is ₹ 4000 at B (4, 4), *i.e.* when 4 units of item M and 4 units of item N are produced. **Ans.**

32. A coin is biased so that the head is three times as likely to occur as tail. If the coin is tossed twice, find the probability distribution of number of tails. Hence find the mean of the number of tails. [4]

OR

Suppose that 5 men out of 100 and 25 women out of 1000 are good orators. Assuming that there are equal number of men and women, find the probability of choosing a good orator. [4]

Ans. According to question :

$$P(H) = 3P(T)$$

But $P(H) + P(T) = 1$

$\Rightarrow$ $3P(T) + P(T) = 1$

$\Rightarrow$ $4P(T) = 1$

$\Rightarrow$ $P(T) = \frac{1}{4}$

So, $P(H) = \frac{3}{4}$

Let X denotes number of tails obtained when the coin is tossed twice. Then, X can be 0, 1 or 2.

$$P(X=0) = P(HH) = \left(\frac{3}{4}\right)^2 = \frac{9}{16}$$

$$P(X=1) = P(HT \text{ or } TH) = \frac{3}{4}\times\frac{1}{4}+\frac{1}{4}\times\frac{3}{4} = \frac{6}{16}$$

$$P(X=2) = P(TT) = \frac{1}{4}\times\frac{1}{4} = \frac{1}{16}$$

∴ Probability distribution table is

X_i	0	1	2
$P(X_i)$	$\frac{9}{16}$	$\frac{6}{16}$	$\frac{1}{16}$

Now, $$\text{Mean} = E(X) = \sum_{i=1}^{n} X_i P(X_i)$$

$$= 0\times\frac{9}{16}+1\times\frac{6}{16}+2\times\frac{1}{16}$$

$$= \frac{8}{16} = \frac{1}{2}$$

∴ $\text{Mean} = \frac{1}{2}$ **Ans.**

OR

Let E_1 be the event of selecting a man and E_2 be the event of selecting a woman.
Let A be the event of selecting a good orator.

Then $$P(E_1) = P(E_2) = \frac{1}{2}$$

{∵ No. of men and women are equal}

$$P(A/E_1) = \frac{5}{100} = \frac{1}{20}$$

$$P(A/E_2) = \frac{25}{1000} = \frac{1}{40}$$

∴ By total theorem of probability :

$P(A) = P(E_1)\,P(A/E_1) + P(E_2)\,P(A/E_2)$

$$= \frac{1}{2}\times\frac{1}{20}+\frac{1}{2}\times\frac{1}{40}$$

$$= \frac{1}{40}+\frac{1}{80} = \frac{3}{80}$$ **Ans.**

SECTION – D

Q. Nos. 33 to 36 carry 6 marks each.

33. Using properties of determinants prove that :

$$\begin{vmatrix} a-b & b+c & a \\ b-c & c+a & b \\ c-a & a+b & c \end{vmatrix} = a^3+b^3+c^3-3abc.$$ **[6]**

OR

If $A = \begin{bmatrix} 1 & 3 & 2 \\ 2 & 0 & -1 \\ 1 & 2 & 3 \end{bmatrix}$, **then show that** $A^3 - 4A^2 - 3A + 11I = 0$. **Hence find** A^{-1}.

Sol. $$\text{L.H.S.} = \begin{vmatrix} a-b & b+c & a \\ b-c & c+a & b \\ c-a & a+b & c \end{vmatrix}$$

Using $C_2 \to C_2 + C_3$ and then taking $(a+b+c)$ common from C_2,

$$= (a+b+c)\begin{vmatrix} a-b & 1 & a \\ b-c & 1 & b \\ c-a & 1 & c \end{vmatrix}$$

Applying $C_1 \to C_1 - C_3$,

$$= (a+b+c)\begin{vmatrix} -b & 1 & a \\ -c & 1 & b \\ -a & 1 & c \end{vmatrix}$$

Applying $R_3 \to R_3 - R_1$ and $R_2 \to R_2 - R_1$,

$$(a+b+c)\begin{vmatrix} -b & 1 & a \\ b-c & 0 & b-a \\ b-a & 0 & c-a \end{vmatrix}$$

Expanding along C_2,

$= (a+b+c)(-1)[(b-c)(c-a)-(b-a)^2]$

$= (a+b+c)(-1)[(bc-ab-c^2+ca-b^2-a^2+2ab)]$

$= (a+b+c)(a^2+b^2+c^2-ac-ab-ca)$

$= a^3+b^3+c^3-3abc = \text{R.H.S.}$ **Hence Proved.**

OR

Given, $$A = \begin{bmatrix} 1 & 3 & 2 \\ 2 & 0 & -1 \\ 1 & 2 & 3 \end{bmatrix}$$

∴ $$A^2 = \begin{bmatrix} 1 & 3 & 2 \\ 2 & 0 & -1 \\ 1 & 2 & 3 \end{bmatrix}\begin{bmatrix} 1 & 3 & 2 \\ 2 & 0 & -1 \\ 1 & 2 & 3 \end{bmatrix}$$

$$= \begin{bmatrix} 9 & 7 & 5 \\ 1 & 4 & 1 \\ 8 & 9 & 9 \end{bmatrix}$$

Also, $A^3 = AA^2 = \begin{bmatrix} 1 & 3 & 2 \\ 2 & 0 & -1 \\ 1 & 2 & 3 \end{bmatrix} \begin{bmatrix} 9 & 7 & 5 \\ 1 & 4 & 1 \\ 8 & 9 & 9 \end{bmatrix}$

$$= \begin{bmatrix} 28 & 37 & 26 \\ 10 & 5 & 1 \\ 35 & 42 & 34 \end{bmatrix}$$

Now, L.H.S. $= A^3 - 4A^2 - 3A + 11I$

$$= \begin{bmatrix} 28 & 37 & 26 \\ 10 & 5 & 1 \\ 35 & 42 & 34 \end{bmatrix} - 4\begin{bmatrix} 9 & 7 & 5 \\ 1 & 4 & 1 \\ 8 & 9 & 9 \end{bmatrix} - 3\begin{bmatrix} 1 & 3 & 2 \\ 2 & 0 & -1 \\ 1 & 2 & 3 \end{bmatrix} + 11\begin{bmatrix} 1 & 0 & 0 \\ 0 & 1 & 0 \\ 0 & 0 & 1 \end{bmatrix}$$

$$= \begin{bmatrix} 28-36-3+11 & 37-28-9 & 26-20-6 \\ 10-4-6 & 5-16+11 & 1-4+3 \\ 35-32-3 & 42-36-6 & 34-36-9+11 \end{bmatrix}$$

$$= \begin{bmatrix} 0 & 0 & 0 \\ 0 & 0 & 0 \\ 0 & 0 & 0 \end{bmatrix} = 0 = \text{R.H.S.}$$ **Hence Proved.**

Now, $A^3 - 4A^2 - 3A + 11I = 0$

Pre-multiplying by A^{-1}

$(A^{-1}A)A^2 - 4(A^{-1}A)A - 3A^{-1}(A) + 11A^{-1}I = A^{-1}0$

$\Rightarrow \quad IA^2 - 4IA - 3I + 11A^{-1} = 0$

$\Rightarrow \quad A^{-1} = \frac{1}{11}\{4A + 3I - A^2\}$

$$= \frac{1}{11}\left\{4\begin{bmatrix} 1 & 3 & 2 \\ 2 & 0 & -1 \\ 1 & 2 & 3 \end{bmatrix} + 3\begin{bmatrix} 1 & 0 & 0 \\ 0 & 1 & 0 \\ 0 & 0 & 1 \end{bmatrix} - \begin{bmatrix} 9 & 7 & 5 \\ 1 & 4 & 1 \\ 8 & 9 & 9 \end{bmatrix}\right\}$$

$$= \frac{1}{11}\begin{bmatrix} 4+3-9 & 12-7 & 8-5 \\ 8-1 & 3-4 & -4-1 \\ 4-8 & 8-9 & 12+3-9 \end{bmatrix}$$

$\Rightarrow \quad A^{-1} = \frac{1}{11}\begin{bmatrix} -2 & 5 & 3 \\ 7 & -1 & -5 \\ -4 & -1 & 6 \end{bmatrix}$ **Ans.**

34. Find the intervals on which the function $f(x) = (x-1)^3 (x-2)^2$ is
(a) strictly increasing, (b) strictly decreasing. **[6]**

OR

Find the dimensions of the rectangle of perimeter 36 cm which will sweep out a volume as large as possible, when revolved about one of its side. Also, find the maximum volume. **[6]**

Sol. Given, $f(x) = (x-1)^3 (x-2)^2$

$\therefore \; f'(x) = (x-1)^3 2(x-2) + (x-2)^2 3(x-1)^2$

$\Rightarrow f'(x) = (x-1)^2 (x-2) [2(x-1) + 3(x-2)]$

$= (x-1)^2 (x-2)(5x-8)$

Putting $\quad f'(x) = 0$

$\Rightarrow \quad x = 1, 2, \frac{8}{5}$

Intervals	Sign of $f'(x) = (x-1)^2 (x-2)(5x-8)$	Strictly increasing/ decreasing
$(-\infty, 1)$	+	Strictly increasing
$\left(1, \frac{8}{5}\right)$	+	Strictly increasing
$\left(\frac{8}{5}, 2\right)$	–	Strictly decreasing
$(2, \infty)$	+	Strictly increasing

(a) f is strictly increasing in $(-\infty, 1) \cup \left(1, \frac{8}{5}\right) \cup (2, \infty)$ **Ans.**

(b) f is strictly decreasing in $\left(\frac{8}{5}, 2\right)$ **Ans.**

OR

Let the length of rectangle be x cm and the breadth of rectangle be y cm.

$\therefore \quad$ Perimeter $= 2(l + b)$

$\Rightarrow \quad 2(x + y) = 36$

$\Rightarrow \quad x + y = 18$

$\Rightarrow \quad x = 18 - y \quad$ ...(i)

When the rectangle is revolved about its length then the dimensions of cylinder so formed are :

$h = x$ cm

$r = y$ cm

$\therefore$ Volume of cylinder is

$V = \pi r^2 h$

$\Rightarrow \quad V = \pi y^2 x$

$\Rightarrow \quad V = \pi y^2(18 - y)$ [Using (i)]

$\Rightarrow \qquad V = \pi(18y^2 - y^3)$

$\Rightarrow \qquad \dfrac{dV}{dy} = \pi(36y - 3y^2)$

Put $\qquad \dfrac{dV}{dy} = 0$

$\Rightarrow \qquad 36y - 3y^2 = 0$

$\Rightarrow \qquad 3y(12 - y) = 0$

$\Rightarrow \qquad y = 0$ (Rejected), $y = 12$

Now, $\qquad \dfrac{d^2V}{dy^2} = \pi(36 - 6y)$

$= 6\pi(6 - y)$

When $y = 12$, $\qquad \dfrac{d^2V}{dy^2} = 6\pi(6 - 12)$

$= 6\pi(-6) < 0$

$\therefore$ Volume is maximum when $y = 12$ cm.

From (i),

$x = 18 - y = 18 - 12 = 6$ cm

$\therefore$ For maximum volume, sides of rectangle are 6 cm and 12 cm.

Maximum volume, $V = \pi y^2 x$

$= \pi(12)^2 6$

$= \pi \times 144 \times 6$

$= 864\pi \text{ cm}^3$ **Ans.**

35. Find the area of the region lying in the first quadrant and enclosed by the x-axis, the line $y = x$ and the circle $x^2 + y^2 = 32$. [6]

Sol.

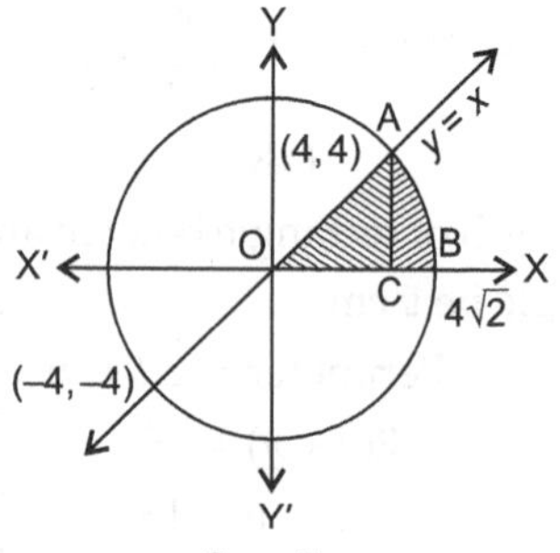

Given, $\qquad x^2 + y^2 = 32$

$\Rightarrow \qquad x^2 + y^2 = (4\sqrt{2})^2 \qquad$ ...(i)

$\therefore$ Centre of this circle is (0, 0)

and $\qquad$ Radius $= 4\sqrt{2}$

Given line is $\qquad y = x \qquad$ ...(ii)

From (i) and (ii),

$x^2 + x^2 = 32$

$\Rightarrow \qquad 2x^2 = 32$

$\Rightarrow \qquad x^2 = 16$

$\Rightarrow \qquad x = \pm 4$

$\therefore$ From (ii), $\qquad y = \pm 4$

Hence points of intersection of circle and line are (4, 4) and (– 4, – 4).

Required area = $ar(AOC) + ar(ABC)$

$= \int_0^4 x\,dx + \int_4^{4\sqrt{2}} \sqrt{(4\sqrt{2})^2 - x^2 dx}$

$= \left[\dfrac{x^2}{2}\right]_0^4 + \left[\dfrac{x}{2}\sqrt{32 - x^2} + \dfrac{32}{2}\sin^{-1}\left(\dfrac{x}{4\sqrt{2}}\right)\right]_4^{4\sqrt{2}}$

$= \left(\dfrac{16}{2} - 0\right) + \left[(0 + 16\sin^{-1}(1)) - \left(2\sqrt{16} + 16\sin^{-1}\left(\dfrac{1}{\sqrt{2}}\right)\right)\right]$

$= 8 + \left[16\left(\dfrac{\pi}{2}\right) - \left(8 + 16\left(\dfrac{\pi}{4}\right)\right)\right]$

$= 8 + (8\pi - 8 - 4\pi)$

$= 8 + 4\pi - 8$

$= 4\pi$ square units **Ans.**

36. Show that the lines $\vec{r} = \vec{a} + \lambda\vec{b}$ and $\vec{r} = \vec{b} + \mu\vec{a}$ are coplanar and the plane containing them is given by $\vec{r}\cdot(\vec{a} \times \vec{b}) = 0$. [6]

Sol. Given lines are :

$\vec{r} = \vec{a} + \lambda\vec{b} \qquad$ ...(i)

$\vec{r} = \vec{b} + \mu\vec{a} \qquad$ ...(ii)

We know that the lines $\vec{r} = \vec{a_1} + \lambda\vec{b_1}$, $\vec{r} = \vec{a_2} + \mu\vec{b_2}$ are coplanar only if

$\left(\vec{a_2} - \vec{a_1}\right)\cdot\left(\vec{b_1} \times \vec{b_2}\right) = 0$

Here, $\vec{a_1} = \vec{a}$, $\vec{b_1} = \vec{b}$, $\vec{a_2} = \vec{b}$, $\vec{b_2} = \vec{a}$.

$\therefore \left(\vec{a_2} - \vec{a_1}\right)\cdot\left(\vec{b_1} \times \vec{b_2}\right) = \left(\vec{b} - \vec{a}\right)\cdot\left(\vec{b} \times \vec{a}\right)$

$= \vec{b}\cdot\left(\vec{b} \times \vec{a}\right) - \vec{a}\cdot\left(\vec{b} \times \vec{a}\right)$

$= \left[\vec{b}\ \vec{b}\ \vec{a}\right] - \left[\vec{a}\ \vec{b}\ \vec{a}\right]$

$= 0 - 0$

{$\because$ When two of three vectors are identical, then their scalar triple product is 0}

$= 0$

Hence given lines are coplanar.

Also equation of plane containing the given lines is :

$$\left(\vec{r}-\vec{a}\right)\cdot\left(\vec{b_1}\times\vec{b_2}\right)=0\left[\because\left(\vec{r}-\vec{a}\right).\vec{n}=0\right]$$

$$\Rightarrow \left(\vec{r}-\vec{a}\right)\cdot\left(\vec{b}\times\vec{a}\right)=0$$

$$\Rightarrow \vec{r}.\left(\vec{b}\times\vec{a}\right)-\vec{a}.\left(\vec{b}\times\vec{a}\right)=0$$

$$\Rightarrow \vec{r}.\left(\vec{b}\times\vec{a}\right)=0$$

$$\Rightarrow -\vec{r}.\left(\vec{a}\times\vec{b}\right)=0 \quad \left[\because\left[\vec{a}\,\vec{b}\,\vec{a}\right]=0\right]$$

$$\Rightarrow \vec{r}.\left(\vec{a}\times\vec{b}\right)=0$$ **Hence Proved.**

●●

Note : Except for the following questions, all the remaining questions have been asked in the previous set.

SECTION – A

8. Let A = $\begin{bmatrix}200 & 50\\10 & 2\end{bmatrix}$ and B = $\begin{bmatrix}50 & 40\\2 & 3\end{bmatrix}$, then |AB| is equal to

(a) 460 **(b) 2000**

(c) 3000 **(d) –7000** [1]

Sol.
$$|AB| = |A|\,|B|$$
$$= (400-500)(150-80)$$
$$= (-100)(70)$$
$$= -7000$$

∴ Option (d) is correct.

9. Let $\vec{a}=\hat{i}-2\hat{j}+3\hat{k}$. If $\vec{b}$ is a vector such that $\vec{a}.\vec{b}=|\vec{b}|^2$ and $|\vec{a}-\vec{b}|=\sqrt{7}$, then $|\vec{b}|$ equals

(a) 7 **(b) 14**

(c) $\sqrt{7}$ **(d) 21** [1]

Sol. $\left|\vec{a}-\vec{b}\right|^2=\left|\vec{a}\right|^2+\left|\vec{b}\right|^2-2\vec{a}.\vec{b}$

$$\Rightarrow (\sqrt{7})^2=(1+4+9)+\left|\vec{b}\right|^2-2\left|\vec{b}\right|^2$$

$$\Rightarrow 7-14=-\left|\vec{b}\right|^2$$

$$\Rightarrow \left|\vec{b}\right|^2=7$$

$$\therefore \left|\vec{b}\right|=\sqrt{7}$$

∴ Option (c) is correct.

10. Three dice are thrown simultaneously. The probability of obtaining a total score of 5 is

(a) $\frac{5}{216}$ **(b) $\frac{1}{6}$**

(c) $\frac{1}{36}$ **(d) $\frac{1}{49}$** [1]

Sol. When three dice are thrown :

$$n(S)=6\times6\times6=216$$

Let A be the event of getting total score 5.

Then A = {(1, 1, 3), (1, 3, 1), (3, 1, 1), (2, 2, 1), (2, 1, 2), (1, 2, 2)}

$$\therefore\quad n(A)=6$$

$$\therefore\quad P(A)=\frac{n(A)}{n(S)}$$

$$=\frac{6}{216}=\frac{1}{36}$$

∴ Option (c) is correct.

15. If $f(x)=2|x|+3|\sin x|+6$, then the right hand derivative of $f(x)$ at $x=0$ is ________. [1]

Sol. 5

Explanation: $f(x)=2|x|+3|\sin x|+6$

Right hand derivative of f at $x=0$ is :

$$Rf'(0)=\lim_{h\to0}\frac{f(0+h)-f(0)}{h}$$, where h is a very small positive real number.

$$=\lim_{h\to0}\frac{(2|h|+3|\sin h|+6)-(2|0|+3\sin|0|+6)}{h}$$

$$=\lim_{h\to0}\frac{(2h+3\sin h+6)-(0+0+6)}{h}$$

$$=\lim_{h\to0}\frac{2h+3\sin h}{h}$$

$$=\lim_{h\to0}\frac{2h}{h}+3\lim_{h\to0}\frac{\sin h}{h}=2+3(1)=5$$ **Ans.**

19. Find $\int\sin^5\left(\frac{x}{2}\right).\cos\left(\frac{x}{2}\right)dx$ [1]

Sol. Let, $I=\int\sin^5\left(\frac{x}{2}\right).\cos\left(\frac{x}{2}\right)dx$

Let $\sin\frac{x}{2} = t$

$\Rightarrow \cos\left(\frac{x}{2}\right)dx = 2dt$

$\therefore I = 2\int t^5 dt$

$\Rightarrow I = \frac{2t^6}{6} + c$

$= \frac{1}{3}\sin^6\left(\frac{x}{2}\right) + c$ **Ans.**

20. If A = $\begin{bmatrix} 1 & 0 \\ 1 & 1 \end{bmatrix}$, then find A³. **[1]**

Sol. Given, $A = \begin{bmatrix} 1 & 0 \\ 1 & 1 \end{bmatrix}$

$A^2 = \begin{bmatrix} 1 & 0 \\ 1 & 1 \end{bmatrix}\begin{bmatrix} 1 & 0 \\ 1 & 1 \end{bmatrix}$

$= \begin{bmatrix} 1 & 0 \\ 2 & 1 \end{bmatrix}$

And, $A^3 = A.A^2$

$= \begin{bmatrix} 1 & 0 \\ 1 & 1 \end{bmatrix}\begin{bmatrix} 1 & 0 \\ 2 & 1 \end{bmatrix}$

$\Rightarrow A^3 = \begin{bmatrix} 1 & 0 \\ 3 & 1 \end{bmatrix}$ **Ans.**

SECTION – B

25. Find the derivative of $x^{\log x}$ w.r.t. log x. **[2]**

Sol. Let $u = x^{\log x}$

Taking log on both sides, we get

$\log u = \log x^{\log x}$

$\Rightarrow \log u = \log x . \log x$

$\Rightarrow \log u = (\log x)^2$

Diff. both sides w.r.t x,

$\Rightarrow \frac{1}{u}\frac{du}{dx} = \frac{2\log x}{x}$

$\Rightarrow \frac{du}{dx} = \frac{2u\log x}{x}$

$\Rightarrow \frac{du}{dx} = \frac{2x^{\log x}\log x}{x}$

Also, let $v = \log x$

$\therefore \frac{dv}{dx} = \frac{1}{x}$

$\therefore \frac{du}{dv} = \frac{du}{dx}\times\frac{dx}{dv} = \frac{2x^{\log x}\log x}{x}\times\frac{x}{1}$

$\Rightarrow \frac{du}{dv} = 2x^{\log x}.\log x$ **Ans.**

26. Find the distance between the parallel planes $2x + y + 2z = 8$ and $4x + 2y + 4z + 5 = 0$ **[2]**

Sol. Given lines are :

$2x + y + 2z = 8$ ...(i)

$4x + 2y + 4z + 5 = 0$ ...(ii)

Multiplying equation (i) by 2,

$4x + 2y + 4z - 16 = 0$

Distance between given parallel lines

$= \frac{|D_1 - D_2|}{\sqrt{A^2 + B^2 + c^2}}$

$= \frac{|-16-5|}{\sqrt{4^2+2^2+4^2}} = \frac{21}{\sqrt{36}}$

$= \frac{21}{6} = \frac{7}{2}$ units **Ans.**

SECTION – C

31. Show that the function f : R → R defined by $f(x) = \frac{x}{x^2+1}$ ∀ $x \in$ R is neither one-one nor onto. **[4]**

Sol. Given, $f(x) = \frac{x}{x^2+1}$

$\because f(2) = \frac{2}{2^2+1} = \frac{2}{5}$

and $f\left(\frac{1}{2}\right) = \frac{\frac{1}{2}}{\left(\frac{1}{2}\right)^2+1} = \frac{\frac{1}{2}}{\frac{1}{4}+1} = \frac{\frac{1}{2}}{\frac{5}{4}}$

$= \frac{1}{2}\times\frac{4}{5} = \frac{2}{5}$

$\therefore f(2) = f\left(\frac{1}{2}\right)$

But $2 \neq \frac{1}{2}$

$\therefore$ f is not one-one.

For onto :

Let $y \in$ R

Such that $f(x) = y$

$\Rightarrow \frac{x}{x^2+1} = y$

$$\Rightarrow \quad x^2y + y = x$$

$$\Rightarrow \quad x^2y - x + y = 0$$

$$\Rightarrow \quad x = \frac{-(-1) \pm \sqrt{1-4y^2}}{2(y)}$$

$$\Rightarrow \quad x = \frac{1 \pm \sqrt{1-4y^2}}{2y}$$

Here x is not real when $1 - 4y^2 < 0$

$$\Rightarrow \quad 1 < 4y^2$$

$$\Rightarrow \quad y > \left(\frac{1}{2}\right)^2$$

$$\Rightarrow \quad y < \frac{-1}{2}$$

or $\quad y > \frac{1}{2}$

$\therefore$ for $\quad y \in \left(-\infty, \frac{-1}{2}\right) \cup \left(\frac{1}{2}, \infty\right)$

There exists no $x \in$ R such that $f(x) = y$.

Hence f is not an onto function. **Hence Proved.**

32. Evaluate : $\int_{-1}^{2} |x^3 - x|\, dx$ **[4]**

Sol. Let, $\quad I = \int_{-1}^{2} |x^3 - x|\, dx$

Putting $\quad x^3 - x = 0$

$$\Rightarrow \quad x(x^2 - 1) = 0$$

$$\Rightarrow \quad x(x-1)(x+1) = 0$$

$$\Rightarrow \quad x = 0, x = 1, x = -1$$

$$\therefore I = \int_{-1}^{0} |x^3 - x|\, dx + \int_{0}^{1} |x^3 - x| + \int_{1}^{2} |x^3 - x|\, dx$$

$$\therefore \quad I = I_1 + I_2 + I_3$$

$$I_1 = \int_{-1}^{0} |x^3 - x|\, dx = \int_{-1}^{0} [x^3 - x]\, dx$$

$[\because x^3 - x \geq 0 \ \forall \ x \leq 0]$

$$= \left[\frac{x^4}{4} - \frac{x^2}{2}\right]_{-1}^{0} = 0 - \left(\frac{1}{4} - \frac{1}{2}\right) = \frac{1}{4}$$

$$I_2 = \int_{0}^{1} |x^3 - x|\, dx = -\int_{0}^{1} (x^3 - x)\, dx$$

$$= -\left[\frac{x^4}{4} - \frac{x^2}{2}\right]_{0}^{1}$$

$$= -\left[\left(\frac{1}{4} - \frac{1}{2}\right) - 0\right] = \frac{1}{4}$$

$$I_3 = \int_{1}^{2} |x^3 - x|\, dx$$

$$= \int_{1}^{2} (x^3 - x)\, dx = \left[\frac{x^4}{4} - \frac{x^2}{2}\right]_{1}^{2}$$

$$= \left[(4-2) - \left(\frac{1}{4} - \frac{1}{2}\right)\right] = \left[2 + \frac{1}{4}\right] = \frac{9}{4}$$

Now, $\quad I = I_1 + I_2 + I_3$

$$= \frac{1}{4} + \frac{1}{4} + \frac{9}{4} = \frac{11}{4}$$ **Ans.**

SECTION – D

36. Using integration find the area of the region :
$\{(x, y) : 0 \leq y \leq x^2, 0 \leq y \leq x, 0 \leq x \leq 2\}$ **[6]**

Sol. Given,

$\{(x, y) : 0 \leq y \leq x^2, 0 \leq y \leq x, 0 \leq x \leq 2\}$

$\Rightarrow \quad y \leq x^2 \quad$ ...(i)

$y \leq x \quad$ ...(ii)

$x \leq 2 \quad$ ...(iii)

$x \geq 0 \quad$ ...(iv)

$y \geq 0 \quad$ ...(v)

Considering inequalities as equation :

$y = x^2, y = x, x = 2, x = 0, y = 0$

Solving $y = x^2$ and $y = x$

$$\Rightarrow \quad x^2 = x$$

$$\Rightarrow \quad x(x-1) = 0$$

$$\Rightarrow \quad x = 0, x = 1$$

$$\therefore \quad y = 0, y = 1$$

$\therefore$ Points of intersection of curve (i) and the (ii) are (0, 0) and (1, 1)

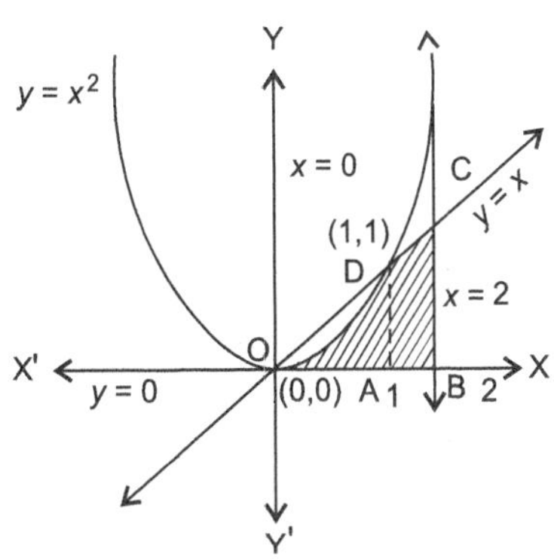

$\therefore$ Required area $= \int_0^1 (y \text{ of the parabola})\, dx + \int_1^2 (y \text{ of the line}) dx$

$\therefore$ Required area $= \int_0^1 x^2 dx + \int_1^2 x\, dx$

$$= \left[\frac{x^3}{3}\right]_0^1 + \left[\frac{x^2}{2}\right]_1^2$$

$$= \left(\frac{1}{3} - 0\right) + \left(\frac{4}{2} - \frac{1}{2}\right)$$

$$= \frac{1}{3} + \left(2 - \frac{1}{2}\right)$$

$$= \frac{1}{3} + \frac{3}{2} = \frac{2+9}{6}$$

$= \frac{11}{6}$ square units. **Ans.**

Outside Delhi [Set-III]

Note : Except for the following questions, all the remaining questions have been asked in the previous set.

SECTION – A

8. The value of $\tan^{-1}\left[\frac{1}{2}\cos^{-1}\left(\frac{\sqrt{5}}{3}\right)\right]$ is

(a) $\frac{3+\sqrt{5}}{2}$ **(b)** $\frac{3-\sqrt{5}}{2}$

(c) $\frac{-3+\sqrt{5}}{2}$ **(d)** $\frac{-3-\sqrt{5}}{2}$ **[1]**

Sol. $\tan^{-1}\left[\frac{1}{2}\cos^{-1}\left(\frac{\sqrt{5}}{3}\right)\right]$

This question is wrong.

It should be $\tan\left[\frac{1}{2}\cos^{-1}\left(\frac{\sqrt{5}}{3}\right)\right]$

Now, let $y = \tan\left[\frac{1}{2}\cos^{-1}\left(\frac{\sqrt{5}}{3}\right)\right]$

Putting, $x = \cos^{-1}\left(\frac{\sqrt{5}}{3}\right)$

$\Rightarrow \cos x = \frac{\sqrt{5}}{3}$ ...(i)

Now, $y = \tan\left(\frac{1}{2} \times x\right)$

$y = \tan\frac{x}{2}$

$\therefore y = \sqrt{\frac{1-\cos x}{1+\cos x}}$ (using property)

$\therefore$ [From (i)] $y = \sqrt{\frac{1-\sqrt{5}/3}{1+\sqrt{5}/3}}$

$\therefore y = \sqrt{\frac{3-\sqrt{5}}{3+\sqrt{5}}} = \sqrt{\frac{(3-\sqrt{5})(3-\sqrt{5})}{(3+\sqrt{5})(3-\sqrt{5})}}$

$\therefore y = \sqrt{\frac{(3-\sqrt{5})^2}{9-5}} = \frac{3-\sqrt{5}}{\sqrt{4}} = \frac{3-\sqrt{5}}{2}$

$\therefore y = \frac{3-\sqrt{5}}{2}$

i.e., $\tan\left[\frac{1}{2}\cos^{-1}\left(\frac{\sqrt{5}}{3}\right)\right] = \frac{3-\sqrt{5}}{2}$

$\therefore$ Option (b) is correct.

9. If A = $\begin{bmatrix} a & 0 & 0 \\ 0 & a & 0 \\ 0 & 0 & a \end{bmatrix}$, then det (adj A) equals

(a) a^{27} **(b)** a^9

(c) a^6 **(d)** a^2 **[1]**

Sol. $|A| = a^3$

$\therefore |\text{adj A}| = |A|^2 = (a^3)^2 = a^6$

$\therefore$ Option (c) is correct.

10. The line $\frac{x-2}{3} = \frac{y-3}{4} = \frac{z-4}{5}$ is parallel to the plane

(a) $2x + 3y + 4z = 0$ **(b)** $3x + 4y - 5z = 7$

(c) $2x + y - 2z = 0$ **(d)** $x - y + z = 2$ **[1]**

Sol. Option (b) or (c) is correct.

As for (b) → $3.3 + 4 \times 4 - 5 \times 5 = 0$

or (c) → $2.3 + 4 \times 1 - 2 \times 5 = 0$

15. If $f(x) = x\,|x|$, then $f'(x) =$ ________ . [1]

Sol. $2x, x \geq 0, -2x, x < 0$

Explanation:
$$f(x) = x|x| = \begin{cases} x^2, & \text{if } x \geq 0 \\ -x^2, & \text{if } x < 0 \end{cases}$$

$$\therefore \quad f'(x) = \begin{cases} 2x, & \text{if } x \geq 0 \\ -2x, & \text{if } x < 0 \end{cases}$$

19. Find $\int \frac{1}{x(1+x^2)} dx$

Sol. Let, $I = \int \frac{1}{x(1+x^2)} dx$

$$= \int \frac{x}{x^2(x^2+1)} dx$$

Let $x^2 = t$

$$\Rightarrow x\,dx = \frac{dt}{2}$$

$$\therefore \quad I = \frac{1}{2}\int \frac{dt}{t(t+1)} = \frac{1}{2}\int \frac{(t+1)-t}{t(t+1)} dt$$

$$= \frac{1}{2}\int \left(\frac{1}{t} - \frac{1}{t+1}\right) dt$$

$$= \frac{1}{2}[\log|t| - \log|t+1|] + c$$

$$= \frac{1}{2}\log\left|\frac{t}{t+1}\right| + c = \frac{1}{2}\log\left|\frac{x^2}{x^2+1}\right| + c$$ **Ans.**

20. If [x] denotes the greatest integer function, then find $\int_0^{3/2} [x^2]\, dx$. [1]

Sol.
$$\int_0^{3/2} [x^2]dx = \int_0^1 [x^2]dx + \int_1^{\sqrt{2}} [x^2]dx + \int_{\sqrt{2}}^{3/2} [x^2]dx$$

$$= \int_0^1 0\,dx + \int_1^{\sqrt{2}} 1\,dx + \int_{\sqrt{2}}^{3/2} 2\,dx = 0 + [x]_1^{\sqrt{2}} + 2[x]_{\sqrt{2}}^{3/2}$$

$$= \sqrt{2} - 1 + 2\left(\frac{3}{2} - \sqrt{2}\right) = \sqrt{2} - 1 + 3 - 2\sqrt{2}$$

$$= 2 - \sqrt{2}$$ **Ans.**

SECTION–B

25. Show that the function $f(x) = \frac{x}{3} + \frac{3}{x}$ decreases in the intervals $(-3, 0) \cup (0, 3)$. [2]

Sol. Given,
$$f(x) = \frac{x}{3} + \frac{3}{x}$$

$$\Rightarrow \quad f'(x) = \frac{1}{3} - \frac{3}{x^2}$$

$$\Rightarrow \quad f'(x) = \frac{x^2 - 9}{3x^2}$$

Now, f is decreasing when $f'(x) < 0$

$$\Rightarrow \quad \frac{x^2-9}{3x^2} < 0$$

$$\Rightarrow \quad x^2 < 9, x \neq 0$$

$$\Rightarrow \quad x^2 < 3^2, x \neq 0$$

$$\Rightarrow \quad -3 < x < 3, x \neq 0$$

⇒ so, $f(x)$ decreases in $(-3, 0) \cup (0, 3)$

Hence Proved.

26. Three distinct numbers are chosen randomly from the first 50 natural numbers. Find the probability that all the three numbers are divisible by both 2 and 3. [2]

Sol. Here, $n(S) = {}^{50}C_3$

Numbers divisible by both 2 and 3 from first 50 natural numbers are 6, 12, 18, 24, 30, 36, 42, 48

i.e., 8 numbers

$$\therefore \text{ Required Probability} = \frac{{}^8C_3}{{}^{50}C_3}$$

$$= \frac{8 \times 7 \times 6}{50 \times 49 \times 48} = \frac{1}{350}$$ **Ans.**

SECTION–C

31. Evaluate $\int_0^1 \sqrt{3 - 2x - x^2}\, dx$ [4]

Sol.
$$\sqrt{3-2x-x^2}\,dx = \int_0^1 \sqrt{-[x^2+2x-3]}\,dx$$

$$= \int_0^1 \sqrt{-[(x+1)^2 - 2^2]}\,dx = \int_0^1 \sqrt{2^2 - (x+1)^2}\,dx$$

$$= \left[\frac{x+1}{2}\sqrt{3-2x-x^2}+\frac{2^2}{2}\sin^{-1}\left(\frac{x+1}{2}\right)\right]_0^1$$

$$= \left[\frac{2}{2}\sqrt{0}+2\sin^{-1}(1)\right]-\left[\frac{1}{2}\sqrt{3}+2\sin^{-1}\left(\frac{1}{2}\right)\right]$$

$$= 2\left(\frac{\pi}{2}\right)-\frac{\sqrt{3}}{2}-2\left(\frac{\pi}{6}\right) = \pi-\frac{\sqrt{3}}{2}-\frac{\pi}{3} = \frac{2\pi}{3}-\frac{\sqrt{3}}{2}$$

Ans.

32. Find the general solution of the differential equation $\frac{dy}{dx}+\frac{1}{x}=\frac{e^y}{x}$. [4]

Sol. Given,

$$\frac{dy}{dx}+\frac{1}{x}=\frac{e^y}{x}$$

$$\Rightarrow \frac{dy}{dx} = \frac{e^y-1}{x}$$

$$\Rightarrow \int\frac{dy}{e^y-1} = \int\frac{dx}{x}$$

$$\Rightarrow \int\frac{e^y}{e^y(e^y-1)}dy = \int\frac{dx}{x}$$

Let $e^y = t$

$$\Rightarrow e^y dy = dt$$

$$\Rightarrow \int\frac{dt}{t(t-1)} = \int\frac{dx}{x}$$

$$\Rightarrow \int\frac{t-(t-1)}{t(t-1)}dt = \int\frac{dx}{x}$$

$$\Rightarrow \int\frac{1}{t-1}dt-\int\frac{1}{t}dt = \int\frac{dx}{x}$$

$$\Rightarrow \log|t-1|-\log|t| = \log|x|+\log|C|$$

$$\Rightarrow \log\left|\frac{t-1}{t}\right| = \log|Cx|$$

$$\Rightarrow \frac{t-1}{t} = Cx$$

$$\Rightarrow e^y-1 = Cxe^y$$

$$\Rightarrow 1-e^{-y} = Cx$$

Ans.

SECTION – D

36. Find the image of the point (–1, 3, 4) in the plane $x-2y=0$. [6]

Sol.

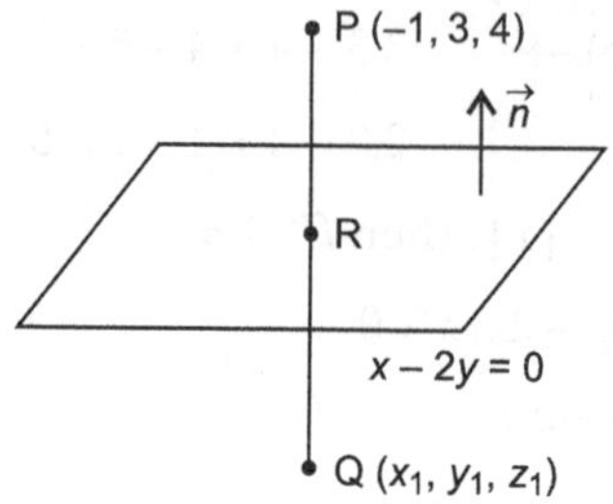

Let the image of the point P(–1, 3, 4) in the plane $x-2y=0$ be Q (x_1, y_1, z_1).

$\therefore$ Direction ratios of $\vec{PQ}$ are

$a_1 = x_1+1, b_1 = y_1-3, c_1 = z_1-4$

Also, direction ratios of $\vec{n}$ are

$a_2 = 1, b_2 = -2, c_2 = 0$

Since $\vec{PQ} \parallel \vec{n}$

$$\Rightarrow \frac{a_1}{a_2} = \frac{b_1}{b_2} = \frac{c_1}{c_2}$$

$$\Rightarrow \frac{x_1+1}{1} = \frac{y_1-3}{-2} = \frac{z_1-4}{0} = \lambda \text{ (say)}$$

$$\Rightarrow x_1 = \lambda-1, y_1 = 3-2\lambda, z_1 = 4$$

$$\therefore \text{Point Q} = (\lambda-1, 3-2\lambda, 4)$$

$$\text{Mid-point of PQ} = R\left(\frac{\lambda-1-1}{2}, \frac{3-2\lambda+3}{2}, \frac{4+4}{2}\right)$$

$$= R\left(\frac{\lambda-2}{2}, 3-\lambda, 4\right)$$

$\because$ Point R lies on the plane $x-2y=0$.

$$\therefore \frac{\lambda-2}{2}-2(3-\lambda) = 0$$

$$\Rightarrow \frac{\lambda-2}{2} = 6-2\lambda$$

$$\Rightarrow \lambda-2 = 12-4\lambda$$

$$\Rightarrow 5\lambda = 14$$

$$\Rightarrow \lambda = \frac{14}{5}$$

$$\therefore Q = \left(\frac{14}{5}-1, 3-\frac{28}{5}, 4\right)$$

$$= \left(\frac{9}{5}, \frac{-13}{5}, 4\right)$$

Hence, the image of the point (–1, 3, 4) in the plane $x-2y=0$ is $\left(\frac{9}{5}, -\frac{13}{4}, 4\right)$ **Ans.**

●●

MATHEMATICS

Delhi [Set-I]

2020

Time allowed : 3 hours

Maximum marks : 80

SECTION–A

Question numbers 1 to 10 are multiple choice questions of 1 mark each. Select the correct option :

1. If A is a square matrix of order 3, such that A (adj A) = 10 I, then |adj A| is equal to :

(a) 1 (b) 10

(c) 100 (d) 101 [1]

Sol. Given : $A (\text{adj } A) = 10I$

We know, $A.\text{adj } A = |A|.I$

$\therefore \quad |A| = 10$

$\because \quad |\text{adj } A| = |A|^{n-1}$

$\therefore \quad |\text{adj } A| = |A|^{3-1} = 10^2 = 100$

∴ Option (c) is correct.

2. If A is a 3 × 3 matrix such that |A| = 8, then |3A| equals:

(a) 8 (b) 24

(c) 72 (d) 216 [1]

Sol. Given : $|A| = 8$

We know, $|kA| = k^3|A|$, where k is constant

$\therefore \quad |3A| = 3^3|A|$

$= 27 \times 8 = 216$

∴ Option (d) is correct.

3. If $y = Ae^{5x} + Be^{-5x}$, then $\frac{d^2y}{dx^2}$ is equal to:

(a) 25 y (b) 5 y

(c) – 25 y (d) 15 y [1]

Sol. Given : $y = Ae^{5x} + Be^{-5x}$

$\therefore \quad \frac{dy}{dx} = 5Ae^{5x} - 5Be^{-5x}$

$\Rightarrow \quad \frac{d^2y}{dx^2} = 25Ae^{5x} + 25Be^{-5x}$

$\Rightarrow \quad \frac{d^2y}{dx^2} = 25y$

∴ Option (a) is correct.

4. $\int x^2 e^{x^3} dx$ equals :

(a) $\frac{1}{3}e^{x^3}+C$ (b) $\frac{1}{3}e^{x^4}+C$

(c) $\frac{1}{2}e^{x^3}+C$ (d) $\frac{1}{2}e^{x^2}+C$ [1]

Sol. Let $I = \int x^2 e^{x^3} dx$

Put $x^3 = t$

$\Rightarrow \quad 3x^2dx = dt$

$\Rightarrow \quad x^2dx = \frac{1}{3}dt$

$\therefore \quad I = \frac{1}{3}\int e^t dt$

$I = \frac{1}{3}e^t + C$

$I = \frac{1}{3}e^{x^3} + C$

∴ Option (a) is correct.

5. If $\hat{i}, \hat{j}, \hat{k}$ are unit vectors along three mutually perpendicular directions, then

(a) $\hat{i}.\hat{j} = 1$ (b) $\hat{i}\times\hat{j} = 1$

(c) $\hat{i}.\hat{k} = 0$ (d) $\hat{i}\times\hat{k} = 0$ [1]

Sol. Option (c) is correct.

6. ABCD is a rhombus whose diagonals intersect at E. Then $\vec{EA} + \vec{EB} + \vec{EC} + \vec{ED}$ equals :

(a) $\vec{0}$ (b) $\vec{AD}$

(c) $2\vec{BC}$ (d) $2\vec{AD}$ [1]

Sol. Option (a) is correct.

7. The lines $\frac{x-2}{1} = \frac{y-3}{1} = \frac{4-z}{k}$ and $\frac{x-1}{k} = \frac{y-4}{2} = \frac{z-5}{-2}$ are mutually perpendicular if the value of k is :

(a) $-\frac{2}{3}$ (b) $\frac{2}{3}$

(c) – 2 (d) 2 [1]

Sol. First line is,

$$\frac{x-2}{1} = \frac{y-3}{1} = \frac{z-4}{-k}$$

∴ Direction ratios of first line = < 1, 1, – k >

Second line is,

$$\frac{x-1}{k} = \frac{y-4}{2} = \frac{z-5}{-2}$$

∴ Direction ratios of second line = < k, 2, – 2>

We know, two lines are perpendicular to each other, if the dot product of their direction ratios is zero.

i.e., $1 \times k + 1 \times 2 + (-k) \times (-2) = 0$

$$\Rightarrow \quad k + 2 + 2k = 0$$

$$\Rightarrow \quad 3k = -2$$

$$\Rightarrow \quad k = -\frac{2}{3}$$

∴ Option (a) is correct.

8. The graph of the inequality $2x + 3y > 6$ is

(a) half plane that contains the origin

(b) half plane that neither contains the origin nor the points of the line $2x + 3y = 6$

(c) whole XOY–plane excluding the points on the line $2x + 3y = 6$

(d) entire XOY plane. **[1]**

Sol. Option (b) is correct.

9. A card is picked at random from a pack of 52 playing cards. Given that the picked card is a queen, the probability of this card to be a card of spade is :

(a) $\frac{1}{3}$ **(b)** $\frac{4}{13}$

(c) $\frac{1}{4}$ **(d)** $\frac{1}{2}$ **[1]**

Sol. Let event A = Picked card is a queen. and event B = card of spade is picked.

$$\therefore \quad P(A) = \frac{4}{52}, \ P(B) = \frac{13}{52}$$

and, $P(A \cap B) = \frac{1}{52}$ [$\because A \cap B$ = Queen of spade]

Now, required probability,

$$P(B/A) = \frac{P(A \cap B)}{P(A)} = \frac{1/52}{4/52} = \frac{1}{4}$$

∴ Option (c) is correct.

10. A die is thrown once. Let A be the event that the number obtained is greater than 3. Let B be the event that the number obtained is less than 5. Then P(A∪B) is :

(a) $\frac{2}{5}$ **(b)** $\frac{3}{5}$

(c) 0 **(d)** 1 **[1]**

Sol.
$$P(A) = \frac{3}{6}$$

$$P(B) = \frac{4}{6}$$

and $$P(A \cap B) = \frac{1}{6}$$

We know, $P(A \cup B) = P(A) + P(B) - P(A \cap B)$

$$= \frac{3}{6} + \frac{4}{6} - \frac{1}{6}$$

$$= \frac{6}{6} = 1$$

∴ Option (d) is correct.

Fill in the blanks in Questions from 11 to 15.

11. A relation in a set A is called ________ relation, if each element of A is related to itself. **[1]**

Sol. reflexive

12. If $A + B = \begin{bmatrix} 1 & 0 \\ 1 & 1 \end{bmatrix}$ and $A - 2B = \begin{bmatrix} -1 & 1 \\ 0 & -1 \end{bmatrix}$, then A = ________. **[1]**

Sol. $\begin{bmatrix} \frac{1}{3} & \frac{1}{3} \\ \frac{2}{3} & \frac{1}{3} \end{bmatrix}$

Explanation:

Given : $$A + B = \begin{bmatrix} 1 & 0 \\ 1 & 1 \end{bmatrix} \quad ...(i)$$

and $$A - 2B = \begin{bmatrix} -1 & 1 \\ 0 & -1 \end{bmatrix} \quad ...(ii)$$

$$\Rightarrow \quad A - 2\left\{\begin{bmatrix} 1 & 0 \\ 1 & 1 \end{bmatrix} - A\right\} = \begin{bmatrix} -1 & 1 \\ 0 & -1 \end{bmatrix} \quad \text{[From (i)]}$$

$$\Rightarrow \quad A - \begin{bmatrix} 2 & 0 \\ 2 & 2 \end{bmatrix} + 2A = \begin{bmatrix} -1 & 1 \\ 0 & -1 \end{bmatrix}$$

$$3A = \begin{bmatrix} -1 & 1 \\ 0 & -1 \end{bmatrix} + \begin{bmatrix} 2 & 0 \\ 2 & 2 \end{bmatrix}$$

$$3A = \begin{bmatrix} 1 & 1 \\ 2 & 1 \end{bmatrix}$$

$$A = \frac{1}{3}\begin{bmatrix} 1 & 1 \\ 2 & 1 \end{bmatrix} = \begin{bmatrix} 1/3 & 1/3 \\ 2/3 & 1/3 \end{bmatrix}$$

13. The least value of the function $f(x) = ax + \frac{b}{x}$ $(a > 0, b > 0, x > 0)$ is ________. **[1]**

Sol. $2\sqrt{ab}$

Explanation:

$$\because \quad f(x) = ax + \frac{b}{x}$$

$$\therefore \quad f'(x) = a - \frac{b}{x^2}$$

For maximum and minimum function,

Put $f'(x) = 0$

$\Rightarrow \quad a - \frac{b}{x^2} = 0$

$\Rightarrow \quad ax^2 - b = 0$

$\Rightarrow \quad ax^2 = b$

$\Rightarrow \quad x = \sqrt{\frac{b}{a}} \quad [\because x > 0]$

Now, $f''(x) = 0 + \frac{2b}{x^3}$

$= \frac{2b}{(\sqrt{b/a})^3} > 0$

$\therefore f(x)$ is minimum at $x = \sqrt{\frac{b}{a}}$

$\therefore$ The least value of $f(x)$ is,

$$f\left(\sqrt{\frac{b}{a}}\right) = a\sqrt{\frac{b}{a}} + \frac{b}{\sqrt{\frac{b}{a}}}$$

$$= \sqrt{ab} + \sqrt{ab} = 2\sqrt{ab}$$

14. The integrating factor of the differential equation $x\frac{dy}{dx} + 2y = x^2$ is _______ . [1]

OR

The degree of the differential equation $1 + \left(\frac{dy}{dx}\right)^2 = x$ is _______ .

Sol. x^2

Explanation:

$\because \quad x\frac{dy}{dx} + 2y = x^2$

$\Rightarrow \quad \frac{dy}{dx} + \frac{2y}{x} = x$

Here, $P = \frac{2}{x}, \ Q = x$

$\therefore \quad \text{I.F.} = e^{\int P dx} = e^{\int \frac{2}{x} dx}$

$= e^{2\log x} = e^{\log x^2}$

$= x^2$

OR

2

15. The vector equation of a line which passes through the points (3, 4, – 7) and (1, –1, 6) is _______ . [1]

OR

The line of shortest distance between two skew lines is _______ to both the lines.

Sol. $(3\hat{i} + 4\hat{j} - 7\hat{k}) + \lambda(-2\hat{i} - 5\hat{j} + 13\hat{k})$

Explanation:

$\because$ A = (3, 4, – 7) and B = (1, – 1, 6)

$\therefore$ Direction ratios of AB = < – 2, –5, 13 >

$\therefore$ Equation of required line is,

$$\vec{r} = (3\hat{i} + 4\hat{j} - 7\hat{k}) + \lambda(-2\hat{i} - 5\hat{j} + 13\hat{k})$$

$$\vec{r} = (3\hat{i} + 4\hat{j} - 7\hat{k}) - \lambda(2\hat{i} + 5\hat{j} - 13\hat{k})$$

OR

Perpendicular

Q. Nos. 16 to 20 are of very short answer type questions.

16. Find the value of $\sin^{-1}\left[\sin\left(-\frac{17\pi}{8}\right)\right]$. [1]

Sol. We know that

$$\sin^{-1}(\sin x) = x$$

$$\text{So, } \sin^{-1}\left[\sin\left(\frac{-17\pi}{8}\right)\right] = \sin^{-1}\left[-\sin\frac{17\pi}{8}\right]$$

$$= \sin^{-1}\left[-\sin\left(2\pi + \frac{\pi}{8}\right)\right]$$

$$= \sin^{-1}\left(-\sin\frac{\pi}{8}\right)$$

$$= \sin^{-1}\left[\sin\left(-\frac{\pi}{8}\right)\right]$$

$$= -\frac{\pi}{8}$$ **Ans.**

17. For A = $\begin{bmatrix} 3 & -4 \\ 1 & -1 \end{bmatrix}$ write A^{-1}. [1]

Sol. Given : $A = \begin{bmatrix} 3 & -4 \\ 1 & -1 \end{bmatrix}$

$\therefore \quad |A| = 3 \times (-1) - (-4) \times 1$

$= -3 + 4 = 1$

and $\quad \text{adj } A = \begin{bmatrix} -1 & 4 \\ -1 & 3 \end{bmatrix}$

$\because \quad A^{-1} = \frac{1}{|A|}\text{adj} A$

$\therefore \quad A^{-1} = \frac{1}{1}\begin{bmatrix} -1 & 4 \\ -1 & 3 \end{bmatrix} = \begin{bmatrix} -1 & 4 \\ -1 & 3 \end{bmatrix}$ **Ans.**

18. If the function f defined as :

$$f(x) = \begin{cases} \dfrac{x^2-9}{x-3}, & x \neq 3 \\ k, & x = 3 \end{cases}$$

is continuous at $x = 3$, find the value of k. **[1]**

Sol. Clearly, $f(x) = \dfrac{x^2-9}{x-3} = x+3$

and $f(3) = k$

Now, for R.H.L.,

$$\lim_{x \to 3^+} f(x) = \lim_{h \to 0}(3+h)+3 = 6$$

and, for L.H.L.,

$$\lim_{x \to 3^-} f(x) = \lim_{h \to 0}(3-h)+3 = 6$$

$\because f(x)$ is continuous at $x = 3$

$\therefore$ RHL = LHL = $f(3)$

$\therefore$ $6 = k$

$\therefore$ $k = 6$ **Ans.**

19. If $f(x) = x^4 - 10$, then find the approximate value of $f(2.1)$. **[1]**

OR

Find the slope of the tangent to the curve $y = 2\sin^2(3x)$ at $x = \dfrac{\pi}{6}$.

Sol. Given : $f(x) = x^4 - 10$

$\therefore$ $f'(x) = 4x^3$

Let $x = 2$ and $h = 0.1$

We know that

$$f'(x) = \frac{f(x+h)-f(x)}{h}$$

or $h.f'(x) = f(x+h) - f(x)$

$\Rightarrow$ $0.1 \times \{4(2)^3\} = f(2.1) - (2^4 - 10)$

$\Rightarrow$ $3.2 = f(2.1) - 6$

$\Rightarrow$ $f(2.1) = 9.2$ **Ans.**

OR

Given, equation of curve is,

$$y = 2\sin^2(3x)$$

$\therefore$ $\dfrac{dy}{dx} = 4\sin 3x \cos 3x.3$

$\Rightarrow$ $\dfrac{dy}{dx} = 12\sin 3x.\cos 3x$

$\therefore$ $\left[\dfrac{dy}{dx}\right]_{x=\frac{\pi}{6}} = 12\sin\dfrac{\pi}{2}\cos\dfrac{\pi}{2} = 0$

$\therefore$ Slope of tangent = 0 **Ans.**

20. Find the value of $\int_1^4 |x-5|\,dx$ **[1]**

Sol. Let $I = \int_1^4 |x-5|\,dx$

$\Rightarrow$ $I = \int_1^4 -(x-5)\,dx$

$\Rightarrow$ $I = \dfrac{-1}{2}[x^2]_1^4 + 5[x]_1^4$

$\Rightarrow$ $I = \dfrac{-1}{2}(16-1) + 5(4-1)$

$\Rightarrow$ $I = \dfrac{-15}{2} + 15 = \dfrac{15}{2}$ **Ans.**

SECTION-B

Q. Nos. 21 to 26 carry 2 marks each.

21. If $f(x) = \dfrac{4x+3}{6x-4}$, $x \neq \dfrac{2}{3}$, then show that $(fof)(x) = x$, for all $x \neq \dfrac{2}{3}$ Also, write inverse of f. **[2]**

OR

Check if the relation R in the set R of real numbers defined as R = {$(a, b) : a < b$} is (i) symmetric, (ii) transitive

Sol. Given: $f(x) = \dfrac{4x+3}{6x-4}$

Now, $(fof)\,x = f(f(x))$

$$= \frac{4\left(\dfrac{4x+3}{6x-4}\right)+3}{6\left(\dfrac{4x+3}{6x-4}\right)-4}$$

$$= \frac{16x+12+18x-12}{24x+18-24x+16}$$

$$= \frac{34x}{34} = x$$

$\therefore$ $(fof)x = x$ **Hence Proved.**

To find inverse of $f(x)$,

Let $f(x) = y$

$\therefore$ $y = \dfrac{4x+3}{6x-4}$

$\Rightarrow$ $6xy - 4y = 4x + 3$

$\Rightarrow$ $6xy - 4x = 4y + 3$

$\Rightarrow \quad x(6y-4) = 4y+3$

$\Rightarrow \quad x = \dfrac{4y+3}{6y-4}$

$= f^{-1}(y) \quad \left[\begin{array}{l} \because f(x) = y \\ \therefore x = f^{-1}(y) \end{array}\right]$

$\therefore \quad f^{-1}(x) = \dfrac{4x+3}{6x-4}, x \neq \dfrac{2}{3}$ **Ans.**

OR

Given : R = {(a, b) : a < b}

Let A be the set of real numbers.

(i) For symmetric : Let $a, b \in A$

Then, $(a, b) \in R$ such that $a < b$

$\Rightarrow$ $b < a$ which is false

$\Rightarrow \quad (b, a) \notin R$

$\therefore$ R is not symmetric. **Ans.**

(ii) For transitive : Let $a, b, c \in A$

Then, $(a, b) \in R$ and $(b, c) \in R$

$\Rightarrow \quad a < b$ and $b < c$

$\Rightarrow \quad a < c$

$\Rightarrow \quad (a, c) \in R$

$\therefore$ R is transitive. **Ans.**

22. Find $\int \dfrac{x}{x^2+3x+2} dx$ **[2]**

Sol. Let $I = \int \dfrac{x}{x^2+3x+2} dx$

$= \int \dfrac{x}{(x+1)(x+2)} dx$

Let $\dfrac{x}{(x+1)(x+2)} = \dfrac{A}{x+1} + \dfrac{B}{x+2}$

$\Rightarrow \quad x = A(x+2) + B(x+1)$

$\Rightarrow \quad x = (A+B)x + (2A+B)$

Comparing coefficients of x both sides, we get

$1 = A + B$ ...(i)

Comparing constant function both sides, we get

$0 = 2A + B$ ...(ii)

Solving equations (i) and (ii), we get

$A = -1$ and $B = 2$

$\therefore \quad I = -\int \dfrac{dx}{x+1} + 2\int \dfrac{dx}{x+2}$

$\Rightarrow \quad I = -\log|x+1| + 2\log|x+2| + C$

$\Rightarrow \quad I = -\log|x+1| + \log|x+2|^2 + C$

$\Rightarrow \quad I = \log\left|\dfrac{(x+2)^2}{x+1}\right| + C$ **Ans.**

23. If $x = a\cos\theta$; $y = b\sin\theta$, then find $\dfrac{d^2y}{dx^2}$. **[2]**

OR

Find the differential of $\sin^2 x$ w.r.t. $e^{\cos x}$.

Sol. Given: $x = a\cos\theta$ and $y = b\sin\theta$

$\therefore \quad \dfrac{dx}{d\theta} = a(-\sin\theta) = -a\sin\theta$

and $\quad \dfrac{dy}{d\theta} = b(\cos\theta) = b\cos\theta$

$\therefore \quad \dfrac{dy}{dx} = \dfrac{dy}{d\theta} \times \dfrac{d\theta}{dx} = \dfrac{dy}{d\theta} \times \dfrac{1}{dx/d\theta}$

$= b\cos\theta \times \dfrac{1}{(-a\sin\theta)}$

$= -\dfrac{b}{a}\cot\theta$

Now, $\quad \dfrac{d^2y}{dx^2} = -\dfrac{b}{a}(-\text{cosec}^2\theta)\dfrac{d\theta}{dx}$

$= \dfrac{b}{a}\text{cosec}^2\theta \times \dfrac{1}{(-a\sin\theta)}$

$= -\dfrac{b}{a^2}\text{cosec}^3\theta$ **Ans.**

OR

Let $\quad u = \sin^2 x$

$\therefore \quad \dfrac{du}{dx} = 2\sin x\cos x$

Also, $\quad v = e^{\cos x}$

$\therefore \quad \dfrac{dv}{dx} = -\sin x\, e^{\cos x}$

Now, $\quad \dfrac{du}{dv} = \dfrac{du}{dx} \times \dfrac{dx}{dv}$

$\dfrac{du}{dv} = \dfrac{2\sin x\cos x}{-\sin x\, e^{\cos x}}$

$= -2\cos x\, e^{-\cos x}$ **Ans.**

24. Evaluate $\int_1^2 \left[\dfrac{1}{x} - \dfrac{1}{2x^2}\right] e^{2x} dx$ **[2]**

Sol. Let $I = \int_1^2 \left[\dfrac{1}{x} - \dfrac{1}{2x^2}\right] e^{2x} dx$

$$\Rightarrow I = \int_1^2 \frac{1}{x} e^{2x} dx - \int_1^2 \frac{1}{2x^2} e^{2x} dx$$

$$\Rightarrow I = \left[\frac{1}{x}\frac{e^{2x}}{2} - \int\left(\frac{-1}{x^2}\right)\frac{e^{2x}}{2} dx\right]_1^2 - \int_1^2 \frac{1}{2x^2} e^{2x} dx$$

[Using by parts]

$$\Rightarrow I = \frac{1}{2}\left[\frac{1}{x} e^{2x}\right]_1^2 + \int_1^2 \frac{1}{2x^2} e^{2x} dx - \int_1^2 \frac{1}{2x^2} e^{2x} dx$$

$$\Rightarrow I = \frac{1}{2}\left[\frac{1}{2} e^4 - 1e^2\right]$$

$$\Rightarrow I = \frac{e^4}{4} - e^2/2$$ **Ans.**

25. Find the value of $\int_0^1 x(1-x)^n dx$. **[2]**

Sol. Let $I = \int_0^1 x(1-x)^n dx$

$$\Rightarrow I = \int_0^1 (1-x)x^n dx \quad \left[\because \int_0^a f(x)dx = \int_0^a f(a-x)dx\right]$$

$$\Rightarrow I = \int_0^1 x^n dx - \int_0^1 x^{n+1} dx$$

$$\Rightarrow I = \frac{1}{n+1}\left[x^{n+1}\right]_0^1 - \frac{1}{n+2}\left[x^{n+2}\right]_0^1$$

$$\Rightarrow I = \frac{1}{n+1}[1-0] - \frac{1}{n+2}[1-0]$$

$$\Rightarrow I = \frac{1}{n+1} - \frac{1}{n+2}$$

$$\Rightarrow I = \frac{n+2-n-1}{(n+1)(n+2)} = \frac{1}{(n+1)(n+2)}$$ **Ans.**

26. Given two independent events A and B such that P(A) = 0.3 and P(B) = 0.6, find P(A′∩ B′). **[2]**

Sol. Given : P(A) = 0.3 and P(B) = 0.6

∵ A and B are independent events

Then,

P(A′∩ B′) = P(A′) P(B′)
= (1 – P(A)) (1– P(B))
= (1 – 0.3) (1 – 0.6)
= 0.7 × 0.4
= 0.28 **Ans.**

SECTION-C

Q. Nos. 27 to 32 carry 4 marks each.

27. Solve for x : $\sin^{-1}(1-x) - 2\sin^{-1}(x) = \frac{\pi}{2}$ **[4]**

Sol. Given :

$$\sin^{-1}(1-x) - 2\sin^{-1}(x) = \frac{\pi}{2}$$

$$\Rightarrow \sin^{-1}(1-x) = \frac{\pi}{2} + 2\sin^{-1}(x)$$

$$\Rightarrow 1-x = \sin\left(\frac{\pi}{2} + 2\sin^{-1}x\right)$$

$$\Rightarrow 1-x = \cos(2\sin^{-1}x)$$

$$\left[\because \sin\left(\frac{\pi}{2}+\theta\right) = \cos\theta\right]$$

Let $\sin^{-1}x = \theta$

$\Rightarrow x = \sin\theta$

$\therefore \cos(2\sin^{-1}x) = \cos(2\theta)$

$\Rightarrow = 1 - 2\sin^2\theta$

$= 1 - 2(x)^2$

$= 1 - 2x^2$

$\therefore 1 - x = 1 - 2x^2$

$\Rightarrow 2x^2 - x = 0$

$\Rightarrow x(2x-1) = 0$

$\Rightarrow x = 0$ or $\frac{1}{2}$

For $x = \frac{1}{2}$,

$$\sin^{-1}(1-x) - 2\sin^{-1}x = \sin^{-1}\frac{1}{2} - 2\sin^{-1}\frac{1}{2}$$

$$= -\sin^{-1}\frac{1}{2}$$

$$= -\frac{\pi}{2} \neq \frac{\pi}{2}$$

$\therefore x = 0$ **Ans.**

28. If $y = (\log x)^x + x^{\log x}$, then find $\frac{dy}{dx}$. **[4]**

Sol. Given : $y = (\log x)^x + x^{\log x}$

Let $(\log x)^x = u$

and $x^{\log x} = v$

$\therefore y = u + v$

and $\frac{dy}{dx} = \frac{du}{dx} + \frac{dv}{dx}$

Now, $u = (\log x)^x$

Taking log on both sides, we get

$$\log u = \log (\log x)^x$$

$$\Rightarrow \quad \log u = x \log (\log x)$$

Differentiate both sides w.r.t. x, we get

$$\frac{1}{u}\times\frac{du}{dx} = x\times\frac{1}{\log x}\times\frac{1}{x}+\log(\log x)$$

$$\Rightarrow \quad \frac{du}{dx} = u\left[\frac{1}{\log x}+\log(\log x)\right]$$

$$\Rightarrow \quad \frac{du}{dx} = (\log x)^x\left[\frac{1}{\log x}+\log(\log x)\right]$$

and $\quad v = x^{\log x}$

Taking log on both sides, we get

$$\log v = \log(x)^{\log x}$$

$$\Rightarrow \quad \log v = \log x . \log x$$

Differentiate both sides w.r.t. x we get

$$\frac{1}{v}\frac{dv}{dx} = \frac{\log x}{x}+\frac{\log x}{x}$$

$$\Rightarrow \quad \frac{dv}{dx} = v\left(\frac{2\log x}{x}\right)$$

$$\Rightarrow \quad \frac{dv}{dx} = x^{\log x}\left(\frac{2\log x}{x}\right)$$

Now, $\quad \frac{dy}{dx} = \frac{du}{dx}+\frac{dv}{dx}$

$$\Rightarrow \quad \frac{dy}{dx} = (\log x)^x\left[\frac{1}{\log x}+\log(\log x)\right] + x^{\log x}\left[\frac{2\log x}{x}\right]$$

Ans.

29. Solve the differential equation :

$$x \sin\left(\frac{y}{x}\right)\frac{dy}{dx}+x-y \sin\left(\frac{y}{x}\right)=0$$

Given that $x = 1$ when $y = \frac{\pi}{2}$. **[4]**

Sol. Given, differential equation is,

$$x\sin\left(\frac{y}{x}\right)\frac{dy}{dx}+x-y\sin\left(\frac{y}{x}\right) = 0$$

$$\Rightarrow \quad x\sin\left(\frac{y}{x}\right)\frac{dy}{dx} = y\sin\left(\frac{y}{x}\right)-x$$

or $$\frac{dy}{dx} = \frac{y\sin\left(\frac{y}{x}\right)-x}{x\sin\left(\frac{y}{x}\right)}$$

Put $\quad y = vx$

$$\therefore \quad \frac{dy}{dx} = v+x\frac{dv}{dx}$$

$$\Rightarrow \quad v+x\frac{dv}{dx} = \frac{(vx)\sin\left(\frac{vx}{x}\right)-x}{x\sin\left(\frac{vx}{x}\right)}$$

$$\Rightarrow \quad v+x\frac{dv}{dx} = \frac{v\sin v-1}{\sin v}$$

$$\Rightarrow \quad \frac{xdv}{dx} = \frac{v\sin v-1}{\sin v}-v$$

$$= \frac{v\sin v-1-v\sin v}{\sin v}$$

$$\Rightarrow \quad x\frac{dv}{dx} = \frac{-1}{\sin v}$$

$$\Rightarrow \quad \sin v\, dv = -\frac{dx}{x}$$

Integrating both sides, we get

$$\int\sin v\, dv = -\int\frac{dx}{x}$$

$$\Rightarrow \quad -\cos v = -\log x + C$$

$$\Rightarrow \quad -\cos\left(\frac{y}{x}\right) = -\log x + C \quad (\because y = vx)$$

It is given that,

$$x = 1 \text{ when } y = \frac{\pi}{2}$$

$$\therefore \quad -\cos\left(\frac{\pi}{2}\right) = -\log 1 + C$$

$$\Rightarrow \quad C = 0$$

$$\therefore \quad -\cos\left(\frac{y}{x}\right) = -\log x$$

$$\Rightarrow \quad \cos\left(\frac{y}{x}\right) = \log x$$

or $$\log x-\cos\left(\frac{y}{x}\right) = 0$$ **Ans.**

30. If $\vec{a} = \hat{i}+2\hat{j}+3\hat{k}$ and $\vec{b} = 2\hat{i}+4\hat{j}-5\hat{k}$ represent two adjacent sides of a parallelogram, find unit vectors parallel to the diagonals of the parallelogram. **[4]**

OR

Using vectors, find the area of the triangle ABC with vertices A (1, 2, 3), B (2, –1, 4) and C (4, 5, – 1).

Sol. Given : $\vec{a} = \hat{i}+2\hat{j}+3\hat{k}$

and $\vec{b} = 2\hat{i}+4\hat{j}-5\hat{k}$

Let $\vec{d_1}$ and $\vec{d_2}$ be the diagonal vectors of the parallelogram.

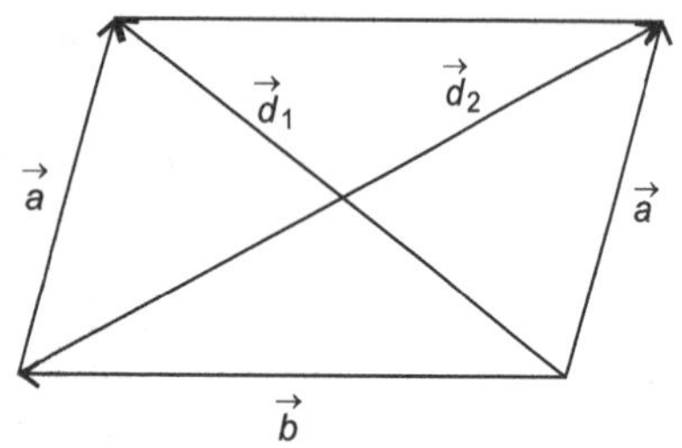

$\therefore \quad \vec{d_1} = \vec{a}+\vec{b}$

and $\vec{d_2} = \vec{a}-\vec{b}$

So, $\vec{d_1} = (\vec{i}+2\vec{j}+3\vec{k})+(2\vec{i}+4\vec{j}-5\vec{k})$

$= 3\vec{i}+6\vec{j}-2\vec{k}$

and $\vec{d_2} = (\vec{i}+2\vec{j}+3\vec{k})-(2\vec{i}+4\vec{j}-5\vec{k})$

$= -\vec{i}-2\vec{j}+8\vec{k}$

$\therefore$ Unit vector parallel to $\vec{d_1} = \dfrac{\vec{d_1}}{|\vec{d_1}|}$

$= \dfrac{3\hat{i}+6\hat{j}-2\hat{k}}{\sqrt{3^2+6^2+(-2)^2}}$

$= \dfrac{3}{7}\hat{i}+\dfrac{6}{7}\hat{j}-\dfrac{2}{7}\hat{k}$

and, unit vector parallel to $\vec{d_2} = \dfrac{\vec{d_2}}{|\vec{d_2}|}$

$= \dfrac{-\hat{i}-2\hat{j}+8\hat{k}}{\sqrt{(-1)^2+(-2)^2+8^2}}$

$= -\dfrac{1}{\sqrt{69}}\hat{i}-\dfrac{2}{\sqrt{69}}\hat{j}+\dfrac{8}{\sqrt{69}}\hat{k}$

Hence, the unit vectors are $\dfrac{3}{7}\hat{i}+\dfrac{6}{7}\hat{j}-\dfrac{2}{7}\hat{k}$ and $-\dfrac{1}{\sqrt{69}}\hat{i}-\dfrac{2}{\sqrt{69}}\hat{j}+\dfrac{8}{\sqrt{69}}\hat{k}$. **Ans.**

OR

Given : Coordinates of vertices of ΔABC are A(1, 2, 3), B (2, – 1, 4) and C(4, 5, – 1)

$\therefore \quad \vec{AB} = \hat{i}-3\hat{j}+\hat{k}$

and $\vec{AC} = 3\hat{i}+3\hat{j}-4\hat{k}$

We know that,

$ar(\Delta ABC) = \dfrac{1}{2}\left|\vec{AB}\times\vec{AC}\right|$

Now, $\vec{AB}\times\vec{AC} = \begin{vmatrix} \vec{i} & \vec{j} & \vec{k} \\ 1 & -3 & 1 \\ 3 & 3 & -4 \end{vmatrix}$

$= \vec{i}(12-3)-\vec{j}(-4-3)+\vec{k}(3+9)$

$= 9\vec{i}+7\vec{j}+12\vec{k}$

$\therefore \quad \left|\vec{AB}\times\vec{AC}\right| = \sqrt{9^2+7^2+(12)^2} = \sqrt{274}$

So, $ar(\Delta ABC) = \dfrac{1}{2}\sqrt{274}$ sq. units **Ans.**

31. A company manufactures two types of novelty souvenirs made of plywood. Souvenirs of type A requires 5 minutes each for cutting and 10 minutes each for assembling. Souvenirs of type B require 8 minutes each for cutting and 8 minutes each for assembling. Given that total time for cutting is 3 hours 20 minutes and for assembling 4 hours. The profit for type A souvenir is ₹ 100 each and for type B souvenir, profit is ₹ 120 each. How many souvenirs of each type should the company manufacture in order to maximize the profit ? Formulate the problem as an LPP and solve it graphically. [4]

Sol. Let the company manufactures x souvenirs of type A and y souvenirs of type B.

Then the objective function is,

Maximize Z = $100x + 120y$

Subject to constraints

$5x + 8y \leq 3$ h 20 min

$\Rightarrow \quad 5x + 8y \leq 200$ ...(i)

and $10x + 8y \leq 4$ h

$$\Rightarrow \quad 10x + 8y \le 240$$

$$\Rightarrow \quad 5x + 4y \le 120 \quad ...(ii)$$

Now, $5x + 8y = 200$

x	40	0
y	0	25

$5x + 4y = 120$

x	24	0
y	0	30

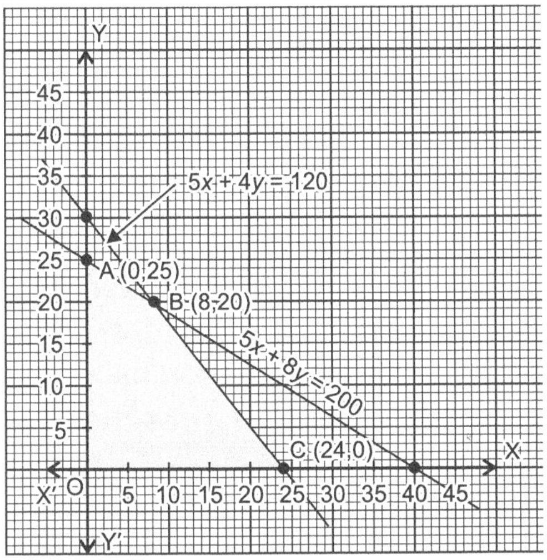

The feasible region is the shaded region OABC .

Points	Objective function Z = 100x + 120y
O (0, 0)	Z = 0
A (0, 25)	Z = 100(0) + 120(25) = 3000
B (8, 20)	Z = 100 (8) + 120 (20) = 3200 (Maximum)
C (24, 0)	Z = 100 (24) + 120 (0) = 2400

Hence, the profit will be maximum if 8 souvenirs of type A and 20 souvenirs of type B are manufactured. **Ans.**

32. Three rotten apples are mixed with seven fresh apples. Find the probability distribution of the number of rotten apples, if three apples are drawn one by one with replacement. Find the mean of the number of rotten apples. [4]

OR

In a shop X, 30 tins of ghee of type A and 40 tins of ghee of type B which look alike, are kept for sale. While in shop Y, similar 50 tins of ghee of type A and 60 tins of ghee of type B are there. One tin of ghee is purchased from one of the randomly selected shop and is found to be of type B. Find the probability that it is purchased from shop Y.

Sol. Let A be the event of taking out a rotten apple and B be the event of taking out a fresh apple.

$$\therefore \quad P(A) = \frac{3}{10}$$

and $$P(B) = \frac{7}{10}$$

Let x_i be the number of rotten apples taken out in the three chances.

So, $i = 0, 1, 2, 3$

$$\therefore \quad P(x=0) = P(B) \times P(B) \times P(B)$$

$$= \frac{7}{10}\times\frac{7}{10}\times\frac{7}{10} = \frac{343}{1000}$$

$$P(x = 1) = P(A) \times P(B) \times P(B) + P(B) \times P(A) \times P(B) + P(B) \times P(B) \times P(A)$$

$$= \frac{3}{10}\times\frac{7}{10}\times\frac{7}{10}+\frac{7}{10}\times\frac{3}{10}\times\frac{7}{10}+\frac{7}{10}\times\frac{7}{10}\times\frac{3}{10}$$

$$= \frac{147}{1000}+\frac{147}{1000}+\frac{147}{1000} = \frac{441}{1000}$$

$$P(x = 2) = P(A) \times P(A) \times P(B) + P(A) \times P(B) \times P(A) + P(B) \times P(A) \times P(A)$$

$$= \frac{3}{10}\times\frac{3}{10}\times\frac{7}{10}+\frac{3}{10}\times\frac{7}{10}\times\frac{3}{10}+\frac{7}{10}\times\frac{3}{10}\times\frac{3}{10}$$

$$= \frac{63}{1000}+\frac{63}{1000}+\frac{63}{1000} = \frac{189}{1000}$$

$$P(x = 3) = P(A) \times P(A) \times P(A)$$

$$= \frac{3}{10}\times\frac{3}{10}\times\frac{3}{10} = \frac{27}{1000}$$

$\therefore$ The probability distribution is,

x_i	0	1	2	3
$P(x_i)$	$\frac{343}{1000}$	$\frac{441}{1000}$	$\frac{189}{1000}$	$\frac{27}{1000}$

Now, $$\text{Mean} = \sum_{i=0}^{3} x_i P(x_i)$$

$$= 0\times\frac{343}{1000}+1\times\frac{441}{1000}+2\times\frac{189}{1000}+3\times\frac{27}{1000}$$

$$= 0+\frac{441}{1000}+\frac{378}{1000}+\frac{81}{1000}$$

$$= \frac{900}{1000} = 0.9$$

Hence, the mean of number of rotten apples is 0.9. **Ans.**

OR

Let H_1 and H_2 be the events of purchasing tins of ghee from shop X and Y respectively.

$\therefore \quad P(H_1) = P(H_2) = \frac{1}{2}$

Also, let E be the event that ghee of type B is purchased.

$\therefore \quad P\left(\frac{E}{H_1}\right) = \frac{40}{70} = \frac{4}{7}$

and $\quad P\left(\frac{E}{H_2}\right) = \frac{60}{110} = \frac{6}{11}$

By Bayes' theorem, required probability is,

$$P\left(\frac{H_2}{E}\right) = \frac{P\left(\frac{E}{H_2}\right)P(H_2)}{P\left(\frac{E}{H_1}\right)P(H_1) + P\left(\frac{E}{H_2}\right)P(H_2)}$$

$$= \frac{\frac{6}{11}\times\frac{1}{2}}{\frac{4}{7}\times\frac{1}{2}+\frac{6}{11}\times\frac{1}{2}}$$

$$= \frac{\frac{6}{11}}{\frac{44+42}{77}} = \frac{6}{11}\times\frac{77}{86} = \frac{21}{43}$$

Ans.

SECTION-D

Q. 33 to 36 carry 6 marks each.

33. Find the vector and cartesian equations of the line which is perpendicular to the lines with equations

$$\frac{x+2}{1} = \frac{y-3}{2} = \frac{z+1}{4} \text{ and } \frac{x-1}{2} = \frac{y-2}{3} = \frac{z-3}{4}$$

and passes through the point (1, 1, 1). Also find the angle between the given lines. [6]

Sol. Given : Equation of given lines are,

$$\frac{x+2}{1} = \frac{y-3}{2} = \frac{z+1}{4}$$

and $\quad \frac{x-1}{2} = \frac{y-2}{3} = \frac{z-3}{4}$

$\therefore$ Direction ratios of first line = < 1, 2, 4 >

Direction ratios of second line = < 2, 3, 4 >

Let the direction ratios of required line be $<a, b, c>$.

$\because$ The required line is perpendicular to the given lines,

$\therefore \quad a + 2b + 4c = 0 \quad$ ...(i)

and $\quad 2a + 3b + 4c = 0 \quad$ ...(ii)

Solving equations (i) and (ii) by cross multiplication method, we get

$$\frac{a}{8-12} = \frac{-b}{4-8} = \frac{c}{3-4} = k\text{(say)}$$

$\Rightarrow a = -4k,\ b = 4k$ and $c = -k$

$\therefore$ Direction ratios of required line = <–4, 4, –1>

Now, vector equation of line passing through (x, y, z) and having direction ratios $< a, b, c >$ is given as,

$$(x\hat{i}+y\hat{j}+z\hat{k}) + \lambda(a\hat{i}+b\hat{j}+c\hat{k}) = 0$$

$$\Rightarrow \quad (\hat{i}+\hat{j}+\hat{k}) + \lambda(-4\hat{i}+4\hat{j}-\hat{k}) = 0 \quad \textbf{Ans.}$$

And, its cartesian form is,

$$\frac{x-1}{-4} = \frac{y-1}{4} = \frac{z-1}{-1} \quad \textbf{Ans.}$$

Now, angle between two lines is given as,

$$\cos\theta = \frac{a_1a_2 + b_1b_2 + c_1c_2}{\sqrt{a_1^2+b_1^2+c_1^2}\sqrt{a_2^2+b_2^2+c_2^2}}$$

Here, $\quad < a_1, b_1, c_1 > = < 1, 2, 4 >$

and $\quad < a_2, b_2, c_2 > = < 2, 3, 4 >$

$$\Rightarrow \quad \cos\theta = \frac{1\times2+2\times3+4\times4}{\sqrt{1^2+2^2+4^2}\sqrt{2^2+3^2+4^2}}$$

$$= \frac{2+6+16}{\sqrt{21}\sqrt{29}} = \frac{24}{\sqrt{609}}$$

$$\Rightarrow \quad \theta = \cos^{-1}\left(\frac{24}{\sqrt{609}}\right)$$

Hence, the angle between the given lines is $\cos^{-1}\left(\frac{24}{\sqrt{609}}\right)$. **Ans.**

34. Using integration find the area of the region bounded between the two circles $x^2 + y^2 = 9$ and $(x - 3)^2 + y^2 = 9$. [6]

OR

Evaluate the following integral as the limit of sums $\int_1^4 (x^2 - x)\, dx$.

Sol. Given :

Equation of first circle is,

$$x^2 + y^2 = 9 \quad \text{...(i)}$$

$\therefore \quad$ Its centre = (0, 0)

and $\quad$ Radius = 3 units

Also, equation of second circle is,

$$(x - 3)^2 + y^2 = 9 \quad \text{...(ii)}$$

$\therefore \quad$ Its centre = (3, 0)

and $\quad$ Radius = 3 units

From equations (i) and (ii), we get

$$(x-3)^2 + 9 - x^2 = 9$$

$$\Rightarrow \quad x^2 - 6x + 9 - x^2 = 0$$

$$\Rightarrow \quad -6x + 9 = 0$$

$$\Rightarrow \quad x = \frac{3}{2}$$

Putting $x = \frac{3}{2}$ in equation (i), we get

$$\left(\frac{3}{2}\right)^2 + y^2 = 9$$

$$\Rightarrow \quad y^2 = 9 - \frac{9}{4} = \frac{36-9}{4} = \frac{27}{4}$$

$$\Rightarrow \quad y = \pm\frac{3\sqrt{3}}{2}$$

$\therefore$ The points of intersection of two circles are $\left(\frac{3}{2}, \frac{3\sqrt{3}}{2}\right)$ and $\left(\frac{3}{2}, \frac{-3\sqrt{3}}{2}\right)$.

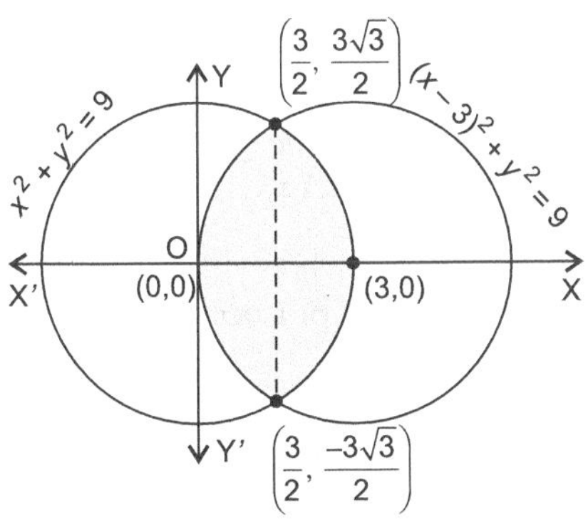

Now, area of bounded region

$$= 2\left[\int_0^{3/2} (y \text{ of } 2^{\text{nd}} \text{ circle})\,dx + \int_{3/2}^{3} (y \text{ of } 1^{\text{st}} \text{ circle})\,dx\right]$$

$$\Rightarrow I = 2\left[\int_0^{3/2} \sqrt{3^2-(x-3)^2}\,dx + \int_{3/2}^{3} \sqrt{3^2-x^2}\,dx\right]$$

$$\Rightarrow I = 2\,[I_1 + I_2]$$

Now, $\quad I_1 = \int_0^{3/2} \sqrt{3^2-(x-3)^2}\,dx$

$$= \left[\frac{x-3}{2}\sqrt{3^2-(x-3)^2} + \frac{9}{2}\sin^{-1}\left(\frac{x-3}{3}\right)\right]_0^{3/2}$$

$$= \frac{1}{2}\left(\frac{3}{2}-3\right)\sqrt{3^2-\left(\frac{3}{2}-3\right)^2} + \frac{9}{2}\sin^{-1}\left\{\frac{1}{3}\left(\frac{3}{2}-3\right)\right\}$$

$$- \left\{\left(\frac{0-3}{2}\right)\sqrt{3^2-(0-3)^2} + \frac{9}{2}\sin^{-1}\left(\frac{0-3}{3}\right)\right\}$$

$$= \frac{1}{2}\left(\frac{-3}{2}\right)\sqrt{9-\frac{9}{4}} + \frac{9}{2}\sin^{-1}\left(\frac{-1}{2}\right) - \left(\frac{-3}{2}\right)\sqrt{9-9}$$

$$-\frac{9}{2}\sin^{-1}(-1)$$

$$= -\frac{3}{4}\cdot\frac{3\sqrt{3}}{2} + \frac{9}{2}\left(-\frac{\pi}{6}\right) - \frac{9}{2}\left(-\frac{\pi}{2}\right)$$

$$= \frac{-9\sqrt{3}}{8} - \frac{3\pi}{4} + \frac{9\pi}{4}$$

$$= \frac{3\pi}{2} - \frac{9\sqrt{3}}{8} = \frac{1}{2}\left[3\pi - \frac{9\sqrt{3}}{4}\right]$$

Similarly,

$$I_2 = \int_{3/2}^{3} \sqrt{3^2-x^2}\,dx$$

$$\Rightarrow I_2 = \left[\frac{x}{2}\sqrt{9-x^2} + \frac{9}{2}\sin^{-1}\left(\frac{x}{3}\right)\right]_{3/2}^{3}$$

$$= \frac{1}{2}\left[\{0+9\sin^{-1}(1)\} - \left\{\frac{3}{2}\sqrt{9-\frac{9}{4}} + 9\sin^{-1}\left(\frac{\frac{3}{2}}{3}\right)\right\}\right]$$

$$\Rightarrow \quad I_2 = \frac{1}{2}\left[\frac{9\pi}{2} - \left\{\frac{3}{2}\sqrt{\frac{27}{4}} + 9\sin^{-1}\frac{1}{2}\right\}\right]$$

$$\Rightarrow \quad I_2 = \frac{1}{2}\left[\frac{9\pi}{2} - \frac{9\sqrt{3}}{4} - \frac{9\pi}{6}\right]$$

$$\Rightarrow \quad I_2 = \frac{1}{2}\left[\frac{9\pi}{2} - \frac{9\sqrt{3}}{4} - \frac{3\pi}{2}\right]$$

$$\Rightarrow \quad I_2 = \frac{1}{2}\left[\frac{6\pi}{2} - \frac{9\sqrt{3}}{4}\right]$$

$$\Rightarrow \quad I_2 = \frac{1}{2}\left[3\pi - \frac{9\sqrt{3}}{4}\right]$$

$$\because \quad I = 2(I_1 + I_2)$$

$$\Rightarrow \quad I = 2\times\frac{1}{2}\left[3\pi - \frac{9\sqrt{3}}{4} + 3\pi - \frac{9\sqrt{3}}{4}\right]$$

$$\Rightarrow \quad I = \left(6\pi - \frac{18}{4}\sqrt{3}\right)$$

$$= \left(6\pi - \frac{9\sqrt{3}}{2}\right) \text{ sq. units}$$

Ans.

OR

Let $\quad I = \int_1^4 (x^2 - x)\,dx$

Here, $a = 1$, $b = 4$ and $nh = 4 - 1 = 3$

Let $f(x) = x^2 - x$

Clearly,

$$f(1) = (1)^2 - 1$$
$$f(1 + h) = (1 + h)^2 - (1 + h)$$
$$f(1 + 2h) = (1 + 2h)^2 - (1 + 2h)$$
$$f[1 + (n - 1)h] = [1 + (n - 1)h]^2 - [1 + (n - 1)h]$$

$\therefore$ Required sum

$$= \lim_{x\to 0} h[f(1)+f(1+h)+...+f(1+(n-1)h)]$$

$$= \lim_{h\to 0} h\ [(1^2 - 1) + \{(1 + h)^2 - (1 + h)\} + \{(1 + 2h)^2 - (1 + 2h)\} + ... + \{(1 + (n - 1)h)^2 - (1 + (n - 1)h)\}]$$

$$= \lim_{h\to 0} h[(1 - 1) + (1 + 2h + h^2 - 1 - h) + (1 + 4h + 4h^2 - 1 - 2h) + ... + \{1 + 2(n - 1)h + \{(n - 1)h\}^2 - 1 - (n - 1)h\}]$$

$$= \lim_{h\to 0} h[0 + (h^2 + h) + (4h^2 + 2h) + ...\{(n - 1)^2h^2 + (n - 1)h\}]$$

$$= \lim_{h\to 0} h\ [(h^2 + 4h^2 + 9h^2 + ... + (n - 1)^2\ h^2) + (h + 2h + 3h + ...(n - 1)h)]$$

$$= \lim_{h\to 0} h\ [h^2(1 + 2^2 + 3^2 + ... (n - 1)^2) + h\ (1 + 2 + 3 + ... (n - 1))]$$

$$= \lim_{h\to 0} h\left[h^2\frac{n(n-1)(2n-1)}{6}+h\frac{n(n-1)}{2}\right]$$

$$= \lim_{h\to 0}\left[\frac{h^3n(n-1)(2n-1)}{6}+\frac{h^2n(n-1)}{2}\right]$$

$$= \lim_{h\to 0}\left[\frac{nh(nh-h)(2nh-h)}{6}+\frac{nh(nh-h)}{2}\right]$$

$$= \frac{3(3-0)(6-0)}{6}+\frac{3(3-0)}{2}$$

$$= 9+\frac{9}{2}=\frac{27}{2}$$ **Ans.**

35. Find the minimum value of $(ax + by)$, where $xy = c^2$. **[6]**

Sol. Given : $xy = c^2$

$\Rightarrow$ $y = \dfrac{c^2}{x}$ ...(i)

Let $t = ax + by$

$\Rightarrow$ $t = ax+\dfrac{bc^2}{x}$ [Using (i)]

Now, $\dfrac{dt}{dx} = a-\dfrac{bc^2}{x^2}$

For minimum value of t,

$$\frac{dt}{dx} = 0$$

$\Rightarrow$ $a-\dfrac{bc^2}{x^2} = 0$

$\Rightarrow$ $ax^2 - bc^2 = 0$

$\Rightarrow$ $x = \pm c\sqrt{\dfrac{b}{a}}$

Now,

$$\frac{d^2t}{dx^2} = 0-(bc^2)\left(\frac{-2}{x^3}\right)=\frac{2bc^2}{x^3}$$

$\Rightarrow$ $\left[\dfrac{d^2t}{dx^2}\right]_{x=c\sqrt{\frac{b}{a}}} = \dfrac{2bc^2}{c^3\dfrac{b\sqrt{b}}{a\sqrt{a}}}=\dfrac{2a}{c}\sqrt{\dfrac{a}{b}}$

and $\left[\dfrac{d^2t}{dx^2}\right]_{x=-c\sqrt{\frac{b}{a}}} = \dfrac{2bc^2}{-c^3\dfrac{b\sqrt{b}}{a\sqrt{a}}}=-\dfrac{2a}{c}\sqrt{\dfrac{a}{b}}$

$\therefore$ Minimum value of t occurs at $x = c\sqrt{\dfrac{b}{a}}$

$\therefore$ Minimum value of $t = a\left(c\sqrt{\dfrac{b}{a}}\right)+b\left(\dfrac{c^2}{c\sqrt{\dfrac{b}{a}}}\right)$

$$= c\sqrt{ab}+c\sqrt{ab}$$

$$= 2c\sqrt{ab}$$ **Ans.**

36. If a, b, c are p^{th}, q^{th} and r^{th} terms respectively of a G.P., then prove that $\begin{vmatrix}\log a & p & 1\\ \log b & q & 1\\ \log c & r & 1\end{vmatrix}=0$ **[6]**

OR

If A = $\begin{bmatrix}2 & -3 & 5\\ 3 & 2 & -4\\ 1 & 1 & -2\end{bmatrix}$, then find A^{-1}.

Using A^{-1}, solve the following system of equations :

$$2x - 3y + 5z = 11$$
$$3x + 2y - 4x = -5$$
$$x + y - 2z = -3$$

Sol. Given : a, b, c are p^{th}, q^{th} and r^{th} terms of G.P., respectively.

Let A be the first term and R be the common ratio of the G.P.

$$\therefore \quad a_p = a$$
$$\Rightarrow \quad AR^{p-1} = a$$
$$\Rightarrow \quad \log a = \log(AR^{p-1})$$
$$= \log A + \log R^{p-1}$$

Thus, $\log a = \log A + (p-1)\log R$

Similarly, $\log b = \log A + (q-1)\log R$

And, $\log c = \log A + (r-1)\log R$

Now,

$$\text{L.H.S.} = \begin{vmatrix} \log a & p & 1 \\ \log b & q & 1 \\ \log c & r & 1 \end{vmatrix}$$

$$= \begin{vmatrix} \log A+(p-1)\log R & p & 1 \\ \log A+(q-1)\log R & q & 1 \\ \log A+(r-1)\log R & r & 1 \end{vmatrix}$$

Applying $C_1 \to C_1 - \log A\, C_3$, we get

$$\text{L.H.S.} = \begin{vmatrix} (p-1)\log R & p & 1 \\ (q-1)\log R & q & 1 \\ (r-1)\log R & r & 1 \end{vmatrix}$$

Taking out log R common from C_1,

$$\text{L.H.S.} = \log R \begin{vmatrix} p-1 & p & 1 \\ q-1 & q & 1 \\ r-1 & r & 1 \end{vmatrix}$$

Applying $C_1 \to C_1 + C_3$, we get

$$\text{L.H.S.} = \log R \begin{vmatrix} p & p & 1 \\ q & q & 1 \\ r & r & 1 \end{vmatrix}$$

$\because C_1$ and C_2 are identical.

$$\therefore \quad \begin{vmatrix} p & p & 1 \\ q & q & 1 \\ r & r & 1 \end{vmatrix} = 0$$

$\therefore$ L.H.S. $= \log R \times 0 = 0 =$ R.H.S. **Hence Proved.**

OR

Given : $A = \begin{bmatrix} 2 & -3 & 5 \\ 3 & 2 & -4 \\ 1 & 1 & -2 \end{bmatrix}$

$$\therefore \quad |A| = 2(-4+4) + 3(-6+4) + 5(3-2)$$
$$= 0 + (-6) + 5$$
$$= -1 \neq 0$$

$\therefore$ Matrix A is invertible, *i.e.*, A^{-1} exists.

Now, cofactors of matrix A are,

$$C_{11} = 0, C_{21} = -1, C_{31} = 2$$
$$C_{12} = 2, C_{22} = -9, C_{32} = 23$$
$$C_{13} = 1, C_{23} = -5, C_{33} = 13$$

$$\therefore \text{ Cofactor matrix} = \begin{bmatrix} 0 & 2 & 1 \\ -1 & -9 & -5 \\ 2 & 23 & 13 \end{bmatrix}$$

$$\because \quad \text{adj } A = \begin{bmatrix} 0 & 2 & 1 \\ -1 & -9 & -5 \\ 2 & 23 & 13 \end{bmatrix}^T$$

$$= \begin{bmatrix} 0 & -1 & 2 \\ 2 & -9 & 23 \\ 1 & -5 & 13 \end{bmatrix}$$

Now, $$A^{-1} = \frac{1}{|A|}\text{adj } A$$

$$A^{-1} = \frac{1}{(-1)}\begin{bmatrix} 0 & -1 & 2 \\ 2 & -9 & 23 \\ 1 & -5 & 13 \end{bmatrix}$$

$$A^{-1} = \begin{bmatrix} 0 & 1 & -2 \\ -2 & 9 & -23 \\ -1 & 5 & -13 \end{bmatrix}$$ **Ans.**

Now, given equations are

$$2x - 3y + 5z = 11$$
$$3x + 2y - 4z = -5$$
$$x + y - 2z = -3$$

Let $$A = \begin{bmatrix} 2 & -3 & 5 \\ 3 & 2 & -4 \\ 1 & 1 & -2 \end{bmatrix}$$

$$X = \begin{bmatrix} x \\ y \\ z \end{bmatrix}$$

and $$B = \begin{bmatrix} 11 \\ -5 \\ -3 \end{bmatrix}$$

$$\therefore \quad AX = B$$

or $$X = A^{-1}B \quad \text{...(i)}$$

$$\Rightarrow \begin{bmatrix} x \\ y \\ z \end{bmatrix} = \begin{bmatrix} 0 & 1 & -2 \\ -2 & 9 & -23 \\ -1 & 5 & -13 \end{bmatrix} \begin{bmatrix} 11 \\ -5 \\ -3 \end{bmatrix}$$

$$\Rightarrow \begin{bmatrix} x \\ y \\ z \end{bmatrix} = \begin{bmatrix} 0-5+6 \\ -22-45+69 \\ -11-25+39 \end{bmatrix}$$

$$\begin{bmatrix} x \\ y \\ z \end{bmatrix} = \begin{bmatrix} 1 \\ 2 \\ 3 \end{bmatrix}$$

$\therefore x = 1, y = 2$ and $z = 3$ **Ans.**

Delhi [Set-II]

Note : Except for the following questions, all the remaining questions have been asked in the previous set.

SECTION-A

1. If $[x \quad 1]\begin{bmatrix} 1 & 0 \\ -2 & 0 \end{bmatrix} = 0$, then x equals

(a) 0 **(b) –2**

(c) –1 **(d) 2** [1]

Sol. Given :

$$[x \quad 1]\begin{bmatrix} 1 & 0 \\ -2 & 0 \end{bmatrix} = 0$$

$$\Rightarrow [x-2 \quad 0] = [0 \quad 0]$$

$$\Rightarrow x - 2 = 0$$

$$\Rightarrow x = 2$$

$\therefore$ Option (d) is correct.

2. $\int 4^x 3^x \, dx$ equals

(a) $\dfrac{12^x}{\log 12} + C$ **(b) $\dfrac{4^x}{\log 4} + C$**

(c) $\left(\dfrac{4^x . 3^x}{\log 4 . \log 3}\right) + C$ **(d) $\dfrac{3^x}{\log 3} + C$** [1]

Sol. Let, $I = \int 4^x 3^x \, dx$

$$\Rightarrow I = \int 12^x \, dx$$

$$\Rightarrow I = \frac{12^x}{\log 12} + C$$

$$\left[\because \int a^x \, dx = \frac{a^x}{\log x} + C\right]$$

$\therefore$ Option (a) is correct.

3. A number is chosen randomly from numbers 1 to 60. The probability that the chosen number is multiple of 2 or 5 is

(a) $\dfrac{2}{5}$ **(b) $\dfrac{3}{5}$**

(c) $\dfrac{7}{10}$ **(d) $\dfrac{9}{10}$** [1]

Sol. Let A = Numbers multiple of 2

and B = Numbers multiple of 5

$$\therefore P(A \cup B) = P(A) + P(B) - P(A \cap B)$$

$$= \frac{30}{60} + \frac{12}{60} - \frac{6}{60}$$

$$= \frac{36}{60} = \frac{3}{5}$$

$\therefore$ Option (b) is correct.

11. A relation R on a set A is called ________, if $(a_1, a_2) \in$ R and $(a_2, a_3) \in$ R implies that $(a_1, a_3) \in$ R for $a_1, a_2, a_3 \in$ A. [1]

Sol. Transitive

16. Evaluate : $\sin\left[\dfrac{\pi}{3} - \sin^{-1}\left(\dfrac{-1}{2}\right)\right]$ [1]

Sol. $\sin\left[\dfrac{\pi}{3} - \sin^{-1}\left(\dfrac{-1}{2}\right)\right]$

$= \sin\left[\dfrac{\pi}{3} + \sin^{-1}\dfrac{1}{2}\right]$ $[\because \sin^{-1}(-x) = -\sin x]$

$= \sin\left(\dfrac{\pi}{3} + \dfrac{\pi}{6}\right) = \sin\dfrac{\pi}{2} = 1$ **Ans.**

17. Using differential, find the approximate value of $\sqrt{36.6}$ upto 2 decimal places. [1]

OR

Find the slope of tangent to the curvey = $2\cos^2(3x)$ at $x = \dfrac{\pi}{6}$.

Sol. Let $y = \sqrt{x}$, $x = 36$ and $\Delta x = 0.6$

Now, $y = \sqrt{x}$

$$\Rightarrow \quad \frac{dy}{dx} = \frac{1}{2\sqrt{x}}$$

$$\because \quad \Delta y = \frac{dy}{dx}\Delta x$$

$$= \left(\frac{1}{2\sqrt{x}}\right)_{x=36} \times 0.6$$

$$= \frac{1}{2\sqrt{36}} \times 0.6$$

$$= \frac{1}{2\times 6} \times 0.6 = 0.05$$

Now, $\Delta y = f(x + \Delta x) - f(x)$

$$\Rightarrow \quad 0.05 = \sqrt{36+0.6} - \sqrt{36}$$

$$\Rightarrow \quad 0.05 = \sqrt{36.6} - 6$$

$$\Rightarrow \quad \sqrt{36.6} = 6.05$$ **Ans.**

OR

Given : $y = 2\cos^2(3x)$

$$\Rightarrow \quad \frac{dy}{dx} = 2[2\cos 3x.(-\sin 3x).3]$$

$$= -12\cos 3x \sin 3x$$

$$= -6\times 2\cos 3x \sin 3x$$

$$= -6\sin 6x$$

$$\Rightarrow \quad \left(\frac{dy}{dx}\right)_{x=\frac{\pi}{6}} = -6\sin \pi = 0$$

$\because$ Slope of tangent $= \dfrac{dy}{dx}$

$\therefore$ Slope of tangent $= 0$ **Ans.**

SECTION–B

21. Find $\int \dfrac{x+1}{x(1-2x)}dx$. **[2]**

Sol. Let $I = \int \dfrac{x+1}{x(1-2x)}dx$

Also, let $\dfrac{x+1}{x(1-2x)} = \dfrac{A}{x} + \dfrac{B}{1-2x}$

$$\Rightarrow \quad \frac{x+1}{x(1-2x)} = \frac{A(1-2x)+Bx}{x(1-2x)}$$

$$\Rightarrow \quad x + 1 = x(-2A + B) + A$$

Comparing coefficient of x both sides, we get

$$1 = -2A + B \quad ...(i)$$

Comparing constant term both sides, we get

$$A = 1$$

$\therefore$ From equation (i),

$$1 = -2(1) + B$$

$$\Rightarrow \quad B = 3$$

$$\therefore \quad \int \frac{x+1}{x(1-2x)}dx = \int \frac{1}{x}dx + \int \frac{3}{1-2x}dx$$

$$= \log|x| + \frac{3\log|1-2x|}{-2} + C$$

$$= \log|x| - \frac{3}{2}\log|1-2x| + C$$ **Ans.**

22. Evaluate $\int \dfrac{x\sin^{-1}(x^2)}{\sqrt{1-x^4}}dx$. **[2]**

Sol. Let $I = \int \dfrac{x\sin^{-1}(x^2)}{\sqrt{1-x^4}}dx$

Let $\sin^{-1}(x^2) = t$

$$\therefore \quad \frac{1}{\sqrt{1-x^4}}(2x)dx = dt$$

$$\Rightarrow \quad \frac{x}{\sqrt{1-x^4}}dx = \frac{dt}{2}$$

$$\therefore \quad I = \int \frac{tdt}{2} = \frac{1}{2}\int tdt$$

$$I = \frac{1}{2}\cdot\frac{t^2}{2} + c$$

$$I = \frac{1}{4}(\sin^{-1}x^2)^2 + C$$ **Ans.**

SECTION–C

27. Prove that $\tan\left[2\tan^{-1}\dfrac{1}{2} - \cot^{-1}3\right] = \dfrac{9}{13}$ **[4]**

Sol. L.H.S. $= \tan\left[2\tan^{-1}\dfrac{1}{2} - \cot^{-1}3\right]$

$$= \tan\left[\tan^{-1}\left(\frac{2\times\frac{1}{2}}{1-\left(\frac{1}{2}\right)^2}\right) - \tan^{-1}\frac{1}{3}\right]$$

$$\left[\because 2\tan^{-1}x = \tan^{-1}\left(\frac{2x}{1-x^2}\right) \text{ and } \cot^{-1}x = \tan^{-1}\frac{1}{x}\right]$$

$$= \tan\left[\tan^{-1}\left(\frac{1}{1-\frac{1}{4}}\right) - \tan^{-1}\frac{1}{3}\right]$$

$$= \tan\left[\tan^{-1}\frac{4}{3} - \tan^{-1}\frac{1}{3}\right]$$

$$= \tan\left[\tan^{-1}\left(\frac{\frac{4}{3}-\frac{1}{3}}{1+\frac{4}{3}\times\frac{1}{3}}\right)\right] = \tan\left[\tan^{-1}\left(\frac{1}{\frac{9+4}{9}}\right)\right]$$

$$= \tan\left[\tan^{-1}\left(\frac{9}{13}\right)\right] = \frac{9}{13} = \text{R.H.S.}$$ **Hence Proved.**

28. If $y = (\cos x)^x + \tan^{-1}\sqrt{x}$, find $\frac{dy}{dx}$. **[4]**

Sol. Given : $y = (\cos x)^x + \tan^{-1}\sqrt{x}$

Let $u = (\cos x)^x$

and $v = \tan^{-1}\sqrt{x}$

$\therefore$ $y = u + v$

$$\Rightarrow \quad \frac{dy}{dx} = \frac{du}{dx} + \frac{dv}{dx} \quad ...(i)$$

Now, $u = (\cos x)^x$

Taking log on both sides, we get

$$\log u = \log(\cos x)^x$$

$$\Rightarrow \quad \log u = x \log(\cos x)$$

Differentiating both sides w.r.t. x, we get

$$\frac{1}{u}\frac{du}{dx} = \frac{x(-\sin x)}{\cos x} + \log(\cos x) \times 1$$

$$\Rightarrow \quad \frac{du}{dx} = u\,[-x\tan x + \log(\cos x)]$$

$$\Rightarrow \quad \frac{du}{dx} = (\cos x)^x\,[-x\tan x + \log(\cos x)]$$

Similarly, $v = \tan^{-1}\sqrt{x}$

Differentiating both sides w.r.t. x, we get

$$\frac{dv}{dx} = \frac{1}{\left(1+(\sqrt{x})^2\right)} \times \frac{1}{2\sqrt{x}}$$

$$\Rightarrow \quad \frac{dv}{dx} = \frac{1}{2\sqrt{x}(1+x)}$$

Putting the values of $\frac{du}{dx}$ and $\frac{dv}{dx}$ in (i), we get

$$\frac{dy}{dx} = (\cos x)^x\,[-x\tan x + \log\cos x] + \frac{1}{2\sqrt{x}(1+x)}$$

Ans.

SECTION–D

36. Find the point on the curve $y^2 = 4x$ which is nearest to the point (2, 1). **[6]**

Sol. Let A(x, y) be the required point on the curve $y^2 = 4x$.

$$\therefore \quad \text{Distance} = \sqrt{(2-x)^2 + (1-y)^2}$$

$$\Rightarrow \quad D = \sqrt{(2-x)^2 + (1-y)^2}$$

$$\Rightarrow \quad D(y) = \sqrt{\left(2-\frac{y^2}{4}\right)^2 + (1-y)^2}$$

$$\left[\because y^2 = 4x \quad \therefore x = \frac{y^2}{4}\right]$$

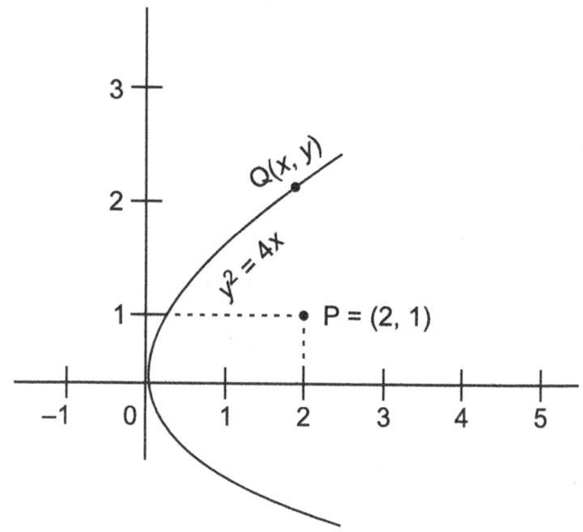

Squaring both sides, we get

$$D^2(y) = \left(2-\frac{y^2}{4}\right)^2 + (1-y)^2$$

Let $Z = D^2(y)$

$$\therefore \quad Z = \left(2-\frac{y^2}{4}\right)^2 + (1-y)^2$$

Now, $$\frac{dZ}{dy} = 2\left(2-\frac{y^2}{4}\right)\left(\frac{-2y}{4}\right) + 2(1-y)(-1)$$

$$\Rightarrow \quad \frac{dZ}{dy} = -y\left(2-\frac{y^2}{4}\right) - 2(1-y)$$

$$\Rightarrow \quad \frac{dZ}{dy} = -2y + \frac{y^3}{4} - 2 + 2y$$

$$\Rightarrow \quad \frac{dZ}{dy} = \frac{y^3}{4} - 2$$

For maximum and minimum value,

Put $\frac{dZ}{dy} = 0$

$$\Rightarrow \quad \frac{y^3}{4} - 2 = 0$$

$$\Rightarrow \quad \frac{y^3}{4} = 2$$

$$\Rightarrow \quad y^3 = 8$$

$$\Rightarrow \quad y = 2$$

Now, $$\frac{d^2 Z}{dy^2} = \frac{3y^2}{4}$$

$$\left[\frac{d^2Z}{dy^2}\right]_{y=2} = \frac{3(2)^2}{4} = \frac{12}{4} = 3 > 0$$

So, Z is minimum at $y = 2$.

$\because \quad y^2 = 4x$

$\Rightarrow \quad (2)^2 = 4x$

$\Rightarrow \quad x = 1$

$\therefore$ Required point on the curve is (1, 2). **Ans.**

Delhi [Set-III]

Note : Except for the following questions, all the remaining questions have been asked in the previous sets.

SECTION-A

1. If A is a skew symmetric matrix of order 3, then the value of |A| is :

(a) 3 **(b) 0**

(c) 9 **(d) 27** **[1]**

Sol. Since, A is a skew symmetric matrix,

$\therefore \quad A^T = -A$

$\Rightarrow \quad |A^T| = |-A|$

$\Rightarrow \quad |A| = |-1|^3 |A|$

$\Rightarrow \quad |A| = -|A|$

$\Rightarrow \quad 2|A| = 0$

$\Rightarrow \quad |A| = 0$

$\therefore$ Option (b) is correct.

6. If $y = \log_e\left(\frac{x^2}{e^2}\right)$, then $\frac{d^2y}{dx^2}$ equals :

(a) $-\frac{1}{x}$ **(b)** $-\frac{1}{x^2}$

(c) $\frac{2}{x^2}$ **(d)** $-\frac{2}{x^2}$ **[1]**

Sol. Given : $$y = \log\left(\frac{x^2}{e^2}\right)$$

$$\therefore \quad \frac{dy}{dx} = \frac{1}{\frac{x^2}{e^2}} \times \frac{1}{e^2}(2x)$$

$$\Rightarrow \quad \frac{dy}{dx} = \frac{2}{x}$$

$$\Rightarrow \quad \frac{d^2y}{dx^2} = \frac{-2}{x^2}$$

$\therefore$ Option (d) is correct.

9. The distance of the origin (0, 0, 0) from the plane $-2x + 6y - 3z = -7$ is :

(a) 1 unit **(b) $\sqrt{2}$ units**

(c) $2\sqrt{2}$ units **(d) 3 units** **[1]**

Sol. Given : Equation of plane is,

$$-2x + 6y - 3z = -7$$

$$\Rightarrow \quad -2x + 6y - 3z + 7 = 0$$

We know distance of point (x_1, y_1, z_1) from a plane $Ax + By + Cz + D = 0$ is given as

$$\text{Distance} = \left|\frac{Ax_1 + By_1 + Cz_1 + D}{\sqrt{A^2+B^2+C^2}}\right|$$

$$\therefore \quad \text{Distance} = \left|\frac{-2(0)+6(0)+(-3)(0)+7}{\sqrt{4+36+9}}\right|$$

$$= \left|\frac{7}{\sqrt{49}}\right| = \left|\frac{7}{7}\right| = 1 \text{ unit}$$

$\therefore$ Option (a) is correct.

11. If A and B are square matrices each of order 3 and |A| = 5, |B| = 3, then the value of |3AB| is ________ . **[1]**

Sol. 405

Explanation : $\because \quad |A| = 5, |B| = 3$

$$\therefore \quad |3\,AB| = 3^3 |A||B|$$

$$= 27 \times 5 \times 3 = 405$$

16. Find the cofactors of all the elements of $\begin{bmatrix} 1 & -2 \\ 4 & 3 \end{bmatrix}$. **[1]**

Sol. Let $$A = \begin{bmatrix} 1 & -2 \\ 4 & 3 \end{bmatrix}$$

$\therefore$ Co-factors of A are

$C_{11} = 3, C_{21} = 2, C_{12} = -4$ and $C_{22} = 1$ **Ans.**

17. Let $f(x) = x|x|$, for all $x \in$ R check its differentiability at $x = 0$. **[1]**

Sol. Given,

$$f(x) = x|x|$$

$$\therefore \quad f(x) = \begin{cases} +x(x), & x \geq 0 \\ -x(x), & x < 0 \end{cases}$$

$$\Rightarrow \quad f(x) = \begin{cases} x^2, & x \geq 0 \\ -x^2, & x < 0 \end{cases}$$

Now,

$$\text{R.H.D.} = \lim_{h \to 0} \frac{f(0+h) - f(0)}{h}$$

$$= \lim_{h \to 0} \frac{(0+h)^2 - (0)^2}{h}$$

$$= \lim_{h \to 0} \frac{h^2}{h} = 0$$

$$\text{L.H.D.} = \lim_{h \to 0} \frac{f(0-h) - f(0)}{-h}$$

$$= \lim_{h \to 0} \frac{-(0-h)^2 - (0)^2}{-h}$$

$$= \lim_{h \to 0} \frac{-h^2}{-h} = 0$$

$\because$ L.H.D. = R.H.D

So, $f(x)$ is differentiable at $x = 0$. **Ans.**

SECTION – B

21. Find $\int \frac{x+1}{(x+2)(x+3)} dx$. **[2]**

Sol. Let $\quad I = \int \frac{x+1}{(x+2)(x+3)}$

Let $\quad \frac{x+1}{(x+2)(x+3)} = \frac{A}{x+2} + \frac{B}{x+3}$

$\Rightarrow \quad x + 1 = A(x + 3) + B(x + 2)$

$\Rightarrow \quad x + 1 = (A + B)x + (3A + 2B)$

Comparing coefficients of x both sides, we get

$1 = A + B$ ...(i)

Comparing constant terms both sides, we get

$1 = 3A + 2B$ ...(ii)

Solving equations (i) and (ii),

$3A + 2(1 - A) = 0$

$\Rightarrow \quad 3A + 2 - 2A = 0$

$\Rightarrow \quad A + 2 = 0$

$\Rightarrow \quad A = -2$

Putting $A = -2$ in equation (i), we get

$B = 3$

$$\therefore \quad I = -2\int \frac{dx}{x+2} + 3\int \frac{dx}{x+3}$$

$$\Rightarrow \quad I = -2 \log |x + 2| + 3 \log |x + 3| + C$$

$$= \log |x + 2|^{-2} + 3 \log |x + 3|^3 + C$$

$$= \log \left|\frac{1}{x+2}\right|^2 + \log |x+3|^3 + C$$

$$\Rightarrow \quad I = \log \left|\frac{(x+3)^3}{(x+2)^2}\right| + C$$ **Ans.**

26. Find the value of $\int_0^1 \tan^{-1}\left(\frac{1-2x}{1+x-x^2}\right) dx$. **[2]**

Sol. Let $\quad I = \int_0^1 \tan^{-1}\left(\frac{1-2x}{1+x-x^2}\right) dx$

$$\Rightarrow \quad I = \int_0^1 \tan^{-1}\left(\frac{(1-x)-x}{1+x(1-x)}\right) dx$$

$$\Rightarrow \quad I = \int_0^1 \left[\tan^{-1}(1-x) - \tan^{-1} x\right] dx$$

$$\left[\because \tan^{-1} A - \tan^{-1} B = \tan^{-1}\left(\frac{A-B}{1+AB}\right)\right]$$

$$\Rightarrow \quad I = \int_0^1 \tan^{-1}(1-x)\, dx - \int_0^1 \tan^{-1} x\, dx$$

$$\Rightarrow \quad I = \int_0^1 \tan^{-1}[1-(1-x)]\, dx - \int_0^1 \tan^{-1} x\, dx$$

$$\left[\because \int_0^a f(x)dx = \int_0^a f(a-x)dx\right]$$

$$\Rightarrow \quad I = \int_0^1 \tan^{-1} x\, dx - \int_0^1 \tan^{-1} x\, dx$$

$\Rightarrow \quad I = 0.$ **Ans.**

SECTION - C

27. Solve the equation for x:

$$\sin^{-1}\left(\frac{5}{x}\right) + \sin^{-1}\left(\frac{12}{x}\right) = \frac{\pi}{2} (x \neq 0)$$ **[4]**

Sol. Given : $\sin^{-1}\left(\frac{5}{x}\right) + \sin^{-1}\left(\frac{12}{x}\right) = \frac{\pi}{2}$

$$\Rightarrow \sin^{-1}\left\{\frac{5}{x}\sqrt{1-\left(\frac{12}{x}\right)^2}+\frac{12}{x}\sqrt{1-\left(\frac{5}{x}\right)^2}\right\} = \frac{\pi}{2}$$

$$\left[\because \sin^{-1}x+\sin^{-1}y = \sin^{-1}\left(x\sqrt{1-y^2}+y\sqrt{1-x^2}\right)\right]$$

$$\Rightarrow \frac{5}{x}\sqrt{\frac{x^2-144}{x^2}}+\frac{12}{x}\sqrt{\frac{x^2-25}{x^2}} = \sin\frac{\pi}{2}$$

$$\Rightarrow \frac{5}{x}\frac{\sqrt{x^2-144}}{x}+\frac{12}{x}\frac{\sqrt{x^2-25}}{x} = 1$$

$$\Rightarrow \frac{5\sqrt{x^2-144}}{x^2}+\frac{12\sqrt{x^2-25}}{x^2} = 1$$

$$\Rightarrow 5\sqrt{x^2-144}+12\sqrt{x^2-25} = x^2$$

$$\Rightarrow 5\sqrt{x^2-144} = x^2 - 12\sqrt{x^2-25}$$

Squaring both sides, we get

$$\Rightarrow 25(x^2-144) = x^4 + 144(x^2-25) - 2 \times 12x^2\sqrt{x^2-25}$$

$$\Rightarrow 25x^2 - 3600 = x^4 + 144x^2 - 3600 - 24x^2\sqrt{x^2-25}$$

$$\Rightarrow 25x^2 - 144x^2 - x^4 = -24x^2\sqrt{x^2-25}$$

$$\Rightarrow -119x^2 - x^4 = -24x^2\sqrt{x^2-25}$$

$$\Rightarrow -x^2(119+x^2) = -24x^2\sqrt{x^2-25}$$

$\Rightarrow x = 0$ (Rejected, as $x \neq 0$)

or $\quad 119 + x^2 = -24\sqrt{x^2-25}$

Again squaring both sides, we get

$$14161 + x^4 + 238x^2 = 576(x^2-25)$$

$$\Rightarrow x^4 + 238x^2 + 14161 = 576x^2 - 14400$$

$$\Rightarrow x^4 - 338x^2 + 28561 = 0$$

$$\Rightarrow x^2 = \frac{338 \pm \sqrt{114244-114244}}{2}$$

$$\Rightarrow x^2 = \frac{338}{2}$$

$$\Rightarrow x^2 = 169$$

$$\Rightarrow x = \pm 13$$

$$\therefore x = 13$$ **Ans.**

28. Find the general solution of the differential equation :

$ye^{x/y}\,dx = (xe^{x/y} + y^2)\,dy,\ y \neq 0$ **[4]**

Sol. Given : differential equation is,

$$ye^{\frac{x}{y}}dx = \left(xe^{\frac{x}{y}}+y^2\right)dy$$

$$\Rightarrow \frac{dx}{dy} = \frac{xe^{\frac{x}{y}}+y^2}{ye^{\frac{x}{y}}}$$

Put $\quad x = vy$

$$\Rightarrow \frac{dx}{dy} = v + y\frac{dv}{dy}$$

Now, $$v + y\frac{dv}{dy} = \frac{(vy)e^{vy/y}+y^2}{ye^{vy/y}}$$

$$\Rightarrow v + y\frac{dv}{dy} = \frac{vye^v+y^2}{ye^v}$$

$$\Rightarrow v + y\frac{dv}{dy} = \frac{ve^v+y}{e^v}$$

$$\Rightarrow y\frac{dv}{dy} = \frac{ve^v+y}{e^v} - v$$

$$\Rightarrow y\frac{dv}{dy} = \frac{ve^v+y-ve^v}{e^v}$$

$$\Rightarrow y\frac{dv}{dy} = \frac{y}{e^v}$$

$$\Rightarrow \frac{dv}{dy} = \frac{1}{e^v}$$

$$\Rightarrow e^v dv = dy$$

Integrating both sides,

$$\Rightarrow \int e^v dv = \int dy$$

$$\Rightarrow e^v = y + C$$

$$\Rightarrow e^{x/y} = y + C$$

or $\quad y = e^{x/y} + K$

which is the required solution of given differential equation. **Ans.**

SECTION - D

33. Find the distance of the point P (3, 4, 4) from the point, where the line joining the points A(3, – 4, – 5) and B (2, – 3, 1) intersects the plane $2x + y + z = 7$. [6]

Sol. Equation of line passing through points A(3, –4, – 5) and B(2, –3, 1) is,

$$\frac{x-3}{2-3} = \frac{y+4}{-3-(-4)} = \frac{z+5}{1-(-5)}$$

$$\Rightarrow \quad \frac{x-3}{-1} = \frac{y+4}{1} = \frac{z+5}{6}$$

Let $$\frac{x-3}{-1} = \frac{y+4}{1} = \frac{z+5}{6} = k$$

$\therefore$ $x = 3 - k,\ y = k - 4$ and $z = 6k - 5$

$\therefore$ Coordinates of any point on the line are M$(3 - k, k - 4, 6k - 5)$

Since, point M lies on the plane $2x + y + z - 7 = 0$

$\therefore 2(3 - k) + (k - 4) + (6k - 5) - 7 = 0$

$\Rightarrow \quad 6 - 2k + k - 4 + 6k - 5 - 7 = 0$

$\Rightarrow \quad 5k - 10 = 0$

$\Rightarrow \quad k = 2$

$\therefore$ Coordinates of M $= (3 - 2, 2 - 4, 6(2) - 5)$

$= (1, -2, 7)$

Now, distance between points P(3, 4, 4) and M(1, – 2, 7) is,

$$PM = \sqrt{(1-3)^2 + (-2-4)^2 + (7-4)^2}$$

$$= \sqrt{(-2)^2 + (-6)^2 + (3)^2}$$

$$= \sqrt{4+36+9} = \sqrt{49}$$

$= 7$ units **Ans.**

●●